AF725229

DRUG TRANSPORTERS

BICENTENNIAL 1807 WILEY 2007 BICENTENNIAL

The Wiley Bicentennial–Knowledge for Generations

Each generation has its unique needs and aspirations. When Charles Wiley first opened his small printing shop in lower Manhattan in 1807, it was a generation of boundless potential searching for an identity. And we were there, helping to define a new American literary tradition. Over half a century later, in the midst of the Second Industrial Revolution, it was a generation focused on building the future. Once again, we were there, supplying the critical scientific, technical, and engineering knowledge that helped frame the world. Throughout the 20th Century, and into the new millennium, nations began to reach out beyond their own borders and a new international community was born. Wiley was there, expanding its operations around the world to enable a global exchange of ideas, opinions, and know-how.

For 200 years, Wiley has been an integral part of each generation's journey, enabling the flow of information and understanding necessary to meet their needs and fulfill their aspirations. Today, bold new technologies are changing the way we live and learn. Wiley will be there, providing you the must-have knowledge you need to imagine new worlds, new possibilities, and new opportunities.

Generations come and go, but you can always count on Wiley to provide you the knowledge you need, when and where you need it!

William J. Pesce
President and Chief Executive Officer

Peter Booth Wiley
Chairman of the Board

DRUG TRANSPORTERS

Molecular Characterization and Role in Drug Disposition

Edited by

GUOFENG YOU
Rutgers University

MARILYN E. MORRIS
University at Buffalo
State University of New York

BICENTENNIAL 1807 WILEY 2007 BICENTENNIAL

WILEY-INTERSCIENCE

A JOHN WILEY & SONS, INC., PUBLICATION

Copyright © 2007 by John Wiley & Sons, Inc. All rights reserved.

Published by John Wiley & Sons, Inc., Hoboken, New Jersey.
Published simultaneously in Canada.

No part of this publication may be reproduced, stored in a retrieval system, or transmitted in any form or by any means, electronic, mechanical, photocopying, recording, scanning, or otherwise, except as permitted under Section 107 or 108 of the 1976 United States Copyright Act, without either the prior written permission of the Publisher, or authorization through payment of the appropriate per-copy fee to the Copyright Clearance Center, Inc., 222 Rosewood Drive, Danvers, MA 01923, (978) 750-8400, fax (978) 750-4470, or on the web at www.copyright.com. Requests to the Publisher for permission should be addressed to the Permissions Department, John Wiley & Sons, Inc., 111 River Street, Hoboken, NJ 07030, (201) 748–6011, fax (201) 748-6008 or online at http://www.wiley.com/go/permission.

Limit of Liability/Disclaimer of Warranty: While the publisher and author have used their best efforts in preparing this book, they make no representations or warranties with respect to the accuracy or completeness of the contents of this book and specifically disclaim any implied warranties of merchantability or fitness for a particular purpose. No warranty may be created or extended by sales representatives or written sales materials. The advice and strategies contained herein may not be suitable for your situation. You should consult with a professional where appropriate. Neither the publisher nor author shall be liable for any loss of profit or any other commercial damages, including but not limited to special, incidental, consequential, or other damages.

For general information on our other products and services or for technical support, please contact our Customer Care Department within the United States at (800) 762-2974, outside the United States at (317) 572-3993 or fax (317) 572-4002.

Wiley also publishes its books in a variety of electronic formats. Some content that appears in print may not be available in electronic formats. For more information about Wiley products, visit our web site at www.wiley.com

Wiley Bicentennial Logo: Richard J. Pacifico

Library of Congress Cataloging-in-Publication Data:

Drug transporters : molecular characterization and role in drug disposition [edited by] Guofeng You, Marilyn E. Morris.
p. ; cm.
Includes bibliographical references.
ISBN 978-0-471-78491-3
1. Carrier proteins. 2. Drug carriers (Pharmacy) I. You, Guofeng. II. Morris, Marilyn Emily.
[DNLM: 1. Biological Transport–physiology. 2. Membrane Transport Proteins–physiology.
3. Pharmacokinetics.
QU 120 D794 2007]
QP552.C34D78 2007
572′.69–dc22

2006038880

Printed in the United States of America

10 9 8 7 6 5 4 3 2 1

CONTENTS

CONTRIBUTORS

Pieter Annaert, Department of Pharmaceutical Sciences, Catholic University of Leuven, Leuven, Belgium

Naohiko Anzai, Department of Pharmacology and Toxicology, Kyorin University School of Medicine, Tokyo, Japan

Karlo Babakhanian, Department of Pharmaceutical Sciences, Leslie Dan Faculty of Pharmacy, University of Toronto, Toronto, Ontario, Canada

Susan E. Bates, Center for Cancer Research, National Cancer Institute, Bethesda, Maryland

Jos H. Beijnen, Department of Medical Oncology, Netherlands Cancer Institute, Amsterdam, The Netherlands; and Department of Pharmaceutical Sciences, Utrecht University, Utrecht, The Netherlands

Reina Bendayan, Department of Pharmaceutical Sciences, Leslie Dan Faculty of Pharmacy, University of Toronto, Toronto, Ontario, Canada

Rajinder K. Bhardwaj, Bristol-Myers Squibb Co., Wallingford, Connecticut

Vibha Bhatnagar, Departments of Medicine, and Family and Preventive Medicine, and the San Diego Veterans Administration Medical Center, University of California San Diego, La Jolla, California

Kim L. R. Brouwer, Division of Pharmacotherapy and Experimental Therapeutics, School of Pharmacy, University of North Carolina, Chapel Hill, North Carolina

Kevin T. Bush, Department of Pediatrics, University of California San Diego, La Jolla, California

Stephen M. Carl, Department of Industrial and Physical Pharmacy, College of Pharmacy, Nursing and Health Sciences, Purdue University, West Lafayette, Indiana

Xingguo Cheng, Department of Pharmacology, Toxicology and Therapeutics, University of Kansas Medical Center, Kansas City, Kansas

Leslie W. Chinn, Department of Biopharmaceutical Sciences and Center for Human Genetics, University of California–San Francisco, San Francisco, California

John Deeken, Center for Cancer Research, National Cancer Institute, Bethesda, Maryland

Lester R. Drewes, Department of Biochemistry and Molecular Biology, University of Minnesota, Duluth, Minnesota

Ihsan Elimrani, IBD Research Laboratory, McGill University Research Institute, Montreal, Quebec, Canada

Jyrki J. Eloranta, Division of Clinical Pharmacology and Toxicology, University Hospital Zurich, Zurich, Switzerland

Hitoshi Endou, Department of Pharmacology and Toxicology, Kyorin University School of Medicine, Tokyo, Japan

Satish A. Eraly, Department of Pediatrics, University of California San Diego, La Jolla, California

Kathleen M. Giacomini, Department of Pharmaceutical Sciences, University of California–San Francisco, San Francisco, California

Jason M. Gow, Department of Biopharmaceutical Sciences and Center for Human Genetics, University of California–San Francisco, San Francisco, California

Olafur Gudmundsson, Bristol-Myers Squibb Research Institute, Discovery Pharmaceutics, Princeton, New Jersey

Dea Herrera-Ruiz, Facultad de Farmacia, Universidad Autónoma del Estado de Morelos, Cuernavaca, México

Mei Hong, Department of Pharmaceutics, Ernest Mario School of Pharmacy, Rutgers, the State University of New Jersey, Piscataway, New Jersey

Gregory Kaler, Department of Pediatrics, University of California San Diego, La Jolla, California

Yukio Kato, Division of Pharmaceutical Sciences, Graduate School of Natural Science and Technology, Kanazawa University, Kakuma-machi, Kanazawa, Japan

Dietrich Keppler, Division of Tumor Biochemistry, German Cancer Research Center, Heidelberg, Germany

Richard B. Kim, Department of Medicine, University of Western Ontario, London, Ontario, Canada

Curtis D. Klaassen, Department of Pharmacology, Toxicology and Therapeutics, University of Kansas Medical Center, Kansas City, Kansas

Gregory T. Knipp, Department of Industrial and Physical Pharmacy, College of Pharmacy, Nursing and Health Sciences, Purdue University, West Lafayette, Indiana

Deanna L. Kroetz, Department of Biopharmaceutical Sciences and Center for Human Genetics, University of California–San Francisco, San Francisco, CA, California

Yoshiyuki Kubo, Division of Pharmaceutical Sciences, Graduate School of Natural Science and Technology, Kanazawa University, Kakuma-machi, Kanazawa, Japan

Gerd A. Kullak-Ublick, Division of Clinical Pharmacology and Toxicology, University Hospital Zurich, Zurich, Switzerland

Jin Kyung Lee, Division of Pharmacotherapy and Experimental Therapeutics, School of Pharmacy, University of North Carolina, Chapel Hill, North Carolina

Shanjun Liu, Department of Pharmaceutical Sciences, Leslie Dan Faculty of Pharmacy, University of Toronto, Toronto, Ontario, Canada

Kazuya Maeda, Department of Molecular Pharmacokinetics, Graduate School of Pharmaceutical Sciences, University of Tokyo, Tokyo, 113-0033 Japan

Serena Marchetti, Department of Medical Oncology, Netherlands Cancer Institute, Amsterdam, The Netherlands; and Department of Internal Medicine, Postgraduate School in Oncology, University of Florence, Azienda Ospedaliero-Universitaria Careggi, Florence, Italy

Roberto Mazzanti, Department of Internal Medicine, Postgraduate School in Oncology, University of Florence, Azienda Ospedaliero-Universitaria Careggi, Florence, Italy

Grant Mitchell, Division of Medical Genetics, Hospital Sainte-Justine, Montreal, Quebec, Canada

Marilyn E. Morris, Department of Pharmaceutical Sciences, School of Pharmacy and Pharmaceutical Sciences, University at Buffalo, State University of New York, Amherst, New York

Megha Nagle, Department of Pediatrics, University of California San Diego, La Jolla, California

Anne T. Nies, Division of Tumor Biochemistry, German Cancer Research Center, Heidelberg, Germany

Sanjay K. Nigam, Departments of Pediatrics, Medicine, and Cellular and Molecular Medicine, University of California San Diego, La Jolla, California

K. Sandy Pang, Department of Pharmaceutical Sciences, Leslie Dan Faculty of Pharmacy, University of Toronto, Toronto, Ontario, Canada

Orsolya Polgar, Center for Cancer Research, National Cancer Institute, Bethesda, Maryland

Ijaz A. Qureshi, Division of Medical Genetics, Hospital Sainte-Justine, Montreal, Quebec, Canada

Maria Rius, Division of Tumor Biochemistry, German Cancer Research Center, Heidelberg, Germany

Robert W. Robey, Center for Cancer Research, National Cancer Institute, Bethesda, Maryland

Patrick T. Ronaldson, Department of Pharmaceutical Sciences, Leslie Dan Faculty of Pharmacy, University of Toronto, Toronto, Ontario, Canada

Jan H. M. Schellens, Department of Medical Oncology, Netherlands Cancer Institute, Amsterdam, The Netherlands; and Department of Pharmaceutical Sciences, Utrecht University, Utrecht, The Netherlands

Ernest Seidman, IBD Research Laboratory, McGill University Research Institute, Montreal, Quebec, Canada

Frances J. Sharom, Department of Molecular and Cellular Biology, University of Guelph, Guelph, Ontario, Canada

Patrick J. Sinko, Department of Pharmaceutics, Ernest Mario School of Pharmacy, Rutgers, the State University of New Jersey, Piscataway, New Jersey

Justin A. Spanier, Department of Biochemistry and Molecular Biology, University of Minnesota, Duluth, Minnesota

Yaming Su, Department of Pharmaceutics, Ernest Mario School of Pharmacy, Rutgers, the State University of New Jersey, Piscataway, New Jersey

Yuichi Sugiyama, Department of Molecular Pharmacokinetics, Graduate School of Pharmaceutical Sciences, University of Tokyo, Tokyo, Japan

Huadong Sun, Department of Pharmaceutical Sciences, Leslie Dan Faculty of Pharmacy, University of Toronto, Toronto, Ontario, Canada

Brandon Swift, Division of Pharmacotherapy and Experimental Therapeutics, School of Pharmacy, University of North Carolina, Chapel Hill, North Carolina

Rommel G. Tirona, Department of Physiology and Pharmacology, Department of Medicine, University of Western Ontario, London, Ontario, Canada

Kenneth K. W. To, Center for Cancer Research, National Cancer Institute, Bethesda, Maryland

David M. Truong, Department of Pediatrics, University of California San Diego, La Jolla, California

Akira Tsuji, Division of Pharmaceutical Sciences, Graduate School of Natural Science and Technology, Kanazawa University, Kakuma-machi, Kanazawa, Japan

Thomas J. Urban, Department of Biopharmaceutical Sciences, University of California–San Francisco, San Francisco, California

Joanne Wang, Department of Pharmaceutics, School of Pharmacy, University of Washington, Seattle, Washington

Xiaodong Wang, Department of Pharmaceutical Sciences, School of Pharmacy and Pharmaceutical Sciences, University at Buffalo, State University of New York, Amherst, New York

Wei Wu, Department of Pediatrics, University of California San Diego, La Jolla, California

Li Xia, Department of Pharmaceutics, School of Pharmacy, University of Washington, Seattle, Washington

Guofeng You, Department of Pharmaceutics, Ernest Mario School of Pharmacy, Rutgers, the State University of New Jersey, Piscataway, New Jersey

Mingyan Zhou, Department of Pharmaceutics, School of Pharmacy, University of Washington, Seattle, Washington

PREFACE

Transporters are membrane proteins that span cellular membranes and are the gatekeepers for all cells and organelles, controlling the intake and efflux of crucial endogenous substrates such as sugars, amino acids, nucleotides, and inorganic ions. The specificity of many transporters is not, however, limited to their physiological substrates, and for some, their physiological substrates remain undiscovered. Xenobiotics (i.e., drugs, dietary and environmental compounds) have the potential to be recognized by transporters, which crucially influence the absorption, distribution, and elimination of drugs in the body.

Due to their hydrophobic nature and relatively low abundance, the molecular identification of transporters had been a difficult task until the development of the expression cloning technique for transporters in the early 1990s. This powerful approach, combined with recent genome analysis, has facilitated the identification and characterization of numerous transporters that are important in drug disposition.

Given the considerable advances in the identification of these transporters, a textbook covering basic transport mechanisms to specific descriptions of transporter families, including substrate and inhibitor specificity, subcellular and tissue localization, mechanisms governing transport, species differences, the clinical implications of these transporters in human physiology and disease, and their role in drug distribution, elimination, and interactions in drug therapy, is both timely and necessary. Such a book has not been available, so our aim is twofold: the first half of the book provides an overview of the relevant drug transporters useful for both beginning and experienced scientists and researchers. The second half of the book presents the principles of drug transport and its associated techniques in sufficient detail to enable nonspecialist readers to understand them. Such readers include graduate students in the pharmacological or physiological sciences and academic or industrial scientists in related fields of study. It is anticipated that this book will be used as a textbook in graduate courses in drug/membrane transport and as a desk reference for researchers working in the transporter field as well as in the areas of drug metabolism and pharmacokinetics in the pharmaceutical industry.

Credit for this comprehensive textbook belongs to the many dedicated scholars who contributed chapters in their area of expertise. To all we express our deepest gratitude and respect. We also acknowledge the contributions of Jonathan Rose and the many professionals at John Wiley who worked with us to ensure the best book possible. Finally, we extend our heartfelt thanks to our families for their constant support and encouragement.

Guofeng You
Marilyn E. Morris

1

OVERVIEW OF DRUG TRANSPORTER FAMILIES

GUOFENG YOU
Ernest Mario School of Pharmacy, Rutgers, the State University of New Jersey, Piscataway, New Jersey

MARILYN E. MORRIS
School of Pharmacy and Pharmaceutical Sciences, University at Buffalo, State University of New York, Amherst, New York

1.1. WHAT ARE DRUG TRANSPORTERS?

Transporters are membrane proteins whose primary function is to facilitate the flux of molecules into and out of cells. Drug transporters did not evolve to transport specific

Drug Transporters: Molecular Characterization and Role in Drug Disposition, Edited by Guofeng You and Marilyn E. Morris
Copyright © 2007 John Wiley & Sons, Inc.

drugs. Instead, their primary functions are to transport nutrients or endogenous substrates, such as sugars, amino acids, nucleotides, and vitamins, or to protect the body from dietary and environmental toxins. However, the specificity of these transporters is not strictly restricted to their physiological substrates. Drugs that bear significant structural similarity to the physiological substrates have the potential to be recognized and transported by these transporters. As a consequence, these transporters also play significant roles in determining the bioavailability, therapeutic efficacy, and pharmacokinetics of a variety of drugs. Nevertheless, because drugs may compete with the physiological substrates of these transporters, they are also likely to interfere with the transport of endogenous substrates and consequently produce deleterious effects on body homeostasis.

1.2. STRUCTURE AND MODEL OF DRUG TRANSPORTERS

Because of the involvement of transporters in all facets of drug absorption, excretion, and toxicity, characterization of transporter structure can provide a scientific basis for understanding drug delivery and disposition, as well as the molecular mechanisms of drug interactions and interindividual and interspecies differences. However, compared to soluble proteins, the atomic resolution crystal structures of membrane transporters have been extremely difficult to obtain, for several reasons. First is the amphipathic nature of the surface of the transporters, with a hydrophobic area in contact with membrane phospholipids and polar surface areas in contact with the aqueous phases on both sides of the membrane; second is the low abundance of many transporters in the membrane, making it impossible to overexpress them, a prerequisite for structural studies; and third is the inherent conformational flexibility of the transporters, making it difficult to obtain stable crystals.

Due to these difficulties, high-resolution three-dimensional structures have been obtained for only a limited number of transporters. For other transporters, three-dimensional structures have been achieved through homology modeling. In this approach, similar folding patterns between any protein and one for which the crystal structure is known enable the construction of a fairly accurate three-dimensional protein model of the unknown structure using the related crystal structure as a template and modern computational techniques. Three-dimensional structures have revealed that transporters have α-helical structures of the membrane-spanning domains, and some of the helices have irregular shapes with kinks and bends. Certain transporters undergo substantial movements during the substrate translocation process. Construction of three-dimensional transporter models have provided insight into functional mechanisms and molecular structures and enabled formulation of new hypotheses regarding transporter structure and function, which may be validated experimentally.

1.3. TRANSPORT MECHANISMS

Not only do different transporters reside in the membrane with different three-dimensional structures, but they also transport their substrates through different transport mechanisms. According to their transport mechanisms, transporters can be divided into passive and active transporters: *passive transporters*, also called *facilitated transporters*, allow molecules to move across cell membrane down their electrochemical gradients. Such a spontaneous process decreases free energy and increases entropy in a system and therefore does not consume any chemical energy. In contrast to facilitated transporters, active transporters typically move molecules against their electrochemical gradients; such a process is entropically unfavorable and therefore needs coupling of the hydrolysis of adenosine triphosphate (ATP) as an energy source. This coupling can be either primary or secondary. In primary active transport, transporters that move molecules against their electrical or chemical gradient, hydrolyze ATP. In the secondary active transport, transporters utilize ion gradients, such as sodium or proton gradients, across the membrane produced by the primary active transporters and transport substrates against an electrochemical difference.

1.4. POLARIZED EXPRESSION OF DRUG TRANSPORTERS IN BARRIER EPITHELIUM

Most drug transporters are expressed in tissues with barrier functions such as the liver, kidney, intestine, placenta, and brain. Cells at the border of these barriers are usually polarized. For example, enterocytes of intestine and proximal tubule cells of kidney have an apical domain facing the lumen and a basolateral domain facing the blood circulation; hepatocytes polarize into a canalicular membrane facing the bile duct and a sinusoidal membrane facing the blood circulation; syncytiotrophoblasts of placenta have an apical domain facing maternal blood and a basolateral domain facing the fetus. Brain capillary endothelial cells, which function as the blood–brain barrier, also polarize into apical and basolateral membranes. In most cases, the expression of drug transporters is highly restricted to one side (i.e., apical or basolateral domain) of polarized cells. Such polarized expression of the transporters is essential for the concerted transport of drugs in the same direction. One of the best studied examples of concerted transport is the kidney. Kidney proximal tubule cells play a critical role in the body clearance of drugs. These drugs are first taken up from the blood into the proximal tubule cells by transporters at the basolateral membrane. Once inside the cells, these drugs are then transported out of the cells into the tubule lumen by transporters at the apical membrane and subsequently eliminated in the urine. The alliance between transporters at both the basolateral membrane and the apical membrane of the kidney proximal tubule cells ensures clearance of the drugs from the body.

1.5. CLASSIFICATIONS OF DRUG TRANSPORTERS

Drug transporters can be classified in a number of different manners, including as efflux transporters versus influx transporters, secretory transporters versus absorptive transporters, and ATP-binding cassette (ABC) transporters versus solute carrier (SLC) transporters.

1.5.1. Definition of Efflux and Influx Transporters

Drug transporters can be categorized as efflux or influx transporters according to the direction in which they transport substrate across the cell membranes. This classification is often observed in the literature where drug transport studies are performed at the cellular level. With this definition, transporters that pump the substrates out of the cells are called *efflux transporters*, whereas transporters that transfer substrates into cells are called *influx transporters*.

1.5.2. Definition of Absorptive and Secretory Transporters

The other way of classifying drug transporters is from a pharmacodynamic or pharmacokinetic point of view. In such a classification, the transporter that transfers its substrates into the systemic blood circulation is called an *absorptive transporter*, whereas the transporter that excretes its substrates from the blood circulation into bile, urine, or the gut lumen is known as a *secretory transporter*. However, when absorptive or secretory transporters in the brain blood–brain barrier and placenta are discussed, the definition needs to be modified. The brain and fetus have traditionally been considered as two "isolated" compartments in the human body. In drug therapy, many strategies have been utilized to achieve either enhanced or reduced penetration of drugs into these two compartments. Conventionally, the transporters facilitating drug penetration into the brain or fetus are referred to as absorptive transporters.

1.5.3. Relationship Between Influx–Efflux and Absorptive–Secretory Transporters

An absorptive transporter does not necessarily mean that it influxes a substrate. Similarly, a secretory transporter does not have to be an efflux pump. For example, the organic anion transporter OAT1, present at the basolateral membrane of the kidney proximal tubule, is an influx transporter based on its role of taking up drugs from the blood into the proximal tubule cells for their subsequent exit across the apical membrane into the urine for elimination. However, considering its overall role of removing drugs out of the blood circulation into the urine, OAT1 is a secretory transporter. Intestinally expressed organic anion–transporting polypeptide-A (OATP-A) is localized on the apical domain of enterocytes. It can take up (i.e., influx) into the enterocytes orally administered drugs for their subsequent exit across the basolateral membrane into the bloodstream, so OATP-A is considered an absorptive transporter. Therefore, influx transporters can function as either absorptive or secretory transporters, depending on the tissue and on the membrane domain where they are expressed.

1.5.4. ABC and SLC Transporters

Most drug transporters can also be molecularly and mechanistically classified as a member of the ABC or the SLC transporter family (Table 1.1). *ABC* (ATP-binding cassette) *transporters* are a family of membrane transport proteins that require ATP

TABLE 1.1. Classification of Drug Transporters

Transporter Family	Family Member	Gene Name	Human Chromosome Locus	Ref.[a]
Organic cation transporter (OCT)	hOCT1	SLC22A1	6q26	1
	hOCT2	SLC22A2	6q26	1
	hOCT3	SLC22A3	6q26–q27	2
Organic cation/carnitine transporter (OCTN)	OCTN1	SLC22A4	5q31.1	3
	OCTN2	SLC22A5	5q31	4
	OCTN3	SLC22A21	5q31	5
	CT2	SLC22A16	6q22.1	6
Organic anion transporter (OAT)	OAT1	SLC22A6	11q13.1–q13.2	7
	OAT2	SLC22A7	6p21.2–p21.1	8
	OAT3	SLC22A8	11q11.7	9
	OAT4	SLC22A11	11q13.1	10
	OAT5	SLC22A10	11q12.3	11
	OAT6	SLC22A20	unknown	—
	URAT1	SLC22A12	11q13.1	10
Organic anion transporter polypeptides (OATPs)	OATP1C1	SLCO1C1	12p12.2	12
	OATP1B1	SLCO1B1	12p12.2	13
	OATP1A2	SLCO1A2	12p12	12
	OATP1B3	SLCO1B3	12p12	14
	OATP2A1	SLCO2A1	3q21	15
	OATP2B1	SLCO2B1	11q13	16
	OATP3A1	SLCO3A1	15q26	17
	OATP4A1	SLCO4A1	20q13.33	17
	OATP4C1	SLCO4C1	5q21.2	18
	OATP5A1	SLCO5A1	8q13.3	19
	OATP6A1	SLCO6A1	5q21.1	20
Peptide transporter (PEPT)	PEPT1	SLC15A1	13q33–q34	21
	PEPT2	SLC15A2	3q21.1	22
	PHT1	SLC15A4	12q24.32	23
	PHT2	SLC15A3	11q12.2	24
Monocarboxylate transporters (MCTs, sMCTs)	MCT1	SLC16A1	1p12	25
	MCT2	SLC16A7	12q13	26
	MCT3	SLC16A8	22q12.3–q13.2	27
	MCT4	SLC16A3	17q25	28
	SMCT1	SLC5A8	12q23	29
	SMCT2	SLC5A12	11p14	24

(Continued)

TABLE 1.1. ***(Continued)***

Transporter Family	Family Member	Gene Name	Human Chromosome Locus	Ref.[a]
Nucleoside transporters (CNTs, ENTs)	CNT1	SLC28A1	15q25–26	30
	CNT2	SLC28A2	15q15	31
	CNT3	SLC28A3	9q22.2	32
	ENT1	SLC29A1	6p21.1–p21.2	33
	ENT2	SLC29A2	11q13	34
	ENT3	SLC29A3	10q22.1	35
	ENT4	SLC29A4	7p22.1	36
Bile acid transporters	NTCP	SLC10A1	14q24.1	37
	ASBT	SLC10A2	13q33	38
	BSEP	ABCB11	2q24	39
	OST-α	—	3q29	40
	OST-β	—	15q22.31	40
Multidrug resistance protein (MDR)	MDR1	ABCB1	7q21.1	41
Multidrug resistance–associated protein (MRP)	MRP1	ABCC1	16p13.1	42
	MRP2	ABCC2	10q24	43
	MRP3	ABCC3	17q22	44
	MRP4	ABCC4	13q32	45
	MRP5	ABCC5	3q27	46
	MRP6	ABCC6	16p13.1	47
	MRP7	ABCC10	6p21.1	48
	MRP8	ABCC11	16q12.1	49
	MRP9	ABCC12	16q12.1	49
Breast cancer resistance protein (BCRP)	BCRP1	ABCG2	4q22	50

[a]References where the human chromosome locus can be found.

hydrolysis for the transport of substrates across membranes. Therefore, ABC transporters are primary active transporters. The protein family derives its name from the ATP-binding domain found on the protein. The best studied drug transporters which are classified as ABC transporters are multidrug resistance protein (MDR), multidrug resistance–associated protein (MRP), and breast cancer resistance protein (BCRP).

Some *SLC* (solute carrier) *transporters* utilize an electrochemical potential difference in the substrate transported and are therefore classified as *facilitated transporters*; other SLC transporters utilize an ion gradient such as a sodium or proton gradient across the membrane produced by primary active transporters and transport substrates against an electrochemical difference. These transporters are classified as *secondary active transporters*. In contrast to ABC transporters, SLC transporters do not possess ATP-binding sites. Most drug transporters belong to the SLC family of transporters.

REFERENCES

1. Koehler MR, Wissinger B, Gorboulev V, Koepsell H, Schmid M, 1997: The two human organic cation transporter genes SLC22A1 and SLC22A2 are located on chromosome 6q26. Cytogenet Cell Genet 79: 198–200.
2. Verhaagh S, Schweifer N, Barlow DP, Zwart R, 1999: Cloning of the mouse and human solute carrier 22a3 (Slc22a3/SLC22A3) identifies a conserved cluster of three organic cation transporters on mouse chromosome 17 and human 6q26–q27. Genomics 55: 209–218.
3. Peltekova VD, Wintle RF, Rubin LA, Amos CI, Huang Q, Gu X, Newman B, van Oene M, Cescon D, Greenberg G, et al., 2004: Functional variants of OCTN cation transporter genes are associated with Crohn disease. Nature Genet 36: 471–475.
4. Shoji Y, Koizumi A, Kayo T, Ohata T, Takahashi T, Harada K, Takada G, 1998: Evidence for linkage of human primary systemic carnitine deficiency with D5S436: a novel gene locus on chromosome 5q. Am J Hum Genet 63: 101–108.
5. Lamhonwah AM, Skaug J, Scherer SW, Tein I, 2003: A third human carnitine/organic cation transporter (OCTN3) as a candidate for the 5q31 Crohn's disease locus (IBD5). Biochem Biophys Res Commun Jan 31;301(1): 98–101.
6. Enomoto A, Wempe MF, Tsuchida H, Shin HJ, Cha SH, Anzai N, Goto A, Sakamoto A, Niwa T, Kanai Y, et al., 2002: Molecular identification of a novel carnitine transporter specific to human testis: insights into the mechanism of carnitine recognition. J Biol Chem 277: 36262–36271.
7. Bahn A, Prawitt D, Buttler D, Reid G, Enklaar T, Wolff NA, Ebbinghaus C, Hillemann A, Schulten HJ, Gunawan B, et. al., 2000: Genomic structure and in vivo expression of the human organic anion transporter 1 (hOAT1) gene. Biochem Biophys Res Commun 275: 623–630.
8. Kok LD, Siu SS, Fung KP, Tsui SK, Lee CY, Waye MM, 2000: Assignment of liver-specific organic anion transporter (SLC22A7) to human chromosome 6 bands p21.2–p21.1 using radiation hybrids. Cytogenet Cell Genet 88: 76–77.
9. Race JE, Grassl SM, Williams WJ, Holtzman EJ, 1999: Molecular cloning and characterization of two novel human renal organic anion transporters (hOAT1 and hOAT3). Biochem Biophys Res Commun Feb 16;255(2): 508–514.
10. Enomoto A, Kimura H, Chairoungdua A, Shigeta Y, Jutabha P, Cha SH, Hosoyamada M, Takeda M, Sekine T, Igarashi T, et al., 2002: Molecular identification of a renal urate-anion exchanger that regulates blood urate levels. Nature 417: 447–452.
11. http://www.gene.ucl.ac.uk/nomenclature/data/get_data.php?hgnc_id=HGNC:18057.
12. Kullak-Ublick GA, Beuers U, Meier PJ, Domdey H, Paumgartner G, 1996: Assignment of the human organic anion transporting polypeptide (OATP) gene to chromosome 12p12 by fluorescence in situ hybridization. J Hepatol Dec;25(6): 985–987.
13. Jung D, Hagenbuch B, Gresh L, Pontoglio M, Meier PJ, Kullak-Ublick GA, 2001: Characterization of the human OATP-C (SLC21A6) gene promoter and regulation of liver-specific OATP genes by hepatocyte nuclear factor 1. J Biol Chem 276: 37206–37214.
14. Konig J, Cui Y, Nies AT, Keppler D, 2000: Localization and genomic organization of a new hepatocellular organic anion transporting polypeptide. J Biol Chem 275: 23161–23168.
15. Lu R, Schuster VL, 1998: Molecular cloning of the gene for the human prostaglandin transporter hPGT: gene organization, promoter activity, and chromosomal localization. Biochem Biophys Res Commun 246: 805–812.

16. Nagase T, Ishikawa K, Suyama M, Kikuno R, Hirosawa M, Miyajima N, Tanaka A, Kotani H, Nomura N, Ohara O, 1998: Prediction of the coding sequences of unidentified human genes. XII. The complete sequences of 100 new cDNA clones from brain which code for large proteins in vitro. DNA Res 5: 355–364.
17. Tamai I, Nezu J, Uchino H, Sai Y, Oku A, Shimane M, Tsuji A, 2000: Molecular identification and characterization of novel members of the human organic anion transporter (OATP) family. Biochem Biophys Res Commun 273: 251–260.
18. Mikkaichi T, Suzuki T, Onogawa T, Tanemoto M, Mizutamari H, Okada M, Chaki T, Masuda S, Tokui T, Eto N, et al., 2004: Isolation and characterization of a digoxin transporter and its rat homologue expressed in the kidney. Proc Natl Acad Sci 101: 3569–3574.
19. http://www.gene.ucl.ac.uk/nomenclature/data/get_data.php?hgnc_id=19046.
20. http://www.gene.ucl.ac.uk/nomenclature/data/get_data.php?hgnc_id=HGNC:23613.
21. Liang R, Fei YJ, Prasad PD, Ramamoorthy S, Han H, Yang-Feng TL, Hediger MA, Ganapathy V, Leibach FH, 1995: Human intestinal H(+)/peptide cotransporter: cloning, functional expression, and chromosomal localization. J Biol Chem 270: 6456–6463.
22. Ramamoorthy S, Liu W, Ma YY, Yang-Feng TL, Ganapathy V, Leibach FH, 1995: Proton/peptide cotransporter (PEPT 2) from human kidney: functional characterization and chromosomal. Biochim Biophys Acta Nov 22;1240(1): 1–4.
23. Daniel H, Kottra G, 2004: The proton oligopeptide cotransporter family SLC15 in physiology and pharmacology. Pflugers Arch Feb;447(5): 610–618.
24. Taylor TD, Noguchi H, Totoki Y, Toyoda A, Kuroki Y, Dewar K, Lloyd C, Itoh T, Takeda T, Kim DW, She X, Barlow KF, Bloom T, Bruford E, Chang JL, Cuomo CA, Eichler E, FitzGerald MG, Jaffe DB, LaButti K, Nicol R, Park HS, Seaman C, Sougnez C, Yang X, Zimmer AR, Zody MC, Birren BW, Nusbaum C, Fujiyama A, Hattori M, Rogers J, Lander ES, Sakaki Y, 2006: Human chromosome 11 DNA sequence and analysis including novel gene identification. Nature Mar 23;440(7083): 497–500.
25. Garcia CK, Li X, Luna J, Francke U, 1994: cDNA cloning of the human monocarboxylate transporter 1 and chromosomal localization of the SLC16A1 locus to 1p13.2–p12. Genomics 23: 500–503.
26. Lin RY, Vera JC, Chaganti RSK, Golde DW, 1998: Human monocarboxylate transporter 2 (MCT2) is a high affinity pyruvate transporter. J Biol Chem 273: 28959–28965.
27. Yoon H, Donoso LA, Philp NJ, 1999: Cloning of the human monocarboxylate transporter MCT3 gene: localization to chromosome 22q12.3–q13.2. Genomics Sep 15;60(3): 366–370.
28. Halestrap AP, Meredith D, 2004: The SLC16 gene family—from monocarboxylate transporters (MCTs) to aromatic amino acid transporters and beyond. Pflugers Arch Feb;447(5): 619–628.
29. Rodriguez AM, Perron B, Lacroix L, Caillou B, Leblanc G, Schlumberger M, Bidart JM, Pourcher T, 2002: Identification and characterization of a putative human iodide transporter located at the apical membrane of thyrocytes. J Clin Endocrinol Metab. 87:3500–3503.
30. Ritzel MWL, Yao SYM, Huang MY, Elliott JF, Cass CE, Young JD, 1997: Molecular cloning and functional expression of cDNAs encoding a human Na(+)-nucleoside cotransporter (hCNT1). Am J Physiol 272: C707–C714.
31. Wang J, Su SF, Dresser MJ, Schaner ME, Washington CB, 1997: Giacomini, K. M.: Na(+)-dependent purine nucleoside transporter from human kidney: cloning and functional characterization. Am J Physiol 273: F1058–F1065.

32. Ritzel MWL, Ng AML, Yao SYM, Graham K, Loewen SK, Smith KM, Ritzel RG, Mowles DA, Carpenter P, Chen XZ, et al., 2001: Molecular identification and characterization of novel human and mouse concentrative Na(+)-nucleoside cotransporter proteins (hCNT3 and mCNT3) broadly selective for purine and pyrimidine nucleosides (system cib). J Biol Chem 276: 2914–2927.

33. Coe IR, Griffiths M, Young JD, Baldwin SA, Cass CE, 1997: Assignment of the human equilibrative nucleoside transporter (hENT1) to 6p21.1–p21.2. Genomics 45: 459–460.

34. Williams JB, Rexer B, Sirripurapu S, John S, Goldstein R, Phillips JA III, Haley LL, Sait SNJ, Shows TB, Smith CM, Gerhard DS, 1997: The human HNP36 gene is localized to chromosome 11q13 and produces alternative transcripts that are not mutated in multiple endocrine neoplasia, type 1 (MEN I) syndrome. Genomics 42: 325–330.

35. Baldwin SA, Beal PR, Yao SY, King AE, Cass CE, Young JD, 2004: The equilibrative nucleoside transporter family, SLC29. Pflugers Arch Feb;447(5): 735–743.

36. Engel K, Zhou M, Wang J, 2004: Identification and characterization of a novel monoamine transporter in the human brain. J Biol Chem 279: 50042–50049.

37. Shiao T, Iwahashi M, Fortune J, Quattrochi L, Bowman S, Wick M, Qadri I, Simon FR, 2000: Structural and functional characterization of liver cell–specific activity of the human sodium/taurocholate cotransporter. Genomics 69: 203–213.

38. Wong MH, Rao PN, Pettenati MJ, Dawson PA, 1996: Localization of the ileal sodium–bile acid cotransporter gene (SLC10A2) to human chromosome 13q33. Genomics 33: 538–540.

39. Strautnieks SS, Bull LN, Knisely AS, Kocoshis SA, Dahl N, Arnell H, Sokal E, Dahan K, Childs S, Ling V, et al., 1998: A gene encoding a liver-specific ABC transporter is mutated in progressive familial intrahepatic cholestasis. Nat Genet Nov;20(3): 233–238.

40. Seward DJ, Koh AS, Boyer JL, Ballatori N, 2003: Functional complementation between a novel mammalian polygenic transport complex and an evolutionarily ancient organic solute transporter, OSTalpha-OSTbeta. J Biol Chem Jul 25;278(30): 27473–27482.

41. Trent JM, Witkowski CM,1987: Clarification of the chromosomal assignment of the human P-glycoprotein/mdr1 gene: possible coincidence with the cystic fibrosis and c-met oncogene. Cancer Genet Cytogenet May;26(1): 187–190.

42. Cole SPC, Bhardwaj G, Gerlach JH, Mackie JE, Grant CE, Almquist KC, Stewart AJ, Kurz EU, Duncan AMV, Deeley RG, 1992: Overexpression of a transporter gene in a multidrug-resistant human lung cancer cell line. Science 258: 1650–1654.

43. Taniguchi K, Wada M, Kohno K, Nakamura T, Kawabe T, Kawakami M, Kagotani K, Okumura K, Akiyama S, Kuwano M, 1996: A human canalicular multispecific organic anion transporter (cMOAT) gene is overexpressed in cisplatin-resistant human cancer cell lines with decreased drug accumulation. Cancer Res. 56: 4124–4129.

44. Uchiumi T, Hinoshita E, Haga S, Nakamura T, Tanaka T, Toh S, Furukawa M, Kawabe T, Wada M, Kagotani K, et al., 1998: Isolation of a novel human canalicular multispecific organic anion transporter, cMOAT2/MRP3, and its expression in cisplatin-resistant cancer cells with decreased ATP-dependent drug transport. Biochem Biophys Res Commun 252: 103–110.

45. Lee K, Belinsky MG, Bell DW, Testa JR, Kruh GD, 1998: Isolation of MOAT-B, a widely expressed multidrug resistance-associated protein/canalicular multispecific organic anion transporter–related transporter. Cancer Res 58: 2741–2747.

46. Suzuki T, Nishio K, Sasaki H, Kurokawa H, Saito-Ohara F, Ikeuchi T, Tanabe S, Terada M, Saijo N, 1997: cDNA cloning of a short type of multidrug resistance protein homologue,

SMRP, from a human lung cancer cell line. Biochem Biophys Res Commun 238: 790–794.

47. Kuss BJ, O'Neill GM, Eyre H, Doggett NA, Callen DF, Davey RA, 1998: ARA, a novel ABC transporter, is located at 16p13.1, is deleted in inv(16) leukemias, and is shown to be expressed in primitive hematopoietic precursors. Genomics 51: 455–458.
48. Allikmets R, Gerrard B, Hutchinson A, Dean M, 1996: Characterization of the human ABC superfamily: isolation and mapping of 21 new genes using the expressed sequence tags database. Hum Mol Genet 5: 1649–1655.
49. Tammur J, Prades C, Arnould I, Rzhetsky A, Hutchinson A, Adachi M, Schuetz JD, Swoboda KJ, Ptacek LJ, Rosier M, Dean M, Allikmets R, 2001: Two new genes from the human ATP-binding cassette transporter superfamily, ABCC11 and ABCC12, tandemly duplicated on chromosome 16q12. Gene 273: 89–96.
50. Allikmets R, Schriml LM, Hutchinson A, Romano-Spica V, Dean M, 1998: A human placenta-specific ATP-binding cassette gene (ABCP) on chromosome 4q22 that is involved in multidrug resistance. Cancer Res 58: 5337–5339.

2

ORGANIC CATION TRANSPORTERS

THOMAS J. URBAN AND KATHLEEN M. GIACOMINI
University of California San Francisco, San Francisco, California

Drug Transporters: Molecular Characterization and Role in Drug Disposition, Edited by Guofeng You and Marilyn E. Morris
Copyright © 2007 John Wiley & Sons, Inc.

2.1. INTRODUCTION

Renal elimination, including both passive glomerular filtration and active tubular secretion, is a major route of clearance for many drugs and drug metabolites. Active secretion of drugs by the kidney takes place primarily in the renal proximal tubule, where drug transporters facilitate the flux of drug molecules from the blood to the tubular lumen for excretion. As early as 1947, it was recognized that transport systems exist that aid in the elimination of a variety of structurally diverse organic cations, including drugs and toxic xenobiotics, as well as their metabolites.[1] Prior to the molecular cloning of organic cation transporters, it had been shown that uptake of low-molecular-weight organic cations into the renal proximal tubule (the first step in active secretion of organic cations) is facilitated by a polyspecific membrane potential–sensitive transport system. Efflux of organic cations into the tubular lumen, the second step in active tubular secretion, was shown to occur via an H^+- or cation-exchange mechanism. Additionally, uptake of organic cations into other tissues, such as hepatic uptake, or transport across the blood–brain barrier, was known to display properties of protein-mediated transport across the plasma membrane (e.g., saturability, and sensitivity to temperature, membrane potential, and small-molecule inhibitors).

The first member of the organic cation transporter (OCT) family [rat OCT1 (rOCT1)] was identified in rat by expression cloning from rat kidney.[2] This was followed quickly by homology cloning of other OCT isoforms in rodents and human.[3–13] Of the membrane transporters cloned to date, three have been identified as members of the basolateral (membrane potential–driven) organic cation transporter system: OCT1 (*SLC22A1*), OCT2 (*SLC22A2*), and OCT3 (*SLC22A3*). All are members of the *SLC22A* family of solute carrier (SLC) transporters, which includes the novel organic cation transporters OCTN1 and OCTN2 (*SLC22A4* and *SLC22A5*, discussed in detail in Chapter 3) and the organic anion transporters OAT1–4 (*SLC22A6*, *SLC22A7*, *SLC22A8*, and *SLC22A11*, discussed in detail in Chapter 4), as well as several transporters that are less well characterized. The molecular identification of the organic cation transporters (OCTs) has led to more detailed studies of the transport mechanism, substrate specificity, and tissue distribution of these transporters, and to a greater understanding of their roles in organic cation disposition and elimination.

2.2. STRUCTURAL–FUNCTIONAL CHARACTERISTICS AND TRANSPORT MECHANISM

In terms of structural and functional characteristics, all three of the OCTs share several common features: (1) a similar transmembrane topology, (2) a shared group of preferred substrates, and (3) a common transport mechanism.[14–20] The OCTs also share a common genomic structure of 11 coding exons. The genes encoding the three OCTs are located in close proximity on human chromosome 6q26, and presumably arose via gene duplication from a single ancestral OCT gene.

The predicted secondary structure of the OCTs, based on sequence or hydropathy analysis, consists of 12 transmembrane domains (TMDs) with cytoplasmic amino

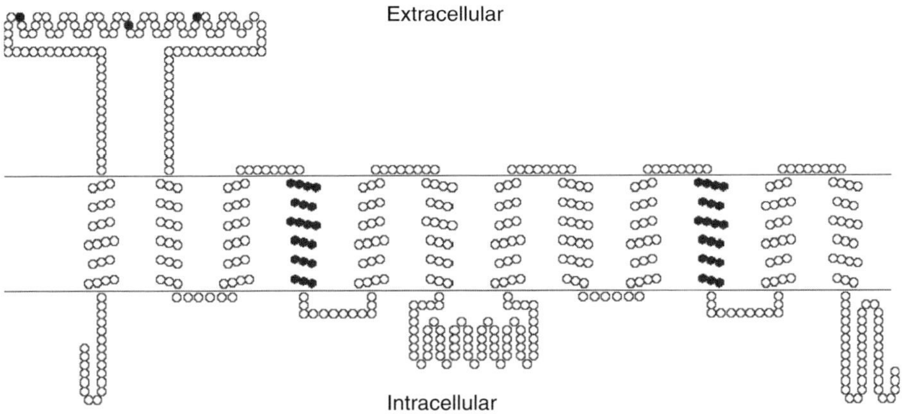

FIGURE 2.1. Secondary structure predicted for OCT1. OCT1 shares a similar transmembrane topology to the other OCTs, consisting of 12 transmembrane domains (TMDs), and a large extracellular loop between TMD1 and TMD2 which contains several putative N-glycosylation sites. TMD4 and TMD10 have been shown to contribute to substrate recognition by OCT1.

and carboxy termini. The OCTs share a large extracellular loop between TMDs 1 and 2 which contains potential N-glycosylation sites, as well as a large intracellular loop between TMDs 6 and 7 which includes predicted phosphorylation sites. The potential importance of N-glycosylation and phosphorylation for OCT activity is discussed below. Figure 2.1 shows the transmembrane topology of OCT1, which is typical of the OCTs. The transmembrane domains are believed to be important for substrate recognition by the OCTs; specifically, recent evidence suggests that the fourth and tenth transmembrane domains are critically involved in substrate recognition by the OCTs,[21,22] and differences between isoforms in terms of substrate specificity (see Section 2.3) may be related to differences in these critical regions. However, it is important to note that given the broad substrate selectivity of the OCTs, the key domains or residues involved in substrate recognition may be different from one substrate to another, even within the same protein. For example, in mutational analysis studies of rat OCT1, mutation of two residues in the fourth TMD (Trp218Tyr and Tyr222Leu) resulted in increased affinity for both tetraethylammonium (TEA) and 1-methyl-4-phenylpyridinium (MPP^+), whereas a third mutant (Thr226Ala) had increased affinity for MPP^+ but no change in affinity for TEA.[21] This suggests that OCT1, and probably all of the OCTs, contain multiple overlapping but nonidentical recognition sites for the various structurally diverse substrates.

Transport of organic cations by the OCTs occurs by facilitated diffusion and is driven by the inside-negative membrane potential.[14–20] Positively charged cations are taken up into cells according to the electrochemical gradient (see Figure 2.2), and this process is membrane potential sensitive [i.e., artificially reducing the membrane potential (as through replacement of extracellular Na^+ with K^+, or treatment with ionophores such as valinomycin) reduces the rate of transport by OCTs].[2,9]

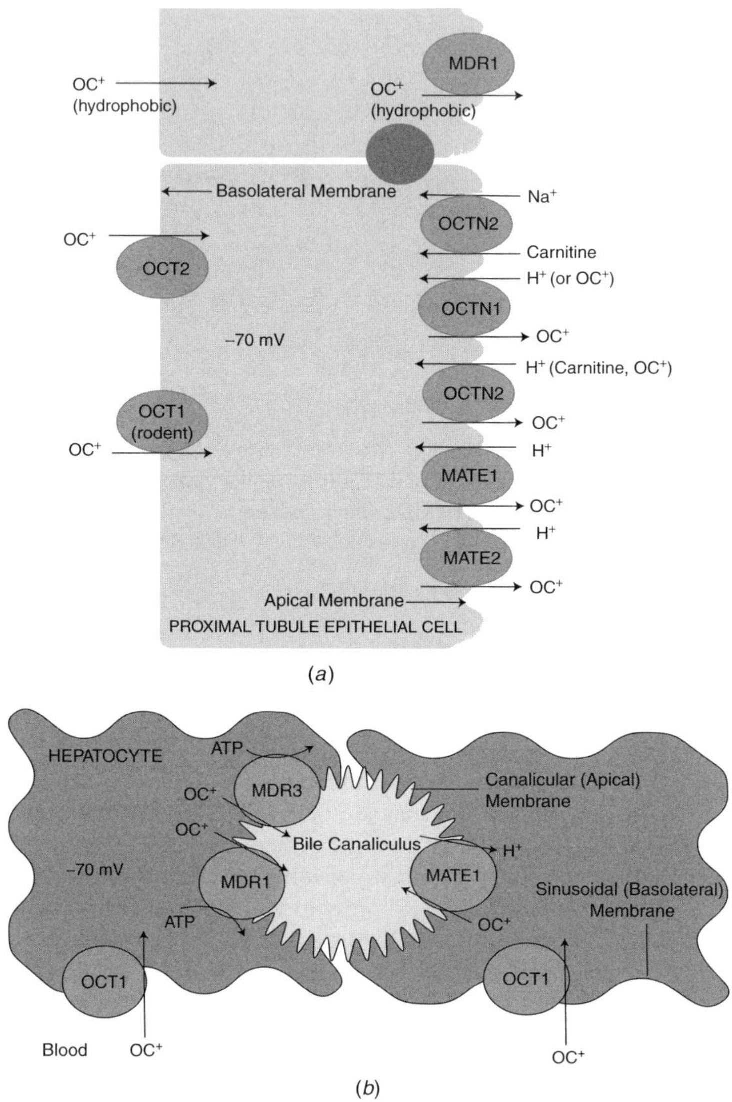

FIGURE 2.2. Model of organic cation transport in excretory organs. (*a*) Kidney: In humans, OCT2 is the primary transporter responsible for uptake of hydrophilic organic cations (OC^+) across the basolateral membrane, in a process driven by the inside-negative membrane potential. In rodents, OCT1 and OCT2 are expressed to a similar degree in kidney. Efflux of organic cations across the luminal (apical) membrane for excretion occurs by an organic cation-H^+ exchange mechanism, and several transporters are suspected to be involved in apical OC^+ transport. Membrane-permeable organic cations may not require facilitative transport across the basolateral membrane, and these hydrophobic cations may be substrates for the apical efflux pump MDR1. (*b*) Liver: OCT1 is by far the most abundant organic cation transporter in the liver, where it serves as a basolateral, potential-driven uptake transporter. After entry into the hepatocyte, cationic drugs may be metabolized or excreted into bile via the primary active efflux transporters MDR1 and MDR3, or via the organic cation-H^+ exchanger MATE1.

Electrophysiological studies using *Xenopus laevis* oocytes microinjected with OCT1, OCT2, or OCT3 cRNA show directly that transport of organic cations by OCTs is electrogenic (i.e., addition of substrate to the extracellular media results in inward currents).[4,10,23–29] OCT activity is insensitive to changes in the Na^+ gradient.[17] Transport of weak bases by OCTs has also been shown to be sensitive to extracellular pH; however, this is explained by the decrease in fractional ionization of the substrate at high pH and does not indicate that transport of organic cations is driven by proton exchange.[30] OCT transport can also occur by cation–cation exchange. Preloading OCT1-injected *X. laevis* oocytes with unlabeled TEA enhances uptake of [^{3}H]MPP^+,[26] and similar results have been found for OCT2.[31]

2.3. SUBSTRATE SELECTIVITY

Common substrates of all OCTs include low-molecular-weight relatively hydrophilic organic cations such as the prototypical cation TEA, the neurotoxin MPP^+, and the endogenous compound *N*-methylnicotinamide (NMN).[18,20,32] Several clinically important drugs have been shown to interact with all of the OCTs, including the antidiabetic drug metformin,[18,33] and the peptic ulcer drug famotidine,[34] demonstrating the broad potential for influence of OCTs on drug disposition and drug action. Figure 2.3 shows the structures of several compounds that interact with OCTs. Table 2.1 lists known substrates and inhibitors of human OCTs.[18,20,30,32–38]

Included among OCT substrates are endogenous compounds such as the biogenic amine neurotransmitters. Dopamine, epinephrine, norepinephrine, histamine, and serotonin have all been shown to interact with one or more OCT isoforms.[20,32] Of particular interest in this regard is OCT3, which was cloned as the extraneuronal monoamine transporter (EMT) and is thought to comprise the uptake-2 system of catecholamine transport, which aids in clearance of catecholamines in extraneuronal tissues.[39] Despite the particular preference of OCT3 for biogenic amines, it is suggested that all three of the cloned OCT transporters combine to contribute to extraneuronal clearance of amine neurotransmitters.[32,40]

Although the OCT family shows broad overlap in substrate specificity, there are examples of relatively isoform-specific substrates and inhibitors. Examples include corticosterone, which has approximately 40-fold greater affinity than Oct1 for rat Oct2, and conversely, mepiperphenidol and *O*-methylisoprenaline, which show a 70-fold greater affinity for rOct1 than for rOct2.[23] OCT3 appears to have the most unique substrate selectivity among the OCT family, with a preference for endogenous monoamines such as dopamine and norepinephrine.

Notably, although most known substrates of the OCTs are cations, some OCT substrates are anionic or neutral compounds at physiological pH. For example, the anions prostaglandin E_2 and prostaglandin $F_{2\alpha}$ have been shown to be substrates for both OCT1 and OCT2,[41] and the neutral steroid β-estradiol interacts with OCT3 with high affinity.[39] Thus, a net positive charge does not appear to be an absolute requirement for interaction with the OCTs.

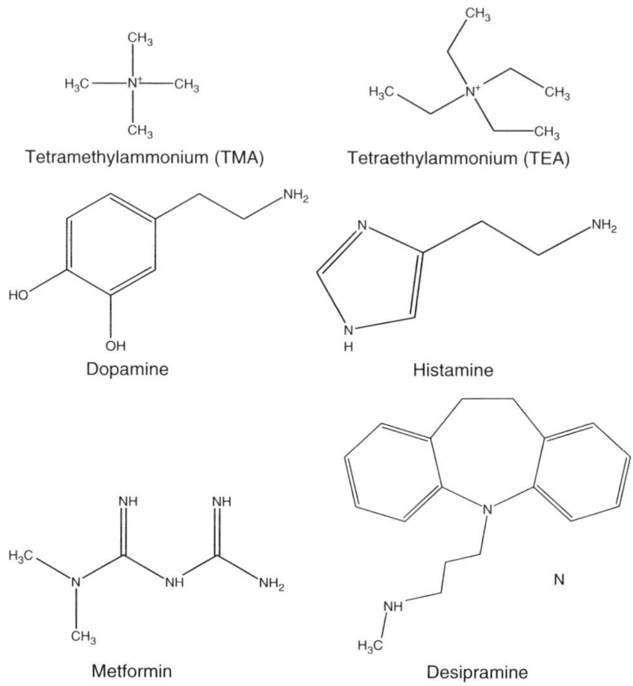

FIGURE 2.3. Representative substrates of organic cation transporters. OCTs are capable of transporting a variety of structurally diverse substrates, including tetraalkylammonium compounds (TMA, TEA), endogenous compounds (dopamine, histamine) and clinically important drugs (desipramine, metformin).

Importantly, OCT orthologs from different species may exhibit differences in substrate specificity. For example, OCT1 orthologs from human, rabbit, mouse, and rat all transport the low-molecular-weight tetraalkylammonium compounds tetramethylammonium (TMA) and TEA. In contrast to human and rabbit, rat and mouse Oct1 do not transport the larger tetrapropylammonium (TPA) or tetrabutylammonium (TBA),[26] suggesting that molecular mass or hydrophobicity may effect differences in recognition of OCT substrates across species.

2.4. TISSUE DISTRIBUTION

Despite the similarities in structure and function, there are important differences in the tissue distribution of mRNA transcripts of the OCTs in human tissues. OCT1 is expressed predominantly in human liver, with lower levels of expression in other tissues.[4,9] OCT1 is thus characterized as a *liver-specific* OCT. Analogously, OCT2, which is expressed almost exclusively in kidney, is described as the *kidney-specific OCT*.[4,42] OCT3, also referred to as the *extraneuronal monoamine transporter* (EMT), has a broad tissue distribution, with moderate levels of mRNA found in the aorta,

TABLE 2.1. Substrates and Inhibitors of Human OCT Isoforms[a]

	OCT1		OCT2		OCT3		
Compound	Inhibitor	Substrate	Inhibitor	Substrate	Inhibitor	Substrate	Refs.
Acebutolol	+	N.D.	N.D.	N.D.	N.D.	N.D.	18,20
Acyclovir	+	+	N.D.	N.D.	N.D.	N.D.	18
Agmatine	+	+	+	+	+	+	32
Amantadine	+	N.D.	+	+	N.D.	N.D.	20
Aquinavir	+	N.D.	N.D.	N.D.	N.D.	N.D.	18
Choline	+	N.D.	+	+	N.D.	–	20,32
Cimetidine	+	N.D.	+	+	+	+	18,20,32
Clonidine	+	–	N.D.	N.D.	+	–	18,20
Cocaine	N.D.	N.D.	+	N.D.	N.D.	N.D.	18
Corticosterone	+	N.D.	+	N.D.	+	N.D.	20,23
Creatinine	+	N.D.	N.D.	N.D.	N.D.	–	20,32,36
Cyanine-863	N.D.	N.D.	+	N.D.	N.D.	N.D.	20
Debrisoquine	N.D.	N.D.	+	+	N.D.	N.D.	18
Decynium-22	+	N.D.	+	N.D.	+	N.D.	20,37
Desipramine	+	N.D.	+	N.D.	+	N.D.	18,20
Disopyramide	+	N.D.	N.D.	N.D.	N.D.	N.D.	18,20
Disprocynium-24	N.D.	N.D.	N.D.	N.D.	+	N.D.	20
Dopamine	+	N.D.	+	+	+	+	20,32,35
Epinephrine	N.D.	N.D.	N.D.	N.D.	+	+	32
β-Estradiol	+	N.D.	+	N.D.	+	N.D.	37
Famotidine	+	+	+	–	+	–	34
Ganciclovir	+	+	N.D.	N.D.	N.D.	N.D.	38
Guanidine	N.D.	N.D.	N.D.	N.D.	+	–	36
Histamine	+	–	+	+	+	+	20
Indinavir	+	N.D.	N.D.	N.D.	N.D.	N.D.	18,20
Memantine	N.D.	N.D.	+	+	N.D.	N.D.	18,20
Mepiperphenidol	N.D.	N.D.	+	N.D.	N.D.	N.D.	18,20

(*Continued*)

TABLE 2.1. (*Continued*)

Compound	OCT1		OCT2		OCT3		Refs.
	Inhibitor	Substrate	Inhibitor	Substrate	Inhibitor	Substrate	
Metformin	+	+	+	+	N.D.	N.D.	33
O-Methylisoprenaline	+	N.D.	+	N.D.	+	N.D.	18
Midazolam	+	N.D.	N.D.	N.D.	N.D.	N.D.	18,20
MPP^+	+	+	+	+	+	+	20,32
Nelfinavir	+	N.D.	N.D.	N.D.	N.D.	N.D.	18,20
NMN	+	+	+	+	+	N.D.	20,32
Norepinephrine	N.D.	N.D.	+	+	+	+	20,32
Phenformin	+	N.D.	+	N.D.	N.D.	N.D.	18
Phenoxybenzamine	+	N.D.	+	N.D.	+	N.D.	18
Prazosin	+	N.D.	+	N.D.	+	N.D.	18
Procainamide	+	N.D.	+	N.D.	+	N.D.	18,20
Progesterone	+	N.D.	+	N.D.	+	N.D.	37
Quinidine	+	N.D.	+	N.D.	N.D.	N.D.	20,30
Quinine	+	N.D.	+	+	N.D.	N.D.	20,30
Ranitidine	+	+	+	+	+	−	34
Ritonavir	+	N.D.	N.D.	N.D.	N.D.	N.D.	18,20
Saquinavir	+	N.D.	N.D.	N.D.	N.D.	N.D.	20
Serotonin	N.D.	N.D.	+	+	+	+	20,32
Tetrabutylammonium	+	+	+	N.D.	N.D.	N.D.	20,69
Tetraethylammonium	+	+	+	+	+	+	20,32
Tetraheptylammonium	+	N.D.	N.D.	N.D.	N.D.	N.D.	20
Tetramethylammonium	+	+	+	N.D.	N.D.	N.D.	20
Tetrapropylammonium	+	+	+	N.D.	N.D.	N.D.	20,69
Tyramine	N.D.	N.D.	N.D.	N.D.	+	+	20,32
Vecuronium	+	N.D.	N.D.	N.D.	N.D.	N.D.	18,20
Verapamil	+	N.D.	+	N.D.	N.D.	N.D.	18,20

[a] Includes compounds that have been tested as either substrates or inhibitors of any OCT isoform. Compounds shown to be substrates or inhibitors are designated as "+," and those shown positively not to be substrates or inhibitors are designated as "−." Compounds that have not been tested as either substrate or inhibitor are listed as "N.D." (not determined) for that isoform.

TABLE 2.2. Tissue Distribution of Human OCT Isoforms

	OCT1	OCT2	OCT3
Liver	+++	−	+
Kidney	−	+++	+
Lung	+	−	+
Trachea	−	−	−
Heart	+	−	++
Skeletal Muscle	++	−	++
Placenta	+	−	++
Pancreas	−	−	−
Brain	−	−	+
Spinal cord	−	−	+
Adrenal gland	+	−	−
Testis	−	−	−
Ovary	+	−	+
Fetal liver	+	−	−
Fetal lung	+	−	−
Fetal brain	+	−	−

[a]The expression level is estimated relative to other tissues for each isoform, based on Northern blot analysis.[4,9,12,42]

skeletal muscle, prostate, adrenal gland, salivary gland, liver, term placenta, and fetal lung.[12] See Table 2.2 for details on tissue distribution of each OCT isoform.

Both OCT1 and OCT2 have been localized to the basolateral membrane in renal proximal tubular epithelial cells in rat[43–45] as well as in cell culture models of kidney (GFP-rOCT2-transfected MDCK cells).[46] In human proximal tubules, OCT2 expression is restricted to the basolateral membrane; however, OCT1 is expressed at very low levels in human kidney.[47] Despite early controversy regarding OCT2 localization,[4,5,48] the emerging consensus is that OCT2 is indeed a basolateral organic cation uptake transporter in kidney.[15] The pharmacological significance of OCT2 thus appears primarily to be its role in active secretion of organic cations in the kidney, facilitating the first step (uptake of organic cations across the basolateral membrane) of active secretion. Efflux of organic cations across the apical membrane is thought to occur by a separate transporter-mediated process, either by the primary active efflux pump, MDR1 (which generally transports bulky hydrophobic cationic compounds) or by an organic cation-H^+ exchange mechanism for small, hydrophilic organic cations (see Figure 2.2).[14,15,20,49] The molecular identity of the organic cation-H^+ exchanger is currently unclear; the novel organic cation transporters (OCTN1 and OCTN2) and the more recently discovered multidrug and toxin exclusion transporters (MATE1 and MATE2) are likely candidates, as all of these are highly expressed in kidney and have been shown to transport the typical cation TEA in a manner consistent with H^+ exchange.[14,15,20,49–51] It is possible that all of these transporters participate in apical organic cation efflux, with differential importance depending on the substrate under question. Among the apical organic cation transporters, only OCTN2 has been shown to influence renal organic cation (TEA) secretion in vivo[52]; however, appropriate animal models to study the in vivo role of OCTN1, MATE1, and MATE2 in renal organic cation secretion do not yet exist.

In liver, the primary site of OCT1 expression, OCT1 was found to be expressed at the sinusoidal (basolateral) membrane by immunohistochemistry and Western blot.[53] This suggests that OCT1 acts as a sinusoidal uptake transporter in liver, aiding in the transfer of organic cations from blood into the hepatocyte for elimination by metabolism or biliary excretion. In addition, for drugs that target the liver for their pharmacological activity, OCT1 may be a limiting step in drug access to hepatocytes and may therefore influence drug action.

It is important to note that species differences exist in the tissue-specific expression of OCTs. Most notably, whereas in human kidney OCT2 is by far the most abundant OCT in terms of mRNA expression, rodent kidney expresses nearly similar levels of OCT1 and OCT2. As with substrate-specificity differences, differences between species in terms of tissue distribution of OCTs are important to consider when interpreting the results of studies performed in animal models.

2.5. REGULATION OF OCT EXPRESSION AND ACTIVITY

As mentioned previously, the large intracellular loop between transmembrane domains 6 and 7 contains several predicted phosphorylation sites that are conserved among the several OCT isoforms. These include target sequences for protein kinase A (PKA), protein kinase C (PKC), protein kinase G (PKG), and tyrosine kinase. There exists ample evidence for the modulation of activity of the OCTs by activation of these kinases, presumably through phosphorylation of the OCTs at the protein level. Several N-glycosylation sites in the large extracellular loop between transmembrane domains 1 and 2 are also potential sites of protein-level regulation. Additionally, hormonal regulation, which is thought to reflect effects on transcription, has been described for OCT2. Regulation of OCT expression is also modulated during development and in certain pathophysiological states. For an extensive review of the regulation of organic cation transport, see ref. 54. Current knowledge of the regulation of the OCTs will be discussed individually for each isoform.

2.5.1. OCT1

Early studies of rat OCT1 showed that stimulation of either PKA or PKC results in a significant increase in organic cation transport, as measured by uptake of the fluorescent compound 4-[4-(dimethylamino)-styryl]-*N*-methylpyridinium (ASP^+) in HEK-293 cells stably transfected with rOCT1[55] PKC stimulation also results in increased affinity for TEA, TPA, and quinine. The effect of PKC stimulation on rOCT1 activity is thought to occur by phosphorylation of the rOCT1 protein, resulting in a conformational change that enhances substrate binding. rOCT1 activity is significantly reduced by treatment with aminogenistein, a specific inhibitor of $p56^{lck}$ tyrosine kinase, suggesting that this kinase is an endogenous activator of OCT1; however, inhibitors of other tyrosine kinases do not affect OCT1 function.

In contrast to rOCT1, stimulation of PKA results in decreased activity of human OCT1 in two different expression systems.[56] In the same studies, PKC activation has no appreciable effect on activity of human OCT1. However, as with rOCT1, human OCT1 appears to be positively regulated by the $p56^{lck}$ tyrosine kinase, as evidenced by reduced hOCT1 activity after treatment with aminogenistein. For both rOCT1 and hOCT1, PKG signaling does not appear to influence OCT1 activity. Human OCT1 has further been shown to be regulated by the Ca^{2+}–calmodulin complex. Specifically, inhibition of Ca^{2+}- and calmodulin-dependent protein kinase II (CaMKII) results in reduced ASP^+ transport by hOCT1, suggesting that this kinase consitutively stimulates hOCT1 activity. In both PKA stimulation and CaMKII inhibition, the reduction in transport activity by hOCT1 is the result of decreased affinity for the substrates tested. These results suggest, again, that the observed effects of protein kinases are due to phosphorylation of the OCT1 protein, causing a conformational change that results in either decreased or increased substrate-binding affinity.

2.5.2. OCT2

Studies on regulation of basolateral organic cation transport have been performed in isolated human proximal tubules, in which OCT2 is known to be the major OCT isoform based on mRNA expression and relative affinity for various substrates. These studies have shown that activation of PKA, PKC, or PKG leads to reduced uptake of the fluorescence organic cation 4-(4-(dimethylamino)styryl)-*N*-methylpyridinium (ASP^+) into renal proximal tubules, suggesting an inhibitory effect of these kinases on OCT2 activity.[57] Inhibition of PKC by calphostin C reverses the inhibition of OCT2 activity, demonstrating the specificity of the PKC-mediated pathway. However, when heterologously expressed in HEK293 cells, hOCT2 was shown to be inhibited only by PKA activation, not by PKC.[58] These conflicting results may be explained by differences between these experimental systems in terms of the expression of genes involved in the PKC pathway.[54]

Further studies using heterologously expressed OCT2 have identified a role for G protein–coupled receptor (GPCR) signaling in regulation of OCT2. The muscarinic receptor agonist carbachol causes a dose-dependent decrease in OCT2 activity (i.e., uptake of ASP^+) that is not explained by direct inhibition of the OCT2 protein by carbachol.[58] Similarly, activation of the phospholipase C (PLC) pathway through purinergic receptor agonism by ATP leads to dose-dependent decreases in OCT2 activity. As both of these pathways result in increased intracellular Ca^{2+} concentrations, it has been hypothesized that the mechanism for OCT2 regulation by GPCRs involves the Ca^{2+}–calmodulin complex. Reduction of intracellular Ca^{2+} concentration or treatment with a Ca^{2+} chelator results similarly in decreased OCT2 activity, and inhibition of calmodulin and its downstream targets CaMKII and myosin light-chain kinase (MLCK) produces a similar effect. Although the effect of muscarinic receptor stimulation on OCT2 activity is partially explained by its effects on intracellular Ca^{2+}, additional studies have shown that the majority of this effect is due to activation of phospatidylinositol 3-kinase (PI3K), as inhibition of PI3K almost completely prevents the inhibitory effect of carbachol on OCT2-mediated transport. These studies suggest that PKA, PI3K, and the Ca^{2+}–calmodulin complex are endogenous regulators of OCT2; however, in contrast to the effect of protein kinases on OCT1, the inhibition of transport activity is not explained by effects on substrate-binding affinity. It is therefore thought that these proteins regulate other aspects of OCT2, such as maturation and trafficking of OCT2 to the plasma membrane, or stimulation of endocytosis.[54]

Glycosylation of OCT2 may be an additional mechanism for regulation of organic cation transport. The rabbit OCT2 isoform was shown to contain three N-glycosylation sites (Asn71, Asn96, and Asn112), each of which was found to be necessary for optimal transport activity by mutational analysis.[59] The Asn112Gln mutant resulted in a significant decrease in membrane expression (to approximately 20% of wild-type) in stably transfected Chinese hamster ovary (CHO) cells. Asn71Gln and Asn96Gln did not alter OCT2 surface expression, but the unglycosylated triple mutant (Asn71Gln/Asn96Gln/Asn112Gln) was retained intracellularly. Affinity for TEA was increased approximately twofold by mutation of each of the three glycosylation sites. Maximal transport rate was decreased fivefold by the Asn112Gln mutation (explained by the reduced surface expression of Asn112Gln) and threefold by Asn96Gln (presumably, due to a lower transporter turnover number). Thus, N-glycosylation may regulate maturation and trafficking of OCT2 as well as substrate recognition and transporter turnover.

There is also evidence for hormonal regulation of OCT2. Functional studies of rat renal cortical slices showed higher basolateral transport of TEA in slices from male rats versus female rats, and rOCT2 mRNA and protein expression was also shown to be higher in male rats than in female rats.[60] Further, treatment of rats with testosterone increased rOCT2 and protein expression in the kidney of both male and female rats.[61] In Madin–Darby canine kidney (MDCK) cells, treatment with dexamethasone, hydrocortisone, or testosterone increased both OCT2 mRNA and protein levels as well as uptake of TEA across the basolateral membrane.[62] Conversely, treatment with estradiol led to decreased OCT2 expression and activity in male rats and in MDCK cells. The physiological significance of hormonal regulation of OCT2 is unclear but may be related to gender-specific differences in clearance of endogenous metabolites.[32]

2.5.3. OCT3

Although OCT3 contains several conserved target sequences for PKA, PKC, and PKG, studies of heterologously expressed OCT3 have shown that activation of these pathways has no apparent effect on OCT3 activity.[63] However, inhibition of the p44/42 mitogen-activated protein kinase (MAPK) has been shown to decrease MPP^+ transport by OCT3, suggesting that MAPK is a positive regulator of OCT3. Similarly, inhibition of calmodulin as well as of the downstream effector CaMKII results in decreased OCT3 activity. Selective inhibition of the calmodulin-dependent phosphodiesterase PDE1 also results in decreased OCT3 activity, suggesting PDE1 stimulation as an additional mechanism for Ca^{2+}- and calmodulin-dependent regulation of OCT3.

2.6. ANIMAL MODELS

Targeted genetic disruption of the organic cation transporters in mice (i.e., OCT knockout mice) provided the first direct evidence of the significance of organic cation transporters to drug disposition in vivo.

2.6.1. Oct1 Knockout Mice

Oct1 −/− knockout mice were shown to be viable and fertile, showing no obvious physiological abnormalities compared with their wild-type littermates, suggesting that Oct1 is dispensible for normal physiology.[64] However, significant differences were observed in the disposition of organic cations in these mice. For example, when administered the typical organic cation, TEA, *Oct1 −/−* mice showed significantly (approximately fivefold) reduced uptake of TEA into the liver, the site of highest OCT1 expression. Biliary excretion of TEA was 2.5-fold lower in *Oct1 −/−* mice, which was explained by the reduced hepatic uptake of TEA. Additionally, direct intestinal excretion of TEA was reduced by approximately 50%. Renal excretion of TEA was paradoxically increased (~50%) in *Oct1 −/−* mice; however, this was explained by the lack of OCT1 in the liver of *Oct1 −/−* mice; that is, decreased hepatic uptake of TEA in these mice resulted in increased availability to the kidney for renal excretion. Further, the coexpression of Oct1 and Oct2 in rodent kidney may have led to a confounding effect of Oct2 activity in *Oct1 −/−* mouse kidney.

In addition to the pharmacokinetic differences observed for TEA, *Oct1 −/−* mice showed similar decreases in hepatic uptake of other OCT1 substrates, including the neurotoxin MPP^+ (~60% reduced) and *meta*-iodobenzylguanidine (MIBG) (~75% reduced).[64] Cimetidine, an OCT1 inhibitor (but not a substrate), did not show significant differences in hepatic uptake in *Oct1 −/−* mice compared with wild-type mice. Similar results were found after intravenous injection of [^{14}C]choline, which is an Oct1 substrate; however, it was suggested that rapid metabolism of the radiolabeled compound to species that are not Oct1 substrates, as well as redundancy (i.e., other transporters capable of transporting choline), may have masked any possible effect of Oct1 on the pharmacokinetics of this compound.

Further studies of *Oct1 −/−* mice have focused on the antidiabetic drug metformin, which exerts its pharmacological effects in the liver. *Oct1 −/−* mice showed a greater than 30-fold decrease in metformin uptake into liver compared with wild-type littermates.[65] Further studies investigated the role of Oct1 in the development of metformin-induced lactic acidosis, a leading toxicity from this drug. A significant increase in serum lactic acid concentration was observed after administration of metformin to wild-type mice, but only slight elevations in serum lactate were seen in *Oct1 −/−* mice, despite similar pharmacokinetic profiles between genotype groups.[66] Taken together, these results suggest that Oct1-mediated metformin transport is a limiting step in metformin uptake into liver, and that the lactic acidosis induced by metformin is related to the availability of the drug to this target organ.

2.6.2. Oct2 Knockout Mice

Similar to *Oct1 –/–* mice, *Oct2 –/–* mice were viable and fertile, showing no apparent physiological defects.[67] However, unlike the *Oct1 –/–* mice, pharmacokinetics of intravenous [^{14}C]TEA were quite similar in *Oct2 –/–* mice and wild-type mice, with the exception of a slight decrease in distribution of TEA into brain in Oct2-deficient mice. Renal clearance of TEA in *Oct2 –/–* mice was not different from that in wild-type mice, with active secretion being the primary mechanism of clearance. The explanation for these findings is that in mouse kidney, Oct1 expression is sufficient to maintain the capacity for active secretion of organic cations even in the absence of Oct2. Thus, to develop a mouse model of renal disposition of organic cations, it was necessary to generate an *Oct1/2 –/–* double-knockout mouse.

2.6.3. Oct1/Oct2 Knockout Mice

Generation of *Oct1/2 –/–* double-knockout mice revealed that the absence of both genes, as with each of the single knockouts, was compatible with normal physiology (i.e., normal viability, fertility, and life span were observed, with no apparent physiological abnormalities).[67] However, unlike the single knockouts, *Oct1/2 –/–* mice show significant impairment in the active tubular secretion of organic cations in the kidney. Specifically, renal tubular secretion of TEA was effectively abolished in *Oct1/2 –/–* mice, with renal clearance approximating glomerular filtration. This resulted in significantly elevated plasma levels of TEA in these mice compared with wild-type or *Oct1 –/–* single-knockout mice. After steady-state infusion of TEA, plasma levels were elevated approximately sixfold in the *Oct1/2 –/–* double-knockout mice compared to *Oct1 –/–*, *Oct2 –/–*, or wild-type mice.

2.6.4. Oct3 Knockout Mice

As mentioned earlier, OCT3 is believed to be primarily responsible for the transport of endogenous monoamines in extraneuronal tissues, also known as the *uptake-2 transport system*. Zwart et al. generated an Oct3-deficient mouse in order to determine the importance of this transporter in the peripheral disposition of monoamine neurotransmitters.[68] Although the *Oct3 –/–* mice were viable and fertile and showed no deficiency in dopamine or norepinephrine metabolism, they did show impaired uptake-2 activity as evidenced by a $>70\%$ reduced uptake of the synthetic monoamine MPP^+ into heart in *Oct3 –/–* mice compared to wild-type mice. Additionally, it was shown that in *Oct3 +/–* heterozygous females impregnated in a heterozygous cross, intravenous injection of MPP^+ resulted in a threefold reduction of MPP^+ uptake in *Oct3 –/–* embryos versus wild-type embryos, whereas the accumulation of MPP^+ in the placenta was not different between groups. These results suggest that OCT3 acts as an uptake transporter for monoamines at the fetoplacental interface. Apart from the distribution of MPP^+ into heart and placental transfer of MPP^+, no significant phenotypic differences have been observed in *Oct3 –/–* mice.

2.7. GENETIC VARIATION

Much attention has recently been focused on the pharmacogenetics of drug transporters (i.e., the effect of natural human genetic variation in drug transporter genes on drug disposition and drug response). Research in this area has taken primarily a sequence-based approach, beginning with identification of genetic variants likely to produce a functional effect (e.g., amino acid substitutions, or insertions or deletions in the coding region), followed in some cases by functional studies of the variant proteins in heterologous expression systems.

2.7.1. OCT1

Several functionally significant polymorphisms in OCT1 have been described. In a sample of 57 healthy Caucasians, 25 genetic variants were discovered, of which eight resulted in a change in protein sequence.[69] Of these, five (Arg61Cys, Cys88Arg, Phe160Leu, Gly401Ser, and Met420del) were tested for function by measurement of transport activity in *X. laevis* oocytes. Cys88Arg and Gly401Ser showed virtually complete loss of activity toward MPP^+, while Arg61Cys retained approximately 30% of wild-type activity. In contrast to MPP^+, uptake of [^{3}H]serotonin by Cys88Arg and Gly401Ser was detectable (~10% of wild-type), suggesting that these variants influence the substrate selectivity of OCT1.

Shu et al. discovered 15 amino acid sequence–altering variants of OCT1 that were identified in a large sample of ethnically diverse healthy subjects, and tested these for function by uptake of MPP^+ in *X. laevis* oocytes.[70] Variants with significantly reduced or complete loss of function included the Arg61Cys and Gly401Ser described by Kerb et al.[69], as well as Pro341Leu, Gly220Val, and Gly465Arg. An additional variant, Ser14Phe, showed a significant increase in activity; interestingly, this variant corresponds to the likely ancestral allele, as the consensus mammalian OCT1 sequence includes phenylalanine at this position. It was noted that all of the variants with reduced function occurred at evolutionarily conserved amino acid residues, and that variants with reduced function tended to be amino acid substitutions that result in a large chemical change. Of the variants with significant functional differences from the reference OCT1, five (Ser14Phe, Arg61Cys, Pro341Leu, Gly401Ser, and Gly465Arg) occurred at >1% allele frequency in at least one ethnic group and are attractive candidates for association with drug response phenotypes for OCT1 substrate drugs.

Functional studies of several additional OCT1 variants have been performed by two independent groups using microinjected *X. laevis* oocytes[71] or transiently transfected HEK293 cells.[72] Three amino acid sequence variants, Pro283Leu, Arg287Gly, and the previously described Pro341Leu, showed significantly reduced transport activity, with Pro283Leu and Arg287Gly having no appreciable activity toward the model substrates TEA and MPP^+. The reduction in activity of these variants was not explained by reduction in protein expression, as immunofluorescence studies showed all three variants to have membrane expression levels similar to that of the wild-type OCT1.[72]

2.7.2. OCT2

Human genetic variants of OCT2 have been investigated comprehensively by resequencing of the coding region in a large ethnically diverse sample.[35] Eight of the variants identified in this screen result in amino acid substitutions, and one single-nucleotide insertion (134-135insA) leads to a premature stop codon at amino acid position 48. Four of the nonsynonymous variants (Met165Ile, Ala270Ser, Arg400Cys, and Lys432Gln) were polymorphic, with ethnic-specific allele frequencies $\geq 1\%$. The remaining four nonsynonymous variants (Pro54Ser, Phe161Leu, Met165Val, and Ala297Gly) as well as the insertion variant were found on only one of 494 chromosomes screened.

Of the four polymorphic OCT2 protein sequence variants (Met165Ile, Ala270Ser, Arg400Cys, and Lys432Gln), all retained function as measured by uptake of the prototypical organic cation substrate, MPP^+, in *X. laevis* oocytes expressing the variant transporters. However, quantitative differences in MPP^+ uptake activity and kinetic differences in interactions with organic cations were observed for the common variants, with Lys432Gln having twofold increased affinity for MPP^+, and the Met165Ile and Arg400Cys variants having lower maximal transport rates (V_{max}) for MPP^+ than the reference OCT2.[35] In inhibition kinetics studies, Ala270Ser showed an increased K_i value for TBA, while Arg400Cys and Lys432Gln had lower K_i values for TBA inhibition than did the reference OCT2. When the rare variants of OCT2 were expressed in the same system, all of the amino acid substitutions were found to have no effect on OCT2 function.[30] However, as expected, the single nucleotide insertion (frameshift) variant showed a complete loss of function.

Additional single-nucleotide polymorphisms (SNPs) in the OCT2 gene were identified by direct sequencing of genomic DNA from 48 unrelated Japanese persons[73] and 116 arrhythmic Japanese patients.[74] In the former study, 27 SNPs and five deletion polymorphisms in the OCT2 gene were identified.[73] Two synonymous variants, also reported by Leabman et al.,[35] were identified at amino acid positions 130 and 150. The remaining SNPs and deletion polymorphisms appeared in introns, 3′ untranslated regions, and 3′ flanking regions of the OCT2 gene. In the latter study, 33 genetic variants, including 14 novel ones, were found.[74] Two nonsynonymous variants were identified (Thr199Ile and Thr201Met) and other SNPs and insertion and deletion polymorphisms were located in exons, the 3′-untranslated region of exon 11, introns, and the 3′-flanking region. The functional effects of these polymorphisms have not been investigated.

In addition to direct sequencing of the OCT2 gene, other lines of evidence exist for a role of genetic variation in OCT2 in variable drug response. Based on data from published literature, an estimate of the genetic component contributing to variation in the renal clearance of metformin, which undergoes transporter-mediated secretion, was found to be particularly high (>90%).[75] This finding suggests that variation in the renal clearance of metformin has a strong genetic component and that genetic variation in OCT2 may explain a large part of this pharmacokinetic variability.

2.7.3. OCT3

In a study of genetic variation in 26 membrane transporter genes in a large sample ($n =$ 247) of ethnically diverse human subjects, Leabman et al. discovered 14 nucleotide substitutions in OCT3.[76] The survey region included the exons and flanking intronic region. Of the 14 variants identified, five were located in the coding exonic regions, and three of these (Thr44Met, Ala116Ser, and Thr400Ile) resulted in a change in the amino acid sequence. Only Ala116Ser was polymorphic, with an allele frequency of 1.7% in the African-American subset of the sample. These variants have not been tested for effects on OCT3 function.

Genetic variation in OCT3 was investigated independently by resequencing the core promoter, exons, and 3′ untranslated region of the gene in 100 healthy persons of European descent.[77] Six nucleotide substitutions and one single base pair deletion were discovered, including −29A>G in the OCT3 promoter and three synonymous substitutions in the coding region. The functional consequences of these variants have not been studied. The authors suggest that the synonymous 1233G>A substitution may generate a cryptic 3′-splice acceptor site.

Additional data on genetic variation in organic cation transporter genes are available on online public databases. The Pharmacogenetics and Genomics Knowledge Base (PharmGKB, www.pharmgkb.org) is a repository for data and information related to all areas of pharmacogenetics. PharmGKB includes well-annotated pages describing genetic variation in OCTs, identified by the Pharmacogenetics of Membrane Transporters project (PMT, www.pharmacogenetics.ucsf.edu) in a large collection of DNA samples from ethnically diverse populations. The NCBI single-nucleotide polymorphism database (dbSNP, www.ncbi.nlm.nih.gov/SNP/) also includes data on OCT variants. The HapMap project (www.hapmap.org), which aims to take advantage of linkage between SNPs to increase power in genetic association studies, may prove to be a useful tool for studying the genetics of response to OCT substrates.

2.8. CONCLUSIONS

The OCTs are a fairly well-studied family of multispecific organic cation transporters, with potential influence for a large number of endogenous and pharmacological compounds. The tissue distribution and subcellular localization of these transporters suggest that their primary role is in the excretion of toxic xenobiotic and endogenous organic cations. OCT1 is a hepatic sinusoidal uptake transporter and appears to be most important for drug distribution into liver, where it aids in presentation of substrate drugs to hepatic metabolizing enzymes or in biliary excretion. OCT2, the kidney-specific organic cation transporter, is primarily responsible for basolateral uptake of organic cations into renal proximal tubules and acts as the first step in active tubular secretion of its substrates. OCT3 has a more diffuse pattern of expression, as well as a relatively unique substrate selectivity profile, and is thought to be important for extraneuronal clearance of monoamine neurotransmitters as well as uptake of monoamines into the heart and across the placenta. Knockout mouse models of the OCTs have provided crucial information regarding the importance of these transporters to xenobiotic and endobiotic disposition in vivo. The existence of genetic variants of OCTs with altered function in cellular assays suggests that genetic variation in OCTs may contribute to interindividual variability in drug disposition or drug response; further study of the importance of these variants to human drug disposition may soon allow for clinical pharmacogenetic testing and improvements in rational drug therapy.

Acknowledgments

This work was supported by National Institutes of Health grants GM36780 and GM61390.

REFERENCES

1. Rennick BR. 1981. Renal tubule transport of organic cations. Am J Physiol 240(2):F83–F89.
2. Gründemann D, Gorboulev V, Gambaryan S, Veyhl M, Koepsell H. 1994. Drug excretion mediated by a new prototype of polyspecific transporter. Nature 372(6506):549–552.
3. Okuda M, Saito H, Urakami Y, Takano M, Inui K. 1996. cDNA cloning and functional expression of a novel rat kidney organic cation transporter, OCT2. Biochem Biophys Res Commun 224(2):500–507.
4. Gorboulev V, Ulzheimer JC, Akhoundova A, Ulzheimer-Teuber I, Karbach U, Quester S, Baumann C, Lang F, Busch AE, Koepsell H. 1997. Cloning and characterization of two human polyspecific organic cation transporters. DNA Cell Biol 16(7):871–881.
5. Gründemann D, Babin-Ebell J, Martel F, Ording N, Schmidt A, Schömig E. 1997. Primary structure and functional expression of the apical organic cation transporter from kidney epithelial LLC-PK1 cells. J Biol Chem 272(16):10408–10413.
6. Mooslehner KA, Allen ND. 1999. Cloning of the mouse organic cation transporter 2 gene, Slc22a2, from an enhancer-trap transgene integration locus. Mamm Genome 10(3):218–224.
7. Schweifer N, Barlow DP. 1996. The Lx1 gene maps to mouse chromosome 17 and codes for a protein that is homologous to glucose and polyspecific transmembrane transporters. Mamm Genome 7(10):735–740.
8. Terashita S, Dresser MJ, Zhang L, Gray AT, Yost SC, Giacomini KM. 1998. Molecular cloning and functional expression of a rabbit renal organic cation transporter. Biochim Biophys Acta 1369(1):1–6.
9. Zhang L, Dresser MJ, Gray AT, Yost SC, Terashita S, Giacomini KM. 1997. Cloning and functional expression of a human liver organic cation transporter. Mol Pharmacol 51(6):913–921.
10. Kekuda R, Prasad PD, Wu X, Wang H, Fei YJ, Leibach FH, Ganapathy V. 1998. Cloning and functional characterization of a potential-sensitive, polyspecific organic cation transporter (OCT3) most abundantly expressed in placenta. J Biol Chem 273(26):15971–15979.
11. Gründemann D, Schechinger B, Rappold GA, Schömig E. 1998. Molecular identification of the corticosterone-sensitive extraneuronal catecholamine transporter. Nat Neurosci 1(5):349–351.

12. Verhaagh S, Schweifer N, Barlow DP, Zwart R. 1999. Cloning of the mouse and human solute carrier 22a3 (Slc22a3/SLC22A3) identifies a conserved cluster of three organic cation transporters on mouse chromosome 17 and human 6q26–q27. Genomics 55(2):209–218.
13. Wu X, Huang W, Ganapathy ME, Wang H, Kekuda R, Conway SJ, Leibach FH, Ganapathy V. 2000. Structure, function, and regional distribution of the organic cation transporter OCT3 in the kidney. Am J Physiol 279(3):F449–F458.
14. Wright SH. 2005. Role of organic cation transporters in the renal handling of therapeutic agents and xenobiotics. Toxicol Appl Pharmacol 204(3):309–319.
15. Wright SH, Dantzler WH. 2004. Molecular and cellular physiology of renal organic cation and anion transport. Physiol Rev 84(3):987–1049.
16. Koepsell H, Gorboulev V, Arndt P. 1999. Molecular pharmacology of organic cation transporters in kidney. J Membr Biol 167(2):103–117.
17. Koepsell H, Schmitt BM, Gorboulev V. 2003. Organic cation transporters. Rev Physiol Biochem Pharmacol 150:36–90.
18. Koepsell H. 2004. Polyspecific organic cation transporters: their functions and interactions with drugs. Trends Pharmacol Sci 25(7):375–381.
19. Koepsell H, Endou H. 2004. The SLC22 drug transporter family. Pflugers Arch 447(5):666–676.
20. Dresser MJ, Leabman MK, Giacomini KM. 2001. Transporters involved in the elimination of drugs in the kidney: organic anion transporters and organic cation transporters. J Pharm Sci 90(4):397–421.
21. Popp C, Gorboulev V, Muller TD, Gorbunov D, Shatskaya N, Koepsell H. 2005. Amino acids critical for substrate affinity of rat organic cation transporter 1 line the substrate binding region in a model derived from the tertiary structure of lactose permease. Mol Pharmacol 67(5):1600–1611.
22. Gorboulev V, Shatskaya N, Volk C, Koepsell H. 2005. Subtype-specific affinity for corticosterone of rat organic cation transporters rOCT1 and rOCT2 depends on three amino acids within the substrate binding region. Mol Pharmacol 67(5):1612–1619.
23. Arndt P, Volk C, Gorboulev V, Budiman T, Popp C, Ulzheimer-Teuber I, Akhoundova A, Koppatz S, Bamberg E, Nagel G, Koepsell H. 2001. Interaction of cations, anions, and weak base quinine with rat renal cation transporter rOCT2 compared with rOCT1. Am J Physiol Renal Physiol 281(3):F454–F468.
24. Budiman T, Bamberg E, Koepsell H, Nagel G. 2000. Mechanism of electrogenic cation transport by the cloned organic cation transporter 2 from rat. J Biol Chem 275(38):29413–29420.
25. Busch AE, Quester S, Ulzheimer JC, Waldegger S, Gorboulev V, Arndt P, Lang F, Koepsell H. 1996. Electrogenic properties and substrate specificity of the polyspecific rat cation transporter rOCT1. J Biol Chem 271(51):32599–32604.
26. Dresser MJ, Gray AT, Giacomini KM. 2000. Kinetic and selectivity differences between rodent, rabbit, and human organic cation transporters (OCT1). J Pharmacol Exp Ther 292(3):1146–1152.
27. Gorboulev V, Volk C, Arndt P, Akhoundova A, Koepsell H. 1999. Selectivity of the polyspecific cation transporter rOCT1 is changed by mutation of aspartate 475 to glutamate. Mol Pharmacol 56(6):1254–1261.

28. Nagel G, Volk C, Friedrich T, Ulzheimer JC, Bamberg E, Koepsell H. 1997. A reevaluation of substrate specificity of the rat cation transporter rOCT1. J Biol Chem 272(51):31953–31956.
29. Sweet DH, Pritchard JB. 1999. rOCT2 is a basolateral potential-driven carrier, not an organic cation/proton exchanger. Am J Physiol 277(6 Pt 2):F890–F898.
30. Fujita T, Urban TJ, Leabman MK, Fujita K, Giacomini KM. 2006. Transport of drugs in the kidney by the human organic cation transporter, OCT2 and its genetic variants. J Pharm Sci 95(1):25–36.
31. Busch AE, Karbach U, Miska D, Gorboulev V, Akhoundova A, Volk C, Arndt P, Ulzheimer JC, Sonders MS, Baumann C, et al. 1998. Human neurons express the polyspecific cation transporter hOCT2, which translocates monoamine neurotransmitters, amantadine, and memantine. Mol Pharmacol 54(2):342–352.
32. Jonker JW, Schinkel AH. 2004. Pharmacological and physiological functions of the polyspecific organic cation transporters: OCT1, 2, and 3 (SLC22A1-3). J Pharmacol Exp Ther 308(1):2–9.
33. Kimura N, Masuda S, Tanihara Y, Ueo H, Okuda M, Katsura T, Inui K. 2005. Metformin is a superior substrate for renal organic cation transporter OCT2 rather than hepatic OCT1. Drug Metab Pharmacokinet 20(5):379–386.
34. Bourdet DL, Pritchard JB, Thakker DR. 2005. Differential substrate and inhibitory activities of ranitidine and famotidine toward human organic cation transporter 1 (hOCT1; SLC22A1), hOCT2 (SLC22A2), and hOCT3 (SLC22A3). J Pharmacol Exp Ther 315(3): 1288–1297.
35. Leabman MK, Huang CC, Kawamoto M, Johns SJ, Stryke D, Ferrin TE, DeYoung J, Taylor T, Clark AG, Herskowitz I, Giacomini KM. 2002. Polymorphisms in a human kidney xenobiotic transporter, OCT2, exhibit altered function. Pharmacogenetics 12(5):395–405.
36. Gründemann D, Liebich G, Kiefer N, Köster S, Schömig E. 1999. Selective substrates for non-neuronal monoamine transporters. Mol Pharmacol 56(1):1–10.
37. Hayer-Zillgen M, Bruss M, Bonisch H. 2002. Expression and pharmacological profile of the human organic cation transporters hOCT1, hOCT2 and hOCT3. Br J Pharmacol 136(6):829–836.
38. Takeda M, Khamdang S, Narikawa S, Kimura H, Kobayashi Y, Yamamoto T, Cha SH, Sekine T, Endou H. 2002. Human organic anion transporters and human organic cation transporters mediate renal antiviral transport. J Pharmacol Exp Ther 300(3):918–924.
39. Wu X, Kekuda R, Huang W, Fei YJ, Leibach FH, Chen J, Conway SJ, Ganapathy V. 1998. Identity of the organic cation transporter OCT3 as the extraneuronal monoamine transporter (uptake2) and evidence for the expression of the transporter in the brain. J Biol Chem 273(49):32776–32786.
40. Eisenhofer G. 2001. The role of neuronal and extraneuronal plasma membrane transporters in the inactivation of peripheral catecholamines. Pharmacol Ther 91(1):35–62.
41. Kimura H, Takeda M, Narikawa S, Enomoto A, Ichida K, Endou H. 2002. Human organic anion transporters and human organic cation transporters mediate renal transport of prostaglandins. J Pharmacol Exp Ther 301(1):293–298.
42. Urakami Y, Akazawa M, Saito H, Okuda M, Inui K. 2002. cDNA cloning, functional characterization, and tissue distribution of an alternatively spliced variant of organic cation transporter hOCT2 predominantly expressed in the human kidney. J Am Soc Nephrol 13(7):1703–1710.

43. Urakami Y, Okuda M, Masuda S, Saito H, Inui KI. 1998. Functional characteristics and membrane localization of rat multispecific organic cation transporters, OCT1 and OCT2, mediating tubular secretion of cationic drugs. J Pharmacol Exp Ther 287(2):800–805.

44. Karbach U, Kricke J, Meyer-Wentrup F, Gorboulev V, Volk C, Loffing-Cueni D, Kaissling B, Bachmann S, Koepsell H. 2000. Localization of organic cation transporters OCT1 and OCT2 in rat kidney. Am J Physiol Renal Physiol 279(4):F679–F687.

45. Sugawara-Yokoo M, Urakami Y, Koyama H, Fujikura K, Masuda S, Saito H, Naruse T, Inui K, Takata K. 2000. Differential localization of organic cation transporters rOCT1 and rOCT2 in the basolateral membrane of rat kidney proximal tubules. Histochem Cell Biol 114(3):175–180.

46. Sweet DH, Miller DS, Pritchard JB. 2000. Basolateral localization of organic cation transporter 2 in intact renal proximal tubules. Am J Physiol Renal Physiol 279(5):F826–F834.

47. Motohashi H, Sakurai Y, Saito H, Masuda S, Urakami Y, Goto M, Fukatsu A, Ogawa O, Inui K. 2002. Gene expression levels and immunolocalization of organic ion transporters in the human kidney. J Am Soc Nephrol 13(4):866–874.

48. Dudley AJ, Bleasby K, Brown CD. 2000. The organic cation transporter OCT2 mediates the uptake of beta-adrenoceptor antagonists across the apical membrane of renal LLC-PK(1) cell monolayers. Br J Pharmacol 131(1):71–79.

49. Lee W, Kim RB. 2004. Transporters and renal drug elimination. Annu Rev Pharmacol Toxicol 44:137–166.

50. Masuda S, Terada T, Yonezawa A, Tanihara Y, Kishimoto K, Katsura T, Ogawa O, Inui K. 2006. Identification and functional characterization of a new human kidney-specific H^+/organic cation antiporter, kidney-specific multidrug and toxin extrusion 2. J Am Soc Nephrol 17(8):2127–2135.

51. Otsuka M, Matsumoto T, Morimoto R, Arioka S, Omote H, Moriyama Y. 2005. A human transporter protein that mediates the final excretion step for toxic organic cations. Proc Natl Acad Sci U S A 102(50):17923–17928.

52. Ohashi R, Tamai I, Nezu Ji J, Nikaido H, Hashimoto N, Oku A, Sai Y, Shimane M, Tsuji A. 2001. Molecular and physiological evidence for multifunctionality of carnitine/organic cation transporter OCTN2. Mol Pharmacol 59(2):358–366.

53. Meyer-Wentrup F, Karbach U, Gorboulev V, Arndt P, Koepsell H. 1998. Membrane localization of the electrogenic cation transporter rOCT1 in rat liver. Biochem Biophys Res Commun 248(3):673–678.

54. Ciarimboli G, Schlatter E. 2005. Regulation of organic cation transport. Pflugers Arch 449(5):423–441.

55. Mehrens T, Lelleck S, Cetinkaya I, Knollmann M, Hohage H, Gorboulev V, Boknik P, Koepsell H, Schlatter E. 2000. The affinity of the organic cation transporter rOCT1 is increased by protein kinase C–dependent phosphorylation. J Am Soc Nephrol 11(7):1216–1224.

56. Ciarimboli G, Struwe K, Arndt P, Gorboulev V, Koepsell H, Schlatter E, Hirsch JR. 2004. Regulation of the human organic cation transporter hOCT1. J Cell Physiol 201(3):420–428.

57. Pietig G, Mehrens T, Hirsch JR, Cetinkaya I, Piechota H, Schlatter E. 2001. Properties and regulation of organic cation transport in freshly isolated human proximal tubules. J Biol Chem 276(36):33741–33746.

58. Cetinkaya I, Ciarimboli G, Yalcinkaya G, Mehrens T, Velic A, Hirsch JR, Gorboulev V, Koepsell H, Schlatter E. 2003. Regulation of human organic cation transporter hOCT2 by

PKA, PI3K, and calmodulin-dependent kinases. Am J Physiol Renal Physiol 284(2):F293–F302.

59. Pelis RM, Suhre WM, Wright SH. 2006. Functional influence of N-glycosylation in OCT2-mediated tetraethylammonium transport. Am J Physiol Renal Physiol 290(5):F1118–F1126.

60. Urakami Y, Nakamura N, Takahashi K, Okuda M, Saito H, Hashimoto Y, Inui K. 1999. Gender differences in expression of organic cation transporter OCT2 in rat kidney. FEBS Lett 461(3):339–342.

61. Slitt AL, Cherrington NJ, Hartley DP, Leazer TM, Klaassen CD. 2002. Tissue distribution and renal developmental changes in rat organic cation transporter mRNA levels. Drug Metab Dispos 30(2):212–219.

62. Shu Y, Bello CL, Mangravite LM, Feng B, Giacomini KM. 2001. Functional characteristics and steroid hormone–mediated regulation of an organic cation transporter in Madin-Darby canine kidney cells. J Pharmacol Exp Ther 299(1):392–398.

63. Martel F, Keating E, Calhau C, Grundemann D, Schömig E, Azevedo I. 2001. Regulation of human extraneuronal monoamine transporter (hEMT) expressed in HEK293 cells by intracellular second messenger systems. Naunyn-Schmiedebergs Arch Pharmacol 364(6):487–495.

64. Jonker JW, Wagenaar E, Mol CA, Buitelaar M, Koepsell H, Smit JW, Schinkel AH. 2001. Reduced hepatic uptake and intestinal excretion of organic cations in mice with a targeted disruption of the organic cation transporter 1 (Oct1 [Slc22a1]) gene. Mol Cell Biol 21(16):5471–5477.

65. Wang DS, Jonker JW, Kato Y, Kusuhara H, Schinkel AH, Sugiyama Y. 2002. Involvement of organic cation transporter 1 in hepatic and intestinal distribution of metformin. J Pharmacol Exp Ther 302(2):510–515.

66. Wang DS, Kusuhara H, Kato Y, Jonker JW, Schinkel AH, Sugiyama Y. 2003. Involvement of organic cation transporter 1 in the lactic acidosis caused by metformin. Mol Pharmacol 63(4):844–848.

67. Jonker JW, Wagenaar E, Van Eijl S, Schinkel AH. 2003. Deficiency in the organic cation transporters 1 and 2 (Oct1/Oct2 [Slc22a1/Slc22a2]) in mice abolishes renal secretion of organic cations. Mol Cell Biol 23(21):7902–7908.

68. Zwart R, Verhaagh S, Buitelaar M, Popp-Snijders C, Barlow DP. 2001. Impaired activity of the extraneuronal monoamine transporter system known as uptake-2 in Orct3/Slc22a3-deficient mice. Mol Cell Biol 21(13):4188–4196.

69. Kerb R, Brinkmann U, Chatskaia N, Gorbunov D, Gorboulev V, Mornhinweg E, Keil A, Eichelbaum M, Koepsell H. 2002. Identification of genetic variants of the human organic cation transporter hOCT1 and their functional consequences. Pharmacogenetics 12(8):591–595.

70. Shu Y, Leabman MK, Feng B, Mangravite LM, Huang CC, Stryke D, Kawamoto M, Johns SJ, DeYoung J, Carlson E, et al. 2003. Evolutionary conservation predicts function of variants of the human organic cation transporter, OCT1. Proc Natl Acad Sci U S A 100(10):5902–5907.

71. Sakata T, Anzai N, Shin HJ, Noshiro R, Hirata T, Yokoyama H, Kanai Y, Endou H. 2004. Novel single nucleotide polymorphisms of organic cation transporter 1 (SLC22A1) affecting transport functions. Biochem Biophys Res Commun 313(3):789–793.

72. Takeuchi A, Motohashi H, Okuda M, Inui K. 2003. Decreased function of genetic variants, Pro283Leu and Arg287Gly, in human organic cation transporter hOCT1. Drug Metab Pharmacokinet 18(6):409–412.

73. Saito S, Iida A, Sekine A, Ogawa C, Kawauchi S, Higuchi S, Nakamura Y. 2002. Catalog of 238 variations among six human genes encoding solute carriers (hSLCs) in the Japanese population. J Hum Genet 47(11):576–584.

74. Fukushima-Uesaka H, Maekawa K, Ozawa S, Komamura K, Ueno K, Shibakawa M, Kamakura S, Kitakaze M, Tomoike H, Saito Y, Sawada J. 2004. Fourteen novel single nucleotide polymorphisms in the SLC22A2 gene encoding human organic cation transporter (OCT2). Drug Metab Pharmacokinet 19(3):239–244.

75. Leabman MK, Giacomini KM. 2003. Estimating the contribution of genes and environment to variation in renal drug clearance. Pharmacogenetics 13(9):581–584.

76. Leabman MK, Huang CC, DeYoung J, Carlson EJ, Taylor TR, de la Cruz M, Johns SJ, Stryke D, Kawamoto M, Urban TJ, et al. 2003. Natural variation in human membrane transporter genes reveals evolutionary and functional constraints. Proc Natl Acad Sci U S A 100(10):5896–5901.

77. Lazar A, Gründemann D, Berkels R, Taubert D, Zimmermann T, Schömig E. 2003. Genetic variability of the extraneuronal monoamine transporter EMT (SLC22A3). J Hum Genet 48(5):226–230.

3

ORGANIC CATION/CARNITINE TRANSPORTERS

IJAZ A. QURESHI
Hospital Sainte-Justine, Montreal, Quebec, Canada

IHSAN ELIMRANI AND ERNEST SEIDMAN
McGill University Research Institute, Montreal, Quebec, Canada

GRANT MITCHELL
Hospital Sainte-Justine, Montreal, Quebec, Canada

Drug Transporters: Molecular Characterization and Role in Drug Disposition, Edited by Guofeng You and Marilyn E. Morris
Copyright © 2007 John Wiley & Sons, Inc.

3.1. INTRODUCTION

Carnitine (3-hydroxy-4-*N*-trimethylaminobutyric acid), a zwitterion, is an essential cofactor in the metabolism of lipids and consequently in the production of cellular energy. L-Carnitine, the active form, has important physiological roles, including involvement in the β-oxidation of fatty acids, facilitation of the transport of long-chain fatty acids across the mitochondrial inner membrane as acylcarnitine esters, and modulation of intracellular coenzyme A homeostasis.[1] It is involved in transfer of the products of peroxisomal β-oxidation to mitochondria. In humans, L-carnitine is obtained from dietary sources, primarily meat and dairy products, and produced endogenously from the amino acids lysine and methionine, with both sources contributing to plasma and tissue levels.

The major site of carnitine absorption is the small intestine, but the mechanism of intestinal uptake of L-carnitine is unclear. Previous studies using Caco-2 cells demonstrated that the transport of L-carnitine involved a carrier-mediated system.[2] Results from our laboratory showed that L-carnitine uptake in differentiated Caco-2 cells was mediated primarily by the organic cation/carnitine transporter OCTN2, located on the brush border membrane.[3] Carnitine is eliminated as free carnitine or acylcarnitine almost exclusively by the kidney, which plays a crucial role in homeostatic regulation of carnitine concentrations in body fluids. Carnitine reabsorption appears to be regulated such that normal plasma carnitine concentration is maintained without significant loss of carnitine in the urine. Although there is evidence of different polarized carnitine transporters in the kidney, the brush border membrane transporter is probably OCTN2, as indicated by a strong reactivity with the anti-OCTN2 polyclonal antibody.[4] There is considerable evidence that various organic cation/carnitine transporters may be involved in carnitine distribution, and organic cation absorption and elimination in several tissues.[5]

The organic cation/carnitine transporters are a subgroup of the SLC22 (solute carrier) drug transporter family, which belongs to the major facilitator superfamily.[6] This superfamily comprises uniporters, symporters, and antiporters from bacteria, lower eukaryotes, plants, and mammals. In human species and mammals, they participate in the absorption and/or excretion of drugs, xenobiotics, and endogenous compounds in the intestine, liver, and/or kidney, and perform homeostatic functions in brain and heart. Various subgroups of the SLC22 drug transporter family are discussed in detail in other chapters of this book. In the present chapter, the discussion focuses on the known members of the organic cation/carnitine transporter (OCTN) subgroup and related carnitine transporters. Emphasis is on the clinical importance of OCTN transporters. Primary systemic carnitine deficiency (SCD) and secondary L-carnitine deficiency syndromes known to be related to OCTN levels and expression lead to a reduced use of fatty acids in energy production. SCD is an autosomal recessive disorder caused by mutations in the gene encoding for the Na^+/L-carnitine transporter OCTN2.[1] Recently, two variants in the OCTN1 and OCTN2 genes have been shown to exist in a haplotype, which is associated with a susceptibility to Crohn's disease.[7]

3.2. BIOCHEMISTRY AND PHYSIOLOGY OF OCTNs

As mentioned above, the organic cation/carnitine transporters are members of the SLC22 polyspecific organic cation transporter family. Homology screening led to the identification of four members of the SLC22 family that translocate the zwitterion carnitine together with Na^+ and/or organic cations. They transport multiple and different substrates in addition to L-carnitine and various acylcarnitine/carnitine esters. This subgroup comprises the transporters OCTN1-3 and hCT2. These transporters may function as an organic cation uniporter or H^+/organic cation antiporter (e.g., OCTN1) or as uniporters for organic cations or Na^+/carnitine cotransporters (e.g., OCTN2). Since many of these transporters are expressed in the intestine, liver, kidney, and other tissues, the OCTNs play a pivotal role in drug and carnitine coabsorption and excretion. In the following sections we deal with the biochemical and physiological characteristics of each of these transporters.

3.2.1. OCTN 1

Structure–Function Relationships OCTN1 was cloned from human fetal kidney, rat, and mouse in 1997.[8–10] Human OCTN1 has 551 amino acids with 11 putative transmembrane domains. It is a multispecific, bidirectional, and pH-dependent organic cation transporter. OCTN1 is expressed in the kidneys, skeletal muscle, placenta, prostate, heart, and lungs.[8] The human gene locus for hOCTN1 is reported to be 5q23.3. A recent study on the tissue distribution and ontogeny of Octn1 in mice indicates that mRNA expression is highest in the kidney, followed by the small intestine,[12] a pattern very similar to the mRNA expression levels of Octn2. Mouse octn1, having 553 amino acids, can mediate carnitine transport in a Na^+-dependent manner,[9] illustrating an apparent species difference in specificity for the same transporter type.

Ligand Specificity and Drug Substrates Human OCTN1 (hOCTN1) transports the cations tetraethylammonium (TEA), quinidine, pyrilamine, and verapamil and the zwitterion carnitine.[11] Many other cations interact with the zwitterion cephaloridine and the anions ofloxacin and levofloxacin. The data show that hOCTN1 is a polyspecific transporter that may have a preference for organic cations. hOCTN1 mediates TEA transport in either direction with an apparent K_m value of 0.2 to 0.4 mM in the influx mode.[8,11] hOCTN1 is independent of an inwardly directed H^+ gradient and of the membrane potential, whereas an inwardly directed H^+ gradient stimulates TEA efflux.[11] hOCTN1 may work as an electroneutral H^+/organic cation antiporter that mediates the cellular efflux of organic cations. OCTN1 and OCTN2 genes have a similar structure and are located in tandem. The sequence of OCTN1 is conserved to a degree similar to that of OCTN2, which is consistent with functional roles. OCTN1 is functional and its properties are distinct from those of OCTN2.[8] It is a multispecific and bidirectional organic cation transporter and interacts with a variety of organic cations. The transport of TEA by OCTN1 is Na^+ independent and pH dependent. OCTN1 has a lower affinity for carnitine than does OCTN2, and depending on the species, it differs in its specificity for carnitine transport. In rat, this transporter does not mediate Na^+-coupled carnitine transport to a significant extent,[10] but in mice the carnitine transport is Na^+ dependent.[9]

Tissue Distribution and Cellular Membrane Localization Wu et al.[10] demonstrated by in situ hybridization that rat Octn1 mRNA was expressed in a wide variety of rat tissues, including the liver, intestine, kidney, brain, heart, and placenta. Large species differences exist with regard to localization and function of OCTN1. OCTN1 could not be detected in liver from adult humans, whereas in rats, that is the site with the strongest expression.[10] Koepsell et al.[5] suggested that hOCTN1 has a luminal location in brush border membranes of renal proximal tubules, where it may be engaged in cation excretion in different species. Since it transports in both directions, OCTN1 may also participate in the reabsorption of organic cations. Also, carnitine uptake was Na^+ dependent for mouse Octn1 but not for rat Octn1.[9,11] Loss of function mutations in hOCTN1 would predictably increase the nephrotoxic potential of cationic drugs that enter proximal tubules. Pharmaceutical up-regulation of hOCTN1 might improve the renal excretion of cationic drugs or xenobiotics without increasing their intracellular concentrations.

Xuan et al.[13] have used anti-mOctn1 antibody to demonstrate the expression of OCTN1 in sperm. They have also shown a low-affinity carnitine uptake in sperm with a K_m of 412 μM, which is consistent with values demonstrated previously for OCTN1 in different tissues and species. Evidence has been presented of the existence in sperm of all three OCTN species. It has been suggested that OCTN1 may serve an important role in the intracellular shuttling of acylcarnitines in sperm mitochondria.

3.2.2. OCTN 2

Structure–Function Relationships Among the organic cation transporters, OCTN2 is identified as the most important carnitine transporter. Koepsell et al.[5] have reviewed extensively the biochemistry and physiology of carnitine transport mediated by OCTN2 and other OCTN transporters. OCTN2 is a Na^+/carnitine cotransporter with a high affinity for carnitine but can function alternatively as a polyspecific and Na^+-independent cation uniporter.[14–17] Na^+-dependent transport of L-carnitine by OCTN2 is electrogenic and to some degree stereospecific.[16,18]

Ligand Specificity and Drug Substrates Several uptake measurements have shown that human OCTN2 has an apparent K_m of 4 to 5 μM for L-carnitine,[14,16] and it exhibits half-maximal concentrations for Na^+ activation between 2 and 19 μM.[16,18,19] Therefore, the molecule responsible for the high-affinity Na^+/L-carnitine cotransport activity is likely to be represented by OCTN2. This activity is observed both in plasma membrane vesicles from skeletal muscle and in brush border membrane vesicles from the kidney and intestine.[20–23] Short-chain acyl esters of carnitine are also transported by hOCTN2 with high affinity in the presence of Na^+ with a K_m around 8.5 μM.[18] hOCTN2 can also transport cephaloridine, a zwitterionic β-lactam antibiotic[24] in the presence of Na^+. In addition, as suggested by electrophysiological studies, hOCTN2 is probably able to transport L-lysine and L-methionine under the same transport conditions. Furthermore, human, rat, and mouse OCTN2 are able to translocate or interact with various organic cations (e.g., TEA, pyrilamine, quinidine, verapamil, choline) in a Na^+- independent fashion.[16–18,25]

Ohashi et al.[26] have determined that the TEA uptake by hOCTN2 has a K_m of around 0.3 μM. It has also been established that carnitine uptake by hOCTN2 was competitively inhibited by TEA.[19,26] Other studies have shown that in the presence of Na^+, carnitine uptake via OCTN2 is inhibited by other organic cations or weak bases [e.g., nicotine, MPTP [1-methyl 4-phenyl 1, 2, 3, 6-tetrahydropyridine], procainamide, quinine, metamphetamine, emetine, clonidine, cimetidine], zwitterions (cephaloridine, cefepime, cefoselis), and noncharged compounds (e.g., corticosterone, aldosterone).[16,18,24,27] The IC_{50} values for cephaloridine, cephepime, and cefoselis were 0.23, 1.7, and 6.4 mM, respectively. The removal of Na^+ decreases the affinity for cephaloridine but not for cefepime and cefoselis.[24] Therefore, it is likely that the binding of carnitine and cations occurs within one binding pocket.

Tissue Distribution and Cellular Membrane Localization Our recent studies have demonstrated the expression and localization of this transporter in human placenta and Caco-2 cells.[3,28] In an earlier study we had shown the presence of OCTN2 on the apical membrane of small intestine and renal tubules.[1] In addition to intestine, OCTN2 exhibited abundant expression in the stomach, colon, and rectum.[29] OCTN2, also present in the brush border membrane of renal proximal tubules in the rat and mouse,[9] participates in absorption and secretion of organic cations in the small intestine and kidney. Mouse OCTN2 can translocate organic cations in either direction, as exemplified by *trans*-stimulation of carnitine uptake by TEA as well as of TEA efflux by carnitine.[25] Japanese researchers have demonstrated significant contributions of OCTN2 to the renal excretion of TEA by injecting radiolabeled TEA into control mice and juvenile visceral steatosis (JVS) mice that suffer from a homozygous loss-of-function mutation of OCTN2.[25,30–32] Secretion of TEA was reduced by 50% in JVS mice compared to controls, and accumulation of TEA in the kidney was increased by 150%.[25] Xuan et al.[13] have suggested that OCTN2 may be responsible for carnitine transport across the spermatozoan plasma membrane. They detected the expression of a 63-kDa OCTN2 protein in sperm which showed a high-affinity carnitine transport, with a K_m value of 3.4 μM. This is very similar to the K_m values of 2 to 6 μM observed for high-affinity carnitine transport in kidney, skeletal muscle, heart, placenta, and cultured skin fibroblasts. Kobayashi et al.[33] have demonstrated that the high-affinity carnitine transporter OCTN2, which is localized at the basolateral membrane of epididymal epithelial cells, mediates carnitine supply into those cells from the systemic circulation as the first step of permeation from blood to spermatozoa. Another recent study[34] reported the expression and localization of OCTN2 in the α- but not the β-cells of mouse pancreas. The authors hypothesize that other carnitine transporters may be implicated in carnitine transport in β-cells.

3.2.3. OCTN 3

Structure–Function Relationships Originally found in mice,[9] OCTN3 has 564 amino acids in its structure. When expressed in HEK-293 cells, it exhibits Na^+-independent carnitine uptake. Among the mouse tissues examined, OCTN3 was expressed strongly in the testis, weakly in the kidney, but not at all in the intestine. However, Duran et al. subsequently demonstrated the presence of OCTN3 on the basolateral membrane of mouse enterocytes.[35] Mouse Octn3 translocates L-carnitine with an apparent K_m of 3 μM, an affinity seven times higher than that of mouse Octn2.

Ligand Specificity and Drug Substrates The transporter has been investigated to date in mouse, chicken, and in human sperm. In contrast to OCTN2 from different species, including mouse, OCTN3 transports carnitine independent of Na^+. Given that carnitine uptake by OCTN3 was not inhibited by 0.5 mM choline and inhibited only 54% by 0.5 mM TEA, OCTN3 appears less relevant quantitatively for organic cation transport than do OCTN1 and OCTN2.[5] In a study by Duran et al.,[35] chicken enterocytes and basolateral membrane vesicles were found to express OCTN3. L-Carnitine uptake was inhibited by TEA and betaine. It was Na^+ independent, *trans*-stimulated by intravesicular L-carnitine, and *cis*-inhibited by TEA and cold L-carnitine. Both uptake and efflux of carnitine were inhibited by verapamil and unaffected by extraxcellular pH. OCTN3 is unique in its limited tissue distribution and Na^+-independent carnitine transport.

Tissue Distribution and Cellular Membrane Localization Duran et al.[35] measured L-carnitine transport in enterocytes and basolateral membrane vesicles isolated from chicken intestinal epithelia. Reverse transcriptase polymerase chain reaction with specific primers for the mouse Octn3 transporter revealed the existence of OCTN3 mRNA in mouse intestine, which was confirmed by in situ hybridization studies. Immunohistochemical analysis showed that OCTN3 protein was associated primarily with the basolateral membranes of rat and chicken enterocytes, whereas OCTN2 was detected at the apical membrane. Mouse Octn3 is strongly expressed in human testis.[9] The expression of a human OCTN3 protein localized to liver peroxisomes with an intermediate affinity K_m of 20 μM for carnitine has also been reported.[36] Xuan et al.[13] confirmed the expression of OCTN3 in human sperm with an anti-mOctn3 antibody.

3.2.4. CT 2

Structure–Function Relationships CT2 is a high-affinity transporter that has been identified in human tissues.[37] Apart from several amino acids in the N-terminal part of the protein, CT2 is identical to the gene product OCT6 described in hematopoietic cells.[38]

Ligand Specificity and Drug Substrates CT2 has a higher substrate selectivity than OCTN1 and OCTN2, insofar as it interacts with carnitine and betaine but not with TEA and several organic anions.[37]

Tissue Distribution and Cellular Membrane Localization CT2 is expressed in the testis by Sertoli cells and by epithelial cells of the epididymal ducts.[37] CT2 mRNA is further expressed in fetal liver, bone marrow, leukocytes, and some leukemic cell lines.[38] CT2 translocates carnitine in either direction across the plasma membrane and may thus contribute to the transepithelial transport of carnitine in the epididymus. In human mammary gland, Kwok et al.[39] have demonstrated that Flipt2/OCT6, a splice variant of CT2, can coexist with ATB(0^+) in a network of multiple SLC organic cation/nutrient transporters in drug transfer. Among other SLC22 family transporters, they found hOCTN1 and hOCTN2, and besides hOCT1 and hOCT3 in the mammary gland epithelial cell line, MCF12A. Modeling analysis predicted multiplicity of uptake mechanisms with the high-affinity systems characterized by a K_m of 5.1 μM for carnitine and 1.6 mM for TEA, respectively, apparently similar to the hOCTN2 parameter reported for carnitine and that of EMT/hOCT3 for TEA. Verapamil, cimetidine, carbamazepine. quinidine, and desipramine inhibited carnitine uptake but required supratherapeutic concentrations, suggesting the robustness of carnitine uptake systems against xenobiotic challenge.

3.2.5. Comparison with Other Carnitine Transporters

ATB(0^+) Transporter Nakanishi et al.[40] identified a second energy-coupled carnitine transporter, called *amino acid transporter system* B(0^+) [ATB(0^+)]. It is primarily an amino acid transporter[40] and plays an important role in the intestine. ATB(0^+) is also expressed in the lung, mammary gland, and eye. It is a Na^+- and Cl^--coupled transport system for neutral and cationic amino acids. Many of the amino acids and amino acid derivatives that serve as substrates for ATB(0^+) are therapeutic agents (e.g., D-serine, carnitine, and nitric oxide synthase inhibitors). Recent studies have shown that the potential of ATB(0^+) as a drug delivery system may be greater than previously envisaged.[41] Taylor[42] speculated that the mature intestine may have an active transport and a passive carnitine diffusion mechanism, representing contributions from OCTN2 and ATB(0^+), respectively. We investigated the influence of Cl^- on intestinal carnitine uptake by replacing it from the uptake buffer with NaH_2PO_4, and noted no effect.[3] We also tested known amino acid substrates of ATB(0^+) and noted that neither leucine nor tryptophan inhibited carnitine transport in Caco-2 cells. In the mammary gland epithelia, ATB(0^+) can exist in a network of various transporters, with hOCTN1, hOCTN2, and Flipt2/OCT6 (a splice variant of CT2).[39] Berezowski et al.[43] have shown that OCTN2 and ATB(0^+) could be involved in carnitine transport in both the apical and basolateral membranes of brain capillary epithelial cells.

3.3. CLINICAL IMPLICATIONS

The physiological significance of organic cation/carnitine transporters in humans is highlighted by the identification of the hereditary disorder caused by mutations in the gene encoding for OCTN2 protein, an autosomal recessive disease known as *primary systemic carnitine deficiency* (SCD).[1] More commonly, secondary carnitine deficiency syndromes are generated by the inhibition of carnitine transport by drugs that use the same transporters.[44]

3.3.1. Primary Systemic Carnitine Deficiency

Patients affected by homozygous mutations in OCTN2 gene are severely symptomatic due to the resulting cardiomyopathy, progressive skeletal weakness, nonketotic hypoglycemia, and hyperammonemia. Many mutations in the OCTN2 protein have been identified, as, reviewed by our laboratory.[1] The impairment of fatty acid oxidation, particularly during fasting, can also have negative effects on skeletal, muscular, cardiac, and liver function in heterozygotes.[45]

Mutations and Single Nucleotide Polymorphisms Lahjouji et al.[1] summarized the OCTN2 mutations described up until 2001. These are classified as nonsense and missense mutations, presenting the specific genotype, ethnicity, gender, and age of the patients with consanguinity and the clinical manifestations. Affected patients are from different ethnic origins, and the majority of cases are present during infancy. Both missense and nonsense mutations are known. The missense mutations can also be classified into those that abolish carnitine transport entirely and those that reduce transport to residual activity. A list of single-nucleotide polymorphisms is also provided.[1]

Juvenile Visceral Steatosis Mouse The JVS phenotype is inherited in an autosomal recessive manner. Localization of the mutation has been determined with molecular genetic studies using detailed chromosome linkage analysis and positional cloning as within a 1.6 cM (centiMorgan) region on mouse chromosome 11. This corresponds to the OCTN2 region on human chromosome 5q31.[5] Lu et al.[31] have isolated the mouse Octn2 gene and identified the mutation in the JVS mouse to be L352R. This mutation is located in the middle of the putative seventh transmembrane domain of Octn2 in the JVS mouse.[32] The hydrophobic residue (L) in a membrane-spanning region is predicted to disturb topology that is essential for function. Yokogawa et al.[46] have compared the characteristics of L-carnitine transport in isolated hepatocytes from wild-type and JVS mice. The uptake of carnitine in controls was saturable and showed two distinct components, high and low affinity. The high-affinity uptake, which showed properties similar to those of Octn2, was absent in JVS hepatocytes. All these results indicate that JVS mice represent a valid animal model for human primary carnitine deficiency. Functional and molecular studies have established that the JVS mice phenotype is caused by mutations in the high-affinity OCTN2.

3.3.2. Drug Inhibitors and Secondary Carnitine Deficiency

Tein,[44] in an excellent review, has discussed the inhibition of carnitine uptake by a large number of xenobiotics that are in extensive clinical use, including lipophilic organic cations (quinidine, verapamil, and emetine) and zwitterionic compounds such as β-lactam antibiotics (cephaloridine). The anionic compounds valproic acid (VPA) and probenecid were found to be moderate inhibitors. Certain compounds (TEA, quinidine, verapamil, cephaloridine, valproic acid) have been shown to be directly transported by hOCTN2 in transfected cells. Pivalate and its prodrugs are also known to cause secondary carnitine deficiency, but their interaction with OCTN transporters is not well documented.[47] Given the key role of hOCTN2 in the transport of acylcarnitines[48] and other cationic compounds, and the fact that it is inhibited by a wide variety of xenobiotics, OCTN2 may be of considerable pharmacological and toxicological importance to humans. Since several drugs and carnitine compete for the same binding sites on OCTN2, it is plausible that an excess of cationic substrates could cause secondary carnitine deficiency.

Valproate Toxicity Carnitine deficiency associated with valproate therapy is considered multifactorial.[44] An important mechanism is the inhibition of tissue carnitine uptake into tissues, including decreased renal tubular reabsorption of free carnitine. The inhibition may be due to competition for OCTN2 between free carnitine and acylcarnitines, including valproylcarnitine esters and short-chain acylcarnitines.

β-Lactam Antibiotic Toxicity Ganapathy et al.[24] have investigated the interaction of several β-lactam antibiotics with OCTN2 using human cell lines that express the transporter constitutively as well as using cloned human and rat OCTN2 expressed heterologously in human cell lines. The β-lactam antibiotics cephaloridine, cefoselis, cefepime, and cefluprenam were found to inhibit OCTN2-mediated carnitine transport. These antibiotics possess quaternary nitrogen, as does carnitine. The interaction of cephaloridine with OCTN2 is Na^+ dependent, whereas the interaction of cefoselis and cefepime with OCTN2 is largely Na^+ independent. These studies showed that OCTN2 plays a crucial role in the pharmacokinetics and therapeutic efficacy of certain β-lactam antibiotics and that cephaloridine-induced carnitine deficiency is probably due to the inhibition of carnitine reabsorption in the kidney.

3.3.3. OCTNs and Autoimmune Diseases

Schreiber et al.[49] recently reviewed the importance of genetic risk factors involved in inflammatory bowel disease (IBD). A genome-wide search revealed that chromosome 5 contains a gene for early-onset Crohn's disease (CD) termed *IBD5*.[50] Further studies revealed that the mutations at 5q31 conferring CD susceptibility code for OCTN1 and OCTN2.[7] This critical region also encompasses the principal cytokine gene cluster, which includes many plausible candidate genes for IBD, such as interleukins 4, 5, and 13. A silent substitution in the coding region of OCTN2 ($p = 0.004$) and a missense substitution (Thr to Ile) in OCTN1 ($p < 0.003$) conferred increased risk for CD.[7] Linkage disequilibrium studies on the IBD5 locus and genotype–phenotype analysis have confirmed the association of IBD5 risk haplotype with CD and revealed that this association was especially strong in patients with perianal disease.[51] Recent studies by Peltekova et al.[52] have reported two variants in the organic cation/carnitine transport cluster at 5q31 which form a haplotype associated with susceptibility to CD. These variants modify transcription and transporter functions of OCTN1 and OCTN2 and increase risk for CD with an odds ratio between 3.4 and 5.1 in homozygotes.[52] The OCTN mutations also interact with CARD15, another gene associated with CD, to increase the risk of CD, with an odds ratio between 7.2 and 10.5 compared to the control population.[52]

Lamhonwah et al.[53] speculated that the human OCTN3 protein, whose corresponding gene is not yet cloned, may also be involved in the etiology of CD. In a recent study, the same group[54] have hypothesized that an amino acid epitope is shared by the OCTN1 variant with *Campylobacter jejuni* and *Mycobacterium paratuberculosis*, which would cross-react in the presence of enterocolitis and cause an impairment of mitochondrial β-oxidation to initiate IBD. Martinez et al.[55] have examined the association of the OCTN genes with CD in a case-controlled study in a Spanish cohort. Their data support the hypothesis of certain polymorphisms in the SLC22A4 and SLC22A5 as genetic markers of susceptibility/protection for CD.

Recently, Waller et al.[56] reported that OCTN variants were as strongly associated with ulcerative colitis (UC) as they were with CD. OCTN variants were in tight linkage disequilibrium with the extended IBD5 risk haplotypes. OCTN substrates including carnitine, butyrylcarnitine, and propionylcarnitine have been documented to have an important effect in the gut.[57] The colonic epithelium is principally nourished by short-chain fatty acids, particularly butyrate.[58] Butyrate, produced in the colon by bacterial breakdown of dietary fiber, has beneficial effects on the colon, and butyrate enemas decrease colon inflammation in IBD.[59] Many mechanisms for butyrate's suppression of inflammation have been described.[60–62] Whatever the mechanism, OCTNs may be implicated, as butyrate is transported as butyrylcarnitine by OCTN1 and OCTN2.[9,18] The link with autoimmune disorders has been further supported by studies showing that OCTN genes are also associated with rheumatoid arthritis.[63]

3.4. REGULATION OF CARNITINE TRANSPORTERS

There are very few reports regarding the regulation of OCTN transporters. Stephens et al.[64] showed that plasma insulin levels have a regulatory effect on OCTN2 transcription. This finding is in accordance with the hypothesis that insulin can augment Na^+-dependent skeletal muscle carnitine uptake, secondary to its action of increasing sarcolemmal Na^+/K^+-ATPase pump activity, and thus intracellular Na^+ flux.[65] The Na^+-dependent active transport of carnitine into human skeletal muscle is mediated via the high-affinity OCTN2. An additional novel finding from this study was that the combination of hypercarnitinemia and hyperinsulinemia increased OCTN2 mRNA expression, whereas hypercarnitinemia alone appeared to have no effect. Whether the increase in OCTN mRNA expression was the result of the increase in intracellular total carnitine content or the elevated serum insulin concentration requires further investigation. Nevertheless, these findings suggest that OCTN2 is regulated at the transcriptional level, which presents another possible target for increasing muscle carnitine stores.

Immunohistochemical studies of renal brush border membrane vesicles by Kato et al.[66] have revealed that PDZK1 protein and OCTN2 are colocalized in renal brush border membranes. Double transfection of OCTN2 with PDZK1 stimulated the uptake by OCTN2 of its endogenous substrate carnitine, and this increase could be accounted for by the six-fold increase in transport capacity. Such an increase was not observed for OCTN2 when the last four amino acids were deleted, indicating an interaction of PDZK1 with the C-terminus of OCTN2. Garcia-Miranda[67] presented evidence to support the hypothesis that there is a decline in Na^+-dependent L-carnitine in the jejunum and ileum of rats after maturation. Neither delayed weaning nor L-carnitine supplementation prevented the down-regulation of Na^+/L-carnitine transport activity. These results demonstrate that intestinal Na^+-dependent L-carnitine uptake activity is under genetic regulation at the transcriptional level.

3.5. CONCLUSIONS

This review provides a summary of current knowledge about the physiology and clinical significance of the OCTN subfamily of organic cation transporters with carnitine/cation cotransport function. All three carnitine transporters (OCTN1, 2, and 3), with different characteristics as to affinity, specificity, and tissue distribution, for example, have wide-ranging clinical importance. There is also evidence to support the presence of additional carnitine transporters, whose identification, localization, and functional characterization must be explored further for a complete understanding of cation/carnitine interactions.

The most physiologically important member of the OCTN group may be considered to be OCTN2, due to its high affinity for carnitine and expression in many tissues. A multitude of mutations and single-nucleotide polymorphisms of OCTN2 cause systemic carnitine deficiency syndrome. Recent genetic studies have implicated the IBD5 locus at 5q31, which codes for OCTN1 and OCTN2, in the etiology of inflammatory bowel disease and rheumatoid arthritis. This region may give rise to variants that modify transcription and transporter functions of OCTN1 and OCTN2 to increase susceptibility to CD and UC. Other researchers have speculated that OCTN3 may also be involved in the etiology of IBD.[53] Recent studies also highlight the pharmacological importance of OCTN transporters, since they transport many drugs, such as TEA, valproate, verapamil, pyrilamine, and β-lactam antibiotics. Short-chain carnitine esters are also potential drugs transported by OCTN transporters, which can be designed to correct the anomalies of carnitine transport and resultant mitochondrial energy dysfunction, which are at the root of many clinical syndromes. The hormonal regulation of OCTN transporters is another developing area of research which may provide guidelines to an understanding of the carnitine/cation cotransport and how to stimulate their influx into a specific organ and to inhibit their efflux.

Acknowledgments

This work was supported by a research grant from the Dairy Farmers of Canada.

REFERENCES

1. Lahjouji K, Mitchell GA, Qureshi IA. 2001. Carnitine transport by organic cation transporters and systemic carnitine deficiency. Mol Genet Metab 73:287–297.
2. McCloud E, Ma TY, Grant KE, Mathis RK, Said HM. 1996. Uptake of L-carnitine by a human intestinal epithelial cell line, Caco-2. Gastroenterology 111:1534–1540.
3. Elimrani I, Lahjouji K, Seidman E, Roy MJ, Mitchell GA, Qureshi I. 2003. Expression and localization of organic cation/carnitine transporter OCTN2 in Caco-2 cells. Am J Physiol Gastrointest Liver Physiol 284:G863–G871.
4. Lahjouji K, Malo C, Mitchell GA, Qureshi IA. 2002. L-Carnitine transport in mouse renal and intestinal brush-border and basolateral membrane vesicles. Biochim Biophys Acta 1558:82–93.
5. Koepsell H, Schmitt BM, Gorboulev V. 2003. Organic cation transporters. Rev Physiol Biochem Pharmacol 150:36–90.
6. Koepsell H, Endou H. 2004. The SLC22 drug transporter family. Pflugers Arch 447:666–676.
7. Rioux JD, Daly MJ, Silverberg MS, Lindblad K, Steinhart H, Cohen Z, Delmonte T, Kocher K, Miller K, Guschwan S, et al. 2001. Genetic variation in the 5q31 cytokine gene cluster confers susceptibility to Crohn disease. Nat Genet 29:223–228.
8. Tamai I, Yabuuchi H, Nezu J, Sai Y, Oku A, Shimane M, Tsuji A. 1997. Cloning and characterization of a novel human pH-dependent organic cation transporter, OCTN1. FEBS Lett 419:107–111.
9. Tamai I, Ohashi R, Nezu JI, Sai Y, Kobayashi D, Oku A, Shimane M, Tsuji A. 2000. Molecular and functional characterization of organic cation/carnitine transporter family in mice. J Biol Chem 275:40064–40072.
10. Wu X, George RL, Huang W, Wang H, Conway SJ, Leibach FH, Ganapathy V. 2000. Structural and functional characteristics and tissue distribution pattern of rat OCTN1, an organic cation transporter, cloned from placenta. Biochim Biophys Acta 1466:315–327.
11. Yabuuchi H, Tamai I, Nezu J, Sakamoto K, Oku A, Shimane M, Sai Y, Tsuji A. 1999. Novel membrane transporter OCTN1 mediates multispecific, bidirectional, and pH-dependent transport of organic cations. J Pharmacol Exp Ther 289:768–773.
12. Alnouti Y, Petrick JS, Klaassen CD. 2006. Tissue distribution and ontogeny of organic cation transporters in mice. Drug Metab Dispos 34:477–482.
13. Xuan W, Lamhonwah AM, Librach C, Jarvi K, Tein I. 2003. Characterization of organic cation/carnitine transporter family in human sperm. Biochem Biophys Res Commun 306:121–128.
14. Tamai I, Ohashi R, Nezu J, Yabuuchi H, Oku A, Shimane M, Sai Y, Tsuji A. 1998. Molecular and functional identification of sodium ion–dependent, high affinity human carnitine transporter OCTN2. J Biol Chem 73:20378–20382.
15. Tamai I, China K, Sai Y, Kobayashi D, Nezu J, Kawahara E, Tsuji A. 2001. Na(+)-coupled transport of L-carnitine via high-affinity carnitine transporter OCTN2 and its subcellular localization in kidney. Biochim Biophys Acta 1512:273–284.
16. Wagner CA, Lükewille U, Kaltenbach S, Moschen I, Bröer A, Risler T, Bröer S, Lang F. 2000. Functional and pharmacological characterization of human Na(+)-carnitine cotransporter hOCTN2. Am J Physiol Renal Physiol 279:F584–F591.

17. Wu X, Huang W, Prasad PD, Seth P, Rajan DP, Leibach FH, Chen J, Conway SJ, Ganapathy V. 1999. Functional characteristics and tissue distribution pattern of organic cation transporter 2 (OCTN2), an organic cation/carnitine transporter. J Pharmacol Exp Ther 290:1482–1492.

18. Ohashi R, Tamai I, Yabuuchi H, Nezu JI, Oku A, Sai Y, Shimane M, Tsuji A. 1999. Na(+)-dependent carnitine transport by organic cation transporter (OCTN2): its pharmacological and toxicological relevance. J Pharmacol Exp Ther 291:778–784.

19. Seth P, Wu X, Huang W, Leibach FH, Ganapathy V. 1999. Mutations in novel organic cation transporter (OCTN2), an organic cation/carnitine transporter, with differential effects on the organic cation transport function and the carnitine transport function. J Biol Chem 274:33388–33392.

20. Berardi S, Stieger B, Hagenbuch B, Carafoli E, Krahenbuhl S. 2000. Characterization of L-carnitine transport into rat skeletal muscle plasma membrane vesicles. Eur J Biochem 267:1985–1994.

21. Prasad PD, Huang W, Ramamoorthy S, Carter AL, Leibach FH, Ganapathy V. 1996. Sodium-dependent carnitine transport in human placental choriocarcinoma cells. Biochim Biophys Acta 1284:109–117.

22. Roque AS, Prasad PD, Bhatia JS, Leibach FH, Ganapathy V. 1996. Sodium-dependent high-affinity binding of carnitine to human placental brush border membranes. Biochim Biophys Acta 1282:274–282.

23. Stieger B, O'Neill B, Krahenbuhl S. 1995. Characterization of L-carnitine transport by rat kidney brush-border-membrane vesicles. Biochem J 309: 643–647.

24. Ganapathy ME, Huang W, Rajan DP, Carter AL, Sugawara M, Iseki K, Leibach FH, Ganapathy V. 2000. Beta-lactam antibiotics as substrates for OCTN2, an organic cation/carnitine transporter. J Biol Chem 275:1699–1707.

25. Ohashi R, Tamai I, Nezu Ji J, Nikaido H, Hashimoto N, Oku A, Sai Y, Shimane M, Tsuji A. 2001. Molecular and physiological evidence for multifunctionality of carnitine/organic cation transporter OCTN2. Mol Pharmacol 59:358–366.

26. Ohashi R, Tamai I, Inano A, Katsura M, Sai Y, Nezu J, Tsuji A. 2002. Studies on functional sites of organic cation/carnitine transporter OCTN2 (SLC22A5) using a Ser467Cys mutant protein. J Pharmacol Exp Ther 302:1286–1294.

27. Wu X, Prasad PD, Leibach FH, Ganapathy V. 1998. cDNA sequence, transport function, and genomic organization of human OCTN2, a new member of the organic cation transporter family. Biochem Biophys Res Commun 246:589–595.

28. Lahjouji K, Elimrani I, Lafond J, Leduc L, Qureshi IA, Mitchell GA. 2004. L-Carnitine transport in human placental brush-border membranes is mediated by the sodium-dependent organic cation transporter OCTN2. Am J Physiol Cell Physiol 287:C263–C269.

29. Terada T, Shimada Y, Pan X, Kishimoto K, Sakurai T, Doi R, Onodera H, Katsura T, Imamura M, Inui K. 2005. Expression profiles of various transporters for oligopeptides, amino acids and organic ions along the human digestive tract. Biochem Pharmacol l70:1756–1763.

30. Hashimoto N, Suzuki F, Tamai I, Nikaido H, Kuwajima M, Hayakawa J, Tsuji A. 1998. Gene-dose effect on carnitine transport activity in embryonic fibroblasts of JVS mice as a model of human carnitine transporter deficiency. Biochem Pharmacol 55:1729–1732.

31. Lu K, Nishimori H, Nakamura Y, Shima K, Kuwajima M. 1998. A missense mutation of mouse OCTN2, a sodium-dependent carnitine cotransporter, in the juvenile visceral steatosis mouse. Biochem Biophys Res Commun 252:590–594.

32. Nezu J, Tamai I, Oku A, Ohashi R, Yabuuchi H, Hashimoto N, Nikaido H, Sai Y, Koizumi A, Shoji Y et al. 1999. Primary systemic carnitine deficiency is caused by mutations in a gene encoding sodium ion–dependent carnitine transporter. Nat Genet 21:91–94.
33. Kobayashi D, Irokawa M, Maeda T, Tsuji A, Tamai I. 2005. Carnitine/organic cation transporter OCTN2-mediated transport of carnitine in primary-cultured epididymal epithelial cells. Reproduction 130:931–937.
34. Kai S, Yakushiji K, Yamauchi M, Ito C, Kuwajima M, Osada Y, Toshimori K. 2005. Expression of novel organic cation/carnitine transporter (OCTN2) in the mouse pancreas. Tissue Cell 37:309–315.
35. Duran JM, Peral MJ, Calonge ML, Ilundain AA. 2005. OCTN3: A Na^+-independent L-carnitine transporter in enterocytes basolateral membrane. J Cell Physiol 202:929–935.
36. Lamhonwah AM, Ackerley CA, Tilups A, Edwards VD, Wanders RJ, Tein I. 2005. OCTN3 is a mammalian peroxisomal membrane carnitine transporter. Biochem Biophys Res Commun 338:1966–1972.
37. Enomoto A, Wempe MF, Tsuchida H, Shin HJ, Cha SH, Anzai N, Goto A, Sakamoto A, Niwa T, Kanai Y, Anders MW, Endou H. 2002. Molecular identification of a novel carnitine transporter specific to human testis. Insights into the mechanism of carnitine recognition. J Biol Chem 277:36262–36271.
38. Gong S, Lu X, Xu Y, Swiderski CF, Jordan CT, Moscow JA. 2002. Identification of OCT6 as a novel organic cation transporter preferentially expressed in hematopoietic cells and leukemias. Exp Hematol 30:1162–1169.
39. Kwok B, Yamauchi A, Rajesan R, Chan L, Dhillon U, Gao W, Xu H, Wang B, Takahashi S, Semple J, et al. 2006. Carnitine/xenobiotics transporters in the human mammary gland epithelia, MCF12A. Am J Physiol Regul Integr Comp Physiol 290:R793–802.
40. Nakanishi T, Hatanaka T, Huang W, Prasad PD, Leibach FH, Ganapathy ME, Ganapathy V. 2001. Na^+- and Cl^--coupled active transport of carnitine by the amino acid transporter ATB($0,^+$) from mouse colon expressed in HRPE cells and *Xenopus* oocytes. J Physiol 532:297–304.
41. Ganapathy ME, Ganapathy V. 2005. Amino Acid Transporter ATB$0,^+$ as a delivery system for drugs and prodrugs. Curr Drug Targets Immune Endocrinol Metab Disord 5:357–364.
42. Taylor PM. 2001. Absorbing competition for carnitine. J Physiol 532:283.
43. Berezowski V, Miecz D, Marszalek M, Bröer A, Bröer S, Cecchelli R, Nalecz KA. 2004. Involvement of OCTN2 and B$0,^+$ in the transport of carnitine through an in vitro model of the blood–brain barrier. J Neurochem 91:860–1872.
44. Tein I. 2003. Carnitine transport: pathophysiology and metabolism of known molecular defects. J Inherit Metab Dis 26:147–169.
45. Lahjouji K, Elimrani I, Wu J, Mitchell GA, Qureshi IA. 2002. A heterozygote phenotype is present in the jvs +/− mutant mouse livers. Mol Genet Metab 76:76–80.
46. Yokogawa K, Yonekawa M, Tamai I, Ohashi R, Tatsumi Y, Higashi Y, Nomura M, Hashimoto N, Nikaido H, Hayakawa J, et al. 1999. Loss of wild-type carrier-mediated L-carnitine transport activity in hepatocytes of juvenile visceral steatosis mice. Hepatology 30:997–1001.
47. Brass EP. 2002. Pivalate-generating prodrugs and carnitine homeostasis in man. Pharmacol Rev 54:589–598.

48. Stanley CA, Hale DE, Berry GT, Deleeuw S, Boxer J, Bonnefont JP. 1992. Brief report: a deficiency of carnitine–acylcarnitine translocase in the inner mitochondrial membrane. N Engl J Med 327:19–23.

49. Schreiber S, Hanpe J, Nikolaus S, Foelsch UR. 2004. Exploration of the genetic aetiology of inflammatory bowel disease: implications for diagnosis and therapy. Aliment Pharmacol Ther 20:1–8.

50. Rioux JD, Silverberg MS, Daly MJ, Steinhart AH, McLeod RS, Griffiths AM, Green T, Brettin TS, Stone V, Bull SB, et al. 2000. Genomewide search in Canadian families with inflammatory bowel disease reveals two novel susceptibility loci. Am J Hum Genet 66:1863–1870.

51. Armuzzi A, Ahmad T, Ling KL, de Silva A, Cullen S, van Heel D, Orchard TR, Welsh KI, Marshall SE, Jewell DP. 2003. Genotype–phenotype analysis of the Crohn's disease susceptibility haplotype on chromosome 5q31. Gut 52:1133–1139.

52. Peltekova VD, Wintle RF, Rubin LA, Amos CI, Huang Q, Gu X, Newman B, van Oene M, Cescon D, Greenberg G, et al. 2004. Functional variants of OCTN cation transporter genes are associated with Crohn disease. Nat Genet 36:471–475.

53. Lamhonwah AM, Skaug J, Scherer SW, Tein I. 2003. A third human carnitine/organic cation transporter (OCTN3) as a candidate for the 5q31 Crohn's disease locus (IBD5). Biochem Biophys Res Commun 30:98–101.

54. Lamhonwah AM, Ackerley C, Onizuka R, Tilups A, Lamhonwah D, Chung C, Tao KS, Tellier R, Tein I. 2005. Epitope shared by functional variant of organic cation/carnitine transporter, OCTN1, *Campylobacter jejuni* and *Mycobacterium paratuberculosis* may underlie susceptibility to Crohn's disease at 5q31. Biochem Biophys Res Commun 337:1165–1175.

55. Martinez A, Del Carmen Martin M, Mendoza JL, Taxonera C, Diaz-Rubio M, de la Concha EG, Urcelay E. 2006. Association of the organic cation transporter OCTN genes with Crohn's disease in the Spanish population. Eur J Hum Genet 14:222–226.

56. Waller S, Tremelling M, Bredin F, Godfrey L, Howson J, Parkes M. 2005. Evidence for association of OCTN genes and IBD5 with ulcerative colitis. Gut 55:809–814.

57. Lerner A, Gruener N, Iancu TC. 1993. Serum carnitine concentrations in coeliac disease. Gut 34:933–935.

58. Ruemmele FM, Dionne S, Qureshi I, Sarma DS, Levy E, Seidman EG. 1999. Butyrate mediates Caco-2 cell apoptosis via up-regulation of pro-apoptotic BAK and inducing caspase-3 mediated cleavage of poly-(ADP-ribose) polymerase (PARP). Cell Death Differ 6:729–735.

59. Roediger WE. 1992. Oxidative and synthetic functions of *n*-butyrate in colonocytes. Dis Colon Rectum 35:511–512.

60. Segain JP, Raingeard de la Bletiere D, Bourreille A, Leray V, Gervois N, Rosales C, Ferrier L, Bonnet C, Blottiere HM, Galmiche JP. 2000. Butyrate inhibits inflammatory responses through NFkappaB inhibition: implications for Crohn's disease. Gut 47:397–403.

61. Huang N, Katz JP, Martin DR, Wu GD. 1997. Inhibition of IL-8 gene expression in Caco-2 cells by compounds which induce histone hyperacetylation. Cytokine 9:27–36.

62. Berger J, Moller DE. 2002. The mechanisms of action of PPARs. Annu Rev Med 53:409–435.

63. Tokuhiro S, Yamada R, Chang X, Suzuki A, Kochi Y, Sawada T, Suzuki M, Nagasaki M, Ohtsuki M, Ono M, et al. 2003. An intronic SNP in a RUNX1 binding site of SLC22A4,

encoding an organic cation transporter, is associated with rheumatoid arthritis. Nat Genet 35:341–348.

64. Stephens FB, Constantin-Teodosiu D, Laithwaite D, Simpson EJ, Greenhaff PL. 2006. Insulin stimulates L-carnitine accumulation in human skeletal muscle. FASEB J 20:377–379.
65. Clausen T. 2003. Na^{+}-K^{+} pump regulation and skeletal muscle contractility. Physiol Rev 83:1269–1324.
66. Kato Y, Sai Y, Yoshida K, Watanabe C, Hirata T, Tsuji A. 2005. PDZK1 directly regulates the function of organic cation/carnitine transporter OCTN2. Mol Pharmacol 67:734–743.
67. Garcia-Miranda P, Duran JM, Peral MJ, Ilundain AA. 2005. Developmental maturation and segmental distribution of rat small intestinal L-carnitine uptake. J Membr Biol Jul:206:9–16.

4

ORGANIC ANION TRANSPORTERS

MEGHA NAGLE, DAVID M. TRUONG, VIBHA BHATNAGAR, GREGORY KALER, KEVIN T. BUSH, WEI WU, SATISH A. ERALY, AND SANJAY K. NIGAM

University of California San Diego, La Jolla, California

4.1. INTRODUCTION

The "classical" organic anion (OA) or organic acid secretory system is thought to be involved fundamentally in the excretion of common drugs, toxins, and endogenous metabolites into the urine, thereby maintaining the equilibrium of anionic metabolites

Drug Transporters: Molecular Characterization and Role in Drug Disposition, Edited by Guofeng You and Marilyn E. Morris
Copyright © 2007 John Wiley & Sons, Inc.

in the body. Organic anion transporters (OATs), the proteins that underlie the OA secretory system, typify secondary (and/or tertiary) active membrane transporters, which translocate substrates across the cell membrane by utilizing a transmembrane electrochemical gradient of the substrate itself or of another solute.[1] To date, the OATs identified include OAT1 (originally, NKT), OAT2 (originally, NLT), OAT3 (originally, Roct), OAT4, OAT5, mouse OAT6, and URAT1 (originally, RST) (Table 4.1). as well as a number of orphan family members.[2–7]

OATs belong to the largest family of secondary active membrane transporters: the major facilitator superfamily (MFS), which is conserved from bacteria to mammals.[8–10] MFS transporters function as transmembrane uniporters, symporters, and antiporters, and transport a wide range of hydrophilic and amphiphilic substrates,[11] including inorganic ions (Na^+, Cl^-, HCO_3^-, etc.), endogenous metabolites (amino acids, sugars, neurotransmitters, etc.) and xenobiotics (drugs and toxins). However, OATs specialize in transporting organic anions, and most of the accumulated knowledge concerning OAT function has been found by studying prototypical substrates such as *p*-aminohippurate and estrone 3-sulfate. Within this superfamily, the OATs are members of the solute carrier family SLC22a, which includes the structurally and functionally related organic cation transporters (OCTs), organic carnitine transporters (OCTNs), unknown soluble transporters (USTs), and flylike putative transporters (Flipts) (http://www.bioparadigms.org/slc/intro.asp).[12]

OATs are heavily expressed in excretory organs such as the kidney, suggesting a critical role in maintaining endogenous homeostasis. Of note, the OATs are implicated in several clinical disorders and are important modulators of drug efficacy and toxicity, due to their ability to transport a large number of the most commonly prescribed pharmaceuticals, such as angiotensin-converting enzyme (ACE) inhibitors, loop and thiazide diuretics, β-lactam antibiotics, methotrexate, and nonsteroidal antiinflammatory drugs.[13]

In recent years there has been considerable progress in understanding the molecular basis of the renal secretory process, owing to the continued identification of organic anion transporters that mediate basolateral and apical organic anion uptake. Furthermore, these transporters have also been found to be present in numerous epithelial tissues other than the kidney, including the olfactory mucosa, liver, choroid plexus, retina, and placenta,[14] which potentially implicates OATs more broadly in secretory processes. Moreover, the sequencing of the human genome has uncovered the chromosomal organization of OAT genes, enabling a deeper picture of the OATs at the molecular, evolutionary, and regulatory levels.

4.2. GENOMICS OF OATs

Aside from their obvious structural and functional conformity (Figure 4.1), the OATs also manifest similar tissue expression patterns.[15] OATs are expressed predominantly in the kidney; however, functionally similar genes such as OAT1 and OAT3 are also both found in choroid plexus, and hOAT5 and UST3 are both found in the liver. Therefore, there may be regulatory elements that function to link expression of these genes in certain tissues.

TABLE 4.1. OAT Family Members

Gene Name (Gene symbol)[a]	Original Name and Reference	Prototypical Substrate(s)	Transport Mechanism[b]	Human Tissue Distribution	Membrane Localization	Species Identified	Gender Difference	Human Gene Locus
OAT1/NKT (SLC22A6)	NKT; ref. 2	*p*-Aminohippurate	OA–DC exchange	Kidney, brain, eyes	Basolateral	Human, mouse, rat, pig, flounder	(rat) male > female	11q12.3
OAT2/NLT (SLC22A7)	NLT; ref. 3	*p*-Aminohippurate	unknown	Kidney, liver	unknown	Human, mouse, rat	(mouse liver) female > male	6q21.1-2
OAT3/Roct (SLC22A8)	Roct; ref. 4	Estrone 3-sulfate	OA–DC exchange	Kidney, brain, eyes, liver, adrenal glands	Basolateral	Human, mouse, rat, pig, rabbit	(rat) male > female	11q12.3
OAT4 (SLC22A11)	Ref. 14	Estrone 3-sulfate	OA–DC exchange	Kidney, placenta, mouse brain and mouse testis	Apical	Human	unknown	11q13.1
OAT5 (SLC22A19)	Ref. 84	Ochratoxin A	unknown	Kidney	unknown	Mouse, rat	unknown	11q12.3
OAT6 (SLC22A20)	Ref. 75	Estrone 3-sulfate	unknown	Mouse olfactory mucosa	unknown	Mouse	unknown	unknown
URAT1/RST (SLC22A12)	RST; ref. 85	Urate	Urate–anion exchange	Kidney	Apical	Human, mouse	(mouse) male > female	11q13.1

[a] SLC, solute carrier family.
[b] OA, organic anion; DC, dicarboxylate.

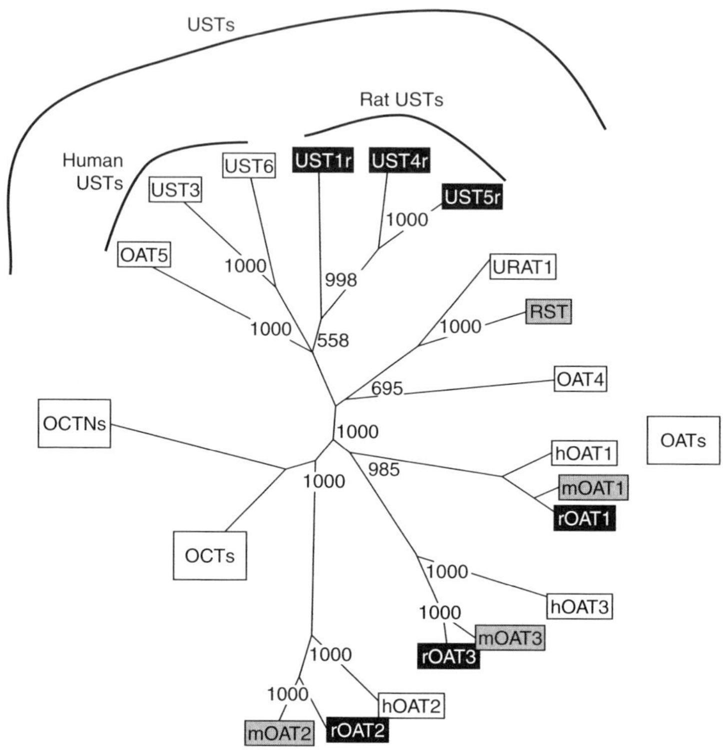

FIGURE 4.1. Schematic illustrating the SLC22 family generated by aligning the sequences of rat, mouse, and human OATs (black, gray, and white boxes, respectively), and representative OCTs and OCTNs, and using the alignment output to generate a dendrogram. (From ref. 82.)

A relatively unique feature of the SLC22a family genes is that many members tend to appear as pairs or clusters in the genome. Six of the eight known OATs are found in three tightly linked pairs (i.e., as adjoining neighbors without other genes interposed between them): OAT1–OAT3, OAT4–RST/URAT1, and UST3–hOAT5.[15] The dendrogram of the OAT family in Figure 4.2 reveals that these physical pairs are also closely related phylogenetic pairs. Both *OAT1* and *OAT3* are found on human chromosome 11q12.3 in a tandem repeat, separated by a mere 8 kilobase pairs; *OAT4* and *URAT1* are found on 11q13.1 in a tandem repeat, separated by 20 kb. These observations suggest that the pairing of OAT genes (which are expressed primarily in the kidney) might exist to facilitate the coordinated transcription (coregulation) of pairmembers.

In the postgenomic era, tools for identifying potential regulatory elements and transcription binding sites have been utilized to explore potential control regions[16–19] and determine consensus sequence binding sites for transcription factors integral to biological processes during kidney development (e.g., PAX1, PBX, WT1, TCF, and

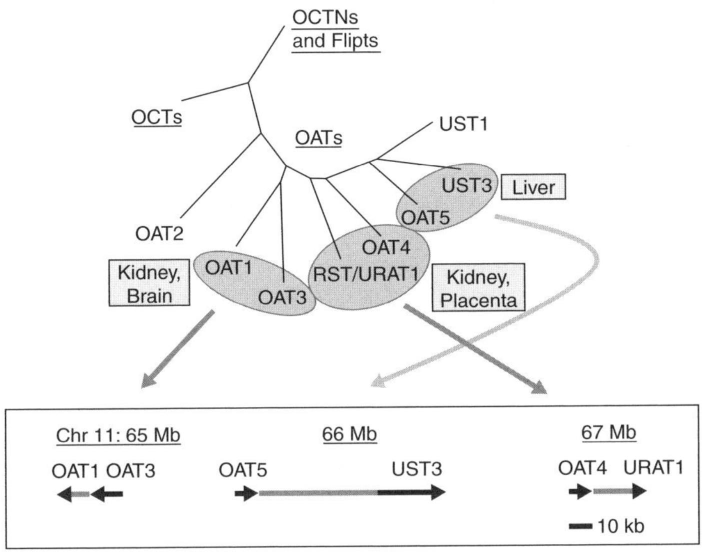

FIGURE 4.2. Pairing, phylogeny, and tissue distribution of the OATs. Ovals indicate gene pairing, and the adjacent text boxes indicate their tissue distributions. The gene pairs are represented schematically in the rectangle below the dendrogram (to the scale in the lower right corner of the figure). Genes are drawn as arrows that indicate transcription direction and the intergenic regions between paired genes as line segments. The approximate chromosomal locations of the gene pairs are indicated (in megabases from the p telomere of chromosome 11). (From ref. 13.)

HNF1).[20] For example, some putative transcription factor binding sites present in OAT noncoding regions may play a role in proximal tubule differentiation. Furthermore, experiments have shown that liver-enriched hepatocyte nuclear factor HNF-4α (nuclear receptor 2A1) can bind to and transactivate the *SLC22A7/OAT2* promoter.[21]

Although we are only at the early stages of understanding regulation of the expression of these transporters, earlier physiological studies have provided some insight into the regulation of OATs. Postnatal maturation of OA transport has been found to be influenced by a number of factors, including substrate availability, age, and gender. For example, increased OA uptake into renal cortical slices can be induced by prior exposure to organic anions. Since the increase could be prevented by the protein synthesis inhibitor, cycloheximide, substrate exposure is believed to induce the synthesis of either the OATs themselves or of other proteins required for OA uptake.[22,23] OA transport has also been shown to be under endocrine regulation. Transport of the prototypical OAT substrate, *p*-aminohippurate (PAH), can be stimulated by treatment with dexamethasone or thyroid hormones, particularly in young rats.[24,25] Such regulation may be one potential mechanism by which organic anion transport is activated during a critical period for OAT synthesis: neonatal development.

4.3. ONTOGENY OF OATs

Identification of the specific proteins mediating organic ion secretion at the molecular level has allowed for analysis of the pre- and postnatal ontogeny of OATs, which may have clinical implications for the dosing of drugs to premature infants and full-term newborns. Analysis of the postnatal expression of these transporters showed that OAT1 expression in the postnatal kidney correlated with the changes in OA transport observed previously in physiological studies[26] and that the expression levels for the OATs increased progressively postnatally in the rat.[27] However, significant differences were observed between male and female rats.[27] For example, while mRNA levels of OAT1 increased progressively in male rats, its expression peaked in female rats at postpartum day 30 with a subsequent decline such that by postpartum day 45, expression levels of OAT1 were significantly lower in female rats than in male rats.[27] OAT2, on the other hand, was minimally expressed in both male and female rats for the first 30 days of postnatal development, but then increased in female rats only, through day 45. Finally, OAT3 was found to increase progressively over the observed course of postnatal development in both male and female rats. Taken together with the results of the physiological studies described above, these data (despite any gender differences in expression) support the notion that the postnatal maturation of renal OA transport observed may depend on increased expression of the transporters.

Despite numerous physiological studies, few data exist on the embryonic ontogeny of these critically important transporters. When OAT1 was first identified by our group (as NKT), its expression was determined in the embryonic kidney.[2] Subsequently, using a combination of Northern blot analysis and in situ hybridization, the spatiotemporal expression patterns of the organic anion transporters (OAT1, OAT2, and OAT3) during murine development were examined.[28] Interestingly, the expression patterns for all of these genes in the kidney were remarkably similar. The mRNA encoding these proteins was detectable in the kidney at days 14 to 16 of embryogenesis, and the expression levels continued to rise through embryogenesis, with the highest level detectable (by in situ hybridization and Northern blot analysis) in adult kidney.[28] Similar patterns of expression were observed in a separate study of the ontogeny of OAT1 in the embryonic rat kidney.[26] The roughly cotemporal expression of these transporters in the developing proximal tubule of the kidney suggests the existence of a common transcriptional regulatory pathway for the expression of OATs.[28] Recently, detailed QPCR examination of the prenatal renal ontogeny of the OATs in the rat was reported (Figure 4.3), in which OAT1 and OAT3, as well as markers of proximal tubular differentiation, Na^+/glucose cotransporter (SGLT1) and Na^+/phosphate cotransporter (NaP2), were found to increase progressively over the embryonic period examined (ed13-ed18).[29] Transporter expression and function was also examined in two commonly used models of nephrogenesis: (1) culture of the whole embryonic kidney[29] and (2) coculture of the uninduced metanephric mesenchyme (an embryonic progenitor tissue of kidney which gives rise to the nephrons of the kidney) with a heterologous inducing tissue (i.e., embryonic spinal cord).[30] Both of these culture systems are invaluable tools for the study of kidney development since much of early nephrogenesis is reproduced in vitro with remarkable fidelity (Figure 4.4).[31] The

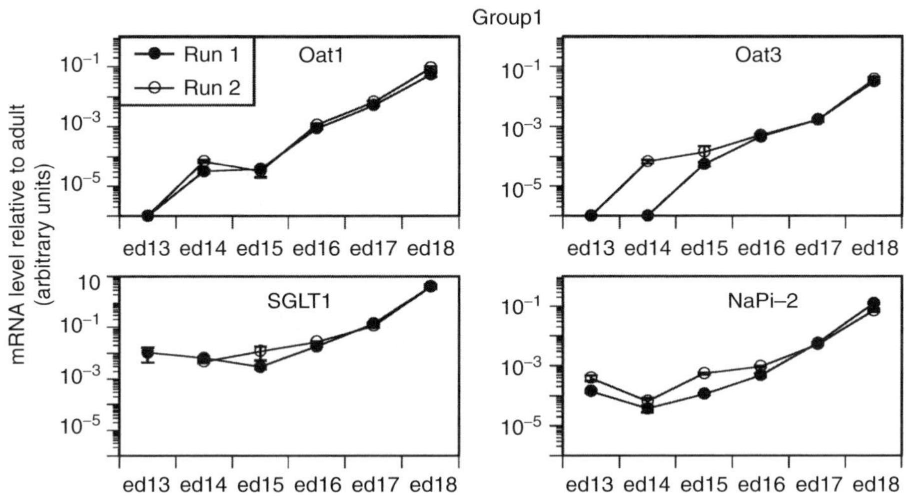

FIGURE 4.3. Relative gene expression levels of OAT genes in embryonic kidney as determined by QPCR, illustrating the up-regulation of SLC family genes as rat kidney maturation progresses. Data were normalized to the expression of glyceraldehyde-3-phosphate dehydrogenase in the same sample and are reported as the ratio of the level of normalized gene expression in the sample of interest to the level of gene expression in adult kidney (mean values ± S.D.). (From ref. 29.)

expression patterns of OAT1, OAT3, SGLT1, and NaPi2 were found to closely resemble those observed in vivo.[29] Moreover, using fluorescein uptake as an in vitro functional assay of OAT expression, inhibition of fluorescein accumulation by probenecid was observed within the tubular structures of the cultured organs.[29] Together with the expression studies, these findings indicate not only that the OATs are expressed functionally during kidney development and might be markers of proximal tubule differentiation, but also raise the possibility that these organ culture systems may represent convenient and reliable in vitro models for study of the developmental induction of the OATs. Given the increasingly lower gestational ages at which infants are being delivered, these types of studies may eventually shed light on the ability of kidneys in premature babies to handle drugs and toxins.

4.4. STRUCTURE AND FUNCTION OF OATs

Although OATs are thought to be similar to other MFS transporters, having 12 α-helical transmembrane domains (TMDs) as well as functional similarity, the crystal structure of OATs is not known, and hence their transport mechanism remains unclear. However, close examination of the primary structure reveals integral clues; weak sequence homology between the two six-helix halves suggests that MFS proteins originated from the duplication of a six-TMD ancestor.[8,32] There are two large

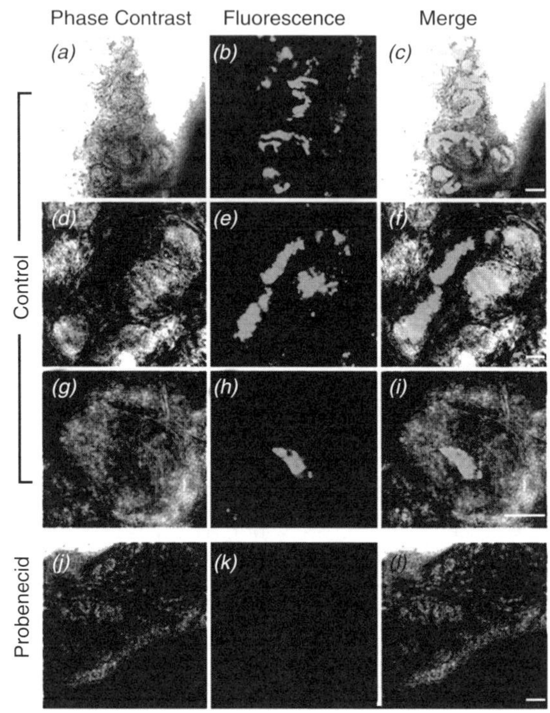

FIGURE 4.4. Confocal microscopic images showing fluorescein accumulation in induced metanephric mesenchyme (MM) due to expression of organic anion transporters. Control; (*a–i*) or presence of 2 mM probenecid (probenecid; *j–l*). (*a, d, g, j*) Phase-contrast photomicrographs of the induced epithelial tubular structures. (*b, e, h, k*) Corresponding fluorescence photomicrograph of the tissue shown in the preceding panel. (*c, f, i, l*) Merged image of the two preceding photomicrographs, indicating the accumulation of fluorescein in the induced MM. (*a–c*) Low-magnification examination of a group of tubular structures in the induced mesenchyme in the absence of probenecid; bar = 100 μM. (*d–f*) High-magnification examination of tubular structures in the mesenchyme induced in the absence of probenecid; bar = 20 μM. (*g–i*) Higher-magnification examination of a tubular structure in the induced mesenchyme in the absence of probenecid; bar = 20 μM. Note the accumulation of fluorescein, to a concentration greater than the medium, in what appears to be a fluid-filled space (presumptive lumen). (*j–l*) Low-magnification examination of a group of tubular structures in the mesenchyme induced in the presence of 2 mM probenecid; bar = 100 μM. Note the absence of concentrative fluorescein accumulation. (From ref. 29.) (*See insert for color representation of figure.*)

interconnecting loops in OAT1, one between TMD 1/2, and another between TMD 6/7[13,33–35] (Figure 4.5). The first large loop is extracellular and contains multiple consensus N-glycosylation sites. The role of these sites in the regulation of transport function has not been determined, although glycosylation of OAT1 was shown to be required for protein trafficking to the membrane.[36] The second large loop is intracellular and contains several canonical protein kinase C (PKC) phosphorylation sites.[35,37]

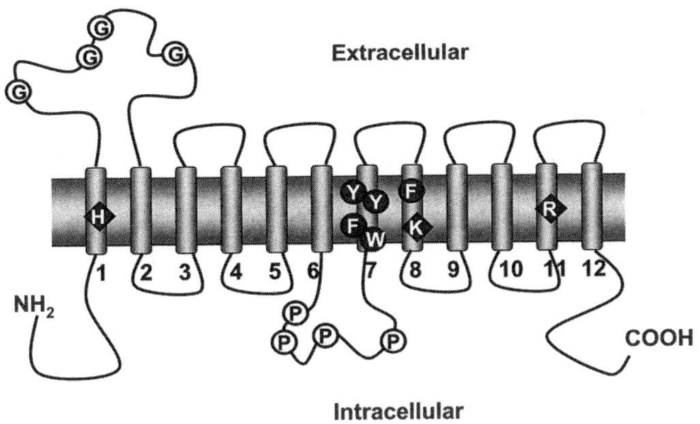

FIGURE 4.5. Schematic illustrating the transmembrane topology of organic anion transporters. Hydropathy analyses indicate that the OATs comprise 12 TMDs (numbered in the figure). The large loops between TMDs 1 and 2 (extracellular) and TMDs 6 and 7 (intracellular) contain several consensus glycosylation (G) and PKC phosphorylation (P) sites, respectively. (From ref. 13.)

Despite the fact that PKC down-regulates OAT function, the phosphorylation sites of the second loop are probably not involved directly in this down-regulation.[35,38–41]

OATs are multispecific transporters that bind a vast array of substrates, with binding being based on general substrate physicochemical properties (charge, hydrophobicity, hydrogen-bonding ability) rather than distinctive molecular characteristics.[42,43] However, there are no direct data on the molecular interactions of substrate ions with the amino acid residues lining the channel of the OATs which might indicate the exact properties that determine substrate binding and transport. Mutagenesis studies have been conducted targeting highly conserved amino acid residues believed to be essential for transport function. These studies have identified conserved aromatic and cationic residues, which appear to interact, respectively, with hydrophobic and anionic moieties of substrates.[44–46] For example, in rat OAT3, five conserved aromatic residues (located in TMDs 7 and 8) and three conserved basic residues (in TMDs 1, 8, and 11) are required for transport activity, as identified by expression studies in *Xenopus* oocytes.[45] Importantly, in the OCTs (cation-transporting homologs of the OATs), acidic or neutral residues are located in the positions corresponding to the critical basic residues in the OATs. Therefore, these conserved basic residues are believed to determine the substrate charge specificity of the OATs. Remarkably, a rat OAT3 double mutant with the Lys370 and Arg454 residues substituted by one neutral and one acidic residue (K370A/R454D) has been reported to change its substrate orientation from anions to cations.[44]

Although the exact mechanism for substrate transport is not known, considerable data indicate that it is based on ion exchange across the membranes of the tubular cell. Basolateral transport is driven by the concentration gradient of intracellular dicarboxylates, mostly α-ketoglutarate (αKG), which are exchanged for extracellular anions

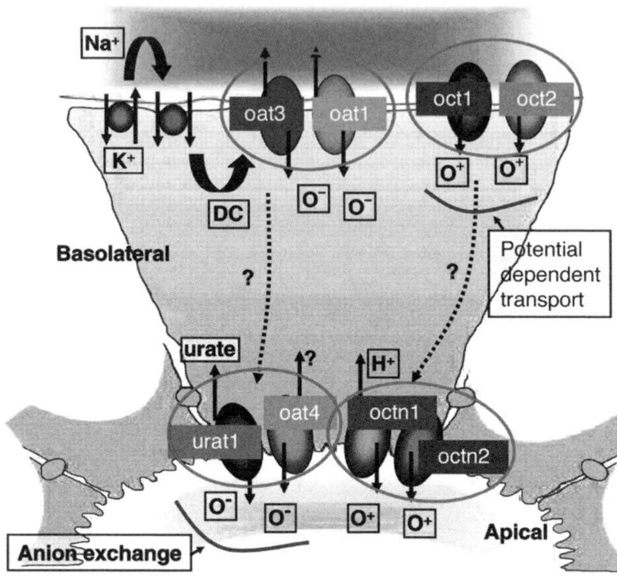

FIGURE 4.6. Mechanisms of proximal tubular uptake and efflux of organic ions mediated by organic anion and cation transporters (OATs, OCTs, OCTNs). Transporters whose encoding genes are chromosomally paired are enclosed in ovals. (From ref. 82.) (*See insert for color representation of figure.*)

in an adenosine triphosphate (ATP)-independent mechanism. As a requisite, the high concentration gradient of dicarboxylates is maintained by the sodium–dicarboxylate cotransporter, which in turn is driven by the inwardly directed sodium gradient across the basolateral membrane generated by Na^+/K^+-ATPase (Figure 4.6). Basolateral entry of anions is followed by apical efflux through the brush border membrane, probably through a sodium-independent system whose exact mechanism remains unclear.

Even though the structure of the OATs has not been elucidated, the crystal structure has been determined recently for three related MFS proteins: the glycerol 3-phosphate (G3P) transporter from the *Escherichia coli* inner membrane (a G3P/P_i antiporter, GlpT), the *E. coli* lactose permease (a lactose/H^+ symporter, LacY), and the oxalate transporter from *Oxalobacter formigenes* (an oxalate–formate antiporter, OxlT).[47–54] These transporters have similar topology, suggesting similar structural design for other MFS proteins, including the OATs. Therefore, the resolved structure of these bacterial transporters, with substrate-binding sites located at the interface between the N- and C-terminal halves of the protein, may prove useful as a template for structural modeling of the OATs and other SLC22 transporters.

4.4.1. Substrate Translocation

Using MFS transporter mechanisms as models for OATs, one might postulate a single-binding-site alternating access mechanism such as that suggested for the MFS transporters[55,56] and elaborated in more detail in thermodynamic and kinetic studies.[57–59] In this mechanism the transporter switches between two alternating conformations, inward facing (C_i) and outward facing (C_o), thus allowing translocation of substrate across the membrane. Recently, a more detailed mechanism of the conformational change has been proposed based on the crystal structure of a bacterial antiporter, GlpT.[53] According to this mechanism (Figure 4.7), substrate binding within the transporter binding site reduces the energy barrier between the inward- and outward-facing conformations (C_i and C_o) and thus facilitates a transition between them, involving a swiveling movement of the two six-TMD halves of the transporter relative to each other ("rocker-switch" mechanism). Analysis of the density map of another bacterial transporter, OxlT, suggests the existence of a third, "closed" conformation of the transporter (C_c) with the substrate-binding site isolated from both cytoplasm and periplasm, as an intermediate state in the transition between the C_i and C_o conformations.[51]

Importantly, in anion-transporting MFS proteins, including OATs, the binding site supposedly comprises (is formed by) two positively charged amino acid residues. These two key residues are, for example, Arg45 (located in the α-helix H1) and Arg269 (H7) in GlpT, Arg46 (H1) and Arg275 (H7) in UhpT (sugar phosphate transporter of *E. coli*), Arg272 (H8) and Lys355 (H11) in OxlT, and Lys370 (H8) and Arg454 (H11) in rOAT3.[44,49,50,52,60] Based on the structure of GlpT, the two key positively charged residues in the substrate-binding site were suggested to form hydrogen bonds simultaneously with a negatively charged substrate ion.[52,53] The formation of such a "bridge" complex (e.g., Arg45 $\cdots$ P_i $\cdots$ Arg269 in GlpT) upon binding a substrate ion to a transporter in the conformation open to one side of the membrane was suggested as a mechanism to pull the N- and C-terminal domains closer to each other and make further tilting easier (Figure 4.8).

4.5. ANIMAL MODELS

As noted earlier, the evidence accumulated indicates that OAT1 and OAT3 manifest functional properties and anatomical localization consistent with a critical role in the basolateral uptake step of renal secretion of organic anions by the classical pathway. However, most of this evidence has derived from in vitro studies, so that the actual in vivo function of OATs, in the context of the whole kidney and/or the entire organism, has remained uncertain. Recently, mice containing null alleles for OAT1 or OAT3 were generated using homologous recombination. Studies of OA transport in these knockout mice have begun to define the role of OAT1 and OAT3 in vivo[61,62] and their potential role in nephrotoxicity and drug–drug interactions (Figure 4.9). Some of these results are discussed below.

4.5.1. Analysis of Transport Ex Vivo

Although OAT1 and OAT3 knockout mice are viable, fertile, and appear grossly normal, they do manifest specific functional defects in renal organic anion handling

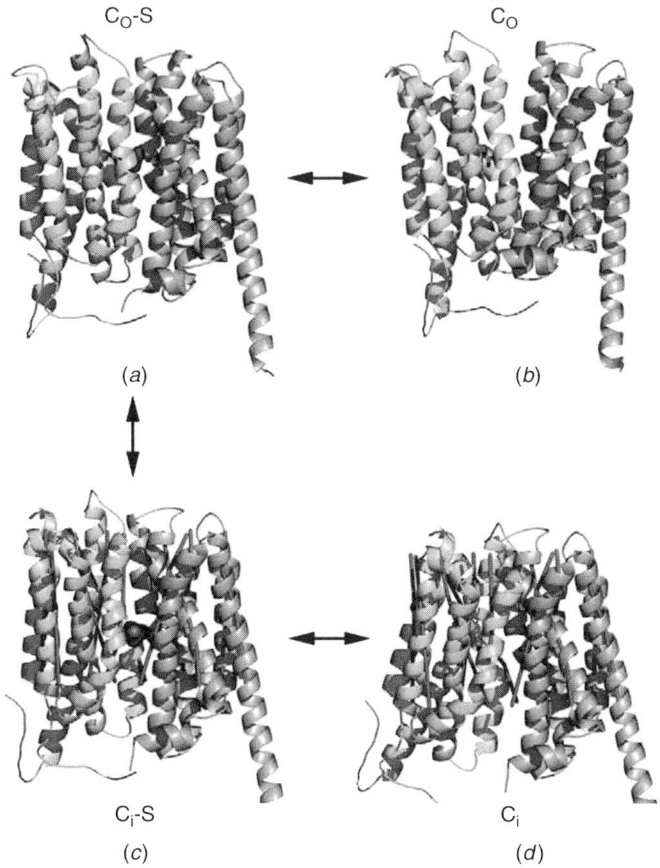

FIGURE 4.7. Proposed conformational changes that accompany substrate translocation by GlpT. The crystal structure (*d*) represents the C_i conformation of the protein. The C_i–P_i conformation (*c*) was generated by fitting the GlpT model into the 6.5-Å map of the substrate-bound form of OxlT,[50] the latter being represented by elongated rods. By rotating the two halves of the GlpT model separately in opposite directions along an axis at their interface and parallel to the membrane, it was found that a 6° rotation of each domain can generate a structure that fits the OxlT map reasonably well. The C_o–P_i conformation (*a*) was produced by a 10° rotation of each domain that is sufficient to close the pore on the cytosolic side of the molecule and to open a pore on the periplasmic side. Finally, the C_o conformation (*b*) was generated by a 16° rotation. (From ref. 53.) (*See insert for color representation of figure.*)

which were detectable upon measurement of the transport of particular OAT substrates. Uptake of organic anions was determined in isolated renal cortical slices from wild-type and knockout mice in the absence and presence of the OAT inhibitor probenecid. To determine the OAT-specific component of transport, the difference between uptake in the absence versus in the presence of inhibitor was estimated. As expected, the OAT-specific component of the uptake of the prototypic organic anion

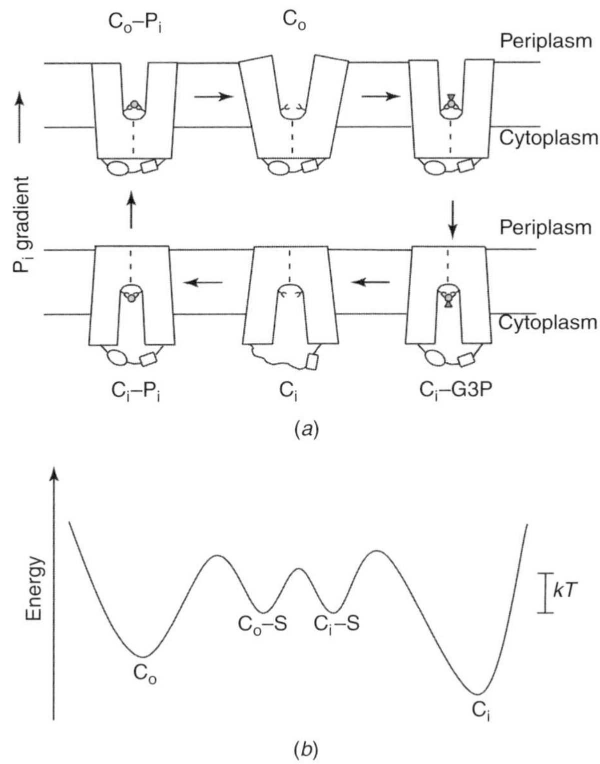

FIGURE 4.8. Proposed single binding site, alternating access mechanism with a rocker-switch type of movement. (*a*) Reaction cycle of substrate translocation. The figure shows the positions of Arg45 and Arg269. P_i is represented by a small disk, and the G3P molecule is represented by a small disk and a triangle. (*b*) Schematic drawing to illustrate the free-energy levels of various conformations of GlpT in the translocation cycle. S denotes a substrate. [(*a*) From ref. 52; (*b*) from ref. 53.]

p-aminohippurate (PAH) was largely abolished in renal slices from OAT1 knockout animals, whereas uptake of estrone sulfate (ES), a well-characterized substrate of OAT3 (but not of OAT1),[63–65] was not reduced significantly. Conversely, renal slices from OAT3 knockout mice manifested nearly complete loss of ES uptake but only partial loss of PAH uptake. These findings suggest that the bulk of renal basolateral transport of PAH is mediated by OAT1, while that of ES is mediated by OAT3. In the OAT3 knockout mouse, organic anion transport in choroid plexus was also found to be defective.[61]

4.5.2. Analysis of Secretion In Vivo

The data above demonstrated the presence of specific deficits in the capacity for basolateral uptake of organic anions into the renal cortex of OAT knockout mice. To

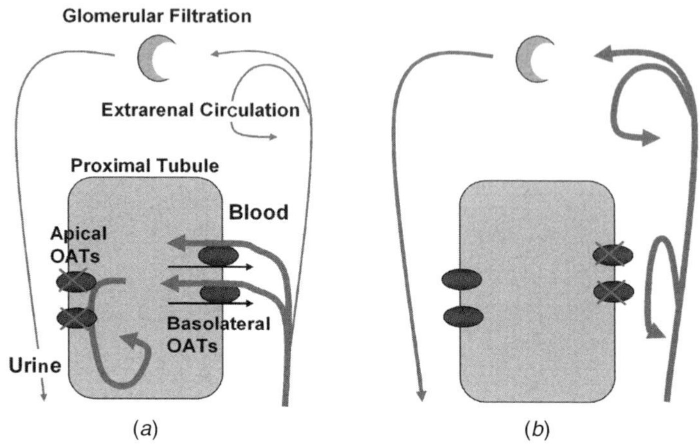

FIGURE 4.9. Knockout of organic anion transporter pairs as models for defects in renal drug handling. (*a*) Knockout of apical organic anion transporters (OATs) might result in increased nephrotoxicity due to "unopposed" transport of potential nephrotoxins into the proximal tubular cell by basolateral OATs. (*b*) Conversely, knockout of basolateral OATs could lead to increased extrarenal toxicity due to delayed clearance. (From ref. 83.)

determine whether these transport deficits translated to diminished renal secretion, the urinary excretion of organic anions in OAT1 knockout and wild-type mice was compared by performing clearance experiments under anesthesia. Although there was no significant decrement in ES clearance in the OAT1 knockout mouse (as expected given the ex vivo results above), the renal clearance of PAH was severely reduced (Figure 4.10). Since there were no differences in glomerular filtration between the genotypes (as determined by measurement of inulin clearance), all of this reduction was attributable to loss of net secretion. A long body of physiological work supports the notion that PAH is the prototype of the highly diverse set of organic anions that are handled in common by the classical pathway. Therefore, our finding that OAT1 is responsible for the bulk of renal secretion of PAH in vivo suggests an integral role in the functioning of this pathway, and thus in the excretion of the numerous clinically important compounds counted among its substrates. The naturetic response to the diuretic furosemide was also blunted in OAT1 knockout mice. Perhaps most interesting was the accumulation of a number of endogenous metabolites in the serum of these animals. Several of these compounds were found to inhibit PAH transport by OAT1 in *Xenopus* oocytes, suggesting that they may be "true" OAT substrates in vivo.

4.6. HUMAN VARIATION AND CLINICAL IMPLICATIONS

Because the OATs are at crucial cellular interfaces for various excretory organs (such as the liver and kidneys), variation or polymorphisms in these genes may account for clinically important differences in drug efficacy and renal drug handling of commonly

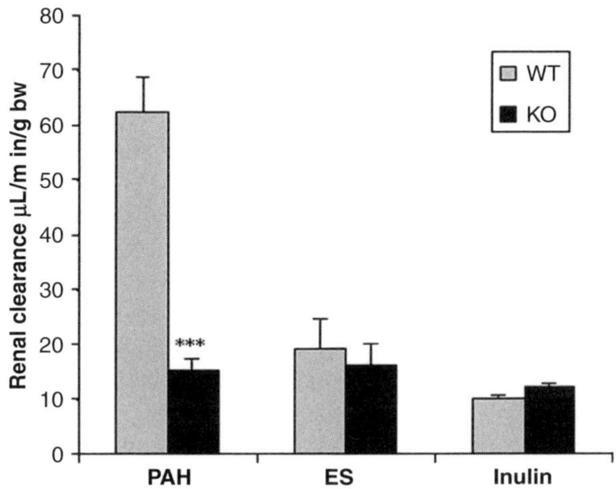

FIGURE 4.10. Renal clearance of organic anions in wild-type and *OAT1* knockout mice. Establishment of stable plasma levels of PAH, ES, and inulin (via continuous intravenous infusion) and then measurement of urinary excretion in wild-type mice (WT, gray bars) and *OAT1* knockout mice (KO, black bars) allowed calculation of renal clearance (ratio of rate of urinary excretion to plasma concentration). The data represent the mean ± S.E. of measurements in four to six mice. Renal clearance of PAH, but not of ES, was significantly lower in the knockout mice than in the wild-type mice. Clearance of inulin, representing a measure of glomerular filtration, was similar in wild-type and knockout mice. ***, $p < 0.001$. (From ref. 62.)

prescribed pharmaceuticals [e.g., ACE inhibitors, methotrexate, and nonsteroidal anti-inflammatory drugs (NSAIDs)] as well as elimination of key metabolites. Therefore, polymorphisms in the OAT genes may account partially for differing disease prevalence and risk of adverse drug reactions (ADRs) among different human subpopulations. ADRs are of particular interest because there are over 2 million incidents and approximately 100,000 deaths related to ADRs in the United States each year. Differences in the expression or function of OATs may alter the elimination of many pharmaceutical agents significantly, thereby increasing the risk of a significant ADR.

In particular, there has been considerable interest in single-nucleotide polymorphisms (SNPs) in this family of genes and their potential role in drug handling by the kidney, thereby affecting drug concentrations and half-lives in the serum. These considerations also apply to transport across all the barrier epithelia in which OATs are expressed: renal clearance (OAT1, OAT2, OAT3, OAT4, and URAT1), transport across the blood–brain barrier via the choroid plexus (OAT1 and OAT3), transport across the placenta (OAT4), and transport across the olfactory mucosa (mouse OAT6).

It has been shown that certain coding region polymorphisms affect drug transport by OATs. For example, certain OAT1 and OAT3 SNPs are known to affect function in *Xenopus* oocyte expression systems. The effects of these coding region SNPs on in vivo drug handling remain unclear.[44,45,66,67] However, coding region polymorphisms seem relatively uncommon,[68] raising the possibility that noncoding region

SNPs, which could affect OAT transcriptional activity, may be clinically important. Such polymorphisms have recently been identified. The pairing of many OATs in the genome also raises the possibility that a polymorphism in the transcriptional region of one OAT could also affect the expression of the paired OAT.[69] Moreover, given the redundancy of OATs as demonstrated by both transport studies and in knockout mice, it is probably most useful to consider multiple OAT SNPs together. Furthermore, since net transport of organic anions across an epithelial barrier (e.g., the renal proximal tubule) depends on both apical and basolateral transport mechanisms, it is probably important to consider the entire set of apical and basolateral transporters in order to obtain a more accurate perspective on how various polymorphisms affect such transport.[69] Thus, the combinatorial effects of certain SNPs in different OATs on organic anion transport in vivo could be quite complex, but perhaps a more clinically relevant approximation.

However, genomic variation (either at the level of expression or function) may have clinical implications (change in drug efficacy or toxicity) only with certain disease states (an environment–gene interaction). For example, expression of OAT3 is thought to be directly related to the clearance of cefazolin, a commonly prescribed antibiotic. This relationship, however, was only observed in patients with mesangial proliferative glomerulonephritis, in whom decreased expression of OAT3 mRNA was strongly associated with decreased renal clearance of cefozolin.[70] This implies that certain patients are at a higher risk of developing cefazolin-related drug toxicity (e.g., hepatitis), depending on individual differences in expression of the OAT3 gene. Therefore, genomic variation may be especially important in stressed or disease-specific environments.

There is also a growing body of research that is delineating the importance of OATs in organ systems other than the kidney, and these are mentioned here only briefly. Efflux of xenobiotics and organic anion substrates across the blood–brain barrier, for example, is thought to occur in a three-step process in the epithelial cells of the choroid plexus, similar to the transport of substrates across the renal proximal tubule cell: apical uptake, transport across the cell, and basolateral transport into the blood.[71] Consistent with this model, OAT3 is located on the apical surface of epithelial cells of the choroid plexus, and knockout mice demonstrate decreased organic anion transport, suggesting an important role for OATs in protecting the central nervous system from toxic endogenous and exogenous substrates.[72] Similarly, the placenta protects the fetus from exposure to toxic substrates and allows the delivery of pharmaceutical drugs (such as zidovudine prophylaxis against HIV) by similar mechanisms.[73] OAT4, for example, has been localized to the placenta and is known to transport many drugs, such as zidovudine, antibiotics, and antihypertensives.[74] The discovery of the olfactory OAT, OAT6, also raises the possibility that SNPs in this gene could be important in determining the efficacy of nasally administered drugs as well as in olfaction.[75,76]

4.6.1. Other Clinical Aspects

OAT1 and OAT3, expressed primarily on the basolateral surface of renal proximal tubular cells, are responsible for the uptake from the blood into the cell, whereas

OAT4 and RST/URAT1, located on the apical surface, are considered candidate genes involved in efflux from the cell into the proximal tubule and/or substrate reabsorption into the proximal tubular cell.[13] One disease thought to be a result of altered organic anion clearance is gout. Gout can be associated with uric acid crystal deposition in the kidney (resulting in nephropathy) or within joints (resulting in an acutely painful inflammatory arthritis). Uric acid is a product of purine metabolism and undergoes net reabsorption in the renal proximal tubule. URAT1, the human homolog of the gene first identified as RST,[5] is thought to transport uric acid across the apical proximal tubular cell. Accordingly, case–control and cohort studies have suggested that loss of function polymorphisms on URAT1 are associated with *hypouricemia*. One study found decreased renal uric acid associated with a human polymorphism close to the N-terminus (C426T),[77] and uncovered another human polymorphism, 313A (a deletion from 313D to 333P), that had no significant uric acid transport in *Xenopus* ooyctes.[78] Nevertheless, the molecular basis of renal urate handling in vivo remains poorly understood. A detailed understanding of molecular mechanisms underlying uric acid clearance could result in novel pharmaceutical treatments for gout that target uric acid transporters.[13,79]

Competition for OAT-binding sites may explain certain types of toxic reactions. Most of the drugs in plasma bind competitively to OATs at the level of the substrate-binding site and influence the pharmacokinetics of other drugs. Since this transporter system has the capacity to recognize and bind a variety of endogenous substrates, OATs are likely to be involved in certain toxic or nephrotoxic drug reactions (Table 4.2).[80] For example, ochratoxin A, a mycotoxin that causes Balkan nephropathy, has been shown experimentally to be accumulated via OAT1 in the renal proximal tubular cells.[35] Another example is the nephrotoxicity due to the antiviral drug

TABLE 4.2. Examples of Nephrotoxic and Neurotoxic Agents Demonstrated to Interact with OATs

Nonsteroidal anti-inflammatory drugs	Uremic toxins	Antivirals
Acetaminophen	Hippuric acid	Acyclovir
Diclofenac	Indoleacetic acid	Adefovir
Ibuprofen	Indoxyl sulfate	Azidothymidine
Indomethacin	Chemotherapeutics	Cidofovir
Ketoprofen	Methotrexate	Ganciclovir
Naproxen	Heavy Metals	Mycotoxins
Phenacetin	Cadmium	Ochratoxin A
Piroxicam	Mercury	Neurotransmitter metabolites
Salicylate	Chlorinated phenoxyacetates	3,4-Dihydroxymandelic acid
Antibiotics	2,4-Dichlorophenoxyacetic acid	3,4-Dihydroxyphenylacetic acid (DOPAC)
Cephalosporins	Chlorinated haloalkenes	Miscellaneous
Penems	1,2-Dichlorovinyl-L-cysteines	Homovanillic acid (HVA)
Penicillins		Hydroxyindoleacetic acid (5-HIAA)

Source: ref. 81.

cidofovir (used to treat ophthalmic cytomegalovirus infection and thought to be active against orthopox viruses such as smallpox), which is known to be mediated by active secretion of cidofovir in the renal tubules. This effect has shown to be counteracted by coadministration of probenecid, a potent inhibitor of organic anion secretion in the proximal renal tubular cells.[64] Methotrexate clearance has been reported to be reduced with NSAIDs acting as competitive substrates for the binding sites of renal organic anion transporters.[35] The net effect of this competition is an increase in plasma concentration of methotrexate and the manifestation of severe toxicity, in the form of bone marrow suppression and damage to intestinal epithelium. Alternatively, the multispecificity of OATs can be used to prolong the action of drugs by competitive inhibition of secretion at the OAT binding site. Such inhibition decreases renal excretion and enhances retention of drugs, an example being the common practice of coadministering probenecid to prolong the action of β-lactam antibiotics. In summary, it may be appropriate to consider the classic secretory transporter system as a target for the treatment of various drug-related side effects and that modification at this cellular level could open up a new era of therapeutic strategies for clinically significant diseases.

4.7. CONCLUSIONS

Since the identification of NKT (later called OAT1), OATs have been shown to play key roles in mediating the renal absorption and excretion of drugs, xenobiotics, and endogenous metabolites. Thus, It has become readily apparent that targeting of this family of transporters, and an understanding of their transport mechanisms, is critical for elucidating mechanisms of drug handling and nephrotoxicity. Given that most of the OAT family members have overlapping substrate specificities and tissue distribution, and both primary sequence and structural homology, it would not be that surprising if they exhibited complex, coupled regulation, as if a clustered unit. Widespread genetic profiling of humans and a more comprehensive list of substrate affinities may ultimately lead to individualized patient care and usher in an era where renal drug, xenobiotic handling, and certain types of toxicity are seen in light of OAT transport capacity. Further work will focus on the structural basis of substrate transport as well as the creation of a list of transported endogenous and exogenous molecules that could potentially uncover the true physiological functions of the OATs and other SLC22 family members. Finally, the complex regulatory mechanisms likely to underlie OAT family transcription and cell surface regulation must be understood in vivo, and to do this, animal models, including single and multiple OAT knockouts, must be created. These approaches should lead the field closer to a more integrated understanding of renal drug handling and toxicity.

Acknowledgments

S.K.N. was supported by grants R01 AI057695 and HD40011, S.A.E. by DK064839.

REFERENCES

1. Mitchell P. 1963. Molecule, group and electron translocation through the natural membranes. Biochem Soc Symp 22:141–168.
2. Lopez-Nieto CE, You G, Bush KT, Barros EJ, Beier DR, Nigam SK. 1997. Molecular cloning and characterization of NKT, a gene product related to the organic cation transporter family that is almost exclusively expressed in the kidney. J Biol Chem 272(10): 6471–6478.
3. Simonson GD, Vincent AC, Roberg KJ, Huang Y, Iwanij V. 1994. Molecular cloning and characterization of a novel liver-specific transport protein. J Cell Sci 107(4):1065–1072.
4. Brady KP, Dushkin H, Fornzler D, Koike T, Magner F, Her H, Gullans S, Segre GV, Green RM, Beier DR. 1999. A novel putative transporter maps to the osteosclerosis (oc) mutation and is not expressed in the oc mutant mouse. Genomics 56(3):254–261.
5. Mori K, Ogawa Y, Ebihara K, Aoki T, Tamura N, Sugawara A, Kuwahara T, Ozaki S, Mukoyama M, Tashiro K. 1997. Kidney-specific expression of a novel mouse organic cation transporter-like protein. FEBS Lett 417(3):371.
6. Wolff NA, Werner A, Burckhardt S, Burckhardt G. 1997. Expression cloning and characterization of a renal organic anion transporter from winter flounder. FEBS Lett 417(3):287–291.
7. Sweet DH, Wolff NA, Pritchard JB. 1997. Expression cloning and characterization of ROAT 1. The basolateral organic anion transporter in rat kidney. J Biol Chem 272(48):30088–30095.
8. Pao SS, Paulsen IT, Saier MH Jr. 1998. Major facilitator superfamily. Microbiol Mol Biol Rev 62(1):1–34.
9. Reizer J, Finley K, Kakuda D, MacLeod CL, Reizer A, Saier MH Jr. 1993. Mammalian integral membrane receptors are homologous to facilitators and antiporters of yeast, fungi, and eubacteria. Protein Sci 2(1):20–30.
10. Ren Q, Kang KH, Paulsen IT. 2004. TransportDB: a relational database of cellular membrane transport systems. Nucleic Acids Res 32(database issue):D284–D288.
11. Saier MH Jr, Beatty JT, Goffeau A, Harley KT, Heijne WH, Huang SC, Jack DL, Jahn PS, Lew K, Liu J, et al. 1999. The major facilitator superfamily. J Mol Microbiol Biotechnol 1(2):257–279.
12. Koepsell H, Endou H. 2004. The SLC22 drug transporter family. Pflugers Arch 447(5):666–676.
13. Eraly SA, Bush KT, Sampogna RV, Bhatnagar V, Nigam SK. 2004. The molecular pharmacology of organic anion transporters: From DNA to FDA? Mol Pharmacol 65(3):479–487.
14. Cha SH, Sekine T, Kusuhara H, Yu E, Kim JY, Kim DK, Sugiyama Y, Kanai Y, Endou H. 2000. Molecular cloning and characterization of multispecific organic anion transporter 4 expressed in the placenta. J Biol Chem 275(6):4507–4512.
15. Eraly SA, Hamilton BA, Nigam SK. 2003. Organic anion and cation transporters occur in pairs of similar and similarly expressed genes. Biochem Biophys Res Commun 300(2):333–342.
16. Fickett JW, Wasserman WW. 2000. Discovery and modeling of transcriptional regulatory regions. Curr Opin Biotechnol 11(1):19–24.

17. Frech K, Quandt K, Werner T. 1997. Finding protein-binding sites in DNA sequences: the next generation. Trends Biochem Sci 22(3):103–104.
18. Hardison RC. 2000. Conserved noncoding sequences are reliable guides to regulatory elements. Trends Genet 16(9):369–372.
19. Pennacchio LA, Olivier M, Hubacek JA, Cohen JC, Cox DR, Fruchart J-C, Krauss RM, Rubin EM. 2001. An apolipoprotein influencing triglycerides in humans and mice revealed by comparative sequencing. Science 294(5540):169–173.
20. Eraly SA, Nigam SK. 2002. Novel human cDNAs homologous to *Drosophila* Orct and mammalian carnitine transporters. Biochem Biophys Res Commun 297(5):1159–1166.
21. Popowski K, Eloranta JJ, Saborowski M, Fried M, Meier PJ, Kullak-Ublick GA. 2005. The human organic anion transporter 2 gene is transactivated by hepatocyte nuclear factor-4 alpha and suppressed by bile acids. Mol Pharmacol 67(5):1629–1638.
22. Hirsch GH, Hook JB. 1969. Stimulation of *p*-aminohippurate transport by slices of rat renal cortex following in vivo administration of triiodothyronine. Proc Soc Exp Biol Med 131(2):513–517.
23. Stopp M, Hartwich R, Braunlich H. 1978. Enhancement of *p*-aminohippurate accumulation in renal cortical slices after repeated administrations of various organic anionic drugs to rats of different ages. Experientia 34(11):1493–1494.
24. Braunlich H. 1988. Hormonal control of postnatal development of renal tubular transport of weak organic acids. Pediatr Nephrol 2(1):151–155.
25. Braunlich H. 1988. Hormonal regulation of postnatal development of renal tubular transport processes. Physiol Bohemoslov 37(4):347–350.
26. Nakajima N, Sekine T, Cha SH, Tojo A, Hosoyamada M, Kanai Y, Yan K, Awa S, Endou H. 2000. Developmental changes in multispecific organic anion transporter 1 expression in the rat kidney. Kidney Int 57(4):1608–1616.
27. Buist SC, Cherrington NJ, Choudhuri S, Hartley DP, Klaassen CD. 2002. Gender-specific and developmental influences on the expression of rat organic anion transporters. J Pharmacol Exp Ther 301(1):145–151.
28. Pavlova A, Sakurai H, Leclercq B, Beier DR, Yu AS, Nigam SK. 2000. Developmentally regulated expression of organic ion transporters NKT (OAT1), OCT1, NLT (OAT2), and Roct. Am J Physiol Renal Physiol 278(4):F635–F643.
29. Sweet DH, Eraly SA, Vaughn DA, Bush KT, Nigam SK. 2006. Organic anion and cation transporter expression and function during embryonic kidney development and in organ culture models. Kidney Int 69(5):837–845.
30. Grobstein C. 1967. Mechanisms of organogenetic tissue interaction. Natl Cancer Inst Monogr 26:279–299.
31. Steer DL, Bush KT, Meyer TN, Schwesinger C, Nigam SK. 2002. A strategy for in vitro propagation of rat nephrons. Kidney Int 62(6):1958–1965.
32. Maiden MC, Davis EO, Baldwin SA, Moore DC, Henderson PJ. 1987. Mammalian and bacterial sugar transport proteins are homologous. Nature 325(6105):641–643.
33. Burckhardt G, Wolff NA. 2000. Structure of renal organic anion and cation transporters. Am J Physiol Renal Physiol 278(6):F853–F866.
34. Sweet DH, Pritchard JB. 1999. The molecular biology of renal organic anion and organic cation transporters. Cell Biochem Biophys 31(1):89–118.

35. You G. 2002. Structure, function, and regulation of renal organic anion transporters. Med Res Rev 22(6):602–616.

36. Kuze K, Graves P, Leahy A, Wilson P, Stuhlmann H, You G. 1999. Heterologous expression and functional characterization of a mouse renal organic anion transporter in mammalian cells. J Biol Chem 274(3):1519–1524.

37. Sweet DH, Bush KT, Nigam SK. 2001. The organic anion transporter family: from physiology to ontogeny and the clinic. Am J Physiol Renal Physiol 281(2):F197–F205.

38. Berkhin EB, Humphreys MH. 2001. Regulation of renal tubular secretion of organic compounds. Kidney Int 59(1):17–30.

39. Terlouw SA, Masereeuw R, Russel FG. 2003. Modulatory effects of hormones, drugs, and toxic events on renal organic anion transport. Biochem Pharmacol 65(9):1393–1405.

40. Wolff NA, Thies K, Kuhnke N, Reid G, Friedrich B, Lang F, Burckhardt G. 2003. Protein kinase C activation downregulates human organic anion transporter 1–mediated transport through carrier internalization. J Am Soc Nephrol 14(8):1959–1968.

41. You G, Kuze K, Kohanski RA, Amsler K, Henderson S. 2000. Regulation of mOAT-mediated organic anion transport by okadaic acid and protein kinase C in LLC-PK(1) cells. J Biol Chem 275(14):10278–10284.

42. Ullrich KJ. 1997. Renal transporters for organic anions and organic cations. Structural requirements for substrates. J Membr Biol 158(2):95–107.

43. Ullrich KJ. 1999. Affinity of drugs to the different renal transporters for organic anions and organic cations. Pharm Biotechnol 12:159–179.

44. Feng B, Dresser MJ, Shu Y, Johns SJ, Giacomini KM. 2001. Arginine 454 and lysine 370 are essential for the anion specificity of the organic anion transporter, rOAT3. Biochemistry 40(18):5511–5520.

45. Feng B, Shu Y, Giacomini KM. 2002. Role of aromatic transmembrane residues of the organic anion transporter, rOAT3, in substrate recognition. Biochemistry 41(28):8941–8947.

46. Wolff NA, Grunwald B, Friedrich B, Lang F, Godehardt S, Burckhardt G. 2001. Cationic amino acids involved in dicarboxylate binding of the flounder renal organic anion transporter. J Am Soc Nephrol 12(10):2012–2018.

47. Abramson J, Kaback HR, Iwata S. 2004. Structural comparison of lactose permease and the glycerol-3-phosphate antiporter: members of the major facilitator superfamily. Curr Opin Struct Biol 14(4):413–419.

48. Abramson J, Smirnova I, Kasho V, Verner G, Kaback HR, Iwata S. 2003. Structure and mechanism of the lactose permease of *Escherichia coli*. Science 301(5633):610–615.

49. Hirai T, Heymann JA, Maloney PC, Subramaniam S. 2003. Structural model for 12-helix transporters belonging to the major facilitator superfamily. J Bacteriol 185(5):1712–1718.

50. Hirai T, Heymann JA, Shi D, Sarker R, Maloney PC, Subramaniam S. 2002. Three-dimensional structure of a bacterial oxalate transporter. Nat Struct Biol 9(8):597–600.

51. Hirai T, Subramaniam S. 2004. Structure and transport mechanism of the bacterial oxalate transporter OxlT. Biophys J 87(5):3600–3607.

52. Huang Y, Lemieux MJ, Song J, Auer M, Wang DN. 2003. Structure and mechanism of the glycerol-3-phosphate transporter from *Escherichia coli*. Science 301(5633):616–620.

53. Lemieux MJ, Huang Y, Wang DN. 2004. The structural basis of substrate translocation by the *Escherichia coli* glycerol-3-phosphate transporter: a member of the major facilitator superfamily. Curr Opin Struct Biol 14(4):405–412.

54. Lemieux MJ, Song J, Kim MJ, Huang Y, Villa A, Auer M, Li XD, Wang DN. 2003. Three-dimensional crystallization of the *Escherichia coli* glycerol-3-phosphate transporter: a member of the major facilitator superfamily. Protein Sci 12(12):2748–2756.

55. Patlak C. 1957. Contributions to the theory of active transport: II. The gate type non-carrier mechanism and generalizations concerning tracer flow, efficiency, and measurement of energy expenditure. Bull Math Biophys 19:209–235.

56. Vidaver GA. 1966. Inhibition of parallel flux and augmentation of counter flux shown by transport models not involving a mobile carrier. J Theor Biol 10(2):301–306.

57. Jencks WP. 1980. The utilization of binding energy in coupled vectorial processes. Adv Enzymol Relat Areas Mol Biol 51:75–106.

58. Tanford C. 1983. Mechanism of free energy coupling in active transport. Annu Rev Biochem 52:379–409.

59. West IC. 1997. Ligand conduction and the gated-pore mechanism of transmembrane transport. Biochim Biophys Acta 1331(3):213–234.

60. Fann M, Davies AH, Varadhachary A, Kuroda T, Sevier C, Tsuchiya T, Maloney PC. 1998. Identification of two essential arginine residues in UhpT, the sugar phosphate antiporter of *Escherichia coli*. J Membr Biol 164(2):187–195.

61. Sweet DH, Miller DS, Pritchard JB, Fujiwara Y, Beier DR, Nigam SK. 2002. Impaired organic anion transport in kidney and choroid plexus of organic anion transporter 3 (Oat3 (Slc22a8)) knockout mice. J Biol Chem 277(30):26934–26943.

62. Eraly SA, Vallon V, Vaughn DA, Gangoiti JA, Richter K, Nagle M, Monte JC, Rieg T, Truong DM, Long JM, Barshop BA, Kaler G, Nigam SK. 2006. Decreased renal organic anion secretion and plasma accumulation of endogenous organic anions in OAT1 knock-out mice. J Biol Chem 281(8):5072–5083.

63. Burckhardt BC, Burckhardt G. 2003. Transport of organic anions across the basolateral membrane of proximal tubule cells. Rev Physiol Biochem Pharmacol 146:95–158.

64. Russel FG, Masereeuw R, van Aubel RA. 2002. Molecular aspects of renal anionic drug transport. Annu Rev Physiol 64:563–594.

65. Dresser MJ, Leabman MK, Giacomini KM. 2001. Transporters involved in the elimination of drugs in the kidney: organic anion transporters and organic cation transporters. J Pharm Sci 90(4):397–421.

66. Bleasby K, Hall LA, Perry JL, Mohrenweiser HW, Pritchard JB. 2005. Functional consequences of single nucleotide polymorphisms in the human organic anion transporter hOAT1 (SLC22A6). J Pharmacol Exp Ther 314(2):923–931.

67. Matsuda A, Hirota T, Akahoshi M, Shimizu M, Tamari M, Miyatake A, Takahashi A, Nakashima K, Takahashi N, Obara K, et al. 2005. Coding SNP in tenascin-C Fn-III-D domain associates with adult asthma. Hum Mol Genet 14(19):2779–2786.

68. Xu G, Bhatnagar V, Wen G, Hamilton BA, Eraly SA, Nigam SK. 2005. Analyses of coding region polymorphisms in apical and basolateral human organic anion transporter (OAT) genes [OAT1 (NKT), OAT2, OAT3, OAT4, URAT (RST)]. Kidney Int 68(4):1491–1499.

69. Bhatnagar V, Xu G, Hamilton BA, Truong DM, Eraly SA, Wu W, Nigam SK. 2006. Analyses of 5′ regulatory region polymorphisms in human SLC22A6 (OAT1) and SLC22A8 (OAT3). J Hum Genet 51(6):575–580.

70. Sakurai Y, Motohashi H, Ogasawara K, Terada T, Masuda S, Katsura T, Mori N, Matsuura M, Doi T, Fukatsu A, Inui K. 2005. Pharmacokinetic significance of renal OAT3 (SLC22A8) for anionic drug elimination in patients with mesangial proliferative glomerulonephritis. Pharm Res 22(12):2016–2022.
71. Miller DS. 2004. Confocal imaging of xenobiotic transport across the choroid plexus. Adv Drug Deliv Rev 56(12):1811–1824.
72. Kusuhara H, Sugiyama Y. 2004. Efflux transport systems for organic anions and cations at the blood–CSF barrier. Drug Transfer Choroid Plexus Multiplicity Substr Specif Transport 56(12):1741.
73. Syme MR, Paxton JW, Keelan JA. 2004. Drug transfer and metabolism by the human placenta. Clin Pharmacokinet 43(8):487–514.
74. Zhou F, Illsley NP, You G. 2006. Functional characterization of a human organic anion transporter hOAT4 in placental BeWo cells. Eur J Pharm Sci 27(5):518–523.
75. Monte JC, Nagle MA, Eraly SA, Nigam SK. 2004. Identification of a novel murine organic anion transporter family member, OAT6, expressed in olfactory mucosa. Biochem Biophys Res Commun 323(2):429–436.
76. Schnabolk GW, Youngblood GL, Sweet DH. 2006. Transport of estrone sulfate by the novel organic anion transporter Oat6 (Slc22a20). Am J Physiol Renal Physiol 291(2):314–321.
77. Graessler J, Graessler A, Unger S, Kopprasch S, Tausche AK, Kuhlisch E, Schroeder HE. 2006. Association of the human urate transporter 1 with reduced renal uric acid excretion and hyperuricemia in a German Caucasian population. Arthritis Rheum 54(1):292–300.
78. Iwai N, Mino Y, Hosoyamada M, Tago N, Kokubo Y, Endou H. 2004. A high prevalence of renal hypouricemia caused by inactive SLC22A12 in Japanese. Kidney Int 66(3):935–944.
79. Hediger MA, Johnson RJ, Miyazaki H, Endou H. 2005. Molecular physiology of urate transport. Physiology (Bethesda) 20:125–133.
80. Tahara H, Kusuhara H, Maeda K, Koepsell H, Fuse E, Sugiyama Y. 2006. Inhibition of oat3-mediated renal uptake as a mechanism for drug–drug interaction between fexofenadine and probenecid. Drug Metab Dispos 34(5):743–747.
81. Sweet DH. 2005. Organic anion transporter (Slc22a) family members as mediators of toxicity. Toxicol Appl Pharmacol 204(3):198.
82. Eraly SA, Monte JC, Nigam SK. 2004. Novel slc22 transporter homologs in fly, worm, and human clarify the phylogeny of organic anion and cation transporters. Physiol Genom 18(1):12–24.
83. Eraly SA, Roland C, Bhatnagar V, Nigam SK. 2003. Novel aspects of renal organic anion transporters. Curr Opin Nephrol Hypertension 12(5):551–558.
84. Young blood GL, Sweet DH. 2004. Identification and functional assessment of the novel murine organic anion transporter Oat 5 (Slc 22a 19) expressed in kidney. Am J Physiol Renal Physiol 287(2):236–244.
85. Mori K, Ogawa Y, Ebihara K, Aoki T, Tamara N, Sugawaa A, Kuwahara T, Ozaki S, Mukoyama M, Tashiro K. 1997. Kidney-specific expression of a novel mouse organic cation transporter-like protein. FEBS Lett 417(3):371.

5

ORGANIC ANION–TRANSPORTING POLYPEPTIDES

ROMMEL G. TIRONA AND RICHARD B. KIM

University of Western Ontario, London, Ontario, Canada

Drug Transporters: Molecular Characterization and Role in Drug Disposition, Edited by Guofeng You and Marilyn E. Morris
Copyright © 2007 John Wiley & Sons, Inc.

5.1. OVERVIEW OF THE OATP SUPERFAMILY

5.1.1. Introduction

The study of drug transport across biological membranes has gathered considerable attention over the last decade as a result of rapid advances in genetic and molecular biological tools that promoted the elucidation of the responsible transporter proteins. Among the most widely investigated transport proteins are the organic anion–transporting polypeptides (OATPs). These proteins are characteristically known for their sodium-independent transport of substrates with broad specificity. Interest in the OATPs continues to grow due to the cumulative evidence gathered from in vitro and in vivo studies that demonstrate important roles for these transporters in determining intestinal drug absorption, hepatic and renal clearance, and tissue distribution.

In this chapter we begin with a review of the OATP superfamily nomenclature, followed by sections covering molecular characteristics, aspects of gene expression and regulation, transport substrates and inhibitors, and relevance to pharmacology and physiology. Specific attention is focused on members of the human OATP family, although illustrations with nonhuman Oatps are noted where relevant. Readers are also directed toward several general reviews[1–3] and chapters in this book that cover in greater detail elements of OATPs in various species.

5.1.2. Nomenclature

For several years after the discovery of the first Oatp,[4] the rapid identification of new members of the transporter superfamily in various species along with the various naming systems was a source of confusion for those who were not closely following the field. It remained difficult to determine the relatedness and identities of the Oatps within and between species until the new classification and nomenclature was developed by Hagenbuch and Meier[2] and approved by the HUGO Gene Nomenclature Committee. In this system, Oatps are classified according to conventions originally established for the cytochrome P450 (CYP) enzyme superfamily,[5] which are based on amino acid sequence identities. For the Oatps, the italicized gene symbol begins with *Slco* while the encoded proteins are named with the root Oatp. For human genes and proteins, capitalized letters are used (e.g., *SLCO*/OATP), whereas for rodents only the initial letter is in capitals with the rest being in lowercase letters (e.g., *Slco*/Oatp). Oatps within the same family that share $\geq 40\%$ amino acid sequence identities have root names followed by Arabic numbers, of which six families are known in humans. Subfamilies that share $\geq 60\%$ amino acid sequence identities are indicated by a letter following the family designation. For instance, OATP family 1 (OATP1) has three subfamilies (A, B, and C). Specific transporter proteins are then given numerical designation after the subfamily heading. Hence OATP1B1 is the first named member belonging to family 1, subfamily B. The final numeral is named chronologically and allows for unambiguous identification of Oatp transporters between species. It was envisioned that this consensus nomenclature would promote standardization and clarity in transporter science. However, despite its introduction, the new nomenclature has not prevented some inadvertent comparisons of nonhomologous Oatps among humans and rodents.[6,7]

Allelic variants of *OATPs* are often named based on haplotype analysis and given a chronologically determined numerical designation after the "star" symbol "*" (e.g., OATP1B1*15), where typically, *1 represents the reference allele. At present, an allelic nomenclature committee for *OATPs* has not been established similar to one organized for the CYP enzymes.[8] Hence, a standardized *OATP* allelic nomenclature as assigned by a formal committee is required in the promotion of human transporter genetics.

5.2. MOLECULAR CHARACTERISTICS OF OATPs

5.2.1. Gene Structure

The human OATPs are encoded by the *SLCO* genes located on a number of different chromosomes (Table 5.1). Members of the *SLCO1* family are found in a gene locus on chromosome 12, including a pseudogene related to the *SCLO1B* subfamily.[9] The *SLCO* genes span from 30 to 310 kb in length and consist of 10 to 18 exons (Table 5.1). In silico analysis (www.genecards.org) predicts that many of the human OATPs are expressed as splice variants (Table 5.1), but experimental verification, tissue distribution, and functional assessment of such isoforms remains lacking. Functional splice variants of rodent Oatp1a3 have been described.[10,11]

5.2.2. Protein Structure

The OATPs are predicted to represent integral membrane proteins that contain 12 transmembrane (TM) helices that harbor the characteristic superfamily signature amino acid sequence D-X-RW-(I,V)-GAWW-X-G-(F,L)-L.[1] Amino and carboxy termini are oriented to the cytoplasmic spaces. Predicted and confirmed N-glycosylation sites are found, many conserved between transporters, in extracellular loops 2 and 5 (see below). Little is known regarding the tertiary structures of OATPs, although more recent studies are beginning to address this aspect of OATP biology. In silico, structural modeling studies with OATP1B3 and OATP2B1 have suggested that the OATPs share features of the major facilitator superfamily (MFS).[3] In these models, OATP1B3 is predicted to possess a central pseudo twofold symmetry axis perpendicular to the membrane plane and a central pore. The pore formed by TM helices 1, 2, 4, 5, 7, 8, 10, and 11 contains conserved basic/polar residues thought to be important to substrate binding and transport mechanisms.[3] Experimental support of the functional importance of these pore residues would be of interest. In addition, modeling of the large extracellular loop 5 in OATP1B3 revealed similarities to Kazal-type serine protease inhibitors and predicted internal disulfide bonds of the present cysteine residues.[3] The relevance of the 10 cysteine residues in the fifth extracellular loop of OATP2B1 was examined by mutational analysis.[12] Indeed, mutation of any of the cysteine residues or deletion of the loop itself caused mistrafficking

TABLE 5.1. Genetic and Protein Characteristics of Human OATPs

Gene				mRNA			Protein		
Name	Chromosome	Size (kb)	Exons	Reference Accession	Size (bp)	Splice Variants	Name	Previous Names	Size (aa)
SLCO1A2	12p12	126	17	NM_021094	2919	yes	OATP1A2	OATP, OATP-A	670
SLCO1B1	12p12	109	14	NM_006446	2830		OATP1B1	OATP2, LST-1, OATP-C	691
SLCO1B3	12p12	101	13	NM_019844	2646		OATP1B3	OATP8	702
SLCO1C1	12p12	58	18	NM_017435	3381	yes	OATP1C1	OATP-F	712
SLCO2A1	3q21	97	14	NM_005630	4040	yes	OATP2A1	hPGT	643
SLCO2B1	11q13	55	16	NM_007256	4108	yes	OATP2B1	OATP-B	709
SLCO3A1	15q26	310	16	NM_013272	2740	yes	OATP3A1	OATP-D	710
SLCO4A1	20q13.1	30	16	NM_016354	2796	yes	OATP4A1	OATP-E	722
SLCO4C1	5q21	63	13	NM_180991	5334		OATP4C1	OATP-H	724
SLCO5A1	8q13.1	163	10	NM_030958	3692		OATP5A1	OATP-J	848
SLCO6A1	5q21	127	14	NM_173488	2515		OATP6A1	OATP-I, GST	719

of the protein to the cell surface. Moreover, each cysteine residue was found to be disulfide bridged. Given that the electrostatic potential of extracellular loop 5 is not basic,[3] it is not predicted to have functional interactions with solutes, highlighting the important functional role of this domain in membrane insertion. Defining the quaternary structure of OATPs has received little attention despite the perception that members of this family form homo- or heterooligomers. For instance, even under reducing conditions, high-molecular-weight bands suggesting multimers are present after SDS–PAGE analysis of OATP1A2 protein heterologously expressed in mammalian cells.[13] Cross-linking experiments with OATP2B1 showed that amino groups between two OATP2B1 molecules would have to be minimally 12 Å apart to be consistent with the observed homo-cross-linking found in cells overexpressing the protein.[12] Attempts to determine whether mouse Oatp1a1 and Oatp1a4 heterodimerize in liver using immunopreciptation failed to show direct association between the proteins.[14]

5.2.3. Transport Mechanisms

It is generally considered that transport via OATPs occurs in a bidirectional fashion dictated by the solute gradients across the membrane. The mechanisms underlying solute transport by the OATPs have been investigated in some detail, especially in the context of bile acid physiology. Attention has focused largely on understanding driving forces, especially since early in vitro studies intriguingly found a lack of stimulation in OATP transport activity by an inwardly directed sodium gradient.[15–17] typical of that which is found for bile acid uptake into hepatocytes. Studies with rat Oatp1a1 and Oatp1a4 first demonstrated that solute uptake into cells was energized by countertransport with either bicarbonate[18] or reduced glutathione (GSH).[19,20] The stoichiometry for GSH/bile acid exchange for Oatp1 is 1 : 1.[19] However, Oatp1-mediated GSH efflux was not dependent on obligate exchange with solute such as bile acids.[21] Interestingly, human OATP2B1 was found to possess pH-dependent transport properties that were solute selective.[17,22] Extracellular acidification promoted solute uptake, a property of OATP2B1 that bears relevance to the environment in which the transporter is expressed on the apical membrane of enterocytes. pH-stimulated transport by OATP2B1 is in contrast to the insensitivity of rat Oatp1 activity to proton gradients.[23] A predominant bile acid efflux function for OATP1B3 in liver has been proposed. The fascinating finding that bile acid transport by OATP1B3 and not OATP1B1 occurs by a GSH cotransport mechanism raises the possibility that the functional transporter affords protection to hepatocytes by limiting the accumulation of toxic intracellular solutes.[24] A 2 : 1 GSH/bile acid cotransport stoichiometry for OATP1B3-mediated transport was observed. OATP2A1, the prostaglandin transporter PGT, appears to energize solute uptake by outward exchange with lactate.[25] Overall, it is becoming evident that the various OATPs can be defined by different transport mechanisms. Structural modeling of OATP1B3 and OATP2B1 has suggested that OATPs transport solutes across membranes through a rocker-switch type of mechanism,[3] but the location of solute-binding sites and molecular mechanisms of the transport process are unclear. How the transport mechanisms relate with physiological roles of OATPs remains poorly understood.

TABLE 5.2. Tissue Expression of Human OATPs

Transporter (Refs.)	Liver		Kidney		Small Intestine		Large Intestine		Brain		Placenta		Testes	
	mRNA	Protein	mRNA	Protein	mRNA	Protein	mRNA	Protein	mRNA	Protein	mRNA	Protein	mRNA	Protein
OATP1A2 (13, 15, 30, 41, 114)	+	+	+	+	+	+			+	+	+			
OATP1B1 (16, 115, 116)	+	+												
OATP1B3 (26, 55, 117)	+	+			+						+			
OATP1C1 (9,30)									+				+	+
OATP2A1 (118–120)	+		+	+	+		+		+		+		+	
OATP2B1 (22,27–32, 116)	+	+	+		+	+	+		+	+	+	+		
OATP3A1 (116, 121)			+						+				+	
OATP4A1 (9, 30, 116, 122)	+		+		+		+		+		+	+	+	
OATP4C1 (123)			+	+										
OATP5A1														
OATP6A1 (106)													+	+

5.3. EXPRESSION AND REGULATION OF OATPs

5.3.1. Tissue Distribution

Several members of the OATP superfamily, including OATP1A2, OATP2A1, OATP2B1, OATP3A1, and OATP4A1, are expressed widely among human tissues (Table 5.2). Others appear to have tissue-specific expression, such as OATP1B1 and OATP1B3 in liver, OATP4C1 in kidney, and OATP6A1 in testes (Table 5.2). For some time, mRNA analysis had been the only source of information about OATP gene expression in certain tissues, but now with the continued development of specific antibodies, protein localization studies have begun to shed light on the potential physiological and pharmacological roles for this transporter family. Hence for a tissue such as liver, OATP1B1, OATP1B3, and OATP2B1 are the main members of the transporter family expressed on the basolateral surface of hepatocytes,[16,26,27] while OATP1A2 is localized on the apical membrane of cholangiocytes.[13] The spectrum of tissues expressing OATP2B1 protein is now quite large and includes liver,[27] small intestine,[22] brain,[28] placenta,[29] eye,[30] skin,[31] and mammary gland.[32] By contrast, little is known about the expression (and function) of OATP5A1. Immunohistochemical analyses have revealed that the polarized expression of many OATPs in epithelial cells is tissue dependent (Table 5.2). For example, OATP2B1 is expressed on the apical membrane of enterocytes, whereas basolateral localization is found in hepatocytes and syncytiotrophoblasts. The reasons that account for differences in polarized expression of OATPs in tissues remain unknown.

Heart		Skeletal Muscle		Lung		Eye		Skin		Breast		
mRNA	Protein	mRNA	Protein	mRNA	Protein	mRNA	Protein	mRNA	Protein	mRNA	Protein	Polarization
				+		+	+	+				AP (enterocyte, cholangiocyte, renal distal tubule), BL (ciliary body epithelia),
												BL (hepatocytes)
												BL (hepatocyte)
+		+				+	+					
+				+		+						
+				+		+	+	+	+	+	+	AP (enterocyte), BL (hepatocyte), BL (syncytiotrophoblast), BL (ciliary body epithelia)
+				+		+	+	+		+		BL (ciliary body epithelia)
		+		+		+	+			+		AP (syncytiotrophoblast), BL (ciliary body epithelia)
												BL (rat renal proximal tubule)

5.3.2. Postranslational Regulation

Phosphorylation Rapid regulation of OATP activity is thought to occur through protein phosphorylation signaling. Evidence supporting this mechanism initially came from studies that demonstrated Oatp transport activity in cultured rat hepatocytes was quickly down-regulated by treatment with extracellular ATP.[33] This reduction in transport activity by ATP was not due to rapid internalization of rat Oatp1a1 and was associated with direct serine phosphorylation of the transporter,[34] suggesting a role for protein kinases. Indeed, a protein kinase C (PKC) activator suppressed rat Oatp1a1 and Oatp1a4 transport when expressed in frog eggs.[35] The identities of the phosphopeptides were recently elucidated after immunopurification of Oatp1a1 from rat liver and mass-spectrometric analysis of tryptic peptides.[36] Two adjacent serine residues near the carboxy terminus of Oatp1a1 undergo ordered phosphorylation, raising the possibility that phospho-signaling functionally alters the interaction of Oatp1a1 with PDZ adapter proteins.[36] More studies are required to understand the molecular determinants of reduced Oatp transport activity by phosphorylation as well as the physiological relevance.

N-Glycosylation Transport function and cellular localization of OATPs appears to be regulated by N-linked glycosylation. Studies with rat Oatp1a1 in glycosylation-deficient yeast expression systems and in glycosylation-competent Xenopus *laevis* oocytes revealed that glycosylation of the four extracellular asparagine residues is required for the transport function.[37] Studies with polymorphic or mutated variants of OATPs also support a role for N-glycosylation in regulating transport function. For example, OATP1B1 variants found on the cell surface in a mammalian heterologous expression system were mainly glycosylated proteins, whereas those confined intracellularly were underglycosylated.[38] Moreover, the degree of expression of glycosylated plasma membrane–localized OATP1B1 mirrored transport activity. In human liver, glycosylated and unglycosylated OATP1B1 are present[16] and there exists significant variability in the expression of both forms among individuals,[39] but the functional significance of these findings is unclear. A polymorphic variant of OATP1A2 (*6) results in an asparagine-to-isoleucine transversion in amino acid 135 of the second extracellular loop.[13] Consistent with an important role of N-glycosylation, the OATP1A2*6 variant is poorly glycosylated, retained intracellularly, and exhibits significantly reduced transport function in vitro. It is also interesting to note that the glycosylation state of OATP1A2 appears to be tissue dependent since the transporter has a larger apparent molecular mass in liver than in brain capillary endothelial cells.[40,41] Finally, investigations with OATP2B1 showed that directed mutations of cysteine residues in extracellular loop IX–X was associated with reduced cell surface expression and transport function, due perhaps to incomplete glycosylation.[42]

Adapter Protein Interactions With the exception of OATP2A1, which resides largely in intracellular spaces, the OATPs are localized on plasma membranes. Members of the PDZ domain-containing proteins have been shown to regulate plasma membrane sorting as well as to modulate transport function by direct interactions with C-terminal amino acids of various solute transporters.[43] Several human OATPs, including OATP1A2 and OATP2B1, contain potential PDZ consensus sequences.[14] Indeed, protein interaction studies have demonstrated that OATP1A2, OATP3A1, and OATP1C1 bind directly to members of the PDZ proteins, including PDZK1, IKEPP, NHERF1, and NHERF2.[44] A convincing role for PDZ proteins on OATP functions was demonstrated elegantly in the cellular distribution and transport function of Oatp1a1 in PDZK1 knockout mice.[14] Although such mice expressed significant amounts of hepatic Oatp1a1, the transporter was largely localized intracellularly. This lack of functional membrane transporter expression translated in vivo to reduced solute (bromosulfophthalein) clearance after intravenous administration. It remains to be determined what role PDZ proteins play in the regulation of human OATP transporter expression and function at both the molecular and physiological levels. Furthermore, it would be of interest to understand whether other protein interactions are required for cell surface sorting of OATPs, particularly for those transporters that do not possess consensus PDZ domains.

5.3.3. Transcriptional Regulation

Basal and adaptive expression of OATPs in various tissues is controlled, in part, by transcriptional mechanisms. At present, much of what is known about transcriptional regulation of OATPs arises from work aimed to understand transporter expression in liver and hence focused on the roles of hepatic transcription factors.

Hepatocyte nuclear factor 1α (HNF1α) is a transcription factor important to the maintenance of liver phenotype among other organs. Absence of Hnf1α in mice causes hepatic dysfunction resulting from disregulation of bile acid and cholesterol homeostasis.[45,46] The alterations in bile acid transport are associated with changes in the hepatic expression of Oatps such as Oatp1a1, Oatp1a5, Oatp1b2, and Oatp2b1, which are markedly down-regulated.[46,47] The human *SLCO1B1* and *SLCO1B3* genes contain functional HNF1α response elements in their proximal promoters,[48] a finding that is consistent with their liver-selective expression. Various pathological conditions such as inflammation are known to down-regulate the expression of HNF1α, leading to decreased expression of hepatic OATPs.[49]

The inducible expression of hepatic OATPs appears to result from bile acid–mediated activation of the farnesoid X receptor (FXR).[50] The *SLCO1B3* gene is directly transactivated by FXR, to cause an up-regulation in OATP1B3 expression in a hepatically derived cell line.[51] Together with the notion that OATP1B3 acts as a hepatic bile acid efflux transporter and its expression is maintained in cholestatic liver disease, it appears that transactivation of *SLCO1B3* by FXR serves as a means of cytoprotection in the face of elevated bile acid exposure.[24]

OATP1B3 expression is decreased in hepatocellular carcinoma (HCC), coincident with increased levels of the transcription factor hepatocyte nuclear factor 3β (HNF3β).[52] It appears that HNF3β interacts with negative response elements in the *SLCO1B3* promoter.[52,53] to suppress OATP1B3 expression. The *SLCO1B3* promoter is also transactivated by the growth hormone- and prolactin-activated transcription factor Stat5, as determined by cell-based reporter gene assays.[54] The physiological implications of these findings remain to be clarified. Finally, OATP1B3 is expressed exclusively in perivenous hepatocytes,[26,39,55] suggesting local transcriptional control of gene expression. It has been proposed that the diminished oxygen tension in the perivenous liver stimulates the function of hypoxia inducible factor 1 (HIF1) locally in the transactivation of a presumed response element in the first intron of the *SLCO1B3* gene.[55] Experimental confirmation of such a transcriptional mechanism is required.

The pregnane X receptor (PXR) is a promiscuous, ligand-activated transcription factor important to the inductive response to xenobiotics.[56] Given that a broad array of hepatic drug detoxification genes are regulated by PXR,[57] it is not unexpected that the expression of OATPs may be induced through this signaling pathway. Thus, rat *Slco1a4* gene expression has been shown to be directly regulated by pxr,[58] and hepatocyte Oatp1a4 levels are strongly induced in rats treated by the rodent pxr agonist prenenolone carbonitrile.[59] In cultured human hepatocytes, OATP1B1 is modestly induced by treatment with rifampin, suggesting that the *SLCO1B1* gene may be under regulation by PXR.[60] Currently, it is unknown whether drug-mediated OATP1B1 induction occurs in vivo. Studies in breast carcinoma have demonstrated high levels of OATP1A2 in tumor, which contrasts with the absence of expression in adjacent tissue.[61] Moreover, the expression of OATP1A2 correlated with that of PXR, suggestive of a potential mechanism for elevated transporter expression.[61] A direct role of PXR in regulating OATP1A2 remains to be determined in addition to its relevance to breast cancer pathogenesis and drug interactions.

5.4. OATP SUBSTRATES AND INHIBITORS

5.4.1. Substrates

As a whole, members of the OATP superfamily are broadly selective transporters interacting with solutes with diverse characteristics (Table 5.3). OATP substrates are relatively large and range in size from 334 Da (benzylpenicillin) to 1143 Da (cholecystokinin octapeptide, CCK-8). Some common features of OATP substrates are steroidal or peptidic (linear or cyclic) structural templates. Generally, solutes transported by OATPs are negatively charged, but there are several examples of neutral (digoxin) and cationic (*N*-methylquinine) substrates. Several drug classes are susceptible to transport by OATPs and those include 3-hydroxy-3-methylglutarylcoenzyme A (HMG-CoA) reductase inhibitors (statins), angiotensin II receptor antagonists, angiotensin converting enzyme inhibitors, and cardiac glycosides. The endogenous substrates for OATPs are hormones such as thyroxine and steroid conjugates, bile acids, bilirubin, and prostaglandins. In the absence of structural data, the understanding of molecular determinants of substrates in their interactions with the OATPs has been learned on the basis of three-dimensional quantitative structure–activity relationship and pharmacophore modeling approaches.[62,63] These studies support the requirement of a hydrophobic region and hydrogen acceptor and donors for OATP substrates.

5.4.2. Substrate Specificity of OATPs

Many OATPs share common substrates. For example, estrone 3-sulfate (E_1S) can be considered somewhat of a pan-substrate for OATPs. But it is quite apparent that there are clear differences among the OATPs in substrate specificity (Table 5.3). Certainly, CCK-8 seems to be solely transported by OATP1B3. Among the members of the superfamily, OATP1A2 possesses perhaps the broadest spectrum of solutes in that compounds of acidic, basic, and neutral character are substrates. Most other OATPs appear to have a predilection for acidic, amphipathic solutes. Why certain compounds have propensities for transport by specific OATPs remains an area of much interest, particularly since isoform-specific substrates could be used as chemical tools to interrogate the activity of OATPs in vivo and complex in vitro systems such as primary cells.

5.4.3. Inhibitors

As with substrates, inhibitors of OATPs can be useful tools to study the pharmacological and physiological impact of this transporter superfamily in vivo and in vitro. Selective inhibitors of OATPs have been sought, especially to understand the relative contribution of each hepatic OATP in drug clearance. However, the use of isoform-specific substrates to identify selective OATP inhibitors has not yet been

TABLE 5.3. Substrates and Inhibitors of Human OATPs[a]

Transporter	Substrates	Nonsubstrates	Inhibitors
OATP1A2	Fexofenadine[124], BSP[15,27,125], T_3[27], T_4[27], E_2G[27], E_1S[27,125], GCA[15], TCA[15,27], DHEAS[27], deltophin[27], DPDPE[27], BQ-123[27], oubain[27,125], PGE_2[27], *N*-methylquinine[27], rosuvastatin[39], pitavastatin[126], MTX[127], microcystin-LR[128]	Digoxin[27], LTC_4[27]	Grapefruit juice[94], orange juice[94], apple juice[94]
OATP1B1	E_1S[27], benzylpenicillin, PGE_2[27], E_2G[27], BSP[27], T_3[27], T_4[27], E_2G[27], E_1S[27], GCA, TCA[27], DHEAS[27], DPDPE[27], BQ-123[27], cerivastatin[73,129], atorvastatin[73,130], rosuvastatin[39], pitavastatin[126], caspofungin[131], phalloidin[132,133], troglitazone-sulfate[134], rifampin[66], bilirubin[96,97], bilirubin-glucuronides[96], arsenic[135], atrasentan[81], valsartan[136], olmesartan[137], enalapril[138], MTX[38], temocaprilat[75], DADLE[64], microcystin-LR[128], SN-38[74]	Digoxin[27], oubain[27], *N*-methylquinine[27], deltorphin[27]	CyA[66,67], FK-506[67,132], rapamycin[132], glycirrizic acid[139], glibenclamide[67], ketoconazole[140], gemfibrozil[129], gemfibrozil-glucuronide[129], ciprofibrate[141], bezafibrate[67,141], clarithromycin[67], erythromycin[67], indinavir[66], nelfinavir[66], ritonavir[66], saquinavir[66], probenacid[67], rifamycin SV[142], digoxin[67], verapamil[67], warfarin[67], MK-571[141], biochanin A[143], Genistein[143], epigallocatechin-3-gallate[143], hyperforin[66]
OATP1B3	Fexofenadine[144], BSP[27], T3[27], T4[27], E_2G[27], E_1S[27], GCA[27], TCA[27], TUDC[145], GUDC[145], DHEAS[27], deltophin[27], DPDPE[27], BQ-123[27], oubain[27], PGE_2, digoxin[27], rosuvastatin[39], valsartan[136], pitavastatin[126], fluo-3[146], docetaxel[140], paclitaxel[140], CCK-8[147], phalloidin[132,133], rifampin[66,146], MTX, bilirubin[97], repaglinide, telmisartan[65], olmesartan[137], enalapril[138], temocaprilat[75], microcystin-LR[128]	*N*-Methylquinine[27], PGE_2[27], caspofungin[131], folate	Ketoconazole[140], glycirrizic acid[139], rifamycin SV[142]
OATP1C1	BSP[9], T_3[9], T_4[9], E_2G[9], E_1S[9]	GCA[9], TCA[9], DHEAS[9], deltophin[9], DPDPE[9], BQ-123[9], digoxin[9], oubain[9], LTC_4[9], PGE_2[9], *N*-Methylquinine[9], MTX[9], folate[9]	

(Continued)

TABLE 5.3. (*Continued*)

Transporter	Substrates	Nonsubstrates	Inhibitors
OATP2A1	PGE_2[118], PGE_1[118], $PGF_{2\alpha}$[118], PGD_2[118], TXB_2[118]	Iloprost[118]	Furosemide[118], TGBz T34[148]
OATP2B1	E_1S[116], benzylpenicillin[116], PGE_2[116], BSP[27], DHEAS[27], pravastatin[22], fluvastatin, rosuvastatin[39], glybenclamide[149], fexofenadine[17]	E_2G[116], GCA[27], TCA[27], oubain[27], digoxin[27], LTC_4[27], PGE_2[27], T_3[27], T_4[27], deltophin[27], DPDPE[27], BQ-123[27]	Benzoate[22], nicotinate[22], phthalate[22], PAH[150], indomethacin[150], TCA[150], cimetidine[150], salicylate[150], valproate[150], rifamycin SV[142], grapefruit juice[149], orange juice[149], naringin[149], naringenin[149], quercetin[149], bergamottin[149], dihydroxybergamottin[149], tangeritin[149], nobelitin[149], bilberry[151], echinacea[151], green tea[151], banaba[151], grape seed[151], ginkgo[151], soybean[151], mulberry[151], black cohosh[151], and Siberian ginseng[151]
OATP3A1	PGE_1[121], PGE_2[116,121], $PGF_{2\alpha}$[121], benzylpenicillin[116], E_1S[116]	PGD_2[121], TBX[121], iloprost[121], MTX[121], TCA[116], E_2G[116]	PGD_2[121], PAH[121]
OATP4A1	T_3[152], T_4[152], TCA[152], E_2G[116], benzylpenicillin[116], PGE_2[116], E_1S[116]	PAH[152], PGE_1[152], PGD_2[152], PGF2α[152]	BSP[152]
OATP4C1	Digoxin[147], oubain[147], T_3[147], T_4[147], MTX[147], cAMP[147]	TCA[147], E_2G[147], PGE_2[147], PAH[147], pravastatin[147], temocaprilat[147], ASA[147], salicylate[147], urate[147], acyclovir[147], ochratoxin[147], benzylpenicillin[147], cGMP[147], TEA[147]	Digitoxin[147], Digitoxigenin[147]
OATP5A1	?	?	?
OATP6A1	rGST1 (T_3, T_4, DHEAS, TCA)[106]		rGST1 (β-estradiol, testosterone)[106]

[a] BSP, bromosulfophthalein; CCK-8, cholecystokinin octapeptide; CyA, cyclosporine A; DADLE, [D-Ala(2), D-Leu(5)]-enkephalin; DPDPE, [D-penicillamine[2,5]] enkephalin; E_1S, estrone sulfate; E_2G, estradiol 17β-glucuronide; DHEAS, dehydroepiandrostane sulfate; GCA, glycocholate; GUDC, glycoursodeoxycholate; LTC_4, leukotriene C_4; MTX, methotrexate; NMQ, *N*-methylquinidine; PAH, *p*-aminohipurrate; PGD_2, prostaglandin D_2; PGE_1, prostaglandin E_1; PGE_2, prostaglandin E_2; $PGF_{2\alpha}$, prostaglandin $F_{2\alpha}$; T_3, triiodothyronine; T_4, thyroxine; TBX, thromboxane A_2; TCA, taurocholate; TUDC, tauroursodeoxycholate.

a successful strategy. An example CCK-8, the selective OATP1B3 substrate, is not only an inhibitor of its transporter but also inhibits the other major hepatic transporter, OATP1B1.[64] One compound that has been used as a relatively selective OATP1B1 inhibitor to study contribution of various transporters in hepatic drug uptake is E_1S. It appears that E_1S is a far more potent inhibitor of OATP1B1 than OATP1B3.[65] Furthermore, there has been interest in understanding whether certain clinically relevant drug interactions occur because of OATP inhibition. Many compounds inhibit the transport activity of the OATPs (Table 5.3), but their inhibitory constants (K_i) range widely. For instance, drugs such as cyclosporine A, rifampin, and ritonavir have submicromolar inhibitory constants for OATP1B1-mediated uptake and are considered potent in vitro inhibitors.[66] K_i values for OATP-mediated uptake should be used in context with estimated hepatic drug concentrations to predict clinical drug-interaction potential.[67]

5.5. PHARMACOLOGY AND PHYSIOLOGY OF OATPs

5.5.1. Clinical Pharmacology

The interplay between drug metabolism and transport that occurs locally within tissues determines overall drug absorption, distribution, and elimination. For many drugs, particularly those capable of significant diffusion across membranes, the major impact of drug metabolism overshadows the influence of facilitated membrane transport on drug disposition. The dynamic, reversible nature of drug transport, as opposed to drug metabolism, has made it challenging for investigators to understand its role in pharmacokinetics. Like other transporters, the importance of the OATPs in drug disposition is most apparent for compounds that are metabolically inert or those whose rates of metabolism are affected significantly by transporter-mediated delivery to eliminating enzymes. But similar to the field of drug metabolism, the clinical relevance of drug transport by the OATPs has begun to emerge from studies that examine the interindividual variability in drug response caused by genetics and drug interactions. At present these studies have drawn our attention to roles of OATPs in intestinal and liver drug disposition, while little has been revealed regarding OATP functions at the blood–brain–barrier and kidney.

Pharmacogenetics As a potential determinant of interindividual variability in drug disposition, there has been significant effort in cataloguing polymorphisms in *SLCO* genes as well as their frequencies in ethnic populations (Table 5.4). Arguably, the consequences of polymorphisms in the hepatic uptake transporter OATP1B1 have received most attention. The cumulative research has shown convincingly that there exists a few relatively common polymorphisms in *SLCO1B1* that are associated with altered oral drug exposure. One single-nucleotide polymorphism (SNP) is 388A>G in codon 130, which converts an asparagine residue to aspartate, in the second extracellular loop of OATP1B1.[38] The frequencies of the *SLCO1B1* 388G allele (*1b) in Caucasians, African Americans, and Asians is approximately 40, 75, and 60%, respectively.[38,68–72]

TABLE 5.4. Polymorphisms in Selected OATPs

Gene	Polymorphism	Position	Amino Acid Changes	In Vitro Function	Refs.
SLCO1A2	G-916A	5′-flanking	—	?	13,127,153
	G-843A	5′-flanking	—	?	
	T-526C	5′-flanking	—	?	
	G-172A	5′-flanking	—	?	
	−188 ins A	5′-flanking	—	?	
	T38C	exon	I13T	↔	
	A382T	exon	N128Y	↔	
	A404T	exon	N135I	↓	
	C502T	exon	R168C	↓	
	A516C	exon	E172D	↓	
	G559A	exon	A187T	↓	
	C830A	exon	T277N	↔	
	A833del	exon	N28del	↓	
	A841G	exon	I281V	↔	
	T968C	exon	L323P	↔	
	A1063G	exon	I355V	↔	
	C2003G	exon	T778S	↔	
SLCO1B1	G-11187A	5′-flanking	—	?	38,68,69,71,73, 86,153–155
	T-11100G	5′-flanking	—	?	
	A-10499C	5′-flanking	—	?	
	A-2569G	5′-flanking	—	?	
	C-2361T	5′-flanking	—	?	
	T-2239G	5′-flanking	—	?	
	A-1718G	5′-flanking	—	?	
	T-1175G	5′-flanking	—	?	
	G-806A	5′-flanking	—	?	
	T217C	exon	F73L	↓	
	T245C	exon	V82A	↓	
	A388G	exon	N130D	↔	
	G411A	exon	S137S	↔	
	A452G	exon	N151S	?	
	G455A	exon	R152K	↔	
	C463A	exon	P155T	↔	
	A467G	exon	E156G	↓	
	T521C	exon	V174A	↓	
	C571T	exon	L191L	↔	
	C597T	exon	F199F	↔	
	G721A	exon	D241N	↔	
	T1058C	exon	I1353T	↓	
	A1294G	exon	N432D	↔	
	A1385G	exon	D462G	↔	
	G1463C	exon	G488A	↓	
	A1964G	exon	D655G	↓	
	A2000G	exon	E667G	↔	

TABLE 5.4. (*Continued*)

Gene	Polymorphism	Position	Amino Acid Changes	In Vitro Function	Refs.
SLCO1B3	−1590 to	5′-flanking	—	?	92,93,153
	−1587 del	5′-flanking	—	?	
	G-1413C	5′-flanking	—	?	
	A-1345G	5′-flanking	—	?	
	−28 to −11 del	5′-flanking	—	?	
	−7 to −4 del	exon	S112A	↑	
	T334G	exon	M233I	↔	
	G699A	exon	L424L	↔	
	A1272G	exon	A519A	↔	
	A1557G	exon	G522C	↔↓	
	G1564T	exon	G611G	↔	
	G1833A				
SLCO2B1	A644T	exon	D215V	?	68
	C663T	exon	S221S	↔	
	T1175C	exon	I392T	?	
	C1457T	exon	S486F	?	

Another common SNP is 521T>C in codon 174, which changes an amino acid residue located in the TM IV from valine to alanine.[38] This allele (*SLCO1B1*5*) has frequencies of approximately 15, 2, and 15% in Caucasians, African Americans, and Asians, respectively.[38,68–72] The 388G and 521C SNPs are in linkage disequilibrium and form the *SLCO1B1*15* haplotype.[68] Pharmacokinetic studies with the HMGCo-A reductase inhibitor pravastatin performed by different investigators strongly indicate that people with the *SLCO1B1*5* or *15 genotype have increased drug exposure in comparison to those carrying the reference allele *SLCO1B1*1a* (388A, 521T) (Table 5.5). Presumably, differences in the bioavailability of pravastatin among people with different *SLCO1B1* reflect the varying degrees of hepatic first-pass effect. These findings are consistent with in vitro studies showing that the OATP1B1 521T (V174A) variant routes poorly to the plasma membrane of expressing cells and has decreased transport function toward a variety of substrates.[38,39,66,73,74] A few studies suggest that *SLCO1B1*1b* is a high-transport-activity genotype since pravastatin levels are lower in subjects harboring this variation than in those with the reference allele.[70,75] Interestingly, in vitro studies have not observed higher transport activity of the OATP1B1*1b protein.[38,66,73] An examination of a liver bank showed a lack of influence of *SLCO1B1* genotype on the total hepatic protein expression of OATP1B1.[39] Further studies are required to clarify the mechanisms responsible for these in vivo observations. Apart from pravastatin, the pharmacokinetics of a growing number of drugs appear to be dependent on the *SLCO1B1* 521T > C genotype, including pitavastatin,[76] rosuvastatin,[77] repaglinide,[78] nateglinide,[79] fexofenadine,[80] atrasentan,[81] valsartan,[75] irinotecan,[82] and ezetimibe[83] (Table 5.5). Intriguingly, there appear to be differences in the impact of *SLCO1B1* genotype on pravastatin pharmacokinetics

TABLE 5.5. SLCO1B1 Genotype and Pharmacokinetics

Drug	*SLCO1B1* Genotype	Ethnicity	PK Effect in Comparison with Reference Genotype[a]	Ref.
Pravastatin	*15/*15	Asian	AUC↑ 187%	69
	*1a/*5	Caucasian	AUC↑ 143%	70
	*1b/*1b	Caucasian	AUC↓ 40%	70
	*17/*17	Caucasian	AUC↑ 130%	71
	*1b/*1b	Asian	AUC↓ 35%	75
	*5, *15, *17 variant haplotype	Caucasian	AUC↑ 110%	89
	*15/*15	Caucasian, African American	AUC ↑ 92 %	unpublished
Rosuvastatin	521CC	Caucasian	AUC↑ 217%	77
Pitavastatin		Asian	AUC↑ %	76
Repaglinide	521CC	Caucasian	AUC↑ 188%	78
Nateglinide	521CC	Asian	AUC↑ 108%	79
Atrasentan	Low-activity genotype	Caucasian, non-caucasian	AUC↑ 73%	81
Valsartan	*1b/*1b	Asian	AUC↓ 27%	75
Fexofenadine	521CC	Caucasian	AUC↑ 127%	80
Irinotecan	*15 carriers	Asian	AUC↑ 182%	82
Ezetimibe-glucuronide	*15 carriers	Caucasian	AUC↑ 305%	83

[a]Reference genotype, *SLCO1B1**1a*.

in children with heterozygous familial hypercholesterolemia.[84] Children with the *SLCO1B1* 521C genotype had lower pravastatin plasma levels than those with 521T. This trend goes in the opposite direction to that seen in adults and highlights the complexities of understanding *SLCO1B1* genotype–phenotype relationships.

With the understanding that HMG-CoA reductase inhibitor drug levels are somewhat determined by genetics, there has been interest in assessing whether *SLCO1B1* genotype also predicts risk for muscle toxicity and cholesterol lowering effect. It is reported that that increased drug dose is a risk factor for HMG-CoA reductase inhibitor-mediated myopathies including severe rhabdomyolysis.[85] suggesting that enhanced drug exposure resulting from *SLCO1B1* genetics similarly elevates risk for such side effects.[86] In the case of atorvastatin, *SCLO1B1* polymorphisms were not different in patients who did or did not experience myopathy.[87] Because OATP1B1 presents HMG-CoA reductase inhibitor drugs to their target in hepatocytes, investigators have examined the role of transporter genetics and the pharmacological effects. In one study, patients with the *SLCO1B1* 521C genotype had reduced lipid lowering effect by statin drugs than those carrying 521T.[88] By contrast, there was a lack of influence of *SLCO1B1* genotype to the lipid lowering of pravastatin in two studies despite that in one study the drug levels were clearly different among subject groups.[89,90] Larger studies will be required to fully clarify the role of *SLCO1B1* polymorphisms in HMG-CoA reductase inhibitor toxicity and efficacy.

Genetic polymorphisms in *SLCO1A2* have been identified and variant proteins have been characterized in vitro.[13,91] (Table 5.5). Few variants are common with most occurring at a frequency <10%. One relatively uncommon variant with a frequency of <5%, OATP1A1*3 (516A>C, Glu172Asp), has reduced transport activity in vitro as a result from a cell surface trafficking defect.[13] Similarly, genetic variations in *SLCO1B3* have been identified.[92,93] (Table 5.5). The 1564G>T (Gly522Cys) polymorphic variant which is relatively rare (<2% allelic frequency) in *SLCO1B3* when expressed in vitro, has reduced transport function. The influence of *SLCO1A2* and *SLCO1B3* genetic polymorphisms on drug disposition in vivo remains to be determined.

Drug Interactions The importance of OATPs in drug disposition has also been made evident from studies describing drug interactions (Table 5.6). In one study, coadministration of grapefruit juice with fexofenadine resulted in 63% decreased plasma exposure of the antihistiminic drug.[94] Fexofenadine is a drug that is not appreciably metabolized and is subject to facilitated membrane transport. That grapefruit juice was able to effectively inhibit fexofenadine transport mediated by OATP1A2 but not P-glycoprotein suggests that modulation of intestinal OATP1A2 activity was the mechanism of the drug interaction. Another dramatic drug interaction involving OATPs is the demonstration that cyclosporine A (CyA) coadministration increased the plasma exposure of rosuvastatin by sevenfold.[95] Again, rosuvastatin is a drug that is not metabolized significantly. Since CyA can potently inhibit OATP1B1-mediated transport of rosuvastatin,[39,95] the mechanism for this drug interaction may be inhibition of hepatic first-pass effect.

TABLE 5.6. Drug Interactions Implicating a Role for OATPs in the Mechanism

OATP Implicated	Drug Affected	Interacting Substance	PK Impact on Affected Drug	Ref.
OATP1A2	Fexofenadine	Grapefruit juice	AUC↓ 63%	94
	Fexofenadine	Orange juice	AUC↓ 70%	94
	Talinolol	Grapefruit juice	AUC↓ 44%	156
OATP1B1	Pravastatin	Orange juice	AUC↑ 152%	157
	Pravastatin	Cyclosporine A	AUC↑	Pravachol product monograph
	Pravastatin	Gemfibrozil	AUC↑ 202%	158
	Rosuvastatin	Cyclosporine A	AUC↑ 710%	95
	Rosuvastatin	Gemfibrozil	AUC↑ 188%	159
	Cerivastatin	Cyclosporine A	AUC↑ 3 to 5-fold	160
	Cerivastatin	Gemfibrozil	AUC↑ 559%	161

5.5.2. Physiological and Pathophysiological Roles

Bilirubin Homeostasis The liver serves to remove albumin-bound bilirubin from the circulation for further metabolism through glucuronidation and eventual excretion of conjugates in bile. The uptake of bilirubin from blood into hepatocytes appears to occur via OATP-mediated facilitated transport. In different experimental systems, unconjugated and conjugated bilirubin was transported by OATP1B1[96,97] and OATP1B3.[97] It has also been well known that administration of organic anions that are OATP substrates such as rifampin can cause unconjugated hyperbilirubinemia.[98,99] However, there is contradicting evidence to suggest that OATP1B1 alone does not transport unconjugated bilirubin[100] and that other unrelated proteins are involved in hepatic bilirubin uptake.[101] In any case, what has become increasingly apparent is the link between hyperbilirubinemia and *SLCO1B1* genotype. An increased risk (odds ratio of 3) of neonatal jaundice was observed in newborns carrying the *SLCO1B1* 388G genotype but not the 521C gentoype.[102] This result suggests that OATP1B1 variants with the aspartate in codon 130 (388G) are defective in bilirubin uptake in neonates. However, in adults, higher unconjugated bilirubin levels are observed in persons with the *SLCO1B1* 521C but not the 388G genotype,[103] underscoring that the effect of *SLCO1B1* genotype is age dependent. In patients with the benign form of mild unconjugated hyperbilirubinemia, Gilbert's syndrome, there was a greater likelihood for carrying either *SLCO1B1* 388G or 521C polymorphism.[104] These findings are consistent with the observation that people with Gilbert's syndrome have a hepatic organic anion transport defect.[105] Although there seems ample evidence to support a role for OATPs in bilirubin homeostasis, further research is required to clarify some of the issues and to determine mechanisms.

Other Physiological Roles Overall, there is a poor understanding of the physiological roles of OATPs. It is only possible to speculate the molecular physiology based on the respective tissue distributions and the spectrum of endogenous substrates. Hence, high affinity transport of T_3 and T_4 by OATP1C1 coupled with selective expression in brain and testes suggests an organ-specific role of this transporter in thyroid hormone physiology.[9] Similarly, a testicular-specific expression OATP6A1, along with substrate selectivity toward androgen conjugates and thyroid hormones, suggests a role in male reproductive physiology.[106] For OATP2A1, a transport specificity for prostaglandins together with broad tissue distribution indicates a role in the autocrine and paracrine regulation of eicosanoid signaling.[107] Finally, there is compelling evidence to support a role for OATP1B1 and OATP1B3 in bile acid homeostasis and enterohepatic recirculation based on several clues: direct transcriptional regulation by FXR[51] and down-regulation in expression during cholestatic disease.[108]

Molecular Pathophysiology The role of OATPs in the pathogenesis of disease has been considered and requires further study. For instance, OATP1B3 is highly expressed in certain colon cancers,[55] which draws an intriguing relationship between bile acids and disease risk and progression with bile acid transport. In breast cancer, an overexpression of estrogen-transporting OATPs may be involved in the pathogenesis of the disease.[61,109,110] The OATPs are also implicated to play central roles in programmed cell death. The rapid loss of intracellular glutathione by a specific efflux process preceding apoptotic cells death[111] has been ascribed to an OATP-like transporter.[112] Treatment of cells induced to apoptosis by known substrates and inhibitors of OATPs appears to modulate the progression to cell death.[112,113]

5.6. CONCLUSIONS

The OATPs are a superfamily of drug and endobiotic transporters whose importance to pharmacology and physiology is now becoming better established. Yet despite over a decade of study, much remains to be understood. It is of particular relevance to drug discovery and clinical pharmacology to better understand the OATP structure–function–transport energetics relationships among substrates, inhibitors and coupled ions for use in the development of drugs with improved tissue targeting and optimized pharmacokinetic and safety profiles. A better comprehension of the functional expression and significance of OATPs throughout human tissues, especially brain and kidney, would provide tools for predicting drug response and clearance. Understandably, we need to direct our attention to deciphering the physiology and pathophysiology of OATPs, especially for those transporters that are least characterized. All these questions will ultimately be answered through the combination of studies at the genetic and molecular level with animal models, including those involving genetic manipulation, and clinical studies aimed at understanding variability in drug response and toxicity.

REFERENCES

1. Hagenbuch B, Meier PJ. 2003. The superfamily of organic anion transporting polypeptides. Biochim Biophys Acta 1609(1):1–18.
2. Hagenbuch B, Meier PJ. 2004. Organic anion transporting polypeptides of the OATP/SLC21 family: phylogenetic classification as OATP/SLCO superfamily, new nomenclature and molecular/functional properties. Pflugers Arch 447(5):653–665.
3. Meier-Abt F, Mokrab Y, Mizuguchi K. 2005. Organic anion transporting polypeptides of the OATP/SLCO superfamily: identification of new members in nonmammalian species, comparative modeling and a potential transport mode. J Membr Biol 208(3):213–227.
4. Jacquemin E, Hagenbuch B, Stieger B, Wolkoff AW, Meier PJ. 1994. Expression cloning of a rat liver Na(+)-independent organic anion transporter. Proc Natl Acad Sci U S A 91(1):133–137.
5. Nelson DR, Koymans L, Kamataki T, Stegeman JJ, Feyereisen R, Waxman DJ, Waterman MR, Gotoh O, Coon MJ, Estabrook RW, et al. 1996. P450 superfamily: update on new sequences, gene mapping, accession numbers and nomenclature. Pharmacogenetics 6(1):1–42.

6. Stedman CA, Liddle C, Coulter SA, Sonoda J, Alvarez JG, Moore DD, Evans RM, Downes M. 2005. Nuclear receptors constitutive androstane receptor and pregnane X receptor ameliorate cholestatic liver injury. Proc Natl Acad Sci U S A 102(6):2063–2068.
7. Sonoda J, Chong LW, Downes M, Barish GD, Coulter S, Liddle C, Lee CH, Evans RM. 2005. Pregnane X receptor prevents hepatorenal toxicity from cholesterol metabolites. Proc Natl Acad Sci U S A 102(6):2198–2203.
8. Ingelman-Sundberg M, Daly AK, Oscarson M, Nebert DW. 2000. Human cytochrome P450 (CYP) genes: recommendations for the nomenclature of alleles. Pharmacogenetics 10(1):91–93.
9. Pizzagalli F, Hagenbuch B, Stieger B, Klenk U, Folkers G, Meier PJ. 2002. Identification of a novel human organic anion transporting polypeptide as a high affinity thyroxine transporter. Mol Endocrinol 16(10):2283–2296.
10. Saito H, Masuda S, Inui K. 1996. Cloning and functional characterization of a novel rat organic anion transporter mediating basolateral uptake of methotrexate in the kidney. J Biol Chem 271(34):20719–20725.
11. Masuda S, Ibaramoto K, Takeuchi A, Saito H, Hashimoto Y, Inui KI. 1999. Cloning and functional characterization of a new multispecific organic anion transporter, OAT-K2, in rat kidney. Mol Pharmacol 55(4):743–752.
12. Hanggi E, Freimoser Grundschober A, Leuthold S, Meier PJ, St-Pierre MV. 2006. Functional analysis of the extracellular cysteine residues in the human organic anion transporting polypeptide, Oatp2b1. Mol Pharmacol 70(3):806–817.
13. Lee W, Glaeser H, Smith LH, Roberts RL, Moeckel GW, Gervasini G, Leake BF, Kim RB. 2005. Polymorphisms in human organic anion–transporting polypeptide 1A2 (OATP1A2): implications for altered drug disposition and central nervous system drug entry. J Biol Chem 280(10):9610–9617.
14. Wang P, Wang JJ, Xiao Y, Murray JW, Novikoff PM, Angeletti RH, Orr GA, Lan D, Silver DL, Wolkoff AW. 2005. Interaction with PDZK1 is required for expression of organic anion transporting protein 1A1 on the hepatocyte surface. J Biol Chem 280(34):30143–30149.
15. Kullak-Ublick GA, Hagenbuch B, Stieger B, Schteingart CD, Hofmann AF, Wolkoff AW, Meier PJ. 1995. Molecular and functional characterization of an organic anion transporting polypeptide cloned from human liver. Gastroenterology 109(4):1274–1282.
16. Konig J, Cui Y, Nies AT, Keppler D. 2000. A novel human organic anion transporting polypeptide localized to the basolateral hepatocyte membrane. Am J Physiol Gastrointest Liver Physiol 278(1):G156–G164.
17. Nozawa T, Imai K, Nezu J, Tsuji A, Tamai I. 2004. Functional characterization of pH-sensitive organic anion transporting polypeptide OATP-B in human. J Pharmacol Exp Ther 308(2):438–445.
18. Satlin LM, Amin V, Wolkoff AW. 1997. Organic anion transporting polypeptide mediates organic anion/HCO_3-exchange. J Biol Chem 272(42):26340–26345.
19. Li L, Lee TK, Meier PJ, Ballatori N. 1998. Identification of glutathione as a driving force and leukotriene C4 as a substrate for oatp1, the hepatic sinusoidal organic solute transporter. J Biol Chem 273(26):16184–16191.
20. Li L, Meier PJ, Ballatori N. 2000. Oatp2 mediates bidirectional organic solute transport: a role for intracellular glutathione. Mol Pharmacol 58(2):335–340.

21. Mittur A, Wolkoff AW, Kaplowitz N. 2002. The thiol sensitivity of glutathione transport in sidedness-sorted basolateral liver plasma membrane and in Oatp1-expressing HeLa cell membrane. Mol Pharmacol 61(2):425–435.
22. Kobayashi D, Nozawa T, Imai K, Nezu J, Tsuji A, Tamai I. 2003. Involvement of human organic anion transporting polypeptide OATP-B (SLC21A9) in pH-dependent transport across intestinal apical membrane. J Pharmacol Exp Ther 306(2):703–708.
23. Marin JJ, Mangas D, Martinez-Diez MC, El-Mir MY, Briz O, Serrano MA. 2003. Sensitivity of bile acid transport by organic anion–transporting polypeptides to intracellular pH. Biochim Biophys Acta 1611(1–2):249–257.
24. Briz O, Romero MR, Martinez-Becerra P, Macias RI, Perez MJ, Jimenez F, San Martin FG, Marin JJ. 2006. Oatp8/1B3-mediated cotransport of bile acids and glutathione: An export pathway for organic anions from hepatocytes? J Biol Chem 281(41):30326–30335.
25. Chan BS, Endo S, Kanai N, Schuster VL. 2002. Identification of lactate as a driving force for prostanoid transport by prostaglandin transporter PGT. Am J Physiol Renal Physiol 282(6):F1097–F1102.
26. König J, Cui Y, Nies AT, Keppler D. 2000. Localization and genomic organization of a new hepatocellular organic anion transporting polypeptide. J Biol Chem 275(30):23161–23168.
27. Kullak-Ublick GA, Ismair MG, Stieger B, Landmann L, Huber R, Pizzagalli F, Fattinger K, Meier PJ, Hagenbuch B. 2001. Organic anion–transporting polypeptide B (OATP-B) and its functional comparison with three other OATPs of human liver. Gastroenterology 120(2):525–533.
28. Bronger H, König J, Kopplow K, Steiner HH, Ahmadi R, Herold-Mende C, Keppler D, Nies AT. 2005. ABCC drug efflux pumps and organic anion uptake transporters in human gliomas and the blood–tumor barrier. Cancer Res 65(24):11419–11428.
29. St-Pierre MV, Hagenbuch B, Ugele B, Meier PJ, Stallmach T. 2002. Characterization of an organic anion–transporting polypeptide (OATP-B) in human placenta. J Clin Endocrinol Metab 87(4):1856–1863.
30. Gao B, Huber RD, Wenzel A, Vavricka SR, Ismair MG, Reme C, Meier PJ. 2005. Localization of organic anion transporting polypeptides in the rat and human ciliary body epithelium. Exp Eye Res 80(1):61–72.
31. Schiffer R, Neis M, Holler D, Rodriguez F, Geier A, Gartung C, Lammert F, Dreuw A, Zwadlo-Klarwasser G, Merk H, Jugert F, Baron JM. 2003. Active influx transport is mediated by members of the organic anion transporting polypeptide family in human epidermal keratinocytes. J Invest Dermatol 120(2):285–291.
32. Pizzagalli F, Varga Z, Huber RD, Folkers G, Meier PJ, St-Pierre MV. 2003. Identification of steroid sulfate transport processes in the human mammary gland. J Clin Endocrinol Metab 88(8):3902–3912.
33. Campbell CG, Spray DC, Wolkoff AW. 1993. Extracellular ATP4- modulates organic anion transport by rat hepatocytes. J Biol Chem 268(21):15399–15404.
34. Glavy JS, Wu SM, Wang PJ, Orr GA, Wolkoff AW. 2000. Down-regulation by extracellular ATP of rat hepatocyte organic anion transport is mediated by serine phosphorylation of oatp1. J Biol Chem 275(2):1479–1484.
35. Guo GL, Klaassen CD. 2001. Protein kinase C suppresses rat organic anion transporting polypeptide 1- and 2-mediated uptake. J Pharmacol Exp Ther 299(2):551–557.

36. Xiao Y, Nieves E, Angeletti RH, Orr GA, Wolkoff AW. 2006. Rat organic anion transporting protein 1A1 (Oatp1a1): purification and phosphopeptide assignment. Biochemistry 45(10):3357–3369.

37. Lee TK, Koh AS, Cui Z, Pierce RH, Ballatori N. 2003. N-Glycosylation controls functional activity of Oatp1, an organic anion transporter. Am J Physiol Gastrointest Liver Physiol 285(2):G371–G381.

38. Tirona RG, Leake BF, Merino G, Kim RB. 2001. Polymorphisms in OATP-C. Identification of multiple allelic variants associated with altered transport activity among European- and African-Americans. J Biol Chem 276(38):35669–35675.

39. Ho RH, Tirona RG, Leake BF, Glaeser H, Lee W, Lemke CJ, Wang Y, Kim RB. 2006. Drug and bile acid transporters in rosuvastatin hepatic uptake: function, expression, and pharmacogenetics. Gastroenterology 130(6):1793–1806.

40. Kullak-Ublick GA, Glasa J, Boker C, Oswald M, Grutzner U, Hagenbuch B, Stieger B, Meier PJ, Beuers U, Kramer W, et al. 1997. Chlorambucil–taurocholate is transported by bile acid carriers expressed in human hepatocellular carcinomas. Gastroenterology 113(4):1295–1305.

41. Gao B, Hagenbuch B, Kullak-Ublick GA, Benke D, Aguzzi A, Meier PJ. 2000. Organic anion-transporting polypeptides mediate transport of opioid peptides across blood–brain barrier. J Pharmacol Exp Ther 294(1):73–79.

42. Hanggi E, Grundschober AF, Leuthold S, Meier PJ, St-Pierre MV. 2006. Functional analysis of the extracellular cysteine residues in the human organic anion transporting polypeptide, OATP2B1. Mol Pharmacol 70(3):806–817.

43. Sugiura T, Kato Y, Tsuji A. 2006. Role of SLC xenobiotic transporters and their regulatory mechanisms PDZ proteins in drug delivery and disposition. J Control Release 116(2):238–246.

44. Kato Y, Yoshida K, Watanabe C, Sai Y, Tsuji A. 2004. Screening of the interaction between xenobiotic transporters and PDZ proteins. Pharm Res 21(10):1886–1894.

45. Pontoglio M, Barra J, Hadchouel M, Doyen A, Kress C, Bach JP, Babinet C, Yaniv M. 1996. Hepatocyte nuclear factor 1 inactivation results in hepatic dysfunction, phenylketonuria, and renal Fanconi syndrome. Cell 84(4):575–585.

46. Shih DQ, Bussen M, Sehayek E, Ananthanarayanan M, Shneider BL, Suchy FJ, Shefer S, Bollileni JS, Gonzalez FJ, Breslow JL, Stoffel M. 2001. Hepatocyte nuclear factor-1alpha is an essential regulator of bile acid and plasma cholesterol metabolism. Nat Genet 27(4):375–382.

47. Maher JM, Slitt AL, Callaghan TN, Cheng X, Cheung C, Gonzalez FJ, Klaassen CD. 2006. Alterations in transporter expression in liver, kidney, and duodenum after targeted disruption of the transcription factor HNF1alpha. Biochem Pharmacol 72(4):512–522.

48. Jung D, Hagenbuch B, Gresh L, Pontoglio M, Meier PJ, Kullak-Ublick GA. 2001. Characterization of the human OATP-C (SLC21A6) gene promoter and regulation of liver-specific OATP genes by hepatocyte nuclear factor 1 alpha. J Biol Chem 276(40):37206–37214.

49. Li N, Klaassen CD. 2004. Role of liver-enriched transcription factors in the down-regulation of organic anion transporting polypeptide 4 (oatp4; oatplb2; slc21a10) by lipopolysaccharide. Mol Pharmacol 66(3):694–701.

50. Parks DJ, Blanchard SG, Bledsoe RK, Chandra G, Consler TG, Kliewer SA, Stimmel JB, Willson TM, Zavacki AM, Moore DD, Lehmann JM. 1999. Bile acids: natural ligands for an orphan nuclear receptor. Science 284(5418):1365–1368.

51. Jung D, Podvinec M, Meyer UA, Mangelsdorf DJ, Fried M, Meier PJ, Kullak-Ublick GA. 2002. Human organic anion transporting polypeptide 8 promoter is transactivated by the farnesoid X receptor/bile acid receptor. Gastroenterology 122(7):1954–1966.
52. Vavricka SR, Jung D, Fried M, Grutzner U, Meier PJ, Kullak-Ublick GA. 2004. The human organic anion transporting polypeptide 8 (SLCO1B3) gene is transcriptionally repressed by hepatocyte nuclear factor 3beta in hepatocellular carcinoma. J Hepatol 40(2):212–218.
53. Ohtsuka H, Abe T, Onogawa T, Kondo N, Sato T, Oshio H, Mizutamari H, Mikkaichi T, Oikawa M, Rikiyama T, et al. 2006. Farnesoid X receptor, hepatocyte nuclear factors 1alpha and 3beta are essential for transcriptional activation of the liver-specific organic anion transporter-2 gene. J Gastroenterol 41(4):369–377.
54. Wood M, Ananthanarayanan M, Jones B, Wooton-Kee R, Hoffman T, Suchy FJ, Vore M. 2005. Hormonal regulation of hepatic organic anion transporting polypeptides. Mol Pharmacol 68(1):218–225.
55. Abe T, Unno M, Onogawa T, Tokui T, Kondo TN, Nakagomi R, Adachi H, Fujiwara K, Okabe M, Suzuki T, et al. 2001. LST-2, a human liver-specific organic anion transporter, determines methotrexate sensitivity in gastrointestinal cancers. Gastroenterology 120(7):1689–1699.
56. Lehmann JM, McKee DD, Watson MA, Willson TM, Moore JT, Kliewer SA. 1998. The human orphan nuclear receptor PXR is activated by compounds that regulate CYP3A4 gene expression and cause drug interactions. J Clin Invest 102(5):1016–1023.
57. Tirona RG, Kim RB. 2005. Nuclear receptors and drug disposition gene regulation. J Pharm Sci 94(6):1169–1186.
58. Guo GL, Staudinger J, Ogura K, Klaassen CD. 2002. Induction of rat organic anion transporting polypeptide 2 by pregnenolone-16alpha-carbonitrile is via interaction with pregnane X receptor. Mol Pharmacol 61(4):832–839.
59. Guo GL, Choudhuri S, Klaassen CD. 2002. Induction profile of rat organic anion transporting polypeptide 2 (oatp2) by prototypical drug-metabolizing enzyme inducers that activate gene expression through ligand-activated transcription factor pathways. J Pharmacol Exp Ther 300(1):206–212.
60. Jigorel E, Le Vee M, Boursier-Neyret C, Parmentier Y, Fardel O. 2006. Differential regulation of sinusoidal and canalicular hepatic drug transporter expression by xenobiotics activating drug-sensing receptors in primary human hepatocytes. Drug Metab Dispos 34(10):1756–1763.
61. Miki Y, Suzuki T, Kitada K, Yabuki N, Shibuya R, Moriya T, Ishida T, Ohuchi N, Blumberg B, Sasano H. 2006. Expression of the steroid and xenobiotic receptor and its possible target gene, organic anion transporting polypeptide-A, in human breast carcinoma. Cancer Res 66(1):535–542.
62. Yarim M, Moro S, Huber R, Meier PJ, Kaseda C, Kashima T, Hagenbuch B, Folkers G. 2005. Application of QSAR analysis to organic anion transporting polypeptide 1a5 (Oatp1a5) substrates. Bioorg Med Chem 13(2):463–471.
63. Chang C, Pang KS, Swaan PW, Ekins S. 2005. Comparative pharmacophore modeling of organic anion transporting polypeptides: a meta-analysis of rat Oatp1a1 and human OATP1B1. J Pharmacol Exp Ther 314(2):533–541.
64. Nozawa T, Tamai I, Sai Y, Nezu J, Tsuji A. 2003. Contribution of organic anion transporting polypeptide OATP-C to hepatic elimination of the opioid pentapeptide analogue [D-Ala2, D-Leu5]-enkephalin. J Pharm Pharmacol 55(7):1013–1020.

65. Ishiguro N, Maeda K, Kishimoto W, Saito A, Harada A, Ebner T, Roth W, Igarashi T, Sugiyama Y. 2006. Predominant contribution of OATP1B3 to the hepatic uptake of telmisartan, an angiotensin II receptor antagonist, in humans. Drug Metab Dispos 34:1109–1115.
66. Tirona RG, Leake BF, Wolkoff AW, Kim RB. 2003. Human organic anion transporting polypeptide-C (SLC21A6) is a major determinant of rifampin-mediated pregnane X receptor activation. J Pharmacol Exp Ther 304(1):223–228.
67. Hirano M, Maeda K, Shitara Y, Sugiyama Y. 2006. Drug–drug interaction between pitavastatin and various drugs via OATP1B1. Drug Metab Dispos 34(7):1229–1236.
68. Nozawa T, Nakajima M, Tamai I, Noda K, Nezu J, Sai Y, Tsuji A, Yokoi T. 2002. Genetic polymorphisms of human organic anion transporters OATP-C (SLC21A6) and OATP-B (SLC21A9): allele frequencies in the Japanese population and functional analysis. J Pharmacol Exp Ther 302(2):804–813.
69. Nishizato Y, Ieiri I, Suzuki H, Kimura M, Kawabata K, Hirota T, Takane H, Irie S, Kusuhara H, Urasaki Y, et al. 2003. Polymorphisms of OATP-C (SLC21A6) and OAT3 (SLC22A8) genes: consequences for pravastatin pharmacokinetics. Clin Pharmacol Ther 73(6):554–565.
70. Mwinyi J, Johne A, Bauer S, Roots I, Gerloff T. 2004. Evidence for inverse effects of OATP-C (SLC21A6) 5 and 1b haplotypes on pravastatin kinetics. Clin Pharmacol Ther 75(5):415–421.
71. Niemi M, Schäffeler E, Lang T, Fromm MF, Neuvonen M, Kyrklund C, Backman JT, Kerb R, Schwab M, Neuvonen PJ, et al. 2004. High plasma pravastatin concentrations are associated with single nucleotide polymorphisms and haplotypes of organic anion transporting polypeptide-C (OATP-C, SLCO1B1). Pharmacogenetics 14(7):429–440.
72. Lee E, Ryan S, Birmingham B, Zalikowski J, March R, Ambrose H, Moore R, Lee C, Chen Y, Schenck D. 2005. Rosuvastatin pharmacokinetics and pharmacogenetics in Caucasian and Asian subjects residing in the same environment. Clin Pharmacol Ther 78(4):330–341.
73. Kameyama Y, Yamashita K, Kobayashi K, Hosokawa M, Chiba K. 2005. Functional characterization of SLCO1B1 (OATP-C) variants, SLCO1B1*5, SLCO1B1*15 and SLCO1B1*15+C1007G, by using transient expression systems of HeLa and HEK293 cells. Pharmacogenet Genom 15(7):513–522.
74. Nozawa T, Minami H, Sugiura S, Tsuji A, Tamai I. 2005. Role of organic anion transporter OATP1B1 (OATP-C) in hepatic uptake of irinotecan and its active metabolite, 7-ethyl-10-hydroxycamptothecin: in vitro evidence and effect of single nucleotide polymorphisms. Drug Metab Dispos 33(3):434–439.
75. Maeda K, Ieiri I, Yasuda K, Fujino A, Fujiwara H, Otsubo K, Hirano M, Watanabe T, Kitamura Y, Kusuhara H, Sugiyama Y. 2006. Effects of organic anion transporting polypeptide 1B1 haplotype on pharmacokinetics of pravastatin, valsartan, and temocapril. Clin Pharmacol Ther 79(5):427–439.
76. Chung JY, Cho JY, Yu KS, Kim JR, Oh DS, Jung HR, Lim KS, Moon KH, Shin SG, Jang IJ. 2005. Effect of OATP1B1 (SLCO1B1) variant alleles on the pharmacokinetics of pitavastatin in healthy volunteers. Clin Pharmacol Ther 78(4):342–350.
77. Lee E, Ryan S, Birmingham B, Zalikowski J, March R, Ambrose H, Moore R, Lee C, Chen Y, Schneck D. 2005. Rosuvastatin pharmacokinetics and pharmacogenetics in white and Asian subjects residing in the same environment. Clin Pharmacol Ther 78(4):330–341.

78. Niemi M, Backman JT, Kajosaari LI, Leathart JB, Neuvonen M, Daly AK, Eichelbaum M, Kivisto KT, Neuvonen PJ. 2005. Polymorphic organic anion transporting polypeptide 1B1 is a major determinant of repaglinide pharmacokinetics. Clin Pharmacol Ther 77(6):468–478.

79. Zhang W, He YJ, Han CT, Liu ZQ, Li Q, Fan L, Tan ZR, Zhang WX, Yu BN, Wang D, Hu DL, Zhou HH. 2006. Effect of SLCO1B1 genetic polymorphism on the pharmacokinetics of nateglinide. Br J Clin Pharmacol 62(5):567–572.

80. Niemi M, Kivisto KT, Hofmann U, Schwab M, Eichelbaum M, Fromm MF. 2005. Fexofenadine pharmacokinetics are associated with a polymorphism of the SLCO1B1 gene (encoding OATP1B1). Br J Clin Pharmacol 59(5):602–604.

81. Katz DA, Carr R, Grimm DR, Xiong H, Holley-Shanks R, Mueller T, Leake B, Wang Q, Han L, Wang PG, et al. 2006. Organic anion transporting polypeptide 1B1 activity classified by SLCO1B1 genotype influences atrasentan pharmacokinetics. Clin Pharmacol Ther 79(3):186–196.

82. Xiang X, Rao Jada S, Hua Li H, Fan L, San Tham L, Ing Wong C, Chin Lee S, Lim R, Yu Zhou Q, Cher Goh B, Huat Tan E, Chowbay B. 2006. Pharmacogenetics of SLCO1B1 gene and the impact of *1b and *15 haplotypes on irinotecan disposition in Asian cancer patients. Pharmacogenet Genom 16(9):683–691.

83. Oswald S, Scheuch E, Cascorbi I, Siegmund W. 2006. A LC-MS/MS method to quantify the novel cholesterol lowering drug ezetimibe in human serum, urine and feces in healthy subjects genotyped for SLCO1B1. J Chromatogr B Anal Technol Biomed Life Sci 830(1):143–150.

84. Hedman M, Antikainen M, Holmberg C, Neuvonen M, Eichelbaum M, Kivisto KT, Neuvonen PJ, Niemi M. 2006. Pharmacokinetics and response to pravastatin in paediatric patients with familial hypercholesterolaemia and in paediatric cardiac transplant recipients in relation to polymorphisms of the SLCO1B1 and ABCB1 genes. Br J Clin Pharmacol 61(6):706–715.

85. Thompson PD, Clarkson P, Karas RH. 2003. Statin-associated myopathy. JAMA 289 (13):1681–1690.

86. Morimoto K, Oishi T, Ueda S, Ueda M, Hosokawa M, Chiba K. 2004. A novel variant allele of OATP-C (SLCO1B1) found in a Japanese patient with pravastatin-induced myopathy. Drug Metab Pharmacokinet 19(6):453–455.

87. Hermann M, Bogsrud MP, Molden E, Asberg A, Mohebi BU, Ose L, Retterstol K. 2006. Exposure of atorvastatin is unchanged but lactone and acid metabolites are increased several-fold in patients with atorvastatin-induced myopathy. Clin Pharmacol Ther 79(6):532–539.

88. Tachibana-Iimori R, Tabara Y, Kusuhara H, Kohara K, Kawamoto R, Nakura J, Tokunaga K, Kondo I, Sugiyama Y, Miki T. 2004. Effect of genetic polymorphism of OATP-C (SLCO1B1) on lipid-lowering response to HMG-CoA reductase inhibitors. Drug Metab Pharmacokinet 19(5):375–380.

89. Igel M, Arnold KA, Niemi M, Hofmann U, Schwab M, Lutjohann D, von Bergmann K, Eichelbaum M, Kivisto KT. 2006. Impact of the SLCO1B1 polymorphism on the pharmacokinetics and lipid-lowering efficacy of multiple-dose pravastatin. Clin Pharmacol Ther 79(5):419–426.

90. Gerloff T, Schaefer M, Mwinyi J, Johne A, Sudhop T, Lutjohann D, Roots I, von Bergmann K. 2006. Influence of the SLCO1B1*1b and *5 haplotypes on pravastatin's cholesterol

lowering capabilities and basal sterol serum levels. Naunyn Schmiedebergs Arch Pharmacol 373(1):45–50.

91. Badagnani I, Castro RA, Taylor TR, Brett CM, Huang CC, Stryke D, Kawamoto M, Johns SJ, Ferrin TE, Carlson EJ, et al. 2006. Interaction of methotrexate with organic-anion transporting polypeptide 1A2 and its genetic variants. J Pharmacol Exp Ther 318(2):521–529.
92. Letschert K, Keppler D, König J. 2004. Mutations in the SLCO1B3 gene affecting the substrate specificity of the hepatocellular uptake transporter OATP1B3 (OATP8). Pharmacogenetics 14(7):441–452.
93. Tsujimoto M, Hirata S, Dan Y, Ohtani H, Sawada Y. 2006. Polymorphisms and linkage disequilibrium of the OATP8 (OATP1B3) gene in Japanese subjects. Drug Metab Pharmacokinet 21(2):165–169.
94. Dresser GK, Bailey DG, Leake BF, Schwarz UI, Dawson PA, Freeman DJ, Kim RB. 2002. Fruit juices inhibit organic anion transporting polypeptide-mediated drug uptake to decrease the oral availability of fexofenadine. Clin Pharmacol Ther 71(1):11–20.
95. Simonson SG, Raza A, Martin PD, Mitchell PD, Jarcho JA, Brown CD, Windass AS, Schneck DW. 2004. Rosuvastatin pharmacokinetics in heart transplant recipients administered an antirejection regimen including cyclosporine. Clin Pharmacol Ther 76(2):167–177.
96. Cui Y, Konig J, Leier I, Buchholz U, Keppler D. 2001. Hepatic uptake of bilirubin and its conjugates by the human organic anion transporter SLC21A6. J Biol Chem 276(13):9626–9630.
97. Briz O, Serrano MA, MacIas RI, Gonzalez-Gallego J, Marin JJ. 2003. Role of organic anion-transporting polypeptides, OATP-A, OATP-C and OATP-8, in the human placenta–maternal liver tandem excretory pathway for foetal bilirubin. Biochem J 371(Pt 3):897–905.
98. Acocella G, Nicolis FB, Tenconi LT. 1965. The effect of an intravenous infusion of rifamycin SV on the excretion of bilirubin, bromsulphalein, and indocyanine green in man. Gastroenterology 49(5):521–525.
99. Campbell SD, de Morais SM, Xu JJ. 2004. Inhibition of human organic anion transporting polypeptide OATP 1B1 as a mechanism of drug-induced hyperbilirubinemia. Chem Biol Interact 150(2):179–187.
100. Wang P, Kim RB, Chowdhury JR, Wolkoff AW. 2003. The human organic anion transport protein SLC21A6 is not sufficient for bilirubin transport. J Biol Chem 278(23):20695–20699.
101. Passamonti S, Terdoslavich M, Margon A, Cocolo A, Medic N, Micali F, Decorti G, Franko M. 2005. Uptake of bilirubin into HepG2 cells assayed by thermal lens spectroscopy. Function of bilitranslocase. FEBS J 272(21):5522–5535.
102. Huang MJ, Kua KE, Teng HC, Tang KS, Weng HW, Huang CS. 2004. Risk factors for severe hyperbilirubinemia in neonates. Pediatr Res 56(5):682–689.
103. Ieiri I, Suzuki H, Kimura M, Takane H, Nishizato Y, Irie S, Urae A, Kawabata K, Higuchi S, Otsubo K, Sugiyama Y. 2004. Influence of common variants in the pharmacokinetic genes (OATP-C, UGT1A1, and MRP2) on serum bilirubin levels in healthy subjects. Hepatol Res 30(2):91–95.
104. Huang CS, Huang MJ, Lin MS, Yang SS, Teng HC, Tang KS. 2005. Genetic factors related to unconjugated hyperbilirubinemia amongst adults. Pharmacogenet Genom 15(1):43–50.

105. Gentile S, Persico M, Tiribelli C. 1990. Abnormal hepatic uptake of low doses of sulfobromophthalein in Gilbert's syndrome: the role of reduced affinity of the plasma membrane carrier of organic anions. Hepatology 12(2):213–217.
106. Suzuki T, Onogawa T, Asano N, Mizutamari H, Mikkaichi T, Tanemoto M, Abe M, Satoh F, Unno M, Nunoki K, et al. 2003. Identification and characterization of novel rat and human gonad–specific organic anion transporters. Mol Endocrinol 17(7):1203–1215.
107. Nomura T, Chang HY, Lu R, Hankin J, Murphy RC, Schuster VL. 2005. Prostaglandin signaling in the renal collecting duct: release, reuptake, and oxidation in the same cell. J Biol Chem 280(31):28424–28429.
108. Zollner G, Fickert P, Zenz R, Fuchsbichler A, Stumptner C, Kenner L, Ferenci P, Stauber RE, Krejs GJ, Denk H, Zatloukal K, Trauner M. 2001. Hepatobiliary transporter expression in percutaneous liver biopsies of patients with cholestatic liver diseases. Hepatology 33(3):633–646.
109. Nozawa T, Suzuki M, Takahashi K, Yabuuchi H, Maeda T, Tsuji A, Tamai I. 2004. Involvement of estrone-3-sulfate transporters in proliferation of hormone-dependent breast cancer cells. J Pharmacol Exp Ther 311(3):1032–1037.
110. Nozawa T, Suzuki M, Yabuuchi H, Irokawa M, Tsuji A, Tamai I. 2005. Suppression of cell proliferation by inhibition of estrone-3-sulfate transporter in estrogen-dependent breast cancer cells. Pharm Res 22(10):1634–1641.
111. van den Dobbelsteen DJ, Nobel CS, Schlegel J, Cotgreave IA, Orrenius S, Slater AF. 1996. Rapid and specific efflux of reduced glutathione during apoptosis induced by anti-Fas/APO-1 antibody. J Biol Chem 271(26):15420–15427.
112. Franco R, Cidlowski JA. 2006. SLCO/OATP-like transport of glutathione in FasL-induced apoptosis: Glutathione efflux is coupled to an organic anion exchange and is necessary for the progression of the execution phase of apoptosis. J Biol Chem 281(40):29542–29557.
113. Ghibelli L, Fanelli C, Rotilio G, Lafavia E, Coppola S, Colussi C, Civitareale P, Ciriolo MR. 1998. Rescue of cells from apoptosis by inhibition of active GSH extrusion. FASEB J 12(6):479–486.
114. Patel P, Weerasekera N, Hitchins M, Boyd CA, Johnston DG, Williamson C. 2003. Semi quantitative expression analysis of MDR3, FIC1, BSEP, OATP-A, OATP-C,OATP-D, OATP-E and NTCP gene transcripts in 1st and 3rd trimester human placenta. Placenta 24(1):39–44.
115. Hsiang B, Zhu Y, Wang Z, Wu Y, Sasseville V, Yang WP, Kirchgessner TG. 1999. A novel human hepatic organic anion transporting polypeptide (OATP2): identification of a liver-specific human organic anion transporting polypeptide and identification of rat and human hydroxymethylglutaryl-CoA reductase inhibitor transporters. J Biol Chem 274(52):37161–37168.
116. Tamai I, Nezu J, Uchino H, Sai Y, Oku A, Shimane M, Tsuji A. 2000. Molecular identification and characterization of novel members of the human organic anion transporter (OATP) family. Biochem Biophys Res Commun 273(1):251–260.
117. Marin JJ, Macias RI, Briz O, Perez MJ, Serrano MA. 2005. Molecular bases of the excretion of fetal bile acids and pigments through the fetal liver–placenta–maternal liver pathway. Ann Hepatol 4(2):70–76.
118. Lu R, Kanai N, Bao Y, Schuster VL. 1996. Cloning, in vitro expression, and tissue distribution of a human prostaglandin transporter cDNA(hPGT). J Clin Invest 98(5):1142–1149.

119. Schuster VL, Lu R, Coca-Prados M. 1997. The prostaglandin transporter is widely expressed in ocular tissues. Surv Ophthalmol 41 (Suppl 2):S41–S45.

120. Bao Y, Pucci ML, Chan BS, Lu R, Ito S, Schuster VL. 2002. Prostaglandin transporter PGT is expressed in cell types that synthesize and release prostanoids. Am J Physiol Renal Physiol 282(6):F1103–F1110.

121. Adachi H, Suzuki T, Abe M, Asano N, Mizutamari H, Tanemoto M, Nishio T, Onogawa T, Toyohara T, Kasai S, et al. 2003. Molecular characterization of human and rat organic anion transporter OATP-D. Am J Physiol Renal Physiol 285(6):F1188–F1197.

122. Sato K, Sugawara J, Sato T, Mizutamari H, Suzuki T, Ito A, Mikkaichi T, Onogawa T, Tanemoto M, Unno M, Abe T, Okamura K. 2003. Expression of organic anion transporting polypeptide E (OATP-E) in human placenta. Placenta 24(2–3):144–148.

123. Mikkaichi T, Suzuki T, Onogawa T, Tanemoto M, Mizutamari H, Okada M, Chaki T, Masuda S, Tokui T, Eto N, et al. 2004. Isolation and characterization of a digoxin transporter and its rat homologue expressed in the kidney. Proc Natl Acad Sci U S A 101(10):3569–3574.

124. Cvetkovic M, Leake B, Fromm MF, Wilkinson GR, Kim RB. 1999. OATP and P-glycoprotein transporters mediate the cellular uptake and excretion of fexofenadine. Drug Metab Dispos 27(8):866–871.

125. Bossuyt X, Muller M, Meier PJ. 1996. Multispecific amphipathic substrate transport by an organic anion transporter of human liver. J Hepatol 25(5):733–738.

126. Fujino H, Saito T, Ogawa S, Kojima J. 2005. Transporter-mediated influx and efflux mechanisms of pitavastatin, a new inhibitor of HMG-CoA reductase. J Pharm Pharmacol 57(10):1305–1311.

127. Badagnani I, Castro RA, Taylor TR, Brett CM, Huang CC, Stryke D, Kawamoto M, Johns SJ, Ferrin TE, Carlson EJ, et al. 2006. Interaction of methotrexate with OATP1A2 and its genetic variants. J Pharmacol Exp Ther 318(2):521–529.

128. Fischer WJ, Altheimer S, Cattori V, Meier PJ, Dietrich DR, Hagenbuch B. 2005. Organic anion transporting polypeptides expressed in liver and brain mediate uptake of microcystin. Toxicol Appl Pharmacol 203(3):257–263.

129. Shitara Y, Hirano M, Sato H, Sugiyama Y. 2004. Gemfibrozil and its glucuronide inhibit the organic anion transporting polypeptide 2 (OATP2/OATP1B1:SLC21A6)-mediated hepatic uptake and CYP2C8-mediated metabolism of cerivastatin: analysis of the mechanism of the clinically relevant drug–drug interaction between cerivastatin and gemfibrozil. J Pharmacol Exp Ther 311(1):228–236.

130. Lau YY, Okochi H, Huang Y, Benet LZ. 2006. Multiple transporters affect the disposition of atorvastatin and its two active hydroxy metabolites: application of in vitro and ex situ systems. J Pharmacol Exp Ther 316(2):762–771.

131. Sandhu P, Lee W, Xu X, Leake BF, Yamazaki M, Stone JA, Lin JH, Pearson PG, Kim RB. 2005. Hepatic uptake of the novel antifungal agent caspofungin. Drug Metab Dispos 33(5):676–682.

132. Fehrenbach T, Cui Y, Faulstich H, Keppler D. 2003. Characterization of the transport of the bicyclic peptide phalloidin by human hepatic transport proteins. Naunyn-Schmiedebergs Arch Pharmacol 368(5):415–420.

133. Meier-Abt F, Faulstich H, Hagenbuch B. 2004. Identification of phalloidin uptake systems of rat and human liver. Biochim Biophys Acta 1664(1):64–69.

134. Nozawa T, Sugiura S, Nakajima M, Goto A, Yokoi T, Nezu J, Tsuji A, Tamai I. 2004. Involvement of organic anion transporting polypeptides in the transport of troglitazone

sulfate: implications for understanding troglitazone hepatotoxicity. Drug Metab Dispos 13(4):525–253.

135. Lu WJ, Tamai I, Nezu JI, Lai ML, Huang JD. 2006. Organic anion transporting polypeptide-C mediates arsenic uptake in HEK-293 cells. J Biomed Sci 13(4):525–533.
136. Yamashiro W, Maeda K, Hirouchi M, Adachi Y, Hu Z, Sugiyama Y. 2006. Involvement of transporters in the hepatic uptake and biliary excretion of valsartan, a selective antagonist of the angiotensin II AT1-receptor, in humans. Drug Metab Dispos 34(7):1247–1254.
137. Nakagomi-Hagihara R, Nakai D, Kawai K, Yoshigae Y, Tokui T, Abe T, Ikeda T. 2006. OATP1B1, OATP1B3, and mrp2 are involved in hepatobiliary transport of olmesartan, a novel angiotensin II blocker. Drug Metab Dispos 34(5):862–869.
138. Liu L, Cui Y, Chung AY, Shitara Y, Sugiyama Y, Keppler D, Pang KS. 2006. Vectorial transport of enalapril by Oatp1a1/Mrp2 and OATP1B1 and OATP1B3/MRP2 in rat and human livers. J Pharmacol Exp Ther 318(1):395–402.
139. Ismair MG, Stanca C, Ha HR, Renner EL, Meier PJ, Kullak-Ublick GA. 2003. Interactions of glycyrrhizin with organic anion transporting polypeptides of rat and human liver. Hepatol Res 26(4):343–347.
140. Smith NF, Acharya MR, Desai N, Figg WD, Sparreboom A. 2005. Identification of OATP1B3 as a high-affinity hepatocellular transporter of paclitaxel. Cancer Biol Ther 4(8):815–818.
141. Yamazaki M, Li B, Louie SW, Pudvah NT, Stocco R, Wong W, Abramovitz M, Demartis A, Laufer R, Hochman JH, Prueksaritanont T, Lin JH. 2005. Effects of fibrates on human organic anion–transporting polypeptide 1B1-, multidrug resistance protein 2- and P-glycoprotein-mediated transport. Xenobiotica 35(7):737–753.
142. Vavricka SR, van Montfoort J, Ha HR, Meier PJ, Fattinger K. 2002. Interactions of rifamycin SV and rifampicin with organic anion uptake systems of human liver. Hepatology 36(1):164–172.
143. Wang X, Wolkoff AW, Morris ME. 2005. Flavonoids as a novel class of human organic anion–transporting polypeptide OATP1B1 (OATP-C) modulators. Drug Metab Dispos 33(11):1666–1672.
144. Shimizu M, Fuse K, Okudaira K, Nishigaki R, Maeda K, Kusuhara H, Sugiyama Y. 2005. Contribution of OATP (organic anion–transporting polypeptide) family transporters to the hepatic uptake of fexofenadine in humans. Drug Metab Dispos 33(10):1477–1481.
145. Maeda K, Kambara M, Tian Y, Hofmann AF, Sugiyama Y. 2006. Uptake of ursodeoxycholate and its conjugates by human hepatocytes: role of Na(+)-taurocholate cotransporting polypeptide (NTCP), organic anion transporting polypeptide (OATP) 1B1 (OATP-C), and oatp1B3 (OATP8). Mol Pharmacol 3(1):70–77.
146. Cui Y, König J, Keppler D. 2001. Vectorial transport by double-transfected cells expressing the human uptake transporter SLC21A8 and the apical export pump ABCC2. Mol Pharmacol 60(5):934–943.
147. Ismair MG, Stieger B, Cattori V, Hagenbuch B, Fried M, Meier PJ, Kullak-Ublick GA. 2001. Hepatic uptake of cholecystokinin octapeptide by organic anion-transporting polypeptides OATP4 and OATP8 of rat and human liver. Gastroenterology 121(5):1185–1190.
148. Chi Y, Khersonsky SM, Chang YT, Schuster VL. 2006. Identification of a new class of prostaglandin transporter inhibitors and characterization of their biological effects on prostaglandin E_2 transport. J Pharmacol Exp Ther 316(3):1346–1350.

149. Satoh H, Yamashita F, Tsujimoto M, Murakami H, Koyabu N, Ohtani H, Sawada Y. 2005. Citrus juices inhibit the function of human organic anion–transporting polypeptide OATP-B. Drug Metab Dispos 33(4):518–523.

150. Sai Y, Kaneko Y, Ito S, Mitsuoka K, Kato Y, Tamai I, Artursson P, Tsuji A. 2006. Predominant Contribution of organic anion transporting polypeptide OATP-B (OATP2B1) to apical uptake of estrone-3-sulfate by human intestinal Caco-2 cells. Drug Metab Dispos 34(8):1423–1431.

151. Fuchikami H, Satoh H, Tsujimoto M, Ohdo S, Ohtani H, Sawada Y. 2006. Effects of herbal extracts on the function of human organic anion–transporting polypeptide OATP-B. Drug Metab Dispos 34(4):577–582.

152. Fujiwara K, Adachi H, Nishio T, Unno M, Tokui T, Okabe M, Onogawa T, Suzuki T, Asano N, Tanemoto M, et al. 2001. Identification of thyroid hormone transporters in humans: different molecules are involved in a tissue-specific manner. Endocrinology 142(5):2005–2012.

153. Iida A, Saito S, Sekine A, Mishima C, Kondo K, Kitamura Y, Harigae S, Osawa S, Nakamura Y. 2001. Catalog of 258 single-nucleotide polymorphisms (SNPs) in genes encoding three organic anion transporters, three organic anion-transporting polypeptides, and three NADH:ubiquinone oxidoreductase flavoproteins. J Hum Genet 46(11):668–683.

154. Iwai M, Suzuki H, Ieiri I, Otsubo K, Sugiyama Y. 2004. Functional analysis of single nucleotide polymorphisms of hepatic organic anion transporter OATP1B1 (OATP-C). Pharmacogenetics 14(11):749–757.

155. Michalski C, Cui Y, Nies AT, Nuessler AK, Neuhaus P, Zanger UM, Klein K, Eichelbaum M, Keppler D, König J. 2002. A naturally occurring mutation in the SLC21A6 gene causing impaired membrane localization of the hepatocyte uptake transporter. J Biol Chem 277(45):43058–43063.

156. Schwarz UI, Seemann D, Oertel R, Miehlke S, Kuhlisch E, Fromm MF, Kim RB, Bailey DG, Kirch W. 2005. Grapefruit juice ingestion significantly reduces talinolol bioavailability. Clin Pharmacol Ther 77(4):291–301.

157. Koitabashi Y, Kumai T, Matsumoto N, Watanabe M, Sekine S, Yanagida Y, Kobayashi S. 2006. Orange juice increased the bioavailability of pravastatin, 3-hydroxy-3-methylglutaryl CoA reductase inhibitor, in rats and healthy human subjects. Life Sci 78(24):2852–2859.

158. Kyrklund C, Backman JT, Neuvonen M, Neuvonen PJ. 2003. Gemfibrozil increases plasma pravastatin concentrations and reduces pravastatin renal clearance. Clin Pharmacol Ther 73(6):538–544.

159. Schneck DW, Birmingham BK, Zalikowski JA, Mitchell PD, Wang Y, Martin PD, Lasseter KC, Brown CD, Windass AS, Raza A. 2004. The effect of gemfibrozil on the pharmacokinetics of rosuvastatin. Clin Pharmacol Ther 75(5):455–463.

160. Muck W, Mai I, Fritsche L, Ochmann K, Rohde G, Unger S, Johne A, Bauer S, Budde K, Roots I, Neumayer HH, Kuhlmann J. 1999. Increase in cerivastatin systemic exposure after single and multiple dosing in cyclosporine-treated kidney transplant recipients. Clin Pharmacol Ther 65(3):251–261.

161. Backman JT, Kyrklund C, Neuvonen M, Neuvonen PJ. 2002. Gemfibrozil greatly increases plasma concentrations of cerivastatin. Clin Pharmacol Ther 72(6):685–691.

6

MAMMALIAN OLIGOPEPTIDE TRANSPORTERS

STEPHEN M. CARL
Purdue University, West Lafayette, Indiana

DEA HERRERA-RUIZ
Universidad Autónoma del Estado de Morelos, Cuernavaca, México

RAJINDER K. BHARDWAJ
Bristol-Myers Squibb Co., Wallingford, Connecticut

OLAFUR GUDMUNDSSON
Bristol-Myers Squibb Research Institute, Princeton, New Jersey

GREGORY T. KNIPP
Purdue University, West Lafayette, Indiana

Drug Transporters: Molecular Characterization and Role in Drug Disposition, Edited by Guofeng You and Marilyn E. Morris
Copyright © 2007 John Wiley & Sons, Inc.

6.1. INTRODUCTION

The underlying mechanisms of protein and peptide membrane trafficking, absorption, and secretion have traditionally been under debate. In the past, the virtual absence of peptides in portal blood following a meal was sufficient evidence to conclude that luminal protein digestion had to proceed all the way to free amino acids before absorption could occur. Coupled with an apparent high capacity of the small intestine to absorb free amino acids through various mechanisms, researchers largely ignored the possibility of additional absorptive systems, either intestinally or otherwise. Limitations associated with gastrointestinal protein and peptide drug delivery substantiated this hypothesis. However, this hypothesis exhibits a number of flaws in that it does not explain how certain dipeptides, such as carnosine (β-alanylhistidine) and anserine (β-alanyl-1-methylhistidine), could be absorbed intestinally intact.[1] Traditionally, transport of these dipeptides was explained by receptor-mediated endocytotic processes. Upon further investigation, it has been determined that other active transport mechanisms exist for di- and tripeptide transport, similar to those observed for a majority of nutrients and xenobiotics.

In fact, the bulk of nitrogen absorption occurs largely through the epithelial apical absorption of intact di- and tripeptides and to a much lesser extent, free amino acids.[1] The lack of intact di- and tripeptides in portal blood after a meal is dependent on the activity of intracellular, cytosolic peptidases, which exhibit little affinity for derived or nonclassical amino acids, such as β-alanine.[1] The basal secretion of digested substrates in the form of free amino acids into blood is indicative of the overall efficiency of these absorptive mediators. Still, the need for concentrative transport mechanisms to mediate peptide absorption is not readily apparent given that peptide concentrations in the intestinal lumen are most certainly higher than plasma levels. However, considering potential physicochemical barriers to peptide permeation and the significance of peptide absorption to cellular energy and overall nutrient requirements, it seems logical that these important substrates should be actively absorbed.

The luminal morphology of the gastrointestinal tract varies significantly from relatively no folding in the esophagus to folds of Kerckring, villi, and microvilli in the small intestine.[2,3] Polarized columnar epithelial cells of the small intestinal villi are the primary mediators for gastrointestinal absorption of orally administered drugs and nutrients. The presence of tight junctional complexes and high nutrient selectivity of these cells impart further barrier functionality to the small intestine, excluding many compounds that may be harmful to the organism or detrimental to cellular function.[2] Based on these barriers, the physicochemical nature of a compound dictates the route and extent of its intestinal absorption.

Compounds traverse biological barriers through either paracellular or transcellular routes.[4–6] Paracellular absorption occurs by diffusion of dissolved solute between cells through the tight junctional complex, or zonula occludens, and the tortuous intercellular pathways.[4,5] The paracellular pathway is quite restrictive depending on the pore size and charge of the tight junctions as well as the cell barrier's porosity. In contrast, the transcellular route comprises several potentially parallel pathways for drug permeation, including passive transcellular diffusion, ion channels, facilitated diffusion, active transport, and endocytosis.[7] The perceived mechanism of barrier permeation is highly dependent on a number of physicochemical properties of the drug, or xenobiotic, such as its net overall charge, hydrophilicity, shape/conformation, size, and molecular weight.[4,5] Transcellular transport of a compound is also highly dependent on a number of physiological factors, some of which are discussed below.

Although the passive paracellular diffusion of hydrophilic amino acids and peptides is possible, size restrictions, secondary structural considerations, and charge–charge interactions through this pathway largely preclude it from serving as a primary mechanism of permeation of many small peptides, oligopeptides, and protein-based compounds. Due to these physiological restrictions, a majority of amino acids and peptides require that absorption occur via the transcellular route. However, the physicochemical properties of amino acids and peptides, including hydrophilicity and charge, generally favor a hydrophilic permeation route (i.e., paracellular diffusion) and prevent their passive transcellular diffusion through the hydrophobic cellular membrane. Given these restrictions as well as those observations concerning the intestinal absorption of undigested dipeptides, such as carnosine and anserine, researchers have realized that the bulk of di- and tripeptide absorption occurs through the function of active, concentrative transporter proteins.

Numerous classes of transporter proteins have been identified to date, each with different and sometimes overlapping substrate specificities, capacities and affinities as well as specific tissue, cellular, and temporal expression patterns. Not surprisingly, the physicochemical properties of a compound dictate its interactions with transporter proteins. Considering that transport is a multifaceted process, variability due to overlap of the substrate selectivity of transporters in passive diffusion due to the lipophilic character and potential solubility differences may result in fluctuations in the observed net membrane transport for a substrate. Notwithstanding other potential sources of variability, the net observable transport of amino acids and di- and tripeptides is mediated by a number of different transporter families. Di- and tripeptide transport is, however, generally believed to be attributable to the activity of the proton-coupled oligopeptide transporter [POT; SLC15 (solute carrier)] superfamily of transporters.

Several comprehensive reviews can be found describing the common characteristics of the oligopeptide transporter proteins.[8–13] To date, four mammalian members of the POT superfamily have been identified and described functionally (Table 6.1). The currently known peptide transporters include peptide transporters 1 and 2, PepT1 (SLC15A1) and PepT2 (SLC15A2), and peptide/histidine

TABLE 6.1. Mammalian Members of the POT Family of Transporters

Transporter Name	Human Gene	Accession Number	Species	Length (a.a.)	PKC/PKA Sites[a]	Human Tissue Expression
PepT1	SLC15A1	AAA17721	*O. cuniculus*	707	1/1	Intestine, liver, kidney, pancreas
		NP_005064	*H. sapiens*	708	2/0	
		NP_476462	*R. norvegicus*	710	1/1	
		NP_444309	*M. musculus*	709	1/1	
		NP_001028071	*Mac mulatta*	708	N.D.	
		NP_001003036	*C. familiaris*	708	N.D.	
		NP_001009758	*O. aries*	707	4/3	
		AAO43094	*Sus scrofa*	708	N.D.	
PepT2	SLC15A2	P46029	*O. cuniculus*	729	4/0	Kidney, lung, brain, mammary gland
		NP_066568	*H. sapiens*	729	5/0	
		NP_113860	*R. norvegicus*	729	3/3	
		NP_067276	*M. musculus*	740	N.D.	
		NP_001028125	*Mac mulatta*	729		
PHT1	SLC15A4	NP_663623	*H. sapiens*	577	11/2	Brain, intestine, retina, placenta
		NP_653359	*R. norvegicus*	572	11/0	
		NP_598656	*M. musculus*	574	N.D.	
PHT2	SLC15A3	NP_057666	*H. sapiens*	581	N.D.	Lung, spleen, thymus
		NP_647557	*R. norvegicus*	582	4/2	
		NP_075531	*M. musculus*	578	N.D.	

[a]N.D., not determined.

transporters 1 and 2, PHT1 (SLC15A4) and PHT2 (SLC15A3). Interestingly, the human intestinal peptide transporter, HPT1 (CDH17), is a member of the cadherin family, which to our knowledge has been identified as the only active di- and tripeptide transporter that is not a member of the POT superfamily.[13,14]

Human PepT1, the most widely studied of the POT family members, was first cloned from a rabbit intestinal cDNA library.[15] Later studies determined that PepT1 was a low-affinity high-capacity transport system for di- and tripeptides, with little to no affinity for amino acids.[16,17] Subsequent identification and cloning of the high-affinity low-capacity PepT2 transporter from a human kidney cDNA library [18,19] suggested that these two transporter systems act concertedly to facilitate the absorption and conservation of di- and tripeptides. Interestingly, this hypothesis does not account for the relative activities of either PHT1 or PHT2, which have only recently been described.[20,21] Discerning the importance of a single transporter from the functional effects of other transporters illustrates one of the many problems inherent in the study of drug transport (i.e., overlapping specificity of these systems may confound the applicability of transport data to actual in vivo physiological systems). Also, current methodology limits the feasibility of studying multiple transporter systems simultaneously, further hindering our efforts to understand the relevance of a particular transporter to overall transcellular flux. A fundamental understanding of a drug transporter's net effect on absorption is essential when analyzing a drug's physiological behavior and response [pharmacokinetics/pharmacodynamics (PK/PD)] after peroral administration. Difficulties in characterizing the intestinal transepithelial transport of drugs underscores the need for a complete understanding of the biophysical and biochemical barriers that could potentially alter a drug's PK activity (i.e., absorption, distribution, metabolism, and excretion). Members of the POT superfamily of transporter proteins facilitate the absorption and secretion of a wide range of di- and tripeptides and of peptidomimetic pharmaceutical agents.

Figure 6.1 shows a representative schematic of the potential transporters and permeation routes available for peptide and peptide-based drug absorption and their potential intracellular fates when traversing a cellular barrier (e.g., the intestinal epithelium). The concepts discussed below further clarify many of the points highlighted in this conceptualized diagram. In this chapter we discuss their relative importance to overall physiology and the PK/PD of potential pharmaceutical substrates. Also, the functional roles of POT members to drug screening and rational drug design are presented from an industrial perspective.

6.2. MOLECULAR AND STRUCTURAL CHARACTERISTICS

Members of the POT superfamily share a common topological map consisting of 12 putative α-helical transmembrane domains with intracellularly localized N- and C-termini.[13,22,23] Two characteristic protein signatures of the POT family members have been identified, known as the PTR2 family signatures: (1) [GA]–[GAS]–[LIVMFYWA]–[LIVM]–[GAS]–D–x–[LIVMFYWT]–[LIVMFYW]–G–x(3)–[TAV]–[IV]–x(3)–[GSTAV]–x–[LIVMF]–x(3)–[GA], and (2) [FYT]–x(2)–[LMFY]–[FYV]–[LIVMFYWA]–x–[IVG]–N–[LIVMAG]–G–[GSA]–[LIMF].[24] A

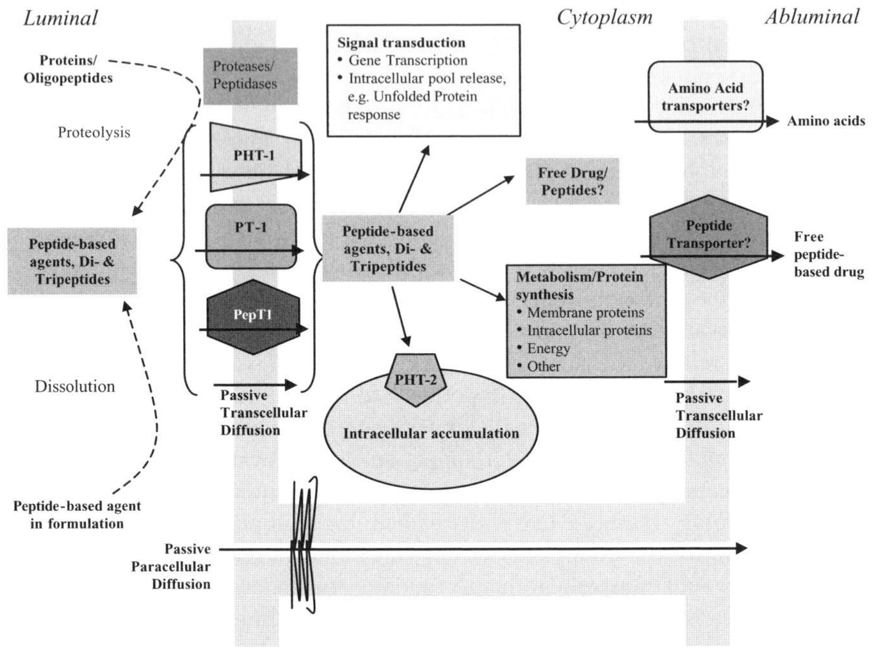

FIGURE 6.1. Potential parallel or competing pathways available for oligopeptide and peptide-based drug permeation and their potential intracellular fates across cellular barriers. Depicted are PepT1, PHT1, PHT2, and PT1 as the concentrative oligopeptide transporters. PepT2 is not expressed in intestinal epithelial cells and is therefore not illustrated here.

third consensus sequence has been proposed by Fei et al., (GTGGIKPXV).[25] Based on their phylogenetic analysis, Saier et al. have also proposed three different signature sequences associated with the POT superfamily.[26] Despite these identified consensus sequences, there is little identity between the PepT and PHT members of this superfamily. For instance, while the human PepT1 (hPepT1) shares 83.5% identity with its rat homolog, its identity with the human PHT1 (hPHT1) is just 20.4%.[13] Interestingly, hPepT1 shares just 26.0% identity with the intracellularly localized hPHT2.[13]

As mentioned previously, PepT1 is arguably the most widely studied member of the POT superfamily of transporters. However, relatively little information is known concerning its tissue and cellular localizations as well as species variations. The cloned human PepT1 cDNA sequence encodes a 708-amino acid protein with an estimated molecular weight of 79 kDa and an isoelectrical point of 8.6.[27] PepT1 expression has been demonstrated in several animal species,[15,30–33] with each isoform exhibiting high homology with other species. Although hPepT1 shows high homology with its various orthologs, studies have demonstrated different tissue expressions of PepT1 among different species.[34] These expressional differences have not been limited to PepT1, as rPHT1 was not expressed in the gastrointestinal tract, in contrast to its human homolog.[20,35] Furthermore, correlation of the expression data from one physiological system may not correlate with another. Consequently, great care should be taken in discerning too much from the data and extrapolating to other physiological models.

PepT1 is expressed primarily in small intestinal epithelial cells, the S1 (pars convoluta) segment of renal proximal tubules, hepatic bile duct epithelial cells, and the pancreas.[13,35–38] Localization studies have also demonstrated that PepT1 is expressed to a lesser extent in the kidney, placenta, and prostate, while in the small intestine, expression is maximal in the duodenum.[35] Ogihara et al. further demonstrated PepT1 localization to the apical plasma membrane of enterocytes in rats.[39] Interestingly, PepT1 expression decreases along the descending small intestinal segments, with ileal expression appearing to be lower than jejunal.[35] Recently, Terada et al. demonstrated that PepT1 is also relatively highly expressed in the stomach of cancer patients, although interindividual differences were apparent.[40] Interestingly, histological results suggest that gastric PepT1 may originate from intestinal metaplasia, which is characterized by the transdifferentiation of gastric epithelial cells to an intestinal phenotype.[40] The functional and subsequent regulatory significance of this expression remains to be elucidated.

Additional studies have demonstrated the intracellular localization of PepT1 isoforms to lysosomal compartments of renal tubular cells.[41,42] The functional significance of this finding remains to be fully elucidated, although it is surmised that PepT1 may prevent accumulation of di- and tripeptides in renal lysosomes by transporting substrates to the cytosolic compartment for subsequent hydrolysis.[41] Of note is the fact that the apical expression of PepT1 has been well established in both prenatal and mature animals,[43,44] although the cellular localization does vary and is highly regulated.

Interestingly, PepT2 is more widely expressed than PepT1, with primary expression in the S2 and S3 segments of the apical membranes in renal proximal tubular cells (pars recta), brain astrocytes, epithelial cells of choroids plexus, retina, mammary gland, and bronchial epithelial cells.[36,45–48] The cloned human PepT2 cDNA is 2190 bp long, encoding a predicted protein of 729 amino acids that shares approximately 50% identity and 70% similarity with its corresponding PepT1 ortholog.[18] The gene encoding PepT2 has been mapped to chromosome 3q13.3–q21.[49] In vitro translation of rabbit PepT2 cRNA resulted in an unglyosylated 83-kDa product and a core-glycosylated 107-kDa product.[50] PepT2 mRNA expression has been identified in human,[18] rat,[51] mouse,[52] and rabbit.[50]

PepT2 was first identified due to the functional assessment that the renal peptide transport system was similar but not identical to its intestinal counterpart (PepT1).[18] In fact, Liu et al. identified PepT2 by screening a human kidney cDNA library with a probe derived from the rabbit intestinal PepT1 cDNA.[18] Since its first identification and characterization, considerable research has been focused on elucidating the functional expression and activity of PepT2 in the kidney. Moreover, recent studies have hypothesized that its primary role is the active reabsorption (secretion) of renally eliminated small oligopeptides.[53] While the expression of both PepT1 and PepT2 have been demonstrated in the proximal tubules of microdissected rat nephrons, all other renal sections were negative for their expression.[36] The positive expression of PepT1 and PepT2 in proximal tubules correlated with in vivo uptake of a fluorescent dipeptide derivative probe (D-Ala-Lys-AMCA) in the inner cortex and outer stripe in rat kidney.[47] Interestingly, the uptake of the substrate was competitively inhibited by cefadroxil and Gly-Gln, indicative of PepT2-mediated absorption.[47] However, PepT2 is much more widely distributed than just the kidney and as such must play a more important role in peptide absorption and accumulation in other tissues as well.

In particular, PepT2 transcripts,[45] protein,[54,55] and functional activity[56,57] have been shown in choroid plexus and brain, as nicely reviewed by Smith et al.[58] As such, this transporter is believed to play a significant role in neuropeptide homeostasis and the efflux of peptides and peptidomimetic xenobiotics from cerebrospinal fluid. Immunoblotting analysis determined the neural PepT2 tissue localization, with protein expression apparent in cerebral cortex, olfactory bulb, basal ganglia, cerebellum, and hindbrain sections of adult brain.[59] The strongest signals were found in the cerebral cortex, while PepT1 protein expression was not found in brain.[59] Interestingly, expression levels were maximal in the fetus and declined with age, indicative of a complex regulatory mechanism to be discussed later. Also, the protein was expressed exclusively on the apical membrane (CSF-facing), indicative of its role in peptide homeostasis and transport.[59]

Shen et al. demonstrated that PepT2 is the primary POT member responsible for Gly-Sar uptake in the choroid plexus of PepT2 knockout mice.[60] However, the lack of an observable pathological phenotype indicates that other systems are able to compensate for the loss of activity.[60] Furthermore, while Gly-Sar is a model substrate for PepT transporters, recent studies in our laboratory have demonstrated that it is not a substrate for the human PHT1 isoform in transiently transfected COS-7 cells.[61] The rat PHT1 isoform was first suggested to be expressed in the apical membranes of the rat choroid plexus,[20] although our results may not reflect species differences in the function of the PHT1 isoforms. Taken together, these data suggest that redundancies may exist for the transport of di- and tripeptides at the choroid plexus and presumably, other biological barriers.

Interestingly, although many studies have demonstrated that PepT2 is not expressed in the gastrointestinal tract, Rühl et al. investigated peptide transport activity in the neuromuscular layers of whole-mount preparations from mouse, rat, and guinea pig stomach and small and large intestines.[62] Surprisingly, D-Ala-Lys-AMCA specifically accumulated in both ganglionic layers of the enteric nervous system, and its accumulation was inhibited by Gly-Sar, D-Phe-Ala, Gly-Gln, and cefadroxil, but not free histidine or benzylpenicillin (a PepT2 substrate).[62] Immunohistochemical analyses demonstrated that the dipeptide uptake was localized to enteric glial cells [as demonstrated by 100% uptake in of glial fibrillary acidic protein ($GFAP^+$) and $S100^+$ cells] and periganglionic tissue-resident macrophages.[62] The researchers surmised that PepT2-mediated dipeptide transport in enteric glia could contribute to the clearance of neuropeptides in the enteric nervous system.[62]

The findings described above underscore the importance of POT members with respect to peptide homeostasis, uptake, and accumulation. Recent evidence indicates that enteric glia play a significant role in modulation of gastrointestinal functions and may be involved in signaling processes of the enteric nervous system.[62] Loss of enteric glial cells in genetically modified animals resulted in neuronal degeneration or changes in the neurochemical coding of enteric neurons, emphasizing the importance of enteric glia for proper neuronal maintenance in the enteric nervous system, as stressed by Rühl et al.[62] The researchers also surmised that PepT2 expression in the enteric glia is unrelated to nutritive peptide absorption, due to its spatial localization and regional distribution throughout the intestine. However, the functional significance of PepT2 expression and function in the enteric neuronal system remains to be elucidated.

Recently, two putative human PHT (hPHT1 and hPHT2) transporters have been identified with expression observed in several human tissues.[63,64] The hPHT1 mRNA sequence is approximately 2.7 kb long, encoding a translated 577-amino acid protein with an estimated molecular weight of 62 kDa and a predicted pI value of 9.2.[65] Four N-linked glycosylation sites were predicted along with several protein phosphorylation sites.[13] Interestingly, though, PHT1 was first cloned in the rat and was the first POT transporter identified in brain.[20] The rat cDNA was 2751 bp, with an open reading frame of 1719 bp that encoded a protein of 572 amino acids.[20] The rPHT1 exhibits approximately 86.5% identity to its human ortholog[13] and was found to be expressed primarily throughout the whole brain, in hippocampus, choroid plexus, cerebellum, and pontine nucleus.[20] The rat PHT1 isoform was also found in both neuronal and small nonneuronal cells by in situ hybridization, while the protein was expressed abundantly in brain and retina, with lower expression in lung and spleen, as determined by Northern blotting.[20] Rat PHT1 expression was not observed in the pancreas, kidney, intestine, liver, heart, and skeletal muscle.[20]

In contrast to its rat ortholog, hPHT1 was found to be expressed at low levels in the human gastrointestinal tract.[35] Interestingly, mRNA expression was also demonstrated in brain, colon, heart, kidney, leukocytes, liver, lung, ovary, pancreas, placenta, prostate, skeletal muscle, small intestine, spleen, testis, and thymus, which were confirmed by Southern blotting.[35] Furthermore, mRNA expression was also demonstrated in retinal pigment epithelium.[66] The functional significance of hPHT1 expression in these tissues remains to be elucidated. However, our laboratory has recently demonstrated hPHT1 expression in the enterocytes of intestinal tissue segments, as well as hPHT1 functional activity in transiently transfected COS-7 cells.[61] The contribution of hPHT1 to overall peptide transport remains to be elucidated.

Similar to hPHT1, hPHT2 has not been widely studied and little is known concerning its physiological significance. Formerly named PTR3, PHT2 was first isolated from the human placenta and has an open reading frame of 1.7 kb, encoding a protein with 581 amino acids with an estimated molecular mass of 64.6 kDa.[65] The rat PHT2 (rPHT2) ortholog has been partially evaluated, having been cloned from a rat brain cDNA.[21] The encoding cDNA was 1979 bp long with an open reading frame of 1748 bp (including the termination codon) and encoded a protein of 582 amino acids that exhibited 49% identity to PHT1 and 80% homology to the human orthologs.[21] Three N-linked glycosylation sites on the rPHT2 protein are predicted and protein phosphorylation sites (PKA and PKC) were identified.[13]

Human PHT2 was found to be widely expressed in various tissues, with mRNA expression demonstrated throughout the gastrointestinal tract, with increased expression in the colon.[35] Furthermore, hPHT2 mRNA expression was shown in the brain, colon, heart, kidney, leukocytes, liver, lung, ovary, pancreas, placenta, skeletal muscle, small intestine, spleen, testis, and thymus, which was all confirmed by Southern blot analysis.[35] Interestingly, Sakata et al. demonstrated rPHT2 mRNA expression in lung, spleen, and thymus, with lower expression in brain, liver, adrenal gland, and heart by RT-PCR.[21] Strong expression was also determined by in situ hybridization in immunocytes, specifically eosinophils, macrophages, and other phagocytes. Further confounding the elucidation of PHT2's function were observations by Sakata et al. suggesting an intracellular localization of PHT2 in the Golgi, lysosomes, autophagosomes, and vacuoles of HEK-293T and baby hamster kidney (BHK) cells.[21] Interestingly, some cells had apparent PHT2 localization on the outer nuclear membrane.[21]

Unfortunately, one cannot ascertain the functional significance of these transporters by investigating their tissue and cellular localizations. Further, peptide transport may also be potentially mediated by a number of other transporter proteins that exhibit overlapping specificity for POT member substrates or play a role in mediating the energy forces driving transport. The primary role of PepT1, PepT2, and PHT1 transporters seems to be the active absorption and accumulation of peptide nutrients. The role of PepT2 may be extended to the clearance of neuropeptides, while the role of PHT2 remains to be elucidated. Further complicating matters is as yet unidentified but functionally characterized basolateral transporters that facilitate the active transcellular flux of these substrates. Interestingly, similar to differences in PepT1 and PepT2 tissue localizations, the intestinal and renal basolateral transporters are also different.[67] In addition, non-POT members may also serve as a confounding factor in understanding peptide transport.

First described in 1994 by Dantzig et al., hPT1 shares only 16% identity and 41% similarity with the hPepT1 amino acid sequence.[13,14] The PT1 cDNA coding region is approximately 2.5 kb long and encodes a 120-kDa protein consisting of 832 amino acids.[14] HPT1 was first identified as a result of a monoclonal antibody that inhibited the transport of β-lactam antibiotics in Caco-2 cells.[14] Chinese hamster ovary (CHO) cells subsequently transfected with HPT1 demonstrated dipeptide and β-lactam transport that was not competitively inhibited by amino acids.[68] Interestingly, hPT1 function has not been fully characterized, although a rat isoform has been shown to be regulated by dietary protein content.[69] Although the molecular, structural, and functional characterization of PT1 has yet to be completed (presumably due to patent issues), the concept of a cadherin family member mediating peptide transport is an important finding and is indicative of the potential overlap in substrate affinities/selectivity between families.

Table 6.2 provides a brief summary of the molecular and functional characteristics of the established human oligopeptide transporters and splice variants. There are other reports of nonclassical peptide transporter behavior, but those transporters will not be elaborated on here. In short, considerably more research is required to fully elucidate all of the functional peptide transporter systems.

TABLE 6.2. Summary of the Molecular and Functional Characteristics of Established Human Oligopeptides Transporters and Splice Variants

Transporter	PepT1	PepT1-RF	Pg-PepT1	PepT2	PHT1	PHT2	PT-1
Family	POT	PepT1-splice variant	PepT1-splice variant	POT	POT	POT	Cadherin
Protein length	708 a.a.	208 a.a.	150 a.a.	729 a.a.	572 a.a.	581 a.a.	832 a.a
Transmembrane domains	12	6	3	12	12	12	10
Established substrates	Di- and tripeptides, peptide-based agents	N.A.: alters pH sensitivity of PepT1	N.A.: forms heterodimer with hPepT1 to decrease Gly-Sar uptake	Di- and tripeptides, peptide-based agents	Histidine, di- and tripeptides, valacyclovir	Histidine, carnosine, di- and tripeptides	Di- and tripeptides, peptide-based agents

6.3. FUNCTIONAL PROPERTIES

6.3.1. Mechanism of Transport

POT transporters use a proton gradient and the membrane potential as the main driving forces, as reviewed previously[9,17,70,71], to mediate the uptake of peptides or peptide-based drugs. In bacteria, yeast, and plant cells, the proton gradient driving force is supplied primarily by membrane ATPases,[17] while in mammalian cells it is generally provided by electroneutral proton–cation exchangers (e.g., Na^+/H^+ antiporters).[72,73] For example, PepT1-mediated intestinal absorption of peptide-based substrates occurs in an asymmetrical manner, with the proton gradient generally considered to be the predominant force influencing transport in the upper small intestine.

Further studies have demonstrated that charged substrates present different binding affinities based on their degree of ionization. Di- and tripeptides, and peptidomimetic compounds with no net charge at the site of absorption are best transferred across the cell membranes by PepT-like transporters.[74–78] PepT substrates share the same substrate-binding site regardless of substrate charge.[79,80] Proton coupling occurs in the H^+-binding site of PepT1, where a H^+ is bound prior to anionic or neutral substrate uptake but is not required for cationic substrates.[71] Irie et al. have developed a computational model to illustrate the H^+-coupled substrate transport of neutral and charged molecules, establishing a PepT1 mechanistic model based on two assumptions: (1) H^+ binds not only to the H^+-binding site but also to the substrate-binding site, and (2) H^+ at the substrate-binding site inhibits the interaction of neutral and cationic substrates but is necessary for that of anionic substrates.[81]

6.3.2. Molecular Requirements for Substrate Recognition and Transport

Some molecular requirements on the structure of PepT-like transporters have been recognized as essential for substrate recognition and transport. Site-directed mutagenesis analyses of single amino acids located within different transmembrane domains (TMD2, TMD4, and TMD5) have shown that Y56, Y64, Y167, N171, and S174 residues modify or inactivate PepT-like transport activity and/or substrate binding completely.[82–85] Other studies have shown that mutations on W294 and E595 reduced Gly-Sar uptake significantly.[82]

Sequence alignments of PepTs have shown the presence of conserved histidyl residues (H57, H121, and H260). Experimental evidence suggests that the histidyl residue H57 is involved in H^+-binding and is essential for peptide transport activity.[86] H121 is involved in substrate recognition, while the role of H260 remains to be elucidated.[84] In hPepT2, H87 has been shown to be absolutely essential to maintain transport activity.[87]

Döring et al. demonstrated that the N-terminal region of the protein (TMD1 to TMD9) confers all the phenotypic characteristics, while studies utilizing PepT1/PepT2 chimeras suggest that the first 400 residues (TMD1 to TMD6) contain the substrate-binding pocket and the region that determines pH dependence.[88–90] Supporting these findings, Döring et al. further demonstrated that a section between TMD2 and TMD3 (amino acid residues 60 to 91) plays a significant role in the pH-dependent transport process.[89] Furthermore, Fei et al. demonstrated that TMD7 to TMD9 play a role in defining substrate affinity.[25] Döring et al. also determined that the first 59 amino acid residues of both PepT1 and PepT2 contribute significantly to substrate affinity.[89] The functional significance of the C-terminal region of oligopeptide transporters is unknown at this time; however, portions of the C-terminus may aid in transporter trafficking, membrane insertion, and/or transporter regulation, as evidenced by its inclusion in the functionally active rat pineal gland PepT1 (pgPepT1) splice variant.[91]

Studies using the substituted cysteine accessibility method (SCAM) have provided more detail concerning PepT structure.[85,92] TMD5, the most conserved region across species, displays solvent accessibility along its entire length, lining the putative aqueous channel akin to related antiporters.[85] Similarly, TMD7 also exhibits solvent accessibility. Preliminary molecular modeling, along with some experimental evidence, has suggested that the extracellular end of TMD7 may shift following substrate binding, providing the basis for channel opening and substrate translocation.[92] However, additional studies are required to clearly elucidate the molecular PepT structure. Taken together, these findings suggest that the N-terminal TMDs form a porelike structure, while TMD7 to TMD9 compose the substrate-binding pocket.[12]

6.3.3. General Substrate Specificities

The three-dimensional structure of any of the POT proteins has not been elucidated. As such, the structural requirements for molecular recognition have been based on uptake or transport experiments, as well as on computational modeling. Several studies have summarized the primary chemical moieties that a peptidomimetic molecule should contain to function as a PepT substrate or inhibitor.[11,93] For years it was believed that a peptide bond was essential for a molecule to have affinity for PepT, however, nowadays it has been well established that this is not the case.[94–97] Furthermore, an entire range of molecules, with varying functional moieties, are also transported by PepT1 with affinity constants similar to those of some dipeptides.[98] Distinctively, substrates or inhibitors with high PepT1 affinity (<0.5 mM) have been recognized as molecules containing the following characteristics: (1) L-amino acids, (2) an acidic or hydrophobic moiety at the C-terminus, (3) a weakly basic group in α-position at the N-terminus, (4) a ketomethylene or acid amide bond, and in the case of a molecule with a peptide bond, (5) to have a trans conformation.[99] Bailey et al. have corroborated experimentally experimentally some of the structural characteristics already proposed and identified some structural requirements necessary to increase PepT1 affinity for a D-enantiomer over its corresponding L-enantiomer.[100] Furthermore, analysis of the binding and transport characteristics of PepT1 have led to the development of several molecular models attempting to establish a PepT–substrate template.[93,101–105]

Comparatively, PepT1 is considered a low-affinity (K_m of 200 μM to 10 mM), high-capacity transport system, while PepT2 has a higher substrate affinity (K_m of 5 to 500 μM) with a lower transport capacity. Interestingly, PepT2 affinity has been attributed to the requirement of an α- or β-aminocarbonyl moiety for substrate recognition.[90] Moreover, the general structural requirements for PepT2 substrate or inhibitor recognition correspond with those described for PepT1.[106,107] However, Those et al. also reported that the spatial localization of a side-chain amino-protecting group in a dipeptide containing a diaminocarbonic acid, and its intramolecular distance from the alpha C-atom, are integral features for transforming a PepT2 substrate into a PepT2 inhibitor.[107]

The molecular and structural characteristics for PHT-like transporters have not yet been determined. Given their recent identification and cloning, it is not surprising that only a few studies have been published regarding their functional properties.[13,20,21,61] In fact, additional studies are required to fully characterize the functional aspects of PHT-mediated transport. As such, additional studies should be designed systematically to provide the required information to analyze the essential structural elements for substrate or inhibitor recognition.

6.3.4. Established Endogenous and Exogenous Substrates

The established physiological substrates of PepT-like transporters are di- and tripeptides,[16] which potentially include different combinations of the 20 L-α-amino acids. These combinations correspond to approximately 400 dipeptides and 8000 tripeptides, not including respective combinations of nonclassical or derived amino acids, such as hydroxyproline, trimethyllysine, and ornithine, homocysteine, methylhistidine, and hydroxylysine. These numbers also do not include the presence of D-enantiomers of the 20 classical amino acids, whose incorporation at the di- and tripeptide N-terminal results in substrates with good affinities and high rates of transport.[108] These observations have resulted in the conclusion that PepT-like proteins are capable of mediating the transport of a huge variety of molecules, with a large diversity of physicochemical characteristics.

Several peptidomimetic drugs, prodrugs, and nonpeptidic compounds have been established as exogenous substrates of the PepT-like transporters. Pharmacologically active compounds that are demonstrated substrates of PepT-like transporters include β-lactam antibiotics (as reviewed previously[12,13]); ACE inhibitors such as captopril, enalapril, and fosinopril[50,109]; bestatin[51]; prodrugs of L-dopa[110–112]; prodrugs of various nucleosides, such as valacyclovir[113] and valganciclovir[114]; prodrugs of some bisphosphonates, such as L-Pro-L-Phe-alendonate and L-Pro-L-Phe-pamidronate[115]; and antibacterial peptide analogs such as alafosfalin.[116] The magnitude of the interaction of β-lactam antibiotics with PepT1 and PepT2 has been analyzed recently in the context of old and new cephalosporins and penicillins, establishing a ranking of affinity constants that may help establishing structure–function studies to further analyze structural substrate/inhibitor requirements.[117]

The peptide/histidine transporters PHT1 and PHT2 have demonstrated dipeptide transport activity, both with high affinity for histidine, and PHT1 for carnosine as well.[20,21] Rat PHT1 when expressed in *Xenopus* oocytes exhibited a high-affinity proton-dependent histidine uptake that was inhibited by several di- and tripeptides but not by other amino acids.[20] Studies in our laboratory suggest that hPHT1 mediates the transport not only of carnosine and L-histidine, but also valacyclovir, in a proton-dependent sodium-independent manner.[61] Interestingly, these studies also concluded that unlike PepT1 and PepT2, the dipeptide glycylsarcosine (Gly-Sar) is not a substrate for hPHT1. Taken together, these data are indicative of both the overlapping specificity, in terms of valacyclovir, as well as the functionally distinct specificity of these transporters. Nevertheless, there is still much to elucidate concerning the affinity and substrate specificity of these peptide transporters.

6.4. REGULATION

Oligopeptide transporter research has predominantly been focused on delineating the functional characteristics of each transporter isoform. Although considerable research is still needed in this area, particularly with undercharacterized isoforms such as PHT1, our focus must now shift toward developing a greater appreciation of their molecular characteristics. In particular, understanding the underlying mechanisms of oligopeptide transporter regulation by various physiological and exogenous stimuli is integral to comprehending their role as mediators of nutritional and pharmacological substrate absorption. Clearly, alterations in oligopeptide transporter expression could potentially result in significant changes that affect the PK/PD phenomena of peptide-based drug substrates. PepT1 has been the most widely studied isoform and has been shown to be regulated via a number of different mechanisms under varying conditions. Studies on the regulation of other POT members (PepT2, PHT1, and PHT2) and PT1 are scarce at best. Factors such as diet, development phase, hormonal influences, pathological conditions (diabetes, intestinal disorders), and various pharmacological agents have all been demonstrated to regulate PepT1 expression and are also discussed in various reviews.[13,118–121] Some of these factors are discussed in greater detail below. However, it is evident that additional studies are required to better elucidate the pre- and posttranscriptional changes that the oligopeptide transporters undergo following various stimuli.

6.4.1. Dietary Regulation

Changes in PepT1 expression due to dietary influences are relatively well established. Erickson et al. demonstrated that a high-casein diet induced a parallel 1.5- to twofold increase in both the PepT1 and rat PT1 mRNA expressions in the middle and distal regions of the rat small intestine when contrasted with low-casein diet control.[69] No apparent change was observed in the proximal region of the rat small intestine, leading to the suggestion that the middle and distal regions may be the most sensitive to dietary fluctuations. Shiraga et al. further investigated the molecular mechanisms of PepT1 expressional changes in the Caco-2 cell model.[122] Incubation of cells with dipeptides (Gly-Sar, Gly-Phe, Phe-Val, Lys-Phe, and Asp-Lys) and individual amino acids (phenylalanine, arginine, and lysine) resulted in activation of the PepT-1 gene promoter. Subsequent induction of the PepT1 mRNA was caused by transcriptional activation through an AP-1 binding site and an amino acid–response element present at the −295 and −277 nucleotides relative to the transcription start site in this region.

Taken together, these expressional changes represent an important evolutionary adaptation to maximize nutrient absorption in response to substrate availability.

Thamotharan et al. also utilized Caco-2 cells to demonstrate that pretreatment with Gly-Sar and cefadroxil for 24 hours increased the V_{max} value of glycylglutamine (Gly-Gln) transport twofold without any significant change in its K_m.[123] Subsequent Western blot analysis revealed a greater than twofold increase in the protein expression of PepT1 in the apical membrane, and a threefold increase in PepT1 mRNA was also observed. Interestingly, an inhibitor of the trans-Golgi network, brefeldin, significantly increased PepT1 mRNA with no demonstrable effect on Gly-Gln transport, suggesting that dipeptides induce PepT1 transcription.[123] Moreover, Walker et al. demonstrated that previous exposure with Gly-Gln (4 mM) for 3 to 4 days in Caco-2 cells resulted in a twofold increase in the V_{max} of Gly-Sar transport, without any change in its K_m.[124] Results suggest that Gly-Gln treatment increased the cellular mRNA level and subsequent PepT1 membrane expression by twofold.

Thamotharan et al. demonstrated that a brief fast or starvation can also up-regulate PepT1 expression on both the mRNA and protein levels.[125] Brush border membrane vesicles prepared from jejunum of fed and 1-day-fasted state rats exhibited a twofold increase of Gly-Gln (V_{max}) uptake upon fasting without a subsequent change observed in the K_m.[125] In addition, PepT1 protein and mRNA increased threefold in both the intestinal brush border membrane and intestinal mucosa, respectively, suggesting a transcriptional regulatory mechanism. In addition, Ogihara et al. demonstrated that PepT1 expression at the microvillous plasma membrane is regulated by diet, whereas the transporter protein gradient along the crypt–villus axis was maintained between the fasting and fed states.[39] Another study comparing rats starved for 4 days and semistarved (50% of freely fed control) for 10 days resulted in a 179% and a 161% increase in jejunal PepT1 mRNA expression of the starved and semistarved groups, respectively, compared to the freely fed control group.[126] These results suggest that PepT1 gene expression can also be modulated by malnutrition.

For better insight into this regulatory mechanism, Naruhashi et al. correlated the functional transport of a model PepT1 substrate, cefadroxil, in both starved and fed rats with PepT1 mRNA expression.[127] Briefly, the transport of cefadroxil was quantified across the small intestines of both groups via an Ussing chamber, and PepT1 mRNA expression was determined using quantitative reverse transcriptase polymerase chain reaction. A good correlation was observed between mRNA expression levels and the permeability coefficients ($r^2 = 0.859$), suggesting that cefadroxil transport is directly proportional to PepT1 expression. Although higher in the proximal regions, PepT1 mRNA expression in the starved rats was increased in all small intestinal segments. Similarly, cefadroxil transport was also higher in the proximal small intestinal regions of starved rats in contrast to the fed rats, indicative of its correlation to PepT1 mRNA expression.[128] Furthermore, Howard et al. investigated the effect of luminal nutrients on the expression of PepT1 in starved and fed rats.[128] Removal of luminal nutrients increased ileal PepT1 mRNA expression compared with orally fed rats. These results allude to the importance of a continuous amino acid supply for the maintenance of proper gut mucosal function. Collectively, these studies suggest that up regulation of PepT1 expression occurs whether fasting is brief or prolonged and may be indicative of a nutrient-sensing feedback mechanism.

6.4.2. Developmental Regulation

Studies have also demonstrated developmental stage–dependent differences in oligopeptide transporter expression and dipeptide absorption during peri- and post-natal periods in rabbits and guinea pigs.[129,130] More recently, Miyamoto et al. demonstrated a dramatically higher intestinal PepT1 mRNA expression in 10-day-old rats, which rapidly decreased and then leveled off reaching a plateau at adult expression levels by day 28 after birth.[131] Further investigation revealed that rat PepT1 mRNA and protein levels in the duodenum, jejunum, and ileum attained maximal levels by days 3 to 5 after birth. Interestingly, these mRNA and protein levels proceeded to fall rapidly to 11 to 13% of maximal expression by day 14, but then rose again to 23 to 58% of maximal expression by day 24. Higher expression of PepT1 at weaning, especially in the ileum, may be related to an adaptive response to a reduced protein diet or nourishment at this time period.[43] The age-dependent PepT1 and PepT2 expression patterns observed in the developing rat kidney were similar to that observed for PepT1 in the small intestine and colon. However, the expression of PepT1 and PepT2 was found to be more consistent in rat kidney than in the intestine. The results suggest that nutritional state may play a less prominent role in regulating the renal expression of these transporters or that an autocrine effect from PepT1 substrates is involved in its regulatory control.[43]

In a recent immunocytochemical study, PepT1 expression was measured using an antibody specific to the C-terminal sequence of PepT1 in rat duodenum at different stages of growth.[132] PepT1 was found to be distributed exclusively in the apical brush border membrane of enterocytes from both prenatal and mature animals. Interestingly, PepT1 expression extends to the subapical cytoplasm, basal cytoplasm, and basolateral membrane of enterocytes immediately after birth.[132] Since the nature of oligopeptide transporter expression at the basolateral membrane has not been defined clearly, atypical PepT1 distribution at birth could be due to enteroendocrine mechanisms following first exposure to enteral (i.e., luminal) nutrition or humoral signals generated during parturition.

6.4.3. Regulation by Circadian Rhythms

Recently, Pan et al. demonstrated the effect of diurnal rhythm on PepT1 expression utilizing an in situ intestinal loop model and everted rat intestines, maintained in a 12-hour photoperiod with free access to food.[133] Kinetic studies demonstrated enhanced Gly-Sar uptake and PepT1 protein and mRNA expression in dark phase (8:00 PM; beginning of the dark phase)–treated rats comparative to light phase (8:00 AM; beginning of the light phase)–treated rats in a 24-hour study cycle.[133] A subsequent study determined that fasting conditions may disrupt the effect of diurnal rhythm on intestinal PepT1 expression and transport activity.[134] In this PK study of ceftibuten oral absorption, higher plasma concentrations were observed at 8:00 P.M. (dark phase) in contrast to 8:00 A.M. (light phase) in fed rats, but no differences in 4-day-fasted rats.

Recently, a PepT1 gene product has been identified in the rat pineal gland (pg-PepT1), which encodes 150-amino acid protein encompassing the three C-terminal membrane-spanning domains of intestinal PepT1 and three additional N-terminal residues.[91] PgPepT1 forms a functional PepT1-like transporter through oligomerization. Interestingly, pgPepT1 mRNA and protein (approximately 16 kDa) levels were found to be >100-fold higher at night, suggestive of a diurnal regulatory mechanism. The neural pathway that controls pineal melatonin production has been suggested to play an important role in regulating the expression of pgPepT1. Thus, a peptide transporter in the pineal gland could play an important role in circadian pineal physiology and contribute to clearance of active or degraded neuropeptides.[91]

6.4.4. Disease State–Dependent Regulation

Many disease state–dependent mechanisms that may alter oligopeptide transporter expression and function could be conceived. Inflammatory disorders, particularly those associated with the release of proinflammatory peptides, are examples of such disease state–dependent variability. A recent study using colonic mucosal samples from diseased (chronic ulcerative colitis, microscopic colitis) and control patients (normal colon) revealed aberrant PepT1 expression under inflammatory conditions.[135] Further studies utilizing coloniclike cells (HT29-Cl.19A) stably transfected with hPepT1 tagged on the N-terminus with green fluorescence protein, with and without inflammatory peptides, indicated that in some states of chronic inflammation, hPepT1 may be expressed anomalously.[135] Recently, Vavricka et al. investigated the effects of tumor necrosis factor alpha (TNFα), interferon-γ (IFNγ), and inflammatory cytokines and mediators [interleukins (IL)-1β, IL-2, IL-8, IL-10] on PepT1 regulation in Caco-2/bbe monolayers and in mice.[136] From these studies it appeared that TNFα and IFNγ play the most prominent roles in activating PepT1 protein expression in the colon. These findings were supported further by the observation of higher Gly-Sar uptake in TNFα- and IFNγ-treated cells when contrasted to controls.[136] Interestingly, Barbot et al. revealed that transcriptional up-regulation of PepT1 occurred in the brush border membrane of enterocytes in *Cryptosporidium parvum* (an amino acid malnutrition disease)–infected rats from day 4 to day 50.[137]

Intestinal transplantation is a surgical procedure performed in patients suffering from short bowel syndrome with complications of parenteral nutrition and irreversible intestinal failure.[138] Syngeneic small intestinal transplantation in rats resulted in increased PepT1 protein expression levels, indicating that transporter expression may be useful to evaluate intestinal graft function in addition to the current histopathological examination of the mucosa.[139] Studies conducted in nephrectomized rats also revealed increased renal high-affinity peptide transport activity and PepT2 expression, while PepT1 levels were undetectable.[140] Two weeks after surgery, Gly-Sar uptake rate (V_{max}) in renal brush border membrane vesicles was significantly increased compared to sham-operated controls.

Utilizing Western and Northern blotting techniques, Gangopadhyay et al. further illustrated that an increase in PepT1 protein and mRNA expression occurred in streptozotocin-induced diabetic rats.[141] Diabetes also increased PepT1 activity in the brush border membrane of the intestinal mucosa, as demonstrated by an increased V_{max} for Gly-Gln. Furthermore, both PepT1 protein and mRNA expression were up-regulated in the brush border membrane of renal tubules. Interestingly, Watanabe et al.

compared the intestinal absorption and pharmacokinetics of cephalexin in type 1 diabetes (streptozotocin-induced) and hyperinsulinemic type 2 diabetic GK and Zucker-fa/fa (Zucker) rats.[142] Pharmacokinetic parameters k_a, $C_{\max}$, and $AUC^{0\rightarrow\infty}$ were significantly higher in type 2 diabetic GK and Zucker rats compared to control rats, whereas the type 1 diabetic streptozotocin-induced rats demonstrated no significant differences from the control. Expression studies supported the PK findings demonstrating increased protein expression in both type 2 diabetic rat models. In contrast, the intestinal PepT1 mRNA levels were not significantly different between controls and any of the study groups. These findings suggest that the PepT1-mediated intestinal absorption of drugs may be enhanced in hyperinsulinemic type 2 diabetic rats. When these results are compared to those determined by Gangopadhyay et al.,[141] the streptozotocin-induced diabetic results are contradictory. This contradiction may be attributed to differences in the experimental conditions, potentially including dietary differences.

6.4.5. Hormonal Regulation

The interplay between insulin and leptin regulation of glucose and lipid homeostasis is well known and forms the basis for the complexities of the metabolic syndrome-associated phenotypes. As discussed above, recent findings in both diabetic and Zucker rats, may indicate that these hormones may also have a concerted effect on protein/peptide homeostasis in part through the modulation of PepT1 expression. In terms of a classical biochemistry view, it would appear obvious and essential that the three major sources of nutrients (lipid, carbohydrates, and protein) would be subject to coordinate regulation.

Toward this aim, recent studies performed in Caco-2 cells demonstrated that within 30 to 60 minutes of adding physiologically relevant insulin concentrations (5 mM), enhanced Gly-Gln uptake was observed and attributed to increased PepT1 expression.[143] In support of this conclusion, kinetic studies revealed a twofold increase in $V_{\max}$ with no significant change in K_m, which was reduced by the tyrosine kinase inhibitor genistein. Additionally, the stimulatory effect of insulin was observed even after treatment with brefeldin, which disrupts the Golgi apparatus, thereby modifying the cells' ability to process newly synthesized PepT1 protein for insertion into the plasma membrane.[143] Furthermore, when colchicine (disrupts the translocation of proteins targeted for membrane insertion) was added 20 minutes before the addition of insulin, it abolished the insulin stimulatory effect completely. Moreover, increased PepT1 protein expression in the apical membrane was demonstrated; however, no corresponding changes in PepT1 gene expression were observed. Nielsen et al. further demonstrated in Caco-2 cell monolayers that short-term insulin pretreatment resulted in a Gly-Sar uptake rate ($V_{\max}$) that was significantly higher than that of the controls, and K_m remained constant.[144] Moreover, the substantial increase in Gly-Sar uptake was not accompanied by changes in hPepT1 mRNA or by measurable changes in the cytosolic pH.

Watanabe et al. further investigated the effects of insulin on cephalexin transport and PepT1 mRNA and protein expression in Caco-2 cells.[145] Consistent with other studies, a 10-minute insulin pretreatment significantly increased the cephalexin uptake and PepT1 protein expression on the apical membrane. Again, no changes were observed in the PepT1 mRNA expression levels. These findings suggest that insulin binds to its receptor on the membranes, promoting PepT1 translocation from the intracellular pool to the apical membrane surface. Collectively, these results indicate that insulin signals an increased PepT1 translocation from a preformed cytoplasmic pool to the apical membrane as opposed to de novo synthesis, which is similar to events observed with the unfolded protein response.[143,146] These results are acute, as opposed to the studies in diabetic rat models that are reflective of a chronic response, further suggesting an alternative pathway by which insulin might increase PepT1 transcription and translation.

In addition to insulin, other hormonal PepT1 regulators have been identified and studied, such as epidermal growth factor (EGF) and leptin. In contrast to insulin, EGF was found to inhibit PepT1 by decreasing the number of transporter molecules on the apical membrane.[147] Kinetic studies demonstrated a significant decrease in V_{max} on the apical uptake of Gly-Sar, whereas K_m remained constant. In contrast, basolateral Gly-Sar uptake demonstrated no changes in either V_{max} or K_m after EGF treatment.[147] Inhibition of transcellular Gly-Sar flux was observed after 5 days of EGF treatment and reached a maximum after 15 days, indicating that a decrease in peptide transport occurs after long-term EGF treatment. Interestingly, brief exposure of EGF has been demonstrated to enhance the functional activity of PepT1.[147] A dose-dependent increase in the apical Gly-Sar uptake with no change in K_m was observed in Caco-2 cells after 5 minutes of EGF exposure. Furthermore, short-term EGF treatment did not alter PepT1 mRNA expression or the intracellular pH. Unfortunately, the actual mechanism by which short-term EGF treatment increases PepT1 functional expression remains unclear.[147] Down-regulation of PepT2 mRNA was also observed after long-term treatment with EGF in the rat proximal tubule cell line SKPT0193 cl.2 (SKPT),[148] implicating either a transcriptional regulatory mechanism through an EGF-responsive element or regulation of PepT2 mRNA stability. However, further studies are required to discern if the EGF regulatory mechanisms for PepT1 and PepT2 share common features.

Buyse et al. explored hormonal PepT1 regulation further by demonstrating that leptin secreted by the stomach may reach the small intestinal lumen to mediate transporter expression.[149] The addition of leptin to intestinal perfusate increased the absorption of Gly-Sar across the rat jejunum in in vivo perfusion studies.[149] In addition, uptake studies conducted utilizing Caco-2 cells demonstrated that leptin significantly increased the V_{max} of Gly-Sar and cephalexin without changes in K_m values. Further, leptin addition to the apical side of Caco-2 cells resulted in a 60% increase in membrane PepT1 protein, with a corresponding decrease in intracellular PepT1. Moreover, enhanced Gly-Sar uptake in response to leptin was abolished with colchicine pretreatment but unaffected by brefeldin.[149] Taken together, these results suggest increased trafficking of PepT1 from the intracellular pool to the apical membrane as a possible mechanism to enhance the transcellular flux of substrates across Caco-2 cells.

Interestingly, gastrointestinal hormones and/or insulin elicit their regulatory effects through specific binding to their respective surface receptors, while thyroid hormones may regulate the expression of oligopeptide transporters by entering the cells directly.[150,151] Significant inhibition of Gly-Sar uptake in Caco-2 cells was observed after pretreatment with 3,5,3′-L-triiodothyronine (T3; 100 nM) for 4 days.[151] Kinetic analysis revealed a twofold decrease in the $V_{\max}$ values with no apparent change witnessed in the K_m. Conversely, Western blot analysis illustrated a 70% reduction in the apical membrane expression of PepT1 protein versus untreated cells. Since decreased gene expression was also observed, it was suggested that T3 inhibition of dipeptide transport is due to decreases in the transcription and/or stability of PepT1 mRNA.[151] However, there is no evidence suggestive of a thyroid hormone-responsive element in the promoter region of the PepT1 gene. Alternatively, it is possible that after binding to its nuclear receptor, T3 could affect transcription through an indirect mechanism(s). Lu and Klaassen recently demonstrated increased PepT1 and PepT2 mRNA expressions in male rat kidneys after undergoing thyroidectomies.[152] These changes are suggestive of nutrient conservation during malnutrition, as evidenced by increased renal reabsorption of di- and tripeptides due to up-regulation of these two transporters.

6.4.6. Regulation by Pharmaceutical Agents

Various pharmaceutical agents, such as clonidine,[153] pentazocine,[154] and cyclophosphamide,[155] have also been demonstrated to affect the activity of oligopeptide transporters. Berlioz et al. studied the receptors responsible for the stimulatory effect of clonidine utilizing Caco-2 cells that were engineered to stably express α_{2A}-adrenergic receptors (Caco-2 3B) at a density similar to that found in the mucosa.[153] Interestingly, higher cephalexin transport was observed in Caco-2 3B cells pretreated with clonidine than in wild-type (expressing PepT1 but not having substantial α_2-adrenergic receptors) and HT29 19A clones (a clone that expresses α_2-adrenergic receptors but does not express detectable levels of PepT1). Kinetic analysis revealed an increase $V_{\max}$ without altering K_m in these cells and that the stimulatory effect of clonidine was blocked completely by Gly-Sar and α_{2A}-adrenergic receptor antagonists. Furthermore, disrupting the microtubule integrity by colchicine abolished the clonidine effect, whereas no effect was observed by Na^+/H^+ exchanger blocker (amiloride), suggesting the role of microtubular integrity on the effect of α_2-agonists. Collectively, these results suggest that clonidine can directly activate α_2-adrenergic receptors located on epithelial cells and thus modulate PepT1 expression by increasing translocation to the apical membrane.[153]

Pentazocine pretreatment for 24 hours resulted in increased Gly-Sar uptake in Caco-2 cells in a concentration- and time-dependent manner.[154] Kinetic analyses have indicated that pentazocine enhanced the $V_{\max}$ for Gly-Sar uptake without significantly altering K_m in Caco-2 cells. Increased PepT1 mRNA expression was observed, suggesting that pentazocine induces PepT1 transcription in Caco-2 cells, leading to increased PepT1 protein expression at the cell membrane.[154] In contrast, treatment with the immunosuppressive agents tacrolimus and cyclosporine A decreased PepT1 activity through an unknown mechanism, as demonstrated by reduction in $V_{\max}$ of Gly-Sar uptake without any discernable effect on K_m.[139] Similarly, pretreatment with steroid hormones (progesterone and norethisterone at 3, 10, and 30 mM) for 24 hours resulted in a significant decrease in the apical-to-basal transport of cephalexin in Caco-2 cell monolayers.[156]

6.4.7. Single-Nucleotide Polymorphisms

Single-nucleotide polymorphisms (SNPs) of transporter genes have also been investigated as possible factors influencing interindividual variations in PK and PD profiles. Coding region SNPs (cSNPs) of transporter genes may induce amino acid mutations, resulting in changes in transporter intrinsic activity by altering the protein's affinity to substrates (K_m) and/or translocation ability or capacity (V_{max}).[157] Despite increasing efforts to understand the molecular and functional characteristics of oligopeptide transporters, little information is available on their respective genetic polymorphisms. Zhang et al. reported nine nonsynonymous and four synonymous cSNPs in the PepT1 gene in a population of 44 persons of different ethnic backgrounds.[158] Plasmids were constructed by site directed mutagenesis for each of the nine nonsynonymous variants and transiently transfected into HeLa cells for functional characterization. Transport kinetics were studied utilizing Gly-Sar as a model substrate; and PepT1 expression was determined by Western blot and immunocytochemical analyses. Results demonstrated that among the nine nonsynonymous SNPs, P586L resulted in significantly reduced transport capacity (V_{max}) and decreased PepT1 protein expression levels.[158]

Following its initial cloning from a human cDNA kidney library, a single PepT2 transcript was identified and accepted as a consensus sequence.[18] Currently, five nonsynonymous cSNPs (R57H, L350F, P409S, R509K, and M704L) have been identified (http://www.ncbi.nlm.nih.gov/SNP/).[159] However, the functional activities of only two of these variants, R57H and P409S, have been characterized functionally in transiently transfected HEK293 cells.[159] Uptake studies performed using these transients and contrasted with wild-types revealed elimination of Gly-Sar uptake in the presence of the R57H variant. Interestingly, the P409S SNP had little effect on function since comparable activity was observed in the transient comparative to the wild-type hPepT2. Western blot analysis and immunofluorescence in *Xenopus* oocytes expressing the PepT2 R57H variant verified protein expression at the membrane surface, indicating that the loss of transport activity was not due to changes in expression, but rather, in it functional characteristics.

6.4.8. Splice Variants

The importance of splice variants in general cannot be underestimated, as they increase protein diversity by allowing multiple, sometimes functionally distinct proteins to be encoded by the same gene. However, elucidating the function of each splice variant has been a tremendous task, especially given the focus on the parent isoform. Inui and colleagues first identified a novel pH-sensing regulatory protein cofactor which modulates the transport activity of hPepT1.[160] The cDNA was cloned from human duodenum having 1704 bp that encoded a predicted protein of 208 amino acid residues.

Functional studies were performed in *Xenopus* oocytes injected with water, hPepT1 cRNA, the 208 hPepT1 splice variant cRNA, and hPepT1 cRNA combined with the hPepT1 splice variant cRNA. Interestingly, results demonstrated a lack of transport activity for Gly-Sar in hPepT1-RF-injected oocytes; however, cotransfection of the hPepT1 splice variant with hPepT1 affected the pH sensitivity of peptide uptake in the double transfect, suggesting a pH-regulatory function for the splice variant and leading to its hPepT1-RF name.

Studies by Sadee and co-workers reported multiple splice variants in the hPHT1 gene.[63] In several regions of the hPHT1 sequence deduced, multiple expressed sequence tag (EST) overlap was observed, suggesting possible sequence variations in the human population. Expression analysis in several tissues and by EST alignments suggests the presence of possible hPHT1 splice variants, as indicated by an insert or gap in two regions of the *hPHT1* coding sequence. Indeed, our laboratory identified a previously unreported hPHT1 splice variant from Caco-2 cells utilizing the primers designed for the cloning of the hPHT1 full-length sequence.[65] This PHT1 cDNA contained a gap of 58 bp between bases 842 and 900 from the start codon. This gap induced a shift in the coding frame, resulting in a predicted protein sequence of 295 amino acids.[65] Interestingly, this hPHT1 splice variant comprised the first six conserved hPHT1 TMDs, similar to the hPepT1-RF,[161] raising potential questions concerning modularity in the hPHT1 gene coding information and its functional relevancy.[13]

Recently, Pinsonneault et al. reported haplotype analysis of the PepT2 gene to determine the effect of variants on PepT2 expression and function.[162] Among these variants, PepT2*1 and *2 were shown to be significantly present in 247 human genomic DNA samples from different ethnic groups. A haplotype is the combination of several sequence variants on a single chromosome at a specific locus and may more accurately reflect genetic diversity in the human population. Wild-type hPepT2 and its variants were transfected into CHO cells and their function was investigated after 24 hours. Results indicated significantly different K_m values for Gly-Sar uptake for each variant. Furthermore, varying levels of hPepT2*1 and *2 mRNA were observed in nine heterozygous kidney tissue samples, suggesting the presence of *cis*-acting polymorphisms affecting transcription or mRNA processing.

6.5. PHARMACEUTICAL DRUG SCREENING

Permeation of peptide-based compounds across cellular barriers can occur by several routes (Figure 6.1). One needs to be mindful that each of these routes of permeation can apply to different peptide-based compounds; however, the physicochemical properties of the compound and the physiological significance of each route will determine the relative importance in controlling the compound's net absorption. In screening compound permeation across cellular barriers, delineation of the key rate-determining steps for controlling absorption needs to be identified. Methods to delineate the functional relevancy of each route are now being enhanced further by focusing the kinetics

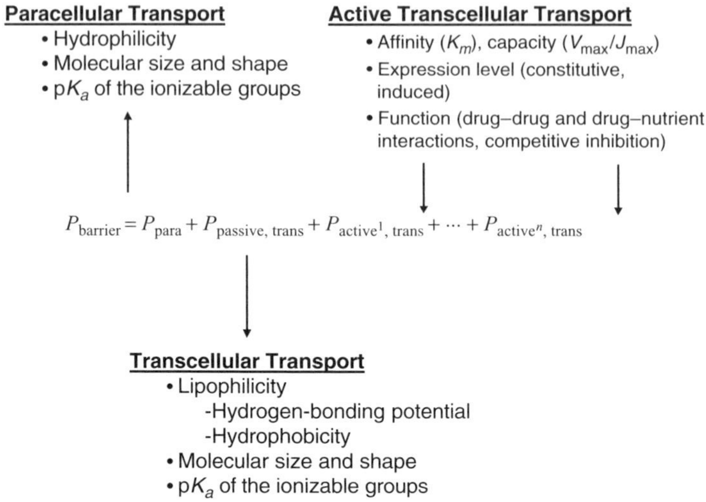

FIGURE 6.2. Some of the rate-limiting mass transfer resistances/characteristics of the parallel paths that oligopeptide-based compounds typically encounter as they traverse cell barriers.The effective permeability of the physiological barrier is a function of all competing transport pathways.

on potential paracellular and transcellular events. Adson et al.[4] demonstrated that the delineation of the paracellular and transcellular mass transfer resistances from observed cell-based assay permeability studies was possible. Several studies conducted by Burton and colleagues further demonstrated that hydrogen-bonding potential may provide a better means for predicting the potential passive transcellular diffusion of peptides.[163–166] Based on these observations, several studies conducted with larger oligopeptides demonstrated that conformation and lipophilicity may dictate the extent of passive paracellular and transcellular peptide transport across Caco-2 and bovine brain microvessel endothelial cell (BBMEC) barriers. [167,168] These studies focused on passive diffusion; however, permeation of smaller di- and tripeptides and peptide-based analogs of similar size may also have affinity for oligopeptides transporters that can be rate controlling for their absorption. A summary of the characteristics for each parallel pathway of transport based on an extension of these and other findings is presented in Figure 6.2.

Although it is clearly evident that oligopeptide transporters play a major physiological role in the absorption of various nutrients and peptidelike xenobiotics, the pharmaceutical industry has yet to fully realize and exploit the potential advantages associated with targeted delivery to these transporters. Probably the most readily apparent application of oligopeptide transporter research is the marketability of Valtrex (valacyclovir) comparative to Zovirax (acyclovir). Comparing clinical trial data (below), the addition of a valyl moiety to the ether end of acyclovir increased the bioavailability from approximately 15% with Zovirax[169] to 54% with Valtrex.[170] Although this increase in acyclovir bioavailability is significant, this case is the exception and not the rule, as it is arguably difficult to directly synthesize compounds with high affinity for a particular transporter and acceptable biological activity. As such, the focus of industry has been to screen compounds for their potential transporter affinities or effective permeability across cellular monolayers and choose lead compounds that are balanced in their physicochemical properties and potential transporter interactions. However, given the relative youth of the transporter field, many scientists have had problems interpreting when a transporter may be important. Although this is largely dependent on the compound and whether or not it is a substrate for a transporter, decision making in this area apparently remains vague. Some have suggested altering the Biopharmaceutics Classification System (BCS) to include elimination criteria to provide criteria to help us understand when transporters will significantly alter the pharmacokinetic profile of a new drug.[171] However, one still needs to know if the compound of interest is a substrate for a particular transporter. This is normally done by permeability screening assays.

Traditionally, the Caco-2 cell line, which is derived from a colonic adenocarcinoma, is used to measure relative gastrointestinal permeability of compounds. However, given varying growth conditions, selective pressures, and intrinsic cellular heterogeneity, it may be difficult to reproduce permeability data accurately between laboratories and individuals. For instance, even increasing the number of passages within the same laboratory changes the expression of numerous transporters in Caco-2 cells.[172] Given the widespread use of these screening models, these differences in transporter expression may not properly translate to the actual physiological condition resulting in selection of lead candidates based on erroneous data.

These obvious concerns have led a number of laboratories to look at the contribution of individual transporters through either transient or stable transfection of the transporter of interest into different cell lines and then measuring transport kinetics.[173–175] Our laboratory developed stably transfected Madin–Darby canine kidney (MDCK)/hPepT1-V5&His clonal cell lines expressing varying levels of epitope-tagged hPepT1 protein to quantify the relationship between hPepT1 expression and its functional kinetics.[175] The utility of this cell line is that it may serve as a useful in vitro model for screening and quantifying peptide and peptide-like drug transport as a function of hPepT1 expression in drug discovery.[176]

Unfortunately, though, even these altered cell lines are limited in that they can model physiology only under extremely controlled conditions. As mentioned above, oligopeptide transporters alone are regulated by a whole host of different mechanisms. Of particular importance, dietary protein increases the gut intralumenal amino acid and oligopeptide concentrations, ultimately resulting in increased POT expression patterns. Given that POT up-regulation is virtually immediate, we (and others) surmise that the increased membrane expression is due to mobilization from intracellular pools of unfolded protein (unfolded protein response). Although the initial thought is that these findings can be utilized to increase the bioavailability of orally administered peptidomimetic drugs through concomitant prescription of a high-protein diet, this presents other potential issues, especially with respect to food–drug interactions.

Studies suggest that dipeptides may regulate their own transport by modifying PepT1 gene expression via two possible mechanisms: (1) increasing the stabilization of mRNA encoding the gene, and (2) increasing the gene transcription.[123] In fact, the effect of dietary protein intake on oligopeptide transporter regulation is probably influenced by a combination of these mechanisms. In short, dietary protein increases the gut intralumenal amino acid and oligopeptide concentrations, ultimately resulting in altered oligopeptide transporter expression patterns. With respect to a compound that exhibits a narrow therapeutic index that happens to be a good substrate for a POT transporter, the concerted up-regulation of the protein in response to food could dramatically increase drug absorption and subsequent blood concentrations. Such an incidence could result in toxic drug effects. Although some may think that such an occurrence could be averted through simply assessing PK parameters in response to food, subtle hormonal regulatory effects, such as up-regulation due to insulin, could present adverse clinical events. Unfortunately, many people do not consider that a small, sugary snack could seriously affect their drug therapy. However, if such a drug were a POT substrate, the resulting insulin secretion in response to the snack could result in serious adverse effects. Although we are not aware of any compounds or situations where these problems have been manifested, we feel that screening tools should be improved to better assess the regulatory mechanisms associated with POT member up/down-regulation to address such potential issues.

Case Study: Targeting Peptide Transporters for Increased Oral Absorption—Valcyclovir Valacyclovir is an L-valine ester prodrug of acyclovir that is used for the treatment of herpes, varicella zoster (VZV), and cytomegaloviruses (CMV). Valacyclovir was developed to increase the oral absorption and plasma levels of acyclovir. Increased plasma concentrations of acyclovir are important in maintaining antiviral activity, especially in immunocompromised patients and in treatment of less sensitive viruses, such as VZV and CMV.[177] Suboptimal exposures can lead to more resistant viral strains. To achieve high enough exposures, acyclovir needs to be dosed intravenously or in multiple high peroral doses.[178] In the design of valacyclovir, the following criteria were met: It was as safe as acyclovir, efficiently converted, and gave exposures after oral administration that were comparable to plasma levels of intravenously dosed acyclovir. Several reviews describe the development, pharmacokinetics, and efficacy of valacyclovir.[179,180]

Initially, two prodrugs of acyclovir with modifications on the purine ring were synthesized. The development of these prodrugs was discontinued due to incomplete conversion and to chronic toxicity that was worse than that of acyclovir.[177,178] The toxicity was suspected to be due to phosphorylation of the prodrug.[179] Subsequently, 18 amino acid esters of acyclovir were synthesized.[177] The bioavailability was measured following oral dosing to rats. Of the 18 prodrugs, 10 exhibited greater recovery of acyclovir in the urine than acyclovir itself when dosed orally. Of the 10, the L-valyl ester, valacyclovir, performed the best, with 63% excreted as acyclovir in the urine. Valacyclovir was also not phosphorylated, reducing the risk of the potential toxicity observed with the initial prodrugs utilized.

Further studies were established to determine if the high absorption of valacyclovir was due to carrier-mediated transport. The absorption of acyclovir and valacyclovir was studied in cynomologus monkey intestinal brush border membrane vesicles, where the influx of valacyclovir into the vesicles was six- to 10-fold higher than the influx of acyclovir.[194] Additional studies in Caco-2 cells showed that transport of valacylovir was seven times higher than acyclovir transport. In rats, L-amino acid ester analogs of acyclovir show better absorption than D or D-L analogs, indicating that the transport was also stereoselective.[177]

Several studies in Caco-2 and transfected cell lines have shown that valacylovir is transported by rat and human PepT1, even though the prodrug does not have a peptide bond.[176,181–187] Recent work has, however, shown that several additional transporters may be involved in valacyclovir transport in humans, and that PepT1 may not be the predominant transporter of this prodrug in humans. Most recently it has been suggested that valacyclovir transport by PHT1 and hPT1 may also be contributing absorption pathways.[61,188]

The absolute bioavailability of acyclovir when 100 mg of oral valacyclovir prodrug is dosed to healthy human subjects is 54%, compared to only 15 to 20% (200- to 600-mg doses) after acyclovir was dosed orally.[189,190] Larger variability was observed after dosing of acyclovir than following valacyclovir dosing. The bioavailability of acyclovir from orally dosed valacylovir is similar in rats and cynomolgus monkeys.[191,192] Over 99% of valacyclovir that is not absorbed is converted to acyclovir.

When valacyclovir is administered orally, there is a slightly less than dose proportional increase in acyclovir exposure and an increase in $T_{\max}$ with increasing doses. The slightly reduced absorption of valacyclovir, with increasing dose is not likely, due to saturable conversion to acyclovir, because of low urinary recovery of valacyclovir, and because the valacyclovir/acyclovir ratio remains the same with increasing dose. The reduced absorption may be due to saturation of absorption sites along the gastrointestinal tract.[193,194]

Although valacyclovir is more soluble than acyclovir (174 mg/mL versus 1.3 mg/mL),[177] solubility is unlikely to limit absorption of either compound. This is supported by the fact that many of the other amino acid ester analogs that have been studied also exhibit improved solubility over acyclovir. However, these analogs have very diverse bioavailability, probably due to differences in their carrier-mediated transport.[177]

6.6. CONCLUSIONS

Figures 6.1 and 6.2 summarize the parallel pathways of permeation and some of the important factors associated with the mass transfer resistances encountered with each permeation route for oligopeptides passive and active transport across cell barriers. It is clear that the passive routes of permeation are very restrictive and limited in the capacity for proper significant absorption of peptide-based drugs, particularly across the gastrointestinal epithelium. Based on this, increasing attention has been focused on developing agents that may be able to traverse these barriers through increasing their affinity and capacity for peptide transporters. Until recently, most of the attention for facilitating the active oral absorption of peptide-based drugs has been focused on the PepT1 isoform. Recent evidence has suggested that the presence of other peptide transporters, PHT isoforms and PT1, may require greater consideration. Assessing and delineating the functional relevancy and overlap of these isoforms are needed to properly understand their impact on peptide uptake and absorption.

Although attention to proper assessment of the functional relevancy of other oligopeptides transporters is merited, there are several universal considerations that will also hold true in all cases. The most important consideration is the impact of diet on the function and expression of the transporters, which has been studied with PepT1 and to a lesser extent with PepT2. Hormonal regulation is another key element that may be altered in certain disease states, in particular metabolic syndrome, which could significantly alter the membrane expression and the clinical impact of oligopeptides transporters. Further studies on the functional relevancy of the oligopeptide transporter splice variants and SNPs are also warranted, especially when one considers the two PepT1 splice variants that do convey functional relevancy. Many other factors do exist, including the fact that we are still discovering new oligopeptides transporter isoforms which will require significant efforts for their proper elucidation. In short, it is apparent that considerably more research is required to characterize all of the functional peptide transporter systems to recognize their full clinical impact.

Acknowledgments

The authors would like to acknowledge support provided by the National Institute of General Medical Sciences (NIGMS; RO1-GM65448), the College of Pharmacy, Nursing and Health Sciences, and the Department of Industrial and Physical Pharmacy at Purdue University.

REFERENCES

1. Devlin, T.M. (2002). *Textbook of Biochemistry with Clinical Correlations*, 5th ed., Wiley-Liss, New York, pp. 1097–1100.
2. Washington, N., Washington, C., and Wilson, C.G. (2001). *Physiological Pharmaceutics: Barriers to Drug Absorption*, 2nd ed., Taylor & Francis, New York, pp. 109–142.
3. Tortora, G.J., and Grabowski, S.R. (1993). *Principles of Anatomy and Physiology*, Harper-Collins College Publishers, New York, p. 768.
4. Adson, A., Burton, P.S., Raub, T.J., Barsuhn, C.L., Audus, K.L., and Ho, N.F. (1995). Passive diffusion of weak organic electrolytes across Caco-2 cell monolayers: uncoupling the contributions of hydrodynamic, transcellular, and paracellular barriers. *J. Pharm. Sci.* 84(10):1197–1204.
5. Knipp, G.T., Ho, N.F., Barsuhn, C.L., and Borchardt, R.T. (1997). Paracellular diffusion in Caco-2 cell monolayers: effect of perturbation on the transport of hydrophilic compounds that vary in charge and size. *J. Pharm. Sci.* 86(10):1105–1110.

6. Sorensen, M., Steenberg, B., Knipp, G.T., Wang, W., Steffansen, B., Frokjaer, S., and Borchardt, R.T. (1997). The effect of beta-turn structure on the permeation of peptides across monolayers of bovine brain microvessel endothelial cells. *Pharm. Res.* 14(10):1341–1348.
7. Oh, D.-M., and Amidon, G.L. (1999). Overview of Membrane Transport, in *Membrane Transorters as Drug Targets*, edited by G.L. Amidon and W. Sadée, Kluwer Academic/Plenum Publishers, New York, pp. 1–27.
8. Graul, R.C., and Sadée, W. (1997). Sequence alignments of the H(+)-dependent oligopeptide transporter family PTR: inferences on structure and function of the intestinal PEPT1 transporter. *Pharm. Res.* 14:388–400.
9. Nussberger, S., Steel, A., and Hediger, M. (1997). Structure and pharmacology of proton-linked peptide transporters. *J. Control. Release* 46:31–38.
10. Yang, C.Y., Dantzing, A.H., and Pidgeon, C. (1999). Intestinal peptide transport of systems and oral drug availability. *Pharm. Res.* 16:1331–1343.
11. Meredith, D., and Boyd, C.A.R. (2000). Structure and function of eukaryotic peptide transporters. *Cell. Mol. Life Sci.* 57:754–778.
12. Rubio-Aliaga, I., and Daniel, H. (2002). Mammalian peptide transporters as targets for drug delivery. *Trends Pharm. Sci.* 23(9):434–440.
13. Herrera-Ruiz, D., and Knipp, G.T. (2003). Current perspectives on established and putative mammalian oligopeptide transporters. *J. Pharm. Sci.* 92(4):691–714.
14. Dantzig, A.H., Hoskins, J.A., Tabas, L.B., Bright, S., Shepard, R.L., Jenkins, I.L., Duckworth, D.C., Sportsman, J.R., Mackensen, D., Rosteck, P.R., Jr., and Skatrud, P.L. (1994). Association of intestinal peptide transport with a protein related to the cadherin superfamily. *Science* 264:430–433.
15. Fei, Y.J., Kanai, Y., Nussberger, S., Ganapathy, V., Leibach, F.H., Romero, M.F., Singh, S.K., Boron, W.F., and Hediger, M.A. (1994). Expression cloning of a mammalian proton-coupled oligopeptide transporter. *Nature* 368:563–566.
16. Leibach, F.H., and Ganapathy, V. (1996). Peptide transport in the intestine and the kidney. *Annu. Rev. Nutr.* 16:99–119.
17. Daniel, H. (1996). Function and molecular structure of brush border membrane peptide/H^+ symporters. *J. Membr. Biol.* 154:197–203.
18. Liu, W., Liang, R., Ramamoorthy, S., Fei, Y.J., Ganapathy, M.E., Hediger, M.A., Ganapathy, V., and Leibach, F.H. (1995). Molecular cloning of PEPT2, a new member of the H(+)/peptide cotransporter family, from human kidney. *Biochim. Biophys. Acta* 1235:461–466.
19. Daniel, H., and Herget, M. (1997). Cellular and molecular mechanisms of renal peptide transport. *Am. J. Physiol. Renal Physiol.* 273:F1–F8.
20. Yamashita, T., Shimada, S., Guo, W., Sato, K., Kohmura, E., Hayakawa, T., Takagi, T., and Tohyama, M. (1997). Cloning and functional expression of a brain peptide/histidine transporter. *J. Biol. Chem.* 272:10205–10211.
21. Sakata, K., Yamashita, T., Maeda, M., Moriyama, Y., Shimada, S., and Tohyama, M. (2001). Cloning of a lymphatic peptide/histidine transporter. *Biochem. J.* 356:53–60.
22. Covitz, Y.K.-M., Amidon, G.L., and Sadée, W. (1998). Membrane topology of the human dipeptides transporter, hPepT1, determined by epitope insertions. *Biochemistry* 37:15214–15221.
23. Lee, V.H.L. (2000). Membrane transporters. *Eur. J. Pharm. Sci.* 11:S41–S50.

24. Steiner, H.Y., Naider, F., and Becker, J.M. (1995). The PTR family: a new group of peptide transporters. *Mol. Microbiol.* 16:825–834.
25. Fei, Y.J., Ganapathy, V., and Leibach, F.H. (1998). Molecular and structural features of the proton-coupled oligopeptide transporter superfamily. *Prog. Nucleic Acid Res. Mol. Biol.* 58:239–261.
26. Saier, M.H., Jr., Eng, B.H., Fard, S., Garg, J., Haggerty, D.A., Hutchinson, W.J., Jack, D.L., Lai, E.C., Liu, H.J., Nusinew, D.P., et al. (1999). Phylogenetic characteristics of novel transport protein families revealed by genome analyses. *Biochim. Biophys. Acta* 1422:1–56.
27. Liang, R., Fei, Y.J., Prasad, P.D., Ramamoorthy, S., Han, H., Yang-Feng, T.L., Hediger, M.A., Ganapathy, V., and Leibach, F.H. (1995). Human intestinal H^+/peptide cotransporter. Cloning, functional expression, and chromosomal localization. *J. Biol. Chem.* 270:6456–6463.
28. Saito, H., Okuda, M., Terada, T., Sasaki, S., and Inri, K.-I. (1995). Cloning and characterization of a rat H^+/peptide cotransporter mediating absorption of beta-lactam antibiotics in the intestine and kidney. *J. Pharmacol. Exp. Ther.* 275:1631–1637.
29. Chen, H., Wong, E.A., and Webb, K.E. (1999). Tissue distribution of a peptide transporter mRNA in sheep, daily cows, pigs, and chickens. *J. Anim. Sci* 77:1277–1283.
30. Fei, Y.J., Sugawara, M., Liu, J.C., Li, H.W., Ganapathy, V., Ganapathy, M.E., and Leibach, F.H. (2000). cDNA structure, genomic organization, and promoter analysis of the mouse intestinal peptide transporter PepT1. *Biochim. Biophys. Acta* 1492:145–154.
31. Pan, Y.X., Wong, E.A., Bloomquist, J.R., and Webb, K.E., Jr. (2001). Expression of a cloned ovine gastrointestinal peptide transporter (oPepT1) in *Xenopus* oocytes induces uptake of oligopeptides in vitro. *J. Nutr* 131:1264–1270.
32. Klang, J.E., Burnworth, L.A., Pan, Y.X., Webb, K.E., Jr., and Wong, E.A. (2005). Functional characterization of a cloned pig intestinal peptide transporter (pPepT1). *J. Anim. Sci.* 83(1):172–181.
33. Van, L., Pan, Y.X., Bloomquist, J.R., Webb, K.E., Jr., and Wong, E.A. (2005). Developmental regulation of a turkey intestinal peptide transporter (PepT1). *Poult. Sci.* 84(1):75–82.
34. Zhang, E.Y., Emerick, R.M., Pak, Y.A., Wrighton, S.A., and Hillgren, K.M. (2004). Comparison of human and monkey peptide transporters: PEPT1 and PEPT2. *Mol. Pharmacol.* 1(3):201–210.
35. Herrera-Ruiz, D., Wang, Q., Gudmundsson, O.S., Cook, T.J., Smith, R.L., Faria, T.N., and Knipp, G.T. (2001). Spatial expression patterns of peptide transporters in the human and rat gastrointestinal tracts, Caco-2 in vitro cell culture, and multiple human tissues. *AAPS PharmSci* 3:article 9.
36. Smith, D.E., Pavlova, A., Berger, U.V., Hediger, M.A., Yang, T., Huang, Y., and Schnermann, J.B. (1998). Tubular localization and tissue distribution of peptide transporters in rat kidney. *Pharm. Res.* 15(8):1244–1249.
37. Shen, H., Smith, D.E., Yang, T., Huang, Y.G., Schnermann, J.B., and Brosius F.C., 3rd (1999). Localization of PEPT1 and PEPT2 proton coupled oligopeptide transporter mRNA and protein in rat kidney. *Am. J. Physiol. Renal Physiol.* 276:F658–F665.
38. Knutter, I., Rubio-Aliaga, I., Boll, M., Hause, G., Daniel, H., Neubert, K., and Brandsch, M. (2002). H^+-peptide cotransport in the human bile duct epithelium cell line SK-ChA-1. *Am. J. Physiol. Gastrointest. Liver Physiol.* 283(1):G222–G229.

39. Ogihara, H., Suzuki, T., Nagamachi, Y., Inui, K., and Takata, K. (1999). Peptide transporter in the rat small intestine: ultrastructural localization and the effect of starvation and administration of amino acids. *Histochem. J.* 31:169–174.
40. Terada, T., Shimada, Y., Pan, X., Kishimoto, K., Sakurai, T., Doi, R., Onodera, H., Katsura, T., Imamura, M., and Inui, K. (2005). Expression profiles of various transporters for oligopeptides, amino acids and organic ions along the human digestive tract. *Biochem. Pharmacol.* 70(12):1756–63.
41. Zhou, X., Thamotharan, M., Gangopadhyay, A., Serdikoff, C., and Adibi, S.A. (2000). Characterization of an oligopeptide transporter in renal lysosomes. *Biochim. Biophys. Acta* 1466:372–378.
42. Sun, D., Landowski, C.P., Chu, X., Wallsten, R., Fleisher, D., and Amidon, G.L. (2001). Drug inhibition of Gly-Sar uptake and hPepT1 localization using hPepT1-GFP fusion protein. *AAPS PharmSci* 3:article 2.
43. Shen, H., Smith, D.E., and Brosius, F.C. (2001). Developmental expression of PEPT1 and PEPT2 in rat small intestine, colon, and kidney. *Pediatr. Res.* 49:789–795.
44. Rome, S., Barbot, L., Windsor, E., Kapel, N., Tricottet, V., Huneau, J.F., Reynes, M., Gobert, J.G., and Tomé, D. (2002). The regionalization of PepT1, NBAT and EAAC1 transporters in the small intestine of rats are uncharged from birth to adulthood. *J. Nutr.* 132:1009–1011.
45. Berger, U.V., and Hediger, M.A. (1999). Distribution of peptide transporter PEPT2 mRNA in the rat nervous system. *Anat. Embryol.* 199:439–449.
46. Groneberg, D.A., Nickolaus, M., Springer, J., Döring, F., Daniel, H., and Fischer, A. (2001). Localization of the peptide transporter PEPT2 in the lung: implications for pulmonary oligopeptide uptake. *Am. J. Pathol.* 158(2):707–714.
47. Groneberg, D.A., Döring, F., Theis, S., Nickolaus, M., Fischer, A., and Daniel, H. (2002). Peptide transport in the mammary gland: expression and distribution of PEPT2 mRNA and protein. *Am. J. Physiol. Endocrinol. Metab.* 282(5):E1172–E1179.
48. Izzedine, H., Launay-Vacher, V., and Deray, G. (2005). Renal tubular transporters and antiviral drugs: an update. *AIDS* 19(5):455–462.
49. Ramamoorthy, S., Liu, W., Ma, Y.Y., Yang-Feng, T.L., Ganapathy, V., and Leibach, F.H. (1995). Proton/peptide cotransporter (PEPT 2) from human kidney: functional characterization and chromosomal localization. *Biochim. Biophys. Acta* 1240(1): 1–4.
50. Boll, M., Herget, M., Wagener, M., Weber, W.M., Markovich, D., Biber, J., Clauss, W., Murer, H., and Daniel, H. (1996). Expression cloning and functional characterization of the kidney cortex high-affinity proton-coupled peptide transporter. *Proc. Natl. Acad. Sci. U. S. A.* 93(1):284–289.
51. Saito, H., Terada, T., Okuda, M., Sasaki, S., and Inui, K. (1996). Molecular cloning and tissue distribution of rat peptide transporter PEPT2. *Biochim. Biophys. Acta* 1280(2):173–177.
52. Rubio-Aliaga, I., Boll, M., and Daniel, H. (2000). Cloning and characterization of the gene encoding the mouse peptide transporter PEPT2. *Biochem. Biophys. Res. Commun.* 276(2):734–741.
53. Inui, K.-I., Masuda, S., and Saito, H. (2000). Cellular and molecular aspects of drug transport in the kidney. *Kidney Int.* 58:944–958.

54. Shu, C., Shen, H., Keep, R.F., and Smith, D.E. (2002). Role of PEPT2 in peptide/mimetic trafficking at the blood–CSF barrier: studies in rat choroid plexus epithelial cells in primary culture. *J. Pharmacol. Exp. Ther.* 301:820–829.

55. Novotny, A., Xiang, J., Stummer, W., Teuscher, N.S., Smith, D.E., and Keep, R.F. (2000). Mechanisms of 5-aminolevulinic acid uptake at the choroid plexus. *J. Neurochem.* 75:321–328.

56. Teuscher, N.S., Novotny, A., Keep, R.F., and Smith, D.E. (2000). Functional evidence for presence of PEPT2 in rat choroid plexus: studies with glycylsarcosine. *J. Pharmacol. Exp. Ther.* 294(2):494–499.

57. Teuscher, N.S., Keep, R.F., and Smith, D.E. (2001). PEPT2-mediated uptake of neuropeptides in rat choroid plexus. *Pharm. Res.* 18(6):807–813.

58. Smith, D.E., Johanson, C.E., and Keep, R.F. (2004). Peptide and peptide analog transport systems at the blood–CSF barrier. *Adv. Drug. Deliv. Rev.* 56:1765–1791.

59. Shen, H., Smith, D.E., Keep, R.F., and Brosius, F.C., 3rd (2004). Immunolocalization of the proton-coupled oligopeptide transporter PEPT2 in developing rat brain. *Mol. Pharmacol.* 1(4):248–256.

60. Shen, H., Smith, D.E., Keep, R.F., Xiang, J., and Brosius, F.C., 3rd (2003). Targeted disruption of the PEPT2 gene markedly reduces dipeptide uptake in choroid plexus. *J. Biol. Chem.* 278(7):4786–4791.

61. Bhardwaj, R.K., Herrera-Ruiz, D., Eltoukhy, N., Saad, M., and Knipp, G.T. (2006). The functional evaluation of human peptide/histidine transporter 1 (hPHT1) in transiently transfected COS-7 cells. *Eur. J. Pharm. Sci.* 27(5):533–542.

62. Rühl, A., Hoppe, S., Frey, I., Daniel, H., and Schemann, M. (2005). Functional expression of the peptide transporter PEPT2 in the mammalian enteric nervous system. *J. Comp. Neurol.* 490(1):1–11.

63. Botka, C.W., Wittig, T.W., Graul, R.C., Nielsen, C.U., Higaka, K., Amidon, G.L., and Sadée, W. (2000). Human proton/oligopeptide transporter (POT) genes: identification of putative human genes using bioinformatics. *AAPS PharmSci* 2:article 16.

64. Knipp, G.T. and Herrera-Ruiz, D. (2004). Nucleic acid encoding the human peptide histidine transporter 1 and methods of use thereof. U.S. patent 6,683,169; filling date: 5/31/01, isuued: 1/27/04.

65. Herrera-Ruiz, D. (2002). Investigations into the molecular and transport characteristics of established and putative peptide transporters. Thesis, Rutgers, The State University of New Jersey, New Brunswick, NJ.

66. Ocheltree, S.M., Keep, R.F., Shen, H., Yang, D., Hughes, B.A., and Smith, D.E. (2003). Preliminary investigation into the expression of proton-coupled oligopeptide transporters in neural retina and retinal pigment epithelium (RPE): lack of functional activity in RPE plasma membranes. *Pharm. Res.* 20(9):1364–1372.

67. Terada, T., Sawada, K., Tatsuya, I., Saito, H., Hashimoto, Y., and Inui, K.-I. (2000). Functional expression of novel peptide transporter in renal basolateral membranes. *Am J. Physiol. Renal Physiol.* 279:F851–F857.

68. Yang, C.Y. (1998). Studies on the human intestinal di-/tripeptide transporter HPT-1 as a potential carrier for orally administered drugs. Thesis, Purdue University. West Lafayette, IN.

69. Erickson, R.H., Gum, J.R., Jr., Lindstrom, M.M., McKean, D., and Kim, Y.S. (1995). Regional expression and dietary regulation of rat small intestinal peptide and amino acid transporter mRNAs. *Biochem. Biophys. Res. Commun.* 216:249–257.

70. Adibi, S.A. (1997). The oligopeptide transporter (PepT1) in human intestine: biology and function. *Gastroenterology* 113:322–340.
71. Nussberger, S., Steel, A., Trott, D., Romero, M., Boron, W.F., and Hediger, M. (1997). Symmetry of H^+ binding to the intra- and extracellular side of the H^+-coupled oligopeptide cotransporter PepT1. *J. Biol. Chem.* 272:7777–7785.
72. Kennedy, D.J., Leibach, F.H., Ganapathy, V., and Thwaites, D.T. (2002). Optimal absorptive transport of the dipeptide glycylsarcosine is dependent on functional Na^+/H^+ exchange activity. *Pflugers Arch. Eur. J. Physiol.* 445(1):139–146.
73. Thwaites, D.T., Kennedy, D.J., Raldua, D., Anderson, C.M., Mendoza, M.E., Bladen, C.L., and Simmons, N.L. (2002). H/dipeptide absorption across the human intestinal epithelium is controlled indirectly via a functional Na/H exchanger. *Gastroenterology* 122(5):1322–1333.
74. Temple, C.S., Bronk, J.K., Bailey, P.D., and Boyd, C.A. (1995). Substrate charge dependence of stoichiometry shows membrane potential is driving force for proton–peptide cotransport in rat renal cortex. *Pflugers Arch. Eur. J. Physiol.* 430:825–829.
75. Temple, C.S., Bailey, P.D., Bronk, J.R., and Boyd, C.A.R. (1996). A model for the kinetics of neutral and anionic dipeptide–proton cotransport by the apical membrane of rat kidney cortex. *J. Physiol.* 494(3):795–808.
76. Amasheh, S., Wenzel, U., Boll, M., Dom, D., Weber, W., and Clauss, W.D.H. (1997). Transport of charged dipeptides by the intestinal H^+/peptide symporter PepT1 expressed in *Xenopus laevis* oocytes. *J. Membr. Biol.* 155:247–256.
77. Balimane, P.V., and Sinko, P.J. (2000). Effect of ionization on the variable uptake of valacyclovir via the human intestinal peptide transporter (hPepT1) in CHO cells. *Biopharm. Drug Dispos.* 21:165–174.
78. Kottra, G., Stamfort, A., and Daniel, H. (2002). PEPT1 as a paradigm for membrane carriers that mediate electrogenic bidirectional transport of anionic, cationic and neutral substrates. *J. Biol. Chem.* 277(36):32683–32691.
79. Wenzel, U., Gebert, I., Weintraut, H., Weber, W.-M., Clauss, W., and Daniel, H. (1996). Transport characteristics of differently charged cephalosporin antibiotics in oocytes expressing the cloned intestinal peptide transporter PepT1 and in human intestinal Caco-2 cells. *J. Pharmacol. Exp. Ther.* 277:831–839.
80. Mackenzie, B., Loo, D.D.F., Fei, Y.-F., Liu, W., Ganapathy, V., Leibach, F.H., and Wright, E.M. (1996). Mechanisms of the human intestinal H^+-coupled oligopeptide transporter hPEPT1. *J. Biol. Chem.* 271(10):5430–5437.
81. Irie, M., Terada, T., Katsura, T., Matsuoka, S., and Inui, K.-I. (2005). Computational modelling of H^+-coupled peptide transport via human PEPT1. *J. Physiol.* 565:429–439.
82. Bolger, M.B., Haworth, I.S., Yeung, A.K., Ann, D., von Grafenstein, H., Hamm-Alvarez, S., Okamoto, C.T., Kim, K.J., Basu, S.K., Wu, S., and Lee, V.H. (1998). Structure, function, and molecular modeling approaches to the study of the intestinal dipeptides transporter PepT1. *J. Pharm. Sci.* 87:1286–1291.
83. Yeung, A.K., Basu, S.K., Wu, S., Chu, C., Okamoto, C.T., Hamm-Alvarez, S., von Grafenstein, H., Shen, W.C., Kim, K.J., Bolger, M.B., et al. (1998). Molecular identification of a role for tyrosine 167 in the function of the human intestinal proton-coupled dipeptide transporter (hPepT1). *Biochem. Biophys. Res. Commun.* 250:103–107.

84. Chen, X.-Z., Steel, A., and Hediger, M.A. (2000). Functional roles of histidine and tyrosine residues in the H^+-peptide transporter PepT1. *Biochem. Biophys. Res. Commun.* 272:726–730.
85. Kulkarni, A.A, Haworth, I.S., and Lee, V.H. (2003). Transmembrane segment 5 of the dipeptide transporter hPepT1 forms a part of the substrate translocation pathway. *Biochem. Biophys. Res. Commun.* 306:177–185.
86. Uchiyama, T., Kulkarni, A.A., Davies, D.L., and Lee, V.H. (2003). Biophysical evidence for His57 as a proton-binding site in the mammalian intestinal transporter hPepT1. *Pharm. Res.* 20(12):1911–1916.
87. Fei, Y.J., Liu, W., Prasad, P.D., Kekuda, R., Oblak, T.G., Ganapathy, V., and Leibach, F.H. (1997). Identification of the histidyl residue obligatory for the catalytic activity of the human H^+/peptide cotransporters PepT1 and PepT2. *Biochemistry* 14:452–460.
88. Döring, F., Dorn, D., Bachfisher, U., Amasheh, S., Herget, M., and Daniel, H. (1996). Functional analysis of a chimeric mammalian peptide transporter derived from the intestinal and renal isoforms. *J. Physiol.* 497:773–779.
89. Döring, F., Martini, C., Walter, J., and Daniel, H. (2002). Importance of a small N-terminal region in mammalian peptide transporters for substrate affinity and function. *J. Membr. Biol.* 186:55–62.
90. Terada, T., Saito, H., Sawada, K., Hashimoto, Y., and Inui, K.-I. (2000). N-Terminal halves of rat H^+/peptide transporters are responsible for their substrate recognition. *Pharm. Res.* 17:15–20.
91. Gaildrat, P., Moller, M., Mukda, S., Humphries, A., Carter, D.A., Ganapathy, V., and Klein, D.C. (2005). A novel pineal-specific product of the oligopeptide transporter PepT1 gene: circadian expression mediated by cAMP activation of an intronic promoter. *J. Biol. Chem.* 280(17):16851–16860.
92. Kulkarni, A.A., Haworth, I.S., Uchiyama, T., and Lee, V.H. (2003). Analysis of transmembrane segment 7 of the dipeptide transporter hPepT1 by cysteine-scanning mutagenesis. *J. Biol. Chem.* 278(51):51833–51840.
93. Bailey, P.D, Boyd, C.A, Bronk, J.R., Collier, I.D., Meredith, D., Morgan, K.M., and Temple, C.S. (2000). How to make drugs orally active: a substrate model for the peptide transporter PepT1. *Angew Chem. Int. Ed.* 39:505–508.
94. Brandsch, M., Thunecke, F., Kullertz, G., Schutkowski, M., Fischer, G., and Neubert, K. (1998). Evidence for the absolute conformational specificity of the intestinal H^+/peptide symporter, PEPT1. *J. Biol. Chem.* 273(7):3861–3864.
95. Enjoh, M., Hashimoto, K., Arai, S., and Shimizu, M. (1996). Inhibitory effect of arphamenine A on intestinal dipeptide transport. *Biosci. Biotechnol. Biochem.* 60(11):1893–1895.
96. Temple, C., Stewart, A., Meredith, D., Listr, N.A., Morgan, K.M., Collier, I.D., Vaughan-Jones, R.D., Boyd, C.A.R., Bailey, P., and Bronk, J.R. (1998). Peptide mimics as substrates for the intestinal peptide transporter. *J. Biol. Chem.* 273:20–22.
97. Vabeno, J., Lejon, T., Nielsen, C.U., Steffansen, B., Chen, W., Ouyang, H., Borchardt, R.T., and Luthman, K. (2004). Phe-Gly dipeptidomimetics designed for the di-/tripeptide transporters PEPT1 and PEPT2: synthesis and biological investigations. *J. Med. Chem.* 47(4):1060–1069.
98. Döring, F., Will, J., Amasheh, S., Clauss, W., Ahlbrecht, H., and Daniel, H. (1998). Minimal molecular determinants of substrates for recognition by the intestinal peptide transporter. *J. Biol. Chem.* 273:23211–23218.

99. Brandsch, M., Knutter, I., and Leibach, F.H. (2004). The intestinal H^+/peptide symporter PEPT1: structure–affinity relationships. *Eur. J. Pharm. Sci.* 21(1):53–60.

100. Bailey, P.D., Boyd, C.A., Collier, I.D., George, J.G., Kellett, G.L., Meredith, D., Morgan, K.M., Pettecrew, R., Price, R.A., and Pritchard, R.G. (2005). Conformational and spacial preferences for substrates of PepT1. *Chem. Commun. (Camb.)* 42:5352–5354.

101. Swaan, P.W., and Tukker, J.J. (1997). Molecular determinants of recognition for the intestinal peptide carrier. *J. Pharm. Sci.* 86:596–602.

102. Zhang, E.Y., Knipp, G.T., Ekins, S., and Swaan, P.W. (2002). Structural biology and function of solute transporters: implications for identifying and designing substrates. *Drug Metab. Rev.* 34(4):709–750.

103. Zhang, E.Y., Mitch, A.P., Cheng, C., Ekis, S., and Swaan, P.W. (2002). Modeling of active transport systems. *Adv. Drug Deliv. Rev.* 54:329–354.

104. Gebauer, S., Knutter, I., Hartrodt, B., Brandsch, M., Neubert, K., and Thondorf, I. (2003). Three-dimensional quantitative structure–activity relationship analyses of peptide substrates of the mammalian H^+/peptide cotransporter PEPT1. *J. Med. Chem.* 46(26):5725–5734.

105. Biegel, A., Gebauer, S., Hartrodt, B., Brandsch, M., Neubert, K., and Thondorf, I. (2005). Three-dimensional quantitative structure–activity relationship analyses of beta-lactam antibiotics and tripeptides as substrates of the mammalian H^+/peptide cotransporter PEPT1. *J. Med. Chem.* 48(13):4410–4409.

106. Theis, S., Hartrodt, B., Kottra, G., Neubert, K., and Daniel, H. (2002). Defining minimal structural features in substrates of the H^+/peptide cotransporter PepT2 using novel amino acid and dipeptide derivatives. *Mol. Pharmacol.* 61:214–221.

107. Theis, S., Knütter, I., Hartrodt, B., Brandsch, M., Kottra, G., Neubert, K., and Daniel, H. (2002). Synthesis and characterization of high affinity inhibitors of the H^+/peptide transporter PepT2. *J. Biol. Chem.* 277:7287–7292.

108. Eriksson, A.H., Elm, P.L., Begtrup, M., Nielsen, R., Steffansen, B., and Brodin, B. (2005). hPEPT1 affinity and translocation of selected Gln-Sar and Glu-Sar dipeptide derivatives. *Mol. Pharmacol.* 2(3):242–249.

109. Shu, C., Shen, H., Hopfer, U., and Smith, D.E. (2001). Mechanism of intestinal absorption and renal reabsorption of an orally active ace inhibitor: uptake and transport of fosinopril in cell cultures. *Drug Metab. Dispos.* 29(10):1307–1315.

110. Tamai, I., Nakanishi, T., Hayashi, K., Terao, T., Sai, Y., Shiraga, T., Miyamoto, K., Takeda, E., Higashida, H., and Tsuji, A. (1997). The predominant contribution of oligopeptide transporter PepT1 to intestinal absorption of beta-lactam antibiotics in the rat small intestine. *J. Pharm. Pharmacol.* 49:796–801.

111. Tsuji, A., Tamai, I., Nakanishi, M., and Amidon, G.L. (1990). Mechanism of absorption of the dipeptide alpha-methyldopa-phe in intestinal brush-border membrane vesicles. *Pharm. Res.* 7(3):308–309.

112. Wang, H.P., Lu, H.H., Lee, J.S., Cheng, C.Y., Mah, J.R., Ku, C.Y., Hsu, W., Yen, C.F., Lin, C.J., and Kuo, H.S. (1996). Intestinal absorption studies on peptide mimetic alpha-methyldopa prodrugs. *J. Pharm. Pharmacol.* 48(3):270–276.

113. Ganapathy, M.E., Huang, W., Wang, H., Ganapathy, V., and Leibach, F.H. (1998). Valacyclovir: a substrate for the intestinal and renal peptide transporters PepT1 and PepT2. *Biochem. Biophys. Res. Commun.* 246:470–475.

114. Sugawara, M., Huang, W., Fei, Y.J., Leibach, F.H., Ganapathy, V., and Ganapathy, M.E. (2000). Transport of valganciclovir, a ganciclovir prodrug, via peptide transporters PEPT1 and PEPT2. *J. Pharm. Sci.* 89(6):781–789.

115. Ezra, A., Hoffman, A., Breuer, E., Alferiev, I.S., Monkkonen, J., El Hanany-Rozen, N., Weiss, G., Stepensky, D., Gati, I., Cohen, H., et al. (2000). A peptide prodrug approach for improving bisphosphonate oral absorption. *J. Med. Chem.* 43(20):3641–3652.

116. Neumann, J., Bruch, M., Gebauer, S., and Brandsch, M. (2004). Transport of the phosphonodipeptide alafosfalin by the H^+/peptide cotransporters PEPT1 and PEPT2 in intestinal and renal epithelial cells. *Eur. J. Biochem.* 271(10):2012–2017.

117. Luckner, P., and Brandsch, M. (2005). Interaction of 31 beta-lactam antibiotics with the H^+/peptide symporter PEPT2: analysis of affinity constants and comparison with PEPT1. *Eur. J. Pharm. Biopharm.* 59(1):17–24.

118. Adibi, S.A. (2003). Regulation of expression of the intestinal oligopeptide transporter (Pept-1) in health and disease. *Am. J. Physiol. Gastrointest. Liver Physiol* 285(5):G779–G788.

119. Nielsen, C.U., and Brodin, B. (2003). Di/tri-peptide transporters as drug delivery targets: regulation of transport under physiological and patho-physiological conditions. *Curr. Drug Targets* 4(5):373–388.

120. Daniel, H. (2004). Molecular and integrative physiology of intestinal peptide transport. *Annu. Rev. Physiol.* 66:361–384.

121. Terada, T., and Inui, K. (2004). Peptide transporters: structure, function, regulation and application for drug delivery. *Curr. Drug Metab.* 5(1):85–94.

122. Shiraga, T., Miyamoto, K.I., Tanaka, H., Yamamoto, H., Taketani, Y., Morita, K., Tamai, I., Tsuji, A., and Takeda, E. (1999). Cellular and molecular mechanisms of dietary regulation on rat intestinal H^+/peptide transporter PepT1. *Gastroenterology* 116:354–362.

123. Thamotharan, M., Bawani, S.Z., Zhou, X., and Adibi, S.A. (1998). Mechanism of dipeptide stimulation of its own transport in a human intestinal cell line. *Proc. Assoc. Am. Physicians* 110:361–368.

124. Walker, D., Thwaites, D.T., Simmons, N.L., Gilbert, H.J., and Hirst, B.H. (1998). Substrate upregulation of the human small intestinal peptide transporter, hPepT1. *J. Physiol.* 507:697–706.

125. Thamotharan, M., Bawani, S.Z., Zhou, X., and Adibi, S.A. (1999). Functional and molecular expression of intestinal oligopeptide transporter (Pept-1) after a brief fast. *Metabolism* 48:681–684.

126. Ihara, T., Tsujikawa, T., Fujiyama, Y., and Bamba, T. (2000). Regulation of PepT1 peptide transporter expression in the rat small intestine under malnourished conditions. *Digestion* 61(1):59–67.

127. Naruhashi, K., Sai, Y., Tamai, I., Suzuki, N., and Tsuji, A. (2002). PepT1 mRNA expression is induced by starvation and its level correlates with absorptive transport of cefadroxil longitudinally in the rat intestine. *Pharm. Res.* 19(10):1417–1423.

128. Howard, A., Goodlad, R.A., Walters, J.R., Ford, D., and Hirst, B.H. (2004). Increased expression of specific intestinal amino acid and peptide transporter mRNA in rats fed by TPN is reversed by GLP-2. *J. Nutr.* 134(11):2957–2964.

129. Himukai, M., Konno, T., and Hoshi, T. (1980). Age-dependent change in intestinal absorption of dipeptides and their constituent amino acids in the guinea pig. *Pediatr. Res.* 14:1272–1275.

130. Guandalini, S., and Rubino, A. (1982). Development of dipeptide transport in the intestinal mucosa of rabbits. *Pediatr. Res.* 16:99–103.

131. Miyamoto, K., Shiraga, T., Morita, K., Yamamoto, H., Haga, H., Taketani, Y., Tamai, I., Sai, Y., Tsuji, A., and Takeda, E. (1996). Sequence, tissue distribution and developmental changes in rat intestinal oligopeptide transporter. *Biochim. Biophys. Acta* 1305:34–38.

132. Hussain, I., Kellett, G.L., Affleck, J., Shepherd, E.J., and Boyd, C.A.R. (2002). Expression and cellular distribution during development of the peptide transporter (PepT1) in the small intestinal epithelium of the rat. *Cell Tissue Res.* 307:139–142.

133. Pan, X., Terada, T., Irie, M., Saito, H., and Inui, K. (2002). Diurnal rhythm of H^+-peptide cotransporter in rat small intestine. *Am. J. Physiol. Gastrointest. Liver Physiol.* 283(1):G57–G64.

134. Pan, X., Terada, T., Okuda, M., and Inui, K. (2003). Altered diurnal rhythm of intestinal peptide transporter by fasting and its effects on the pharmacokinetics of ceftibuten. *J. Pharmacol. Exp. Ther.* 307(2):626–632.

135. Merlin, D., Si-Tahar, M., Sitaraman, S.V., Eastburn, K., Williams, I., Liu, X., Hediger, M.A., and Madara, J.L. (2001). Colonic epithelial hPepT1 expression occurs in inflammatory bowel disease: transport of bacterial peptides influences expression of MHC class 1 molecules. *Gastroenterology* 120(7):1666–1679.

136. Vavricka, S.R., Musch, M.W., Fujiya, M., Kles, K., Chang, L., Eloranta, J.J., Kullak-Ublick, G.A., Drabik, K., Merlin, D., and Chang, E.B. (2005). Tumor necrosis factor-alpha and interferon-gamma increase PepT1 expression and activity in the human colon carcinoma cell line Caco-2/bbe and in mouse intestine. *Pflugers Arch. Eur. J. Physiol.* 19:1–10.

137. Barbot, L., Windsor, E., Rome, S., Tricottet, V., Reynes, M., Topouchian, A., Huneau, J.F., Gobert, J.G., Tomé, D., and Kapel, N. (2003). Intestinal peptide transporter PepT1 is overexpressed during acute cryptosporidiosis in suckling rats as a result of both malnutrition and experimental parasite infection. *Parasitol. Res.* 89(5):364–370.

138. Goulet, O., Lacaille, F., Jan, D., and Ricour, C. (2000). Intestinal transplantation: indications, results and strategy. *Curr. Opin. Clin. Nutr. Metab. Care* 3(5):329–338.

139. Motohashi, H., Masuda, S., Katsura, T., Saito, H., Sakamoto, S., Uemoto, S., Tanaka, K., and Inui, K. (2001). Expression of peptide transporter following intestinal transplantation in the rat. *J. Surg. Res.* 99(2):294–300.

140. Takahashi, K., Masuda, S., Nakamura, N., Saito, H., Futami, T., Doi, T., and Inui, K. (2001). Upregulation of H(+)-peptide cotransporter PEPT2 in rat remnant kidney. *Am. J. Physiol. Renal Physiol.* 281(6):F1109–F1116.

141. Gangopadhyay, A., Thamotharan, M., and Adibi, S.A. (2002). Regulation of oligopeptide transporter (Pept-1) in experimental diabetes. *Am. J. Physiol. Gastrointest. Liver Physiol.* 283:G133–G138.

142. Watanabe, K., Terada, K., and Sato, J. (2003). Intestinal absorption of cephalexin in diabetes mellitus model rats. *Eur. J. Pharm. Sci.* 19(2–3):91–98.

143. Thamotharan, M., Bawani, S.Z., Zhou, X., and Adibi, S.A. (1999). Hormonal regulation of oligopeptide transporter pept-1 in a human intestinal cell line. *Am. J. Physiol. Cell Physiol.* 276(4):C821–C826.

144. Nielsen, C.U., Amstrup, J., Nielsen, R., Steffansen, B., Frokjaer, S., and Brodin, B. (2003). Epidermal growth factor and insulin short-term increase hPepT1-mediated glycylsarcosine uptake in Caco-2 cells. *Acta. Physiol. Scand.* 178(2):139–148.

145. Watanabe, K., Terada, K., Jinriki, T., and Sato, J. (2004). Effect of insulin on cephalexin uptake and transepithelial transport in the human intestinal cell line Caco-2. *Eur. J. Pharm. Sci.* 21(1):87–95.

146. Fernandez, J., Bode, B., Koromilas, A., Diehl, J.A., Krukovets, I., Snider, M.D., and Hatzoglou, M. (2002). Translation mediated by the internal ribosome entry site of the cat-1 mRNA is regulated by glucose availability in a PERK kinase-dependent manner. *J. Biol. Chem.* 277(14):11780–11787.

147. Nielsen, C.U., Amstrup, J., Steffansen, B., Frokjaer, S., and Brodin, B. (2001). Epidermal growth factor inhibits glycylsarcosine transport and hPepT1 expression in a human intestinal cell line. *Am. J. Physiol. Gastrointest. Liver Physiol.* 281(1):G191–G199.

148. Bravo, S.A., Nielsen, C.U., Amstrup, J., Frokjaer, S., and Brodin, B. (2004). Epidermal growth factor decreases PEPT2 transport capacity and expression in the rat kidney proximal tubule cell line SKPT0193 cl.2. *Am. J. Physiol. Renal Physiol.* 286(2):F385–F393.

149. Buyse, M., Berlioz, F., Guilmeau, S., Tsocas, A., Voisin, T., Péranzi, G., Merlin, D., Laburthe, M., Lewin, M.J.M., Rozé, C., and Bado, A. (2001). PepT1-mediated epithelial transport of dipeptides and cephalexin is enhanced by luminal leptin in the small intestine. *J. Clin. Invest.* 108:1483–1494.

150. Yen, P.M., and Chin W.W. (1994). New advances in understanding the molecular mechanisms of thyroid hormone action. *Trends Endocrinol. Metab.* 5:65–72.

151. Ashida, K., Katsura, T, Motohashi, H., Saito, H., and Inui, K. (2002). Thyroid hormone regulates the activity and expression of the peptide transporter PEPT1 in Caco-2 cells. *Am. J. Physiol. Gastrointest. Liver Physiol.* 282:G617–G623.

152. Lu, H., and Klaassen, C. (2006). Tissue distribution and thyroid hormone regulation of Pept1 and Pept2 mRNA in rodents. *Peptides* 27(4):850–857.

153. Berlioz, F., Maoret, J.J., Paris, H., Laburthe, M., Farinotti, R., and Rozé, C. (2000). α_2-Adrenergic receptors stimulate oligopeptide transport in a human intestinal cell line. *J. Pharmacol. Exp. Ther.* 294:466–472.

154. Fujita, T., Majikawa, Y., Umehisa, S., Okada, N., Yamamoto, A., Ganapathy, V., and Leibach, F.H. (1999). σ-Receptor ligand-induced up-regulation of the H^+/peptide transporter PEPT1 in the human intestinal cell line Caco-2. *Biochem. Biophys. Res. Commun.* 261:242–246.

155. Satoh, J., Tsujikawa, T., Fujiyama, Y., and Bamba, T. (2003). Nutritional benefits of enteral alanyl-glutamine supplementation on rat small intestinal damage induced by cyclophosphamide. *J. Gastroenterol. Hepatol.* 18(6):719–725.

156. Watanabe, K., Jinriki, T., and Sato, J. (2006). Effects of progesterone and norethisterone on cephalexin transport and peptide transporter PEPT1 expression in human intestinal cell line Caco-2. *Biol. Pharm. Bull.* 29(1):90–95.

157. Evans, W.E., and Relling, M.V. (1999). Pharmacogenomics: translating functional genomics into rational therapeutics. *Science* 286:487–491.

158. Zhang, E.Y., Fu, D.J., Pak, Y.A., Stewart, T., Mukhopadhyay, N., Wrighton, S.A., and Hillgren, K.M. (2004). Genetic polymorphisms in human proton–dependent dipeptide transporter PEPT1: implications for the functional role of Pro586. *J. Pharmacol. Exp. Ther.* 310(2):437–445.

159. Terada, T., Irie, M., Okuda, M., and Inui, K. (2004). Genetic variant Arg57His in human H^+/peptide cotransporter 2 causes a complete loss of transport function. *Biochem. Biophys. Res. Commun.* 316(2):416–420.

160. Saito, H., Motohashi, H., Mukai, M., and Inui, K. (1997). Cloning and characterization of a pH-sensing regulatory factor that modulates transport activity of the human H^+/peptide cotransporter, PEPT1. *Biochem. Biophys. Res. Commun.* 237(3):577–582.

161. Urtti, A., Sadée, W., and Johns, S.J. (2001). Genomic structure of proton-coupled oligopeptide transporter hPepT1 and pH-sensing regulatory splice variant. *AAPS Pharm-Sci* 3:article 6.

162. Pinsonneault, J., Nielsen, C.U., and Sadée, W. (2004). Genetic variants of the human H^+/dipeptide transporter PEPT2: analysis of haplotype functions. *J. Pharmacol. Exp. Ther* 311(3):1088–1096.

163. Conradi, R.A., Hilgers, A.R., Ho, N.F., and Burton, P.S. (1992). The influence of peptide structure on transport across Caco-2 cells: II. Peptide bond modification which results in improved permeability. *Pharm. Res.* 9(3):435–439.

164. Kim, D.C., Burton, P.S., and Borchardt, R.T. (1993). A correlation between the permeability characteristics of a series of peptides using an in vitro cell culture model (Caco-2) and those using an in situ perfused rat ileum model of the intestinal mucosa. *Pharm. Res.* 10(12):1710–1714.

165. Conradi, R.A., Hilgers, A.R., Burton, P.S., and Hester, J.B. (1994). Epithelial cell permeability of a series of peptidic HIV protease inhibitors: aminoterminal substituent effects. *J. Drug Target.* 2(2):167–171.

166. Chikhale, E.G., Ng, K.Y., Burton, P.S., and Borchardt, R.T. (1994). Hydrogen bonding potential as a determinant of the in vitro and in situ blood–brain barrier permeability of peptides. *Pharm. Res.* 11(3):412–419.

167. Knipp, G.T., Vander Velde, D.G., Siahaan, T.J., and Borchardt, R.T. (1997). The effect of beta-turn structure on the passive diffusion of peptides across Caco-2 cell monolayers. *Pharm. Res.* 14(10):1332–1340.

168. Sorensen, M., Steenberg, B., Knipp, G.T., Wang, W., Steffansen, B., Frokjaer, S., and Borchardt, R.T. (1997). The effect of beta-turn structure on the permeation of peptides across monolayers of bovine brain microvessel endothelial cells. *Pharm. Res.* 14(10):1341–1348.

169. Prescribing information for Zovirax®. Available online at http://us.gsk.com/products/assets/us_zovirax.pdf. Accessed April 24, 2006.

170. Prescribing information for Valtrex®. Available online at http://us.gsk.com/products/assets/us_valtrex.pdf. Accessed April 24, 2006.

171. Wu, C.Y., and Benet, L.Z. (2005). Predicting drug disposition via application of BCS: transport/absorption/elimination interplay and development of a biopharmaceutics drug disposition classification system. *Pharm. Res.* 22(1):11–23.

172. Yu, H., Cook, T.J., and Sinko, P.J. (1997). Evidence for diminished functional expression of intestinal transporters in Caco-2 cell monolayers at high passages. *Pharm. Res.* 14(6):757–762.

173. Han, H.K., Oh, D.M., and Amidon, G.L. (1998). Cellular uptake mechanism of amino acid ester prodrugs in Caco-2/hPEPT1 cells overexpressing a human peptide transporter. *Pharm. Res.* 5(9):1382–1386.

174. Chu, C., Okamoto, C.T., Hamm-Alvarez, S.F., and Lee, V.H. (2004). Stable transfection of MDCK cells with epitope-tagged human PepT1. *Pharm. Res.* 21(11):1970–1973.

175. Herrera-Ruiz, D., Faria, T.N., Bhardwaj, R.K., Timoszyk, J., Gudmundsson, O.S., Moench, P., Wall, D.A., Smith, R.L., and Knipp, G.T. (2004). A novel hPepT1 stably

transfected cell line: establishing a correlation between expression and function. *Mol. Pharm.* 1(2):136–144.

176. Bhardwaj, R.K., Herrera-Ruiz, D., Sinko, P.J., Gudmundsson, O.S., and Knipp, G.T. (2005). Delineation of HPEPT1 meditated uptake and transport of substrates with varying transporter affinities utilizing stably transfected HPEPT1/MDCK clones and Caco-2 cells. *J. Pharmacol. Exp. Ther.* 314:1093–1100.
177. Beauchamp, L.M., Orr, G.F., de Miranda, P., Burnette, T., and Krenitsky, T.A. (1992). Amino acid ester prodrugs of acyclovir. *Antiviral Chem. Chemother.* 3(3):157–164.
178. de Miranda, P., and Burnette, T.C. (1994). Metabolic fate and pharmacokinetics of the acyclovir prodrug valaciclovir in cynomolgus monkeys. *Drug Metab. Dispos.* 22(1):55–59.
179. Beutner, K.R. (1995). Valacyclovir: a review of its antiviral activity, pharmacokinetic properties, and clinical efficacy. *Antiviral Res.* 28(4):281–290.
180. Crooks, R.J., and Murray, A. (1994). Valaciclovir: a review of a promising new antiherpes agent. *Antiviral Chem. Chemother.* 5(1):31–37.
181. Balimane, P.V., Tamai, I., Guo, A., Nakanishi, T., Kitada, H., Leibach, F.H., Tsuji, A., and Sinko, P.J. (1998). Direct evidence for peptide transporter (PepT1)–mediated uptake of a nonpeptide prodrug, valacyclovir. *Biochem. Biophys. Res. Commun.* 250(2):246–251.
182. de Vrueh, R.L.A., Smith, P.L., and Lee, C.-P. (1998). Transport of L-valine-acyclovir via the oligopeptide transporter in the human intestinal cell line, Caco-2. *J. Pharmacol. Exp. Ther.* 286(3):1166–1170.
183. Ganapathy, M.E., Huang, W., Wang, H., Ganapathy, V., and Leibach, F.H. (1998). Valacyclovir: a substrate for the intestinal and renal peptide transporters PEPT1 and PEPT2. *Biochem. Biophys. Res. Commun.* 246(2):470–475.
184. Han, H.-K., de Vrueh, R.L.A., Rhie, J.K., Covitz, K.-M.Y., Smith, P.L., Lee, C.-P., Oh, D.-M., Sadée, W., and Amidon, G.L. (1998). 5′-Amino acid esters of antiviral nucleosides, acyclovir, and AZT are absorbed by the intestinal PEPT1 peptide transporter. *Pharm. Res.* 15(8):1154–1159.
185. Guo, A., Hu, P., Balimane, P.V., Leibach, F.H., and Sinko, P.J. (1999). Interactions of a nonpeptidic drug, valacyclovir, with the human intestinal peptide transporter (hPEPT1) expressed in a mammalian cell line. *J. Pharmacol. Exp. Ther.* 289(1):448–454.
186. Sinko, P.J., and Balimane, P.V. (1998). Carrier-mediated intestinal absorption of valacyclovir, the L-valyl ester prodrug of acyclovir. 1: Interactions with peptides, organic anions and organic cations in rats. *Biopharm. Drug Dispos.* 19(4):209–217.
187. Sawada, K., Terada, T., Saito, H., Hashimoto, Y., and Inui, K.-I. (1999). Recognition of L-Amino acid ester compounds by rat peptide transporters PEPT1 and PEPT2. *J. Pharmacol. Exp. Ther.* 291(2):705–709.
188. Landowski, C.P., Sun, D., Foster, D.R., Menon, S.S., Barnett, J.L., Welage, L.S., Ramachandran, C., and Amidon, G.L. (2003). Gene expression in the human intestine and correlation with oral valacyclovir pharmacokinetic parameters. *J. Pharmacol. Exp. Ther.* 306(2):778–786.
189. de Miranda, P., and Blum, M.R. (1983). Pharmacokinetics of acyclovir after intravenous and oral administration. *J. Antimicrob. Chemother.* 12(suppl. B):29–37.
190. Soul-Lawton, J., Seaber, E., On, N., Wootton, R., Rolan, P., and Posner, J. (1995). Absolute bioavailability and metabolic disposition of valaciclovir, the L-valyl ester of acyclovir, following oral administration to humans. *Antimicrob. Agents Chemother.* 39(12):2759–2764.

191. de Miranda, P., and Good, S.S. (1992). Species differences in the metabolism and disposition of antiviral nucleoside analogs: 1. Acyclovir. *Antiviral Chem. Chemother.* 3(1):1–8.

192. Burnette, T.C., and de Miranda, P. (1994). Metabolic disposition of the acyclovir prodrug valaciclovir in the rat. *Drug Metab. Dispos.* 22(1):60–64.

193. Weller, S., Blum, M.R., Doucette, M., Burnette, T., Cederberg, D.M., de Miranda, P., and Smiley, M.L. (1993). Pharmacokinetics of the acyclovir pro-drug valaciclovir after escalating single- and multiple-dose administration to normal volunteers. *Clin. Pharmacol. Ther.* 54(6):595–605.

194. Perry, C.M., and Faulds, D. (1996). Valaciclovir: A review of its antiviral activity, pharmacokinetic properties and therapeutic efficacy in herpesvirus infections. *Drugs* 52(5):754–772.

7

MONOCARBOXYLATE TRANSPORTERS

JUSTIN A. SPANIER AND LESTER R. DREWES

University of Minnesota, Duluth, Minnesota

Drug Transporters: Molecular Characterization and Role in Drug Disposition, Edited by Guofeng You and Marilyn E. Morris
Copyright © 2007 John Wiley & Sons, Inc.

7.1. INTRODUCTION

7.1.1. SLC16 Gene Family: MCTs

The solute carrier 16 (SLC16) gene family is composed of 14 sequence-related isoforms, known as the *monocarboxylate transporters* (MCTs).[1] DNA and protein sequence analysis indicates that the members of the monocarboxylate transporter family possess many similar sequence motifs, consistent with the structure of proteins embedded in a lipid bilayer membrane. The topology predicted includes the presence of 12 transmembrane domains, intracellular N- and C-termini, and a large intracellular loop between transmembrane domains 6 and 7. The identification of the first MCT came after the discovery of a mutant protein in Chinese hamster ovary cells that preferentially transported mevalonate.[2] Subsequent cloning and expression of the wild-type gene in a breast cancer cell line showed that pyruvate was the preferred substrate.[3] Of the now 14 identified SLC16 gene family members, only MCT1 to MCT4, MCT8, and TAT1 have been functionally characterized and their substrates identified. Within the known MCT family members only MCT1–4 are known to transport monocarboxylates such as lactate, pyruvate, butyrate, and the ketone bodies β-hydroxybutyrate and acetoacetate.[4] In contrast, MCT8, formerly XPCT, is known to facilitate the proton-independent transport of thyroid hormones T3 and T4,[5] and TAT1 is a transporter of neutral amino acids. MCT6 was recently shown to transport the anion drug bumetanide in a pH- and membrane potential–sensitive manner, although its physiological substrate was not identified.[6] Phylogenetic analysis reveals that MCT1–4 have a high degree of sequence similarity (Figure 7.1). Therefore, it is suggestive that MCT1–4 are the only MCTs likely to transport monocarboxylates. Additionally, it is possible that due to their relatedness to TAT1 and MCT8, MCT9 and MCT14 may be involved in amino acid transport (Figure 7.1). The mechanism of transport for MCT1–4 is electroneutral because a single proton is translocated across the plasma membrane, along with each monocarboxylate anion. The process begins with the binding of a proton followed by binding of the monocarboxylate anion. The two substrates are then translocated across the plasma membrane and released in the reverse order.[7]

Through site-directed mutagenesis, many clues have been uncovered as to the mechanism of monocarboxylate transport by rat MCT1. Several critical amino acids have been identified that are important for both substrate recognition and transport activity. Arginine at position 306 (R306) is conserved among almost all MCT family members and is probably involved in binding of the monocarboxylate anion.[8] D302 and E369 are thought to be involved in H^+ binding.[8] It is hypothesized that the binding of the monocarboxylate anion to R306 may cause the H^+ to move from D302 to E369. The binding of the H^+ to E369 then causes a conformational change that results in the release of the monocarboxylate anion on the opposite side of the membrane.[9] In addition to the residues that may interact physically with the substrates, there are several other residues that are critical for transport activity. Mutation of R143, G153, and F360 results in a loss of transport activity or inability to be inserted into the plasma membrane.[9] The loss of MCT1 expression at the cell surface suggests

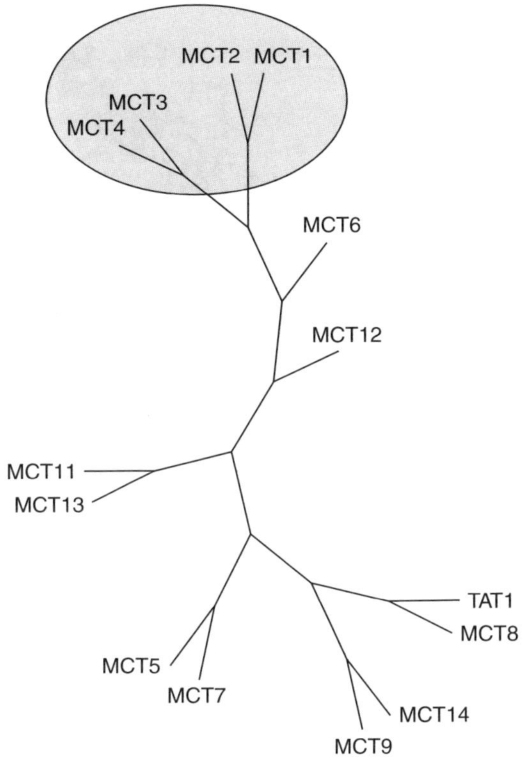

FIGURE 7.1. Phylogentic relationship predicted among all SLC family members. A Newick string containing all 14 protein sequences was generated using the ClustalW algorithm and plotted as a radial tree using T-REX software (freely available from http://www.labunix.uqam.ca/~makarenv/trex.html).

that these residues are critical for the proper association of MCT1 with its ancillary protein CD147. Finally, experiments suggest that R143 and K142 may be involved in the stereospecificity of MCT1.[9] The hypothetical protein structure of MCT1 (Figure 7.2) based on the crystal structure of the *Escherichia coli* lactose permease protein illustrates the position of the critical amino acids. It is evident that the predicted positions of the amino acids necessary for substrate binding do indeed face the putative translocation pore.

7.1.2. SLC5 Gene Family: SMCTs

The recent identification of a mutation in a sodium-dependent butyrate transporter associated with colon cancer led to the discovery of a set of sodium-dependent MCTs that transport monocarboxylates with a high degree of specificity but do not exhibit the typical gene or protein structure common to the SLC16 transporter family.[10–12] These transporters, denoted *sodium-coupled monocarboxylate transporters* (SMCTs),

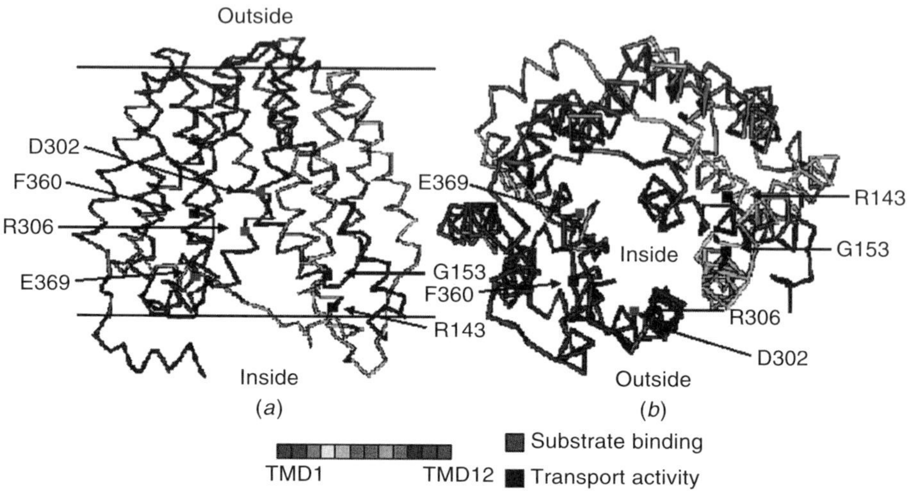

FIGURE 7.2. *(a)* A side view of the hypothetical structure of MCT1 embedded in the plasma membrane is depicted with solid lines indicating the edge of the lipid bilayer. *(b)* MCT1 is shown as in *(a)*, but viewed from the exofacial membrane surface. Each transmembrane domain (TMD) is color coded from the N- (red) to the c-terminus (purple). The positions predicted for the critical amino acids necessary for substrate binding (red) and transport activity (black) are highlighted and numbered according to their position in the protein sequence. This structure was deduced by threading the MCT1 protein sequence onto the crystal structure of the *E. coli* lactose permease A chain using the Cn3D software available from NCBI. *(See insert for color representation of figure.)*

comprise two of 12 identified members of the solute carrier 5 gene family and are denoted SLC5A8 and SLC5A12. This family, also known as the *sodium/glucose cotransporter gene family*, contains carriers whose substrates include glucose, myoinositol, iodide, choline, and B-complex vitamins.[13] Hydropathy analysis of SMCT protein sequences predicts the presence of 13 transmembrane domains with extracellular N- and intracellular C-termini. The link between colon cancer and SMCT1 gene suppression make this protein particularly important from a pharmacological perspective for clinical and therapeutic purposes. Unlike many MCTs, which generally have relatively broad tissue expression patterns, the protein expression of SMCTs is highly tissue specific. This leads to the supposition that these transporters may have evolved as a mechanism to cope with the unique metabolic requirements associated with specific tissues. Aside from tissue distribution, there are many other properties of SMCTs that differ from those of MCTs. As the name implies, the transport mechanism for SMCTs involves the transport of monocarboxylates along with sodium ions in an electrogenic fashion. That is, the symporter cotransports multiple (between two and four) sodium ions along with a monocarboxylate anion down a concentration gradient.[10] Finally, the substrate affinity of SMCTs for monocarboxylates differs greatly from that reported for MCTs (Table 7. 1).

TABLE 7.1. MCT and SMCT Transporter Kinetics

Transporter	Source	Expression System	Substrate	K_m(mM)
MCT1	Human[52]	Oocyte	L- Lactate	6.0
			Pyruvate	2.5
	Rat[21]	Oocyte	L-Lactate	3.5
			Pyruvate	1.0
			β-Hydroxybutyrate	12.5
			Acetoacetate	5.5
	Hamster[113]	sf9 (insect cell)	L-Lactate	8.3
			Pyruvate	3.1
MCT2	Human[52]	Oocyte	L-Lactate	6.5
			Pyruvate	0.03
	Rat[53]	Oocyte	L-Lactate	0.74
			Pyruvate	0.08
			β-Hydroxybutyrate	1.2
			Acetoacetate	0.8
	Hamster[113]	sf9	L-Lactate	8.7
			Pyruvate	0.8
MCT3	Chicken[54]	Yeast	L-Lactate	5.8
MCT4	Human[55]	Oocyte	L-Lactate	28
			Pyruvate	150
			β-Hydroxybutyrate	130
			Acetoacetate	210
	Rat[22]	Oocyte	L-Lactate	34
SMCT1	Mouse[10]	Oocyte	L-Lactate	0.23
			D-Lactate	0.72
			Butyrate	0.08
			Acetate	2.46
			Propionate	0.12
			Nicotinic acid	0.30
SMCT2	Mouse[12]	Oocyte	L-Lactate	[a]
			D-Lactate	[a]
			Butyrate	[a]
			Propionate	[a]

[a] K_m values could not be determined because the transporter could not be saturated.

7.2. MCT AND SMCT TRANSPORTER EXPRESSION

7.2.1. Tissue Distribution and Subcellular Localization of MCTs

In humans, MCT1 is expressed nearly ubiquitously in almost every tissue in the body and serves as the carrier for lactate flux across the plasma membrane of most cells. In the case of polarized cells such as the choroid plexus and intestinal epithelium, MCT1 is coexpressed with another MCT, but trafficking mechanisms segregate the MCTs to opposite sides of the cell (apical and basolateral). This asymmetric distribution may contribute to vectoral flux of substrates across the cellular barrier.

In contrast to MCT1, MCT2–4 are known to play a much more tissue-specific role in the transport of monocarboxylates. Analysis of MCT2 mRNA and protein expression in a variety of species has shown that major species-specific variations in tissue distribution exist. MCT2 expression in rodents is evident in a variety of tissues, including brain, liver, kidney, and testes.[14] Although the major MCT present in rodent neurons is MCT2, its expression in human brain appears to be minimal. Both expressed sequence tag (EST) expression and Northern blot analysis of human brain has found only very low or trace expression of MCT2.[15] However, lactate uptake by neurons has been suggested, with some controversy, as a major source of energy substrates for neuronal activation.[16,17] Therefore, if a high-affinity lactate transporter is required by adult human neurons, its identity and characteristics are yet to be determined.

MCT3 is unique among MCTs in that it is reportedly expressed in the retinal pigment epithelium (RPE), although the human EST database (Unigene) suggests a somewhat broader expression profile (National Center for Biotechnology Information). MCT3 is located at the basolateral membrane, where it functions in tandem with MCT1 on the apical side.[18] The presence of the lower-affinity MCT3 ($K_m \sim 5.8$ mM versus $K_m \sim 3.5$ mM) in the basolateral membrane of the RPE cell creates a more effective mechanism for lactate efflux out of the retina and into the bloodstream.

MCT4 works synergistically with MCT1 and is limited in its tissue expression. In humans MCT4 expression is limited to highly glycolytic tissues and has been detected in skeletal muscle, heart, and liver.[19] In human skeletal muscle MCT4 expression is highly variable among individuals and is localized primarily to type II glycolytic muscle fibers, while MCT1 expression varies little between individuals and is present in both type I and type II muscle fibers.[20] This coexpression of MCT1 with other MCTs probably results from the unique physical and biochemical properties associated with each MCT family member.[19] For example, the K_m of MCT1 for lactate is about 3.5 mM,[21] while that of MCT4 is significantly higher ($K_m \sim 17$ to 34 mM).[22] Because the concentration of lactate in the blood is normally about 1.5 mM, this extreme difference in substrate affinity results in MCT4 acting more like an assistant to MCT1 to further enhance lactate efflux from glycolytic muscle fibers.[19]

In the human colon, there exist symbiotic bacteria that breakdown undigested carbohydrates to produce significant amounts of short-chain fatty acids (SCFAs) such as butyrate, propionate, and acetate. These SCFAs serve as major precursors in generating glucose, amino acids, and ketone bodies.[23] SCFAs also provide a major fuel source for the colonocytes, with butyrate being the most preferred substrate for oxidation.[24] The colonocytes that line the lumen of the human colon exhibit polarized expression of MCTs, where MCT1 is expressed on the apical surface and MCT4 is localized to the basolateral surface.[25] These transporters are critical for proper maintenance of the colon epithelium because SCFAs such as butyrate in the colon have been shown to regulate many essential cellular functions, including cell growth, proliferation, and differentiation.[26] The presence of MCT1 on the apical side and the low-affinity MCT4 on the basolateral surface of the colon epithelium provides a means by which SCFAs can be concentrated in the colonic epithelial cell for catabolism and regulatory processes. Additionally, the polarized expression of these transporters restricts transfer of monocarboxylates from the basolateral surface out into the gut lumen.

7.2.2. Tissue Distribution and Subcellular Localization of SMCTs

Adding to the complexity of monocarboxylate transport in the intestinal tract, researchers have recently confirmed the presence of SMCT1 and SMCT2 in the human and mouse intestine.[11,12] Interestingly, these two transporters have very different expression patterns along the entire length of the intestinal tract. In mouse, SMCT1 was found to have the most abundant transcript expression in the distal part of the small intestine, cecum, and colon, whereas SMCT2 mRNA expression was localized exclusively to the proximal region of the small intestine.[10,12] The presence of SMCT2 in the proximal (sterile portion) of the small intestine implies that it may play a major role in the absorption of monocarboxylates that stem from dietary sources rather than from bacterial sources as in the more distal portions of the intestine.[12] In addition to assisting the H^+- linked MCTs in nutrient uptake, the presence of SMCTs in the colon also presents a unique means by which the intestinal epithelial cell can import D-lactate from the lumen. Although humans cannot produce D-lactate, it is the metabolic by-product of many bacteria in the intestinal tract and can serve as an energy source for mammalian tissues.[10]

In situ hybridization studies have also identified unique expression patterns of SMCTs in the mouse kidney. SMCT2 expression was shown to be greatest in the cortical regions of the proximal tubule epithelium, where concentrations of lactate are highest, while SMCT1 transcript expression is high in the cortex and moderate in medullary regions of the tubule epithelium, where lactate concentrations are low.[12,27] Furthermore, the identification of SMCTs in the nephritic proximal tubule epithelium is significant because no known H^+-linked MCT has been shown to be expressed on the apical surface of the proximal or distal region of the tubule epithelium. The SMCTs would then facilitate the movement of monocarboxylates into and out of the nephron tubule lumen. It is known that ~95% of blood lactate is recovered in the kidney.[28] It makes sense, then, that both high- and low-affinity sodium-coupled cotransporters are expressed in the proximal tubule, whereas only the high-affinity transporter is expressed in the distal nephron. During intense exercise the concentration of lactate in blood can become elevated. The presence of the low-affinity SMCT will ensure efficient capture of lactate from the blood filtrate, while the high-affinity transporter will capture the remaining lactate during its passage through the distal nephron.

7.3. REGULATION OF GENE EXPRESSION AND ACTIVITY

7.3.1. Regulation of Gene Expression

Most studies on the regulation of MCT gene expression focus on MCT1. Experimental evidence suggests that MCT1 function may be regulated at both the transcription and translational levels. The availability of substrate has been demonstrated to influence

MCT1 gene expression; however, the details of this mechanism are unknown. It has been shown that suckling rat pups, where the diet is composed primarily of fats from the mother's milk, have increased levels of MCT1 protein in brain endothelial cells compared to adult rats.[29] Additionally, adult rats fed a high-fat diet, which results in high levels of plasma ketones, have increased expression of MCT1 at the blood–brain barrier.[30] Furthermore, butyrate greatly enhances MCT1 transcript and protein expression in a human intestinal cell line.[31] Exercise training in humans also dramatically affects MCT expression in skeletal muscle. During periods of intense muscle activity in humans, lactate production by glycolytic muscle cells is greatly enhanced. MCT1 in the plasma membrane reportedly increases by as much as 76%, while MCT4 protein content increased by 34% after 8 weeks of exercise training.[20] Additionally, in the heart, where lactate is a major fuel source for oxidative phosphorylation, up-regulation of MCTs is even more dramatic following periods of moderate- and high-intensity exercise.[19] Although it is clear from these results that high levels of lactate and circulating ketone bodies can induce MCT1 gene transcription, the exact pathway that is involved remains elusive.

Some clues as to the mechanism of MCT1 transcriptional regulation are emerging. With the recent release of the human genome sequence and characterization of the human MCT1 promoter, researchers have been able to identify several transcription factor–binding sites within the promoter region. These conserved sequences within the human MCT1 promoter include binding sites for upstream stimulatory factors (USFs), nuclear factor-κB (NFκB), stimulating protein 1 (SP1), activator protein 1 (AP1), and activator protein 2 (AP2).[32] Of these putative MCT1 transcriptional regulators, only USFI and USFII have been shown in vitro to regulate MCT1 gene expression. Overexpression of USFI and USFII in a human intestinal cell line (Caco-2) led to a 34% and 84% decrease in MCT1 promoter activity, respectively.[33] In another study, $CoCl_2$, an oxidative phosphorylation inhibitor and putative activator of the transcription factor HIF-1α, caused an increase in MCT1 mRNA in cultured rat brain endothelial cells (RBE4).[4] Additionally, in a more recent study it was hypothesized that increased MCT1 mRNA expression in brain may be mediated by HIF1α in a spontaneous hypertensive rat model after an ischemic insult.[34] In contrast to these reports, a recent analysis of the human MCT1 and MCT4 promoters indicated that only MCT4 is responsive to hypoxia through a direct interaction with HIF1α.[35] Although the MCT1 promoters of different species contain many of the same transcription factor–binding sites, the sequences vary greatly. Therefore, it is possible that some transcriptional regulator elements among species are conserved, whereas others, such as response to hypoxia, are not.

Evidence supporting hormonal regulation of MCT1 transcription has also been reported. In a thyroid cell line (FRTL-5) MCT1 transcription was found to be regulated by thyroid-stimulating hormone (TSH) through cAMP-dependent pathways.[36] In a human intestinal cell line (Caco-2), addition of leptin to the apical surface leads to a small but significant increase in MCT1 protein in the apical membrane.[37] This increase in MCT1 expression induced by leptin was attributed to increased MCT1 mRNA production and enhanced translocation of the transporter protein to the plasma membrane from intracellular pools.[37] Further studies with Caco-2 cells have shown a fivefold increase in MCT1 protein expression following long-term exposure to phorbol ester and down-regulation of protein kinase C (PKC).[38]

7.3.2. Intracellular Trafficking

Many multispanning integral membrane transport proteins, ion channels, and receptors are associated with single-spanning glycosylated proteins that facilitate their translocation to the plasma membrane. These ancillary proteins not only are required for delivery of the transporter to the plasma membrane, but in some cases they have been found to modulate transporter activity.[39–41] Two glycoproteins in the immunoglobulin superfamily facilitate the translocation of MCT1–4 to the plasma membrane. CD147 (also known as basigin, EMMPRIN, HT7, and OX-47) is the ancillary protein to MCT1, MCT3, and MCT4,[42,43] while gp70 (embigin) is the ancillary protein for MCT2.[44] In erythrocytes, where CD147 is absent, gp70 appears able to serve as the ancillary protein to MCT1.[45] These two glycoproteins have the same basic structure in that they contain variably glycosylated extracellular IgG-like domains along with a highly conserved single-pass transmembrane segment and short intracellular C-terminus. It is hypothesized that the interaction between MCTs and their accessory proteins involves electrostatic interactions between a glutamic acid residue in the transmembrane segment of either CD147 or gp70 and an arginine residue conserved in transmembrane domain 8 of MCT1.[46] Recently, this hypothesis was challenged by showing that mutation of the Glu-221 of CD147 (E221A) failed to disrupt the interaction between MCT1 and CD147.[47] The exact nature of the interaction between CD147 and MCTs is unresolved.

In polarized epithelial cells, MCTs are often targeted to either the apical or basolateral membranes via signals within the transporter itself or through sorting signals within the ancillary protein. It was recently reported that the C-terminus of CD147 contains a basolateral sorting signal for translocation of MCTs in MDCK cells.[47,48] Interestingly, in the retinal pigment epithelium (RPE), where both apical and basolateral targeting of CD147 occurs, the CD147 sorting signal is not recognized and results in the delivery of the CD147/MCT1 heterodimer to the apical surface.[47] This may also be true in the intestinal epithelium, where MCT1 is targeted to the apical side of the plasma membrane. Although there is a lack of identified sorting signals, MCT1, MCT3, and MCT4 were found to contain strong basolateral sorting signals that are recognized in both MDCK and RPE cells. The fact that various MCT isoforms have evolved different mechanisms to achieve polarized expression within epithelial cells has greatly enhanced the ability of the cell to regulate the transport of monocarboxylates to meet the specific needs of individual cell types. The interaction of MCTs with their accessory proteins is multifunctional. Not only is the interaction essential for proper membrane trafficking, but the direct protein–protein interaction must be retained in the membrane to maintain functional transport activity. Disruption of the disulfide bridges in CD147 with *p*-chloromercuribenzene sulfonate (pCMBS), which causes the protein to be more loosely associated with MCTs, abolishes transport activity in *Xenopus* oocytes.[44]

7.3.3. Regulation of Transporter Activity

Studies regarding the modulation of transport kinetics of MCTs within the plasma membrane provide important evidence regarding the regulation of MCT1 activity. Activation of cAMP-dependent pathways in rat brain endothelial cells (RBE4) caused a 40 to 60% reduction in lactate transport.[49] This reduction in lactate transport was attributed to loss of transporter function in the membrane by either internalization of the transporter or inactivation of the transporter by protein modification. These results suggest that phosphorylation of MCT1, CD147, or a closely associated protein is likely to be involved in the cAMP response.[49] Indeed, MCT1 contains potential phosphorylation sites, but to date there is no conclusive evidence for direct phosphorylation. Aside from CD147, carbonic anhydrase isoform II (CAII) is the only other protein that has been shown to interact directly with MCT1. Addition of CAII to *Xenopus* oocytes expressing MCT1 increased transporter activity three- to fivefold.[50] Interestingly, it was shown that the C-terminus of MCT1 was not important for binding with CAII to occur, but it was essential for modulation of transporter activity. In addition to direct protein–protein interactions with MCTs, it is probable that protein interactions with CD147 may also modulate transporter function. There is evidence for a "CD147–CD98hc supercomplex" containing several transporters and cell surface glycoproteins that are tightly associated with each other in the plasma membrane.[51] This complex includes CD147, MCT1, CD98hc, the amino acid transporters LAT1 and ASCT2, and the glycoprotein EpCAM. This complex is coordinately expressed at the transcriptional level in tumor cells, and due to their close association in the membrane, it is likely that functional modulation of one of these proteins could alter the transport activity of other proteins in the complex.[51]

7.4. CLINICAL APPLICATIONS AND IMPLICATIONS FOR DRUG DELIVERY

7.4.1. Substrates

Natural Substrates MCT1–4 and SMCTs 1 and 2 are known to transport a variety of physiologically relevant substrates, including lactate, pyruvate, and butyrate. Each of these monocarboxylate transporters exhibits unique kinetic properties. The affinity for monocarboxylates by MCT1 and MCT4 is very similar across species, while the affinity for monocarboxylates by MCT2 varies greatly between rodents and humans (Table 7. 1). Rat MCT2 was characterized as the high-affinity pyruvate carrier because of its low K_m value, 0.08 mM.[52] MCT2 also has a relatively high affinity for lactate, but major differences exist between human, rat, and hamster transporters. While rat MCT2 has a K_m of 0.74 mM for lactate, human and hamster MCT2 have a roughly 10-fold higher K_m of 6.5 and 8.7 mM, respectively.[52,53] The differences in transporter kinetics between species can make it difficult to form conclusions about the physiological role of MCTs in rodents compared to that of humans. Cross-species studies of MCT kinetics are absent for MCT3, SMCT1, and SMCT2. Therefore, species comparisons are not possible.

Multiple monocarboxylates have been tested as substrates for MCT1–4 (Table 7.1). While MCT1 has a moderate affinity for lactate and pyruvate, it appears to have a slightly higher affinity for pyruvate in all species that have been studied. Few data are available on MCT3 kinetics, but the chicken isoform appears to have a similar affinity for lactate as MCT1.[54] In both rat and human, MCT4 is a low-affinity lactate transporter (K_m 28 to 34 mM), where its proposed function is to extrude lactate from highly glycolytic cells, such as skeletal muscle fibers and astrocytes.[22,55]

In contrast to MCT1–4, SMCT1 transports monocarboxylates with a very high affinity, with butyrate being the preferred substrate.[10] Another unique property of SMCT1 is that it has similar transport affinities for both L- and D-lactate. Although SMCT1 and SMCT2 transport similar substrates, the K_m values for SMCT2 could not be computed, due to the inability to saturate the transporter.[12] In addition to lactate, pyruvate, and SCFA, both SMCT1 and MCT1 contribute to the uptake of the water-soluble B-complex vitamin nicotinic acid in the intestinal epithelium.[56,57] Additionally, SMCT1 was originally identified as a iodide transporter, but recent studies suggest that the transporter acts as an iodide channel that is gated by monocarboxylates.[58,59]

Clinically Relevant Drugs Although drugs administered orally can enter the body through passive diffusion in the intestinal tract, transport involving specific membrane carriers in the intestinal epithelium may contribute significantly. The expression of MCTs in the human gut and elsewhere throughout the body makes this carrier family particularly important for the enhanced uptake of exogenous and clinically relevant substrates from the gut lumen and for their subsequent delivery into target tissues. Additionally, the presence of MCTs in many barrier systems, including the blood–brain barrier, blood–testis barrier, and blood–placental barrier, provide a means by which drugs can be targeted to tissues where passive diffusion is unlikely to occur. Exogenous substrates that are transported via MCTs are typically anions of small, weak, monovalent organic acids that are either hydrophobic or hydrophilic.[4] Facilitated diffusion of weak organic acids such as acetate, propionate, benzoate, and nicotinate have been demonstrated both in in vitro and in vivo and are likely to be mediated by MCTs.[60–63]

Although it has been proposed that nonsteroidal anti-inflammatory compounds (NSAIs) such as salicylic acid and ibuprofen are transported by MCTs,[64,65] studies in a human trophoblast cell line show that MCT-mediated transport of these compounds is unlikely.[66] However, even though NSAIs may not be transported, these compounds appear to interact with MCTs and inhibit the transport of endogenous monocarboxylates. Several β-lactam antibiotics, including phenethicillin, propicillin, and carindacillin, are reportedly transported by MCTs.[67] Cellular uptake of another β-lactam drug, cefdinir, is in part by MCT activity.[68] Transport of the 3-hydroxy-3-methylglutaryl-coenzyme A (HMG-CoA) reductase inhibitors simvastatin and atorvastatin across the blood–brain barrier (BBB) is also mediated by MCTs.[69–71] The use of valproic acid, an antiepileptic agent, during pregnancy has been associated with causing fetal malformations, and its transport across the intestinal epithelium and blood–placental barrier is hypothesized to occur via MCTs.[72] Additionally, XP13512, a recently developed gabapentin prodrug used for treatment of seizures, is reported to be transported with high affinity by both MCTs and the sodium-dependent multivitamin transporter (SMVT) in Caco-2 cells.[73] The hypoglycemic agent nateglinide has also been demonstrated to be transported in part by MCTs, but other significant transport mechanisms exist.[74] The ineffectiveness of the anticancer drug 6-mercaptopurine (6-MP) in the central nervous system of patients with acute lymphoblastic leukemia has been attributed to a high rate of 6-MP efflux across the BBB by MCTs.[75]

Many compounds that are derived from dietary sources can have powerful antioxidant properties. Some of these naturally occurring compounds, such as quercetin, artepillin, caffeic acid, and their metabolites, produced by bacterial metabolism are known to be transported by MCTs expressed in Caco-2 cells, and thus by inference, across the intestinal epithelium.[76–79] Although many clinically relevant compounds are transported by MCTs, a few drugs influence MCT1 gene and protein expression but do not serve as substrates. The anti-cancer drug flutamide, used in the treatment of prostate cancer, has been shown to cause a significant decrease in MCT1 and MCT2 mRNA expression and causes long-term deficits in lactate transport in rats exposed to the drug during the fetal life stage.[80] Furthermore, the β_2-agonist clenbuterol, a bronchodilator with potent anabolic properties, was found to cause significant decreases in MCT1 protein expression in both the soleus and extensor digitorum longus (EDL) muscles in rats after 4 weeks of drug exposure.[81]

7.4.2. Inhibitors

There is a wide range of reversible and irreversible inhibitors of monocarboxylate transport. These inhibitors can be classified into four general categories. First, substituted aromatic monocarboxylates such as phenylpyruvate and derivatives of α-cyanocinnamate (α-CHC) serve as potent competitive reversible MCT inhibitors. Although α-CHC is commonly used as a MCT inhibitor, it is a more potent inhibitor of the mitochondrial pyruvate carrier.[82] Therefore, this inhibitor blocks oxidative phosphorylation and causes the toxic buildup of lactic acid. Second, pyrrolopyrimidinedione (PPD) derivatives serve as highly specific MCT inhibitors.[83,84] These compounds are known to have potent immunosuppressant effects by specifically preventing T-lymphocyte activation through inhibition of lactate transport.[83] Third, there are amphiphilic compounds that can also serve as competitive reversible MCT inhibitors. Compounds in this group include the anion transport inhibitors 5-nitro-2-(3-phenylpropylamino)benzoate (NPPB) and niflumate. Additionally, the naturally occurring flavonoids quercetin and phloretin have been shown to bind competitively with MCTs. Aside from its antioxidant properties, quercetin also suppresses the stress response by inhibiting the expression of heat shock proteins.[85] Phloretin has been shown to increase membrane fluidity and inhibit the transport of many organic acids, inorganic ions, and nonelectrolytes.[86] Finally, stilbene disulfonates such as 4,4′-diisothiocyanostilbene-2,2′-disulfonate (DIDS) and 4,4′-dibenzamidostilbene-2,2′-disulfonate (DBDS) serve as very potent, membrane-impermeant, competitive, and reversible inhibitors of MCTs.[87] Prolonged exposure (>1 hour) to DIDS, however, can lead to irreversible inhibition due to the modification of amino groups at the cell surface. Therefore, DBDS may be a more practical inhibitor in this category because it does not contain reactive isothiocyanate groups and may be less toxic to cells during long-term exposure.[87] A recent report suggests that bumetanide is transported by MCT6.[6] However, because the physiological substrate for MCT6 is not known, it is conjecture whether this drug inhibits the normal function of this transporter.

In addition to the MCT inhibitors that interact directly with the transporter itself, pCMBS irreversibly inhibits MCT transport through modification of CD147. Inhibition of monocarboxylate transport by pCMBS results from the modification of disulfide bridges within the IgG-like domains of CD147, thus disrupting the interaction between the transporter and CD147 at the cell surface.[44] DEPC may also inhibit transport by MCTs through an indirect mechanism.[54] Whether any of the compounds within these four inhibitor categories act on SMCTs is unknown, but the identification of SMCT-specific inhibitors will surely provide valuable information as to the overall contribution of these transporters to monocarboxylate transport.

Of all the inhibitors described previously, only the PPDs appear to be MCT specific; the remainder are less specific and may inhibit other transporter families.[83] The PPDs are known to be highly specific for MCT1 and to a lesser extent for MCT2. Although most of these inhibitors are not isoform specific, their affinities for the various MCT isoforms vary significantly (Table 7.2). Most of the MCT inhibitors identified are competitive, and therefore the high-affinity MCT2 is most sensitive, while the low-affinity MCT4 is least sensitive to the classical MCT inhibitors. The most potent inhibitors

TABLE 7.2. Inhibitors of MCTs

Transporter	Source	Expression System	Substrate	IC_{50}(μM)[a]
MCT1	Rat[53]	Oocyte	α-CHC	425
			Phloretin	28
			Benzobromaron	22
			Quercetin	14
			DIDS	1100[22]
			DBDS	N.A.
			pCMBS	N.A.
MCT2	Rat[53]	Oocyte	α-CHC	24
			Phloretin	14
			Benzobromaron	9.0
			Quercetin	5.0
			DIDS	N.A.
			DBDS	N.A.
MCT3	Rat[54]	Yeast	DEPC	400
MCT4	Rat[22]	Oocyte	α-CHC	350
			Phloretin	40
			DIDS	500
MCT6	Human[6]	Oocyte	Bumetanide	0.08[b]

[a] N.A., not available.
[b] K_m value (mM).

are the flavonoids phloretin and quercetin. The hystidyl-selective reagent diethyl pyrocarbonate (DEPC) is the only compound known to inhibit MCT3. Treatment with hydroxylamine did not reverse the inhibition; therefore, the authors concluded that DEPC inhibition was caused by modification of lysine or arginine residues.[54]

Interestingly, MCT3 was found to be insensitive to the classical MCT inhibitors, such as phloretin, α-CHC, and pCMBS.[54] The reason for the insensitivity of MCT3 to the classical MCT inhibitors is unclear but may be explained in part by the use of a yeast expression system that did not include coexpression of the accessory protein CD147. In another study, coexpression with CD147 was necessary for functional expression of the transporter (MCT1) in yeast.[88] These results suggest that it is likely that another protein within yeast, which is insensitive to pCMBS, facilitated the translocation of MCT3 to the plasma membrane. Further studies will be needed to validate MCT3 inhibitors and clarify its trafficking, localization, structure, and transport mechanism in yeast.

7.4.3. Implications in Physiology and Disease Treatment

Cancer Biology Neoplastic tissues exhibit high metabolic rates that correlate with increased cellular growth and proliferation. Interestingly, tumor cells have increased glucose transport capabilities and rely predominantly on glycolysis for adenine triphosphate (ATP) production even in the presence of oxygen. Thus, tumor cells are major producers of lactic and pyruvic acids, physiological substrates for the MCTs and SMCTs. To cope with the increased production of lactate and H^+, tumors may up-regulate MCT expression at the cell surface in order to maintain a homeostatic pH.[89] Additionally, the MCT accessory protein, CD147, is an inducer of matrix metalloproteinases, enzymes essential for tissue basement membrane digestion. Thus, increased MCT expression may lead to increased expression of CD147 and aid in tumor invasion and metastasis.[90] High levels of MCT1 and MCT2 expression have been detected in many human epithelial tumor cell lines.[52] Furthermore, it was demonstrated that MCT1 protein expression increased in some melanoma cell lines where they play a significant role in the regulation of intracellular pH and may be critical for tumor cell survival.[91] Additionally, there is increased MCT1 protein expression in human high-grade glial neoplasms compared to normal brain.[92] The ability of both MCTs and their accessory proteins to promote tumor cell growth and survival makes them attractive drug targets for cancer therapy. It has been suggested that inhibition of lactic acid export from tumor cells by inhibiting MCTs may be an effective anticancer strategy. The theory is that failure of MCT function will cause intracellular pH to fall and disable cellular metabolism and processes. However, most MCT inhibitors are not sufficiently specific and may be toxic to normal and tumor cells. An exception may be the recently reported PPDs that appear to block immune cell responses by an MCT-specific mechanism but are not generally toxic.[83] Further studies to evaluate the effect of this new class of MCT inhibitors on tumor cells are warranted.

In contrast to the up-regulation of MCTs in many tumor cells, MCT and SMCT expression often becomes down-regulated in colorectal tumors. The short-chain fatty acid butyrate, produced through bacterial fermentation of dietary fiber in the colon, is important for the regulation of many genes involved in proliferation and differentiation as well as apoptosis within colonocytes.[93] Butyrate is thought to regulate gene expression primarily by inhibiting histone deacetylases.[59] Alterations in butyrate transport in the colonic epithelium have been shown to have significant effects on cellular development. Thus, the proper expression of MCTs and SMCTs in the colon is critical in maintaining tissue homeostasis. Diminished MCT expression and consequently decreased uptake of butyrate can lead to the promotion of tumor growth.[94] During the transition from normalcy to malignancy MCT1 mRNA and protein expression are significantly down-regulated in the human colon, while MCT2 and MCT4 remain relatively low in abundance.[95] Whether the down-regulation of MCT1 expression is caused by a decrease in substrate availability or a malfunction in the signaling pathways involved in MCT1 gene expression is unknown, but this alteration in gene expression certainly may contribute to the malignant phenotype. Although MCT1 may play a significant role in butyrate transport in the human intestine, the role of SMCTs may be equally significant. Because the transmembrane gradient for Na^+ is much larger than for H^+, the driving force for butyrate uptake by colonocytes is much greater through SMCT1 than through MCT1.[59] Given these facts, it is not surprising that SMCT1 is a potent tumor suppressor in the colon.[96] SMCT1 gene expression is silenced in approximately 60% of colorectal cancers.[96] Additionally, when SMCT1 expression is restored in vitro, colony formation is decreased by at least 75% in some colon cancer cell lines.[59] The mechanism of SMCT1 suppression in colorectal tumors appears to result from aberrant methylation of exon 1, which occurs during the very early stages of tumor development.[96] In addition to observations of colorectal tumors, SMCT1 suppression has also been seen in human tumors from the stomach, thyroid gland, and brain.[59] Butyrate is found in very low concentrations in the blood, so its availability in tissues other than the colon is limited. The role of SMCT1 in tissues where butyrate is not available requires further investigation.

Cerebral Vascular Disease and Stroke It has long been recognized that brain oxygen deprivation stimulates glycolysis, lactate accumulation, and a decline in pH to levels incompatible with those of normal enzyme function and pathways of energy metabolism.[97,98] However, what is less well recognized is that in contrast to muscle and other tissues that readily eliminate anaerobically produced lactate and H^+ ions by efflux into the blood, the adult brain significantly lacks this capacity.[99]

Higher plasma glucose levels (and higher brain lactic acid levels) have also been associated with greater ischemic damage in adult animals.[100,101] Consistent with this observation is the report that fasting (and lower plasma glucose) results in smaller ischemic infarcts.[102] However, paradoxically, higher plasma glucose (and more substrate for maintaining energy levels) is also correlated with a reduction in ischemic damage in suckling rat pups.[103] Notably, lactic acid efflux was not determined in these studies, and it may play an important role. In suckling rats, MCT1 protein levels at the BBB are elevated 25-fold compared to adults.[29] Moreover, a ketogenic diet, although conferring stroke neuroprotection, also elevates brain vascular MCT1.[30] This adaptation for processing monocarboxylic acids may enable the brain to eliminate efficiently the excess lactic acid produced by glycolysis during and following a stroke (hypoxia/ischemia). A corollary to this hypothesis is that lactate and hydrogen ions from the brain interstitial space are transported into the blood by MCT1 in brain endothelial cells. Interestingly, fasting induces ketonemia and confers neuroprotection in rats subject to hypoxic insult[102,104] while also enhancing MCT function.[105] Elevated MCT1 expression correlates with a reduced risk of ischemic damage, and MCT1 may play a key role in this neuroprotection. Complementary to increased numbers of MCT1 proteins is the possibility that increased functional activity of existing transporters by intracellular regulatory mechanisms also may play a key role in facilitating lactic acid processing and efflux from the brain.[49–51]

Diabetes Diabetes (types I and II) is a disease involving disturbances in insulin signaling and disruption of glucose metabolism. Altered glucose transport and oxidation generates products such as pyruvate and lactate that utilize MCTs for cellular transport. Therefore, a significant role of MCTs in diabetes may be hypothesized. It is reported that elevated levels of plasma lactate may be a significant risk factor for the development of type II diabetes.[106,107] Skeletal muscle contributes significantly to insulin-dependent glucose uptake and is a major source of lactate in human body.

In both type I and type II diabetes, MCT expression is altered severely. MCT1 protein content in skeletal muscle in men with type II diabetes was about 35% lower than in healthy men and increased to normal levels following moderate exercise training.[108] In contrast, MCT4 protein was not significantly different between the two groups, but after exercise training, MCT4 content increased in healthy men and remained unchanged in men with type II diabetes.[108] This would indicate that the expression of MCT1 and MCT4 in muscle is altered in diabetes and that their regulation in response to exercise is modified.

In rats with streptozotocin (STZ)-induced type I diabetes, it was found that MCT1 protein was reduced in heart, and MCT1 and MCT4 protein were reduced in skeletal muscle.[109] Consistent with previous observations, this decrease in MCT1 and MCT4 protein was elevated back to normal levels in both heart and skeletal muscle with intense exercise training.[109] This suggests that it may take a high level of exercise training to elevate MCT4 protein levels in patients with diabetes. In addition to skeletal muscle, adipocytes are also important sites of lactate production, where MCT1 is expressed to facilitate lactate efflux. In a STZ-induced type I diabetes rat model it was shown that MCT1 protein expression in adipocytes is decreased by almost 80% that of control rats.[110] It is apparent from these results that MCT expression is severely altered in both skeletal muscle and adipocytes in rats and humans with diabetes. Whether these changes in MCT expression contribute to disease progression or are a result of altered glucose metabolism (or a combination of the two) is unknown. Pharmacological manipulation of MCT expression and/or activity to maintain normal plasma lactate levels may aid in the prevention and treatment of diabetes.

7.4.4. Drug Design

Currently, only structural comparisons of substrates have been performed to examine the minimum requirements necessary for MCT-mediated transport. Generally,

MCT1–4 prefer monocarboxylates containing two to four unbranched, aliphatic carbon atoms, with the highest affinity for compounds containing three carbons. Single carbon atoms containing monocarboxylates, such as formate and bicarbonate, are poor substrates.[1] Substitutions in the C2 and C3 positions are tolerated (excluding amino and amido groups), with carbonyl or hydroxyl substitutions at C2 generally being preferred (pyruvate and lactate). MCTs are stereoselective for L-lactate but not for 2-chloropropionate and β-hydroxybutyrate.[14]

Monocarboxylate compounds containing long-branched aliphatic or aromatic side chains are also acceptable substrates for MCTs. Some of these compounds can bind to the transporter with high affinity but are not easily translocated and thus function as inhibitors. The monocarboxylate group does not appear to be essential because compounds such as the PPD derivatives, quercetin, phloretin, and 6-MP also serve as substrates for MCTs. Additionally, the PPD derivatives may serve as good templates for drug design because they are isoform specific.

X-ray crystallography provides a means by which the structures of the translocation pore of membrane transporters may be deduced.[111,112] Thus far no MCT protein has succumbed to any such structural analysis. However, a hypothetical three-dimensional structure of MCT proteins can be obtained by threading the amino acid sequence onto the previously determined three-dimensional structures of similar 12-transmembrane-domain-containing proteins (lactose permease, glycerol-3-phosphate transporter) (Figure 7. 1). In the near future it may be possible to apply software to describe the dimensions, electronic configuration, and putative amino acids that define the substrate binding and translocation features of MCTs.

Few studies have been conducted concerning the substrates and inhibitors of SMCTs, therefore, it is difficult to determine its overall substrate preference. It can be assumed, however, that because MCT1–4 and SMCTs transport similar natural substrates, they are both likely to be involved in transport of the same clinically relevant substrates. From the available kinetic studies it appears that both SMCT1 and SMCT2 have a higher affinity for unsubstituted monocarboxylates (butyrate). Additionally, as the chain length increases from two to four carbons, the affinity increases. Furthermore, the ability of SMCTs to transport both L- and D-lactate with high affinity may provide a means by which synthetic compounds can be targeted specifically for SMCT-mediated transport.

Acknowledgments

This work was supported in part by the University of Minnesota–Duluth Department of Biology and grants from the American Heart Association and National Institutes of Health (NS37764).

REFERENCES

1. Halestrap AP, Meredith D. 2004. The SLC16 gene family: from monocarboxylate transporters (MCTs) to aromatic amino acid transporters and beyond. Pflugers Arch 447(5):619–628.

2. Kim CM, Goldstein JL, Brown MS. 1992. cDNA cloning of MEV, a mutant protein that facilitates cellular uptake of mevalonate, and identification of the point mutation responsible for its gain of function. J Biol Chem 267(32):23113–23121.
3. Garcia CK, Goldstein JL, Pathak RK, Anderson RG, Brown MS. 1994. Molecular characterization of a membrane transporter for lactate, pyruvate, and other monocarboxylates: implications for the Cori cycle. Cell 76(5):865–873.
4. Enerson BE, Drewes LR. 2003. Molecular features, regulation, and function of monocarboxylate transporters: implications for drug delivery. J Pharm Sci 92(8):1531–1544.
5. Friesema EC, Ganguly S, Abdalla A, Manning Fox JE, Halestrap AP, Visser TJ. 2003. Identification of monocarboxylate transporter 8 as a specific thyroid hormone transporter. J Biol Chem 278(41):40128–40135.
6. Murakami Y, Kohyama N, Kobayashi Y, Ohbayashi M, Ohtani H, Sawada Y, Yamamoto T. 2005. Functional characterization of human monocarboxylate transporter 6 (SLC16A5). Drug Metab Dispos 33(12):1845–1851.
7. De Bruijne AW, Vreeburg H, Van Steveninck J. 1983. Kinetic analysis of L-lactate transport in human erythrocytes via the monocarboxylate-specific carrier system. Biochim Biophys Acta 732(3):562–568.
8. Rahman B, Schneider HP, Bröer A, Deitmer JW, Bröer S. 1999. Helix 8 and helix 10 are involved in substrate recognition in the rat monocarboxylate transporter MCT1. Biochemistry 38(35):11577–11584.
9. Galic S, Schneider HP, Bröer A, Deitmer JW, Bröer S. 2003. The loop between helix 4 and helix 5 in the monocarboxylate transporter MCT1 is important for substrate selection and protein stability. Biochem J 376(Pt 2):413–422.
10. Miyauchi S, Gopal E, Fei YJ, Ganapathy V. 2004. Functional identification of SLC5A8, a tumor suppressor down-regulated in colon cancer, as a Na(+)-coupled transporter for short-chain fatty acids. J Biol Chem 279(14):13293–13296.
11. Coady MJ, Chang MH, Charron FM, Plata C, Wallendorff B, Sah JF, Markowitz SD, Romero MF, Lapointe JY. 2004. The human tumour suppressor gene SLC5A8 expresses a Na^+-monocarboxylate cotransporter. J Physiol 557(Pt 3):719–731.
12. Srinivas SR, Gopal E, Zhuang L, Itagaki S, Martin PM, Fei YJ, Ganapathy V, Prasad PD. 2005. Cloning and functional identification of slc5a12 as a sodium-coupled low-affinity transporter for monocarboxylates (SMCT2). Biochem J 392(Pt 3):655–664.
13. Wright EM, Turk E. 2004. The sodium/glucose cotransport family SLC5. Pflugers Arch 447(5):510–518.
14. Halestrap AP, Price NT. 1999. The proton-linked monocarboxylate transporter (MCT) family: structure, function and regulation. Biochem J 343(Pt 2):281–299.
15. Price NT, Jackson VN, Halestrap AP. 1998. Cloning and sequencing of four new mammalian monocarboxylate transporter (MCT) homologues confirms the existence of a transporter family with an ancient past. Biochem J 329(Pt 2):321–328.
16. 16. Pellerin L, Magistretti PJ. 1994. Glutamate uptake into astrocytes stimulates aerobic glycolysis: a mechanism coupling neuronal activity to glucose utilization. Proc Natl Acad Sci U S A 91(22):10625–10629.
17. Gladden LB. 2004. Lactate metabolism: a new paradigm for the third millennium. J Physiol 558(Pt 1):5–30.

18. Philp NJ, Yoon H, Grollman EF. 1998. Monocarboxylate transporter MCT1 is located in the apical membrane and MCT3 in the basal membrane of rat RPE. Am J Physiol 274(6 Pt 2):R1824–R1828.

19. Bonen A. 2001. The expression of lactate transporters (MCT1 and MCT4) in heart and muscle. Eur J Appl Physiol 86(1):6–11.

20. Pilegaard H, Domino K, Noland T, Juel C, Hellsten Y, Halestrap AP, Bangsbo J. 1999. Effect of high-intensity exercise training on lactate/H^+ transport capacity in human skeletal muscle. Am J Physiol 276(2 Pt 1):E255–E261.

21. Bröer S, Schneider HP, Bröer A, Rahman B, Hamprecht B, Deitmer JW. 1998. Characterization of the monocarboxylate transporter 1 expressed in *Xenopus laevis* oocytes by changes in cytosolic pH. Biochem J 333(Pt 1):167–174.

22. Dimmer KS, Friedrich B, Lang F, Deitmer JW, Bröer S. 2000. The low-affinity monocarboxylate transporter MCT4 is adapted to the export of lactate in highly glycolytic cells. Biochem J 350(Pt 1):219–227.

23. Roediger WE. 1980. Role of anaerobic bacteria in the metabolic welfare of the colonic mucosa in man. Gut 21(9):793–798.

24. Clausen MR, Mortensen PB. 1994. Kinetic studies on the metabolism of short-chain fatty acids and glucose by isolated rat colonocytes. Gastroenterology 106(2):423–432.

25. Gill RK, Saksena S, Alrefai WA, Sarwar Z, Goldstein JL, Carroll RE, Ramaswamy K, Dudeja PK. 2005. Expression and membrane localization of MCT isoforms along the length of the human intestine. Am J Physiol Cell Physiol 289(4):C846–C852.

26. Treem WR, Ahsan N, Shoup M, Hyams JS. 1994. Fecal short-chain fatty acids in children with inflammatory bowel disease. J Pediatr Gastroenterol Nutr 18(2):159–164.

27. Gopal E, Fei YJ, Sugawara M, Miyauchi S, Zhuang L, Martin P, Smith SB, Prasad PD, Ganapathy V. 2004. Expression of slc5a8 in kidney and its role in Na(+)-coupled transport of lactate. J Biol Chem 279(43):44522–44532.

28. Craig F. 1946. Renal tubular reabsorption, metabolic utilization, and isomeric fractionation of lactic acid in the dog. Am J Physiol 146:146–159.

29. Leino RL, Gerhart DZ, Drewes LR. 1999. Monocarboxylate transporter (MCT1) abundance in brains of suckling and adult rats: a quantitative electron microscopic immunogold study. Brain Res Dev Brain Res 113(1–2):47–54.

30. Leino RL, Gerhart DZ, Duelli R, Enerson BE, Drewes LR. 2001. Diet-induced ketosis increases monocarboxylate transporter (MCT1) levels in rat brain. Neurochem Int 38(6):519–527.

31. Cuff MA, Lambert DW, Shirazi-Beechey SP. 2002. Substrate-induced regulation of the human colonic monocarboxylate transporter, MCT1. J Physiol 539(Pt 2):361–371.

32. Cuff MA, Shirazi-Beechey SP. 2002. The human monocarboxylate transporter, MCT1: genomic organization and promoter analysis. Biochem Biophys Res Commun 292(4):1048–1056.

33. Hadjiagapiou C, Borthakur A, Dahdal RY, Gill RK, Malakooti J, Ramaswamy K, Dudeja PK. 2005. Role of USF1 and USF2 as potential repressor proteins for human intestinal monocarboxylate transporter 1 promoter. Am J Physiol Gastrointest Liver Physiol 288(6):G1118–G1126.

34. Zhang F, Vannucci SJ, Philp NJ, Simpson IA. 2005. Monocarboxylate transporter expression in the spontaneous hypertensive rat: effect of stroke. J Neurosci Res 79(1–2):139–145.

35. Ullah MS, Davies AJ, Halestrap AP. 2006. The plasma membrane lactate transporter MCT4, but not MCT1, is up-regulated by hypoxia through a HIF-1alpha dependent mechanism. J Biol Chem 281(14):9030–9037.

36. Fanelli A, Grollman EF, Wang D, Philp NJ. 2003. MCT1 and its accessory protein CD147 are differentially regulated by TSH in rat thyroid cells. Am J Physiol Endocrinol Metab 285(6):E1223–E1229.

37. Buyse M, Sitaraman SV, Liu X, Bado A, Merlin D. 2002. Luminal leptin enhances CD147/MCT-1-mediated uptake of butyrate in the human intestinal cell line Caco2-BBE. J Biol Chem 277(31):28182–28190.

38. Alrefai WA, Tyagi S, Gill R, Saksena S, Hadjiagapiou C, Mansour F, Ramaswamy K, Dudeja PK. 2004. Regulation of butyrate uptake in Caco-2 cells by phorbol 12-myristate 13-acetate. Am J Physiol Gastrointest Liver Physiol 286(2):G197–203.

39. Kafitz KW, Rose CR, Thoenen H, Konnerth A. 1999. Neurotrophin-evoked rapid excitation through TrkB receptors. Nature 401(6756):918–921.

40. McLatchie LM, Fraser NJ, Main MJ, Wise A, Brown J, Thompson N, Solari R, Lee MG, Foord SM. 1998. RAMPs regulate the transport and ligand specificity of the calcitonin-receptor-like receptor. Nature 393(6683):333–339.

41. Palacin M, Kanai Y. 2004. The ancillary proteins of HATs: SLC3 family of amino acid transporters. Pflugers Arch 447(5):490–494.

42. Kirk P, Wilson MC, Heddle C, Brown MH, Barclay AN, Halestrap AP. 2000. CD147 is tightly associated with lactate transporters MCT1 and MCT4 and facilitates their cell surface expression. EMBO J 19(15):3896–3904.

43. Philp NJ, Ochrietor JD, Rudoy C, Muramatsu T, Linser PJ. 2003. Loss of MCT1, MCT3, and MCT4 expression in the retinal pigment epithelium and neural retina of the 5A11/basigin-null mouse. Invest Ophthalmol Vis Sci 44(3):1305–1311.

44. Wilson MC, Meredith D, Fox JE, Manoharan C, Davies AJ, Halestrap AP. 2005. Basigin (CD147) is the target for organomercurial inhibition of monocarboxylate transporter isoforms 1 and 4: the ancillary protein for the insensitive MCT2 is EMBIGIN (gp70). J Biol Chem 280(29):27213–27221.

45. Poole RC, Halestrap AP. 1997. Interaction of the erythrocyte lactate transporter (monocarboxylate transporter 1) with an integral 70-kDa membrane glycoprotein of the immunoglobulin superfamily. J Biol Chem 272(23):14624–14628.

46. Wilson MC, Meredith D, Halestrap AP. 2002. Fluorescence resonance energy transfer studies on the interaction between the lactate transporter MCT1 and CD147 provide information on the topology and stoichiometry of the complex in situ. J Biol Chem 277(5):3666–3672.

47. Deora AA, Philp N, Hu J, Bok D, Rodriguez-Boulan E. 2005. Mechanisms regulating tissue-specific polarity of monocarboxylate transporters and their chaperone CD147 in kidney and retinal epithelia. Proc Natl Acad Sci U S A 102(45):16245–16250.

48. Deora AA, Gravotta D, Kreitzer G, Hu J, Bok D, Rodriguez-Boulan E. 2004. The basolateral targeting signal of CD147 (EMMPRIN) consists of a single leucine and is not recognized by retinal pigment epithelium. Mol Biol Cell 15(9):4148–4165.

49. Smith JP, Drewes LR. 2006. Modulation of monocarboxylic acid transporter-1 kinetic function by the cAMP signaling pathway in rat brain endothelial cells. J Biol Chem 281(4):2053–2060.

50. Becker HM, Hirnet D, Fecher-Trost C, Sultemeyer D, Deitmer JW. 2005. Transport activity of MCT1 expressed in *Xenopus* oocytes is increased by interaction with carbonic anhydrase. J Biol Chem 280(48):39882–39889.

51. Xu D, Hemler ME. 2005. Metabolic activation-related CD147–CD98 complex. Mol Cell Proteom 4(8):1061–1071.

52. Lin RY, Vera JC, Chaganti RS, Golde DW. 1998. Human monocarboxylate transporter 2 (MCT2) is a high affinity pyruvate transporter. J Biol Chem 273(44):28959–28965.

53. Bröer S, Bröer A, Schneider HP, Stegen C, Halestrap AP, Deitmer JW. 1999. Characterization of the high-affinity monocarboxylate transporter MCT2 in *Xenopus laevis* oocytes. Biochem J 341(Pt 3):529–535.

54. Grollman EF, Philp NJ, McPhie P, Ward RD, Sauer B. 2000. Determination of transport kinetics of chick MCT3 monocarboxylate transporter from retinal pigment epithelium by expression in genetically modified yeast. Biochemistry 39(31):9351–9357.

55. Manning Fox JE, Meredith D, Halestrap AP. 2000. Characterization of human monocarboxylate transporter 4 substantiates its role in lactic acid efflux from skeletal muscle. J Physiol 529(Pt 2):285–293.

56. Gopal E, Fei YJ, Miyauchi S, Zhuang L, Prasad PD, Ganapathy V. 2005. Sodium-coupled and electrogenic transport of B-complex vitamin nicotinic acid by slc5a8, a member of the Na/glucose co-transporter gene family. Biochem J 388(Pt 1):309–316.

57. Takanaga H, Maeda H, Yabuuchi H, Tamai I, Higashida H, Tsuji A. 1996. Nicotinic acid transport mediated by pH-dependent anion antiporter and proton cotransporter in rabbit intestinal brush-border membrane. J Pharm Pharmacol 48(10):1073–1077.

58. Rodriguez AM, Perron B, Lacroix L, Caillou B, Leblanc G, Schlumberger M, Bidart JM, Pourcher T. 2002. Identification and characterization of a putative human iodide transporter located at the apical membrane of thyrocytes. J Clin Endocrinol Metab 87(7):3500–3503.

59. Gupta N, Martin PM, Prasad PD, Ganapathy V. 2005. SLC5A8 (SMCT1)-mediated transport of butyrate forms the basis for the tumor suppressive function of the transporter. Life Sci 78(21):2419–2425.

60. Oldendorf WH. 1973. Carrier-mediated blood–brain barrier transport of short-chain monocarboxylic organic acids. Am J Physiol 224(6):1450–1453.

61. Kang YS, Terasaki T, Tsuji A. 1990. Acidic drug transport in vivo through the blood–brain barrier. A role of the transport carrier for monocarboxylic acids. J Pharmacobiodyn 13(2):158–163.

62. Tsuji A, Takanaga H, Tamai I, Terasaki T. 1994. Transcellular transport of benzoic acid across Caco-2 cells by a pH-dependent and carrier-mediated transport mechanism. Pharm Res 11(1):30–37.

63. Kido Y, Tamai I, Okamoto M, Suzuki F, Tsuji A. 2000. Functional clarification of MCT1-mediated transport of monocarboxylic acids at the blood–brain barrier using in vitro cultured cells and in vivo BUI studies. Pharm Res 17(1):55–62.

64. Tamai I, Sai Y, Ono A, Kido Y, Yabuuchi H, Takanaga H, Satoh E, Ogihara T, Amano O, Izeki S, Tsuji A. 1999. Immunohistochemical and functional characterization of pH-dependent intestinal absorption of weak organic acids by the monocarboxylic acid transporter MCT1. J Pharm Pharmacol 51(10):1113–1121.

65. Terasaki T, Takakuwa S, Moritani S, Tsuji A. 1991. Transport of monocarboxylic acids at the blood–brain barrier: studies with monolayers of primary cultured bovine brain capillary endothelial cells. J Pharmacol Exp Ther 258(3):932–937.

66. Emoto A, Ushigome F, Koyabu N, Kajiya H, Okabe K, Satoh S, Tsukimori K, Nakano H, Ohtani H, Sawada Y. 2002. H(+)-linked transport of salicylic acid, an NSAID, in the human trophoblast cell line BeWo. Am J Physiol Cell Physiol 282(5):C1064–C1075.

67. Itoh TTM, Li Y-H, Yamada H. 1998. Transport of phenethicillin into rat intestinal brush border membrane vesicles: role of the monocarboxylic acid transporter system. Int J Pharm 172:103–112.

68. Tsuji A, Tamai I, Nakanishi M, Terasaki T, Hamano S. 1993. Intestinal brush-border transport of the oral cephalosporin antibiotic, cefdinir, mediated by dipeptide and monocarboxylic acid transport systems in rabbits. J Pharm Pharmacol 45(11):996–998.

69. Wu X, Whitfield LR, Stewart BH. 2000. Atorvastatin transport in the Caco-2 cell model: contributions of P-glycoprotein and the proton–monocarboxylic acid co-transporter. Pharm Res 17(2):209–215.

70. Tsuji A, Saheki A, Tamai I, Terasaki T. 1993. Transport mechanism of 3-hydroxy-3-methylglutaryl coenzyme A reductase inhibitors at the blood–brain barrier. J Pharmacol Exp Ther 267(3):1085–1090.

71. Saheki A, Terasaki T, Tamai I, Tsuji A. 1994. In vivo and in vitro blood–brain barrier transport of 3-hydroxy-3-methylglutaryl coenzyme A (HMG-CoA) reductase inhibitors. Pharm Res 11(2):305–311.

72. Utoguchi N, Audus KL. 2000. Carrier-mediated transport of valproic acid in BeWo cells, a human trophoblast cell line. Int J Pharm 195(1–2):115–124.

73. Cundy KC, Annamalai T, Bu L, De Vera J, Estrela J, Luo W, Shirsat P, Torneros A, Yao F, Zou J, Barrett RW, Gallop MA. 2004. XP13512 [(+/−)-1-([(alpha-isobutanoyloxyethoxy)carbonyl] aminomethyl)-1-cyclohexane acetic acid], a novel gabapentin prodrug: II. Improved oral bioavailability, dose proportionality, and colonic absorption compared with gabapentin in rats and monkeys. J Pharmacol Exp Ther 311(1): 324–333.

74. Okamura A, Emoto A, Koyabu N, Ohtani H, Sawada Y. 2002. Transport and uptake of nateglinide in Caco-2 cells and its inhibitory effect on human monocarboxylate transporter MCT1. Br J Pharmacol 137(3):391–399.

75. Deguchi Y, Yokoyama Y, Sakamoto T, Hayashi H, Naito T, Yamada S, Kimura R. 2000. Brain distribution of 6-mercaptopurine is regulated by the efflux transport system in the blood–brain barrier. Life Sci 66(7):649–662.

76. Konishi Y. 2005. Transepithelial transport of artepillin C in intestinal Caco-2 cell monolayers. Biochim Biophys Acta 1713(2):138–144.

77. Konishi Y. 2005. Transepithelial transport of microbial metabolites of quercetin in intestinal Caco-2 cell monolayers. J Agric Food Chem 53(3):601–607.

78. Konishi Y, Kobayashi S. 2004. Transepithelial transport of chlorogenic acid, caffeic acid, and their colonic metabolites in intestinal Caco-2 cell monolayers. J Agric Food Chem 52(9):2518–2526.

79. Konishi Y, Kobayashi S. 2004. Microbial metabolites of ingested caffeic acid are absorbed by the monocarboxylic acid transporter (MCT) in intestinal Caco-2 cell monolayers. J Agric Food Chem 52(21):6418–6424.

80. Goddard I, Florin A, Mauduit C, Tabone E, Contard P, Bars R, Chuzel F, Benahmed M.

2003. Alteration of lactate production and transport in the adult rat testis exposed in utero to flutamide. Mol Cell Endocrinol 206(1–2):137–146.

81. Kitaura T, Tsunekawa N, Hatta H. 2001. Decreased monocarboxylate transporter 1 in rat soleus and EDL muscles exposed to clenbuterol. J Appl Physiol 91(1):85–90.
82. Halestrap AP, Denton RM. 1974. Specific inhibition of pyruvate transport in rat liver mitochondria and human erythrocytes by alpha-cyano-4-hydroxycinnamate. Biochem J 138(2):313–316.
83. Murray CM, Hutchinson R, Bantick JR, Belfield GP, Benjamin AD, Brazma D, Bundick RV, Cook ID, Craggs RI, Edwards S, et al. 2005. Monocarboxylate transporter MCT1 is a target for immunosuppression. Nat Chem Biol 1(7):371–376.
84. Michne WF, Schroeder JD, Guiles JW, Treasurywala AM, Weigelt CA, Stansberry MF, McAvoy E, Shah CR, Baine Y, Sawutz DG, et al. 1995. Novel inhibitors of the nuclear factor of activated T cells (NFAT)–mediated transcription of beta-galactosidase: potential immunosuppressive and antiinflammatory agents. J Med Chem 38(14):2557–2569.
85. Nagai N, Nakai A, Nagata K. 1995. Quercetin suppresses heat shock response by down regulation of HSF1. Biochem Biophys Res Commun 208(3):1099–1105.
86. Deuticke B, Lutkemeier P, Poser B. 1991. Influence of phloretin and alcohols on barrier defects in the erythrocyte membrane caused by oxidative injury and electroporation. Biochim Biophys Acta 1067(2):111–122.
87. Poole RC, Halestrap AP. 1991. Reversible and irreversible inhibition, by stilbenedisulphonates, of lactate transport into rat erythrocytes: identification of some new high-affinity inhibitors. Biochem J 275(Pt 2):307–312.
88. Makuc J, Cappellaro C, Boles E. 2004. Co-expression of a mammalian accessory trafficking protein enables functional expression of the rat MCT1 monocarboxylate transporter in *Saccharomyces cerevisiae*. FEMS Yeast Res 4(8):795–801.
89. Izumi H, Torigoe T, Ishiguchi H, Uramoto H, Yoshida Y, Tanabe M, Ise T, Murakami T, Yoshida T, Nomoto M, Kohno K. 2003. Cellular pH regulators: potentially promising molecular targets for cancer chemotherapy. Cancer Treat Rev 29(6):541–549.
90. Toole BP. 2003. Emmprin (CD147), a cell surface regulator of matrix metalloproteinase production and function. Curr Top Dev Biol 54:371–389.
91. Wahl ML, Owen JA, Burd R, Herlands RA, Nogami SS, Rodeck U, Berd D, Leeper DB, Owen CS. 2002. Regulation of intracellular pH in human melanoma: potential therapeutic implications. Mol Cancer Ther 1(8):617–628.
92. Froberg MK, Gerhart DZ, Enerson BE, Manivel C, Guzman-Paz M, Seacotte N, Drewes LR. 2001. Expression of monocarboxylate transporter MCT1 in normal and neoplastic human CNS tissues. Neuroreport 12(4):761–765.
93. Daly K, Cuff MA, Fung F, Shirazi-Beechey SP. 2005. The importance of colonic butyrate transport to the regulation of genes associated with colonic tissue homoeostasis. Biochem Soc Trans 33(Pt 4):733–735.
94. Singh B, Halestrap AP, Paraskeva C. 1997. Butyrate can act as a stimulator of growth or inducer of apoptosis in human colonic epithelial cell lines depending on the presence of alternative energy sources. Carcinogenesis 18(6):1265–1270.
95. Lambert DW, Wood IS, Ellis A, Shirazi-Beechey SP. 2002. Molecular changes in the expression of human colonic nutrient transporters during the transition from normality to malignancy. Br J Cancer 86(8):1262–1269.

96. Li H, Myeroff L, Smiraglia D, Romero MF, Pretlow TP, Kasturi L, Lutterbaugh J, Rerko RM, Casey G, Issa JP, et al. 2003. SLC5A8, a sodium transporter, is a tumor suppressor gene silenced by methylation in human colon aberrant crypt foci and cancers. Proc Natl Acad Sci U S A 100(14):8412–8417.
97. Siesjo BK. 1978. *Brain Energy Metabolism.* New York: Wiley, pp. 398–526.
98. De Vivo DC. 2001. Cerebral energy failure. Curr Neurol Neurosci Rep 1(3):203–204.
99. Drewes LR, Gilboe DD. 1973. Glycolysis and the permeation of glucose and lactate in the isolated, perfused dog brain during anoxia and postanoxic recovery. J Biol Chem 248(7):2489–2496.
100. Kalimo H, Rehncrona S, Soderfeldt B, Olsson Y, Siesjo BK. 1981. Brain lactic acidosis and ischemic cell damage: 2. Histopathology. J Cereb Blood Flow Metab 1(3):313–327.
101. Pulsinelli WA, Waldman S, Rawlinson D, Plum F. 1982. Moderate hyperglycemia augments ischemic brain damage: a neuropathologic study in the rat. Neurology 32(11):1239–1246.
102. Marie C, Bralet AM, Gueldry S, Bralet J. 1990. Fasting prior to transient cerebral ischemia reduces delayed neuronal necrosis. Metab Brain Dis 5(2):65–75.
103. Vannucci RC, Brucklacher RM, Vannucci SJ. 1996. The effect of hyperglycemia on cerebral metabolism during hypoxia–ischemia in the immature rat. J Cereb Blood Flow Metab 16(5):1026–1033.
104. Go KG, Prenen GH, Korf J. 1988. Protective effect of fasting upon cerebral hypoxic–ischemic injury. Metab Brain Dis 3(4):257–263.
105. Gjedde A, Crone C. 1975. Induction processes in blood–brain transfer of ketone bodies during starvation. Am J Physiol 229(5):1165–1169.
106. Reaven GM, Hollenbeck C, Jeng CY, Wu MS, Chen YD. 1988. Measurement of plasma glucose, free fatty acid, lactate, and insulin for 24 h in patients with NIDDM. Diabetes 37(8):1020–1024.
107. Ohlson LO, Larsson B, Bjorntorp P, Eriksson H, Svardsudd K, Welin L, Tibblin G, Wilhelmsen L. 1988. Risk factors for type 2 (non-insulin-dependent) diabetes mellitus. Thirteen and one-half years of follow-up of the participants in a study of Swedish men born in 1913. Diabetologia 31(11):798–805.
108. Juel C, Holten MK, Dela F. 2004. Effects of strength training on muscle lactate release and MCT1 and MCT4 content in healthy and type 2 diabetic humans. J Physiol 556(Pt 1):297–304.
109. Enoki T, Yoshida Y, Hatta H, Bonen A. 2003. Exercise training alleviates MCT1 and MCT4 reductions in heart and skeletal muscles of STZ-induced diabetic rats. J Appl Physiol 94(6):2433–2438.
110. Hajduch E, Heyes RR, Watt PW, Hundal HS. 2000. Lactate transport in rat adipocytes: identification of monocarboxylate transporter 1 (MCT1) and its modulation during streptozotocin-induced diabetes. FEBS Lett 479(3):89–92.
111. Abramson J, Smirnova I, Kasho V, Verner G, Kaback HR, Iwata S. 2003. Structure and mechanism of the lactose permease of *Escherichia coli*. Science 301(5633):610–615.
112. Lemieux MJ, Song J, Kim MJ, Huang Y, Villa A, Auer M, Li XD, Wang DN. 2003. Three-dimensional crystallization of the *Escherichia coli* glycerol-3-phosphate transporter: a member of the major facilitator superfamily. Protein Sci 12(12):2748–2756.
113. Garcia CK, Brown MS, Pathak RK, Goldstein JL. 1995. cDNA cloning of MCT2, a second monocarboxylate transporter expressed in different cells than MCT1. J Biol Chem 270(4):1843–1849.

8

NUCLEOSIDE TRANSPORTERS: CNTs AND ENTs

LI XIA, MINGYAN ZHOU, AND JOANNE WANG
University of Washington, Seattle, Washington

8.1. INTRODUCTION

Endogenous nucleosides (Figure 8.1) are important precursors of nucleic acid synthesis, a process fundamental to the growth and metabolism in all living systems. The purine nucleoside adenosine is also a key signaling molecule that exerts profound

Drug Transporters: Molecular Characterization and Role in Drug Disposition, Edited by Guofeng You and Marilyn E. Morris
Copyright © 2007 John Wiley & Sons, Inc.

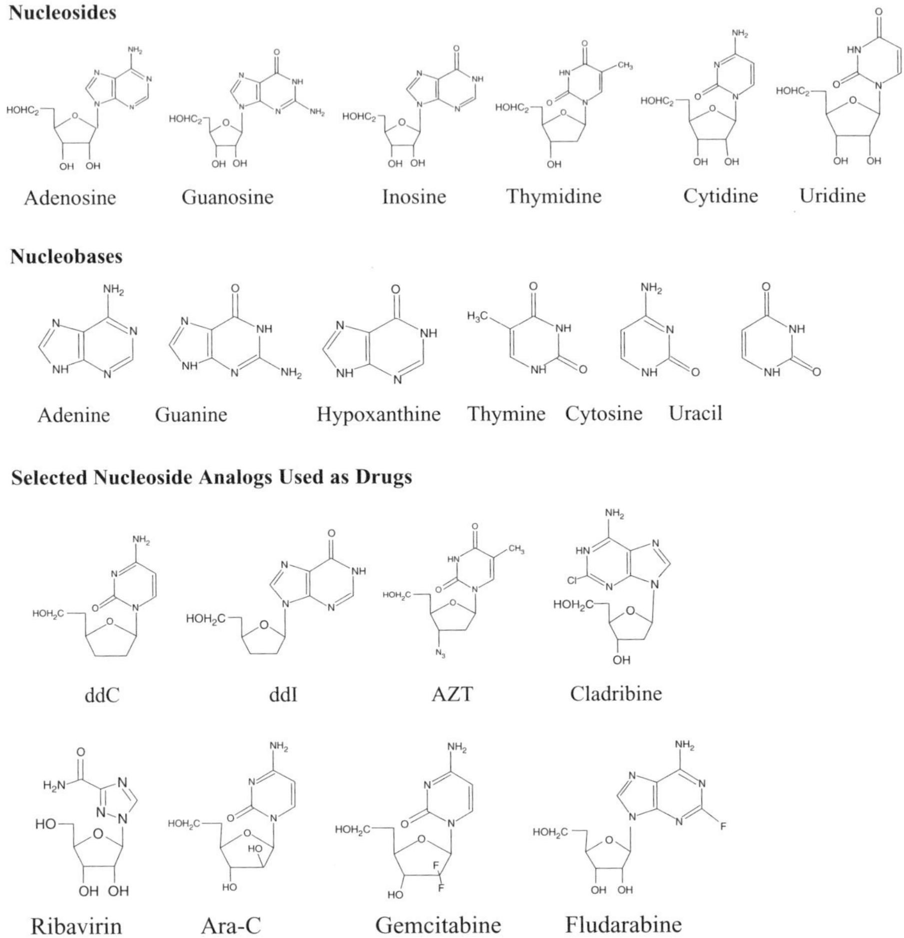

FIGURE 8.1. Structures of selected nucleosides and nucleoside analogs.

effects on many tissues and organs by activating specific purinergic receptors.[1,2] Nucleosides are hydrophilic and have low membrane permeability. To facilitate the movement of nucleosides across cell membranes, mammalian cells have evolved two major classes of transporters: concentrative nucleoside transporters (CNTs) and equilibrative nucleoside transporters (ENTs). At the cellular level, these transporters play key roles in salvaging nucleosides for DNA and RNA syntheses and in regulating adenosine signaling at the receptor sites. From a whole-organism point of view, nucleoside transporters expressed in absorptive and excretory organs are important for maintaining total body homeostasis of nucleosides. The pharmacological significance of nucleoside transporters originates from the wide use of nucleoside analogs in the treatment of cancer, viral infections, and other pathophysiological conditions. The structures of a few clinically used nucleoside analogs are shown in Figure 8.1. Many of the therapeutic nucleoside analogs rely on nucleoside transporters to enter or exit cells. Consequently, the expression and functional characteristics of these transporters in target or nontarget cells, as well as in absorptive and excretory organs, will have an important impact on the efficacy and toxicity of therapeutic nucleoside analogs.

Historically, nucleoside transport activities were first studied by functional and kinetic analyses in mammalian cells and tissue preparations.[1] These early studies established that there exist multiple nucleoside transporters, which can be classified into two major systems: equilibrative and concentrative.[1] The *equilibrative system* consists of some low-affinity facilitated carrier proteins that are further divided into two subtypes, *es* and *ei*, based on their sensitivity to a highly specific inhibitor, nitrobenzylthioinosine [nitrobenzylmercaptopurine riboside (NBMPR)]. The *es* (equilibrative inhibitor sensitive) subtype is potently inhibited by NBMPR (K_i, 0.1 to 10 nM), whereas the *ei* (equilibrative inhibitor insensitive) subtype is not affected by NBMPR at concentrations below 1 μM.[1] The *es* and *ei* transport both purine and pyrimidine nucleosides and are widely distributed in different tissues and cell types. The *concentrative system* consists of some high-affinity Na^+-dependent transporters that mediate uphill transport of nucleosides by coupling to the physiologic Na^+ gradient. Several subtypes with distinct substrate specificity were recognized, with three of them (*cit, cif, cib*) well characterized.[1] The *cit* (N2) subtype is pyrimidine selective but also transports adenosine. The *cif* (N1) subtype is mainly purine selective but also transports uridine. The *cib* (N3) subtype is broadly selective, transporting both purine and pyrimidine nucleosides. The Na^+-dependent transporters appear to have relatively limited tissue distributions and are typically found in epithelial tissues such as the intestine, kidney, liver, and choroid plexus.[1]

Molecular cloning studies in the past decade led to the isolation of genes encoding individual membrane proteins underlying the foregoing nucleoside transport activities in human and other mammalian species.[3–5] They belong to two structurally unrelated solute carrier gene families, SLC28 and SLC29. The SLC28 family consists of three members, CNT1 (SLC28A1), CNT2 (SLC28A2), and CNT3 (SLC28A3), which correspond to the previously characterized *cit*, *cif*, and *cib* subtypes respectively. The SLC29 family consists of four members, ENT1 (SLC29A1), ENT2 (SLC29A2), ENT3 (SLC29A3), and ENT4 (SLC29A4). ENT1 underlies the *es* activity, whereas ENT2 is responsible for the *ei* activity. ENT3 and ENT4 are novel members identified from the human genome project. ENT3 is an intracellular low-affinity nucleoside transporter,[6] and ENT4 exhibits novel substrate specificity by functioning as a polyspecific organic cation transporter.[7,8] An overview of individual CNT and ENT family members is presented in Table 8.1.

8.2. CONCENTRATIVE NUCLEOSIDE TRANSPORTERS (SLC28)

8.2.1. Family Members

CNT1 (SLC28A1) The first cloned Na^+-dependent nucleoside transporter, rCNT1, was isolated from a rat jejunal cDNA library in 1994 by expression cloning in *Xenopus laevis* oocytes.[9] Its human homolog, hCNT1, was subsequently cloned from human kidney.[10] The *hCNT1* gene is localized to chromosome 15q25–26. hCNT1 (649 a.a.) and rCNT1 (648 a.a.) are 85% identical in protein sequence, and are predicted to

TABLE 8.1. Molecular and Functional Characteristics of Human Nucleoside Transporters

Transporter	Molecular Features	Substrates	Transport Mode	Cellular Location	Tissue Distribution
CNT family					
hCNT1	Gene name: *SLC28A1* Gene locus: 15q25–26 Protein length: 649 a.a. TMs predicted: 13 TMs	Nucleosides: uridine, thymidine, cytidine, adenosine Nucleoside analogs: gemcitabine, cytarabine, 5′-deoxy-5-fluorouridine, 2′-deoxycytidine, ddC, 3TC, AZT	Concentrative Na^+–nucleoside cotransporter 1 Na^+ : 1 nucleoside	Plasma membrane Apical membrane	Liver, kidney, intestine, brain
hCNT2 (hSPNT1)	Gene name: *SLC28A2* Gene locus: 15q15 Protein length: 658 a.a. TMs predicted: 13 TMs	Nucleosides: adenosine, guanosine, inosine, uridine Nucleoside analogs: ribavirin clofarabine 5-fluorouridine	Concentrative Na^+–nucleoside cotransporter 1 Na^+ : 1 nucleoside	Plasma membrane Apical localization in kidney and intestine cells	Kidney, heart, liver, skeletal muscle, pancreas, placenta, brain, cervix, prostate, small intestine, rectum, colon, lung
hCNT3	Gene name: *SLC28A3* Gene locus: 9q22.2 Protein length: 691 a.a. TMs predicted: 13 TMs	Nucleosides: adenosine, guanosine, inosine, uridine, thymidine, cytidine Nucleoside analogs: 5-fluorouridine, 5-fluoro-2′-deoxyuridine, clofarabine, fludarabine, zebularine, cladribine, gemcitabine, AZT, ddC, ddI, ribavirin, 3-deazauridien	Concentrative Na^+–nucleoside cotransporter 2 Na^+ : 1 nucleoside	Plasma membrane	Mammary gland, pancreas, bone marrow, trachea, intestine, liver, lung, placenta, prostate, testis, brain, heart

ENT family					
hENT1	Gene name: *SLC29A1* Gene locus: 6p21.1–p21.2 Protein length: 456 a.a. TMs predicted: 11 TMs	Nucleosides: adenosine, guanosine, inosine, uridine, thymidine, cytidine Nucleoside analogs: cladribine, gemcitabine, fludarabine, cytarabine, ribavirin, clofarabine	Facilitated carrier	Plasma membrane, basolateral localization in MDCK Some expression also found in mitochondria	Ubiquitous
hENT2	Gene name: *SLC29A2* Gene locus: 11q13 Protein length: 456 a.a. TMs predicted: 11 TMs	Nucleosides: adenosine, inosine, uridine, thymidine, guanosine, cytidine Nucleobase: adenine, hypoxanthine, guanine, uracil, thymine, cytosine Nucleoside analogs: gemcitabine, clofarabine AZT, ddI, ddC	Facilitated carrier	Plasma membrane Basolateral localization in MDCK	Widely distributed (skeletal muscle, heart, pancreas, brain, kidney, placenta, prostate, small intestine, lung, thymus, etc.)

(Continued)

TABLE 8.1. ***(Continued)***

Transporter	Molecular Features	Substrates	Transport Mode	Cellular Location	Tissue Distribution
ENT family (*Continued*)					
hENT3	Gene name: *SLC29A3* Gene locus: 10q22.1 Protein length: 475 a.a. TMs predicted: 11 TMs	Nucleosides: adenosine, inosine, guanosine, thymidine, uridine Nucleobase: adenine Nucleoside analogs: cladribine, cordycepin, tubercidine, fludarabine, zebularine, AZT, ddC, ddI	Not determined pH-sensitive	Intracellular, partially colocalization with late endosomes/ lysosomes	Widely distributed (placenta, uterus, ovary, spleen, lymph node, bone marrow, brain, heart)
hENT4 (PMAT)	Gene name: *SLC29A4* Gene locus: 7p22.1 Protein length: 530 a.a. TMs predicted: 11 TMs	Biogenic amines: serotonin, dopamine, norepinephrine, histamine, epinephrine, tyramine Nucleosides: adenosine Other organic cations: MPP^+, TEA	Na^+-independent electrogenic	Plasma membrane	Brain, skeletal muscle, kidney, heart, liver

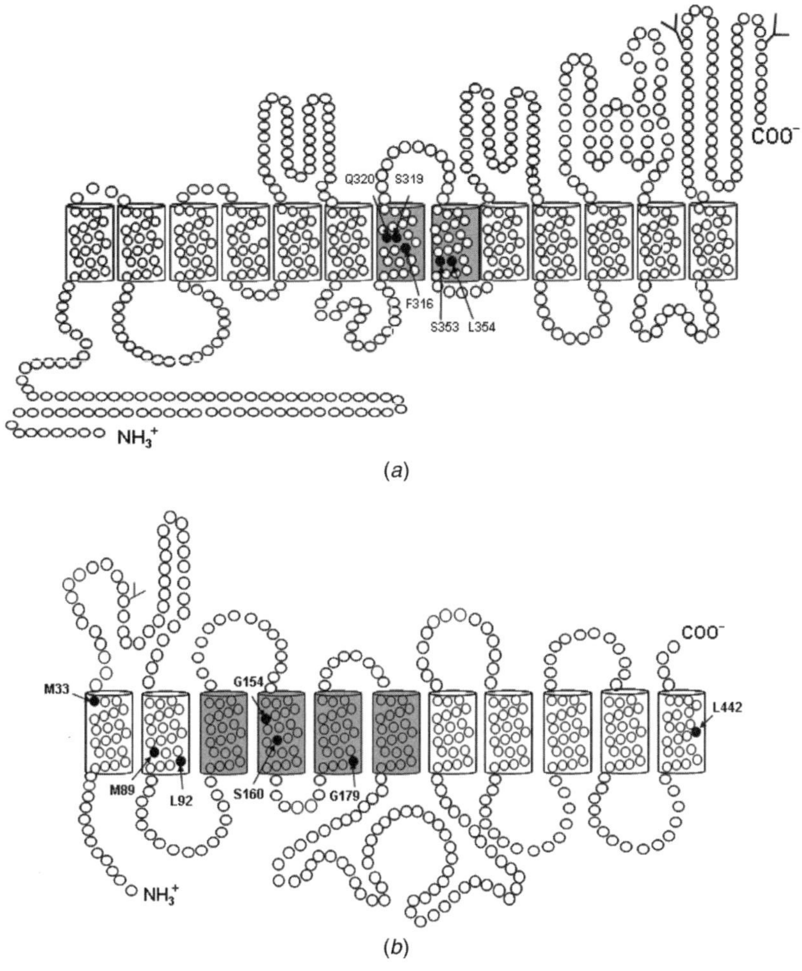

FIGURE 8.2. Secondary structures predicted for hCNT1 (*a*) and hENT1 (*b*). The N-glycosylation sites are indicated as "Y." Transmembrane domains important for substrate–inhibitor interactions are shaded in gray. Residues shown as solid circles have been implicated in interaction with substrates or inhibitors.

possess 13 transmembrane (TM) α-helices (Figure 8.2*a*). A CNT1 ortholog has also been cloned from pig.[11]

In humans, hCNT1 mRNA is expressed predominantly in the liver and kidney and exhibits much lower expression in other tissues.[12] There appear to be large interindividual differences of hCNT1 mRNA expression level in the kidney.[12] In rat, rCNT1 mRNA was found in the jejunum, kidney, liver, and various regions of the brain.[13,14] Western blot further demonstrated the expression of rCNT1 protein in the small intestine, kidney, and liver.[15] Fluorescence protein-tagged constructs of hCNT1 and rCNT1 were expressed on the apical membranes of polarized LLC-PK1 and MDCK cells.[16,17] In rat tissues, immunohistochemistry studies showed that rCNT1 is localized predominantly to the apical membranes of the rat jejunum, renal cortical tubules, and liver parenchymal cells.[15] In rat hepatocytes it was suggested that rCNT1 is targeted to the apical (canalicular) membranes via a transcytotic pathway.[18]

Expression of hCNT1 and rCNT1 in *Xenopus* oocytes and transfected cell lines revealed transport properties that are consistent with those of the pyrimidine-selective subtype *cit*.[19–22] hCNT1 and rCNT1 selectively transport naturally occurring pyrimidine nucleosides (uridine, thymidine, cytidine) as well as the purine nucleoside, adenosine, in a Na^+-dependent manner with high affinities (K_m values in the low-micromolar range).[5] However, both hCNT1 and rCNT1 transport adenosine with much lower capacities (V_{max}) than pyrimidine nucleosides,[10,21,23] suggesting that these transporters may function mainly as a pyrimidine-specific nucleoside transporter in vivo. hCNT1 and rCNT1 interact with a number of pyrimidine and adenosine analogs.[5] hCNT1 transports the anticancer nucleoside analogs gemcitabine (2′,2′-difluorodeoxycytidine) and cytarabine [1-(β-D-arabinofuranosyl)cytosine (AraC)].[24,25] 5′-Deoxy-5-fluorouridine (5′-DFUR), an intermediate metabolite of the anticancer drug capecitabine, is a good hCNT1 substrate.[26] hCNT1 also accepts a number of antiviral nucleoside analogs, including zidovudine [3′-azido-3′-deoxythymidine (AZT)], lamivudine [2′,3′-dideoxy-3′-thiacytidine (3TC)], and zalcitabine [2′,3′-dideoxycytidine (ddC)].[10,25] Similarly, rCNT1 accepts a number of pyrimidine and adenosine analogs as substrates, such as 5-fluoro-2′-deoxyuridine, 5-fluorouridine, gemcitabine, AZT, ddC, cladribine, and cytarabine.[20,21,27,28]

CNT2 (SLC28A2) rCNT2 cDNA was first isolated from rat liver by expression cloning and was initially named SPNT (sodium-dependent purine nucleoside transporter).[29] The human homolog hCNT2 (also termed hSPNT1) was subsequently cloned by homology cloning from human kidney.[30,31] The *hCNT2* gene is localized to chromosome 15q15. hCNT2 protein is 72% identical to hCNT1 and shares a similar 13 TM membrane topology. Orthologs of CNT2 were also cloned from rabbit (rbCNT2) and mouse (mCNT2).[32,33]

Northern blot studies revealed that hCNT2 is widely distributed in heart, liver, skeletal muscle, kidney, intestine, pancreas, placenta, brain, and lung.[12,30] hCNT2 is also expressed in normal human leukocytes and a number of neoplastic tissues and cancer cell lines.[12,34] rCNT2 is expressed in rat liver, jejunum, spleen, heart, skeletal muscle, and various regions of the brain.[29,35] Interestingly, rCNT2 transcripts were not detected by Northern blot analysis in the rat kidney.[29] rCNT2 was also found in rat brain endothelial and choroid plexus epithelial cells,[35,36] which form the blood–brain barrier (BBB) and blood–cerebrospinal fluid (CSF) barrier respectively. Consistent with previous functional data demonstrating *cif* activities in brush border membranes of kidney or intestinal epithelia,[1,37,38] green fluorescence protein (GFP)-tagged rCNT2 is predominantly targeted to the apical membranes of polarized MDCK and LLC-PK cells.[16] Interestingly, in liver parenchymal cells, an antibody study by Duflot et al. showed that rCNT2 protein is highly expressed in the basolateral (sinusoidal) membrane and minimally presents at the apical (canalicular) membrane,[18] suggesting that membrane trafficking of rCNT2 in hepatocytes may be different from that in renal or intestinal epithelial cells.

The transport characteristics of CNT2 cloned from human, rat, mouse, and rabbit have been studied in *Xenopus* oocytes and transfected cell lines.[21,29,30,32,33,35,39,40] Substrate specificities of these recombinant CNT2 proteins are consistent with the properties of the classic purine nucleoside-selective subtype *cif*. hCNT2 transports purine nucleosides (adenosine, guanosine, and inosine) and the pyrimidine nucleoside uridine with high affinities and exhibits relatively higher affinities for purine nucleosides (K_m values ranging from 8 to 14 μM for adenosine and inosine) than for uridine (K_m ranging from 80 to 116 μM).[5,30,31] rCNT2, rbCNT2, and mCNT2 exhibited similar transport properties to hCNT2, selectively transporting purine nucleosides and uridine.[21,30,33] CNT2 transporters interact with a number of nucleoside analogs.[5] Ribavirin, a broad-spectrum antiviral nucleoside analog used in the treatment of hepatitis C, and clofarabine, an investigational antileukemia drug, were shown to be transported by hCNT2.[41–43] 5-Fluorouridine is transported by hCNT2 with kinetics similar to those of uridine.[44] Other nucleoside analogs, including gemcitabine, AZT, ddC, ddA, AraA, 5-fluoro-2′-deoxyuridine (FdU), and 5-iodo-2′-deoxyuridine (IdU), are not substrates of hCNT2.[31,39,45] There appear to be significant species differences in substrate specificity between the human and rat CNT2 in interacting with nucleoside analogs. For example, ddI is either not a substrate or is poorly transported by hCNT2.[40,45] In contrast, rCNT2 transports ddI ($K_m = 29.2$ μM).[46] The anticancer drug cladribine (2-CdA) appears to be a good substrate for rCNT2 but is not transported by hCNT2.[45]

CNT3 (SLC28A3) hCNT3 was cloned from human mammary gland and differentiated human myeloid HL-60 cells.[47] hCNT3 is about 47 to 48% identical in amino acid sequence to hCNT1 and hCNT2, and 57% identical to hfCNT, a broadly selective CNT from the ancient marine prevertebrate hagfish.[48] The *hCNT3* gene is localized to chromosome 9q22.2. The mouse ortholog mCNT3 was isolated from liver, and the protein shares 78% identity with hCNT3. Both hCNT3 and mCNT3 are predicted to possess 13 TMs similar to CNT1 and CNT2.

The distribution pattern of hCNT3 was investigated in a human multiple tissue expression RNA array.[47] The highest levels are found in mammary gland, pancreas, bone marrow, trachea, and intestine. Lower levels are found in liver, lung, placenta, prostate, testis, and other tissues, including some regions of the brain and heart. hCNT3 is also found in fetal tissues and various cultured cell lines, including K562, HeLa, and HL-60. A recent study also revealed low expression of hCNT3 in leukocytes and several leukemia cell lines.[49] The subcellular localization of hCNT3 in polarized epithelial cells has not been investigated directly.

When expressed in *Xenopus* oocytes, hCNT3 and mCNT3 exhibit *cib* subtype nucleoside transport properties, with broad selectivities for both purine and pyrimidine nucleosides. hCNT3 transports purine and pyrimidine nucleosides with reported K_m values ranging from 15 to 53 μM in one study[47] and 3 to 11 μM in another.[50] Compared to hCNT1 and hCNT2, hCNT3 is capable of transporting a broader range of synthetic nucleoside analogs. hCNT3 efficiently transports both pyrimidine (5-fluorouridine,

5-fluoro-2′-deoxyuridine, zebularine, and gemcitabine) and purine (cladribine, clofarabine, and fludarabine) nucleoside analogs.[47,43] Low uptake is also observed for AZT, ddC, and ddI.[47] Recent electrophysiology studies demonstrated further that ribavirin and 3-deazauridine are also hCNT3 substrates.[51] So far, specific high-affinity inhibitors have not been described for hCNT3 and other CNTs.

8.2.2. Transport Mechanisms

Mammalian CNTs are Na^+-coupled transporters which utilize the inwardly directed Na^+ gradient established by Na^+/K^+-ATPase to translocate substrates into cells. Early radio tracer uptake studies in cell and tissue preparations suggested that the Na^+/nucleoside coupling stoichiometry is 1 : 1 for *cit* (CNT1) and *cif* (CNT2); and 2 : 1 for *cib* (CNT3).[1] Recent electrophysiological studies have confirmed these coupling stoichiometry for hCNT1 and hCNT3.[51–53] Electrophysiological analysis of hCNT1 also suggested an ordered simultaneous transport model in which Na^+ first binds to the transporter in order to enhance subsequent binding of nucleoside.[52] Interestingly, while ion coupling of hCNT1 and hCNT2 is Na^+ specific, hCNT3 exhibits unique broad cation specificity and can use H^+ and Li^+ to substitute Na^+ for coupling.[53] Chimeric studies between hCNT1 and hCNT3 suggested that the site for Na^+ ion interaction may reside in the C-terminal half.[53] CNT homologs have also been found in bacteria, yeast, nematodes, insects, and hagfish.[4,54] CNTs isolated from *Escherichia coli*, *Candida albicans*, and *Caenorhabditis elegans* use a proton gradient as the driving force, whereas the hagfish hfCNT is coupled to Na^+.[53]

From the substrate point of view, CNT1–3 specifically transport nucleosides and their analogs. CNT1–3 do not transport nucleobases (i.e., free purine or pyrimidine bases) or nucleotides. A number of recent studies have examined the structure–activity relationships of the CNTs.[19,55–58] Besides isoform-dependent specificity toward the purine or pyrimidine moieties, several common features of the ribose structure are important for CNT–substrate interaction. Generally speaking, the 3′-hydroxyl group of the ribose ring is essential for substrate interaction with the CNTs, and modifications at the C(3′) position are usually not tolerated. CNTs are less sensitive to modifications at the 5′-hydroxyl group and generally tolerate changes at the 2′-hydroxyl group.

From the protein standpoint, mammalian CNTs are proposed to have 13 transmembranes (TMs) with an intracellular N-terminus and an extracellular C-terminus based on antibody and glycosylation studies (Figure 8.2A).[15] hCNT1, hCNT2, and hCNT3 all contain potential N-glycosylation sites and are likely to exist in the glycosylated form.[5,50] Mutational analysis of the three putative N-linked glycosylation sites in rCNT2 suggests that glycosylation does not affect transporter function or membrane sorting.[59] Protein chimera and site-directed mutagenesis studies have been used to examine the structure–function relationships of CNTs. Domain-swapping studies in rCNT1 and rCNT2 first identified TMs 7 and 8 as the essential sites for substrate recognition and discrimination.[60] Site-directed mutagenesis studies further identified Ser318 on TM 7 of rCNT1 as essential for its pyrimidine selectivity.[61] Replacing Ser318 with glycine transformed rCNT2 into a broadly selective CNT3-like transporter.[61,62] A Gln319Met mutation augmented this effect by enhancing the apparent affinity for purine nucleosides.[61] Using a similar approach, Loewen et al. confirmed these findings in hCNT1 and hCNT2, and further identified Ser319/Gln320 and Ser353/Leu354 as essential determinants for substrate selectivity.[63] In addition, mutation of a conserved residue Phe316 in TM7 of hCNT1 to tyrosine or alanine increased transporter sensitivity to guanosine.[64] Together, these studies demonstrated that TMs 7 and 8 are essential components of the substrate binding and translocation channel, and that substrate selectivity of the CNTs may be determined by a few key amino acid residues located on TMs 7 and 8. Cysteine-accessibility analysis of hCNT3 suggests that TMs 11 and 12 may also participate in forming the nucleoside translocation pathway.[65] In contrast, TMs 1 to 3 appear not essential for CNT function. When TMs 1 to 3 were truncated from rCNT1 or hCNT1, the resulting transporters retained significant Na^+-dependent transport activity.[15] NupC, a H^+-coupled CNT from *E. coli*, lacks TMs 1 to 3, further suggesting that TMs 1 to 3 may not be required for CNT function.[66]

8.3. EQUILIBRATIVE NUCLEOSIDE TRANSPORTERS (SLC29)

8.3.1. Family Members

ENT1 (SLC29A1) Purification and N-terminal sequencing of the prototype *es* transporter from human erythrocytes led to the cloning of hENT1 in 1997.[67] Human ENT1 is a 456-residue glycoprotein with 11 predicted TMs (Figure 8.2B). The *hENT1* gene is localized to chromosome 6p21.1–21.2. ENT1 orthologs were cloned from rat and mouse.[68–70] rENT1 (457 a.a.) and mENT1.1 (460 a.a.) are about 78 to 79% identical to hENT1.[68–70] A splice variant, mENT1.2, has been reported for mouse.[70] Studies at both the mRNA and protein levels have revealed that ENT1 is almost ubiquitously distributed in human and rodent tissues, although its abundance varies among tissues.[12,67,70] High expression of hENT1 protein or mRNA has been demonstrated and confirmed in many tissues, including erythrocytes, placenta, brain, heart, liver, lung, and colon.[12,71] However, striking intertissue and interindividual differences in the expression levels of hENT1 were observed.[12] GFP- or YFP (yellow fluorescence protein)-tagged hENT1 is localized predominantly on the basolateral membrane of differentiated MDCK cells, whereas a small amount of hENT1 also appears on the apical membrane.[17,72] Besides plasma membrane, Lai et al. showed that hENT1, but not mENT1, is also present in mitochondria, which may contribute to enhanced mitochondrial toxicity of certain antiviral nucleoside analogs.[73] A PEXN motif located in the first extracellular loop of hENT1 was recently identified as an important signal for mitochondrial targeting of hENT1.[74]

Recombinant human, rat, and mouse ENT1 transporters transport both purine and pyrimidine nucleosides and exhibit typical *es*-type activities. The K_m values of hENT1 toward endogenous nucleosides range from 50 μM for adenosine to 680 μM for cytidine.[75] Transport is Na^+ independent and inhibited by nanomolar NBMPR. hENT1 does not interact with nucleobases.[67,75] hENT1 interacts with many nucleoside analogs widely used in the treatment of cancer and viral infections. Cladribine, gemcitabine, fludarabine, cytarabine, clofarabine, and ribavirin have been demonstrated to be transported by hENT1.[28,43,76,77] However, hENT1 does not transport AZT and only weakly transports ddC and ddI.[78] Coronary vasodilators such as dipyridamole, dilazep, and draflazine are potent hENT1 inhibitors (K_i values in nanomolar concentrations).[79] These compounds bind to the ENT1 protein but are not transported into the cells. There are significant species differences among ENT1 in interacting with coronary drugs. Rat ENT1 is much less sensitive to dipyridamole and dilazep than human ENT1, and mouse ENT1 is only intermediately sensitive to dipyridamole.[69,79] In addition to NBMPR and coronary vasodilators, a number of protein kinase inhibitors, including the Bcr-Abl tyrosine kinase inhibitor STI-571 (Gleevec) used for treatment of chronic myelogenous leukemia, were shown to be effective in inhibiting ENT1 and/or ENT2, with IC_{50} values ranging from 60 to 1000 nM.[80] Interestingly, Carrier et al. recently showed that plant-derived cannabinoids are potent inhibitors of ENT1 (K_i < 250 nM).[81]

ENT2 (SLC29A2) The human ENT2 transporter was cloned from HeLa cells and human placenta by two independent groups.[82,83] hENT2 (456 a.a.) is 46% identical in amino acid sequence to hENT1. The most homologous regions between hENT1 and hENT2 are the transmembrane segments, while the least are the hydrophilic termini and loops. The human *hENT2* gene is localized to chromosome 11q13. Two mRNA-spliced variants, encoding nonfunctional truncated transporter proteins, were reported.[72,83] The rodent orthologs rENT2 and mENT2 were also cloned.[68,69] The 456-residue mENT2 and rENT2 are 88% identical to hENT2. hENT2 mRNA is particularly abundant in skeletal muscle, but is also expressed in a wide range of tissues, including brain, heart, placenta, thymus, pancreas, prostate, and kidney.[12,83] In MDCK cells, GFP-tagged hENT2 is localized exclusively to the basolateral membrane. A C-terminal dileucine repeat is implicated in the surface expression of hENT2.[72]

Recombinant human and rat ENT2 proteins transport a broad range of purine and pyrimidine nucleosides, and they confer NBMPR-insensitive Na^+-independent nucleoside transport typical to the *ei* system. Although both hENT1 and hENT2 transporters are broadly selective for purine and pyrimidine nucleosides, hENT2 exhibits 7.7- and 19.3-fold lower affinities for guanosine and cytidine, respectively, but a four fold higher affinity for inosine.[75] More important, ENT2 transports a wide range of purine and pyrimidine nucleobases, whereas ENT1 does not.[84] Specifically, hENT2 and rENT2 transport hypoxanthine, adenine, guanine, uracil, and thymine with low affinities (K_m ranging from 0.7 to 2.6 mM).[84] hENT2, but not rENT2, also transports cytosine. h/rENT2 transports AZT, ddC, and ddI with much higher efficiency than h/rENT1. Gemcitabine and clofarabine have been shown to be transported by hENT2.[28,43,79] A number of uridine and cytidine analogs also interact with hENT2.[85]

ENT3 (SLC29A3) The third member of the ENT family, ENT3, was first identified and cloned from human and mouse due to sequence similarity to ENT1 and ENT2.[86] The *hENT3* gene is localized to chromosome 10q22.1. The 475-residue hENT3 and mENT3 proteins are 73% identical in amino acid sequence and exhibit ~30 to 33% identities to ENT1 and ENT2. Structurally, ENT3 differs from ENT1 and ENT2 in possessing a long hydrophilic N-terminal region that contains a dileucine motif (DE)XXXL(LI) for endosomal/lysosomal targeting.[6] ENT3 transcripts and proteins are widely distributed in human and rodent tissues. In humans the highest levels of mRNA were found in the placenta, uterus, ovary, spleen, lymph node, and bone marrow; the lowest levels were found in brain and heart.[6] In contrast to ENT1 and ENT2, the endogenous as well as GFP-tagged hENT3 proteins were found to reside predominantly intracellularly in HeLa cells and to colocalize partially with lysosomal markers. Truncation of the N-terminal region or mutation of the dileucine motif to alanine (hENT3AA) relocated hENT3 to the plasma membrane in HeLa cells and in *Xenopus* oocytes.[6]

Characterization of a cell surface–expressed mutant hENT3AA suggested that hENT3 is a broadly selective and low-affinity nucleoside transporter that transports purine and pyrimidine nucleosides as well as the nucleobase adenine.[6] hENT3AA-mediated transport is Na^+ independent but stimulated strongly by low pH, with the optimum pH being 5.5.[6] At pH 5.5, hENT3AA exhibits about 40- and 10-fold lower affinities for adenosine and uridine than for hENT1. Nucleoside analogs, including cladribine, cordycepin (3′-deoxyadenosine), tubercidin (7-deazaadenosine), fludarabine, zebularine, ddI, ddC, and AZT, are transported by hENT3AA. hENT3AA is relatively insensitive to the classical nucleoside transport inhibitors, such as NBMPR, dipyridamole, and dilazep.

ENT4 (SLC29A4) The fourth member of the human SLC29 family, hENT4, was first cloned and characterized in our laboratory.[7] Gene orthologs of ENT4 are also found in mouse and rat.[7,87] hENT4 is a 530-residue protein which displays about 86% identity to the 528-residue mouse homolog mENT4. The mENT4 and hENT4 proteins share low but significant sequence identity (about 18 to 20%) to ENT1-3 and possess a similar 11 TM membrane topology. However, hENT4 has a longer hydrophilic N-terminus and exhibits a low overall sequence homology to other ENTs. Phylogenetic analysis suggests that the mammalian ENT4 lineage is evolutionarily distinct from the ENT1/2/3 lineage(s).[87] The gene encoding hENT4 is located at chromosome 7p22.1. Our Northern blot analysis showed that hENT4 mRNA is expressed preferentially in the human brain but is also found in skeletal muscle, kidney, heart, and liver.[7] In human and mouse brain, hENT4 transcripts are widely distributed in different regions of the CNS (Wang et al., unpublished data).[7] YFP tagging of hENT4 in MDCK cells demonstrated that this protein is localized primarily to the plasma membrane.

Although ENT4 was originally suggested to be a nucleoside transporter,[3] extensive functional analysis carried out in this laboratory showed that except for a moderate activity toward adenosine, hENT4 minimally interacts with other nucleosides, nucleobases, or their analogs.[7] Instead, it mediates Na^+-independent, electrogenic transport of monoamine neurotransmitters such as serotonin and dopamine (K_m = 114 and 329 μM, respectively).[7] Functional analysis of the mouse ENT4 ortholog also confirmed that the monoamines are the physiologic substrates of ENT4 (Wang et al., unpublished data). Therefore, we previously proposed an alternative functional name, plasma membrane monoamine transporter (PMAT), for ENT4. Our recent analysis showed that hENT4 also transports cationic xenobiotics such as MPP^+ and TEA, and shares striking functional similarities to a genetically unrelated drug transporter family, organic cation transporters (OCTs) (reviewed in Chapter 2).[8] Because ENT4 does not typically function as a nucleoside transporter, we will not include it for further discussion.

8.3.2. Transport Mechanisms

ENT1 and ENT2 are facilitated carriers that transport substrates down their concentration gradients. However, in many cell types, cellular uptake of nucleosides and nucleoside analogs are tightly coupled to intracellular metabolism (e.g., phosphorylation). Rapid enzymatic conversion of nucleosides to metabolites (e.g., nucleotides) can thus provide a metabolic "driving force" to promote cellular uptake.[1] Efflux occurs only when intracellular concentrations of free nucleoside exceed those outside the cell. The transport mechanism of the lysosome-localized ENT3 is not known yet. ENT3 activity is strongly stimulated by proton, exhibiting maximal activity at pH 5.5 and no activity at pH 8.0. It is unclear whether the pH dependence of ENT3 reflects a proton–nucleoside cotransport mechanism, or an evolutionary adaptation of ENT3 to the acidic environment of lysosomes. Mammalian ENT homologs are also found in fungi, protozoans, nematodes, insects, and plants, but not in bacteria.[87] Interestingly, ENT members from *Leishmania donovani* function as electrogenic proton cotransporters,[88] suggesting that ENT-type transporters are not always "equilibrative" and can be electrogenic and concentrative in certain species.

Like the CNTs, the 3′-OH of the ribose moiety is critical for substrate interaction with ENT1 and ENT2, whereas the 2′ and 5′ hydroxyl groups are less important.[58,85,89] Halogen modifications on most positions at the base are generally accepted. In general, hENT2 is more tolerant than hENT1 to modification at the ribose ring and even transports nucleobases that lack the ribose moiety. Analysis of a series of uridine analogs with hENT1 and hENT2 suggests that the C(2′)-OH is a structural determinant for uridine-hENT1 but not for uridine-hENT2 interactions.[89] Moreover, hENT2 displayed more tolerance than hENT1 to removal of C(5′)-OH. The changes in binding energies between transporter proteins and the various uridine analogs suggest that hENT1 may form strong interactions with C(3′)-OH and moderate interactions with C(2′)-OH and C(5′)-OH of uridine, whereas hENT2 may form strong interactions with C(3′)-OH, weak interactions with C(5′)-OH, and no interaction with C(2′)-OH.[89]

Glycosylation scanning and antibody studies have confirmed the originally proposed 11 TM topology of hENT1 (Figure 8.2B).[90] hENT1 is N-glycosylated at a single site and hENT2 at two sites in the large extracellular loop linking TMs 1 and 2.[90] Glycosylation is not required for the transport activity of hENT1 but may slightly affect the binding affinity to transport inhibitors such as NBMPR.[91] Glycosylation does not change hENT2 function but is required for efficient targeting of hENT2 protein to the plasma membrane.[92] Investigation of human and rat ENT1/2 has begun to identify functionally important domains of these transporters. The human and rat ENT1 transporters exhibit similar binding affinities toward NBMPR, but there is more than a 50-fold difference between hENT1 and rENT1 in interacting with dipyridamole.[79] Using chimeric constructs of hENT1 and rENT1, Sundaram et al. first demonstrated that TMs 1 to 6 of hENT1 are required for interaction with dipyridamole and dilazep, with TMs 3 to 6 being the major site of interaction.[79] Studies with chimeras between rENT1 and rENT2 indicate that TMs 3 to 6 also represent the major site of NBMPR interaction.[93] Additional work revealed that transplantation of a segment encompassing TMs 5 and 6 of rENT2, which transports nucleobases, could enable rENT1 to transport nucleobases.[84] Consistent with these studies demonstrating the importance of TMs 3 to 6, covalent modification with *p*-chloromercuriphenyl sulfonate of Cys140 in TM 4 of rENT2 resulted in the loss of transport activity, and this modification could be protected by coincubation with the substrate uridine.[94] Mutagenesis of hENT1 and hENT2, which differ in their sensitivities to dipyridamole and dilazep, revealed that residue Met33, located at the end of the predicted TM 1, controls sensitivity to these inhibitors.[95,96] Site-directed mutagenesis of Gly154 and Gly179 within TM 4 and 5 of hENT1 resulted in impaired transport activity and reduced sensitivity to NBMPR, suggesting potential roles for these residues in substrate recognition and NBMPR binding.[97,99] Ser160 and Met89 of hENT1 may play a dominant role in conferring sensitivity to dipyridamole and adenosine/guanosine affinity.[98] In addition, Leu92 of hENT1 has been implicated in interacting with inosine and guanosine, and with NBMPR and dilazep.[99] Although it is well accepted that TMs 1 to 6 contain the functional domains of ENTs, a recent study demonstrated that some residues in TM 11 of hENT1 (e.g., Leu 442) are also important for interactions with substrates and inhibitors.[100]

8.4. REGULATION OF NUCLEOSIDE TRANSPORTERS

Nucleoside transport regulation is cell-type dependent, involves multiple mechanisms, and occurs at both transcriptional and posttranscriptional levels. Pathways that stimulate cell proliferation and differentiation influence the type and abundance of nucleoside transporters in cells and tissues.[1,5,34,101] A number of hormones and chemicals, through interaction with cellular signaling pathways, regulate CNTs and ENTs.[1,5,34,101]

Regulation of nucleoside transport activities by cell differentiation has been studied in a number of cell culture models.[1,5,34,101] For example, marked changes in nucleoside transport activity during differentiation were observed in human HL-60 promyelocytic leukemia cells.[102–105] Undifferentiated HL-60 cells exhibit hENT1- and hENT2-mediated *es* and *ei* transport activities predominantly, with a small amount of Na^+-dependent activity. Upon induction of differentiation by the phorbol ester PMA or dimethyl sulfoxide (DMSO), the cells exhibited a decrease in equilibrative Na^+-independent nucleoside transport that was accompanied by an increase in concentrative Na^+-dependent transport of both pyrimidine and purine nucleosides.[103–105] The decreased Na^+-independent transport was attributed to *es* and *ei* transporters with a decrease in V_{max} and no change in affinity. For the *es* transporter, the decrease in nucleoside uptake was paralleled by a decline in NBMPR binding sites.[103,104] The increased Na^+-dependent transport activity was consistent with CNT3. Recent studies showed that hCNT3 mRNA transcripts were substantially increased in differentiated vs. undifferentiated HL-60 cells.[47] An upstream PMA response element was found in the promoter region of the hCNT3 gene, indicating that PMA may directly activate hENT3 transcription in HL-60 cells.[47]

A number of cellular signaling pathways have been implicated in the regulation of nucleoside transport. Lipopolysaccharides (LPSs) and phorbol esters (e.g., PMA), which activate B cells, up-regulated CNTs and down-regulated ENTs in lymphocytes from the bone marrow and in a human B-lymphoblast cell line.[106–109] The induction of the CNT activity and the down-regulation of the *es* system by LPS and PMA were PKC dependent, and activation of CNT-type transporters also involved tumor necrosis factor-α and nitric oxide signaling pathways in activated human B lymphocytes.[106–109] Protein and mRNA levels for hENT1 were reduced and transport activity was decreased in human endothelial cells with exposure to D-glucose, a condition that leads to activation of purinergic receptors.[110] A number of studies reported that limited oxygen availability (hypoxia) down-regulated ENT1 mRNA and protein in several cell types.[111–115] This down regulation of ENT1 is beneficial because it increases extracellular concentrations of adenosine, which initiates a series of innate protective mechanisms under stressful conditions through activation of cell surface receptors.[111,113,114] Recent studies suggest that down-regulation of ENT1 is mediated by the hypoxia inducible factor 1 (HIF-1). The hENT1 promoter region contains HIF-1 binding sites, and binding to HIF-1 is responsible for the transcriptional repression of ENT1 observed during hypoxia.[114]

Besides transcriptional regulation, nucleoside transporters are also regulated post-translationally. For example, in cultured MCF-7 breast cancer cells and HeLa cells, acute stimulation of PKC caused a rapid increase in hENT1-associated nucleoside uptake, which appeared to involve PKC δ and/or ε, but not α, β, or γ.[116] This up-regulation of hENT1 activity appears not to involve transporter recruitment or de novo synthesis, but it is not yet clear whether changes in the phosphorylation state of the transporter are involved.[116]

Regulation of nucleoside transporters has also been studied in primary cell cultures and in vivo animal models. In rat liver, expression of rCNT1, rCNT2, and rENT1 was shown to be affected by developmental stages, growth hormones, partial hepatectomy, interleukin-6, and bile acids.[117–121] In rat small intestine, where CNT1 and CNT2 are coexpressed in the brush border membranes, fasting of the animals resulted in an increased expression of rCNT1 protein with a concomitant increase in Na^+-dependent thymidine and gemcitabine uptake activity.[122] Changes of mRNA levels of CNTs and ENTs were also observed in various tissues from diabetic rats and human umbilical artery smooth muscle cells from subjects with gestational diabetes.[123,124]

8.5. PHYSIOLOGICAL ROLES OF NUCLEOSIDE TRANSPORTERS

Mammalian cells acquire purines and pyrimidines via two pathways, de novo biosynthesis and salvage pathways. The primary physiological function of ENTs and CNTs in mammalian cells is to facilitate cellular uptake of natural nucleosides, derived from the diet or produced by tissues such as the liver, for nucleotide synthesis in the salvage pathways.[1,3–5] This function is particularly important for cells that lack de novo biosynthetic pathways, such as leukocytes, erythrocytes, bone marrow, and certain brain cells.[1] The wide distributions of ENT1 and ENT2 in various cell types and the ability of these transporters to transport both purine and pyrimidine nucleosides suggest that they play a primary role in nucleoside salvage pathways. The ability of ENT2 to transport nucleobases indicates that this transporter may also contribute to cellular uptake of hypoxanthine, a major salvageable purine nucleobase in animals.[125,126] In addition, ENTs expressed in intracellular compartments may play a role in organelle transport of nucleosides. It has been shown that some hENT1 is also expressed in mitochondria[73] and thus may transport nucleoside into this organelle for mitochondrial DNA synthesis. ENT3, an intracellular transporter expressed in lysosomes, has been postulated to export nucleosides produced by RNA degradation from the lysosomes.[6] The CNTs are particularly abundant in epithelial cells of the intestine, kidney, and liver. Expressed in the apical membranes, CNTs are thought to work in concert with the basolaterally localized ENTs to mediate vectorial transepithelial flux of nucleosides and therefore regulate total body homeostasis of nucleosides.[4,5]

Another important physiological function of nucleoside transporters is related to the purine nucleoside adenosine, a key signaling molecule that exerts profound effects on many tissues and organs. Through binding to cell surface receptors (A_1, A_{2A}, A_{2B}, and A_3), adenosine regulates a myriad of physiological processes, such as coronary blood flow, vascular tone regulation, neurotransmitter release, platelet aggregation, immunosuppression during cellular stress, and inflammation.[1–3,5] By influencing the concentration of adenosine available to adenosine receptors, nucleoside transporters, particular ENT1, play important regulatory roles in adenosine signaling. For example, studies with rENT1 antibody revealed abundant expression of rat ENT1 protein in the sinoatrial node, cardiac atrial, and ventricular cells, suggesting an important role of this transporter in regulating the cardiovascular activity of adenosine.[127] In the human brain, hENT1 and A1 receptors colocalize in various brain regions, indicating an important role of hENT1 in the control of neuromodulatory action of adenosine in the CNS.[71] Interestingly, studies of knockout mice lacking the *ENT1* gene suggested that ENT1 has a physiological role in ethanol-associated behaviors via A1 adenosine receptor.[128] The $ENT1^{-/-}$ mice appeared phenotypically normal but consumed twice as much alcohol as the wild-type mice.[128] Finally, recent studies have demonstrated that ENT1 is involved in the anti-inflammatory effects of adenosine under stressful conditions such as hypoxia.[111,112,114] Down-regulation of ENT1 appears to serve as an innate protective mechanism during hypoxia.

8.6. CLINICAL SIGNIFICANCE OF NUCLEOSIDE TRANSPORTERS

The clinical significance of nucleoside transporters can be viewed in several aspects. First, a number of anticancer nucleoside analogs rely on nucleoside transporters to enter cells to reach their cellular targets. As such, the expression level of nucleoside transporters on the target cells is an important determinant for intracellular drug bioavailability, and consequently, responsiveness to therapy. Second, if a drug is a substrate for nucleoside transporters, the distribution of these transporters in various tissues and organs, particularly the absorptive and excretory organs, may influence its pharmacokinetic and toxicological properties. Finally, nucleoside transporters are potential drug targets. By modulating extracellular concentrations of adenosine, nucleoside transport inhibitors can regulate a variety of physiological processes, which can potentially lead to therapeutic benefits.

Nucleoside analogs are an important class of drugs used in the treatment of various forms of leukemia (e.g., cytarabine, fludarabine, cladribine) as well as in solid tumors (e.g., gemcitabine). All of the anticancer nucleoside analogs share a similar mechanism of action. After entering the cells, nucleoside analogs undergo sequential phosphorylation to form the corresponding nucleoside triphosphates, which exert cytotoxicity by interfering with DNA and/or RNA synthesis and metabolisms.[5] Many anticancer nucleoside analogs are hydrophilic and rely on nucleoside transporters, especially ENT1, to enter cells. In this case the number of functional transporters on the target cells becomes an important determinant for intracellular drug bioavailability and, consequently, responsiveness to therapy. Down-regulation of transporter expression or selection of transporter-deficient cells may contribute to clinical resistance to cytotoxic nucleoside analogs.[129] Indeed, many in vitro studies have demonstrated that nucleoside transporters are necessary for many nucleoside analogs to enter cells, and deficiency in transport can result in resistance to cytotoxic nucleoside analogs such as cytarabine, 5-fluorouridine, 5-fluorodeoxyuridine, and gemcitabine.[24,26,44,129] For example, Mackey et al. have shown in a panel of cancer cell lines that treatment of cells with ENT inhibitors such as NBMPR or dipyridamole increases resistance to gemcitabine 39- to 1800-fold.[24] Consistent with the in vitro observations, several recent clinical studies have implicated a role of hENT1 in clinical resistance to cancer chemotherapy. Galmarini et al. retrospectively analyzed the effects of hENT1 mRNA expression on efficacy of cytarabine treatment of acute myelogenous leukemia in blast cells of 123 patients treated with cytarabine. Decreased expression of hENT1 was associated with an increased risk of early relapse.[130] Stam et al. studied 18 infants and 24 children with acute lymphoblastic leukemia to determine why infants were sensitive to cytarabine. Their results showed that leukemic blasts from infants were threefold more sensitive to cytarabine than blasts from children.[131] Decreased mRNA levels of deoxycytidine kinase (dCK) but increased levels of hENT1 were observed in infants, which was thought to be responsible for the increased sensitivity of infant acute lymphocytic leukemia. In the case of gemcitabine, Spratlin et al. first examined hENT1 and hCNT3 expression in tumor biopsies from 21 pancreatic cancer patients treated with gemcitabine. Their analysis indicated that patients with detectable hENT1 immunostaining had a significantly longer survival time after gemcitabine chemotherapy than did patients without detectable hENT1.[132] More recently, Giovannetti et al. characterized the expression pattern of genes involved in gemcitabine activity in pancreas tumor specimens from 102 gemcitabine-treated patients and its correlation with treatment outcome.[133] hENT1 expression significantly correlated with clinical outcome; patients with high levels of hENT1 had a significantly longer overall survival.[133] These encouraging results suggest that measurement of transporter abundance may provide a predictive tool for guiding the appropriate use of anticancer drugs in individual patients for treatment optimization.

Nucleoside transporters are found in absorptive and excretory organs and thus may influence the systemic pharmacokinetics of nucleoside analogs. For example, in the epithelia of the intestine and kidney, the CNTs and the ENTs are localized to the apical and basolateral membranes, respectively, and are thought to mediate absorption, reabsorption, or elimination of nucleosides and nucleoside analogs. CNTs and ENTs are also asymmetrically expressed in the blood–brain barrier and choroids plexus.[36] How quantitatively these transporters contribute to the bioavailability, clearance, and tissue-specific disposition of therapeutic nucleoside analogs is largely unknown in vivo. Pharmacokinetic studies using transporter-deficient animal models may help to elucidate the roles of CNTs and ENTs in determining the absorption, distribution, and excretion of nucleoside analogs.

Nucleoside transporters themselves can serve as drug targets. As stated above, the endogenous nucleoside adenosine functions as a signaling molecule and regulates a number of physiological processes via binding to specific cell surface receptors.[1,2] Nucleoside transporter–mediated cellular uptake represents a major mechanism for the termination of adenosine signaling. By enhancing local adenosine concentrations, ENT inhibitors can potentiate the adenosine effect and produce therapeutic benefits. Coronary vasodilators such as dipyridamole, dilazep, and draflazine are potent hENT1 inhibitors that can substantially increase and prolong the cardiovascular effects of adenosine.[2] Dipyridamole is used clinically as a vasodilator, and draflazine has been shown to exert cardioprotective effect in humans.[134] Adenosine is an endogenous neuroprotective agent. By binding to A1 and A2 receptors, adenosine decreases excitatory amino acid neurotransmission, inhibits inflammation, and promotes vasodilation.[135] Transport inhibitors are thus potentially valuable in ameliorating ischemia- or hypoxia-induced neuronal injury. The neuroprotective effect of propentofylline in acute stroke patients was thought, at least in part, to be due to its ability to inhibit adenosine transport.[135] Recently, Carrier et al. suggested that the immunosuppressive effects of cannabinoids may also be due to their ability to enhance adenosine signaling through inhibition of ENT1.[81]

8.7. POLYMORPHISMS OF NUCLEOSIDE TRANSPORTERS

There are larger interindividual variations in nucleoside transporter expression levels and activities in normal and malignant human tissues and cells.[12,136,137] This variation has been implicated in interindividual difference in response to nucleoside analog therapies.[4,129] Genetic polymorphisms in CNT and ENT genes have been hypothesized to be one of the factors underlying the variations in the expression or activity of nucleoside transporters in humans.[4,129] Recently, Giacomini and colleagues have investigated genetic variations in the coding regions and flanking intronic regions of CNTs and ENTs in a collection of 270 DNA samples from ethnically diverse U.S. populations. Their studies suggest that there are a number of functionally significant protein-altering single-nucleotide polymorphisms (SNPs) in the *hCNT1* gene coding region.[138,139] In particular, a rare variant (Ser546Pro) and a single base pair deletion variant (1153del) with an allele frequency of 3% in the African-American population were found to be nonfunctional. A common CNT1 variant (Val189Ile) with a frequency of 26% had a reduced affinity for the anticancer nucleoside analog gemcitabine.[139] In contrast, few functionally significant coding-region SNPs were identified for hCNT2, hCNT3, hENT1, and hENT2, suggesting that coding region SNPs of these transporters are unlikely to contribute to interindividual differences in response to nucleoside analog drugs.[42,140–142] Recently, Myers et al. examined 1.6 kb upstream of the transcription initiation site of *hENT1* in a limited DNA sample collection.[143] Three SNPs (−1345C > G, −1050G > A, and −706G > C) were found, and analysis of four naturally occurring haplotypes suggests that promoter region haplotypes may affect hENT1 gene expression.[143] The regulatory-region SNPs might offer additional clues to the considerable variability in nucleoside transporter abundance observed in normal and tumor tissues.

8.8. CONCLUSIONS

In recent years, significant progress has been made in the studies of nucleoside transporters. Molecular cloning of the CNT and ENT genes has promoted studies that have greatly enhanced our understanding of the substrate and inhibitor specificities, tissue and cellular localization, structure–function relationships, and regulation mechanisms of nucleoside transporters. Knowledge obtained from these studies will pave the way for rational design and development of nucleoside analog drugs with improved pharmacokinetic and pharmacodynamic properties. The first nucleoside transporter-deficient animal has recently been generated for ENT1.[128] Detailed phenotyping and pharmacological analysis in nucleoside-transporter knockout animal models will provide insights into the physiological, pharmacological, and toxicological functions of nucleoside transporters in vivo, in the context of complex intracellular metabolism and the presence of other transporters (e.g., OATs, MRPs) that also transport some nucleoside analogs and/or their active metabolites. Recently, studies have been launched to correlate nucleoside transporter expression levels with clinical outcomes of therapeutic nucleosides. Pharmacogenomic studies have been undertaken to elucidate the mechanisms leading to inter-individual variations in response to nucleoside drugs. Knowledge gained from these studies will undoubtedly benefit the design, development, and clinical optimization of new and existing nucleoside analogs.

Acknowledgments

This work is supported by NIH grant GM66233.

REFERENCES

1. Griffith DA, Jarvis SM. 1996. Nucleoside and nucleobase transport systems of mammalian cells. Biochim Biophys Acta 1286(3):153–181.
2. Thorn JA, Jarvis SM. 1996. Adenosine transporters. Gen Pharmacol 27(4):613–620.
3. Baldwin SA, Beal PR, Yao SY, King AE, Cass CE, Young JD. 2004. The equilibrative nucleoside transporter family, SLC29. Pflugers Arch 447(5):735–743.

4. Gray JH, Owen RP, Giacomini KM. 2004. The concentrative nucleoside transporter family, SLC28. Pfluegers Arch 447(5):728–734.
5. Kong W, Engel K, Wang J. 2004. Mammalian nucleoside transporters. Curr Drug Metab 5(1):63–84.
6. Baldwin SA, Yao SY, Hyde RJ, Ng AM, Foppolo S, Barnes K, Ritzel MW, Cass CE, Young JD. 2005. Functional characterization of novel human and mouse equilibrative nucleoside transporters (hENT3 and mENT3) located in intracellular membranes. J Biol Chem 280(16):15880–15887.
7. Engel K, Zhou M, Wang J. 2004. Identification and characterization of a novel monoamine transporter in the human brain. J Biol Chem 279(48):50042–50049.
8. Engel K, Wang J. 2005. Interaction of organic cations with a newly identified plasma membrane monoamine transporter. Mol Pharmacol 68(5):1397–1407.
9. Huang QQ, Yao SY, Ritzel MW, Paterson AR, Cass CE, Young JD. 1994. Cloning and functional expression of a complementary DNA encoding a mammalian nucleoside transport protein. J Biol Chem 269(27):17757–17760.
10. Ritzel MW, Yao SY, Huang MY, Elliott JF, Cass CE, Young JD. 1997. Molecular cloning and functional expression of cDNAs encoding a human Na^+–nucleoside cotransporter (hCNT1). Am J Physiol 272(2 Pt 1):C707–C714.
11. Pajor AM. 1998. Sequence of a pyrimidine-selective Na^+/nucleoside cotransporter from pig kidney, pkCNT1. Biochim Biophys Acta 1415(1):266–269.
12. Pennycooke M, Chaudary N, Shuralyova I, Zhang Y, Coe IR. 2001. Differential expression of human nucleoside transporters in normal and tumor tissue. Biochem Biophys Res Commun 280(3):951–959.
13. Cass CE, Young JD, Baldwin SA, Cabrita MA, Graham KA, Griffiths M, Jennings LL, Mackey JR, Ng AM, Ritzel MW, Vickers MF, Yao SY. 1999. Nucleoside transporters of mammalian cells. Pharm Biotechnol 12:313–352.
14. Anderson CM, Baldwin SA, Young JD, Cass CE, Parkinson FE. 1999. Distribution of mRNA encoding a nitrobenzylthioinosine-insensitive nucleoside transporter (ENT2) in rat brain. Brain Res Mol Brain Res 70(2):293–297.
15. Hamilton SR, Yao SY, Ingram JC, Hadden DA, Ritzel MW, Gallagher MP, Henderson PJ, Cass CE, Young JD, Baldwin SA. 2001. Subcellular distribution and membrane topology of the mammalian concentrative Na^+–nucleoside cotransporter rCNT1. J Biol Chem 276(30):27981–27988.
16. Mangravite LM, Lipschutz JH, Mostov KE, Giacomini KM. 2001. Localization of GFP-tagged concentrative nucleoside transporters in a renal polarized epithelial cell line. Am J Physiol Renal Physiol 280(5):F879–F885.
17. Lai Y, Bakken AH, Unadkat JD. 2002. Simultaneous expression of hCNT1-CFP and hENT1-YFP in Madin–Darby canine kidney cells. Localization and vectorial transport studies. J Biol Chem 277(40):37711–37717.
18. Duflot S, Calvo M, Casado FJ, Enrich C, Pastor-Anglada M. 2002. Concentrative nucleoside transporter (rCNT1) is targeted to the apical membrane through the hepatic transcytotic pathway. Exp Cell Res 281(1):77–85.
19. Dresser MJ, Gerstin KM, Gray AT, Loo DD, Giacomini KM. 2000. Electrophysiological analysis of the substrate selectivity of a sodium-coupled nucleoside transporter (rCNT1) expressed in *Xenopus laevis* oocytes. Drug Metab Dispos 28(9):1135–1140.

20. Crawford CR, Cass CE, Young JD, Belt JA. 1998. Stable expression of a recombinant sodium-dependent, pyrimidine-selective nucleoside transporter (CNT1) in a transport-deficient mouse leukemia cell line. Biochem Cell Biol 76(5):843–851.
21. Yao SY, Ng AM, Ritzel MW, Gati WP, Cass CE, Young JD. 1996. Transport of adenosine by recombinant purine- and pyrimidine-selective sodium/nucleoside cotransporters from rat jejunum expressed in *Xenopus laevis* oocytes. Mol Pharmacol 50(6):1529–1535.
22. Fang X, Parkinson FE, Mowles DA, Young JD, Cass CE. 1996. Functional characterization of a recombinant sodium-dependent nucleoside transporter with selectivity for pyrimidine nucleosides (cNT1rat) by transient expression in cultured mammalian cells. Biochem J 317(Pt 2):457–465.
23. Vickers MF, Young JD, Baldwin SA, Ellison MJ, Cass CE. 2001. Functional production of mammalian concentrative nucleoside transporters in *Saccharomyces cerevisiae*. Mol Membr Biol 18(1):73–79.
24. Mackey JR, Mani RS, Selner M, Mowles D, Young JD, Belt JA, Crawford CR, Cass CE. 1998. Functional nucleoside transporters are required for gemcitabine influx and manifestation of toxicity in cancer cell lines. Cancer Res 58(19):4349–4357.
25. Graham KA, Leithoff J, Coe IR, Mowles D, Mackey JR, Young JD, Cass CE. 2000. Differential transport of cytosine-containing nucleosides by recombinant human concentrative nucleoside transporter protein hCNT1. Nucleosides Nucleotides Nucleic Acids 19(1–2):415–434.
26. Mata JF, Garcia-Manteiga JM, Lostao MP, Fernandez-Veledo S, Guillen-Gomez E, Larrayoz IM, Lloberas J, Casado FJ, Pastor-Anglada M. 2001. Role of the human concentrative nucleoside transporter (hCNT1) in the cytotoxic action of 5[prime]-deoxy-5-fluorouridine, an active intermediate metabolite of capecitabine, a novel oral anticancer drug. Mol Pharmacol 59(6):1542–1548.
27. Wang J, Schaner ME, Thomassen S, Su SF, Piquette-Miller M, Giacomini KM. 1997. Functional and molecular characteristics of Na(+)-dependent nucleoside transporters. Pharm Res 14(11):1524–1532.
28. Mackey JR, Yao SY, Smith KM, Karpinski E, Baldwin SA, Cass CE, Young JD. 1999. Gemcitabine transport in *Xenopus* oocytes expressing recombinant plasma membrane mammalian nucleoside transporters. J Natl Cancer Inst 91(21):1876–1881.
29. Che M, Ortiz DF, Arias IM. 1995. Primary structure and functional expression of a cDNA encoding the bile canalicular, purine-specific Na(+)-nucleoside cotransporter. J Biol Chem 270(23):13596–13599.
30. Wang J, Su SF, Dresser MJ, Schaner ME, Washington CB, Giacomini KM. 1997. Na(+)-dependent purine nucleoside transporter from human kidney: cloning and functional characterization. Am J Physiol 273(6 Pt 2):F1058–F1065.
31. Ritzel MW, Yao SY, Ng AM, Mackey JR, Cass CE, Young JD. 1998. Molecular cloning, functional expression and chromosomal localization of a cDNA encoding a human Na^+/nucleoside cotransporter (hCNT2) selective for purine nucleosides and uridine. Mol Membr Biol 15(4):203–211.
32. Gerstin KM, Dresser MJ, Wang J, Giacomini KM. 2000. Molecular cloning of a Na^+-dependent nucleoside transporter from rabbit intestine. Pharm Res 17(8):906–910.
33. Patel DH, Crawford CR, Naeve CW, Belt JA. 2000. Cloning, genomic organization and chromosomal localization of the gene encoding the murine sodium-dependent, purine-selective, concentrative nucleoside transporter (CNT2). Gene 242(1–2):51–58.

34. Pastor-Anglada M, Casado FJ, Valdes R, Mata J, Garcia-Manteiga J, Molina M. 2001. Complex regulation of nucleoside transporter expression in epithelial and immune system cells. Mol Membr Biol 18(1):81–85.
35. Li JY, Boado RJ, Pardridge WM. 2001. Cloned blood–brain barrier adenosine transporter is identical to the rat concentrative Na^+ nucleoside cotransporter CNT2. J Cereb Blood Flow Metab 21(8):929–936.
36. Redzic ZB, Biringer J, Barnes K, Baldwin SA, Al-Sarraf H, Nicola PA, Young JD, Cass CE, Barrand MA, Hladky SB. 2005. Polarized distribution of nucleoside transporters in rat brain endothelial and choroid plexus epithelial cells. J Neurochem 94(5):1420–1426.
37. Gutierrez MM, Brett CM, Ott RJ, Hui AC, Giacomini KM. 1992. Nucleoside transport in brush border membrane vesicles from human kidney. Biochim Biophys Acta 1105(1):1–9.
38. Patil SD, Ngo LY, Unadkat JD. 2000. Structure-inhibitory profiles of nucleosides for the human intestinal N1 and N2 Na^+–nucleoside transporters. Cancer Chemother Pharmacol 46(5):394–402.
39. Schaner ME, Wang J, Zevin S, Gerstin KM, Giacomini KM. 1997. Transient expression of a purine-selective nucleoside transporter (SPNTint) in a human cell line (HeLa). Pharm Res 14(10):1316–1321.
40. Schaner ME, Wang J, Zhang L, Su SF, Gerstin KM, Giacomini KM. 1999. Functional characterization of a human purine-selective, Na^+-dependent nucleoside transporter (hSPNT1) in a mammalian expression system. J Pharmacol Exp Ther 289(3):1487–1491.
41. Patil SD, Ngo LY, Glue P, Unadkat JD. 1998. Intestinal absorption of ribavirin is preferentially mediated by the Na^+–nucleoside purine (N1) transporter. Pharm Res 15(6):950–952.
42. Owen RP, Gray JH, Taylor TR, Carlson EJ, Huang CC, Kawamoto M, Johns SJ, Stryke D, Ferrin TE, Giacomini KM. 2005. Genetic analysis and functional characterization of polymorphisms in the human concentrative nucleoside transporter, CNT2. Pharmacogenet Genom 15(2):83–90.
43. King KM, Damaraju VL, Vickers MF, Yao SY, Lang T, Tackaberry TE, Mowles DA, Ng AM, Young JD, Cass CE. 2006. A comparison of the transportability, and its role in cytotoxicity, of clofarabine, cladribine, and fludarabine by recombinant human nucleoside transporters produced in three model expression systems. Mol Pharmacol 69(1):346–353.
44. Lang TT, Selner M, Young JD, Cass CE. 2001. Acquisition of human concentrative nucleoside transporter 2 (hcnt2) activity by gene transfer confers sensitivity to fluoropyrimidine nucleosides in drug-resistant leukemia cells. Mol Pharmacol 60(5):1143–1152.
45. Gerstin KM, Dresser MJ, Giacomini KM. 2002. Specificity of human and rat orthologs of the concentrative nucleoside transporter, SPNT. Am J Physiol Renal Physiol 283(2):F344–F349.
46. Li JY, Boado RJ, Pardridge WM. 2001. Differential kinetics of transport of 2′,3′-dideoxyinosine and adenosine via concentrative Na^+ nucleoside transporter CNT2 cloned from rat blood–brain barrier. J Pharmacol Exp Ther 299(2):735–740.
47. Ritzel MW, Ng AM, Yao SY, Graham K, Loewen SK, Smith KM, Ritzel RG, Mowles DA, Carpenter P, Chen XZ, et al. 2001. Molecular identification and characterization of novel human and mouse concentrative Na^+-nucleoside cotransporter proteins (hCNT3 and mCNT3) broadly selective for purine and pyrimidine nucleosides (system cib). J Biol Chem 276(4):2914–2927.

48. Yao SY, Ng AM, Loewen SK, Cass CE, Baldwin SA, Young JD. 2002. An ancient prevertebrate Na^+-nucleoside cotransporter (hfCNT) from the Pacific hagfish (*Eptatretus stouti*). Am J Physiol Cell Physiol 283(1):C155–C168.

49. Molina-Arcas M, Bellosillo B, Casado FJ, Montserrat E, Gil J, Colomer D, Pastor-Anglada M. 2003. Fludarabine uptake mechanisms in B-cell chronic lymphocytic leukemia. Blood 101(6):2328–2334.

50. Toan SV, To KK, Leung GP, de Souza MO, Ward JL, Tse CM. 2003. Genomic organization and functional characterization of the human concentrative nucleoside transporter-3 isoform (hCNT3) expressed in mammalian cells. Pflugers Arch 447(2):195–204.

51. Hu H, Endres CJ, Chang C, Umapathy NS, Lee EW, Fei YJ, Itagaki S, Swaan PW, Ganapathy V, Unadkat JD. 2006. Electrophysiological characterization and modeling of the structure activity relationship of the human concentrative nucleoside transporter 3 (hCNT3). Mol Pharmacol 69(5):1542–1553.

52. Smith KM, Ng AM, Yao SY, Labedz KA, Knaus EE, Wiebe LI, Cass CE, Baldwin SA, Chen XZ, Karpinski E, Young JD. 2004. Electrophysiological characterization of a recombinant human Na^+-coupled nucleoside transporter (hCNT1) produced in *Xenopus* oocytes. J Physiol 558(Pt 3):807–823.

53. Smith KM, Slugoski MD, Loewen SK, Ng AM, Yao SY, Chen XZ, Karpinski E, Cass CE, Baldwin SA, Young JD. 2005. The broadly selective human Na^+/nucleoside cotransporter (hCNT3) exhibits novel cation-coupled nucleoside transport characteristics. J Biol Chem 280(27):25436–25449.

54. Baldwin SA, Mackey JR, Cass CE, Young JD. 1999. Nucleoside transporters: molecular biology and implications for therapeutic development. Mol Med Today 5(5):216–224.

55. Zhang J, Visser F, Vickers MF, Lang T, Robins MJ, Nielsen LP, Nowak I, Baldwin SA, Young JD, Cass CE. 2003. Uridine binding motifs of human concentrative nucleoside transporters 1 and 3 produced in *Saccharomyces cerevisiae*. Mol Pharmacol 64(6):1512–1520.

56. Zhang J, Smith KM, Tackaberry T, Visser F, Robins MJ, Nielsen LP, Nowak I, Karpinski E, Baldwin SA, Young JD, Cass CE. 2005. Uridine binding and transportability determinants of human concentrative nucleoside transporters. Mol Pharmacol 68(3):830–839.

57. Lang TT, Young JD, Cass CE. 2004. Interactions of nucleoside analogs, caffeine, and nicotine with human concentrative nucleoside transporters 1 and 2 stably produced in a transport-defective human cell line. Mol Pharmacol 65(4):925–933.

58. Chang C, Swaan PW, Ngo LY, Lum PY, Patil SD, Unadkat JD. 2004. Molecular requirements of the human nucleoside transporters hCNT1, hCNT2, and hENT1. Mol Pharmacol 65(3):558–570.

59. Mangravite LM, Giacomini KM. 2003. Sorting of rat SPNT in renal epithelium is independent of N-glycosylation. Pharm Res 20(2):319–323.

60. Wang J, Giacomini KM. 1997. Molecular determinants of substrate selectivity in Na^+-dependent nucleoside transporters. J Biol Chem 272(46):28845–28848.

61. Wang J, Giacomini KM. 1999. Serine 318 is essential for the pyrimidine selectivity of the N2 Na^+-nucleoside transporter. J Biol Chem 274(4):2298–2302.

62. Wang J, Giacomini KM. 1999. Characterization of a bioengineered chimeric Na^+–nucleoside transporter. Mol Pharmacol 55(2):234–240.

63. Loewen SK, Ng AM, Yao SY, Cass CE, Baldwin SA, Young JD. 1999. Identification of amino acid residues responsible for the pyrimidine and purine nucleoside specificities of

human concentrative Na(+) nucleoside cotransporters hCNT1 and hCNT2. J Biol Chem 274(35):24475–24484.

64. Lai Y, Lee EW, Ton CC, Vijay S, Zhang H, Unadkat JD. 2005. Conserved residues F316 and G476 in the concentrative nucleoside transporter 1 (hCNT1) affect guanosine sensitivity and membrane expression, respectively. Am J Physiol Cell Physiol 288(1):C39–C45.
65. Zhang J, Tackaberry T, Ritzel MW, Raborn T, Barron G, Baldwin SA, Young JD, Cass CE. 2006. Cysteine-accessibility analysis of transmembrane domains 11–13 of human concentrative nucleoside transporter 3. Biochem J 394(Pt 2):389–398.
66. Loewen SK, Yao SY, Slugoski MD, Mohabir NN, Turner RJ, Mackey JR, Weiner JH, Gallagher MP, Henderson PJ, Baldwin SA, Cass CE, Young JD. 2004. Transport of physiological nucleosides and anti-viral and anti-neoplastic nucleoside drugs by recombinant *Escherichia coli* nucleoside–H(+) cotransporter (NupC) produced in *Xenopus laevis* oocytes. Mol Membr Biol 21(1):1–10.
67. Griffiths M, Beaumont N, Yao SY, Sundaram M, Boumah CE, Davies A, Kwong FY, Coe I, Cass CE, Young JD, Baldwin SA. 1997. Cloning of a human nucleoside transporter implicated in the cellular uptake of adenosine and chemotherapeutic drugs. Nat Med 3(1):89–93.
68. Yao SY, Ng AM, Muzyka WR, Griffiths M, Cass CE, Baldwin SA, Young JD. 1997. Molecular cloning and functional characterization of nitrobenzylthioinosine (NBMPR)-sensitive (es) and NBMPR-insensitive (ei) equilibrative nucleoside transporter proteins (rENT1 and rENT2) from rat tissues. J Biol Chem 272(45):28423–28430.
69. Kiss A, Farah K, Kim J, Garriock RJ, Drysdale TA, Hammond JR. 2000. Molecular cloning and functional characterization of inhibitor-sensitive (mENT1) and inhibitor-resistant (mENT2) equilibrative nucleoside transporters from mouse brain. Biochem J 352(Pt 2):363–372.
70. Handa M, Choi DS, Caldeiro RM, Messing RO, Gordon AS, Diamond I. 2001. Cloning of a novel isoform of the mouse NBMPR-sensitive equilibrative nucleoside transporter (ENT1) lacking a putative phosphorylation site. Gene 262(1–2):301–307.
71. Jennings LL, Hao C, Cabrita MA, Vickers MF, Baldwin SA, Young JD, Cass CE. 2001. Distinct regional distribution of human equilibrative nucleoside transporter proteins 1 and 2 (hENT1 and hENT2) in the central nervous system. Neuropharmacology 40(5):722–731.
72. Mangravite LM, Xiao G, Giacomini KM. 2003. Localization of human equilibrative nucleoside transporters, hENT1 and hENT2, in renal epithelial cells. Am J Physiol Renal Physiol 14:14.
73. Lai Y, Tse CM, Unadkat JD. 2004. Mitochondrial expression of the human equilibrative nucleoside transporter 1 (hENT1) results in enhanced mitochondrial toxicity of antiviral drugs. J Biol Chem 279(6):4490–4497.
74. Lee EW, Lai Y, Zhang H, Unadkat JD. 2006. Identification of the mitochondrial targeting signal of the human equilibrative nucleoside transporter 1 (hENT1): implications for interspecies differences in mitochondrial toxicity of fialuridine. J Biol Chem 281(24):16700–16706.
75. Ward JL, Sherali A, Mo ZP, Tse CM. 2000. Kinetic and pharmacological properties of cloned human equilibrative nucleoside transporters, ENT1 and ENT2, stably expressed in nucleoside transporter-deficient PK15 cells. Ent2 exhibits a low affinity for guanosine and cytidine but a high affinity for inosine. J Biol Chem 275(12):8375–8381.

76. Jarvis SM, Thorn JA, Glue P. 1998. Ribavirin uptake by human erythrocytes and the involvement of nitrobenzylthioinosine-sensitive (es)-nucleoside transporters. Br J Pharmacol 123(8):1587–1592.

77. Clarke ML, Damaraju VL, Zhang J, Mowles D, Tackaberry T, Lang T, Smith KM, Young JD, Tomkinson B, Cass CE. 2006. The role of human nucleoside transporters in cellular uptake of 4′-thio-{beta}-D-arabinofuranosylcytosine and {beta}-D-arabinosylcytosine. Mol Pharmacol 70(1):303–310.

78. Yao SY, Ng AM, Sundaram M, Cass CE, Baldwin SA, Young JD. 2001. Transport of antiviral 3′-deoxy-nucleoside drugs by recombinant human and rat equilibrative, nitrobenzylthioinosine (NBMPR)-insensitive (ENT2) nucleoside transporter proteins produced in *Xenopus* oocytes. Mol Membr Biol 18(2):161–167.

79. Sundaram M, Yao SY, Ng AM, Griffiths M, Cass CE, Baldwin SA, Young JD. 1998. Chimeric constructs between human and rat equilibrative nucleoside transporters (hENT1 and rENT1) reveal hENT1 structural domains interacting with coronary vasoactive drugs. J Biol Chem 273(34):21519–21525.

80. Huang M, Wang Y, Cogut SB, Mitchell BS, Graves LM. 2003. Inhibition of nucleoside transport by protein kinase inhibitors. J Pharmacol Exp Ther 304(2):753–760.

81. Carrier EJ, Auchampach JA, Hillard CJ. 2006. Inhibition of an equilibrative nucleoside transporter by cannabidiol: a mechanism of cannabinoid immunosuppression. Proc Natl Acad Sci U S A 103(20):7895–7900.

82. Griffiths M, Yao SY, Abidi F, Phillips SE, Cass CE, Young JD, Baldwin SA. 1997. Molecular cloning and characterization of a nitrobenzylthioinosine-insensitive (ei) equilibrative nucleoside transporter from human placenta. Biochem J 328(Pt 3):739–743.

83. Crawford CR, Patel DH, Naeve C, Belt JA. 1998. Cloning of the human equilibrative, nitrobenzylmercaptopurine riboside (NBMPR)-insensitive nucleoside transporter ei by functional expression in a transport-deficient cell line. J Biol Chem 273(9):5288–5293.

84. Yao SY, Ng AM, Vickers MF, Sundaram M, Cass CE, Baldwin SA, Young JD. 2002. Functional and molecular characterization of nucleobase transport by recombinant human and rat equilibrative nucleoside transporters 1 and 2. Chimeric constructs reveal a role for the ENT2 helix 5–6 region in nucleobase translocation. J Biol Chem 277(28):24938–24948.

85. Vickers MF, Kumar R, Visser F, Zhang J, Charania J, Raborn RT, Baldwin SA, Young JD, Cass CE. 2002. Comparison of the interaction of uridine, cytidine, and other pyrimidine nucleoside analogues with recombinant human equilibrative nucleoside transporter 2 (hENT2) produced in *Saccharomyces cerevisiae*. Biochem Cell Biol 80(5):639–644.

86. Hyde RJ, Cass CE, Young JD, Baldwin SA. 2001. The ENT family of eukaryote nucleoside and nucleobase transporters: recent advances in the investigation of structure/function relationships and the identification of novel isoforms. Mol Membr Biol 18(1):53–63.

87. Acimovic Y, Coe IR. 2002. Molecular evolution of the equilibrative nucleoside transporter family: identification of novel family members in prokaryotes and eukaryotes. Mol Biol Evol 19(12):2199–2210.

88. Stein A, Vaseduvan G, Carter NS, Ullman B, Landfear SM, Kavanaugh MP. 2003. Equilibrative nucleoside transporter family members from *Leishmania donovani* are electrogenic proton symporters. J Biol Chem 278(37):35127–35134.

89. Vickers MF, Zhang J, Visser F, Tackaberry T, Robins MJ, Nielsen LP, Nowak I, Baldwin SA, Young JD, Cass CE. 2004. Uridine recognition motifs of human equilibrative

nucleoside transporters 1 and 2 produced in *Saccharomyces cerevisiae*. Nucleosides Nucleotides Nucleic Acids 23(1–2):361–373.

90. Sundaram M, Yao SY, Ingram JC, Berry ZA, Abidi F, Cass CE, Baldwin SA, Young JD. 2001. Topology of a human equilibrative, nitrobenzylthioinosine (NBMPR)-sensitive nucleoside transporter (hENT1) implicated in the cellular uptake of adenosine and anti-cancer drugs. J Biol Chem 276(48):45270–45275.
91. Vickers MF, Mani RS, Sundaram M, Hogue DL, Young JD, Baldwin SA, Cass CE. 1999. Functional production and reconstitution of the human equilibrative nucleoside transporter (hENT1) in *Saccharomyces cerevisiae*. Interaction of inhibitors of nucleoside transport with recombinant hENT1 and a glycosylation-defective derivative (hENT1/N48Q). Biochem J 339(Pt 1):21–32.
92. Ward JL, Leung GP, Toan SV, Tse CM. 2003. Functional analysis of site-directed glycosylation mutants of the human equilibrative nucleoside transporter-2. Arch Biochem Biophys 411(1):19–26.
93. Sundaram M, Yao SY, Ng AM, Cass CE, Baldwin SA, Young JD. 2001. Equilibrative nucleoside transporters: mapping regions of interaction for the substrate analogue nitrobenzylthioinosine (NBMPR) using rat chimeric proteins. Biochemistry 40(27): 8146–8151.
94. Yao SY, Sundaram M, Chomey EG, Cass CE, Baldwin SA, Young JD. 2001. Identification of Cys140 in helix 4 as an exofacial cysteine residue within the substrate-translocation channel of rat equilibrative nitrobenzylthioinosine (NBMPR)-insensitive nucleoside transporter rENT2. Biochem J 353(Pt 2):387–393.
95. Visser F, Vickers MF, Ng AM, Baldwin SA, Young JD, Cass CE. 2002. Mutation of residue 33 of human equilibrative nucleoside transporters 1 and 2 alters sensitivity to inhibition of transport by dilazep and dipyridamole. J Biol Chem 277(1):395–401.
96. Visser F, Zhang J, Raborn RT, Baldwin SA, Young JD, Cass CE. 2005. Residue 33 of human equilibrative nucleoside transporter 2 is a functionally important component of both the dipyridamole and nucleoside binding sites. Mol Pharmacol 67(4):1291–1298.
97. SenGupta DJ, Lum PY, Lai Y, Shubochkina E, Bakken AH, Schneider G, Unadkat JD. 2002. A single glycine mutation in the equilibrative nucleoside transporter gene, hENT1, alters nucleoside transport activity and sensitivity to nitrobenzylthioinosine. Biochemistry 41(5):1512–1519.
98. Endres CJ, Unadkat JD. 2005. Residues Met89 and Ser160 in the human equilibrative nucleoside transporter 1 affect its affinity for adenosine, guanosine, *S*6-(4-nitrobenzyl)mercaptopurine riboside, and dipyridamole. Mol Pharmacol 67(3):837–844.
99. Endres CJ, SenGupta DJ, Unadkat JD. 2004. Mutation of leucine-92 selectively reduces the apparent affinity of inosine, guanosine, NBMPR [*S*6-(4-nitrobenzyl)mercaptopurine riboside] and dilazep for the human equilibrative nucleoside transporter, hENT1. Biochem J 380(Pt 1):131–137.
100. Visser F, Baldwin SA, Isaac RE, Young JD, Cass CE. 2005. Identification and mutational analysis of amino acid residues involved in dipyridamole interactions with human and *Caenorhabditis elegans* equilibrative nucleoside transporters. J Biol Chem 280(12):11025–11034.
101. Cabrita MA, Baldwin SA, Young JD, Cass CE. 2002. Molecular biology and regulation of nucleoside and nucleobase transporter proteins in eukaryotes and prokaryotes. Biochem Cell Biol 80(5):623–638.

102. Takimoto T, Kubota M, Tsuruta S, Kitoh T, Tanizawa A, Akiyama Y, Mikawa H. 1989. Cell cycle related change of Ara-C transport in HL-60 cells after differentiation induction. FEBS Lett 247(2):173–176.
103. Lee CW, Sokoloski JA, Sartorelli AC, Handschumacher RE. 1994. Differentiation of HL-60 cells by dimethylsulfoxide activates a Na(+)-dependent nucleoside transport system. In Vivo 8(5):795–801.
104. Lee CW. 1994. Decrease in equilibrative uridine transport during monocytic differentiation of HL-60 leukaemia: involvement of protein kinase C. Biochem J 300(Pt 2):407–412.
105. Lee CW, Sokoloski JA, Sartorelli AC, Handschumacher RE. 1991. Induction of the differentiation of HL-60 cells by phorbol 12-myristate 13-acetate activates a Na(+)-dependent uridine-transport system. Involvement of protein kinase C. Biochem J 274(Pt 1):85–90.
106. Soler C, Felipe A, Mata JF, Casado FJ, Celada A, Pastor-Anglada M. 1998. Regulation of nucleoside transport by lipopolysaccharide, phorbol esters, and tumor necrosis factor-alpha in human B-lymphocytes. J Biol Chem 273(41):26939–26945.
107. Soler C, Felipe A, Casado FJ, Celada A, Pastor-Anglada M. 2000. Nitric oxide regulates nucleoside transport in activated B lymphocytes. J Leukoc Biol 67(3):345–349.
108. Soler C, Garcia-Manteiga J, Valdes R, Xaus J, Comalada M, Casado FJ, Pastor-Anglada M, Celada A, Felipe A. 2001. Macrophages require different nucleoside transport systems for proliferation and activation. FASEB J 15(11):1979–1988.
109. Soler C, Valdes R, Garcia-Manteiga J, Xaus J, Comalada M, Casado FJ, Modolell M, Nicholson B, MacLeod C, Felipe A, Celada A, Pastor-Anglada M. 2001. Lipopolysaccharide-induced apoptosis of macrophages determines the up-regulation of concentrative nucleoside transporters Cnt1 and Cnt2 through tumor necrosis factor-alpha-dependent and -independent mechanisms. J Biol Chem 276(32):30043–30049.
110. Parodi J, Flores C, Aguayo C, Rudolph MI, Casanello P, Sobrevia L. 2002. Inhibition of nitrobenzylthioinosine-sensitive adenosine transport by elevated D-glucose involves activation of P2Y2 purinoceptors in human umbilical vein endothelial cells. Circ Res 90(5):570–577.
111. Casanello P, Torres A, Sanhueza F, Gonzalez M, Farias M, Gallardo V, Pastor-Anglada M, San Martin R, Sobrevia L. 2005. Equilibrative nucleoside transporter 1 expression is downregulated by hypoxia in human umbilical vein endothelium. Circ Res 97(1):16–24.
112. Chaudary N, Naydenova Z, Shuralyova I, Coe IR. 2004. Hypoxia regulates the adenosine transporter, mENT1, in the murine cardiomyocyte cell line, HL-1. Cardiovasc Res 61(4):780–788.
113. Chaudary N, Naydenova Z, Shuralyova I, Coe IR. 2004. The adenosine transporter, mENT1, is a target for adenosine receptor signaling and protein kinase Cepsilon in hypoxic and pharmacological preconditioning in the mouse cardiomyocyte cell line, HL-1. J Pharmacol Exp Ther 310(3):1190–1198.
114. Eltzschig HK, Abdulla P, Hoffman E, Hamilton KE, Daniels D, Schonfeld C, Loffler M, Reyes G, Duszenko M, Karhausen J, et al. 2005. HIF-1-dependent repression of equilibrative nucleoside transporter (ENT) in hypoxia. J Exp Med 202(11):1493–1505.
115. Kobayashi S, Zimmermann H, Millhorn DE. 2000. Chronic hypoxia enhances adenosine release in rat PC12 cells by altering adenosine metabolism and membrane transport. J Neurochem 74(2):621–632.
116. Coe I, Zhang Y, McKenzie T, Naydenova Z. 2002. PKC regulation of the human equilibrative nucleoside transporter, hENT1. FEBS Lett 517(1–3):201–205.

117. Del Santo B, Valdes R, Mata J, Felipe A, Casado FJ, Pastor-Anglada M. 1998. Differential expression and regulation of nucleoside transport systems in rat liver parenchymal and hepatoma cells. Hepatology 28(6):1504–1511.

118. Felipe A, Ferrer-Martinez A, Casado FJ, Pastor-Anglada M. 1997. Expression of sodium-dependent purine nucleoside carrier (SPNT) mRNA correlates with nucleoside transport activity in rat liver. Biochem Biophys Res Commun 233(2):572–575.

119. Del Santo B, Tarafa G, Felipe A, Casado FJ, Pastor-Anglada M. 2001. Developmental regulation of the concentrative nucleoside transporters CNT1 and CNT2 in rat liver. J Hepatol 34(6):873–880.

120. Fernandez-Veledo S, Huber-Ruano I, Aymerich I, Duflot S, Casado FJ, Pastor-Anglada M. 2006. Bile acids alter the subcellular localization of CNT2 (concentrative nucleoside cotransporter) and increase CNT2-related transport activity in liver parenchymal cells. Biochem J 395(2):337–344.

121. Fernandez-Veledo S, Valdes R, Wallenius V, Casado FJ, Pastor-Anglada M. 2004. Up-regulation of the high-affinity pyrimidine-preferring nucleoside transporter concentrative nucleoside transporter 1 by tumor necrosis factor-alpha and interleukin-6 in liver parenchymal cells. J Hepatol 41(4):538–544.

122. Valdes R, Ortega MA, Casado FJ, Felipe A, Gil A, Sanchez-Pozo A, Pastor-Anglada M. 2000. Nutritional regulation of nucleoside transporter expression in rat small intestine. Gastroenterology 119(6):1623–1630.

123. Pawelczyk T, Podgorska M, Sakowicz M. 2003. The effect of insulin on expression level of nucleoside transporters in diabetic rats. Mol Pharmacol 63(1):81–88.

124. Aguayo C, Sobrevia L. 2000. Nitric oxide, cGMP and cAMP modulate nitrobenzylthioinosine-sensitive adenosine transport in human umbilical artery smooth muscle cells from subjects with gestational diabetes. Exp Physiol 85(4):399–409.

125. MacDonnell P, Huff K, Grouse L, Guroff G. 1980. Brain nucleic acids. In *Biochemistry of Brain* (Kumar S, ed.), Pergamon Press, New York, pp. 211–240.

126. Kong W, Wang J. 2003. Hypoxanthine transport in human glioblastoma cells and effect on cell susceptibility to methotrexate. Pharm Res 20(11):1804–1811.

127. Musa H, Dobrzynski H, Berry Z, Abidi F, Cass CE, Young JD, Baldwin SA, Boyett MR. 2002. Immunocytochemical demonstration of the equilibrative nucleoside transporter rENT1 in rat sinoatrial node. J Histochem Cytochem 50(3):305–309.

128. Choi DS, Cascini MG, Mailliard W, Young H, Paredes P, McMahon T, Diamond I, Bonci A, Messing RO. 2004. The type 1 equilibrative nucleoside transporter regulates ethanol intoxication and preference. Nat Neurosci 7(8):855–861.

129. Damaraju VL, Damaraju S, Young JD, Baldwin SA, Mackey J, Sawyer MB, Cass CE. 2003. Nucleoside anticancer drugs: the role of nucleoside transporters in resistance to cancer chemotherapy. Oncogene 22(47):7524–7536.

130. Galmarini CM, Thomas X, Calvo F, Rousselot P, Rabilloud M, El Jaffari A, Cros E, Dumontet C. 2002. In vivo mechanisms of resistance to cytarabine in acute myeloid leukaemia. Br J Haematol 117(4):860–868.

131. Stam RW, den Boer ML, Meijerink JP, Ebus ME, Peters GJ, Noordhuis P, Janka-Schaub GE, Armstrong SA, Korsmeyer SJ, Pieters R. 2003. Differential mRNA expression of Ara-C-metabolizing enzymes explains Ara-C sensitivity in MLL gene-rearranged infant acute lymphoblastic leukemia. Blood 101(4):1270–1276.

132. Spratlin J, Sangha R, Glubrecht D, Dabbagh L, Young JD, Dumontet C, Cass C, Lai R, Mackey JR. 2004. The absence of human equilibrative nucleoside transporter 1 is associated with reduced survival in patients with gemcitabine-treated pancreas adenocarcinoma. Clin Cancer Res 10(20):6956–6961.

133. Giovannetti E, Del Tacca M, Mey V, Funel N, Nannizzi S, Ricci S, Orlandini C, Boggi U, Campani D, Del Chiaro M, et al. 2006. Transcription analysis of human equilibrative nucleoside transporter-1 predicts survival in pancreas cancer patients treated with gemcitabine. Cancer Res 66(7):3928–3935.

134. Rongen GA, Smits P, Ver Donck K, Willemsen JJ, De Abreu RA, Van Belle H, Thien T. 1995. Hemodynamic and neurohumoral effects of various grades of selective adenosine transport inhibition in humans. Implications for its future role in cardioprotection. J Clin Invest 95(2):658–668.

135. Parkinson FE, Rudolphi KA, Fredholm BB. 1994. Propentofylline: a nucleoside transport inhibitor with neuroprotective effects in cerebral ischemia. Gen Pharmacol 25(6):1053–1058.

136. Reiman T, Clarke ML, Dabbagh L, Vsianska M, Coupland RW, Belch AR, Baldwin SA, Young JD, Cass CE, Mackey JR. 2002. Differential expression of human equilibrative nucleoside transporter 1 (hENT1) protein in the Reed–Sternberg cells of Hodgkin's disease. Leuk Lymphoma 43(7):1435–1440.

137. Mackey JR, Jennings LL, Clarke ML, Santos CL, Dabbagh L, Vsianska M, Koski SL, Coupland RW, Baldwin SA, Young JD, Cass CE. 2002. Immunohistochemical variation of human equilibrative nucleoside transporter 1 protein in primary breast cancers. Clin Cancer Res 8(1):110–116.

138. Leabman MK, Huang CC, DeYoung J, Carlson EJ, Taylor TR, de la Cruz M, Johns SJ, Stryke D, Kawamoto M, Urban TJ, et al. 2003. Pharmacogenetics of Membrane Transporters. I: Natural variation in human membrane transporter genes reveals evolutionary and functional constraints. Proc Natl Acad Sci U S A 100(10):5896–5901.

139. Gray JH, Mangravite LM, Owen RP, Urban TJ, Chan W, Carlson EJ, Huang CC, Kawamoto M, Johns SJ, Stryke D, Ferrin TE, Giacomini KM. 2004. Functional and genetic diversity in the concentrative nucleoside transporter, CNT1, in human populations. Mol Pharmacol 65(3):512–519.

140. Owen RP, Lagpacan LL, Taylor TR, de la Cruz M, Huang CC, Kawamoto M, Johns SJ, Stryke D, Ferrin TE, Giacomini KM. 2006. Functional characterization and haplotype analysis of polymorphisms in the human equilibrative nucleoside transporter, ENT2. Drug Metab Dispos 34(1):12–15.

141. Badagnani I, Chan W, Castro RA, Brett CM, Huang CC, Stryke D, Kawamoto M, Johns SJ, Ferrin TE, Carlson EJ, Burchard EG, Giacomini KM. 2005. Functional analysis of genetic variants in the human concentrative nucleoside transporter 3 (CNT3; SLC28A3). Pharmacogenom J 5(3):157–165.

142. Osato DH, Huang CC, Kawamoto M, Johns SJ, Stryke D, Wang J, Ferrin TE, Herskowitz I, Giacomini KM. 2003. Functional characterization in yeast of genetic variants in the human equilibrative nucleoside transporter, ENT1. Pharmacogenetics 13(5):297–301.

143. Myers SN, Goyal RK, Roy JD, Fairfull LD, Wilson JW, Ferrell RE. 2006. Functional single nucleotide polymorphism haplotypes in the human equilibrative nucleoside transporter 1. Pharmacogenet Genom 16(5):315–320.

9

BILE ACID TRANSPORTERS

JYRKI J. ELORANTA AND GERD A. KULLAK-UBLICK
University Hospital Zurich, Division of Clinical Pharmacology and Toxicology, Zurich, Switzerland

Drug Transporters: Molecular Characterization and Role in Drug Disposition, Edited by Guofeng You and Marilyn E. Morris
Copyright © 2007 John Wiley & Sons, Inc.

9.1. OVERVIEW OF THE ENTEROHEPATIC CIRCULATION OF BILE ACIDS

Bile acids are amphipathic physiological detergents that play essential roles in promoting absorption, excretion, and transport of cholesterol, lipids, lipophilic nutrients, and other hydrophobic compounds in the liver and the intestine.[1] The two primary bile acids in humans are cholic acid (CA) and chenodeoxycholic acid (CDCA). In the intestinal bacterial flora these can be converted to secondary bile acids, deoxycholic acid (DCA) and lithocholic acid (LCA), respectively.

Bile acids are synthesized from cholesterol in the liver and stored in the gallbladder, from which they are postprandially released into the small intestine. From the small intestine the bile acids are recycled back to the liver via portal blood. A bile acid molecule may shuttle up to a dozen times between the liver and the intestine during one day, each time crossing the membrane domains of hepatocytes and enterocytes. This enterohepatic circulation of bile acids is highly efficient in healthy human individuals: Less than 10% of the total bile acid pool escapes ileal reabsorption, and thus enters the colon. In the colon, secondary bile acids are formed which may be either passively absorbed into colonocytes or lost through fecal excretion. This small loss into feces is compensated for by hepatic de novo synthesis of bile acids from cholesterol, catalyzed by cascades of cytochrome P450 (CYP) enzymes located in endoplasmic reticulum, in mitochondria, or in cytosol.[2] Thus, conversion of cholesterol into bile acids also provides an important route for elimination of cholesterol from the human body.

Disruption of the biliary secretion of bile acids may result in cholestasis, in which intrahepatic and systemic accumulation of bile acids causes cytotoxicity. In the liver this may eventually progress to fibrosis and cirrhosis.

9.2. CHIEF TRANSPORTERS IN THE ENTEROHEPATIC CIRCULATION OF BILE ACIDS

Enterohepatic circulation of bile acids is mediated by specific transporters, most of which are integral plasma membrane proteins, expressed in hepatocytes and enterocytes in a polarized manner.[3] The chief transporters involved in bile acid cycling and bile formation are presented in Figure 9.1 and discussed in more detail below.

Following their synthesis, bile acids are excreted from hepatocytes into bile canaliculi. This represents a major driving force for generation of bile flow and is a rate-limiting step in overall bile acid transport. The efflux of monovalent bile acids from hepatocytes occurs mainly via the canalicular bile salt export pump (BSEP, *ABCB11*), which belongs to the superfamily of ATP-binding cassette (ABC) transporters.[4] The ABC transporters couple the energy released from ATP hydrolysis to their transport activity.[5] Most eukaryotic ABC proteins share several evolutionarily conserved domains and typically contain two hydrophobic transmembrane domains, each of which spans the membrane six times. While BSEP is responsible for the efflux of monovalent bile acids from hepatocytes into bile, the multidrug resistance–associated protein 2 (MRP2, *ABCC2*) exports divalent and sulfated and/or glucuronidated bile acids as

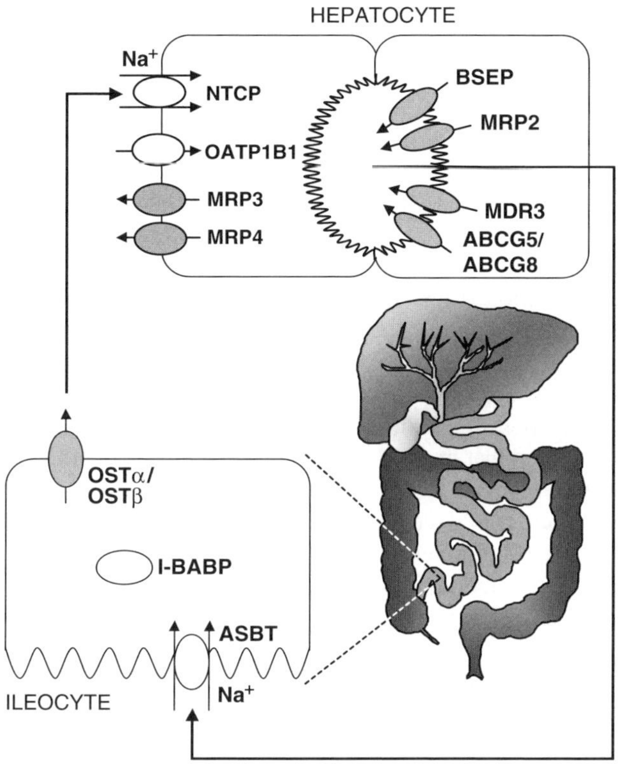

FIGURE 9.1. Transporters involved in the enterohepatic circulation of bile acids and bile formation. Bile acids are taken up into the ileocyte from the intestinal lumen by the sodium-dependent transporter ASBT, and putatively trafficked through the ileocyte by I-BABP. Bile acids are effluxed from the ileocyte to portal venous blood by the action of the OSTα/OSTβ heterodimer. At the basolateral membrane of the hepatocyte the main bile acid uptake system is NTCP, which transports bile acids from portal blood in a sodium-dependent manner. OATP1B1 may also contribute to hepatic bile acid uptake in a sodium-independent manner. In normal conditions, very little, if any, bile acids are effluxed back to portal blood at the basolateral membrane of hepatocytes. However, in states of cholestasis, expression of the bile acid spillover pumps MRP3 and MRP4 is increased, and they may mediate efflux of bile acids into systemic circulation. The main efflux system for bile acids from hepatocytes into bile is BSEP at the canalicular hepatocyte membrane. In addition, MRP2 may also export divalent and sulphated or glucuronidated bile acids into bile. MDR3 and the ABCG5/ABCG8 heterodimer transport phospholipids and cholesterol, respectively, from hepatocytes into bile.

well as other conjugated anions, such as chemotherapeutic agents and antibiotics, into bile.[6] MRP2 is also a member of the ABC transporter family which is located at the canalicular membrane of hepatocytes.

In addition to BSEP and MRP2, other transporters localized at the canalicular membrane of hepatocytes are involved in bile formation. Bile acids are the most prominent (60 to 70%) solid component of bile, but hepatic bile also contains phospholipids, notably phosphatidylcholine, and cholesterol. Phospholipids are translocated from the inner to the outer leaflet of the canalicular membranes of hepatocytes by yet another ABC transporter, the multidrug resistance protein 3 (MDR3, *ABCB4*).[5] The transport system for cholesterol at the canalicular membrane of hepatocytes is the protein heterodimer ABCG5/ABCG8.[7] The efflux of lipids mediated by MDR3 and ABCG5/ABCG8 promotes the formation of mixed micelles containing bile acids, cholesterol, and phospholipids. This process not only solubilizes cholesterol but also helps to protect cholangiocytes against the harmful detergent effects of bile acids in the biliary tree.

After their passage to the intestinal lumen, bile acids are efficiently taken up into ileocytes via the apical sodium-dependent bile acid transporter (ASBT, *SLC10A2*).[8] Characteristic of the SLC10 (solute carrier) family of transporters,[9] ASBT contains seven transmembrane domains, and its bile acid uptake activity is electrogenically coupled with cotransport of sodium.

It has been proposed that transcellular shuttling of bile acids from the apical membrane domain of enterocytes to the basolateral membrane may be facilitated by the small cytosolic protein called the ileal bile acid–binding protein (I-BABP).[10,11] It has been suggested that I-BABP interacts physically with ASBT,[12] although the functional significance of this interaction remains unclear.

At the basolateral membrane domain of ileal enterocytes, bile acids are extruded into portal blood by the heterodimeric organic solute transporter OSTα/OSTβ.[13] The OSTα protein is predicted to contain seven membrane-spanning domains, whereas the smaller subunit OSTβ has a single transmembrane domain. Coexpression of OSTα and OSTβ polypeptides is required for correct localization of the heterodimer at the cell membrane and for transport activity. Consistent with its role as the intestinal bile acid efflux transporter, distribution of the OSTα/OSTβ heterodimer along the intestine closely mirrors that of the bile acid uptake system ASBT.

Finally, to complete the enterohepatic circulation, bile acids are extracted from the portal blood circulation by the liver. The sodium–taurocholate cotransporting polypeptide (NTCP, *SLC10A1*) at the sinusoidal membrane of hepatocytes is the chief uptake system for bile acids from portal blood into the parenchymal cells of the liver.[14] NTCP belongs to the same sodium-dependent SLC10 transporter family as ASBT, and is similarly likely to contain seven transmembrane domains and a cytoplasmic carboxy terminus.[9,15] The amino acid identity between the human NTCP and ASBT is approximately 35%.

While NTCP is responsible for the sodium-dependent uptake of bile acids into hepatocytes, certain members of the organic anion transporter (OATP, *SLCO*) family, notably OATP1B1 (gene symbol *SLCO1B1*; previously known as OATP-C/OATP2/*SLC21A6*), may also contribute to bile acid extraction from the portal blood at the basolateral membrane of hepatocytes in a sodium-independent manner.[16]

Under normal physiological conditions, only negligible amounts of bile acids are effluxed back to portal blood at the basolateral hepatocyte membrane. However, in cholestatic states, the expression of the two basolateral bile acid overflow systems of the ABC transporter family, multidrug resistance-associated proteins 3 (MRP3, *ABCC3*) and 4 (MRP4, *ABCC4*), is increased.[17–19] Thus, in cholestasis, MRP3 and

MRP4 may transport substantial amounts of bile acids from hepatocytes back into the systemic circulation for subsequent renal excretion. In addition to MRP3 and MRP4, OSTα and OSTβ are also expressed at the basolateral membranes of hepatocytes in humans.[20] It is thus conceivable that OSTα/OSTβ may also contribute to alternative bile acid efflux during cholestasis.

Similar to I-BABP in enterocytes, proteins that are involved putatively in intracellular trafficking of bile acids from the basolateral to the canalicular membrane have been identified in hepatocytes.[21,22] One such intracellular protein capable of binding bile acids with a high affinity in the human liver is the hepatic bile acid–binding protein (HBAB), which may thus assist in the rapid transcellular vectorial transport of bile acids in hepatocytes.[23]

9.3. ENTEROHEPATIC BILE ACID TRANSPORTERS IN LIVER DISEASE

Chronic cholestatic liver diseases such as primary biliary cirrhosis and primary sclerosing cholangitis are characterized by an impairment of bile formation or of bile flow. Altered expression or function of bile acid transporters can be either a cause or a consequence of cholestasis, thus leading to hepatotoxicity due to accumulation of bile acids and cholephilic toxins in hepatocytes. Among the genes encoding transporters that are involved in bile acid transport or bile formation are several that have been identified or proposed as disease genes in the pathogenesis of cholestasis.

Progressive familial intrahepatic cholestasis type 2 (PFIC2) is caused by mutations in the *ABCB11* gene, which encodes BSEP.[24,25] These mutations in the *ABCB11* gene lead to a rapidly progressive hepatic dysfunction in early infancy. In such patients the biliary bile salt levels can be reduced to less than 1% that of normal subjects. In a recent case report, specific *ABCB11* mutations identified in an adolescent cholestatic patient correlated with reduced BSEP protein expression in vivo and decreased bile acid transport activity in vitro.[26] Another case report suggested that heterozygous BSEP deficiency may predispose to transient neonatal cholestasis.[27] Furthermore, defective or altered function or expression of BSEP may contribute to certain types of drug-induced cholestasis[28] and may be associated with intrahepatic cholestasis of pregnancy.[29]

Defective MDR3 expression has also been associated with the inherited liver disease PFIC, type 3.[30,31] PFIC3 is characterized by high bile acid concentrations and elevated γ-glutamyl transpeptidase activity in serum. Several PFIC3-associated mutations in the *ABCB4* gene may lead to either absent or severely decreased MDR3 expression at the canalicular membrane of hepatocytes. Similar to BSEP, there is increasing evidence suggesting that deficiency of impaired activity of MDR3 may be involved in cholestasis induced by drugs such as oral contraceptives.[32,33]

Inherited mutations in the *ABCC2* gene encoding the canalicular transporter MRP2 are linked to the Dubin–Johnson syndrome, characterized by reduced efflux of conjugated bilirubin into bile.[34–37] Some of these mutations have been reported to result in an absence of the MRP2 protein from the canalicular membrane of hepatocytes. In contrast to the PFIC syndromes, hepatic function is preserved in the Dubin–Johnson syndrome.

Mutations in the *SLC10A2* gene encoding ASBT have been identified that can cause primary bile acid malabsorption, a rare disorder of the intestine characterized by congenital diarrhoea, steatorrhea and reduced plasma cholesterol levels.[38] The ASBT variants carrying these mutations exhibit severely reduced bile acid transport activity in vitro.

No mutations in the *SLC10A1* gene encoding NTCP leading to clinically manifest defects in hepatic bile acid uptake have been characterized thus far. However, a recent study identified ethnicity-dependent single-nucleotide polymorphisms in the *SLC10A1* gene that were associated with a considerable decrease in transport function in vitro.[39] Thus, genetic heterogeneity in the *SLC10A1* gene may play a role in the etiology of hypercholanemia. Furthermore, certain human diseases, such as advanced stage primary biliary cirrhosis[40] and cholestatic alcoholic hepatitis,[41] are associated with reduced NTCP expression. However, this change in NTCP expression may be a consequence of cholestatic liver injury rather than a cause of it.

9.4. CONTROL OF BILE ACID TRANSPORT AND METABOLISM

In addition to their role as physiological detergents, bile acids possess crucial regulatory properties which allow them to control their own transport and metabolism within the enterohepatic circulation through multiple feedforward and feedback mechanisms. Hepatocytes and enterocytes possess numerous signaling pathways that are activated or modulated by bile acids, and ultimately serve to maintain intracellular concentrations of potentially toxic bile acids at a constant level.

An important mechanism toward controlling bile acid levels within cells is to adjust the cellular uptake or efflux of bile acids by regulating the expression and/or activity of uptake and efflux proteins, as discussed in detail below. It should be noted, however, that additional mechanisms are also operational in preventing intracellular bile acid concentrations from reaching toxic levels. One such mechanism is to regulate the de novo synthesis of bile acids according to the existing intracellular bile acid content. To reduce bile acid synthesis, the expression levels of the key CYP enzymes involved in de novo bile acid synthesis (i.e., CYP7A1, CYP8B1, and CYP27A1) are suppressed.[42] Furthermore, expression levels of several phase II enzymes that in addition to their role in drug detoxification may convert bile acids into less toxic and more hydrophilic derivatives are induced in response to elevated levels of bile acids.[43] These metabolizing enzymes include uridine 5′-diphosphate-glucuronosyltransferase 2B4 (UGT2B4) and dehydroepiandrosterone sulfotransferase (SULT2A1).

In this review we focus on the mechanisms that regulate the expression of bile acid transporters at the transcriptional level. However, it should be remembered that the activity of bile acid transporters is also known to be regulated at other levels, particularly through posttranslational protein modification and protein–protein interactions.[44,45] The relative importance of transcriptional and posttranslational events in controlling bile acid transport activity in either normal physiology or pathophysiology remains largely unelucidated. Both are likely to be highly important. It seems likely that the mechanisms involving modification at the protein level could elicit the most rapid changes in transporter activity, whereas transcriptional changes may be responsible for more intermediate- and longer-term regulation of transport.

9.5. NUCLEAR RECEPTORS AS TRANSCRIPTIONAL REGULATORS OF BILE ACID HOMEOSTASIS

Nuclear and steroid receptors form a large family of transcriptional regulators, with over 100 members in all metazoan organisms and almost 50 members in humans.[46] Most nuclear/steroid receptors share a conserved overall structural design: a ligand-independent activation function at the amino terminus, a central conserved DNA-binding domain, and a carboxy-terminal region containing regions mediating ligand binding, dimerization, and ligand-dependent transactivation. Most nuclear/steroid receptors bind to their DNA response elements as either hetero- or homodimers, which is reflected in their preferred DNA-binding motifs typically containing two hexameric half sites. These hexamers, the general consensus sequence for which is AGGTCA, can be arranged as direct (DR), inverted (IR), or everted (ER) repeats, separated by a variable and receptor-specific number of base pairs.

The full transcriptional activity of most, but not all, nuclear/steroid receptors depends on a physical interaction by an agonist with their ligand-binding pocket. These ligands are typically small lipophilic molecules such as hormones, fatty acids, oxysterols, or bile acids. Their binding induces a conformational shift in the carboxy termini of the receptors, allowing their interaction with transcriptional coactivators.[47] These coactivators may act by modifying histones or other promoter-associated proteins or by altering local chromatin structure in a way that increases the rate of transcriptional initiation. Conversely, in the absence of an agonistic ligand, or when bound to an antagonistic ligand, the carboxy termini of nuclear and steroid receptors associate with transcriptional corepressors that render the proximal promoters less permissive for transcription. The dependence of the transcriptional activity of most nuclear and steroid receptors on specific ligands allows them to monitor intracellular environment and to elicit rapid transcriptional responses to changes in the concentrations of specific compounds.

9.5.1. FXR: The Master Regulator of Bile Acid Transport and Metabolism

The chief sensor of intracellular bile acid levels and the main executor of bile acid–induced transcriptional programs is the nuclear receptor farnesoid X receptor (FXR).[48] Bile acids interact directly with the ligand-binding domain of FXR. In transactivation and coactivator recruitment assays, CDCA is the most efficient FXR activator, followed by DCA and CA.[49–51] LCA alone can weakly activate FXR; however, it strongly antagonizes CDCA-mediated stimulation of FXR.[52] This apparent antagonism of FXR function may contribute to LCA-induced cholestasis.

Bile acids are not the only ligands that interact with FXR directly. Recently, it has been suggested that the oxysterol 22(*R*)-hydroxycholesterol, an intermediate in the synthesis of bile acids and steroid hormones, can also interact directly with the ligand-binding pocket of FXR and mediate gene activation.[53] Traditionally, oxysterols have been considered to be agonistic ligands for another member of the nuclear receptor family, the liver X receptor (LXR), which functions as a chief regulator of cholesterol homeostasis. Furthermore, androsterone, a testosterone metabolite, can interact directly with the FXR ligand-binding domain and enhance transcriptional activity of FXR through coactivator recruitment.[54] It is possible that different ligands induce different conformational states of FXR, leading to distinct patterns of gene regulation.[55] In this review we only discuss the significance of bile acids as FXR ligands and activators of FXR-induced transcriptional programs.

In accordance with its function as a bile acid receptor, FXR is expressed abundantly in the tissues exposed to bile acids: liver, intestine, and kidneys. The consensus DNA-binding motif for FXR is in the format of an inverted repeat-1 (IR-1, inverted hexameric repeat separated by one base pair),[56] to which it binds as a heterodimer with another nuclear receptor, the retinoid X receptor (RXR). However, other configurations, such as IR-0 and ER-8, may allow binding of FXR in the context of specific promoters.[57,58]

Supporting its physiological role in controlling bile acid homeostasis, FXR-null mice exhibit a phenotype similar to that of persons suffering from Byler disease, an inherited cholestatic liver disorder.[59] Upon feeding with cholic acid, these mice lacking FXR exhibit severely increased hepatotoxicity compared to the wild-type counterparts. In further support of the importance of FXR in human bile acid homeostasis, certain hereditary forms of cholestasis are associated with decreased FXR activity.[60]

Encouraging reports showing hepatoprotective effects by synthetic FXR agonists in rodent models of cholestasis[61,62] suggest that specific FXR ligands could also prove to be useful in the treatment of cholestatic liver diseases in humans. One major concern regarding FXR-based therapy is that this nuclear receptor is involved in other metabolic processes in addition to bile acid homeostasis. Thus, its activation even with specific ligands may affect these unrelated processes in an undesired manner. For example, negative feedback suppression of bile acid–synthesizing enzymes by bile acids is mediated by a complex cascade involving FXR, leading to reduced conversion of cholesterol to bile acids. In addition to controlling bile acid transporter and synthesis genes, FXR is crucially involved in glucose homeostasis via regulation of the genes encoding gluconeogenic enzymes.[63,64] In addition, FXR activated by bile acids has recently been shown to be an important mediator of liver regeneration.[65] It was hypothesized that FXR activation by increased levels of bile acids would be a signal of reduced hepatic capacity and function. Thus, until the liver reaches the sufficient number of new hepatocytes to cope with the bile acid levels, FXR continues to trigger their proliferation.

9.5.2. Role of PXR and VDR as Bile Acid Sensors

FXR is not the only nuclear receptor that can use bile acids as ligands that modulate transcriptional activity. In addition, the pregnane X receptor (PXR) and the vitamin D receptor (VDR) can mediate transcriptional responses to certain bile acids. Together with its xenosensor partner, constitutive androstane receptor (CAR), PXR is a nuclear receptor that typically utilizes drugs and xenobiotics as its ligands.[66] In response to these ligands, PXR induces the expression of genes encoding proteins involved in drug detoxification and elimination pathways. In addition to xenobiotics, certain bile acids, such as the highly toxic LCA, can serve as agonistic ligands for PXR.[67,68] Indeed, activation of PXR can protect mouse livers against LCA-mediated injury.[67] Double-knockout mice lacking both FXR and PXR exhibit more severe disturbances of bile acid metabolism than mice lacking only one of the nuclear receptors, demonstrating that both contribute to bile acid homeostasis.[69] PXR is a master regulator of the gene encoding the CYP3A4 enzyme,[70] which, in addition to its role in detoxifying xenobiotics, also metabolizes bile acids to less toxic and more easily excreted derivatives. Thus, by being both activators of the *CYP3A4* gene and substrates of the CYP3A4 enzyme, bile acids can initiate a hepatoprotective feedforward loop via PXR. Similar to PXR, VDR can utilize the toxic secondary bile acid LCA as an agonistic ligand.[71] Whereas VDR is expressed only weakly in hepatocytes, it is present abundantly in enterocytes. It is thus likely to play a more prominent role in the intestine in response to LCA. In addition to PXR, VDR can also transactivate the *CYP3A4* gene. Thus, VDR activated by elevated levels of LCA in the intestine may, in a feedforward manner, contribute to enhanced expression of the CYP3A4 enzyme, which is capable of detoxifying LCA.

Common to all three nuclear receptors that can utilize bile acids as ligands (i.e., FXR, PXR, VDR) is that they all bind to their respective DNA response elements as heterodimers with the nuclear receptor RXR, with very few exceptions. Positive or negative effects on the transcriptional activity of RXR-containing heterodimers by the RXR ligand 9-*cis*-retinoic acid appear to depend on the exact promoter context.[72,73]

9.5.3. Bile Acid–Induced Transcriptional Repressor SHP

FXR can also negatively regulate the rate of transcription from specific promoters. In rare cases, FXR is known to repress its target genes, such as those encoding the human apolipoprotein A-I[74] and human apolipoprotein C-III,[75] through direct binding to the promoter. However, FXR more commonly mediates negative transcriptional responses to elevated levels of bile acids indirectly, through inducing expression of its target gene encoding the transcriptional repressor called *small heterodimer partner*, SHP. SHP is an atypical member of the nuclear receptor family; it does not contain a DNA-binding domain and does not depend on a ligand for its activity. Instead, it interacts directly with a variety of DNA-bound transcriptional activators, interfering with their transcriptional activity. Although SHP can suppress the activity of transcription factors from certain other families, it most commonly targets other nuclear receptors and steroid receptors (reviewed in ref. 76). Via the SHP pathway, bile acids can influence the activities of a wider range of transcription factors than only those nuclear receptors that are modulated by them directly.

Several mechanisms have been suggested for SHP-mediated suppression of transactivator proteins. SHP may compete with transcriptional coactivators over the same or overlapping interaction surface on transactivators that are bound to the promoter elements.[77] Alternatively, SHP may interfere with the binding of transactivators to their DNA response elements.[78] Additionally, SHP may recruit transcriptional corepressors to its target promoters, thus contributing to the reduced transcriptional rate of a given promoter.[79]

The importance of SHP in the control of bile acid homeostasis is indicated by the fact that SHP-null mice exhibit an imbalance in bile acid metabolism and abnormal responses when challenged with diets rich in bile acids.[80,81]

9.6. FXR-DEPENDENT MECHANISMS THAT REGULATE HUMAN BILE ACID TRANSPORTER GENES

We discuss next the FXR-dependent effects of bile acids on the expression of transporter genes in hepatocytes and enterocytes. These concepts are summarized in Figure 9.2. It should be emphasized that although we focus on FXR-dependent effects by

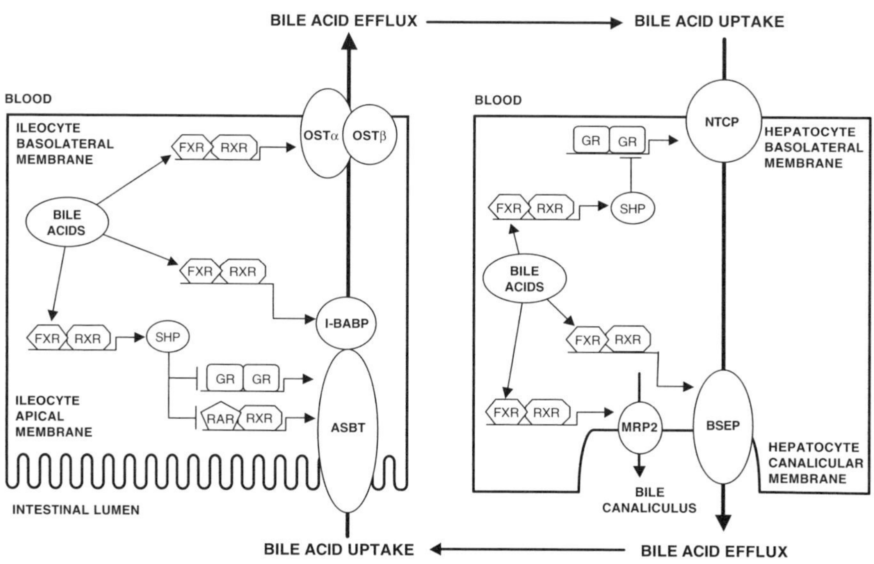

FIGURE 9.2. Bile acid–induced FXR-dependent transcriptional mechanisms that regulate the genes encoding enterohepatic bile acid transporters. FXR activated by bile acids induces the expression of the intestinal (OSTα/OSTβ) and hepatic (BSEP, MRP2) bile acid efflux systems, as well as of the intestinal intracellular bile acid transporter (I-BABP), via direct binding to its response elements in the respective regulatory promoter regions. FXR binds DNA as a heterodimer with the nuclear receptor RXR. Decreased expression of the intestinal (ASBT) and hepatic (NTCP) bile acid uptake transporters occurs by a mechanism that involves FXR indirectly. In this cascade, bile acid–activated FXR induces the expression of the transcriptional repressor SHP, which subsequently interferes with the activity of the transactivator proteins that regulate the expression of the genes encoding bile acid–uptake systems. In the case of the human *SLC10A1* (NTCP) and *SLC10A2* (ASBT) promoters, the transactivator targeted by SHP is the steroid receptor GR. In the context of the human *SLC10A2* promoter, the nuclear receptor heterodimer RAR-RXR has been suggested as an alternative or parallel target for SHP-mediated transcriptional suppression.

bile acids in this review, bile acids also elicit signaling pathways that lead to changes in specific gene expression which are independent of FXR or other nuclear receptors. These alternative bile acid–stimulated pathways include signaling through mitogen-activated protein kinases[82,83] and through the G-protein-coupled receptor TGR5.[84] Parallel bile acid–stimulated signaling pathways may ensure that the desired transcriptional responses are achieved.

9.6.1. Positive Feedforward Control of Bile Acid Efflux Systems by Bile Acids

In response to bile acids, FXR induces BSEP expression via direct interaction of FXR-RXR heterodimers with an IR-1 element located in the proximal promoter of the *ABCB11* gene.[85–87] Thus, excessive levels of bile acids stimulate hepatocanalicular clearance of bile acids. The FXR-RXR binding element is conserved between the human and rodent *ABCB11/Abcb11* promoters, supporting its functional importance. In agreement with FXR being a crucial activator of the *ABCB11* gene, the BSEP expression is reduced in mice lacking the *FXR* gene.[59]

The *ABCC2* gene encoding the MRP2 transporter provides an illustrative example of the complex interactions between the metabolic nuclear receptors. Both the human and rodent *ABCC2/Abcc2* promoters can be activated by either FXR, PXR, or CAR in the presence of their respective ligands or activators.[58] Interestingly, each of these three nuclear receptors can interact with the same atypical ER-8 element present in the regulatory region of the *ABCC2* gene. The relative significance or the degree of redundancy between the three transcription factors in the regulation of MRP2 expression is not yet clear.

FXR also transactivates the *ABCB4* gene encoding MDR3, the phospholipid transporter at the canalicular membrane of hepatocytes.[85] Thus, bile acids may, in a coordinated manner and via activation of FXR, induce the excretion of both bile acids (BSEP, MRP2) and phospholipids (MDR3) into bile. In agreement with this, FXR-null mice are prone to developing cholesterol gallstones, caused by the deficiency in the excretion of bile acids and phospholipids.[59]

Expression of the genes encoding the alternative bile acid export pumps at the basolateral hepatocyte membranes, MRP3 (*ABCC3*) and MRP4 (*ABCC4*), appears to be induced in cholestasis in an FXR-independent manner, at least in bile duct–ligated or bile acid–fed mice.[86,87] It may be that the other bile acid–responsive nuclear receptor, PXR, is responsible for the bile acid–mediated induction of the *ABCC3* and *ABCC4* genes, or can at least compensate in the absence of FXR.[88] In agreement with the cholestatic mouse models, it has been reported that in human cholestatic liver disease, expression of MRP3 and MRP4 is similarly elevated.[40,89,90]

In analogy to the *ABCB11* gene encoding the hepatic efflux system BSEP, the two genes encoding the heterodimeric efflux system in the intestine, OSTα and OSTβ, are induced by bile acids through direct binding of FXR-RXR heterodimers to the two human *OST* promoters.[91,92] While the *OST*β promoter appears to have a single IR-1-like binding site for FXR-RXR, the human *OST*α promoter contains two adjacent IR-1-like FXR response elements, both of which are functional and required for full response to bile acids. Physiological evidence in support of the induction of OST expression by bile acids is provided by a recent study showing that both mRNA and protein levels of OSTα and OSTβ are increased in cholestatic liver tissue of primary biliary cirrhosis patients.[93] In further agreement with the proposed role of FXR in inducing the *OST* genes, their baseline expression is reduced and their bile acid–mediated induction is abolished in FXR-null mice.[92–94] Accordingly, the location and sequence of the IR-1-like FXR response elements are largely conserved in both human and mouse *OSTα* and *OSTβ* genes.

Bile acid–activated FXR also stimulates expression of the gene encoding the intracellular bile acid transporter I-BABP in the ileum.[95] Supporting this, I-BABP expression is reduced dramatically in FXR-deficient mice.[59] FXR response elements of the IR-1 type have been identified in both the human and rodent *I-BABP/I-babp* promoters. The bile acid–activated FXR can thus up-regulate the expression of both the membrane-bound bile acid uptake system OSTα/OSTβ and the intracellular bile acid transporter I-BABP in a coordinated manner within enterocytes.

9.6.2. Negative Feedback Control of Bile Acid Uptake Systems by Bile Acids

In rodent models of cholestasis, such as bile duct ligation or bile acid feeding, expression of the hepatic bile acid uptake system Ntcp is suppressed at both the protein and mRNA levels.[96,97] Furthermore, certain human cholestatic states, such as advanced-stage primary biliary cirrhosis[40] and cholestatic alcoholic hepatitis,[41] are similarly associated with reduced NTCP expression. It has recently been suggested that the mechanism of down-regulation of human NTCP expression in response to bile acids involves the FXR-SHP cascade. FXR-induced SHP targets the transactivator glucocorticoid receptor (GR), which interacts with its response element just upstream of the transcription initiation site on the human *SLC10A1* promoter.[98]

FXR also suppresses transcription of the *SLCO1B1* gene encoding the sodium-independent bile acid transporter OATP1B1 at the basolateral membrane of human hepatocytes.[99] FXR-mediated repression of the *SLCO1B1* promoter takes place through a multistep regulatory cascade which involves bile acid–mediated repression of the gene encoding the liver-enriched homeodomain factor hepatocyte nuclear factor 1α (HNF1α). HNF1α, in turn, is a strong DNA-binding transactivator of the *OATP1B1* promoter. The regulatory region of the *HNF1α* gene contains a consensus DNA-binding response element for the nuclear receptor HNF4α, the transcriptional activity of which can be targeted by negative interference by FXR-induced SHP.

Similarly to NTCP, treatment of cultured cells with bile acids suppresses expression of the intestinal bile acid uptake transporter ASBT, another member of the SLC10 transporter family in humans.[100] In accordance with this, *SLC10A2* gene expression is reduced in patients with obstructive cholestasis compared to healthy controls.[101] Two transactivators of the human *SLC10A2* promoter, the GR[98,102] and the nuclear receptor heterodimer retinoic acid receptor (RAR)-RXR,[100] have been suggested as targets for negative interference by the bile acid–induced transcriptional repressor SHP. It remains to be determined whether these two transactivators are targeted simultaneously by SHP to ensure efficient down-regulation of the *SLC10A2* promoter by bile acids, or whether the two pathways are targeted under different circumstances. Interestingly, while both the *SLC10A1* (NTCP) and *SLC10A2* (ASBT) genes are transactivated by

GR, the relative locations and configurations of the GR response elements within the two promoters are not conserved.

9.7. CROSSTALK BETWEEN THE TRANSCRIPTIONAL CONTROL OF BILE ACID AND DRUG TRANSPORTERS

Transport and metabolism of exogenous compounds, such as drugs, nutrients, and environmental xenobiotics, bear many similarities to those of endogenous bile acids. Indeed, drugs may undergo enterohepatic circulation similarly to bile acids. Several transporters recognize both bile acids and drugs as substrates. For example, although their significance for overall bile acid transport in normal physiology remains uncertain, the two transporters of the human OATP1B subfamily, OATP1B1 and OATP1B3, can transport bile acids in addition to drugs such as methotrexate and rifampicin.[103] The transcriptional regulatory circuits controlling the expression of drug transporter genes often contain feedforward and feedback loops reminiscent of the mechanisms outlined for regulation of bile acid transport above (reviewed in ref. 76). It is becoming increasingly apparent that in addition to reprogramming of genes involved in bile acid homeostasis, changes in intracellular concentrations of bile acids also influence the expression levels of hepatic and intestinal drug transporters. Thus, changes in intracellular bile acid levels may affect the efficiency of drug extraction and elimination.

HNF4α is a liver-enriched nuclear receptor with a critical role in maintaining the hepatic pattern of gene expression.[104] It binds to the DNA response elements on its target promoters, preferably of the DR-1 or DR-2 configuration, as homodimers. The ligand-binding domain of HNF4α has been suggested to be bound constitutively by endogenous fatty acids[105,106]; thus, its activity may not be modulated readily by exogenous ligands. In addition to its previous roles in regulation of glucose and cholesterol metabolism, HNF4α has recently emerged as a regulator of hepatic drug transport in humans. The genes encoding two major hepatic basolateral drug transporters in humans, organic anion transporter 2 (OAT2; *SLC22A7*)[107] and organic cation transporter 1 (OCT1;*SLC22A1*),[108] are transactivated by HNF4α. OAT2 mediates uptake of drugs such as salicylates and cephalosporins from sinusoidal blood, whereas OCT1 transports dopamine, metformin, and verapamil, among other substrates.

In the regulatory regions of the human *SLC22A7* gene, HNF4α transactivation is mediated by a single DR-1 element. The human *SLC22A1* promoter, in turn, contains two adjacent lower-affinity DR-2 elements, both of which are required for maximal transactivation by HNF4α. In both promoter contexts, the bile acid–induced transcriptional repressor SHP interferes with transactivation by HNF4α. It thus appears that in conditions of elevated intracellular bile acid concentrations, the expression of two major drug uptake systems at the basolateral hepatic membrane is reduced. This could limit the amount of xenobiotics that enters the hepatocyte for subsequent metabolism when intracellular levels of toxic bile acids are already elevated. Also, the possibility of decreased hepatic extraction of drugs that are substrates of OAT2 and OCT1 should be taken into account when such drugs are administered to patients suffering from cholestasis.

9.8. CONCLUSIONS

Enterohepatic bile acid transporters are crucial in maintaining bile acid homeostasis in healthy persons. Their importance for hepatic and intestinal function is emphasized by numerous recent demonstrations that either hereditary or acquired disturbances in the activity and/or expression of bile acid transporters are associated with cholestatic disease conditions. Bile acids regulate the expression of transporters that mediate the enterohepatic circulation of both bile acids and drugs via complex feedforward and feedback mechanisms. The main orchestrator of these transcriptional circuits is FXR, which, in response to bile acids, controls the expression levels of bile acid transporters at all membrane domains of enterohepatic circulation.

Increased knowledge of the transcriptional mechanisms governing changes in bile acid transporter gene expression has improved our understanding of the pathophysiology of cholestatic liver diseases. Additionally, this information is likely to provide us with novel tools for designing therapeutic strategies to combat diseases of the liver and the intestine.

REFERENCES

1. Hofmann, A.F. (1999) The continuing importance of bile acids in liver and intestinal disease. *Arch Intern Med*, *159*, 2647–2658.
2. Russell, D.W. (2003) The enzymes, regulation, and genetics of bile acid synthesis. *Annu Rev Biochem*, *72*, 137–174.
3. Meier, P.J. and Stieger, B. (2002) Bile salt transporters. *Annu Rev Physiol*, *64*, 635–661.
4. Arrese, M. and Ananthanarayanan, M. (2004) The bile salt export pump: molecular properties, function and regulation. *Pflugers Arch*, *449*, 123–131.
5. Borst, P. and Elferink, R.O. (2002) Mammalian ABC transporters in health and disease. *Annu Rev Biochem*, *71*, 537–592.
6. Keppler, D. and König, J. (2000) Hepatic secretion of conjugated drugs and endogenous substances. *Semin Liver Dis*, *20*, 265–272.
7. Yu, L., Hammer, R.E., Li-Hawkins, J., Von Bergmann, K., Lutjohann, D., Cohen, J.C. and Hobbs, H.H. (2002) Disruption of Abcg5 and Abcg8 in mice reveals their crucial role in biliary cholesterol secretion. *Proc Natl Acad Sci U S A*, *99*, 16237–16242.
8. Wong, M.H., Oelkers, P., Craddock, A.L. and Dawson, P.A. (1994) Expression cloning and characterization of the hamster ileal sodium-dependent bile acid transporter. *J Biol Chem*, *269*, 1340–1347.
9. Hagenbuch, B. and Dawson, P. (2004) The sodium bile salt cotransport family SLC10. *Pflugers Arch*, *447*, 566–570.
10. Gong, Y.Z., Everett, E.T., Schwartz, D.A., Norris, J.S. and Wilson, F.A. (1994) Molecular cloning, tissue distribution, and expression of a 14-kDa bile acid–binding protein from rat ileal cytosol. *Proc Natl Acad Sci U S A*, *91*, 4741–4745.
11. Kramer, W., Stengelin, S., Baringhaus, K.H., Enhsen, A., Heuer, H., Becker, W., Corsiero, D., Girbig, F., Noll, R. and Weyland, C. (1999) Substrate specificity of the ileal and the hepatic Na(+)/bile acid cotransporters of the rabbit: I. Transport studies with membrane vesicles and cell lines expressing the cloned transporters. *J Lipid Res*, *40*, 1604–1617.

12. Kramer, W., Wess, G., Bewersdorf, U., Corsiero, D., Girbig, F., Weyland, C., Stengelin, S., Enhsen, A., Bock, K., Kleine, H., Le Dreau, M.A. and Schafer, H.L. (1997) Topological photoaffinity labeling of the rabbit ileal Na^+/bile-salt-cotransport system. *Eur J Biochem*, *249*, 456–464.

13. Dawson, P.A., Hubbert, M., Haywood, J., Craddock, A.L., Zerangue, N., Christian, W.V. and Ballatori, N. (2005) The heteromeric organic solute transporter alpha-beta, Ostalpha-Ostbeta, is an ileal basolateral bile acid transporter. *J Biol Chem*, *280*, 6960–6968.

14. Hagenbuch, B., Stieger, B., Foguet, M., Lubbert, H. and Meier, P.J. (1991) Functional expression cloning and characterization of the hepatocyte Na^+/bile acid cotransport system. *Proc Natl Acad Sci U S A*, *88*, 10629–10633.

15. Mareninova, O., Shin, J.M., Vagin, O., Turdikulova, S., Hallen, S. and Sachs, G. (2005) Topography of the membrane domain of the liver Na^+-dependent bile acid transporter. *Biochemistry*, *44*, 13702–13712.

16. Hagenbuch, B. and Meier, P.J. (2003) The superfamily of organic anion transporting polypeptides. *Biochim Biophys Acta*, *1609*, 1–18.

17. Donner, M.G. and Keppler, D. (2001) Up-regulation of basolateral multidrug resistance protein 3 (Mrp3) in cholestatic rat liver. *Hepatology*, *34*, 351–359.

18. Soroka, C.J., Lee, J.M., Azzaroli, F. and Boyer, J.L. (2001) Cellular localization and up-regulation of multidrug resistance–associated protein 3 in hepatocytes and cholangiocytes during obstructive cholestasis in rat liver. *Hepatology*, *33*, 783–791.

19. Denk, G.U., Soroka, C.J., Takeyama, Y., Chen, W.S., Schuetz, J.D. and Boyer, J.L. (2004) Multidrug resistance-associated protein 4 is up-regulated in liver but down-regulated in kidney in obstructive cholestasis in the rat. *J Hepatol*, *40*, 585–591.

20. Ballatori, N., Christian, W.V., Lee, J.Y., Dawson, P.A., Soroka, C.J., Boyer, J.L., Madejczyk, M.S. and Li, N. (2005) OSTalpha-OSTbeta: a major basolateral bile acid and steroid transporter in human intestinal, renal, and biliary epithelia. *Hepatology*, *42*, 1270–1279.

21. Stolz, A., Takikawa, H., Ookhtens, M. and Kaplowitz, N. (1989) The role of cytoplasmic proteins in hepatic bile acid transport. *Annu Rev Physiol*, *51*, 161–176.

22. Agellon, L.B. and Torchia, E.C. (2000) Intracellular transport of bile acids. *Biochim Biophys Acta*, *1486*, 198–209.

23. Stolz, A., Hammond, L., Lou, H., Takikawa, H., Ronk, M. and Shively, J.E. (1993) cDNA cloning and expression of the human hepatic bile acid–binding protein: a member of the monomeric reductase gene family. *J Biol Chem*, *268*, 10448–10457.

24. Strautnieks, S.S., Bull, L.N., Knisely, A.S., Kocoshis, S.A., Dahl, N., Arnell, H., Sokal, E., Dahan, K., Childs, S., Ling, V. et al. (1998) A gene encoding a liver-specific ABC transporter is mutated in progressive familial intrahepatic cholestasis. *Nat Genet*, *20*, 233–238.

25. Jansen, P.L., Strautnieks, S.S., Jacquemin, E., Hadchouel, M., Sokal, E.M., Hooiveld, G.J., Köning, J.H., De Jager-Krikken, A., Kuipers, F., Stellaard, F. et al. (1999) Hepatocanalicular bile salt export pump deficiency in patients with progressive familial intrahepatic cholestasis. *Gastroenterology*, *117*, 1370–1379.

26. Noe, J., Kullak-Ublick, G.A., Jochum, W., Stieger, B., Kerb, R., Haberl, M., Mullhaupt, B., Meier, P.J. and Pauli-Magnus, C. (2005) Impaired expression and function of the bile salt export pump due to three novel ABCB11 mutations in intrahepatic cholestasis. *J Hepatol*, *43*, 536–543.

27. Hermeziu, B., Sanlaville, D., Girard, M., Leonard, C., Lyonnet, S. and Jacquemin, E. (2006) Heterozygous bile salt export pump deficiency: a possible genetic predisposition to transient neonatal cholestasis. *J Pediatr Gastroenterol Nutr*, *42*, 114–116.

28. Stieger, B., Fattinger, K., Madon, J., Kullak-Ublick, G.A. and Meier, P.J. (2000) Drug- and estrogen-induced cholestasis through inhibition of the hepatocellular bile salt export pump (Bsep) of rat liver. *Gastroenterology*, *118*, 422–430.

29. Eloranta, M.L., Hakli, T., Hiltunen, M., Helisalmi, S., Punnonen, K. and Heinonen, S. (2003) Association of single nucleotide polymorphisms of the bile salt export pump gene with intrahepatic cholestasis of pregnancy. *Scand J Gastroenterol*, *38*, 648–652.

30. De Vree, J.M., Jacquemin, E., Sturm, E., Cresteil, D., Bosma, P.J., Aten, J., Deleuze, J.F., Desrochers, M., Burdelski, M., Bernard, O. et al. (1998) Mutations in the MDR3 gene cause progressive familial intrahepatic cholestasis. *Proc Natl Acad Sci U S A*, *95*, 282–287.

31. Jacquemin, E., De Vree, J.M., Cresteil, D., Sokal, E.M., Sturm, E., Dumont, M., Scheffer, G.L., Paul, M., Burdelski, M., Bosma, P.J. et al. (2001) The wide spectrum of multidrug resistance 3 deficiency: from neonatal cholestasis to cirrhosis of adulthood. *Gastroenterology*, *120*, 1448–1458.

32. Rosmorduc, O., Hermelin, B. and Poupon, R. (2001) MDR3 gene defect in adults with symptomatic intrahepatic and gallbladder cholesterol cholelithiasis. *Gastroenterology*, *120*, 1459–1467.

33. Pauli-Magnus, C., Lang, T., Meier, Y., Zodan-Marin, T., Jung, D., Breymann, C., Zimmermann, R., Kenngott, S., Beuers, U., Reichel, C. et al. (2004) Sequence analysis of bile salt export pump (ABCB11) and multidrug resistance P-glycoprotein 3 (ABCB4, MDR3) in patients with intrahepatic cholestasis of pregnancy. *Pharmacogenetics*, *14*, 91–102.

34. Kartenbeck, J., Leuschner, U., Mayer, R. and Keppler, D. (1996) Absence of the canalicular isoform of the MRP gene-encoded conjugate export pump from the hepatocytes in Dubin–Johnson syndrome. *Hepatology*, *23*, 1061–1066.

35. Paulusma, C.C., Kool, M., Bosma, P.J., Scheffer, G.L., ter Borg, F., Scheper, R.J., Tytgat, G.N., Borst, P., Baas, F. and Oude Elferink, R.P. (1997) A mutation in the human canalicular multispecific organic anion transporter gene causes the Dubin–Johnson syndrome. *Hepatology*, *25*, 1539–1542.

36. Tsujii, H., Konig, J., Rost, D., Stöckel, B., Leuschner, U. and Keppler, D. (1999) Exon–intron organization of the human multidrug-resistance protein 2 (MRP2) gene mutated in Dubin–Johnson syndrome. *Gastroenterology*, *117*, 653–660.

37. Keitel, V., Kartenbeck, J., Nies, A.T., Spring, H., Brom, M. and Keppler, D. (2000) Impaired protein maturation of the conjugate export pump multidrug resistance protein 2 as a consequence of a deletion mutation in Dubin–Johnson syndrome. *Hepatology*, *32*, 1317–1328.

38. Oelkers, P., Kirby, L.C., Heubi, J.E. and Dawson, P.A. (1997) Primary bile acid malabsorption caused by mutations in the ileal sodium-dependent bile acid transporter gene (SLC10A2). *J Clin Invest*, *99*, 1880–1887.

39. Ho, R.H., Leake, B.F., Roberts, R.L., Lee, W. and Kim, R.B. (2004) Ethnicity-dependent polymorphism in Na^+–taurocholate cotransporting polypeptide (SLC10A1) reveals a domain critical for bile acid substrate recognition. *J Biol Chem*, *279*, 7213–7222.

40. Zollner, G., Fickert, P., Silbert, D., Fuchsbichler, A., Marschall, H.U., Zatloukal, K., Denk, H. and Trauner, M. (2003) Adaptive changes in hepatobiliary transporter expression in primary biliary cirrhosis. *J Hepatol*, *38*, 717–727.

41. Zollner, G., Fickert, P., Zenz, R., Fuchsbichler, A., Stumptner, C., Kenner, L., Ferenci, P., Stauber, R.E., Krejs, G.J., Denk, H., Zatloukal, K. and Trauner, M. (2001) Hepatobiliary transporter expression in percutaneous liver biopsies of patients with cholestatic liver diseases. *Hepatology*, *33*, 633–646.
42. Chiang, J.Y. (2004) Regulation of bile acid synthesis: pathways, nuclear receptors, and mechanisms. *J Hepatol*, *40*, 539–551.
43. Zamek-Gliszczynski, M.J., Hoffmaster, K.A., Nezasa, K., Tallman, M.N. and Brouwer, K.L. (2006) Integration of hepatic drug transporters and phase II metabolizing enzymes: mechanisms of hepatic excretion of sulfate, glucuronide, and glutathione metabolites. *Eur J Pharm Sci*, *27*, 447–486.
44. Haussinger, D., Schmitt, M., Weiergraber, O. and Kubitz, R. (2000) Short-term regulation of canalicular transport. *Semin Liver Dis*, *20*, 307–321.
45. Kullak-Ublick, G.A., Stieger, B. and Meier, P.J. (2004) Enterohepatic bile salt transporters in normal physiology and liver disease. *Gastroenterology*, *126*, 322–342.
46. Berkenstam, A. and Gustafsson, J.A. (2005) Nuclear receptors and their relevance to diseases related to lipid metabolism. *Curr Opin Pharmacol*, *5*, 171–176.
47. Xu, L., Glass, C.K. and Rosenfeld, M.G. (1999) Coactivator and corepressor complexes in nuclear receptor function. *Curr Opin Genet Dev*, *9*, 140–147.
48. Kalaany, N.Y. and Mangelsdorf, D.J. (2006) LXRS and FXR: the yin and yang of cholesterol and fat metabolism. *Annu Rev Physiol*, *68*, 159–191.
49. Makishima, M., Okamoto, A.Y., Repa, J.J., Tu, H., Learned, R.M., Luk, A., Hull, M.V., Lustig, K.D., Mangelsdorf, D.J. and Shan, B. (1999) Identification of a nuclear receptor for bile acids. *Science*, *284*, 1362–1365.
50. Parks, D.J., Blanchard, S.G., Bledsoe, R.K., Chandra, G., Consler, T.G., Kliewer, S.A., Stimmel, J.B., Willson, T.M., Zavacki, A.M., Moore, D.D. and Lehmann, J.M. (1999) Bile acids: natural ligands for an orphan nuclear receptor. *Science*, *284*, 1365–1368.
51. Lew, J.L., Zhao, A., Yu, J., Huang, L., De Pedro, N., Pelaez, F., Wright, S.D. and Cui, J. (2004) The farnesoid X receptor controls gene expression in a ligand- and promoter-selective fashion. *J Biol Chem*, *279*, 8856–8861.
52. Yu, J., Lo, J.L., Huang, L., Zhao, A., Metzger, E., Adams, A., Meinke, P.T., Wright, S.D. and Cui, J. (2002) Lithocholic acid decreases expression of bile salt export pump through farnesoid X receptor antagonist activity. *J Biol Chem*, *277*, 31441–31447.
53. Deng, R., Yang, D., Yang, J. and Yan, B. (2006) Oxysterol 22(*R*)-hydroxycholesterol induces the expression of the bile salt export pump through nuclear receptor farsenoid X receptor but not liver X receptor. *J Pharmacol Exp Ther*, *317*, 317–325.
54. Wang, S., Lai, K., Moy, F.J., Bhat, A., Hartman, H.B. and Evans, M.J. (2006) The nuclear hormone receptor farnesoid X receptor (FXR) is activated by androsterone*. *Endocrinology*, *147*, 4025–4033.
55. Downes, M., Verdecia, M.A., Roecker, A.J., Hughes, R., Hogenesch, J.B., Kast-Woelbern, H.R., Bowman, M.E., Ferrer, J.L., Anisfeld, A.M., Edwards, P.A. et al. (2003) A chemical, genetic, and structural analysis of the nuclear bile acid receptor FXR. *Mol Cell*, *11*, 1079–1092.
56. Laffitte, B.A., Kast, H.R., Nguyen, C.M., Zavacki, A.M., Moore, D.D. and Edwards, P.A. (2000) Identification of the DNA binding specificity and potential target genes for the farnesoid X-activated receptor. *J Biol Chem*, *275*, 10638–10647.

57. Song, C.S., Echchgadda, I., Baek, B.S., Ahn, S.C., Oh, T., Roy, A.K. and Chatterjee, B. (2001) Dehydroepiandrosterone sulfotransferase gene induction by bile acid–activated farnesoid X receptor. *J Biol Chem*, *276*, 42549–42556.

58. Kast, H.R., Goodwin, B., Tarr, P.T., Jones, S.A., Anisfeld, A.M., Stoltz, C.M., Tontonoz, P., Kliewer, S., Willson, T.M. and Edwards, P.A. (2002) Regulation of multidrug resistance-associated protein 2 (ABCC2) by the nuclear receptors pregnane X receptor, farnesoid X–activated receptor, and constitutive androstane receptor. *J Biol Chem*, *277*, 2908–2915.

59. Sinal, C.J., Tohkin, M., Miyata, M., Ward, J.M., Lambert, G. and Gonzalez, F.J. (2000) Targeted disruption of the nuclear receptor FXR/BAR impairs bile acid and lipid homeostasis. *Cell*, *102*, 731–744.

60. Chen, F., Ananthanarayanan, M., Emre, S., Neimark, E., Bull, L.N., Knisely, A.S., Strautnieks, S.S., Thompson, R.J., Magid, M.S., Gordon, R. et al. (2004) Progressive familial intrahepatic cholestasis, type 1, is associated with decreased farnesoid X receptor activity. *Gastroenterology*, *126*, 756–764.

61. Liu, Y., Binz, J., Numerick, M.J., Dennis, S., Luo, G., Desai, B., MacKenzie, K.I., Mansfield, T.A., Kliewer, S.A., Goodwin, B. and Jones, S.A. (2003) Hepatoprotection by the farnesoid X receptor agonist GW4064 in rat models of intra- and extrahepatic cholestasis. *J Clin Invest*, *112*, 1678–1687.

62. Fiorucci, S., Clerici, C., Antonelli, E., Orlandi, S., Goodwin, B., Sadeghpour, B.M., Sabatino, G., Russo, G., Castellani, D., Willson, T.M. et al. (2005) Protective effects of 6-ethyl chenodeoxycholic acid, a farnesoid X receptor ligand, in estrogen-induced cholestasis. *J Pharmacol Exp Ther*, *313*, 604–612.

63. Stayrook, K.R., Bramlett, K.S., Savkur, R.S., Ficorilli, J., Cook, T., Christe, M.E., Michael, L.F. and Burris, T.P. (2005) Regulation of carbohydrate metabolism by the farnesoid X receptor. *Endocrinology*, *146*, 984–991.

64. Ma, K., Saha, P.K., Chan, L. and Moore, D.D. (2006) Farnesoid X receptor is essential for normal glucose homeostasis. *J Clin Invest*, *116*, 1102–1109.

65. Huang, W., Ma, K., Zhang, J., Qatanani, M., Cuvillier, J., Liu, J., Dong, B., Huang, X. and Moore, D.D. (2006) Nuclear receptor–dependent bile acid signaling is required for normal liver regeneration. *Science*, *312*, 233–236.

66. Handschin, C. and Meyer, U.A. (2005) Regulatory network of lipid-sensing nuclear receptors: roles for CAR, PXR, LXR, and FXR. *Arch Biochem Biophys*, *433*, 387–396.

67. Staudinger, J.L., Goodwin, B., Jones, S.A., Hawkins-Brown, D., MacKenzie, K.I., LaTour, A., Liu, Y., Klaassen, C.D., Brown, K.K., Reinhard, J. et al. (2001) The nuclear receptor PXR is a lithocholic acid sensor that protects against liver toxicity. *Proc Natl Acad Sci U S A*, *98*, 3369–3374.

68. Xie, W., Radominska-Pandya, A., Shi, Y., Simon, C.M., Nelson, M.C., Ong, E.S., Waxman, D.J. and Evans, R.M. (2001) An essential role for nuclear receptors SXR/PXR in detoxification of cholestatic bile acids. *Proc Natl Acad Sci U S A*, *98*, 3375–3380.

69. Guo, G.L., Lambert, G., Negishi, M., Ward, J.M., Brewer, H.B., Jr., Kliewer, S.A., Gonzalez, F.J. and Sinal, C.J. (2003) Complementary roles of farnesoid X receptor, pregnane X receptor, and constitutive androstane receptor in protection against bile acid toxicity. *J Biol Chem*, *278*, 45062–45071.

70. Lehmann, J.M., McKee, D.D., Watson, M.A., Willson, T.M., Moore, J.T. and Kliewer, S.A. (1998) The human orphan nuclear receptor PXR is activated by compounds that

regulate CYP3A4 gene expression and cause drug interactions. *J Clin Invest*, *102*, 1016–1023.

71. Makishima, M., Lu, T.T., Xie, W., Whitfield, G.K., Domoto, H., Evans, R.M., Haussler, M.R. and Mangelsdorf, D.J. (2002) Vitamin D receptor as an intestinal bile acid sensor. *Science*, *296*, 1313–1316.
72. Kassam, A., Miao, B., Young, P.R. and Mukherjee, R. (2003) Retinoid X receptor (RXR) agonist-induced antagonism of farnesoid X receptor (FXR) activity due to absence of coactivator recruitment and decreased DNA binding. *J Biol Chem*, *278*, 10028–10032.
73. Tzameli, I., Chua, S.S., Cheskis, B. and Moore, D.D. (2003) Complex effects of rexinoids on ligand dependent activation or inhibition of the xenobiotic receptor, CAR. *Nucl Recept*, *1*, 2.
74. Claudel, T., Sturm, E., Duez, H., Torra, I.P., Sirvent, A., Kosykh, V., Fruchart, J.C., Dallongeville, J., Hum, D.W., Kuipers, F. and Staels, B. (2002) Bile acid–activated nuclear receptor FXR suppresses apolipoprotein A-I transcription via a negative FXR response element. *J Clin Invest*, *109*, 961–971.
75. Claudel, T., Inoue, Y., Barbier, O., Duran-Sandoval, D., Kosykh, V., Fruchart, J., Fruchart, J.C., Gonzalez, F.J. and Staels, B. (2003) Farnesoid X receptor agonists suppress hepatic apolipoprotein CIII expression. *Gastroenterology*, *125*, 544–555.
76. Eloranta, J.J. and Kullak-Ublick, G.A. (2005) Coordinate transcriptional regulation of bile acid homeostasis and drug metabolism. *Arch Biochem Biophys*, *433*, 397–412.
77. Borgius, L.J., Steffensen, K.R., Gustafsson, J.A. and Treuter, E. (2002) Glucocorticoid signaling is perturbed by the atypical orphan receptor and corepressor SHP. *J Biol Chem*, *277*, 49761–49766.
78. Kim, J.Y., Kim, H.J., Kim, K.T., Park, Y.Y., Seong, H.A., Park, K.C., Lee, I.K., Ha, H., Shong, M., Park, S.C. and Choi, H.S. (2004) Orphan nuclear receptor small heterodimer partner represses hepatocyte nuclear factor 3/Foxa transactivation via inhibition of its DNA binding. *Mol Endocrinol*, *18*, 2880–2894.
79. Kemper, J.K., Kim, H., Miao, J., Bhalla, S. and Bae, Y. (2004) Role of an mSin3A-Swi/Snf chromatin remodeling complex in the feedback repression of bile acid biosynthesis by SHP. *Mol Cell Biol*, *24*, 7707–7719.
80. Wang, L., Lee, Y.K., Bundman, D., Han, Y., Thevananther, S., Kim, C.S., Chua, S.S., Wei, P., Heyman, R.A., Karin, M. and Moore, D.D. (2002) Redundant pathways for negative feedback regulation of bile acid production. *Dev Cell*, *2*, 721–731.
81. Kerr, T.A., Saeki, S., Schneider, M., Schaefer, K., Berdy, S., Redder, T., Shan, B., Russell, D.W. and Schwarz, M. (2002) Loss of nuclear receptor SHP impairs but does not eliminate negative feedback regulation of bile acid synthesis. *Dev Cell*, *2*, 713–720.
82. Gupta, S., Natarajan, R., Payne, S.G., Studer, E.J., Spiegel, S., Dent, P. and Hylemon, P.B. (2004) Deoxycholic acid activates the c-Jun N-terminal kinase pathway via FAS receptor activation in primary hepatocytes: role of acidic sphingomyelinase-mediated ceramide generation in FAS receptor activation. *J Biol Chem*, *279*, 5821–5828.
83. Qiao, L., Han, S.I., Fang, Y., Park, J.S., Gupta, S., Gilfor, D., Amorino, G., Valerie, K., Sealy, L., Engelhardt, J.F. et al. (2003) Bile acid regulation of C/EBPbeta, CREB, and c-Jun function, via the extracellular signal-regulated kinase and c-Jun NH2-terminal kinase pathways, modulates the apoptotic response of hepatocytes. *Mol Cell Biol*, *23*, 3052–3066.

84. Kawamata, Y., Fujii, R., Hosoya, M., Harada, M., Yoshida, H., Miwa, M., Fukusumi, S., Habata, Y., Itoh, T., Shintani, Y. et al. (2003) A G protein–coupled receptor responsive to bile acids. *J Biol Chem*, *278*, 9435–9440.

85. Huang, L., Zhao, A., Lew, J.L., Zhang, T., Hrywna, Y., Thompson, J.R., de Pedro, N., Royo, I., Blevins, R.A., Pelaez, F. et al. (2003) Farnesoid X receptor activates transcription of the phospholipid pump MDR3. *J Biol Chem*, *278*, 51085–51090.

86. Wagner, M., Fickert, P., Zollner, G., Fuchsbichler, A., Silbert, D., Tsybrovskyy, O., Zatloukal, K., Guo, G.L., Schuetz, J.D., Gonzalez, F.J. et al. (2003) Role of farnesoid X receptor in determining hepatic ABC transporter expression and liver injury in bile duct–ligated mice. *Gastroenterology*, *125*, 825–838.

87. Zollner, G., Fickert, P., Fuchsbichler, A., Silbert, D., Wagner, M., Arbeiter, S., Gonzalez, F.J., Marschall, H.U., Zatloukal, K., Denk, H. and Trauner, M. (2003) Role of nuclear bile acid receptor, FXR, in adaptive ABC transporter regulation by cholic and ursodeoxycholic acid in mouse liver, kidney and intestine. *J Hepatol*, *39*, 480–488.

88. Schuetz, E.G., Strom, S., Yasuda, K., Lecureur, V., Assem, M., Brimer, C., Lamba, J., Kim, R.B., Ramachandran, V., Komoroski, B.J. et al. (2001) Disrupted bile acid homeostasis reveals an unexpected interaction among nuclear hormone receptors, transporters, and cytochrome P450. *J Biol Chem*, *276*, 39411–39418.

89. Shoda, J., Kano, M., Oda, K., Kamiya, J., Nimura, Y., Suzuki, H., Sugiyama, Y., Miyazaki, H., Todoroki, T., Stengelin, S. et al. (2001) The expression levels of plasma membrane transporters in the cholestatic liver of patients undergoing biliary drainage and their association with the impairment of biliary secretory function. *Am J Gastroenterol*, *96*, 3368–3378.

90. Keitel, V., Burdelski, M., Warskulat, U., Kuhlkamp, T., Keppler, D., Haussinger, D. and Kubitz, R. (2005) Expression and localization of hepatobiliary transport proteins in progressive familial intrahepatic cholestasis. *Hepatology*, *41*, 1160–1172.

91. Landrier, J.F., Eloranta, J.J., Vavricka, S.R. and Kullak-Ublick, G.A. (2006) The nuclear receptor for bile acids, FXR, transactivates human organic solute transporter-alpha and -beta genes. *Am J Physiol Gastrointest Liver Physiol*, *290*, G476–G485.

92. Lee, H., Zhang, Y., Lee, F.Y., Nelson, S.F., Gonzalez, F.J. and Edwards, P.A. (2006) FXR regulates organic solute transporters alpha and beta in the adrenal gland, kidney, and intestine. *J Lipid Res*, *47*, 201–214.

93. Boyer, J.L., Trauner, M., Mennone, A., Soroka, C.J., Cai, S.Y., Moustafa, T., Zollner, G., Lee, J.Y. and Ballatori, N. (2006) Upregulation of a basolateral FXR-dependent bile acid efflux transporter OSTalpha-OSTbeta in cholestasis in humans and rodents. *Am J Physiol Gastrointest Liver Physiol*, *290*, G1124–G1130.

94. Zollner, G., Wagner, M., Moustafa, T., Fickert, P., Silbert, D., Gumhold, J., Fuchsbichler, A., Halilbasic, E., Denk, H., Marschall, H.U. and Trauner, M. (2006) Coordinated induction of bile acid detoxification and alternative elimination in mice: role of FXR-regulated organic solute transporter-alpha/beta in the adaptive response to bile acids. *Am J Physiol Gastrointest Liver Physiol*, *290*, G923–G932.

95. Grober, J., Zaghini, I., Fujii, H., Jones, S.A., Kliewer, S.A., Willson, T.M., Ono, T. and Besnard, P. (1999) Identification of a bile acid–responsive element in the human ileal bile acid–binding protein gene. Involvement of the farnesoid X receptor/9-*cis*-retinoic acid receptor heterodimer. *J Biol Chem*, *274*, 29749–29754.

96. Gartung, C., Schuele, S., Schlosser, S.F. and Boyer, J.L. (1997) Expression of the rat liver Na^+/taurocholate cotransporter is regulated in vivo by retention of biliary constituents but not their depletion. *Hepatology*, *25*, 284–290.

97. Fickert, P., Zollner, G., Fuchsbichler, A., Stumptner, C., Pojer, C., Zenz, R., Lammert, F., Stieger, B., Meier, P.J., Zatloukal, K., Denk, H. and Trauner, M. (2001) Effects of ursodeoxycholic and cholic acid feeding on hepatocellular transporter expression in mouse liver. *Gastroenterology*, *121*, 170–183.

98. Eloranta, J.J., Jung, D. and Kullak-Ublick, G.A. (2006) The human Na^+-taurocholate cotransporting polypeptide gene is activated by glucocorticoid receptor and peroxisome proliferator-activated receptor-gamma coactivator-1alpha, and suppressed by bile acids via a small heterodimer partner-dependent mechanism. *Mol Endocrinol*, *20*, 65–79.

99. Jung, D. and Kullak-Ublick, G.A. (2003) Hepatocyte nuclear factor 1 alpha: a key mediator of the effect of bile acids on gene expression. *Hepatology*, *37*, 622–631.

100. Neimark, E., Chen, F., Li, X. and Shneider, B.L. (2004) Bile acid–induced negative feedback regulation of the human ileal bile acid transporter. *Hepatology*, *40*, 149–156.

101. Hruz, P., Zimmermann, C., Gutmann, H., Degen, L., Beuers, U., Terracciano, L., Drewe, J. and Beglinger, C. (2006) Adaptive regulation of the ileal apical sodium dependent bile acid transporter (ASBT) in patients with obstructive cholestasis. *Gut*, *55*, 395–402.

102. Jung, D., Fantin, A.C., Scheurer, U., Fried, M. and Kullak-Ublick, G.A. (2004) Human ileal bile acid transporter gene ASBT (SLC10A2) is transactivated by the glucocorticoid receptor. *Gut*, *53*, 78–84.

103. Hagenbuch, B. and Meier, P.J. (2004) Organic anion transporting polypeptides of the OATP/ SLC21 family: phylogenetic classification as OATP/ SLCO superfamily, new nomenclature and molecular/functional properties. *Pflugers Arch*, *447*, 653–665.

104. Hayhurst, G.P., Lee, Y.H., Lambert, G., Ward, J.M. and Gonzalez, F.J. (2001) Hepatocyte nuclear factor 4alpha (nuclear receptor 2A1) is essential for maintenance of hepatic gene expression and lipid homeostasis. *Mol Cell Biol*, *21*, 1393–1403.

105. Dhe-Paganon, S., Duda, K., Iwamoto, M., Chi, Y.I. and Shoelson, S.E. (2002) Crystal structure of the HNF4 alpha ligand binding domain in complex with endogenous fatty acid ligand. *J Biol Chem*, *277*, 37973–37976.

106. Wisely, G.B., Miller, A.B., Davis, R.G., Thornquest, A.D., Jr., Johnson, R., Spitzer, T., Sefler, A., Shearer, B., Moore, J.T., Willson, T.M. and Williams, S.P. (2002) Hepatocyte nuclear factor 4 is a transcription factor that constitutively binds fatty acids. *Structure*, *10*, 1225–1234.

107. Popowski, K., Eloranta, J.J., Saborowski, M., Fried, M., Meier, P.J. and Kullak-Ublick, G.A. (2005) The human organic anion transporter 2 gene is transactivated by hepatocyte nuclear factor-4 alpha and suppressed by bile acids. *Mol Pharmacol*, *67*, 1629–1638.

108. Saborowski, M., Kullak-Ublick, G.A. and Eloranta, J.J. (2006) The human organic cation transporter-1 gene is transactivated by hepatocyte nuclear factor-4alpha. *J Pharmacol Exp Ther*, *317*, 778–785.

10

MULTIDRUG RESISTANCE PROTEIN: P-GLYCOPROTEIN

FRANCES J. SHAROM

University of Guelph, Guelph, Ontario, Canada

Drug Transporters: Molecular Characterization and Role in Drug Disposition, Edited by Guofeng You and Marilyn E. Morris
Copyright © 2007 John Wiley & Sons, Inc.

10.1. P-GLYCOPROTEIN GENE FAMILY

P-Glycoprotein (Pgp) was one of the first members of the adenosine triphosphate (ATP)-binding cassette (ABC) superfamily to be studied. Overexpression of Pgp was linked to multidrug resistance (MDR) in mammalian cell lines and human cancers, evoking intense interest first from molecular and cell biologists, and later, when purified Pgp became available, from biochemists and biophysicists. Today, this fascinating protein, which is proposed to operate as an ATP-powered drug efflux pump, remains one of the most studied membrane transporters. Pgp genes from human, mouse, and Chinese hamster, among others, have been cloned and sequenced, and homologs have been identified in other species, including *Drosophila melanogaster* and *Caenorhabditis elegans*.[1,2] Pgp in higher mammals forms a small gene family, with two isoforms expressed in humans and three isoforms in rodents. The class I and III isoforms (human *MDR1/ABCB1*, mouse *mdr1/Abcb1a* and *mdr3/Abcb1*) are drug transporters, while the class II isoforms (human *MDR2/3/ABCB4*, mouse *mdr2/Abcb4*) carry out export of phosphatidylcholine (PC) into the bile.[3] The two human genes arose from a duplication event and are adjacent to each other on the chromosome. The drug-transporting isoform shares 78% amino acid sequence identity with the PC-exporting isoform, suggesting that they share similar structures and mechanisms of action. For the rest of this chapter, the term *Pgp* is used to indicate the ABCB1 gene product.

10.2. TISSUE DISTRIBUTION OF P-GLYCOPROTEIN

Early studies of Pgp distribution in human[4] and rodent[5] tissues showed that the protein is expressed at low levels in most tissues but is found in much higher amounts at the apical surface of epithelial cells lining the colon, small intestine, pancreatic ductules, bile ductules and kidney proximal tubules, and the adrenal gland. Thus, epithelial cells with excretory roles generally express Pgp. The transporter is also located in the endothelial cells of the blood–brain barrier,[6] the blood–testis barrier,[7] and the blood–mammary tissue barrier,[8] and has recently been found to play a role in the blood–inner ear barrier, where it is expressed in the capillary endothelial cells of the cochlea and vestibule.[9] Thus the role of Pgp in the blood–brain and blood–tissue barriers is likely to protect these organs from toxic compounds that gain entry into the circulatory system. Pgp is expressed at high levels at the luminal surface of secretory epithelial cells in the pregnant endometrium,[10] as well as the placenta,[11] where it may provide protection for the fetus.[12] The protein is also found on the surface of hematopoietic cells, where its function remains enigmatic. The ABCB4 protein is expressed at high levels on the bile canalicular membrane of hepatocytes, in accordance with its proposed role in transport of PC into the bile.[13]

10.3. ROLE OF P-GLYCOPROTEIN IN HUMAN PHYSIOLOGY

The tissue localization of Pgp suggests that the protein plays a physiological role in the protection of susceptible organs such as the brain, testis, and inner ear from toxic xenobiotics, the secretion of metabolites and xenobiotics into bile, urine, and the lumen of the gastrointestinal tract, and possibly the transport of hormones from the adrenal gland and the uterine epithelium. These ideas have been strongly supported by studies on transgenic knockout mice lacking one or both of the genes encoding the drug-transporting Pgps Abcb1a and Abcb1b. Both single- and double-knockout mice are fertile, viable, and phenotypically indistinguishable from wild-type mice under normal conditions. So Pgp does not appear to fulfill any essential physiological functions. However, Pgp knockout mice showed radical changes in the way that they handled a challenge with many drugs.[14] mdr3 knockout mice displayed a disrupted blood–brain barrier and were 100-fold more sensitive to the pesticide ivermectin, which was neurotoxic to the animals.[15] This Pgp isoform appears to play the major role in preventing accumulation of drugs in the brain.[15,16] The double-knockout mouse has proved useful in evaluating the effect of Pgp-mediated transport on drugs that are targeted to the central nervous system.[17] Certain dogs of the collie lineage[18] and several other dog breeds[19,20] have a naturally occurring lack of Pgp due to a frameshift mutation in the MDR1 gene and are also hypersensitive to ivermectin. To date, no human null alleles have been reported, despite widespread use of drugs that are Pgp substrates.

Pgp in the intestinal epithelium plays an important role in the extrusion of many drugs from the blood into the intestinal lumen, and in preventing drugs in the intestinal lumen from entering the bloodstream. Pgp activity can therefore reduce the absorption and oral bioavailability of those drugs that are transport substrates.

One important goal in clinical medicine has been the development of techniques for in vivo functional imaging of Pgp-mediated drug transport in normal tissues and tumors and its inhibition by specific Pgp modulators. The radiopharmaceutical technetium-99m-sestamibi (^{99m}Tc-MIBI) has been validated as a Pgp transport substrate. Scintigraphic studies of human subjects showed rapid clearance of the radiotracer from normal liver and kidneys in vivo; however, it was retained selectively in these organs after administration of the Pgp modulator, PSC833.[21] Later studies have shown the prognostic value of this approach in different types of tumors, including breast and lung cancer, sarcoma, and lymphoma.[22] The activity of Pgp at the human blood brain barrier has also been imaged using positron emission tomography using ^{11}C-labeled verapamil or carvedilol (Pgp transport substrates).[23]

10.4. P-GLYCOPROTEIN SUBSTRATES AND MODULATORS

Pgp has the ability to interact with literally hundreds of structurally diverse *substrates* (see Table 10.1), which are generally nonpolar, weakly amphipathic compounds, and include natural products, anticancer drugs, steroids, fluorescent dyes, linear and cyclic peptides, and ionophores. The unusual promiscuity of the transporter has made it difficult to find "nonsubstrates." Potential physiological substrates for Pgp could include peptides, steroid hormones, lipids, and small cytokines, such as interleukin-2, intereukin-4, and interferon-γ. However, there is little information on the extent to which endogenous compounds are transported by Pgp in vivo. Identification of

TABLE 10.1. Pgp Substrates and Modulators

Substrates		Modulators
Vinca alkaloids	*Detergents*	*Ca^{2+} channel blockers*
vinblastine	Triton X-100	verapamil
vincristine	nonylphenol ethoxylate	nifedipine
Anthracyclines	*Fluorescent dyes*	azidopine
doxorubicin	rhodamine 123	dexniguldipine
daunorubicin	tetramethylrosamine	*Calmodulin antagonists*
Taxanes	Hoechst 33342	trifluoperazine
paclitaxel	LDS-751	chloropromazine
docetaxel	calcein acetoxymethyl ester	*trans*-flupenthixol
Epipodophyllotoxins	*Linear/cyclic peptides*	*Cyclic peptides*
etoposide	ALLN	cyclosporin A
teniposide	NAc-LLY-amide	PSC833
Steroids	leupeptin	*Steroids*
aldosterone	pepstatin A	progesterone
dexamethasone	*Ionophores*	tamoxifen
HIV protease inhibitors	gramicidin D	cortisol
indinavir	nonactin	*Miscellaneous*
saquinavir	beauvericin	GF120918
nelfinavir	*Cytotoxic agents*	LY335979
ritonavir	colchicine	XR9576
Analgesics	actinomycin D	OC144-093
morphine	mitoxantrone	disulfiram
Cardiac glycosides	*Miscellaneous*	quinidine
digoxin	loperamide	chloroquine
Antihelminthics	cimetidine	reserpine
ivermectin		amiodarone
		terfenadine

a specific compound as a Pgp substrate is often indirect, although more specific spectroscopic approaches now allow measurement of binding affinity.[24] Direct measurement of Pgp-mediated transport has been carried out for only a small fraction of these substrates. Work with reconstituted Pgp has shown that it is an active transporter, generating a substrate concentration gradient across the membrane.[25,26] In intact cells, the drug concentration in the cytosol is kept low enough to circumvent cytotoxicity, and they thus become multidrug resistant.

A second class of compounds exists which interact with Pgp, the *modulators* (also known as MDR chemosensitizers, reversers, or inhibitors; see Table 10.1). Modulators are able to reverse MDR in intact cells in vitro, by interfering with the ability of Pgp to efflux drug and thus generate a drug concentration gradient. The ability to block the action of Pgp selectively is of importance clinically, whether the goal is to achieve more efficacious cancer chemotherapy, improve drug bioavailability and uptake in the intestine, or deliver drugs to the brain. Numerous pharmacologic agents have been identified as Pgp modulators, many by serendipity or trial and error (see Table 10.1).

Modulators are as diverse structurally as substrates.[27] They appear to interact with the same binding site(s) as drugs and compete with them for transport. Many modulators (e.g., verapamil, cyclosporin A, *trans*-flupenthixol) are themselves transported by the protein. Cells are generally not resistant to killing by modulators, but they are killed by MDR drugs in combination with modulators. The way in which modulators exert their action at the molecular level is still not well understood.

10.5. P-GLYCOPROTEIN STRUCTURE

Like many other ABC proteins,[28,29] Pgp comprises two membrane-bound domains, each made up of six transmembrane (TM) helices and two cytoplasmic nucleotide-binding (NB) domains which bind and hydrolyze ATP (Figure 10.1*a*). The topology of Pgp was established using molecular biological methods such as Cys mutations and insertion of glycosylation sites.[30,31] Earlier studies using various heterologous expression systems suggested alternative topologies in which putative TM segments were displaced outside the membrane, however, it seems likely that these arrangements were the result of misfolding and do not reflect the true topology of the transporter in vivo.[32] The TM regions from both halves of Pgp form the drug-binding region of the protein,[33] and drugs enter this binding pocket from the lipid bilayer.[34]

High-resolution x-ray crystal structures of two ABC proteins, the catalytic domains of the DNA repair enzyme Rad50cd[35] and the vitamin B12 importer BtuCD,[36] showed that the two NB domains were in close contact to form a dimeric structure. Two molecules of ATP were bound at the dimer interface, with each binding site comprising the Walker A and B motifs of the *cis*-NB domain and the LSGGQ signature C motif of the *trans*-NB domain. This "sandwich dimer" structure has also been observed for the isolated NB domain of the ABC protein MJ0796, which forms a stable dimer when the ATPase activity of the protein is inactivated by the mutation E171Q.[37] It seems likely that this dimerization process plays a critical role in the catalytic cycle of the ABC proteins and may be closely tied to the power stroke.[29]

No high resolution x-ray crystal structure is available for Pgp. Early work by Rosenberg et al. using electron microscopy (EM) single-particle image analysis of purified Pgp produced a very low resolution structure which suggested the existence of a large 5-nm-diameter central pore in the protein.[38] This pore was closed at the cytoplasmic face of the membrane, forming an aqueous chamber within the membrane from which entry points to the membrane lipid were observed. Two widely separated 3-nm lobes on the cytoplasmic side of the membrane were thought to represent the NB domains. This structure was at odds with both biochemical studies, which suggested kinetic cooperativity between the two catalytic sites, and the high-resolution x-ray crystal structures of other ABC proteins described above, which showed close physical association of the two NB domains. Fluorescence resonance energy transfer (FRET) studies in which two different fluorescent probes were covalently linked to each Walker A motif Cys residue also indicated that the positioning of the two NB domains is compatible with the sandwich dimer model (Figure 10.1*b*),[39] and Urbatsch et al. found that the two Walker A Cys residues could readily cross-link spontaneously.[40]

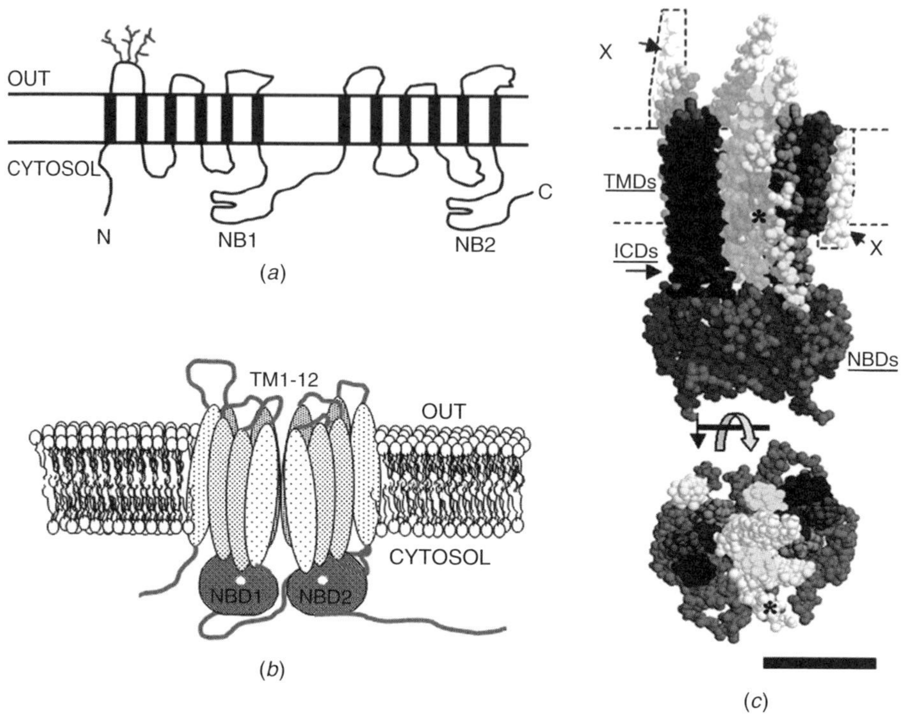

FIGURE 10.1. Topology and structure of Pgp. (*a*) Pgp is proposed to consist of two equivalent halves, each with six TM segments and an NB domain on the cytosolic side. Both the N- and C-terminus are cytosolic. (*b*) The low-resolution structural model of Pgp generated using several different FRET measurements of the distances separating key regions of the protein. (*c*) Medium-resolution structural model of Pgp obtained from cryo-EM studies. Top: a side view of the protein is shown with the NB domains at the bottom. The 12 putative TM α-helices are arranged in a pseudosymmetrical relationship. Bottom: view of Pgp looking down on the TM helices from the extracellular side of the membrane. The dashed lines indicate the putative boundary of a 4.5-nm-thick lipid bilayer (scale bar = 5 nm), [(*b*) From ref. 158; (*c*) adapted from ref. 45, with permission.]

In addition, Loo et al. showed that Cys residues in the Walker A motifs could be cross-linked at low temperatures to Cys residues in the LSGGQ motif, indicating that the signature sequences in one NB domain are adjacent to the Walker A site in the other NB domain.[41] Later work by Rosenberg and co-workers showed that nucleotide binding causes a repacking of the TM regions of Pgp,[42] which could open the central pore to allow access of hydrophobic drugs directly from the lipid bilayer.[43] It was proposed from this reorganization that ATP binding, not hydrolysis, drives the conformational changes associated with transport.[42] The vanadate-trapped complex of Pgp, which is thought to resemble the catalytic transition state structurally, displayed a third distinct conformation of the protein, suggesting that rotation of TM α-helices had taken place.[42] Mouse Pgp has also been studied by EM and image analysis of two-dimensional crystals of purified protein in a lipid bilayer.[44] The resulting low-resolution projection structure (22Å) was compact and suggested that the two cytoplasmic NB domains interact closely.

More recently, a higher-resolution EM structure was obtained for human Pgp which shows close association of the NB domains[45] and bears a much greater resemblance to the mouse Pgp structure (Figure 10.1*c*), so it seems likely that the NB domains indeed form the sandwich dimer observed for other ABC proteins. This structure also clearly showed the existence of 12 TM segments, supporting the proposed topology of the protein, but the resolution was not high enough to discern further details. The packing arrangement of the TM helices of Pgp has been explored systematically by Clarke and co-workers, who introduced Cys residues into defined positions within a Cys-less Pgp construct and then carried out cross-linking studies.[46] The pattern observed suggested that TM6 is close to TM10, 11, and 12, whereas TM12 is close to TM4, 5, and 6. Recent work showed that the ends of TM2 and TM11 are close together on the cytoplasmic side of the membrane,[47] as are the cytoplasmic ends of TM5 and TM8.[48]

10.6. SUBCELLULAR SYSTEMS FOR STUDYING P-GLYCOPROTEIN

Much early work on the molecular basis of MDR was carried out on intact cells selected for MDR by growth in high concentrations of drugs, such as colchicine and vinblastine. However, the difficulties involved in dissecting such a complex system soon led to attempts to use simpler subcellular systems to study the MDR phenomenon. Native plasma membrane vesicles isolated from MDR cells expressing high levels of Pgp have proved to be very useful. Compared to membrane preparations from the drug-sensitive parent cell line, they often display much higher levels of ATPase activity, which are attributable to the presence of large amounts of Pgp in the plasma membrane.[49,50] In addition, membrane vesicles were found to be labeled by photoaffinity analogs of both MDR drugs[51] and nucleotides,[52] providing some of the first biochemical evidence that Pgp binds these molecules. Since then, membrane vesicles have been used for sophisticated kinetic studies of substrate binding using radiolabeled drugs.[53]

Plasma membrane vesicles have also proved useful in studies of Pgp-mediated drug transport. Most vesicle preparations consist of a mixture of right-side-out and inside-out vesicles,[54] and if they are well sealed, the latter population can transport drug from the external medium into the vesicle lumen when provided with ATP. When using a vesicle system where other membrane-bound ATPases are present, it is often necessary to add an ATP-regenerating system, such as creatine kinase and creatine phosphate, to prevent rapid depletion of ATP in the external solution. Substrate uptake into the vesicle interior can be measured in one of two ways. If drug is available in radioactive form (e.g., [^{3}H]colchicine, [^{3}H]vinblastine, [^{125}I]peptide), it is added to the vesicle preparation at time zero, together with ATP and a regenerating system, and vesicles are removed at various times (typically, ranging up to 30 minutes) and collected by rapid filtration.[54] Drug uptake into the vesicles increases with time, usually reaching a plateau value which represents a steady state. This steady state is a result of two competing processes: active transport of drug by Pgp into the vesicle lumen (up a concentration gradient) and passive diffusion of the hydrophobic drug out of the vesicle (down a concentration gradient). Addition of excess unlabeled drug to the vesicle exterior once the steady state has been reached results in very rapid exchange with labeled drug in the vesicle interior.[54] Ruetz and Gros expressed all three mouse Pgps in the yeast mutant strain sec 6-4, which accumulates large numbers of secretory vesicles because of a trafficking defect.[55] These vesicles contained sufficient Pgp for characterization of the drug transport process using a rapid filtration approach.

Caution should be taken when using the fixed time-point rapid filtration approach since transport can become nonlinear within 1 minute, making estimation of the initial rates of Pgp-mediated transport difficult. In these situations, maximal uptake of drug is measured instead, however, steady-state uptake values cannot be treated as kinetic data and do not allow, for example, determination of K_m or $V_{\max}$ for the drug transport process. In addition, this approach consumes relatively large amounts of membrane vesicles and radiolabeled drug. Fluorescence approaches have been developed that circumvent these problems and allow continuous real-time monitoring of Pgp-mediated drug transport in native membrane vesicle systems. For example, fluorescence quenching of daunorubicin transported into the interior of DNA-loaded plasma membrane vesicles allowed kinetic characterization of Pgp-mediated drug transport.[56]

Biochemical characterization of Pgp requires purification of the protein in a functional state. This has been accomplished by several research groups, using a variety of drug-selected MDR cell lines and cells transfected with the MDR1 gene, as the source of protein.[57–61] In general, expression of Pgp in heterologous systems (*Escherichia coli*, baculovirus-infected insect cells, and yeast) has been fraught with difficulties and has not led to the widespread use of this approach. The use of *E. coli* as a host cell for expression was shown to lead to misfolding of the protein.[32] Overexpression in the yeast *Pichia pastoris* is the exception and has led to the purification of milligram amounts of both wild-type and mutant Pgps.[62] This system has also proved very useful for overexpression of other ABC transport proteins.[63] Purified Pgp has been characterized with respect to both its ATPase and drug transport activities (see below), and various biophysical studies have been carried out to assess its structure and conformation, using circular dichroism (CD) spectroscopy,[64] fluorescence spectroscopy,[24] and EM.[44] Pgp has been reconstituted successfully into proteoliposomes, so that both its ATPase and drug transport functions are retained.[25,26,58,65–68]

10.7. ATP BINDING AND HYDROLYSIS BY P-GLYCOPROTEIN

ATP hydrolysis supplies the energy for active drug transport. In most ATP-driven transporters, ATP hydrolysis is tightly coupled to substrate transport, so that it is hydrolyzed only when substrate is transported concurrently. However, Pgp is unusual in displaying a high level of constitutive (basal) ATPase activity, which is observed in the absence of added drugs for plasma membrane vesicles from MDR cells[49,50] and insect cells overexpressing recombinant Pgp,[69,70] and purified Pgp.[57–59] Constitutive

ATPase activity has since been reported for other eukaryotic ABC proteins, including MRP1 (ABCC1), CFTR (ABCC7), ABCA1, ABCR (ABCA4), and several bacterial ABC transporters. Purified Pgp has a maximal basal ATPase activity as high as 3 to 5 μmol/min per milligram of protein, depending on the presence of detergent, lipids, and drugs.[62,71]

The K_m for ATP hydrolysis by membrane-bound and purified Pgp reported by several laboratories is quite high (in the range 0.4 to 0.8 mM), indicating that Pgp has a relatively low nucleotide affinity compared to other transporters. A divalent cation is necessary for ATP hydrolysis. Physiologically, this ion is Mg^{2+}, although both Mn^{2+} and Co^{2+} can support ATP hydrolysis at lower rates.[72] Several inhibitors of Pgp ATPase activity have been identified, including *ortho*-vanadate and various sulfhydryl-modifying agents, including maleimides, 7-chloro-4-nitrobenzo-2-oxa-1,3-diazole (NBD-Cl), *p*-chloromercuribenzenesulfonate, $HgCl_2$, and others. Sulfhydryl reagents covalently modify two Cys residues, one in each Walker A motif (Cys 431 and 1074 in human Pgp)[73] and thereby inhibit catalysis, although ATP binding still takes place.[74] These Cys are not required for ATPase activity, since a Cys-less Pgp protein is still active,[30] and the loss of activity when they are modified probably results from steric interference.

The basal ATPase activity of Pgp is modulated by drug substrates and modulators in a complex and puzzling fashion. Three different patterns have been observed. Many drugs display a biphasic pattern, with stimulation of ATPase activity at low concentrations and varying degrees of inhibition at higher concentrations. Some compounds have been observed to only stimulate activity; for example, many linear peptides, cyclic peptides, and ionophores stimulate Pgp ATPase activity up to 2.5-fold.[75] On the other hand, some substrates appear to produce only inhibition of activity. The molecular basis of these differences in ATPase modulation is not known. The biphasic pattern might arise from the presence of a "stimulatory" drug-binding site and an overlapping "inhibitory" drug-binding site,[76] but why such an arrangement would be intrinsic to the mechanism of Pgp is not clear. To complicate matters further, extremely variable results have been seen from one research lab to another. For example, vinblastine stimulated the ATPase activity of human Pgp[77] but inhibited the ATPase activity of hamster Pgp.[57–59] Modulation of Pgp ATPase activity by drugs and modulators is also highly dependent on the detergent used to isolate the protein or the surrounding lipid environment.[62,65,78]

The ATPase activity of Pgp is inhibited rapidly and completely by the P_i analog, *ortho*-vanadate (V_i), in the presence of ATP. V_i is trapped after a single catalytic turnover in only one NB domain,[72] as the complex $ADP{\cdot}V_i{\cdot}M^{2+}$, where M^{2+} is a divalent cation, usually Mg^{2+}. The trapped complex can also form from ADP and V_i, but at a lower rate. The V_i-trapped complex displays no ATPase activity, suggesting that both catalytic sites must be functional for ATP hydrolysis to take place. Based on these observations, Senior et al. proposed that Pgp operates by an alternating sites mechanism, whereby only one catalytic site can be in the transition-state conformation at any time, and the two sites alternate in catalysis.[79] Based on studies of myosin and other nucleotide-utilizing proteins, the V_i-trapped complex is believed to resemble the catalytic transition state structurally.[80] However, the V_i-trapped complex of Pgp is very stable; V_i and ADP dissociate slowly from the catalytic site, and ATPase activity is regained.[72]

Nucleotide binding to Pgp is of relatively low affinity, making it difficult to measure by classical techniques. Binding of both unmodified nucleotides[74,81] and fluorescent TNP-labeled nucleotides[71] to purified Pgp has been quantitated using fluorescence spectroscopic approaches.[24] Recently, an electron paramagnetic resonance (EPR) spectroscopy study also examined binding of a spin-labeled ATP analog.[82] These studies were consistent in showing a K_d value for ATP or ADP binding in the range 0.2 to 0.4 mM. TNP-labeled nucleotides bind with higher affinity (K_d of 30 to 40 μM), probably because the nitrophenyl ring engages in additional interactions with hydrophobic residues in the nucleotide-binding site.[71] The stoichiometry of ATP binding is normally 2 (i.e., both catalytic sites are occupied)[82,83] In the V_i-trapped complex, the second untrapped catalytic site can still bind ATP with the same affinity.[83]

10.8. DRUG BINDING TO P-GLYCOPROTEIN

Several different approaches have been used to characterize the binding of drugs and modulators to Pgp. Photoaffinity labeling by analogs of substrates and modulators, such as [^{3}H]azidopine and [^{125}I]iodoarylazidoprazosin, has been used widely to study the drug-binding properties of Pgp.[84,85] Competition experiments with photoactive substrate analogs have given an indication of binding affinity and demonstrated interactions between substrates and modulators. However, labeling stoichiometry is often very low, complicating interpretation of the results. In addition, kinetic analysis of binding, and quantitation of dissociation constants, is not possible.

Direct binding studies using radioactive drugs and modulators have been carried out using native plasma membrane vesicles containing Pgp.[53,86–89] Such an approach is technically difficult because of the high levels of nonspecific background binding obtained with hydrophobic drugs, which arises from nonspecific partitioning into the membrane. Detailed kinetic analysis led to the estimation of K_d values for binding, and rates of association and dissociation could also be quantitated. Complex allosteric interactions were found between multiple drug-binding sites.[53]

Fluorescence quenching approaches have been developed to monitor binding and obtain quantitative estimates of K_d for binding of drugs, and modulators to purified Pgp.[24,90] These techniques can measure equilibrium binding without the need to separate Pgp-bound drug from free drug. The first approach used Pgp labeled at the two Walker A motif Cys residues with 2-(4′-maleimidylanilino)naphthalene-6-sulfonic acid (MIANS).[74] Saturable quenching of MIANS fluorescence was obtained with nucleotides, drugs, and modulators, and fitting of the data led to an estimate of the K_d value for binding. More recently, saturable quenching of the intrinsic Trp fluorescence of purified Pgp was observed with nucleotides, drugs, and modulators, and again led to quantitation of the binding affinity.[81] Values of K_d for a large number of different drugs and modulators range from 37 nM for paclitaxel (a high-affinity substrate) to 158 μM for colchicine (a low-affinity substrate).[24,91] Thus, the substrate-binding affinity of Pgp covers a range of 10^4.

10.9. P-GLYCOPROTEIN-MEDIATED DRUG TRANSPORT

The transport activity of Pgp can be studied in intact cells or in simpler subcellular systems such as plasma membrane vesicles and reconstituted proteoliposomes. In general, it has proved difficult to characterize the transport properties of Pgp in complex intact cell systems. However, one exception to this has been the use of polarized epithelial cells (such as MDCK, LLC-PK1, or Caco-2 cells) grown as monolayers on permeable filters that allow separate access to the basal and apical compartments.[92] Transfection of Pgp results in expression of the protein at the apical surface, and quantitative measurements of basal-to-apical and apical-to-basal fluxes of a drug can be made.[93–95] This approach can be very useful for direct determination of whether a drug is transported by Pgp, and showed that many MDR modulators are themselves transport substrates.[94,96–98] However, these cell lines may also show endogenous expression of drug transporters, although at low levels, which may complicate interpretation of experimental data.

Plasma membrane vesicles from MDR cells have been used extensively for measurements of Pgp-mediated drug transport. Inside-out vesicles (present in variable amounts in plasma membrane preparations) transport drug into the lumen when supplied with ATP and an ATP-regenerating system.[54,56,99–103] Radiolabeled drugs such as [^{3}H]vinblastine, [^{3}H]daunorubicin, or [^{3}H]colchicine are usually employed. Some early work purporting to measure drug-binding to membrane vesicles did not differentiate between binding and transport, since ATP was included in the samples (at the time it was not known if ATP was required for drug-binding). These studies probably measured drug transport rather than ATP-dependent drug-binding. Osmotic sensitivity is a useful test to differentiate between transport and binding, and has been used for both plasma membrane vesicles[54,99,100] and reconstituted systems.[54] One additional difficulty is the high background levels of drug often observed for hydrophobic drugs such as vinblastine.[54]

In general, drug transport into plasma membrane vesicles or proteoliposomes is saturable at high drug concentrations and requires ATP hydrolysis; nonhydrolyzable analogs do not support transport. A drug concentration gradient is generated across the membrane, which can usually only be estimated indirectly.[54] Drugs and modulators block transport of other drugs with varying degrees of effectiveness. Reconstituted proteoliposomes containing fully or partially purified Pgp have also been used to characterize drug transport. An ATP-regenerating system is often not required, since other membrane-bound enzymes do not deplete ATP. Similar approaches using radiolabeled substrates have been used to monitor transport in proteoliposomes. In this more defined system, the magnitude of the drug concentration gradient was estimated more precisely; Pgp built up a five- to six-fold gradient of colchicine[25] and NAc-LLY-amide.[26]

Real-time fluorescence assays can monitor the Pgp-mediated transport of fluorescent substrates continuously. H33342 is highly fluorescent when partitioned into the membrane but loses fluorescence after export into the aqueous solution, allowing the initial rate of movement of the dye out of plasma membrane vesicle to be quantitated in real time.[104] The same system was used to demonstrate H33342 transport by purified Pgp reconstituted into liposomes in an inside-out orientation.[66] Tetramethylrosamine (TMR), which loses its fluorescence when transported into the interior of reconstituted vesicles containing Pgp, was used to measure the initial rate of transport over times as short as 30 seconds.[105] Kinetic analysis of TMR transport showed that Pgp obeyed Michaelis–Menten kinetics with respect to both ATP ($K_m = 0.48$ mM, close to the K_m for ATP hydrolysis) and TMR ($K_m = 0.3$ μM).

The stoichiometry of ATP hydrolysis relative to the number of drug molecules transported by Pgp is a controversial issue that has still not been resolved, mainly because of the high basal levels of ATP hydrolysis. Sharom et al. estimated that three or four additional molecules of ATP were hydrolyzed for every molecule of colchicine transported,[25] while Ambudkar et al. reported that 2.8 ATP molecules were hydrolyzed for every molecule of vinblastine transported.[106] Shapiro and Ling estimated that the apparent rate of transport of H33342 from the membrane was 50-fold lower than the rate of ATP hydrolysis.[66] It seems likely that the true turnover rate of Pgp transport will always be underestimated by conventional transport experiments, since the rate of *net* drug accumulation inside a vesicle or proteoliposome is measured. Lipophilic drugs that are moved from the membrane into the lumen immediately repartition into the bilayer, where they are then reexported, resulting in futile cycling of the transporter that does not result in a net increase in drug moved across the membrane. A transport system using the ionophore valinomycin (a Pgp substrate) with bound $^{86}Rb^+$ circumvented this problem.[107] Pgp mediated the ATP-dependent uptake of valinomycin–$^{86}Rb^+$ complex into the proteoliposome lumen, where the radioactive cation accumulated to a concentration of 8 mM since it is not lipid-soluble. When the ATPase and transport activities of Pgp were measured under the same conditions, comparable rates of valinomycin transport and ATP hydrolysis were found, with 0.5 to 0.8 ionophore molecule transported per ATP molecule hydrolyzed.[67]

10.10. SUBSTRATE SPECIFICITY OF P-GLYCOPROTEIN AND NATURE OF THE DRUG-BINDING SITE

Pgp displays a remarkable ability to interact with, and transport, a large variety of compounds, ranging from chemotherapeutic drugs to peptides. Most preferred substrates are amphipathic and relatively hydrophobic, although some are not (colchicine, for example, is quite water soluble). Pgp substrates range in size from large complex molecules such as paclitaxel and vinblastine to smaller drugs such as daunorubicin and doxorubicin. Pgp also interacts with linear and cyclic peptides and ionophores, including gramicidin D, valinomycin, *N*-acetyl-leucyl-leucyl-norleucinal (ALLN), leupeptin, pepstatin A, and several bioactive peptides.[75,108] Small tripeptides such as NAc-LLY-amide are excellent transport substrates.[26] Even nonionic detergents such as Triton X-100 and nonylphenol ethoxylates interact with Pgp.[91,109,110] Many substrates, but not all, contain planar aromatic rings and positively charged tertiary N atoms. No highly conserved recognition elements have been found in Pgp substrates and modulators.

Many attempts have been made over the years to develop a quantitative structure–activity relationship (QSAR) for Pgp substrates and modulators, to link various properties of these molecules (physical, chemical, or structural) with their biological activity.[111] One problem in achieving this goal has been the wide variety of biological assays (many indirect) used to infer interaction with Pgp, the use of limited series of structurally related compounds, and the lack of consistency in the molecular descriptors used. Very few studies have measured binding affinity directly, and various surrogates, such as inhibition of drug transport or stimulation of ATPase activity, have been used instead. An additional complication has been the likely existence of at least two binding regions within the drug-binding pocket of the protein, which interact with each other allosterically.[112] Hydrophobic peptides appear to differ from other Pgp substrates (e.g., they are much smaller than a typical substrate and often have no aromatic rings or tertiary N atoms), and most studies on common pharmacophores have not considered them.

Seelig defined a set of structural elements that are required for interaction of a compound with Pgp,[113] consisting of two or three electron donors (hydrogen-bond acceptors) arranged in a fixed spatial separation. Any molecule containing at least one of these structural units was predicted to be a Pgp substrate, and binding affinity was predicted to increase with hydrogen-bond strength. The TM domains of Pgp contain a high fraction of amino acids with side chains capable of acting as hydrogen-bond donors to interact with substrates. A more recent three-dimensional quantitative structure–activity relationship study supported these ideas and suggested that interaction of the substrate with one or more sites within the protein plays a key role in efflux.[114] Substrate recognition was proposed to be based on the dimensions of the drug molecule, and the presence of two types of recognition elements, two hydrophobic groups 16.5Å apart and two hydrogen-bond acceptors 11.5Å apart. Another three-dimensional (QSAR) approach was used to generate a Pgp pharmacophore consisting of one hydrogen-bond acceptor, one aromatic ring, and two hydrophobic units.[115] Pajeva and Wiese proposed a pharmacophore model consisting of two hydrophobic units, three hydrogen-bond acceptors, and one hydrogen-bond donor.[116] They concluded that drug-binding affinity depends on the number of points involved simultaneously in the interaction with Pgp and proposed that different drugs can be involved in different binding modes with these points. QSAR studies have also been carried out for Pgp modulators,[117,118] and attempts have been made to classify them based on their structures.[119] Artificial neural networks have also been used,[120] with the aim of employing such analysis as a predictive tool to identify new MDR-reversing agents.

Many questions remain about how Pgp can bind and transport so many structurally diverse compounds. Biochemical studies have been used to argue for a single common drug-binding site, or two or more separate sites. Based on ATPase inhibition studies, it was proposed that drugs, peptides, and modulators all competed for a common drug-binding site,[121] whereas another group concluded that two separate pharmacophores existed.[122] Photoaffinity labeling studies suggested that Pgp contained two nonidentical drug-binding sites, one in each half of the protein.[123] Binding studies using radiolabeled drug supported the existence of multiple binding sites, which displayed complex allosteric interactions and could switch between high- and low-affinity conformations.[53] Shapiro and Ling demonstrated the existence of two "functional" transport sites within Pgp; the H-site showed preference for the drug Hoechst 33342 (H33342), whereas the R-site showed preference for rhodamine 123

(R123).[112] The two sites interacted with each other allosterically, such that H-site and R-site drugs mutually stimulated each other's transport, whereas two H-site drugs inhibited each other's transport, as did two R-site drugs. Later work suggested the existence of a third allosterically linked drug transport site.[124]

Soluble bacterial transcription factors that bind multiple drugs (i.e., QacR, BmrR, and MarR) have provided intriguing insights into how a single drug-binding site can accommodate many structurally diverse compounds.[125] Crystallographic studies of QacR complexes with six drugs showed that the protein contains a large, flexible binding pocket, rich in aromatic amino acids, but also containing some polar residues.[126] Van der Waals and hydrophobic interactions play a major role in drug-binding, augmented by electrostatic interactions between charged groups on the drug and charged amino acid side chains. The size and flexibility of the binding pocket allow drugs with different structures to establish interactions with different subsets of residues. Two distinct but partially overlapping binding pockets were observed. Later studies showed that two drugs could bind to the protein simultaneously.[127] Structural studies of the human xenobiotic nuclear receptor, PXR, showed that the same drug can bind within a large hydrophobic cavity in three different orientations, each stabilized by a different complement of polar side chains.[128]

Multidrug transport proteins such as Pgp probably bind their substrates using principles similar to those observed for soluble multidrug-binding proteins.[129] The crystal structure of the bacterial RND-family multidrug efflux pump AcrB, binding four structurally diverse drugs, showed that three ligand molecules bind simultaneously to a large central cavity, primarily by hydrophobic, aromatic stacking, and van der Waals interactions.[130] Each drug binds to AcrB using a different subset of amino acid residues. Studies using Cys mutants and thiol-reactive substrate analogs support the idea of a common hydrophobic pocket within Pgp and show that residues from multiple TM segments contribute to the binding region.[131–134] Cys cross-linking experiments showed that the packing of the TM segments of Pgp is altered when drugs bind: in a different way for each substrate.[135] This "induced-fit" type of mechanism can explain how the binding pocket accommodates such a broad range of structurally diverse compounds.

Like the transcriptional regulator proteins, the drug-binding pocket of Pgp appears to be able to accommodate more than one compound simultaneously. Based on their cross-linking data, Loo et al. proposed that a thiol-reactive substrate and a second drug molecule could occupy different regions of the binding pocket simultaneously.[136] More recently, fluorescence approaches showed that LDS-751 and R123 could both bind to the R-site of Pgp at the same time, interacting in a noncompetitive fashion.[137] The dimensions of the drug-binding pocket, determined using a thiol-reactive cross-linking substrate, also suggest that it is large enough to accommodate two substrates at the same time.[138]

Several approaches have been used to locate and characterize the regions of Pgp that form the drug-binding pocket. Labeling of the protein with various photoactive drug analogs, followed by chemical or proteolytic cleavage and identification of the labeled peptides showed that several TM segments in both halves of Pgp were involved in substrate-binding.[139–142] Different regions of the protein were labeled by different drug analogs, suggesting that they did not all bind at exactly the same location.

Mutagenesis studies indicated that residues in TM 4, 5, and 6 in the N-terminal half of Pgp and TM 9, 10, 11, and 12 in the C-terminal half were involved in forming the drug-binding pocket.[143] Loo and Clarke systematically inserted single Cys residues by site-directed mutagenesis into 252 positions in the TM segments, and then reacted them with either a thiol-reactive substrate or a drug analog.[131,132,144] Overall, the drug-binding pocket is envisioned as funnel-shaped, narrower at the cytoplasmic side of the membrane,[145] where TM2/TM11 and TM5/TM8 come together.[47,48] They concluded that the drug-binding pocket is found at the interface between the two TM "halves" of Pgp. This was confirmed by Pleban et al., who used propafenone-type substrate photoaffinity ligands, in conjunction with matrix-assisted laser desorption/ionization time-of-flight mass spectrometry, to define the substrate-binding site(s) of Pgp.[146] They observed the highest labeling in TM 3, 5, 8, and 11, and when this pattern was projected onto a three-dimensional homology model of Pgp, labeling was found to occur at the interface formed by TM3 and 11 in one half of the protein, and TM5 and 8 in the other half.

Pawagi et al. proposed that aromatic amino acid residues may play an important role in the binding and transport path for drug substrates.[147] Studies of intrinsic Pgp fluorescence also showed that Trp residues were highly quenched by binding of certain drugs, and FRET from Trp to substrate molecules took place with high efficiency,[81,148] suggesting that Trp residues are located close to the sites of drug-binding.

10.11. P-GLYCOPROTEIN AS A HYDROPHOBIC VACUUM CLEANER OR DRUG FLIPPASE

Pgp substrates are typically hydrophobic and are expected to partition into the membrane. The substrate-binding sites of Pgp appear to be contained within its TM regions, and drugs gain access to these sites after partitioning into the lipid bilayer (Figure 10.2*a*).[34] The idea that the transporter acts as a "vacuum cleaner" for hydrophobic molecules present in the membrane was first suggested by Higgins and Gottesman[149] and has found widespread acceptance. In intact cells, Pgp substrates entering the cell from the extracellular medium are intercepted at the plasma membrane and extruded to the exterior without entering the cytosol.[150] Lipid bilayers are amphipathic multilayered structures and do not behave like an organic solvent such as octanol in terms of drug partitioning. After entering the membrane, Pgp substrates (which are generally amphipathic in nature) are not distributed uniformly in the hydrophobic core of the lipid bilayer but tend to concentrate in the interfacial regions of the membrane.[151]

Several studies suggested that the drug-binding pocket of Pgp is probably located within the cytoplasmic leaflet of the membrane. Drugs appear to gain access to this binding site after moving to the cytoplasmic leaflet by spontaneous "flip-flop" from the outer leaflet, which can be a slow process for many compounds that are Pgp substrates.[152,153] Transport by reconstituted Pgp of the fluorescent dyes H33342 and LDS-751 suggested that they were probably extracted from the cytoplasmic leaflet of the membrane.[154,155] The positively charged compound *N*-methyldexniguldipine, which is unable to flip-flop to the inner leaflet, could interact with Pgp if added to cell fragments and membrane vesicles where the cytoplasmic membrane face is accessible,

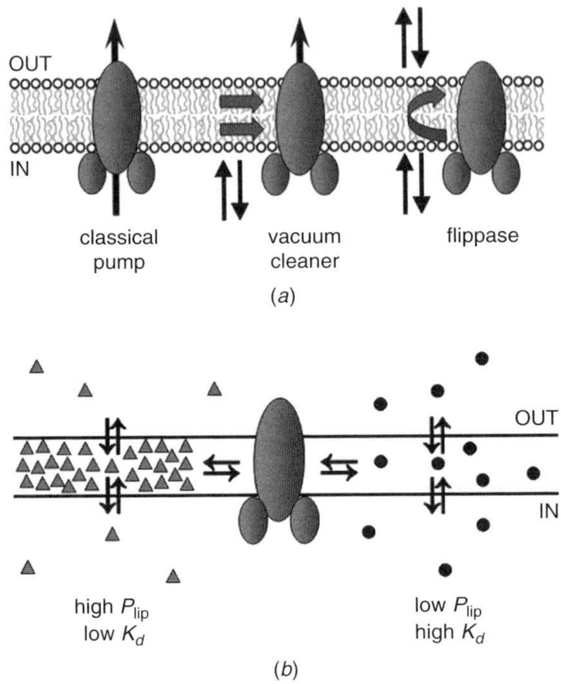

FIGURE 10.2. (*a*) Classical pump, vacuum cleaner, and flippase models of Pgp action. Classical pumps, such as lactose permease, transport a polar substrate from the aqueous phase on one side of the membrane to the aqueous phase on the other side. The substrate does not come into contact with the lipid bilayer and moves through a hydrophilic path formed by the TM regions of the protein. In the vacuum cleaner model, drugs (both substrates and modulators) partition into the lipid bilayer and interact with Pgp within the membrane. They are subsequently effluxed into the aqueous phase on the extracellular side of the membrane. In the flippase model, drugs partition into the membrane, interact with the drug-binding pocket in Pgp (which is located within the cytoplasmic leaflet) and are then translocated, or flipped, to the outer membrane leaflet. Drugs will be present at a higher concentration in the outer leaflet compared to the inner leaflet, and an experimentally measurable drug concentration gradient is generated when drugs partition rapidly from the two membrane leaflets into the aqueous phase on each side of the membrane. (*b*) The effect of membrane partitioning on drug-binding to Pgp. The binding affinity of Pgp for a particular substrate or modulator (K_d) is related to the lipid–water partition coefficient of the drug (P_{lip}). A drug with a high value of P_{lip} (left side of the figure) will accumulate to a high concentration within the membrane. This will favor binding to Pgp and result in a low apparent K_d value. In contrast, a drug with a low value of P_{lip} (right side of the figure) will have a lower membrane concentration and a high apparent K_d value.

but not if added to intact cells.[156] Similarly, some peptide modulators cannot interact with Pgp in intact cells if supplied in the extracellular medium but can do so in membrane vesicles, where they can reach the cytoplasmic leaflet.[157] More recently, a FRET approach estimated the distance separating bound H33342 from a fluorescent probe linked covalently to the catalytic sites, and the results clearly place the drug-binding site within the cytoplasmic membrane leaflet.[148] The binding site for another fluorescent substrate, LDS-751, was also localized by FRET to the cytoplasmic half of the bilayer, although closer to the membrane surface in the interfacial region.[158] H33342 binds to the H-site and LDS-751 to the R-site, so it appears that both of these functional drug transport sites are in Pgp domains within the cytoplasmic membrane leaflet.

It was proposed that Pgp may operate as a drug flippase, moving hydrophobic drug molecules from the inner to the outer leaflet of the membrane (Figure 10.2*a*).[149] Given the high level of amino acid similarity between Pgp and its close relative the MDR3/ABCB4 protein, which functions as a PC-specific phospholipid flippase, this suggestion seems reasonable. The location of the drug-binding pocket in the cytoplasmic leaflet of the membrane is also compatible with this idea. If Pgp maintains a higher drug concentration in the outer leaflet than in the inner leaflet, equilibration of drug between the membrane and the aqueous phase on each side would result in the observed drug concentration gradient. Such partitioning and equilibration of nonpolar drugs between lipid bilayers and water is a very fast process, limited only by diffusion.[159] In fact, it is not possible to distinguish experimentally between a transport process in which drugs are moved from the inner to the outer leaflet, followed by rapid partitioning into the aqueous phases on each side, and one in which drugs are moved from the inner leaflet directly to the extracellular aqueous phase, followed by repartitioning of drug into the outer leaflet.

Several studies have indicated that Pgp can move fluorescent NBD-labeled phospholipid derivatives from the inner to the outer leaflet of the plasma membrane in intact cells overexpressing the protein,[160,161] and glycosphinglipids (GSLs) have also been considered as substrates.[162,163] Since then, reconstituted Pgp in lipid bilayer vesicles has been shown to act as an outwardly-directed flippase for NBD-labeled phospholipids and simple GSLs such as glucosyl-, galactosyl-, and lactosylceramide.[164,165] The lipid translocation process shares many biochemical features with drug transport: it requires ATP hydrolysis, it is inhibited by V_i, and drug substrates compete with flippase activity.[164,165] Thus, both drugs and membrane lipids appear to follow the same transport route through the Pgp molecule, increasing the likelihood that drug transport takes place via a flippase-like mechanism. It is possible that Pgp plays a physiological role in flipping glucosylceramide from the cytoplasmic leaflet to the luminal leaflet of the Golgi apparatus, which is a required step in the biosynthesis of complex GSL.[165,166]

10.12. ROLE OF THE LIPID BILAYER IN P-GLYCOPROTEIN FUNCTION

The hydrophobic vacuum cleaner model proposes that drugs and modulators partition into the membrane before interacting with the transporter. Pgp substrates generally have high lipid:water partition coefficients[152,167] and accumulate within the membrane to high concentrations (Figure 10.2*b*). Pgp would thus experience a much higher drug concentration than that nominally added to the aqueous phase, by 300- to

20,000-fold.[152,167] The role of the lipid bilayer is thus to concentrate the drug; Pgp itself may have a relatively low intrinsic substrate-binding affinity. The kinetics of transport of the lipophilic dye H33342 out of the membrane were measured using a fluorescence approach.[104] The transport rate was directly proportional to the amount of H33342 in the lipid phase and inversely proportional to its concentration in the aqueous phase, thus demonstrating that Pgp removes the dye from the lipid bilayer.

The mode of action of Pgp modulators also appears to be intimately linked to the presence of the membrane and the behavior of drugs within it. Modulators show the same structural features as substrates, interact with the drug-binding pocket, and (in many cases) are also transported by Pgp, yet cells are not resistant to them, and they reverse drug resistance. The behavior of modulators has been linked to their rate of diffusion across the membrane.[168] Pgp substrates were found to cross a lipid bilayer relatively slowly (e.g., R123 had a half-life of 3 minutes), while the transbilayer movement rate of several MDR modulators was too fast to be detected.[169] Amphipathic drugs and modulators localize in the interfacial regions of the bilayer, and appear to cross membranes by a flip-flop mechanism.[152] Thus, it was proposed that drugs and modulators are handled similarly by Pgp; they are transported, with hydrolysis of ATP. Compounds that have been pumped out can reenter the outer leaflet and flip-flop back into the inner leaflet (i.e., diffuse across the membrane) before interacting with Pgp again and being reexported. For substrates, the rate of membrane reentry is slow enough for efflux via Pgp to keep pace, and a drug gradient is established, causing resistance. For modulators, the rate of membrane reentry is so rapid that Pgp cannot keep pace and essentially operates in a futile cycle; the transport turnover and rate of ATP hydrolysis are high, but no concentration gradient is generated; thus, cells are not resistant to modulators. This model[159] suggests that effective modulators are high-affinity substrates with a high transbilayer diffusion rate.

Reconstitution of Pgp into bilayers of defined lipids has been an important tool in exploring the role of specific lipids, and the membrane in general, on its activity. Differential scanning calorimetry (DSC) studies indicated that reconstituted Pgp perturbed a large number of membrane phospholipids, even at relatively high lipid/protein ratios.[170] When Pgp was reconstituted into proteoliposomes composed of three different PCs, drug substrates displayed different partition coefficients into these bilayers.[167] The affinity of drug binding measured using fluorescence quenching was highly correlated with the lipid : water partition coefficient, so that K_d decreased as the partition coefficient increased (Figure 10.2*b*). The concentration of the drug in the membrane is thus important for high-affinity interaction of drugs with Pgp.

The ATPase activity of Pgp is also modulated by the lipid environment surrounding the protein. Addition of various phospholipids to purified Pgp resulted in concentration-dependent increases or decreases in activity and protected the protein from thermal inactivation.[78] The values of both K_m for ATP hydrolysis and K_d for ATP binding were different above and below the melting temperature of the lipid bilayer, and the activation and energies for ATP hydrolysis in the gel and liquid-crystalline phases of the bilayer were also significantly different.[171] Thus, both ATP binding and ATP hydrolysis by Pgp are affected by the phase state of the membrane in which it is reconstituted, possibly because the NB domains may interact with the bilayer surface.

The profile for ATPase activation and inhibition by drugs and modulators changes when Pgp is moved from detergent solution into a lipid bilayer[25] and also varies with the nature of the host lipid in which the protein is reconstituted.[65] This suggests that coupling between drug-binding sites and NB domains is affected by the lipid environment of the protein.

Pgp-mediated drug transport is also affected in an interesting way by the fluidity of the membrane. Changes in the fluidity of canalicular membrane vesicles altered Pgp-mediated transport of daunorubicin and vinblastine.[172] When lipid fluidity was increased using membrane fluidizers, drug transport was significantly inhibited, suggesting that the physical state of the membrane affects Pgp transport function. This idea was explored further using Pgp reconstituted into proteoliposomes composed of two synthetic PCs with different melting points.[105] A real-time fluorescence assay used to measure the initial rate of transport found a highly unusual biphasic temperature dependence: a high rate of transport in the rigid gel phase, the maximum transport rate at the melting temperature of the bilayer, and a lower transport rate in the fluid liquid-crystalline phase. This pattern suggests that the rate of drug transport by Pgp may be dominated by partitioning of drug into the bilayer, which shows similar temperature dependence.

10.13. MECHANISM OF ACTION OF P-GLYCOPROTEIN

Much remains to be understood about how Pgp transports (or flips) drugs and how coupled ATP hydrolysis powers transport. Transport can be broken down into several steps: entry of substrates into the binding pocket within the cytoplasmic leaflet, conformational changes in Pgp driven by ATP binding/hydrolysis, and release of drug to either the outer leaflet or the extracellular aqueous phase. Many different experimental approaches, including various biochemical and spectroscopic techniques, have provided evidence that conformational changes take place during the catalytic cycle of Pgp and other ABC proteins.[173] It is assumed that release of drug from Pgp involves reorientation of the drug-binding site from the cytosolic side of the membrane (or the inner membrane leaflet) to the extracellular side (or the outer membrane leaflet), accompanied by a switch from high to low drug-binding affinity. Superimposed on the transport cycle is the ATP hydrolysis cycle, which involves ATP binding, formation of the nucleotide sandwich dimer, ATP hydrolysis, dissociation of P_i, and dissociation of ADP. A recent review discusses the drug translocation mechanism of Pgp in detail.[174]

Substrates may diffuse from the lipid bilayer into the drug-binding pocket through "gates" formed by TM segments at either end of the pocket.[175] The nature of the local environment within the drug-binding pocket is still controversial. Loo et al. tested whether Cys residues within the drug-binding pocket of Pgp were able to react with charged thiol-reactive compounds and concluded that the drug-binding pocket is accessible to water.[176] In contrast, the fluorescence properties of drugs bound to purified Pgp clearly indicate that the local environment of the binding site is very hydrophobic, with a polarity lower than that of chloroform.[158] Several drugs (e.g., H33342, LDS-751) show large increases in the intensity of their fluorescence emission and a substantial blue shift in their emission wavelength on binding to Pgp, both hallmarks of a hydrophobic local environment.[148,158]

Conformational communication must exist between the drug-binding pocket and the NB domains, so that substrate binding activates ATP hydrolysis and initiates the transport cycle. Binding of drugs caused large changes in the fluorescence of MIANS groups covalently linked within the catalytic site of the NB domains, thus confirming this idea. More recently, the effect of drug-binding on cross-linking between Cys residues in the signature C and Walker A motifs was tested. Drug-binding in the TM regions induced long-range conformational changes in both NB domains, to decrease or increase the distance between these two sequence motifs.[177] TM2/TM11[47] and TM5/TM8[48] are close together and probably enclose the drug-binding pocket, which is located at the interface between the TM halves at the cytoplasmic side of Pgp. These regions may form the "hinges" required for conformational changes during the transport cycle. Covalent linkage of a drug analog to position 306 led to permanent activation of Pgp ATPase activity, suggesting that this region of the protein may be part of the signal that switches on ATP hydrolysis when the drugs occupy the binding pocket.[178]

A photoaffinity labeling study reported that the V_i-trapped complex of Pgp, which is thought to resemble the catalytic transition state, has drastically reduced affinity for drug substrates.[179] It was proposed that following ATP hydrolysis, drug is moved from a high- to a low-affinity binding site, thus promoting release from Pgp on the other side of the membrane.[180] However, this has been contradicted by quantitative fluorescence quenching measurements of drug and nucleotide binding to the V_i-trapped complex, which showed that it can bind several different drugs with high-affinity.[181] High-affinity substrate-binding was also observed for the V_i-trapped complex of the ABC protein TAP1/TAP2.[182] Based on these results, a concerted transport mechanism was proposed rather than a multistep reaction.[181] In this model, release of drug from Pgp during the catalytic cycle is proposed to occur simultaneously with ATP hydrolysis, and precedes formation of the V_i-trapped complex.

Drug release was proposed to occur as a result of rehydration of the substrate when it enters the drug-binding pocket,[176] which was originally envisioned as a large cavity filled by water,[38] and it was suggested that hydration may prevent drug from repartitioning into the lipid bilayer. However, recent EM structures of Pgp do not show a large water-filled cavity.[44,45] In addition, hydration of drug when it is released into an aqueous environment cannot prevent it from reentering the membrane, since a hydrated hydrophobic molecule is an unfavorable situation entropically, and it will very rapidly partition into the lipid bilayer.

ATP provides the energy for the *power stroke*, which consists of conformational change(s) that drive drug transport. Two proposals exist that propose different origins for the power stroke of Pgp.[183] Based on observations that drug-binding altered ATP binding affinity,[88] Higgins and Linton proposed the ATP switch model, in which dimerization of the NB domains driven by ATP binding is the source of the power stroke.[184] Tight ATP binding following drug-binding is proposed to drive formation of the sandwich dimer, transmitting conformational changes to the drug-binding domains that result in drug transport. ATP hydrolysis is then used to separate the NB domains and reset the transporter. However, this model is controversial, since several groups have reported that drug-binding does not affect ATP binding,[49,70,77] and quantitation of ATP binding affinity showed only small changes upon drug binding.[83] Senior et al. proposed that transport is driven by relaxation of a high-energy intermediate formed during ATP hydrolysis, which thus provides the power stroke.[79] One molecule of ATP was proposed to drive the transport of one drug molecule. Sauna and Ambudkar have proposed an alternative model in which two molecules of ATP are hydrolyzed per cycle.[185] In this model, drug and ATP binding do not influence each other: hydrolysis of one ATP molecule drives drug transport, and hydrolysis of a second ATP molecule resets the transporter. This model is also unsatisfactory. There has been no independent verification of the proposed requirement for two rounds of ATP hydrolysis per drug molecule transported. Sauna et al. reported that Pgps with mutations in the Walker B Glu residues (E556Q and E1201Q) failed to undergo the second round of ATP hydrolysis required to reset the transport cycle.[186] However, this was contradicted by Senior and co-workers, who found that these mutants could undergo multiple catalytic turnovers. Rapid kinetic studies that dissect out various steps in the transport cycle, and define their kinetic and thermodynamic constants, may be required to fully understand the mechanism of action of Pgp.

10.14. ROLE OF P-GLYCOPROTEIN IN DRUG THERAPY

Pgp substrates include many drugs that are used in the treatment of common human diseases. The protein consequently plays a central role in drug absorption and disposition in vivo and is an important determinant in the pharmacokinetic profile of many drugs and ultimately, the clinical response.[187,188] Pgp substrates include anticancer drugs, HIV protease inhibitors, analgesics, calcium channel blockers, immunosuppressive agents, cardiac glycosides, antihelminthics, antibiotics, and H_2-receptor antagonists, to name just a few (see Table 10.1).

High levels of Pgp are found in the luminal membrane of the capillary endothelial cells, where it immediately pumps drugs back into the blood. The presence of Pgp strongly reduces the brain accumulation of many different drugs, and in knockout mice, penetration of substrates into the brain is increased 10- to 100-fold. Pgp prevents the penetration of HIV protease inhibitors into the brain, limiting treatment efficacy. Anticancer drugs directed to brain tumors are also prevented from reaching their desired site of action.

Pgp appears to be a major player in limiting absorption of drugs from the intestinal lumen. Studies in knockout mice showed that the bioavailability of orally administered paclitaxel, a drug known for its poor solubility, increased from 11% to 35% in animals lacking Pgp.[189] Paclitaxel and other drugs are also excreted directly from the blood circulation into the intestinal lumen. However, not all Pgp substrates show compromised drug absorption. For example, digoxin, HIV protease inhibitors, verapamil, and quinidine all show high oral bioavailability despite being good Pgp substrates.[187] Thus, Pgp may not be as quantitatively important as first thought in drug absorption. It is possible that high drug concentrations in the intestinal lumen saturate the transporter; as well, Pgp-mediated efflux may have a limited effect on bioavailability if passive diffusion rates are high. Because of the increased likelihood that a drug will fail to be effective in animal and human trials if it is a Pgp substrate, many pharmaceutical companies have added interactions with Pgp to their drug discovery screening processes in an attempt at early identification of these compounds. This is especially important for drugs targeted to the central nervous system.

Blockade of Pgp with modulators can have dramatic effects on systemic drug disposition by decreasing drug elimination through the intestine, bile, and urine. Initially, the focus was on using modulators with anticancer drugs to improve the efficacy of chemotherapy treatment,[190] but later it was realized that modulators could be useful in altering the pharmacological behavior of many drugs, to improve their delivery. Modulators may enhance intestinal drug absorption and increase drug penetration through biologically important protective barriers such as the blood–brain, blood–cerebrospinal fluid, and maternal–fetal barriers. Delivery of drugs to the brain, either to treat epilepsy and other central nervous system diseases, AIDS, or brain tumors such as gliomas might therefore be increased by addition of an effective modulator. This has been shown to be feasible in a mouse model using highly effective modulators such as PSC833 and GF120918.[191–193] The future development of more effective Pgp modulators may make enhanced drug delivery to the brain a realistic clinical goal.

First-generation modulators (e.g., verapamil, quinidine) were poor Pgp inhibitors, requiring high plasma levels to reverse MDR, which could not be obtained without unacceptable patient toxicity. In addition, these drugs were used clinically to treat other medical conditions and produced pharmacological side effects. Several advanced MDR-reversing agents are in various stages of development.[194] Second- and third-generation MDR inhibitors with good clinical potential include PSC833, GF120918, XR9576, LY335979, VX-710, and OC 144-093.

Several Pgp modulators also inhibit one or more cytochrome P450 enzymes (e.g., CYP3A4) that function to metabolize drugs. Thus, it has been widely observed that treatment with Pgp modulators decreases drug clearance, resulting in increased toxicity of coadministered drugs. Plasma drug levels remain higher for longer, increasing the "area under the curve" (AUC) and often necessitating a substantial reduction in drug dose to avoid toxicity. More selective third-generation Pgp modulators, such as LY335979 and XR9576, do not inhibit the CYP enzymes and show only small increases in AUC, so that dose reduction is not needed. Understanding how Pgp modulators affect the toxicity and pharmacokinetics of other drugs is important for the design of clinical trials of MDR modulation.

10.15. MODULATION OF P-GLYCOPROTEIN IN CANCER TREATMENT

A major reason for the failure of chemotherapy treatment to cure human cancers is the ability of tumor cells to become resistant to several anticancer drugs simultaneously. Many mechanisms are known to contribute to MDR in tumor cells, of which the presence of multidrug efflux pumps is only one. Three ABC family members, Pgp, MRP1 (ABCC1), and BCRP (ABCG2), are likely to be the major drug efflux pumps expressed in human cancers.[195] Tumor cells are notoriously heterogeneous, and correlations between drug resistance and the expression of efflux pumps have been difficult to establish. Some tumors express Pgp before chemotherapy treatment (e.g., colorectal and renal cancers), while in others, expression increases after exposure to MDR drugs (e.g., leukemias, lymphomas, myeloma, and breast and ovarian carcinomas). In general, patients with Pgp-positive tumors respond less well to chemotherapy and have a poorer outlook and long-term survival. There is strong evidence linking Pgp expression with poor response to chemotherapy in acute myelogenous leukemia (AML).[196,197]

Studies to validate the role of MDR reversal in the treatment of various malignancies are under way; there have been some partial successes, and many failures. However, there is still no consensus on the useful of MDR modulators in treating human cancers, and the controversy is likely to continue.[198,199] Four contributing factors make the results of many clinical trials with modulators uninterpretable. First, there is a need to establish whether a patient's tumor contains Pgp and whether the level is clinically significant. Second, many modulator clinical trials have used first- and second-generation compounds that are poorly effective at the clinically achievable dose. This limitation will hopefully be overcome by new, more potent and specific third-generation Pgp modulators. A third factor is that modulators affect the disposition of other drugs, either by decreasing drug elimination via Pgp or by inhibiting drug metabolism via cytochrome P450. Cancer patients treated with both chemotherapy drugs and a modulator are thus exposed to higher levels of anticancer drug, which confounds interpretation of the results. In some trials, the dose of anticancer drug was lowered to avoid toxicity and allow direct comparison of results from the two study arms. Finally, tumors have multiple, often redundant mechanisms of cellular resistance to drugs.[200] Not only do tumor cells have other defense mechanisms at their disposal, they can also express other multidrug efflux pumps. Thus, the potential contribution of Pgp to drug resistance in a tumor is very difficult to assess. Modulation of Pgp in tumors is likely to be accompanied by altered Pgp function in normal tissues.[201] However, in some trials, tumor regression was obtained without apparent increases in normal tissue toxicity. There have been suggestions that MDR modulation may delay the emergence of clinical drug resistance. Thus, administration of modulators in the earlier stages of cancer may prevent drug resistance.

10.16. REGULATION OF P-GLYCOPROTEIN EXPRESSION

Cells adapt to the presence of toxic xenobiotics in their environment by up-regulation of drug efflux pumps, such as Pgp, which provides them with a long-term survival advantage. The MDR1 gene is activated, and a stable MDR phenotype induced, after short-term exposure of cells to a variety of environmental insults. This response is of fundamental importance in the case of emergence of MDR in tumor cells exposed to anticancer drugs. MDR1 expression may be up-regulated by two mechanisms: an increase in the amount of MDR1 message by transcriptional regulation, and stabilization of the mRNA. A considerable amount is now known about the transcriptional regulation of ABC proteins, including the MDR1 gene.[202–204] Transcription of a particular gene is determined by various response elements present within the promoter sequence, their accessibility, and the transcription factors available to interact with them, which depend on both the intracellular milieu and extracellular signals. The multiprotein complexes that assemble on the promoter sequence are also dynamic in nature, and influenced by chromatin structure. There appear to be multiple interacting pathways for activation of MDR1. A redundant network of MDR1 regulation ensures the rapid emergence of resistance in cells subjected to chemical stress. By more fully understanding the molecular mechanisms by which the MDR1 gene is activated, it may be possible to intervene clinically to prevent its transcriptional activation.

Most MDR1 transcripts arise from downstream promoter sequences located in the middle of exon 1,[205] which lacks a TATA box. An inverted CCAAT box interacts with the trimeric transcription factor NF-Y and the Sp family transcription factors Sp1 and Sp3. In general, MDR1 transcription is up-regulated as part of a general cellular "stress" response to stimuli such as heat shock, exposure to anticancer drugs and carcinogens, serum deprivation, inflammation, hypoxia, and ionizing radiation. The activation of several signaling pathways, including the protein kinase C and mitogen-activated protein kinase cascades, an increase in intracellular Ca^{2+}, and induction of NFκB, can up-regulate MDR1 expression.[204] Chemical modification of chromatin may affect gene expression, and the MDR1 promoter is regulated negatively by methylation. Posttranscriptional mechanisms also appear to play a role in regulating MDR1 expression, and the stability of MDR1 mRNA is increased in cells subjected to various stresses.

10.17. P-GLYCOPROTEIN GENE POLYMORPHISMS AND THEIR IMPLICATIONS IN DRUG THERAPY AND DISEASE

Changes in Pgp expression and function would be expected to alter the absorption, plasma concentration, tissue distribution, and excretion of its drug substrates. Pgp polymorphisms might thus influence the outcome of drug treatment. Variations in the nucleotide sequence of the Pgp gene can affect both expression and function of the transporter. The first polymorphism to be reported in the human MDR1 gene was the G2677T variant, which results in the amino acid change A893S. Since then, about 30 single-nucleotide polymorphisms (SNPs) have been discovered by sequencing the MDR1 gene in large numbers of people of different ethnic origin.[206–209] The most common variants have probably been identified, although it is possible that some rare polymorphisms remain to be detected. There are considerable differences in the frequency of these variant alleles in different populations of Caucasian, African, and Asian origin.[210] Distinct haplotypes exist, with considerable heterogeneity found within a single ethnic group; however, all ethnic groups appear to possess the three most common haplotypes, which were found in >70% of the total population. Some SNPs result in a change in the amino acid coding sequence of the protein (nonsynonymous), whereas others do not (synonymous).

Polymorphisms have been reported to alter both the expression and the function of the transporter. For example, the synonymous C3435T variant (exon 26) appears to result in reduced Pgp expression levels, leading to increased oral absorption of digoxin and higher plasma drug levels. These results, however, were later contradicted by those of other groups. A recent metaanalysis suggested that the C3435T SNP has no effect on the expression of MDR1 mRNA or the pharmacokinetics of digoxin.[211] Conflicting data have been reported on the effects of other alleles using various drug substrates, and the controversy seems likely to continue. The differential effects of Pgp polymorphisms on Pgp expression and drug disposition will probably not be resolved until progress is made in standardizing parameters such as sample size and makeup, environmental factors, and the assays used for Pgp protein and mRNA quantification. MDR1 haplotypes, rather than individual SNPs, are also more likely to affect the pharmacokinetics of MDR1 substrates. Two common Pgp polymorphisms (G2677T/A and C3435T) may play a role in the differential response to the cholesterol-lowering statin drugs.[212] When haplotypes were also considered, a subgroup of female patients was identified who showed a remarkable response to treatment, which was not linked to a single polymorphism.

Pgp variants carrying spontaneous mutations have been found in cultured cell lines. The first to be reported was the G195V substitution, which confers increased resistance to colchicine but has little effect on resistance to several other drugs.[213] Deletion of F335 was reported in another cell line,[214] which also showed altered resistance to a variety of drugs. The effect of several polymorphic sequence variations common in human populations on Pgp drug transport function has been investigated in transfected mammalian cells in vitro. Little difference in cell surface expression and transport function was noted between any of the variants and the wild-type protein.[215,216] On the other hand, the G1199A polymorphism, which results in a S400N change, changed the efflux and trans-epithelial flux of a fluorescent substrate, and altered cellular resistance to some drugs but not to others.[217] Thus, it seems likely that a number of Pgp polymorphisms may influence the disposition and therapeutic efficacy of selected drugs.

Given the role played by Pgp in protecting tissues and organs from toxicants, it would not be surprising to find that polymorphisms play a role in human susceptibility to various disease states. mdr1 knockout mice spontaneously develop a form of colitis that can be prevented by antibiotic treatment,[218] suggesting that Pgp functions as a defense against bacteria or toxins in the intestine. Confirming this idea, inflammatory bowel diseases (Crohn's disease and ulcerative colitis) are linked to the missense variant A893S/T,[219] and patients with ulcerative colitis (but not Crohn's disease) have a higher frequency of the C3435T genotype, which results in lowered Pgp expression in the intestine.[220] Anti-HIV drugs are known to be Pgp substrates, so a link between treatment efficacy and Pgp polymorphisms would not be unexpected. Although several common polymorphisms had no apparent effect on susceptibility to infection,[221] they were reported to influence drug treatment[222,223]; however, this was contradicted by another study.[224]

Variant Pgp alleles can also affect cancer susceptibility. The genotypic frequency of the C3435T SNP was not altered in colorectal tumor cells from a total patient population compared to controls[225]; however, when an under-50 patient population was examined, carriers of the 3435TT genotype or 3435T allele were at substantially higher risk of developing the disease.[226] Evidence also suggests that Pgp polymorphisms influence the risk of developing renal epithelial tumors; C3435T and C3435TT carriers are again at higher risk.[227]

An association was reported between the response of epilepsy patients to drug treatment and the C3435T polymorphism in the *MDR1* gene.[228] Patients with seizures that were not controlled by drugs were more likely to be homozygous for the C-variant allele, which is associated with higher Pgp transport function, suggesting that the drugs have a lower efficiency of penetration across the blood–brain barrier in this group. However, two later studies failed to confirm these results.[229,230] The anti-Parkinson drug budipine is exported actively out of the brain by Pgp in mice,[231] and Parkinson's disease susceptibility has been linked to Pgp polymorphisms in Chinese populations, where a MDR1 haplotype containing the SNPs 2677T and 3435T was found to protect against the disease.[232]

10.18. CONCLUSIONS

P-Glycoprotein is a drug transporter of the ABC superfamily that functions as an ATP-powered drug efflux pump. Rapid progress has been made in recent years in understanding the three-dimensional structure and ATP hydrolysis cycle of this protein, and many tools are now available for its study at the molecular level. Although the transporter can interact with hundreds of nonpolar, weakly amphipathic compounds with no apparent structural similarity, progress is being made in developing a pharmacophore model to describe its binding regions. The protein appears to interact with its multiple substrates via a large flexible drug-binding pocket, to which drugs gain access from the bilayer, leading to the suggestion that it is a "vacuum cleaner" for hydrophobic compounds that concentrate within the membrane. The drug transport mechanism of Pgp remains ill defined and may involve "flipping" of substrates from the inner to the outer membrane leaflet. The primary physiological role of the protein appears to be protection of sensitive organs and tissues from xenobiotic toxicity. Many drugs used in clinical therapy are P-glycoprotein substrates, and the transporter is now increasingly recognized to play a central role in the absorption and disposition of many drugs, including chemotherapeutic agents. Other compounds, known as modulators, that block the drug efflux function of Pgp are under development and may have clinical applications in the future. Nucleotide polymorphisms in the Pgp gene that may affect its regulation and expression have been identified in human populations. The effect of these variants on drug response and disease susceptibility is an important focus of future research.

REFERENCES

1. Croop JM. 1993. P-glycoprotein structure and evolutionary homologies. Cytotechnology 12:1–32.
2. Lincke CR, Broeks A, The I, Plasterk RH, Borst P. 1993. The expression of two P-glycoprotein (pgp) genes in transgenic *Caenorhabditis elegans* is confined to intestinal cells. EMBO J 12:1615–1620.
3. Ruetz S, Gros P. 1994. Phosphatidylcholine translocase: a physiological role for the mdr2 gene. Cell 77:1071–1081.

4. Thiebaut F, Tsuruo T, Hamada H, Gottesman MM, Pastan I, Willingham MC. 1987. Cellular localization of the multidrug-resistance gene product P-glycoprotein in normal human tissues. Proc Natl Acad Sci U S A 84:7735–7738.
5. Croop JM, Raymond M, Haber D, Devault A, Arceci RJ, Gros P, Housman DE. 1989. The three mouse multidrug resistance (mdr) genes are expressed in a tissue-specific manner in normal mouse tissues. Mol Cell Biol 9:1346–1350.
6. Beaulieu E, Demeule M, Ghitescu L, Béliveau R. 1997. P-glycoprotein is strongly expressed in the luminal membranes of the endothelium of blood vessels in the brain. Biochem J 326:539–544.
7. Melaine N, Liénard MO, Dorval I, Le Goascogne C, Lejeune H, Jégou B. 2002. Multidrug resistance genes and P-glycoprotein in the testis of the rat, mouse, guinea pig, and human. Biol Reprod 67:1699–1707.
8. Edwards JE, Alcorn J, Savolainen J, Anderson BD, McNamara PJ. 2005. Role of P-glycoprotein in distribution of nelfinavir across the blood–mammary tissue barrier and blood–brain barrier. Antimicrob Agents Chemother 49:1626–1628.
9. Saito T, Zhang ZJ, Tsuzuki H, Ohtsubo T, Yamada T, Yamamoto T, Saito H. 1997. Expression of P-glycoprotein in inner ear capillary endothelial cells of the guinea pig with special reference to blood–inner ear barrier. Brain Res 767:388–392.
10. Arceci RJ, Croop JM, Horwitz SB, Housman D. 1988. The gene encoding multidrug resistance is induced and expressed at high levels during pregnancy in the secretory epithelium of the uterus. Proc Natl Acad Sci U S A 85:4350–4354.
11. Gil S, Saura R, Forestier F, Farinotti R. 2005. P-glycoprotein expression of the human placenta during pregnancy. Placenta 26:268–270.
12. Kalabis GM, Kostaki A, Andrews MH, Petropoulos S, Gibb W, Matthews SG. 2005. Multidrug resistance phosphoglycoprotein (ABCB1) in the mouse placenta: fetal protection. Biol Reprod 73:591–597.
13. Smit JJ, Schinkel AH, Mol CA, Majoor D, Mooi WJ, Jongsma AP, Lincke CR, Borst P. 1994. Tissue distribution of the human MDR3 P-glycoprotein. Lab Invest 71:638–649.
14. Schinkel AH. 1998. Pharmacological insights from P-glycoprotein knockout mice. Int J Clin Pharmacol Ther 36:9–13.
15. Schinkel AH, Smit JJ, van Tellingen O, Beijnen JH, Wagenaar E, van Deemter L, Mol CA, van der Valk MA, Robanus-Maandag EC, Te RH. 1994. Disruption of the mouse mdr1a P-glycoprotein gene leads to a deficiency in the blood–brain barrier and to increased sensitivity to drugs. Cell 77:491–502.
16. Jette L, Pouliot JF, Murphy GF, Béliveau R. 1995. Isoform I (mdr3) is the major form of P-glycoprotein expressed in mouse brain capillaries: evidence for cross-reactivity of antibody C219 with an unrelated protein. Biochem J 305:761–766.
17. Doran A, Obach RS, Smith BJ, Hosea NA, Becker S, Callegari E, Chen C, Chen X, Choo E, Cianfrogna J, et al. 2005. The impact of P-glycoprotein on the disposition of drugs targeted for indications of the central nervous system: evaluation using the MDR1A/1B knockout mouse model. Drug Metab Dispos 33:165–174.
18. Roulet A, Puel O, Gesta S, Lepage JF, Drag M, Soll M, Alvinerie M, Pineau T. 2003. MDR1-deficient genotype in collie dogs hypersensitive to the P-glycoprotein substrate ivermectin. Eur J Pharmacol 460:85–91.
19. Nelson OL, Carsten E, Bentjen SA, Mealey KL. 2003. Ivermectin toxicity in an Australian shepherd dog with the MDR1 mutation associated with ivermectin sensitivity in collies. J Vet Intern Med 17:354–356.

20. Neff MW, Robertson KR, Wong AK, Safra N, Broman KW, Slatkin M, Mealey KL, Pedersen NC. 2004. Breed distribution and history of canine mdr1-1delta, a pharmacogenetic mutation that marks the emergence of breeds from the collie lineage. Proc Natl Acad Sci U S A 101:11725–11730.
21. Luker GD, Fracasso PM, Dobkin J, Piwnica-Worms D. 1997. Modulation of the multidrug resistance P-glycoprotein: detection with technetium-99m-sestamibi in vivo. J Nucl Med 38:369–372.
22. Del Vecchio S, Zannetti A, Aloj L, Salvatore M. 2003. MIBI as prognostic factor in breast cancer. J Nucl Med 47:46–50.
23. Sasongko L, Link JM, Muzi M, Mankoff DA, Yang XD, Collier AC, Shoner SC, Unadkat JD. 2005. Imaging P-glycoprotein transport activity at the human blood–brain barrier with positron emission tomography. Clin Pharmacol Ther 77:503–514.
24. Sharom FJ, Liu R, Qu Q, Romsicki Y. 2001. Exploring the structure and function of the P-glycoprotein multidrug transporter using fluorescence spectroscopic tools. Seminars Cell Dev Biol 12:257–266.
25. Sharom FJ, Yu X, Doige CA. 1993. Functional reconstitution of drug transport and ATPase activity in proteoliposomes containing partially purified P-glycoprotein. J Biol Chem 268:24197–24202.
26. Sharom FJ, Yu X, DiDiodato G, Chu JWK. 1996. Synthetic hydrophobic peptides are substrates for P-glycoprotein and stimulate drug transport. Biochem J 320:421–428.
27. Robert J, Jarry C. 2003. Multidrug resistance reversal agents. J Med Chem 46:4805–4817.
28. Schmitt L, Tampé R. 2002. Structure and mechanism of ABC transporters. Curr Opin Struct Biol 12:754–760.
29. Jones PM, George AM. 2004. The ABC transporter structure and mechanism: perspectives on recent research. Cell Mol Life Sci 61:682–699.
30. Loo TW, Clarke DM. 1995. Membrane topology of a cysteine-less mutant of human P-glycoprotein. J Biol Chem 270:843–848.
31. Kast C, Canfield V, Levenson R, Gros P. 1996. Transmembrane organization of mouse P-glycoprotein determined by epitope insertion and immunofluorescence. J Biol Chem 271:9240–9248.
32. Linton KJ, Higgins CF. 2002. P-glycoprotein misfolds in *Escherichia coli*: evidence against alternating-topology models of the transport cycle. Mol Membr Biol 19:51–58.
33. Loo TW, Clarke DM. 1999. The transmembrane domains of the human multidrug resistance P-glycoprotein are sufficient to mediate drug-binding and trafficking to the cell surface. J Biol Chem 274:24759–24765.
34. Raviv Y, Pollard HB, Bruggemann EP, Pastan I, Gottesman MM. 1990. Photosensitized labeling of a functional multidrug transporter in living drug-resistant tumor cells. J Biol Chem 265:3975–3980.
35. Hopfner KP, Karcher A, Shin DS, Craig L, Arthur LM, Carney JP, Tainer JA. 2000. Structural biology of Rad50 ATPase: ATP-driven conformational control in DNA double-strand break repair and the ABC-ATPase superfamily. Cell 101:789–800.
36. Locher KP, Lee AT, Rees DC. 2002. The *E. coli* BtuCD structure: a framework for ABC transporter architecture and mechanism. Science 296:1091–1098.
37. Smith PC, Karpowich N, Millen L, Moody JE, Rosen J, Thomas PJ, Hunt JF. 2002. ATP binding to the motor domain from an ABC transporter drives formation of a nucleotide sandwich dimer. Mol Cell 10:139–149.

38. Rosenberg MF, Callaghan R, Ford RC, Higgins CF. 1997. Structure of the multidrug resistance P-glycoprotein to 2.5 nm resolution determined by electron microscopy and image analysis. J Biol Chem 272:10685–10694.
39. Qu Q, Sharom FJ. 2001. FRET analysis indicates that the two ATPase active sites of the P-glycoprotein multidrug transporter are closely associated. Biochemistry 40:1413–1422.
40. Urbatsch IL, Gimi K, Wilke-Mounts S, Lerner-Marmarosh N, Rousseau ME, Gros P, Senior AE. 2001. Cysteines 431 and 1074 are responsible for inhibitory disulfide cross-linking between the two nucleotide-binding sites in human P-glycoprotein. J Biol Chem 276:26980–26987.
41. Loo TW, Bartlett MC, Clarke DM. 2002. The "LSGGQ" motif in each nucleotide-binding domain of human P-glycoprotein is adjacent to the opposing Walker A sequence. J Biol Chem 277:41303–41306.
42. Rosenberg MF, Velarde G, Ford RC, Martin C, Berridge G, Kerr ID, Callaghan R, Schmidlin A, Wooding C, Linton KJ, Higgins CF. 2001. Repacking of the transmembrane domains of P-glycoprotein during the transport ATPase cycle. EMBO J 20:5615–5625.
43. Rosenberg MF, Kamis AB, Callaghan R, Higgins CF, Ford RC. 2003. Three-dimensional structures of the mammalian multidrug resistance P-glycoprotein demonstrate major conformational changes in the transmembrane domains upon nucleotide binding. J Biol Chem 278:8294–8299.
44. Lee JY, Urbatsch IL, Senior AE, Wilkens S. 2002. Projection structure of P-glycoprotein by electron microscopy: evidence for a closed conformation of the nucleotide binding domains. J Biol Chem 277:40125–40131.
45. Rosenberg MF, Callaghan R, Modok S, Higgins CF, Ford RC. 2005. Three-dimensional structure of P-glycoprotein: the transmembrane regions adopt an asymmetric configuration in the nucleotide-bound state. J Biol Chem 280:2857–2862.
46. Loo TW, Clarke DM. 2000. The packing of the transmembrane segments of human multidrug resistance P-glycoprotein is revealed by disulfide cross-linking analysis. J Biol Chem 275:5253–5256.
47. Loo TW, Bartlett MC, Clarke DM. 2004. Val^{133} and Cys^{137} in transmembrane segment 2 are close to Arg^{935} and Gly^{939} in transmembrane segment 11 of human P-glycoprotein. J Biol Chem 279:18232–18238.
48. Loo TW, Bartlett MC, Clarke DM. 2004. Disulfide cross-linking analysis shows that transmembrane segments 5 and 8 of human P-glycoprotein are close together on the cytoplasmic side of the membrane. J Biol Chem 279:7692–7697.
49. al-Shawi MK, Senior AE. 1993. Characterization of the adenosine triphosphatase activity of Chinese hamster P-glycoprotein. J Biol Chem 268:4197–4206.
50. Doige CA, Yu X, Sharom FJ. 1992. ATPase activity of partially purified P-glycoprotein from multidrug-resistant Chinese hamster ovary cells. Biochim Biophys Acta 1109:149–160.
51. Safa AR, Glover CJ, Meyers MB, Biedler JL, Felsted RL. 1986. Vinblastine photoaffinity labeling of a high molecular weight surface membrane glycoprotein specific for multidrug-resistant cells. J Biol Chem 261:6137–6140.
52. Cornwell MM, Tsuruo T, Gottesman MM, Pastan I. 1987. ATP-binding properties of P-glycoprotein from multidrug-resistant KB cells. FASEB J 1:51–54.

53. Martin C, Berridge G, Higgins CF, Mistry P, Charlton P, Callaghan R. 2000. Communication between multiple drug-binding sites on P-glycoprotein. Mol Pharmacol 58:624–632.
54. Doige CA, Sharom FJ. 1992. Transport properties of P-glycoprotein in plasma membrane vesicles from multidrug-resistant Chinese hamster ovary cells. Biochim Biophys Acta 1109:161–171.
55. Ruetz S, Gros P. 1994. Functional expression of P-glycoproteins in secretory vesicles. J Biol Chem 269:12277–12284.
56. Guiral M, Viratelle O, Westerhoff HV, Lankelma J. 1994. Cooperative P-glycoprotein mediated daunorubicin transport into DNA-loaded plasma membrane vesicles. FEBS Lett 346:141–145.
57. Urbatsch IL, al-Shawi MK, Senior AE. 1994. Characterization of the ATPase activity of purified Chinese hamster P-glycoprotein. Biochemistry 33:7069–7076.
58. Shapiro AB, Ling V. 1994. ATPase activity of purified and reconstituted P-glycoprotein from Chinese hamster ovary cells. J Biol Chem 269:3745–3754.
59. Sharom FJ, Yu X, Chu JWK, Doige CA. 1995. Characterization of the ATPase activity of P-glycoprotein from multidrug-resistant Chinese hamster ovary cells. Biochem J 308:381–390.
60. Ambudkar SV, Lelong IH, Zhang J, Cardarelli C. 1998. Purification and reconstitution of human P-glycoprotein. Methods Enzymol 292:492–504.
61. Loo TW, Clarke DM. 1995. Rapid purification of human P-glycoprotein mutants expressed transiently in HEK 293 cells by nickel-chelate chromatography and characterization of their drug-stimulated ATPase activities. J Biol Chem 270:21449–21452.
62. Lerner-Marmarosh N, Gimi K, Urbatsch IL, Gros P, Senior AE. 1999. Large scale purification of detergent-soluble P-glycoprotein from *Pichia pastoris* cells and characterization of nucleotide binding properties of wild-type, Walker A, and Walker B mutant proteins. J Biol Chem 274:34711–34718.
63. Cai J, Gros P. 2003. Overexpression, purification, and functional characterization of ATP-binding cassette transporters in the yeast, *Pichia pastoris*. Biochim Biophys Acta 1610:63–76.
64. Dong M, Ladavière L, Penin F, Deléage G, Baggetto LG. 1998. Secondary structure of P-glycoprotein investigated by circular dichroism and amino acid sequence analysis. Biochim Biophys Acta 1371:317–334.
65. Urbatsch IL, Senior AE. 1995. Effects of lipids on ATPase activity of purified Chinese hamster P-glycoprotein. Arch Biochem Biophys 316:135–140.
66. Shapiro AB, Ling V. 1995. Reconstitution of drug transport by purified P-glycoprotein. J Biol Chem 270:16167–16175.
67. Eytan GD, Regev R, Assaraf YG. 1996. Functional reconstitution of P-glycoprotein reveals an apparent near stoichiometric drug transport to ATP hydrolysis. J Biol Chem 271:3172–3178.
68. Ambudkar SV. 1995. Purification and reconstitution of functional human P-glycoprotein. J Bioenerg Biomembr 27:23–29.
69. Rao US. 1995. Mutation of glycine 185 to valine alters the ATPase function of the human P-glycoprotein expressed in Sf9 cells. J Biol Chem 270:6686–6690.
70. Sarkadi B, Price EM, Boucher RC, Germann UA, Scarborough GA. 1992. Expression of the human multidrug resistance cDNA in insect cells generates a high activity drug-stimulated membrane ATPase. J Biol Chem 267:4854–4858.

71. Liu R, Sharom FJ. 1997. Fluorescence studies on the nucleotide binding domains of the P-glycoprotein multidrug transporter. Biochemistry 36:2836–2843.

72. Urbatsch IL, Sankaran B, Weber J, Senior AE. 1995. P-glycoprotein is stably inhibited by vanadate-induced trapping of nucleotide at a single catalytic site. J Biol Chem 270:19383–19390.

73. Loo TW, Clarke DM. 1995. Covalent modification of human P-glycoprotein mutants containing a single cysteine in either nucleotide-binding fold abolishes drug-stimulated ATPase activity. J Biol Chem 270:22957–22961.

74. Liu R, Sharom FJ. 1996. Site-directed fluorescence labeling of P-glycoprotein on cysteine residues in the nucleotide binding domains. Biochemistry 35:11865–11873.

75. Sharom FJ, DiDiodato G, Yu X, Ashbourne KJ. 1995. Interaction of the P-glycoprotein multidrug transporter with peptides and ionophores. J Biol Chem 270:10334–10341.

76. Gottesman MM, Pastan I, Ambudkar SV. 1996. P-glycoprotein and multidrug resistance. Curr Opin Genet Dev 6:610–617.

77. Ambudkar SV, Lelong IH, Zhang J, Cardarelli CO, Gottesman MM, Pastan I. 1992. Partial purification and reconstitution of the human multidrug-resistance pump: characterization of the drug-stimulatable ATP hydrolysis. Proc Natl Acad Sci U S A 89:8472–8476.

78. Doige CA, Yu X, Sharom FJ. 1993. The effects of lipids and detergents on ATPase-active P-glycoprotein. Biochim Biophys Acta 1146:65–72.

79. Senior AE, al-Shawi MK, Urbatsch IL. 1995. The catalytic cycle of P-glycoprotein. FEBS Lett 377:285–289.

80. Smith CA, Rayment I. 1996. X-ray structure of the magnesium(II)·ADP·vanadate complex of the *Dictyostelium discoideum* myosin motor domain to 1.9 Å resolution. Biochemistry 35:5404–5417.

81. Liu R, Siemiarczuk A, Sharom FJ. 2000. Intrinsic fluorescence of the P-glycoprotein multidrug transporter: sensitivity of tryptophan residues to binding of drugs and nucleotides. Biochemistry 39:14927–14938.

82. Delannoy S, Urbatsch IL, Tombline G, Senior AE, Vogel PD. 2005. Nucleotide binding to the multidrug resistance P-glycoprotein as studied by ESR spectroscopy. Biochemistry 44:14010–14019.

83. Qu Q, Russell PL, Sharom FJ. 2003. Stoichiometry and affinity of nucleotide binding to P-glycoprotein during the catalytic cycle. Biochemistry 42:1170–1177.

84. Safa AR. 1998. Photoaffinity labels for characterizing drug interaction sites of P-glycoprotein. Methods Enzymol 292:289–307.

85. Safa AR. 1999. Photoaffinity analogs for multidrug resistance-related transporters and their use in identifying chemosensitizers. Drug Resist Updates 2:371–381.

86. Martin C, Berridge G, Higgins CF, Callaghan R. 1997. The multi-drug resistance reversal agent SR33557 and modulation of vinca alkaloid binding to P-glycoprotein by an allosteric interaction. Br J Pharmacol 122:765–771.

87. Taylor JC, Ferry DR, Higgins CF, Callaghan R. 1999. The equilibrium and kinetic drug-binding properties of the mouse P-gp1a and P-gp1b P-glycoproteins are similar. Br J Cancer 81:783–789.

88. Martin C, Berridge G, Mistry P, Higgins C, Charlton P, Callaghan R. 2000. drug-binding sites on P-glycoprotein are altered by ATP binding prior to nucleotide hydrolysis. Biochemistry 39:11901–11906.

89. Martin C, Higgins CF, Callaghan R. 2001. The vinblastine binding site adopts high- and low-affinity conformations during a transport cycle of P-glycoprotein. Biochemistry 40:15733–15742.

90. Sharom FJ, Russell PL, Qu Q, Lu P. 2003. Fluorescence techniques for studying membrane transport proteins: the P-glycoprotein multidrug transporter. Methods Mol Biol 227: 109–128.

91. Sharom FJ, Liu R, Romsicki Y, Lu P. 1999. Insights into the structure and substrate interactions of the P-glycoprotein multidrug transporter from spectroscopic studies. Biochim Biophys Acta 1461:327–345.

92. Tanigawara Y, Okamura N, Hirai M, Yasuhara M, Ueda K, Kioka N, Komano T, Hori R. 1992. Transport of digoxin by human P-glycoprotein expressed in a porcine kidney epithelial cell line (LLC-PK1). J Pharmacol Exp Ther 263:840–845.

93. Hunter J, Jepson MA, Tsuruo T, Simmons NL, Hirst BH. 1993. Functional expression of P-glycoprotein in apical membranes of human intestinal Caco-2 cells: kinetics of vinblastine secretion and interaction with modulators. J Biol Chem 268:14991–14997.

94. Saeki T, Ueda K, Tanigawara Y, Hori R, Komano T. 1993. Human P-glycoprotein transports cyclosporin A and FK506. J Biol Chem 268:6077–6080.

95. Ueda K, Okamura N, Hirai M, Tanigawara Y, Saeki T, Kioka N, Komano T, Hori R. 1992. Human P-glycoprotein transports cortisol, aldosterone, and dexamethasone, but not progesterone. J Biol Chem 267:24248–24252.

96. Saeki T, Ueda K, Tanigawara Y, Hori R, Komano T. 1993. P-glycoprotein-mediated transcellular transport of MDR-reversing agents. FEBS Lett 324:99–102.

97. Ueda K, Saeki T, Hirai M, Tanigawara Y, Tanaka K, Okamura M, Yasuhara M, Hori R, Inui K, Komano T. 1994. Human P-glycoprotein as a multi-drug transporter analyzed by using transepithelial transport system. Jpn J Physiol 44 (Suppl 2):S67–S71.

98. Hirai M, Tanaka K, Shimizu T, Tanigawara Y, Yasuhara M, Hori R, Kakehi Y, Yoshida O, Ueda K, Komano T. 1995. Cepharanthin, a multidrug resistant modifier, is a substrate for P-glycoprotein. J Pharmacol Exp Ther 275:73–78.

99. Horio M, Gottesman MM, Pastan I. 1988. ATP-dependent transport of vinblastine in vesicles from human multidrug-resistant cells. Proc Natl Acad Sci U S A 85:3580–3584.

100. Kamimoto Y, Gatmaitan Z, Hsu J, Arias IM. 1989. The function of Gp170, the multidrug resistance gene product, in rat liver canalicular membrane vesicles. J Biol Chem 264:11693–11698.

101. Tamai I, Safa AR. 1990. Competitive interaction of cyclosporins with the vinca alkaloid-binding site of P-glycoprotein in multidrug-resistant cells. J Biol Chem 265:16509–16513.

102. Horio M, Lovelace E, Pastan I, Gottesman MM. 1991. Agents which reverse multidrug-resistance are inhibitors of [^{3}H]vinblastine transport by isolated vesicles. Biochim Biophys Acta 1061:106–110.

103. Piwnica-Worms D, Rao VV, Kronauge JF, Croop JM. 1995. Characterization of multidrug resistance P-glycoprotein transport function with an organotechnetium cation. Biochemistry 34:12210–12220.

104. Shapiro AB, Corder AB, Ling V. 1997. P-glycoprotein-mediated Hoechst 33342 transport out of the lipid bilayer. Eur J Biochem 250:115–121.

105. Lu P, Liu R, Sharom FJ. 2001. Drug transport by reconstituted P-glycoprotein in proteoliposomes: effect of substrates and modulators, and dependence on bilayer phase state. Eur J Biochem 268:1687–1697.

106. Ambudkar SV, Cardarelli CO, Pashinsky I, Stein WD. 1997. Relation between the turnover number for vinblastine transport and for vinblastine-stimulated ATP hydrolysis by human P-glycoprotein. J Biol Chem 272:21160–21166.

107. Eytan GD, Borgnia MJ, Regev R, Assaraf YG. 1994. Transport of polypeptide ionophores into proteoliposomes reconstituted with rat liver P-glycoprotein. J Biol Chem 269:26058–26065.

108. Sarkadi B, Müller M, Homolya L, Hollo Z, Seprodi J, Germann UA, Gottesman MM, Price EM, Boucher RC. 1994. Interaction of bioactive hydrophobic peptides with the human multidrug transporter. FASEB J 8:766–770.

109. Charuk JH, Grey AA, Reithmeier RA. 1998. Identification of the synthetic surfactant nonylphenol ethoxylate: a P-glycoprotein substrate in human urine. Am J Physiol 274:F1127–F1139.

110. Loo TW, Clarke DM. 1998. Nonylphenol ethoxylates, but not nonylphenol, are substrates of the human multidrug resistance P-glycoprotein. Biochem Biophys Res Commun 247:478–480.

111. Stouch TR, Gudmundsson A. 2002. Progress in understanding the structure–activity relationships of P-glycoprotein. Adv Drug Deliv Rev 54:315–328.

112. Shapiro AB, Ling V. 1997. Positively cooperative sites for drug transport by P-glycoprotein with distinct drug specificities. Eur J Biochem 250:130–137.

113. Seelig A. 1998. A general pattern for substrate recognition by P-glycoprotein. Eur J Biochem 251:252–261.

114. Cianchetta G, Singleton RW, Zhang M, Wildgoose M, Giesing D, Fravolini A, Cruciani G, Vaz RJ. 2005. A pharmacophore hypothesis for P-glycoprotein substrate recognition using GRIND-based three-dimensional-QSAR. J Med Chem 48:2927–2935.

115. Ekins S, Kim RB, Leake BF, Dantzig AH, Schuetz EG, Lan LB, Yasuda K, Shepard RL, Winter MA, Schuetz JD, Wikel JH, Wrighton SA. 2002. Application of three-dimensional quantitative structure–activity relationships of P-glycoprotein inhibitors and substrates. Mol Pharmacol 61:974–981.

116. Pajeva IK, Wiese M. 2002. Pharmacophore model of drugs involved in P-glycoprotein multidrug resistance: explanation of structural variety (hypothesis). J Med Chem 45: 5671–5686.

117. Klopman G, Shi LM, Ramu A. 1997. Quantitative structure–activity relationship of multidrug resistance reversal agents. Mol Pharmacol 52:323–334.

118. Wiese M, Pajeva IK. 2001. structure–activity relationships of multidrug resistance reversers. Curr Med Chem 8:685–713.

119. Bakken GA, Jurs PC. 2000. Classification of multidrug-resistance reversal agents using structure-based descriptors and linear discriminant analysis. J Med Chem 43:4534–4541.

120. Tmej C, Chiba P, Schaper KJ, Ecker G, Fleischhacker W. 1999. Artificial neural networks as versatile tools for prediction of MDR-modulatory activity. Adv Exp Med Biol 457: 95–105.

121. Borgnia MJ, Eytan GD, Assaraf YG. 1996. Competition of hydrophobic peptides, cytotoxic drugs, and chemosensitizers on a common P-glycoprotein pharmacophore as revealed by its ATPase activity. J Biol Chem 271:3163–3171.

122. Garrigues A, Loiseau N, Delaforge M, Ferté J, Garrigos M, André F, Orlowski S. 2002. Characterization of two pharmacophores on the multidrug transporter P-glycoprotein. Mol Pharmacol 62:1288–1298.

123. Dey S, Ramachandra M, Pastan I, Gottesman MM, Ambudkar SV. 1997. Evidence for two nonidentical drug-interaction sites in the human P-glycoprotein. Proc Natl Acad Sci U S A 94:10594–10599.
124. Shapiro AB, Fox K, Lam P, Ling V. 1999. Stimulation of P-glycoprotein-mediated drug transport by prazosin and progesterone. Evidence for a third drug-binding site. Eur J Biochem 259:841–850.
125. Schumacher MA, Brennan RG. 2002. Structural mechanisms of multidrug recognition and regulation by bacterial multidrug transcription factors. Mol Microbiol 45:885–893.
126. Schumacher MA, Miller MC, Grkovic S, Brown MH, Skurray RA, Brennan RG. 2001. Structural mechanisms of QacR induction and multidrug recognition. Science 294:2158–2163.
127. Schumacher MA, Miller MC, Brennan RG. 2004. Structural mechanism of the simultaneous binding of two drugs to a multidrug-binding protein. EMBO J 23:2923–2930.
128. Watkins RE, Wisely GB, Moore LB, Collins JL, Lambert MH, Williams SP, Willson TM, Kliewer SA, Redinbo MR. 2001. The human nuclear xenobiotic receptor PXR: structural determinants of directed promiscuity. Science 292:2329–2333.
129. Neyfakh AA. 2002. Mystery of multidrug transporters: the answer can be simple. Mol Microbiol 44:1123–1130.
130. Yu EW, McDermott G, Zgurskaya HI, Nikaido H, Koshland DE Jr. 2003. Structural basis of multiple drug-binding capacity of the AcrB multidrug efflux pump. Science 300: 976–980.
131. Loo TW, Clarke DM. 1999. Identification of residues in the drug-binding domain of human P-glycoprotein: analysis of transmembrane segment 11 by cysteine-scanning mutagenesis and inhibition by dibromobimane. J Biol Chem 274:35388–35392.
132. Loo TW, Clarke DM. 1997. Identification of residues in the drug-binding site of human P-glycoprotein using a thiol-reactive substrate. J Biol Chem 272:31945–31948.
133. Loo TW, Clarke DM. 2001. Defining the drug-binding site in the human multidrug resistance P-glycoprotein using a methanethiosulfonate analog of verapamil, MTS-verapamil. J Biol Chem 276:14972–14979.
134. Loo TW, Clarke DM. 2002. Location of the rhodamine-binding site in the human multidrug resistance P-glycoprotein. J Biol Chem 277:44332–44338.
135. Loo TW, Bartlett MC, Clarke DM. 2003. Substrate-induced conformational, changes in the transmembrane segments of human P-glycoprotein: direct evidence for the substrate-induced fit mechanism for drug-binding. J Biol Chem 278:13603–13606.
136. Loo TW, Bartlett MC, Clarke DM. 2003. Simultaneous binding of two different drugs in the binding pocket of the human multidrug resistance P-glycoprotein. J Biol Chem 278:39706–39710.
137. Lugo MR, Sharom FJ. 2005. Interaction of LDS-751 and rhodamine 123 with P-glycoprotein: evidence for simultaneous binding of both drugs. Biochemistry 44:14020–14029.
138. Loo TW, Clarke DM. 2001. Determining the dimensions of the drug-binding domain of human P-glycoprotein using thiol cross-linking compounds as molecular rulers. J Biol Chem 276:36877–36880.
139. Bruggemann EP, Currier SJ, Gottesman MM, Pastan I. 1992. Characterization of the azidopine and vinblastine binding site of P-glycoprotein. J Biol Chem 267:21020–21026.

140. Greenberger LM. 1993. Major photoaffinity drug labeling sites for iodoaryl azidoprazosin in P-glycoprotein are within, or immediately C-terminal to, transmembrane domains 6 and 12. J Biol Chem 268:11417–11425.

141. Demmer A, Thole H, Kubesch P, Brandt T, Raida M, Fislage R, Tummler B. 1997. Localization of the iodomycin binding site in hamster P-glycoprotein. J Biol Chem 272:20913–20919.

142. Demeule M, Laplante A, Murphy GF, Wenger RM, Béliveau R. 1998. Identification of the cyclosporin-binding site in P-glycoprotein. Biochemistry 37:18110–18118.

143. Loo TW, Clarke DM. 1999. Molecular dissection of the human multidrug resistance P-glycoprotein. Biochem Cell Biol 77:11–23.

144. Loo TW, Clarke DM. 2000. Identification of residues within the drug-binding domain of the human multidrug resistance P-glycoprotein by cysteine-scanning mutagenesis and reaction with dibromobimane. J Biol Chem 275:39272–39278.

145. Loo TW, Clarke DM. 2005. Recent progress in understanding the mechanism of P-glycoprotein-mediated drug efflux. J Membr Biol 206:173–185.

146. Pleban K, Kopp S, Csaszar E, Peer M, Hrebicek T, Rizzi A, Ecker GF, Chiba P. 2005. P-glycoprotein substrate-binding domains are located at the transmembrane domain/transmembrane domain interfaces: a combined photoaffinity labeling-protein homology modeling approach. Mol Pharmacol 67:365–374.

147. Pawagi AB, Wang J, Silverman M, Reithmeier RA, Deber CM. 1994. Transmembrane aromatic amino acid distribution in P-glycoprotein: a functional role in broad substrate specificity. J Mol Biol 235:554–564.

148. Qu Q, Sharom FJ. 2002. Proximity of bound Hoechst 33342 to the ATPase catalytic sites places the drug-binding site of P-glycoprotein within the cytoplasmic membrane leaflet. Biochemistry 41:4744–4752.

149. Higgins CF, Gottesman MM. 1992. Is the multidrug transporter a flippase? Trends Biochem Sci 17:18–21.

150. Homolya L, Hollo Z, Germann UA, Pastan I, Gottesman MM, Sarkadi B. 1993. Fluorescent cellular indicators are extruded by the multidrug resistance protein. J Biol Chem 268:21493–21496.

151. Gallois L, Fiallo M, Laigle A, Priebe W, Garnier-Suillerot A. 1996. The overall partitioning of anthracyclines into phosphatidyl-containing model membranes depends neither on the drug charge nor the presence of anionic phospholipids. Eur J Biochem 241:879–887.

152. Regev R, Yeheskely-Hayon D, Katzir H, Eytan GD. 2005. Transport of anthracyclines and mitoxantrone across membranes by a flip-flop mechanism. Biochem Pharmacol 70: 161–169.

153. Regev R, Eytan GD. 1997. Flip-flop of doxorubicin across erythrocyte and lipid membranes. Biochem Pharmacol 54:1151–1158.

154. Shapiro AB, Ling V. 1997. Extraction of Hoechst 33342 from the cytoplasmic leaflet of the plasma membrane by P-glycoprotein. Eur J Biochem 250:122–129.

155. Shapiro AB, Ling V. 1998. Transport of LDS-751 from the cytoplasmic leaflet of the plasma membrane by the rhodamine-123-selective site of P-glycoprotein. Eur J Biochem 254:181–188.

156. Ferry D, Boer R, Callaghan R, Ulrich WR. 2000. Localization of the 1,4-dihydropyridine drug acceptor of P-glycoprotein to a cytoplasmic domain using a permanently charged derivative *N*-methyl dexniguldipine. Int J Clin Pharmacol Ther 38:130–140.

157. Sharom FJ, Lu P, Liu R, Yu X. 1998. Linear and cyclic peptides as substrates and modulators of P-glycoprotein: peptide binding and effects on drug transport and accumulation. Biochem J 333:621–630.

158. Lugo MR, Sharom FJ. 2005. Interaction of LDS-751 with P-glycoprotein and mapping of the location of the R drug binding site. Biochemistry 44:643–655.

159. Eytan GD. 2005. Mechanism of multidrug resistance in relation to passive membrane permeation. Biomed Pharmacother 59:90–97.

160. van Helvoort A, Smith AJ, Sprong H, Fritzsche I, Schinkel AH, Borst P, van Meer G. 1996. MDR1 P-glycoprotein is a lipid translocase of broad specificity, while MDR3 P-glycoprotein specifically translocates phosphatidylcholine. Cell 87:507–517.

161. Bosch I, Dunussi-Joannopoulos K, Wu RL, Furlong ST, Croop J. 1997. Phosphatidylcholine and phosphatidylethanolamine behave as substrates of the human MDR1 P-glycoprotein. Biochemistry 36:5685–5694.

162. van Meer G, Sillence D, Sprong H, Kälin N, Raggers R. 1999. Transport of (glyco)sphingolipids in and between cellular membranes: multidrug transporters and lateral domains. Biosci Rep 19:327–333.

163. de Rosa MF, Sillence D, Ackerley C, Lingwood C. 2004. Role of multiple drug resistance protein 1 in neutral but not acidic glycosphingolipid biosynthesis. J Biol Chem 279:7867–7876.

164. Romsicki Y, Sharom FJ. 2001. Phospholipid flippase activity of the reconstituted P-glycoprotein multidrug transporter. Biochemistry 40:6937–6947.

165. Eckford PD, Sharom FJ. 2005. The reconstituted P-glycoprotein multidrug transporter is a flippase for glucosylceramide and other simple glycosphingolipids. Biochem J 389: 517–526.

166. Lala P, Ito S, Lingwood CA. 2000. Retroviral transfection of Madin–Darby canine kidney cells with human *MDR1* results in a major increase in globotriaosylceramide and 10^5- to 10^6-fold increased cell sensitivity to verocytotoxin: role of P-glycoprotein in glycolipid synthesis. J Biol Chem 275:6246–6251.

167. Romsicki Y, Sharom FJ. 1999. The membrane lipid environment modulates drug interactions with the P-glycoprotein multidrug transporter. Biochemistry 38:6887–6896.

168. Eytan GD, Kuchel PW. 1999. Mechanism of action of P-glycoprotein in relation to passive membrane permeation. Int Rev Cytol 190:175–250.

169. Eytan GD, Regev R, Oren G, Assaraf YG. 1996. The role of passive transbilayer drug movement in multidrug resistance and its modulation. J Biol Chem 271:12897–12902.

170. Romsicki Y, Sharom FJ. 1997. Interaction of P-glycoprotein with defined phospholipid bilayers: a differential scanning calorimetric study. Biochemistry 36:9807–9815.

171. Romsicki Y, Sharom FJ. 1998. The ATPase and ATP binding functions of P-glycoprotein: modulation by interaction with defined phospholipids. Eur J Biochem 256:170–178.

172. Sinicrope FA, Dudeja PK, Bissonnette BM, Safa AR, Brasitus TA. 1992. Modulation of P-glycoprotein-mediated drug transport by alterations in lipid fluidity of rat liver canalicular membrane vesicles. J Biol Chem 267:24995–25002.

173. Sharom FJ. 2003. Probing of conformational changes, catalytic cycle and ABC transporter function. In Holland IB, Kuchler K, Higgins C, Cole SP, editors, *ABC Proteins: From Bacteria to Man*. London: Academic Press. pp. 107–133.

174. Callaghan R, Ford RC, Kerr ID. 2006. The translocation mechanism of P-glycoprotein. FEBS Lett 580:1056–1063.

175. Loo TW, Clarke DM. 2005. Do drug substrates enter the common drug-binding pocket of P-glycoprotein through "gates"? Biochem Biophys Res Commun 329:419–422.

176. Loo TW, Bartlett MC, Clarke DM. 2004. The drug-binding pocket of the human multidrug resistance P-glycoprotein is accessible to the aqueous medium. Biochemistry 43:12081–12089.

177. Loo TW, Bartlett MC, Clarke DM. 2003. drug-binding in human P-glycoprotein causes conformational changes in both nucleotide-binding domains. J Biol Chem 278:1575–1578.

178. Loo TW, Bartlett MC, Clarke DM. 2003. Permanent activation of the human P-glycoprotein by covalent modification of a residue in the drug-binding site. J Biol Chem 278:20449–20452.

179. Ramachandra M, Ambudkar SV, Chen D, Hrycyna CA, Dey S, Gottesman MM, Pastan I. 1998. Human P-glycoprotein exhibits reduced affinity for substrates during a catalytic transition state. Biochemistry 37:5010–5019.

180. Sauna ZE, Ambudkar SV. 2001. Characterization of the catalytic cycle of ATP hydrolysis by human P-glycoprotein. The two ATP hydrolysis events in a single catalytic cycle are kinetically similar but affect different functional outcomes. J Biol Chem 276:11653–11661.

181. Qu Q, Chu JW, Sharom FJ. 2003. Transition state P-glycoprotein binds drugs and modulators with unchanged affinity, suggesting a concerted transport mechanism. Biochemistry 42:1345–1353.

182. Chen M, Abele R, Tampé R. 2003. Peptides induce ATP hydrolysis at both subunits of the transporter associated with antigen processing. J Biol Chem 278:29686–29692.

183. Ambudkar SV, Kim IW, Sauna ZE. 2006. The power of the pump: mechanisms of action of P-glycoprotein (ABCB1). Eur J Pharm Sci 27:392–400.

184. Higgins CF, Linton KJ. 2004. The ATP switch model for ABC transporters. Nat Struc Biol 11:918–926.

185. Sauna ZE, Smith MM, Müller M, Kerr KM, Ambudkar SV. 2001. The mechanism of action of multidrug-resistance-linked P-glycoprotein. J Bioenerg Biomembr 33:481–491.

186. Sauna ZE, Müller M, Peng XH, Ambudkar SV. 2002. Importance of the conserved Walker B glutamate residues, 556 and 1201, for the completion of the catalytic cycle of ATP hydrolysis by human P-glycoprotein (ABCB1). Biochemistry 41:13989–14000.

187. Lin JH, Yamazaki M. 2003. Clinical relevance of P-glycoprotein in drug therapy. Drug Metab Rev 35:417–454.

188. Lin JH, Yamazaki M. 2003. Role of P-glycoprotein in pharmacokinetics: clinical implications. Clin Pharmacokinet 42:59–98.

189. Sparreboom A, van Asperen J, Mayer U, Schinkel AH, Smit JW, Meijer DK, Borst P, Nooijen WJ, Beijnen JH, van Tellingen O. 1997. Limited oral bioavailability and active epithelial excretion of paclitaxel (Taxol) caused by P-glycoprotein in the intestine. Proc Natl Acad Sci U S A 94:2031–2035.

190. Sikic BI, Fisher GA, Lum BL, Halsey J, Beketic-Oreskovic L, Chen G. 1997. Modulation and prevention of multidrug resistance by inhibitors of P-glycoprotein. Cancer Chemother Pharmacol 40 (Suppl):S13–S19.

191. Choo EF, Leake B, Wandel C, Imamura H, Wood AJJ, Wilkinson GR, Kim RB. 2000. Pharmacological inhibition of P-glycoprotein transport enhances the distribution of HIV-1 protease inhibitors into brain and testes. Drug Metab Dispos 28:655–660.

192. Mayer U, Wagenaar E, Dorobek B, Beijnen JH, Borst P, Schinkel AH. 1997. Full blockade of intestinal P-glycoprotein and extensive inhibition of blood–brain barrier P-glycoprotein by oral treatment of mice with PSC833. J Clin Invest 100:2430–2436.

193. Hendrikse NH, Schinkel AH, De Vries EGE, Fluks E, van der Graaf WTA, Willemsen ATM, Vaalburg W, Franssen EJF. 1998. Complete in vivo reversal of P-glycoprotein pump function in the blood–brain barrier visualized with positron emission tomography. Br J Pharmacol 124:1413–1418.

194. Tan B, Piwnica-Worms D, Ratner L. 2000. Multidrug resistance transporters and modulation. Curr Opin Oncol 12:450–458.

195. Litman T, Druley TE, Stein WD, Bates SE. 2001. From MDR to MXR: new understanding of multidrug resistance systems, their properties and clinical significance. Cell Mol Life Sci 58:931–959.

196. Leith CP, Kopecky KJ, Godwin J, McConnell T, Slovak ML, Chen IM, Head DR, Appelbaum FR, Willman CL. 1997. Acute myeloid leukemia in the elderly: assessment of multidrug resistance (MDR1) and cytogenetics distinguishes biologic subgroups with remarkably distinct responses to standard chemotherapy. A Southwest Oncology Group study. Blood 89:3323–3329.

197. Leith CP, Kopecky KJ, Chen IM, Eijdems L, Slovak ML, McConnell TS, Head DR, Weick J, Grever MR, Appelbaum FR, Willman CL. 1999. Frequency and clinical significance of the expression of the multidrug resistance proteins MDR1/P-glycoprotein, MRP1, and LRP in acute myeloid leukemia: a Southwest Oncology Group Study. Blood 94:1086–1099.

198. Garraway LA, Chabner B. 2002. MDRI inhibition: less resistance or less relevance? Eur J Cancer [A] 38:2337–2340.

199. Polgar O, Bates SE. 2005. ABC transporters in the balance: Is there a role in multidrug resistance? Biochem Soc Trans 33:241–245.

200. Gottesman MM. 2002. Mechanisms of cancer drug resistance. Annu Rev Med 53:615–627.

201. Chen CC, Meadows B, Regis J, Kalafsky G, Fojo T, Carrasquillo JA, Bates SE. 1997. Detection of in vivo P-glycoprotein inhibition by PSC 833 using Tc-99m sestamibi. Clin Cancer Res 3:545–552.

202. Scotto KW. 2003. Transcriptional regulation of ABC drug transporters. Oncogene 22:7496–7511.

203. Labialle S, Gayet L, Marthinet E, Rigal D, Baggetto LG. 2002. Transcriptional regulators of the human multidrug resistance 1 gene: recent views. Biochem Pharmacol 64:943–948.

204. Shtil AA, Azare J. 2005. Redundancy of biological regulation as the basis of emergence of multidrug resistance. Int Rev Cytol 246:1–29.

205. Chin JE, Soffir R, Noonan KE, Choi K, Roninson IB. 1989. Structure and expression of the human MDR (P-glycoprotein) gene family. Mol Cell Biol 9:3808–3820.

206. Schwab M, Eichelbaum M, Fromm MF. 2003. Genetic polymorphisms of the human MDR1 drug transporter. Annu Rev Pharmacol Toxicol 43:285–307.

207. Fromm MF. 2002. The influence of *MDR1* polymorphisms on P-glycoprotein expression and function in humans. Adv Drug Deliv Rev 54:1295–1310.

208. Pauli-Magnus C, Kroetz DL. 2004. Functional implications of genetic polymorphisms in the multidrug resistance gene *MDR1* (*ABCB1*). Pharm Res 21:904–913.

209. Marzolini C, Paus E, Buclin T, Kim RB. 2004. Polymorphisms in human *MDR1* (P-glycoprotein): recent advances and clinical relevance. Clin Pharmacol Ther 75:13–33.

210. Chelule PK, Gordon M, Palanee T, Page T, Mosam A, Derm FC, Coovadia HM, Cassol S. 2003. *MDR1* and *CYP3A4* polymorphisms among African, Indian, and white populations in KwaZulu-Natal, South Africa. Clin Pharmacol Ther 74:195–196.

211. Chowbay B, Li HH, David M, Cheung YB, Lee EJD. 2005. Meta-analysis of the influence of MDR1 C3435T polymorphism on digoxin pharmacokinetics and MDR1 gene expression. Br J Clin Pharmacol 60:159–171.

212. Kajinami K, Brousseau ME, Ordovas JM, Schaefer EJ. 2004. Polymorphisms in the multidrug resistance-1 (*MDR1*) gene influence the response to atorvastatin treatment in a gender-specific manner. Am J Cardiol 93:1046–1050.

213. Kioka N, Tsubota J, Kakehi Y, Komano T, Gottesman MM, Pastan I, Ueda K. 1989. P-glycoprotein gene (MDR1) cDNA from human adrenal: normal P-glycoprotein carries Gly185 with an altered pattern of multidrug resistance. Biochem Biophys Res Commun 162:224–231.

214. Chen G, Duran GE, Steger KA, Lacayo NJ, Jaffrezou JP, Dumontet C, Sikic BI. 1997. Multidrug-resistant human sarcoma cells with a mutant P-glycoprotein, altered phenotype, and resistance to cyclosporins. J Biol Chem 272:5974–5982.

215. Kimchi-Sarfaty C, Gribar JJ, Gottesman MM. 2002. Functional characterization of coding polymorphisms in the human MDR1 gene using a vaccinia virus expression system. Mol Pharmacol 62:1–6.

216. Morita N, Yasumori T, Nakayama K. 2003. Human MDR1 polymorphism: G2677T/A and C3435T have no effect on MDR1 transport activities. Biochem Pharmacol 65:1843–1852.

217. Woodahl EL, Yang ZP, Bui T, Shen DD, Ho RJY. 2004. Multidrug resistance gene G1199A polymorphism alters efflux transport activity of P-glycoprotein. J Pharmacol Exp Ther 310:1199–1207.

218. Maggio-Price L, Bielefeldt-Ohmann H, Treuting P, Iritani BM, Zeng WP, Nicks A, Tsang M, Shows D, Morrissey P, Viney JL. 2005. Dual infection with *Helicobacter bilis* and *Helicobacter hepaticus* in P-glycoprotein-deficient mdr1a−/− mice results in colitis that progresses to dysplasia. Am J Pathol 166:1793–1806.

219. Brant SR, Panhuysen CIM, Nicolae D, Reddy DM, Bonen DK, Karaliukas R, Zhang LL, Swanson E, Datta LW, Moran T, et al. 2003. *MDR1* Ala893 polymorphism is associated with inflammatory bowel disease. Am J Hum Genet 73:1282–1292.

220. Schwab M, Schäffeler E, Marx C, Fromm MF, Kaskas B, Metzler J, Stange E, Herfarth H, Schoelmerich J, Gregor M, et al. 2003. Association between the C3435T *MDR1* gene polymorphism and susceptibility for ulcerative colitis. Gastroenterology 124:26–33.

221. Bleiber G, May M, Suarez C, Martinez R, Marzolini C, Egger M, Telenti A, Swiss HIV CS. 2004. *MDR1* genetic polymorphism does not modify either cell permissiveness to HIV-1 or disease progression before treatment. J Infect Dis 189:583–586.

222. Fellay J, Marzolini C, Meaden ER, Back DJ, Buclin T, Chave JP, Decosterd LA, Furrer H, Opravil M, Pantaleo G, et al. 2002. Response to antiretroviral treatment in HIV-1-infected individuals with allelic variants of the multidrug resistance transporter 1: a pharmacogenetics study. Lancet 359:30–36.

223. Brumme ZL, Dong WW, Chan KJ, Hogg RS, Montaner JS, O'Shaughnessy MV, Harrigan PR. 2003. Influence of polymorphisms within the CX3CR1 and MDR-1 genes on initial antiretroviral therapy response. AIDS 17:201–208.

224. Nasi M, Borghi V, Pinti M, Bellodi C, Lugli E, Maffei S, Troiano L, Richeldi L, Mussini C, Esposito R, Cossarizza A. 2003. MDR1 C3435T genetic polymorphism does not influence the response to antiretroviral therapy in drug-naive HIV-positive patients. AIDS 17:1696–1698.

225. Humeny A, Rödel F, Rödel C, Sauer R, Füzesi L, Becker CM, Efferth T. 2003. *MDR1* single nucleotide polymorphism C3435T in normal colorectal tissue and colorectal carcinomas detected by MALDI-TOF mass spectrometry. Anticancer Res 23:2735–2740.

226. Kurzawski M, Drozdzik M, Suchy J, Kurzawski G, Bialecka M, Gornik W, Lubinski J. 2005. Polymorphism in the P-glycoprotein drug transporter MDR1 gene in colon cancer patients. Eur J Clin Pharmacol 61:389–394.

227. Siegsmund M, Brinkmann U, Schäffeler E, Weirich G, Schwab M, Eichelbaum M, Fritz P, Burk O, Decker J, Alken P, et al. 2002. Association of the P-glycoprotein transporter $MDR1^{C3435T}$ polymorphism with the susceptibility to renal epithelial tumors. J Am Soc Nephrol 13:1847–1854.

228. Siddiqui A, Kerb R, Weale ME, Brinkmann U, Smith A, Goldstein DB, Wood NW, Sisodiya SM. 2003. Association of multidrug resistance in epilepsy with a polymorphism in the drug-transporter gene ABCB1. N Engl J Med 348:1442–1448.

229. Sills GJ, Mohanraj R, Butler E, McCrindle S, Collier L, Wilson EA, Brodie MJ. 2005. Lack of association between the C3435T polymorphism in the human multidrug resistance (MDR1) gene and response to antiepileptic drug treatment. Epilepsia 46:643–647.

230. Tan NCK, Heron SE, Scheffer IE, Pelekanos JT, McMahon JM, Vears DF, Mulley JC, Berkovic SF. 2004. Failure to confirm association of a polymorphism in ABCB1 with multidrug-resistant epilepsy. Neurology 63:1090–1092.

231. Uhr M, Ebinger M, Rosenhagen MC, Grauer MT. 2005. The anti-Parkinson drug budipine is exported actively out of the brain by P-glycoprotein in mice. Neurosci Lett 383:73–76.

232. Tan EK, Chan DKY, Ng PW, Woo J, Teo YY, Tang K, Wong LP, Chong SS, Tan C, Shen H, Zhao Y, Lee. CGL 2005. Effect of MDR1 haplotype on risk of Parkinson disease. Arch Neurol 62:460–464.

11

MULTIDRUG RESISTANCE PROTEINS OF THE ABCC SUBFAMILY

ANNE T. NIES, MARIA RIUS, AND DIETRICH KEPPLER

German Cancer Research Center, Heidelberg, Germany

Drug Transporters: Molecular Characterization and Role in Drug Disposition, Edited by Guofeng You and Marilyn E. Morris
Copyright © 2007 John Wiley & Sons, Inc.

11.1. INTRODUCTION

The absorption, distribution, and elimination of drugs is controlled decisively by integral plasma membrane proteins.[1] The family of adenosine triphosphate (ATP)-binding cassette (ABC) transporters includes a number of members that are located in the plasma membrane and mediate the ATP-dependent efflux of endogenous and xenobiotic substances.[2] Based on amino acid sequence similarity and phylogeny, the 48 known human ABC transporters are grouped into seven subfamilies designated A through G.[2] Members of three ABC subfamilies have been recognized to play a key role in drug efflux from cells: the MDR1 P-glycoprotein (ABCB1) of the ABCB subfamily (discussed in Chapter 10), the breast cancer resistance protein (BCRP, ABCG2) of the ABCG subfamily (discussed in Chapter 12), and the multidrug resistance proteins (MRPs) of the ABCC subfamily (discussed in this chapter).

The human ABCC subfamily consists of 12 members: the nine MRPs as well as the cystic fibrosis transmembrane conductance regulator (CFTR), and the two sulfonylurea receptors SUR1 and SUR2. According to current topology prediction programs, ABCC proteins have two cytoplasmic nucleotide-binding domains (NBDs) and, depending on the isoform, two or three membrane-spanning domains (MSDs). The binding of ATP at the NBDs and subsequent ATP hydrolysis is required for ABCC/MRP-mediated transport of substances across the plasma membrane.[3] ABCCs/MRPs mediate the efflux of many endogenous and xenobiotic lipophilic organic anions, and several have overlapping substrate specificities; some are also involved in conferring resistance to a wide variety of chemotherapeutic and antiviral agents.[3,4] Their predominant localization in intestinal and renal epithelia, in hepatocytes, and in blood–tissue barriers signifies an important role of ABCC/MRP transporters in drug absorption, distribution, and elimination, as well as in protecting tissues against the entry of xenobiotic toxins.[5]

Numerous naturally occurring sequence variants leading to amino acid substitutions in the ABCCs/MRPs have been identified. These nonsynonymous sequence variants are of considerable clinical interest because they may cause interindividual variation in drug response, but only some of them have been functionally characterized so far. Sequence variants resulting in the loss of a functional ABCC2 and ABCC6 are the molecular basis of the two hereditary disorders: Dubin–Johnson syndrome and pseudoxanthoma elasticum, respectively. The function of ABCC2 as an apical efflux pump for bilirubin glucuronosides and other anionic conjugates is well established.[6] Absence of a functional ABCC2 protein from the hepatocyte canalicular membrane[7,8] hence leads to conjugated hyperbilirubinemia, characteristic of Dubin–Johnson syndrome.[9,10] In contrast, the physiological function of ABCC6 is still unknown and it is currently not possible to explain how the loss of ABCC6 function causes the connective tissue disorder pseudoxanthoma elasticum.[11]

In this chapter we focus on the molecular properties of the ABCC/MRP efflux pumps, their functional characteristics, and their tissue distribution. The functional consequences of naturally occurring sequence variants are discussed as well. Aspects of transcriptional and posttranscriptional regulation have been summarized in recent reviews.[3,12,13]

11.2. DISCOVERY OF THE ABCC/MRP EFFLUX PUMPS AND GENOMIC ORGANIZATION OF THE *ABCC* GENES

Tumor cell lines that have been made resistant to anticancer drugs by exposing them in vitro to increasing concentrations of a cytotoxic agent served as a valuable tool in the identification of the first ABCC/MRP efflux pump. Studies with the doxorubicin-resistant cell lines HL60/ADR and H69AR revealed that both cell lines showed a multidrug resistance phenotype without overexpressing MDR1 P-glycoprotein.[14,15] By photoaffinity labeling with 8-azido-ATP, a 190-kDa protein was identified to be overexpressed in the HL60/ADR cells and it was hypothesized to be a nucleotide-binding protein distinct from MDR1 P-glycoprotein, causing the multidrug resistance phenotype.[16] In 1992, Cole et al. succeeded in the molecular identification of this 190-kDa protein by cloning of the respective cDNA from H69AR cells[17]; the encoded protein was initially termed multidrug resistance–associated protein (MRP)[17] and is now known as ABCC1 or multidrug resistance protein 1, MRP1 (Table 11.1). The amino acid sequence identity of human ABCC1 with MDR1 P-glycoprotein, which belongs to the ABCB subfamily,[2] is only 24%, underlining the fact that the ABCB and the ABCC subfamilies are very distinct from each other. The ABCC subfamily also includes the ATP-gated chloride channel CFTR (ABCC7)[30] and the ATP-dependent sulfonylurea receptors ABCC8 (SUR1) and ABCC9 (SUR2), which are potassium channel regulators[31,32] (Table 11.1).

TABLE 11.1. Nomenclature and Properties of the Human *ABCC* Genes and ABCC Proteins

Symbol	Alternative Name	Gene Accession Number[a]	Chromosomal Localization[a]	Gene Size (bases)[a]	Number of Exons[a]	Protein Accession Number[b]	Amino Acids[b]	Amino Acid Identity[b]	Topology
ABCC1	MRP1	NM_004996	16p13.1	192,839	31, ref. 18	NP_004987	1531	100	MSD0-MSD1-NBD1-MSD2-NBD2
ABCC2	MRP2	NM_000392	10q24	69,460	32, refs. 19,20	NP_000383	1545	50	MSD0-MSD1-NBD1-MSD2-NBD2
ABCC3	MRP3	NM_003786	17q22	56,835	31, ref. 21	NP_003777	1527	58	MSD0-MSD1-NBD1-MSD2-NBD2
ABCC4	MRP4	NM_005845	13q32	281,593	31, refs. 21,22	NP_005836	1325	41	MSD1-NBD1-MSD2-NBD2
ABCC5	MRP5	NM_005688	3q27	98,005	30, ref. 23	NP_005679	1437	38	MSD1-NBD1-MSD2-NBD2
ABCC6	MRP6	NM_001171	16p13.1	73,324	31, ref. 24	NP_001162	1503	46	MSD0-MSD1-NBD1-MSD2-NBD2
ABCC10	MRP7	NM_033450	6p21.1	22,871	22, ref. 23	NP_258261	1464	35	MSD0-MSD1-NBD1-MSD2-NBD2
ABCC11	MRP8	NM_033151	16q12.1	80,483	29, ref. 25 30, ref. 26 31, ref. 27	NP_149163	1382	33	MSD1-NBD1-MSD2-NBD2
ABCC12	MRP9	NM_033226	16q12.1	63,797	29, refs. 25,26	NP_150229	1359	36	MSD1-NBD1-MSD2-NBD2
CFTR	ABCC7	NM_000492	7q31.2	188,698	27, ref. 28	NP_000483	1480	30	MSD1-NBD1-MSD2-NBD2
ABCC8	SUR1	NM_000352	11p15.1	84,016	39, ref. 29	NP_000343	1581	36	MSD0-MSD1-NBD1-MSD2-NBD2
ABCC9	SUR2A SUR2B	NM_005691 NM_020297	12p12.1 12p12.1	140,189 140,189	38, ref. 29 38, ref. 29	NP_005682 NP_064693	1549 1549	35 36	MSD0-MSD1-NBD1-MSD2-NBD2

[a]Gene accession number and chromosomal localization compiled from the "Gene" database at http://www.ncbi.nlm.nih.gov, gene size from the database at http://www.genecards.org, and the number of exons from the references indicated.

[b]Protein accession number and amino acid number compiled from the "Protein" database at http://www.ncbi.nlm.nih.gov, and amino acid identity analyzed by the GAP tool of the HUSAR program package at http://genome.dkfz-heidelberg.de.

Subsequent to the cloning of ABCC1, additional members have been identified that are all encoded by distinct genes and are often located on different chromosomes (Table 11.1). The first evidence for the existence of more than one ABCC/MRP efflux pump came from the cloning and sequencing of a novel 347-bp cDNA fragment from normal rat liver, which was related to but not identical to rat *Abcc1* cDNA and was not expressed in transport-deficient GY/TR$^-$ mutant rats.[33] These mutant rats had long been known to lack the ATP-dependent biliary excretion of anionic conjugates.[33–40] The 347-bp cDNA fragment was subsequently shown to be part of the full-length cDNA encoding rat Abcc2/Mrp2,[41] which was identified as an ABCC1-related protein located in the canalicular membrane of normal hepatocytes but not of hepatocytes from transport-deficient mutant rats.[8,41,42] Because of this initial localization in the hepatocyte canalicular membrane and its function as an efflux pump for anionic conjugates, Abcc2 was formerly termed canalicular Mrp (cMrp)[41] or canalicular multispecific organic anion transporter (cMoat).[42] The screening of expressed sequence tag (EST) databases[26,27,43,44] or of the high-throughput genomic sequence database[25] revealed the presence of additional ABCC/MRP efflux pumps and led to the cloning and molecular characterization of the full-length cDNAs encoding human ABCC3,[45–48] ABCC4,[22,49–53] ABCC5,[45,54–56] ABCC6,[24,57,58] ABCC10,[23,59] ABCC11,[25–27,60] and ABCC12[25,26,61] by several laboratories. The recently identified *ABCC13* gene encodes a truncated and likely functionally inactive protein.[62,63]

All human *ABCC* genes have been analyzed with respect to their genomic organization (Table 11.1). Interestingly, four of the nine *ABCC/MRP* genes are located on chromosome 16: *ABCC1* and *ABCC6* adjacent to each other in a tail-to-tail orientation at region p13.1[24] and *ABCC11* and *ABCC12* adjacent to each other in a tail-to-head orientation at region q12.1[25,26] indicating that each "gene tandem" originated by gene duplication. A comparison of the genomic organization of human *ABCC1*, *ABCC2*, and *ABCC3* showed remarkable similarities among all three genes with respect to exon number and exon size; moreover, *ABCC1*, *ABCC2*, and *ABCC3* have 21 identical splice sites based on an amino acid alignment of the three respective proteins, suggesting a close evolutionary relationship of these genes as well.[64]

11.3. MOLECULAR CHARACTERIZATION

11.3.1. Domain Structure of the ABCC Proteins

Many studies have been performed to elucidate the structure of the ABCC proteins, especially of ABCC1 and ABCC2. The ABCC transporters share with other members of the ABC superfamily a typical core structure: the hydrophobic membrane-spanning domains MSD1 and MSD2, each followed by a hydrophilic and highly conserved intracellular nucleotide-binding domain: NBD1 and NBD2, respectively (Figure 11.1*a*). Hydropathy profiles and limited proteolysis experiments suggested that ABCC1 contains an additional hydrophobic amino terminal MSD0 of approximately 200 amino acids, which precedes the typical core structure.[65,66] Glycosylation-site mutants and epitope insertion revealed the presence of five additional transmembrane segments

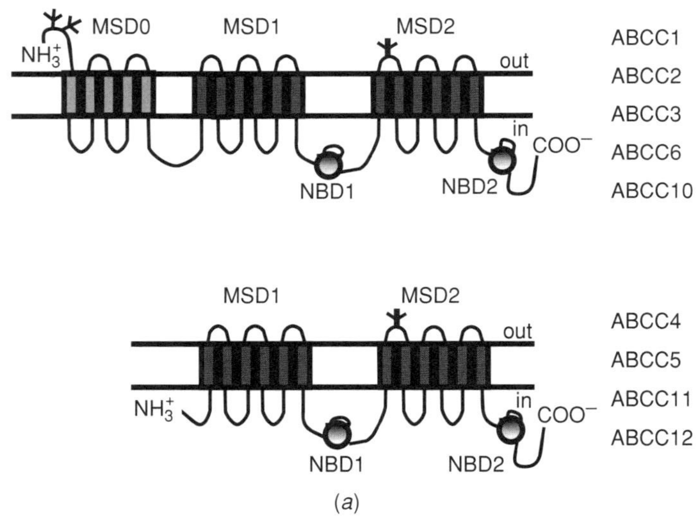

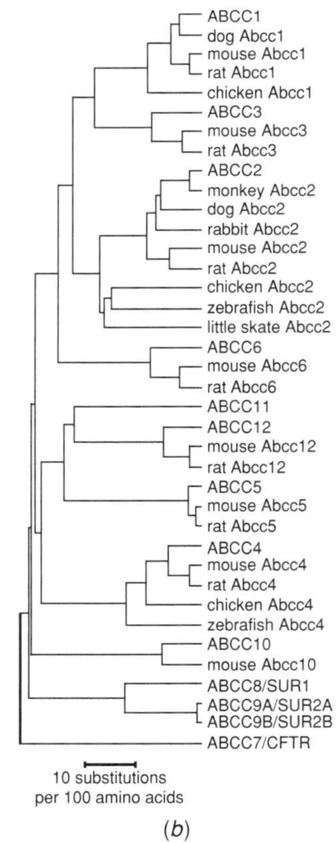

FIGURE 11.1 Topology of the human ABBC subfamily members and phylogenetic tree of full-length amino acid sequences from ABCC orthologs. (*a*) Schematic representation of the ABC core structure with two membrane-spanning domains MSD1 and MSD2, each followed by the nucleotide-binding domains, NBD1 and NBD2, respectively, shared by ABCC4, ABCC5, ABCC11, and ABCC12. An additional amino terminal MSD0, which precedes the ABC core structure, is predicted for ABCC1-3, ABCC6, and ABCC10. The number of transmembrane segments in MSD2 of ABCC2 remains unclear, varying between four and six depending on the algorithms used. (*b*) Phylogenetic tree reconstructed from the pairwise evolutionary distance between aligned ABCC/Abcc sequences. The analysis of distance was performed by the PATH tool, and the phylogenetic tree was reconstructed by the GrowTree tool of the HUSAR program package at http://genome.dkfz-heidelberg.de.

with an extracellularly located amino terminus for ABCC1.[67–69] Computational analysis of rat Abcc2 also predicted an extracellular localization of the amino terminus.[41] Further, an antibody against the amino-terminal sequence of human ABCC2 confirmed the extracellular localization of the amino terminus by immunofluorescence microscopy.[70] On the basis of sequence similarity, it is expected that ABCC3, ABCC6, ABCC8 (SUR1), ABCC9 (SUR2), and ABCC10 share this topological feature with an additional extracellular amino-terminal MSD0.[47,48,58,59,71] In contrast, a topology model with two MSDs and an intracellular amino terminus is predicted for ABCC4, ABCC5, ABCC7, ABCC11, and ABCC12 in a manner similar to that for MDR1 P-glycoprotein.[25,72,73]

Several *N*-glycosylation sites have been identified in the ABCC protein sequences, especially for ABCC1 and ABCC2[67]; however, site-directed mutagenesis of these specific amino acids did not alter sorting, transport properties, and substrate specificity of ABCC1.[67,74]

11.3.2. Amino Acid Sequence Identities of Human ABCC/MRP Paralogs

The deduced amino acid sequence lengths of the nine human ABCCs/MRPs range between 1325 amino acids for ABCC4 and 1545 amino acids for ABCC2 (Table 11.1). Within the complete human ABCC subfamily, the sulfonylurea receptors ABCC8 (SUR1) and ABCC9 (SUR2A and SUR2B) comprise 1581 and 1549 amino acids, respectively, thus being the longest proteins of the ABCC subfamily. In comparison with ABCC1, ABCC3 shares 58% amino acid identity and is the member closest to ABCC1 (Table 11.1). In contrast, CFTR (ABCC7) shares only 30% amino acid identity with ABCC1 and thus has the lowest amino acid sequence identity. The ABCC/MRP members that have the same topology as ABCC1 are also those most closely related to ABCC1 (Table 11.1).

11.3.3. ABCC/MRP Orthologs in Other Species and Phylogenetic Analysis

A number of Abcc/Mrp orthologs has been identified in several mammalian species and to a minor extent in other vertebrates. At present, seven rat orthologs (Abcc1 to Abcc6 and Abcc12) and eight mouse orthologs (Abcc1–6, Abcc10, and Abcc12) have been described (Figure 11.1*b*). In addition to humans, rat and mouse are thus the mammalian species with the largest number of known full-length Abcc/Mrp sequences. Phylogenetic analysis of the multiple amino acid sequence alignments shows the high degree of homology among human, rat, and mouse ABCC/MRP orthologs (Figure 11.1*b*). For ABCC1, the full-length sequences from five mammalian species are currently known, including those from human, rat, mouse, dog, and cow, with amino acid sequence identities ranging between 87 and 91%. ABCC2/Abcc2 full-length sequences have been identified in human, rat, mouse, rabbit, dog, and rhesus monkey, in addition to chicken, zebrafish, and little skate, with amino acid sequence identities varying between 57% for the little skate Abcc2 and 96% for the rhesus monkey Abcc2 compared to the human ABCC2. In addition, the ABCC4 full-length sequences from human, rat, mouse, chicken, and zebrafish are known and their amino acid sequence identities vary from 87% for rat and mouse Abcc4 to 70% for zebrafish Abcc4 in comparison to the human ABCC4. Furthermore, members of the ABCC subfamily have been also described in plants, such as in *Arabidopsis* with 15 ABCCs and in rice with 17 ABCCs.[75]

11.4. SUBSTRATE SPECIFICITY AND MULTIDRUG RESISTANCE PROFILES OF HUMAN ABCCs/MRPs

The ABCCs/MRPs are ATP-dependent unidirectional efflux pumps for conjugated and unconjugated organic anions. Many of these pumps have been identified to confer drug resistance to natural-product drugs as well as to purine and pyrimidine 5′-nucleoside monophosphate analogs. The expression of a defined ABCC in a cell system is an important tool for the specific characterization of its substrate specificity and multidrug resistance profile. This has been achieved using cells, including polarized cell lines, such as Madin–Darby canine kidney (MDCK), human HepG2, and pig LLC-PK_1 cells, as well as nonpolarized cell lines, such as human HEK293, human HeLa, hamster V79, and insect Sf9 cells. The substrate specificity can be characterized in the intact cell system. However, the studies in intact cell systems have a number of limitations [e.g., an endogenous uptake transporter for the compound under study must be present (section 11.6)]. In contrast, studies using inside-out membrane vesicles provide direct evidence for ABCC-mediated ATP-dependent transport of compounds, the only restriction being that compounds must be radiolabeled or fluorescent.[76] In the following sections we provide an overview of the substrate specificity as well as the multidrug resistance profiles of the human ABCC/MRP efflux pumps (Table 11.2).

11.4.1. ABCC1

In 1994, the cysteinyl leukotriene LTC_4 was identified as the first physiological substrate for ABCC1.[77,78] This finding was the result of the search for the molecular identity of the ATP-dependent efflux pump that mediates the release of LTC_4 from mastocytoma cells.[80] In addition, several cell systems allowed the identification of ABCC1 and its substrate specificity, especially ABCC1-overexpressing drug-selected cell lines, such as the doxorubicin-selected small cell lung cancer cell lines H69/AR[15] and GLC4/ADR,[81] as well as the doxorubicin-selected human leukemia HL60/ADR cells.[14] In addition, HeLa cells transfected with the recombinant *ABCC1* allowed direct evidence supporting the results obtained with the drug-selected cell lines.[78] After the discovery of LTC_4 as an ABCC1 substrate, further compounds were identified as ABCC1 substrates, including LTD_4, LTE_4, *S*-glutathionyl 2,4-dinitrobenzene (DNP-SG), 17β-glucuronosyl estradiol ($E_2 17\beta G$), lithocholyltaurine 3-sulfate, 6α-glucuronosylhyodeoxycholate, oxidized glutathione (GSSG), and bilirubin glucuronosides.[78,82–84] In addition, glutathione conjugates from ethacrynic acid, prostaglandin A_1, melphalan, chlorambucil, and aflatoxin B_1 are also transported

TABLE 11.2. Substrates and Drug Resistance Profiles of Human ABCC/MRP Transporters

Substrate[a]	K_m (μM)	Refs.	Drug Resistance[b]	Refs.
ABCC1				
Leukotriene C_4	0.1	77,78,89	Doxorubicin	99,100
Leukotriene D_4		78,82	Daunorubicin	99
Leukotriene E_4		78,82	Epirubicin	99
N-Acetyl leukotriene E_4		82	Vincristine	99,100
S-Glutathionyl aflatoxin B1	0.2	87	Vinblastine	99
S-Glutathionyl 2,4-dinitrobenzene	3.6	77,79,82	Etoposide	99
S-Glutathionyl prostaglandin A_2		86	Arsenite	99
S-Glutathionyl ethacrynic acid	28	85	Antimony	99
S-Glutathionyl *N*-ethylmaleimide		101	Taxol	99
Chlorambucil			Colchicine	99
Monochloro monoglutathionyl		88		
Monohydroxy monoglutathionyl		88		
Bisglutathionyl		88		
Melphalan				
Monochloro monoglutathionyl		82,88		
Monohydroxy monoglutathionyl		88		
Glutathione disulfide (GSSG)	93	83		
GSH (+ verapamil)[c]	83	92		
Vincristine + GSH[d]		89–91		
Daunorubicin + GSH[d]		91		
Aflatoxin B1 + GSH[d]		87		
Estrone 3-sulfate (+ GSH)[c]	0.7	95		
Estrone 3-sulfate	4.2	95		
β-*O*-Glucuronosyl NNAL (+ GSH)[c]	39	96		
Bilirubin				
Monoglucuronosyl		102		
Bisglucuronosyl		102		
17β-Glucuronosyl estradiol	b1.5	82		
	2.5	84		
6α-Glucuronosyl hyodeoxycholate		82		
Glucuronosyl etoposide		82		
Lithocholyltaurine 3-sulfate		82		
Folate		76		
Methotrexae	50	97		
Fluo-3	12	103		
p-Aminohippurate	372	98		
ABCC2				
Leukotriene C_4	1	70	Cisplatin	70,108
S-Glutathionyl 2,4-dinitrobenzene	6.5	105	Etoposide	70
S-Glutathionyl ethacrynic acid		105	Vincristine	70
Bilirubin			Doxorubicin	70
Monoglucuronosyl	0.7	104	Epirubicin	70
Bisglucuronosyl	0.9	104	Methotrexate	97

(*Continued*)

TABLE 11.2. ***(Continued)***

Substrate[a]	K_m (μM)	Refs.	Drug Resistance[b]	Refs.
17β-Glucuronosyl estradiol	7.2	70		
p-Aminohippurate	880	98		
Ochratoxin A		98		
Cholecystokinin peptide	8.1	107		
Estrone 3-sulfate		106		
ABCC3				
Leukotriene C_4	5.3	109	Etoposide	48,112
S-Glutathionyl 2,4-dinitrobenzene	5.7	109	Teniposide	48,112
17β-Glucuronosyl estradiol	25.6	109–111	Methotrexate	112
Cholylglycine	248	109		
Methotrexate	776	109,113		
Folate	1960	113		
Leucovorin	1740	113		
Dehydroepiandrosterone 3-sulfate	46	110		
Bilirubin				
Monoglucuronosyl		110		
Bisglucuronosyl		110		
ABCC4				
cGMP[e]	9.7	51,116,118	PMEA[e]	50,114,115
cAMP[e]	44.5	51,116,118	Azidothymidine	114
17β-Glucuronosyl estradiol	30.3	51,116	Methotrexate	49
Folate	170	119	6-Thioguanine	50,116
Methotrexate	220	51,119	6-Mercaptopurine	116
Leucovorin	640	119	Ganciclovir	117
Dehydroepiandrosterone 3-sulfate	2	120	Topotecan	126
Prostaglandin E_1	2.1	121		
Prostaglandin E_2	3.5	121,122		
Prostaglandin $F_{2\alpha}$	12.6	122		
Thromboxane B_2	9.9	122		
Urate	1500	124		
p-Aminohippurate	160	125		
Cholyltaurine (+ GSH)[d]	7.7	53		
Cholyltaurine (+ *S*-methylglutathione)[d]	3.8	53		
GSH (+ cholyltaurine)[d]	2700	53		
S-methyl-glutathione (+ cholyltaurine)[d]	1200	53		
Cholate (+ GSH)[d]	14.8	123		
Cholylglycine (+ GSH)[d]	25.8	123		
Deoxycholylglycine (+ GSH)[d]	6.7	123		

TABLE 11.2. ***(Continued)***

Substrate[a]	K_m (μM)	Refs.	Drug Resistance[b]	Refs.
Chenodeoxycholylglycine (+ GSH)[d]	5.9	123		
Chenodeoxycholyltaurine (+ GSH)[d]	3.6	123		
Ursodeoxycholylglycine (+ GSH)[d]	12.5	123		
Ursodeoxycholyltaurine (+ GSH)[d]	7.8	123		
ADP		118		
ABCC5				
cGMP[e]		55	Cadmium chloride	54
cAMP[e]		55	Potassium antimonyl	54
5-Fluoro-2′-deoxyuridine monophosphate	1100	127	6-Mercaptopurine	56
5-Fluorouridine monophosphate		127	6-Thioguanine	56,127
2′-Deoxyuridine monophosphate		127	PMEA[e]	56
Folate	1000	128	5-Fluorouracil	127
Methotrexate	1300	128	Methotrexate	127,128
Diglutamylated methotrexate	700	128		
ABCC6				
BQ-123		57,58	Etoposide	57
Leukotriene C_4	~0.6	57,58	Teniposide	57
S-Glutathionyl *N*-ethylmaleimide	~282	58		
S-Glutathionyl 2,4-dinitrobenzene		57		
ABCC10				
17β-Glucuronosyl estradiol	58	129	Docetaxel	130
Leukotriene C_4		129	Paclitaxel	130
			Vincristine	130
			Vinblastine	130
ABCC11				
5-Fluoro-2′-deoxyuridine monophosphate		131	5-Fluorouracil	131
Dehydroepiandrosterone 3-sulfate	13	60,132	5-Fluoro-2′-deoxyuridine	131
Estrone 3-sulfate	>150	60,132	5-Fluoro-5′-deoxyuridine	131
Leukotriene C_4		132	2′,3′-Dideoxycytidine	131

(Continued)

TABLE 11.2. ***(Continued)***

Substrate[a]	K_m (μM)	Refs.	Drug Resistance[b]	Refs.
S-Glutathionyl 2,4-dinitrobenzene		132	PMEA[e]	131
17β-Glucuronosyl estradiol	63	132		
17β-Glucuronosyl estradiol + dehydroepiandrosterone 3-sulfate[c]		132		
cGMP[e]		132		
cAMP[e]		132		
Cholylglycine		132		
Cholyltaurine		132		
Folate		132		
Methotrexate	957	132		

[a]Compounds listed have been identified as substrates by measurement of the ATP-dependent transport into inside-out membrane vesicles prepared from ABCC-expressing cells.
[b]ABCC-expressing cells confer resistance against the listed compounds by analysis of drug sensitivity.
[c]ATP-dependent stimulated transport of the first compound listed by the second compound listed.
[d]ATP-dependent cotransport of both compounds; for substrate concentrations, see refs. 53 and 123; K_m values refer to the compound not in parentheses.
[e]Abbreviations: cAMP, adenosine 3′,5′-cyclic monophosphate; cGMP, guanosine 3′,5′-cyclic monophosphate; PMEA, 9-(2-phosphonylmethoxyethyl)adenine.

by ABCC1.[82,85–88] All these substrates share the property of being organic amphiphilic anions and conjugates with glutathione, glucuronate, or sulfate. The glutathione *S*-conjugate LTC_4 is the substrate with the highest affinity for ABCC1 identified so far, with a K_m value of 97 nM.[78] As mentioned above, GSSG is a physiological substrate for ABCC1 with a relatively low affinity, suggesting a role of ABCC1 in the cellular defense against oxidative stress by decreasing concentrations of GSSG.[83] Moreover, ABCC1 mediates ATP-dependent transport of the cytostatic drugs vincristine and daunorubicin in the presence of reduced glutathione (GSH), which functions as a cosubstrate.[89–91] GSH itself is not or only poorly transported by ABCC1[83,89–91]; however, xenobiotics such as verapamil (a calcium channel blocker and an inhibitor of MDR1 P-glycoprotein) and several dietary flavonoids, including apigenin, stimulate GSH transport without being transported themselves.[92–94] In addition, the conjugated estrogen estrone 3-sulfate (E_13S) and the tobacco-specific carcinogen 4-(methylnitrosamino)-1-(3-pyridyl)-butanol (NNAL) were efficiently transported by ABCC1 only in the presence of GSH, without any evidence of cotransport.[95,96] Thus, GSH can play different roles in the ABCC1-mediated transport: first, as a cosubstrate together with hydrophobic compounds such as the *Vinca* alkaloids vincristine and vinblastine; second, as a substrate, but only in the presence of xenobiotics such as verapamil and dietary flavonoids; and third, as enhancer of ABCC1-mediated transport of glucuronated and sulfated conjugates, without being cotransported itself. These studies suggest that ABCC1 contains a bipartite binding site for hydrophobic and anionic moieties that allows for binding of conjugated substrates such as LTC_4, but allows also for cooperative binding of a hydrophobic compound and GSH.[89,90,92]

Several unconjugated amphiphilic anions are also transported by ABCC1 independently of GSH, such as folate and its antimetabolite methotrexate[76,97] as well as the pentaanionic fluorescent dye fluo-3 and the monoanionic *p*-aminohippurate (PAH).[98] Several inhibitors of ABCC1-mediated transport have been described, although their specificity for ABCC1 is not very well defined and they may also inhibit other ABCC transporters. Structural analogs of cysteinyl leukotrienes, mostly developed as leukotriene D_4 receptor antagonists such as the quinoline derivative MK571, are potent inhibitors of ABCC1-mediated ATP-dependent LTC_4 transport, with a K_i value of 0.6 μM for MK571.[78] However, MK571 also inhibits the ATP-dependent transport mediated by other ABCC members (i.e., ABCC2 and ABCC4).[41,53] Cyclosporin A inhibits the ABCC1-mediated ATP-dependent LTC_4 transport, with a K_i value of 5 μM[78] and the ABCC2-mediated ATP-dependent transport of monoglucuronosyl bilirubin with a K_i value of 21 μM.[104] However, cyclosporin A is a known MDR1 P-glycoprotein inhibitor.[133] Its nonimmunosuppressive derivative PSC833, also an inhibitor of MDR1 P-glycoprotein, is a relatively weak inhibitor of ABCC1-mediated transport.[78] Several tricyclic isoxazoles have been described to be potent and specific inhibitors of the ABCC1-mediated ATP-dependent transport in a GSH-dependent manner, in particular LY475776 and LY402913.[134–137] At present, it is known that these compounds do not inhibit ABCC2 and ABCC3, and can therefore for the time being, be considered as ABCC1 specific.

11.4.2. ABCC2

Before the molecular identification of ABCC2, the function and substrate specificity of rat Abcc2 were studied by comparison of normal and hyperbilirubinemic mutant rats: the Eisai hyperbilirubinemic (EHBR) rats and the GY/TR^- mutant rats.[8,36,138] These mutant rats are deficient in the secretion of anionic conjugates into bile[37–40] because they lack a functional Abcc2 in the hepatocyte canalicular membrane.[8,41,42] ATP-dependent transport measurements using inside-out-oriented canalicular membrane vesicles from these rats[33–36,139] contributed retrospectively to our knowledge on the substrate specificity of Abcc2.[8,140,141] In addition, the ABCC2 substrate specificity has been explored more extensively in inside-out membrane vesicles from cells stably expressing recombinant human ABCC2[70,105,141] and with ABCC2 purified to homogeneity.[142] Its substrate specificity comprises several endogenous substrates that are also prototypic for ABCC1, especially LTC_4 as a high-affinity substrate,[70] as well as other conjugates with glutathione, glucuronate, or sulfate, such as DNP-SG, *S*-glutathionyl ethacrynate, mono- and bisglucuronosyl bilirubin, $E_2 17\beta G$, lithocholyltaurine sulfate, chenodeoxycholyltaurine sulfate, and $E_1 3S$.[70,104–106] In addition, several unconjugated anionic substances have been identified as ABCC2 substrates using inside-out membrane vesicles, such as bromosulfophtalein (BSP), PAH, ochratoxin A, the sulfated cholecystokinin octapeptide CCK-8, and methotrexate.[97,98,107,143] Although the substrate specificity and transport efficiency of ABCC2 and ABCC1 are similar, differences in the kinetic properties have been established, for example the K_m values of ABCC2 for LTC_4 and $E_2 17\beta G$ are 10- and fivefold higher, respectively, than those for ABCC1.[70] In contrast, ABCC2 shows a higher affinity for mono- and bisglucuronosyl bilirubin than does ABCC1.[104] Furthermore, ABCC2 mediates low-affinity transport of GSH and also of GSSG.[139,144,145] Thus, ABCC2 has a decisive role in the hepatobiliary and renal elimination of many anionic conjugates of endogenous and xenobiotic substances.

ABCC2 is also able to confer drug resistance. The earliest studies on multidrug resistance were based on Northern blot and RNase protection analyses of several resistant cell lines in comparison to nonresistant parental cell lines.[44,146] Direct evidence of the ABCC2-dependent drug resistance profile was obtained with MDCK and HEK293 cells expressing recombinant ABCC2. These cells were resistant to cisplatin, etoposide, vincristine, doxorubicin, and epirubicin.[70] The resistance to etoposide and vincristine was partially reversed by inhibition of GSH synthesis. However, no direct evidence exists at present for ABCC2-mediated cotransport of GSH with vincristine or etoposide. ABCC2-mediated resistance to cisplatin was confirmed in ABCC2-expressing LLC-PK_1 cells,[108] and resistance to methotrexate was demonstrated in transfected human ovarian carcinoma cells.[97]

11.4.3. ABCC3

Similar to ABCC1 and ABCC2, human ABCC3 transports a broad range of endogenous, mostly conjugated organic anions. Studies with rat Abcc3 demonstrated transport of $E_2 17\beta G$ with a K_m value of 67 μM as well as of methotrexate, whereas transport of LTC_4 and DNP-SG was not detected.[147] Several glucuronate conjugates were able to inhibit the Abcc3-mediated $E_2 17\beta G$ transport.[147] However, recombinant human ABCC3 was able to transport LTC_4 and DNP-SG with K_m values of 5.3 and 5.7 μM, respectively, in addition to $E_2 17\beta G$ and methotrexate.[109] In contrast to ABCC1 and ABCC2, glutathione *S*-conjugates are poor substrates for ABCC3. Another report confirmed $E_2 17\beta G$, LTC_4, and DNP-SG as substrates for ABCC3 when expressed in Sf9 insect cells[148]; additionally, no transport of GSH or dehydroepiandrosterone 3-sulfate (DHEAS) was detected.[148] When ABCC3 was expressed in MDCK cells, DHEAS as well as LTC_4 and $E_2 17\beta G$ were transported in an ATP-dependent manner by ABCC3.[110] Furthermore, mono- and bisglucuronosyl bilirubin were also identified as important physiological substrates for ABCC3.[110]

Rat Abcc3 is able to transport bile salts (e.g., cholyltaurine, cholylglycine, sulfatochenodeoxycholyltaurine, and sulfatolithocholyltaurine) with high affinity.[149] However, subsequent studies showed that human ABCC3 mediates cholylglycine transport only with a low affinity and no significant cholyltaurine transport,[109,111,150] indicating species differences with regard to substrate specificity and affinity, at least for bile salts.[111] The recent development and anlysis of *Abcc3*$^{-/-}$ knockout mice revealed the physiological importance of Abcc3 as a glucuronoside transporter, mediating especially the efflux of bilirubin glucuronosides but not that of bile salts.[151]

The multidrug resistance profile of ABCC3 has been analyzed in several cell systems.[48,112,113,148] ABCC3 confers resistance to methotrexate and to the epipodophyllotoxins etoposide and teniposide.[48,112,148] Inhibition of GSH biosynthesis, with consequently decreased intracellular GSH concentration, did not affect the ABCC3-mediated resistance to etoposide or teniposide.[148] Furthermore, no increased GSH efflux or reduced intracellular GSH concentration were detected in

ABCC3-expressing cells.[48,112] The resistance to methotrexate was further characterized using membrane vesicles demonstrating ABCC3-mediated ATP-dependent transport of folate, methotrexate, and N^5-formyltetrahydrofolate (leucovorin), but not of the polyglutamylated metabolites of methotrexate.[113] Unlike ABCC1 and ABCC2, ABCC3 does not mediate significant resistance to *Vinca* alkaloids, doxorubicin, cisplatin, and paclitaxel.[112,148]

11.4.4. ABCC4

Functional characterization identified ABCC4 as an ATP-dependent organic anion transporter of broad substrate specificity. The first substrates of ABCC4 identified were nucleoside monophosphate analogs that act as inhibitors of HIV reverse transcriptase and are used as antiretroviral drugs.[114] The human T-lymphoid cell line CEM, selected for resistance to nucleoside monophosphate analogs, showed amplification of the *ABCC4* gene and increased ABCC4 expression associated with increased efflux of 9-(2-phosphonylmethoxyethyl)adenine (PMEA, a nucleoside phosphonate analog), azidothymidine (AZT) monophosphate, as well as other nucleoside monophosphate analogs.[114] These results were confirmed with cells expressing recombinant ABCC4. These cells, additionally, showed increased resistance to purine analogs such as 6-thioguanine and 6-mercaptopurine, as well as to the antiviral agent ganciclovir, used in chemo- and immunotherapy.[50,115–117] However, the multidrug resistance profile of ABCC4 does not include natural-product drugs such as anthracyclines, etoposide, *Vinca* alkaloids, and paclitaxel.[115]

Several studies demonstrated that ABCC4 mediates transport of the cyclic nucleotides guanosine 3′,5′-monophosphate (cGMP) and adenosine 3′,5′-monophosphate (cAMP).[50,51,116] Lai and Tan proposed that ABCC4-mediated transport of cyclic nucleotides is influenced by GSH.[50] Inhibition of GSH synthesis resulted in a decreased efflux of cAMP from the cells, even when synthesis of cAMP was stimulated with forskolin.[50] However, no direct demonstration of this stimulation has been reported in inside-out membrane vesicles. In addition to the cyclic nucleotides, ABCC4 also regulates ADP storage in dense granules from human platelets.[118] In platelet membrane vesicles, ABCC4-mediated transport of ADP was orthovanadate-sensitive, and transport of cGMP and cAMP was ATP-dependent.[118]

Transport of the physiological substrate folate and its antimetabolites methotrexate and methotrexate polyglutamate was also detected in ABCC4-transfected cells.[51,119] ABCC4 is also an efflux pump for the endogenous conjugates $E_2 17\beta G$[51,116] and DHEAS.[120] Furthermore, ABCC4 mediates ATP-dependent transport of the prostaglandins PGE_1 and PGE_2.[121] Indirect evidence based on inhibition of PGE_2 transport suggested that other prostanoids may be substrates as well.[121] Independent studies demonstrated that ABCC4 mediates transport not only of PGE_2 but also of $PGF_{2\alpha}$ and thromboxane B_2.[122] Thus, ABCC4 can be considered as an export pump for prostanoids.[122] Nonsteroidal anti-inflammatory drugs inhibited ABCC4-mediated $E_2 17\beta G$ transport.[121]

At present, direct evidence of GSH transport by ABCC4 has been reported only by our group.[53,123] In vesicles from ABCC4-expressing V79 hamster fibroblasts, ATP-dependent transport of monoanionic bile salts was detected only in the presence of

GSH or GSH derivatives (*S*-methylglutathione and ophthalmate), and ATP-dependent transport of GSH required the presence of monoanionic bile salts.[53] This cotransport is an obligatory coefflux of both cosubstrates.[123] Thus, ABCC4 transports together with GSH a wide range of bile salts, including unconjugated cholate as well as taurine- and glycine-conjugated bile salts.[123] Because of its basolateral localization in hepatocytes,[53] ABCC4 can mediate the efflux of GSH and bile salts from the hepatocytes into blood.

Membrane vesicles from Sf9 insect cells expressing ABCC4 showed transport of urate, the end product of purine catabolism.[124] In this cell system, ABCC4-mediated cGMP transport was stimulated by urate by a complex allosteric interaction.[124] Transport of PAH was also mediated by ABCC4, and to a lesser extent by ABCC2, indicating that ABCC4 is the decisive export pump for PAH in kidney proximal tubules. [125] The localization of ABCC4 in the apical membrane of proximal tubule epithelial cells[51] suggests that urate and PAH are excreted into urine via ABCC4.

Topotecan is apparently another ABCC4 substrate.[126,152] Abcc4-deficient mice showed enhanced accumulation of topotecan in brain tissue and cerebrospinal fluid.[152] Moreover, cells expressing recombinant murine Abcc4 conferred resistance to topotecan, and Abcc4-mediated $E_2 17\beta G$ transport was inhibited by topotecan.[152] Cells expressing recombinant human ABCC4 also showed reduced accumulation of topotecan.[126] However, direct demonstration of topotecan as a substrate using membrane vesicle transport studies is currently lacking.

11.4.5. ABCC5

The initial functional studies were performed in intact cells expressing recombinant human ABCC5.[54,56] In these cells, ABCC5 was able to mediate efflux of the anionic dye fluorescein diacetate in an ATP-dependent and GSH-independent manner,[54] as well as of DNP-SG and of GSH.[56] Low-level resistance mediated by ABCC5 was reported against cadmium chloride and potassium antimonyl tartrate,[54] against the purine analogs 6-mercaptopurine and 6-thioguanine, and against PMEA.[56]

Studies in inside-out membrane vesicles demonstrated ATP-dependent transport of cGMP and cAMP by ABCC5, describing for the first time an ABCC/MRP substrate with a phosphate residue as the negatively charged moiety.[55] No significant transport was detected for the glutathione and glucuronate conjugates LTC_4 and $E_2 17\beta G$, respectively, as well as for GSSG.[55] This finding of ABCC5 as a cyclic nucleotide export pump was subsequently confirmed in studies with intact cells.[153] The cyclic nucleotide transport by ABCC5 was inhibited by several phosphodiesterase inhibitors. Some of these compounds are structurally related to cGMP and able to enhance intracellular cGMP concentrations by blocking degradation as well as export of cGMP.[55] Thus, ABCC5, together with ABCC4, may play an important role in the regulation of the tissue levels of cGMP. However, the affinity of ABCC5 to cGMP and the potency of inhibition by phosphodiesterase inhibitors seem to be less than observed originally by Jedlitschky et al.[55,127,153]

In cytotoxicity assays, ABCC5 conferred high-level resistance to 5-fluorouracil and 6-thioguanine, and to a lesser extent, to methotrexate.[127] Measurements of ATP-dependent transport in vesicles demonstrated ABCC5-mediated transport of 5-fluorouracil metabolites (i.e., 5-fluoro-2′-deoxyuridine monophosphate and 5-fluorouridine monophosphate) and of 2′-deoxyuridine monophosphate, but not of 5-fluorouridine or 5-fluoro-5′-deoxyuridine.[127] Transport of 5-fluoro-2′-deoxyuridine monophosphate by ABCC5 was inhibited by several organic anions, such as probenecid and MK571, as well as by cyclic nucleotides and phosphodiesterase inhibitors.[127] Other studies showed that ABCC5 confers resistance to methotrexate and methotrexate polyglutamate.[128] ATP-dependent transport of folate, methotrexate, and the diglutamylated form of methotrexate was also detected in ABCC5-containing membrane vesicles with high K_m values of 1, 1.3, and 0.7 mM, respectively.[128]

11.4.6. ABCC6

Studies with inside-out membrane vesicles from Sf9 insect cells expressing rat Abcc6 showed ATP-dependent transport of the anionic cyclopentapeptide and endothelin receptor antagonist BQ-123.[154] However, Abcc6 did not mediate transport of the sulfated conjugates E_13S and DHEAS, the bile acid cholylglycine, the prostaglandins PGE_1 and PGE_2, thyroxine and triiodothyroxine, PAH, several hydrophobic drugs, and aminophospholipids.[154] Further studies on human ABCC6 expressed in Sf9 insect cells revealed the glutathione *S*-conjugates LTC_4 and NEM-SG, as well as BQ-123, as substrates for ABCC6.[58] The estimated K_m values for LTC_4 and NEM-SG were 0.6 and 282 μM, respectively, both thus having a low affinity for ABCC6; the K_m value for BQ-123 could not be determined, due to its low transport rate.[58] The ATP-dependent transport of NEM-SG was inhibited by the organic anions probenecid, benzbromarone, and indomethacin.[58] In accordance with these findings, Belinsky et al. also reported ABCC6-mediated transport of BQ-123, LTC_4, and DNP-SG, and no transport of $E_217\beta G$.[57] Moreover, ABCC6 conferred a low level of resistance to etoposide, teniposide, doxorubicin, and daunorubicin.[57] However, it is currently not clear how the substrates identified so far relate to the pathology of pseudoxanthoma elasticum, where ABCC6 is hereditarily deficient (Section 11.7.2).

11.4.7. ABCC10

Little is known about the substrate specificity of ABCC10. Functional studies in membrane vesicles from HEK293 cells expressing recombinant ABCC10 showed ATP-dependent transport of $E_217\beta G$ with a K_m value of about 58 μM.[129] The ABCC10-mediated transport of $E_217\beta G$ was inhibited by LTC_4, MK571, cyclosporin A, and conjugated bile salts; however, only a low rate of transport of LTC_4 was detected.[129] Several other organic anions, such as DNP-SG, glycocholate, methotrexate, folate, and cyclic nucleotides, were not transported by ABCC10.[129] ABCC10 was reported to confer high levels of resistance to docetaxel and low levels of resistance to paclitaxel and *Vinca* alkaloids.[130]

11.4.8. ABCC11 and ABCC12

Initial studies on ABCC11 proposed this transporter as a cyclic nucleotide efflux pump.[131] Intact LLC-PK$_1$ cells expressing recombinant human ABCC11 showed enhanced cellular efflux of cAMP after stimulation of the intracellular cAMP synthesis by forskolin, an activator of adenylate cyclase.[131] Similar results were observed for the efflux of cGMP. These results were confirmed by transport studies in inside-out membrane vesicles in a subsequent study, in which the substrate specificity of ABCC11 was analyzed extensively.[132] In addition to cAMP and cGMP, ABCC11 is able to transport LTC$_4$ and DNP-SG, E$_2$17βG, the monoanionic bile salts cholylglycine and cholyltaurine, folate and its antimetabolite methotrexate, and the steroid sulfates E$_1$3S and DHEAS.[132] The steroid transport was restricted to the sulfoconjugates, whereas unconjugated steroids were not transported. The ABCC11-mediated transport of E$_1$3S and DHEAS was further confirmed in vesicles from MDCK cells expressing recombinant ABCC11.[60] Transport of E$_2$17βG was stimulated by DHEAS, but transport of DHEAS was not stimulated by E$_2$17βG, suggesting a nonreciprocal interaction of both substances.[132] Several compounds were able to inhibit ABCC11-mediated DHEAS transport, including some well-known CFTR inhibitors, such as the arylaminobenzoates diphenylamine-2-carboxylate and 5-nitro-2-(3-phenylpropylamino)benzoate.[60] The ABCC11-transfected LLC-PK$_1$ cells were resistant to a variety of nucleoside and nucleotide analogs, including the fluoropyrimidines 5-fluorouracil, 5-fluoro-2′-deoxyuridine, and 5-fluoro-5′-deoxyuridine, as well as 2′,3′-dideoxycytidine and PMEA.[131] ATP-dependent transport of 5-fluoro-2′-deoxyuridine monophosphate was detected in ABCC11-expressing cells; in contrast, no transport of 5-fluorouracil or 5-fluoro-2′-deoxyuridine was detected.[131] For ABCC12, no functional characterization and no multidrug resistance profile have been reported so far.

11.5. LOCALIZATION OF ABCC/MRP EFFLUX PUMPS IN NORMAL HUMAN TISSUES AND IN HUMAN CANCERS

By screening the human EST databases ("Unigene" database at http://www.ncbi.nlm.nih.gov) one can assess the mRNA expression profile of each ABCC/MRP efflux pump. Whereas *ABCC1, ABCC3, ABCC4, ABCC5,* and *ABCC10* transcripts are found in almost all of the 46 tissues analyzed, *ABCC2* and *ABCC6* mRNA expression is restricted to about 12 and that of *ABCC11* and *ABCC12* to seven and five of the investigated tissues, respectively. Because the protein levels need not be proportional to the levels of the corresponding mRNAs, it is essential to analyze the ABCC/MRP protein expression profiles as well. Moreover, ABCC/MRP protein expression is often cell-type specific, and localization of the transporters in distinct cell types of a given tissue is therefore equally important. Protein expression and localization studies in normal human tissues and in human cancers have been performed primarily for ABCC1 to ABCC6 (Sections 11.5.1 to 11.5.6). Increasing knowledge of the function and localization will advance the understanding of ABCC/MRP-mediated substance transport in various tissues.

For functional characterization, ABCCs/MRPs are often recombinantly expressed in polarized MDCK cells. These transfected cells are used either for the isolation of plasma membrane vesicles for transport studies (Section 11.4) or for the measurement of vectorial transport of substances in intact cells (Section 11.6). In these polarized MDCK cells, human ABCCs/MRPs acquire a domain-specific plasma membrane localization. Whereas ABCC2[70,105] and ABCC11[60] are localized in the apical membrane, ABCC1,[86] ABCC3,[47,48] ABCC4,[50] ABCC5,[56] and ABCC6[155] are routed to the basolateral membrane (Figure 11.2*a*). Hepatocytes are polarized cells that express endogenously at least four different ABCC/MRP efflux pumps: that is, ABCC2 localized in the apical (canalicular) membrane[8,41,156] and ABCC3,[47,48] ABCC4,[53,157,158] and ABCC6[154,159–161] localized in the basolateral (sinusoidal) membrane (Figure 11.2*b*).

The molecular mechanisms of domain-specific ABCC/MRP targeting are incompletely understood. Instead of a distinct linear sequence motif, which had been suggested in one study for the apical localization of ABCC2,[162] targeting signals appear to be composed of several motifs within different parts of the respective ABCC/MRP that come together only in the intact protein. For instance, MSD0 is required for proper localization, at least of ABCC1 and ABCC2.[163,164] Other regions also contribute to the correct targeting of ABCC1 and ABCC2.[165–167] The interaction of ABCCs/MRPs with other proteins may be required as well, as indicated by studies with radixin knockout mice.[168] Radixin belongs to the ezrin/radixin/moesin family and cross-links actin with several integral membrane proteins.[169] The Abcc2 protein is selectively absent from the hepatocyte canalicular membrane of these radixin knockout mice, which are therefore characterized by conjugated hyperbilirubinemia.[168] Canalicular localization of ABCC2 in human hepatocytes may also depend on the interaction with radixin.[170] Another class of scaffolding proteins comprises the PSD95/Dlg/ZO-1 (PDZ) proteins that bind to specific consensus sequences at the C-termini of integral membrane proteins. Some PDZ proteins interact with ABCC2 in vitro.[171,172] Whether an interaction is also required in vivo is not clear, because Abcc2 is properly localized in the apical membrane of kidney proximal tubule cells of Pdzk1 knockout mice.[173]

11.5.1. ABCC1

The ABCC1 protein is detectable in a number of human cells and tissues with the highest levels in lung, testis, kidney, and macrophages.[17,174] Normal human hepatocytes lack detectable amounts of ABCC1,[159,175] but ABCC1 seems to be up-regulated in proliferating hepatocyte-derived cells.[175] Immunohistochemical and immunofluorescence analyses have shown that ABCC1 is localized predominantly in cells of blood–tissue barriers which form pharmacological sanctuaries in the body. ABCC1 is, for instance, present in the basolateral membrane of Sertoli cells of the blood–testis barrier,[176] of choroid plexus cells of the blood–cerebrospinal fluid barrier,[177] in bronchial epithelium,[178,179] and in the apical syncytiotrophoblast membrane of the placenta.[180] Whether ABCC1 also contributes to the function of the blood–brain barrier is still controversial.[177,181–183] ABCC1 was localized in the luminal membrane of human[182] brain capillary endothelial cells in one study, but not in others.[177,183] Studies with Abcc1 knockout mice were also not conclusive, because

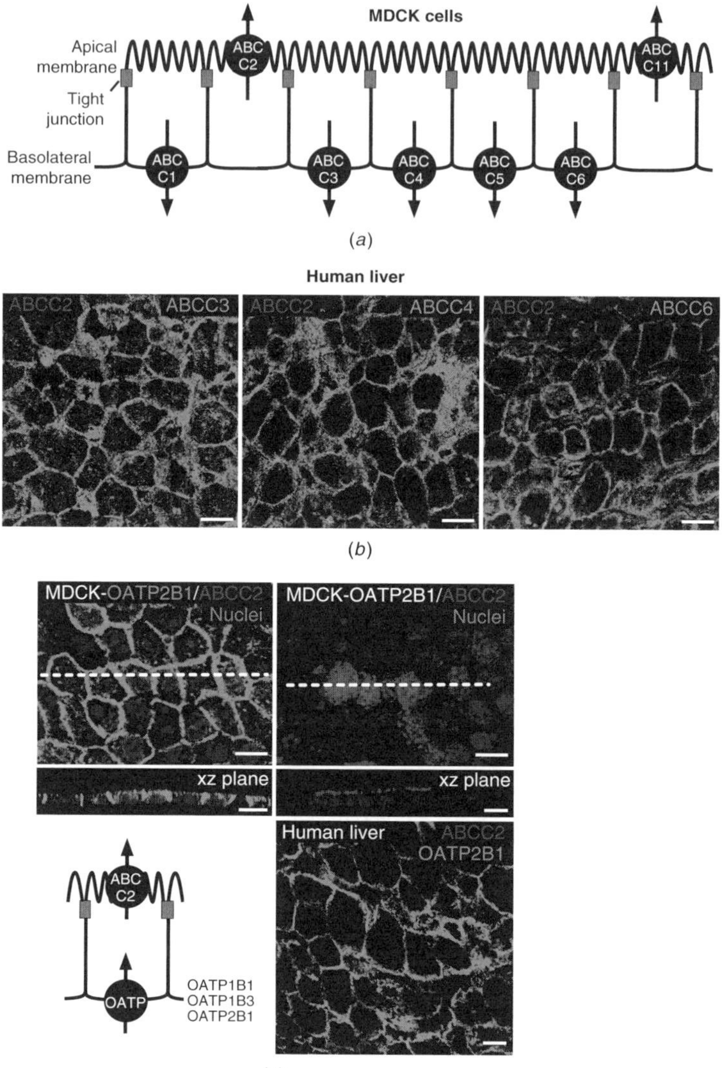

FIGURE 11.2. Subcellular localization of ABCC/MRP efflux pumps. (*a*) Schematic representation of polarized Madin–Darby canine kidney (MDCK) cells recombinantly expressing human ABCC/MRP efflux pumps, which acquire a domain-specific localization either in the apical membrane (ABCC2, ABCC11) or in the basolateral membrane (ABCC3, ABCC4, ABCC5, ABCC6). (*b*) Confocal laser scanning micrographs of ABCC/MRP efflux pumps in human hepatocytes. At least four different ABCC/MRP transporters have been identified in human hepatocytes [i.e., ABCC2 (red) in the canalicular (apical) membrane and ABCC3, ABCC4, and ABCC6 (green) in the sinusoidal (basolateral) membrane]. Bars, 20 μm. (*c*) Confocal laser scanning micrographs of MDCK cells simultaneously expressing recombinant human OATP2B1 (green) as an uptake transporter and ABCC2 (red) as an efflux pump for organic anions. The lines indicate where the optical xz-sections had been taken. These double-transfected cells serve as valuable tools to study the vectorial transport of organic anions that undergo hepatobiliary elimination. In human hepatocytes, OATP2B1 (green) is located in the sinusoidal membrane and ABCC2 (red) in the canalicular membrane. Bars, 10 μm. (*See insert for color representation of figure*.)

some studies suggested an involvement of Abcc1 in the efflux of substances from the brain into blood,[184] whereas others did not.[185,186] In human and murine kidney, ABCC1/Abcc1 is present in the epithelial cells of the loop of Henle and urinary collecting ducts.[176,187] In colon, highest ABCC1 levels are found in the Paneth cells of the crypts, but not in the epithelial cells.[187] This ABCC1 tissue distribution and localization pattern, which is almost mutually exclusive to that of ABCC2 (Section 11.5.2), indicates that ABCC1 has an important function in the detoxification of substances from those cells and tissues that do not express ABCC2.

11.5.2. ABCC2

Subsequent to its first demonstration in the canalicular membrane of rat and human hepatocytes,[8,41,156] ABCC2/Abcc2 was also identified in the apical membrane of polarized cells of rat and human kidney proximal tubules,[188,189] human small intestine,[190,191] colon,[141,191] gallbladder,[192] bronchi,[141,191] and placenta.[180,193] The exclusive apical localization in these polarized cell types underscores the function of ABCC2 in the terminal excretion and detoxification of endogenous and xenobiotic organic anions. In contrast, ABCC2 is not present in some blood–tissue barriers such as the blood–testis barrier[191,194] and the blood–brain barrier in human cerebral cortex.[182,191,195] However, epilepsy in humans and rats apparently leads to synthesis of ABCC2/Abcc2 in brain capillaries of the hippocampus.[196,197] This up-regulation may contribute to the frequently observed resistance to antiepileptic drug treatment. Some studies also describe Abcc2 in brain capillaries of normal rats[198,199] and a strain-specific Abcc2 expression in murine brain.[181] ABCC2 is not synthesized in detectable amounts in a number of human tissues (e.g., in skin, exocrine pancreas, female reproductive system, lymphatic system, cardiovascular system, and connective tissue).[191]

11.5.3. ABCC3

ABCC3 is localized in the basolateral membrane of hepatocytes,[47,48] cholangiocytes,[48] and polarized cells from the gallbladder,[192] colon, pancreas, kidney, spleen, and adrenal cortex.[200] Hepatic ABCC3 expression is inducible but appears to be constitutive in other organs. This was observed initially in mutant rats with chronic conjugated hyperbilirubinemia,[201] which are unable to secrete bilirubin glucuronosides into bile because they lack a functionally active Abcc2[37,40] (Section 11.7.1). In humans, ABCC3 levels may vary up to 80-fold among individuals,[202] and several factors may play a role in hepatic ABCC3 induction. Hereditary and acquired liver disorders, including Dubin–Johnson syndrome,[47] progressive familial intrahepatic cholestasis type 3,[200] icteric primary biliary cirrhosis,[203] and obstructive cholestasis,[204] lead to increased ABCC3 protein levels. These liver disorders are associated with elevated serum concentrations of bilirubin glucuronosides, which are normally secreted into bile by the apical conjugate efflux pump ABCC2.[205] The identification of ABCC3 as an efflux pump for bilirubin glucuronosides,[110] together with its localization in the hepatocyte basolateral membrane,[47,48] supports the concept that basolateral efflux via ABCC3 accounts for the conjugated hyperbilirubinemia observed in certain liver disorders and thereby compensates when the biliary elimination of organic anions via ABCC2 is impaired.

11.5.4. ABCC4

In contrast to the strict apical localization of ABCC2, ABCC4 can acquire an apical as well as a basolateral membrane localization, depending on the tissue and the cell type. The first immunofluorescence analyses localized ABCC4 in the basolateral membrane of glandular epithelial cells of the prostate.[115] Subsequently, ABCC4/Abcc4 was demonstrated in the apical membrane of proximal tubule epithelial cells of human[51] and rat kidney.[157] In human and murine brain capillary endothelial cells, ABCC4/Abcc4 is also localized in the apical (i.e., luminal) membrane.[152,182] A basolateral localization is observed in human, rat, and mouse hepatocytes,[53,157,158] pancreatic duct epithelial cells,[206] and in choroid plexus epithelial cells,[152] as well as after recombinant expression in polarized MDCK cells.[50] Other cell types and tissues that express the ABCC4 protein include astrocytes,[182] erythrocytes,[207] platelets,[118] adrenal gland,[120] epithelial cells of seminal vesicles and ureter,[122] and smooth muscle cells in the urogenital tract.[122]

11.5.5. ABCC5

Using two different antibodies, ABCC5 was detected in the epithelial cells of the urethra.[208] Smooth muscle cells and endothelial cells in heart[209] and the urogenital tract[208] are also positive for ABCC5. In placenta, ABCC5 is present in the basal membrane of syncytiotrophoblasts and in and around fetal vessels.[210] In brain, ABCC5 has been detected in astrocytes and pyramidal neurons[182] as well as in the blood–brain barrier, where it was found in the luminal (i.e., apical) membrane of brain capillary endothelial cells, either in cryosections of human brain[182] or in cultured bovine cells.[195] ABCC5 is localized in the basolateral membrane after recombinant expression in polarized MDCK cells.[56] Similar to ABCC4, ABCC5 appears to acquire either an apical or a basolateral localization, depending on the cell type.

11.5.6. ABCC6

Knowledge of ABCC6 tissue distribution is especially important, because lack of a functional ABCC6 protein is the molecular basis of pseudoxanthoma elasticum (Section 11.7.2). ABCC6 is expressed predominantly in the liver and kidney, where it localizes in the basolateral membrane of rat[154] and human hepatocytes[159–161] and of proximal tubule epithelial cells.[160,161] It is currently controversial whether ABCC6 is present in other cell types and tissues as well.[160,161]

11.5.7. ABCC10 to ABCC12

So far, immunolocalization analyses have been performed only for ABCC11, which was localized to the axonal membrane of neurons in human cerebral cortex as well as to the apical membrane of HepG2 and MDCK cells expressing recombinant ABCC11.[60]

11.5.8. ABCC/MRP Efflux Pumps in Human Cancers

Most of the ABCC/MRP efflux pumps have been shown in animal and in vitro studies to confer resistance to a wide variety of anticancer and antiviral drugs (Section 11.4 and Table 11.2). Despite this established contribution to the development of cellular drug resistance, the role of ABCC/MRP efflux pumps in clinical drug resistance is not well defined and remains under investigation.[211,212] A large number of studies shows the overexpression of one or more *ABCC/MRP* mRNAs in various human cancers and in cell lines derived from human cancers. Several studies have correlated this overexpression with clinicopathological data. High levels of *ABCC1* mRNA are, for instance, a prognostic factor for a poor outcome of retinoblastoma[213] and neuroblastoma.[214] Fewer studies have analyzed clinical specimens with respect to ABCC/MRP protein levels and to the immunolocalization of these efflux pumps (Table 11.3). In several trials, focusing primarily on ABCC1, the protein levels were determined before and after chemotherapy or were correlated with outcome. The ABCC1 protein level appears to be a significant negative indicator of response to chemotherapy and overall survival in some cancers (e.g., in non-small cell lung cancer[240,241] and in breast cancer).[216,242,243] Other studies have correlated the ABCC1 protein levels with cancer stage and invasiveness of prostate cancer[244] and with stage[245] and overall survival[246] in acute myeloblastic leukemia. The determination of the ABCC/MRP protein expression profile in a given tumor sample, together with knowledge of the drug resistance profile of each ABCC/MRP transporter as well as the detection and localization of uptake transporters, will advance the development of personalized chemotherapy regimens for patients.

11.6. DOUBLE-TRANSFECTED CELLS AS A MODEL SYSTEM TO STUDY THE VECTORIAL TRANSPORT OF SUBSTANCES

Vectorial transport is a key step in the hepatobiliary and renal elimination of endogenous substances and xenobiotics. Double-transfected cells expressing a basolateral uptake transporter and an apical ATP-dependent efflux pump simultaneously are now considered valuable tools for the study of this vectorial transport in vitro.[143,247,248] In addition, vectorial transport studies are of interest for pharmacokinetic analyses in drug discovery, maximizing drug efficacy and reducing interference of the transport of physiological endogenous substrates by drug candidates at the sites of their uptake and elimination.

For vectorial transport studies in vitro, it is required that the cells grow in a polarized fashion, forming a basolateral and an apical plasma membrane domain. This can be achieved by culturing them on semipermeable filter supports. The human colon carcinoma Caco-2 cell line, for instance, gains an epithelial cell polarity when cultured on certain filter membranes, with ABCC2 being localized in the apical membrane.[249] The opossum kidney OK cells and the porcine kidney LLC-PK_1 cells are established cell culture models of renal epithelial vectorial transport.[250] MDCK cells stably

TABLE 11.3. Immunodetection of ABCC/MRP Efflux Pumps in Clinical Specimens of Selected Human Solid Tumors

Type of Tumor	ABCC1	ABCC2	ABCC3	ABCC4	ABCC5	ABCC6	Ref.
Breast carcinoma	2/11[a]						215
	80/100						216
		25/49					191
Colorectal	4/13						215
carcinoma	30/30						217
	56/139						218
		39/50					191
Esophagus	12/12						215
squamous cell	54/86						218
carcinoma							
Gastric carcinoma	35/103						218
	11/20	8/13					219
							191
Gastrointestinal	9/26						220
stromal tumors	13/21						221
Glioma	16/23						222
	12/20						223
	12/17						224
	14/15						225
	0/50	0/46	0/46	36/61	35/58	0/46	226
	13/25		7/25	11/25			227
Hepatocellular	3/6	33/38	9/9				228
carcinoma							
Non-small cell	28/59						229
lung cancer	23/27						215
	5/27						230
	13/109						178
		14/32					191
Melanoma	0/10						215
	13/21						231
Ovarian carcinoma	9/15						215
	13/58						232
	17/21						233
	39/60	16/17					234
	51/115	19/115					191
							235
Renal carcinoma	2/6						215
		18/19					189
		9/22					191
Retinoblastoma	8/16	5/16		0/16			236
Soft tissue sarcoma	0/14						215
	37/86	27/86	31/86				237
Testicular	56/56						238
carcinoma							
Thyroid carcinoma		9/12					239

[a]The first number is that of samples positive for the respective ABCC/MRP protein; the second number is the total number of samples investigated.

expressing combinations of uptake transporters and efflux pumps have been established to study vectorial transport under more defined conditions.[143,247,248]

11.6.1. Establishment and Characterization of Double-Transfected Cells

Several double-transfected MDCK cell lines have been established that express an uptake transporter in the basolateral membrane and an ATP-dependent efflux pump in the apical membrane.[106,143,247] In the particular case of studies on hepatobiliary elimination, members of the organic anion-transporting polypeptide (OATP) family, comprising OATP1B1,[251–253] OATP1B3,[254] and OATP2B1[255,256] as well as the sodium-dependent bile salt transporter NTCP,[12] the organic cation transporter OCT1,[257] and the organic anion transporter OAT2,[257] are of interest because they contribute to the uptake of endogenous substances and drugs across the sinusoidal (basolateral) membrane of hepatocytes. Although several members of the ABCC subfamily are expressed in hepatocytes, only ABCC2 contributes to the efflux across the canalicular (apical) hepatocyte membrane. However, other ATP-dependent efflux pumps are also localized in the apical membrane of the hepatocyte and mediate transport of endogenous and xenobiotic substances, including ABCB1 (MDR1 P-glycoprotein), ABCG2 (BCRP), and ABCB11 (BSEP, bile salt export pump).

The first double-transfected cells for vectorial transport were MDCK cells, which expressed recombinant human OATP1B3 in the basolateral membrane and ABCC2 in the apical membrane.[143] Subsequently, double-transfected MDCK cells expressing OATP1B1 and ABCC2 or OATP2B1 and ABCC2 were established and characterized[106,247] (Figure 11.2*c*). Additional transporter combinations in transfected cells have been established: NTCP and ABCB11,[258,259] OATP1B1 and ABCB1, or OATP1B1 and ABCG2.[260]

11.6.2. Functional Characterization of Double-Transfected Cells: Vectorial Transport

For functional characterization, double-transfected cells are grown polarized on semipermeable filter supports, and transcellular transport is measured either from the basolateral to the apical side or from the apical to the basolateral side. In the case of cells expressing ABCC2 and an uptake transporter, a vectorial transport resulting from the combined uptake and efflux is obtained only from the basolateral to the apical side, not in the opposite direction. This has been shown for BSP, a prototypic amphiphilic organic anion and a substrate for OATP1B1, OATP1B3, OATP2B1, as well as ABCC2,[106,143] and also for DHEAS, $E_2 17\beta G$, pravastatin, LTC_4, CCK-8, and $E_1 3S$.[106,107,143,247] All these compounds are substrates for the efflux pump ABCC2 and for at least one of the three OATPs expressed in human hepatocytes. When cells express an uptake transporter exclusively, the vectorial transport may be negligible and only intracellular substrate accumulation may be detected.[143]

Vectorial transport of organic anions by double-transfected cells also permits the analysis of transport inhibitors.[143] A selective inhibition of the efflux pump should result in a decreased vectorial transport from the basolateral to the apical side and in intracellular substrate accumulation. For instance, 1-chloro-2,4-dinitrobenzene is transported into the cell by basolateral uptake and conjugated with GSH to form DNP-SG, which then inhibits the ABCC2-mediated efflux transport of BSP.[143] However, inhibitors may not be selective for only one transport process, as exemplified by cyclosporin A and the quinoline derivative MK571, which inhibit both the OATP1B3-mediated uptake and the ABCC2-mediated efflux of CCK-8.[107]

11.7. SEQUENCE VARIANTS OF HUMAN *ABCC/MRP* GENES AND THE HEREDITARY DEFICIENCIES OF ABCC2 IN DUBIN–JOHNSON SYNDROME AND OF ABCC6 IN PSEUDOXANTHOMA ELASTICUM

When one compares a pair of human chromosomes, a single nucleotide variant occurs approximately every 1200 bp.[261] By March 2007, about 12 million of these single-nucleotide polymorphisms (SNPs) had been compiled in the SNP database of the National Center for Biotechnology Information (NCBI; http://www.ncbi.nlm.nih.gov/SNP). Moreover, the International HapMap Consortium has generated a haplotype map of the human genome by identifying more than 1 million SNPs from 269 people of four different populations.[262] In this chapter we adhere to the recommendations of the Human Genome Variation Society (http://www.hgvs.org and ref. 263) that a nucleotide or amino acid change should be designated as a *sequence variant* rather than as a polymorphism or mutation, because the term *mutation* is used ambiguously to indicate either a change or a disease-causing (pathogenic) change. Similarly, the term *polymorphism* is used both to describe a non-disease-causing (benign) change or a change found at a frequency of 1% or higher in the population.

For the nine human *ABCC/MRP* genes, several thousand single-nucleotide variants are listed in the NCBI-SNP database. Whereas many sequence variants are present in the introns and may have no phenotypic consequences at all, others are located in the 5′- and 3′-flanking regions and may lead to an altered expression level of the respective ABCC/MRP protein. Sequence variants within the exons, also designated as *coding SNPs* (cSNPs), may result in amino acid substitutions. These nonsynonymous or missense variants are of considerable interest because they may affect the transport function of the ABCC/MRP efflux pumps. The complete lack of a functional protein may eventually lead to disease. Within the human ABCC subfamily, sequence variants with functional consequences have been identified for ABCC2 (Dubin–Johnson syndrome, Section 11.7.1), ABCC6 (pseudoxanthoma elasticum, Section 11.7.2), ABCC7 (cystic fibrosis, reviewed in ref. 264), ABCC8 (persistent hyperinsulinemic hypoglycemia of infancy, reviewed in ref. 265), and ABCC9 (dilated cardiomyopathy)[266]. Because the *ABCC11* and *ABCC12* genes have been mapped to a region harboring the genes for paroxysmal kinesigenic choreoathetosis, both may be candidate genes affected in this or in related neurological disorders.[25]

At present, major research efforts are being taken to identify non-disease-associated ABCC/MRP variants and to analyze their potential functional consequences. Because ABCCs/MRPs function as drug efflux pumps (Section 11.4), these variants may account for interindividual variation (e.g., in multidrug resistance, drug disposition, and adverse drug effects).[1] One approach is to associate the frequency of a sequence variant with the frequency of a specific trait in the population.[267] Using this type of linkage analysis, a sequence variant of ABCC11 has recently been identified that is associated with wet or dry earwax in humans.[268] Among others, the online tool PolyPhen (polymorphism phenotyping, http://genetics.bwh.harvard.edu/pph) represents an approach to assessing the potential effects of single amino acid substitutions. Based on multiple sequence alignments and information from known three-dimensional protein structures, this algorithm predicts with some probability whether an amino acid substitution has an impact on protein structure and function.[269] However, neither association studies nor *in silico* predictions can substitute for the experimental analysis of each amino acid variant to proof functional changes of the respective ABCC/MRP efflux pump. Accordingly, several ABCC1 and ABCC3 variants have been characterized functionally by expression in mammalian cells (Section 11.7.3).

11.7.1. Hereditary ABCC2 Sequence Variants Causing Dubin–Johnson Syndrome

An increasing number of sequence variants in the human *ABCC2* gene have been identified in patients with Dubin–Johnson syndrome (Table 11.4) since the initial demonstration that the absence of a functionally active ABCC2 protein from the hepatocyte canalicular membrane is the molecular basis of this hereditary disorder.[7,8] Dubin–Johnson syndrome was originally described in 1954 and characterized by the presence of a dark liver and by conjugated hyperbilirubinemia,[9,10] because bilirubin conjugates are effluxed from hepatocytes into blood via the basolateral ABCC3, which compensates for the deficiency in ABCC2-mediated biliary elimination.[205] The liver of persons affected by Dubin–Johnson syndrome appears dark blue or black, due to deposition of a dark pigment in the pericanalicular area of the hepatocytes.[280] The disorder is inherited in an autosomal recessive mode,[9,10,281] and its incidence ranges from 1 : 1300 among Iranian Jews[281] and 1 : 300,000 in the Japanese population.[276]

Many Dubin–Johnson syndrome-associated variants (Table 11.5) are single nucleotide changes, resulting in premature stop codons,[156,270,274,282] amino acid substitutions,[20,272,274,275,277,279] or alternative splicing.[20,271,275,276] The NCBI-SNP database also comprises one entry (rs17222547:C>A, dbSNP build 126) that predicts a premature stop codon at codon 967; however, this variant has not yet been identified in patients with Dubin–Johnson syndrome. Other Dubin–Johnson syndrome–associated variants include a 6-nucleotide deletion leading to the loss of two amino acids from NBD2,[19] large deletions causing frame shifts and subsequent premature stop codons,[273,274] and an in-frame deletion/insertion.[273] Premature stop codons may lead to rapid degradation of the variant *ABCC2* mRNA due to *nonsense-mediated decay*, a mechanism that recognizes during translation whether a stop codon precedes the last splice site.[283] Actually, a truncated ABCC2 protein has not yet been detected.[7,19,156] Other Dubin–Johnson syndrome–associated sequence variants result in the synthesis of immature ABCC2 proteins that are recognized by the cellular

TABLE 11.4. Nucleotide Sequence Variants in the Human *ABCC2* Gene Identified in Patients with Dubin–Johnson Syndrome

Type of Variant	Location	Nucleotide Change[a]	Consequence Predicted[a]	Affected Region in ABCC2 Protein[b]	First Reported (Ref.)
Nonsense	Exon 3	c.298 C>T[c]	p.R100X	Truncation in MSD0	270
Splicing	Exon 8	c.1031+4A>G	Complex splicing	Truncation in MSD1	271
Missense	Exon 9	c.1177 C>T	p.R393W	MSD1	272
In-frame indel	Exon 10	c.1256_1272indelCT[c]	p.E419_M424indelA	Disruption of MSD1	273
Missense	Exon 10	c.1321 C>A[c]	p.L441M	MSD1	274
Splicing	Exon 13	c.1815+2T>A	Exon 13 skipping	Disruption of MSD1	275
Splicing	Exon 15	c.1967+2T>C	p.K635X	Truncation in NBD1	276
Missense	Exon 16	c.2026 G>C[c]	p.G676R	NBD1	277
Missense	Exon 17	c.2125 T>C	p.W709R	NBD1	278
Missense	Exon 18	c.2302 C>T	p.R768W	NBD1	275
Splicing	Exon 18	c.2439+2T>C	Exon 18 skipping	Disruption of NBD1	20
Deletion	Exon 21	c.2748_2838del[c]	p.S916fsX	Truncation in MSD2	274
Nonsense	Exon 23	c.3196 C>T	p.R1066X	Truncation in MSD2	156
Deletion	Exon 24	c.3399_3400del[c]	p.Y1134CfsX	Truncation in MSD2	274
Missense	Exon 25	c.3449 G>A	p.R1150H	MSD2	279
Missense	Exon 25	c.3517 A>T	p.I1173F	MSD2	279
Deletion	Exon 26	c.3615_3843del[c]	p.W1206fsX	Truncation in MSD2	274
Nonsense	Exon 27	c.3825 C>G[c]	p.Y1275X	Truncation in MSD2	274
Nonsense	Exon 28	c.3928 C>T[c]	p.R1310X	Truncation in NBD2	282
Missense	Exon 29	c.4054 G>C[c]	p.E1352Q	NBD2	274
Missense	Exon 29	c.4145 A>G[c]	p.Q1382R	NBD2	20
Deletion	Exon 30	c.4175_4180del	p.R1392_M1393del	NBD2	19
Deletion	Exon 30	c.4292_4293del[c]	p.T1431RfsX	Truncation in NBD2	273

[a] As recommended by the Human Genome Variation Society (http://www.hgvs.org/mutnomen) and by ref. 263, nucleotide position +1 is the A of the ATG of the translation initiation codon in the *ABCC2* cDNA sequence, "c." describes a nucleotide change in relation to the *ABCC2* cDNA sequence (NM_000392), "p." describes a change in relation to the ABCC2 protein sequence deduced (NP_000383), "X" denotes a premature stop codon, "fs" denotes a frame shift, and "indel" denotes insertion/deletion.
[b] See Figure 11.1 for regions.
[c] Heterozygous variant.

TABLE 11.5. Naturally Occurring Nonsynonymous Amino Acid Substitutions in Human ABCC1 and ABCC3

Amino Acid Substitution[a]	Effect Predicted by PolyPhen[b]	Experimentally Proven Functional Consequence	NCBI-SNP ID[c] and/or Ref.
ABCC1			
p.C43S	Benign	Impaired plasma membrane localization, decreased doxorubicin resistance[330]	296
p.T73I	Benign	None apparent[328]	296
p.S92F	Possibly damaging	None apparent[328]	rs8187844
p.T117M	Benign	None apparent[328]	325
p.R230Q	Benign	None apparent[328]	rs8187848
p.V353M	Benign		rs8187852
p.R433S	Benign	Decreased transport, increased doxorubicin resistance[326]	326
p.R633Q	Benign	None apparent[328]	306
p.G671V	Probably damaging	None apparent[331]	306
p.R723Q	Benign	None apparent[328]	296, rs4148356
p.T844M	Benign		327
p.A989T	Benign	Decreased 17β-glucuronosyl estradiol transport[328]	328
p.C1047S	Benign	None apparent[328]	rs13337489
p.R1058Q	Probably damaging	None apparent[328]	296
p.L1247V	Benign		295
p.T1401M	Benign		rs8057331
p.S1512L	Benign	None apparent[328]	325
ABCC3			
p.G11D	Possibly damaging		rs11568609
p.K13N	Benign		202
p.H68Y	Benign		202
p.S314I	Possibly damaging		rs11568580
p.S346F	Benign		202, rs11568605
p.Q513K	Benign		202
p.T527R	Benign		rs1003354
p.A528G	Benign		rs1003355
p.L548Q	Possibly damaging		110
p.S607N	Benign		rs11568608
p.K718M	Possibly damaging		rs11568584
p.R1286G	Benign		rs11568593
p.R1297H	Probably damaging	None apparent[110]	110,202, rs11568591
p.R1348C	Benign		rs11568588
p.L1362V	Benign		rs1051625
p.Q1365R	Benign		rs11568590

(*Continued*)

TABLE 11.5. ***(Continued)***

Amino Acid Substitution[a]	Effect Predicted by PolyPhen[b]	Experimentally Proven Functional Consequence	NCBI-SNP ID[c] and/or Ref.
p.R1381S	Probably damaging		rs11568597
p.A1398V	Benign		rs11549764
p.G1423R	Probably damaging		202
p.A1513D	Benign		rs11656685

[a] As recommended by the Human Genome Variation Society (http://www.hgvs.org/mutnomen) and by ref. 263, "p." describes a change in relation to the deduced ABCC1 (NP_004987) or ABCC3 (NP_003777) protein sequence.
[b] PolyPhen online tool for assessing potential effects of amino acid substitutions, http://genetics.bwh.harvard.edu/pph.
[c] Single-nucleotide polymorphism database, http://www.ncbi.nlm.gov/SNP; data based on NCBI dbSNP build 126, regular updates available.

quality control machinery, retained in the endoplasmic reticulum, and finally degraded by proteasomes.[279,284–286] Some ABCC2 variants mature properly and are apically localized but functionally inactive.[279,286]

Abcc2 sequence variants have also been identified in two well-characterized hyperbilirubinemic rat strains, the GY/TR⁻ rat[37] and the Eisai hyperbilirubinemic rat (EHBR),[39,40] that are deficient in the secretion of anionic conjugates into bile. Based on the molecular identification and cloning of rat Abcc2 as the apical conjugate efflux pump, which is deficient in these rat strains,[41,42,287] both are considered to be animal models of human Dubin–Johnson syndrome. Distinct sequence variants in the rat *Abcc2* gene leading to premature stop codons [i.e., at codon 401 (GY/TR[42]) and at codon 855 (EHBR[287])] result in the absence of the Abcc2 protein from the hepatocyte canalicular membrane.[33,41,42] Using an adenoviral expression system, the human *ABCC2* gene was introduced into the Abcc2-deficient EHBRs and resulted in synthesis of human ABCC2 being localized in the hepatocyte canalicular membrane.[288] Human ABCC2 was apparently functional in the biliary efflux of bilirubin glucuronosides, because the conjugated hyperbilirubinemia, usually observed in the EHBRs,[39,40] was largely reduced.[288] These rats may thus be used to study human ABCC2 function by hepatobiliary elimination studies. Recently, several groups have generated and examined Abcc2 knockout mice[289–291] that appear to be healthy and fertile, similar to Abcc2-deficient rats. Abcc2-deficient mice cross-bred with mice lacking other Abc transporters will be useful to investigate the consequences when several of these efflux pumps are deficient simultaneously.

Because ABCC2 contributes to the oral bioavailability of drugs as well as to the hepatobiliary and renal elimination of drugs and anionic drug conjugates, ABCC2 variants with altered transport function may affect drug disposition in the body and account for adverse drug reactions.[141,292] A number of naturally occurring nonsynonymous sequence variants that do not cause Dubin–Johnson syndrome have been identified whose functional consequences are currently unknown (refs. 6, 21, and 292–295 and NCBI-SNP database). Most of them have low allele frequencies of 0.01 to

0.03 in Japanese[293,296] and Caucasians,[297] except for variant p.V417I, which has an allele frequency of 0.12 to 0.21. Several variants were introduced into the *ABCC2* cDNA by site-directed mutagenesis, and the variant proteins were functionally analyzed after their expression in mammalian cells. Some ABCC2 variants showed a reduced methotrexate transport activity (p.R412G[298]) or resulted in a reduced ABCC2 protein level (p.S789F, p.A1450T[299]). Variant p.V417I did not alter transport function,[299] but its mRNA expression was reduced in preterm placenta.[193] ABCC2 variants may also account for interindividual variation of transporter expression in normal liver.[300]

In addition to recreating naturally occurring sequence variants, amino acids were also intentionally changed in human and rat ABCC2/Abcc2. A nonconservative substitution of Trp1254 in the human ABCC2 protein resulted in the loss of methotrexate transport activity.[301] Charged amino acids within some transmembrane-spanning helices of rat Abcc2 play an important role in substrate recognition and determination of substrate specificity.[302,303] Interestingly, the substitution of Arg586 with Leu or Ile and of Arg1096 with Lys, Leu, or Met resulted in the acquisition of bile salt transport activity by rat Abcc2.[304]

11.7.2. Hereditary ABCC6 Sequence Variants Causing Pseudoxanthoma Elasticum

It was a notable discovery that the hereditary deficiency of ABCC6 is the molecular basis of pseudoxanthoma elasticum,[305–307] also known as Groenblad–Strandberg syndrome.[308,309] This autosomal recessive disorder leads to calcification of elastic fibers in the skin, arteries, and retina and manifests in skin lesions, vision loss, and cardiovascular complications.[310–312] The prevalence is estimated to be 1 : 100,000.[313] Due to a high carrier frequency and consanguinity, pseudo dominance is also observed.[313]

Approximately 100 different *ABCC6* sequence variants have been identified in patients with pseudoxanthoma elasticum,[305–307,314–323] most of which are missense (55%), nonsense (15%), and small deletion variants (15%).[11] *ABCC6* is currently the only gene considered to be defective in pseudoxanthoma elasticum, but its role in manifestation of the disease is largely unknown. Two major issues still need to be investigated. First, the physiological ABCC6 substrate is not known. Although some ABCC6 substrates have been identified (Section 11.4), they can currently not explain the pathology of pseudoxanthoma elasticum. Recently generated Abcc6 knockout mice[324] have a pathology similar to that observed in patients affected by pseudoxanthoma elasticum. These mice may be valuable in identifying additional ABCC6 substrates. Moreover, candidate substrates may come from the analysis of blood samples from patients in comparison to nonaffected individuals.[11] Second, the ABCC6 protein is mainly expressed in hepatocytes and proximal tubule epithelial cells of the kidney,[159–161] whereas it is still controversial whether it is present in tissues affected by pseudoxanthoma elasticum (i.e., the skin, retina, and arteries).[160,161] Therefore, it is currently not clear whether substances have to be effluxed from liver and kidney and need to be provided to skin, retina, and arteries, or whether pseudoxanthoma elasticum results from substances accumulating in skin, retina, and arteries.

11.7.3. Naturally Occurring Sequence Variants of ABCC1, ABCC3, ABCC4, and ABCC5

Many nonsynonymous sequence variants are listed in the NCBI-SNP database and have been identified by several groups in ABCC1,[21,295,296,306,325–328] ABCC3,[110,202] ABCC4,[21] and ABCC5.[21,209,295] To date, no human genetic disorders are known that can be attributed to a deficient transport function of any of these efflux pumps. Knockout mice may help to identify human clinical conditions to which a deficient ABCC transporter might contribute. For instance, Abcc1-deficient mice are hypersensitive to etoposide,[176] Abcc3-deficient mice show altered morphine pharmacokinetics,[329] and Abcc4-deficient mice are hypersensitive to topotecan.[152]

The expression of ABCC/MRP variants in mammalian cells is another valuable tool to identify functional changes caused by amino acid substitutions. Whereas none of the currently known ABCC4[21] and ABCC5[21,209,295] variants have been characterized functionally in vitro, several studies have been performed with ABCC1 and ABCC3 variants. Table 11.5 lists currently known nonsynonymous variants of human ABCC1 and ABCC3, their potential effects as predicted by the PolyPhen tool, and functional consequences of those variants that have been characterized in vitro. Most of the amino acid substitutions analyzed were predicted to be benign, and many ABCC1 variants were proven to have transport properties similar to those of the wild-type ABCC1.[328] Interestingly, *in silico* prediction and experimental results were discrepant for several variants. ABCC1-p.C43S and ABCC1-p.R433S, considered as benign by the PolyPhen algorithm, resulted in impaired plasma membrane localization and decreased doxorubicin resistance[330] and to decreased transport of several organic anions and increased resistance to doxorubicin,[326] respectively. On the contrary, because they are within NBD1 (Gly671 in ABCC1) or NBD2 (Arg1297 in ABCC3), substitution of amino acids that are highly conserved, was expected to be damaging, but functional consequences were detected for neither the ABCC1 variant p.G617V[331] nor for the ABCC3 variant p.R1297H.[110] Although the ABCC1-p.G671V variant is apparently functionally active in vitro, it may become deleterious in combination with other, not yet known factors. This is suggested by two studies which show that the variant is potentially under negative selection in four different populations[327] and that it is associated with doxorubicin-induced cardiomyopathy.[332] Taken together, the analysis of phenotypic consequences of ABCC/MRP variants is a challenging task and it is likely that many variants may have no clinical impact at all.

As for ABCC2, sequence variants have been introduced intentionally into the *ABCC1* and *ABCC3* cDNAs. Their functional analysis has provided important insights into the nature of the substrate recognition sites and the mechanism of transport (recently reviewed in ref. 333).

11.8. CONCLUSIONS

Multidrug resistance proteins of the ABCC subfamily mediate the unidirectional efflux of many endogenous and xenobiotic organic anions, including drugs, across the plasma membrane. The ABCC/MRP transporters differ markedly from MDR1

P-glycoprotein, in amino acid sequence and substrate specificity, the first mammalian ATP-dependent drug efflux pump that had been identified. The ABCCs/MRPs vary among each other with respect to substrate specificity, tissue distribution, and domain-specific localization in polarized cells. Several ABCC/MRP subfamily members are capable of conferring multidrug resistance. ABCC1, the first cloned ABCC/MRP subfamily member, has been crystallized and resolved structurally, with a 22-Å limit.[334] A higher resolution is expected in the future, which will give more insight into how the transporter works at the molecular level.

The hereditary deficiency of ABCC2 leads to Dubin–Johnson syndrome, which is characterized by conjugated hyperbilirubinemia, because bilirubin glucuronosides cannot be secreted into bile. Although the hereditary deficiency of ABCC6 has been identified as the molecular origin of pseudoxanthoma elasticum, a disease-relevant physiological substrate of ABCC6 is currently unknown. It remains to be resolved why the loss of a functional ABCC6 protein in hepatocytes and proximal tubule epithelial cells of the kidney causes pathology in skin, eyes, and arteries.

The predominant localization of ABCCs/MRPs in liver, kidney, intestine, and blood–tissue barriers shows that these transporters are important drug efflux pumps determining the oral bioavailability of drugs, their distribution within the body, and their elimination. In this context, naturally occurring sequence variants (i.e., single-nucleotide polymorphisms) are of considerable clinical interest, because they may affect drug disposition. However, functional analysis of the many known nonsynonymous variants is just at the beginning and remains a challenging task for future research.

Acknowledgments

The studies in the authors' laboratory were supported by the German Cancer Research Center, Heidelberg, a collaboration between the German Cancer Research Center and Pfizer Research Laboratories (Gorton, Connecticut), the German Research Foundation (DFG), and the Wilhelm Sander-Stiftung, München. We acknowledge the contributions to this work from past and present members of our laboratory, particularly Markus Büchler, Jörg König, Gabriele Jedlitschky, Inka Leier, Yunhai Cui, Verena Keitel, Daniel Rost, Thomas Schaub, Wolfgang Hagmann, Hiroyki Tsujii, Young-Min Lee, Miriam Bortfeld, Manuela Brom, Johanna Hummel-Eisenbeiss, Elke Herrmann, and Daniela Keller, as well as the collaboration with Jürgen Kartenbeck and Herbert Spring from the Cell Biology Division of the German Cancer Research Center.

REFERENCES

1. Giacomini KM, Sugiyama Y. 2006. Membrane transporters and drug response. In: Brunton LL, Lazo JS, Parker RL, editors. *Goodman & Gilman's. The Pharmacological Basis of Therapeutics*. New York: McGraw-Hill, pp 41–70.
2. Dean M, Annilo T. 2005. Evolution of the ATP-binding cassette (ABC) transporter superfamily in vertebrates. Annu Rev Genom Hum Genet 6:123–142.

3. Haimeur A, Conseil G, Deeley RG, Cole SP. 2004. The MRP-related and BCRP/ABCG2 multidrug resistance proteins: biology, substrate specificity and regulation. Curr Drug Metab 5:21–53.
4. Kruh GD, Belinsky MG. 2003. The MRP family of drug efflux pumps. Oncogene 22: 7537–7552.
5. Leslie EM, Deeley RG, Cole SP. 2005. Multidrug resistance proteins: role of P-glycoprotein, MRP1, MRP2, and BCRP (ABCG2) in tissue defense. Toxicol Appl Pharmacol 204:216–237.
6. Nies AT, Keppler D. 2007. The apical conjugate efflux pump ABCC2 (MRP2). Pflugers Arch 453:643–659.
7. Kartenbeck J, Leuschner U, Mayer R, Keppler D. 1996. Absence of the canalicular isoform of the *MRP* gene-encoded conjugate export pump from the hepatocytes in Dubin–Johnson syndrome. Hepatology 23:1061–1066.
8. Keppler D, Kartenbeck J. 1996. The canalicular conjugate export pump encoded by the *cmrp/cmoat* gene. In: Boyer JL, Ockner RK, editors. *Progress in Liver Diseases*. Philadelphia; PA: Saunders, pp 55–67.
9. Dubin IN, Johnson FB. 1954. Chronic idiopathic jaundice with unidentified pigment in liver cells: a new clinicopathologic entity with report of 12 cases. Medicine 33:155–179.
10. Sprinz H, Nelson RS. 1954. Persistent nonhemolytic hyperbilirubinemia associated with lipochrome-like pigment in liver cells; report of four cases. Ann Intern Med 41:952–962.
11. Bergen AA, Plomp AS, Hu X, de Jong PT, Gorgels TG. 2007. ABCC6 and pseudoxanthoma elasticum. Pflugers Arch 453:685–691.
12. Kullak-Ublick GA, Stieger B, Meier PJ. 2004. Enterohepatic bile salt transporters in normal physiology and liver disease. Gastroenterology 126:322–342.
13. van de Water FM, Masereeuw R, Russel FG. 2005. Function and regulation of multidrug resistance proteins (MRPs) in the renal elimination of organic anions. Drug Metab Rev 37:443–471.
14. McGrath T, Center MS. 1987. Adriamycin resistance in HL60 cells in the absence of detectable P-glycoprotein. Biochem Biophys Res Commun 145:1171–1176.
15. Mirski SE, Gerlach JH, Cole SP. 1987. Multidrug resistance in a human small cell lung cancer cell line selected in adriamycin. Cancer Res 47:2594–2598.
16. McGrath T, Latoud C, Arnold ST, Safa AR, Felsted RL, Center MS. 1989. Mechanisms of multidrug resistance in HL60 cells. Analysis of resistance associated membrane proteins and levels of mdr gene expression. Biochem Pharmacol 38:3611–3619.
17. Cole SPC, Bhardwaj G, Gerlach JH, Mackie JE, Grant CE, Almquist KC, Stewart AJ, Kurz EU, Duncan AMV, Deeley RG. 1992. Overexpression of a transporter gene in a multidrug-resistant human lung cancer cell line. Science 258:1650–1654.
18. Grant CE, Kurz EU, Cole SPC, Deeley RG. 1997. Analysis of the intron-exon organization of the human multidrug-resistance protein gene (*MRP*) and alternative splicing of its mRNA. Genomics 45:368–378.
19. Tsujii H, König J, Rost D, Stöckel B, Leuschner U, Keppler D. 1999. Exon-intron organization of the human multidrug-resistance protein 2 (MRP2) gene mutated in Dubin–Johnson syndrome. Gastroenterology 117:653–660.
20. Toh S, Wada M, Uchiumi T, Inokuchi A, Makino Y, Horie Y, Adachi Y, Sakisaka S, Kuwano M. 1999. Genomic structure of the canalicular multispecific organic

anion-transporter gene (MRP2/cMOAT) and mutations in the ATP-binding-cassette region in Dubin–Johnson syndrome. Am J Hum Genet 64:739–746.

21. Saito S, Iida A, Sekine A, Miura Y, Ogawa C, Kawauchi S, Higuchi S, Nakamura Y. 2002. Identification of 779 genetic variations in eight genes encoding members of the ATP-binding cassette, subfamily C (ABCC/MRP/CFTR). J Hum Genet 47:147–171.
22. Lamba JK, Adachi M, Sun D, Tammur J, Schuetz EG, Allikmets R, Schuetz JD. 2003. Nonsense mediated decay downregulates conserved alternatively spliced ABCC4 transcripts bearing nonsense codons. Hum Mol Genet 12:99–109.
23. Kao HH, Chang MS, Cheng JF, Huang JD. 2003. Genomic structure, gene expression, and promoter analysis of human multidrug resistance–associated protein 7. J Biomed Sci 10:98–110.
24. Kool M, van der Linden M, de Haas M, Baas F, Borst P. 1999. Expression of human MRP6, a homologue of the multidrug resistance protein gene MRP1, in tissues and cancer cells. Cancer Res 59:175–182.
25. Tammur J, Prades C, Arnould I, Rzhetsky A, Hutchinson A, Adachi M, Schuetz JD, Swoboda KJ, Ptacek LJ, Rosier M, Dean M, Allikmets R. 2001. Two new genes from the human ATP-binding cassette transporter superfamily, ABCC11 and ABCC12, tandemly duplicated on chromosome 16q12. Gene 273:89–96.
26. Yabuuchi H, Shimizu H, Takayanagi S, Ishikawa T. 2001. Multiple splicing variants of two new human ATP-binding cassette transporters, ABCC11 and ABCC12. Biochem Biophys Res Commun 288:933–939.
27. Bera TK, Lee S, Salvatore G, Lee B, Pastan I. 2001. MRP8, a new member of ABC transporter superfamily, identified by EST database mining and gene prediction program, is highly expressed in breast cancer. Mol Med 7:509–516.
28. Zielenski J, Rozmahel R, Bozon D, Kerem B, Grzelczak Z, Riordan JR, Rommens J, Tsui LC. 1991. Genomic DNA sequence of the cystic fibrosis transmembrane conductance regulator (CFTR) gene. Genomics 10:214–228.
29. Aguilar-Bryan L, Clement JP, Gonzalez G, Kunjilwar K, Babenko A, Bryan J. 1998. Toward understanding the assembly and structure of KATP channels. Physiol Rev 78:227–245.
30. Riordan JR, Rommens JM, Kerem B, Alon N, Rozmahel R, Grzelczak Z, Zielenski J, Lok S, Plavsic N, Chou JL. 1989. Identification of the cystic fibrosis gene: cloning and characterization of complementary DNA. Science 245:1066–1073.
31. Aguilar-Bryan L, Nichols CG, Wechsler SW, Clement JP, Boyd AE III, Gonzalez G, Herrera-Sosa H, Nguy K, Bryan J, Nelson DA. 1995. Cloning of the beta cell high-affinity sulfonylurea receptor: a regulator of insulin secretion. Science 268:423–426.
32. Inagaki N, Gonoi T, Clement JP, Wang CZ, Aguilar-Bryan L, Bryan J, Seino S. 1996. A family of sulfonylurea receptors determines the pharmacological properties of ATP-sensitive K^+ channels. Neuron 16:1011–1017.
33. Mayer R, Kartenbeck J, Büchler M, Jedlitschky G, Leier I, Keppler D. 1995. Expression of the *MRP* gene-encoded conjugate export pump in liver and its selective absence from the canalicular membrane in transport-deficient mutant hepatocytes. J Cell Biol 131:137–150.
34. Ishikawa T, Müller M, Klünemann C, Schaub T, Keppler D. 1990. ATP-dependent primary active transport of cysteinyl leukotrienes across liver canalicular membrane: Role of the

ATP-dependent transport system for glutathione S-conjugates. J Biol Chem 265:19279–19286.

35. Kitamura T, Jansen P, Hardenbrook C, Kamimoto Y, Gatmaitan Z, Arias IM. 1990. Defective ATP-dependent bile canalicular transport of organic anions in mutant (TR^-) rats with conjugated hyperbilirubinemia. Proc Natl Acad Sci U S A 87:3557–3561.
36. Oude Elferink RPJ, Meijer DKF, Kuipers F, Jansen PLM, Groen AK, Groothuis GMM. 1995. Hepatobiliary secretion of organic compounds: molecular mechanisms of membrane transport. Biochim Biophys Acta 1241:215–268.
37. Jansen PLM, Peters WHM, Lamers WH. 1985. Hereditary chronic conjugated hyperbilirubinemia in mutants rats caused by defective hepatic anion transport. Hepatology 5:573–579.
38. Huber M, Guhlmann A, Jansen PL, Keppler D. 1987. Hereditary defect of hepatobiliary cysteinyl leukotriene elimination in mutant rats with defective hepatic anion excretion. Hepatology 7:224–228.
39. Takikawa H, Sano N, Narita T, Uchida Y, Yamanaka M, Horie T, Mikami T, Tagaya O. 1991. Biliary excretion of bile acid conjugates in a hyperbilirubinemic mutant Sprague–Dawley rat. Hepatology 14:352–360.
40. Hosokawa S, Tagaya O, Mikami T, Nozaki Y, Kawaguchi A, Yamatsu K, Shamoto M. 1992. A new rat mutant with chronic conjugated hyperbilirubinemia and renal glomerular lesions. Lab Anim Sci 42:27–34.
41. Büchler M, König J, Brom M, Kartenbeck J, Spring H, Horie T, Keppler D. 1996. cDNA cloning of the hepatocyte canalicular isoform of the multidrug resistance protein, cMRP, reveals a novel conjugate export pump deficient in hyperbilirubinemic mutant rats. J Biol Chem 271:15091–15098.
42. Paulusma CC, Bosma PJ, Zaman GJR, Bakker CTM, Otter M, Scheffer GL, Scheper RJ, Borst P, Oude Elferink RPJ. 1996. Congenital jaundice in rats with a mutation in a multidrug resistance associated-protein gene. Science 271:1126–1127.
43. Allikmets R, Gerrard B, Hutchinson A, Dean M. 1996. Characterization of the human ABC superfamily: isolation and mapping of 21 new genes using the expressed sequence tags database. Hum Mol Genet 5:1649–1655.
44. Kool M, de Haas M, Scheffer GL, Scheper RJ, van Eijk MJ, Juijn JA, Baas F, Borst P. 1997. Analysis of expression of cMOAT (MRP2), MRP3, MRP4, and MRP5, homologues of the multidrug resistance–associated protein gene (MRP1), in human cancer cell lines. Cancer Res 57:3537–3547.
45. Belinsky MG, Bain LJ, Balsara BB, Testa JJ, Kruh GD. 1998. Characterization of MOAT-C and MOAT-D, new members of the MRP/cMOAT subfamily of transporter proteins. J Natl Cancer Inst 90:1735–1741.
46. Kiuchi Y, Suzuki H, Hirohashi T, Tyson CA, Sugiyama Y. 1998. cDNA cloning and inducible expression of human multidrug resistance associated protein 3 (MRP3). FEBS Lett 433:149–152.
47. König J, Rost D, Cui Y, Keppler D. 1999. Characterization of the human multidrug resistance protein isoform MRP3 localized to the basolateral hepatocyte membrane. Hepatology 29:1156–1163.
48. Kool M, van der Linden M, de Haas M, Scheffer GL, de Vree JM, Smith AJ, Jansen G, Peters GJ, Ponne N, Scheper RJ, et al. 1999. MRP3, an organic anion transporter able to transport anti-cancer drugs. Proc Natl Acad Sci U S A 96:6914–6919.

49. Lee K, Belinsky MG, Bell DW, Testa JR, Kruh GD. 1998. Isolation of MOAT-B, a widely expressed multidrug resistance–associated protein/canalicular multispecific organic anion transporter-related transporter. Cancer Res 58:2741–2747.

50. Lai L, Tan TM. 2002. Role of glutathione in the multidrug resistance protein 4 (MRP4/ABCC4)–mediated efflux of cAMP and resistance to purine analogues. Biochem J 361:497–503.

51. van Aubel RA, Smeets PH, Peters JG, Bindels RJ, Russel FG. 2002. The MRP4/ABCC4 gene encodes a novel apical organic anion transporter in human kidney proximal tubules: putative efflux pump for urinary cAMP and cGMP. J Am Soc Nephrol 13:595–603.

52. Wielinga PR, Reid G, Challa EE, van der Heijden I, van Deemter L, de Haas M, Mol C, Kuil AJ, Groeneveld E, Schuetz JD, et al. 2002. Thiopurine metabolism and identification of the thiopurine metabolites transported by MRP4 and MRP5 overexpressed in human embryonic kidney cells. Mol Pharmacol 62:1321–1331.

53. Rius M, Nies AT, Hummel-Eisenbeiss J, Jedlitschky G, Keppler D. 2003. Cotransport of reduced glutathione with bile salts by MRP4 (ABCC4) localized to the basolateral hepatocyte membrane. Hepatology 38:374–384.

54. McAleer MA, Breen MA, White NL, Matthews N. 1999. pABC11 (also known as MOAT-C and MRP5), a member of the ABC family of proteins, has anion transporter activity but does not confer multidrug resistance when overexpressed in human embryonic kidney 293 cells. J Biol Chem 274:23541–23558.

55. Jedlitschky G, Burchell B, Keppler D. 2000. The multidrug resistance protein 5 functions as an ATP-dependent export pump for cyclic nucleotides. J Biol Chem 275:30069–30074.

56. Wijnholds J, Mol CA, van Deemter L, de Haas M, Scheffer GL, Baas F, Beijnen JH, Scheper RJ, Hatse S, De Clercq E, Balzarini J, Borst P. 2000. Multidrug-resistance protein 5 is a multispecific organic anion transporter able to transport nucleotide analogs. Proc Natl Acad Sci U S A 97:7476–7481.

57. Belinsky MG, Chen ZS, Shchaveleva I, Zeng H, Kruh GD. 2002. Characterization of the drug resistance and transport properties of multidrug resistance protein 6 (MRP6, ABCC6). Cancer Res 62:6172–6177.

58. Ilias A, Urban Z, Seidl TL, Le Saux O, Sinko E, Boyd CD, Sarkadi B, Varadi A. 2002. Loss of ATP-dependent transport activity in pseudoxanthoma elasticum–associated mutants of human ABCC6 (MRP6). J Biol Chem 277:16860–16867.

59. Hopper E, Belinsky MG, Zeng H, Tosolini A, Testa JR, Kruh GD. 2001. Analysis of the structure and expression pattern of MRP7 (ABCC10), a new member of the MRP subfamily. Cancer Lett 162:181–191.

60. Bortfeld M, Rius M, König J, Herold-Mende C, Nies AT, Keppler D. 2006. Human multidrug resistance protein ABCC11 (MRP8), an apical efflux pump for steroid sulfates, is an axonal protein of the central and peripheral nervous system. Neuroscience 137:1247–1257.

61. Bera TK, Iavarone C, Kumar V, Lee S, Lee B, Pastan I. 2002. MRP9, an unusual truncated member of the ABC transporter superfamily, is highly expressed in breast cancer. Proc Natl Acad Sci U S A 99:6997–7002.

62. Yabuuchi H, Takayanagi S, Yoshinaga K, Taniguchi N, Aburatani H, Ishikawa T. 2002. ABCC13, an unusual truncated ABC transporter, is highly expressed in fetal human liver. Biochem Biophys Res Commun 299:410–417.

63. Annilo T, Dean M. 2004. Degeneration of an ATP-binding cassette transporter gene, ABCC13, in different mammalian lineages. Genomics 84:34–46.
64. Keppler D, Kamisako T, Leier I, Cui Y, Nies AT, Tsujii H, König J. 2000. Localization, substrate specificity, and drug resistance conferred by conjugate export pumps of the MRP family. Adv Enzyme Regul 40:339–349.
65. Stride BD, Valdimarsson G, Gerlach JH, Wilson GM, Cole SP, Deeley RG. 1996. Structure and expression of the messenger RNA encoding the murine multidrug resistance protein, an ATP-binding cassette transporter. Mol Pharmacol 49:962–971.
66. Bakos E, Hegedus T, Hollo Z, Welker E, Tusnady GE, Zaman GJ, Flens MJ, Varadi A, Sarkadi B. 1996. Membrane topology and glycosylation of the human multidrug resistance–associated protein. J Biol Chem 271:12322–12326.
67. Hipfner DR, Almquist KC, Leslie EM, Gerlach JH, Grant CE, Deeley RG, Cole SP. 1997. Membrane topology of the multidrug resistance protein (MRP): a study of glycosylation-site mutants reveals an extracytosolic NH2 terminus. J Biol Chem 272:23623–23630.
68. Kast C, Gros P. 1997. Topology mapping of the amino-terminal half of multidrug resistance–associated protein by epitope insertion and immunofluorescence. J Biol Chem 272:26479–26487.
69. Kast C, Gros P. 1998. Epitope insertion favors a six transmembrane domain model for the carboxy-terminal portion of the multidrug resistance–associated protein. Biochemistry 37:2305–2313.
70. Cui Y, König J, Buchholz U, Spring H, Leier I, Keppler D. 1999. Drug resistance and ATP-dependent conjugate transport mediated by the apical multidrug resistance protein, MRP2, permanently expressed in human and canine cells. Mol Pharmacol 55:929–937.
71. Conti LR, Radeke CM, Shyng SL, Vandenberg CA. 2001. Transmembrane topology of the sulfonylurea receptor SUR1. J Biol Chem 276:41270–41278.
72. Borst P, Evers R, Kool M, Wijnholds J. 2000. A family of drug transporters: the multidrug resistance–associated proteins. J Natl Cancer Inst 92:1295–1302.
73. Chang XB, Hou YX, Jensen TJ, Riordan JR. 1994. Mapping of cystic fibrosis transmembrane conductance regulator membrane topology by glycosylation site insertion. J Biol Chem 269:18572–18575.
74. Müller M, Yong M, Peng XH, Petre B, Arora S, Ambudkar SV. 2002. Evidence for the role of glycosylation in accessibility of the extracellular domains of human MRP1 (ABCC1). Biochemistry 41:10123–10132.
75. Klein M, Burla B, Martinoia E. 2006. The multidrug resistance–associated protein (MRP/ABCC) subfamily of ATP-binding cassette transporters in plants. FEBS Lett 580:1112–1122.
76. Keppler D, Jedlitschky G, Leier I. 1998. Transport function and substrate specificity of multidrug resistance protein. Methods Enzymol 292:607–616.
77. Jedlitschky G, Leier I, Buchholz U, Center M, Keppler D. 1994. ATP-dependent transport of glutathione S-conjugates by the multidrug resistance–associated protein. Cancer Res 54:4833–4836.
78. Leier I, Jedlitschky G, Buchholz U, Cole SP, Deeley RG, Keppler D. 1994. The MRP gene encodes an ATP-dependent export pump for leukotriene C4 and structurally related conjugates. J Biol Chem 269:27807–27810.
79. Müller M, Meijer C, Zaman GJ, Borst P, Scheper RJ, Mulder NH, de Vries EG, Jansen PL. 1994. Overexpression of the gene encoding the multidrug resistance–associated protein

results in increased ATP-dependent glutathione S-conjugate transport. Proc Natl Acad Sci U S A 91:13033–13037.

80. Leier I, Jedlitschky G, Buchholz U, Keppler D. 1994. Characterization of the ATP-dependent leukotriene C4 export carrier in mastocytoma cells. Eur J Biochem 220:599–606.
81. Zijlstra JG, de Vries EG, Mulder NH. 1987. Multifactorial drug resistance in an adriamycin-resistant human small cell lung carcinoma cell line. Cancer Res 47:1780–1784.
82. Jedlitschky G, Leier I, Buchholz U, Barnouin K, Kurz G, Keppler D. 1996. Transport of glutathione, glucuronate, and sulfate conjugates by the MRP gene–encoded conjugate export pump. Cancer Res 56:988–994.
83. Leier I, Jedlitschky G, Buchholz U, Center M, Cole SP, Deeley RG, Keppler D. 1996. ATP-dependent glutathione disulphide transport mediated by the MRP gene-encoded conjugate export pump. Biochem J 314(Pt 2):433–437.
84. Loe DW, Almquist KC, Cole SP, Deeley RG. 1996. ATP-dependent 17 beta-estradiol 17-(beta-D-glucuronide) transport by multidrug resistance protein (MRP). Inhibition by cholestatic steroids. J Biol Chem 271:9683–9689.
85. Zaman GJ, Cnubben NH, van Bladeren PJ, Evers R, Borst P. 1996. Transport of the glutathione conjugate of ethacrynic acid by the human multidrug resistance protein MRP. FEBS Lett 391:126–130.
86. Evers R, Cnubben NH, Wijnholds J, van Deemter L, van Bladeren PJ, Borst P. 1997. Transport of glutathione prostaglandin A conjugates by the multidrug resistance protein 1. FEBS Lett 419:112–116.
87. Loe DW, Stewart RK, Massey TE, Deeley RG, Cole SP. 1997. ATP-dependent transport of aflatoxin B1 and its glutathione conjugates by the product of the multidrug resistance protein (MRP) gene. Mol Pharmacol 51:1034–1041.
88. Barnouin K, Leier I, Jedlitschky G, Pourtier-Manzanedo A, König J, Lehmann WD, Keppler D. 1998. Multidrug resistance protein-mediated transport of chlorambucil and melphalan conjugated to glutathione. Br J Cancer 77:201–209.
89. Loe DW, Almquist KC, Deeley RG, Cole SP. 1996. Multidrug resistance protein (MRP)-mediated transport of leukotriene C4 and chemotherapeutic agents in membrane vesicles. Demonstration of glutathione-dependent vincristine transport. J Biol Chem 271:9675–9682.
90. Loe DW, Deeley RG, Cole SP. 1998. Characterization of vincristine transport by the M(r) 190,000 multidrug resistance protein (MRP): evidence for cotransport with reduced glutathione. Cancer Res 58:5130–5136.
91. Renes J, de Vries EG, Nienhuis EF, Jansen PL, Müller M. 1999. ATP- and glutathione-dependent transport of chemotherapeutic drugs by the multidrug resistance protein MRP1. Br J Pharmacol 126:681–688.
92. Loe DW, Deeley RG, Cole SP. 2000. Verapamil stimulates glutathione transport by the 190-kDa multidrug resistance protein 1 (MRP1). J Pharmacol Exp Ther 293:530–538.
93. Leslie EM, Mao Q, Oleschuk CJ, Deeley RG, Cole SP. 2001. Modulation of multidrug resistance protein 1 (MRP1/ABCC1) transport and atpase activities by interaction with dietary flavonoids. Mol Pharmacol 59:1171–1180.
94. Leslie EM, Deeley RG, Cole SP. 2003. Bioflavonoid stimulation of glutathione transport by the 190-kDa multidrug resistance protein 1 (MRP1). Drug Metab Dispos 31:11–15.

95. Qian YM, Song WC, Cui H, Cole SP, Deeley RG. 2001. Glutathione stimulates sulfated estrogen transport by multidrug resistance protein 1. J Biol Chem 276:6404–6411.
96. Leslie EM, Ito K, Upadhyaya P, Hecht SS, Deeley RG, Cole SP. 2001. Transport of the beta-O-glucuronide conjugate of the tobacco-specific carcinogen 4-(methylnitrosamino)-1-(3-pyridyl)-1-butanol (NNAL) by the multidrug resistance protein 1 (MRP1): requirement for glutathione or a non-sulfur-containing analog. J Biol Chem 276:27846–27854.
97. Hooijberg JH, Broxterman HJ, Kool M, Assaraf YG, Peters GJ, Noordhuis P, Scheper RJ, Borst P, Pinedo HM, Jansen G. 1999. Antifolate resistance mediated by the multidrug resistance proteins MRP1 and MRP2. Cancer Res 59:2532–2535.
98. Leier I, Hummel-Eisenbeiss J, Cui Y, Keppler D. 2000. ATP-dependent *para*-aminohippurate transport by apical multidrug resistance protein MRP2. Kidney Int 57:1636–1642.
99. Cole SP, Sparks KE, Fraser K, Loe DW, Grant CE, Wilson GM, Deeley RG. 1994. Pharmacological characterization of multidrug resistant MRP-transfected human tumor cells. Cancer Res 54:5902–5910.
100. Grant CE, Valdimarsson G, Hipfner DR, Almquist KC, Cole SP, Deeley RG. 1994. Overexpression of multidrug resistance-associated protein (MRP) increases resistance to natural product drugs. Cancer Res 54:357–361.
101. Bakos E, Evers R, Szakacs G, Tusnady GE, Welker E, Szabo K, de Haas M, van Deemter L, Borst P, Varadi A, Sarkadi B. 1998. Functional multidrug resistance protein (MRP1) lacking the N-terminal transmembrane domain. J Biol Chem 273:32167–32175.
102. Jedlitschky G, Leier I, Buchholz U, Hummel-Eisenbeiss J, Burchell B, Keppler D. 1997. ATP-dependent transport of bilirubin glucuronides by the multidrug resistance protein MRP1 and its hepatocyte canalicular isoform MRP2. Biochem J 327(Pt 1):305–310.
103. Keppler D, Cui Y, König J, Leier I, Nies A. 1999. Export pumps for anionic conjugates encoded by MRP genes. Adv Enzyme Regul 39:237–246.
104. Kamisako T, Leier I, Cui Y, König J, Buchholz U, Hummel-Eisenbeiss J, Keppler D. 1999. Transport of monoglucuronosyl and bisglucuronosyl bilirubin by recombinant human and rat multidrug resistance protein 2. Hepatology 30:485–490.
105. Evers R, Kool M, van Deemter L, Janssen H, Calafat J, Oomen LCJM, Paulusma CC, Oude Elferink RPJ, Baas F, Schinkel AH, Borst P. 1998. Drug export activity of the human canalicular multispecific organic anion transporter in polarized kidney MDCK cells expressing *cMOAT* (*MRP2)* cDNA. J Clin Invest 101:1310–1319.
106. Kopplow K, Letschert K, König J, Walter B, Keppler D. 2005. Human hepatobiliary transport of organic anions analyzed by quadruple-transfected cells. Mol Pharmacol 68:1031–1038.
107. Letschert K, Komatsu M, Hummel-Eisenbeiss J, Keppler D. 2005. Vectorial transport of the peptide CCK-8 by double-transfected MDCKII cells stably expressing the organic anion transporter OATP1B3 (OATP8) and the export pump ABCC2. J Pharmacol Exp Ther 313:549–556.
108. Chen ZS, Kawabe T, Ono M, Aoki S, Sumizawa T, Furukawa T, Uchiumi T, Wada M, Kuwano M, Akiyama SI. 1999. Effect of multidrug resistance-reversing agents on transporting activity of human canalicular multispecific organic anion transporter. Mol Pharmacol 56:1219–1228.
109. Zeng H, Liu G, Rea PA, Kruh GD. 2000. Transport of amphipathic anions by human multidrug resistance protein 3. Cancer Res 60:4779–4784.

110. Lee YM, Cui Y, König J, Risch A, Jäger B, Drings P, Bartsch H, Keppler D, Nies AT. 2004. Identification and functional characterization of the natural variant MRP3-Arg1297His of human multidrug resistance protein 3 (MRP3/ABCC3). Pharmacogenetics 14:213–223.

111. Akita H, Suzuki H, Hirohashi T, Takikawa H, Sugiyama Y. 2002. Transport activity of human MRP3 expressed in Sf9 cells: comparative studies with rat MRP3. Pharm Res 19:34–41.

112. Zeng H, Bain LJ, Belinsky MG, Kruh GD. 1999. Expression of multidrug resistance protein-3 (multispecific organic anion transporter-D) in human embryonic kidney 293 cells confers resistance to anticancer agents. Cancer Res 59:5964–5967.

113. Zeng H, Chen ZS, Belinsky MG, Rea PA, Kruh GD. 2001. Transport of methotrexate (MTX) and folates by multidrug resistance protein (MRP) 3 and MRP1: effect of polyglutamylation on MTX transport. Cancer Res 61:7225–7232.

114. Schuetz JD, Connelly MC, Sun D, Paibir SG, Flynn PM, Srinivas RV, Kumar A, Fridland A. 1999. MRP4: A previously unidentified factor in resistance to nucleoside-based antiviral drugs. Nat Med 5:1048–1051.

115. Lee K, Klein-Szanto AJ, Kruh GD. 2000. Analysis of the MRP4 drug resistance profile in transfected NIH3T3 cells. J Natl Cancer Inst 92:1934–1940.

116. Chen ZS, Lee K, Kruh GD. 2001. Transport of cyclic nucleotides and estradiol 17-beta-D-glucuronide by multidrug resistance protein 4: resistance to 6-mercaptopurine and 6-thioguanine. J Biol Chem 276:33747–33754.

117. Adachi M, Sampath J, Lan LB, Sun D, Hargrove P, Flatley R, Tatum A, Edwards MZ, Wezeman M, Matherly L, Drake R, Schuetz J. 2002. Expression of MRP4 confers resistance to ganciclovir and compromises bystander cell killing. J Biol Chem 277:38998–39004.

118. Jedlitschky G, Tirschmann K, Lubenow LE, Nieuwenhuis HK, Akkerman JW, Greinacher A, Kroemer HK. 2004. The nucleotide transporter MRP4 (ABCC4) is highly expressed in human platelets and present in dense granules, indicating a role in mediator storage. Blood 104:3603–3610.

119. Chen ZS, Lee K, Walther S, Raftogianis RB, Kuwano M, Zeng H, Kruh GD. 2002. Analysis of methotrexate and folate transport by multidrug resistance protein 4 (ABCC4): MRP4 is a component of the methotrexate efflux system. Cancer Res 62:3144–3150.

120. Zelcer N, Reid G, Wielinga P, Kuil A, van der Heijden, I, Schuetz JD, Borst P. 2003. Steroid and bile acid conjugates are substrates of human multidrug-resistance protein (MRP) 4 (ATP-binding cassette C4). Biochem J 371:361–367.

121. Reid G, Wielinga P, Zelcer N, van der Heijden, I, Kuil A, de Haas M, Wijnholds J, Borst P. 2003. The human multidrug resistance protein MRP4 functions as a prostaglandin efflux transporter and is inhibited by nonsteroidal antiinflammatory drugs. Proc Natl Acad Sci U S A 100:9244–9249.

122. Rius M, Thon WF, Keppler D, Nies AT. 2005. Prostanoid transport by multidrug resistance protein 4 (MRP4/ABCC4) localized in tissues of the human urogenital tract. J Urol 174:2409–2414.

123. Rius M, Hummel-Eisenbeiss J, Hofmann AF, Keppler D. 2006. Substrate specificity of human ABCC4 (MRP4)-mediated cotransport of bile acids and reduced glutathione. Am J Physiol Gastrointest Liver Physiol 290:G640–G649.

124. van Aubel RA, Smeets PH, van den Heuvel JJ, Russel FG. 2005. Human organic anion transporter MRP4 (ABCC4) is an efflux pump for the purine end metabolite urate

with multiple allosteric substrate binding sites. Am J Physiol Renal Physiol 288:F327–F333.

125. Smeets PH, van Aubel RA, Wouterse AC, van den Heuvel JJ, Russel FG. 2004. Contribution of multidrug resistance protein 2 (MRP2/ABCC2) to the renal excretion of *p*-aminohippurate (PAH) and identification of MRP4 (ABCC4) as a novel PAH transporter. J Am Soc Nephrol 15:2828–2835.

126. Tian Q, Zhang J, Chan SY, Tan TM, Duan W, Huang M, Zhu YZ, Chan E, Yu Q, Nie YQ, et al. 2006. Topotecan is a substrate for multidrug resistance associated protein 4. Curr Drug Metab 7:105–118.

127. Pratt S, Shepard RL, Kandasamy RA, Johnston PA, Perry W III, Dantzig AH. 2005. The multidrug resistance protein 5 (ABCC5) confers resistance to 5-fluorouracil and transports its monophosphorylated metabolites. Mol Cancer Ther 4:855–863.

128. Wielinga P, Hooijberg JH, Gunnarsdottir S, Kathmann I, Reid G, Zelcer N, van der Born K, de Haas M, van der Heijden, I, Kaspers G, et al. 2005. The human multidrug resistance protein MRP5 transports folates and can mediate cellular resistance against antifolates. Cancer Res 65:4425–4430.

129. Chen ZS, Hopper-Borge E, Belinsky MG, Shchaveleva I, Kotova E, Kruh GD. 2003. Characterization of the transport properties of human multidrug resistance protein 7 (MRP7, ABCC10). Mol Pharmacol 63:351–358.

130. Hopper-Borge E, Chen ZS, Shchaveleva I, Belinsky MG, Kruh GD. 2004. Analysis of the drug resistance profile of multidrug resistance protein 7 (ABCC10): resistance to docetaxel. Cancer Res 64:4927–4930.

131. Guo Y, Kotova E, Chen ZS, Lee K, Hopper-Borge E, Belinsky MG, Kruh GD. 2003. MRP8, ATP-binding cassette C11 (ABCC11), is a cyclic nucleotide efflux pump and a resistance factor for fluoropyrimidines 2′,3′-dideoxycytidine and 9′-(2′-phosphonylmethoxyethyl)adenine. J Biol Chem 278:29509–29514.

132. Chen ZS, Guo Y, Belinsky MG, Kotova E, Kruh GD. 2005. Transport of bile acids, sulfated steroids, estradiol 17-beta-D-glucuronide, and leukotriene C4 by human multidrug resistance protein 8 (ABCC11). Mol Pharmacol 67:545–557.

133. Barrand MA, Rhodes T, Center MS, Twentyman PR. 1993. Chemosensitisation and drug accumulation effects of cyclosporin A, PSC-833 and verapamil in human MDR large cell lung cancer cells expressing a 190k membrane protein distinct from P-glycoprotein. Eur J Cancer 29A:408–415.

134. Norman BH, Dantzig AH, Kroin JS, Law KL, Tabas LB, Shepard RL, Palkowitz AD, Hauser KL, Winter MA, Sluka JP, Starling JJ. 1999. Reversal of resistance in multidrug resistance protein (MRP1)–overexpressing cells by LY329146. Bioorg Med Chem Lett 9:3381–3386.

135. Norman BH, Gruber JM, Hollinshead SP, Wilson JW, Starling JJ, Law KL, Self TD, Tabas LB, Williams DC, Paul DC, Wagner MM, Dantzig AH. 2002. Tricyclic isoxazoles are novel inhibitors of the multidrug resistance protein (MRP1). Bioorg Med Chem Lett 12:883–886.

136. Mao Q, Qiu W, Weigl KE, Lander PA, Tabas LB, Shepard RL, Dantzig AH, Deeley RG, Cole SP. 2002. GSH-dependent photolabeling of multidrug resistance protein MRP1 (ABCC1) by [^{125}I]LY475776: evidence of a major binding site in the COOH-proximal membrane spanning domain. J Biol Chem 277:28690–28699.

137. Qian YM, Grant CE, Westlake CJ, Zhang DW, Lander PA, Shepard RL, Dantzig AH, Cole SP, Deeley RG. 2002. Photolabeling of human and murine multidrug resistance protein 1 with the high affinity inhibitor [^{125}I]LY475776 and azidophenacyl-[^{35}S]glutathione. J Biol Chem 277:35225–35231.

138. Akita H, Suzuki H, Ito K, Kinoshita S, Sato N, Takikawa H, Sugiyama Y. 2001. Characterization of bile acid transport mediated by multidrug resistance associated protein 2 and bile salt export pump. Biochim Biophys Acta 1511:7–16.

139. Fernandez-Checa JC, Takikawa H, Horie T, Ookhtens M, Kaplowitz N. 1992. Canalicular transport of reduced glutathione in normal and mutant Eisai hyperbilirubinemic rats. J Biol Chem 267:1667–1673.

140. König J, Nies AT, Cui Y, Leier I, Keppler D. 1999. Conjugate export pumps of the multidrug resistance protein (MRP) family: localization, substrate specificity, and MRP2-mediated drug resistance. Biochim Biophys Acta 1461:377–394.

141. König J, Nies AT, Cui Y, Keppler D. 2003. MRP2, the apical export pump for anionic conjugates. In: Holland IB, Kuchler K, Higgins C, Cole SPC, editors. *ABC Proteins: From Bacteria to Man.* London: Academic Press, pp 423–443.

142. Hagmann W, Schubert J, König J, Keppler D. 2002. Reconstitution of transport-active multidrug resistance protein 2 (MRP2; ABCC2) in proteoliposomes. Biol Chem 383:1001–1009.

143. Cui Y, König J, Keppler D. 2001. Vectorial transport by double-transfected cells expressing the human uptake transporter SLC21A8 and the apical export pump ABCC2. Mol Pharmacol 60:934–943.

144. Paulusma CC, van Geer MA, Evers R, Heijn M, Ottenhoff R, Borst P, Oude Elferink RP. 1999. Canalicular multispecific organic anion transporter/multidrug resistance protein 2 mediates low-affinity transport of reduced glutathione. Biochem J 338(Pt 2):393–401.

145. Evers R, de Haas M, Sparidans R, Beijnen J, Wielinga PR, Lankelma J, Borst P. 2000. Vinblastine and sulfinpyrazone export by the multidrug resistance protein MRP2 is associated with glutathione export. Br J Cancer 83:375–383.

146. Taniguchi K, Wada M, Kohno K, Nakamura T, Kawabe T, Kawakami M, Kagotani K, Okumura K, Akiyama S, Kuwano M. 1996. A human canalicular multispecific organic anion transporter (cMOAT) overexpressed in cisplatin-resistant human cancer cell lines with decreased drug accumulation. Cancer Res 56:4124–4129.

147. Hirohashi T, Suzuki H, Sugiyama Y. 1999. Characterization of the transport properties of cloned rat multidrug resistance–associated protein 3 (MRP3). J Biol Chem 274:15181–15185.

148. Zelcer N, Saeki T, Reid G, Beijnen JH, Borst P. 2001. Characterization of drug transport by the human multidrug resistance protein 3 (ABCC3). J Biol Chem 276:46400–46407.

149. Hirohashi T, Suzuki H, Takikawa H, Sugiyama Y. 2000. ATP-dependent transport of bile salts by rat multidrug resistance–associated protein 3 (Mrp3). J Biol Chem 275:2905–2910.

150. Zelcer N, Saeki T, Bot I, Kuil A, Borst P. 2003. Transport of bile acids in multidrug-resistance-protein 3-overexpressing cells co-transfected with the ileal Na^+-dependent bile-acid transporter. Biochem J 369:23–30.

151. Belinsky MG, Dawson PA, Shchaveleva I, Bain LJ, Wang R, Ling V, Chen ZS, Grinberg A, Westphal H, Klein-Szanto A, Lerro A, Kruh GD. 2005. Analysis of the in vivo functions of Mrp3. Mol Pharmacol 68:160–168.

152. Leggas M, Adachi M, Scheffer GL, Sun D, Wielinga P, Du G, Mercer KE, Zhuang Y, Panetta JC, Johnston B, et al. 2004. Mrp4 confers resistance to topotecan and protects the brain from chemotherapy. Mol Cell Biol 24:7612–7621.

153. Wielinga PR, van der Heijden, I, Reid G, Beijnen JH, Wijnholds J, Borst P. 2003. Characterization of the MRP4- and MRP5-mediated transport of cyclic nucleotides from intact cells. J Biol Chem 278:17664–17671.

154. Madon J, Hagenbuch B, Landmann L, Meier PJ, Stieger B. 2000. Transport function and hepatocellular localization of mrp6 in rat liver. Mol Pharmacol 57:634–641.

155. Sinko E, Ilias A, Ujhelly O, Homolya L, Scheffer GL, Bergen AA, Sarkadi B, Varadi A. 2003. Subcellular localization and N-glycosylation of human ABCC6, expressed in MDCKII cells. Biochem Biophys Res Commun 308:263–269.

156. Paulusma CC, Kool M, Bosma PJ, Scheffer GL, ter Borg F, Scheper RJ, Tytgat GNJ, Borst P, Baas F, Oude Elferink RPJ. 1997. A mutation in the human canalicular multispecific organic anion transporter gene causes the Dubin–Johnson syndrome. Hepatology 25:1539–1542.

157. Denk GU, Soroka CJ, Takeyama Y, Chen YQ, Schuetz JD, Boyer JL. 2004. Multidrug resistance–associated protein 4 is up-regulated in liver but down-regulated in kidney in obstructive cholestasis in the rat. J Hepatol 40:585–591.

158. Wagner M, Fickert P, Zollner G, Fuchsbichler A, Silbert D, Tsybrovskyy O, Zatloukal K, Guo GL, Schuetz JD, Gonzalez FJ, et al. 2003. Role of farnesoid X receptor in determining hepatic ABC transporter expression and liver injury in bile duct–ligated mice. Gastroenterology 125:825–838.

159. Keppler D, König J, Nies AT. 2001. Conjugate export pumps of the multidrug resistance protein (MRP) family in liver. In: Arias IM, Boyer JL, Chisari FV, Fausto N, Schachter D, Shafritz DA, editors. *The Liver: Biology and Pathobiology*. New York: Lippincott Williams & Wilkins, pp 373–382.

160. Scheffer GL, Hu X, Pijnenborg AC, Wijnholds J, Bergen AA, Scheper RJ. 2002. MRP6 (ABCC6) detection in normal human tissues and tumors. Lab Invest 82:515–518.

161. Beck K, Hayashi K, Dang K, Hayashi M, Boyd CD. 2005. Analysis of ABCC6 (MRP6) in normal human tissues. Histochem Cell Biol 123:517–528.

162. Harris MJ, Kuwano M, Webb M, Board PG. 2001. Identification of the apical membrane-targeting signal of the multidrug resistance-associated protein 2 (MRP2/MOAT). J Biol Chem 276:20876–20881.

163. Fernandez SB, Hollo Z, Kern A, Bakos E, Fischer PA, Borst P, Evers R. 2002. Role of the N-terminal transmembrane region of the multidrug resistance protein MRP2 in routing to the apical membrane in MDCKII cells. J Biol Chem 277:31048–31055.

164. Westlake CJ, Cole SP, Deeley RG. 2005. Role of the NH2 terminal membrane spanning domain of multidrug resistance protein 1/ABCC1 in protein processing and trafficking. Mol Biol Cell 16:2483–2492.

165. Konno T, Ebihara T, Hisaeda K, Uchiumi T, Nakamura T, Shirakusa T, Kuwano M, Wada M. 2003. Identification of domains participating in the substrate specificity and subcellular localization of the multidrug resistance proteins MRP1 and MRP2. J Biol Chem 278:22908–22917.

166. Nies AT, König J, Cui Y, Brom M, Spring H, Keppler D. 2002. Structural requirements for the apical sorting of human multidrug resistance protein 2 (ABCC2). Eur J Biochem 269:1866–1876.

167. Westlake CJ, Payen L, Gao M, Cole SP, Deeley RG. 2004. Identification and characterization of functionally important elements in the multidrug resistance protein 1 COOH-terminal region. J Biol Chem 279:53571–53583.

168. Kikuchi S, Hata M, Fukumoto K, Yamane Y, Matsui T, Tamura A, Yonemura S, Yamagishi H, Keppler D, Tsukita S. 2002. Radixin deficiency causes conjugated hyperbilirubinemia with loss of Mrp2 from bile canalicular membranes. Nat Genet 31:320–325.

169. Bretscher A, Edwards K, Fehon RG. 2002. ERM proteins and merlin: integrators at the cell cortex. Nat Rev Mol Cell Biol 3:586–599.

170. Kojima H, Nies AT, König J, Hagmann W, Spring H, Uemura M, Fukui H, Keppler D. 2003. Changes in the expression and localization of hepatocellular transporters and radixin in primary biliary cirrhosis. J Hepatol 39:693–702.

171. Kocher O, Comella N, Gilchrist A, Pal R, Tognazzi K, Brown LF, Knoll JH. 1999. PDZK1, a novel PDZ domain-containing protein up-regulated in carcinomas and mapped to chromosome 1q21, interacts with cMOAT (MRP2), the multidrug resistance–associated protein. Lab Invest 79:1161–1170.

172. Hegedus T, Sessler T, Scott R, Thelin W, Bakos E, Varadi A, Szabo K, Homolya L, Milgram SL, Sarkadi B. 2003. C-terminal phosphorylation of MRP2 modulates its interaction with PDZ proteins. Biochem Biophys Res Commun 302:454–461.

173. Kocher O, Pal R, Roberts M, Cirovic C, Gilchrist A. 2003. Targeted disruption of the PDZK1 gene by homologous recombination. Mol Cell Biol 23:1175–1180.

174. Flens MJ, Zaman GJR, van der Valk P, Izquierdo MA, Schroeijers AB, Scheffer GL, van der Groep P, de Haas M, Meijer CJLM, Scheper RJ. 1996. Tissue distribution of the multidrug resistance protein. Am J Pathol 148:1237–1247.

175. Roelofsen H, Vos TA, Schippers IJ, Kuipers F, Koning H, Moshage H, Jansen PLM, Müller M. 1997. Increased levels of the multidrug resistance protein in lateral membranes of proliferating hepatocyte-derived cells. Gastroenterology 112:511–521.

176. Wijnholds J, Scheffer GL, van der Valk M, van der Valk P, Beijnen JH, Scheper RJ, Borst P. 1998. Multidrug resistance protein 1 protects the oropharyngeal mucosal layer and the testicular tubules against drug-induced damage. J Exp Med 188:797–808.

177. Rao VV, Dahlheimer JL, Bardgett ME, Snyder AZ, Finch RA, Sartorelli AC, Piwnica-Worms D. 1999. Choroid plexus epithelial expression of MDR1 P glycoprotein and multidrug resistance–associated protein contribute to the blood–cerebrospinal fluid–drug permeability barrier. Proc Natl Acad Sci U S A 96:3900–3905.

178. Wright SR, Boag AH, Valdimarsson G, Hipfner DR, Campling BG, Cole SP, Deeley RG. 1998. Immunohistochemical detection of multidrug resistance protein in human lung cancer and normal lung. Clin Cancer Res 4:2279–2289.

179. Brechot JM, Hurbain I, Fajac A, Daty N, Bernaudin JF. 1998. Different pattern of MRP localization in ciliated and basal cells from human bronchial epithelium. J Histochem Cytochem 46:513–517.

180. St Pierre MV, Serrano MA, Macias RI, Dubs U, Hoechli M, Lauper U, Meier PJ, Marin JJ. 2000. Expression of members of the multidrug resistance protein family in human term placenta. Am J Physiol Regul Integr Comp Physiol 279:R1495–R1503.

181. Soontornmalai A, Vlaming ML, Fritschy JM. 2006. Differential, strain-specific cellular and subcellular distribution of multidrug transporters in murine choroid plexus and blood–brain barrier. Neuroscience 138:159–169.

182. Nies AT, Jedlitschky G, König J, Herold-Mende C, Steiner HH, Schmitt HP, Keppler D. 2004. Expression and immunolocalization of the multidrug resistance proteins, MRP1-MRP6 (ABCC1–ABCC6), in human brain. Neuroscience 129:349–360.

183. Aronica E, Gorter JA, Jansen GH, van Veelen CW, van Rijen PC, Leenstra S, Ramkema M, Scheffer GL, Scheper RJ, Troost D. 2003. Expression and cellular distribution of multidrug transporter proteins in two major causes of medically intractable epilepsy: focal cortical dysplasia and glioneuronal tumors. Neuroscience 118:417–429.

184. Sugiyama D, Kusuhara H, Lee YJ, Sugiyama Y. 2003. Involvement of multidrug resistance associated protein 1 (Mrp1) in the efflux transport of 17beta estradiol-D-17beta-glucuronide (E217betaG) across the blood–brain barrier. Pharm Res 20:1394–1400.

185. Sun H, Johnson DR, Finch RA, Sartorelli AC, Miller DW, Elmquist WF. 2001. Transport of fluorescein in MDCKII-MRP1 transfected cells and mrp1-knockout mice. Biochem Biophys Res Commun 284:863–889.

186. Cisternino S, Rousselle C, Lorico A, Rappa G, Scherrmann JM. 2003. Apparent lack of Mrp1-mediated efflux at the luminal side of mouse blood–brain barrier endothelial cells. Pharm Res 20:904–909.

187. Peng KC, Cluzeaud F, Bens M, van Huyen JP, Wioland MA, Lacave R, Vandewalle A. 1999. Tissue and cell distribution of the multidrug resistance-associated protein (MRP) in mouse intestine and kidney. J Histochem Cytochem 47:757–768.

188. Schaub TP, Kartenbeck J, König J, Vogel O, Witzgall R, Kriz W, Keppler D. 1997. Expression of the conjugate export pump encoded by the *mrp2* gene in the apical membrane of kidney proximal tubules. J Am Soc Nephrol 8:1213–1221.

189. Schaub TP, Kartenbeck J, König J, Spring H, Dörsam H, Stähler G, Störkel S, Thon WF, Keppler D. 1999. Expression of the MRP2 gene-encoded conjugate export pump in human kidney proximal tubules and in renal-cell carcinoma. J Am Soc Nephrol 10:1159–1169.

190. Fromm MF, Kauffmann HM, Fritz P, Burk O, Kroemer HK, Warzok RW, Eichelbaum M, Siegmund W, Schrenk D. 2000. The effect of rifampin treatment on intestinal expression of human MRP transporters. Am J Pathol 157:1575–1580.

191. Sandusky GE, Mintze KS, Pratt SE, Dantzig AH. 2002. Expression of multidrug resistance–associated protein 2 (MRP2) in normal human tissues and carcinomas using tissue microarrays. Histopathology 41:65–74.

192. Rost D, König J, Weiss G, Klar E, Stremmel W, Keppler D. 2001. Expression and localization of the multidrug resistance proteins MRP2 and MRP3 in human gallbladder epithelia. Gastroenterology 121:1203–1208.

193. Meyer zu Schwabedissen HE, Jedlitschky G, Gratz M, Haenisch S, Linnemann K, Fusch C, Cascorbi I, Kroemer HK. 2005. Variable expression of MRP2 (ABCC2) in human placenta: influence of gestational age and cellular differentiation. Drug Metab Dispos 33:896–904.

194. Bart J, Hollema H, Groen HJ, de Vries EG, Hendrikse NH, Sleijfer DT, Wegman TD, Vaalburg W, van der Graaf WT. 2004. The distribution of drug-efflux pumps, P-gp, BCRP, MRP1 and MRP2, in the normal blood–testis barrier and in primary testicular tumours. Eur J Cancer 40:2064–2070.

195. Zhang Y, Schuetz JD, Elmquist WF, Miller DW. 2004. Plasma membrane localization of multidrug resistance–associated protein homologs in brain capillary endothelial cells. J Pharmacol Exp Ther 311:449–455.

196. Aronica E, Gorter JA, Ramkema M, Redeker S, Ozbas-Gerceker F, van Vliet EA, Scheffer GL, Scheper RJ, van der Kalk P, Baayen JC, Troost D. 2004. Expression and cellular distribution of multidrug resistance–related proteins in the hippocampus of patients with mesial temporal lobe epilepsy. Epilepsia 45:441–451.

197. Hoffmann K, Gastens AM, Volk HA, Loscher W. 2006. Expression of the multidrug transporter MRP2 in the blood–brain barrier after pilocarpine-induced seizures in rats. Epilepsy Res 69:1–14.

198. Miller DS, Nobmann SN, Gutmann H, Toeroek M, Drewe J, Fricker G. 2000. Xenobiotic transport across isolated brain microvessels studied by confocal microscopy. Mol Pharmacol 58:1357–1367.

199. van Vliet EA, Redeker S, Aronica E, Edelbroek PM, Gorter JA. 2005. Expression of multidrug transporters MRP1, MRP2, and BCRP shortly after status epilepticus, during the latent period, and in chronic epileptic rats. Epilepsia 46:1569–1580.

200. Scheffer GL, Kool M, de Haas M, de Vree JM, Pijnenborg AC, Bosman DK, Oude Elferink RPJ, van der Valk P, Borst P, Scheper RJ. 2002. Tissue distribution and induction of human multidrug resistance protein 3. Lab Invest 82:193–201.

201. Hirohashi T, Suzuki H, Ito K, Ogawa K, Kume K, Shimizu T, Sugiyama Y. 1998. Hepatic expression of multidrug resistance–associated protein-like proteins maintained in Eisai hyperbilirubinemic rats. Mol Pharmacol 53:1068–1075.

202. Lang T, Hitzl M, Burk O, Mornhinweg E, Keil A, Kerb R, Klein K, Zanger UM, Eichelbaum M, Fromm MF. 2004. Genetic polymorphisms in the multidrug resistance–associated protein 3 (ABCC3, MRP3) gene and relationship to its mRNA and protein expression in human liver. Pharmacogenetics 14:155–164.

203. Zollner G, Fickert P, Silbert D, Fuchsbichler A, Marschall HU, Zatloukal K, Denk H, Trauner M. 2003. Adaptive changes in hepatobiliary transporter expression in primary biliary cirrhosis. J Hepatol 38:717–727.

204. Shoda J, Kano M, Oda K, Kamiya J, Nimura Y, Suzuki H, Sugiyama Y, Miyazaki H, Todoroki T, Stengelin S, et al. 2001. The expression levels of plasma membrane transporters in the cholestatic liver of patients undergoing biliary drainage and their association with the impairment of biliary secretory function. Am J Gastroenterol 96:3368–3378.

205. Nies AT, Cui Y, König J, Keppler D. 2004. Transport of bilirubin conjugates across hepatocellular membrane domains and the conjugated hyperbilirubinemia of Dubin–Johnson syndrome. In: Trauner M, Jansen PLM, editors. *Molecular Pathogenesis of Cholestasis*. Georgetown; TX: Landes Bioscience, pp 195–210.

206. König J, Hartel M, Nies AT, Martignoni ME, Guo J, Büchler MW, Friess H, Keppler D. 2005. Expression and localization of human multidrug resistance protein (ABCC) family members in pancreatic carcinoma. Int J Cancer 115:359–367.

207. Klokouzas A, Wu CP, van Veen HW, Barrand MA, Hladky SB. 2003. cGMP and glutathione-conjugate transport in human erythrocytes. Eur J Biochem 270:3696–3708.

208. Nies AT, Spring H, Thon WF, Keppler D, Jedlitschky G. 2002. Immunolocalization of multidrug resistance protein 5 (ABCC5) in human genitourinary system. J Urol 167:2271–2275.

209. Dazert P, Meissner K, Vogelgesang S, Heydrich B, Eckel L, Bohm M, Warzok R, Kerb R, Brinkmann U, Schäffeler E, et al. 2003. Expression and localization of the multidrug resistance protein 5 (MRP5/ABCC5), a cellular export pump for cyclic nucleotides, in human heart. Am J Pathol 163:1567–1577.

210. Meyer zu Schwabedissen HE, Grube M, Heydrich B, Linnemann K, Fusch C, Kroemer HK, Jedlitschky G. 2005. Expression, localization, and function of MRP5 (ABCC5), a transporter for cyclic nucleotides, in human placenta and cultured human trophoblasts: effects of gestational age and cellular differentiation. Am J Pathol 166:39–48.
211. Leonard GD, Fojo T, Bates SE. 2003. The role of ABC transporters in clinical practice. Oncologist 8:411–424.
212. Szakacs G, Paterson JK, Ludwig JA, Booth-Genthe C, Gottesman MM. 2006. Targeting multidrug resistance in cancer. Nat Rev Drug Discov 5:219–234.
213. Chan HS, Lu Y, Grogan TM, Haddad G, Hipfner DR, Cole SP, Deeley RG, Ling V, Gallie BL. 1997. Multidrug resistance protein (MRP) expression in retinoblastoma correlates with the rare failure of chemotherapy despite cyclosporine for reversal of P-glycoprotein. Cancer Res 57:2325–2330.
214. Norris MD, Bordow SB, Marshall GM, Haber PS, Cohn SL, Haber M. 1996. Expression of the gene for multidrug-resistance-associated protein and outcome in patients with neuroblastoma. N Engl J Med 334:231–238.
215. Nooter K, Westerman AM, Flens MJ, Zaman GJ, Scheper RJ, van Wingerden KE, Burger H, Oostrum R, Boersma T, Sonneveld P. 1995. Expression of the multidrug resistance–associated protein (MRP) gene in human cancers. Clin Cancer Res 1:1301–1310.
216. Filipits M, Malayeri R, Suchomel RW, Pohl G, Stranzl T, Dekan G, Kaider A, Stiglbauer W, Depisch D, Pirker R. 1999. Expression of the multidrug resistance protein (MRP1) in breast cancer. Anticancer Res 19:5043–5049.
217. Filipits M, Suchomel RW, Dekan G, Stiglbauer W, Haider K, Depisch D, Pirker R. 1997. Expression of the multidrug resistance–associated protein (MRP) gene in colorectal carcinomas. Br J Cancer 75:208–212.
218. Takebayashi Y, Akiyama S, Natsugoe S, Hokita S, Niwa K, Kitazono M, Sumizawa T, Tani A, Furukawa T, Aikou T. 1998. The expression of multidrug resistance protein in human gastrointestinal tract carcinomas. Cancer 82:661–666.
219. Alexander D, Yamamoto T, Kato S, Kasai S. 1999. Histopathological assessment of multidrug resistance in gastric cancer: expression of P-glycoprotein, multidrug resistance–associated protein, and lung-resistance protein. Surg Today 29:401–406.
220. Plaat BE, Hollema H, Molenaar WM, Torn Broers GH, Pijpe J, Mastik MF, Hoekstra HJ, van den Berg E, Scheper RJ, van der Graaf WT. 2000. Soft tissue leiomyosarcomas and malignant gastrointestinal stromal tumors: differences in clinical outcome and expression of multidrug resistance proteins. J Clin Oncol 18:3211–3220.
221. Theou N, Gil S, Devocelle A, Julie C, Lavergne-Slove A, Beauchet A, Callard P, Farinotti R, Le Cesne A, Lemoine A, Faivre-Bonhomme L, Emile JF. 2005. Multidrug resistance proteins in gastrointestinal stromal tumors: site-dependent expression and initial response to imatinib. Clin Cancer Res 11:7593–7598.
222. Abe T, Mori T, Wakabayashi Y, Nakagawa M, Cole SP, Koike K, Kuwano M, Hori S. 1998. Expression of multidrug resistance protein gene in patients with glioma after chemotherapy. J Neurooncol 40:11–18.
223. Mohri M, Nitta H, Yamashita J. 2000. Expression of multidrug resistance-associated protein (MRP) in human gliomas. J Neurooncol 49:105–115.
224. Tews DS, Nissen A, Kulgen C, Gaumann AK. 2000. Drug resistance-associated factors in primary and secondary glioblastomas and their precursor tumors. J Neurooncol 50:227–237.

225. Benyahia B, Huguet S, Decleves X, Mokhtari K, Criniere E, Bernaudin JF, Scherrmann JM, Delattre JY. 2004. Multidrug resistance-associated protein MRP1 expression in human gliomas: chemosensitization to vincristine and etoposide by indomethacin in human glioma cell lines overexpressing MRP1. J Neurooncol 66:65–70.

226. Bronger H, König J, Kopplow K, Steiner HH, Ahmadi R, Herold-Mende C, Keppler D, Nies AT. 2005. ABCC drug efflux pumps and organic anion uptake transporters in human gliomas and the blood–tumor barrier. Cancer Res 65:11419–11428.

227. Calatozzolo C, Gelati M, Ciusani E, Sciacca FL, Pollo B, Cajola L, Marras C, Silvani A, Vitellaro-Zuccarello L, Croci D, Boiardi A, Salmaggi A. 2005. Expression of drug resistance proteins Pgp, MRP1, MRP3, MRP5 and GST-pi in human glioma. J Neurooncol 74:113–121.

228. Nies AT, König J, Pfannschmidt M, Klar E, Hofmann WJ, Keppler D. 2001. Expression of the multidrug resistance proteins MRP2 and MRP3 in human hepatocellular carcinoma. Int J Cancer 94:492–499.

229. Sugawara I, Yamada H, Nakamura H, Sumizawa T, Akiyama S, Masunaga A, Itoyama S. 1995. Preferential expression of the multidrug-resistance-associated protein (MRP) in adenocarcinoma of the lung. Int J Cancer 64:322–325.

230. Kreisholt J, Sorensen M, Jensen PB, Nielsen BS, Andersen CB, Sehested M. 1998. Immunohistochemical detection of DNA topoisomerase IIalpha, P-glycoprotein and multidrug resistance protein (MRP) in small-cell and non-small-cell lung cancer. Br J Cancer 77:1469–1473.

231. Schadendorf D, Makki A, Stahr C, van Dyck A, Wanner R, Scheffer GL, Flens MJ, Scheper R, Henz BM. 1995. Membrane transport proteins associated with drug resistance expressed in human melanoma. Am J Pathol 147:1545–1552.

232. Yokoyama Y, Sato S, Fukushi Y, Sakamoto T, Futagami M, Saito Y. 1999. Significance of multi-drug-resistant proteins in predicting chemotherapy response and prognosis in epithelial ovarian cancer. J Obstet Gynaecol Res 25:387–394.

233. Koshiyama M, Fujii H, Kinezaki M, Morita Y, Nanno H, Yoshida M. 2001. Immunohistochemical expression of topoisomerase IIalpha (Topo IIalpha) and multidrug resistance–associated protein (MRP), plus chemosensitivity testing, as chemotherapeutic indices of ovarian and endometrial carcinomas. Anticancer Res 21:2925–2932.

234. Yakirevich E, Sabo E, Naroditsky I, Sova Y, Lavie O, Resnick MB. 2006. Multidrug resistance-related phenotype and apoptosis-related protein expression in ovarian serous carcinomas. Gynecol Oncol 100:152–159.

235. Arts HJ, Katsaros D, de Vries EG, Massobrio M, Genta F, Danese S, Arisio R, Scheper RJ, Kool M, Scheffer GL, et al. 1999. Drug resistance–associated markers P-glycoprotein, multidrug resistance-associated protein 1, multidrug-resistance–associated protein 2, and lung resistance protein as prognostic factors in ovarian carcinoma. Clin Cancer Res 5:2798–2805.

236. Wilson MW, Fraga CH, Fuller CE, Rodriguez-Galindo C, Mancini J, Hagedorn N, Leggas ML, Stewart CF. 2006. Immunohistochemical detection of multidrug-resistant protein expression in retinoblastoma treated by primary enucleation. Invest Ophthalmol Vis Sci 47:1269–1273.

237. Oda Y, Saito T, Tateishi N, Ohishi Y, Tamiya S, Yamamoto H, Yokoyama R, Uchiumi T, Iwamoto Y, Kuwano M, Tsuneyoshi M. 2005. ATP-binding cassette superfamily transporter gene expression in human soft tissue sarcomas. Int J Cancer 114:854–862.

238. Eid H, Mingfang L, Institoris E, Bodrogi I, Bak M. 2000. MRP expression of testicular cancers and its clinical relevance. Anticancer Res 20:4019–4022.

239. Ruggeri RM, Sciacchitano S, Vitarelli E, Trimarchi F, Barresi G, Trovato M. 2006. Immunoexpression of multidrug-resistance protein 2 and cyclooxygenase 2 in medullary thyroid carcinomas. Arch Pathol Lab Med 130:1014–1019.

240. Oshika Y, Nakamura M, Tokunaga T, Fukushima Y, Abe Y, Ozeki Y, Yamazaki H, Tamaoki N, Ueyama Y. 1998. Multidrug resistance–associated protein and mutant p53 protein expression in non-small cell lung cancer. Mod Pathol 11:1059–1063.

241. Berger W, Setinek U, Hollaus P, Zidek T, Steiner E, Elbling L, Cantonati H, Attems J, Gsur A, Micksche M. 2005. Multidrug resistance markers P-glycoprotein, multidrug resistance protein 1, and lung resistance protein in non-small cell lung cancer: prognostic implications. J Cancer Res Clin Oncol 131:355–363.

242. Nooter K, Brutel DLR, Look MP, van Wingerden KE, Henzen-Logmans SC, Scheper RJ, Flens MJ, Klijn JG, Stoter G, Foekens JA. 1997. The prognostic significance of expression of the multidrug resistance–associated protein (MRP) in primary breast cancer. Br J Cancer 76:486–493.

243. Rudas M, Filipits M, Taucher S, Stranzl T, Steger GG, Jakesz R, Pirker R, Pohl G. 2003. Expression of MRP1, LRP and Pgp in breast carcinoma patients treated with preoperative chemotherapy. Breast Cancer Res Treat 81:149–157.

244. Sullivan GF, Amenta PS, Villanueva JD, Alvarez CJ, Yang JM, Hait WN. 1998. The expression of drug resistance gene products during the progression of human prostate cancer. Clin Cancer Res 4:1393–1403.

245. Poulain S, Lepelley P, Preudhomme C, Cambier N, Cornillon J, Wattel E, Cosson A, Fenaux P. 2000. Expression of the multidrug resistance-associated protein in myelodysplastic syndromes. Br J Haematol 110:591–598.

246. Filipits M, Stranzl T, Pohl G, Suchomel RW, Zochbauer S, Brunner R, Lechner K, Pirker R. 1999. MRP expression in acute myeloid leukemia: an update. Adv Exp Med Biol 457:141–150.

247. Sasaki M, Suzuki H, Ito K, Abe T, Sugiyama Y. 2002. Transcellular transport of organic anions across a double-transfected Madin–Darby canine kidney II cell monolayer expressing both human organic anion-transporting polypeptide (OATP2/SLC21A6) and multidrug resistance-associated protein 2 (MRP2/ABCC2). J Biol Chem 277:6497–6503.

248. Keppler D. 2005. Uptake and efflux transporters for conjugates in human hepatocytes. Methods Enzymol 400:531–542.

249. Walgren RA, Karnaky KJ Jr, Lindenmayer GE, Walle T. 2000. Efflux of dietary flavonoid quercetin 4′-beta-glucoside across human intestinal Caco-2 cell monolayers by apical multidrug resistance-associated protein-2. J Pharmacol Exp Ther 294:830–836.

250. Horio M, Pastan I, Gottesman MM, Handler JS. 1990. Transepithelial transport of vinblastine by kidney-derived cell lines. Application of a new kinetic model to estimate in situ K_m of the pump. Biochim Biophys Acta 1027:116–122.

251. Abe T, Kakyo M, Tokui T, Nakagomi R, Nishio T, Nakai D, Nomura H, Unno M, Suzuki M, Naitoh T, Matsuno S, Yawo H. 1999. Identification of a novel gene family encoding human liver-specific organic anion transporter LST-1. J Biol Chem 274:17159–17163.

252. Hsiang B, Zhu Y, Wang Z, Wu Y, Sasseville V, Yang WP, Kirchgessner TG. 1999. A novel human hepatic organic anion transporting polypeptide (OATP2). Identification of a liver-specific human organic anion transporting polypeptide and identification of rat

and human hydroxymethylglutaryl-CoA reductase inhibitor transporters. J Biol Chem 274:37161–37168.

253. König J, Cui Y, Nies AT, Keppler D. 2000. A novel human organic anion transporting polypeptide localized to the basolateral hepatocyte membrane. Am J Physiol Gastrointest Liver Physiol 278:G156–G164.

254. König J, Cui Y, Nies AT, Keppler D. 2000. Localization and genomic organization of a new hepatocellular organic anion transporting polypeptide. J Biol Chem 275:23161–23168.

255. Kullak-Ublick GA, Ismair MG, Stieger B, Landmann L, Huber R, Pizzagalli F, Fattinger K, Meier PJ, Hagenbuch B. 2001. Organic anion-transporting polypeptide B (OATP-B) and its functional comparison with three other OATPs of human liver. Gastroenterology 120:525–533.

256. Tamai I, Nezu J, Uchino H, Sai Y, Oku A, Shimane M, Tsuji A. 2000. Molecular identification and characterization of novel members of the human organic anion transporter (OATP) family. Biochem Biophys Res Commun 273:251–260.

257. Koepsell H, Endou H. 2004. The SLC22 drug transporter family. Pflugers Arch 447:666–676.

258. Mita S, Suzuki H, Akita H, Stieger B, Meier PJ, Hofmann AF, Sugiyama Y. 2005. Vectorial transport of bile salts across MDCK cells expressing both rat Na^+-taurocholate cotransporting polypeptide and rat bile salt export pump. Am J Physiol Gastrointest Liver Physiol 288:G159–G167.

259. Mita S, Suzuki H, Akita H, Hayashi H, Onuki R, Hofmann AF, Sugiyama Y. 2006. Vectorial transport of unconjugated and conjugated bile salts by monolayers of LLC-PK1 cells doubly transfected with human NTCP and BSEP or with rat Ntcp and Bsep. Am J Physiol Gastrointest Liver Physiol 290:G550–G556.

260. Matsushima S, Maeda K, Kondo C, Hirano M, Sasaki M, Suzuki H, Sugiyama Y. 2005. Identification of the hepatic efflux transporters of organic anions using double-transfected Madin–Darby canine kidney II cells expressing human organic anion-transporting polypeptide 1B1 (OATP1B1)/multidrug resistance–associated protein 2, OATP1B1/multidrug resistance 1, and OATP1B1/breast cancer resistance protein. J Pharmacol Exp Ther 314:1059–1067.

261. Sherry ST, Ward MH, Kholodov M, Baker J, Phan L, Smigielski EM, Sirotkin K. 2001. dbSNP: the NCBI database of genetic variation. Nucleic Acids Res 29:308–311.

262. The International HapMap Consortium. 2005. A haplotype map of the human genome. Nature 437:1299–1320.

263. den Dunnen JT, Antonarakis SE. 2000. Mutation nomenclature extensions and suggestions to describe complex mutations: a discussion. Hum Mutat 15:7–12.

264. Riordan JR. 1999. Cystic fibrosis as a disease of misprocessing of the cystic fibrosis transmembrane conductance regulator glycoprotein. Am J Hum Genet 64:1499–1504.

265. Bryan J, Munoz A, Zhang X, Dufer M, Drews G, Krippeit-Drews P, Aguilar-Bryan L. 2007. ABCC8 and ABCC9: ABC transporters that regulate K(+) channels. Pflugers Arch 453:703–718.

266. Bienengraeber M, Olson TM, Selivanov VA, Kathmann EC, O'Cochlain F, Gao F, Karger AB, Ballew JD, Hodgson DM, Zingman LV, et al. 2004. ABCC9 mutations identified in human dilated cardiomyopathy disrupt catalytic KATP channel gating. Nat Genet 36:382–387.

267. Kruglyak L. 2005. Power tools for human genetics. Nat Genet 37:1299–1300.

268. Yoshiura K, Kinoshita A, Ishida T, Ninokata A, Ishikawa T, Kaname T, Bannai M, Tokunaga K, Sonoda S, Komaki R, et al. 2006. A SNP in the ABCC11 gene is the determinant of human earwax type. Nat Genet 38:324–330.

269. Ramensky V, Bork P, Sunyaev S. 2002. Human non-synonymous SNPs: server and survey. Nucleic Acids Res 30:3894–3900.

270. Shoda J, Suzuki H, Suzuki H, Sugiyama Y, Hirouchi M, Utsunomiya H, Oda K, Kawamoto T, Matsuzaki Y, Tanaka N. 2003. Novel mutations identified in the human multidrug resistance–associated protein 2 (MRP2/ABCC2) gene in a Japanese patient with Dubin–Johnson syndrome. Hepatol Res 27:323–326.

271. Mor-Cohen R, Zivelin A, Rosenberg N, Goldberg I, Seligsohn U. 2005. A novel ancestral splicing mutation in the multidrug resistance protein 2 gene causes Dubin–Johnson syndrome in Ashkenazi Jewish patients. Hepatol Res 31:104–111.

272. Machida I, Wakusawa S, Sanae F, Hayashi H, Kusakabe A, Ninomiya H, Yano M, Yoshioka K. 2005. Mutational analysis of the MRP2 gene and long-term follow-up of Dubin–Johnson syndrome in Japan. J Gastroenterol 40:366–370.

273. Cebecauerova D, Jirasek T, Budisova L, Mandys V, Volf V, Novotna Z, Subhanova I, Hrebicek M, Elleder M, Jirsa M. 2005. Dual hereditary jaundice: simultaneous occurrence of mutations causing Gilbert's and Dubin–Johnson syndrome. Gastroenterology 129:315–320.

274. Lee JH, Chen HL, Chen HL, Ni YH, Hsu HY, Chang MH. 2006. Neonatal Dubin–Johnson syndrome: long-term follow-up and MRP2 mutations study. Pediatr Res 59:584–589.

275. Wada M, Toh S, Taniguchi K, Nakamura T, Uchiumi T, Kohno K, Yoshida I, Kimura A, Sakisaka S, Adachi Y, Kuwano M. 1998. Mutations in the canalicular multispecific organic anion transporter (cMOAT) gene, a novel ABC transporter, in patients with hyperbilirubinemia II/Dubin–Johnson syndrome. Hum Mol Genet 7:203–207.

276. Kajihara S, Hisatomi A, Mizuta T, Hara T, Ozaki I, Wada I, Yamamoto K. 1998. A splice mutation in the human canalicular multispecific organic anion transporter gene causes Dubin–Johnson syndrome. Biochem Biophys Res Commun 253:454–457.

277. Wakusawa S, Machida I, Suzuki S, Hayashi H, Yano M, Yoshioka K. 2003. Identification of a novel 2026G→C mutation of the MRP2 gene in a Japanese patient with Dubin–Johnson syndrome. J Hum Genet 48:425–429.

278. Machida I, Inagaki Y, Suzuki S, Hayashi H, Wakusawa S. 2004. Mutation analysis of the multidrug resistance protein 2 (MRP2) gene in a Japanese patient with Dubin–Johnson syndrome. Hepatol Res 30:86–90.

279. Mor-Cohen R, Zivelin A, Rosenberg N, Shani M, Muallem S, Seligsohn U. 2001. Identification and functional analysis of two novel mutations in the multidrug resistance protein 2 gene in Israeli patients with Dubin–Johnson syndrome. J Biol Chem 276:36923–36930.

280. Roy Chowdhury N, Arias IM, Wolkoff AW, Roy Chowdhury J. 2001. Disorders of bilirubin metabolism. In: Arias IM, Boyer JL, Chisari FV, Fausto N, Schachter D, Shafritz DA, editors. *The Liver: Biology and Pathobiology*. Philadelphia, PA: Lippincott Williams & Wilkins, pp 291–309.

281. Shani M, Seligsohn U, Gilon E, Sheba C, Adam A. 1970. Dubin–Johnson syndrome in Israel: I. Clinical, laboratory, and genetic aspects of 101 cases. J Med 39:549–567.

282. Tate G, Li M, Suzuki T, Mitsuya T. 2002. A new mutation of the ATP-binding cassette, subfamily C, member 2 (ABCC2) gene in a Japanese patient with Dubin–Johnson syndrome. Genes Genet Syst 77:117–121.

283. Thermann R, Neu-Yilik G, Deters A, Frede U, Wehr K, Hagemeier C, Hentze MW, Kulozik AE. 1998. Binary specifications of nonsense codons by splicing and cytoplasmic translation. EMBO J 17:3484–3494.

284. Keitel V, Kartenbeck J, Nies AT, Spring H, Brom M, Keppler D. 2000. Impaired protein maturation of the conjugate export pump multidrug resistance protein 2 as a consequence of a deletion mutation in Dubin–Johnson syndrome. Hepatology 32:1317–1328.

285. Hashimoto K, Uchiumi T, Konno T, Ebihara T, Nakamura T, Wada M, Sakisaka S, Maniwa F, Amachi T, Ueda K, Kuwano M. 2002. Trafficking and functional defects by mutations of the ATP-binding domains in MRP2 in patients with Dubin–Johnson syndrome. Hepatology 36:1236–1245.

286. Keitel V, Nies AT, Brom M, Hummel-Eisenbeiss J, Spring H, Keppler D. 2003. A common Dubin–Johnson syndrome mutation impairs protein maturation and transport activity of MRP2 (ABCC2). Am J Physiol Gastrointest Liver Physiol 284:G165–G174.

287. Ito K, Suzuki H, Hirohashi T, Kume K, Shimizu T, Sugiyama Y. 1997. Molecular cloning of canalicular multispecific organic anion transporter defective in EHBR. Am J Physiol 272:G16–G22.

288. Hirouchi M, Suzuki H, Sugiyama Y. 2005. Treatment of hyperbilirubinemia in Eisai hyperbilirubinemic rat by transfecting human MRP2/ABCC2 gene. Pharm Res 22:661–666.

289. Chu X, Strauss JR, Mariano MA, Li J, Newton DJ, Cai X, Wang RW, Yabut J, Hartley DP, Evans DC, Evers R. 2006. Characterization of mice lacking the multidrug resistance protein Mrp2 (Abcc2). J Pharmacol Exp Ther 317:579–589.

290. Nezasa K, Tian X, Zamek-Gliszczynski MJ, Patel NJ, Raub TJ, Brouwer KL. 2006. Altered hepatobiliary disposition of 5 (and 6)-carboxy-2′,7′-dichlorofluorescein in Abcg2 (Bcrp1) and Abcc2 (Mrp2) knockout mice. Drug Metab Dispos 34:718–723.

291. Vlaming ML, Mohrmann K, Wagenaar E, de Waart DR, Oude Elferink RP, Lagas JS, van Tellingen O, Vainchtein LD, Rosing H, Beijnen JH, Schellens JH, Schinkel AH. 2006. Carcinogen and anti-cancer drug transport by Mrp2 in vivo: studies using Mrp2 (Abcc2) knockout mice. J Pharmacol Exp Ther 318:319–327.

292. Suzuki H, Sugiyama Y. 2002. Single nucleotide polymorphisms in multidrug resistance associated protein 2 (MRP2/ABCC2): its impact on drug disposition. Adv Drug Deliv Rev 54:1311–1331.

293. Itoda M, Saito Y, Soyama A, Saeki M, Murayama N, Ishida S, Sai K, Nagano M, Suzuki H, Sugiyama Y, Ozawa S, Sawada J. 2002. Polymorphisms in the ABCC2 (cMOAT/MRP2) gene found in 72 established cell lines derived from Japanese individuals: an association between single nucleotide polymorphisms in the 5′-untranslated region and exon 28. Drug Metab Dispos 30:363–364.

294. Colombo S, Soranzo N, Rotger M, Sprenger R, Bleiber G, Furrer H, Buclin T, Goldstein D, Decosterd L, Telenti A. 2005. Influence of ABCB1, ABCC1, ABCC2, and ABCG2 haplotypes on the cellular exposure of nelfinavir in vivo. Pharmacogenet Genom 15:599–608.

295. Leschziner G, Zabaneh D, Pirmohamed M, Owen A, Rogers J, Coffey AJ, Balding DJ, Bentley DB, Johnson MR. 2006. Exon sequencing and high resolution haplotype analysis of ABC transporter genes implicated in drug resistance. Pharmacogenet Genom 16:439–450.

296. Ito S, Ieiri I, Tanabe M, Suzuki A, Higuchi S, Otsubo K. 2001. Polymorphisms of the ABC transporter genes, MDR1, MRP1 and MRP2/cMOAT, in healthy Japanese subjects. Pharmacogenetics 11:175–184.

297. Cascorbi I. 2006. Role of pharmacogenetics of ATP-binding cassette transporters in the pharmacokinetics of drugs. Pharmacol Ther DOI:10.1016/j.pharmthera.2006.04.009.

298. Hulot JS, Villard E, Maguy A, Morel V, Mir L, Tostivint I, William-Faltaos D, Fernandez C, Hatem S, Deray G, et al. 2005. A mutation in the drug transporter gene ABCC2 associated with impaired methotrexate elimination. Pharmacogenet Genom 15:277–285.

299. Hirouchi M, Suzuki H, Itoda M, Ozawa S, Sawada J, Ieiri I, Ohtsubo K, Sugiyama Y. 2004. Characterization of the cellular localization, expression level, and function of SNP variants of MRP2/ABCC2. Pharm Res 21:742–748.

300. Meier Y, Pauli-Magnus C, Zanger UM, Klein K, Schaeffeler E, Nussler AK, Nussler N, Eichelbaum M, Meier PJ, Stieger B. 2006. Interindividual variability of canalicular ATP-binding-cassette (ABC)-transporter expression in human liver. Hepatology 44:62–74.

301. Ito K, Oleschuk CJ, Westlake C, Vasa MZ, Deeley RG, Cole SP. 2001. Mutation of Trp1254 in the multispecific organic anion transporter, multidrug resistance protein 2 (MRP2) (ABCC2), alters substrate specificity and results in loss of methotrexate transport activity. J Biol Chem 276:38108–38114.

302. Ito K, Suzuki H, Sugiyama Y. 2001. Charged amino acids in the transmembrane domains are involved in the determination of the substrate specificity of rat Mrp2. Mol Pharmacol 59:1077–1085.

303. Ryu S, Kawabe T, Nada S, Yamaguchi A. 2000. Identification of basic residues involved in drug export function of human multidrug resistance–associated protein 2. J Biol Chem 275:39617–39624.

304. Ito K, Suzuki H, Sugiyama Y. 2001. Single amino acid substitution of rat MRP2 results in acquired transport activity for taurocholate. Am J Physiol 281:G1034–G1043.

305. Bergen AA, Plomp AS, Schuurman AJ, Terry S, Breuning M, Dauwerse H, Swart J, Kool M, van Soest S, Baas F, ten Brink JB, de Jong PT. 2000. Mutations in ABCC6 cause pseudoxanthoma elasticum. Nat Genet 25:228–231.

306. Le Saux O, Urban Z, Tschuch C, Csiszar K, Bacchelli B, Quaglino D, Pasquali-Ronchetti I, Pope FM, Richards A, Terry S, et al. 2000. Mutations in a gene encoding an ABC transporter cause pseudoxanthoma elasticum. Nat Genet 25:223–227.

307. Ringpfeil F, Lebwohl MG, Christiano AM, Uitto J. 2000. Pseudoxanthoma elasticum: mutations in the MRP6 gene encoding a transmembrane ATP-binding cassette (ABC) transporter. Proc Natl Acad Sci U S A 97:6001–6006.

308. Groenblad E. 1929. Angioid streaks: pseudoxanthoma elasticum, vorlaufige mitteilung. Acta Opthalmol 7.

309. Strandberg J. 1929. Pseudoxanthoma elasticum. Z Haut Geschlechtskr 31:689.

310. Yap EY, Gleaton MS, Buettner H. 1992. Visual loss associated with pseudoxanthoma elasticum. Retina 12:315–319.

311. Mendelsohn G, Bulkley BH, Hutchins GM. 1978. Cardiovascular manifestations of Pseudoxanthoma elasticum. Arch Pathol Lab Med 102:298–302.

312. Lebwohl M, Schwartz E, Lemlich G, Lovelace O, Shaikh-Bahai F, Fleischmajer R. 1993. Abnormalities of connective tissue components in lesional and non-lesional tissue of patients with pseudoxanthoma elasticum. Arch Dermatol Res 285:121–126.

313. Plomp AS, Hu X, de Jong PT, Bergen AA. 2004. Does autosomal dominant pseudoxanthoma elasticum exist? Am J Med Genet A 126:403–412.

314. Struk B, Cai L, Zach S, Ji W, Chung J, Lumsden A, Stumm M, Huber M, Schaen L, Kim CA, et al. 2000. Mutations of the gene encoding the transmembrane transporter protein ABC-C6 cause pseudoxanthoma elasticum. J Mol Med 78:282–286.

315. Cai L, Lumsden A, Guenther UP, Neldner SA, Zach S, Knoblauch H, Ramesar R, Hohl D, Callen DF, Neldner KH, et al. 2001. A novel Q378X mutation exists in the transmembrane transporter protein ABCC6 and its pseudogene: implications for mutation analysis in pseudoxanthoma elasticum. J Mol Med 79:536–546.

316. Meloni I, Rubegni P, De Aloe G, Bruttini M, Pianigiani E, Cusano R, Seri M, Mondillo S, Federico A, Bardelli AM, et al. 2001. Pseudoxanthoma elasticum: point mutations in the ABCC6 gene and a large deletion including also ABCC1 and MYH11. Hum Mutat 18:85.

317. Le Saux O, Beck K, Sachsinger C, Silvestri C, Treiber C, Goring HH, Johnson EW, de Paepe A, Pope FM, Pasquali-Ronchetti I, et al. 2001. A spectrum of ABCC6 mutations is responsible for pseudoxanthoma elasticum. Am J Hum Genet 69:749–764.

318. Pulkkinen L, Nakano A, Ringpfeil F, Uitto J. 2001. Identification of ABCC6 pseudogenes on human chromosome 16p: implications for mutation detection in pseudoxanthoma elasticum. Hum Genet 109:356–365.

319. Ringpfeil F, Nakano A, Uitto J, Pulkkinen L. 2001. Compound heterozygosity for a recurrent 16.5-kb Alu-mediated deletion mutation and single-base-pair substitutions in the ABCC6 gene results in pseudoxanthoma elasticum. Am J Hum Genet 68:642–652.

320. Hu X, Plomp A, Wijnholds J, Ten Brink J, van Soest S, van den Born LI, Leys A, Peek R, de Jong PT, Bergen AA. 2003. ABCC6/MRP6 mutations: further insight into the molecular pathology of pseudoxanthoma elasticum. Eur J Hum Genet 11:215–224.

321. Chassaing N, Martin L, Mazereeuw J, Barrie L, Nizard S, Bonafe JL, Calvas P, Hovnanian A. 2004. Novel ABCC6 mutations in pseudoxanthoma elasticum. J Invest Dermatol 122:608–613.

322. Gheduzzi D, Guidetti R, Anzivino C, Tarugi P, Di Leo E, Quaglino D, Ronchetti IP. 2004. ABCC6 mutations in Italian families affected by pseudoxanthoma elasticum (PXE). Hum Mutat 24:438–439.

323. Miksch S, Lumsden A, Guenther UP, Foernzler D, Christen-Zach S, Daugherty C, Ramesar RK, Lebwohl M, Hohl D, Neldner KH, et al. 2005. Molecular genetics of pseudoxanthoma elasticum: type and frequency of mutations in ABCC6. Hum Mutat 26:235–248.

324. Gorgels TG, Hu X, Scheffer GL, van der Wal AC, Toonstra J, de Jong PT, van Kuppevelt TH, Levelt CN, de Wolf A, Loves WJ, et al. 2005. Disruption of Abcc6 in the mouse: novel insight in the pathogenesis of pseudoxanthoma elasticum. Hum Mol Genet 14:1763–1773.

325. Perdu J, Germain DP. 2001. Identification of novel polymorphisms in the pM5 and MRP1 (ABCC1) genes at locus 16p13.1 and exclusion of both genes as responsible for pseudoxanthoma elasticum. Hum Mutat 17:74–75.

326. Conrad S, Kauffmann HM, Ito K, Leslie EM, Deeley RG, Schrenk D, Cole SP. 2002. A naturally occurring mutation in MRP1 results in a selective decrease in organic anion transport and in increased doxorubicin resistance. Pharmacogenetics 12:321–330.

327. Wang Z, Sew PH, Ambrose H, Ryan S, Chong SS, Lee EJ, Lee CG. 2006. Nucleotide sequence analyses of the MRP1 gene in four populations suggest negative selection on its coding region. BMC Genom 7:111.

328. Letourneau IJ, Deeley RG, Cole SP. 2005. Functional characterization of non-synonymous single nucleotide polymorphisms in the gene encoding human multidrug resistance protein 1 (MRP1/ABCC1). Pharmacogenet Genom 15:647–657.

329. Zelcer N, van de Wetering K, de Waart R, Scheffer GL, Marschall HU, Wielinga PR, Kuil A, Kunne C, Smith A, van der Valk M, Wijnholds J, Elferink RO, Borst P. 2006. Mice lacking Mrp3 (Abcc3) have normal bile salt transport, but altered hepatic transport of endogenous glucuronides. J Hepatol 44:768–775.

330. Leslie EM, Letourneau IJ, Deeley RG, Cole SP. 2003. Functional and structural consequences of cysteine substitutions in the NH2 proximal region of the human multidrug resistance protein 1 (MRP1/ABCC1). Biochemistry 42:5214–5224.

331. Conrad S, Kauffmann HM, Ito K, Deeley RG, Cole SP, Schrenk D. 2001. Identification of human multidrug resistance protein 1 (MRP1) mutations and characterization of a G671V substitution. J Hum Genet 46:656–663.

332. Wojnowski L, Kulle B, Schirmer M, Schluter G, Schmidt A, Rosenberger A, Vonhof S, Bickeboller H, Toliat MR, Suk EK, et al. 2005. NAD(P)H oxidase and multidrug resistance protein genetic polymorphisms are associated with doxorubicin-induced cardiotoxicity. Circulation 112:3754–3762.

333. Deeley RG, Westlake C, Cole SP. 2006. Transmembrane transport of endo- and xenobiotics by mammalian ATP-binding cassette multidrug resistance proteins. Physiol Rev 86:849–899.

334. Rosenberg MF, Mao Q, Holzenburg A, Ford RC, Deeley RG, Cole SP. 2001. The structure of the multidrug resistance protein 1 (MRP1/ABCC1): crystallization and single-particle analysis. J Biol Chem 276:16076–16082.

12

BREAST CANCER RESISTANCE PROTEIN

ROBERT W. ROBEY, ORSOLYA POLGAR, JOHN DEEKEN, KENNETH K. W. TO, AND SUSAN E. BATES

National Cancer Institute, Bethesda, Maryland

Drug Transporters: Molecular Characterization and Role in Drug Disposition, Edited by Guofeng You and Marilyn E. Morris
Copyright © 2007 John Wiley & Sons, Inc.

12.1. INTRODUCTION

As with most transporters associated with drug resistance, the second member of the G subfamily of adenosine triphosphate (ATP)-binding cassette transporters, ABCG2, was first identified as an unusual phenotype, as several groups began reporting non-P-glycoprotein-(Pgp) and non-multidrug resistance protein 1 (MRP1)-mediated drug resistance in cell lines selected with mitoxantrone.[1–5] In addition to high levels of mitoxantrone resistance, these cell lines exhibited resistance to doxorubicin and etoposide but lacked resistance to vinblastine or cisplatin. Further studies of the mitoxantrone-selected breast cancer line, MCF-7/MX, revealed ATP-dependent transport of radioactive mitoxantrone.[3] High levels of cross-resistance to mitoxantrone in the MCF-7/MX line were accompanied by resistance to the camptothecin analogs topotecan, 9-aminocamptothecin, CPT-11, and SN-38, the active metabolite of CPT-11.[6] Resistance to the camptothecin analogs was not caused by alterations in topoisomerase I levels.[6] Mounting evidence suggested the existence of an unknown, energy-dependent transport mechanism in these cells.

Around the same time that the mitoxantrone-resistant cells were described, a nearly identical phenotype was noted in a breast cancer cell line developed in the Fojo laboratory by selection with adriamycin in the presence of verapamil to prevent overexpression of Ppg.[7] These cells, MCF-7/AdrVp, also displayed ATP-dependent transport of adriamycin and rhodamine 123 in the absence of Pgp or MRP1.[8] This cell line was used by Ross and colleagues to clone the novel transporter, calling it the breast cancer resistance protein (BCRP), since it was cloned from the MCF-7/AdrVp subline.[9] Shortly after the discovery of Doyle et al.[9], a nearly identical transporter, termed ABCP for "ABC transporter highly expressed in placenta," was reported by Allikmets et al. after a search of expressed sequence tag databases.[10] And our laboratory cloned a cDNA from the mitoxantrone-selected human colon carcinoma cell line S1-M1-80, derived from the S1-M1-3.2 line first reported by Rabindran et al.[11] We named the gene MXR, for "mitoxantrone resistance."[12]

Subsequent to the cloning of BCRP/ABCP/MXR, the Human Genome Nomenclature Committee assigned the name ABCG2, making it the second member of the G subfamily of ABC transporters. This terminology is used throughout the chapter. The G subfamily of transporters is made up of six half-transporters[13]: two involved in cholesterol transport, ABCG1, the human homolog of the *Drosophila* white protein, and ABCG4[14]; one currently found only in rodents, ABCG3, which appears to have an aberrant ATP-binding domain[15]; and two that heterodimerize with each other to form a functional sterol transporter, ABCG5 and ABCG8.[16]

Laboratories around the world have studied ABCG2, reporting an ever-expanding list of substrates and inhibitors. Observations regarding protein structure, function, and regulation of expression have also appeared. In this chapter we describe the highlights of our current understanding. With well over 400 publications involving ABCG2, a complete compendium of work on the transporter is no longer possible in a few pages.

12.2. GENETICS OF ABCG2

The *ABCG2* gene spans over 66 kb and consists of 16 exons and 15 introns. Exons range in size from 60 to 532 bp, with the translational start site found in the second exon, the Walker A site in exon 3, and the ABC signature motif in exon 6. Characterization of the *ABCG2* promoter revealed that it is TATA-less with multiple Sp1, AP1, and AP2 sites, much like the promoters for other members of the ABC transporter superfamily, such as the *MDR1* (*ABCB1*), *MRP1* (*ABCC1*), and *ABCG1* genes, which all lack TATA boxes and have multiple Sp1 sites.[17] Phylogenetic analysis of the amino acid sequence of ABCG2 compared to other ABC transporters indicates that it is distantly related to ABCB1(Pgp) and ABCC1, but with 29.3% identity it is closely related to ABCG1, the human homolog of the *Drosophila* white protein.[9] Initial evaluation of the promoter activity of the *ABCG2* gene was performed using a series of constructs made from the *ABCG2* 5 upstream region in a luciferase reporter assay. A region 312 bp directly upstream from the transcriptional start site conferred basal promoter activity, with possible positive regulatory elements between –1285 and –628 bp and negative regulatory elements identified in the region between –628 and –312 bp.[17]

We mapped the *ABCG2* gene to human chromosome 4q22, between the markers D4S2462 and D4S1557.[18] Fluorescence in situ hybridization (FISH) studies with a bacterial artificial chromosome probe containing *ABCG2* confirmed localization of the gene at 4q21–4q22 in cells with a normal chromosome 4.[18] Cytogenetic studies in cell lines with high levels of ABCG2 expression revealed multiple rearrangements involving chromosome 4. The most common, although not exclusive, mechanism of ABCG2 overexpression in vitro is gene amplification. In the MCF-7/MX and MCF-7/AdVp3000 cell lines, an amplification peak was found by comparative genomic hybridization studies, and FISH studies with a bacterial artificial chromosome confirmed amplification of the *ABCG2* gene.[18] In the S1-M1-80 cell line, however, only a balanced chromosome 4 and 17 translocation was observed. In a series of mitoxantrone-selected SF295 glioblastoma cell lines, we found that ABCG2 was amplified in the form of double-minute chromosomes and that increasing the concentration of the selecting drug promoted reintegration of the double minutes into multiple sites.[19] The absence of coamplification in other chromosomal regions is consistent with the theory that ABCG2 functions as a homodimer or higher-order multimer.

12.3. PROTEIN STRUCTURE AND FUNCTION

ABCG2, composed of 655 amino acids, runs as a 72-kDa protein on SDS-PAGE and is organized into two main structural elements: an N-terminal ATP-binding domain (NBD) and a C-terminal transmembrane domain (TMD). This particular domain organization, also seen in other members of the ABCG subfamily, is reversed when compared to most other human ABC transporters, where the N-terminal transmembrane segments are followed by the C-terminal NBD. It has previously been demonstrated that eukaryotic ABC transporters require at least two nucleotide-binding domains and two transmembrane domains for transport activity.[20] ABCG2 is considered a half-transporter and is thought to homodimerize in order to function, as opposed to other ABC half-transporters, which engage in heterodimeric association to form functional transporters, a well-characterized example of which is the ABCG5/ABCG8 heterodimer.[21] ABCG2 homodimerization is supported by the fact that when expressed in Sf9 insect cells, where a heterodimeric partner is unlikely, the protein is fully functional. Chimeric fusion proteins containing two ABCG2 monomers either with or without a flexible linker peptide were also shown to be functional, endorsing the idea of homodimer formation.[22] In addition, coimmunoprecipitation experiments using two different tags on the ABCG2 monomers also suggested homodimer formation.[23,24] In general, little is known about the structure and dimerization of ABCG2. The most accurate structural information could be obtained from high-resolution crystal structures; unfortunately, no human ABC transporter has been crystallized to date, and the crystal structure of only a few full-length bacterial ABC transporters is available: the lipid A transporter MsbA from *Escherichia coli*,[25] *Vibrio cholera*,[26] and *Salmonella typhimurium*,[27] and the vitamin B_{12} importer BtuCD from *E. coli*.[28] We should note, however, that the MsbA structures were recently retracted.

The transmembrane domain of ABCG2, composed of residues 361 to 655, is predicted to have six transmembrane segments (TMs) and an extracellular loop between transmembrane helices 5 and 6. This extracellular loop contains cysteine 603, which has been demonstrated to form a symmetrical intermolecular disulfide bond in the ABCG2 homodimer.[29] According to studies by multiple groups, this disulfide bridge is not essential for transport and trafficking to the plasma membrane, given that the C603A mutant displayed wild type–like characteristics. On the other hand, it has been suggested that the other two cysteines of this extracellular loop, C592 and C608, form intramolecular disulfide bonds, and that mutating these residues resulted in impaired localization and function.[29]

There are three putative N-linked glycosylation sites in the ABCG2 TMD (residues 418, 557, and 596), of which only asparagine 596 was shown to be glycosylated, and this posttranslational modification seems not to be required for proper localization and function of the transporter.[30] Asparagine 596 is located in the loop between TMs 5 and 6, helping to substantiate the prediction that this part of the protein is localized extracellularly.

Sequence alignments with the previously mentioned bacterial ABC transporters of known structure reveal a very low overall similarity for the ABCG2 TMD. This low sequence similarity, together with the reversed orientation of ABCG2 (whose consequences are not yet known), makes it difficult to speculate on the three-dimensional structure of the ABCG2 TMD. According to a recently generated computer model,[31] a cone-shaped large central cavity is formed when the 12 TMs, six from each monomer, come together with twofold symmetry, similar to that seen in the MsbA homodimer. Multiple points of interaction exist between the two monomers. One such interface might be formed by the GXXXG motif in TM1 of ABCG2. This motif consists of two glycines separated by any three amino acids and has been shown to be involved in the dimerization of other membrane proteins,[32] the most well-characterized example of which is the glycophorin A homodimer.[33] As the helix turns, the two small glycine residues end up on the same surface and could be involved in tight packing of the helices in a dimer or higher-order multimer. Mutating these glycines to leucines had a marked effect on ABCG2 function, consistent with impaired dimerization.[34] However, a role for these residues in substrate binding cannot be excluded without an accurate model. Although in glycophorin A the GXXXG motif interacts with the same motif in the dimerization partner, in view of the above-mentioned twofold symmetry seen in other ABC half-transporters, it seems unlikely that the TM1 segments from two ABCG2 monomers would come together directly to form an interface in the homodimer. Twofold symmetry has also been observed in the packing of transmembrane helices of full-length transporters, an example of which is TMs 5 and 8 of Pgp being close together on the cytoplasmic side of the membrane in cross-linking studies following cysteine-scanning mutagenesis.[35] Thus, an alternative explanation for the role of GXXXG motif in TM1 of ABCG2 could be its involvement in forming a higher-order oligomer, given that in one study, ABCG2 was suggested primarily to form tetramers, with the possibility of dodecamers as the functional unit.[36]

Another region potentially involved in dimerization is a conserved three-amino acid sequence in TM5 (residues 552 to 554). Mutations in the corresponding residues in the *Drosophila* white protein (an ortholog of ABCG2) are thought to disrupt heterodimerization.[37] Further, the ABCG8 G574E mutant, which corresponds to amino acid 551 of ABCG2, was also characterized as interfering with dimerization. However, the ABCG8 G574R mutant, although demonstrating some effect on ABCG8 maturation, did not prevent formation of the ABCG5/ABCG8 heterodimer. In the case of ABCG2, substituting glycine 553 with either leucine or glutamic acid resulted in decreased protein expression, impaired glycosylation, and retention in the endoplasmic recticulum (ER).[38] Furthermore, the G553L mutant was not rescued from the ER when cotransfected with the wild-type protein, suggesting that no mutant or wild-type dimers were formed. Again, these findings are consistent with impairment in dimerization, although other explanations, such as improper folding, cannot be excluded. On the other hand, when cotransfected with wild type, the inactive L554P mutant partially reversed drug resistance in PA317 cells, a result that implied that residue 554 is critical for function; yet, mutating this residue does not prevent dimerization.[24] Altogether, these findings suggest a critical role for this region of the fifth transmembrane helix.

Little is known about which residues of the ABCG2 TMD bind and translocate its substrates. Similar to P-glycoprotein (ABCB1),[39,40] at least two drug-binding sites within a larger drug-binding pocket have been suggested. This model is based on the fact that certain substrates can abolish labeling with the photoaffinity analog IAAP, whereas others have no effect.[41] Arginine 482, predicted to localize in TM3 near the membrane–cytoplasm interface, is speculated to be part of one of these drug-binding sites, given that replacing this residue by any nonpositively charged residue results in wider substrate specificity.[42,43] The R482G and T variants were found in drug-resistant cancer cell lines and described as gain-of-function mutants, with the addition of substrates such as anthracyclines and rhodamine 123 to drugs transported by the wild-type protein.[44] However, no single-nucleotide polymorphisms (SNPs) were identified at this position, and methotrexate, for example, was found to be transported only by the wild-type protein.[45] The latter findings support a hypothesis that an arginine at this position is required to transport a yet unidentified physiologic substrate of ABCG2, giving the arginine a selective advantage, despite the gain-of-function effect observed when replacing it with multiple other residues.

The nucleotide-binding domain of ABCG2 shows greater sequence similarity to other ABC transporters than does TM domain. This segment of the protein contains the characteristic Walker A and Walker B motifs and the C signature sequence that defines ABC transporters. Mutating a conserved lysine to methionine (K86M) in the Walker A motif, as in other ABC transporters,[46] renders the protein catalytically inactive with an intact dimerization and trafficking pattern.[23,47] This mutant, when cotransfected with the wild-type protein, reduced its activity in a dominant negative manner, providing further evidence that ABCG2 functions as a homodimer. Similarly, a mutation (D210N) introduced in the Walker B motif, supposedly involved in magnesium binding, abrogates transport activity despite intact trafficking to the cell surface.[22]

12.4. FACTORS CONTROLLING ABCG2 EXPRESSION

Beyond recent studies suggesting that ABCG2 expression may be regulated by sex hormones or hypoxia, little is known about the molecular mechanisms controlling *ABCG2* expression. An estrogen response element was identified in the *ABCG2* gene by electrophoretic mobility shift assay and luciferase reporter gene assay.[48] Contradictory results have been obtained regarding the regulation of ABCG2 by steroid hormones. Up-regulation by physiologically relevant quantities of estradiol was observed in one study, while in another, estrogen was shown to induce posttranscriptional down-regulation of ABCG2 in estrogen receptor–positive cell lines.[48,49] Wang and colleagues demonstrated that progesterone increased and estradiol decreased ABCG2 expression and function in placental BeWo cells[50]; however, Yasuda et al. found that progesterone decreased ABCG2 expression in the same cells.[51] A study of ABCG2 expression in rat and mouse tissues attributed high ABCG2 expression in the male rat kidney to the suppressive effects of estradiol, while high ABCG2 expression in the mouse liver was due to the inducive effects of testosterone.[52] Merino et al. reported that Abcg2 expression in the liver was higher in male mice than in female mice by a factor of 2 to 3, and the biliary extrusion of the Abcg2 substrates nitrofurantoin and 2-amino-1-methyl-6-phenylimidazo[4,5-b]pyridine (PhIP) were nine- and twofold higher in males.[53] In a small series of human liver tissues examined for ABCG2 expression by Western blot analysis, expression levels were found to be higher in men than in women, although the levels were not quantified.[53] However, since a greater sensitivity in women to toxic agents has not been demonstrated, it remains to be determined whether women have other mechanisms of protection to endogenous and exogenous toxins that compensate for their relatively lower levels of ABCG2 in the liver. Despite the conflicting results, all of the studies mentioned above do point to the involvement of sex hormones in the regulation of ABCG2 expression.

Krishnamurthy and colleagues were the first to demonstrate that hypoxia regulates ABCG2 expression. They observed a hypoxia responsive element in the *ABCG2* promoter and found expression to be regulated by the HIF-1 pathway, suggesting that this mechanism minimized the detrimental effects of excess porphyrins such as heme during hypoxia.[54] Up-regulation of ABCG2 expression was demonstrated in a number of cell lines exposed to hypoxic conditions, and increased transport of Hoechst 33342 was also observed.[54] Stem cells or tumor cells in hypoxic environments may therefore be protected by high levels of ABCG2. Elucidation of the genomic structure of the *ABCG2* gene and a better characterization of the mechanisms controlling *ABCG2* gene expression are warranted.

Interestingly, Akt has been shown to be involved in controlling ABCG2 expression at the cell membrane. Mogi and colleagues first noted that Akt1-null mice displayed a reduced number of SP cells in their bone marrow.[55] When they treated bone marrow cells with the phosphatidylinositol 3-kinase inhibitor LY294002, in effect an Akt inhibitor, they also noted fewer SP cells. Treatment with LY294002 was found to cause a shift of Abcg2 from the membrane to the cytoplasm, resulting in fewer SP cells.[55] Takada and colleagues confirmed the role of Akt in regulating surface expression of ABCG2, reporting decreased surface expression of the protein in ABCG2-transfected LLC-PK1 cells treated with LY294002 compared to untreated cells.[56] The exact mechanism by which Akt controls cell surface ABCG2 localization has yet to be elucidated.

The mechanisms described to date are most likely to be involved in the regulation of ABCG2 expression in normal tissues. In cancer cells, much work remains to be done to determine the mechanism of activation. Using Pgp as a model, gene amplification is unlikely to occur in the clinical setting. However, gene rearrangement and gene capture by aberrant promoters have been described for Pgp and could also be true for ABCG2.[57]

12.5. SUBSTRATES OF ABCG2

As ABCG2 was cloned originally from drug-resistant cancer cells, chemotherapeutic agents were among the first compounds identified as ABCG2 substrates. A high level of resistance to mitoxantrone is a hallmark of cells that overexpress ABCG2, but the range of substrates now reported may well be comparable to that observed for Pgp.[9,12] The mitoxantrone-related fluorescent compound BBR3390 was also one of the earlier identified substrates of the protein.[58] Cross-resistance to anthracyclines, etoposide, teniposide, and the camptothecin derivatives topotecan, irinotecan, and SN-38 (the active metabolite of irinotecan) is observed in ABCG2-overexpressing, drug-selected cancer cells.[3,4,6,7,11,59] The topoisomerase active drugs J-107088 and NB-506 are readily transported by ABCG2.[60] Cancer cells selected in topotecan[61,62] or SN-38[63] have been shown to up-regulate ABCG2 as a mechanism of resistance. Even selection with camptothecins that are relatively poor ABCG2 substrates, such as DX-8951f and BNP-1350, results in ABCG2 overexpression.[64–66]

Overexpression of ABCG2 has also been shown to confer resistance to methotrexate,[67] although no data exist showing up-regulation of ABCG2 when methotrexate is the selecting drug. Our laboratory reported ABCG2 overexpression in flavopiridol-selected breast cancer cells,[68] and Nakanishi et al. found that ABCG2 overexpression in leukemic blasts correlated with resistance to flavopiridol in an ex vivo assay.[69] Tyrosine kinase inhibitors such as CI1033, gefitinib, and imatinib have been shown to be substrates of ABCG2 and probably act as competitive inhibitors.[70–72] Caco-2 cells continuously exposed to imatinib have been shown to up-regulate ABCG2 as a mechanism of resistance.[73]

In early studies with the MCF-7/AdrVp cell line, we reported the ATP-dependent transport of rhodamine 123, a fluorescent compound transported by Pgp.[8] Rhodamine transport was also detected in the mitoxantrone-resistant S1-M1-3.2 cell line described by Rabindran et al. Since rhodamine had long been used to quantify Pgp expression in resistant cells and clinical samples, we evaluated a series of cell lines for ABCG2 expression by flow cytometry using the inhibitor fumitremorgin C (FTC). Whereas MCF-7/AdrVp and S1-M1-3.2 cells readily transported rhodamine, no other ABCG2-overexpressing cell lines did.[74] We discovered that a drug-induced mutation at amino acid 482, originally noted when the gene was cloned[60], correlated with the ability to transport rhodamine.[75] Cells that expressed ABCG2 with a mutation to a glycine or threonine at amino acid 482 could transport rhodamine, while cells that expressed wild-type ABCG2 with an arginine at this site could not.[44] Subsequently, an arginine to methionine mutation was reported in a $CD4^+$ T-cell line selected with adriamycin.[76] Similarly, Allen and colleagues reported that in mouse fibroblast cell lines lacking functional Pgp or MRP1, selection with adriamycin resulted in ABCG2 overexpression.[59] When they sequenced *Abcg2* in the adriamycin-resistant lines, they found that all cells had acquired a mutation at amino acid 482, changing the wild-type arginine in the protein to a methionine or serine.[77] These results highlighted the importance of amino acid 482 in substrate recognition and suggested it to be a “hot spot” for mutation.

Further studies served to highlight the importance of amino acid 482 in substrate recognition. Cells transfected with ABCG2 harboring an amino acid 482 mutation to an N, C, M, S, T, V, A, G, E, W, D, Q and H, but not Y or K results in higher resistance to mitoxantrone and adriamycin, compared to the wild-type arginine.[42] Similarly, a study in Sf9 insect cells that expressed ABCG2 with a 482 mutation to a G, I, M, S, T, D, N, K, Y found that all mutants except R482K, R482Y and wild-type ABCG2 transported rhodamine 123.[43] In contrast, methotrexate transport was only found in cells expressing wild-type ABCG2.[43] Although it is not clear how amino acid 482 affects substrate recognition, it has been suggested that amino acid 482 is in a drug binding site (see Section 12.3). However, as amino acid 482 mutations have not been observed in clinical samples[69,78] these mutations are principally of academic interest, potentially leading to insights regarding protein structure or drug binding.

Fluorescent substrates have been developed for use in detecting ABCG2 in clinical samples by flow cytometry. Mitoxantrone, BODIPY-prazosin, topotecan, and Hoechst 33342 have all been shown to be substrates of ABCG2 in flow cytometry assays.[74,79–82] As all of these compounds are substrates for Pgp as well as ABCG2, we sought to develop fluorescent assays using ABCG2-specific substrates. Based on the earlier observation of Jonker et al.,[83] we evaluated the porphyrin pheophorbide

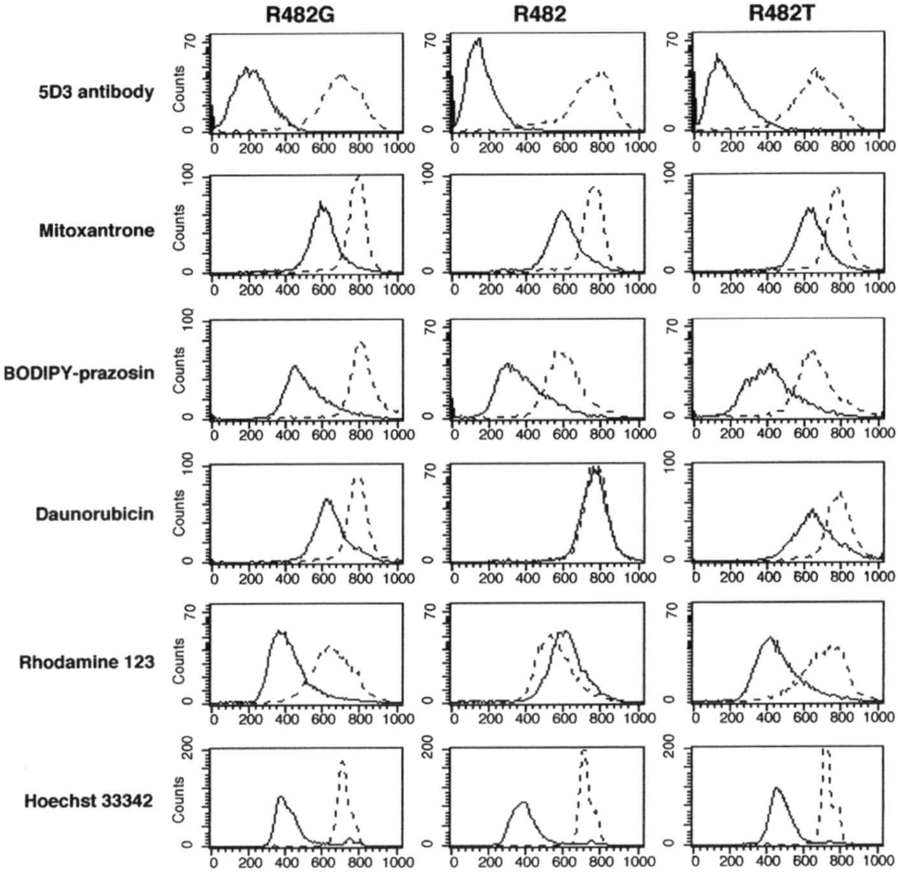

FIGURE 12.1. Effect of amino acid 482 mutations on ABCG2 substrate specificity. HEK293 cells transfected with mutant (R482G, R482T) or wild-type (R482) ABCG2 were incubated in phosphate-buffered saline with 2% BSA with phycoerythrin-labeled negative control antibody (solid line) or phycoerythrin-labeled anti-ABCG2 antibody (clone 5D3, dashed line) for 30 minutes, washed, and subsequently analyzed by flow cytometry to generate the top row of histograms. Cells were also incubated in medium containing 20 μM mitoxantrone (second row of histograms), 250 nM BODIPY-prazosin (third row), 5 μg/mL daunorubicin (fourth row), 0.5 μg/mL rhodamine 123 (fifth row), or 10 μM Hoechst 33342 (bottom row) in the presence (dashed line) or absence (solid line) of 10 μM FTC.

a as a specific ABCG2 substrate.[84] Subsequently, the porphyrins chlorin e6 and pyropheophorbide a methyl ester were also found to be ABCG2-specific substrates.[85] Figure 12.1 demonstrates the effect of amino acid 482 on substrate specificity. HEK293 cells expressing mutant R482G or R482T ABCG2 (left and right columns, respectively) are able to transport mitoxantrone, BODIPY-prazosin, daunorubicin, rhodamine 123, and Hoechst 33342, while cells expressing wild-type ABCG2 (center column) transport only mitoxantrone, BODIPY-prazosin, and Hoechst 33342. All clones display comparable expression of ABCG2 on the cell surface, as shown by the top row of histograms in Figure 12.1.

Emerging evidence points to a potential role of ABCG2 in transporting endogenous steroids. Using a *Lactococcus* expression model, a group led by van Veen found 17β–estradiol to be a substrate for ABCG2.[86] In contrast, Imai et al. used LLC-PK1 cells expressing ABCG2 and reported that sulfated conjugates of estrone and 17β-estradiol were substrates for ABCG2, whereas their respective free estrogens were not.[87] Pavek et al. subsequently showed that estrone and 17β-estradiol as well as other endogenous steroids, such as corticosterone, estradiol, and aldosterone, are not substrates for transport by ABCG2, but found that 17β-estradiol significantly inhibited ABCG2-mediated transport of known substrates.[88] Suzuki et al. demonstrated that estrone3-sulfate (E_1S), as well as dehydroepiandrosterone (DHEAS), were substrates.[89] These conflicting accounts of which endogenous steroids are substrates and even potential inhibitors of ABCG2 still need to be reconciled, including an understanding of whether the cell line, the transfected protein, or experimental conditions account for the differing results.

Since ABCG2 is highly expressed in the gastrointestinal tract, it is thought to have an important role in the oral bioavailability of drugs. Many groups have reported transport of nonchemotherapeutic oral drugs as well as natural products and dietary carcinogens. Several antibiotics, such as nitrofurantoin and erythromycin are reportedly substrates,[53,90,91] as are some antivirals[76,92] and 3-hydroxy-3-methyiglutarylcoenzyme A (HMG-CoA) reductase inhibitors.[93–95] The flavonoids genistein and quercetin have been identified as ABCG2 substrates along with their glucuronides.[96,97] Table 12.1 includes a more complete list of substrates that have been identified.

12.6. INHIBITORS OF ABCG2

The first ABCG2-specific inhibitor to be described was the diketopiperazine fumitremorgin C (FTC). The inhibitory effects of FTC were reported even before ABCG2 was cloned.[11] Isolated from the fermentation broth of *Aspergillus fumigatis*, FTC at a concentration of 5 μM readily inhibited ABCG2-mediated resistance to mitoxantrone, doxorubicin, and topotecan in drug-selected cell lines, but had no effect on resistance mediated by Pgp or MRP1.[11] Resistance in ABCG2-transfected cells was also reversed.[98] Clinical use of FTC was limited due to its known neurotoxicity, thus leading to a search for a less toxic inhibitor, resulting in the subsequent discovery of Ko143.[99] Highly potent, Ko143 exhibited no neurotoxic effects in mice compared to the parent compound, FTC. Other diketopiperazine inhibitors, including indolyl diketopiperazines[100] and tryprostatin a,[101] have also been reported to inhibit ABCG2.

Since there was significant overlap between ABCG2 and Pgp substrates, known inhibitors of Pgp were among the first compounds to be examined for their ability to inhibit ABCG2. Elacridar (GF120918) was found to be an ABCG2 inhibitor in addition to its ability to inhibit Pgp.[102] MRP1-mediated transport was unaffected. Reserpine has also been reported to inhibit Hoechst transport in SP cells.[82] Before the

TABLE 12.1. Substrates and Inhibitors of ABCG2

Substrates	Inhibitors
Chemotherapy agents	Diketopiperazines
Mitoxantrone	Fumitremorgin C[11,98]
BBR3390[58]	Ko143[99]
Anthracyclines[a]	Tryprostatin A[101]
Daunorubicin[a] [7,44,47,77]	Indolyl diketopiperazines[100]
Doxorubicin[a] [7,11,44,47,77]	Steroids and steroidlike compounds
Epirubicin[a] [44]	Corticosterone[88]
Bisantrene[a] [11,44]	Digoxin[88]
Flavopiridol[68,173]	Beclomethasone[88,111]
Etoposide[a] [3,4,7,174]	6-α-Methylprednisolone[88]
Teniposide[174]	Dexamethasone[88]
Camptothecins	Triamcinolone[88]
9-Aminocamptothecin[6,59,175–177]	Mometasone[111]
Topotecan[6,59,61,62,79,177,178]	Ciclesonide[111]
Irinotecan[6,64,177,179]	Antivirals
SN-38[6,63,175,177,180,181]	Ritonavir[92]
SN-38 glucuronide[180,181]	Saquinavir[92]
Diflomotecan[a] (weak)[182]	Nelfinavir[92]
Homocamptothecin[a] (weak)[182]	Immunosupressants
DX-8951f (weak)[64,65]	Tacrolimus[104]
BNP1350 (weak)[66]	Sirolimus[104]
Indolocarbazoles	Cyclosporin A[8,103,104]
J-107088[60]	Pyridines and dihydropyridines
NB-506[60]	Niguldipine[117]
Compound A[183]	Nicardipine[117,118]
UCN-01[83]	Nitrendipine[117,118]
Antifolates	Dipyridamole[118]
Methotrexate[a], methotrexate di[a]- and triglutamate[a] [67,91,184–187]	Nimodipine[118]
	DHP-014[117]
GW1843[a] [186]	Tyrosine kinase inhibitors
Tomudex[a] [186]	Gefitinib[107–109,203]
Tyrosine kinase inhibitors	Imatinib[109,110]
Imatinib[72]	EKI-785[109]
Gefitinib[71,107]	CI1033[70]
CI1033[70]	P-glycoprotein inhibitors
Other compounds	Elacridar (GF120918)[102,177]
Antivirals	Tariquidar (XR9576)[83]
Zidovudine (AZT)[188]	Biricodar[a] (VX-710)[121]
Lamivudine[76]	Flavonoids
HMG-CoA reductase inhibitors	Chrysin[112]
Rosuvastatin[93]	Biochanin A[112]
Pitavastatin[94]	Benzoflavone[204]
Cerivastatin[95]	6-Prenylchrysin[115]
Carcinogens	Tectochrysin[115]
Aflatoxin B1[133]	Naringenin[113]
2-Amino-3-methylimidazo[4,5-*f*]quinoline (IQ)[133]	Quercetin[111,181]
	Acacetin[113]
3-Amino-1,4-dimethyl-5*H*-pyrido[4,3-*b*]indole (Trp-P-1)[133]	Kaempferol[113]
	Silymarin[111]

(*Continued*)

TABLE 12.1. (*Continued*)

Substrates	Inhibitors
2-Amino-1-methyl-6-phenylimidazo(4,5-*b*)pyridine (PhIP)[88,132,138]	Hesperetin[111]
Porphyrins	Daidzein[111]
Pheophorbide a[83,84]	Resveratrol[111]
Pyropheophorbide, a methyl ester[85]	Genistein[113]
Chlorin e6[85]	Naringenin-7-glucoside[113]
Protoporphyrin IX[54,85]	Benzimidazoles
Fluorescent compounds	Pantoprazole[119]
Rhodamine 123[a] [8,44,79]	Omeprazole[119]
Lysotracker Green[a] [44,79]	Oxfendazole[195]
BODIPY-prazosin[44,79]	Estrogens, estrogen agonists, and estrogen antagonists
Hoechst 33342[82,189]	17β-Estradiol[105]
Flavonoids	Estrone[105]
Genistein[113]	Diethylstilbestrol[106]
Quercetin[97,190]	Tamoxifen[106]
Glucuronide, glutathione, and sulfate conjugates	Toremifene[106]
Benzo[*a*]pyrene-3-sulfate (BP3S)[191]	TAG-11[106]
Benzo[*a*]pyrene-3-glucuronide (BP3S)[191]	TAG-139[106]
Estrone 3-sulfate[87,89,95]	Taxanes
4-Methylumbelliferone sulfate[89]	Ortataxel[a] [120]
4-Methylumbelliferone glucuronide[140]	tRA96023[a] [120]
6-Hydroxy-5,7-dimethyl-2-methylamino-4-(3-pyridylmethyl)benzothiazole	tRA98006[a] [122]
glucuronide (E3040)[89,192]	Flavopiridol[68]
Dehydroepiandrosterone sulfate	UCN-01[83]
(DHEAS)[192,193]	Novobiocin[a] [205,206,206]
17β-Estradiol sulfate[87]	Stilbenoids[207]
Acetaminophen sulfate[194]	
Benzimidazoles	
Albendazole sulfoxide[195]	
Oxfendazole[195]	
Pantoprazole[119]	
Antibiotics	
Ciprofloxacin[90]	
Ofloxacin[90]	
Norfloxacin[90]	
Erythromycin[91]	
Dirithromycin[91]	
Rifampicin[91]	
Nitrofurantoin[53]	
Dihydrotestosterone[196]	
Sulfasalazine[197,198]	
Phenethyl isothiocyanate[116,199]	
Dipyridamole[118]	
Ochratoxin A[200]	
GV196771[201]	
Folic acid[184]	
Cimetidine[88]	
ME3229[202]	

[a]Compound affected by the amino acid at position 482.

ABCG2 gene was cloned, rhodamine transport in MCF-7/AdVp cells was found to be inhibited by cyclosporin A.[8] Subsequent to these initial studies with cyclosporin A, the results have been contradicted[41] and confirmed.[103,104] The highly potent Pgp inhibitor tariquidar (XR9576) has been shown to inhibit pheophorbide a transport in ABCG2-transfected cells.[83]

The estrogens estrone and 17β-estradiol were also among the first ABCG2 inhibitors identified.[105] Subsequently, diethylstilbestrol, tamoxifen, and torimefene were found to act as inhibitors.[106] This observation led Sugimoto and colleagues to screen a number of tamoxifen analogs for their ability to inhibit ABCG2, resulting in the discovery of TAG-139, a tamoxifen analog that was fivefold more potent than estrone at inhibiting ABCG2.[106]

Tyrosine kinase inhibitors are an emerging class of compounds that also act as ABCG2 inhibitors. The HER tyrosine kinase inhibitor CI1033 was first demonstrated to inhibit ABCG2-mediated resistance to SN-38 and topotecan in transfected cells.[70] Gefitinib was shown to inhibit ABCG2-mediated drug resistance, as were imatinib and EKI-785.[107–110] It is most likely that these tyrosine kinase inhibitors act as competitive inhibitors since ABCG2 has been shown to transport or confer resistance directly to CI1033, gefitinib, and imatinib.[71,72,107,107]

An increasing number of flavonoids have been shown to inhibit ABCG2. Cooray et al. were among the first to report an interaction between ABCG2 and flavonoids. They identified silymarin, hesperetin, quercetin, and daidzein, as well as the stilbene resveratrol as ABCG2 inhibitors since they increased the intracellular accumulation of mitoxantrone and BODIPY-prazosin in ABCG2-expressing cells.[111] Similarly, Zhang et al. found chrysin and biochanin a to be potent inhibitors of ABCG2-mediated resistance to mitoxantrone.[112] Imai et al. reported that genistein, naringenin, acacetin, and kaempferol reversed resistance to SN-38 and mitoxantrone in ABCG2-transduced K562 cells.[113] Structure–activity studies have also identified novel ABCG2 inhibitors,[114,115] among them 6-prenylchrysin and tectochrysin. Organic isothiocyanates have been identified as inhibitors.[116] Although the fact that flavonoids such as quercetin and genistein are ABCG2 substrates argues that they function as competitive inhibitors, it has been postulated that flavonoids inhibit ABCG2 due to their interaction with its nucleotide-binding domain.[96,114] Since flavonoids occur naturally in food products, these compounds may affect the bioavailability of oral drugs.

As was the case for Pgp, many groups have reported that a number of off-the-shelf compounds are ABCG2 inhibitors. The glucocorticoids beclomethasone, 6α-methylprednisone, triamcinolone, dexamethasone, betamethasone, and prednisone were found to inhibit ABCG2-mediated transport in mouse fibroblast cells but were not themselves transported.[88] Dihydropyridines and pyridines, used as antihypertensives, were found to increase intracellular mitoxantrone accumulation in ABCG2-overexpressing cells.[117] Similarly, Zhang et al. reported that dipyridamole and dihydropyridines were ABCG2 inhibitors, noting that dipyridamole itself was also transported whereas nitrendipine was not.[118] Benzimidazoles have been found to cause significant toxicity when coadministered with methotrexate, and the findings of Breedveld and colleagues suggest that this may be due to their ability to inhibit ABCG2 competitively.[119] A list of reported ABCG2 inhibitors is provided in Table 12.1.

Several studies have noted the impact of amino acid 482 on inhibitor potency. While confirming the effects of reported ABCG2 inhibitors in stably-transfected HEK-293 cells, we noted that novobiocin was effective in reversing ABCG2-mediated transport of fluorescent substrates only in cells transfected with wild-type but not mutant ABCG2.[44] The effects of amino acid changes at position 482 were observed for biricodar, developed as a modulator of both Pgp and MRP1, and the taxane derivatives ortataxel and tRA96023.[120–122] Amino acid 482 mutations have also been shown to alter the ability of some flavonoids to inhibit ABCG2.[115]

The growing number of substrates and inhibitors for ABCG2 is not surprising, considering previous experience with Pgp and its known promiscuous nature. The ability to transport a wide range of unrelated compounds supports its suspected role in protecting cells from xenobiotics, and suggests that like Pgp, drugs bind ABCG2 in a large central cavity rather than in a lock-and-key conformation. The parallel between ABCG2 and Pgp in drug transporter paradigms allowed the rapid identification of inhibitors but probably has hindered the actual clinical development of ABCG2 inhibitors.

12.7. TISSUE LOCALIZATION OF ABCG2

After its initial discovery in resistant cancer cell lines, investigators pursued varying lines of inquiry to determine the location, expression, and physiological role of ABCG2 in vivo. In the initial report identifying ABCG2, Doyle et al. performed Northern blot analysis in commercially prepared human samples representing 16 different tissues.[9] Using cDNA probes to examine expression of ABCG2 mRNA, they found the highest expression in placental tissue. Lower levels were observed in tissue from the brain, prostate, small intestine, testis, ovary, colon, and liver. Transcripts were undetectable in tissues from the heart, lung, skeletal muscle, kidney, pancreas, spleen, thymus, and peripheral blood leukocytes. We later confirmed these results, noting high expression in the central nervous system, liver, adrenal gland, placenta, prostate, testes, and uterus.[123] Lower levels were detected by Northern blot in the small and large intestine, stomach, lung, kidney, and pancreas.[123]

Maliepaard et al. examined ABCG2 expression in normal tissue and cancer cell lines by immunohistochemistry using the BXP-21 and BXP-34 monoclonal antibodies.[124] Again, high expression was found in placental tissue and specifically in the placental syncytiotrophoblast. Strong positive staining was also observed in the colon, and positive staining was seen in the small intestine, biliary canaliculi (but not hepatocytes or bile ductules), breast tissue, venous endothelium, and in capillaries. Comparison was also made of mRNA expression by semiquantitative reverse transcriptase polymerase chain reaction (RT-PCR) with the IHC assay. RT-PCR results generally correlated with IHC staining with BXP-21 and BXP-34 antibodies, except that mRNA levels were higher in tissues with greater blood vessel density (e.g., ovary, cervix, small intestine) and were lower in tissues with a smaller endothelium density (e.g., heart and breast tissue). Studies by Scheffer et al.[125] and Faneyte et al.[126] also found expression in breast tissue to localize to the vascular endothelium rather than normal mammary or breast cancer tissue. Directing a specially generated polyclonal rabbit ABCG2 antibody (termed 405) against an 18-mer peptide near the ATP-binding region together with the commercially available 5D3 monoclonal antibody, we also found high expression of this protein in the syncytiotrophoblast placenta, alveolar pneumocytes in the lung, skin sebaceous glands, small and large intestine, bile canaliculi, and blood vessels, including endothelium from the central nervous system.[123]

12.8. FUNCTION PREDICTED FROM TISSUE DISTRIBUTION

The location and level of expression of ABCG2 highlight the protein's potential roles in the normal host. These include probable roles in protecting stem cells; forming the maternal–fetal barrier, the blood–testis cell barrier, and the blood–brain barrier; and its role in the absorption and efflux of xenobiotics and endogenous metabolic products within the gastrointestinal tract.

12.8.1. Stem Cells

Recent evidence has shown that ABCG2 may play a critical role in the phenotype of stem cells. Hematopoietic stem cells can be recognized as a *side population* (SP), when bone marrow cells are studied via flow cytometry. These SP cells extrude Hoechst 33342 and have surface antigen expression profiles that correspond to a stem cell phenotype.[127] Hoescht 33342 is a known substrate for ABCB1, but after investigators found no differences in the SP cell population in ABCB1-deficient and wild-type mice,[82] Zhou et al. reported that ABCG2 was responsible for the SP phenotype. High levels of ABCG2, which also effluxes Hoechst dye, were found in early murine stem cells, including hematopoietic, muscle, and embryonic stem cells, potentially protecting the cell from exogenous toxins or serving an unknown function in endogenous substrate efflux. As these cells matured, ABCG2 levels subsided, and other efflux mechanisms, such as ABCB1, were up-regulated. Although ABCG2 may serve as part of the stem cell phenotype, it is neither necessary nor sufficient for stem cell development and cellular differentiation, a fact confirmed by the observation that there is no significant developmental pathology associated with *Abcg2*-deficient mice. Since Zhou's publication, ABCG2 has been identified as a potential phenotypic marker for stem cells in multiple tissues, including normal lung[128] and breast tissue.[129]

12.8.2. Placenta

ABCG2 may function in the human placenta to protect the fetus or to transport steroid hormones produced in the placenta.[130] Specifically, E_1S and DHEAS are among the major estrogens produced and secreted by the placenta, and as shown by Suzuki et al. are ABCG2 substrates.[89] However, given its high expression in the syncytiotrophoblast near the apical surface at the chorionic villus, ABCG2 may help form the barrier between the maternal and fetal circulation systems and thus protect the fetus from endogenous and exogenous toxins. Jonker et al. investigated this possibility by exposing *Abcb1*-deficient mice to the known ABCG2 substrate topotecan.[131] When pregnant mice were given topotecan with the ABCG2 inhibitor GF120918, fetal plasma levels of topotecan were twice as high as that measured in maternal plasma. ABCG2 expression is relatively higher in human placental tissue compared to murine indicating that the protein may play an even more important role in protecting human fetal exposure to endogenous and exogenous toxins present in the maternal circulation.

12.8.3. Mammary Gland

As opposed to the role that ABCG2 may play in the placenta, and despite the absence of expression in adult breast epithelial cells, ABCG2 expression in lactating mammary tissue has been found to concentrate substrates, including toxins, into breast milk. Jonker et al. reported that ABCG2 expression is strongly induced in the mammary glands of lactating mice, cows, and humans.[132] Further, levels of the carcinogen and toxin PhIP, as well as topotecan, were highly concentrated in the milk of *Abcg2* wild-type mice, whereas these substances were not present in the milk of *Abcg2*-deficient mice. In addition, the administration of GF120918 reversed the active secretion of topotecan into milk in the wild-type mice. By determining high ratios of substance concentration in breast milk versus maternal plasma, these researchers confirmed that cimetidine and acyclovir were Abcg2 substrates. This same group later verified the role of Abcg2 in concentrating potential substrates in breast milk when they found that the concentrations of heterocyclic amines as well as aflatoxin were over threefold higher in breast milk than in maternal plasma in lactating wild-type versus *Abcg2*-deficient mice.[133]

This functional role in concentrating substrates in lactating mammary epithelium is in contrast to the low expression level of ABCG2 observed in nonlactating breast epithelium. It remains unclear why ABCG2 would be induced in the lactating mammary gland and therefore concentrate potential toxins in breast milk, increasing the potential harm to feeding infants. The physiologic role of an up-regulated ABCG2, as well as the identification of endogenous substrates that might be conveyed from mother to infant by this transporter still need to be discovered.

12.8.4. Testis

High expression of ABCG2 has been reported in normal testis tissue as determined by Northern blot and localized to the Sertoli–Leydig cells by immunohistochemical analysis.[123] Bart et al. found ABCG2 as well as Pgp expressed in the myoid cell layer as well as in endothelial cells of the normal testis.[134] These transporters are expressed on the luminal side of the endothelium as well as the apical side of myoid cells, highlighting their probable function of transporting substrates away from the seminiferous tubules, implying a potential role in protecting developing germ cells from toxins in the male circulation.

12.8.5. Blood–Brain Barrier

Numerous researchers have explored the role of ABCG2 in forming the blood–brain barrier. Using the techniques of RT-PCR, Western blot analysis, and immunofluroescence staining using BXP-21 and BXP-34, Cooray et al. showed ABCG2 expression along with that of ABCB1 in the microvasculature of both normal and malignant tissues from the central nervous system.[135] ABCG2 was expressed on the luminal surface of the endothelium, providing a clue to its role as an efflux pump away from the brain interstitium. Expression of other ABC transporters, including MRP1, MRP3, and MRP5, could not be detected. Zhang et al. investigated *ABCG2* as well as *ABCB1* and *ABCC1* expression in normal and glioblastoma tumors as measured by RT-PCR and showed higher expression of *ABCG2* than of the other two transporters.[136] Cisternino et al. explored the role of Abcg2 in forming the murine blood–brain barrier in *Abcb1*-deficient mice.[137] They found that mRNA levels of *Abcg2* were 700-fold higher in the microvasculature than in the brain cortex. They exposed these Abcb1-deficient and wild-type mice to mitoxantrone with and without GF120918 and found a two- to threefold higher uptake in the mice treated with GF120918. Not all researchers have found Abcg2 playing a critical role in the blood–brain barrier, however. For example, van Herwaarden et al.[138] did not find higher levels of PhIP in the brains of *Abcg2*-deficient mice than in wild-type animals.

12.8.6. Liver and the Gastrointestinal Tract

The presence of ABCG2 in the apical membrane of the small intestine and bile canniculi implies a potential role for the protein in xenobiotic and endogenous substrate efflux from the liver, and substrate absorption along the small intestine. This theory has been confirmed in a variety of settings. Jonker and colleagues administered oral topotecan with and without GF120918 to mice and found significantly higher plasma drug concentrations in animals receiving the inhibitor.[131] Administration of GF120918 decreased plasma clearance, decreased hepatobiliary excretion, and increased reuptake in the small intestine. Taipalensuu et al.[139] examined jejunal mucosa from 13 normal volunteers and compared expression levels of 10 potential drug efflux transporters from the ABC family, including ABCB1 (Pgp), ABCC1, and ABCG2. The highest level detected was for the mRNA encoding ABCG2. Given that this area of the small intestine is quantitatively the most important site for drug absorption, this finding was unexpected, especially given the importance previously ascribed to ABCB1 in clinical pharmacology research.

In one study, van Herwaarden et al.[138] used *Abcg2*-deficient mice to test for the elimination of PhIP and found that wild-type mice eliminated this food carcinogen primarily via fecal excretion, whereas the elimination in *Abcg2*-deficient animals was in the urine. Wild-type and *Abcg2*-deficient animals underwent cannulation of their gall bladders and ligation of their common bile ducts and were then administered intravenous PhIP. Levels of PhIP measured in the small intestine and cecum were significantly higher in wild-type animals, confirming the efflux of the compound into the lumen of the small intestine via Abcg2.

Adachi et al.[140] used both *Abcg2*-deficient mice as well as hereditary hyperbilirubinemic mice deficient in *Abcc2* to examine enterocyte efflux of substrates that have been glucuronidated and sulfated in the liver prior to efflux into the intestinal lumen. They found that Abcg2 had an important role in effluxing these conjugated metabolites, and concluded that Abcc2 had a lesser role in effluxing some glucuronidated conjugates.

12.9. GENETIC POLYMORPHISMS

Due to the potential pharmacologic impact, a number of studies have investigated the impact that SNPs in the *ABCG2* gene have on protein expression and function. A number of SNPs have been identified in coding regions of the gene, and at least four nonsynonymous SNPs have been identified. These SNPs occur at mRNA positions 34 (V12M; exon 2), 421 (Q141K, exon 16), 616 (I206L, exon 6), and 1768 (N590Y, exon 15). The V12M, I206L, and N590Y SNPs have not been found to confer an alteration in protein expression or function.[141,142] However, the nonsynonymous substitution C421A, in which a lysine is substituted for glutamine at amino acid (Q141K), has been shown to have a functional significance. Various researchers have found that this SNP can lead to lower plasma membrane expression,[141,143–145] reduced drug efflux,[141,146] and reduced ATPase activity.[141,146] This polymorphism can lead to lower IC_{50} levels in cell lines exposed to cytotoxic agents that are ABCG2 substrates, including mitoxantrone, irinotecan, and SN-38.[141] Additionally, when patients who had the SNP were exposed to chemotherapy, higher in vivo levels of topotecan and diflomotecan were found,[147,148] coinciding with the role of ABCG2 in biliary and intestinal elimination of xenobiotics.

12.10. ABCG2 EXPRESSION IN CANCER AND ITS ROLE IN DRUG RESISTANCE

Expression of ABCG2 in leukemia is by far the most extensively studied, a summary of which is given in Table 12.2. In acute myelogenous leukemia (AML), conflicting data exist regarding ABCG2 expression. Some groups have reported relatively high ABCG2 expression,[149–151] whereas others reported undetectable levels.[152–154] Similarly, ABCG2 has been found to be predictive of response to chemotherapy in some studies[155–157] but not in others.[154] In acute lymphoblastic leukemia, a similar situation exists, with ABCG2 reported to be predictive[158,159] and unpredictive[160] of response. To clarify the contribution of ABCG2 to resistance in leukemia, more studies with larger patient populations and validated methods are necessary. As discussed below,

TABLE 12.2. Summary of Investigations Measuring ABCG2 Expression or Function in Leukemia

Author	Cancer	Method[a]	Samples	Conclusion
Ross et al.[149]	Leukemia	RT-PCR	20 AML samples 1 ALL sample	Relatively high ABCG2 levels in seven samples (33%). No correlation with ABCB1 expression.
Sargent et al.[208]	AML	IHC: BXP-34 antibody	Blast cells from 20 patients with AML	Six of 22 (27%) samples had >10% of cells positive.
van Der Kolk et al.[152]	AML	FC: BXP-21, mitoxantrone transport	20 paired de novo and relapsed/refractory AML	Low levels of ABCG2 expression in AML. No up-regulation of ABCG2 expression or activity at relapse. High ABCG2 expression with immature phenotype.
Sauerbrey et al.[160]	Childhood ALL	RT-PCR	47 de novo samples 20 relapsed samples	Lower ABCG2 expression in T-cell lineages. No correlation between expression and response.
Abbott et al.[154]	AML	RT-PCR	40 de novo samples	High levels of ABCG2 uncommon in AML, but high levels may be due to small subset of cells. No correlation between ABCG2 expression and response.
Steinbach et al.[155]	Childhood AML	RT-PCR	59 untreated AML 9 relapsed cases	Median ABCG2 levels 10-fold higher in patients who did not achieve remission compared to responders. ABCG2 levels higher in first relapse than at diagnosis. Expression related to prognosis.
van den Huevel-Eibrink et al.[209]	AML	RT-PCR	20 paired de novo and relapsed/refractory AML samples	ABCG2 levels higher in relapsed/refractory samples than at diagnosis. Significant coexpression of ABCG2 and ABCB1.

(Continued)

TABLE 12.2. (*Continued*)

Author	Cancer	Method[a]	Samples	Conclusion
van der Pol et al.[153]	AML	FC: BODIPY-prazosin transport	45 paired samples	ABCG2 function undetectable in almost all samples. No changes observed after therapy.
Nakanishi et al.[173]	Acute leukemia	RT-PCR	21 blast samples	ABCG2 expression range of 200-fold. ABCG2 expression correlated with resistance to flavopiridol-induced apoptosis.
Plasschaert et al.[210]	ALL	FC: mitoxantrone transport	46 de novo samples	ABCG2 expression in B-lineage 2.4-fold higher than T-lineage. Higher gene expression correlated with higher function. No aa482 mutations.
Stam et al.[158]	Childhood ALL	RT-PCR	13 samples from infants 13 samples from noninfants	ABCG2 expression 2.4-fold less in samples from infants. Expression correlated with Ara-C resistance.
Galimberti et al.[150]	AML	RT-PCR	51 AML samples	56% of samples were positive; 48.2% had intermediate levels. ABCG2 and ABCB1 expression correlated.
Suvannasankha et al.[151]	AML	RT-PCR FC: BXP-21, BXP-34, 5D3, mitoxanrone transport	31 pretreatment blast samples	ABCG2 expression in all but one case between 8226 and 8226/MR20 levels. ABCG2 expression and function confined to small subpopulations. No correlation between protein expression and function.

Suvannasankha et al.[159]	ALL	RT-PCR FC: BXP-21, BXP-34, 5D3, mitoxantrone transport	30 pretreatment blast samples	BXP-21, BXP-34, and 5D3 and positivity in 43, 37, and 47% of cases, respectively. Function in 70% of cases. No correlation between gene expression or protein expression and function. BXP-21 staining predictive of PFS. No aa482 mutations.
Benderra et al.[156]	AML	RT-PCR	149 de novo samples	ABCG2 expression predictive of complete remission, 4-year DFS, and 4-year overall survival. Predictive in patients receiving daunorubicin and mitoxantrone but not idarubicin.
Uggla et al.[157]	AML	RT-PCR	40 AML samples	No difference in ABCG2 expression between responders and nonresponders. In responders, higher expression was related to shorter overall survival.

[a]IHC, immunohistochemistry; FC, flow cytometry.

these discrepancies are likely to be due to the lack of standardized methods and the lack of a uniform reference point.

Two studies reported contradictory results when measuring ABCG2 expression in solid tumors by immunohistochemistry. Scheffer et al. examined protein expression in 34 paraffin-embedded untreated tumor samples from various sites and found positive staining in only one small intestine adenocarcinoma sample.[125] In a study of 150 paraffin-embedded untreated tumor samples, Diestra and colleagues reported frequent ABCG2 expression. Highest levels were found in melanoma and tumors of the digestive tract, endometrium, and lung.[161] Both cytoplasmic and membrane staining was observed, raising some concern about specificity. However, this is the most complete immunohistochemical study to date.

In a study of 43 untreated breast carcinoma samples, Kanzaki et al. reported low *ABCG2* gene expression as measured by RT-PCR, and no correlation with clinical outcome.[162] Similarly, Faneyte et al. reported no detectable ABCG2 expression by immunohistochemistry and no correlation between gene expression and survival.[126] Burger and colleagues, however, found that *ABCG2* and *ABCB1* gene expression correlated with response in anthracycline-treated patients, although expression did not correlate with progression-free survival.[163] Since ABCG2 does not transport anthracyclines significantly, it is difficult to discern why expression would correlate with poor response.

Two studies have focused on ABCG2 expression in testicular tumors. Zurita et al. examined protein expression in 56 paraffin-embedded tumor samples from patients diagnosed with advanced testicular germ cell tumors. Strong and intermediate expression was observed in 86 and 7% of samples, respectively.[164] They found no correlation between ABCG2 expression and clinical outcome when patients were treated with platinum-based chemotherapy.[164] The second study found that seminoma and non-seminoma testicular tumor samples from untreated patients expressed ABCG2 when assessed by immunohistochemistry using the BXP-21 antibody, whereas testicular lymphoma samples did not.[134]

Studies demonstrating an interaction between tyrosine kinase inhibitors such as gefitinib and ABCG2 led Theou and colleagues to examine ABCG2 via Western blot in tumor samples obtained from 21 patients with gastrointestinal stromal tumors and three patients with leiomyosarcomas.[165] They found no expression of ABCG2 in any of the samples; however, detection of protein expression by immunoblotting is not a sensitive method and may not detect low but clinically relevant levels of ABCG2.

Yoh et al. examined ABCG2 expression in 72 formalin-fixed paraffin-embedded samples obtained from patients diagnosed with non-small cell lung carcinoma before chemotherapy.[166] Of the 72 samples, 33 (46%) were ABCG2 positive. ABCG2 expression correlated with shorter progression-free survival and overall survival when patients were treated with platinum-based chemotherapy. This finding needs to be reproduced and verified with a different method, since platinum compounds are not ABCG2 substrates.

It is likely that the discordant results reported for ABCG2 expression in cancer are due to the lack of standardized methods of detection. As demonstrated by a 1994 workshop devoted to determining standardized methods for detection of Pgp, different methods of detection offer different levels of sensitivity and specificity, and nonstandard terminology results in varying conclusions.[167] Some of the discordance observed in a study by Suvannasankha et al. may be attributed to these issues.[151] ABCG2 expression was measured in cell lines and 31 AML samples by RT-PCR and by flow cytometry examining mitoxantrone efflux or staining with the antibodies BXP-21, BXP-34, or 5D3. They reported no correlation between ABCG2 expression measured by RT-PCR and mitoxantrone transport or between antibody staining with any of the antibodies and mitoxantrone transport in ALL.[151] While it was suggested that the discordant results are due to the complex biology of ABCG2, it is also likely that the assays have differing levels of sensitivity and specificity. Mitoxantrone is a minor substrate for other ABC transporters, so flow cytometry studies should use more specific ABCG2 substrates. Lack of antibody sensitivity or specificity may explain the discordant results of Scheffer et al. and Diestra et al. The BXP-34 antibody used in the study by Scheffer et al. may not be sensitive enough to detect lower levels of ABCG2 expression than those detected by Diestra et al. using the BXP-21 antibody. Standardized methods for detection of ABCG2 will aid in the determination of the contribution of ABCG2 to resistance in cancer. Further, it is critical that a common point of reference be used. Either a consistent panel of normal tissues should be included with every immunohistochemical study, or a reference cell line should be included in studies of gene expression or functional measurements. MCF-7 cells may be a good choice as a reference line, since expression of ABCG2 is low but readily detectable in functional assays.[74,83] Using a cell line with endogenous levels—also more likely to be clinically relevant—offers the chance to obtain reproducible results from laboratory to laboratory.

12.11. CANCER STEM CELLS

Since hematopoietic stem cells are often found in the SP fraction, where intracellular Hoechst 33342 fluorescence is low due to ABCG2 expression, many groups have identified cancer stem cells in cell lines and solid tumors via this method. In neuroblastomas from 15 of 23 patients, Hirschmann-Jax et al. found SP cells that expressed ABCG2 and generated both SP and non-SP progeny.[168] These cells also demonstrated increased resistance to chemotherapeutic agents. An SP population has also been found in a number of established cancer cell lines[169] as well as in retinoblastoma[170] and gastrointestinal cell lines.[171] Studies such as these have led to the belief that SP cells, characterized by ABCG2 expression, represent a population of cancer stem cells.

Such a conclusion must be viewed with caution. First, the SP fraction, even in bone marrow, is known not to represent a pure stem cell population, and the method for generating the SP fraction greatly affects the types of cells included in it.[172] Additionally, since Hoechst 33342 is a substrate for both ABCG2 and Pgp, the SP fraction will include cells overexpressing endogenous levels of either of these proteins. Further, ABCG2 may be part of the normal phenotype. For example, NCI-H460 lung carcinoma cells have high levels of *ABCG2* mRNA and transport Hoechst 33342, but

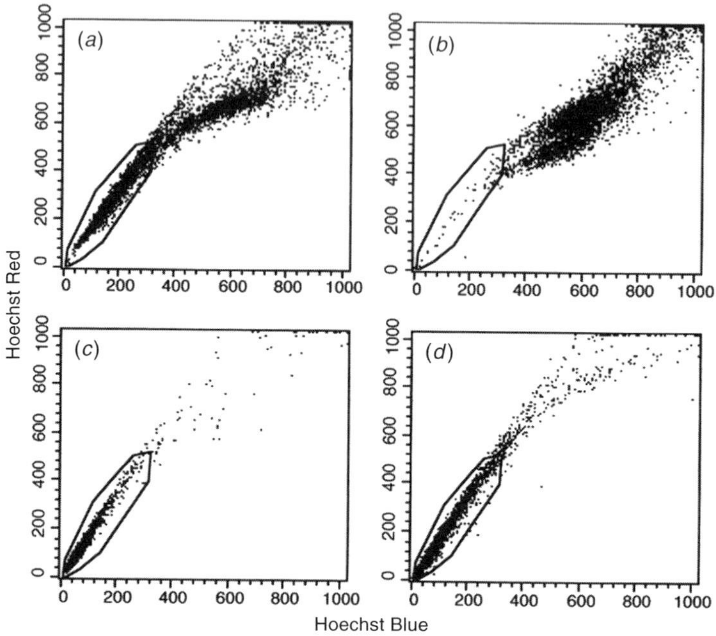

FIGURE 12.2. The stem cell phenotype does not always identify cancer stem cells. NCI-H460 cells (*a*, *b*), Pgp-overexpressing SW620 Ad300 cells (*c*), or ABCG2-overexpressing MCF-7 FLV100 cells (*d*) were incubated in 2.5 μg/mL Hoechst 33342 for 30 minutes, washed, then incubated in Hoechst-free medium for an additional 60 minutes. For (*b*), cells were incubated in Hoechst in the presence of 10 μM FTC to prevent ABCG2-mediated Hoechst transport.

this is because the entire population, rather than a subset of cells, expresses ABCG2. As shown in Figure 12.2*a*, when NCI-H460 cells are incubated with Hoechst 33342, half of the population of cells is in the SP gate, and this phenomenon is reversed by the addition of the ABCG2-specific inhibitor FTC (Figure 12.2*b*). When Pgp-positive SW620 Ad300 cells are incubated with Hoechst, generating the plot shown in Figure 12.2*c*, they give the appearance of an SP population, as is the case for ABCG2-overexpressing MCF-7 FLV500 cells (Figure 12.2*d*). Drug-resistant cells that overexpress ABCG2 are not considered stem cells. Although ABCG2 may be expressed in stem cells as a protection from xenobiotics, it cannot define a stem cell in the normal or malignant state.

12.12. CONCLUSIONS

Identification of ABCG2 in drug-resistant cancer cells opened a line of investigation that has led to multiple new insights into normal human physiology and xenobiotic protection. However, the original hope—that a new therapeutic target for drug resistance reversal has been identified—has not yet been tested or validated in clinical oncology. Rapid discoveries in protein biology will hopefully be followed rapidly by studies that evaluate the role of ABCG2 in cancer drug resistance. The first step is the development of validated, reproducible assays that can determine the expression of ABCG2 in newly diagnosed and drug-resistant cancers.

REFERENCES

1. Taylor, CW, Dalton, WS, Parrish, PR, Gleason, MC, Bellamy, WT, Thompson, FH, Roe, DJ, and Trent, JM. 1991. Different mechanisms of decreased drug accumulation in doxorubicin and mitoxantrone resistant variants of the MCF7 human breast cancer cell line. Br J Cancer 63:923–929.
2. Dietel, M, Arps, H, Lage, H, and Niendorf, A. 1990. Membrane vesicle formation due to acquired mitoxantrone resistance in human gastric carcinoma cell line EPG85-257. Cancer Res 50:6100–6106.
3. Nakagawa, M, Schneider, E, Dixon, KH, Horton, J, Kelley, K, Morrow, C, and Cowan, KH. 1992. Reduced intracellular drug accumulation in the absence of P-glycoprotein (*mdr*1) overexpression in mitoxantrone-resistant human MCF-7 human breast cancer cells. Cancer Res 52:6175–6181.
4. Kellner, U, Hutchinson, L, Seidel, A, Lage, H, Danks, MK, Dietel, M, and Kaufmann, SH. 1997. Decreased drug accumulation in a mitoxantrone-resistant gastric carcinoma cell line in the absence of P-glycoprotein. Int J Cancer 71:817–824.
5. Futscher, BW, Abbaszadegan, MR, Doman, F, and Dalton, WS. 1994. Analysis of MRP mRNA in mitoxantrone-selected, multidrug resistant human tumor cells. Biochem Pharmacol 47:1601–1606.
6. Yang, CJ, Horton, JK, Cowan, KH, and Schneider, E. 1995. Cross-resistance to camptothecin analogues in a mitoxantrone-resistant human breast carcinoma cell line is not due to DNA topoisomerase I alterations. Cancer Res 55:4004–4009.
7. Chen, Y-N, Mickley, LA, Schwartz, AM, Acton, EM, Hwang, J, and Fojo, AT . 1990. Characterization of adriamycin-resistant human breast cancer cells which display overexpression of a novel resistance-related membrane protein. J Biol Chem 265:10073–10080.
8. Lee, JS, Scala, S, Matsumoto, Y, Dickstein, B, Robey, R, Zhan, Z, Altenberg, G, and Bates, SE. 1997. Reduced drug accumulation and multidrug resistance in human breast cancer cells without associated P-glycoprotein or MRP overexpression. J Cell Biochem 65:513–526.
9. Doyle, LA, Yang, W, Abruzzo, LV, Krogmann, T, Gao, Y, Rishi, AK, and Ross, DD. 1998. A multidrug resistance transporter from human MCF-7 breast cancer cells. Proc Natl Acad Sci U S A 95:15665–15670.
10. Allikmets, R, Schriml, LM, Hutchinson, A, Romano-Spica, V, and Dean, M. 1998. A human placenta-specific ATP-binding cassette gene (ABCP) on chromosome 4q22 that is involved in multidrug resistance. Cancer Res 58:5337–5339.
11. Rabindran, SK, He, H, Singh, M, Brown, E, Collins, KI, Annable, T, and Greenberger, LM. 1998. Reversal of a novel multidrug resistance mechanism in human colon carcinoma cells by fumitremorgin C. Cancer Res 58:5850–5858.
12. Miyake, K, Mickley, L, Litman, T, Zhan, Z, Robey, R, Cristensen, B, Brangi, M, Greenberger, L, Dean, M, Fojo, T, and Bates, SE. 1999. Molecular cloning of cDNAs which

are highly overexpressed in mitoxantrone-resistant cells: demonstration of homology to ABC transport genes. Cancer Res 59:8–13.

13. Dean, M, Rzhetsky, A, and Allikmets, R. 2001. The human ATP-binding cassette (ABC) transporter superfamily. Genome Res 11:1156–1166.
14. Wang, N, Lan, D, Chen, W, Matsuura, F, and Tall, AR. 2004. ATP-binding cassette transporters G1 and G4 mediate cellular cholesterol efflux to high-density lipoproteins. Proc Natl Acad Sci U S A 101:9774–9779.
15. Mickley, L, Jain, P, Miyake, K, Schriml, LM, Rao, K, Fojo, T, Bates, S, and Dean, M. 2001. An ATP-binding cassette gene (ABCG3) closely related to the multidrug transporter ABCG2 (MXR/ABCP) has an unusual ATP-binding domain. Mamm Genome 12:86–88.
16. Yu, L, Li-Hawkins, J, Hammer, RE, Berge, KE, Horton, JD, Cohen, JC, and Hobbs, HH. 2002. Overexpression of ABCG5 and ABCG8 promotes biliary cholesterol secretion and reduces fractional absorption of dietary cholesterol. J Clin Invest 110:671–680.
17. Bailey-Dell, KJ, Hassel, B, Doyle, LA, and Ross, DD. 2001. Promoter characterization and genomic organization of the human breast cancer resistance protein (ATP-binding cassette transporter G2) gene. Biochim Biophys Acta 1520:234–241.
18. Knutsen, T, Rao, VK, Ried, T, Mickley, L, Schneider, E, Miyake, K, Ghadimi, BM, Padilla-Nash, H, Pack, S, Greenberger, L, et al. 2000. Amplification of 4q21-q22 and the MXR gene in independently derived mitoxantrone-resistant cell lines. Genes Chromosomes Cancer 27:110–116.
19. Rao, VK, Wangsa, D, Robey, RW, Huff, L, Honjo, Y, Hung, J, Knutsen, T, Ried, T, and Bates, SE. 2005. Characterization of ABCG2 gene amplification manifesting as extrachromosomal DNA in mitoxantrone-selected SF295 human glioblastoma cells. Cancer Genet Cytogenet 160:126–133.
20. Berkower, C, and Michaelis, S. 1991. Mutational analysis of the yeast a-factor transporter STE6, a member of the ATP binding cassette (ABC) protein superfamily. EMBO J 10:3777–3785.
21. Graf, GA, Yu, L, Li, WP, Gerard, R, Tuma, PL, Cohen, JC, and Hobbs, HH. 2003. ABCG5 and ABCG8 are obligate heterodimers for protein trafficking and biliary cholesterol excretion. J Biol Chem 278:48275–48282.
22. Bhatia, A, Schafer, HJ, and Hrycyna, CA. 2005. Oligomerization of the human ABC transporter ABCG2: evaluation of the native protein and chimeric dimers. Biochemistry 44:10893–10904.
23. Henriksen, U, Gether, U, and Litman, T. 2005. Effect of Walker A mutation (K86M) on oligomerization and surface targeting of the multidrug resistance transporter ABCG2. J Cell Sci 118:1417–1426.
24. Kage, K, Tsukahara, S, Sugiyama, T, Asada, S, Ishikawa, E, Tsuruo, T, and Sugimoto, Y. 2002. Dominant-negative inhibition of breast cancer resistance protein as drug efflux pump through the inhibition of S-S dependent homodimerization. Int J Cancer 97:626–630.
25. Chang, G, and Roth, CB. 2001. Structure of MsbA from *E. coli*: a homolog of the multidrug resistance ATP binding cassette (ABC) transporters. Science 293:1793–1800.
26. Chang, G. 2003. Structure of MsbA from *Vibrio cholera*: a multidrug resistance ABC transporter homolog in a closed conformation. J Mol Biol 330:419–430.
27. Reyes, CL, and Chang, G. 2005. Structure of the ABC transporter MsbA in complex with ADP. vanadate and lipopolysaccharide. Science 308:1028–1031.

28. Locher, KP, Lee, AT, and Rees, DC. 2002. The E. coli BtuCD structure: a framework for ABC transporter architecture and mechanism. Science 296:1091–1098.
29. Henriksen, U, Fog, JU, Litman, T, and Gether, U. 2005. Identification of intra- and intermolecular disulfide bridges in the multidrug resistance transporter ABCG2. J Biol Chem 280:36926–36934.
30. Diop, NK, and Hrycyna, CA. 2005. N-Linked glycosylation of the human ABC transporter ABCG2 on asparagine 596 is not essential for expression, transport activity, or trafficking to the plasma membrane. Biochemistry 44:5420–5429.
31. Li, YF, Polgar, O, Okada, M, Esser, L, Bates, SE and Xia, D. 2006. Towards understanding the mechanism of action of the multidrug resistance-linked half-ABC transporter ABCG2: a molecular modeling study. J Mol Graph Model 25:837–851.
32. Russ, WP, and Engelman, DM. 2000. The GxxxG motif: a framework for transmembrane helix–helix association. J Mol Biol 296:911–919.
33. Gerber, D, and Shai, Y. 2001. In vivo detection of hetero-association of glycophorin-A and its mutants within the membrane. J Biol Chem 276:31229–3132.
34. Polgar, O, Robey, RW, Morisaki, K, Dean, M, Michejda, C, Sauna, ZE, Ambudkar, SV, Tarasova, N, and Bates, SE. 2004. Mutational analysis of ABCG2: role of the GXXXG motif. Biochemistry 43:9448–9456.
35. Loo, TW, Bartlett, MC, and Clarke, DM. 2004. Disulfide cross-linking analysis shows that transmembrane segments 5 and 8 of human P-glycoprotein are close together on the cytoplasmic side of the membrane. J Biol Chem 279:7692–7697.
36. Xu, J, Liu, Y, Yang, Y, Bates, S, and Zhang, JT. 2004. Characterization of oligomeric human half-ABC transporter ATP-binding cassette G2. J Biol Chem 279:19781–9.
37. Mackenzie, SM, Brooker, MR, Gill, TR, Cox, GB, Howells, AJ, and Ewart, GD. 1999. Mutations in the white gene of *Drosophila melanogaster* affecting ABC transporters that determine eye colouration. Biochim Biophys Acta 1419:173–185.
38. Polgar, O, Ozvegy-Laczka, C, Robey, RW, Morisaki, K, Okada, M, Tamaki, A, Koblos, G, Elkind, NB, Ward, Y, Dean, M, Sarkadi, B, and Bates, SE. 2006. Mutational studies of G553 in TM5 of ABCG2: a residue potentially involved in dimerization. Biochemistry.
39. Dey, S, Ramachandra, M, Pastan, I, Gottesman, MM, and Ambudkar, SV. 1997. Evidence for two nonidentical drug-interaction sites in the human P-glycoprotein. Proc Natl Acad Sci U S A 94:10594–10599.
40. Shapiro, AB, and Ling, V. 1997. Positively cooperative sites for drug transport by P-glycoprotein with distinct drug specificities. Eur J Biochem 250:130–137.
41. Ejendal, KF, and Hrycyna, CA. 2005. Differential sensitivities of the human ATP-binding cassette transporters ABCG2 and P-glycoprotein to cyclosporin A. Mol Pharmacol 67:902–911.
42. Miwa, M, Tsukahara, S, Ishikawa, E, Asada, S, Imai, Y, and Sugimoto, Y. 2003. Single amino acid substitutions in the transmembrane domains of breast cancer resistance protein (BCRP) alter cross resistance patterns in transfectants. Int J Cancer 107:757–763.
43. Ozvegy-Laczka, C, Koblos, G, Sarkadi, B, and Varadi, A. 2005. Single amino acid (482) variants of the ABCG2 multidrug transporter: major differences in transport capacity and substrate recognition. Biochim Biophys Acta 1668:53–63.
44. Robey, RW, Honjo, Y, Morisaki, K, Nadjem, TA, Runge, S, Risbood, M, Poruchynsky, MS, and Bates, SE. 2003. Mutations at amino acid 482 in the ABCG2 gene affect substrate and antagonist specificity. Br J Cancer 89:1971–1978.

45. Chen, ZS, Lee, K, Walther, S, Raftogianis, RB, Kuwano, M, Zeng, H, and Kruh, GD. 2002. Analysis of methotrexate and folate transport by multidrug resistance protein 4 (ABCC4): MRP4 is a component of the methotrexate efflux system. Cancer Res 62:3144–3150.

46. Müller, M, Bakos, E, Welker, E, Varadi, A, Germann, UA, Gottesman, MM, Morse, BS, Roninson, IB, and Sarkadi, B. 1996. Altered drug-stimulated ATPase activity in mutants of the human multidrug resistance protein. J Biol Chem 271:1877–1883.

47. Ozvegy, C, Varadi, A, and Sarkadi, B. 2002. Characterization of drug transport, ATP hydrolysis and nucleotide trapping by the human ABCG2 multidrug transporter: modulation of substrate specificity by a point mutation. J Biol Chem 277:47980–47990.

48. Ee, PL, Kamalakaran, S, Tonetti, D, He, X, Ross, DD, and Beck, WT. 2004. Identification of a novel estrogen response element in the breast cancer resistance protein (ABCG2) gene. Cancer Res 64:1247–1251.

49. Imai, Y, Ishikawa, E, Asada, S, and Sugimoto, Y. 2005. Estrogen-mediated post transcriptional down-regulation of breast cancer resistance protein/ABCG2. Cancer Res 65:596–604.

50. Wang, H, Zhou, L, Gupta, A, Vethanayagam, RR, Zhang, Y, Unadkat, JD, and Mao, Q. 2005. Regulation of BCRP/ABCG2 expression by progesterone and 17{beta}-estradiol in human placental BeWo cells. Am J Physiol Endocrinol Metab 290:E798–807.

51. Yasuda, S, Itagaki, S, Hirano, T, and Iseki, K. 2005. Expression level of ABCG2 in the placenta decreases from the mid stage to the end of gestation. Biosci Biotechnol Biochem 69:1871–1876.

52. Tanaka, Y, Slitt, AL, Leazer, TM, Maher, JM, and Klaassen, CD. 2005. Tissue distribution and hormonal regulation of the breast cancer resistance protein (Bcrp/Abcg2) in rats and mice. Biochem Biophys Res Commun 326:181–187.

53. Merino, G, Jonker, JW, Wagenaar, E, van Herwaarden, AE, and Schinkel, AH. 2005. The breast cancer resistance protein (BCRP/ABCG2) affects pharmacokinetics, hepatobiliary excretion, and milk secretion of the antibiotic nitrofurantoin. Mol Pharmacol 67:1758–1764.

54. Krishnamurthy, P, Ross, DD, Nakanishi, T, Bailey-Dell, K, Zhou, S, Mercer, KE, Sarkadi, B, Sorrentino, BP, and Schuetz, JD. 2004. The stem cell marker Bcrp/ABCG2 enhances hypoxic cell survival through interactions with heme. J Biol Chem 279:24218–24225.

55. Mogi, M, Yang, J, Lambert, JF, Colvin, GA, Shiojima, I, Skurk, C, Summer, R, Fine, A, Quesenberry, PJ, and Walsh, K. 2003. Akt signaling regulates side population cell phenotype via Bcrp1 translocation. J Biol Chem 278:39068–39075.

56. Takada, T, Suzuki, H, Gotoh, Y, and Sugiyama, Y. 2005. Regulation of the cell surface expression of human BCRP/ABCG2 by the phosphorylation state of Akt in polarized cells. Drug Metab Dispos 33:905–909.

57. Mickley, LA, Spengler, BA, Knutsen, TA, Biedler, JL and Fojo, T. 1997. Gene rearrangement: a novel mechanism for *MDR-1* gene activation. J Clin Invest 99:1947–1957.

58. Hazlehurst, LA, Foley, NE, Gleason-Guzman, MC, Hacker, MP, Cress, AE, Greenberger, LW, De Jong, MC, and Dalton, WS. 1999. Multiple mechanisms confer drug resistance to mitoxantrone in the human 8226 myeloma cell line. Cancer Res 59:1021–1028.

59. Allen, JD, Brinkhuis, RF, Wijnholds, J, and Schinkel, AH. 1999. The mouse Bcrp1/Mxr/Abcp gene: amplification and overexpression in cell lines selected for resistance to topotecan, mitoxantrone, or doxorubicin. Cancer Res 59:4237–4241.

60. Komatani, H, Kotani, H, Hara, Y, Nakagawa, R, Matsumoto, M, Arakawa, H, and Nishimura, S. 2001. Identification of breast cancer resistant protein/mitoxantrone resistance/placenta-specific, ATP-binding cassette transporter as a transporter of NB-506 and J-107088, topoisomerase I inhibitors with an indolocarbazole structure. Cancer Res 61:2827–2832.

61. Ma, J, Maliepaard, M, Nooter, K, Loos, WJ, Kolker, HJ, Verweij, J, Stoter, G, and Schellens, JH. 1998. Reduced cellular accumulation of topotecan: a novel mechanism of resistance in a human ovarian cancer cell line. Br J Cancer 77:1645–1652.

62. Maliepaard, M, van Gastelen, MA, de Jong, LA, Pluim, D, van Waardenburg, RC, Ruevekamp-Helmers, MC, Floot, BG, and Schellens, JH. 1999. Overexpression of the BCRP/MXR/ABCP gene in a topotecan-selected ovarian tumor cell line. Cancer Res 59:4559–4563.

63. Kawabata, S, Oka, M, Shiozawa, K, Tsukamoto, K, Nakatomi, K, Soda, H, Fukuda, M, Ikegami, Y, Sugahara, K, Yamada, Y, et al. 2001. Breast cancer resistance protein directly confers SN-38 resistance of lung cancer cells. Biochem Biophys Res Commun 280:1216–1223.

64. Ishii, M, Iwahana, M, Mitsui, I, Minami, M, Imagawa, S, Tohgo, A, and Ejima, A. 2000. Growth inhibitory effect of a new camptothecin analog, DX-8951f, on various drug-resistant sublines including BCRP-mediated camptothecin derivative-resistant variants derived from the human lung cancer cell line PC-6. Anticancer Drugs 11:353–362.

65. van Hattum, AH, Hoogsteen, IJ, Schluper, HM, Maliepaard, M, Scheffer, GL, Scheper, RJ, Kohlhagen, G, Pommier, Y, Pinedo, HM, and Boven, E. 2002. Induction of breast cancer resistance protein by the camptothecin derivative DX-8951f is associated with minor reduction of antitumour activity. Br J Cancer 87:665–672.

66. van Hattum, AH, Schluper, HM, Hausheer, FH, Pinedo, HM, and Boven, E. 2002. Novel camptothecin derivative BNP1350 in experimental human ovarian cancer: determination of efficacy and possible mechanisms of resistance. Int J Cancer 100:22–29.

67. Volk, EL, Farley, KM, Wu, Y, Li, F, Robey, RW, and Schneider, E. 2002. Overexpression of wild-type breast cancer resistance protein mediates methotrexate resistance. Cancer Res 62:5035–5040.

68. Robey, RW, Medina-Perez, WY, Nishiyama, K, Lahusen, T, Miyake, K, Litman, T, Senderowicz, AM, Ross, DD, and Bates, SE. 2001. Overexpression of the ATP-binding cassette half-transporter, ABCG2 (MXR/BCRP/ABCP1), in flavopiridol-resistant human breast cancer cells. Clin Cancer Res 7:145–152.

69. Nakanishi, T, Karp, JE, Tan, M, Doyle, LA, Peters, T, Yang, W, Wei, D, and Ross, DD. 2003. Quantitative analysis of breast cancer resistance protein and cellular resistance to flavopiridol in acute leukemia patients. Clin Cancer Res 9:3320–3328.

70. Erlichman, C, Boerner, SA, Hallgren, CG, Spieker, R, Wang, XY, James, CD, Scheffer, GL, Maliepaard, M, Ross, DD, Bible, KC, and Kaufmann, SH. 2001. The HER tyrosine kinase inhibitor CI1033 enhances cytotoxicity of 7-ethyl-10-hydroxycamptothecin and topotecan by inhibiting breast cancer resistance protein-mediated drug efflux. Cancer Res 61:739–748.

71. Elkind, NB, Szentpetery, Z, Apati, A, Ozvegy-Laczka, C, Varady, G, Ujhelly, O, Szabo, K, Homolya, L, Varadi, A, Buday, L, et al. 2005. Multidrug transporter ABCG2 prevents tumor cell death induced by the epidermal growth factor receptor inhibitor Iressa (ZD1839, Gefitinib). Cancer Res 65:1770–1777.

72. Burger, H, van Tol, H, Boersma, AW, Brok, M, Wiemer, EA, Stoter, G, and Nooter, K. 2004. Imatinib mesylate (STI571) is a substrate for the breast cancer resistance protein (BCRP)/ABCG2 drug pump. Blood 104:2940–2942.

73. Burger, H, van Tol, H, Brok, M, Wiemer, EA, de Bruijn, EA, Guetens, G, de Boeck, G, Sparreboom, A, Verweij, J, and Nooter, K. 2005. Chronic imatinib mesylate exposure leads to reduced intracellular drug accumulation by induction of the ABCG2 (BCRP) and ABCB1 (MDR1) drug transport pumps. Cancer Biol Ther 4:747–752.

74. Robey, RW, Honjo, Y, van de Laar, A, Miyake, K, Regis, JT, Litman, T, and Bates, SE. 2001. A functional assay for detection of the mitoxantrone resistance protein, MXR (ABCG2). Biochim Biophys Acta 1512:171–182.

75. Honjo, Y, Hrycyna, CA, Yan, QW, Medina-Perez, WY, Robey, RW, van de Laar, A, Litman, T, Dean, M, and Bates, SE. 2001. Acquired mutations in the MXR/BCRP/ABCP gene alter substrate specificity in MXR/BCRP/ABCP-overexpressing cells. Cancer Res 61:6635–6639.

76. Wang, X, Furukawa, T, Nitanda, T, Okamoto, M, Sugimoto, Y, Akiyama, S, and Baba, M. 2003. Breast cancer resistance protein (BCRP/ABCG2) induces cellular resistance to HIV-1 nucleoside reverse transcriptase inhibitors. Mol Pharmacol 63:65–72.

77. Allen, JD, Jackson, SC, and Schinkel, AH. 2002. A mutation hot spot in the Bcrp1 (Abcg2) multidrug transporter in mouse cell lines selected for doxorubicin resistance. Cancer Res 62:2294–2299.

78. Plasschaert, SL, van der Kolk, DM, de Bont, ES, Kamps, WA, Morisaki, K, Bates, SE, Scheffer, GL, Scheper, RJ, Vellenga, E, and de Vries, EG. 2003. The role of breast cancer resistance protein in acute lymphoblastic leukemia. Clin Cancer Res 9:5171–5177.

79. Litman, T, Brangi, M, Hudson, E, Fetsch, P, Abati, A, Ross, DD, Miyake, K, Resau, JH, and Bates, SE. 2000. The multidrug-resistant phenotype associated with overexpression of the new ABC half-transporter, MXR (ABCG2). J Cell Sci 113:2011–2021.

80. Minderman, H, Suvannasankha, A, O'Loughlin, KL, Scheffer, GL, Scheper, RJ, Robey, RW, and Baer, MR. 2002. Flow cytometric analysis of breast cancer resistance protein expression and function. Cytometry 48:59–65.

81. Kawabata, S, Oka, M, Soda, H, Shiozawa, K, Nakatomi, K, Tsurutani, J, Nakamura, Y, Doi, S, Kitazaki, T, Sugahara, K, et al. 2003. Expression and functional analyses of breast cancer resistance protein in lung cancer. Clin Cancer Res 9:3052–3057.

82. Zhou, S, Schuetz, JD, Bunting, KD, Colapietro, AM, Sampath, J, Morris, JJ, Lagutina, I, Grosveld, GC, Osawa, M, Nakauchi, H, and Sorrentino, BP. 2001. The ABC transporter Bcrp1/ABCG2 is expressed in a wide variety of stem cells and is a molecular determinant of the side-population phenotype. Nat Med 7:1028–1034.

83. Jonker, JW, Buitelaar, M, Wagenaar, E, van der Valk, MA, Scheffer, GL, Scheper, RJ, Plosch, T, Kuipers, F, Elferink, RP, Rosing, H, Beijnen, JH, and Schinkel, AH. 2002. The breast cancer resistance protein protects against a major chlorophyll-derived dietary phototoxin and protoporphyria. Proc Natl Acad Sci U S A 99:15649–15654.

84. Robey, RW, Steadman, K, Polgar, O, Morisaki, K, Blayney, M, Mistry, P, and Bates, SE. 2004. Pheophorbide a is a specific probe for ABCG2 function and inhibition. Cancer Res 64:1242–1246.

85. Robey, RW, Steadman, K, Polgar, O, and Bates, SE. 2005. ABCG2-mediated transport of photosensitizers: potential impact on photodynamic therapy. Cancer Biol Ther 4:187–194.

86. Janvilisri, T, Venter, H, Shahi, S, Reuter, G, Balakrishnan, L, and van Veen, HW. 2003. Sterol transport by the human breast cancer resistance protein (ABCG2) expressed in *Lactococcus lactis*. J Biol Chem 278:20645–20651.

87. Imai, Y, Asada, S, Tsukahara, S, Ishikawa, E, Tsuruo, T, and Sugimoto, Y. 2003. Breast cancer resistance protein exports sulfated estrogens but not free estrogens. Mol Pharmacol 64:610–618.

88. Pavek, P, Merino, G, Wagenaar, E, Bolscher, E, Novotna, M, Jonker, JW, and Schinkel, AH. 2005. Human breast cancer resistance protein: interactions with steroid drugs, hormones, the dietary carcinogen 2-amino-1-methyl-6-phenylimidazo(4,5-*b*)pyridine, and transport of cimetidine. J Pharmacol Exp Ther 312:144–152.

89. Suzuki, M, Suzuki, H, Sugimoto, Y, and Sugiyama, Y. 2003. ABCG2 transports sulfated conjugates of steroids and xenobiotics. J Biol Chem 278:22644–22649.

90. Merino, G, Alvarez, AI, Pulido, MM, Molina, AJ, Schinkel, AH, and Prieto, JG. 2006. Breast cancer resistance protein (BCRP/ABCG2) transports fluoroquinolone antibiotics and affects their oral availability, pharmacokinetics and milk secretion. Drug Metab Dispos 34:690–695.

91. Janvilisri, T, Shahi, S, Venter, H, Balakrishnan, L, and van Veen, HW. 2005. Arginine-482 is not essential for transport of antibiotics, primary bile acids and unconjugated sterols by the human breast cancer resistance protein (ABCG2). Biochem J 385:419–426.

92. Gupta, A, Zhang, Y, Unadkat, JD, and Mao, Q. 2004. HIV protease inhibitors are inhibitors but not substrates of the human breast cancer resistance protein (BCRP/ABCG2). J Pharmacol Exp Ther 310:334–341.

93. Huang, L, Wang, Y, and Grimm, SW. 2006. ATP-dependent transport of rosuvastatin in membrane vesicles expressing breast cancer resistant protein. Drug Metab Dispos.

94. Hirano, M, Maeda, K, Matsushima, S, Nozaki, Y, Kusuhara, H, and Sugiyama, Y. 2005. Involvement of BCRP (ABCG2) in the biliary excretion of pitavastatin. Mol Pharmacol 68:800–807.

95. Matsushima, S, Maeda, K, Kondo, C, Hirano, M, Sasaki, M, Suzuki, H, and Sugiyama, Y. 2005. Identification of the hepatic efflux transporters of organic anions using double-transfected Madin–Darby canine kidney II cells expressing human organic anion-transporting polypeptide 1B1 (OATP1B1)/multidrug resistance-associated protein 2, OATP1B1/multidrug resistance 1, and OATP1B1/breast cancer resistance protein. J Pharmacol Exp Ther 314:1059–1067.

96. Morris, ME, and Zhang, S. 2006. Flavonoid-drug interactions: effects of flavonoids on ABC transporters. Life Sci 78:2116–2130.

97. Sesink, AL, Arts, IC, de Boer, VC, Breedveld, P, Schellens, JH, Hollman, PC, and Russel, FG. 2005. Breast cancer resistance protein (Bcrp1/Abcg2) limits net intestinal uptake of quercetin in rats by facilitating apical efflux of glucuronides. Mol Pharmacol 67:1999–2006.

98. Rabindran, SK, Ross, DD, Doyle, LA, Yang, W, and Greenberger, LM. 2000. Fumitremorgin C reverses multidrug resistance in cells transfected with the breast cancer resistance protein. Cancer Res 60:47–50.

99. Allen, JD, van Loevezijn, A, Lakhai, JM, van der Valk, M, van Tellingen, O, Reid, G, Schellens, JHM, Koomen, G-J, and Schinkel, AH. 2002. Potent and specific inhibition of the breast cancer resistance protein multidrug transporter in vitro and in mouse intestine by a novel analogue of fumitremorgin C. Mol Cancer Ther 1:417–425.

100. van Loevezijn, A, Allen, JD, Schinkel, AH, and Koomen, GJ. 2001. Inhibition of BCRP-mediated drug efflux by fumitremorgin-type indolyl diketopiperazines. Bioorg Med Chem Lett 11:29–32.

101. Woehlecke, H, Osada, H, Herrmann, A, and Lage, H. 2003. Reversal of breast cancer resistance protein-mediated drug resistance by tryprostatin A. Int J Cancer 107:721–728.

102. de Bruin, M, Miyake, K, Litman, T, Robey, R, and Bates, SE. 1999. Reversal of resistance by GF120918 in cell lines expressing the ABC half-transporter, MXR. Cancer Lett 146:117–126.

103. Qadir, M, O'Loughlin, KL, Fricke, SM, Williamson, NA, Greco, WR, Minderman, H, and Baer, MR. 2005. Cyclosporin A is a broad-spectrum multidrug resistance modulator. Clin Cancer Res 11:2320–2326.

104. Gupta, A, Dai, Y, Vethanayagam, RR, Hebert, MF, Thummel, KE, Unadkat, JD, Ross, DD, and Mao, Q. 2006. Cyclosporin A, tacrolimus and sirolimus are potent inhibitors of the human breast cancer resistance protein (ABCG2) and reverse resistance to mitoxantrone and topotecan. Cancer Chemother Pharmacol 1–10.

105. Imai, Y, Tsukahara, S, Ishikawa, E, Tsuruo, T, and Sugimoto, Y. 2002. Estrone and 17beta-estradiol reverse breast cancer resistance protein-mediated multidrug resistance. Jpn J Cancer Res 93:231–235.

106. Sugimoto, Y, Tsukahara, S, Imai, Y, Sugimoto, Y, Ueda, K, and Tsuruo, T. 2003. Reversal of breast cancer resistance protein-mediated drug resistance by estrogen antagonists and agonists. Mol Cancer Ther 2:105–112.

107. Yanase, K, Tsukahara, S, Asada, S, Ishikawa, E, Imai, Y, and Sugimoto, Y. 2004. Gefitinib reverses breast cancer resistance protein-mediated drug resistance. Mol Cancer Ther 3:1119–1125.

108. Nakamura, Y, Oka, M, Soda, H, Shiozawa, K, Yoshikawa, M, Itoh, A, Ikegami, Y, Tsurutani, J, Nakatomi, K, Kitazaki, T, et al. 2005. Gefitinib ("Iressa," ZD1839), an epidermal growth factor receptor tyrosine kinase inhibitor, reverses breast cancer resistance protein/ABCG2-mediated drug resistance. Cancer Res 65:1541–1546.

109. Ozvegy-Laczka, C, Hegedus, T, Varady, G, Ujhelly, O, Schuetz, JD, Varadi, A, Keri, G, Orfi, L, Nemet, K, and Sarkadi, B. 2004. High-affinity interaction of tyrosine kinase inhibitors with the ABCG2 multidrug transporter. Mol Pharmacol 65:1485–1495.

110. Houghton, PJ, Germain, GS, Harwood, FC, Schuetz, JD, Stewart, CF, Buchdunger, E, and Traxler, P. 2004. Imatinib mesylate is a potent inhibitor of the ABCG2 (BCRP) transporter and reverses resistance to topotecan and SN-38 in vitro. Cancer Res 64:2333–2337.

111. Cooray, HC, Janvilisri, T, van Veen, HW, Hladky, SB, and Barrand, MA. 2004. Interaction of the breast cancer resistance protein with plant polyphenols. Biochem Biophys Res Commun 317:269–275.

112. Zhang, S, Yang, X, and Morris, ME. 2004. Flavonoids are inhibitors of breast cancer resistance protein (ABCG2)-mediated transport. Mol Pharmacol 65:1208–1216.

113. Imai, Y, Tsukahara, S, Asada, S, and Sugimoto, Y. 2004. Phytoestrogens/flavonoids reverse breast cancer resistance protein/ABCG2-mediated multidrug resistance. Cancer Res 64:4346–4352.

114. Zhang, S, Yang, X, Coburn, RA, and Morris, ME. 2005. Structure activity relationships and quantitative structure activity relationships for the flavonoid-mediated inhibition of breast cancer resistance protein. Biochem Pharmacol 70:627–639.

115. Ahmed-Belkacem, A, Pozza, A, Munoz-Martinez, F, Bates, SE, Castanys, S, Gamarro, F, Di Pietro, A, and Perez-Victoria, JM. 2005. Flavonoid structure-activity studies identify 6-prenylchrysin and tectochrysin as potent and specific inhibitors of breast cancer resistance protein ABCG2. Cancer Res 65:4852–4860.

116. Ji, Y, and Morris, ME. 2005. Membrane transport of dietary phenethyl isothiocyanate by ABCG2 (breast cancer resistance protein). Mol Pharm 2:414–419.

117. Zhou, XF, Yang, X, Wang, Q, Coburn, RA, and Morris, ME. 2005. Effects of dihydropyridines and pyridines on multidrug resistance mediated by breast cancer resistance protein: in vitro and in vivo studies. Drug Metab Dispos 33:1220–1228.

118. Zhang, Y, Gupta, A, Wang, H, Zhou, L, Vethanayagam, RR, Unadkat, JD, and Mao, Q. 2005. BCRP transports dipyridamole and is inhibited by calcium channel blockers. Pharm Res 22:2023–2034.

119. Breedveld, P, Zelcer, N, Pluim, D, Sonmezer, O, Tibben, MM, Beijnen, JH, Schinkel, AH, van Tellingen, O, Borst, P, and Schellens, JH. 2004. Mechanism of the pharmacokinetic interaction between methotrexate and benzimidazoles: potential role for breast cancer resistance protein in clinical drug–drug interactions. Cancer Res 64:5804–5811.

120. Minderman, H, Brooks, TA, O'Loughlin, KL, Ojima, I, Bernacki, RJ, and Baer, MR. 2004. Broad-spectrum modulation of ATP-binding cassette transport proteins by the taxane derivatives ortataxel (IDN-5109, BAY 59-8862) and tRA96023. Cancer Chemother Pharmacol 53:363–369.

121. Minderman, H, O'Loughlin, KL, Pendyala, L, and Baer, MR. 2004. VX-710 (biricodar) increases drug retention and enhances chemosensitivity in resistant cells overexpressing P-glycoprotein, multidrug resistance protein, and breast cancer resistance protein. Clin Cancer Res 10:1826–1834.

122. Brooks, TA, Minderman, H, O'Loughlin, KL, Pera, P, Ojima, I, Baer, MR, and Bernacki, RJ. 2003. Taxane-based reversal agents modulate drug resistance mediated by P-glycoprotein, multidrug resistance protein, and breast cancer resistance protein. Mol Cancer Ther 2:1195–1205.

123. Fetsch, PA, Abati, A, Litman, T, Morisaki, K, Honjo, Y, Mittal, K, and Bates, SE. 2005. Localization of the ABCG2 mitoxantrone resistance-associated protein in normal tissues. Cancer Lett 235:84–92.

124. Maliepaard, M, Scheffer, GL, Faneyte, IF, van Gastelen, MA, Pijnenborg, AC, Schinkel, AH, van De Vijver, MJ, Scheper, RJ, and Schellens, JH. 2001. Subcellular localization and distribution of the breast cancer resistance protein transporter in normal human tissues. Cancer Res 61:3458–3464.

125. Scheffer, GL, Maliepaard, M, Pijnenborg, AC, van Gastelen, MA, de Jong, MC, Schroeijers, AB, van der Kolk, DM, Allen, JD, Ross, DD, van der Valk, P, et al. 2000. Breast cancer resistance protein is localized at the plasma membrane in mitoxantrone- and topotecan-resistant cell lines. Cancer Res 60:2589–2593.

126. Faneyte, IF, Kristel, PM, Maliepaard, M, Scheffer, GL, Scheper, RJ, Schellens, JH, and van de Vijver, MJ. 2002. Expression of the breast cancer resistance protein in breast cancer. Clin Cancer Res 8:1068–1074.

127. Goodell, MA, Rosenzweig, M, Kim, H, Marks, DF, DeMaria, M, Paradis, G, Grupp, SA, Sieff, CA, Mulligan, RC, and Johnson, RP. 1997. Dye efflux studies suggest that hematopoietic stem cells expressing low or undetectable levels of CD34 antigen exist in multiple species. Nat Med 3:1337–1345.

128. Summer, R, Kotton, DN, Sun, X, Ma, B, Fitzsimmons, K, and Fine, A. 2003. Side population cells and Bcrp1 expression in lung. Am J Physiol Lung Cell Mol Physiol 285:L97–L104.

129. Alvi, AJ, Clayton, H, Joshi, C, Enver, T, Ashworth, A, Vivanco, MM, Dale, TC, and Smalley, MJ. 2003. Functional and molecular characterisation of mammary side population cells. Breast Cancer Res 5:R1–R8.

130. Doyle, LA, and Ross, DD. 2003. Multidrug resistance mediated by the breast cancer resistance protein BCRP (ABCG2). Oncogene 22:7340–7358.

131. Jonker, JW, Smit, JW, Brinkhuis, RF, Maliepaard, M, Beijnen, JH, Schellens, JH, and Schinkel, AH. 2000. Role of breast cancer resistance protein in the bioavailability and fetal penetration of topotecan. J Natl Cancer Inst 92:1651–1656.

132. Jonker, JW, Merino, G, Musters, S, van Herwaarden, AE, Bolscher, E, Wagenaar, E, Mesman, E, Dale, TC, and Schinkel, AH. 2005. The breast cancer resistance protein BCRP (ABCG2) concentrates drugs and carcinogenic xenotoxins into milk. Nat Med 11:127–129.

133. van Herwaarden, AE, Wagenaar, E, Karnekamp, B, Merino, G, Jonker, JW, and Schinkel, AH. 2006. Breast cancer resistance protein (Bcrp1/Abcg2) reduces systemic exposure of the dietary carcinogens aflatoxin B1, IQ and Trp-P-1 but also mediates their secretion into breast milk. Carcinogenesis 27:123–130.

134. Bart, J, Hollema, H, Groen, HJ, de Vries, EG, Hendrikse, NH, Sleijfer, DT, Wegman, TD, Vaalburg, W, and van der Graaf, WT. 2004. The distribution of drug-efflux pumps, P-gp, BCRP, MRP1 and MRP2, in the normal blood–testis barrier and in primary testicular tumours. Eur J Cancer 40:2064–2070.

135. Cooray, HC, Blackmore, CG, Maskell, L, and Barrand, MA. 2002. Localisation of breast cancer resistance protein in microvessel endothelium of human brain. Neuroreport 13:2059–2063.

136. Zhang, W, Mojsilovic-Petrovic, J, Andrade, MF, Zhang, H, Ball, M, and Stanimirovic, DB. 2003. The expression and functional characterization of ABCG2 in brain endothelial cells and vessels. FASEB J 17:2085–2087.

137. Cisternino, S, Mercier, C, Bourasset, F, Roux, F, and Scherrmann, JM. 2004. Expression, up-regulation, and transport activity of the multidrug-resistance protein Abcg2 at the mouse blood–brain barrier. Cancer Res 64:3296–3301.

138. van Herwaarden, AE, Jonker, JW, Wagenaar, E, Brinkhuis, RF, Schellens, JH, Beijnen, JH, and Schinkel, AH. 2003. The breast cancer resistance protein (Bcrp1/Abcg2) restricts exposure to the dietary carcinogen 2-amino-1-methyl-6-phenylimidazo[4,5-b]pyridine. Cancer Res 63:6447–6452.

139. Taipalensuu, J, Tornblom, H, Lindberg, G, Einarsson, C, Sjoqvist, F, Melhus, H, Garberg, P, Sjostrom, B, Lundgren, B, and Artursson, P. 2001. Correlation of gene expression of ten drug efflux proteins of the ATP-binding cassette transporter family in normal human jejunum and in human intestinal epithelial Caco-2 cell monolayers. J Pharmacol Exp Ther 299:164–170.

140. Adachi, Y, Suzuki, H, Schinkel, AH, and Sugiyama, Y. 2005. Role of breast cancer resistance protein (Bcrp1/Abcg2) in the extrusion of glucuronide and sulfate conjugates from enterocytes to intestinal lumen. Mol Pharmacol 67:923–928.

141. Morisaki, K, Robey, RW, Ozvegy-Laczka, C, Honjo, Y, Polgar, O, Steadman, K, Sarkadi, B, and Bates, SE. 2005. Single nucleotide polymorphisms modify the transporter activity of ABCG2. Cancer Chemother Pharmacol 56:161–172.

142. Vethanayagam, RR, Wang, H, Gupta, A, Zhang, Y, Lewis, F, Unadkat, JD, and Mao, Q. 2005. Functional analysis of the human variants of breast cancer resistance protein: I206L, N590Y, and D620N. Drug Metab Dispos 33:697–705.

143. Imai, Y, Nakane, M, Kage, K, Tsukahara, S, Ishikawa, E, Tsuruo, T, Miki, Y, and Sugimoto, Y. 2002. C421A polymorphism in the human breast cancer resistance protein gene is associated with low expression of Q141K protein and low-level drug resistance. Mol Cancer Ther 1:611–616.

144. Kondo, C, Suzuki, H, Itoda, M, Ozawa, S, Sawada, J, Kobayashi, D, Ieiri, I, Mine, K, Ohtsubo, K, and Sugiyama, Y. 2004. Functional analysis of SNPs variants of BCRP/ABCG2. Pharm Res 21:1895–1903.

145. Kobayashi, D, Ieiri, I, Hirota, T, Takane, H, Maegawa, S, Kigawa, J, Suzuki, H, Nanba, E, Oshimura, M, Terakawa, N, et al. 2005. Functional assessment of ABCG2 (BCRP) gene polymorphisms to protein expression in human placenta. Drug Metab Dispos 33:94–101.

146. Mizuarai, S, Aozasa, N, and Kotani, H. 2004. Single nucleotide polymorphisms result in impaired membrane localization and reduced atpase activity in multidrug transporter ABCG2. Int J Cancer 109:238–246.

147. Sparreboom, A, Loos, WJ, Burger, H, Sissung, TM, Verweij, J, Figg, WD, Nooter, K, and Gelderblom, H. 2005. Effect of ABCG2 genotype on the oral bioavailability of topotecan. Cancer Biol Ther 4:650–658.

148. Sparreboom, A, Gelderblom, H, Marsh, S, Ahluwalia, R, Obach, R, Principe, P, Twelves, C, Verweij, J, and McLeod, HL. 2004. Diflomotecan pharmacokinetics in relation to ABCG2 421C>A genotype. Clin Pharmacol Ther 76:38–44.

149. Ross, DD, Karp, JE, Chen, TT, and Doyle, LA. 2000. Expression of breast cancer resistance protein in blast cells from patients with acute leukemia. Blood 96:365–368.

150. Galimberti, S, Guerrini, F, Palumbo, GA, Consoli, U, Fazzi, R, Morabito, F, Santini, V, and Petrini, M. 2004. Evaluation of BCRP and MDR-1 co-expression by quantitative molecular assessment in AML patients. Leuk Res 28:367–372.

151. Suvannasankha, A, Minderman, H, O'Loughlin, KL, Nakanishi, T, Greco, WR, Ross, DD, and Baer, MR. 2004. Breast cancer resistance protein (BCRP/MXR/ABCG2) in acute myeloid leukemia: discordance between expression and function. Leukemia 18:1252–1257.

152. van der Kolk, DM, Vellenga, E, Scheffer, GL, Muller, M, Bates, SE, Scheper, RJ, and De Vries, EG. 2002. Expression and activity of breast cancer resistance protein (BCRP) in de novo and relapsed acute myeloid leukemia. Blood 99:3763–3770.

153. van der Pol, MA, Broxterman, HJ, Pater, JM, Feller, N, van der Maas, M, Weijers, GW, Scheffer, GL, Allen, JD, Scheper, RJ, van Loevezijn, A, Ossenkoppele, GJ, and Schuurhuis, GJ. 2003. Function of the ABC transporters, P-glycoprotein, multidrug resistance protein and breast cancer resistance protein, in minimal residual disease in acute myeloid leukemia. Haematologica 88:134–147.

154. Abbott, BL, Colapietro, AM, Barnes, Y, Marini, F, Andreeff, M, and Sorrentino, BP. 2002. Low levels of ABCG2 expression in adult AML blast samples. Blood 100:4594–4601.

155. Steinbach, D, Sell, W, Voigt, A, Hermann, J, Zintl, F, and Sauerbrey, A. 2002. BCRP gene expression is associated with a poor response to remission induction therapy in childhood acute myeloid leukemia. Leukemia 16:1443–1447.

156. Benderra, Z, Faussat, AM, Sayada, L, Perrot, JY, Chaoui, D, Marie, JP, and Legrand, O. 2004. Breast cancer resistance protein and P-glycoprotein in 149 adult acute myeloid leukemias. Clin Cancer Res 10:7896–7902.

157. Uggla, B, Stahl, E, Wagsater, D, Paul, C, Karlsson, MG, Sirsjo, A, and Tidefelt, U. 2005. BCRP mRNA expression v. clinical outcome in 40 adult AML patients. Leuk Res 29:141–146.

158. Stam, RW, van den Heuvel-Eibrink, MM, den Boer, ML, Ebus, ME, Janka-Schaub, GE, Allen, JD, and Pieters, R. 2004. Multidrug resistance genes in infant acute lymphoblastic leukemia: Ara-C is not a substrate for the breast cancer resistance protein. Leukemia 18:78–83.

159. Suvannasankha, A, Minderman, H, O'Loughlin, KL, Nakanishi, T, Ford, LA, Greco, WR, Wetzler, M, Ross, DD, and Baer, MR. 2004. Breast cancer resistance protein (BCRP/MXR/ABCG2) in adult acute lymphoblastic leukaemia: frequent expression and possible correlation with shorter disease-free survival. Br J Haematol 127:392–398.

160. Sauerbrey, A, Sell, W, Steinbach, D, Voigt, A, and Zintl, F. 2002. Expression of the BCRP gene (ABCG2/MXR/ABCP) in childhood acute lymphoblastic leukaemia. Br J Haematol 118:147–150.

161. Diestra, JE, Scheffer, GL, Catala, I, Maliepaard, M, Schellens, JH, Scheper, RJ, Germa-Lluch, JR, and Izquierdo, MA. 2002. Frequent expression of the multi-drug resistance-associated protein BCRP/MXR/ABCP/ABCG2 in human tumours detected by the BXP-21 monoclonal antibody in paraffin-embedded material. J Pathol 198:213–219.

162. Kanzaki, A, Toi, M, Nakayama, K, Bando, H, Mutoh, M, Uchida, T, Fukumoto, M, and Takebayashi, Y. 2001. Expression of multidrug resistance-related transporters in human breast carcinoma. Jpn J Cancer Res 92:452–458.

163. Burger, H, Foekens, JA, Look, MP, Meijer-van Gelder, ME, Klijn, JG, Wiemer, EA, Stoter, G, and Nooter, K. 2003. RNA expression of breast cancer resistance protein, lung resistance-related protein, multidrug resistance-associated proteins 1 and 2, and multidrug resistance gene 1 in breast cancer: correlation with chemotherapeutic response. Clin Cancer Res 9:827–836.

164. Zurita, AJ, Diestra, JE, Condom, E, Garcia Del Muro, X, Scheffer, GL, Scheper, RJ, Perez, J, Germa-Lluch, JR, and Izquierdo, MA. 2003. Lung resistance-related protein as a predictor of clinical outcome in advanced testicular germ-cell tumours. Br J Cancer 88:879–886.

165. Theou, N, Gil, S, Devocelle, A, Julie, C, Lavergne-Slove, A, Beauchet, A, Callard, P, Farinotti, R, Le Cesne, A, Lemoine, A, Faivre-Bonhomme, L, and Emile, JF. 2005. Multidrug resistance proteins in gastrointestinal stromal tumors: site-dependent expression and initial response to imatinib. Clin Cancer Res 11:7593–7598.

166. Yoh, K, Ishii, G, Yokose, T, Minegishi, Y, Tsuta, K, Goto, K, Nishiwaki, Y, Kodama, T, Suga, M, and Ochiai, A. 2004. Breast cancer resistance protein impacts clinical outcome in platinum-based chemotherapy for advanced non-small cell lung cancer. Clin Cancer Res 10:1691–1697.

167. Beck, WT, Grogan, TM, Willman, CL, Cordon-Cardo, C, Parham, DM, Kuttesch, JF, Andreeff, M, Bates, SE, Berard, CW, Boyett, JM, et al. 1996. Methods to detect P-glycoprotein–associated multidrug resistance in patients' tumors: Consensus recommendations. Cancer Res 56:3010–3020.

168. Hirschmann-Jax, C, Foster, AE, Wulf, GG, Nuchtern, JG, Jax, TW, Gobel, U, Goodell, MA, and Brenner, MK. 2004. A distinct "side population" of cells with high drug efflux capacity in human tumor cells. Proc Natl Acad Sci U S A 101:14228–14233.

169. Kondo, T, Setoguchi, T, and Taga, T. 2004. Persistence of a small subpopulation of cancer stem-like cells in the C6 glioma cell line. Proc Natl Acad Sci U S A 101:781–786.

170. Seigel, GM, Campbell, LM, Narayan, M, and Gonzalez-Fernandez, F. 2005. Cancer stem cell characteristics in retinoblastoma. Mol Vis 11:729–737.

171. Haraguchi, N, Utsunomiya, T, Inoue, H, Tanaka, F, Mimori, K, Barnard, GF, and Mori, M. 2005. Characterization of a side population of cancer cells from human gastrointestinal system. Stem Cells.

172. Montanaro, F, Liadaki, K, Schienda, J, Flint, A, Gussoni, E, and Kunkel, LM. 2004. Demystifying SP cell purification: viability, yield, and phenotype are defined by isolation parameters. Exp Cell Res 298:144–154.

173. Nakanishi, T, Doyle, LA, Hassel, B, Wei, Y, Bauer, KS, Wu, S, Pumplin, DW, Fang, HB, and Ross, DD. 2003. Functional characterization of human breast cancer resistance protein (BCRP, ABCG2) expressed in the oocytes of *Xenopus laevis*. Mol Pharmacol 64:1452–1462.

174. Allen, JD, Van Dort, SC, Buitelaar, M, van Tellingen, O, and Schinkel, AH. 2003. Mouse breast cancer resistance protein (Bcrp1/Abcg2) mediates etoposide resistance and transport, but etoposide oral availability is limited primarily by P-glycoprotein. Cancer Res 63:1339–1344.

175. Brangi, M, Litman, T, Ciotti, M, Nishiyama, K, Kohlhagen, G, Takimoto, C, Robey, R, Pommier, Y, Fojo, T, and Bates, SE. 1999. Camptothecin resistance: role of the ATP-binding cassette (ABC), mitoxantrone-resistance half-transporter (MXR), and potential for glucuronidation in MXR-expressing cells. Cancer Res 59:5938–5946.

176. Rajendra, R, Gounder, MK, Saleem, A, Schellens, JH, Ross, DD, Bates, SE, Sinko, P, and Rubin, EH. 2003. Differential effects of the breast cancer resistance protein on the cellular accumulation and cytotoxicity of 9-aminocamptothecin and 9-nitrocamptothecin. Cancer Res 63:3228–3233.

177. Maliepaard, M, van Gastelen, MA, Tohgo, A, Hausheer, FH, van Waardenburg, RC, de Jong, LA, Pluim, D, Beijnen, JH, and Schellens, JH. 2001. Circumvention of breast cancer resistance protein (BCRP)–mediated resistance to camptothecins in vitro using non-substrate drugs or the BCRP inhibitor GF120918. Clin Cancer Res 7:935–941.

178. Yang, CH, Schneider, E, Kuo, ML, Volk, EL, Rocchi, E, and Chen, YC. 2000. BCRP/MXR/ABCP expression in topotecan-resistant human breast carcinoma cells. Biochem Pharmacol 60:831–837.

179. Schellens, JH, Maliepaard, M, Scheper, RJ, Scheffer, GL, Jonker, JW, Smit, JW, Beijnen, JH, and Schinkel, AH. 2000. Transport of topoisomerase I inhibitors by the breast cancer resistance protein. Potential clinical implications. Ann N Y Acad Sci 922:188–194.

180. Nakatomi, K, Yoshikawa, M, Oka, M, Ikegami, Y, Hayasaka, S, Sano, K, Shiozawa, K, Kawabata, S, Soda, H, Ishikawa, T, Tanabe, S, and Kohno, S. 2001. Transport of 7-ethyl-10-hydroxycamptothecin (SN-38) by breast cancer resistance protein ABCG2 in human lung cancer cells. Biochem Biophys Res Commun 288:827–832.

181. Yoshikawa, M, Ikegami, Y, Sano, K, Yoshida, H, Mitomo, H, Sawada, S, and Ishikawa, T. 2004. Transport of SN-38 by the wild type of human ABC transporter ABCG2 and its inhibition by quercetin, a natural flavonoid. J Exp Ther Oncol 4:25–35.

182. Bates, SE, Medina-Perez, WY, Kohlhagen, G, Antony, S, Nadjem, T, Robey, RW, and Pommier, Y. 2004. ABCG2 mediates differential resistance to SN-38 and homocamptothecins. J Pharmacol Exp Ther 310:836–842.

183. Nakagawa, R, Hara, Y, Arakawa, H, Nishimura, S, and Komatani, H. 2002. ABCG2 confers resistance to indolocarbazole compounds by ATP-dependent transport. Biochem Biophys Res Commun 299:669–675.

184. Chen, ZS, Robey, RW, Belinsky, MG, Shchaveleva, I, Ren, XQ, Sugimoto, Y, Ross, DD, Bates, SE, and Kruh, GD. 2003. Transport of methotrexate, methotrexate polyglutamates, and 17beta-estradiol 17-(beta-D-glucuronide) by ABCG2: effects of acquired mutations at R482 on methotrexate transport. Cancer Res 63:4048–4054.

185. Volk, EL, and Schneider, E. 2003. Wild-type breast cancer resistance protein (BCRP/ABCG2) is a methotrexate polyglutamate transporter. Cancer Res 63:5538–5543.

186. Shafran, A, Ifergan, I, Bram, E, Jansen, G, Kathmann, I, Peters, GJ, Robey, RW, Bates, SE, and Assaraf, YG. 2005. ABCG2 harboring the Gly482 mutation confers high-level resistance to various hydrophilic antifolates. Cancer Res 65:8414–22.

187. Mitomo, H, Kato, R, Ito, A, Kasamatsu, S, Ikegami, Y, Kii, I, Kudo, A, Kobatake, E, Sumino, Y, and Ishikawa, T. 2003. A functional study on polymorphism of the ATP-binding cassette transporter ABCG2: critical role of arginine-482 in methotrexate transport. Biochem J 373:767–774.

188. Wang, X, Nitanda, T, Shi, M, Okamoto, M, Furukawa, T, Sugimoto, Y, Akiyama, S, and Baba, M. 2004. Induction of cellular resistance to nucleoside reverse transcriptase inhibitors by the wild-type breast cancer resistance protein. Biochem Pharmacol 68:1363-1370.

189. Scharenberg, CW, Harkey, MA, and Torok-Storb, B. 2002. The ABCG2 transporter is an efficient Hoechst 33342 efflux pump and is preferentially expressed by immature human hematopoietic progenitors. Blood 99:507–512.

190. Youdim, KA, Qaiser, MZ, Begley, DJ, Rice-Evans, CA, and Abbott, NJ. 2004. Flavonoid permeability across an in situ model of the blood–brain barrier. Free Radic Biol Med 36:592–604.

191. Ebert, B, Seidel, A, and Lampen, A. 2005. Identification of BCRP as transporter of benzo[a]pyrene conjugates metabolically formed in Caco-2 cells and its induction by Ah-receptor agonists. Carcinogenesis 26:1754–1763.

192. Mizuno, N, Suzuki, M, Kusuhara, H, Suzuki, H, Takeuchi, K, Niwa, T, Jonker, JW, and Sugiyama, Y. 2004. Impaired renal excretion of 6-hydroxy-5,7-dimethyl-2-methylamino-4-(3-pyridylmethyl) benzothiazole (E3040) sulfate in breast cancer resistance protein (BCRP1/ABCG2) knockout mice. Drug Metab Dispos 32:898–901.

193. Lee, YJ, Kusuhara, H, Jonker, JW, Schinkel, AH, and Sugiyama, Y. 2005. Investigation of efflux transport of dehydroepiandrosterone sulfate and mitoxantrone at the mouse blood–brain barrier: a minor role of breast cancer resistance protein. J Pharmacol Exp Ther 312:44–52.

194. Zamek-Gliszczynski, MJ, Hoffmaster, KA, Tian, X, Zhao, R, Polli, JW, Humphreys, JE, Webster, LO, Bridges, AS, Kalvass, JC, and Brouwer, KL. 2005. Multiple mechanisms are involved in the biliary excretion of acetaminophen sulfate in the rat: role of Mrp2 and Bcrp1. Drug Metab Dispos 33:1158–1165.

195. Merino, G, Jonker, JW, Wagenaar, E, Pulido, MM, Molina, AJ, Alvarez, AI, and Schinkel, AH. 2005. Transport of anthelmintic benzimidazole drugs by breast cancer resistance protein (BCRP/ABCG2). Drug Metab Dispos 33:614–618.

196. Huss, WJ, Gray, DR, Greenberg, NM, Mohler, JL, and Smith, GJ. 2005. Breast cancer resistance protein–mediated efflux of androgen in putative benign and malignant prostate stem cells. Cancer Res 65:6640–6650.

197. van der Heijden, J, de Jong, MC, Dijkmans, BA, Lems, WF, Oerlemans, R, Kathmann, I, Schalkwijk, CG, Scheffer, GL, Scheper, RJ, and Jansen, G. 2004. Development of sulfasalazine resistance in human T cells induces expression of the multidrug resistance transporter ABCG2 (BCRP) and augmented production of TNFalpha. Ann Rheum Dis 63:138–143.

198. van der Heijden, J, de Jong, MC, Dijkmans, BA, Lems, WF, Oerlemans, R, Kathmann, I, Scheffer, GL, Scheper, RJ, Assaraf, YG, and Jansen, G. 2004. Acquired resistance of human T cells to sulfasalazine: stability of the resistant phenotype and sensitivity to non-related DMARDs. Ann Rheum Dis 63:131–137.

199. Ji, Y, and Morris, ME. 2004. Effect of organic isothiocyanates on breast cancer resistance protein (ABCG2)-mediated transport. Pharm Res 21:2261–2269.

200. Schrickx, J, Lektarau, Y, and Fink-Gremmels, J. 2005. Ochratoxin A secretion by ATP-dependent membrane transporters in Caco-2 cells. Arch Toxicol 1–7.

201. Polli, JW, Baughman, TM, Humphreys, JE, Jordan, KH, Mote, AL, Webster, LO, Barnaby, RJ, Vitulli, G, Bertolotti, L, Read, KD, and Serabjit-Singh, CJ. 2004. The systemic exposure of an N-methyl-D-aspartate receptor antagonist is limited in mice by the P-glycoprotein and breast cancer resistance protein efflux transporters. Drug Metab Dispos 32:722–726.

202. Kondo, C, Onuki, R, Kusuhara, H, Suzuki, H, Suzuki, M, Okudaira, N, Kojima, M, Ishiwata, K, Jonker, JW, and Sugiyama, Y. 2005. Lack of improvement of oral absorption of ME3277 by prodrug formation is ascribed to the intestinal efflux mediated by breast cancer resistant protein (BCRP/ABCG2). Pharm Res 22:613–618.

203. Braun, AH, Stark, K, Dirsch, O, Hilger, RA, Seeber, S, and Vanhoefer, U. 2005. The epidermal growth factor receptor tyrosine kinase inhibitor gefitinib sensitizes colon cancer cells to irinotecan. Anticancer Drugs 16:1099–1108.

204. Zhang, S, Wang, X, Sagawa, K, and Morris, ME. 2005. Flavonoids chrysin and benzoflavone, potent breast cancer resistance protein inhibitors, have no significant effect on topotecan pharmacokinetics in rats or mdr1a/1b (−/−) mice. Drug Metab Dispos 33:341–348.

205. Yang, CH, Chen, YC, and Kuo, ML. 2003. Novobiocin sensitizes BCRP/MXR/ABCP overexpressing topotecan-resistant human breast carcinoma cells to topotecan and mitoxantrone. Anticancer Res 23:2519–2523.

206. Shiozawa, K, Oka, M, Soda, H, Yoshikawa, M, Ikegami, Y, Tsurutani, J, Nakatomi, K, Nakamura, Y, Doi, S, Kitazaki, T, et al. 2004. Reversal of breast cancer resistance protein (BCRP/ABCG2)–mediated drug resistance by novobiocin, a coumermycin antibiotic. Int J Cancer 108:146–151.

207. Morita, H, Koyama, K, Sugimoto, Y, and Kobayashi, J. 2005. Antimitotic activity and reversal of breast cancer resistance protein-mediated drug resistance by stilbenoids from *Bletilla striata*. Bioorg Med Chem Lett 15:1051–1054.

208. Sargent, JM, Williamson, CJ, Maliepaard, M, Elgie, AW, Scheper, RJ, and Taylor, CG. 2001. Breast cancer resistance protein expression and resistance to daunorubicin in blast cells from patients with acute myeloid leukaemia. Br J Haematol 115:257–262.

209. van den Heuvel-Eibrink, MM, Wiemer, EA, Prins, A, Meijerink, JP, Vossebeld, PJ, van der Holt, B, Pieters, R, and Sonneveld, P. 2002. Increased expression of the breast cancer resistance protein (BCRP) in relapsed or refractory acute myeloid leukemia (AML). Leukemia 16:833–839.

210. Plasschaert, SL, van der Kolk, DM, De Bont, ES, Vellenga, E, Kamps, WA, and De Vries, EG. 2004. Breast cancer resistance protein (BCRP) in acute leukemia. Leuk Lymphoma 45:649–654.

13

DRUG TRANSPORT IN THE LIVER

PIETER ANNAERT
Catholic University of Leuven, Leuven, Belgium

BRANDON SWIFT, JIN KYUNG LEE, AND KIM L. R. BROUWER
University of North Carolina, Chapel Hill, North Carolina

Drug Transporters: Molecular Characterization and Role in Drug Disposition, Edited by Guofeng You and Marilyn E. Morris
Copyright © 2007 John Wiley & Sons, Inc.

13.1. HEPATIC PHYSIOLOGY: LIVER STRUCTURE AND FUNCTION

The liver is one of the major organs involved in drug metabolism and excretion. Hepatocytes, which are the predominant cell type in the liver, are polarized cells with distinct basolateral and apical domains that extend in chordlike arrays throughout the liver (see Figure 13.1). Hepatoctyes contain the principal metabolic and transport machinery responsible for hepatic drug clearance. Transport proteins play an important role in the clearance of drugs from hepatic sinusoidal blood and the excretion of parent compound and/or metabolite(s) across the apical membrane into the bile canaliculus. Although lipophilic compounds may diffuse from sinusoidal blood across the basolateral (sinusoidal and lateral) membrane domain, it is now recognized that many drugs gain access to the hepatocyte via transport proteins. Biliary excretion of drugs and metabolites is an active process that requires adenosine triphosphate (ATP)-dependent transport proteins that pump substrates into the canalicular lumen. ATP-dependent transport proteins also aid in the removal of drugs and metabolites from the hepatocyte by pumping them across the basolateral membrane into sinusoidal blood as it flows toward the central vein.

The field of hepatic transport is an emerging discipline. In the sections that follow, hepatic transport proteins are introduced based on their localization and function as

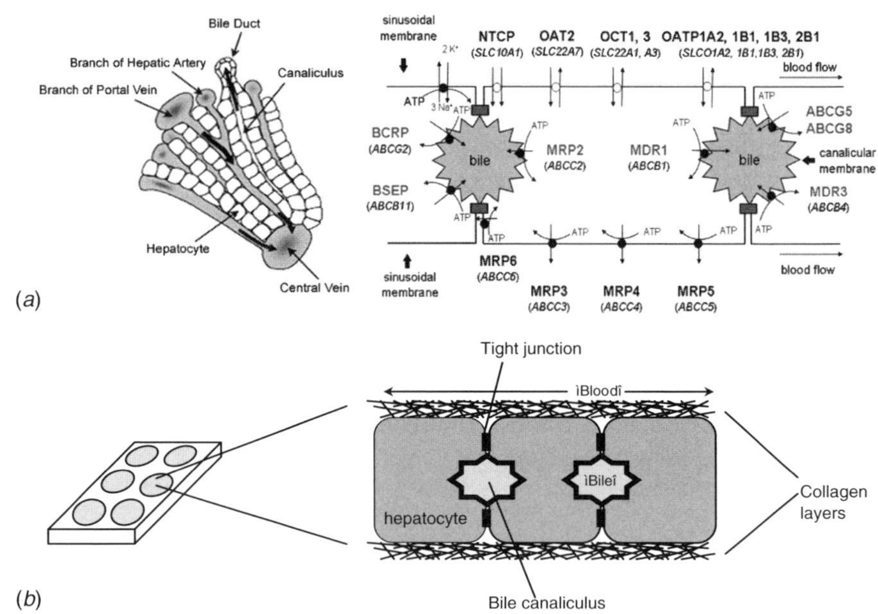

FIGURE 13.1. (*a*) In vivo architecture of the liver and polarized nature of hepatocytes, displaying two separate membrane domains facing blood and bile; (*b*) scheme illustrating the polarized phenotype in vitro when hepatocytes are cultured in a sandwich configuration (between two layers of gelled collagen). [(*a*) In part according to ref. 278, with permission.]

either hepatic uptake or efflux transporters. The nomenclature, substrate specificity, driving force(s) for transport, and clinical significance with respect to disease states and/or drug interactions are summarized. Current information regarding transport protein trafficking, regulation, and the potential role of hepatic transport proteins in drug-induced hepatotoxicity is discussed. One of the greatest challenges in studying hepatobiliary drug transport is selecting the right model system(s) and tools to answer the relevant questions. An overview of approaches currently in use and under development is provided. As scientific knowledge advances in this field, new techniques and tools, and more sophisticated models are anticipated to aid in unraveling the complexities of hepatic drug transport and exploiting these processes to achieve desirable therapeutic outcomes.

13.2. HEPATIC UPTAKE TRANSPORT PROTEINS

The solute carrier (*SLC*) gene family, containing the *SLC10* and *SLC22* subfamilies, together with the *SLCO* gene family represent the predominant transport proteins that have been shown to mediate hepatic uptake of xenobiotics across the sinusoidal hepatocyte membrane (see Figure 13.1). The previous and currently approved nomenclature for these uptake transporters that reside on the hepatic basolateral membrane, and example substrates, are included in Table 13.1.

NTCP *(SLC10A1)* The Na^+-taurocholate cotransporting polypeptide (NTCP), which is expressed exclusively in liver, mediates sodium-dependent bile salt uptake in human liver against a typical 5- to 10-fold concentration gradient and with a 2 : 1 sodium-to-taurocholate (TC) stoichiometry. NTCP accounts for an estimated 80% of total hepatic uptake of conjugated bile acids.[279] Although the substrate specificity of NTCP appears oriented primarily to bile acids, this protein also transports the cholephilic compounds bromosulfophthalein (BSP) and estrone 3-sulfate.[280] Moreover, NTCP recently has been demonstrated to contribute to the hepatic uptake of rosuvastatin.[281] In rats, Ntcp also has been reported to transport thyroid hormones[3] and the mushroom toxin α-amanitin (a cyclic octapeptide).[4] These findings indicate that NTCP substrate specificity extends beyond bile acids and structurally related compounds. Investigation of polymorphisms in NTCP, as reported by Ho et al.,[279] revealed that this transport protein exhibits a region that is critical and specific for bile acid substrate recognition. Indeed, one NTCP variant displayed nearly complete loss of bile acid transport but fully functional transport of estrone sulfate. Interestingly, NTCP also was found to transport certain drugs when covalently bound to taurocholate (i.e., chlorambucil-taurocholate).[5]

OATPs *(SLCO;* Previously, *SLC21A)* The family of organic anion transporting polypeptides (OATPs) plays an essential role in hepatic drug uptake. OATPs often appear to be rate limiting in the hepatobiliary clearance of drugs, thus controlling the hepatic elimination and/or oral bioavailability of various compounds. OATPs are characterized by broad and overlapping substrate specificity and display affinity for

TABLE 13.1. Human Hepatic Uptake Transport Proteins

Protein/ Trivial Names	Gene Symbol	Substrates/References
NTCP	*SLC10A1*	BSP; cholate; estrone 3-sulfate; glycocholate; taurochenodeoxycholate; tauroursodeoxycholate; TC [280], Rosuvastatin [281]
OATP1A2 OATP-A OATP-1 OATP	*SLCO1A2* *(SLC21A3)*	Bile acids; BQ-123; BSP; DHEAS; DPDPE; $E_2$17G; estrone 3-sulfate; *n*-methylquinine; ouabain; prostaglandin E_2; T3; T4 [282], Deltorphin II [206,282], Fexofenadine [139], Microcystin-LR [283], Saquinavir [284]
OATP1B1 OATP-C LST-1 OATP2	*SLCO1B1* *(SLC21A6)*	Atorvastatin [146,208], Benzylpenicillin [209], Bile acids; BQ-123; BSP; DHEAS; DPDPE; $E_2$17G; estrone 3-sulfate; T3; T4 [282], Bilirubin, bilirubin glucuronides [285], LTC_4; prostaglandin E_2 [282,286], Caspofungin [210], Cerivastatin [159,208], Fluvastatin [101], Irinotecan metabolite (SN-38) [211], Methotrexate [286], Microcystin-LR [283], Phalloidin (assessed by dimethyl derivative) [212], Pitavastatin [213,214], Pravastatin [208,215], Repaglinide [166,168], Rifampin [147,216], Rosuvastatin [164,281], Simvastatin [208], Troglitazone-sulfate [217]
OATP1B3 OATP-8 LST-2	*SLCO1B3* *(SLC21A8)*	Bile acids; BQ-123; BSP; Deltorphin II; DHEAS; digoxin; DPDPE; $E_2$17G; estrone 3-sulfate; LTC_4; ouabain; T3; T4 [282], CCK-8 [218], Fexofenadine [219], Fluvastatin [101], Microcystin-LR [283], Monoglucuronosyl bilirubin [285], Paclitaxel [220], Phalloidin (assessed by dimethyl derivative) [212], Pitavastatin [214], Rifampin [216]
OATP2B1 OATP-B	*SLCO2B1* *(SLC21A9)*	BSP; DHEAS; estrone 3-sulfate [282], Fluvastatin [101]
OAT2	*SLC22A7*	5-Fluorouracil; allopurinol; L-ascorbic acid; bumetanide; DHEAS; estrone 3-sulfate; glutarate [11], Erythromycin [178], Methotrexate [10], Prostaglandin E_2 [11,17], Prostaglandin $F_{2\alpha}$ [14], Ranitidine [287], Salicylate (controversial) [12], Tetracycline [221], Theophylline [178], Zidovudine [222]
OCT1	SLC22A1	Acyclovir, ganciclovir [222], Azidoprocainamide methoiodide; *n*-methylquinidine; *n*- methylquinine; tributylmethylammonium [223], Choline [224], Imatinib [225], Metformin [226], MPP^+; *N*-methylnicotinamide; tetraethylammonium [227,228,288], Famotidine, ranitidine [179], Prostaglandin E_2, prostaglandin $F_{2\alpha}$ [17]
OCT3 EMT	SLC22A3	Adrenaline, noradrenaline, tyramine [229], Agmatine, MPP^+; tetraethylammonium [230–232], Atropine, etilefrine [289], Histamine [224]

[a]BQ-123, [cyclo(D-Trp-D-Asp-L-Pro-D-Val-L-Leu)]; DPDPE, [D-penicillamine^{2,5}]-enkephalin; $E_2$17G, estradiol-17β(β-D-glucuronide); T_3, T_4, thyroid hormones; LTC_4, leukotriene C_4; MPP^+, 1-methyl-4-phenylpyridinium.

a spectrum of organic anions, bulky (previously called type II) cations, and neutral steroids. In contrast to NTCP, the OATPs operate in a sodium-independent manner, and some OATP isoforms have been hypothesized to function as glutathione antiporters,[6,7] employing the high intracellular glutathione concentrations as a driving force for hepatic uptake of substrates with high efficiency. Eleven human OATP isoforms and 14 rat Oatp isoforms have been identified thus far. A comparison of substrate specificities of Oatps/OATPs between rat and human was published recently[8] and illustrates the discrepancy between rat and human substrate profiles consistent with the fact that human OATPs are not orthologs of rat Oatps.

OATP1A2, OATP1B1, OATP1B3, and OATP2B1 are the four human OATPs that are now considered to play a substantial role in hepatic uptake of exogenous and endogenous compounds at the liver sinusoidal membrane domain. OATP1B1 and 1B3 are liver-specific, whereas OATP2B1 is widely expressed; OATP1A2 exhibits the highest expression in brain. Despite overlapping substrate specificities (especially between the 1B1 and 1B3 isoforms), distinct substrate affinity profiles have been reported. For instance, while all four human liver-expressed OATPs mediate BSP, dehydroepiandrosterone sulfate (DHEAS), and estrone 3-sulfate transport, bile acids are reportedly transported by OATP1A2, 1B1, and 1B3, but not by OATP2B1. As illustrated in Table 13.1, OATP1B1 clearly displays the largest substrate diversity and appears to have affinity for HMG-CoA reductase inhibitors ("statins"; see Table 13.1). OATP1B1 is the major human liver transport protein that is involved in Na^+-independent bile salt uptake. OATP1B1 also plays a predominant role in hepatic bilirubin uptake, which was demonstrated by correlating drug-associated inhibition of OATP1B1-mediated uptake with the incidence of hyperbilirubinemia associated with these drugs.[9] Interestingly, OATPs also seem to have affinity for several larger-molecular-weight compounds, such as the lipopeptide antifungal agent caspofungin (OATP1B1), the cyclic heptapeptide microcystin-LR (OATP1A2, 1B1, 1A3), and the mushroom toxin phalloidin (OATP1B1 and 1B3). Several substrates were identified as having selective affinity for just one of the four OATP isoforms expressed in human liver [e.g., *n*-methylquinine for OATP1A2; repaglinide and troglitazone sulfate for OATP1B1; and digoxin, cholecystokinin (CCK)-8, and paclitaxel for OATP1B3].

OATs *(SLC22A)* Organic anion transporters (OATs) constitute a family of proteins that mediate transport of primarily negatively charged endogenous and exogenous compounds in exchange for dicarboxylate ions. Since substantial expression in the kidney was demonstrated for all human OATs identified and characterized so far (OAT1–4), the majority of available data regarding OATs has been generated in the field of renal drug transport. However, OAT2 is expressed predominantly in the sinusoidal hepatocyte membrane (with lower expression in basolateral membrane of proximal kidney tubules).[10] In addition to the earlier identification of prostaglandins, zidovudine, and tetracycline as OAT2 substrates, a range of structurally dissimilar compounds (including the drugs taxol, allopurinol and 5-fluorouracil) was recently added to the OAT2 substrate list (see Table 13.1).[11] The affinity of salicylate for OAT2 remains controversial in that it had been identified as a (weak) substrate in two independent studies,[12,13] although it clearly could not be classified as a substrate based on data generated by Kobayashi et al.[11] Human OAT2 has 79% sequence identity to its rat ortholog Oat2,[10] but differences in substrate specificity and even membrane localization (liver vs. kidney) exist.[14,15] Detection of mRNA for two additional OAT transporters (originally referred to as OAT4 and OAT5) with exclusive expression in liver has been reported[10]; however, the substrate specificity of these isoforms (gene products *SLC22A9, SLC22A10*[16]) remains to be elucidated. It should be noted that the latter gene products are different from OAT4 (*SLC22A11*, expressed in placenta and kidney) and mouse Oat5 (*SLC22A19*, expressed in kidney).

OCTs *(SLC22A)* Organic cation transporters (OCTs) are electrogenic uniporters that mediate transport of primarily small (type I) cations in an Na^+-independent fashion, although transport of anionic (e.g., prostaglandins[17]) and uncharged compounds also has been observed (see ref. 18 and references therein). Human OCT1 (*SLC22A1*; expressed exclusively in liver[288]) and OCT3 (*SLC22A3*; broader tissue distribution) are the isoforms expressed at the sinusoidal membrane of the hepatocyte. In addition to small cationic model substrates, various marketed drugs, including the antivirals acyclovir and ganciclovir, as well as the H_2-receptor antagonists famotidine and ranitidine have been identified as OCT1 substrates (see Table 13.1). Substrates for OCT3 appear to be more limited to endogenous substances and neurotransmitters.

13.3. HEPATIC EFFLUX TRANSPORT PROTEINS

Hepatic excretion of xenobiotics from the liver may occur across the basolateral membrane into sinusoidal blood, or across the canalicular (apical) membrane into bile which flows through fine tubular canals between adjacent liver cells (see Figure 13.1). Lists of substrates for human hepatic export proteins are provided in Table 13.2.

13.3.1. Canalicular Transport Proteins

Canalicular transport proteins, responsible for the hepatic excretion of drugs and metabolites, belong to the ATP-binding cassette (ABC) family of transport proteins which mediate ATP-dependent transfer of solutes. Canalicular transport proteins for drugs and metabolites include P-glycoprotein (MDR1, *ABCB1*), the multidrug resistance protein 3 (MDR3, *ABCB4*), the bile salt export pump (BSEP, *ABCB11*), the multidrug resistance–associated protein (MRP2, *ABCC2*), and breast cancer resistance protein (BCRP, *ABCG2*).

P-glycoprotein *(MDR1, ABCB1)* P-glycoprotein (Pgp) was first identified over three decades ago in multidrug-resistant (MDR) tumor cells.[20] Pgp represents the most widely studied ABC transport protein and is responsible for biliary excretion of bulky hydrophobic and cationic substrates: chemotherapeutic agents (e.g., daunorubicin, doxorubicin, etoposide, paclitaxel, vinblastine, vincristine), cardiac

TABLE 13.2. Human Hepatic Efflux Transport Proteins

Protein/ Trivial Names	Gene Symbol	Substrates/References
MRP1 MRP, GS-X	*ABCC1*	Daunorubicin; doxorubicin; etoposide; vincristine [233], Glutathione [38], Methotrexate (MTX) [124]
MRP3 MOAT-D MLP2 cMOAT2	*ABCC3*	Acetaminophen glucuronide [69], $E_2$17G; monovalent and sulfated bile salts; MTX [34], Etoposide [38]
MRP4 MOAT-B	*ABCC4*	Azidothymidine [234], cAMP; cGMP; adefovir [235,236,309], MTX [310], Prostaglandin E_1; prostaglandin E_2 [235]
MRP5 MOAT-C ABC11	*ABCC5*	cAMP; cGMP [236], Adefovir [235]
MRP6 MOAT-E MLP1	*ABCC6*	BQ-123 [307]
BSEP Sister Pgp	*ABCB11*	Conjugated and unconjugated bile acids; TC; Pravastatin [237, 314]
MRP2 CMOAT cMRP	*ABCC2*	Acetaminophen glucuronide; carboxydichlorofluorescein [238], Camptothecin; doxorubicin [239], Cerivastatin; estrone 3-sulfate [240], Cisplatin; vincristine [241], Etoposide [242], Glibenclamide; indomethacin; rifampin [243], Glucuronide, glutathione, and sulfate conjugates; LTC_4 [30], MTX [244], Pravastatin [99]
MDR1 Pgp	*ABCB1*	Amprenavir; indinavir; nelfinavir; ritonavir; saquinavir [311,312], Aldosterone; corticosterone; dexamethasone; digoxin [245], Cyclosporin A; MX [246], Debrisoquine; erythromycin; lovastatin; terfenadine [139], Digoxin; quinidine [313], Doxorubicin; paclitaxel; rhodamine 123 [247,248], Etoposide [249], Fexofenadine [139], Levofloxacin; grepafloxacin [250], Losartan; vinblastine [251], Tacrolimus [252], Talinolol [253], Verapamil; norverapamil [254]
MDR3 PFIC3 Phospholipid flippase MDR2/3	*ABCB4*	Phospholipids [23]
BCRP MXR, ABCP	*ABCG2*	Daunorubicin; doxorubicin; MX; sulfated conjugates glucuronides [21], Topotecan; Irinotecan [32], Prazosin; rhodamine 123 [255], Testosterone; tamoxifen; estradiol [256]

glycosides (e.g., digoxin), rhodamine 123, cyclosporin A, and HIV-1 protease inhibitors (e.g., amprenavir, inidnavir, nelfinavir, ritonavir, saquinavir). Pgp modulators have been evaluated to increase drug sensitivity and oral bioavailability during chemotherapy because overexpression of Pgp on the surface of tumor cells is one cause of multidrug resistance.[305]

MDR3 ***(ABCB4)*** MDR3, a phospholipid flippase, is involved in the biliary secretion of phospholipids and cholephilic compounds. MDR3 plays a crucial role in basic liver physiology in humans and rats; phospholipids and cholesterol are responsible for the formation of micelles that solubilize bile acids in the lumen of the bile canaliculus.[22] Mutations in the MDR3 gene lead to progressive familial intrahepatic cholestasis type 3 (PFIC3), a disease that is characterized by increased γ-glutamyltranspeptidase levels, ductular proliferation, and inflammatory infiltrate that can progress to biliary cirrhosis. MDR3 was reported to have similar substrate specificity to Pgp, but the rate of transport was lower than that of Pgp.[23]

BSEP ***(ABCB11)*** BSEP is the predominant bile salt export protein that mediates the biliary excretion of conjugated and unconjugated bile salts. Recently it was reported that BSEP might also participate in transport of nonbile acid substrates such as pravastatin.[24,25] Patients with progressive familial intrahepatic cholestasis type 2 (PFIC2) have *ABCB11* gene mutations and the absence of BSEP protein expression, leading to hepatocellular injury and necrosis caused by increased intracellular concentrations of detergentlike bile acids.[26] Inhibition of BSEP by several drugs, such as troglitazone, bosentan, cyclosporin A, and rifampicin, has been suggested as one mechanism for drug-induced liver injury.[27,28]

MRP2 ***(ABCC2)*** MRP2 is a major xenobiotic efflux pump on the canalicular membrane. MRP2 plays a key role in the biliary excretion of organic anions, including bilirubin-diglucuronide, glutathione conjugates, sulfated bile salts, and divalent bile salt conjugates, as well as numerous drugs, including sulfopyrazone, indomethacin, penicillin, vinblastine, methotrexate, and telmisartan. Patients with Dubin–Johnson syndrome suffer from defective hepatic biliary excretion due to the absence of MRP2.[29] Animals with hereditary conjugated hyperbilirubinemia [Mrp2-deficient rats (GY/TR$^-$) and Eisai hyperbilirubinemic rats (EHBR)] exhibit increased expression of basolateral Mrp3, a compensatory mechanism that enables the excretion of Mrp2 substrates into the systemic circulation, thus avoiding excessive accumulation of organic anions in the hepatoctye.[30,31]

BCRP ***(ABCG2)*** BCRP is highly expressed in the canalicular membrane of the hepatocyte as well as in the intestine, breast, and placenta. BCRP is a half-transport protein that forms a functional homodimer. BCRP mediates transport of estrone 3-sulfate and various sulfated steroidal compounds. BCRP also is involved in the development of resistance to a variety of anticancer agents, such as SN-38, mitoxantrone (mx), topotecan, and doxorubicin.[21,32]

13.3.2. Basolateral Efflux Transport Proteins

Members of the ATP-dependent MRP subfamily represent a major class of export proteins on the hepatic basolateral membrane. MRP 1, 3, 4, 5, and 6 are involved in cellular transport of both hydrophobic uncharged molecules and hydrophilic anionic compounds (see Table 13.2).[33] MRP1 (*ABCC1*) is responsible for efflux of various organic anions, including glucuronide, glutathione, and sulfate-conjugated drugs. MRP1 expression is very low on the basolateral membrane in healthy liver, but is highly inducible during severe liver injury; MRP1 induction is believed to play a role in liver protection.[306] MRP3 (*ABCC3*) is involved in the hepatic excretion of glucuronide conjugates (e.g., acetaminophen glucuronide), methotrexate, and estradiol-17β-glucuronide ($E_2$17G). The expression level of MRP3 is increased markedly under cholestatic conditions.[34] MRP4 (*ABCC4*) and MRP5 (*ABCC5*) transport the cyclic nucleotides cAMP and cGMP as well as the purine analogs 6-mercaptopurine and 6-thioguanine.[35] MRP4 has been implicated in transport of nucleoside analogs such as zidovudine, lamivudine, and stavudine as well as nonnucleotide substrates (e.g., methotrexate) and reverse transcriptase inhibitors (e.g., azidothymidine). MRP4 is also involved in the transport of sulfate conjugates of bile acids and steroids.[36] While the expression level of MRP5 in healthy liver is relatively low, lipopolysaccharide (LPS) treatment resulted in down-regulation of MRP2 and induction of MRP5, suggesting that MRP5 may participate in hepatic response during cholestasis.[37] MRP6 transports glutathione conjugates and the endothelin receptor antagonist BQ-123.[307] Mutations in *ABCC6* cause pseudoxanthoma elasticum (PXE), an inheritable systemic connective tissue disorder affecting the skin, eyes, and blood vessels.[39]

In addition to the MRPs, other transport proteins on the basolateral membrane may play a role in hepatic basolateral excretion. For example, the OATPs have been hypothesized to function as basolateral export proteins under certain conditions, although the in vivo role of these transport proteins in basolateral hepatic excretion remains to be elucidated.[308]

13.4. INTRACELLULAR TRAFFICKING OF HEPATIC PROTEINS AND XENOBIOTICS

As discussed above, transport proteins residing on the basolateral and apical membranes of the hepatocyte play an important role in the uptake and excretion of endogenous and exogenous compounds. Research to date has focused primarily on characterizing these transport proteins on a functional basis. The mechanism(s) involved in the targeted trafficking of these proteins to the correct plasma membrane domain and the regulation of transport protein trafficking represent areas of ongoing investigation. Membrane transport regulation may be influenced by alterations in transport function and changes in the number or disposition of transport molecules in the membrane. Stimulation by an agonist can increase the number of hepatic transport proteins expressed on the plasma membrane by (1) rapidly recruiting preexisting transport proteins located in intracellular stores (short-term modulation), or (2) synthesis of new protein via transcription and translation (long-term modulation).[41]

Transport proteins stored in intracellular vesicles are targeted to and fuse with the correct plasma domain (i.e., apical or basolateral) in response to appropriate stimuli; when the stimulus is withdrawn or another stimulus is applied, the transporters are removed from the membrane by endocytosis.[41] After endocytosis, the transport proteins remain in intracellular storage vesicles until restimulation, when they are again expressed on the plasma membrane surface. Eventually, these proteins move to a nonrecycling compartment, where they are degraded by lysosomes.[42] Thus, the number of transport proteins present on the cell surface is regulated by de novo synthesis, exocytotic insertion of proteins recruited from the cytoplasm, endocytic retrieval, and protein degradation.[43] A diagram explaining the possible routes and mechanisms for trafficking of hepatic transport proteins is shown in Figure 13.2.

13.4.1. Apical Proteins

Bsep is synthesized in the Golgi apparatus of rat hepatocytes and is sequestered transiently in an intracellular pool en route to the apical plasma membrane.[44] Evidence now suggests that ABC transport proteins (e.g., Pgp and Mdr2) may translocate directly from the Golgi to the canalicular membrane where they are stored in subapical compartments until signaled to the canalicular membrane. Taurocholate and dibutyryl (DB) cAMP modulate trafficking of Bsep, Pgp, and Mdr2; simultaneous administration of DBcAMP and taurocholate had additive effects, suggesting the presence of two separate intracellular pools of proteins, one mobilized by DBcAMP and the other by taurocholate.[45] ABC transporters are trafficked by a microtubule-dependent transcytotic vesicle-transport system which is induced by cAMP, resulting in an increase in transport proteins at the canalicular membrane.[46] Administration of colchicine, which disrupts microtubules, inhibited cAMP- and taurocholate-induced translocation of transport proteins to the membrane.[47] In addition, wortmannin inhibited the taurocholate-induced microtubule transport of Bsep, Mdr1, Mdr2, and Mrp2, suggesting the involvement of phosphoinositide 3-kinase in vesicle trafficking.[48] Bsep also was regulated by the mitogen-activated protein kinase p38 (MAPK) and extracellular signal-regulated kinase 2 (ERK-2). MAP kinase-dependent Bsep translocation from intracellular pools to the canalicular domain was demonstrated in tauroursodeoxycholate-stimulated perfused liver.[49]

Protein kinase C (PKC) plays an important role in the insertion of Mrp2 at the apical membrane. The taurine conjugated ursodeoxycholic acid (TUDCA) was shown to enhance efflux on the apical membrane by stimulating the insertion of Mrp2 via a PKC-dependent mechanism in cholestasis-induced perfused rat livers.[50] Mrp2 localization is also dependent on the presence of radixin (Rdx), a member of the ezrin–radixin–moesin family of proteins that cross-links actin filaments with integral membrane proteins. Kikuchi et al. utilized a Rdx knockout mouse model to demonstrate a decrease in Mrp2 density on the canalicular membrane compared with other proteins, such as Pgp.[51]

Some canalicular transport proteins may traffic in an indirect manner, first to the basolateral membrane, followed by redirection to the canalicular membrane domain. Soroka et al. demonstrated in rat hepatocytes that Mrp2 resides in

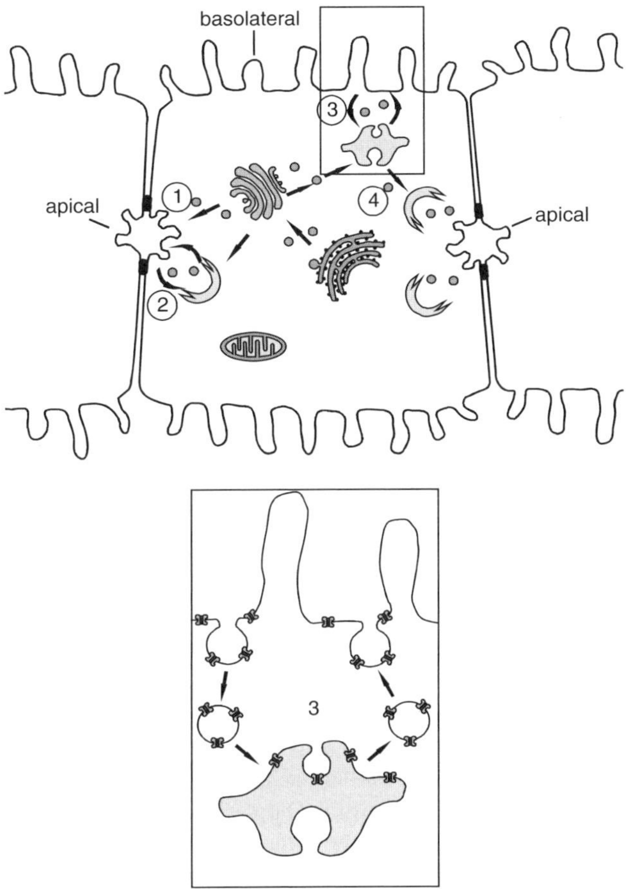

FIGURE 13.2. Intracellular trafficking of transport proteins in the hepatocyte. Newly synthesized proteins are transported from the Golgi network in four possible directions. The first pathway (1) represents direct trafficking of canalicular membrane proteins to the apical membrane, where they are expressed immediately or (2) stored in subapical compartments until some stimulus results in the mobilization of apical proteins to the membrane surface. Basolateral proteins may be trafficked directly to the basolateral membrane (3) or stored in a submembrane endosome until a signaling event causes them to be trafficked to the hepatocyte surface. The insert represents the exocytic insertion and endocytic retrieval of the transport proteins, depending on the intracellular status and signaling of the hepatocyte. The canalicular proteins also may undergo indirect trafficking (4), where they are transported to the basolateral membrane but never inserted into the plasma membrane, and then directed to the apical membrane.

microtubule-associated vesicles together with Bsep and polymeric immunoglobulin A receptor and undergoes vesicular transcytosis.[52] Mrp2 also undergoes trafficking to the canalicular membrane in a microtubule-dependent manner upon stimulation by cAMP.[53]

Osmolarity plays a role in the localization of transport proteins. In perfused rat livers, hypo-osmotic solutions increased taurocholate excretion in bile due to the increased translocation of Bsep to the canalicular membrane. In contrast, a hyper-osmotic environment caused a decrease in biliary excretion of taurocholate and the endocytic retrieval of Bsep.[54] Under hyper-osomotic conditions, Mrp2 resides intracellularly, whereas under hypo-osmotic conditions Mrp2 translocates to the plasma membrane.[55] Lipopolysaccharide exposure also resulted in the retrieval of Mrp2 from the apical membrane.[56]

The role of the N-terminal domain in MRP2 trafficking has been examined.[57] The specific apical localization of MRP2 is due to a sequence in the C-terminal tail not present in basolaterally targeted MRP1.[58] Deletion of three amino acids from the C-terminus of MRP2 results in localization predominantly at the basolateral membrane in MDCK cells. In sandwich-cultured rat hepatocytes, glycosylation plays a role in directing Mrp2 to the canalicular domain.[19]

13.4.2. Basolateral Proteins

Less is known about the intracellular trafficking of basolateral transport proteins. It is probable that transport proteins traffic via microtubule-dependent intracellular pathways to the basolateral domain. The PDZ consensus binding site at the C-terminal domain of OATP proteins plays a role in basolateral membrane localization of this family of proteins; oligomerization of rat Oatp1a1 with PDZK1 allows for proper expression of this isoform at the basolateral membrane surface. In PDZK1 knockout mice, Oatp1a1 protein expression was unaltered, but the protein was localized predominantly in intracellular structures.[59] Another important sequence structure common to all OATPs is a large extracellular loop between transmembrane domains IX and X, with 10 conserved cysteines; this region of the protein plays a role in disulfide linkages and glycosylation which is important for the trafficking and function of OATPs. The absence or substitution of these cysteine residues in OATP2B1 expressed in CHO-K1 cells resulted in a decrease in membrane expression and function.[60] Glycosylation appears to be involved in the targeting of OAT to the basolateral membrane. A common structural feature that is shared among all OATs is a consensus site for N-linked glycosylation in the first extracellular loop within the current secondary structure model. OAT1 remained intracellularly localized following treatment with tunicamycin, an antibiotic that inhibits N-glycosylation.[303]

The phosphorylation state of Ntcp determines whether this protein is localized on the basolateral membrane or intracellularly in the endosomal compartment. cAMP treatment causes Ntcp to translocate from the endosomes to the basolateral membrane.[2] Translocation induced by cAMP involves dephosphorylation of Ntcp by protein phosphatase 2B, which leads to increased retention of the protein in the basolateral membrane.[61,62] cAMP-induced trafficking of Ntcp relies on the phosphoinositide 3-kinase(PI3K)/protein kinase B pathway and intact microfilaments.[63–65]

13.4.3. Intracellular Trafficking of Xenobiotics

Research to date has focused primarily on characterizing the localization and function of hepatic transport proteins that play a pivotal role in the disposition of many exogenous compounds, such as drugs and toxins. The mechanisms of intracellular translocation of xenobiotics have yet to be fully elucidated. Hayes et al. utilized isolated perfused rat livers, isolated rat hepatocyte couplets and the WIF-B cell line to demonstrate that the fluorescent cation daunorubicin sequesters in intracellular compartments; daunorubicin is stored predominantly in vesicles in the pericanalicular region.[66] Stimulation of daunorubicin biliary excretion by taurocholate and DBcAMP indicated that transcytotic vesicular transport is also applicable to drugs. Furthermore, the microtubule destabilizer nocodazole decreased the biliary efflux of daunorubicin.[66] Recently, P-glycoprotein was shown to be localized in the Golgi and in the mitochondria of doxorubicin-resistant K562 cells. This study was the first to demonstrate the presence of functional P-glycoprotein at the organelle site; P-glycoprotein was involved in doxorubicin accumulation inside the organelle and not in efflux.[67] This finding supports the hypothesis that other transport proteins may reside at intracellular sites and play a role in the distribution of xenobiotics within the hepatocyte.

13.5. REGULATION OF HEPATIC DRUG TRANSPORT PROTEINS

As discussed previously, one mechanism of short-term regulation of membrane transport is by altering the number of transport molecules in the membrane, which may be accomplished by trafficking of proteins from intracellular pools. Long-term regulation involves modulation of transport protein expression at the transcriptional and/or translational level. A large superfamily of orphan nuclear receptors plays an important role in the regulation of transporter gene expression in hepatocytes; nuclear receptors mediate developmental as well as physiological responses to both endogenous and exogenous compounds. Upon ligand binding, the orphan nuclear receptor proceeds to bind to its heterodimeric partner (e.g., retinoic acid X receptor, RXR), and this complex modulates the transcription rate. Many xenobiotics interact with nuclear receptors [e.g., pregnane X receptor (PXR), peroximsome proliferator-activated receptor α (PPARα), liver X receptor (LXR), constitutive androstane receptor (CAR), and farnesoid X receptor (FXR)], and the resulting complexes bind to the regulatory region of genes to modulate transport protein expression. Table 13.3 lists the nuclear receptors and the transport proteins with which they interact.

PXR and CAR are involved in coregulation of the basolateral membrane transport proteins Oatp2 and Mrp3 in mice.[68] While treatment of wild-type mice with the PXR-specific ligand pregnenolone-16α-carbonitrile (PCN) resulted in a large increase in Mrp3 RNA, no activation was observed in the PXR knockout mice.[68] Furthermore, it was demonstrated using Wistar Kyoto rats that CAR does not play a key role in phenobarbital-mediated induction of Mrp3.[69] Negative feedback regulation of Ntcp by bile acid–activated FXR in both primary rat hepatocytes and HepG2 cells occurs via induction of shp, a small heterodimeric protein.[70] This pathway provides a mechanism for coordinated down-regulation of bile acid import and synthesis in order to protect the hepatocyte from accumulation of bile acids under cholestatic conditions.[70]

Cytokine-mediated inflammatory cholestasis causes down-regulation of Ntcp and Mrp2 by the RAR (retinoic acid receptor) and RXR heterodimer.[71] *SLCO1B3* gene expression has been reported to be regulated by the bile acid chenodeoxycholic acid (CDCA), which is a ligand of FXR/BAR (farnesoid X receptor/bile acid receptor). This provides another mechanism for decreased hepatic uptake of bile acids during cholestatic stress.[72]

The canalicular protein Bsep contains a promoter region sequence specific for binding of the FXR nuclear receptor, after heterodimerization with RXRα. When this complex is bound by bile acids, it effectively regulates the transcription of several genes involved in bile acid homeostasis.[73] PXR induces CYP3A4 and P-glycoprotein by similar mechanisms in the human colon carcinoma cell line LS174T; the promoter region of the MDR1 gene contains a DR4 motif to which PXR binds.[74] PXR also mediates the induction of MRP1 and MRP2 through redox-active compounds. However, the role of PXR is still unclear because deletion of the binding sites that are hypothesized to mediate the induction process did not result in a significant decrease in induction.[75] Mrp2 mRNA levels were induced following treatment with natural and synthetic FXR ligands; in PXR null and wild-type mice, Mrp2 was induced by the PXR activators, dexamethasone, pregnenolone 16α-carbonitrile, and the CAR agonist phenobarbital.[76] In mice, Mdr2 mRNA and protein are induced by PPARα, which is a nuclear receptor activated by fatty acids and hypolipidaemic fibrates.[77]

13.6. DISEASE STATE ALTERATIONS IN HEPATIC TRANSPORT PROTEINS

Many genetic mutations influence transport protein expression, localization, and/or function. For example, Dubin–Johnson syndrome is a rare autosomal recessive liver disorder in humans characterized by chronic conjugated hyperbilirubinemia that is due to the absence of MRP2 on the canalicular membrane.[78] Similar mutations have been found in the Mrp2 gene in TR^- (transport deficient) Wistar rats and EHBR (Eisai hyperbilirubinemic) Sprague–Dawley rats. One compensatory mechanism for the absence of MRP2/Mrp2 protein in patients with Dubin–Johnson syndrome and EHBR and TR^- rats is increased expression of MRP3/Mrp3 on the basolateral membrane.[30]

Progressive familial intrahepatic cholestasis (PFIC), primarily a childhood disease that affects canalicular bile transport, is classified into three different types. PFIC type 1 is caused by a mutation in the gene encoding the familial intrahepatic cholestasis 1 protein (FIC1). This protein plays a role in maintaining bile salt homeostasis. PFIC types 2 and 3 are caused by mutations in the genes encoding BSEP and the phospholipid translocase (MDR3), respectively. Benign recurrent intrahepatic cholestasis (BRIC), or recurrent familial intrahepatic cholestasis (RFIC), a very rare disease that affects only a few hundred patients worldwide, involves a mutation in the FIC1 gene, similar to PFIC type 1.[79]

There are also many forms of acquired defects in bile salt transport that involve direct inhibition of function or expression of the bile salt transport proteins resulting in cholestasis, including drug-induced cholestasis, sepsis-associated cholestasis,

TABLE 13.3. Regulation of Transport Proteins[a]

Transport Protein	Induce/ Repress	RNA	Protein	Nuclear Receptor	Ligands	Test System	Refs.
Ntcp (r)	↑			RAR	Retinoids (LG364, TTNPB)	HepG2 reporter assay	71
Ntcp (r)	↓			RAR	Cytokines (TNFα, IL-1β)	HepG2 reporter assay	71
OATP1B3	↑			FXR	Bile acids (CDCA)	HepG2 reporter assay	72
MRP1, 2, 3	↑			CAR	Phenobarbital	HepG2, Northern blots	155,257
BSEP	↑	×	×	FXR	Bile acids (CDCA, DCA, LCA, DKCA)	Human tissue immunoblot, HepG2 reporter assay	73,258
Mrp2 (r)	↑			RAR	Retinoids (LG364, TTNPB)	HepG2 reporter assay	71
Mrp2 (r)	↓			RAR	Cytokines (TNFα, IL-1β)	HepG2 reporter assay	71
MRP2	↑	×	×	PXR	Rifampicin, hyperforin, clotrimazol, rifampin	Human hepatocytes, duodenal biopsy, HepG2 reporter assays, RT-PCR, Western blot	75,76,297
MRP2	↑	×		FXR	CDCA and GW4064	HepG2 Northern blot	76
MRP2	↑	×	×	CAR	Phenobarbital	Human hepatocytes	259
MRP3	Slight ↑	×		PXR	Rifampicin	HepG2 RT-PCR	297
MRP3	↑	×		PXR	Clotrimazole, rifampicin, RU486, metyrapone, PCN	HebG2, HuH7	298
MRP4	↑	×		CAR	Phenobarbital	Human hepatocytes, HepG2	260
MRP5	↑	×		PXR	Rifampicin	HepG2 RT-PCR	297
MDR1	↑	×		PXR	Rifampin, rifampicin	LS174T Northern blot, human hepatocytes, LS180 cells Northern blot	74,152
Mrp2 (r)	↑	×	×	PXR	PCN, spironolactone, DEX	Rat hepatocytes, FAO	76,261,262,297,299,300
Mrp2 (r)	↑		×	Nrf2	Ethoxyquin, oltipraz	in vivo	299
Mrp2 (r)	↑	×	×	CAR	Phenobarbital	Rat hepatocytes, FAO	76,259,297,301
Mrp2 (r)	↑	×		FXR	Chenodeoxycholic acid, GW4064	Rat hepatocytes, FAO	76
Mrp2 (r)	Slight ↓		×	PPARα	Clofibric acid, DEHP, PFDA	in vivo	299

(Continued)

TABLE 13.3. (*Continued*)

Transport Protein	Induce/ Repress	RNA	Protein	Nuclear Receptor	Ligands	Test System	Refs.
Mrp3 (r)	↑	×	×	CAR	Phenobarbital, PCB 99, diallyl sulfide, TSO	in vivo	69,155,263,264
Mrp3 (r)	↑	×	×	Nrf2	Oltipraz	in vivo	263,264
Mrp4 (r)	↑	×		Nrf2	Ethoxyquin, Oltipraz	in vivo	302
Mdr1a (r)[b]	↑	×		CAR	Diallyl sulfide	in vivo	265
Mdr1a (r)[b]	↑	×		Nrf2	Oltipraz	in vivo	265
Mdr1b (r)[b]	↑	×		Ahr	Indole 3-carbinol	in vivo	265
Mdr1b (r)[b]	↑	×		CAR	PCB 99	in vivo	265
Bsep (r)	↑	×		PXR	DEX	in vivo	262
Oatp2 (r)	↓		×	AhR	B-Naphthoflavone, indole 3-carbinol, TCDD, PCB 126	in vivo	122
Oatp2 (r)	↑	×	×	CAR	Phenobarbital, diallyl sulfide, PCB 99	in vivo	122,266,267
Oatp2 (r)	↑	×	×	PXR	PCN, spironolactone, DEX	in vivo	122,267
Oatp1a4 (r)	↑		×	PXR	DEX	Rat hepatocytes	300
Ntcp (r)	↓		×	PXR	DEX	Rat hepatocytes	300
Oatp3a1, 2b1 (m)	↑	×		AhR	B-Naphthoflavone, PCB 126, TCDD	in vivo	268
Oatp1a1 (m)	↓	×		CAR	Phenobarbital, diallyl sulfide, TCPOBOP	in vivo	268
Oatp1a4 (m)	↑	×		PXR	PCN, spironolactone	in vivo	268
Oatp1a1, 1b2, 2a1, 2b1 (m)	↓	×		PXR	PCN, spironolactone, DEX	in vivo	268
Oatp1a1, 1b2, 2a1, 2b1 (m)	↓	×		PPARα	Clofibrate, DEHP	in vivo	268
Oatp1a1 (m)	↓	×		Nrf2	Oltipraz, ethoxyquin, butylated hydroxyanisole	in vivo	268

Oatp2b1 (m)	↑	×		Nrf2	Oltipraz, ethoxyquin, butylated hydroxyanisole	in vivo	268
Mrp2, 3, 5, 6 (m)	↑	×		AhR	B-Naphthoflavone, PCB 126, TCDD	in vivo	269
Mrp2, 3, 4, 6, 7 (m)	↑	×	Mrp4	CAR	Phenobarbital, TCPOBOP	in vivo	68,260,269
Mrp 2 (m)	↑	×		PXR	RU486	in vivo	298
Mrp2 (m)	↑	×		CAR	TCPOBOP	Transgenic mice expressing only humaninized CAR, Northern blot	301
Mrp3 (m)	↑	×		PXR	RU486	in vivo	298
Mrp3 (m)	↑	×		PXR	PCN, spironolactone	in vivo	68,269
Mrp3 (m)	↑	×	×	PPARα	Clofibrate, DEHP	in vivo	121,269
Mrp4 (m)	↑	×	×	PPARα	Clofibrate	in vivo	121
Bcrp (m)	↑	×	×	PPARα	Clofibrate	in vivo	121
Mdr1a (m)	↑	×	×	PPARα	Clofibrate	in vivo	121
Mrp2, 3, 4, 5, 6 (m)	↑	×		Nrf2	Oltipraz, ethoxyquin, butylated hydroxyanisole	in vivo	269
Oatp2 (m)	↑	×		PXR	LCA	in vivo	270
Oatp2 (m)	↑	×		PXR	PCN, 2-AAF	in vivo	68,271,272
Oatp2 (m)	↑	×		CAR	Phenobarbital, TCPOBOP	in vivo	68
Bcrp (m)	↑	×		PXR	2-AAF	in vivo	272
Mrp2 (m)	↑	×		PXR	2-AAF	in vivo	272
Mdr2 (m)	↑	×	×	PPARα	Ciprofibrate	in vivo	77

[a]DEHP, di(2-ethylhexyl) phthalate; PFDA, perfluorodecanoic acid; PCN, pregnenolone16α-carbonitrile; TCDD, 2,3,7,8-tetrachlorodibenzo-*p*-dioxin, TCPOBOP, 1,4-bis[2-(3,5-dichloropyridyloxy)]benzene; LCA, lithocholic acid; TSO, *trans*-stilbene oxide; 2-AAF, 2-acetylaminofluorene; CDCDA, chenodeoxycholic acid; DCA, deoxycholic acid; LCA, lithocholic acid; DKCA, 5β-cholanic acid 3,7-dion; FAO, rat hepatoma cell line; LS174T, human colon carcinoma cell line; LS180, colon cancer cell line; PCB, polychlorinated biphenyl; DEX, dexamethasone; 2-AAF, 2-acetylamino fluorene; AhR, aryl hydrocarbon receptor; Nrf2, nuclear factor-E2-related factor2; TTNPB, 4-[E-2-(5,6,7,8-tetrahydro-5,5,8,8-tetramethyl-2-naphthalenyl-1-propenyl] benzoic acid.

[b] Induction not seen with entire class of nuclear receptor inducers.

and intrahepatic cholestasis of pregnancy. Drug-induced cholestasis may involve inhibition of BSEP, which can result in the intracellular accumulation of bile salts and subsequent liver damage. Drugs reported to cause this type of cholestasis include cyclosporin A, rifamycin SV, rifampicin, glibenclamide,[38] troglitazone,[80] and bosentan.[27] Sepsis-associated cholestasis is caused by the effect of bacterial endotoxins on transport systems. Impaired basolateral and canalicular bile acid and organic anion transport has been observed in the isolated perfused rat liver after injection of lipopolysaccharide (LPS).[82] In a similar model using endotoxin-induced cholestasis, LPS induced an early and selective but reversible retrieval of Mrp2 from the apical membrane followed by down-regulation in Mrp2 mRNA.[83] A decrease in both Bsep mRNA and protein accounted for the reduced canalicular secretion of bile acids in endotoxin-treated rats.[84,85] Ntcp expression and activity also were decreased after administration of LPS, tumor necrosis factor α, or IL-1β to rodents.[86] Finally, intrahepatic cholestasis of pregnancy (ICP) usually occurs during the third trimester of pregnancy and resolves spontaneously after delivery. The classic maternal feature is pruritus, and the main biochemical finding is an increase in total serum bile acid concentrations. One important factor that may play a role in predisposing women to ICP is if the child has PFIC3, which is associated with defects in the MDR3 gene.[87] Cholestatic estrogen metabolites also appear to play a major role in the pathogenesis of ICP.[88] Rats treated with ethinyl estradiol exhibited a marked decrease in canalicular transport of taurocholate and dinitrophenylglutathione[88] as well as down-regulation of the uptake transporters Ntcp and Oatp1.[89]

13.7. MODEL SYSTEMS FOR STUDYING HEPATOBILIARY DRUG TRANSPORT

Several model systems have been applied successfully to the study of hepatobiliary drug transport, ranging from models mimicking the transporter-mediated transmembrane movement of drugs (e.g., membrane vesicles) to sophisticated in vivo models (e.g., knockout animals). Significant species differences in transporter affinity and expression profiles[90,91] remain a key consideration in selecting the appropriate model. The relevance of a particular model to human hepatobiliary drug disposition will depend largely on whether the model is derived from human tissue (e.g., human hepatocytes) or whether human transport proteins are expressed. Furthermore, since the rate-limiting step in hepatobiliary disposition differs between drugs, the relative expression levels of drug-metabolizing enzymes and transport proteins must be considered for each model, and data must be interpreted cautiously and in conjunction with complementary model systems when such expression levels do not reflect the in vivo situation. Table 13.4 summarizes applications as well as advantages and disadvantages of various hepatobiliary drug transport models.

The methodology to isolate purified canalicular (cLPM) and basolateral (bLPM) liver plasma membrane vesicles with functional transport properties has been available for more than two decades.[92–94] Hepatic membrane vesicles are useful in assessing the mechanisms of transport (driving forces, electrogenicity) and identifying substrates and inhibitors of transport proteins.[290,291,95–97] Although cLPM and bLPM preparation is labor intensive, it remains an attractive model system for determining differences in the functional activity of transporters obtained from various hepatic tissue sources without the need to isolate intact cells. For example, this approach has been used successfully to compare functional expression of taurocholate transport in the basolateral liver membrane of developing rats compared to adult rats.[98] This model system is less suitable for assessing the role of a specific transport protein in hepatic drug uptake or biliary excretion because multiple transport proteins are expressed in cLPM and bLPM that may mediate transport of a single substrate.

Cloning and expression of (human) transport proteins in polarized mammalian cell lines (predominantly the MDCK cell line) has evolved rapidly to one of the most widely used techniques for generating efficient and flexible in vitro tools for identification of substrates and inhibitors of hepatic transport proteins in humans.[242] Expression of at least two transport proteins (one for uptake and one for excretion) has now become a standard approach used to mimic the in vivo expression of uptake and excretory transporters at the basolateral (sinusoidal) and apical (canalicular) membrane domains.[242,99] Moreover, a correlation was found between in vitro vectorial transport of substrates across Oatp1b2 (Oatp4)- and Mrp2-transfected MDCKII monolayers and in vivo biliary clearance in rats.[100] Recently, Kopplow et al.[101] reported the construction under well-defined conditions of quadruple-transfected MDCKII cells allowing simultaneous screening for OATP1B1/1B3 and 2B1 substrates along with MRP2-mediated excretion at the apical membrane. The use of polarized cell lines mimics the polarized phenotype of the hepatocyte, thus allowing for study of the vectorial transport of nonmetabolized substrates. On the other hand, relative expression levels of multiple cotransfected transport proteins remain difficult to assess, resulting in a rather artificial model exhibiting differences in rate-limiting steps in hepatobiliary drug disposition compared to in vivo. Nonpolarized cell systems such as insect cells and oocytes also have been used for human transporter expression.[102] For example, a characterization of OAT2, which is highly expressed in human liver, was conducted recently with *Xenopus laevis* oocytes, resulting in novel information regarding substrate specificity and transport mechanism.[11] Important advantages of the latter model are that they can be constructed and validated rapidly and that drug transport by a specific transport protein can be characterized without interference from other (transport) processes. Indeed, Saito et al. recently demonstrated that membrane vesicles prepared from BCRP-transfected insect cells provide a high-speed assay to establish a structure–activity relationship for inhibitors.[103] However, the main disadvantage of these model systems is that the relative contribution of the particular transport protein to overall transport processes cannot be determined.[104] In addition, the actual functional activity of a given transport protein is likely to depend on the cellular environment, which is clearly different in a transfected system compared to the normal hepatocyte.[102]

Freshly isolated hepatocytes can be considered the most comprehensive cell-based model for the study of hepatic drug transport. Suspended hepatocytes remain a useful and convenient system for studying drug uptake mechanisms. Unfortunately, hepatocytes lose their cellular polarity rapidly upon isolation,[105] resulting in the absence

TABLE 13.4. Summary of Major Applications and Advantages and Disadvantages of Model Systems Used to Study Hepatobiliary Drug Transport

Model System	Major Applications	Advantages/Disadvantages
Membrane vesicles prepared from transfected systems	High-throughput screening for substrates and inhibitors	+ High throughput – Difficult to assess the relative contribution of multiple transport proteins
Purified membrane vesicles prepared from liver tissue	Mechanistic studies on contribution of specific transporters at particular membrane domain	+ All relevant transporters expressed, no interference from metabolism – Technically challenging
Nonpolarized transfected cells (oocytes, Sf9, HeLa, etc.)	High-throughput screening for substrates and inhibitors	+ High-throughput cell-based model – Nonhepatic cellular context
Polarized transfected cells (MDCK, HEK293, HeLa, etc.)	Vectorial transport across whole cells can be investigated	+ Polarized phenotype mimics hepatocyte polarity + Multiple transfections can be conducted – Use of nonhepatic and/or animal host cells – Difficult to standardize relative expression levels
Isolated hepatocytes	Investigate hepatic uptake kinetics of substrates	+ Most comprehensive cell-based model + Cells are metabolically active – Hepatocytes can only be used within a few hours after isolation – Not suitable for study of canalicular efflux transport
Plated hepatocytes	Mechanistic studies on hepatobiliary disposition, hepatic drug interactions, and transporter regulation at the molecular level	+ Sandwich culture allows for optimal transporter expression + Most hepatic transporter and drug-metabolizing enzymes expressed in culture – Static model (no bile or blood flow) – Gradual loss of hepatic phenotype with culture time (weeks)
Hepatocyte couplets/liver slices	Mechanistic studies on interplay of hepatic metabolism and transport	+ In vivo architecture maintained at cellular level – Viability limited to few hours
Isolated perfused liver	Study mechanisms of hepatic uptake, metabolism, and biliary excretion	+ Intact organ physiology + No interference from other ADME processes – Low throughput, high animal consumption, limited to ~3 hours after liver isolation
In vivo bile excretion study	Most physiologically relevant model	+ Studies may be conducted in freely moving animals – Few examples in humans – Lowest throughput

of the endogenous biochemical context for studying drug transport processes occurring at the basolateral versus canalicular membrane domains. Moreover, hepatocytes cultured on a single layer of an extracellular collagen matrix ("conventional configuration") continue to dedifferentiate rapidly with complete loss of cell polarity and many hepatocyte-specific functions.[106] In contrast, when hepatocytes are cultured between two layers of gelled collagen ("sandwich configuration") (see Figure 13.1), mature hepatocyte morphology and long-term expression of a differentiated phenotype can be maintained.[107–109] This includes re-establishment of cell polarity and functional canalicular networks in culture, thus allowing for the study of hepatobiliary drug metabolism and transport.[110–112] The formation of "excretory domains" that are thought to correspond to canalicular networks also has been observed in hepatocyte-derived cell lines (HepG2, WIF-B), which have been used to study trafficking and regulation of transport proteins.[64,113–115] It should be noted, however, that these cell lines express the hepatic phenotype only partially, and lack expression of many drug-metabolizing enzyme and transport proteins.[116] A recent evaluation of the human hepatoma cell line HepaRG revealed mRNA expression levels and corresponding functional activities of sinusoidal and canalicular transport proteins that were comparable to those observed in primary human hepatocytes.[116] Although neither cell polarity nor expression of the transport proteins at the relevant membrane domains were investigated in this study, HepaRG cells may serve as a promising alternative to primary cultured hepatocytes for in vitro hepatobiliary transport studies. Hepatocyte couplets have been prepared following limited exposure of liver tissue to collagenase and thus provide another model system preserving hepatocyte polarity. Hepatocyte couplets have been used to study hepatobiliary transport mechanisms as well as the physiology underlying bile secretion.[117,118]

The isolated perfused liver (primarily rat) methodology has been used widely to investigate hepatobiliary transport of drugs and xenobiotics. Although time- and animal-consuming, the main advantage of this model system is that the normal liver physiology (including bile flow) is maintained throughout the time of the experiment. Moreover, the ability to sample from both perfusate (blood compartment) and bile makes this model system particularly useful for mechanistic studies (including pharmacokinetic modeling) without potentially confounding factors that exist in vivo. Another attractive feature of the isolated perfused liver method is the ability to assess the effects of in vivo pretreatments (e.g., with inducers) on hepatobiliary transport and metabolism.[119] Livers obtained from animals with a transport protein knocked out or from animals that are natural mutants for a particular transport protein (e.g., TR^- Wistar rats, which have a functionally relevant defect in the *Mrp2* gene) can also be used in perfusion experiments.[69,120]

Genetic models of transport protein deficiencies also have been used for in vivo pharmacokinetic studies to reveal the role of a particular transport protein in drug disposition, including hepatobiliary transport and metabolism. Although compensatory up-regulation of other transport processes as a result of the absence of a pivotal transport system has been reported,[31] preservation of otherwise intact liver physiology and hepatobiliary drug transport and metabolism processes can be considered as the most attractive feature of these in vivo models.

13.8. TOOLS FOR STUDYING HEPATOBILIARY DRUG TRANSPORT

The various tools employed to study transport proteins are often utilized in combination with the many model systems discussed previously. As the nature of science is constantly evolving with new discoveries, so are the tools used to make these discoveries. Many of the tools currently in use have been borrowed from other fields, such as molecular biology or engineering.

Molecular biology techniques often are employed to assess the expression, localization and function of the hepatic transport proteins. For example, Northern blot, Western blot, or real-time polymerase chain reaction (RT-PCR) are utilized to determine relative protein or mRNA levels. These data should be interpreted with caution because quantification of bands on a Western blot using densitometry gives only the relative amount of protein on a single membrane. The same is true for RT-PCR, where quantification of the mRNA expression of transport proteins is not usually correlated with protein level in terms of the relative amount. Studies often examine the regulation of hepatic transport proteins by assessing the induction or repression of protein or mRNA using Western blot or RT-PCR analysis.[68,121,122] Animal studies that employ genetic knockout of a particular transport protein or spontaneous genetic mutations resulting in a transport protein deficiency often use Western or Northern blot to confirm that the transport protein is absent.[120,292] Theoretically, the function of a particular transport protein could be assessed using probe substrates. However, specific probes for most proteins have yet to be identified, making it difficult to determine the role of any single transport protein in the disposition of a given substrate.

The broad overlap in substrate specificity of the hepatic transport proteins also holds true for compounds used as inhibitors. A single compound is often not a specific inhibitor of one particular transport protein, which confounds data interpretation in deciphering the role of a single transport protein in vivo. More recently, the use of small interfering RNA (siRNA) to knock down the expression and function of a transport protein in primary hepatocytes has been demonstrated. Tian et al. utilized siRNA to modulate both Mrp2 and Mrp3 in primary sandwich-cultured rat hepatocytes.[305] RNA interference (RNAi) is reversible, unlike the knockout rodent models or naturally occurring genetic mutations, and this technology could be applied to human-derived cells and/or in vivo. Knockdown of a target gene using RNAi is more potent than with antisense technology.[125] RNAi is also more specific because it targets the mRNA, unlike chemical inhibitors that interact with the transport proteins, which have a high degree of homology and similar structures. More recently, morpholino antisense oligos have been shown to have higher specificity and longer stability than siRNA. Morpholino oligos, named after the morpholine rings in their backbone, contain nucleic acid bases and a nonionic phosphorodiamidate intersubunit linkage that either blocks translation initiation in the cytosol or modifies pre-mRNA splicing in the nucleus.[126,127]

Noninvasive techniques that utilize radionuclide imaging to track the disposition of xenobiotics in vivo also have been used to study biliary clearance of drugs. Compounds tagged with a gamma-emitting radionuclide have been administered in vivo and subsequently tracked through the biological system using an external gamma camera to assess the rate and extent of absorption, distribution, and excretion of a compound. A novel method recently developed to collect biliary secretions utilizes an oroenteric tube with an occlusive balloon. The hepatobiliary imaging agent ^{99m}technetium (Tc) mebrofenin is employed in this clinical protocol to evaluate the degree of gallbladder contraction in response to pharmacological stimulation and to detect any leakage of bile due to partial occlusion of the intestine.[128] Hendrikse et al. used the radiopharmaceuticals ^{99m}Tc-HIDA and ^{99m}Tc-MIBI to assess the function of the hepatic efflux proteins P-glycoprotein, Mrp1, and Mrp2 in vivo.[129] Another imaging technique, positron emission tomography (PET), uses a PET camera to measure the concentration and movement of positron emitters in the living body. An excellent example of the utility of imaging to evaluate transport protein function was performed by Lee et al. using [^{11}C]verapamil to evaluate P-glycoprotein function at the blood–brain barrier in the rhesus monkey.[130] Imaging techniques may emerge as a powerful tool to elucidate drug disposition and predict drug–transport interactions.

Microscopy is a widely used and inexpensive noninvasive tool to study hepatic transport. Both fluorescent and confocal microscopes have been used in conjunction with fluorescent probes or substrates to study the trafficking and localization of fluorescently tagged proteins.[19,66,131,132]

Pharmacokinetic modeling is a useful tool to aid in analyzing and interpreting the data generated from in vitro and in vivo model systems. Such mathematical approaches can provide valuable insights regarding the hepatobiliary disposition of a compound, including the identification of rate-limiting steps and predictions regarding the impact of potential alterations in hepatobiliary disposition. Data generated in various model systems, including the isolated perfused liver, sandwich-cultured hepatocytes, and transfected cell lines, are amenable to pharmacokinetic analyses.[133–136]

13.9. INTERPLAY BETWEEN DRUG METABOLISM AND TRANSPORT

The substantial overlap in substrate specificity, along with the shared response to prototypical inducers between CYP3A4 and Pgp, has been recognized for more than a decade.[137–139] In addition, a thorough screen of various known CYP3A4 inhibitors demonstrated that most CYP3A4 inhibitors were able to inhibit Pgp, albeit generally at higher K_i values.[140] More important, this has led to various studies supporting the concept that CYP3A4 and Pgp act in concert to determine intestinal drug absorption of substrates.[138,141,142] Indeed, inhibition of intestinal Pgp was shown to enhance exposure of common substrates to CYP3A4, thereby increasing overall metabolism during intestinal transepithelial transport across Caco-2 monolayers. Although the cellular orientation of apical transporters relative to CYP enzymes in the liver differs from that in the intestine, hepatic interplay between drug metabolism and biliary excretion also has been demonstrated. For instance, the extent of hepatic metabolism of the dual CYP3A/Pgp substrate tacrolimus was increased due to Pgp inhibition during isolated rat liver perfusion,[143,144] illustrating the controlling influence of efflux transport on drug metabolism in the liver. A similar type of interplay was observed in rat liver for the CYP3A/Pgp substrate digoxin and the Pgp inhibitor quinidine, also resulting in enhanced digoxin metabolism upon coadministration.[145] In the same study, coadministration of digoxin with the Oatp2 inhibitor rifampicin resulted in reduced digoxin metabolism, demonstrating that hepatic uptake transporters can control exposure of hepatic CYP enzymes to drugs. Inhibition of Oatp-mediated hepatic uptake of atorvastatin by rifampicin was proposed as the predominant mechanism to explain the reduction in atorvastatin metabolism following rifampicin coadministration in the isolated perfused rat liver.[146] Uptake transporters regulate access of drugs to nuclear receptor proteins (e.g., PXR, CAR), thereby influencing the intracellular concentrations of drugs that act as nuclear receptor inducers. For example, human OATP1B1 expression is a major determinant of the extent of PXR activation by rifampin.[147] The observation that chemical inhibition of hepatic Pgp results in enhanced substrate exposure to CYP3A leading to increased metabolism is consistent with results obtained with erythromycin in Mdr1a/b (−/−) knockout mice; values of the average area under the curve (AUC) for $^{14}CO_2$ were increased 1.9-fold following the erythromycin breath test in these double-knockout mice compared to control animals.[293]

In addition to the functional interplay between drug-metabolizing enzymes and transporters, the coordinated regulation of these ADME-relevant proteins also has implications for drug disposition. The concomitant in vitro induction of Pgp and CYP3A4 by prototypical inducers is the best known example of coordinated enzyme and transporter regulation.[137,149] This concept is supported by in vivo induction data with protease inhibitors in rats.[150] On the other hand, Matheny et al. showed that in vivo induction of Pgp and CYP3A in rats is tissue and inducer specific.[151] The latter observation indicates that multiple factors rather than just one orphan nuclear receptor (such as PXR) play a role in the regulation of Pgp and CYP3A. Nevertheless, various independent studies support the concept that regulation of CYP3A and Pgp expression occurs via common pathways.[74,152] The nuclear hormone receptor PXR, believed to predominate (hepatic) regulation of CYP3A expression, has been demonstrated to bind at the 5′-upstream region of the MDR1 gene, thereby playing a pivotal role in rifampin-mediated Pgp induction.[74] The multiplicity of PXR activation effects, as well as the overlap with the effects of CAR activation, have been investigated systematically. The battery of genes reported to be under PXR regulation includes drug-metabolizing enzymes as well as transporters such as *Cyp3a11*, *Ugt1a1*, *Abcb1a/b*, and *Slco1a4*.[153] Furthermore, it was demonstrated that 1,7-phenanthroline, which was hypothesized initially to be a “selective” UGT inducer, concomitantly increased Mrp3 expression in rat liver,[294] whereas Oatp1a4 expression was decreased significantly. Coordinate regulation of hepatic drug metabolizing enzymes and transport protein expression is thought to protect the hepatocyte from intracellular accumulation of toxic xenobiotics. Although the role of the orphan nuclear receptor CAR in expression of phase 2 enzymes, including glucuronosyltransferases (UGT1A1), has been reported previously,[154] its involvement in Mrp3 induction remains controversial.[69,155] In this context, the plausible role of multiple nuclear receptors in transporter (and enzyme) regulation, as reported previously for MRP2,[76] should be emphasized.

13.10. DRUG INTERACTIONS IN HEPATOBILIARY TRANSPORT

A collection of recently published clinical data has demonstrated that mechanisms underlying hepatic drug interactions frequently extend beyond the classical involvement of P450-mediated drug metabolism.[139,156,157] Transporter-related drug interactions are expected to gain importance in the drug discovery and development arena, as recent screening programs are aimed at filtering out chemical entities that display high affinities for CYP450 enzymes. As a consequence, there is an increased likelihood during drug discovery to select compounds that are transport inhibitors and/or that depend on transporters for their elimination and/or distribution. Drug-mediated inhibition of transporter activity, as well as induction of transport protein expression, can lead to clinically relevant drug interactions at the hepatobiliary transport level.

A drug–drug interaction in hepatobiliary transport may occur when one drug interferes with the functional activity (e.g., inhibition) and/or expression levels (e.g., induction) of a hepatic transport protein that is critically involved in hepatobiliary elimination of another drug. In addition, since endogenous compounds such as bile acids rely on hepatic transport proteins to maintain normal hepatic physiology (e.g., bile flow), administration of a drug that inhibits the function of key transport proteins may cause important interactions with endogenous substances in the hepatobiliary system. As discussed earlier in this chapter, this may lead to unexpected hepatotoxicity.

A substantial number of drug–drug interactions in hepatobiliary transport reported in recent years occur at the level of hepatic uptake (see Table 13.5). Various studies have revealed interactions of drugs that inhibit the OATP family of transporters with hepatic uptake of HMG-CoA reductase inhibitors ("statins"). For example, concomitant treatment with the CYP3A4 inhibitor cyclosporin A (CsA) elevated cerivastatin plasma concentrations three- to fivefold, despite its dual CYP2C8- and CYP3A4-mediated metabolic pathway.[158] Shitara et al.[159] provided in vitro data supporting the plausible mechanism for this clinically relevant drug–drug interaction by demonstrating saturable OATP1B1-mediated uptake of cerivastatin in human hepatocytes. In addition, the mechanism behind this severe drug interaction was confirmed in vivo in the rat.[160] Earlier studies had demonstrated significant interactions of CsA with statin therapy. Examples include the 6-fold higher HMG-CoA reductase activity in the plasma of patients receiving atorvastatin and CsA,[161] as well as the 5- and 22-fold increase in exposure to pravastatin and lovastatin, respectively, in patients receiving these drugs in combination with CsA.[162] It should be noted, however, that for most clinically relevant drug interactions involving statins, the relative importance of CYP3A4 (or other CYPs)[163] versus OATP1B1[295] inhibition remains to be elucidated. Coadministration of CsA results in 7- and 11-fold increases in rosuvastatin exposure and maximum plasma concentrations, respectively, compared to

TABLE 13.5. Clinically Relevant Drug–Drug Interactions in Hepatobiliary Transport

Inhibiting Drug	Transporter Substrate Affected	Protein(s) Involved, Effect, and Mechanism Proposed	Refs.
Amiodarone	Digoxin	2-fold increased digoxin levels in humans and rats; in the rat, in vitro data support inhibition of Oatp2-mediated digoxin uptake	175–177
Cyclosporin A	Bosentan	30-fold increased bosentan plasma levels in humans; inhibition of OATP-mediated hepatic uptake of bosentan in rat	167
	Cerivastatin	Cerivastatin AUC 4-fold ↑; OATP1B1 inhibition ($K_i = 0.2\ \mu M$)	158,159
	Repaglinide	Repaglinide AUC 4-fold ↑; inhibition of CYP3A4 and OATP1B1	166,168
	Rosuvastatin	Rosuvastatin AUC 7.1-fold ↑; OATP1B1 inhibition ($IC_{50} = 2.2\ \mu M$ at 5 μM rosuvastatin)	164
Erythromycin	Theophylline	25% erythromycin dose reduction advised; erythromycin inhibits OAT2-mediated theophylline transport; enzyme inhibition also may be involved	178,273
Gemfibrozil	Cerivastatin	Cerivastatin AUC 5.6-fold ↑; inhibition of CYP2C8 (major) and OATP1B1 (minor?)	169,170
	Repaglinide	Repaglinide AUC 8.1-fold ↑; CYP2C8 inhibition by gemfibrozil glucuronide; possibly OATP1B1 inhibition	168,274,275
	Rosuvastatin	Rosuvastatin AUC 1.9-fold ↑; OATP1B1 inhibition	276
	Pravastatin	Pravastatin AUC 2-fold ↑; renal Cl pravastatin ↓; possibly hepatic OATP1B1 inhibition, in part	173
	Lovastatin	Lovastatin acid AUC 2.8-fold ↑; mechanism(s) not elucidated	172
	Simvastatin	Simvastatin acid AUC 1.9-fold ↑; mechanism(s) not confirmed; contribution of OATP1B1 to hepatic uptake may be limited	171,208
Quinine/quinidine	Digoxin	Biliary Cl ↓ by 35/42%; Pgp inhibition	180,277
Ritonavir	Digoxin	Digoxin nonrenal Cl ↓ by 48%; Pgp inhibition	182
Verapamil	Digoxin	Biliary Cl ↓ by 43%; Pgp inhibition	181

rosuvastatin administered alone,[164] even though metabolism is a minor route of rosuvastatin elimination.[165] These observations indicate that inhibition of OATP1B1 by CsA likely plays a key role in the drug–drug interaction between rosuvastatin and CsA. Transporter-mediated drug interactions elicited by CsA clearly are not limited to coadministered HMG-CoA reductase inhibitors, since CsA coadministration increased the plasma AUC of the endothelin receptor antagonist bosentan (30-fold) as well as the antidiabetic repaglinide (4-fold).[166,167] In vivo data generated in the rat with bosentan strongly suggest that inhibition of hepatocellular uptake plays an important role in this severe drug interaction. The important role of OATP1B1 in the hepatic uptake of repaglinide has been demonstrated in a separate study.[168] Moreover, this finding also suggests that inhibition of OATP1B1, in addition to CYP2C8, by the fibric acid derivative gemfibrozil may contribute to the drug interaction between gemfibrozil and repaglinide. Gemfibrozil also has been reported to alter the pharmacokinetics of the statin drugs, possibly due in part to inhibition of hepatic OATP1B1. The use of gemfibrozil as comedication during cerivastatin therapy resulted in more than a 5-fold increase in exposure (plasma AUC) of cerivastatin.[169] Shitara et al.[170] conducted in vitro studies to assess the relative importance of the inhibitory effects of gemfibrozil and gemfibrozil glucuronide on CYP2C8-mediated metabolism and on OATP1B1-mediated hepatic uptake of cerivastatin. Both gemfibrozil and gemfibrozil glucuronide were more potent inhibitors of CYP2C8 than OATP1B1, suggesting a major role for CYP2C8 and possibly minor involvement of OATP1B1 in this drug–drug interaction. A less pronounced interaction was reported between gemfibrozil and simvastatin, resulting in about a 1.9-fold increase in simvastatin acid AUC values.[171] Since no inhibition of CYP3A4 by gemfibrozil was observed in vitro, the possible role for OATP inhibition in this drug interaction remains to be elucidated. The same conclusion can be drawn for the drug interactions observed following coadministration of gemfibrozil with pravastatin or lovastatin.[172,173] Interestingly, recent in vitro data indicate that gemfibrozil stands out among other fibric acid derivatives tested for its potency to inhibit OATP1B1.[174]

For some hepatic drug interactions, the use of animal models to elucidate the underlying mechanism(s) of interaction may be justified, despite considerable species differences in transporter affinity profiles and the fact that human OATPs are not orthologs of rat Oatps. For instance, the pharmacokinetic interaction between the antiarrhythmic drug amiodarone and digoxin has been observed in both humans[175] and rats.[176] In vitro data suggest that inhibition of Oatp1a4-mediated hepatic uptake of digoxin by amiodarone probably plays a predominant role in this drug interaction.[177] Less information is available regarding drug interactions mediated by hepatic uptake transporters other than those of the SLCO family. Data generated in OAT2 expressing oocytes strongly support a role for OAT2 in the well-known drug interaction between theophylline and erythromycin in human liver.[178] Since additional in vitro data have demonstrated substantial inhibition of OAT2-mediated tetracycline uptake in oocytes by acetaminophen, erythromycin, chloramphenicol, ibuprofen, bumetanide, and furosemide, it is likely that several other drug interactions involving OAT2-mediated transport remain to be identified.[11] Numerous drugs (including the H_2

antagonist famotidine[179] and several HIV protease inhibitors[296]) have been identified as potent inhibitors of OCT1 and/or OCT3, the two isoforms of the SLC22A family of hepatic uptake transporters, but clinically relevant drug interactions involving these transporters have not been reported.

Inhibition of canalicular transport proteins also has resulted in clinically relevant drug interactions, including a reduction in digoxin biliary clearance (by about 40%) in patients who also were receiving the Pgp inhibitors quinidine or verapamil.[180,181] More recently, inhibition of Pgp by ritonavir was proposed as the mechanism behind the significant decrease in the nonrenal clearance of digoxin in healthy volunteers who were coadministered the HIV protease inhibitor ritonavir.[182]

Apart from the drug–drug interactions in hepatobiliary transport discussed above, various drugs interfere with the hepatic handling of endogenous compounds, including bilirubin, bile acids, and thyroid hormones, due to inhibition of hepatic transport proteins. For instance, Campbell et al. demonstrated that the potency of several compounds (including CsA, rifamycin SV, and the protease inhibitors saquinavir and ritonavir) that inhibited OATP1B1-mediated transport in vitro could be correlated directly to the incidence of hyperbilirubinemia following the use of these drugs in humans.[9] These observations are consistent with the predominant role of OATP1B1 compared to other SLCO gene products in hepatic bilirubin uptake.[183] Several drugs also have been reported to interact with BSEP-mediated bile acid transport, thus causing alterations in the hepatobiliary handling of bile acids, which may lead to hepatotoxicity. CsA, rifamycin, rifampicin, and glibenclamide were reported to *cis*-inhibit BSEP activity, while *trans*-inhibition was observed for $E_2 17G$.[38,184] Inhibition of BSEP also was proposed as one mechanism by which the endothelin receptor antagonist bosentan causes cholestasis in humans and also in rats.[27] This observation was consistent with data generated in a systematic in vitro study examining the relationship between the potency of compounds to inhibit bile acid transport and their cholestatic and/or hepatotoxic potential.[185] In addition to bosentan, CsA, the endothelin-A receptor antagonist CI-1034, glyburide, erythromycin estolate, and troleandomycin were all shown to inhibit transport of taurocholate across the canalicular membrane of human hepatocytes.[185] Moreover, these compounds (except troleandomycin) also were identified as potent inhibitors of the hepatic uptake of taurocholate. CI-1034 potently inhibited the three major human OATP isoforms.[186]

In addition to direct inhibition of transporter activity, altered transporter expression following drug-mediated transporter induction is another important mechanism underlying drug–drug interactions in hepatobiliary transport. Indeed, as discussed in Section 13.5, transporters are under regulatory control of orphan nuclear receptors such as PXR and CAR. Therefore, repeated administration of drugs that are ligands for these orphan nuclear receptors may lead to altered transporter expression and activity (induction/up-regulation or down-regulation). This will cause substantial changes in hepatobiliary elimination kinetics of drugs that are substrates for these transporters. Clinically relevant induction of Pgp has been observed at the intestinal level,[187] and induction of hepatic Pgp in animals has been reported. For instance, rifampicin treatment resulted in a 4- to 13-fold increase in hepatic Pgp in monkeys.[188] Similarly, treatment of rats with the Pgp inducer tamoxifen resulted in about a 12-fold increase in hepatic Mdr1b mRNA accompanied by an increase in biliary excretion of tamoxifen and its metabolites.[189]

In recent years, there have been an increasing number of reports concerning clinically relevant interactions in hepatobiliary drug transport. These findings illustrate the unmet need for sophisticated and representative model systems allowing the early prediction of hepatobiliary drug interaction potential in drug discovery programs.

13.11. POTENTIAL ROLE OF HEPATIC TRANSPORT PROTEINS IN DRUG-INDUCED HEPATOTOXICITY

Drug-induced liver injury has emerged as the most common reason for the withdrawal of Food and Drug Administration–approved drugs from the market.[190] Drug-induced hepatotoxicity may be classified as intrinsic or idiosyncratic. Intrinsic hepatotoxicity is generally predictable, dose-dependent, and characteristic for the drug. In contrast, idiosyncratic toxicity is unpredictable, not dose-related, and has no clear underlying mechanism. Age, gender, and genetic and environmental factors may affect susceptibility to drug-induced liver injury. Early studies focused on bioactivation, covalent adduct formation, immunotoxicity, and disruptions in cellular bioenergetics as mechanisms underlying hepatotoxicity, but more recent data suggest that hepatic transport proteins may be an important site of toxic interactions.

Inhibition of canalicular Bsep has been reported as one mechanism of drug-induced liver injury because impaired Bsep function leads to elevated hepatic concentrations of detergentlike bile acids, which can disrupt key membrane-associated cellular functions.[28,191] Troglitazone (Rezulin) was the first thiazolidinedione used for the treatment of type 2 diabetes prior to its withdrawal from the market in 2000 due to idiosyncratic liver injury.[192] Several mechanisms underlying troglitazone-induced hepatotoxicity have been proposed,[193] including Bsep inhibition by both troglitazone ($K_{i,\text{apparent}} = 1.3\ \mu M$) and its primary metabolite, troglitazone sulfate ($K_{i,\text{apparent}} = 0.23\ \mu M$).[28] Troglitazone causes intrahepatic accumulation of bile acids, which can lead to hepatocellular necrosis and severe liver damage.[194] Following troglitazone administration, Mrp2-deficient rats exhibited a marked delay in the biliary excretion and enhanced urinary excretion of troglitazone sulfate, leading to elevated serum bile acid concentrations relative to wild-type rats.[195] Although troglitazone has been the most studied example of drug-induced liver injury resulting from Bsep inhibition, other drugs, including bosentan, glibenclamide, cyclosporin, and rifampin, also impair Bsep function.[27,38] In addition to *cis*-inhibition of Bsep, $E_2$17G *trans*-inhibits Bsep following biliary excretion by Mrp2.[38] In the absence of Mrp2, $E_2$17G–induced cholestasis is not observed.[38,196]

Hepatic transport proteins may protect the hepatocyte from drug-induced hepatotoxicity by changes in transport protein expression.[4] Intrahepatic and obstructive cholestasis in humans and rats results in down-regulation of Ntcp and Oatp1b2 and up-regulation of Mrp isoforms.[89,123,198,199] In patients with hepatitis or chronic cholestasis, up-regulation of major hepatic export proteins such as MDR1, MRP1, and MRP3 have been reported.[30] Similarly, during liver injury caused by acetaminophen and carbon tetrachloride, which are both well-known hepatic toxicants, the expression of the hepatic transport proteins in mice was modulated[200,201]: Mrp2 protein expression was increased slightly; Mrp3 and Mrp4 expression was increased 3.6- and 16-fold, respectively; Ntcp and Oatps were down-regulated. Collectively, hepatocytes may prevent accumulation of potentially toxic chemicals and protect against subsequent liver injury by limiting influx and enabling efficient efflux of harmful toxicants following down-regulation of uptake transporters and up-regulation of efflux transporters.

In certain cases, impaired hepatic transport may provide protection against drug-induced liver injury. Many electrophiles that form covalent adducts with intracellular macromolecules are detoxified by conjugation with glutathione. Mrp2-deficient rats do not excrete glutathione into bile, resulting in about a two- to threefold increase in hepatic glutathione concentrations.[202] Acetaminophen, a commonly used antipyretic and analgesic agent, induces severe liver injury at high doses, when acetaminophen sulfation and glucuronidation become saturated and oxidation to *N*-acetyl-*p*-benzoquinoneimine, a potent electrophile, becomes a significant metabolic pathway.[203] *N*-acetyl-*p*-benzoquinoneimine may be detoxified by conjugation with glutathione, but as this cofactor becomes depleted, the electrophilic metabolite forms covalent adducts, which can lead to necrosis.[204] Administration of toxic acetaminophen doses (1 g/kg) induced the expected hepatotoxicity in wild-type rats but had no apparent adverse effects in Mrp2-deficient rat livers.[205] Elevated glutathione concentrations in Mrp2-deficient rats appear to play a protective role in acetaminophen-induced hepatotoxicity; a similar effect also would be expected for other electrophiles in Mrp2-deficient rats.[205]

13.12. THE FUTURE OF HEPATIC DRUG TRANSPORT

This is an exciting era in the fast-paced field of hepatic drug transport. During the last decade, the basolateral and apical transport proteins that play a major role in the hepatic uptake and biliary excretion of drugs and metabolites have been identified. Although some work remains in identifying the role of proteins in the hepatic basolateral excretion of drugs/metabolites, it is generally assumed that identification of the key transport proteins responsible for hepatobiliary drug disposition is nearing completion. However, the structural features of a molecule that enhance the probability of interaction (e.g., transport, inhibition) with specific hepatic transport proteins remains to be defined. Molecular modeling approaches hold great promise in answering these fundamental questions. Overlapping substrate specificity for many of these proteins suggests that the likelihood of identifying specific substrates and/or inhibitors, which would serve as useful tools particularly in clinical studies, is remote. Understanding which proteins are likely to be the rate-limiting step in hepatobiliary drug disposition is essential in predicting how genetic differences, drug interactions, and/or disease states will affect systemic, hepatic, and biliary/intestinal exposure to drugs/metabolites. This has important implications for drug efficacy, as well as toxicity, and much work remains to be undertaken in this area. The role of hepatic transport proteins other than those that reside on the basolateral and apical membranes of the hepatocyte (e.g., intracellular trafficking proteins, proteins on the endoplasmic reticulum) and their role in hepatobiliary drug/metabolite disposition, efficacy, and toxicity await exploration. Continued characterization, validation, and refinement of existing model systems are needed. A road map for use with existing models would be particularly helpful in the drug development process to delineate what model(s) should be used to answer which questions, and in what order, and to address how data should be interpreted and used most effectively in drug development. In addition, the development of new techniques, tools, and/or model systems that address specific concerns with existing approaches (e.g., species-dependent differences, sample throughput, physiologic relevance), and that can be used to answer both mechanistic and applied questions regarding hepatobiliary transport, is needed. Such tools will be useful in enhancing our understanding of how drugs and metabolites move through the human hepatocyte to sites of action, biotransformation, and/or excretion. This knowledge is fundamental to the optimization of drug therapy to achieve desirable therapeutic outcomes and to minimize the incidence of side effects, including drug-induced hepatotoxicity.

Acknowledgments

This work was supported in part by the National Institutes of Health grant R0I GM41935.

REFERENCES

1. Shon JH, Yoon YR, Hong WS, Nguyen PM, Lee SS, Choi YG, Cha IJ, Shin JG. 2005. Effect of itraconazole on the pharmacokinetics and pharmacodynamics of fexofenadine in relation to the MDR1 genetic polymorphism. Clin Pharmacol Ther 78(2):191–201.
2. Mukhopadhayay S, Ananthanarayanan M, Stieger B, Meier PJ, Suchy FJ, Anwer MS. 1997. cAMP increases liver Na^+-taurocholate cotransport by translocating transporter to plasma membranes. Am J Physiol 273(4 Pt 1):G842–G848.
3. Friesema EC, Docter R, Moerings EP, Stieger B, Hagenbuch B, Meier PJ, Krenning EP, Hennemann G, Visser TJ. 1999. Identification of thyroid hormone transporters. Biochem Biophys Res Commun 254(2):497–501.
4. Gundala S, Wells LD, Milliano MT, Talkad V, Luxon BA, Neuschwander-Tetri BA. 2004. The hepatocellular bile acid transporter Ntcp facilitates uptake of the lethal mushroom toxin alpha-amanitin. Arch Toxicol 78(2):68–73.
5. Kullak-Ublick GA, Glasa J, Boker C, Oswald M, Grutzner U, Hagenbuch B, Stieger B, Meier PJ, Beuers U, Kramer W, Wess G, Paumgartner G. 1997. Chlorambucil-taurocholate is transported by bile acid carriers expressed in human hepatocellular carcinomas. Gastroenterology 113(4):1295–1305.
6. Briz O, Romero MR, Martinez-Becerra P, Macias RI, Perez MJ, Jimenez F, San Martin FG, Marin JJ. 2006. Oatp8/1B3-mediated cotransport of bile acids and glutathione: an export pathway for organic anions from hepatocytes? J Biol Chem 287(44):30326–30335.

7. Li L, Meier PJ, Ballatori N. 2000. Oatp2 mediates bidirectional organic solute transport: a role for intracellular glutathione. Mol Pharmacol 58(2):335–340.
8. Mikkaichi T, Suzuki T, Tanemoto M, Ito S, Abe T. 2004. The organic anion transporter (OATP) family. Drug Metab Pharmacokinet 19(3):171–179.
9. Campbell SD, de Morais SM, Xu JJ. 2004. Inhibition of human organic anion transporting polypeptide OATP 1B1 as a mechanism of drug-induced hyperbilirubinemia. Chem Biol Interact 150(2):179–187.
10. Sun W, Wu RR, van Poelje PD, Erion MD. 2001. Isolation of a family of organic anion transporters from human liver and kidney. Biochem Biophys Res Commun 283(2):417–422.
11. Kobayashi Y, Ohshiro N, Sakai R, Ohbayashi M, Kohyama N, Yamamoto T. 2005. Transport mechanism and substrate specificity of human organic anion transporter 2 (hOat2 [SLC22A7]). J Pharm Pharmacol 57(5):573–578.
12. Khamdang S, Takeda M, Noshiro R, Narikawa S, Enomoto A, Anzai N, Piyachaturawat P, Endou H. 2002. Interactions of human organic anion transporters and human organic cation transporters with nonsteroidal anti-inflammatory drugs. J Pharmacol Exp Ther 303(2):534–539.
13. Sekine T, Cha SH, Tsuda M, Apiwattanakul N, Nakajima N, Kanai Y, Endou H. 1998. Identification of multispecific organic anion transporter 2 expressed predominantly in the liver. FEBS Lett 429(2):179–182.
14. Enomoto A, Takeda M, Shimoda M, Narikawa S, Kobayashi Y, Kobayashi Y, Yamamoto T, Sekine T, Cha SH, Niwa T, Endou H. 2002. Interaction of human organic anion transporters 2 and 4 with organic anion transport inhibitors. J Pharmacol Exp Ther 301(3):797–802.
15. Kojima R, Sekine T, Kawachi M, Cha SH, Suzuki Y, Endou H. 2002. Immunolocalization of multispecific organic anion transporters, OAT1, OAT2, and OAT3, in rat kidney. J Am Soc Nephrol 13(4):848–857.
16. Koepsell H, Endou H. 2004. The SLC22 drug transporter family. Pflugers Arch 447(5):666–676.
17. Kimura H, Takeda M, Narikawa S, Enomoto A, Ichida K, Endou H. 2002. Human organic anion transporters and human organic cation transporters mediate renal transport of prostaglandins. J Pharmacol Exp Ther 301(1):293–298.
18. Koepsell H, Schmitt BM, Gorboulev V. 2003. Organic cation transporters. Rev Physiol Biochem Pharmacol 150:36–90.
19. Zhang P, Tian X, Chandra P, Brouwer KL. 2005. Role of glycosylation in trafficking of Mrp2 in sandwich-cultured rat hepatocytes. Mol Pharmacol 67(4):1334–1341.
20. Juliano RL, Ling V. 1976. A surface glycoprotein modulating drug permeability in Chinese hamster ovary cell mutants. Biochim Biophys Acta 455(1):152–162.
21. Kawabata S, Oka M, Shiozawa K, Tsukamoto K, Nakatomi K, Soda H, Fukuda M, Ikegami Y, Sugahara K, Yamada Y, et al. 2001. Breast cancer resistance protein directly confers SN-38 resistance of lung cancer cells. Biochem Biophys Res Commun 280(5):1216–1223.
22. Carrella M, Roda E. 1999. Evolving concepts in the pathophysiology of biliary lipid secretion. Ital J Gastroenterol Hepatol 31(7):643–648.
23. Smith AJ, van Helvoort A, van Meer G, Szabo K, Welker E, Szakacs G, Varadi A, Sarkadi B, Borst P. 2000. MDR3 P-glycoprotein, a phosphatidylcholine translocase, transports several cytotoxic drugs and directly interacts with drugs as judged by interference with nucleotide trapping. J Biol Chem 275(31):23530–23539.

24. Kullak-Ublick GA, Stieger B, Hagenbuch B, Meier PJ. 2000. Hepatic transport of bile salts. Semin Liver Dis 20(3):273–292.
25. Hirano M, Maeda K, Hayashi H, Kusuhara H, Sugiyama Y. 2005. Bile salt export pump (BSEP/ABCB11) can transport a nonbile acid substrate, pravastatin. J Pharmacol Exp Ther 314(2):876–882.
26. Kullak-Ublick GA, Stieger B, Meier PJ. 2004. Enterohepatic bile salt transporters in normal physiology and liver disease. Gastroenterology 126(1):322–342.
27. Fattinger K, Funk C, Pantze M, Weber C, Reichen J, Stieger B, Meier PJ. 2001. The endothelin antagonist bosentan inhibits the canalicular bile salt export pump: a potential mechanism for hepatic adverse reactions. Clin Pharmacol Ther 69(4):223–231.
28. Funk C, Pantze M, Jehle L, Ponelle C, Scheuermann G, Lazendic M, Gasser R. 2001. Troglitazone-induced intrahepatic cholestasis by an interference with the hepatobiliary export of bile acids in male and female rats. Correlation with the gender difference in troglitazone sulfate formation and the inhibition of the canalicular bile salt export pump (Bsep) by troglitazone and troglitazone sulfate. Toxicology 167(1):83–98.
29. Tsujii H, Konig J, Rost D, Stockel B, Leuschner U, Keppler D. 1999. Exon–intron organization of the human multidrug-resistance protein 2 (MRP2) gene mutated in Dubin–Johnson syndrome. Gastroenterology 117(3):653–660.
30. Konig J, Rost D, Cui Y, Keppler D. 1999. Characterization of the human multidrug resistance protein isoform MRP3 localized to the basolateral hepatocyte membrane. Hepatology 29(4):1156–1163.
31. Kuroda M, Kobayashi Y, Tanaka Y, Itani T, Mifuji R, Araki J, Kaito M, Adachi Y. 2004. Increased hepatic and renal expressions of multidrug resistance-associated protein 3 in Eisai hyperbilirubinuria rats. J Gastroenterol Hepatol 19(2):146–153.
32. Maliepaard M, van Gastelen MA, de Jong LA, Pluim D, van Waardenburg RC, Ruevekamp-Helmers MC, Floot BG, Schellens JH. 1999. Overexpression of the BCRP/MXR/ABCP gene in a topotecan-selected ovarian tumor cell line. Cancer Res 59(18):4559–4563.
33. Borst P, Evers R, Kool M, Wijnholds J. 1999. The multidrug resistance protein family. Biochim Biophys Acta 1461(2):347–357.
34. Hirohashi T, Suzuki H, Sugiyama Y. 1999. Characterization of the transport properties of cloned rat multidrug resistance-associated protein 3 (MRP3). J Biol Chem 274(21):15181–15185.
35. Sampath J, Adachi M, Hatse S, Naesens L, Balzarini J, Flatley RM, Matherly LH, Schuetz JD. 2002. Role of MRP4 and MRP5 in biology and chemotherapy. AAPS PharmSci 4(3):E14.
36. Zelcer N, Reid G, Wielinga P, Kuil A, van der Heijden I, Schuetz JD, Borst P. 2003. Steroid and bile acid conjugates are substrates of human multidrug-resistance protein (MRP) 4 (ATP-binding cassette C4). Biochem J 371(Pt 2):361–367.
37. Donner MG, Warskulat U, Saha N, Haussinger D. 2004. Enhanced expression of basolateral multidrug resistance protein isoforms Mrp3 and Mrp5 in rat liver by LPS. Biol Chem 385(3–4):331–339.
38. Stieger B, Fattinger K, Madon J, Kullak-Ublick GA, Meier PJ. 2000. Drug- and estrogen-induced cholestasis through inhibition of the hepatocellular bile salt export pump (Bsep) of rat liver. Gastroenterology 118(2):422–430.

39. Perdu J, Germain DP. 2001. Identification of novel polymorphisms in the pM5 and MRP1 (ABCC1) genes at locus 16p13.1 and exclusion of both genes as responsible for pseudoxanthoma elasticum. Hum Mutat 17(1):74–75.
40. Durr D, Stieger B, Kullak-Ublick GA, Rentsch KM, Steinert HC, Meier PJ, Fattinger K. 2000. St John's wort induces intestinal P-glycoprotein/MDR1 and intestinal and hepatic CYP3A4. Clin Pharmacol Ther 68(6):598–604.
41. Bradbury NA, Bridges RJ. 1994. Role of membrane trafficking in plasma membrane solute transport. Am J Physiol 267(1 Pt 1):C1–C24.
42. Gradilone SA, Tietz PS, Splinter PL, Marinelli RA, LaRusso NF. 2005. Expression and subcellular localization of aquaporin water channels in the polarized hepatocyte cell line, WIF-B. BMC Physiol 5:13.
43. Lienhard GE. 1983. Regulation of cellular membrane transport by the exocytotic insertion and endocytic retrieval of transporters. Trend Biochem Sci 8(4):125–127.
44. Kipp H, Arias IM. 2000. Newly synthesized canalicular ABC transporters are directly targeted from the Golgi to the hepatocyte apical domain in rat liver. J Biol Chem 275(21):15917–15925.
45. Kipp H, Pichetshote N, Arias IM. 2001. Transporters on demand: intrahepatic pools of canalicular ATP binding cassette transporters in rat liver. J Biol Chem 276(10):7218–7224.
46. Hayakawa T, Bruck R, Ng OC, Boyer JL. 1990. DBcAMP stimulates vesicle transport and HRP excretion in isolated perfused rat liver. Am J Physiol 259(5 Pt 1):G727–G735.
47. Gatmaitan ZC, Nies AT, Arias IM. 1997. Regulation and translocation of ATP-dependent apical membrane proteins in rat liver. Am J Physiol 272(5 Pt 1):G1041–G1049.
48. Misra S, Ujhazy P, Gatmaitan Z, Varticovski L, Arias IM. 1998. The role of phosphoinositide 3-kinase in taurocholate-induced trafficking of ATP-dependent canalicular transporters in rat liver. J Biol Chem 273(41):26638–26644.
49. Kurz AK, Graf D, Schmitt M, Vom Dahl S, Haussinger D. 2001. Tauroursodesoxycholate-induced choleresis involves p38(MAPK) activation and translocation of the bile salt export pump in rats. Gastroenterology 121(2):407–419.
50. Beuers U, Bilzer M, Chittattu A, Kullak-Ublick GA, Keppler D, Paumgartner G, Dombrowski F. 2001. Tauroursodeoxycholic acid inserts the apical conjugate export pump, Mrp2, into canalicular membranes and stimulates organic anion secretion by protein kinase C-dependent mechanisms in cholestatic rat liver. Hepatology 33(5):1206–1216.
51. Kikuchi S, Hata M, Fukumoto K, Yamane Y, Matsui T, Tamura A, Yonemura S, Yamagishi H, Keppler D, Tsukita S, Tsukita S. 2002. Radixin deficiency causes conjugated hyperbilirubinemia with loss of Mrp2 from bile canalicular membranes. Nat Genet 31(3):320–325.
52. Soroka CJ, Pate MK, Boyer JL. 1999. Canalicular export pumps traffic with polymeric immunoglobulin A receptor on the same microtubule-associated vesicle in rat liver. J Biol Chem 274(37):26416–26424.
53. Roelofsen H, Soroka CJ, Keppler D, Boyer JL. 1998. Cyclic AMP stimulates sorting of the canalicular organic anion transporter (Mrp2/cMoat) to the apical domain in hepatocyte couplets. J Cell Sci 111(Pt 8):1137–1145.
54. Haussinger D, Schmitt M, Weiergraber O, Kubitz R. 2000. Short-term regulation of canalicular transport. Semin Liver Dis 20(3):307–321.
55. Kubitz R, D'Urso D, Keppler D, Haussinger D. 1997. Osmodependent dynamic localization of the multidrug resistance protein 2 in the rat hepatocyte canalicular membrane. Gastroenterology 113(5):1438–1442.

56. Dombrowski F, Kubitz R, Chittattu A, Wettstein M, Saha N, Haussinger D. 2000. Electron-microscopic demonstration of multidrug resistance protein 2 (Mrp2) retrieval from the canalicular membrane in response to hyperosmolarity and lipopolysaccharide. Biochem J 348(Pt 1):183–188.

57. Fernandez SB, Hollo Z, Kern A, Bakos E, Fischer PA, Borst P, Evers R. 2002. Role of the N-terminal transmembrane region of the multidrug resistance protein MRP2 in routing to the apical membrane in MDCKII cells. J Biol Chem 277(34):31048–31055.

58. Harris MJ, Kuwano M, Webb M, Board PG. 2001. Identification of the apical membrane-targeting signal of the multidrug resistance-associated protein 2 (MRP2/MOAT). J Biol Chem 276(24):20876–20881.

59. Wang P, Wang JJ, Xiao Y, Murray JW, Novikoff PM, Angeletti RH, Orr GA, Lan D, Silver DL, Wolkoff AW. 2005. Interaction with PDZK1 is required for expression of organic anion transporting protein 1A1 on the hepatocyte surface. J Biol Chem 280(34):30143–30149.

60. Hanggi E, Freimoser Grundschober A, Leuthold S, Meier PJ, St-Pierre MV. 2006. Functional analysis of the extracellular cysteine residues in the human organic anion transporting polypeptide, Oatp2b1. Mol Pharmacol.

61. Mukhopadhyay S, Ananthanarayanan M, Stieger B, Meier PJ, Suchy FJ, Anwer MS. 1998. Sodium taurocholate cotransporting polypeptide is a serine, threonine phosphoprotein and is dephosphorylated by cyclic adenosine monophosphate. Hepatology 28(6):1629–1636.

62. Mukhopadhyay S, Webster CR, Anwer MS. 1998. Role of protein phosphatases in cyclic AMP-mediated stimulation of hepatic Na+/taurocholate cotransport. J Biol Chem 273(45):30039–30045.

63. Webster CR, Anwer MS. 1999. Role of the PI3K/PKB signaling pathway in cAMP-mediated translocation of rat liver Ntcp. Am J Physiol 277(6 Pt 1):G1165–G1172.

64. Dranoff JA, McClure M, Burgstahler AD, Denson LA, Crawford AR, Crawford JM, Karpen SJ, Nathanson MH. 1999. Short-term regulation of bile acid uptake by microfilament-dependent translocation of rat ntcp to the plasma membrane. Hepatology 30(1):223–229.

65. Webster CR, Srinivasulu U, Ananthanarayanan M, Suchy FJ, Anwer MS. 2002. Protein kinase B/Akt mediates cAMP- and cell swelling-stimulated Na^+/taurocholate cotransport and Ntcp translocation. J Biol Chem 277(32):28578–28583.

66. Hayes JH, Soroka CJ, Rios-Velez L, Boyer JL. 1999. Hepatic sequestration and modulation of the canalicular transport of the organic cation, daunorubicin, in the Rat. Hepatology 29(2):483–493.

67. Munteanu E, Verdier M, Grandjean-Forestier F, Stenger C, Jayat-Vignoles C, Huet S, Robert J, Ratinaud MH. 2006. Mitochondrial localization and activity of P-glycoprotein in doxorubicin-resistant K562 cells. Biochem Pharmacol 71(8):1162–1174.

68. Staudinger JL, Madan A, Carol KM, Parkinson A. 2003. Regulation of drug transporter gene expression by nuclear receptors. Drug Metab Dispos 31(5):523–527.

69. Xiong H, Yoshinari K, Brouwer KLR, Negishi M. 2002. Role of constitutive androstane receptor in the in vivo induction of Mrp3 and CYP2B1/2 by phenobarbital. Drug Metab Dispos 30(8):918–923.

70. Denson LA, Sturm E, Echevarria W, Zimmerman TL, Makishima M, Mangelsdorf DJ, Karpen SJ. 2001. The orphan nuclear receptor, shp, mediates bile acid–induced inhibition of the rat bile acid transporter, ntcp. Gastroenterology 121(1):140–147.

71. Denson LA, Auld KL, Schiek DS, McClure MH, Mangelsdorf DJ, Karpen SJ. 2000. Interleukin-1beta suppresses retinoid transactivation of two hepatic transporter genes involved in bile formation. J Biol Chem 275(12):8835–8843.
72. Jung D, Podvinec M, Meyer UA, Mangelsdorf DJ, Fried M, Meier PJ, Kullak-Ublick GA. 2002. Human organic anion transporting polypeptide 8 promoter is transactivated by the farnesoid X receptor/bile acid receptor. Gastroenterology 122(7):1954–1966.
73. Ananthanarayanan M, Balasubramanian N, Makishima M, Mangelsdorf DJ, Suchy FJ. 2001. Human bile salt export pump promoter is transactivated by the farnesoid X receptor/bile acid receptor. J Biol Chem 276(31):28857–28865.
74. Geick A, Eichelbaum M, Burk O. 2001. Nuclear receptor response elements mediate induction of intestinal MDR1 by rifampin. J Biol Chem 276(18):14581–14587.
75. Kauffmann HM, Pfannschmidt S, Zoller H, Benz A, Vorderstemann B, Webster JI, Schrenk D. 2002. Influence of redox-active compounds and PXR-activators on human MRP1 and MRP2 gene expression. Toxicology 171(2–3):137–146.
76. Kast HR, Goodwin B, Tarr PT, Jones SA, Anisfeld AM, Stoltz CM, Tontonoz P, Kliewer S, Willson TM, Edwards PA. 2002. Regulation of multidrug resistance-associated protein 2 (ABCC2) by the nuclear receptors pregnane X receptor, farnesoid X-activated receptor, and constitutive androstane receptor. J Biol Chem 277(4):2908–2915.
77. Kok T, Bloks VW, Wolters H, Havinga R, Jansen PL, Staels B, Kuipers F. 2003. Peroxisome proliferator-activated receptor alpha (PPARalpha)-mediated regulation of multidrug resistance 2 (Mdr2) expression and function in mice. Biochem J 369(Pt 3):539–547.
78. Paulusma CC, Kool M, Bosma PJ, Scheffer GL, ter Borg F, Scheper RJ, Tytgat GN, Borst P, Baas F, Oude Elferink RP. 1997. A mutation in the human canalicular multispecific organic anion transporter gene causes the Dubin–Johnson syndrome. Hepatology 25(6):1539–1542.
79. Harris MJ, Le Couteur DG, Arias IM. 2005. Progressive familial intrahepatic cholestasis: genetic disorders of biliary transporters. J Gastroenterol Hepatol 20(6):807–817.
80. Funk C, Ponelle C, Scheuermann G, Pantze M. 2001. Cholestatic potential of troglitazone as a possible factor contributing to troglitazone-induced hepatotoxicity: in vivo and in vitro interaction at the canalicular bile salt export pump (Bsep) in the rat. Mol Pharmacol 59(3):627–635.
81. Bolder U, Trang NV, Hagey LR, Schteingart CD, Ton-Nu HT, Cerre C, Elferink RP, Hofmann AF. 1999. Sulindac is excreted into bile by a canalicular bile salt pump and undergoes a cholehepatic circulation in rats. Gastroenterology 117(4):962–971.
82. Bolder U, Ton-Nu HT, Schteingart CD, Frick E, Hofmann AF. 1997. Hepatocyte transport of bile acids and organic anions in endotoxemic rats: impaired uptake and secretion. Gastroenterology 112(1):214–225.
83. Kubitz R, Wettstein M, Warskulat U, Haussinger D. 1999. Regulation of the multidrug resistance protein 2 in the rat liver by lipopolysaccharide and dexamethasone. Gastroenterology 116(2):401–410.
84. Lee JM, Trauner M, Soroka CJ, Stieger B, Meier PJ, Boyer JL. 2000. Expression of the bile salt export pump is maintained after chronic cholestasis in the rat. Gastroenterology 118(1):163–172.
85. Vos TA, Hooiveld GJ, Koning H, Childs S, Meijer DK, Moshage H, Jansen PL, Müller M. 1998. Up-regulation of the multidrug resistance genes, Mrp1 and Mdr1b, and

down-regulation of the organic anion transporter, Mrp2, and the bile salt transporter, Spgp, in endotoxemic rat liver. Hepatology 28(6):1637–1644.

86. Green RM, Beier D, Gollan JL. 1996. Regulation of hepatocyte bile salt transporters by endotoxin and inflammatory cytokines in rodents. Gastroenterology 111(1):193–198.

87. Jacquemin E, de Vree JM, Cresteil D, Sokal EM, Sturm E, Dumont M, Scheffer GL, Paul M, Burdelski M, Bosma PJ, et al. 2001. The wide spectrum of multidrug resistance 3 deficiency: from neonatal cholestasis to cirrhosis of adulthood. Gastroenterology 120(6):1448–1458.

88. Bossard R, Stieger B, O'Neill B, Fricker G, Meier PJ. 1993. Ethinylestradiol treatment induces multiple canalicular membrane transport alterations in rat liver. J Clin Invest 91(6):2714–2720.

89. Simon FR, Fortune J, Iwahashi M, Gartung C, Wolkoff A, Sutherland E. 1996. Ethinyl estradiol cholestasis involves alterations in expression of liver sinusoidal transporters. Am J Physiol 271(6 Pt 1):G1043–G1052.

90. Niinuma K, Kato Y, Suzuki H, Tyson CA, Weizer V, Dabbs JE, Froehlich R, Green CE, Sugiyama Y. 1999. Primary active transport of organic anions on bile canalicular membrane in humans. Am J Physiol 276(5 Pt 1):G1153–G1164.

91. Ninomiya M, Ito K, Horie T. 2005. Functional analysis of dog multidrug resistance-associated protein 2 (Mrp2) in comparison with rat Mrp2. Drug Metab Dispos 33(2):225–232.

92. Blitzer BL, Donovan CB. 1984. A new method for the rapid isolation of basolateral plasma membrane vesicles from rat liver. Characterization, validation, and bile acid transport studies. J Biol Chem 259(14):9295–9301.

93. Kunst M, Sies H, Akerboom TP. 1989. ATP-stimulated uptake of *S*-(2,4-dinitrophenyl) glutathione by plasma membrane vesicles from rat liver. Biochim Biophys Acta 983(1): 123–125.

94. Meier PJ, St Meier-Abt A, Barrett C, Boyer JL. 1984. Mechanisms of taurocholate transport in canalicular and basolateral rat liver plasma membrane vesicles. Evidence for an electrogenic canalicular organic anion carrier. J Biol Chem 259(16):10614–10622.

95. Duffy MC, Blitzer BL, Boyer JL. 1983. Direct determination of the driving forces for taurocholate uptake into rat liver plasma membrane vesicles. J Clin Invest 72(4): 1470–1481.

96. Novak DA, Ryckman FC, Suchy FJ. 1989. Taurocholate transport by basolateral plasma membrane vesicles isolated from human liver. Hepatology 10(4):447–453.

97. Ishizuka H, Konno K, Shiina T, Naganuma H, Nishimura K, Ito K, Suzuki H, Sugiyama Y. 1999. Species differences in the transport activity for organic anions across the bile canalicular membrane. J Pharmacol Exp Ther 290(3):1324–1330.

98. Suchy FJ, Courchene SM, Blitzer BL. 1985. Taurocholate transport by basolateral plasma membrane vesicles isolated from developing rat liver. Am J Physiol 248(6 Pt 1):G648–G654.

99. Sasaki M, Suzuki H, Ito K, Abe T, Sugiyama Y. 2002. Transcellular transport of organic anions across a double-transfected Madin–Darby canine kidney II cell monolayer expressing both human organic anion-transporting polypeptide (OATP2/SLC21A6) and multidrug resistance-associated protein 2 (MRP2/ABCC2). J Biol Chem 277(8):6497–6503.

100. Sasaki M, Suzuki H, Aoki J, Ito K, Meier PJ, Sugiyama Y. 2004. Prediction of in vivo biliary clearance from the in vitro transcellular transport of organic anions across

a double-transfected Madin–Darby canine kidney II monolayer expressing both rat organic anion transporting polypeptide 4 and multidrug resistance associated protein 2. Mol Pharmacol 66(3):450–459.

101. Kopplow K, Letschert K, König J, Walter B, Keppler D. 2005. Human hepatobiliary transport of organic anions analyzed by quadruple-transfected cells. Mol Pharmacol 68(4):1031–1038.

102. Wagner CA, Friedrich B, Setiawan I, Lang F, Bröer S. 2000. The use of *Xenopus laevis* oocytes for the functional characterization of heterologously expressed membrane proteins. Cell Physiol Biochem 10(1–2):1–12.

103. Saito H, Hirano H, Nakagawa H, Fukami T, Oosumi K, Murakami K, Kimura H, Kouchi T, Konomi M, Tao E, et al. 2006. A new strategy of high-speed screening and quantitative SAR analysis to evaluate human ABC transporter ABCG2–drug interactions. J Pharmacol Exp Ther 317(3):1114–1124.

104. Hirano M, Maeda K, Shitara Y, Sugiyama Y. 2004. Contribution of OATP2 (OATP1B1) and OATP8 (OATP1B3) to the hepatic uptake of pitavastatin in humans. J Pharmacol Exp Ther 311(1):139–146.

105. Groothuis GM, Hulstaert CE, Kalicharan D, Hardonk MJ. 1981. Plasma membrane specialization and intracellular polarity of freshly isolated rat hepatocytes. Eur J Cell Biol 26(1):43–51.

106. LeCluyse EL, Bullock PL, Parkinson A. 1996. Strategies for restoration and maintenance of normal hepatic structure and function in long-term cultures of rat hepatocytes. Adv Drug Deliv Rev 22(1–2):133–186.

107. Dunn JC, Yarmush ML, Koebe HG, Tompkins RG. 1989. Hepatocyte function and extracellular matrix geometry: long-term culture in a sandwich configuration. FASEB J 3(2):174–177.

108. Gomez-Lechon MJ, Jover R, Donato T, Ponsoda X, Rodriguez C, Stenzel KG, Klocke R, Paul D, Guillen I, Bort R, Castell JV. 1998. Long-term expression of differentiated functions in hepatocytes cultured in three-dimensional collagen matrix. J Cell Physiol 177(4):553–562.

109. Kudryavtseva EI, Engelhardt NV. 2003. Requirement of 3D extracellular network for maintenance of mature hepatocyte morphology and suppression of alpha-fetoprotein synthesis in vitro. Immunol Lett 90(1):25–31.

110. Liu X, Chism JP, LeCluyse EL, Brouwer KR, Brouwer KLR. 1999. Correlation of biliary excretion in sandwich-cultured rat hepatocytes and in vivo in rats. Drug Metab Dispos 27(6):637–644.

111. Liu X, LeCluyse EL, Brouwer KR, Gan LS, Lemasters JJ, Stieger B, Meier PJ, Brouwer KLR. 1999. Biliary excretion in primary rat hepatocytes cultured in a collagen-sandwich configuration. Am J Physiol 277(1 Pt 1):G12–G21.

112. Liu X, LeCluyse EL, Brouwer KR, Lightfoot RM, Lee JI, Brouwer KLR. 1999. Use of Ca2+ modulation to evaluate biliary excretion in sandwich-cultured rat hepatocytes. J Pharmacol Exp Ther 289(3):1592–1599.

113. Ihrke G, Neufeld EB, Meads T, Shanks MR, Cassio D, Laurent M, Schroer TA, Pagano RE, Hubbard AL. 1993. WIF-B cells: an in vitro model for studies of hepatocyte polarity. J Cell Biol 123(6 Pt 2):1761–1775.

114. Sai Y, Nies AT, Arias IM. 1999. Bile acid secretion and direct targeting of mdr1-green fluorescent protein from Golgi to the canalicular membrane in polarized WIF-B cells. J Cell Sci 112(Pt 24):4535–4545.

115. Zegers MM, Hoekstra D. 1998. Mechanisms and functional features of polarized membrane traffic in epithelial and hepatic cells. Biochem J 336(Pt 2):257–269.

116. Le Vee M, Jigorel E, Glaise D, Gripon P, Guguen-Guillouzo C, Fardel O. 2006. Functional expression of sinusoidal and canalicular hepatic drug transporters in the differentiated human hepatoma HepaRG cell line. Eur J Pharm Sci 28(1–2):109–117.

117. Boyer JL. 1997. Isolated hepatocyte couplets and bile duct units: novel preparations for the in vitro study of bile secretory function. Cell Biol Toxicol 13(4–5):289–300.

118. Coleman R, Wilton JC, Stone V, Chipman JK. 1995. Hepatobiliary function and toxicity in vitro using isolated hepatocyte couplets. Gen Pharmacol 26(7):1445–1453.

119. Patel NJ, Zamek-Gliszczynski MJ, Zhang P, Han YH, Jansen PL, Meier PJ, Stieger B, Brouwer KLR. 2003. Phenobarbital alters hepatic Mrp2 function by direct and indirect interactions. Mol Pharmacol 64(1):154–159.

120. Nezasa K, Tian X, Zamek-Gliszczynski MJ, Patel NJ, Raub TJ, Brouwer KLR. 2006. Altered hepatobiliary disposition of 5 (and 6)-carboxy-2′,7′-dichlorofluorescein in Abcg2 (Bcrp1) and Abcc2 (Mrp2) knockout mice. Drug Metab Dispos 34(4):718–723.

121. Moffit JS, Aleksunes LM, Maher JM, Scheffer GL, Klaassen CD, Manautou JE. 2006. Induction of hepatic transporters multidrug resistance-associated proteins (Mrp) 3 and 4 by clofibrate is regulated by peroxisome proliferator-activated receptor alpha. J Pharmacol Exp Ther 317(2):537–545.

122. Guo GL, Choudhuri S, Klaassen CD. 2002. Induction profile of rat organic anion transporting polypeptide 2 (oatp2) by prototypical drug-metabolizing enzyme inducers that activate gene expression through ligand-activated transcription factor pathways. J Pharmacol Exp Ther 300(1):206–212.

123. Lee J, Azzaroli F, Wang L, Soroka CJ, Gigliozzi A, Setchell KD, Kramer W, Boyer JL. 2001. Adaptive regulation of bile salt transporters in kidney and liver in obstructive cholestasis in the rat. Gastroenterology 121(6):1473–1484.

124. Zeng H, Chen ZS, Belinsky MG, Rea PA, Kruh GD. 2001. Transport of methotrexate (MTX) and folates by multidrug resistance protein (MRP) 3 and MRP1: effect of polyglutamylation on MTX transport. Cancer Res 61(19):7225–7232.

125. Aoki Y, Cioca DP, Oidaira H, Kamiya J, Kiyosawa K. 2003. RNA interference may be more potent than antisense RNA in human cancer cell lines. Clin Exp Pharmacol Physiol 30(1–2):96–102.

126. Summerton J, Weller D. 1997. Morpholino antisense oligomers: design, preparation, and properties. Antisense Nucleic Acid Drug Dev 7(3):187–195.

127. Summerton J. 1999. Morpholino antisense oligomers: the case for an RNase H-independent structural type. Biochim Biophys Acta 1489(1):141–158.

128. Ghibellini G, Johnson BM, Kowalsky RJ, Heizer WD, Brouwer KLR. 2004. A novel method for the determination of biliary clearance in humans. AAPS J 6(4):e33.

129. Hendrikse NH, Kuipers F, Meijer C, Havinga R, Bijleveld CM, van der Graaf WT, Vaalburg W, de Vries EG. 2004. In vivo imaging of hepatobiliary transport function mediated by multidrug resistance associated protein and P-glycoprotein. Cancer Chemother Pharmacol 54(2):131–138.

130. Lee YJ, Maeda J, Kusuhara H, Okauchi T, Inaji M, Nagai Y, Obayashi S, Nakao R, Suzuki K, Sugiyama Y, Suhara T. 2006. In vivo evaluation of P-glycoprotein function at the blood–brain barrier in nonhuman primates using [^{11}C]verapamil. J Pharmacol Exp Ther 316(2):647–653.

131. Marinelli RA, Tietz PS, Larusso NF. 2005. Regulated vesicle trafficking of membrane transporters in hepatic epithelia. J Hepatol 42(4):592–603.
132. Wakabayashi Y, Lippincott-Schwartz J, Arias IM. 2004. Intracellular trafficking of bile salt export pump (ABCB11) in polarized hepatic cells: constitutive cycling between the canalicular membrane and rab11-positive endosomes. Mol Biol Cell 15(7):3485–3496.
133. Turncliff RZ, Hoffmaster KA, Kalvass JC, Pollack GM, Brouwer KLR. 2006. Hepatobiliary disposition of a drug/metabolite pair: comprehensive pharmacokinetic modeling in sandwich-cultured rat hepatocytes. J Pharmacol Exp Ther 318(2):881–889.
134. Tahara H, Kusuhara H, Fuse E, Sugiyama Y. 2005. P-glycoprotein plays a major role in the efflux of fexofenadine in the small intestine and blood–brain barrier, but only a limited role in its biliary excretion. Drug Metab Dispos 33(7):963–968.
135. Zamek-Gliszczynski MJ, Hoffmaster KA, Tian X, Zhao R, Polli JW, Humphreys JE, Webster LO, Bridges AS, Kalvass JC, Brouwer KLR. 2005. Multiple mechanisms are involved in the biliary excretion of acetaminophen sulfate in the rat: role of Mrp2 and Bcrp1. Drug Metab Dispos 33(8):1158–1165.
136. Doherty MM, Poon K, Tsang C, Pang KS. 2006. Transport is not rate-limiting in morphine glucuronidation in the single-pass perfused rat liver preparation. J Pharmacol Exp Ther 317(2):890–900.
137. Wacher VJ, Wu CY, Benet LZ. 1995. Overlapping substrate specificities and tissue distribution of cytochrome P450 3A and P-glycoprotein: implications for drug delivery and activity in cancer chemotherapy. Mol Carcinog 13(3):129–134.
138. Gan LS, Moseley MA, Khosla B, Augustijns PF, Bradshaw TP, Hendren RW, Thakker DR. 1996. CYP3A-like cytochrome P450-mediated metabolism and polarized efflux of cyclosporin A in Caco-2 cells. Drug Metab Dispos 24(3):344–349.
139. Cvetkovic M, Leake B, Fromm MF, Wilkinson GR, Kim RB. 1999. OATP and P-glycoprotein transporters mediate the cellular uptake and excretion of fexofenadine. Drug Metab Dispos 27(8):866–871.
140. Yasuda K, Lan LB, Sanglard D, Furuya K, Schuetz JD, Schuetz EG. 2002. Interaction of cytochrome P450 3A inhibitors with P-glycoprotein. J Pharmacol Exp Ther 303(1):323–332.
141. Raeissi SD, Hidalgo IJ, Segura-Aguilar J, Artursson P. 1999. Interplay between CYP3A-mediated metabolism and polarized efflux of terfenadine and its metabolites in intestinal epithelial Caco-2 (TC7) cell monolayers. Pharm Res 16(5):625–632.
142. Hochman JH, Chiba M, Nishime J, Yamazaki M, Lin JH. 2000. Influence of P-glycoprotein on the transport and metabolism of indinavir in Caco-2 cells expressing cytochrome P-450 3A4. J Pharmacol Exp Ther 292(1):310–318.
143. Wu CY, Benet LZ. 2003. Disposition of tacrolimus in isolated perfused rat liver: influence of troleandomycin, cyclosporine, and gg918. Drug Metab Dispos 31(11):1292–1295.
144. Jeong H, Chiou WL. 2006. Role of P-glycoprotein in the hepatic metabolism of tacrolimus. Xenobiotica 36(1):1–13.
145. Lau YY, Wu CY, Okochi H, Benet LZ. 2004. Ex situ inhibition of hepatic uptake and efflux significantly changes metabolism: hepatic enzyme-transporter interplay. J Pharmacol Exp Ther 308(3):1040–1045.

146. Lau YY, Okochi H, Huang Y, Benet LZ. 2006. Multiple transporters affect the disposition of atorvastatin and its two active hydroxy metabolites: application of in vitro and ex situ systems. J Pharmacol Exp Ther 316(2):762–771.

147. Tirona RG, Leake BF, Wolkoff AW, Kim RB. 2003. Human organic anion transporting polypeptide-C (SLC21A6) is a major determinant of rifampin-mediated pregnane X receptor activation. J Pharmacol Exp Ther 304(1):223–228.

148. Tannergren C, Petri N, Knutson L, Hedeland M, Bondesson U, Lennernas H. 2003. Multiple transport mechanisms involved in the intestinal absorption and first-pass extraction of fexofenadine. Clin Pharmacol Ther 74(5):423–436.

149. Schuetz EG, Beck WT, Schuetz JD. 1996. Modulators and substrates of P-glycoprotein and cytochrome P4503A coordinately up-regulate these proteins in human colon carcinoma cells. Mol Pharmacol 49(2):311–318.

150. Wang Z, Hamman MA, Huang SM, Lesko LJ, Hall SD. 2002. Effect of St John's wort on the pharmacokinetics of fexofenadine. Clin Pharmacol Ther 71(6):414–420.

151. Matheny CJ, Ali RY, Yang X, Pollack GM. 2004. Effect of prototypical inducing agents on P-glycoprotein and CYP3A expression in mouse tissues. Drug Metab Dispos 32(9):1008–1014.

152. Synold TW, Dussault I, Forman BM. 2001. The orphan nuclear receptor SXR coordinately regulates drug metabolism and efflux. Nat Med 7(5):584–590.

153. Maglich JM, Stoltz CM, Goodwin B, Hawkins-Brown D, Moore JT, Kliewer SA. 2002. Nuclear pregnane x receptor and constitutive androstane receptor regulate overlapping but distinct sets of genes involved in xenobiotic detoxification. Mol Pharmacol 62(3):638–646.

154. Sugatani J, Kojima H, Ueda A, Kakizaki S, Yoshinari K, Gong QH, Owens IS, Negishi M, Sueyoshi T. 2001. The phenobarbital response enhancer module in the human bilirubin UDP-glucuronosyltransferase UGT1A1 gene and regulation by the nuclear receptor CAR. Hepatology 33(5):1232–1238.

155. Cherrington NJ, Hartley DP, Li N, Johnson DR, Klaassen CD. 2002. Organ distribution of multidrug resistance proteins 1, 2, and 3 (Mrp1, 2, and 3) mRNA and hepatic induction of Mrp3 by constitutive androstane receptor activators in rats. J Pharmacol Exp Ther 300(1):97–104.

156. Endres CJ, Hsiao P, Chung FS, Unadkat JD. 2005. The role of transporters in drug interactions. Eur J Pharm Sci 27(5):501–517.

157. DuBuske LM. 2005. The role of P-glycoprotein and organic anion-transporting polypeptides in drug interactions. Drug Saf 28(9):789–801.

158. Muck W, Mai I, Fritsche L, Ochmann K, Rohde G, Unger S, Johne A, Bauer S, Budde K, Roots I, Neumayer HH, Kuhlmann J. 1999. Increase in cerivastatin systemic exposure after single and multiple dosing in cyclosporine-treated kidney transplant recipients. Clin Pharmacol Ther 65(3):251–261.

159. Shitara Y, Itoh T, Sato H, Li AP, Sugiyama Y. 2003. Inhibition of transporter-mediated hepatic uptake as a mechanism for drug–drug interaction between cerivastatin and cyclosporin A. J Pharmacol Exp Ther 304(2):610–616.

160. Shitara Y, Hirano M, Adachi Y, Itoh T, Sato H, Sugiyama Y. 2004. In vitro and in vivo correlation of the inhibitory effect of cyclosporin A on the transporter-mediated hepatic uptake of cerivastatin in rats. Drug Metab Dispos 32(12):1468–1475.

161. Asberg A, Hartmann A, Fjeldsa E, Bergan S, Holdaas H. 2001. Bilateral pharmacokinetic interaction between cyclosporine A and atorvastatin in renal transplant recipients. Am J Transplant 1(4):382–386.

162. Olbricht C, Wanner C, Eisenhauer T, Kliem V, Doll R, Boddaert M, O'Grady P, Krekler M, Mangold B, Christians U. 1997. Accumulation of lovastatin, but not pravastatin, in the blood of cyclosporine-treated kidney graft patients after multiple doses. Clin Pharmacol Ther 62(3):311–321.

163. Martin J, Krum H. 2003. Cytochrome P450 drug interactions within the HMG-CoA reductase inhibitor class: Are they clinically relevant? Drug Saf 26(1):13–21.

164. Simonson SG, Raza A, Martin PD, Mitchell PD, Jarcho JA, Brown CD, Windass AS, Schneck DW. 2004. Rosuvastatin pharmacokinetics in heart transplant recipients administered an antirejection regimen including cyclosporine. Clin Pharmacol Ther 76(2):167–177.

165. Martin PD, Warwick MJ, Dane AL, Hill SJ, Giles PB, Phillips PJ, Lenz E. 2003. Metabolism, excretion, and pharmacokinetics of rosuvastatin in healthy adult male volunteers. Clin Ther 25(11):2822–2835.

166. Kajosaari LI, Niemi M, Neuvonen M, Laitila J, Neuvonen PJ, Backman JT. 2005. Cyclosporine markedly raises the plasma concentrations of repaglinide. Clin Pharmacol Ther 78(4):388–399.

167. Treiber A, Schneiter R, Delahaye S, Clozel M. 2004. Inhibition of organic anion transporting polypeptide-mediated hepatic uptake is the major determinant in the pharmacokinetic interaction between bosentan and cyclosporin A in the rat. J Pharmacol Exp Ther 308(3):1121–1129.

168. Niemi M, Backman JT, Kajosaari LI, Leathart JB, Neuvonen M, Daly AK, Eichelbaum M, Kivisto KT, Neuvonen PJ. 2005. Polymorphic organic anion transporting polypeptide 1B1 is a major determinant of repaglinide pharmacokinetics. Clin Pharmacol Ther 77(6):468–478.

169. Backman JT, Kyrklund C, Neuvonen M, Neuvonen PJ. 2002. Gemfibrozil greatly increases plasma concentrations of cerivastatin. Clin Pharmacol Ther 72(6):685–691.

170. Shitara Y, Hirano M, Sato H, Sugiyama Y. 2004. Gemfibrozil and its glucuronide inhibit the organic anion transporting polypeptide 2 (OATP2/OATP1B1:SLC21A6)-mediated hepatic uptake and CYP2C8-mediated metabolism of cerivastatin: analysis of the mechanism of the clinically relevant drug-drug interaction between cerivastatin and gemfibrozil. J Pharmacol Exp Ther 311(1):228–236.

171. Backman JT, Kyrklund C, Kivisto KT, Wang JS, Neuvonen PJ. 2000. Plasma concentrations of active simvastatin acid are increased by gemfibrozil. Clin Pharmacol Ther 68(2):122–129.

172. Kyrklund C, Backman JT, Kivisto KT, Neuvonen M, Laitila J, Neuvonen PJ. 2001. Plasma concentrations of active lovastatin acid are markedly increased by gemfibrozil but not by bezafibrate. Clin Pharmacol Ther 69(5):340–345.

173. Kyrklund C, Backman JT, Neuvonen M, Neuvonen PJ. 2003. Gemfibrozil increases plasma pravastatin concentrations and reduces pravastatin renal clearance. Clin Pharmacol Ther 73(6):538–544.

174. Yamazaki M, Li B, Louie SW, Pudvah NT, Stocco R, Wong W, Abramovitz M, Demartis A, Laufer R, Hochman JH, Prueksaritanont T, Lin JH. 2005. Effects of fibrates on

human organic anion-transporting polypeptide 1B1-, multidrug resistance protein 2- and P-glycoprotein–mediated transport. Xenobiotica 35(7):737–753.

175. Nademanee K, Kannan R, Hendrickson J, Ookhtens M, Kay I, Singh BN. 1984. Amiodarone–digoxin interaction: clinical significance, time course of development, potential pharmacokinetic mechanisms and therapeutic implications. J Am Coll Cardiol 4(1):111–116.
176. Lambert C, Lamontagne D, Hottlet H, du Souich P. 1989. Amiodarone–digoxin interaction in rats. A reduction in hepatic uptake. Drug Metab Dispos 17(6):704–708.
177. Kodawara T, Masuda S, Wakasugi H, Uwai Y, Futami T, Saito H, Abe T, Inu K. 2002. Organic anion transporter oatp2-mediated interaction between digoxin and amiodarone in the rat liver. Pharm Res 19(6):738–743.
178. Kobayashi Y, Sakai R, Ohshiro N, Ohbayashi M, Kohyama N, Yamamoto T. 2005. Possible involvement of organic anion transporter 2 on the interaction of theophylline with erythromycin in the human liver. Drug Metab Dispos 33(5):619–622.
179. Bourdet DL, Pritchard JB, Thakker DR. 2005. Differential substrate and inhibitory activities of ranitidine and famotidine toward human organic cation transporter 1 (hOCT1; SLC22A1), hOCT2 (SLC22A2), and hOCT3 (SLC22A3). J Pharmacol Exp Ther 315(3):1288–1297.
180. Angelin B, Arvidsson A, Dahlqvist R, Hedman A, Schenck-Gustafsson K. 1987. Quinidine reduces biliary clearance of digoxin in man. Eur J Clin Invest 17(3):262–265.
181. Hedman A, Angelin B, Arvidsson A, Beck O, Dahlqvist R, Nilsson B, Olsson M, Schenck-Gustafsson K. 1991. Digoxin–verapamil interaction: reduction of biliary but not renal digoxin clearance in humans. Clin Pharmacol Ther 49(3):256–262.
182. Ding R, Tayrouz Y, Riedel KD, Burhenne J, Weiss J, Mikus G, Haefeli WE. 2004. Substantial pharmacokinetic interaction between digoxin and ritonavir in healthy volunteers. Clin Pharmacol Ther 76(1):73–84.
183. Briz O, Serrano MA, MacIas RI, Gonzalez-Gallego J, Marin JJ. 2003. Role of organic anion-transporting polypeptides, OATP-A, OATP-C and OATP-8, in the human placenta–maternal liver tandem excretory pathway for foetal bilirubin. Biochem J 371(Pt 3): 897–905.
184. Byrne JA, Strautnieks SS, Mieli-Vergani G, Higgins CF, Linton KJ, Thompson RJ. 2002. The human bile salt export pump: characterization of substrate specificity and identification of inhibitors. Gastroenterology 123(5):1649–1658.
185. Kostrubsky VE, Strom SC, Hanson J, Urda E, Rose K, Burliegh J, Zocharski P, Cai H, Sinclair JF, Sahi J. 2003. Evaluation of hepatotoxic potential of drugs by inhibition of bile-acid transport in cultured primary human hepatocytes and intact rats. Toxicol Sci 76(1):220–228.
186. Sahi J, Sinz MW, Campbell S, Mireles R, Zheng X, Rose KA, Raeissi S, Hashim MF, Ye Y, de Morais SM, et al. 2006. Metabolism and transporter-mediated drug–drug interactions of the endothelin-A receptor antagonist CI-1034. Chem Biol Interact 159(2):156–168.
187. Westphal K, Weinbrenner A, Zschiesche M, Franke G, Knoke M, Oertel R, Fritz P, von Richter O, Warzok R, Hachenberg T, et al. 2000. Induction of P-glycoprotein by rifampin increases intestinal secretion of talinolol in human beings: a new type of drug/drug interaction. Clin Pharmacol Ther 68(4):345–355.

188. Gant TW, O'Connor CK, Corbitt R, Thorgeirsson U, Thorgeirsson SS. 1995. In vivo induction of liver P-glycoprotein expression by xenobiotics in monkeys. Toxicol Appl Pharmacol 133(2):269–276.

189. Riley J, Styles J, Verschoyle RD, Stanley LA, White IN, Gant TW. 2000. Association of tamoxifen biliary excretion rate with prior tamoxifen exposure and increased mdr1b expression. Biochem Pharmacol 60(2):233–239.

190. FDA CfBEaR.. 2006. http://www.fda.gov/cber/recalls.htm.ed.

191. Spivey JR, Bronk SF, Gores GJ. 1993. Glycochenodeoxycholate-induced lethal hepatocellular injury in rat hepatocytes. Role of ATP depletion and cytosolic free calcium. J Clin Invest 92(1):17–24.

192. Herrine SK, Choudhary C. 1999. Severe hepatotoxicity associated with troglitazone. Ann Intern Med 130(2):163–164.

193. Smith MT. 2003. Mechanisms of troglitazone hepatotoxicity. Chem Res Toxicol 16(6):679–687.

194. Gores GJ, Miyoshi H, Botla R, Aguilar HI, Bronk SF. 1998. Induction of the mitochondrial permeability transition as a mechanism of liver injury during cholestasis: a potential role for mitochondrial proteases. Biochim Biophys Acta 1366(1–2):167–175.

195. Kostrubsky VE, Vore M, Kindt E, Burliegh J, Rogers K, Peter G, Altrogge D, Sinz MW. 2001. The effect of troglitazone biliary excretion on metabolite distribution and cholestasis in transporter-deficient rats. Drug Metab Dispos 29(12):1561–1566.

196. Sano N, Takikawa H, Yamanaka M. 1993. Estradiol-17 beta-glucuronide-induced cholestasis. Effects of ursodeoxycholate-3-*O*-glucuronide and 3,7-disulfate. J Hepatol 17(2):241–246.

197. Ananthanarayanan M, Ng OC, Boyer JL, Suchy FJ. 1994. Characterization of cloned rat liver Na(+)-bile acid cotransporter using peptide and fusion protein antibodies. Am J Physiol 267(4 Pt 1):G637–G643.

198. Donner MG, Keppler D. 2001. Up-regulation of basolateral multidrug resistance protein 3 (Mrp3) in cholestatic rat liver. Hepatology 34(2):351–359.

199. Zollner G, Fickert P, Silbert D, Fuchsbichler A, Marschall HU, Zatloukal K, Denk H, Trauner M. 2003. Adaptive changes in hepatobiliary transporter expression in primary biliary cirrhosis. J Hepatol 38(6):717–727.

200. Aleksunes LM, Scheffer GL, Jakowski AB, Pruimboom-Brees IM, Manautou JE. 2006. Coordinated expression of multidrug resistance-associated proteins (Mrps) in mouse liver during toxicant-induced injury. Toxicol Sci 89(2):370–379.

201. Aleksunes LM, Slitt AM, Cherrington NJ, Thibodeau MS, Klaassen CD, Manautou JE. 2005. Differential expression of mouse hepatic transporter genes in response to acetaminophen and carbon tetrachloride. Toxicol Sci 83(1):44–52.

202. Dietrich CG, Ottenhoff R, de Waart DR, Oude Elferink RP. 2001. Role of MRP2 and GSH in intrahepatic cycling of toxins. Toxicology 167(1):73–81.

203. Galinsky RE, Levy G. 1981. Dose- and time-dependent elimination of acetaminophen in rats: pharmacokinetic implications of cosubstrate depletion. J Pharmacol Exp Ther 219(1):14–20.

204. Dahlin DC, Miwa GT, Lu AY, Nelson SD. 1984. N-acetyl-p-benzoquinone imine: a cytochrome P450-mediated oxidation product of acetaminophen. Proc Natl Acad Sci U S A 81(5):1327–1331.

205. Silva VM, Thibodeau MS, Chen C, Manautou JE. 2005. Transport deficient (TR-) hyperbilirubinemic rats are resistant to acetaminophen hepatotoxicity. Biochem Pharmacol 70(12):1832–1839.

206. Gao B, Hagenbuch B, Kullak-Ublick GA, Benke D, Aguzzi A, Meier PJ. 2000. Organic anion-transporting polypeptides mediate transport of opioid peptides across blood–brain barrier. J Pharmacol Exp Ther 294(1):73–79.

207. Kubitz R, Sutfels G, Kuhlkamp T, Kolling R, Haussinger D. 2004. Trafficking of the bile salt export pump from the Golgi to the canalicular membrane is regulated by the p38 MAP kinase. Gastroenterology 126(2):541–553.

208. Kameyama Y, Yamashita K, Kobayashi K, Hosokawa M, Chiba K. 2005. Functional characterization of SLCO1B1 (OATP-C) variants, SLCO1B1*5, SLCO1B1*15 and SLCO1B1*15+C1007G, by using transient expression systems of HeLa and HEK293 cells. Pharmacogenet Genom 15(7):513–522.

209. Tamai I, Nezu J, Uchino H, Sai Y, Oku A, Shimane M, Tsuji A. 2000. Molecular identification and characterization of novel members of the human organic anion transporter (OATP) family. Biochem Biophys Res Commun 273(1):251–260.

210. Sandhu P, Lee W, Xu X, Leake BF, Yamazaki M, Stone JA, Lin JH, Pearson PG, Kim RB. 2005. Hepatic uptake of the novel antifungal agent caspofungin. Drug Metab Dispos 33(5):676–682.

211. Nozawa T, Minami H, Sugiura S, Tsuji A, Tamai I. 2005. Role of organic anion transporter OATP1B1 (OATP-C) in hepatic uptake of irinotecan and its active metabolite, 7-ethyl-10-hydroxycamptothecin: in vitro evidence and effect of single nucleotide polymorphisms. Drug Metab Dispos 33(3):434–439.

212. Meier-Abt F, Faulstich H, Hagenbuch B. 2004. Identification of phalloidin uptake systems of rat and human liver. Biochim Biophys Acta 1664(1):64–69.

213. Fujino H, Nakai D, Nakagomi R, Saito M, Tokui T, Kojima J. 2004. Metabolic stability and uptake by human hepatocytes of pitavastatin, a new inhibitor of HMG-CoA reductase. Arzneimittelforschung 54(7):382–388.

214. Hirano M, Maeda K, Shitara Y, Sugiyama Y. 2004. Contribution of OATP2 (OATP1B1) and OATP8 (OATP1B3) to the hepatic uptake of pitavastatin in humans. J Pharmacol Exp Ther 311(1):139–146.

215. Nakai D, Nakagomi R, Furuta Y, Tokui T, Abe T, Ikeda T, Nishimura K. 2001. Human liver-specific organic anion transporter, LST-1, mediates uptake of pravastatin by human hepatocytes. J Pharmacol Exp Ther 297(3):861–867.

216. Vavricka SR, van Montfoort J, Ha HR, Meier PJ, Fattinger K. 2002. Interactions of rifamycin SV and rifampicin with organic anion uptake systems of human liver. Hepatology 36(1):164–172.

217. Nozawa T, Sugiura S, Nakajima M, Goto A, Yokoi T, Nezu J, Tsuji A, Tamai I. 2004. Involvement of organic anion transporting polypeptides in the transport of troglitazone sulfate: implications for understanding troglitazone hepatotoxicity. Drug Metab Dispos 32(3):291–294.

218. Ismair MG, Stieger B, Cattori V, Hagenbuch B, Fried M, Meier PJ, Kullak-Ublick GA. 2001. Hepatic uptake of cholecystokinin octapeptide by organic anion-transporting polypeptides OATP4 and OATP8 of rat and human liver. Gastroenterology 121(5):1185–1190.

219. Shimizu M, Uno T, Sugawara K, Tateishi T. 2006. Effects of itraconazole and diltiazem on the pharmacokinetics of fexofenadine, a substrate of P-glycoprotein. Br J Clin Pharmacol 61(5):538–544.

220. Smith NF, Acharya MR, Desai N, Figg WD, Sparreboom A. 2005. Identification of OATP1B3 as a high-affinity hepatocellular transporter of paclitaxel. Cancer Biol Ther 4(8).

221. Babu E, Takeda M, Narikawa S, Kobayashi Y, Yamamoto T, Cha SH, Sekine T, Sakthisekaran D, Endou H. 2002. Human organic anion transporters mediate the transport of tetracycline. Jpn J Pharmacol 88(1):69–76.

222. Takeda M, Khamdang S, Narikawa S, Kimura H, Kobayashi Y, Yamamoto T, Cha SH, Sekine T, Endou H. 2002. Human organic anion transporters and human organic cation transporters mediate renal antiviral transport. J Pharmacol Exp Ther 300(3):918–924.

223. van Montfoort JE, Muller M, Groothuis GM, Meijer DK, Koepsell H, Meier PJ. 2001. Comparison of "type I" and "type II" organic cation transport by organic cation transporters and organic anion-transporting polypeptides. J Pharmacol Exp Ther 298(1):110–115.

224. Gründemann D, Liebich G, Kiefer N, Koster S, Schömig E. 1999. Selective substrates for non-neuronal monoamine transporters. Mol Pharmacol 56(1):1–10.

225. Thomas J, Wang L, Clark RE, Pirmohamed M. 2004. Active transport of imatinib into and out of cells: implications for drug resistance. Blood 104(12):3739–3745.

226. Kimura N, Masuda S, Tanihara Y, Ueo H, Okuda M, Katsura T, Inui K. 2005. Metformin is a superior substrate for renal organic cation transporter OCT2 rather than hepatic OCT1. Drug Metab Pharmacokinet 20(5):379–386.

227. Gorboulev V, Ulzheimer JC, Akhoundova A, Ulzheimer-Teuber I, Karbach U, Quester S, Baumann C, Lang F, Busch AE, Koepsell H. 1997. Cloning and characterization of two human polyspecific organic cation transporters. DNA Cell Biol 16(7):871–881.

228. Zhang L, Gorset W, Dresser MJ, Giacomini KM. 1999. The interaction of *n*-tetraalkylammonium compounds with a human organic cation transporter, hOCT1. J Pharmacol Exp Ther 288(3):1192–1198.

229. Grundemann D, Schechinger B, Rappold GA, Schomig E. 1998. Molecular identification of the corticosterone-sensitive extraneuronal catecholamine transporter. Nat Neurosci 1(5):349–351.

230. Hayer-Zillgen M, Bruss M, Bonisch H. 2002. Expression and pharmacological profile of the human organic cation transporters hOCT1, hOCT2 and hOCT3. Br J Pharmacol 136(6):829–836.

231. Gründemann D, Hahne C, Berkels R, Schömig E. 2003. Agmatine is efficiently transported by non-neuronal monoamine transporters extraneuronal monoamine transporter (EMT) and organic cation transporter 2 (OCT2). J Pharmacol Exp Ther 304(2):810–817.

232. Wu X, Huang W, Ganapathy ME, Wang H, Kekuda R, Conway SJ, Leibach FH, Ganapathy V. 2000. Structure, function, and regional distribution of the organic cation transporter OCT3 in the kidney. Am J Physiol Renal Physiol 279(3):F449–F458.

233. Cole SP, Sparks KE, Fraser K, Loe DW, Grant CE, Wilson GM, Deeley RG. 1994. Pharmacological characterization of multidrug resistant MRP-transfected human tumor cells. Cancer Res 54(22):5902–5910.

234. Schuetz JD, Connelly MC, Sun D, Paibir SG, Flynn PM, Srinivas RV, Kumar A, Fridland A. 1999. MRP4: a previously unidentified factor in resistance to nucleoside-based antiviral drugs. Nat Med 5(9):1048–1051.

235. Reid G, Wielinga P, Zelcer N, de Haas M, van Deemter L, Wijnholds J, Balzarini J, Borst P. 2003. Characterization of the transport of nucleoside analog drugs by the human multidrug resistance proteins MRP4 and MRP5. Mol Pharmacol 63(5):1094–1103.

236. Jedlitschky G, Burchell B, Keppler D. 2000. The multidrug resistance protein 5 functions as an ATP-dependent export pump for cyclic nucleotides. J Biol Chem 275(39):30069–30074.

237. Gerloff T, Stieger B, Hagenbuch B, Madon J, Landmann L, Roth J, Hofmann AF, Meier PJ. 1998. The sister of P-glycoprotein represents the canalicular bile salt export pump of mammalian liver. J Biol Chem 273(16):10046–10050.

238. Xiong H, Turner KC, Ward ES, Jansen PL, Brouwer KLR. 2000. Altered hepatobiliary disposition of acetaminophen glucuronide in isolated perfused livers from multidrug resistance-associated protein 2-deficient TR(-) rats. J Pharmacol Exp Ther 295(2):512–518.

239. Koike K, Kawabe T, Tanaka T, Toh S, Uchiumi T, Wada M, Akiyama S, Ono M, Kuwano M. 1997. A canalicular multispecific organic anion transporter (cMOAT) antisense cDNA enhances drug sensitivity in human hepatic cancer cells. Cancer Res 57(24):5475–5479.

240. Matsushima S, Maeda K, Kondo C, Hirano M, Sasaki M, Suzuki H, Sugiyama Y. 2005. Identification of the hepatic efflux transporters of organic anions using double-transfected Madin–Darby canine kidney II cells expressing human organic anion-transporting polypeptide 1B1 (OATP1B1)/multidrug resistance-associated protein 2, OATP1B1/multidrug resistance 1, and OATP1B1/breast cancer resistance protein. J Pharmacol Exp Ther 314(3):1059–1067.

241. Kawabe T, Chen ZS, Wada M, Uchiumi T, Ono M, Akiyama S, Kuwano M. 1999. Enhanced transport of anticancer agents and leukotriene C4 by the human canalicular multispecific organic anion transporter (cMOAT/MRP2). FEBS Lett 456(2):327–331.

242. Cui Y, König J, Buchholz JK, Spring H, Leier I, Keppler D. 1999. Drug resistance and ATP-dependent conjugate transport mediated by the apical multidrug resistance protein, MRP2, permanently expressed in human and canine cells. Mol Pharmacol 55(5):929–937.

243. Payen L, Courtois A, Campion JP, Guillouzo A, Fardel O. 2000. Characterization and inhibition by a wide range of xenobiotics of organic anion excretion by primary human hepatocytes. Biochem Pharmacol 60(12):1967–1975.

244. Hooijberg JH, Broxterman HJ, Kool M, Assaraf YG, Peters GJ, Noordhuis P, Scheper RJ, Borst P, Pinedo HM, Jansen G. 1999. Antifolate resistance mediated by the multidrug resistance proteins MRP1 and MRP2. Cancer Res 59(11):2532–2535.

245. Ueda K, Okamura N, Hirai M, Tanigawara Y, Saeki T, Kioka N, Komano T, Hori R. 1992. Human P-glycoprotein transports cortisol, aldosterone, and dexamethasone, but not progesterone. J Biol Chem 267(34):24248–24252.

246. Marie JP, Helou C, Thevenin D, Delmer A, Zittoun R. 1992. In vitro effect of P-glycoprotein (Pgp) modulators on drug sensitivity of leukemic progenitors (CFU-L) in acute myelogenous leukemia (AML). Exp Hematol 20(5):565–568.

247. Advani R, Fisher GA, Lum BL, Hausdorff J, Halsey J, Litchman M, Sikic BI. 2001. A phase I trial of doxorubicin, paclitaxel, and valspodar (PSC 833), a modulator of multidrug resistance. Clin Cancer Res 7(5):1221–1229.

248. van der Sandt IC, Blom-Roosemalen MC, de Boer AG, Breimer DD. 2000. Specificity of doxorubicin versus rhodamine-123 in assessing P-glycoprotein functionality in the LLC-PK1, LLC-PK1:MDR1 and Caco-2 cell lines. Eur J Pharm Sci 11(3):207–214.
249. Lum BL, Fisher GA, Brophy NA, Yahanda AM, Adler KM, Kaubisch S, Halsey J, Sikic BI. 1993. Clinical trials of modulation of multidrug resistance. Pharmacokinetic and pharmacodynamic considerations. Cancer 72(11 Suppl):3502–3514.
250. Yamaguchi H, Yano I, Hashimoto Y, Inui KI. 2000. Secretory mechanisms of grepafloxacin and levofloxacin in the human intestinal cell line caco-2. J Pharmacol Exp Ther 295(1):360–366.
251. Soldner A, Christians U, Susanto M, Wacher VJ, Silverman JA, Benet LZ. 1999. Grapefruit juice activates P-glycoprotein-mediated drug transport. Pharm Res 16(4):478–485.
252. Floren LC, Bekersky I, Benet LZ, Mekki Q, Dressler D, Lee JW, Roberts JP, Hebert MF. 1997. Tacrolimus oral bioavailability doubles with coadministration of ketoconazole. Clin Pharmacol Ther 62(1):41–49.
253. Spahn-Langguth H, Baktir G, Radschuweit A, Okyar A, Terhaag B, Ader P, Hanafy A, Langguth P. 1998. P-glycoprotein transporters and the gastrointestinal tract: evaluation of the potential in vivo relevance of in vitro data employing talinolol as model compound. Int J Clin Pharmacol Ther 36(1):16–24.
254. Pauli-Magnus C, von Richter O, Burk O, Ziegler A, Mettang T, Eichelbaum M, Fromm MF. 2000. Characterization of the major metabolites of verapamil as substrates and inhibitors of P-glycoprotein. J Pharmacol Exp Ther 293(2):376–382.
255. Ozvegy C, Litman T, Szakacs G, Nagy Z, Bates S, Varadi A, Sarkadi B. 2001. Functional characterization of the human multidrug transporter, ABCG2, expressed in insect cells. Biochem Biophys Res Commun 285(1):111–117.
256. Janvilisri T, Venter H, Shahi S, Reuter G, Balakrishnan L, van Veen HW. 2003. Sterol transport by the human breast cancer resistance protein (ABCG2) expressed in *Lactococcus lactis*. J Biol Chem 278(23):20645–20651.
257. Kiuchi Y, Suzuki H, Hirohashi T, Tyson CA, Sugiyama Y. 1998. cDNA cloning and inducible expression of human multidrug resistance associated protein 3 (MRP3). FEBS Lett 433(1–2):149–152.
258. Schuetz EG, Strom S, Yasuda K, Lecureur V, Assem M, Brimer C, Lamba J, Kim RB, Ramachandran V, Komoroski BJ, et al. 2001. Disrupted bile acid homeostasis reveals an unexpected interaction among nuclear hormone receptors, transporters, and cytochrome P450. J Biol Chem 276(42):39411–39418.
259. Courtois A, Payen L, Le Ferrec E, Scheffer GL, Trinquart Y, Guillouzo A, Fardel O. 2002. Differential regulation of multidrug resistance-associated protein 2 (MRP2) and cytochromes P450 2B1/2 and 3A1/2 in phenobarbital-treated hepatocytes. Biochem Pharmacol 63(2):333–341.
260. Assem M, Schuetz EG, Leggas M, Sun D, Yasuda K, Reid G, Zelcer N, Adachi M, Strom S, Evans RM, et al. 2004. Interactions between hepatic Mrp4 and Sult2a as revealed by the constitutive androstane receptor and Mrp4 knockout mice. J Biol Chem 279(21):22250–22257.
261. Courtois A, Payen L, Guillouzo A, Fardel O. 1999. Up-regulation of multidrug resistance-associated protein 2 (MRP2) expression in rat hepatocytes by dexamethasone. FEBS Lett 459(3):381–385.

262. Fardel O, Payen L, Courtois A, Vernhet L, Lecureur V. 2001. Regulation of biliary drug efflux pump expression by hormones and xenobiotics. Toxicology 167(1):37–46.

263. Slitt AL, Cherrington NJ, Maher JM, Klaassen CD. 2003. Induction of multidrug resistance protein 3 in rat liver is associated with altered vectorial excretion of acetaminophen metabolites. Drug Metab Dispos 31(9):1176–1186.

264. Cherrington NJ, Slitt AL, Maher JM, Zhang XX, Zhang J, Huang W, Wan YJ, Moore DD, Klaassen CD. 2003. Induction of multidrug resistance protein 3 (mrp3) in vivo is independent of constitutive androstane receptor. Drug Metab Dispos 31(11):1315–1319.

265. Brady JM, Cherrington NJ, Hartley DP, Buist SC, Li N, Klaassen CD. 2002. Tissue distribution and chemical induction of multiple drug resistance genes in rats. Drug Metab Dispos 30(7):838–844.

266. Hagenbuch N, Reichel C, Stieger B, Cattori V, Fattinger KE, Landmann L, Meier PJ, Kullak-Ublick GA. 2001. Effect of phenobarbital on the expression of bile salt and organic anion transporters of rat liver. J Hepatol 34(6):881–887.

267. Rausch-Derra LC, Hartley DP, Meier PJ, Klaassen CD. 2001. Differential effects of microsomal enzyme-inducing chemicals on the hepatic expression of rat organic anion transporters, OATP1 and OATP2. Hepatology 33(6):1469–1478.

268. Cheng X, Maher J, Dieter MZ, Klaassen CD. 2005. Regulation of mouse organic anion-transporting polypeptides (oatps) in liver by prototypical microsomal enzyme inducers that activate distinct transcription factor pathways. Drug Metab Dispos 33(9):1276–1282.

269. Maher JM, Cheng X, Slitt AL, Dieter MZ, Klaassen CD. 2005. Induction of the multidrug resistance-associated protein family of transporters by chemical activators of receptor-mediated pathways in mouse liver. Drug Metab Dispos 33(7):956–962.

270. Staudinger JL, Goodwin B, Jones SA, Hawkins-Brown D, MacKenzie KI, LaTour A, Liu Y, Klaassen CD, Brown KK, Reinhard J, et al. 2001. The nuclear receptor PXR is a lithocholic acid sensor that protects against liver toxicity. Proc Natl Acad Sci U S A 98(6):3369–3374.

271. Staudinger J, Liu Y, Madan A, Habeebu S, Klaassen CD. 2001. Coordinate regulation of xenobiotic and bile acid homeostasis by pregnane X receptor. Drug Metab Dispos 29(11):1467–1472.

272. Anapolsky A, Teng S, Dixit S, Piquette-Miller M. 2006. The role of pregnane X receptor in 2-acetylaminofluorene-mediated induction of drug transport and -metabolizing enzymes in mice. Drug Metab Dispos 34(3):405–409.

273. Jonkman JH, Upton RA. 1984. Pharmacokinetic drug interactions with theophylline. Clin Pharmacokinet 9(4):309–334.

274. Niemi M, Backman JT, Neuvonen M, Neuvonen PJ. 2003. Effects of gemfibrozil, itraconazole, and their combination on the pharmacokinetics and pharmacodynamics of repaglinide: potentially hazardous interaction between gemfibrozil and repaglinide. Diabetologia 46(3):347–351.

275. Ogilvie BW, Zhang D, Li W, Rodrigues AD, Gipson AE, Holsapple J, Toren P, Parkinson A. 2006. Glucuronidation converts gemfibrozil to a potent, metabolism-dependent inhibitor of CYP2C8: implications for drug-drug interactions. Drug Metab Dispos 34(1):191–197.

276. Schneck DW, Birmingham BK, Zalikowski JA, Mitchell PD, Wang Y, Martin PD, Lasseter KC, Brown CD, Windass AS, Raza A. 2004. The effect of gemfibrozil on the pharmacokinetics of rosuvastatin. Clin Pharmacol Ther 75(5):455–463.

277. Hedman A, Angelin B, Arvidsson A, Dahlqvist R, Nilsson B. 1990. Interactions in the renal and biliary elimination of digoxin: stereoselective difference between quinine and quinidine. Clin Pharmacol Ther 47(1):20–26.

278. Chan LM, Lowes S, Hirst BH. 2004. The ABCs of drug transport in intestine and liver: efflux proteins limiting drug absorption and bioavailability. Eur J Pharm Sci 21: 25–51.

279. Ho RH, Leake BF, Roberts RL, Lee W, Kim RB. 2004. Ethnicity-dependent polymorphism in Na+-taurocholate cotransporting polypeptide (SLC10A1) reveals a domain critical for bile acid substrate recognition. J Biol Chem 279:7213–7222.

280. Meier PJ, Eckhardt U, Schroeder A, Hagenbuch B, Stieger B. 1997. Substrate specificity of sinusoidal bile acid and organic anion uptake systems in rat and human liver. Hepatology 26:1667–1677.

281. Ho RH, Tirona RG, Leake BF, Glaeser H, Lee W, Lemke CJ, Wang Y, Kim RB. 2006. Drug and bile acid transporters in rosuvastatin hepatic uptake: function, expression, and pharmacogenetics. Gastroenterology 130:1793–1806.

282. Kullak-Ublick GA, Ismair MG, Stieger B, Landmann L, Huber R, Pizzagalli F, Fattinger K, Meier PJ, Hagenbuch B. 2001. Organic anion-transporting polypeptide B (OATP-B) and its functional comparison with three other OATPs of human liver. Gastroenterology 120:525–533.

283. Fischer WJ, Altheimer S, Cattori V, Meier PJ, Dietrich DR, Hagenbuch B. 2005. Organic anion transporting polypeptides expressed in liver and brain mediate uptake of microcystin. Toxicol Appl Pharmacol 203:257–263.

284. Su Y, Zhang X, Sinko PJ. 2004. Human organic anion-transporting polypeptide OATP-A (SLC21A3) acts in concert with P-glycoprotein and multidrug resistance protein 2 in the vectorial transport of Saquinavir in Hep G2 cells. Mol Pharm 1:49–56.

285. Cui Y, Konig J, Leier I, Buchholz U, Keppler D. 2001. Hepatic uptake of bilirubin and its conjugates by the human organic anion transporter SLC21A6. J Biol Chem 276: 9626–9630.

286. Abe T, Kakyo M, Tokui T, Nakagomi R, Nishio T, Nakai D, Nomura H, Unno M, Suzuki M, Naitoh T, Matsuno S, Yawo H. 1999. Identification of a novel gene family encoding human liver-specific organic anion transporter LST-1. J Biol Chem 274: 17159–17163.

287. Tahara H, Kusuhara H, Endou H, Koepsell H, Imaoka T, Fuse E, Sugiyama Y. 2005. A species difference in the transport activities of H2 receptor antagonists by rat and human renal organic anion and cation transporters. J Pharmacol Exp Ther 315:337–345.

288. Zhang L, Dresser MJ, Gray AT, Yost SC, Terashita S, Giacomini KM. 1997. Cloning and functional expression of a human liver organic cation transporter. Mol Pharmacol 51:913–921.

289. Muller J, Lips KS, Metzner L, Neubert RH, Koepsell H, Brandsch M. 2005. Drug specificity and intestinal membrane localization of human organic cation transporters (OCT). Biochem Pharmacol 70:1851–1860.

290. Boyer JL, Meier PJ. 1990. Characterizing mechanisms of hepatic bile acid transport utilizing isolated membrane vesicles. Methods Enzymol 192:517–533.

291. Wolters H, Spiering M, Gerding A, Slooff MJ, Kuipers F, Hardonk MJ, Vonk RJ. 1991. Isolation and characterization of canalicular and basolateral plasma membrane fractions from human liver. Biochim Biophys Acta 1069:61–69.

292. Belinsky MG, Dawson PA, Shchaveleva I, Bain LJ, Wang R, Ling V, Chen ZS, Grinberg A, Westphal H, Klein-Szanto A, Lerro A, Kruh GD. 2005. Analysis of the in vivo functions of Mrp3. Mol Pharmacol 68:160–168.

293. Lan LB, Dalton JT, Schuetz EG. 2000. Mdr1 limits CYP3A metabolism in vivo. Mol Pharmacol 58:863–869.

294. Wang S, Hartley DP, Ciccotto SL, Vincent SH, Franklin RB, Kim MS. 2003. Induction of hepatic phase II drug-metabolizing enzymes by 1,7-phenanthroline in rats is accompanied by induction of MRP3. Drug Metab Dispos 31:773–775.

295. Kim RB. 2004. 3-Hydroxy-3-methylglutaryl-coenzyme A reductase inhibitors (statins) and genetic variability (single nucleotide polymorphisms) in a hepatic drug uptake transporter: what's it all about? Clin Pharmacol Ther 75:381–385.

296. Zhang L, Gorset W, Washington CB, Blaschke TF, Kroetz DL, Giacomini KM. 2000. Interactions of HIV protease inhibitors with a human organic cation transporter in a mammalian expression system. Drug Metab Dispos 28:329–334.

297. Schrenk D, Baus PR, Ermel N, Klein C, Vorderstemann B, Kauffmann HM. 2001. Up-regulation of transporters of the MRP family by drugs and toxins. Toxicol Lett 120: 51–57.

298. Teng S, Jekerle V, Piquette-Miller M. 2003. Induction of ABCC3 (MRP3) by pregnane X receptor activators. Drug Metab Dispos 31:1296–1299.

299. Johnson DR, Klaassen CD. 2002. Regulation of rat multidrug resistance protein 2 by classes of prototypical microsomal enzyme inducers that activate distinct transcription pathways. Toxicol Sci 67:182–189.

300. Turncliff RZ, Meier PJ, Brouwer KL. 2004. Effect of dexamethasone treatment on the expression and function of transport proteins in sandwich-cultured rat hepatocytes. Drug Metab Dispos 32:834–839.

301. Huang W, Zhang J, Chua SS, Qatanani M, Han Y, Granata R, Moore DD. 2003. Induction of bilirubin clearance by the constitutive androstane receptor (CAR). Proc Natl Acad Sci USA 100:4156–4161.

302. Chen C, Klaassen CD. 2004. Rat multidrug resistance protein 4 (Mrp4, Abcc4): molecular cloning, organ distribution, postnatal renal expression, and chemical inducibility. Biochem Biophys Res Commun 317:46–53.

303. Tanaka K, Xu W, Zhou F, You G. 2004. Role of glycosylation in the organic anion transporter OAT1. J Biol Chem 279:14961–14966.

304. Tian X, Zamek-Gliszczynski MJ, Zhang P, Brouwer KL. 2004. Modulation of multidrug resistance-associated protein 2 (Mrp2) and Mrp3 expression and function with small interfering RNA in sandwich-cultured rat hepatocytes. Mol Pharmacol 66:1004–1010.

305. Oza AM. 2002. Clinical development of P glycoprotein modulators in oncology. Novartis Found Symp 243:103–115.

306. Ros JE, Libbrecht L, Geuken M, Jansen PL, Roskams TA. 2003. High expression of MDR1, MRP1, and MRP3 in the hepatic progenitor cell compartment and hepatocytes in severe human liver disease. J Pathol 200:553–560.

307. Madon J, Hagenbuch B, Landmann L, Meier PJ, Stieger B. 2000. Transport function and hepatocellular localization of mrp6 in rat liver. Mol Pharmacol 57:634–641.

308. Li L, Meier PJ, Ballatori N. 2000. Oatp2 mediates bidirectional organic solute transport: a role for intracellular glutathione. Mol Pharmacol 58:335–340.

309. Chen ZS, Lee K, Kruh GD. 2001. Transport of cyclic nucleotides and estradiol 17-beta-D-glucuronide by multidrug resistance protein 4. Resistance to 6-mercaptopurine and 6-thioguanine. J Biol Chem 276:33747–33754.

310. Chen ZS, Lee K, Walther S, Raftogianis RB, Kuwano M, Zeng H, Kruh GD. 2002. Analysis of methotrexate and folate transport by multidrug resistance protein 4 (ABCC4): MRP4 is a component of the methotrexate efflux system. Cancer Res 62:3144–3150.

311. Kim RB, Fromm MF, Wandel C, Leake B, Wood AJ, Roden DM, Wilkinson GR. 1998. The drug transporter P-glycoprotein limits oral absorption and brain entry of HIV-1 protease inhibitors. J Clin Invest 101:289–294.

312. Polli JW, Jarrett JL, Studenberg SD, Humphreys JE, Dennis SW, Brouwer KR, Woolley JL. 1999. Role of P-glycoprotein on the CNS disposition of amprenavir (141W94), an HIV protease inhibitor. Pharm Res 16:1206–1212.

313. Fromm MF, Kim RB, Stein CM, Wilkinson GR, Roden DM. 1999. Inhibition of P-glycoprotein-mediated drug transport: A unifying mechanism to explain the interaction between digoxin and quinidine [seecomments]. Circulation 99:552–557.

314. Hirano M, Maeda K, Hayashi H, Kusuhara H, Sugiyama Y. 2005. Bile salt export pump (BSEP/ABCB11) can transport a nonbile acid substrate, pravastatin. J Pharmacol Exp Ther 314, 876–882.

14

DRUG TRANSPORT IN THE BRAIN

PATRICK T. RONALDSON, KARLO BABAKHANIAN, AND REINA BENDAYAN

University of Toronto, Toronto, Ontario, Canada

Drug Transporters: Molecular Characterization and Role in Drug Disposition, Edited by Guofeng You and Marilyn E. Morris
Copyright © 2007 John Wiley & Sons, Inc.

14.1. INTRODUCTION

Pharmacotherapy of central nervous system (CNS) diseases remains difficult, due to limited drug permeation across the blood–brain barrier and blood–cerebrospinal fluid barrier. Therapeutic compounds may cross these barriers by several uptake processes, including transcytosis, receptor-mediated endocytosis, passive diffusion, carrier-mediated (facilitated) transport, and/or active transport.[1] Once across these initial barriers, brain drug accumulation can be restricted further by passive efflux within the cerebrospinal fluid (sink effect), metabolic degradation, and/or active efflux transport. In addition, brain parenchymal cellular compartments (i.e., astrocytes, microglia, oligodendrocytes, and neurons) also play an important role in regulating CNS drug distribution. These cells express several drug transport proteins, which underscore the complexity of xenobiotic disposition within the brain. The objective of this chapter is to summarize the current knowledge on the molecular expression (i.e., gene, protein), cellular localization, and functional activity of drug transporters in the brain.

14.2. PHYSIOLOGY OF THE BRAIN BARRIERS AND BRAIN PARENCHYMA

14.2.1. Blood–Brain Barrier

The blood–brain barrier (BBB) constitutes a remarkable physical and biochemical barrier between the brain and systemic circulation.[2] Structurally, the BBB is composed of a monolayer of nonfenestrated microvessel endothelial cells surrounded by pericytes and perivascular astrocytes. Brain microvessel endothelial cells are joined by tight junctions (i.e., zonulae occludens), which are maintained by trophic factors released from adjacent astrocytes.[3,4] Under physiological conditions, these tight junctions form a continuous, almost impermeable cellular barrier that limits paracellular flux and transport as well as the influx of endogenous and exogenous substances, with the exception of very small lipid-soluble molecules.[5] The high transendothelial electrical resistance (1500 to 2000 $\Omega \cdot cm^2$) of the BBB further restricts the free flow of water and solutes.[6]

Several receptors, ion channels, and influx–efflux transport proteins are expressed prominently at the BBB. Functionally, brain transporters are similar to well-characterized systems in other tissues (e.g., D-glucose transporter, L-amino acid carrier systems, Na^+/K^+-ATPase), although the capacity and rate of transport can vary widely.[7] At the level of brain microvascular endothelium, many of these membrane-bound transport systems are distributed asymmetrically. One example of this asymmetry involves the facilitative glucose transporter, GLUT-1, which is expressed to a fourfold greater degree at the abluminal side of the BBB than at the luminal side.[8]

In addition to these transport systems, endocytosis of macromolecules has also been reported at the BBB.[9] Receptor-mediated and adsorptive endocytotic processes in brain endothelium have been documented for both hormones and plasma proteins.[10]

Examples of cell surface receptors involved in these processes include the endothelial barrier antigen OX-47 (an integral plasma membrane glycoprotein that is involved in cell-to-cell recognition), and endothelial glyocalyx (possible role in vascular permeability and surface charge).[11,12] Transferrin, a plasma protein involved in systemic iron transport, may also be taken up into the brain parenchyma by receptor-mediated endocytosis and transcytosis.[13] In brain microvessel and capillary endothelial cells, specialized plasma membrane microdomains known as caveolae are believed to be involved in endocytosis of various macromolecules, including plasma proteins, immunoglobulins, and metaloproteins.[14] Caveolae are involved in other biological processes, including signal transduction and cholesterol transport.[15,16]

14.2.2. Blood–Cerebrospinal Fluid Barrier

The blood–cerebrospinal fluid (BCSF) barrier is formed by the choroid plexus, which is the major interface between the systemic circulation and the CSF. The BCSF barrier is located at the outer epithelial surface of the choroid plexus, a leaflike highly vascular organ that protrudes into all four cerebral ventricles. It is comprised of fenestrated capillaries that are surrounded by a monolayer of epithelial cells joined together by tight junctions.[17] These tight junctions form the structural basis of the BCSF barrier and seal together adjacent polarized epithelial cells (also known as ependymal cells). Thus, once a solute has crossed the capillary wall, it must also permeate these ependymal cells before entering the CSF.

The primary function of the choroid plexus is to produce the CSF continuously and to maintain its composition. The total volume of CSF (140 mL) is replaced approximately four to five times daily.[18] The continuous flow of CSF through the ventricular system into the subarachnoid space and exiting into the venous system provides a "sink" that reduces the steady-state concentration of a molecule penetrating into the brain and CSF.[19] The sink effect is greater for large-molecular-weight and hydrophilic molecules. The CSF also contains approximately 0.3% of plasma proteins, totaling 15 to 40 mg/mL, depending on sampling site.[20] This is in contrast to the extracellular space of the normal adult brain, which contains no detectable plasma proteins.[21]

Similar to the BBB, the choroid plexus displays polarized expression of various receptors, ion channels, and transport systems that regulate the CSF composition via secretion and reabsorption.[22] The apical membrane expresses the Na^+/K^+-ATPase pump and several channels for monovalent anions and cations.[7] Studies have also demonstrated the expression of facilitated and sodium-dependent carriers for the transport of nonelectrolytes.[18,23,24] The basolateral side is lined with Na^+/H^+ antiporters, Cl^-/HCO_3^- antiporters, facilitated carriers for nonelectrolytes, and carbonic anhydrase.[18,23]

14.2.3. Cellular Compartments of the Brain Parenchyma

The brain parenchyma consists of neurons and the surrounding glial cells. Neurons form the basic structural and functional component of the CNS. The primary function of neurons is to respond to stimuli by conducting electrical signals along conductive processes (e.g., the axon). The conduction of electrical impulses results in the release of neurotransmitters that further regulate (positively and negatively) nearby neuronal responses.[25] This enables the brain to maintain a highly complex communication network.

Glial cells are known to play an instrumental role in the regulation and maintenance of CNS homeostasis and can be classified into two groups: the macroglia and the microglia. The macroglia include both astrocytes and oligodendrocytes, which originate from the ectodermal layer of the gastrula and proliferate throughout life, particularly in response to injury.[26] Astrocytes are the most abundant cell type in the brain. They possess a stellate (i.e., star-shaped) morphology and contain numerous cytoplasmic fibrils, of which glial acidic fibrillary protein (GFAP) is the main constituent.[27] Astrocytes possess numerous functions that aid in maintaining the homeostatic environment of the CNS. These functions include the initiation and regulation of immune and inflammatory events during injury and infection (i.e., production and secretion of cytokines), expression of adhesion molecules for neuronal development, buffering of excess K^+ during periods of neuronal hyperactivity, and secretion of trophic factors required to maintain the integrity of the BBB.[3,28–30] The primary function of the other macroglia cell type, oligodendrocytes, is to form the insulating myelin sheath that surrounds neuronal axons in the CNS. Myelin, an extension of the oligodendrocyte plasma membrane, is a lipid-rich biological membrane that forms multilamellar spirally wrapped sheaths around neuronal axons to increase the resistance for electrical impulses during an action potential.[31]

Microglia, are much smaller in size than macroglia and are the primary immune cells of the brain. Similar to macroglia, microglia may also proliferate in response to injury. Although a few studies have suggested that microglia originate from the neuroectoderm,[32,33] the most widely accepted view is microglia are derived from the hematopoietic lineage of the embryonic mesoderm.[34,35] Microglia are distributed ubiquitously within the CNS, with the basal ganglia and cerebellum possessing considerably greater numbers than the cerebral cortex.[36]

14.3. DRUG TRANSPORTERS IN THE BRAIN

Pharmacological treatment of CNS disorders requires that drugs attain efficacious concentrations in the brain. This therapeutic objective requires that drugs are able to cross the brain barriers successfully (i.e., BBB, BCSF barrier) and in the case of some diseases, also permeate the cellular compartments of the brain parenchyma (i.e., astrocytes, microglia, oligodendrocytes, neurons). Although small, nonionic, lipid-soluble compounds can easily enter the brain by passive diffusion, the CNS permeation of larger, water-soluble, and/or ionic substances is less likely to occur by this mechanism.[37] For many of these compounds, uptake into the brain and extrusion from the brain is governed by drug transport proteins. Many transport proteins that have been shown to be involved in the influx and efflux of drugs [i.e., adenosine triphosphate (ATP)-binding cassette transporters, organic anion and cation transporters, nucleoside transporters, peptide transporters] have been identified both at the brain barriers and

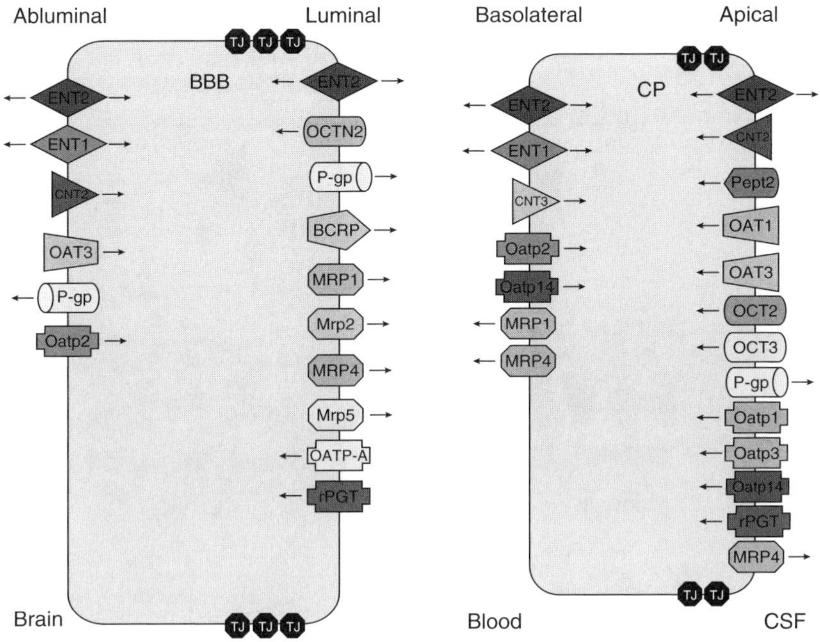

FIGURE 14.1. Xenobiotic transport mechanisms at the brain barriers. The proposed expression of ABC superfamily transporters, organic anion and cation transporters, nucleoside transporters, and peptide transporters is depicted at the BBB endothelium and at the choroid plexus epithelium. Although expression of these transporters has been shown, localization and function remain to be demonstrated. The arrows indicate the direction of substrate transport. References are indicated in the text. (*See insert for color representation of figure.*)

in the cellular compartments of the brain parenchyma. In the following section we summarize the current knowledge on the CNS localization and functional expression of these membrane drug transporters. Localization of various transporters at the brain barriers is presented in Figure 14.1. Similarly, transporter localization in cellular compartments of the brain parenchyma is depicted in Figure 14.2.

14.3.1. ATP-Binding Cassette Drug Transporters

The ATP-binding cassette (ABC) transporter superfamily consists of membrane-bound ATP-driven proteins that extrude from cells xenobiotics and their metabolites. ABC family members are classified according to the presence of various consensus sequences, including two ATP binding motifs (Walker A and Walker B) and the ABC signature C motif (ALSGGQ).[38] To date, there are 48 known human ABC family members belonging to seven different subfamilies. A comprehensive list of currently known mammalian ABC transporters compiled by Michael Müller (Wageningen University, The Netherlands) can be found at http://nutrigene.4t.com/humanabc.htm. Mutations in some of the ABC genes are the underlying cause of genetic disorders such as cystic fibrosis, anemia, Dubin–Johnson syndrome, and retinal degeneration.[39] ABC

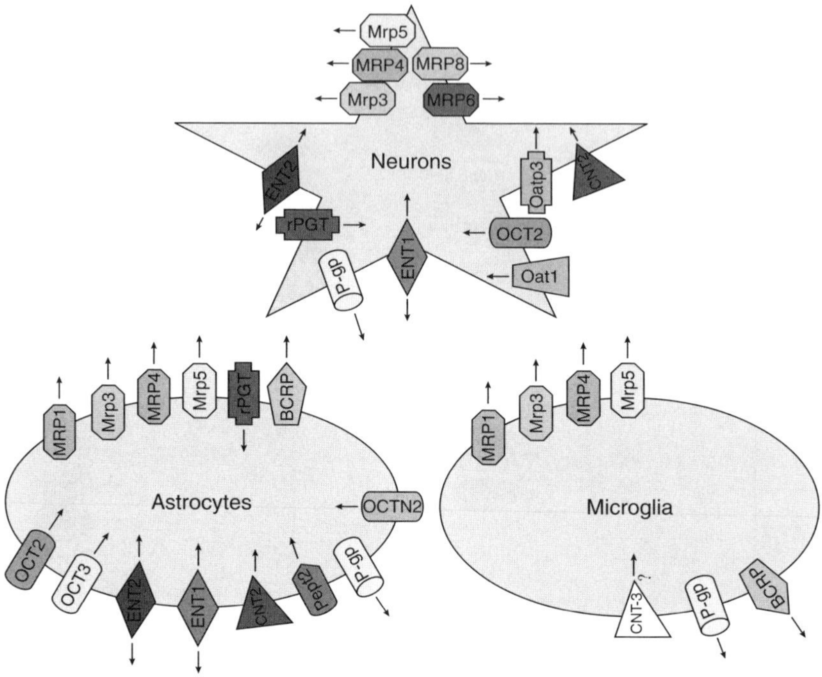

FIGURE 14.2. Xenobiotic transport mechanisms in the brain parenchyma. The proposed expression of ABC superfamily transporters, organic anion and cation transporters, nucleoside transporters, and peptide transporters is depicted in astrocytes, microglia, and neurons. Although expression of these transporters has been shown, localization and function remain to be demonstrated. The localization of membrane drug transporters in oligodendrocytes has not been well characterized and therefore is not included in this figure. The arrows indicate the direction of substrate transport. References are indicated in the text. (*See insert for color representation of figure.*)

transporters such as Pgp, MRP isoforms, and BCRP are also important determinants of drug uptake, distribution, and excretion. These transporters have all been implicated in the development of the multidrug resistance (MDR) phenotype. The MDR phenotype is defined as the simultaneous resistance to several structurally unrelated compounds. MDR does not result from independent genetic mutations which confer resistance to a single xenobiotic.[40]

P-Glycoprotein The ABC superfamily of proteins contains many membrane-bound energy-dependent transporters involved in the cellular efflux of endogeneous and exogeneous compounds. P-glycoprotein (Pgp), a well-characterized ABC transporter discovered in Toronto by Victor Ling,[41,42] is a 170-kDa integral membrane protein encoded by the MDR gene.[40] Two isoforms of the MDR gene, designated MDR1 and MDR2, have been cloned and sequenced in humans.[43,44] In rodents, Pgp is encoded by the mdr1a, mdr1b, and mdr2 genes. While MDR2/mdr2 is expressed primarily in the liver and has been implicated in the translocation of phosphatidylcholine into the bile,[45] overexpression of human MDR1 or rodent mdr1a/mdr1b confers the MDR phenotype.[40,46] In humans, the MDR1 gene product is 1280 amino acids in length and consists of two homologous halves, each made up of six transmembrane domains and one ATP-binding site.[40] Mammalian Pgp may possess between two and four oligosaccharide side chains (i.e., glycosylations) on the first extracellular loop.[40] In addition, phosphorylation is a hallmark of mature Pgp, and these posttranslational modifications are often observed in the linker region between transmembrane domains six and seven.[40]

Pgp was identified initially in Chinese hamster ovary cells selected for resistance to colchicine,[42] where it functions as an energy-dependent efflux pump that can extrude several amphipathic pharmacological agents. Since its discovery, many drugs have been shown to be Pgp substrates, including naturally occurring chemotherapeutic drugs such as anthracyclines (doxorubicin, daunorubicin, mitoxantrone), *Vinca* alkaloids (vincristine, vinblastine), epipodophyllotoxins (etoposide, teniposide), and taxanes (taxol, taxotere) as well as immunosuppressive agents (cyclosporin A and its analog, PSC833), cardiac glycosides (digoxin), antibiotics (rifampin, erythromycin, ciprofloxacin, grepafloxacin), antiallergenics (bepotastine, fexofenadine), beta blockers (talinolol, celiprolol), antiepileptics (phenytoin), steroid hormones (dexamethasone, cortisol), antimycotic agents (ketoconazole, itraconazole), histamine receptor antagonists (cimetidine, ranitidine), and HIV-1 protease inhibitors (saquinavir, indinavir, ritonavir, nelfinavir, amprenavir). Other therapeutic compounds that have also been shown to be Pgp substrates include morphine and atorvastatin.[47] Currently identified Pgp transport inhibitors include calcium channel blockers (verapamil), calmodulin antagonists (trifluoperazine), quinolines (quinidine), cyclosporin A, and the HIV-1 protease inhibitors. Many of these compounds act as both Pgp substrates and inhibitors and may interact with Pgp at more than one binding site.[47] Several studies have shown that Pgp is directly involved in regulating the brain accumulation of these therapeutic agents.[48–53] For example, in vivo studies in mdr1a/1b knockout mice have shown that HIV-1 protease inhibitors (e.g., saquinavir, indinavir, nelfinavir, ritonavir) exhibit four- to 36-fold increases in brain drug accumulation.[54,55]

In the CNS, Pgp expression has been investigated primarily at the brain vascular barriers (i.e., BBB and BCSF barrier). Although Pgp has been localized on the apical side of the choroid plexus epithelia,[56] the location of the protein in the brain microvessel endothelial cells that constitute the blood–brain barrier has been controversial. While several luminal membrane isolation studies, immunohistochemistry, and immunofluorescence laser scanning confocal microscopy studies have localized Pgp to the luminal surface of the mammalian brain endothelium,[57–59] others have identified the localization of Pgp on neighboring astrocyte foot processes.[60,61] Recently, our laboratory has reported Pgp localization on both the luminal and abluminal sides of the brain microvascular endothelium.[62] Furthermore, we have characterized the functional expression of Pgp in a rat microglia cell line (MLS-9),[63] cultured rat astrocytes,[64] and in a rat brain microvessel endothelial cell line and in situ in rat brain capillaries.[65] Recently, Pgp expression has also been reported in human glioma cell lines.[66] In addition to the plasma membrane, the subcellular localization of Pgp in brain cellular compartments has also been investigated. Previous studies have demonstrated the expression of Pgp within caveolae (i.e., plasma membrane microdomains that regulate various cellular processes) in bovine and human brain microvessel endothelial cells.[59,67] Using immunogold immunocytochemistry and immunoprecipitation, we have also reported the localization of Pgp within caveolae and the physical association of caveolin-1 with Pgp in primary cultures of rat astrocytes and in CTX TNA2 cells, an immortalized rat astrocyte cell line.[64] In addition, our laboratory has demonstrated subcellular localization of Pgp along the nuclear envelope and in cytoplasmic non-clathrin- and clathrin-coated vesicles in rat and human brain microvessel endothelial cells.[62,65] as well as in primary cultures of rat astrocytes and in CTX TNA2 cells.[64]

Multidrug Resistance Proteins A second group of ABC transporters that are involved in conferring MDR are the cystic fibrosis transmembrane conductance regulator (CFTR)/multidrug resistance protein (MRP) family (ABC subfamily C). At present, the mammalian MRP family (humans, MRP; rodents, Mrp) contains 13 members, including one ion channel (cystic fibrosis transmembrane regulator gene; CFTR), two surface receptors (sulfonylurea 1 and 2; SUR1 and 2), and a truncated protein that does not mediate transport (ABCC13).[68,69] These proteins are not involved in drug transport and are not discussed in this chapter. The remaining MRP family members, can be subclassified according to their membrane topology.[70] MRP1-3, 6, and 7 contain three transmembrane domains, TMD_0, TMD_1, and TMD_2, which show a $5 + 6 + 6$ configuration in transmembrane helices. Nucleotide-binding domains (NBD) 1 and 2 are located between TMD_1 and TMD_2 and between TMD_2 and the carboxy terminus, respectively. A cytoplasmic linker (L_0) located between the first two TMDs is essential for a functional protein.[71] MRP4, MRP5, MRP8, and possibly MRP9 are considered to be "short" MRPs, as they do not contain TMD_0 but do retain the cytoplasmic linker. Based on these differences, it is not surprising that the various MRPs differ with respect to tissue distribution, substrate specificity, and physiological function.[72]

Similar to Pgp, the MRP/Mrp family of active efflux transport proteins contributes to the MDR phenotype. Since 1992, when the first MRP/Mrp family members were cloned and characterized,[73] several anticancer drugs have been identified to be MRP/Mrp substrates, including vincristine (MRP1, MRP2, MRP6), etoposide (MRP1-3, MRP6), methotrexate (MRP1-3), daunorubicin (MRP1-2), and cisplatin (MRP2) as well as physiological substrates such as glutathione (MRP1/MRP2). MRP1/Mrp1-mediated cellular efflux has been demonstrated for other therapeutic compounds, including HIV-1 protease inhibitors[74–76] and antiepileptic drugs.[77] Increased brain accumulation of phenytoin, an antiepileptic drug, was observed in Mrp2-deficient rats, suggesting that Mrp2 plays a role in determining the CNS penetration of this therapeutic agent.[78] Unlike MRP1-3 and MRP6, MRP4, MRP5, and MRP8 display a unique capacity to transport a variety of monophosphorylated compounds. For example, cells functionally expressing with MRP4 show increased efflux of monophosphorylated nucleotides and nucleotide analogs such as cAMP, cGMP, 9-(2-phosphonylmethoxyethyl)adenine (PMEA), azidothymidine monophosphate, and purine analogs (6-mercaptopurine and 6-thioguanine).[79–82] Taken together with their localization and expression within the CNS, these observations suggest that MRP/Mrp isoforms play a critical role in regulating the brain accumulation and distribution of therapeutic compounds.

Several lines of evidence indicate that all of the functionally characterized MRP isoforms (1 to 8) are expressed in at least one CNS compartment and that they probably play a role in transport of drugs and metabolites. In addition, gene and protein expression in brain tissue has been reported for MRP9; however, functional characterization of this transporter in brain cellular compartments has yet to be examined. It should be noted that there is substantial controversy about the localization and function of certain family members in the CNS. In the following section we provide a summary of current knowledge with respect to MRP substrate profile and expression at the brain barriers and in the cellular components of the brain parenchyma. Our group has recently published a comprehensive review on MRP/Mrp expression and function in the CNS.[83]

At the BBB, many studies have reported the cellular localization and molecular (i.e., gene and protein) expression of MRP/Mrp isoforms. Using RT-PCR analysis, Zhang and colleagues identified the presence of Mrp1 mRNA in both cultured bovine brain microvessel endothelial cells and in capillary-enriched fractions of brain homogenates.[84] More recently, confocal laser scanning microscopy studies have localized MRP1 to the luminal side of the human BBB.[85] At the mammalian BBB, there is also evidence for the expression of Mrp2 (in rat but not detected in cow or human), MRP4/Mrp4 (human, cow, mouse), Mrp5 (cow, mouse), and Mrp6 (cow).[84,86–90] The presence of several different MRP homologs at the BBB may be important in controlling the uptake of organic anions into the brain.

Immunocytochemical studies in rodents have provided evidence for abluminal Mrp1 expression in choroid plexus epithelial cells of the blood–CSF barrier.[56] Studies in triple-knockout mice [e.g., mdr1a (−/−), mdr1b (−/−), Mrp1 (−/−)] showed increased accumulation of etoposide in the CSF compared to double-knockout mice [e.g., mdr1a (−/−), mdr1b (−/−)], which suggests that Mrp1 may act to mediate organic anion efflux into the blood.[56] MRP4/Mrp4 expression has been detected at both luminal and abluminal plasma membranes of choroid plexus epithelial cells, implying that this transporter may be involved in limiting organic anion influx from blood and driving organic anion efflux from brain to blood.[91] High levels of Mrp5 mRNA transcripts have also been observed in rat choroid plexus[92]; however, high-resolution immunofluorescence staining studies were unable to detect Mrp5 expression in murine choroid plexus.[93] Several studies have shown that expression of MRP2/Mrp2.[92] and MRP3/Mrp3.[85,92] at the blood–CSF barrier is negligible.

In the brain parenchyma, MRP/Mrp mRNA expression has been reported in astrocytes, microglia, oligodendrocytes, and neurons. The expression of MRP1/Mrp1, Mrp3, Mrp4, and Mrp5 has been reported in all of these cell types.[85,94–96] Studies in our laboratory have shown Mrp1, Mrp4, and Mrp5 functional expression in a continuous rat microglia cell line (MLS-9).[82,97] Using immunogold cytochemistry at the electron microscope level, Mrp1 was found to localize primarily to the plasma membrane in MLS-9 cells.[76] Although Mrp6 gene and protein expression has not been observed in astrocytes, microglia, and oligodendrocytes,[96,98] a recent study has suggested that human neurons highly express MRP6.[99] In addition, MRP8 has been shown to colocalize with microfilaments in the white matter of human brain, suggesting that it is principally an axonal protein.[100] Although several studies have reported that brain parenchymal cells express negligible levels of MRP2/Mrp2,[92,96,97] Hirrlinger and colleagues detected the expression of Mrp2 mRNA in astrocytes isolated from embryonic rats.[96] These discrepancies may be explained by differences in prenatal versus postnatal expression of Mrp2.[101,102]

Breast Cancer Resistance Protein A third ABC superfamily member that may be involved in xenobiotic efflux is breast cancer resistance protein (BCRP; also known as ABCG2). Unlike Pgp and MRP1, ABCG2 is a "half-transporter" comprised of only six putative transmembrane spanning domains and a single ATP-binding region.[103] Since other half-transporters, such as ABCG5 and ABCG8, have been observed to dimerize and form functional homo- or heterodimers,[104] it has been hypothesized that ABCG2 may also form functional dimers. ABCG2 has significant overlap in substrate-specificity profile with Pgp and has been shown to recognize a vast array of sulfoconjugated organic anions and hydrophobic and amphiphilic compounds.[105] Examples of well-established ABCG2 substrates include fluoroquinolone antibiotics (ciprofloxacin, ofloxacin, norfloxacin), mitoxantrone, camphothecin derivatives (topotecan, irinotecan), dipyridamole, and estradiol-17β-glucuronide.[106–109] ABCG2 may also transport physiologic substrates such as glutathione, steroid hormones, and folic acid.[105,106]

Several recent studies have demonstrated the expression of ABCG2 in the brain, particularly along the luminal side of the BBB.[110,111] Zhang and colleagues reported the gene and protein expression of ABCG2 in both primary cultures of human brain microvessel endothelial cells.[112] Overexpression of human ABCG2 in an immortalized rat brain endothelial cell line resulted in enhanced basolateral-to-apical transport of mitoxantrone, an established ABCG2 substrate.[112] In the porcine brain, ABCG2 appears to be expressed predominantly in brain microvessel endothelial cells and to a lesser extent in choroid plexus epithelial cells and pericytes.[113] Lee and colleagues reported ABCG2 gene and protein expression along the luminal side of mouse brain capillaries.[114] Taken together, these data suggest that ABCG2 may play a role in limiting the brain uptake of several xenobiotics. With respect to brain parenchyma, low expression of ABCG2 has been shown in primary cultures of rat astrocytes and microglia as well as in the MLS-9 cell line[115] and in human fetal astrocytes.[112]

Although expression of ABCG2 in brain cellular compartments has been demonstrated clearly, data on the functional activity of ABCG2 at this site remain inconclusive. ABCG2 transport activity at the BBB has been demonstrated in cultured human and rat brain microvessel endothelial cells, but these cell systems were transfected to overexpress the exogenous protein of interest.[111,112] Studies in primary cultures of human brain endothelial cells and the RBE4 cell line have shown that the accumulation of mitoxantrone, an established ABCG2 substrate, could not be enhanced in the presence of standard ABCG2 inhibitors (e.g., Ko143, FTC), suggesting lack of ABCG2 efflux activity in these cell systems.[115] The same study also reported an absence of ABCG2-mediated efflux of mitoxantrone in primary cultures of rat astrocytes and in the MLS-9 cell line.[115] Data from recent in vivo studies in ABCG2 knockout mice are controversial. Some studies have implied that ABCG2 is not functional at the BBB,[114–116] whereas others have suggested that ABCG2 is in fact active at this brain barrier.[117,118] Further studies are needed to clarify the functional significance of this transporter at the BBB.

14.3.2. Organic Anion–Transporting Polypeptides

Organic anion transporting polypeptides (OATPs in humans; Oatps in rodents) are a group of membrane solute carriers involved in the transport of amphipathic substrates.[119] Although several members of this family are expressed selectively in hepatic tissue, many OATPs/Oatps are present in multiple regions of the rodent and human body, including the CNS. Previous studies have suggested that OATPs/Oatps have overlapping and partially distinct substrate preferences for many solutes, including bile salts, organic dyes, steroid conjugates, thyroid hormones, anionic polypeptides, and numerous xenobiotics.[119]

Hydropathy analysis has predicted that OATP/Oatp membrane topology consists of 12 transmembrane domains. However, the 12-transmembrane-domain model predicted has not been tested experimentally. All OATPs/Oatps share many structural features, including a large extracellular loop between transmembrane domains 9 and 10 which contains many conserved cysteine residues that resemble the zinc finger motifs of DNA-binding proteins.[119] OATP/Oatp family members also possess N-glycosylation sites in extracellular loops 2 and 5. In addition, all OATP/Oatp family members have a conserved amino acid sequence, known as the *OATP superfamily signature*, located at the border between extracellular loop 3 and transmembrane domain 6. Conserved charged amino acids are found at the boundaries of the transmembrane domains as well as within the membrane itself.[119] The role of these conserved amino acids for transmembrane substrate transport has yet to be determined.

OATP/Oatp genes are classified within the solute carrier family 21A (humans: SLC21A; rodent: Slc21a).[119] To date, 11 rat, 8 mouse, and 9 human OATPs/Oatps have been identified[119]; however, not all of these isoforms are expressed in the brain. The information provided below summarizes the molecular (i.e., gene and protein) expression, localization, and substrate profile of those OATP/Oatp isoforms that have been identified in the CNS. Since the human OATPs and rodent Oatps are distinctly different in terms of gene and protein expression and substrate profiles, they are considered separately.

Human OATP Isoforms Previous studies have shown the expression of OATP isoforms in human brain tissue. Immunofluorescence staining of human brain frontal cortex tissue fixed in situ demonstrated the localization of OATP-A along the border of brain capillary and microvessel endothelial cells.[120] Astrocytes and neurons were immunonegative, suggesting that these cells do not express OATP-A.[120] Studies in *Xenopus laevis* oocytes have shown that OATP-A is involved in the uptake of δ-opioid receptor agonists such as D-penicillamine(2,5)-enkephalin and deltorphin II,[120] which implies that this transporter may be prominently involved in the brain uptake of centrally active opioids. The transport of these substrates by OATP-A was inhibited by the opiate antagonists naloxone and naltrindole as well as the μ-opioid receptor agonist Tyr-D-Ala-Gly-*N*-methyl-Phe-glycinol and the endogenous peptide Leu-enkephalin.[120] Other OATP isoforms that have also been detected in human brain include OATP-B,[121] OATP-D,[122] OATP-E,[123] and OATP-F.[124] Using *Xenopus laevis* oocytes and transfected human embryonic kidney cells (HEK293), previous studies have shown that these OATPs may be involved in the transport of both therapeutic agents and endogenous substrates including bromosulfophthalein, pravastatin, sulfated steroids (i.e., estrone 3-sulfate), prostaglandins (PGE_1, PGE_2, $PGF_{2\alpha}$) and thyroid hormones.[121–126] However, the exact cellular localization of these isoforms in the human brain has yet to be elucidated. In addition, mRNA expression of human prostaglandin transporter (hPGT), a member of the SLC21A family of transporters,[119] has been reported in both adult and fetal brain tissue.[127] Although the expression of hPGT in human brain tissue suggests that it may be involved in the CNS transport of prostaglandins, further studies are required to elucidate the exact role of this transporter in the brain.

Rodent Oatp Isoforms Similar to human OATPs, several Oatp isoforms have been detected in tissue isolated from rodent brain. Gene and protein expression of Oatp1 has been observed in neonatal rat choroid plexus.[128] Although fluorescence confocal microscopy studies have shown that this transporter is localized intracellularly in neonatal rats, this same method has shown that Oatp1 is localized primarily to the apical surface of choroid plexus epithelial cells in adult rats.[128] The localization of this transporter suggests that it may be involved in regulating the brain permeation of therapeutic agents such as estradiol 17β-D-glucuronide,[129] fexofenadine,[130] and D-penicillamine(2,5)-enkephalin.[120] Expression of other Oatp isoforms (e.g., Oatp2, Oatp3, Oatp14) has been reported in brain capillary–enriched fractions and/or brain capillary endothelial cells as well as in choroid plexus epithelial cells.[92,131–133] These Oatps may be involved in the brain transport of substrates such as digoxin,[134] fexofenadine,[130] D-penicillamine(2,5)-enkephalin,[135] estrone 3-sulfate,[131] and thyroid hormones.[132]

In terms of the brain parenchyma, a few studies have reported the neuronal expression of both Oatp3 and Oatp9.[129,136] Neuronal expression of Oatp9 is interesting since many of its substrates are involved in inflammatory signaling and regulation (i.e., prostaglandins, leukotriene C4), suggesting that this transporter may play an important role in transporting signals to target cells during brain inflammation.[129] In addition, Gao and colleagues (2000) observed that Oatp2 immunoreactivity did not colocalize with GFAP, which implies that astrocytes do not express Oatp2.[120] Overall, the cellular compartments of the brain parenchyma remain poorly characterized for the expression of Oatp family members.

Immunoblot analysis has detected the expression of prostaglandin transporter (rPGT), an Slc21a family member, in cultured rat cerebral endothelial cells, astrocytes, pericytes, and choroid plexus epithelial cells.[137] Immunocytochemical analysis detected strong rPGT immunoreactivity in the supraoptic and paraventricular nuclei of the hypothalamus as well as in the ependymal cell layer of the third ventricle and in the choroid plexus.[137] The prototypical substrate of rPGT is prostaglandin E2[138]; however, functional studies examining rPGT-mediated transport mechanisms in the brain have not been undertaken.

14.3.3. Organic Anion Transporter Family

Organic anions (i.e., drugs, toxins, as well as endogenous and exogenous molecules) that are sulfated or glucuronidated by phase II metabolism bear transient or permanent negative charge at physiological pH. Such molecules are generally membrane impermeable and are transported by the organic anion transport system. In the kidney, organic anions are secreted by the proximal renal tubules into the urine for elimination. This secretion is divided into two processes: (1) uptake of organic anions from the peritubular blood plasma into the renal tubular cells through the basolateral membrane, and (2) release of organic anions into the tubule lumen through the apical brush border membrane.[139] Apart from the kidney, organic anion transport system is also found in the brain, including the brain capillaries, choroid plexus, and brain parenchyma.[7,140] Their expression in the choroid plexus and BBB, in hand with evidence that steroid hormones, neurotransmitter, and their metabolites, such as 17β-estradiol-D-17β-glucuronide (from 17β-estradiol) and homovanillic acid (from dopamine), are substrates for organic anion transporters suggest that these transporters regulate the composition of brain by controlling the flux of xenobiotics from CSF and ECF into blood.[141]

Organic anion transporters (OATs) belong to family 22 of the solute carrier superfamily (SLC22A).[142] OATs are expressed primarily in epithelial tissues, (i.e., the kidney and liver) but also in placenta, small intestine, choroid plexus, and brain microvasculature that transport xenobiotics.[143–148] Currently, the OAT family comprises OAT1,[149–151] OAT2,[143] OAT3,[145,146] OAT4,[147] OAT5,[152] OAT6,[153] and renal-specific transporter (RST).[144] OATs are categorized into three classes based on their energy requirements: (1) Na^+-dependent OATs, (2) Na^+-independent facilitators or exchangers, and (3) active OATs that require ATP. Na^+-dependent OATs have a narrow substrate specificity and play a major role in the reabsorption of essential anionic substrates into the proximal renal tubules. Na^+-independent and active OATs, on the other hand, possess broad substrate specificity and are involved in the secretion of organic anions in kidney, liver, and brain.[7,154]

Organic anion transporter 1 (Oat1, Slc22a6) is a transporter abundant in the brain and was first cloned from rat and mouse and described as novel kidney transporter.[149–151] Human OAT1 is highly localized at the basolateral membrane of renal proximal tubular cells.[155] One of the hallmarks of this transporter is its ability to exchange extracellular organic anions such as PAH for an intracellular dicarboxylate (α-ketoglutarate).[150] Here, uptake of PAH is coupled indirectly to Na^+/dicarboxylate transport and is tertiary active.[156,157] With respect to localization, OAT1/Oat1 is also weakly expressed in the brain,[150,158–160] placenta, skeletal muscle,[158] and liver.[161] In mouse CNS (choroid plexus, dura mater, root ganglions, and spinal cord), Oat1 mRNA expression is highest at the embryonic stage and decreases toward adulthood, where it is detected only in the meninges.[159] This differential temporal expression suggests a yet undetermined role of this anion transporter during the development of the CNS structures.[162] Murine Oat1 has also been detected in neurons of the cerebral cortex and hippocampus, as well as in the ependymal cell layer of the choroid plexus.[163] Various studies have shown expression of OAT1 at the apical side of choroid plexus epithelial cells.[160,162,164] This in turn suggests OAT1 involvement in removal of anionic drugs and neurotransmitter metabolites such as vanillin mandelate from the CSF to the blood across the CP.[160,162,164]

Organic anion transporter 2 (Oat2; Slc22a7), cloned from rat liver and described as novel liver transporter,[143] was characterized by Sekine and colleagues.[165] OAT2 protein is liver specific and is not expressed in the brain.[139,165] Only Oat2 mRNA has been detected in rat choroid plexus.[92]

Organic anion transporter 3 (Oat3; Slc22a7) was fist cloned from a rat brain cDNA library. It is the most abundant OAT in the brain and is also expressed in liver, kidney, lungs, and eye.[146,166] Human OAT3 cloned from the kidney[167] is also highly expressed in the skeletal muscle and brain.[168] In rodent kidney, Oat3 is localized to the basolateral membrane of proximal tubules.[146,169] In the CNS, immunochemical staining revealed basolateral[141] and faint luminal localization of Oat3 in rat brain capillaries,[170] as well as apical membrane localization in rat[171] and human[164] choroid epithelial cells. Studies in *Xenopus laevis* oocyte expressing Oat3 have shown transport of PAH, estrone sulfate, taurocholate, ochratoxin, benzylpenicillin, cimetidine, and ranitidine but not tetraethylammonium in a Na^+-independent manner.[140,146,171] Furthermore, benzylpenicillin transport from the CSF was also Na^+-independent and inhibited by PAH. Conversely, elimination of cimetidine from the CSF was inhibited by benzylpenicillin and PAH, but not by N^1-methylnicotinamide, a quaternary amine and typical substrate for organic cation transporters.[172,173] Taken together, these results suggest that the organic anion transport properties via the blood–brain and/or BCSF barrier are consistent with transport by OAT3. In studies where PAH and homovanillic acid, a metabolite of dopamine, was injected into the brain, rapid and saturable efflux from the brain was detected via the BBB.[141,174] Furthermore, anionic metabolites of neurotransmitters such as epinephrine, norepinephrine, dopamine, and serotonin potently inhibited the uptake of estrone sulfate via rat Oat3-expressing oocyte, suggesting an involvement of Oat3 in the extrusion and elimination of endogenous organic anions from the brain via the blood–brain and BSCF barrier.[146] In other studies where Oat3 gene was disrupted, there was a significant decrease (~75 %) in fluorescein uptake by murine choroid plexus, but no change in fluorescein-labeled methotrexate uptake by the underlying capillaries of the choroid plexus epithelium.[140] This suggests that Oat3 has apical localization in the choroid plexus cells and that the basolateral exit is not affected by Oat3 loss. Taken together, Oat3 appears to play a vital role in the entry step of substrates into the choroid plexus, leading to elimination of substrates from the CSF.[140]

Organic anion transporter 4 (OAT4; SLC22A11) was cloned from human kidney cDNA library and found to be expressed predominantly in the apical membranes of proximal tubules[175] of the kidney and placenta.[147] RT-PCR studies have also shown OAT4 mRNA in the choroid plexus epithelial cells[176] and brain microvessel endothelial cells.[90] Transport by OAT4 which is bidirectional and Na^+ independent is believed to be involved in reabsorption and secretion of prostaglandins,[177] estrone sulfate, ochratoxin A,[147,175] and other anionic substrates through the kidney and choroid plexus, as well as excretion of toxic substrates from the fetal into maternal circulation in the placenta.[147]

Organic anion transporter 5 (Oat5; Slc22a19) was cloned from rat kidney cDNA library and expressed predominantly at the apical membranes of proximal tubules of the kidney.[178] Organic anion transporter 6 (Oat6; Slc22a20), on the other hand, was cloned from mouse olfactory mucosa.[153] Neither Oat5 nor Oat6 are expressed in the brain.[148,153]

Renal-specific transporter (RST, SLC22A12), the mouse homolog of human urate transporter 1 (URAT1), was cloned from mouse kidney.[144] Both human URAT1 and mouse RST are localized on the apical (brush border) side of the renal proximal tubules and are believed to be involved in the reabsorption of urate in the kidney.[179,180] Although human URAT1 is expressed primarily in the kidney, the mouse homolog protein is highly expressed in the choroid plexus and brain capillary–enriched fraction.[180] Depolarization of membrane voltage stimulates the transport activity of this protein. As suggested by Breen and co-workers, there are membrane voltage–sensitive excretion mechanisms for organic anion and fluorescein-labeled methotrexate in the choroid plexus. Hence, mouse RST, in coordination with OAT3, could potentially play a role in the extrusion of anionic metabolites of neurotransmitters from the CNS.[181]

14.3.4. Organic Cation Transporter Family

Organic molecules with transient or permanent net positive charge are referred to as organic cations and have decreased ability to permeate biological membranes passively. Therefore, entry of cationic molecules across cell membranes requires active transport. Agents varying from (1) endogenous compounds such as acetylcholine, choline, thiamine, N^1-methylnicotinamide, and creatinine, to (2) the biogenic amines dopamine, serotonin, epinephrine, and norepinephrine, to (3) therapeutic agents such as cimetidine, metformin, acyclovir, memantine, and quinidine are among the host of compounds transported by the organic cation transporter system.[182–186] There are also several weak bases, uncharged, and even anionic compounds that are transported by this system.[187]

Organic cation transporters (OCTs), found in various mammalian tissues as well as lower eukaryotes, bacteria, and plants,[188] have been divided into two major groups: oligospecific and polyspecific transporters. *Oligospecific transporters* facilitate the transport of a single main substrate or molecules with closely related structures and include the Na^+-cotransporters for neurotransmitters, high-affinity transporters for thiamine, and vesicular and plasma membrane cotransporters for choline.[183] *Polyspecific transporters*, on the other hand, mediate the transport of organic cations with different molecular structures[186,189,190] and belong to a large transporter family of OCTs, or family 22 of solute carrier superfamily (SLC22A).[191] Two distinct classes of OCT systems have been defined: (1) potential-sensitive organic cation transporters (OCT1-3 in humans and Oct1-3 in rodents)[192–194] localized to the basolateral membrane of various cells, usually involved in the influx of organic cations into a cell, and (2) H^+ gradient-dependent novel organic cation transporters (OCTN1-3 in humans and Octn1-3 in rodents) which mediate the cellular efflux of cationic substrates.[195–201] The combined effect of these two OCT systems results in the transport of cationic substrates across the renal tubular cells from the blood into the luminal fluid, from the blood into the bile across the hepatocyte, or from the fetal into the maternal blood through the placental syncytiotrophoblast, hence eliminating many endogenous amines as well as a wide array of drugs and environmental toxins.[187]

Members of the SLC22A-family share a predicted membrane topology that includes 12 α-helical transmembrane (TM) domains with cytoplasmic N- and C-termini along with a large extracellular loop (between TM1 and TM2) that has several N-glycosylation sites and smaller intercellular loop (between TM6 and TM7). They also possess a C-terminus that has consensus sequences for protein kinase A–, protein kinase C–, and tyrosine kinase–dependent phosphorylation sites.[190,202,203]

Oct1 (Slc22a1) was originally cloned from rat kidney,[192] followed by isolation of the human (OCT1) homolog.[204] Although controversial, very little[92,205] or no[206] OCT1/Oct1 mRNA has been found in the brain (choroid plexus and astrocytes).

Oct2 (Slc22a2) expression is tissue specific and expressed primarily in the kidney.[193,207] Both Oct2 and OCT2 are localized to the basolateral membranes of the proximal renal tubules.[208,209] Other expression sites include human placenta,[193] neurons (human hippocampal pyramidal cells, cerebral cortex, and subcortical nuclei),[210] and rat choroid plexus.[206,211] OCT2 in human neurons has been reported to mediate the transport of endogenous substrates such as dopamine, norepinephrine, serotonin, histamine, choline, and exogenous substrates such as tetraethylammonium (TEA), 1-methyl-4-phenylpyridinium (MPP^+), muscle relaxant memantine, and the antiparkinsonian agent amantidine.[210] Hence, in the CNS, OCT2 may be involved in the uptake of choline into neurons and in choline reabsorption from the cerebrospinal fluid. In other words, OCT2 in brain may help reduce concentrations of basic neurotransmitters and their metabolites and protect the brain from neurotransmitter excitotoxicity.

OCT3 (SLC22A3) has a relatively broad expression profile and is highly expressed in the human placenta, liver, skeletal muscle, kidney, heart, and to a lesser extent, the brain.[212–214] Oct3 mRNA expression in brain is greater than that of Oct1 and Oct2[215]. In situ hybridization studies demonstrated its expression in cerebral cortex and cerebellar and hippocampal neurons of rodents.[213,216] Reverse transcriptase polymerase chain reaction (RT-PCR) analysis of rat superior cervical ganglion and choroid plexus has also detected the Oct3 transcript.[217] Apart from mediating the uptake of TEA, desipramine, and amphetamines, OCT3 also interacts with cationic neurotoxin MPP^+ and the neurotransmitters dopamine and serotonin. Expression of OCT3 in the brain combined with its ability to interact with cationic neurotoxins and neurotransmitters suggests that this polyspecific organic cation transporter may play a significant role in the handling of neuroactive compounds in the brain.[215] Due to the localization of OCT3 on cortical astrocytes and its ability to mediate influx of catecholamines and neurotoxic organic cations such as MPP^+, this transporter has also been termed the extraneuronal monoamine transporter (EMT), known also as the uptake system.[218] However, the true role of OCT3 in the brain is not yet known, partly because distribution of dopamine and serotonin are not highly localized to the regions of the brain where OCT3 is expressed.[215]

Novel Organic Cation Transporters Thus far, three isoforms in rodents (Octn1-3) and two isoforms in humans (OCTN1-2) have been cloned and characterized.[188] These transporters mediate substrate uptake across membranes of renal proximal tubules, fetal liver, trachea, placenta, skeletal muscle, and cardiac muscle. All OCTN members transport organic cations and carnitine,[187,199] and translocation of this endogenous substrate together with Na^+-and/or organic cations is H^+-gradient dependent. L-Carnitine (β-hydroxy-γ-trimethylaminobutyric acid) is a small, highly polar zwitterion at physiological condition that plays an important role in the cotransport of long-chain fatty acids across the inner mitochondrial membrane for β-oxidation and ATP generation. This compound, along with its metabolite, acetyl-L-carnitine, is present in the CNS and has been proposed to have different physiological roles, including the control of acetyl moiety level in the brain parenchyma for the synthesis of acetylcholine.[220] Furthermore, administration of carnitine or acetyl-L-carnitine to Alzheimer's patients has shown improvement in memory functions.[221] These observations suggest that L-carnitine and acetyl-L-carnitine are transported from the circulating blood into the brain across the BBB. In the BBB there are several transport systems for nutrients and xenobiotics to maintain the homeostasis in the CNS.[222] Therefore, it is likely that carnitine and acetyl-L-carnitine are transported into the CNS via specific transporters across the BBB.[223]

OCTN1 (SLC22A4) has been cloned from human and rodent tissue.[197,199,201,224] There is, however, large species difference in localization and function of OCTN1. Humans OCTN1 is expressed primarily in the fetal liver (absent in adult liver), lung, kidney (brush border membrane), bone marrow, trachea, and to a lesser extent in skeletal muscle, placenta, and prostate,[197] while rat Octn1 mRNA is expressed in the spinal cord choroid plexus, hippocampus, cortex, and cerebellum.[92,201,211] Among many others, substrates transported by OCTN1 include carnitine, TEA, quinidine, choline, nicotine, cimetidine, and clonidine.[224] OCTN1 localization has been proposed at the apical membranes of polarized cells to aid in the secretion of organic cations.[211]

OCTN2 (SLC22A5) was first cloned from human placenta,[200] followed by isolation of the mouse and rat homologs.[225,226] OCTN2 functions as a polyspecific and Na^+-independent cation uniporter.[198,219,227,228] Octn2/OCTN2 has also been characterized as high-affinity Na^+-dependent plasmalemmal carnitine electrogenic transporter[227,229] (also known as carnitine transporter 1, CT1), although it also transports TEA, albeit with lower activity compared to that of carnitine. When expressed in HeLa cells, this transporter mediates the transport of a number of classical organic cations, including MPP+.[200] OCTN2 is present in various tissues, including brain, kidney, skeletal muscle, heart, placenta, and others.[198] In CNS, OCTN2 mRNA was detected in neurons from hippocampus, cerebellum, spinal cord, and superior cervical ganglion,[198,211,217,228] although it was not expressed in cerebral cortex neurons.[220] OCTN2 has also been detected in primary cultures of human, rat, mouse, porcine, and bovine brain capillary endothelial cells.[223] Functional loss of the transporter in Octn2$^{-/-}$ mice is associated with decreased systemic and brain concentration of acetyl-L-carnitine[230]; this, along with their expression in brain capillary endothelial cells, has led to the belief that these transporters have luminal localization at the BBB and assist the entry of carnitine into the brain.[223]

Octn3 (Slc22a9) was first cloned from mouse testis and the protein shown to be highly expressed in the testis and liver peroxisomes[231] but weakly in the kidney.[199] Carnitine transporter 2 (CT2, SLC22A16), on the other hand, was cloned and characterized in human testis.[179] However, there is no evidence for the expression of neither of these proteins in the CNS.

14.3.5. Nucleoside Membrane Transport Systems

Nucleotides, phosphate esters of nucleosides, are the building blocks of nucleic acids, DNA and RNA, and are considered one of the most important molecules in the cell. These nucleosides are divided into purines (adenosine and guanosine) and pyrimidines (cytosine, thymidine, and uridine). Nucleotides such as cyclic AMP and cyclic GMP serve as secondary messengers in signal transduction, while ATP serves as an energy store. The nucleotide adenosine is also a ligand for cell surface purinergic P1 receptor involved in physiological responses, including modulation of neural excitotoxicity,[232] coronary vasodilation, and platelet aggregation.[233]

In general, nucleosides are synthesized endogenously via de novo pathway. However, a number of cells and tissues, including bone marrow cells, erythrocytes, leukocytes, and some cells in the brain,[234] lack the enzymes required for de novo nucleoside biosynthesis and therefore rely on salvage of extracellular nucleosides or nucleobases for their intracellular metabolic demands.[235] The brain, is therefore, dependent on a constant supply of nucleosides from the blood and in situ synthesis.[236,237] To achieve this, transporters on various cell membranes facilitate the entry of nucleosides and nucleotides into the cells. It is also of interest to study these transporters because of their involvement (uptake route) in therapeutic applications of many cytotoxic nucleotide derivatives used in the treatment of cancer, stroke, and cardiovascular, parasitic, and viral diseases.[233] Mammalian nucleoside transporters are classified into two structurally and mechanistically unrelated protein families based on their sodium dependency: Na^+-dependent concentrative nucleoside transporters (CNTs), also classified as the SLC28A family, and Na^+-independent equilibrative nucleoside transporters (ENTs) also classified as the SLC29A family.[233,237,238]

Concentrative Nucleoside Transporter Family CNTs are present in both prokaryotes and eukaryotes and mediate the concentrative accumulation of nucleosides inside the cell.[239] Six functionally different transport activities have been described based on their substrate specificity. These are *cit* (pyrimidine nucleosides and adenosine), *cif* (purine nucleosides and uridine), *cib* (purine and pyrimidine nucleosides), *cit*-like (pyrimidine selective but also transports adenosine and guanosine), *cs* (selective for adenosine and its analogs), and *csg* (guanosine selective).[240–243] Three different proteins, CNT1, CNT2, and CNT3, have been described that are insensitive to nitrobenzylmercaptopurine riboside (NBMPR) and are responsible for the *cit*, *cif*, and *cib* activities, respectively. Molecular identities responsible for *cit*-like, *cs*, and *csg* activity, which are NBMPR-sensitive, have not yet been identified.[244] CNT1, -2, and -3 couple uphill transport of nucleosides to downhill transport of sodium ions, and in the case of CNT3, also to downhill transport of protons.[245] With respect to topology, CNTs are integral proteins with 13 transmembrane α-helices with a large putatively extracellular glycosylated extracellular C-terminus.[246] However, unlike ENTs, which are found ubiquitously in many cell types, CNTs are found primarily in specialized cell types, including macrophages,[247] microglia,[248] choroid plexus,[249] leukemia cells,[250] and renal and gastrointestinal epithelia,[241,243] suggesting an important role in absorption, secretion, distribution, and elimination of physiologic nucleosides and nucleoside-analog drugs. CNTs have also been characterized in bacteria, insects, hagfish,[251] *Candida albicans*,[252] and *Caenorhabditis elegans*,[253] and unlike ENTs, many prokaryotes also express this group of transporters.

CNT1 (SLC28A1) is expressed primarily in epithelial tissues, including kidney, small intestine enterocytes, and liver, and is localized to the apical membrane, where it works in concert with ENTs that are localized predominantly in the basolateral membrane to mediate transepithelial nucleoside flux.[254] The Cnt1 gene transcript is also expressed in many regions of the rat brain, including cerebral cortex, cerebellum, hippocampus, striatum, brain stem, superior colliculus, posterior hypothalamus, and choroid plexus.[255,256] However, there is no evidence for the expression of the protein at the BBB,[257] although expression of *cit* nucleoside transport system at the blood–brain and BCSF barriers has been reported.[258] CNT1 mediates the cellular accumulation of various nucleoside analogs used in the treatment of HIV (zidovudine, lamivudine, and zalcitabine) and cancer tumors (cytarabine and gemcitabine).[259]

Messenger RNA for human CNT2 (SLC28A2) has been detected in various tissues, including liver, kidney, spleen, heart, placenta, colon, rectum, pancreas, small intestine, skeletal muscle, and brain.[255,260,261] CNT2 protein is localized at the apical membranes of polarized kidney cells.[262] At the blood–brain and BCSF barriers, CNT2 is limited to the membrane facing interstitial and cerebrospinal fluid, respectively.[263] CNT2 is considered as the BBB adenosine transporter and transports adenosine, guanosine, and certain pyrimidine nucleosides such as uridine.[264] Therefore, this transporter is named sodium-dependent purine nucleoside transporter (SPNT). Localization of this CNT2 at the blood–brain and BCSF barriers suggests likely roles in removing adenosine from brain extracellular fluids.[263] Antiviral compounds such as didanosine and ribavirin, used in the treatment of HIV and hepatitis C, respectively, are also substrates of CNT2.[265]

Human CNT3 (SLC28A3) mRNA has been identified in many tissues, including bone marrow, pancreas, trachea, mammary gland, liver, prostate, heart, brain, and some regions of the intestine.[242] In vitro studies using rabbit choroidal cells have shown *cib*-type transport of nucleosides. In addition, in vivo experiments with isolated rabbit choroidal cells have shown transport of nucleosides from blood to CSF.[266] This system is thus one that facilitates the entry of nucleosides into the brain. Studies undertaken by our group[248] also identified an Na^+-dependent NMBPR-insensitive thymidine uptake by microglia continuous cell line. The *cib*-type transport identified at the basolateral side of rabbit choroid plexus tissue slices is selective for naturally occurring purine and pyrimidine ribo- and deoxyribonucleosides (e.g., guanosine, inosine, formycin B, uridine, cytidine) and for base-modified nucleoside analogs (e.g., 5-fluorouridine, 2-chlororadenosine), but not for synthetic nucleoside analogs (e.g., zidovudine, zalcitabine) substituted on the ribose ring.[249,267] This transporter mediates the cellular uptake of nucleoside analogs used in the treatment of cancer (e.g., cladribine, gemcitabine, 5-fluorouridine, fludarabine, zebularine) as well as HIV (e.g., zidovudine, didanosine, zalcitabine).[243] However, it is interesting that transport of nucleoside analog drugs such as stavudine by the *cib* transport system[268] has been demonstrated in intestinal cells, but this system fails to transport analogs at the blood–brain and BSCF barriers.[267,269]

Equilibrative Nucleoside Transporter Family ENTs are expressed in most human cell types and are widely distributed in eukaryotes but so far appear to be absent in prokaryotes.[239] Four isoforms of ENT, designated ENT1–4 in humans and Ent1–4 in rodents,[270] have been identified and classified into two subtypes, based on their sensitivity to NBMPR. These include the equilibrative sensitive (*es*) type, inhibited by nanomolar concentrations of NBMPR (K_i = 0.1 to 1 nM) and the equilibrative insensitive (*ei*) type inhibited by micromolar concentrations of NBMPR.[237] Transport via ENTs is bidirectional and can mediate uptake or efflux of substrate compounds depending on the nucleoside concentration gradient across the plasma membrane. Furthermore, all ENTs transport adenosine but have different capacities to transport other nucleosides and nucleobases.[270] ENT1 and ENT2, two of the best-characterized subtypes of these transporters, exhibit broad substrate specificity for purine and pyrimidine nucleosides and belong to *es*- and *ei*-type ENTs, respectively. ENT3 is also an *es*-type transporter and displays broad substrate specificity for nucleosides as well as nucleobases.[271] The transporting properties of ENT4 have not yet been fully characterized. As revealed from human ENT1, the archetype for all ENT family members is predicted to possess 11 α-helical transmembrane domains with a cytoplasmic N-terminus, an extracellular C-terminus and a large cytoplasmic loop linking TM6 and TM7.[272] In addition, the extracellular loop between TM1 and TM2 is N-glycosylated.[273]

ENT1 (SLC29A1) is ubiquitously distributed in human and rodent tissues at both the mRNA and protein levels at the plasma and nuclear membranes.[274] In the brain, ENT proteins play an integral role by salvaging nucleosides for biosynthetic pathways and/or controlling exogenous adenosine concentrations that interact with purinergic receptors.[275] Both Ent1 and ENT1 have been located at pyamidal neurons of the hippocampus, granule neurons of the dentate gyrus, Purkinje and granule neurons of the cerebellum, cortical and striatal neurons and astrocytes.[261,276] In humans, ENT1 protein has been localized at frontal and parietal lobes of the cerebral cortex, thalamus, midbrain, and basal ganglia.[275] Colocalization of ENT1 with adenosine A_1 in human brain suggests that this transporter may regulate the neuromodulatory actions of adenosine.[275] Human ENT1 is involved in the transport of such nucleoside analogs as cladribine and cytarabine used in cancer treatment.[277]

ENT2 (SLC29A2) mRNA is highly expressed in human skeletal muscle as well as other tissues, including pancreas, prostate, kidney, heart, thymus, placenta, and brain,[278] and is localized to the basolateral as well as nuclear membranes of canine kidney cells transfected with plasmid encoding the ENT2 gene.[274] Human ENT2 protein is highly expressed in the cerebellum, thalamus, medulla, midbrain, and brainstem regions, particularly the pons, and is expressed weakly in cerebral cortex and basal ganglia.[275] Because of ENT2 localization in brain regions rich in glial cells, involvement of the transporter with nucleoside transport into the glial cells has been speculated.[261,275] Furthermore, Ent2 has been identified as the primary functional nucleoside transporter in rat C6 glioma cells.[279] In contrast to ENT1, ENT2 protein, when expressed in *Xenopus* oocytes, can transport the antiretroviral nucleoside analog zidovudine.[278,280]

ENT3 (SLC29A3) protein, particularly abundant in the placenta, is widely distributed in human and rodent tissues and some neoplasias. ENT3 gene transcript has been detected in brain of both species.[271] Unlike ENT1, -2, and -4, which are found predominantly at the cell surface, ENT3 is localized intercellularly at the lysosomal membranes and thus may function as a organellar transporter.[281] ENT3 has broad substrate specificity, and the antiviral purines and pyrimidine nucleoside analogs didanosine, zalcitabine, and zidovudine are transported efficiently by this protein.[281]

ENT4 (SLC29A4) protein is expressed ubiquitously in human tissue, particularly in the brain, and is a low-affinity nucleoside transporter at the cell surface.[282] ENT4 transports dopamine, serotonin, and neurotoxin MPP^+ efficiently and is therefore called plasma membrane monoamine transporter (PMAT). This transporter has been proposed to maintain homeostasis of monoamine levels, and under certain conditions might supplement the role of high-affinity transporters such as dopamine, serotonin, and norepinephrine transporters in the brain.[282]

14.3.6. Peptide Transporters

The intestine and kidney play key roles in the absorption and reabsorption of dietary nutrients.[283] Several transporters have been identified at these sites that assist in the uptake of peptides from the small intestine and renal tubular lumen into the blood.[284,285] Peptide transporters (PTs) belong to family 15 of the solute carrier superfamily (SLC15A) of proton-dependent oligopeptide transporters (POT) and consist of four members: PEPT1, PEPT2, PHT1, and PHT2.[286,287] They have been reported in mammals, including human, mice, rat, and rabbit,[288,289] as well as in bacteria, fungi, yeast, and plants.[290] All members of the POT family consist of 12 putative α-helical transmembrane domains with cytosolic N- and C-termini. The extracellular loops can possess two to seven glycosylation sites, while the intracellular loops have various protein kinase A and C phosphorylation sites.[287,289] These symporters couple the uphill movement of small peptides with the downhill movement of protons across biological membranes via an inwardly directed electrochemical H^+-gradient and a negative membrane potential.[291–293] Apart from transporting sequence-independent di- and tripeptides, pharmacologically active peptidelike drugs (β-lactam antibiotics, ACE inhibitors, and antiviral nucleoside analogs), neuropeptides (carnosine, an endogenous dipeptide), and endogenous peptidomimetics (5-aminolevulinic acid) are also transported by these PTs.[286,289,294,295] The brain, including cerebral cortex, cerebellum, and choroid plexus,[296,297] also express peptide transporters that are involved in uptake of substrates from the systemic circulation and CSF into the brain cells. Furthermore, uptake by PTs of enzymatic degradation products of neuropeptides that may still be bioactive at the synapse has been proposed.[296–298]

Peptide transporter 1 (PEPT1, SLC15A1) was first cloned from a rabbit intestinal cDNA library[299] and shown to have high capacity and low affinity for di- and tripeptides. In mammals, this transporter is expressed primarily in the apical plasma membrane of enterocytes in the intestine[300] and is responsible for the absorption of small peptides arising from digestion of dietary proteins.[299] No Pept1 protein was found in the rat brain.[301]

Peptide transporter 2 (PEPT2, SLC15A2) was first cloned from a human kidney cDNA library[302] and shown to have low capacity and high affinity for di- and tripeptides.[303] PEPT2 protein is expressed in various tissues, including lung, mammary gland, retina, and brain.[301,304] In the brain, Pept2 is expressed in rat cerebral cortex but the physiological role of the peptide transporter in the cerebral cortex, is yet to be seen.[297] Pept2 transcript has also been identified in rat astrocytes[304,305] and could contribute to brain glutathione metabolism by providing cysteinylglycine derived from extracellular glutathione[298] or for removing neuroactive peptides such as kyotorphin[306] from the extracellular fluid. At the rat choroid plexus, both protein expression[307] and functional activity[308] have been assessed, and localization of the transporter has been identified at the apical membranes of the choroidal epithelium using immunofluorescent confocal microscopy.[309] However, since the BBB is impermeable to 5-aminolevulinic acid (5-ALA), the presence of this compound, which was inhibited by α-amino-containing cephalosporins but not Pept1 inhibitors in the CSF, could be attributed to expression of either Pept2 or other transporters at the basolateral side of the choroidal epithelium.[307] In studies performed by Shu and co-workers, there was a preferential uptake (K_m of GlySar was 59.6 μM for apical and 1400 μM for basolateral uptake) of the model dipeptide into primary culture of rat choroid plexus epithelial cells.[309] Furthermore, plasma concentrations of 5-ALA were much higher than those found in CSF,[307] suggesting the presence of an active transport system for 5-ALA at the choroid plexus epithelium.[307] Taken together, PEPT2 may have a role as an efflux pump for the effective removal of neuropeptides, peptides, and peptidomimetics from CSF into the blood. Pharmacological inhibition of PEPT2 could maximize CNS penetration of therapeutic peptides that are substrates for this transporter.[309]

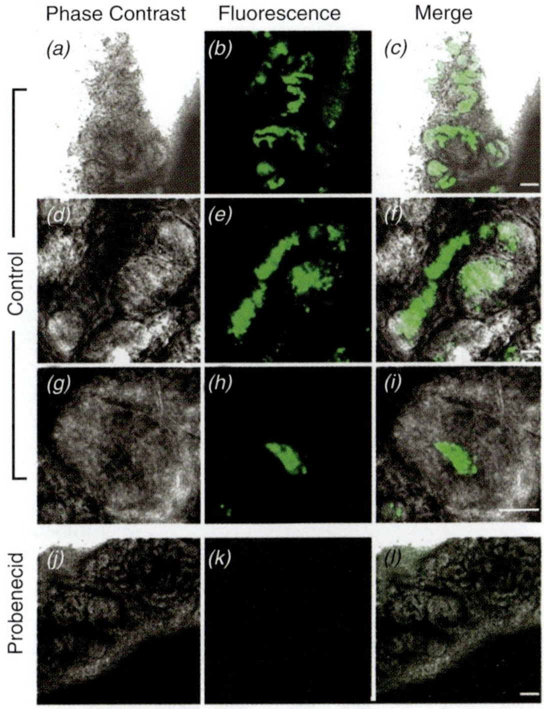

FIGURE 4.4. Confocal microscopic images showing fluorescein accumulation in induced metanephric mesenchyme (MM) due to expression of organic anion transporters. Control; (*a–i*) or presence of 2 mM probenecid (probenecid; *j–l*). (*a, d, g, j*) Phase-contrast photomicrographs of the induced epithelial tubular structures. (*b, e, h, k*) Corresponding fluorescence photomicrograph of the tissue shown in the preceding panel. (*c, f, i, l*) Merged image of the two preceding photomicrographs, indicating the accumulation of fluorescein in the induced MM. (*a–c*) Low-magnification examination of a group of tubular structures in the induced mesenchyme in the absence of probenecid; bar = 100 μM. (*d–f*) High-magnification examination of tubular structures in the mesenchyme induced in the absence of probenecid; bar = 20 μM. (*g–i*) Higher-magnification examination of a tubular structure in the induced mesenchyme in the absence of probenecid; bar = 20 μM. Note the accumulation of fluorescein, to a concentration greater than the medium, in what appears to be a fluid-filled space (presumptive lumen). (*j–l*) Low-magnification examination of a group of tubular structures in the mesenchyme induced in the presence of 2 mM probenecid; bar = 100 μM. Note the absence of concentrative fluorescein accumulation.

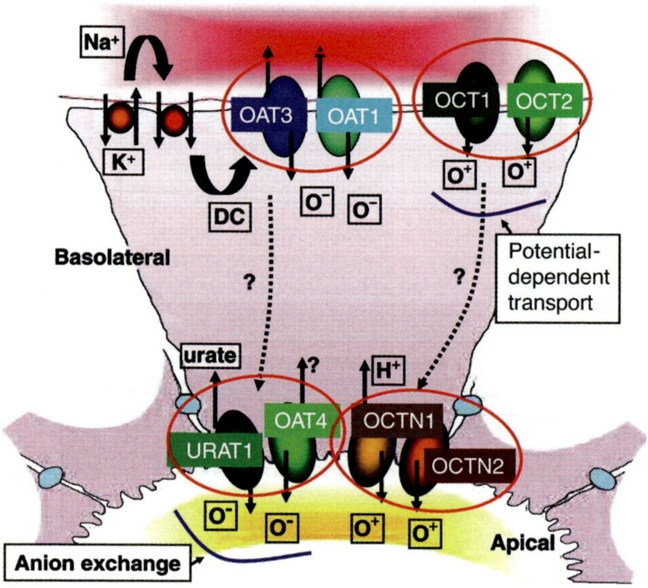

FIGURE 4.6. Mechanisms of proximal tubular uptake and efflux of organic ions mediated by organic anion and cation transporters (OATs, OCTs, OCTNs). Transporters whose encoding genes are chromosomally paired are enclosed in ovals.

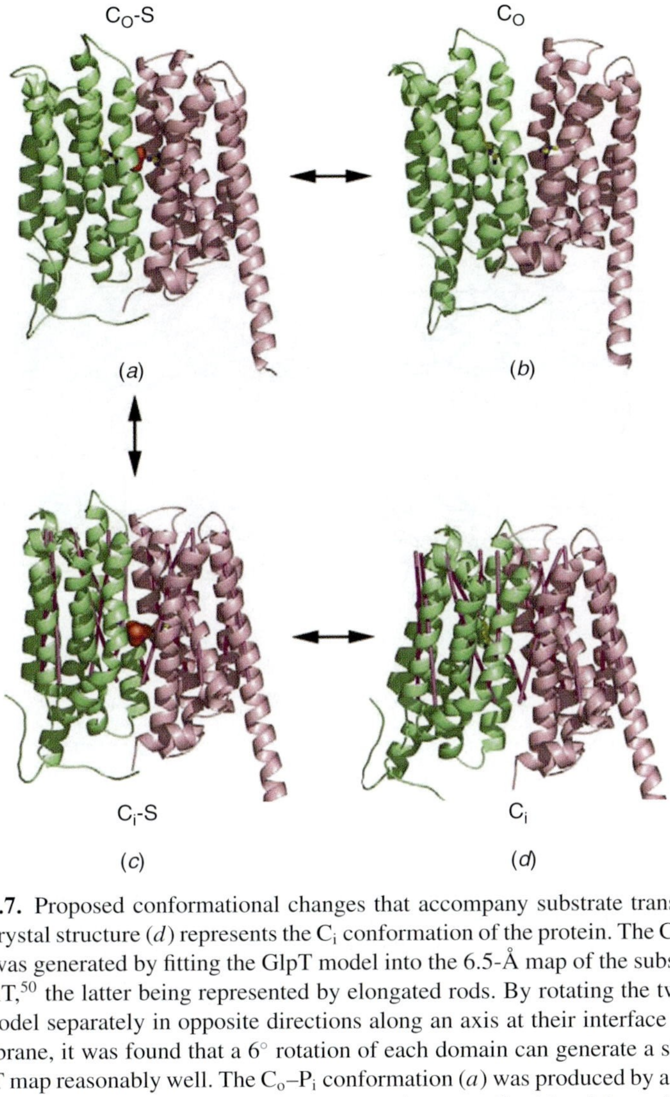

FIGURE 4.7. Proposed conformational changes that accompany substrate translocation by GlpT. The crystal structure (*d*) represents the C_i conformation of the protein. The C_i–P_i conformation (*c*) was generated by fitting the GlpT model into the 6.5-Å map of the substrate-bound form of OxlT,[50] the latter being represented by elongated rods. By rotating the two halves of the GlpT model separately in opposite directions along an axis at their interface and parallel to the membrane, it was found that a 6° rotation of each domain can generate a structure that fits the OxlT map reasonably well. The C_o–P_i conformation (*a*) was produced by a 10° rotation of each domain that is sufficient to close the pore on the cytosolic side of the molecule and to open a pore on the periplasmic side. Finally, the C_o conformation (*b*) was generated by a 16° rotation.

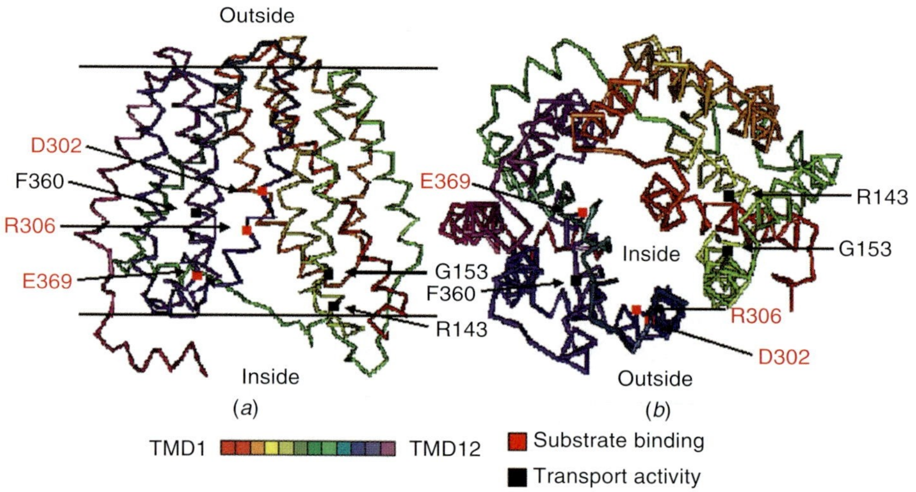

FIGURE 7.2. *(a)* A side view of the hypothetical structure of MCT1 embedded in the plasma membrane is depicted with solid lines indicating the edge of the lipid bilayer. *(b)* MCT1 is shown as in *(a)*, but viewed from the exofacial membrane surface. Each transmembrane domain (TMD) is color coded from the N- (red) to the c-terminus (purple). The positions predicted for the critical amino acids necessary for substrate binding (red) and transport activity (black) are highlighted and numbered according to their position in the protein sequence. This structure was deduced by threading the MCT1 protein sequence onto the crystal structure of the *E. coli* lactose permease A chain using the Cn3D software available from NCBI.

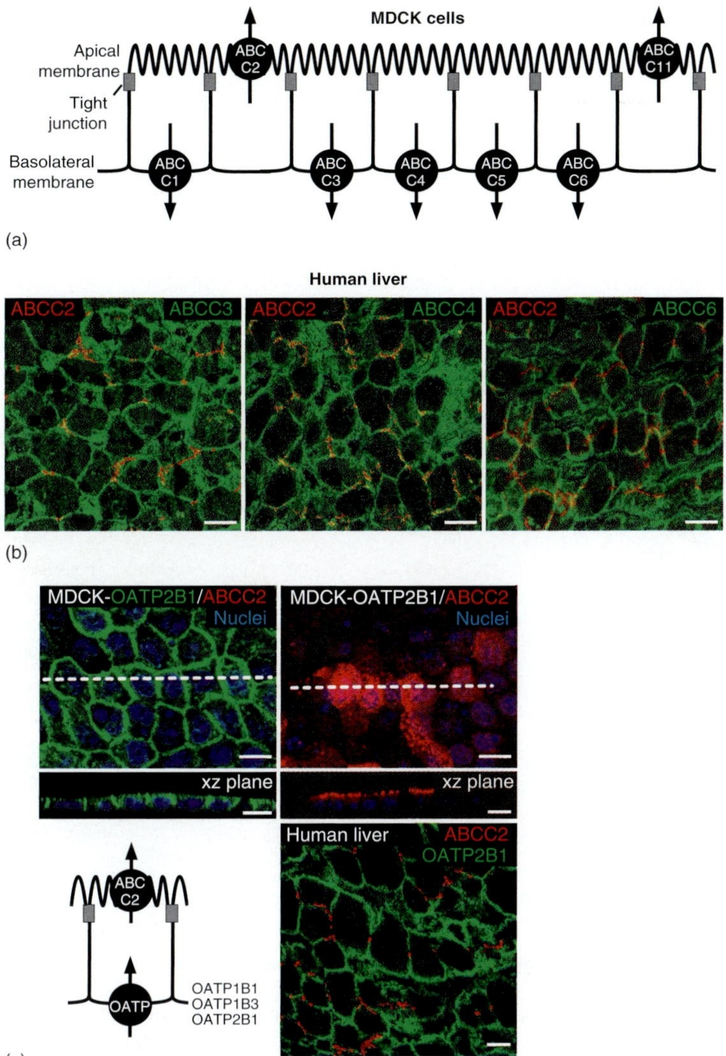

FIGURE 11.2. Subcellular localization of ABCC/MRP efflux pumps. (*a*) Schematic representation of polarized Madin–Darby canine kidney (MDCK) cells recombinantly expressing human ABCC/MRP efflux pumps, which acquire a domain-specific localization either in the apical membrane (ABCC2, ABCC11) or in the basolateral membrane (ABCC3, ABCC4, ABCC5, ABCC6). (*b*) Confocal laser scanning micrographs of ABCC/MRP efflux pumps in human hepatocytes. At least four different ABCC/MRP transporters have been identified in human hepatocytes [i.e., ABCC2 (red) in the canalicular (apical) membrane and ABCC3, ABCC4, and ABCC6 (green) in the sinusoidal (basolateral) membrane]. Bars, 20 μm. (*c*) Confocal laser scanning micrographs of MDCK cells simultaneously expressing recombinant human OATP2B1 (green) as an uptake transporter and ABCC2 (red) as an efflux pump for organic anions. The lines indicate where the optical xz-sections had been taken. These double-transfected cells serve as valuable tools to study the vectorial transport of organic anions that undergo hepatobiliary elimination. In human hepatocytes, OATP2B1 (green) is located in the sinusoidal membrane and ABCC2 (red) in the canalicular membrane. Bars, 10 μm.

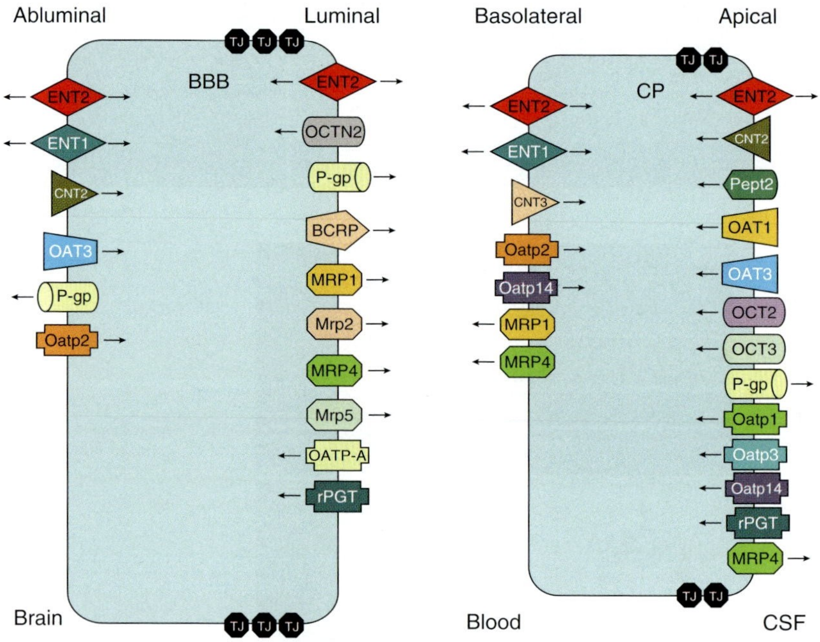

FIGURE 14.1. Xenobiotic transport mechanisms at the brain barriers. The proposed expression of ABC superfamily transporters, organic anion and cation transporters, nucleoside transporters, and peptide transporters is depicted at the BBB endothelium and at the choroid plexus epithelium. Although expression of these transporters has been shown, localization and function remain to be demonstrated. The arrows indicate the direction of substrate transport. References are indicated in the text.

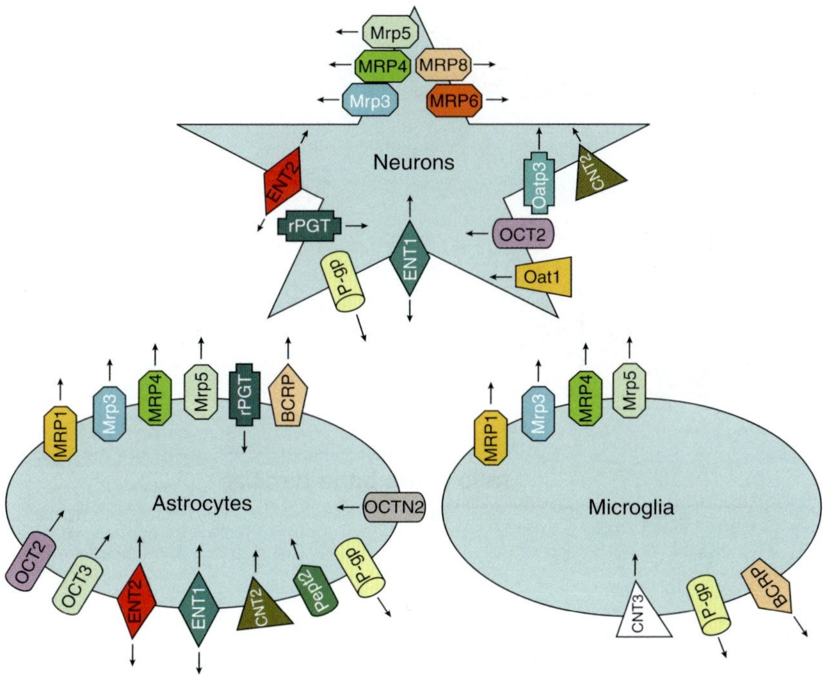

FIGURE 14.2. Xenobiotic transport mechanisms in the brain parenchyma. The proposed expression of ABC superfamily transporters, organic anion and cation transporters, nucleoside transporters, and peptide transporters is depicted in astrocytes, microglia, and neurons. Although expression of these transporters has been shown, localization and function remain to be demonstrated. The localization of membrane drug transporters in oligodendrocytes has not been well characterized and therefore is not included in this figure. The arrows indicate the direction of substrate transport. References are indicated in the text.

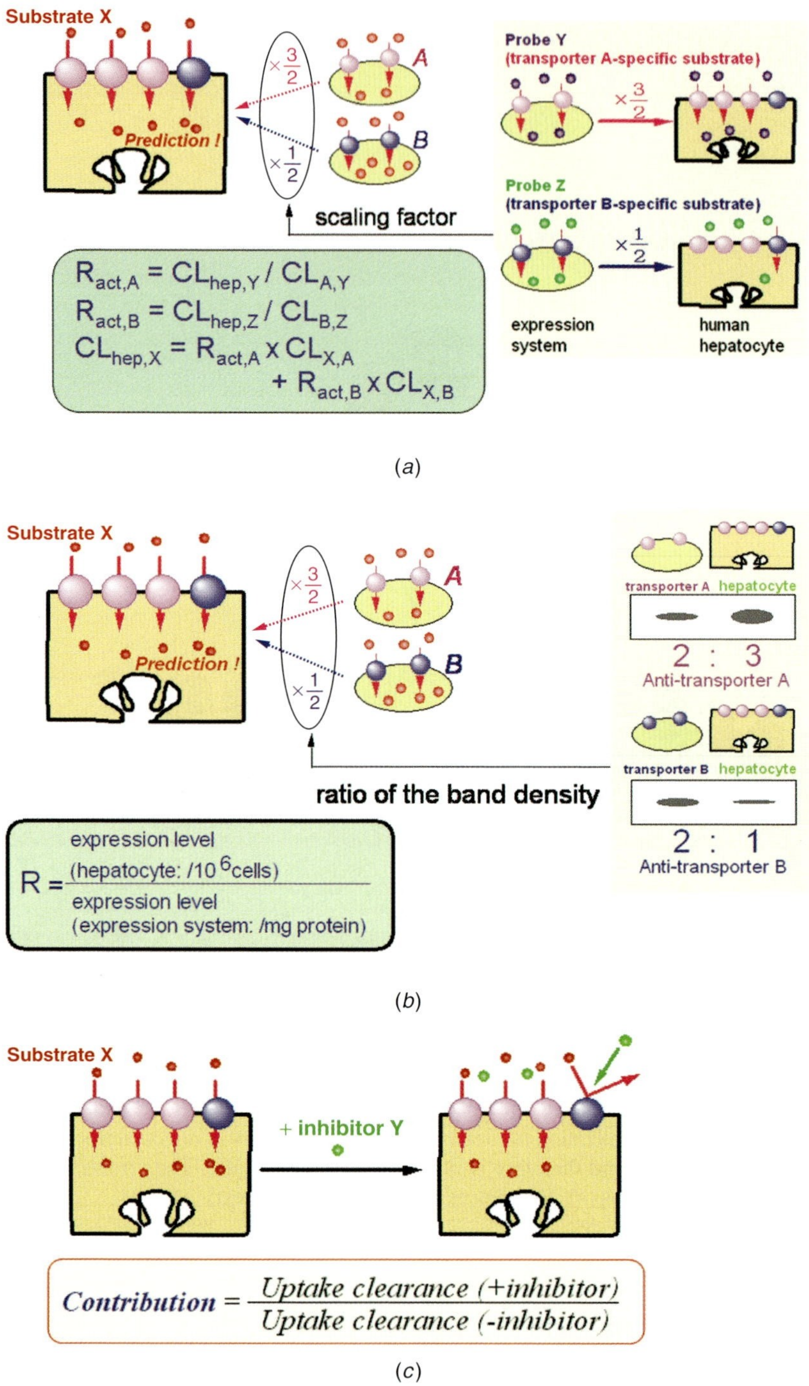

FIGURE 19.2. Schematic diagram of the method for estimating the contribution of each transporter to the overall hepatic uptake: (*a*) using reference compounds; (*b*) using the relative expression levels estimated from Western blot analysis; (*c*) using transporter-specific inhibitors. The details are described in the text.

Both peptide histidine transporter 1 (Pht1, Slc15a4) and peptide histidine transporter 2 (Pht2, Slc15a3) have been cloned from the rat brain cDNA library.[310,311] In situ hybridization studies have also revealed Pht1 localization in several regions of the rat brain, including the hippocampus, cerebellum, and pontine nucleus, and in lower levels of the cerebral cortex, brain stem, thalamus, and hypothalamus.[310] When expressed in *Xenopus* oocytes, rat Pht1 transports histidine and histidine-containing dipeptide, carnosine, in a pH-dependent manner and is inhibited by other di- and tripeptides.[310] On the other hand, in transfected cells, Pht2 acts as a lysosomal histidine and histidylleucine transporter and may play a role in the intracellular trafficking of small peptides.[311] Unfortunately, information on tissue distribution and cellular localization and transport properties of Pht1 and Pht2 is limited.

14.4. RELEVANCE OF DRUG TRANSPORTERS IN DISEASES OF THE CNS

The expression of many different drug transport proteins in the brain implies a highly complex system that regulates the permeation of xenobiotics into the CNS. Many diseases remain refractory to pharmacotherapy, which may be directly related to the expression of drug efflux transporters at the brain barriers and in brain parenchyma. In addition, factors related to the neurological disease and pathological conditions of the CNS may lead to altered drug transporter functional expression. This suggests that drug permeation and distribution in the brain during various disease conditions may be much different than expected. The relationship of drug transport proteins to the pharmacotherapy of neurological disorders such as brain neoplasia, HIV-1 encephalitis, epilepsy, Alzheimer's disease, and Parkinson's disease is discussed in the following section.

14.4.1. Brain Neoplasia

Pharmacological treatment of brain cancer is extremely difficult, possibly due to the limited brain accumulation of chemotherapeutic agents. Many tumors, especially those of neuroepithelial origin (i.e., involving astrocytes, oligodendrocytes, ependymal cells, and the choroid plexus), have been shown to express drug transporters known to regulate the cellular accumulation of antineoplastic drugs. Most current research has focused on the expression and functional activity of ABC superfamily drug efflux transporters in tumor cells and the consequent development of MDR. Tumor cells rapidly develop MDR, which may be either intrinsic or acquired. *Intrinsic* (also known as de novo) *resistance* is present prior to chemotherapy exposure and leads to initial treatment failure. Resistance developed during the course of chemotherapy is known as *acquired MDR*, and these patients often display disease progression even though initially, they responded to treatment.[95] Several lines of evidence support the expression of ABC transporters in brain tumors, although the relationship of transporter expression to tumor grade and cellular origin remains unclear. Pgp expression has been reported in gliomas[312,313] as well as along the tumor microvasculature in both gliomas and glioblastomas.[314,315] In addition, Abe et al. reported MRP1 expression in gliomas isolated from chemotherapy-naive patients.[316] Interestingly, this same study reported that MRP1 expression increased in patients who received aggressive chemotherapy,[316] suggesting that drug treatment may up-regulate the expression of this transporter in brain cancer. Gene and protein expression of several MRP isoforms, including MRP3, MRP4, and MRP5, have been identified in cell lines derived from brain neoplasms.[317] Up-regulation of ABCG2 in brain cancer has also been reported, and it appears to be localized exclusively to the tumor microvasculature.[318] As described earlier in the chapter, many ABC transporter substrates are also commonly used chemotherapeutic agents.[319–321] The expression of ABC transporters in brain tumors may explain, in part, the limited efficacy of pharmacotherapeutic approaches in the treatment of brain cancer.

To date, little is known about the expression of uptake transporters in various types of brain cancer. A recent manuscript has described the mRNA expression of various OATP isoforms, including OATP-A, OATP-B, and OATP-C in isolated human gliomas.[322] The role of these transporters in regulating chemotherapeutic drug distribution in the brain remains to be determined.

14.4.2. HIV-1 Encephalitis

Human immunodeficiency virus type-1 (HIV-1) infection of the brain may lead to HIV-1 encephalitis (HIVE), a chronic neurodegenerative condition characterized by productive viral replication in brain mononuclear phagocytes.[323,324] HIVE may lead to various neurocognitive disorders, including HIV-1-associated dementia (HAD), which is characterized by cognitive and motor dysfunction as well as behavioral abnormalities.[325] The implementation of highly active antiretroviral therapy (HAART) to treat HIV-1 infection has led to a 40 to 50% decrease in the incidence of HAD.[326–328] However, HIV-1-associated neurocognitive disorders, including HAD- and HIV-related sensory neuropathies, continue to be a major cause of morbidity and mortality. In fact, the combined prevalence of HAD-and HIV-related sensory neuropathy is approximately 30 to 50% in patients with advanced HIV-1 infection.[328] Taken together, these observations suggest that HAART may not provide complete protection against the progression of HAD/HIVE.

HIV-1 enters the brain early during the course of infection and infects primarily microglia and to a lesser extent, astrocytes.[324] Once they are infected, these cells may become activated and secrete cytokines (e.g., TNFα, IL-1β, IL-6) and neurotoxins (e.g., arachidonic acid, glutamate, nitric oxide, platelet activating factor, quinolinic acid).[324] Release of these substances may be induced by the binding of HIV-1 viral proteins HIV-1 viral envelope glycoprotein gp120 or HIV-1 transactivator protein (Tat) to chemokine receptors (e.g., CXCR4, CCR5) on the microglia/astrocyte cell surface.[323,324,329] When high concentrations of these substances are present in the brain, neurotoxicity may result due to overactivation of neuronal receptors [e.g., *N*-methyl-D-aspartate receptors (NMDARs)].

Several studies have demonstrated that many neurotoxic mediators released during brain HIV-1 infection, particularly cytokines, can alter drug transporter expression. Studies using isolated rat hepatocytes[330] and isolated rat brain capillaries[331] have shown that cytokines (e.g., TNFα, IL-1β, IL-6) can alter Pgp molecular expression and functional activity. Using a human hepatoma cell line (HepG2), Lee and Piquette-Miller (2003) demonstrated that treatment with IL-1β and IL-6 can significantly increase both MRP1 gene expression and functional activity.[332] Therefore, it is not unexpected that cytokine secretion during HIV-1 infection can lead to altered drug transporter expression and changes in antiretroviral drug permeation and distribution in the CNS. In fact, our laboratory has recently shown increased cytokine secretion in the presence of the HIV-1 viral envelope glycoprotein gp120 and that these cytokines are involved in the regulation of Pgp expression in cultured rat astrocytes.[333]

In addition to the secretion of inflammatory cytokines, other studies have shown that the HIV-1 virus and its related proteins can also alter the functional expression of drug transporters. An early study reported that Pgp expression was up-regulated upon HIV-1 infection of both a T-cell and monocytic cell line.[334] Andreana and colleagues also reported increased Pgp expression in peripheral blood mononuclear cells (PBMCs) isolated from HIV-infected patients.[335] These observations are somewhat controversial, as other studies have shown that various subsets of PBMCs may actually express lower levels of Pgp in HIV-1-infected patients compared to healthy volunteers.[336,337] The fact that soluble viral proteins (e.g., gp120, Tat) are secreted during active HIV-1 infection has generated significant interest. Recently, Tat has been shown to increase in vitro both Pgp and MRP1 expression at the BBB.[338,339] Studies in our laboratory have shown that Pgp expression and functional activity are decreased significantly in cultured astrocytes treated with gp120.[333]

Although implementation of HAART has been successful in reducing systemic viral load in HIV-1-seropositive patients, HIVE remains refractory to drug therapy, possibly due to the poor permeation of antiretroviral drugs into the brain. For example, antiretroviral drugs, particularly HIV-1 protease inhibitors, penetrate the CNS poorly.[340] One possible mechanism for the low brain concentrations of antiretroviral drugs is the expression of ABC transporters (e.g., Pgp, MRPs, ABCG2) at the brain barriers and in brain cellular targets of HIV-1 infection (e.g., microglia, astrocytes). HIV-1 protease inhibitors are well-known transport substrates for both Pgp and MRPs.[50,76,341,342] In addition, nucleotide reverse transcriptase inhibitors (e.g., PMEA) have been shown to interact with Mrp4 and Mrp5 in cultured rat microglia.[82] Zidovudine monophosphate and abacavir have also been shown to be MRP4 substrates,[79,343] whereas stavudine is an MRP5 substrate.[343] Additionally, MRP8 has been shown to be involved in the cellular efflux of PMEA and zalcitabine but not zidovudine or lamivudine.[344] Antiretroviral agents that have been shown to be substrates for ABCG2 include the nucleoside reverse transcriptase inhibitors zidovudine and stavudine.[345,346] In contrast, HIV-1 protease inhibitors (e.g., ritonavir, saquinavir, nelfinavir) have been shown to be inhibitors but not substrates of ABCG2.[347] Other studies suggest that antiretroviral drugs may be responsible for increased Pgp levels in HIV-infected cells. Several groups have reported that HIV-1 protease inhibitors (e.g., saquinavir, ritonavir, nelfinavir, amprenavir) and nonnucleoside reverse transcriptase inhibitors (e.g., efavirenz) may induce significant Pgp expression in peripheral immune and intestinal cells that have been cultured in these drug solutions for a period of time.[337,348–350] In contrast, other researchers have reported that HIV-1 protease inhibitors do not alter Pgp expression and suggest that drug resistance is associated with mutations in the HIV-1 protease enzyme rather than an overexpression of Pgp.[351] Similarly, studies have shown that the induction of drug resistance by selection with increasing concentrations of zidovudine resulted in cells that were resistant to zidovudine but did not express detectable amounts of Pgp.[352] It has been proposed that resistance to zidovudine is associated primarily with mutations in HIV reverse transcriptase and not Pgp overexpression.[352]

To date, a few studies have described uptake systems that may be involved in anti-HIV drug transport in the brain. Our laboratory has reported the functional expression of a Na^+-dependent thymidine transporter[248] as well as an organic cation transporter-like uptake system for zidovudine in cultured rat microglia.[353] Furthermore, nucleoside transporters such as hCNT1 have been implicated in the low-affinity uptake of zidovudine and stavudine.[354]

14.4.3. Epilepsy

Epilepsy is characterized by recurrent seizures and is one of the most common neurological disorders worldwide.[355] Approximately one-third of epileptic patients remain pharmacoresistant despite significant advances in antiepileptic drug therapy.[356] The consequences of uncontrolled epilepsy are severe and include decreased life span, bodily injury, neuropsychological impairment, and social disability.[357] Resistant patients often do not respond to drug therapy with several, if not all, antiepileptics, even though these compounds may act by different pharmacological mechanisms.[358] Interestingly, the blood concentrations of antiepileptic drugs in resistant patients usually fall within the normal therapeutic range, suggesting an intrinsic mechanism in the brain that limits the accumulation of antiepileptic drugs.

Tishler and colleagues were the first group to describe the role of drug transporters in the development of pharmacoresistant epilepsy.[359] They reported overexpression of Pgp in brain tissue isolated from pharmacoresistant patients, which implies that the development of MDR may also be relevant to the treatment of epilepsy. Pgp expression has also been reported in the capillary endothelium in brain tissue of patients with intractable epilepsy.[360] In fact, the brain expression of Pgp may be up-regulated in drug-resistant epilepsy. For example, Rizzi et al. induced seizures in rats by kainate injection and observed increased Pgp expression in the limbic system after repeated seizure activity.[361] Interestingly, exposure to antiepileptic drugs did not increase Pgp expression in rats, suggesting that Pgp upregulation is mediated by factors related to seizure activity itself.[361] MRP1 expression has also been detected in several brain malformations characteristic of refractory epilepsy, including dysplastic neurons, reactive astrocytes, and the glial element of focal cortical dysplasia (i.e., balloon cells).[362,363] In addition, up-regulation of MRP2 and MRP5 mRNA transcripts and protein has been reported in medically intractable epilepsy.[360,364] Interestingly, the protein expression of ABCG2 has also been reported to increase in experimental rat models of status epilepticus.[77] However, the effect of increased ABCG2 expression in the development of resistance to antiepileptic drugs remains to be determined. In addition, studies examining the involvement of various influx transport systems on the brain permeation of antiepileptic drugs have not been described. These transporters (e.g., OATs, OCTs, OATPs, peptide transporters, nucleoside transporters) may represent novel therapeutic targets for enhancing delivery of antiepileptic compounds to the brain.

14.4.4. Alzheimer's Disease

Alzheimer's disease, a progressive neurological disorder, affects approximately 15 million people worldwide and is the most common cause of dementia in the elderly.[365] The pathological features of Alzheimer's disease are neurofibrillary tangles and the formation of neurite plaques, which may be focal or diffuse. The principal component of the neurite plaques is β-amyloid protein, which is believed to induce inflammatory processes characterized by microgliosis and astrocytosis.[366] Activated microglia and astrocytes secrete neurotoxic mediators that may lead to neuronal cell injury and death. Neuronal cell death in Alzheimer's disease may result in several clinical manifestations, including impaired memory as well as cognitive and behavioral dysfunction.[367]

To date, there is very little information on the role of drug transporters in the pathogenesis and treatment of Alzheimer's disease. A few studies have demonstrated that Pgp may play a protective role against the onset of Alzheimer's disease. Using Pgp- enriched plasma membrane vesicles, Lam and colleagues observed the ATP-dependent transport of β-amyloid protein, suggesting that this protein is in fact a Pgp substrate.[368] More recently, an inverse correlation was observed between vascular Pgp expression and deposition of β-amyloid protein in the brain parenchyma, suggesting that Pgp may be involved prominently in the brain clearance of β-amyloid protein.[369] Interestingly, simvastatin and pravastatin, two 3-hydroxy-3-methylglutarylcoenzyme A (HMG-CoA) reductase inhibitors, have been shown to reduce intra- and extracellular levels of β-amyloid peptides in primary cultures of hippocampal neurons, although a definitive correlation with Pgp has not been demonstrated clearly.[370] Taken together, these data imply that the use of therapeutic strategies to increase brain Pgp expression may be useful in the prevention and treatment of Alzheimer's disease.

14.4.5. Parkinson's Disease

Parkinson's disease is a common neurodegenerative disorder characterized by bradykinesia and akinesia (i.e., slowness or inability to initiate voluntary movement), muscle rigidity, and tremor at rest. It has an estimated prevalence ranging between 70 and 170 per 100,000 persons in North America.[371] Parkinson's disease is a syndrome rather than a specific disease and may result from either familial or sporadic etiology. In sporadic Parkinson's disease, environmental factors such as pesticide and well water exposure, rural living, and various drug contaminants (i.e., MPP^+) have been implicated.[372,373] The association of these environmental factors with the onset of Parkinson's disease has led to a significant interest in transporters that may be involved in the uptake and distribution of these toxins in the brain. For example, it has been suggested that genetic polymorphisms in the MDR1 gene may represent a risk factor for Parkinson's disease since Pgp is involved in the active efflux of toxins from the brain. In fact, studies using mdr1a(−/−) mice have shown that Pgp is prominently involved in limiting the accumulation of toxic pesticides in the CNS.[374] Nonetheless, no published research has directly examined the role of drug transporters in either the pathogenesis and/or brain distribution of anti-Parkinsonian drugs.

14.5. CONCLUSIONS

The dynamic and highly controlled environment of the brain is regulated, in part, by the blood–brain and blood–CSF barriers. Within this intricate environment exists the cellular components of the brain parenchyma (i.e., astrocytes, microglia, oligodendrocytes, and neurons). Each brain cellular compartment possesses a specific and selective set of metabolic enzymes, receptor proteins, and secretory factors that serve to maintain the homeostatic environment required for normal brain physiology. In addition, the localization and expression of putative drug transporters in these barriers play a critical role in the influx and efflux of numerous endogeneous and exogeneous substrates, which ultimately exert a significant impact in the overall pharmacokinetic and pharmacodynamic profile of drugs in the brain. The localization and functional expression of influx transporters (e.g., OATs, OCTs, OATPs, NTs, PTs) as well as the efflux transporters (e.g., Pgp, MRPs, ABCG2) at the brain barriers and within the brain parenchyma suggest that complex drug–transporter interactions may occur during the pharmacotherapy of CNS disease. Further investigation is required to characterize drug transporters at the parenchymal barrier sites so as to fully understand their clinical role.

REFERENCES

1. Lee G, Bendayan R. 2004. Functional expression and localization of P-glycoprotein in the central nervous system: relevance to the pathogenesis and treatment of neurological disorders. Pharm Res 21:1313–1330.
2. Lai CH, Kuo KH, Leo JM. 2005. Critical role of actin in modulating BBB permeability. Brain Res Brain Res Rev 50:7–13.
3. Janzer RC, Raff MC. 1987. Astrocytes induce blood–brain barrier properties in endothelial cells. Nature 325:253–257.
4. Abbott NJ. 2002. Astrocyte-endothelial interactions and blood–brain barrier permeability. J Anat 200:629–638.
5. Reese TS, Karnovsky MJ. 1967. Fine structural localization of a blood–brain barrier to exogenous peroxidase. J Cell Biol 34:207–217.
6. Butt AM, Jones HC, Abbott NJ. 1990. Electrical resistance across the blood–brain barrier in anaesthetized rats: a developmental study. J Physiol 429:47–62.

7. Lee G, Dallas S, Hong M, Bendayan R. 2001. Drug transporters in the central nervous system: brain barriers and brain parenchyma considerations. Pharmacol Rev 53: 569–596.
8. Farrell CL, Pardridge WM. 1991. Blood–brain barrier glucose transporter is asymmetrically distributed on brain capillary endothelial lumenal and ablumenal membranes: an electron microscopic immunogold study. Proc Natl Acad Sci U S A 88:5779–5783.
9. Broadwell RD. 1989. Transcytosis of macromolecules through the blood–brain barrier: a cell biological perspective and critical appraisal. Acta Neuropathol (Berl) 79:117–128.
10. Abbott NJ. 2005. Dynamics of CNS barriers: evolution, differentiation, and modulation. Cell Mol Neurobiol 25:5–23.
11. Vorbrodt AW. 1988. Ultrastructural cytochemistry of blood–brain barrier endothelia. Prog Histochem Cytochem 18:1–99.
12. Rippe B, Haraldsson B. 1994. Transport of macromolecules across microvascular walls: the two-pore theory. Physiol Rev 74:163–219.
13. Broadwell RD, Baker-Cairns BJ, Friden PM, Oliver C, Villegas JC. 1996. Transcytosis of protein through the mammalian cerebral epithelium and endothelium. III. Receptor-mediated transcytosis through the blood–brain barrier of blood-borne transferrin and antibody against the transferrin receptor. Exp Neurol 142:47–65.
14. Simionescu M, Gafencu A, Antohe F. 2002. Transcytosis of plasma macromolecules in endothelial cells: a cell biological survey. Microsc Res Tech 57:269–288.
15. Schnitzer JE, Oh P, Pinney E, Allard J. 1994. Filipin-sensitive caveolae-mediated transport in endothelium: reduced transcytosis, scavenger endocytosis, and capillary permeability of select macromolecules. J Cell Biol 127:1217–1232.
16. Robenek MJ, Schlattmann K, Zimmer KP, Plenz G, Troyer D, Robenek H. 2003. Cholesterol transporter caveolin-1 transits the lipid bilayer during intracellular cycling. FASEB J 17:1940–1942.
17. Ghersi-Egea JF, Strazielle N. 2001. Brain drug delivery, drug metabolism, and multidrug resistance at the choroid plexus. Microsc Res Tech 52:83–88.
18. Davson H, Hollingsworth G, Segal MB. 1970. The mechanism of drainage of the cerebrospinal fluid. Brain 93:665–678.
19. Saunders NR, Habgood MD, Dziegielewska KM. 1999. Barrier mechanisms in the brain, I. Adult brain. Clin Exp Pharmacol Physiol 26:11–19.
20. Felgenhauer K. 1974. Protein size and cerebrospinal fluid composition. Klin Wochenschr 52:1158–1164.
21. Azzi G, Bernaudin JF, Bouchaud C, Bellon B, Fleury-Feith J. 1990. Permeability of the normal rat brain, spinal cord and dorsal root ganglia microcirculations to immunoglobulins G. Biol Cell 68:31–36.
22. Spector R, Johanson CE. 1989. The mammalian choroid plexus. Sci Am 261:68–74.
23. Johanson CE, Sweeney SM, Parmelee JT, Epstein MH. 1990. Cotransport of sodium and chloride by the adult mammalian choroid plexus. Am J Physiol 258:C211–C216.
24. Garner C, Brown PD. 1992. Two types of chloride channel in the apical membrane of rat choroid plexus epithelial cells. Brain Res 591:137–145.
25. Ludwig M, Pittman QJ. 2003. Talking back: dendritic neurotransmitter release. Trends Neurosci 26:255–261.

26. Peters A, Josephson K, Vincent SL. 1991. Effects of aging on the neuroglial cells and pericytes within area 17 of the rhesus monkey cerebral cortex. Anat Rec 229:384–398.
27. Walz W. 2000. Controversy surrounding the existence of discrete functional classes of astrocytes in adult gray matter. GLIA 31:95–103.
28. Anderson CM, Swanson RA. 2000. Astrocyte glutamate transport: review of properties, regulation, and physiological functions. GLIA 32:1–14.
29. Bender AS, Reichelt W, Norenberg MD. 2000. Characterization of cystine uptake in cultured astrocytes. Neurochem Int 37:269–276.
30. Speth C, Dierich MP, Sopper S. 2005. HIV-infection of the central nervous system: the tightrope walk of innate immunity. Mol Immunol 42:213–228.
31. Wilson R, Brophy PJ. 1989. Role for the oligodendrocyte cytoskeleton in myelination. J Neurosci Res 22:439–448.
32. Hao C, Richardson A, Fedoroff S. 1991. Macrophage-like cells originate from neuroepithelium in culture: characterization and properties of the macrophage-like cells. Int J Dev Neurosci 9:1–14.
33. Fedoroff S, Zhai R, Novak JP. 1997. Microglia and astroglia have a common progenitor cell. J Neurosci Res 50:477–486.
34. Kaur C, Hao AJ, Wu CH, Ling EA. 2001. Origin of microglia. Microsc Res Tech 54:2–9.
35. Dalmau I, Vela JM, Gonzalez B, Finsen B, Castellano B. 2003. Dynamics of microglia in the developing rat brain. J Comp Neurol 458:144–157.
36. Dickson DW, Mattiace LA, Kure K, Hutchins K, Lyman WD, Brosnan CF. 1991. Microglia in human disease, with an emphasis on acquired immune deficiency syndrome. Lab Invest 64:135–156.
37. Spector R. 1990. Drug transport in the central nervous system: role of carriers. Pharmacology 40:1–7.
38. Leslie EM, Deeley RG, Cole SP. 2001. Toxicological relevance of the multidrug resistance protein 1, MRP1 (ABCC1) and related transporters. Toxicology 167:3–23.
39. Stefkova J, Poledne R, Hubacek JA. 2004. ATP-binding cassette (ABC) transporters in human metabolism and diseases. Physiol Res 53:235–243.
40. Gottesman MM, Hrycyna CA, Schoenlein PV, Germann UA, Pastan I. 1995. Genetic analysis of the multidrug transporter. Annu Rev Genet 29:607–649.
41. Ling V, Thompson LH. 1974. Reduced permeability in CHO cells as a mechanism of resistance to colchicine. J Cell Physiol 83:103–116.
42. Juliano RL, Ling V. 1976. A surface glycoprotein modulating drug permeability in Chinese hamster ovary cell mutants. Biochim Biophys Acta 455:152–162.
43. Chen CJ, Chin JE, Ueda K, Clark DP, Pastan I, Gottesman MM, Roninson IB. 1986. Internal duplication and homology with bacterial transport proteins in the mdr1 (P-glycoprotein) gene from multidrug-resistant human cells. Cell 47:381–389.
44. Roninson IB, Chin JE, Choi KG, Gros P, Housman DE, Fojo A, Shen DW, Gottesman MM, Pastan I. 1986. Isolation of human mdr DNA sequences amplified in multidrug-resistant KB carcinoma cells. Proc Natl Acad Sci U S A 83:4538–4542.
45. Smit JJ, Schinkel AH, Oude Elferink RP, Groen AK, Wagenaar E, van Deemter L, Mol CA, Ottenhoff R, van der Lugt NM, van Roon MA. 1993. Homozygous disruption of the

murine mdr2 P-glycoprotein gene leads to a complete absence of phospholipid from bile and to liver disease. Cell 75:451–462.

46. Ueda K, Cardarelli C, Gottesman MM, Pastan I. 1987. Expression of a full-length cDNA for the human "MDR1" gene confers resistance to colchicine, doxorubicin, and vinblastine. Proc Natl Acad Sci U S A 84:3004–3008.
47. Sun J, He ZG, Cheng G, Wang SJ, Hao XH, Zou MJ. 2004. Multidrug resistance P-glycoprotein: crucial significance in drug disposition and interaction. Med Sci Monit 10:RA5–RA14.
48. Demeule M, Regina A, Jodoin J, Laplante A, Dagenais C, Berthelet F, Moghrabi A, Beliveau R. 2002. Drug transport to the brain: key roles for the efflux pump P-glycoprotein in the blood–brain barrier. Vascul Pharmacol 38:339–348.
49. Cisternino S, Rousselle C, Debray M, Scherrmann JM. 2004. In situ transport of vinblastine and selected P-glycoprotein substrates: implications for drug–drug interactions at the mouse blood–brain barrier. Pharm Res 21:1382–1389.
50. Ronaldson PT, Lee G, Dallas S, Bendayan R. 2004. Involvement of P-glycoprotein in the transport of saquinavir and indinavir in rat brain microvessel endothelial and microglia cell lines. Pharm Res 21:811–818.
51. Tahara H, Kusuhara H, Fuse E, Sugiyama Y. 2005. P-glycoprotein plays a major role in the efflux of fexofenadine in the small intestine and blood–brain barrier, but only a limited role in its biliary excretion. Drug Metab Dispos 33:963–968.
52. Hsiao P, Sasongko L, Link JM, Mankoff DA, Muzi M, Collier AC, Unadkat JD. 2006. Verapamil P-glycoprotein transport across the rat blood–brain barrier: cyclosporine, a concentration inhibition analysis, and comparison with human data. J Pharmacol Exp Ther 317:704–710.
53. Ohashi R, Kamikozawa Y, Sugiura M, Fukuda H, Yabuuchi H, Tamai I. 2006. Effect of P-glycoprotein on intestinal absorption and brain penetration of antiallergic agent bepotastine besilate. Drug Metab Dispos 34:793–799.
54. Kim RB, Fromm MF, Wandel C, Leake B, Wood AJ, Roden DM, Wilkinson GR. 1998. The drug transporter P-glycoprotein limits oral absorption and brain entry of HIV-1 protease inhibitors. J Clin Invest 101:289–294.
55. Washington CB, Wiltshire HR, Man M, Moy T, Harris SR, Worth E, Weigl P, Liang Z, Hall D, Marriott L, Blaschke TF. 2000. The disposition of saquinavir in normal and P-glycoprotein deficient mice, rats, and in cultured cells. Drug Metab Dispos 28:1058–1062.
56. Rao VV, Dahlheimer JL, Bardgett ME, Snyder AZ, Finch RA, Sartorelli AC, Piwnica-Worms D. 1999. Choroid plexus epithelial expression of MDR1 P glycoprotein and multidrug resistance-associated protein contribute to the blood-cerebrospinal-fluid–drug permeability barrier. Proc Natl Acad Sci U S A 96:3900–3905.
57. Beaulieu E, Demeule M, Ghitescu L, Beliveau R. 1997. P-Glycoprotein is strongly expressed in the luminal membranes of the endothelium of blood vessels in the brain. Biochem J 326(Pt 2):539–544.
58. Drion N, Risede P, Cholet N, Chanez C, Scherrmann JM. 1997. Role of P-170 glycoprotein in colchicine brain uptake. J Neurosci Res 49:80–88.
59. Virgintino D, Robertson D, Errede M, Benagiano V, Girolamo F, Maiorano E, Roncali L, Bertossi M. 2002. Expression of P-glycoprotein in human cerebral cortex microvessels. J Histochem Cytochem 50:1671–1676.

60. Golden PL, Pardridge WM. 1999. P-Glycoprotein on astrocyte foot processes of unfixed isolated human brain capillaries. Brain Res 819:143–146.
61. Schlachetzki F, Pardridge WM. 2003. P-Glycoprotein and caveolin-1alpha in endothelium and astrocytes of primate brain. Neuroreport 14:2041–2046.
62. Bendayan R, Ronaldson PT, Gingras D, Bendayan M. 2006. In situ localization of P-glycoprotein (ABCB1) in human and rat brain. J Histochem Cytochem 54:1159–1167.
63. Lee G, Schlichter L, Bendayan M, Bendayan R. 2001. Functional expression of P-glycoprotein in rat brain microglia. J Pharmacol Exp Ther 299:204–212.
64. Ronaldson PT, Bendayan M, Gingras D, Piquette-Miller M, Bendayan R. 2004. Cellular localization and functional expression of P-glycoprotein in rat astrocyte cultures. J Neurochem 89:788–800.
65. Bendayan R, Lee G, Bendayan M. 2002. Functional expression and localization of P-glycoprotein at the blood–brain barrier. Microsc Res Tech 57:365–380.
66. Decleves X, Fajac A, Lehmann-Che J, Tardy M, Mercier C, Hurbain I, Laplanche JL, Bernaudin JF, Scherrmann JM. 2002. Molecular and functional MDR1-Pgp and MRPs expression in human glioblastoma multiforme cell lines. Int J Cancer 98:173–180.
67. Jodoin J, Demeule M, Fenart L, Cecchelli R, Farmer S, Linton KJ, Higgins CF, Beliveau R. 2003. P-Glycoprotein in blood–brain barrier endothelial cells: interaction and oligomerization with caveolins. J Neurochem 87:1010–1023.
68. Borst P, Evers R, Kool M, Wijnholds J. 2000. A family of drug transporters: the multidrug resistance-associated proteins. J Natl Cancer Inst 92:1295–1302.
69. Dallas S, Miller DS, Bendayan R. 2006. Multidrug resistance-associated proteins: expression and function in the central nervous system. Pharmacol Rev 58:140–161.
70. Kruh GD, Belinsky MG. 2003. The MRP family of drug efflux pumps. Oncogene 22:7537–7552.
71. Bakos E, Evers R, Szakacs G, Tusnady GE, Welker E, Szabo K, de Haas M, van Deemter L, Borst P, Varadi A, Sarkadi B. 1998. Functional multidrug resistance protein (MRP1) lacking the N-terminal transmembrane domain. J Biol Chem 273:32167–32175.
72. Dallas S, Miller DS, Bendayan R. 2006. Multidrug resistance-associated proteins: expression and function in the central nervous system. Pharmacol Rev 58:140–161.
73. Cole SP, Bhardwaj G, Gerlach JH, Mackie JE, Grant CE, Almquist KC, Stewart AJ, Kurz EU, Duncan AM, Deeley RG. 1992. Overexpression of a transporter gene in a multidrug-resistant human lung cancer cell line. Science 258:1650–1654.
74. Srinivas RV, Middlemas D, Flynn P, Fridland A. 1998. Human immunodeficiency virus protease inhibitors serve as substrates for multidrug transporter proteins MDR1 and MRP1 but retain antiviral efficacy in cell lines expressing these transporters. Antimicrob Agents Chemother 42:3157–3162.
75. Jones K, Bray PG, Khoo SH, Davey RA, Meaden ER, Ward SA, Back DJ. 2001. P-Glycoprotein and transporter MRP1 reduce HIV protease inhibitor uptake in CD4 cells: potential for accelerated viral drug resistance? AIDS 15:1353–1358.
76. Dallas S, Ronaldson PT, Bendayan M, Bendayan R. 2004. Multidrug resistance protein 1-mediated transport of saquinavir by microglia. Neuroreport 15:1183–1186.
77. van Vliet EA, Redeker S, Aronica E, Edelbroek PM, Gorter JA. 2005. Expression of multidrug transporters MRP1, MRP2, and BCRP shortly after status epilepticus, during the latent period, and in chronic epileptic rats. Epilepsia 46:1569–1580.

78. Potschka H, Fedrowitz M, Loscher W. 2003. Multidrug resistance protein MRP2 contributes to blood–brain barrier function and restricts antiepileptic drug activity. J Pharmacol Exp Ther 306:124–131.

79. Schuetz JD, Connelly MC, Sun D, Paibir SG, Flynn PM, Srinivas RV, Kumar A, Fridland A. 1999. MRP4: a previously unidentified factor in resistance to nucleoside-based antiviral drugs. Nat Med 5:1048–1051.

80. Jedlitschky G, Burchell B, Keppler D. 2000. The multidrug resistance protein 5 functions as an ATP-dependent export pump for cyclic nucleotides. J Biol Chem 275:30069–30074.

81. Chen ZS, Lee K, Kruh GD. 2001. Transport of cyclic nucleotides and estradiol 17-beta-D-glucuronide by multidrug resistance protein 4. Resistance to 6-mercaptopurine and 6-thioguanine. J Biol Chem 276:33747–33754.

82. Dallas S, Schlichter L, Bendayan R. 2004. Multidrug resistance protein (MRP) 4- and MRP 5-mediated efflux of 9-(2-phosphonylmethoxyethyl)adenine by microglia. J Pharmacol Exp Ther 309:1221–1229.

83. Dallas S, Miller DS, Bendayan R. 2006. Multidrug resistance-associated proteins: expression and function in the central nervous system. Pharmacol Rev 58:140–161.

84. Zhang Y, Han H, Elmquist WF, Miller DW. 2000. Expression of various multidrug resistance-associated protein (MRP) homologues in brain microvessel endothelial cells. Brain Res 876:148–153.

85. Nies AT, Jedlitschky G, König J, Herold-Mende C, Steiner HH, Schmitt HP, Keppler D. 2004. Expression and immunolocalization of the multidrug resistance proteins, MRP1-MRP6 (ABCC1-ABCC6), in human brain. Neuroscience 129:349–360.

86. Miller DS, Nobmann SN, Gutmann H, Toeroek M, Drewe J, Fricker G. 2000. Xenobiotic transport across isolated brain microvessels studied by confocal microscopy. Mol Pharmacol 58:1357–1367.

87. Berezowski V, Landry C, Dehouck MP, Cecchelli R, Fenart L. 2004. Contribution of glial cells and pericytes to the mRNA profiles of P-glycoprotein and multidrug resistance-associated proteins in an in vitro model of the blood–brain barrier. Brain Res 1018: 1–9.

88. Leggas M, Adachi M, Scheffer GL, Sun D, Wielinga P, Du G, Mercer KE, Zhuang Y, Panetta JC, Johnston B, et al. 2004. Mrp4 confers resistance to topotecan and protects the brain from chemotherapy. Mol Cell Biol 24:7612–7621.

89. Zhang Y, Schuetz JD, Elmquist WF, Miller DW. 2004. Plasma membrane localization of multidrug resistance–associated protein homologs in brain capillary endothelial cells. J Pharmacol Exp Ther 311:449–455.

90. Kusch-Poddar M, Drewe J, Fux I, Gutmann H. 2005. Evaluation of the immortalized human brain capillary endothelial cell line BB19 as a human cell culture model for the blood–brain barrier. Brain Res 1064:21–31.

91. Dallas S, Miller DS, Bendayan R. 2006. Multidrug resistance-associated Proteins (MRPs): Expression and function in the central nervous system. Pharmacol Rev 58:140–161.

92. Choudhuri S, Cherrington NJ, Li N, Klaassen CD. 2003. Constitutive expression of various xenobiotic and endobiotic transporter mRNAs in the choroid plexus of rats. Drug Metab Dispos 31:1337–1345.

93. Soontornmalai A, Vlaming ML, Fritschy JM. 2006. Differential, strain-specific cellular and subcellular distribution of multidrug transporters in murine choroid plexus and blood–brain barrier. Neuroscience 138:159–169.

94. Decleves X, Regina A, Laplanche JL, Roux F, Boval B, Launay JM, Scherrmann JM. 2000. Functional expression of P-glycoprotein and multidrug resistance–associated protein (Mrp1) in primary cultures of rat astrocytes. J Neurosci Res 60:594–601.

95. Ballerini P, Di Iorio P, Ciccarelli R, Nargi E, D'Alimonte I, Traversa U, Rathbone MP, Caciagli F. 2002. Glial cells express multiple ATP binding cassette proteins which are involved in ATP release. Neuroreport 13:1789–1792.

96. Hirrlinger J, Konig J, Dringen R. 2002. Expression of mRNAs of multidrug resistance proteins (Mrps) in cultured rat astrocytes, oligodendrocytes, microglial cells and neurones. J Neurochem 82:716–719.

97. Dallas S, Zhu X, Baruchel S, Schlichter L, Bendayan R. 2003. Functional expression of the multidrug resistance protein 1 in microglia. J Pharmacol Exp Ther 307:282–290.

98. Hirrlinger J, Moeller H, Kirchhoff F, Dringen R. 2005. Expression of multidrug resistance proteins (Mrps) in astrocytes of the mouse brain: a single cell RT-PCR study. Neurochem Res 30:1237–1244.

99. Beck K, Hayashi K, Dang K, Hayashi M, Boyd CD. 2005. Analysis of ABCC6 (MRP6) in normal human tissues. Histochem Cell Biol 123:517–528.

100. Bortfeld M, Rius M, König J, Herold-Mende C, Nies AT, Keppler D. 2006. Human multidrug resistance protein 8 (MRP8/ABCC11), an apical efflux pump for steroid sulfates, is an axonal protein of the CNS and peripheral nervous system. Neuroscience 137:1247–1257.

101. Kao A, Shiau YC, Tsai SC, Wang JJ, Ho ST. 2002. Technetium-99m methoxy-isobutylisonitrile imaging for parathyroid adenoma: relationship to P-glycoprotein or multidrug resistance-related protein expression. Eur J Nucl Med Mol Imag 29: 1012–1015.

102. Yabuuchi H, Takayanagi S, Yoshinaga K, Taniguchi N, Aburatani H, Ishikawa T. 2002. ABCC13, an unusual truncated ABC transporter, is highly expressed in fetal human liver. Biochem Biophys Res Commun 299:410–417.

103. Doyle LA, Yang W, Abruzzo LV, Krogmann T, Gao Y, Rishi AK, Ross DD. 1998. A multidrug resistance transporter from human MCF-7 breast cancer cells. Proc Natl Acad Sci U S A 95:15665–15670.

104. Graf GA, Li WP, Gerard RD, Gelissen I, White A, Cohen JC, Hobbs HH. 2002. Coexpression of ATP-binding cassette proteins ABCG5 and ABCG8 permits their transport to the apical surface. J Clin Invest 110:659–669.

105. Staud F, Pavek P. 2005. Breast cancer resistance protein (BCRP/ABCG2). Int J Biochem Cell Biol 37:720–725.

106. Mao Q, Unadkat JD. 2005. Role of the breast cancer resistance protein (ABCG2) in drug transport. AAPS J 7:E118–E133.

107. Zhang Y, Gupta A, Wang H, Zhou L, Vethanayagam RR, Unadkat JD, Mao Q. 2005. BCRP transports dipyridamole and is inhibited by calcium channel blockers. Pharm Res 22:2023–2034.

108. Gupta A, Dai Y, Vethanayagam RR, Hebert MF, Thummel KE, Unadkat JD, Ross DD, Mao Q. 2006. Cyclosporin A, tacrolimus and sirolimus are potent inhibitors of the human breast cancer resistance protein (ABCG2) and reverse resistance to mitoxantrone and topotecan. Cancer Chemother Pharmacol 58:374–383.

109. Merino G, Alvarez AI, Pulido MM, Molina AJ, Schinkel AH, Prieto JG. 2006. Breast cancer resistance protein (BCRP/ABCG2) transports fluoroquinolone antibiotics and

affects their oral availability, pharmacokinetics, and milk secretion. Drug Metab Dispos 34:690–695.

110. Cooray HC, Blackmore CG, Maskell L, Barrand MA. 2002. Localisation of breast cancer resistance protein in microvessel endothelium of human brain. Neuroreport 13:2059–2063.

111. Hori S, Ohtsuki S, Tachikawa M, Kimura N, Kondo T, Watanabe M, Nakashima E, Terasaki T. 2004. Functional expression of rat ABCG2 on the luminal side of brain capillaries and its enhancement by astrocyte-derived soluble factor(s). J Neurochem 90:526–536.

112. Zhang W, Mojsilovic-Petrovic J, Andrade MF, Zhang H, Ball M, Stanimirovic DB. 2003. The expression and functional characterization of ABCG2 in brain endothelial cells and vessels. FASEB J 17:2085–2087.

113. Eisenblatter T, Huwel S, Galla HJ. 2003. Characterisation of the brain multidrug resistance protein (BMDP/ABCG2/BCRP) expressed at the blood–brain barrier. Brain Res 971:221–231.

114. Lee YJ, Kusuhara H, Jonker JW, Schinkel AH, Sugiyama Y. 2005. Investigation of efflux transport of dehydroepiandrosterone sulfate and mitoxantrone at the mouse blood–brain barrier: a minor role of breast cancer resistance protein. J Pharmacol Exp Ther 312: 44–52.

115. Lee G, Babakhanian K, Ramaswamy M, Prat A, Wosik K, Bendayan R. 2007. Expression of the ATP-binding cassette membrane transporter, ABCG2, in human and rodent brain microvessel endothelial and glial cell culture systems. Pharm Res. Epub March 23, 2007.

116. van Herwaarden AE, Jonker JW, Wagenaar E, Brinkhuis RF, Schellens JH, Beijnen JH, Schinkel AH. 2003. The breast cancer resistance protein (Bcrp1/Abcg2) restricts exposure to the dietary carcinogen 2-amino-1-methyl-6-phenylimidazo[4,5-*b*]pyridine. Cancer Res 63:6447–6452.

117. Cisternino S, Mercier C, Bourasset F, Roux F, Scherrmann JM. 2004. Expression, up-regulation, and transport activity of the multidrug-resistance protein Abcg2 at the mouse blood–brain barrier. Cancer Res 64:3296–3301.

118. Breedveld P, Pluim D, Cipriani G, Wielinga P, van Tellingen O, Schinkel AH, Schellens JH. 2005. The effect of Bcrp1 (Abcg2) on the in vivo pharmacokinetics and brain penetration of imatinib mesylate (Gleevec): implications for the use of breast cancer resistance protein and P-glycoprotein inhibitors to enable the brain penetration of imatinib in patients. Cancer Res 65:2577–2582.

119. Hagenbuch B, Meier PJ. 2003. The superfamily of organic anion transporting polypeptides. Biochim Biophys Acta 1609:1–18.

120. Gao B, Hagenbuch B, Kullak-Ublick GA, Benke D, Aguzzi A, Meier PJ. 2000. Organic anion–transporting polypeptides mediate transport of opioid peptides across blood–brain barrier. J Pharmacol Exp Ther 294:73–79.

121. Kullak-Ublick GA, Ismair MG, Stieger B, Landmann L, Huber R, Pizzagalli F, Fattinger K, Meier PJ, Hagenbuch B. 2001. Organic anion–transporting polypeptide B (OATP-B) and its functional comparison with three other OATPs of human liver. Gastroenterology 120:525–533.

122. Adachi H, Suzuki T, Abe M, Asano N, Mizutamari H, Tanemoto M, Nishio T, Onogawa T, Toyohara T, Kasai S, et al. 2003. Molecular characterization of human and rat organic anion transporter OATP-D. Am J Physiol Renal Physiol 285:F1188–F1197.

123. Fujiwara K, Adachi H, Nishio T, Unno M, Tokui T, Okabe M, Onogawa T, Suzuki T, Asano N, Tanemoto M, et al. 2001. Identification of thyroid hormone transporters in humans: different molecules are involved in a tissue-specific manner. Endocrinology 142:2005–2012.

124. Pizzagalli F, Hagenbuch B, Stieger B, Klenk U, Folkers G, Meier PJ. 2002. Identification of a novel human organic anion transporting polypeptide as a high affinity thyroxine transporter. Mol Endocrinol 16:2283–2296.

125. Tamai I, Nozawa T, Koshida M, Nezu J, Sai Y, Tsuji A. 2001. Functional characterization of human organic anion transporting polypeptide B (OATP-B) in comparison with liver-specific OATP-C. Pharm Res 18:1262–1269.

126. Kobayashi D, Nozawa T, Imai K, Nezu J, Tsuji A, Tamai I. 2003. Involvement of human organic anion transporting polypeptide OATP-B (SLC21A9) in pH-dependent transport across intestinal apical membrane. J Pharmacol Exp Ther 306:703–708.

127. Lu R, Kanai N, Bao Y, Schuster VL. 1996. Cloning, in vitro expression, and tissue distribution of a human prostaglandin transporter cDNA(hPGT). J Clin Invest 98:1142–1149.

128. Angeletti RH, Bergwerk AJ, Novikoff PM, Wolkoff AW. 1998. Dichotomous development of the organic anion transport protein in liver and choroid plexus. Am J Physiol 275:C882–C887.

129. Nishio T, Adachi H, Nakagomi R, Tokui T, Sato E, Tanemoto M, Fujiwara K, Okabe M, Onogawa T, Suzuki T, et al. 2000. Molecular identification of a rat novel organic anion transporter moat1, which transports prostaglandin D(2), leukotriene C(4), and taurocholate. Biochem Biophys Res Commun 275:831–838.

130. Cvetkovic M, Leake B, Fromm MF, Wilkinson GR, Kim RB. 1999. OATP and P-glycoprotein transporters mediate the cellular uptake and excretion of fexofenadine. Drug Metab Dispos 27:866–871.

131. Ohtsuki S, Takizawa T, Takanaga H, Terasaki N, Kitazawa T, Sasaki M, Abe T, Hosoya K, Terasaki T. 2003. In vitro study of the functional expression of organic anion transporting polypeptide 3 at rat choroid plexus epithelial cells and its involvement in the cerebrospinal fluid-to-blood transport of estrone-3-sulfate. Mol Pharmacol 63:532–537.

132. Sugiyama D, Kusuhara H, Taniguchi H, Ishikawa S, Nozaki Y, Aburatani H, Sugiyama Y. 2003. Functional characterization of rat brain–specific organic anion transporter (Oatp14) at the blood–brain barrier: high affinity transporter for thyroxine. J Biol Chem 278:43489–43495.

133. Taogoshi T, Nomura A, Murakami T, Nagai J, Takano M. 2005. Transport of prostaglandin E1 across the blood–brain barrier in rats. J Pharm Pharmacol 57:61–66.

134. Noé B, Hagenbuch B, Stieger B, Meier PJ. 1997. Isolation of a multispecific organic anion and cardiac glycoside transporter from rat brain. Proc Natl Acad Sci U S A 94:10346–10350.

135. Dagenais C, Ducharme J, Pollack GM. 2001. Uptake and efflux of the peptidic delta-opioid receptor agonist. Neurosci Lett 301:155–158.

136. Abe T, Kakyo M, Sakagami H, Tokui T, Nishio T, Tanemoto M, Nomura H, Hebert SC, Matsuno S, Kondo H, Yawo H. 1998. Molecular characterization and tissue distribution of a new organic anion transporter subtype (oatp3) that transports thyroid hormones and taurocholate and comparison with oatp2. J Biol Chem 273:22395–22401.

137. Kis B, Isse T, Snipes JA, Chen L, Yamashita H, Ueta Y, Busija DW. 2006. Effects of LPS stimulation on the expression of prostaglandin carriers in the cells of the blood–brain and blood–cerebrospinal fluid barriers. J Appl Physiol 100:1392–1399.

138. Ivanov AI, Romanovsky AA. 2004. Prostaglandin E2 as a mediator of fever: synthesis and catabolism. Front Biosci 9:1977–1993.

139. Burckhardt BC, Burckhardt G. 2003. Transport of organic anions across the basolateral membrane of proximal tubule cells. Rev Physiol Biochem Pharmacol 146:95–158.

140. Sweet DH, Miller DS, Pritchard JB, Fujiwara Y, Beier DR, Nigam SK. 2002. Impaired organic anion transport in kidney and choroid plexus of organic anion transporter 3 (Oat3 (Slc22a8)) knockout mice. J Biol Chem 277:26934–26943.

141. Mori S, Takanaga H, Ohtsuki S, Deguchi T, Terasaki T, Kang Y-S, Hosoya K-I. 2003. Rat organic anion transporter 3 (rOAT3) is responsible for brain-to-blood efflux of homovanillic acid at the abluminal membrane of brain capillary endothelial cells. J Cereb Blood Flow Metab 23:432–440.

142. Kusuhara H, Sugiyama Y. 2005. Active efflux across the blood–brain barrier: role of the solute carrier family. NeuroRx 2:73–85.

143. Simonson GD, Vincent AC, Roberg KJ, Huang Y, Iwanij V. 1994. Molecular cloning and characterization of a novel liver-specific transport. J Cell Sci 107:1065–1072.

144. Mori K, Ogawa Y, Ebihara K, Aoki T, Tamura N, Ozaki S, Mukoyama M, Tanaka I, Nakao K, Tashiro K, Sugawara A, Kuwahara T. 1997. Kidney-specific expression of a novel mouse organic cation transporter-like protein. FEBS Lett 417:371–374.

145. Brady KP, Dushkin H, Förnzler D, Magner F, Her H, Beier DR, Gullans S, Koike T, Segre GV, Green RM. 1999. A novel putative transporter maps to the osteosclerosis (oc) mutation and is not expressed in the oc mutant mouse. Genomics 56:254–261.

146. Kusuhara H, Sekine T, Utsunomiya-Tate N, Tsuda M, Cha SH, Kanai Y, Endou H, Sugiyama Y, Kojima R. 1999. Molecular cloning and characterization of a new multispecific organic anion transporter from rat brain. J Biol Chem 274:13675–13680.

147. Cha SH, Sekine T, Kusuhara H, Yu E, Kim JY, Kim DK, Kanai Y, Endou H, Sugiyama Y. 2000. Molecular cloning and characterization of multispecific organic anion transporter 4 expressed in the placenta. J Biol Chem 275:4507–4512.

148. Youngblood GL, Sweet DH. 2004. Identification and functional assessment of the novel murine organic anion transporter Oat5 (Slc22a19) expressed in kidney. Am J Physiol Renal Physiol 287:F236–F244.

149. Lopez-Nieto CE, You G, Bush KT, Barros EJG, Beier DR, Nigam SK. 1997. Molecular cloning and characterization of NKT, a gene product related to the organic cation transporter family that is almost exclusively expressed in the kidney. J Biol Chem 272:6471–6478.

150. Sekine T, Watanabe N, Hosoyamada M, Kanai Y, Endou H. 1997. Expression cloning and characterization of a novel multispecific organic anion transporter. J Biol Chem 272:18526–18529.

151. Sweet DH, Pritchard JB. 1999. The molecular biology of renal organic anion and organic cation transporters. Cell Biochem Biophys 31:89–118.

152. Eraly SA, Nigam SK. 2002. Novel human cDNAs homologous to *Drosophila* Orct and mammalian carnitine transporters. Biochem Biophys Res Commun 297:1159–1166.

153. Monte JC, Nagle MA, Eraly SA, Nigam SK. 2004. Identification of a novel murine organic anion transporter family member, OAT6, expressed in olfactory mucosa. Biochem Biophys Res Commun 323:429–436.

154. Sekine T, Cha SH, Endou H. 2000. The multispecific organic anion transporter (OAT) family. Pflugers Arch Eur J Physiol 440:337–350.

155. Reid G, Wolff NA, Burckhandt G, Dautzenberg FM. 1998. Cloning of a human renal *p*-aminohippurate transporter, hROAT1. Kidney Blood Press Res 21:233–237.

156. Shimada H, Moewes B, Burckhardt G. 1987. Indirect coupling to Na^+ of *p*-aminohippuric acid uptake into rat renal basolateral membrane vesicles. Am J Physiol 253:F795–F801.

157. Pritchard JB. 1988. Coupled transport of *p*-aminohippurate by rat kidney basolateral membrane vesicles. Am J Physiol 255:F597–F604.

158. Hosoyamada M, Sekine T, Kanai Y, Endou H. 1999. Molecular cloning and functional expression of a multispecific organic anion transporter from human kidney. Am J Physiol Renal Physiol 276:F122–F128.

159. Pavlova A, Sakurai H, Leclercq B, Beier DR, Yu ASL, Nigam SK. 2000. Developmentally regulated expression of organic ion transporters NKT(OAT1), OCT1, NLT(OAT2), and Roct. Am J Physiol Renal Physiol 278:F635–F643.

160. Pritchard JB, Sweet DH, Miller DS, Walden R. 1999. Mechanism of organic anion transport across the apical membrane of choroid plexus. J Biol Chem 274:33382–33387.

161. Cihlar T, Lin DC, Fuller MD, Mendel DB, Pritchard JB, Sweet DH. 1999. The antiviral nucleotide analogs cidofovir and adefovir are novel substrates for human and rat renal organic anion transporter 1. Mol Pharmacol 56:570–580.

162. Bahn A, Ebbinghaus C, Ebbinghaus D, Burckhardt G, Hagos Y, Ponimaskin EG, Füzesi L. 2004. Expression studies and functional characterization of renal human organic anion transporter 1 isoforms. Drug Metab Dispos 32:424–430.

163. Bahn A, Ljubojevic M, Lorenz H, Schultz C, Ghebremedhin E, Ugele B, Sabolic I, Burckhardt G, Hagos Y. 2005. Murine renal organic anion transporters mOAT1 and mOAT3 facilitate the transport of neuroactive tryptophan metabolites. Am J Physiol Cell Physiol 289:C1075–C1084.

164. Alebouyeh M, Takeda M, Noshiro R, Hasannejad H, Inatomi J, Khamdang S, Anzai N, Endou H, Onozato ML, Tojo A, Narikawa S, Huang X-L. 2003. Expression of human organic anion transporters in the choroid plexus and their interactions with neurotransmitter metabolites. J Pharm Sci 93:430–436.

165. Sekine T, Tsuda M, Apiwattanakul N, Nakajima N, Kanai Y, Endou H, Cha SH. 1998. Identification of multispecific organic anion transporter 2 expressed predominantly in the liver. FEBS Lett 429:179–182.

166. Buist SCN, Cherrington NJ, Choudhuri S, Hartley DP, Klaassen CD. 2002. Gender-specific and developmental influences on the expression of rat organic anion transporters. J Pharmacol Exp Ther 301:145–151.

167. Race JE, Grassl SM, Williams WJ, Holtzman EJ. 1999. Molecular cloning and characterization of two novel human renal organic anion transporters (hOAT1 and hOAT3). Biochem Biophys Res Commun 255:508–514.

168. Cha SH, Sekine T, Fukushima JI, Kanai Y, Kobayashi Y, Goya T, Endou H. 2001. Identification and characterization of human organic anion transporter 3 expressing predominantly in the kidney. Mol Pharmacol 59:1277–1286.

169. Bakhiya N, Bahn A, Burckhardt G, Wolff NA. 2003. Human organic anion transporter 3 (hOAT3) can operate as an exchanger and mediate secretory urate flux. Cell Physiol Biochem 13:249–256.

170. Kikuchi R, Kusuhara H, Sugiyama D, Sugiyama Y. 2003. Contribution of organic anion transporter 3 (Slc22a8) to the elimination of p-aminohippuric acid and benzylpenicillin across the blood–brain barrier. J Pharmacol Exp Ther 306:51–58.

171. Nagata Y, Kusuhara H, Endou H, Sugiyama Y. 2002. Expression and functional characterization of rat organic anion transporter 3 (rOat3) in the choroid plexus. Mol Pharmacol 61:982–988.

172. Suzuki H, Sawada Y, Sugiyama Y. 1986. Transport of cimetidine by the rat choroid plexus in vitro. J Pharmacol Exp Ther 239:927–935.

173. Suzuki H, Sawada Y, Sugiyama Y, Iga T, Hanano M. 1987. Transport of benzylpenicillin by the rat choroid plexus in vitro. J Pharmacol Exp Ther 242:660–665.

174. Kakee A, Terasaki T, Sugiyama Y. 1997. Selective brain to blood efflux transport of para-aminohippuric acid across the blood–brain barrier: in vivo evidence by use of the brain efflux index method. J Pharmacol Exp Ther 283:1018–1025.

175. Babu E, Takeda M, Narikawa S, Kobayashi Y, Enomoto A, Tojo A, Seok HC, Sekine T, Sakthisekaran D, Endou H. 2002. Role of human organic anion transporter 4 in the transport of ochratoxin A. Biochim Biophys Acta 1590:64–75.

176. Xu G, Bhatnagar V, Wen G, Hamilton BA, Eraly SA, Nigam SK. 2005. Analyses of coding region polymorphisms in apical and basolateral human organic anion transporter (OAT) genes [OAT1 (NKT), OAT2, OAT3, OAT4, URAT (RST)]: rapid communication. Kidney Int 68:1491–1499.

177. Kimura H, Takeda M, Narikawa S, Enomoto A, Ichida K, Endou H. 2002. Human organic anion transporters and human organic cation transporters mediate renal transport of prostaglandins. J Pharmacol Exp Ther 301:293–298.

178. Anzai N, Jutabha P, Enomoto A, Yokoyama H, Hirata T, Shiraya K, He X, Seok HC, Takeda M, Miyazaki H, et al. 2005. Functional characterization of rat organic anion transporter 5 (SLC22a19) at the apical membrane of renal proximal tubules. J Pharmacol Exp Ther 315:534–544.

179. Enomoto A, Kimura H, Chairoungdua A, Shigeta Y, Jutabha P, Cha SH, Hosoyamada M, Takeda M, Sekine T, Igarashi T, et al. 2002. Molecular identification of a renal urate-anion exchanger that regulates blood urate levels. Nature 417:447–452.

180. Imaoka T, Kusuhara H, Hasegawa M, Morita N, Sugiyama Y, Adachi-Akahane S, Endou H. 2004. The renal-specific transporter mediates facilitative transport of organic anions at the brush border membrane of mouse renal tubules. J Am Soc Nephrol 15:2012– 2022.

181. Breen CM, Sykes DB, Fricker G, Miller DS. 2002. Confocal imaging of organic anion transport in intact rat choroid plexus. Am J Physiol Renal Physiol 282:F877–F885.

182. Gründemann D, Liebich G, Kiefer N, Koster S, Schömig E. 1999. Selective substrates for non-neuronal monoamine transporters. Mol Pharmacol 56:1–10.

183. Zhang L, Gorset W, Dresser MJ, Giacomini KM. 1999. The interaction of *n*-tetraalkylammonium compounds with a human organic cation transporter, hOCT1. J Pharmacol Exp Ther 288:1192–1198.

184. Zhang L, Gorset W, Kroetz DL, Giacomini KM, Washington CB, Blaschke TF. 2000. Interactions of HIV protease inhibitors with a human organic cation transporter in a mammalian expression system. Drug Metab Dispos 28:329–334.

185. Arndt P, Volk C, Gorboulev V, Budiman T, Popp C, Ulheimer-Teuber I, Akhoundova A, Koppatz S, Bamberg E, Nagel G, Koepsell H. 2001. Interaction of cations, anions, and weak base quinine with rat renal cation transporter rOCT2 compared with rOCT1. Am J Physiol Renal Physiol 281:F454–F468.

186. Dresser MJ, Leabman MK, Giacomini KM. 2001. Transporters involved in the elimination of drugs in the kidney: Organic anion transporters and organic cation transporters. J Pharm Sci 90:397–421.

187. Koepsell H, Schmitt BM, Gorboulev V. 2003. Organic cation transporters. Rev Physiol Biochem Pharmacol 150:36–90.
188. Koepsell H, Endou H. 2004. The SLC22 drug transporter family. Pflugers Arch Eur J Physiol 447:666–676.
189. Koepsell H. 1998. Organic cation transporters in intestine, kidney, liver, and brain. Annu Rev Physiol 60:243–266.
190. Koepsell H, Gorboulev V, Arndt P. 1999. Molecular pharmacology of organic cation transporters in kidney. J Membr Biol 167:103–117.
191. Jonker JW, Schinkel AH. 2004. Pharmacological and physiological functions of the polyspecific organic cation transporters: OCT1, 2, and 3 (SLC22A1-3). J Pharmacol Exp Ther 308:2–9.
192. Grundemann D, Gorboulev V, Gambaryan S, Veyhl M, Koepsell H. 1994. Drug excretion mediated by a new prototype of polyspecific transporter. Nature 372:549–552.
193. Gorboulev V. 1997. Cloning and characterization of two human polyspecific organic cation transporters. DNA Cell Biol 16:871–881.
194. Gründemann D, Babin-Ebell J, Martel F, Ording N, Schmidt A, Schömig E. 1997. Primary structure and functional expression of the apical organic cation transporter from kidney epithelial LLC-PK1 cells. J Biol Chem 272:10408–10413.
195. Ullrich KJ. 1994. Specificity of transporters for ‘organic anions’ and ‘organic cations’ in the kidney. Biochim Biophys Acta Rev Biomembr 1197:45–62.
196. Sekine T, Kusuhara H, Utsunomiya-Tate N, Tsuda M, Kanai Y, Endou H, Sugiyama Y. 1998. Molecular cloning and characterization of high-affinity carnitine transporter from rat intestine. Biochem Biophys Res Commun 251:586–591.
197. Tamai I, Yabuuchi H, Sai Y, Tsuji A, Nezu J-I, Oku A, Shimane M. 1997. Cloning and characterization of a novel human pH-dependent organic cation transporter, OCTN1. FEBS Lett 419:107–111.
198. Tamai I, Ohashi R, Yabuuchi H, Sai Y, Tsujit A, Nezu J-I, Oku A, Shimane M. 1998. Molecular and functional identification of sodium ion-dependent, high affinity human carnitine transporter OCTN2. J Biol Chem 273:20378–20382.
199. Tamai I, Ohashi R, Nezu J-I, Sai Y, Kobayashi D, Oku A, Shimane M, Tsuji A. 2000. Molecular and functional characterization of organic cation/carnitine transporter family in mice. J Biol Chem 275:40064–40072.
200. Wu X, Leibach FH, Ganapathy V, Prasad PD. 1998. cDNA sequence, transport function, and genomic organization of human OCTN2, a new member of the organic cation transporter family. Biochem Biophys Res Commun 246:589–595.
201. Wu X, Huang W, Wang H, Leibach FH, Ganapathy V, George RL, Conway SJ. 2000. Structural and functional characteristics and tissue distribution pattern of rat OCTN1, an organic cation transporter, cloned from placenta. Biochim Biophys Acta Biomembr 1466:315–327.
202. Pao SS, Paulsen IT, Saier J. 1998. Major facilitator superfamily. Microbiol Mol Biol Rev 62:1–34.
203. Ciarimboli G, Schlatter E. 2005. Regulation of organic cation transport. Pflugers Arch Eur J Physiol 449:423–441.
204. Zhang L, Dresser MJ, Terashita S, Giacomini KM, Gray AT, Yost SC. 1997. Cloning and functional expression of a human liver organic cation transporter. Mol Pharmacol 51:913–921.

205. Inazu M, Takeda H, Matsumiya T. 2005. Molecular and functional characterization of an Na^+-independent choline transporter in rat astrocytes. J Neurochem 94:1427–1437.

206. Sweet DH, Miller DS, Pritchard JB. 2001. Ventricular choline transport: a role for organic cation transporter 2 expressed in choroid plexus. J Biol Chem 276:41611–41619.

207. Okuda M, Saito H, Urakami Y, Takano M, Inui K-I. 1996. cDNA cloning and functional expression of a novel rat kidney organic cation transporter, OCT2. Biochem Biophys Res Commun 224:500–507.

208. Karbach U, Kricke J, Meyer-Wentrup F, Gorboulev V, Volk C, Loffing-Cueni D, Kaissling B, Bachmann S, Koepsell H. 2000. Localization of organic cation transporters OCT1 and OCT2 in rat kidney. Am J Physiol Renal Physiol 279:F679–F687.

209. Motohashi H, Sakurai Y, Saito H, Masuda S, Urakami Y, Goto M, Fukatsu A, Ogawa O, Inui K-I. 2002. Gene expression levels and immunolocalization of organic ion transporters in the human kidney. J Am Soc Nephrol 13:866–874.

210. Busch AE, Baumann C, Waldegger S, Lang F, Karbach U, Miska D, Gorboulev V, Akhoundova A, Volk C, Arndt P, et al. 1998. Human neurons express the polyspecific cation transporter hOCT2, which translocates monoamine neurotransmitters, amantadine, and memantine. Mol Pharmacol 54:342–352.

211. Slitt AL, Cherrington NJ, Hartley DP, Leazer TM, Klaassen CD. 2002. Tissue distribution and renal developmental changes in rat organic cation transporter mRNA levels. Drug Metab Dispos 30:212–219.

212. Gründemann D, Koster S, Kiefer N, Breidert T, Engelhardt M, Spitzenberger F, Obermuller N, Schömig E. 1998. Transport of monoamine transmitters by the organic cation transporter type 2, OCT2. J Biol Chem 273:30915–30920.

213. Kekuda R, Prasad PD, Wu X, Wang H, Fei Y-J, Leibach FH, Ganapathy V. 1998. Cloning and functional characterization of a potential-sensitive, polyspecific organic cation transporter (OCT3) most abundantly expressed in placenta. J Biol Chem 273:15971–15979.

214. Wu X, Huang W, Ganapathy ME, Wang H, Kekuda R, Conway SJ, Leibach FH, Ganapathy V. 2000. Structure, function, and regional distribution of the organic cation transporter OCT3 in the kidney. Am J Physiol Renal Physiol 279:F449–F458.

215. Wu X, Kekuda R, Huang W, Fei Y-J, Leibach FH, Ganapathy V, Chen J, Conway SJ. 1998. Identity of the organic cation transporter OCT3 as the extraneuronal monoamine transporter (uptake2) and evidence for the expression of the transporter in the brain. J Biol Chem 273:32776–32786.

216. Schmitt A, Mossner R, Gossmann A, Fischer IG, Lesch KP, Gorboulev V, Koepsell H, Murphy DL. 2003. Organic cation transporter capable of transporting serotonin is up-regulated in serotonin transporter–deficient mice. J Neurosci Res 71:701–709.

217. Kristufek D, Rudorfer W, Pifl C, Huck S. 2002. Organic cation transporter mRNA and function in the rat superior cervical ganglion. J Physiol 543:117–134.

218. Gründemann D, Schechinger B, Rappold GA, Schömig E. 1998. Molecular identification of the corticosterone-sensitive extraneuronal catecholamine transporter. Nature Neurosci 1:349–351.

219. Tamai I, China K, Sai Y, Kobayashi D, Nezu J-I, Kawahara E, Tsuji A. 2001. Na^+-coupled transport of L-carnitine via high-affinity carnitine transporter OCTN2 and its subcellular localization in kidney. Biochim Biophys Acta Biomembr 1512:273–284.

220. Wawrzenczyk A, Nalecz KA, Nalecz MJ. 1995. Effect of externally added carnitine on the synthesis of acetylcholine in rat cerebral cortex cells. Neurochem Int 26:635–641.

221. Pettegrew JW, Klunk WE, Panchalingam K, Kanfer JN, McClure RJ. 1995. Clinical and neurochemical effects of acetyl-L-carnitine in Alzheimer's disease. Neurobiol Aging 16:1–4.

222. Tamai I, Tsuji A. 2000. Transporter-mediated permeation of drugs across the blood–brain barrier. J Pharm Sci 89:1371–1388.

223. Kido Y, Tamai I, Ohnari A, Sai Y, Kagami T, Nezu J-I, Nikaido H, Hashimoto N, Asano M, Tsuji A. 2001. Functional relevance of carnitine transporter OCTN2 to brain distribution of L-carnitine and acetyl-L-carnitine across the blood–brain barrier. J Neurochem 79:959–969.

224. Yabuuchi H, Tamai I, Sakamoto K, Sai Y, Tsuji A, Nezu J-I, Oku A, Shimane M. 1999. Novel membrane transporter OCTN1 mediates multispecific, bidirectional, and pH-dependent transport of organic cations. J Pharmacol Exp Ther 289:768–773.

225. Brooks H, Krahenbuhl S. 2001. Identification and tissue distribution of two differentially spliced variants of the rat carnitine transporter OCTN2. FEBS Lett 508:175–180.

226. Nezu J-I, Oku A, Shimane M, Tamai I, Ohashi R, Yabuuchi H, Sai Y, Tsuji A, Hashimoto N, Nikaido H, et al. 1999. Primary systemic carnitine deficiency is caused by mutations in a gene encoding sodium ion-dependent carnitine transporter. Nat Genet 21:91–94.

227. Wagner CA, Lükewille U, Kaltenbach S, Moschen I, Bröer A, Risler T, Broer S, Lang F. 2000. Functional and pharmacological characterization of human Na+-carnitine cotransporter hOCTN2. Am J Physiol Renal Physiol 279:F584–F591.

228. Wu X, Huang W, Seth P, Rajan DP, Leibach FH, Ganapathy V, Prasad PD, Chen J, Conway SJ. 1999. Functional characteristics and tissue distribution pattern of organic cation transporter 2 (OCTN2), an organic cation/carnitine transporter. J Pharmacol Exp Ther 290:1482–1492.

229. Ohashi R, Tamai I, Yabuuchi H, Sai Y, Tsuji A, Nezu J-I, Oku A, Shimane M. 1999. Na^+-dependent carnitine transport by organic cation transporter (OCTN2): Its pharmacological and toxicological relevance. J Pharmacol Exp Ther 291:778–784.

230. Inano A, Sai Y, Tsuji A, Tamai I, Nikaido H, Hasimoto N, Asano M. 2003. Acetyl-L-carnitine permeability across the blood–brain barrier and involvement of carnitine transporter OCTN2. Biopharm Drug Dispos 24:357–365.

231. Lamhonwah A-M, Tein I, Ackerley CA, Tilups A, Edwards VD, Wanders RJ. 2005. OCTN3 is a mammalian peroxisomal membrane carnitine transporter. Biochem Biophys Res Commun 338:1966–1972.

232. Von Lubitz DKJE, Carter MF, Beenhakker M, Lin RC, Jacobson KA. 1995. Adenosine: A prototherapeutic concept in neurodegeneration. Ann N Y Acad Sci 765:163–178.

233. Baldwin SA, Mackey JR, Cass CE, Young JD. 1999. Nucleoside transporters: Molecular biology and implications for therapeutic development. Mol Med Today 5:216–224.

234. Griffith DA, Jarvis SM. 1996. Nucleoside and nucleobase transport systems of mammalian cells. Biochim Biophys Acta Rev Biomembr 1286:153–181.

235. Fox IH, Kelley WN. 1978. The role of adenosine and 2′-deoxyadenosine in mammalian cells. Annu Rev Biochem 47:655–686.

236. Tremblay GC, Jimenez U, Crandall DE. 1976. Pyrimidine biosynthesis and its regulation in the developing rat brain. J Neurochem 26:57–64.

237. Kraupp M, Marz R. 1995. Nucleobase and nucleoside transport in mammalian cells. Wien Klin Wochenschr 107:677–680.

238. Cass CE, Young JD, Baldwin SA, Cabrita MA, Graham KA, Griffiths M, Jennings LL, Mackey JR, Ng AM, Ritzel MW, Vickers MF, Yao SY. 1999. Nucleoside transporters of mammalian cells. Pharm Biotechnol 12:313–352.

239. Cabrita MA, Cass CE, Young JD, Baldwin SA. 2002. Molecular biology and regulation of nucleoside and nucleobase transporter proteins in eukaryotes and prokaryotes. Biochem Cell Biol 80:623–638.

240. Ritzel MWL, Yao SYM, Huang M-Y, Young JD, Elliott JF, Cass CE. 1997. Molecular cloning and functional expression of cDNAs encoding a human Na^+-nucleoside cotransporter (hCNT1). Am J Physiol Cell Physiol 272:C707–C714.

241. Yao SYM, Ng AML, Young JD, Ritzel MWL, Mackey JR, Cass CE. 1998. Molecular cloning, functional expression and chromosomal localization of a cDNA encoding a human Na^+/nucleoside cotransporter (hCNT2) selective for purine nucleosides and uridine. Mol Membr Biol 15:203–211.

242. Ritzel MWL, Ng AML, Yao SYM, Graham K, Loewen SK, Smith KM, Hyde RJ, Karpinski E, Cass CE, Baldwin SA, Young JD. 2001. Recent molecular advances in studies of the concentrative Na^+-dependent nucleoside transporter (CNT) family: identification and characterization of novel human and mouse proteins (hCNT3 and mCNT3) broadly selective for purine and pyrimidine nucleosides (system cib). Mol Membr Biol 18:65–72.

243. Ritzel MWL, Ng AML, Yao SYM, Loewen SK, Smith KM, Chen X-Z, Karpinski E, Young JD, Graham K, Mowles DA, et al. 2001. Molecular identification and characterization of novel human and mouse concentrative Na^+-nucleoside cotransporter proteins (hCNT3 and mCNT3) broadly selective for purine and pyrimidine nucleosides (system cib). J Biol Chem 276:2914–2927.

244. Loewen SK, Yao SYM, Slugoski MD, Mohabir NN, Young JD, Turner RJ, Weiner JH, Mackey JR, Cass CE, Gallagher MP, Henderson PJF, Baldwin SA. 2004. Transport of physiological nucleosides and anti-viral and anti-neoplastic nucleoside drugs by recombinant *Escherichia coli* nucleoside-H^+ cotransporter (NupC) produced in *Xenopus laevis* oocytes. Mol Membr Biol 21:1–10.

245. Smith KM, Slugoski MD, Loewen SK, Ng AML, Yao SYM, Chen X-Z, Karpinski E, Young JD, Cass CE, Baldwin SA. 2005. The broadly selective human Na^+/nucleoside cotransporter (hCNT3) exhibits novel cation-coupled nucleoside transport characteristics. J Biol Chem 280:25436–25449.

246. Hamilton SR, Yao SY, Ingram JC, Hadden DA, Ritzel MW, Gallagher MP, Henderson PJ, Cass CE, Young JD, Baldwin SA. 2001. Subcellular distribution and membrane topology of the mammalian concentrative Na^+-nucleoside cotransporter rCNT1. J Biol Chem 276:27981–27988.

247. Plagemann PGW, Aran JM. 1990. Characterization of Na^+-dependent, active nucleoside transport in rat and mouse peritoneal macrophages, a mouse macrophage cell line and normal rat kidney cells. Biochim Biophys Acta Biomembr 1028:289–298.

248. Hong M, Schlichter L, Bendayan R. 2000. A Na(+)-dependent nucleoside transporter in microglia. J Pharmacol Exp Ther 292:366–374.

249. Wu X, Yuan G, Brett CM, Hui AC, Giacomini KM. 1992. Sodium-dependent nucleoside transport in choroid plexus from rabbit. Evidence for a single transporter for purine and pyrimidine nucleosides. J Biol Chem 267:8813–8818.

250. Crawford CR, Ng CYC, Noel LD, Belt JA. 1990. Nucleoside transport in L1210 murine leukemia cells. Evidence for three transporters. J Biol Chem 265:9732–9736.

251. Yao SY, Ng AM, Loewen SK, Cass CE, Baldwin SA, Young JD. 2002. An ancient prevertebrate Na^+-nucleoside cotransporter (hfCNT) from the Pacific hagfish (*Eptatretus stouti*). Am J Physiol Cell Physiol 283:C155–C168.
252. Loewen SK, Ng AML, Mohabir NN, Young JD, Baldwin SA, Cass CE. 2003. Functional characterization of a H^+/nucleoside co-transporter (CaCNT) from *Candida albicans*, a fungal member of the concentrative nucleoside transporter (CNT) family of membrane proteins. Yeast 20:661–675.
253. Xiao G, Wang J, Tangen T, Giacomini KM. 2001. A novel proton-dependent nucleoside transporter, CeCNT3, from *Caenorhabditis elegans*. Mol Pharmacol 59:339–348.
254. Lai Y, Bakken AH, Unadkat JD. 2002. Simultaneous expression of hCNT1-CFP and hENT1-YFP in Madin–Darby canine kidney cells. Localization and vectorial transport studies. J Biol Chem 277:37711–37717.
255. Anderson CM, Xiong W, Parkinson FE, Young JD, Cass CE. 1996. Demonstration of the existence of mRNAs encoding N1/cif and N2/cit sodium/nucleoside cotransporters in rat brain. Mol Brain Res 42:358–361.
256. Lu H, Chen C, Klaassen C. 2004. Tissue distribution of concentrative and equilibrative nucleoside transporters in male and female rats and mice. Drug Metab Dispos 32:1455–1461.
257. Pardridge WM. 2005. The blood–brain barrier: bottleneck in brain drug development. NeuroRx 2:3–14.
258. Thomas SA, Segal MB. 1997. Saturation kinetics, specificity and NBMPR sensitivity of thymidine entry into the central nervous system. Brain Res 760:59–67.
259. Graham KA, Leithoff J, Mowles D, Mackey JR, Cass CE, Coe IR, Young JD. 2000. Differential transport of cytosine-containing nucleosides by recombinant human concentrative nucleoside transporter protein hCNT1. Nucleosides Nucleotides Nucleic Acids 19:415–434.
260. Pennycooke M, Chaudary N, Shuralyova I, Zhang Y, Coe IR. 2001. Differential expression of human nucleoside transporters in normal and tumor tissue. Biochem Biophys Res Commun 280:951–959.
261. Nagai K, Nagasawa K, Fujimoto S. 2005. Transport mechanisms for adenosine and uridine in primary-cultured rat cortical neurons and astrocytes. Biochem Biophys Res Commun 334:1343–1350.
262. Mangravite LM, Giacomini KM, Lipschutz JH, Mostov KE. 2001. Localization of GFP-tagged concentrative nucleoside transporters in a renal polarized epithelial cell line. Am J Physiol Renal Physiol 280:F879–F885.
263. Redzic ZB, Biringer J, Barnes K, Baldwin SA, Al Sarraf H, Nicola PA, Young JD, Cass CE, Barrand MA, Hladky SB. 2005. Polarized distribution of nucleoside transporters in rat brain endothelial and choroid plexus epithelial cells. J Neurochem 94:1420–1426.
264. Cornford EM, Oldendorf WH. 1975. Independent blood–brain barrier transport systems for nucleic acid precursors. Biochim Biophys Acta 394:211–219.
265. Gray JH, Owen RP, Giacomini KM. 2004. The concentrative nucleoside transporter family, SLC28. Pflugers Arch Eur J Physiol 447:728–734.
266. Spector R. 1982. Nucleoside transport in choroid plexus: mechanism and specificity. Arch Biochem Biophys 216:693–703.
267. Wu X, Gutierrez MM, Giacomini KM. 1994. Further characterization of the sodium-dependent nucleoside transporter (N3) in choroid plexus from rabbit. Biochim Biophys Acta Biomembr 1191:190–196.

268. Waclawski AP, Sinko PJ. 1996. Oral absorption of anti-acquired immune deficiency syndrome nucleoside analogues: 2. Carrier-mediated intestinal transport of stavudine in rat and rabbit preparations. J Pharm Sci 85:478–485.

269. Terasaki T, Pardrigde WM. 1988. Restricted transport pf 3′-azido-3′-deoxythymidine and dideoxynucleosides through the blood–brain barrier. J Infect Dis 158:630–632.

270. Baldwin SA, Beal PR, King AE, Yao SYM, Young JD, Cass CE. 2004. The equilibrative nucleoside transporter family, SLC29. Pflugers Arch Eur J Physiol 447:735–743.

271. Podgorska M, Kocbuch K, Pawelczyk T. 2005. Recent advances in studies on biochemical and structural properties of equilibrative and concentrative nucleoside transporters. Acta Biochim Pol 52:749–758.

272. Hyde RJ, Cass CE, Young JD, Baldwin SA. 2001. The ENT family of eukaryote nucleoside and nucleobase transporters: Recent advances in the investigation of structure/function relationships and the identification of novel isoforms. Mol Membr Biol 18:53–63.

273. Sundaram M, Yao SYM, Young JD, Cass CE, Ingram JC, Berry ZA, Abidi F, Baldwin SA. 2001. Topology of a human equilibrative, nitrobenzylthioinosine (NBMPR)-sensitive nucleoside transporter (hENT1) implicated in the cellular uptake of adenosine and anti-cancer drugs. J Biol Chem 276:45270–45275.

274. Mani RS, Marjan JMJ, Graham KA, Cass CE, Young JD, Hammond JR, Baldwin SA. 1998. Demonstration of equilibrative nucleoside transporters (hENT1 and hENT2) in nuclear envelopes of cultured human choriocarcinoma (BeWo) cells by functional reconstitution in proteoliposomes. J Biol Chem 273:30818–30825.

275. Jennings LL, Hao C, Cabrita MA, Vickers MF, Baldwin SA, Young JD, Cass CE. 2001. Distinct regional distribution of human equilibrative nucleoside transporter proteins 1 and 2 (hENT1 and hENT2) in the central nervous system. Neuropharmacology 40:722–731.

276. Anderson CM, Xiong W, Geiger JD, Parkinson FE, Young JD, Cass CE, Baldwin SA. 1999. Distribution of equilibrative, nitrobenzylthioinosine-sensitive nucleoside transporters (ENT1) in brain. J Neurochem 73:867–873.

277. Pastor-Anglada M, Felipe A, Javier Casado F. 1998. Transport and mode of action of nucleoside derivatives used in chemical and antiviral therapies. Trends Pharmacol Sci 19:424–430.

278. Crawford CR, Patel DH, Belt JA, Naeve C. 1998. Cloning of the human equilibrative, nitrobenzylmercaptopurine riboside (NBMPR)-insensitive nucleoside transporter ei by functional expression in a transport-deficient cell line. J Biol Chem 273:5288–5293.

279. Sinclair CJD, LaRiviere CG, Young JD, Cass CE, Baldwin SA, Parkinson FE. 2000. Purine uptake and release in rat C6 glioma cells: nucleoside transport and purine metabolism under ATP-depleting conditions. J Neurochem 75:1528–1538.

280. Yao SYM, Ng AML, Sundaram M, Cass CE, Baldwin SA, Young JD. 2001. Transport of antiviral 3′-deoxy-nucleoside drugs by recombinant human and rat equilibrative, nitrobenzylthioinosine (NBMPR)-insensitive (ENT2) nucleoside transporter proteins produced in *Xenopus* oocytes. Mol Membr Biol 18:161–167.

281. Baldwin SA, Hyde RJ, Foppolo S, Barnes K, Yao SYM, Ng AML, Ritzel MWL, Young JD, Cass CE. 2005. Functional characterization of novel human and mouse equilibrative nucleoside transporters (hENT3 and mENT3) located in intracellular membranes. J Biol Chem 280:15880–15887.

282. Engel K, Zhou M, Wang J. 2004. Identification and characterization of a novel monoamine transporter in the human brain. J Biol Chem 279:50042–50049.

283. Adibi SA. 1997. Renal assimilation of oligopeptides: physiological mechanisms and metabolic importance. Am J Physiol Endocrinol Metab 272:E723–E736.

284. Boll M, Markovich D, Weber W-M, Korte H, Daniel H, Murer H. 1994. Expression cloning of a cDNA from rabbit small intestine related to proton-coupled transport of peptides, β-lactam antibiotics and ACE-inhibitors. Pflugers Arch Eur J Physiol 429:146–149.

285. Boll M, Herget M, Wagener M, Daniel H, Weber WM, Clauss W, Markovich D, Biber J, Murer H. 1996. Expression cloning and functional characterization of the kidney cortex high-affinity proton-coupled peptide transporter. Proc Natl Acad Sci U S A 93:284–289.

286. Herrera-Ruiz D, Knipp GT. 2003. Current perspectives on established and putative mammalian oligopeptide transporters. J Pharm Sci 92:691–714.

287. Daniel H, Kottra G. 2004. The proton oligopeptide cotransporter family SLC15 in physiology and pharmacology. Pflugers Arch Eur J Physiol 447:610–618.

288. Shen H, Smith DE, Yang T, Huang YG, Schneemann JB, Brosius FC III. 1999. Localization of PEPT1 and PEPT2 proton-coupled oligopeptide transporter mRNA and protein in rat kidney. Am J Physiol Renal Physiol 276:F658–F665.

289. Smith DE, Johanson CE, Keep RF. 2004. Peptide and peptide analog transport systems at the blood–CSF barrier. Adv Drug Deliv Rev 56:1765–1791.

290. Meredith D, Boyd CAR. 2000. Structure and function of eukaryotic peptide transporters. Cell Mol Life Sci 57:754–778.

291. Ganapathy V, Leibach FH. 1983. Role of pH gradient and membrane potential in dipeptide transport in intestinal and renal brush-border membrane vesicles from the rabbit. Studies with L-carnosine and glycyl-L-proline. J Biol Chem 258:14189–14192.

292. Brandsch M, Brandsch C, Prasad PD, Ganapathy V, Hopfer U, Leibach FH. 1995. Identification of a renal cell line that constitutively expresses the kidney-specific high-affinity H^+/peptide cotransporter. FASEB 9:1489–1496.

293. Ocheltree SM, Shen H, Hu Y, Smith DE, Keep RF. 2005. Role and relevance of peptide transporter 2 (PEPT2) in the kidney and choroid plexus: in vivo studies with glycylsarcosine in wild-type and PEPT2 knockout mice. J Pharmacol Exp Ther 315:240–247.

294. Rubio-Aliaga I, Daniel H. 2002. Mammalian peptide transporters as targets for drug delivery. Trends Pharmacol Sci 23:434–440.

295. Ganapathy ME, Brandsch M, Prasad PD, Ganapathy V, Leibach FH. 1995. Differential recognition of β-lactam antibiotics by intestinal and renal peptide transporters, PEPT 1 and PEPT 2. J Biol Chem 270:25672–25677.

296. Fujita T, Kishida T, Okada N, Yamamoto A, Ganapathy V, Leibach FH. 1999. Interaction of kyotorphin and brain peptide transporter in synaptosomes prepared from rat cerebellum: implication of high affinity type H^+/peptide transporter PEPT2 mediated transport system. Neurosci Lett 271:117–120.

297. Fujita T, Kishida T, Wada M, Okada N, Yamamoto A, Leibach FH, Ganapathy V. 2004. Functional characterization of brain peptide transporter in rat cerebral cortex: identification of the high-affinity type H^+/peptide transporter PEPT2. Brain Res 997:52–61.

298. Dringen R, Hamprecht B, Bröer S. 1998. The peptide transporter PepT2 mediates the uptake of the glutathione precursor CysGly in astroglia-rich primary cultures. J Neurochem 71:388–393.

299. Fei Y-J, Kanai Y, Nussberger S, Hediger MA, Ganapathy V, Leibach FH, Romero MF, Singh SK, Boron WF. 1994. Expression cloning of a mammalian proton-coupled oligopeptide transporter. Nature 368:563–566.

300. Knutter I, Rubio-Aliaga I, Boll M, Hause G, Daniel H, Neubert K, Brandsch M. 2002. H^+-peptide cotransport in the human bile duct epithelium cell line SK-ChA-1. Am J Physiol Gastrointest Liver Physiol 283:G222–G229.

301. Shen H, Smith DE, Keep RF, Brosius FC 3rd. 2004. Immunolocalization of the proton-coupled oligopeptide transporter PEPT2 in developing rat brain. Mol Pharmacol 1:248–256.

302. Liu W, Liang R, Ramamoorthy S, Fei Y-J, Ganapathy ME, Hediger MA, Gunapathy V, Leibach FH. 1995. Molecular cloning of PEPT 2, a new member of the H^+/peptide cotransporter family, from human kidney. Biochim Biophys Acta Biomembr 1235:461–466.

303. Daniel H, Herget M. 1997. Cellular and molecular mechanisms of renal peptide transport. Am J Physiol Renal Physiol 273:F1–F8.

304. Berger UV, Hediger MA. 1999. Distribution of peptide transporter PEPT2 mRNA in the rat nervous system. Anat Embryol 199:439–449.

305. Xiang J, Chiang P-P, Hu Y, Smith DE, Keep RF. 2006. Role of PEPT2 in glycylsarcosine transport in astrocyte and glioma cultures. Neurosci Lett 396:225–229.

306. Hussain I, Zanic-Grubisic T, Kudo Y, Boyd CAR. 2001. Functional and molecular characterization of a peptide transporter in the rat PC12 neuroendocrine cell line. FEBS Lett 508:350–354.

307. Novotny A, Xiang J, Keep RF, Teuscher NS, Smith DE, Stummer W. 2000. Mechanisms of 5-aminolevulinic acid uptake at the choroid plexus. J Neurochem 75:321–328.

308. Teuscher NS, Smith DE, Novotny A, Keep RF. 2000. Functional evidence for presence of PEPT2 in rat choroid plexus: studies with glycylsarcosine. J Pharmacol Exp Ther 294:494–499.

309. Shu C, Shen H, Teuscher NS, Lorenzi PJ, Keep RF, Smith DE. 2002. Role of PEPT2 in peptide/mimetic trafficking at the blood–cerebrospinal fluid barrier: Studies in rat choroid plexus epithelial cells in primary culture. J Pharmacol Exp Ther 301:820–829.

310. Yamashita T, Takagi T, Shimada S, Guo W, Sato K, Tohyama M, Kohmura E, Hayakawa T. 1997. Cloning and functional expression of a brain peptide/histidine transporter. J Biol Chem 272:10205–10211.

311. Sakata K, Yamashita T, Maeda M, Moriyama Y, Shimada S, Tohyama M. 2001. Cloning of a lymphatic peptide/histidine transporter. Biochem J 356:53–60.

312. von Bossanyi P, Diete S, Dietzmann K, Warich-Kirches M, Kirches E. 1997. Immunohistochemical expression of P-glycoprotein and glutathione S-transferases in cerebral gliomas and response to chemotherapy. Acta Neuropathol (Berl) 94:605–611.

313. Yokogami K, Kawano H, Moriyama T, Uehara H, Sameshima T, Oku T, Goya T, Wakisaka S, Nagamachi S, Jinnouchi S, Tamura S. 1998. Application of SPET using technetium-99m sestamibi in brain tumours and comparison with expression of the MDR-1 gene: Is it possible to predict the response to chemotherapy in patients with gliomas by means of ^{99m}Tc-sestamibi SPET? Eur J Nucl Med 25:401–409.

314. Tanaka Y, Abe Y, Tsugu A, Takamiya Y, Akatsuka A, Tsuruo T, Yamazaki H, Ueyama Y, Sato O, Tamaoki N. 1994. Ultrastructural localization of P-glycoprotein on capillary endothelial cells in human gliomas. Virchows Arch 425:133–138.

315. Korshunov A, Golanov A, Sycheva R, Pronin I. 1999. Prognostic value of tumour associated antigen immunoreactivity and apoptosis in cerebral glioblastomas: an analysis of 168 cases. J Clin Pathol 52:574–580.
316. Abe T, Mori T, Wakabayashi Y, Nakagawa M, Cole SP, Koike K, Kuwano M, Hori S. 1998. Expression of multidrug resistance protein gene in patients with glioma after chemotherapy. J Neurooncol 40:11–18.
317. Decleves X, Fajac A, Lehmann-Che J, Tardy M, Mercier C, Hurbain I, Laplanche JL, Bernaudin JF, Scherrmann JM. 2002. Molecular and functional MDR1-Pgp and MRPs expression in human glioblastoma multiforme cell lines. Int J Cancer 98:173–180.
318. Aronica E, Gorter JA, Redeker S, van Vliet EA, Ramkema M, Scheffer GL, Scheper RJ, van der Valk P, Leenstra S, Baayen JC, Spliet WG, Troost D. 2005. Localization of breast cancer resistance protein (BCRP) in microvessel endothelium of human control and epileptic brain. Epilepsia 46:849–857.
319. Fisher GA, Lum BL, Hausdorff J, Sikic BI. 1996. Pharmacological considerations in the modulation of multidrug resistance. Eur J Cancer 32A:1082–1088.
320. Mohri M, Nitta H, Yamashita J. 2000. Expression of multidrug resistance-associated protein (MRP) in human gliomas. J Neurooncol 49:105–115.
321. van Zanden JJ, de Mul A, Wortelboer HM, Usta M, van Bladeren PJ, Rietjens IM, Cnubben NH. 2005. Reversal of in vitro cellular MRP1 and MRP2 mediated vincristine resistance by the flavonoid myricetin. Biochem Pharmacol 69:1657–1665.
322. Bronger H, König J, Kopplow K, Steiner HH, Ahmadi R, Herold-Mende C, Keppler D, Nies AT. 2005. ABCC drug efflux pumps and organic anion uptake transporters in human gliomas and the blood–tumor barrier. Cancer Res 65:11419–11428.
323. Persidsky Y, Gendelman HE. 2003. Mononuclear phagocyte immunity and the neuropathogenesis of HIV-1 infection. J Leukoc Biol 74:691–701.
324. Kaul M, Zheng J, Okamoto S, Gendelman HE, Lipton SA. 2005. HIV-1 infection and AIDS: consequences for the central nervous system. Cell Death Differ 12(Suppl 1):878–892.
325. Navia BA, Jordan BD, Price RW. 1986. The AIDS dementia complex: I. Clinical features. Ann Neurol 19:517–524.
326. Sacktor N, Lyles RH, Skolasky R, Kleeberger C, Selnes OA, Miller EN, Becker JT, Cohen B, McArthur JC. 2001. HIV-associated neurologic disease incidence changes: Multicenter AIDS Cohort Study, 1990–1998. Neurology 56:257–260.
327. McArthur JC, Haughey N, Gartner S, Conant K, Pardo C, Nath A, Sacktor N. 2003. Human immunodeficiency virus-associated dementia: an evolving disease. J Neurovirol 9:205–221.
328. McArthur JC. 2004. HIV dementia: an evolving disease. J Neuroimmunol 157:3–10.
329. Ruibal-Ares BH, Belmonte L, Bare PC, Parodi CM, Massud I, de Bracco MM. 2004. HIV-1 infection and chemokine receptor modulation. Curr HIV Res 2:39–50.
330. Sukhai M, Yong A, Pak A, Piquette-Miller M. 2001. Decreased expression of P-glycoprotein in interleukin-1beta and interleukin-6 treated rat hepatocytes. Inflamm Res 50:362–370.
331. Hartz AM, Bauer B, Fricker G, Miller DS. 2006. Rapid modulation of P-glycoprotein-mediated transport at the blood–brain barrier by tumor necrosis factor-alpha and lipopolysaccharide. Mol Pharmacol 69:462–470.

332. Lee G, Piquette-Miller M. 2003. Cytokines alter the expression and activity of the multidrug resistance transporters in human hepatoma cell lines; analysis using RT-PCR and cDNA microarrays. J Pharm Sci 92:2152–2163.

333. Ronaldson PT, Bendayan R. 2006. HIV-1 viral envelope glycoprotein gp120 triggers an inflammatory response in cultured rat astrocytes and regulates the functional expression of P-glycoprotein. Mol Pharmacol 70:1087–1098.

334. Gollapudi S, Gupta S. 1990. Human immunodeficiency virus I–induced expression of P-glycoprotein. Biochem Biophys Res Commun 171:1002–1007.

335. Andreana A, Aggarwal S, Gollapudi S, Wien D, Tsuruo T, Gupta S. 1996. Abnormal expression of a 170-kilodalton P-glycoprotein encoded by MDR1 gene, a metabolically active efflux pump, in $CD4^+$ and $CD8^+$ T cells from patients with human immunodeficiency virus type 1 infection. AIDS Res Hum Retrovir 12:1457–1462.

336. Malorni W, Lucia MB, Rainaldi G, Cauda R, Cianfriglia M, Donelli G, Ortona L. 1998. Intracellular expression of P-170 glycoprotein in peripheral blood mononuclear cell subsets from healthy donors and HIV-infected patients. Haematologica 83:13–20.

337. Meaden ER, Hoggard PG, Maher B, Khoo SH, Back DJ. 2001. Expression of P-glycoprotein and multidrug resistance-associated protein in healthy volunteers and HIV-infected patients. AIDS Res Hum Retrovir 17:1329–1332.

338. Hayashi K, Pu H, Andras IE, Eum SY, Yamauchi A, Hennig B, Toborek M. 2006. HIV-TAT protein upregulates expression of multidrug resistance protein 1 in the blood–brain barrier. J Cereb Blood Flow Metab 26:1052–1065.

339. Hayashi K, Pu H, Tian J, Andras IE, Lee YW, Hennig B, Toborek M. 2005. HIV-Tat protein induces P-glycoprotein expression in brain microvascular endothelial cells. J Neurochem 93:1231–1241.

340. Deeks SG, Smith M, Holodniy M, Kahn JO. 1997. HIV-1 protease inhibitors. A review for clinicians. JAMA 277:145–153.

341. Lee CG, Gottesman MM, Cardarelli CO, Ramachandra M, Jeang KT, Ambudkar SV, Pastan I, Dey S. 1998. HIV-1 protease inhibitors are substrates for the MDR1 multidrug transporter. Biochemistry 37:3594–3601.

342. Williams GC, Liu A, Knipp G, Sinko PJ. 2002. Direct evidence that saquinavir is transported by multidrug resistance-associated protein (MRP1) and canalicular multispecific organic anion transporter (MRP2). Antimicrob Agents Chemother 46:3456–3462.

343. Reid G, Wielinga P, Zelcer N, de Haas M, van Deemter L, Wijnholds J, Balzarini J, Borst P. 2003. Characterization of the transport of nucleoside analog drugs by the human multidrug resistance proteins MRP4 and MRP5. Mol Pharmacol 63:1094–1103.

344. Guo Y, Kotova E, Chen ZS, Lee K, Hopper-Borge E, Belinsky MG, Kruh GD. 2003. MRP8, ATP-binding cassette C11 (ABCC11), is a cyclic nucleotide efflux pump and a resistance factor for fluoropyrimidines 2′,3′-dideoxycytidine and 9′-(2′-phosphonylmethoxyethyl)adenine. J Biol Chem 278:29509–29514.

345. Wang X, Furukawa T, Nitanda T, Okamoto M, Sugimoto Y, Akiyama S, Baba M. 2003. Breast cancer resistance protein (BCRP/ABCG2) induces cellular resistance to HIV-1 nucleoside reverse transcriptase inhibitors. Mol Pharmacol 63:65–72.

346. Wang X, Baba M. 2005. The role of breast cancer resistance protein (BCRP/ABCG2) in cellular resistance to HIV-1 nucleoside reverse transcriptase inhibitors. Antivir Chem Chemother 16:213–216.

347. Gupta A, Zhang Y, Unadkat JD, Mao Q. 2004. HIV protease inhibitors are inhibitors but not substrates of the human breast cancer resistance protein (BCRP/ABCG2). J Pharmacol Exp Ther 310:334–341.

348. Huang L, Wring SA, Woolley JL, Brouwer KR, Serabjit-Singh C, Polli JW. 2001. Induction of P-glycoprotein and cytochrome P450 3A by HIV protease inhibitors. Drug Metab Dispos 29:754–760.

349. Perloff MD, von Moltke LL, Marchand JE, Greenblatt DJ. 2001. Ritonavir induces P-glycoprotein expression, multidrug resistance-associated protein (MRP1) expression, and drug transporter-mediated activity in a human intestinal cell line. J Pharm Sci 90:1829–1837.

350. Dupuis ML, Flego M, Molinari A, Cianfriglia M. 2003. Saquinavir induces stable and functional expression of the multidrug transporter P-glycoprotein in human CD4 T-lymphoblastoid CEMrev cells. HIV Med 4:338–345.

351. Chandler B, Almond L, Ford J, Owen A, Hoggard P, Khoo S, Back D. 2003. The effects of protease inhibitors and nonnucleoside reverse transcriptase inhibitors on P-glycoprotein expression in peripheral blood mononuclear cells in vitro. J Acquir Immune Defic Syndr 33:551–556.

352. Yusa K, Oh-hara T, Yamazaki A, Tsukahara S, Satoh W, Tsuruo T. 1990. Cross-resistance to anti-HIV nucleoside analogs in multidrug-resistant human cells. Biochem Biophys Res Commun 169:986–990.

353. Hong M, Schlichter L, Bendayan R. 2001. A novel zidovudine uptake system in microglia. J Pharmacol Exp Ther 296:141–149.

354. Cano-Soldado P, Lorrayoz IM, Molina-Arcas M, Casado FJ, Martinez-Picado J, Lostao MP, Pastor-Anglada M. 2004. Interaction of nucleoside inhibitors of HIV-1 reverse transcriptase with the concentrative nucleoside transporter-1 (SLC28A1). Antivir Ther 9:993–1002.

355. Browne TR, Holmes GL. 2001. Epilepsy. N Engl J Med 344:1145–1151.

356. Kwan P, Brodie MJ. 2000. Epilepsy after the first drug fails: substitution or add-on? Seizure 9:464–468.

357. Loscher W, Potschka H. 2005. Role of drug efflux transporters in the brain for drug disposition and treatment of brain diseases. Prog Neurobiol 76:22–76.

358. Sisodiya SM. 2003. Mechanisms of antiepileptic drug resistance. Curr Opin Neurol 16:197–201.

359. Tishler DM, Weinberg KI, Hinton DR, Barbaro N, Annett GM, Raffel C. 1995. MDR1 gene expression in brain of patients with medically intractable epilepsy. Epilepsia 36: 1–6.

360. Dombrowski SM, Desai SY, Marroni M, Cucullo L, Goodrich K, Bingaman W, Mayberg MR, Bengez L, Janigro D. 2001. Overexpression of multiple drug resistance genes in endothelial cells from patients with refractory epilepsy. Epilepsia 42:1501–1506.

361. Rizzi M, Perego C, Aliprandi M, Richichi C, Ravizza T, Colella D, Veliskova J, Moshe SL, De Simoni MG, Vezzani A. 2003. Glia activation and cytokine increase in rat hippocampus by kainic acid–induced status epilepticus during postnatal development. Neurobiol Dis 14:494–503.

362. Sisodiya SM, Lin WR, Harding BN, Squier MV, Thom M. 2002. Drug resistance in epilepsy: expression of drug resistance proteins in common causes of refractory epilepsy. Brain 125:22–31.

363. Aronica E, Gorter JA, Jansen GH, van Veelen CW, van Rijen PC, Leenstra S, Ramkema M, Scheffer GL, Scheper RJ, Troost D. 2003. Expression and cellular distribution of multidrug transporter proteins in two major causes of medically intractable epilepsy: focal cortical dysplasia and glioneuronal tumors. Neuroscience 118:417–429.

364. Vogelgesang S, Kunert-Keil C, Cascorbi I, Mosyagin I, Schroder E, Runge U, Jedlitschky G, Kroemer HK, Oertel J, Gaab MR, et al. 2004. Expression of multidrug transporters in dysembryoplastic neuroepithelial tumors causing intractable epilepsy. Clin Neuropathol 23:223–231.

365. Benveniste EN, Nguyen VT, O'Keefe GM. 2001. Immunological aspects of microglia: relevance to Alzheimer's disease. Neurochem Int 39:381–391.

366. Giulian D. 1999. Microglia and the immune pathology of Alzheimer disease. Am J Hum Genet 65:13–18.

367. Cummings JL, Vinters HV, Cole GM, Khachaturian ZS. 1998. Alzheimer's disease: etiologies, pathophysiology, cognitive reserve, and treatment opportunities. Neurology 51:S2–S17.

368. Lam FC, Liu R, Lu P, Shapiro AB, Renoir JM, Sharom FJ, Reiner PB. 2001. beta-Amyloid efflux mediated by P-glycoprotein. J Neurochem 76:1121–1128.

369. Vogelgesang S, Warzok RW, Cascorbi I, Kunert-Keil C, Schroeder E, Kroemer HK, Siegmund W, Walker LC, Pahnke J. 2004. The role of P-glycoprotein in cerebral amyloid angiopathy; implications for the early pathogenesis of Alzheimer's disease. Curr Alzheimer Res 1:121–125.

370. Fassbender K, Simons M, Bergmann C, Stroick M, Lutjohann D, Keller P, Runz H, Kuhl S, Bertsch T, von Bergmann K, et al. 2001. Simvastatin strongly reduces levels of Alzheimer's disease beta-amyloid peptides Abeta 42 and Abeta 40 in vitro and in vivo. Proc Natl Acad Sci U S A 98:5856–5861.

371. Ben Shlomo Y, Sieradzan K. 1995. Idiopathic Parkinson's disease: epidemiology, diagnosis and management. Br J Gen Pract 45:261–268.

372. Koller W, Vetere-Overfield B, Gray C, Alexander C, Chin T, Dolezal J, Hassanein R, Tanner C. 1990. Environmental risk factors in Parkinson's disease. Neurology 40:1218–1221.

373. Semchuk KM, Love EJ, Lee RG. 1992. Parkinson's disease and exposure to agricultural work and pesticide chemicals. Neurology 42:1328–1335.

374. Schinkel AH, Wagenaar E, Mol CA, van Deemter L. 1996. P-Glycoprotein in the blood–brain barrier of mice influences the brain penetration and pharmacological activity of many drugs. J Clin Invest 97:2517–2524.

15

DRUG TRANSPORT IN THE KIDNEY

NAOHIKO ANZAI AND HITOSHI ENDOU
Kyorin University School of Medicine, Shinkawa, Mitaka-shi, Tokyo, Japan

Drug Transporters: Molecular Characterization and Role in Drug Disposition, Edited by Guofeng You and Marilyn E. Morris
Copyright © 2007 John Wiley & Sons, Inc.

15.1. INTRODUCTION

A variety of endogenous and exogenous substances that are harmful to the body can be classified into organic anions and cations. Their elimination is therefore essential for the maintenance of homeostasis. Excretory organs such as the kidney, liver, and intestine defend the body against the potentially harmful effects of these compounds by biotransformation into less active metabolites and by the excretory transport process. Particularly in the kidney, drugs and environmental toxicants are eventually excreted into the urine, either in the unchanged form or as biotransformation products. This renal excretion is closely related to the physiological events occurring in nephrons (i.e., filtration, secretion, and reabsorption). More than 70 years ago, phenolsulfophthalein, an anionic dye, was already observed to be highly concentrated in the proximal convoluted tubules, and thus it was used as an indicator of the tubular secretion process.[1] Transport systems responsible for renal tubular secretion of endogenous or exogenous substances have been divided into either organic anion transport systems or organic cation transport systems, based on their preferential substrate selectivity.[2,3] The process of secreting organic anions and cations through the proximal tubular cells is achieved via unidirectional transcellular transport, which involves the uptake of organic ions into the cells from the blood across the basolateral membrane, followed by extrusion across the brush border membrane into the proximal tubular fluid.[4,5]

Generally, the substrates for renal organic anion transport system include a number of chemically heterogeneous weak acids with a carbon backbone and a net negative charge at physiological pH (pK_a < 7). Their structures may be aromatic or aliphatic. Historically, *p*-aminohippurate (PAH) has been used as a prototypical substrate for the renal organic anion transport system.[2,3] PAH is a high-affinity substrate, and it is nearly totally extracted by the renal organic anion transport system during a single passage through the kidney when its serum concentration is low. As a result, the PAH clearance has been used to estimate the renal plasma flow, and the renal organic anion transporter has alternatively been called a PAH transporter. A prominent feature of the PAH transporter is that it interacts with and transports a variety of organic anions with unrelated chemical structures.[3,5,6] Various endogenous organic anions, uremic substances, drugs, and environmental compounds have been assumed to be substrates of the PAH transporter.

Although the substrates for renal organic cation transport systems include numerous structurally divergent cationic compounds, all contain a nitrogen atom and carry a net positive charge at physiological pH. Their structures may be primary, secondary, or tertiary amines or quaternary ammonium salts. Historically, N^1-methylnicotinamide (NMN) and tetraethylammonium (TEA) have been used as model substrates for studying organic cation transport.[3,4] Earlier studies indicate that active secretion and reabsorption of organic cations are both mainly proximal tubular functions. A number of organic cations, particularly when present in low concentrations, are almost completely extracted from blood during a single passage through the kidney. Their clearance can thus approximate the renal plasma flow. The organic cation secretory mechanism is saturable and thereby subject to competitive inhibition.[7] Studies with isolated rat hepatocytes suggest the existence of two different uptake systems for organic cations: type 1 transports small hydrophilic organic cations, and type 2 transports large hydrophobic organic cations.[8,9]

During the last decade, molecular cloning has identified several families of multispecific organic anion and cation transporters, such as organic anion transporter (OAT), organic cation transporter (OCT), organic anion–transporting polypeptide (OATP), sodium-phosphate transporter (NPT), peptide transporter (PEPT), nucleoside transporter (CNT), and the multidrug and toxin extrusion 1 (MATE1). Additional findings have also suggested ATP-dependent organic ion transporters such as MDR1/P-glycoprotein, the multidrug resistance–associated protein (MRP) and the breast cancer resistance protein (BCRP) to act as an efflux pump.

In this chapter we outline the present knowledge of drug transport in the kidney. The first section is an overview of the molecular information on renal drug transporters. It includes the key transporters associated with renal organic anions and cations, peptides, and nucleosides. In the second section we focus on recent advances, particularly the regulatory mechanisms of renal drug transporters, such as gender differences, intracellular signaling, and scaffold proteins. Finally, we discuss the pharmacological and toxicological aspects of renal drug transporters. A number of articles[10,11] and reviews[12–20] concerning organic anion and cation transporters have been published.

15.2. MOLECULAR IDENTIFICATION OF RENAL DRUG TRANSPORTERS

15.2.1. Organic Anion Transporters

Molecular cloning has identified most of the organic anion transporters to belong to the OAT, OATP, NPT, PEPT, CNT, and MRP transporter families and BCRP (Table 15.1 and Figure 15.1). In this part we give an overview of molecular information concerning individual organic anion transporters and their possible roles in renal drug elimination.

Organic Anion Transporter Family SLC22

OAT1 Several research groups have cloned the first member of the OAT family OAT1.[21–23] OAT1 is identical to the previously isolated clone, the novel kidney-specific transporter (NKT).[24] OAT1 mRNA is expressed predominantly in the kidneys and weakly in the brain. In the kidneys, OAT1 protein is localized at the basolateral membrane of proximal tubular cells. The OAT1-mediated uptake of PAH is stimulated by an outwardly directed concentration gradient of dicarboxylates such as α-ketoglutarate, which is consistent with the previous notion that OAT1 is an organic anion–dicarboxylate exchanger.[25] The substrate selectivity of OAT1 is markedly broad, and these substrates include endogenous substances, such as dicarboxylates, cyclic nucleotides, and prostaglandins, and exogenous substances, such as various anionic drugs and environmental compounds.[26] The affinities of OAT1 for these compounds are similar to values reported for the basolateral PAH transporter,[3] and

TABLE 15.1. Drug Transporters Expressed in the Kidneys

Gene Name	Protein Name	Tissue Distribution	Predominant Substrate[a]	Refs.
SLC15: Proton Oligopeptide Cotransporter Family				
SLC15A1	PEPT1	Intestine, kidney	Di- and tripeptides, β–lactam antibiotics, valaciclovir, bestatin	70,71
SLC15A2	PEPT2	Kidney, lung, brain	Di- and tripeptides, β–lactam antibiotics, valaciclovir, bestatin	70,71
SLC17: Vesicular Glutamate Transporter Family				
SLC17A1	NPT1	Kidney, liver	Organic acids, phosphate, chloride	67
SLC21/SLC0: Organic Anion–Transporting Polypepetide Family				
Slco1a1	Oatp1	Brain, colon, kidney, liver, lung, intestine	BSP, TC, $E_2$17βG, LTC_4, DNP-SG, T3, T4, aldosterone, cortisol, ouabain, OTA, temocaprirat	15,58
SLCO1A2	OATP1A2	Brain	Bile salt, PGE_2, fexofenadine	15,60
Slco1a3	Oat-k1	Kidney	TC, $E_2$17βG, ES, DHEAS, folate, T3, T4, MTX	15,61
Slco1a3	Oat-k2	Kidney	TC, $E_2$17βG, ES, DHEAS, folate, T3, T4, MTX	15,62
Slco1a7	Oatp3	Kidney, lung, retina	TC, T3, T4	15,63
SLCO2A1	PGT	Ubiquitous	Eicosanoids, PGs, thromboxane B2, $E_2$17βG	15,60
SLCO2B1	OATP-B	Brain, liver, kidney, heart, placenta	ES, DHEAS, PGE_2, $E_2$17βG, BSP	15,60
SLCO3A1	OATP-D	Ubiquitous	ES, PGE_2, benzylpenicillin	15,60
SLCO4A1	OATP-E	Ubiquitous	TC, T3, T4, rT3, PGE_2, benzylpenicillin	15,60
SLCO4C1	OATP-H	Kidney	T3, digoxin, ouabain	61
Slco1a6	Oatp5	Kidney	Unknown	15,64
SLC22: Organic Cation–Anion Transporter Family				
SLC22A1	OCT1	Liver	TEA, NMN, choline, MPP, dopamine	88,89
SLC22A2	OCT2	Kidney	TEA, NMN, choline, MPP, dopamine	87,93
SLC22A3	OCT3	Liver, skeletal muscle	TEA, guanidine, histamine	95,96
SLC22A4	OCTN1	Kidney, skeletal muscle	TEA, quinidine, verapamil, carnitine	100–102

SLC22A5	OCTN2	Skeletal muscle, kidney	Carnitine, TEA	105,106
SLC22A6	OAT1	Kidney	PAH, PGE_2, cidofovir, acyclovir, ganciclovir, adefovir, MTX, rifampicin, omeprazole, theophylline, OTA, urate	27,28,35,36,46,52
SLC22A7	OAT2	Liver	PGE_2, $PGF_2\alpha$, tetracycline,	35–37
SLC22A8	OAT3	Kidney, brain	ES, cAMP, cGMP, $E_2 17\beta G$, DHEAS, PGE_2, PGF_2a, OTA, MTX, cimetidine, tetracycline, urate	37,40,45,46,48
SLC22A11	OAT4	Kidney, placenta	ES, DHEAS, tetracycline, MTX, OTA, AZT, urate	35,37,44, 46,51,52
SLC22A12	URAT1	Kidney	Urate, nicotinate	53
Slc22a19	Oat5	Kidney	ES, DHEAS, OTA	54–56
SLC28: Concentrative Nucleoside Transporter Family				
SLC28A1	CNT1	Liver, kidney, small Intestine	Pyrimidine nucleoside, azidothymidine	74
SLC28A2	CNT2	Kidney	Purine nucleoside, didanosine	74
SLC??: Multidrug and Toxin Extrusion (MATE) Family				
?	MATE1	Liver, skeletal muscle, kidney	TEA, MPP	113
?	MATE2	Kidney	Unknown	113
			MDR (Superfamily B)	
ABCB1	MDR1	Brain, intestine, kidney	$E_2 17\beta G$, calcein, fluo-3, rhodamine 123, cardiac glycosides, anticancer agents, verapamil, anti-HIV drugs	112
			MRP (Superfamily C)	
ABCC1	MRP1	Ubiquitous	LTC_4, DNP-SG, $E_2 17\beta G$, GSSG, AFB_1-SG, PGA_1-SG, PGA_2-SG, GSH, PAH, MTX, etoposide-G, fluo-3, calcein	188

(*Continued*)

TABLE 15.1. ***(Continued)***

Gene Name	Protein Name	Tissue Distribution	Predominant Substrate[a]	Refs.
ABCC2	MRP2	Brain, intestine, kidney, liver	LTC_4, DNP-SG, $E_2 17\beta G$, folate, glycholate, urate, cAMP, cGMP, PMEA, AZTMP, PAH, MTX, etoposide-G, fluo-3, calcein	188
ABCC3	MRP3	Adrenals, intestine, kidney, liver, pancreas	LTC_4, DNP-SG, $E_2 17\beta G$, folate, glycholate, MTX	188
ABCC4	MRP4	Ubiquitous	$E_2 17\beta G$, urate, cAMP, cGMP, PMEA, AZTMP, MTX, adefovir	188
ABCC5	MRP5	Ubiquitous	GSH, cAMP, cGMP, 6-Mp, adefovir, DNP-SG, CMFDA, BCECF, FDA	188
ABCC6	MRP6	Kidney, liver	LTC_4, NEM-SG, BQ123	188
BCRP (Superfamily G)				
ABCG2	BCRP	Placenta, intestine, kidney (in rodents)	ES, $E_2 17\beta G$, MTX	81

[a]TEA, tetraethylammonium; NMN, N^1-methylnicotinamide; MPP, 1-methyl-4-phenylpyridinium; OTA, ochratoxin A; ES, estrone sulfate; DHEAS, dehydroepiandrosterone sulfate; $E_2 17\beta G$, estradiol-17β-glucuronide; PG, prostaglandin; BSP, bromosulfophtalein; MTX, methotrexate; TC, taurocholate; AFB1-SG, S-(aflatoxin B1)-glutathione; AZTMP, azidothymidine monophosphate; CMFDA, 5-chlomethylfluorescein; 6-MP, 6-mercaptopurine; MTAL, medullary thick ascending limb; NEM-SG, *N*-ethylmaleimide glutathione; PMEA, 9-(2-phosphonylmethoxyethyl)adenine; BQ123 is an endothelin-receptor antagonist, cyclo[Trp-Asp-Pro-Val-Leu].

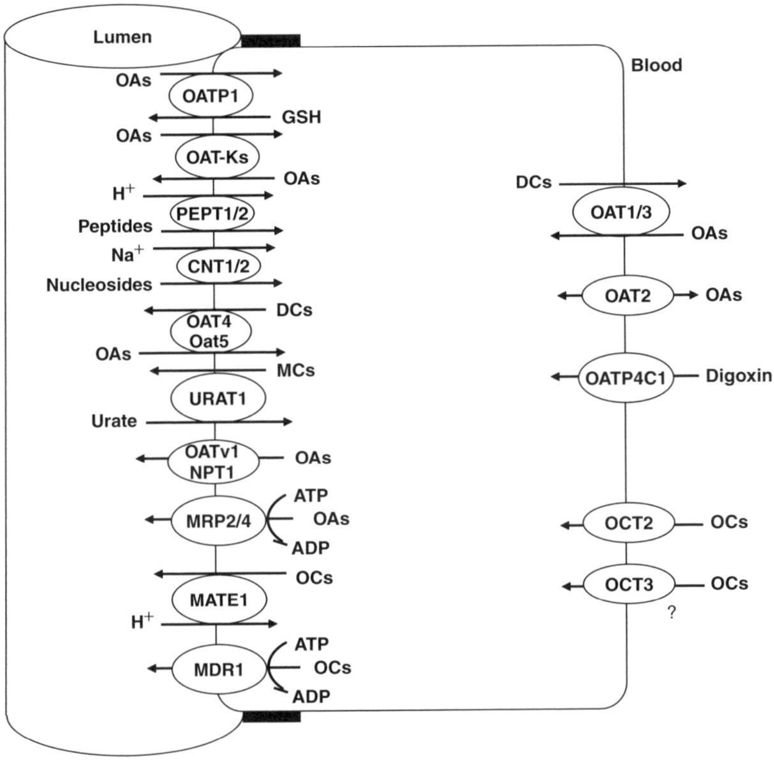

FIGURE 15.1. Proposed model of organic anion and cation transporters in renal proximal tubules. OAs, organic anions; OCs, organic cations; DCs, dicarboxylates; MCs, monocarboxylates.

the functional properties and localization of OAT1 are identical to those of the renal PAH transport system. The alternative splice variants of human OAT1 have been identified[27,28]; OAT1-1 and OAT1-2 appear to be functionally almost identical, whereas no functions have been detected for OAT1-3 and OAT1-4.[29]

OAT2 OAT2 was originally isolated from the rat liver as a novel liver-specific transport protein with an unknown function.[30] Because of its structural similarities to OAT1, OAT2 was functionally characterized.[31] OAT2 is expressed predominantly in the liver and weakly in the kidneys. The typical substrates of OAT2 are salicylate, acetylsalicylate, prostaglandin E_2 (PGE_2), dicarboxylates, PAH, zidovudine (AZT), and tetracycline.[32–37]

OAT3 OAT3 was isolated from the rat,[38] and it seems to us structurally identical to Roct, which has been identified as a transporterlike protein that exhibits a reduced expression in osteosclerosis mice.[39] OAT3 mRNA is expressed in the kidneys, liver, brain, and eye.[38] In the kidneys, OAT3 is localized at the basolateral membrane of the proximal tubular cells.[40,41] In the brain, OAT3 is localized at the luminal membrane of choroid plexus cells[42,43] and in capillary endothelial cells.[44] Similar to OAT1, OAT3 recognizes a broad spectrum of substrates, which mediates the high-affinity transport of PAH, estrone sulfate, ochratoxin A, and various drugs, including the cationic drug cimetidine, in exchange for dicarboxylates inside cells.[35–37,45–48]

OAT4 OAT4 was cloned from human kidneys.[49] OAT4 mRNA is expressed in the kidneys and is localized at the apical membrane of proximal tubular cells. In the placenta, OAT4 is expressed on the fetal side of the syncytiotrophoblast cells.[50] When expressed in *Xenopus* oocytes, OAT4 mediates the Na^+-independent, high-affinity transport of estrone sulfate, dehydroepiandrosterone sulfate, ochratoxin A, and PGE_2, $PGF_2\alpha$, and urate.[35,37,46,49,51,52] A recent study demonstrated that OAT4 functions as an organic anion–dicarboxylate exchanger.[53]

URAT1 URAT1 is expressed exclusively in the kidneys, where it is located in the apical membrane of proximal tubular cells.[54] URAT1 exhibits Na^+-independent uptake of urate and regulates the serum urate level; genetic defects in URAT1 are the predominant causes of idiopathic renal hypouricemia. In a study using URAT1 cRNA-injected *Xenopus* oocytes, the *cis*-inhibitory effect of uricosuric drugs (e.g., probenecid, benzbromarone, sulfinpyrazone, losartan) and the *trans*-stimulatory effect of antiuricosuric drugs (e.g., pyrazinoic acid, the metabolite of the antituberculous agent pyrazinamide) on the URAT1-mediated transport of urate were demonstrated.[54]

Oat5 Oat5 was recently identified from mouse[55] and rat.[56] It is expressed exclusively in the kidneys, and rat Oat5 is localized at the apical side of the proximal tubules.[56] Oat5 mediates the transport of steroid sulfates as well as ochratoxin A.[55–57]

Other than these six clones, Sun et al. has reported two OAT-related clones and designated them as hOAT4 and hOAT5, respectively.[58] However, they did not demonstrate any transport function, and human OAT4, identified by Cha and colleagues,[49] and mouse and rat Oat5,[55,56] are not identical to these two clones.

Organic Anion–Transporting Polypeptide Family SLC21/SLCO The first member of this family, oatp1, was identified from the rat liver by an expression cloning method as a sodium-independent bile acid transporter.[59] Thus far, 11 human isoforms and 14 rat isoforms have been identified in the OATP family.[60] Although some OATPs are involved selectively in the hepatic uptake of bulky and relatively hydrophobic organic anions, most OATPs are expressed in many tissues, such as the blood–brain barrier, choroids plexus, lungs, heart, intestine, kidneys, placenta, and testes.[60] To clarify the confusing and species-dependent "old" nomenclature, a novel nomenclature has recently been assigned to the OATP family (Table 15.1). The OATP superfamily was subdivided into several families (40% amino acid sequence identity) and subfamilies (60% amino acid sequence identity).[60] The OATP family is now divided into six families (OATP1 to OATP6). There are considerable species differences in the OATP family among rodents and humans. Among human OATPs, only OATP4C1 is mainly expressed in the kidneys.[61] Oatp1a3v1 (previous name: OAT-K1)[62] and Oatp1a3v2 (previous name: OAT-K2)[63] are expressed specifically in the rat. Oatp1a1 (previous name: oatp1),[59] Oatp1a5 (previous name: oatp3),[64] Oatp1a6 (previous name: oatp5),[65] and Oatp4c1[61] are expressed in rodent kidneys. The orthologs of these isoforms, except OATP4C1, are absent in humans. Because of the above-mentioned remarkable species differences in OATP, it is difficult to assign distinct physiological roles to each OATP in the kidneys. The role of OATP4C1 in the kidneys is evident. There are several important substances that are preferable substrates for the OATP family, which are excreted primarily via the kidneys. One example is digoxin, a cardiac glycoside. The exit pathway for digoxin at the apical membrane of proximal tubular cells has been assumed to be an ATP-dependent efflux pump, P-glycoprotein (Pgp).[66] However, the basolateral entrance for digoxin was as yet unknown. Recently, OATP4C1 has been revealed to be a digoxin transporter.[61] OATP4C1 is expressed exclusively in the basolateral membrane of proximal tubular cells and mediates the high-affinity transport of digoxin ($K_m = 7.8$ μM) and ouabain ($K_m = 0.38$ μM), as well as thyroid hormones such as triiodothyronine ($K_m = 5.9$ μM). These data suggest that OATP4C1 is a digoxin transporter localized in the basolateral membrane of proximal tubular cells and plays a central role in the renal elimination of digoxin.

Sodium/Phosphate Transporter Type I Family SLC17 Molecular studies have determined that type 1 phosphate transporters (NPT1s; SLC17), a family of proteins initially characterized as phosphate carriers, expressed at the apical membrane of renal proximal tubular cells, mediate the transport of organic anions (Figure 15.1).[67] Mouse and human NPT1 were shown to mediate the transport of various organic anions in a chloride-dependent manner. Moreover, because human NPT1 exhibits an affinity for PAH, corresponding to previous reports using brush border membrane vesicles, NPT1 is also suggested as representing the classical voltage-dependent PAH transporter.[13] However, an influence of the membrane potential on PAH transport was not demonstrated.[68] We isolated a novel transport protein with the properties of voltage-driven organic anion transport from pig kidney cortex by expression cloning in *Xenopus* oocytes.[69] A cDNA encoding a 467-amino acid peptide was designated as OATv1 (voltage-driven organic anion transporter 1). The predicted amino acid sequence of OATv1 exhibited 60 to 65% identity to those of human, rat, rabbit, and mouse NPT1/Npt1. OATv1 mediates the transport of PAH, estrone sulfate, estradiol-17β-glucuronide, and urate. PAH transport via OATv1 was affected by the changes in membrane potential. This transport protein is localized at the apical membrane of renal proximal tubule, which is consistent with the proposed localization of a voltage-driven organic anion transporter. OATv1 has therefore been proposed to play an important role in the excretion of various organic anions, including PAH and urate driven by membrane voltage through the apical membrane of the tubular epithelial cells into the urine.

Peptide Transporter Family SLC15 Peptide transporters are involved in electrogenic, H^+-dependent transport of small peptides as well as various peptidelike drugs,[70,71] such as β-lactam antibiotics, angiotensin-converting enzyme (ACE) inhibitors, and anticancer drugs. Two peptide transporters, designated as PEPT1 and PEPT2, have been cloned.[70,71] PEPT1, a low-affinity/high-capacity transporter, was first cloned from the rabbit intestine and subsequently from rat and human. Rat Pept1 was localized to the apical side of intestinal epithelial cells[72] and in early regions (S1 segments) of apical proximal tubules.[73] PEPT2, a high-affinity/low-capacity transporter, appeared to have different tissue localization than PEPT1, in that PEP2 is highly expressed in the kidney but not in the intestine. Rat Pept2 was localized to the apical side of the proximal tubule in more distal regions (S3 segments).[73] Generally, substrates for Pept2 are similar to those of Pept1, although their affinities are different.

Nucleoside Transporter Family SLC28 Nucleoside transporters in the the concentrative inwardly directed Na^+/nucleoside cotransport system (CNT family; SLC28)[74] and the equilibrative bidirectional facilitators (ENT family; SLC29)[75] play critical roles in nucleoside salvage pathways, where they mediate the first step of nucleotide biosynthesis. In addition, these transporters work in concert to terminate adenosine signaling. CNT family members are crucial determinants of response to a variety of anticancer and antiviral nucleoside analogs, as they modulate the entry of these analogs into target tissues. Furthermore, this family is involved in the absorption and disposition of many nucleoside analogs.

Multidrug Resistance–Associated Protein Family ABCC The MRP family consists of primarily active transporter with ATP-binding cassette motifs. The prototype of this family is Pgp,[66] which extrudes various hydrophobic molecules, particularly antineoplastic compounds, such as vincristine, vinblastine, adriamycin, and daunorubicin, and it confers multidrug resistance on cancer cells.[76] MRP1 and MRP2 were isolated from cancer cells with multidrug resistance that do not express Pgp. In addition to antineoplastic drugs, MRP2 transports glucronides and cysteine conjugates, and it is expressed in the canalicular membrane of hepatocytes.[13] MRP2-deficient mice lack the activity to extrude conjugate anions from the liver, thus resulting in the phenotype of the Dubin–Johnson syndrome.[77] To date, many isoforms have been identified in the MRP family,[13,78] and several of these isoforms are expressed in the apical membrane of proximal tubular cells (Figure 15.1). MRP members in proximal tubular cells supposedly function as an extrusion pump for organic anions from the apical membrane, especially large and hydrophobic organic anions. Regarding the renal physiology and pharmacology, particular attention should be paid to two isoforms: MRP2 and MRP4. MRP2 has been shown to transport PAH, but its affinity for PAH is low ($K_m = 2$ mM). The observation that the renal excretion of PAH in isolated perfused kidneys from Mrp2-deficient rats is not significantly different from those in the kidneys from wild-type rats suggests a modest contribution of MRP2, if any, to the efflux of PAH.[79] In contrast, human MRP4, which is also localized in the apical membrane of proximal tubular cells, transports PAH with a much higher affinity ($K_m = 160\ \mu$M) compared with MRP2. Furthermore, real-time PCR and Western blot analysis showed that the renal cortical expression of MRP4 is approximately fivefold higher than that of MRP2.[79] These data demonstrate that MRP4 plays a certain role in the efflux of PAH and several small hydrophilic organic anions, such as urate, cAMP, and cGMP, into the tubular lumen.[80]

Breast Cancer Resistance Protein/ABCG2 Breast cancer resistance protein (BCRP) is the second isoform of the WHITE subfamily of the human ATP-binding cassette (ABC) transporter superfamily and thus also named as ABCG2.[81] BCRP is called a *half-transporter* because it contains a single N-terminal ATP-binding cassette followed by six putative transmembrane segments and may function as a homodimer or homotetramer. BCRP expression was detected in several hematological malignancies and solid tumors, suggesting its role in clinical drug resistance of cancer.[81] BCRP can accept a variety of organic anions, such as sulfated conjugates of steroids and xenobiotics in addition to the anticancer drugs mitoxantrone, topotecan, and methotrexate.[81] Bcrp mRNA levels were high in rodent kidneys[82] but BCRP seems unlikely to play an important role in renal secretion of drugs in humans as in rodents, since there is little expression of BCRP in human kidney.

15.2.2. Organic Cation Transporters

In this part we give an overview of molecular information concerning individual organic cation transporters (OCT family and P-glycoprotein) and their possible roles in renal drug elimination (Table 15.1 and Figure 15.1).

Organic Cation Transporter Family SLC22

OCT1 OCT1 was cloned from rat kidney and characterized functionally in 1994 using *Xenopus* oocytes.[83] Rat Oct1 (rOct1) mRNA was expressed in the liver, intestine, and kidney. rOct1 protein was localized to the basolateral membrane of the S1 and S2 segments of renal proximal tubules.[84,85] When expressed in oocytes, rOct1 stimulated its TEA uptake and inhibited by diverse organic cations.[83] Electrophysiological study indicated that the rOct1-mediated cation transport is electrogenic. TEA uptake was decreased by acidifying the medium pH, suggesting that rOct1-mediated uptale was pH sensitive. The human OCT1 mRNA expression was observed predominantly in the liver.[86,87] Although it has been shown that hydrophobicity is a major determinant of drug interactions with OCT1,[88] significant differences have been found among the cloned OCT1 transporters from different species in terms of the kinetics and substrate selectivities.[89] These findings suggest that OCT1 may be responsible, in part, for interspecies differences in the disposition of organic cations.

OCT2 Oct2 was isolated from a rat kidney cDNA library by Okuda et al.[90] Rat Oct2 (rOct2) mRNA was expressed mainly in the kidney, but not in the liver, lung, or intestine. rOct2 was localized to the basolateral membrane of the proximal tubules.[91,92] rOct2 has been shown to interact with various cationic compounds, such as *N*-methyl-4-phenylpyridinium (MPP^+), cimetidine, NMN, nicotine, quinine, and quinidine.[85] When expressed in oocytes, rOct2-mediated TEA uptake was suppressed by the replacement of Na^+ with K^+, thus indicating that the rOct2 transport was membrane-potential dependent. Human OCT2 has also been identified.[87] Interestingly, unlike rOct2, hOCT2 was localized to the apical membrane of the distal tubule.[87] A splice variant of hOCT2, OCT2A, has a truncated C-terminus lacking the last three putative transmembrane domains, transported TEA with about 5% of the wild type's maximal rate, but revealed higher affinity for several organic catios.[93] The reason for the species differences observed in the localization of OCT2 in the kidney is not yet known.

OCT3 OCT3 was isolated from a rat placental cDNA library,[94] and their human and mouse homologs were identified successively.[95–97] Rodent Oct3 and human OCT3 mRNA was expressed in various tissues, including the kidney.[94,97] Rat Oct3 exhibited an uptake of TEA and guanidine, which was inhibited by MPP^+. Electrophysiological studies revealed that rOct3-mediated TEA uptake evoked a potential-dependent inward current, which was markedly influenced by the extracellular pH. Rat Oct3 interacted with dopamine, the neurotoxins amphetamine and methamphetamine, as well as a variety of steroids.[98] Although human OCT3 transports diverse organic cations, including catecholamines,[95,96] OCT3 appears to be inhibited selectively by corticosterone and *o*-methylisoprenaline compared to other OCTs.[99] The localization and intrarenal distribution of OCT3 is unknown.

OCTN1 Two other members of the OCT family, OCTN1 and OCTN2, have been cloned based on their homology to OCT. OCTN1 was identified from human fetal liver.[100] Human OCTN1 (hOCTN1) mRNA was expressed abundantly in the kidney, trachea, bone marrow, fetal liver, and several human cancer cell lines.[100] When expressed in HEK293 cells, hOCTN1 mediated the saturable and pH-dependent TEA uptake. TEA efflux mediated by OCTN1 was also dependent on the acidic external medium pH.[101] OCTN1 transported several drugs and endogenous compounds, including quinidine, verapamil, and carnitine.[101] Furthermore, OCTN1 was inhibited by various drugs, such as cimetidine, procainamide, pyrilamine, quinine, cephaloridine, and verapamil.[101] Very recently, Grundemann et al. demonstrated that the key substrate of OCTN1 is ergothioneine, a product biosynthesized by fungi and mycobacteria. They thus proposed the functional name ETT (ET transporter) instead of OCTN1.[102]

OCTN2 OCTN2 was identified from a human placental trophoblast cell line by homology search, and their mouse and rat homologs were isolated successively.[103,104] OCTN2 mRNA was detected strongly in adult human kidney, trachea, spleen, bone marrow, skeletal muscle, heart, and placenta.[105] When expressed in HEK293 cells, hOCTN2 mediated L-carnitine uptake in a sodium-dependent manner, and it also mediated the uptake of TEA and guanidine.[105] Mutations in the OCTN2 gene *SLC22A5* have been linked causally to primary systemic carnitine deficiency, an autosomal recessive disease characterized by low serum and intracellular concentrations of carnitine.[106–109] Two mutations in OCTN2, P478L and L352R, resulted in a complete loss of carnitine transport function, but P478L actually had higher organic cation transport activity than the wild type.[110] These results indicate that the binding sites of OCTN2 for carnitine and organic cations significantly overlap but are not identical.

Based on the search for OAT isoforms, we identified a clone with relatively ubiquitous tissue distribution and named it carnitine transporter 1 (CT1).[111] CT1 mediated the high-affinity transport of L-carnitine ($K_m = 25$ μM) in a partially sodium-dependent manner. Octanoylcarnitine, acetylcarnitine, and γ-butyrobetaine potently inhibited the CT1-mediated carnitine transport.

MDR1/P-Glycoprotein ABCB MDR1 is a member of the ABC transporter superfamily. MDR1 actively extruded from the cells drugs with diverse structures, such as *Vinca* alkaloids, steroids, cyclosporines, tacrolimus, anthracyclines, and miscellaneous hydrophobiccations.[66] Although the cellular drug efflux mediated by Pgp was first identified in cancer cells, Pgp was found to be highly expressed in a number of normal tissues, such as liver, pancreas, kidney, colon, and jejunum. In the kidney, MDR1 was found to be concentrated particularly on the apical surface of epithelial cells of the proximal tubules, where it secretes various drug substrates into the lumen.[112] The finding that a cardiac glycoside digoxin is actively secreted via Pgp in renal proximal tubules is clinically important for transporter-mediated drug interactions.

MATE1 Very recently, Otsuka et al. show that MATE1, a human and mouse ortholog of the multidrug and toxin extrusion family conferring multidrug resistance on bacteria, is expressed primarily in the kidney and liver, where it is localized to the luminal membranes of the renal tubules and bile canaliculi.[113] When expressed in HEK293 cells, MATE1 mediates H^+-coupled electroneutral exchange of TEA and MPP^+. Its substrate specificity is similar to those of renal and hepatic H^+-coupled organic cations export. Thus, MATE1 appears to be the long-searched-for polyspecific organic cation exporter that directly transports toxic organic cations into urine and bile.

15.3. REGULATION OF RENAL DRUG TRANSPORTERS

15.3.1. Gender and Developmental Differences

Gender differences in mRNA and/or protein expression have been reported for OAT1,[114,115] OAT2,[34,115–117] OAT3,[114–116] URAT1,[118] and OCT2,[119,120] thereby suggesting that some OAT members and OCT2 are regulated by sex hormones. The mouse Oat1 mRNA levels were higher in the male kidney than in the female kidney. Rat Oat2 mRNA expression was higher in the female kidney than in the male kidney or liver. In contrast, the mouse Oat2 mRNA levels were highest in both kidneys and low in the male liver. In the male liver, rat Oat3 mRNA expression was detected. The mouse Urat1 mRNA levels were higher in the male kidney than in the female kidney. The rat Oct2 mRNA levels were higher in the male kidney than in the female kidney. Developmental changes of expression have been reported in OATs[115,121,122]; the mRNA expression levels of OAT1, OAT2, and OAT3 increased during postnatal development.

15.3.2. Phosphorylation

All OAT isoforms have several sites for phosphorylation by protein kinase C (PKC) in the large intracellular loop between the sixth and seventh transmembrane domains (TMDs). Several studies have revealed that the activation of PKC decreases the transport activity of OATs.[123–126] This inhibitory effect is also associated with altered substrate selectivity. A reduced OAT-mediated transport activity is rescued by PKC inhibitors.[123,124] Furthermore, recent studies have demonstrated the OAT1 activity to be stimulated by epidermal growth factor (EGF) via mitogen-activated protein kinases (MAPKs).[127] In addition to the sites for phosphorylation by PKC, OAT isoforms have putative sites for phosphorylation by PKA, casein kinase II, or tyrosine kinase. It is not clear whether these protein kinases are involved in the regulation of transporter functions.

Similar to OATs, All OCT isoforms also have several sites for PKC phosphorylation. In HEK293 cells, the rOct1-mediated transport of 4-(4-(dimethylamino)styryl)-*N*-methylpyridinium iodide (ASP) was stimulated by PKC, PKA, and tyrosine kinase activators.[128] In contrast, ASP uptake into hOCT2-expressing HEK293 cells was not affected by PKC activator, whereas it was slightly inhibited by PKA agonist.[129] OCT2 is constitutively activated by Ca^{2+}/calmodulin complex.[129]

15.3.3. Protein–Protein Interaction

Several renal apical transporters possess the PDZ motif at their C-terminus.[13,130] The PDZ motif is one of the important protein–protein interaction modules, and it is composed of three amino acid residues: S/T-X-Φ (where X is any amino acid and Φ is a hydrophobic residue). The renal apical organic anion transporters URAT1, OAT4, PEPT2, and OCTN2 possess the PDZ motif at their C-terminus. Yeast two-hybrid experiments revealed that these transporters interact with the multivalent PDZ domain-containing proteins such as PDZK1 and NHERF1 via their C-terminal PDZ motifs.[131–134] The coexpression of transporters and PDZK1 in HEK293 cells increases their transport activity. This synergic effect is abolished when the C-terminal PDZ motif deletion mutants of transporters are coexpressed with PDZK1. These results indicate that PDZK1 regulates transport activities via interaction with the PDZ motif.

15.3.4. Glycosylation

The glycosylation sites in the first extracellular loop between first and second TMDs are conserved in OATs. Tunicamycin, an inhibitor of asparagine-linked glycosylation, inhibited PAH transport activity in mOat1-transfected COS7 cells.[135] Immunofluorescence revealed that the mOat1 protein remained primarily in the intracellular compartment after tunicamycin treatment.[135] This study indicates that the glycosylation of the mOat1 protein is necessary for proper trafficking of the protein to the plasma membrane. Other experiments have demonstrated that disrupting Asp39 (one of the glycosylated sites) in mice resulted in a complete loss of transport activity of OAT1 without affecting its surface expression.[136] As a result, glycosylation could also be responsible for substrate recognition.

15.4. RENAL HANDLING OF SELECTED DRUGS

15.4.1. Cephalosporin Antibiotics

Cephalosporin antibiotics are suggested not only to be filtered through the glomeruli but also secreted actively by the proximal tubules. Cephalosporins inhibited PAH uptake in rat renal slices[137] and renal plasma membrane vesicles.[138] Cephaloridine, a cephalosporin that possesses both anionic and cationic moieties inhibited PAH transport but not NMN transport in basolateral membrane vesicles.[139] Cephalosporin antibiotics are thus considered to be secreted by the proximal tubule via the PAH transport system.[8] Consistent with these results, we have observed that rOat1 as well as rOat3 interacts with various cephalosporin antibiotics.[140,141]

Using proximal tubular cells stably expressing human OAT1, OAT3, and OAT4 (S_2-hOAT1, S_2-hOAT3, S_2-OAT4), we elucidated the interaction of human OATs with various cephalosporin antibiotics.[142] All of cephalosporin antibiotics used (cephalothin, cefoperazone, cefazolin, ceftriaxone, cephaloridine, cefotaxime, cefadroxil, and cefamandole) significantly inhibited organic anion uptake mediated by hOAT1, hOAT3, and OAT4 in a competitive manner.

15.4.2. Diuretics

Diuretics cause natriuresis and are therefore used to treat patients with volume overload, including hypertension, liver cirrhosis, nephrotic syndrome, and congestive heart failure.[143] Thiazides and loop diuretics exhibit their diuretic effects from the luminal side by inhibiting the Na^+–Cl^- cotransporter of the distal tubule and the Na^+–K^+–$2Cl^-$ cotransporter of the loop of Henle, respectively.[143] In addition, because the binding of diuretics to plasma proteins is generally high (more than 90%), tubular secretion is the main route of urinary excretion of the diuretics. Tubular secretion has thus been thought to play a critical role in the action of loop and thiazide diuretics. Renal tubular secretion of diuretics has been demonstrated in studies dealing with the secretion of bumetanide and furosemide in the isolated perfused rat kidney[144,145] and renal tubular secretion of chlorothiazide and hydrochlorothiazide in the avian kidney.[146] Thiazide and loop diuretics, which both contain a sulfamoyl group (sulfonamide diuretics) as a common chemical characteristic, are weak organic acids. Consistent with this, the involvement of the organic anion transport system in the tubular secretion of diuretics has been suggested in studies including bumetanide inhibition of the PAH transport in rat renal slices[147] and the PAH inhibition of furosemide excretion in the rabbit.[148]

Using S_2-hOATs, we elucidated the interaction of human OATs with various diuretics.[149] Diuretics tested (i.e., thiazides, loop diuretics, and carbonic anhydrase inhibitors) inhibited organic anion uptake mediated by hOAT1, hOAT2, hOAT3, and

OAT4 in a competitive manner. hOAT1 exhibited the highest-affinity interactions for thiazides, whereas hOAT3 did those for loop diuretics.

15.4.3. Nonsteroidal Anti-inflammatory Drugs

Nonsteroidal anti-inflammatory drugs (NSAIDs) have been widely used for their anti-inflammatory and analgesic properties. Previous studies indicated the active accumulation of NSAIDs in the renal proximal tubular cells. The accumulation of indomethacin and salicylate has been demonstrated in rat proximal tubular cells.[150,151] In particular, renal handling of salicylate was studied extensively by micropuncture experiments in vivo,[152] isolated proximal tubules,[151,153] renal cortical slices,[154,155] and a kidney epithelial cell line.[156] In addition, there have been reports on the interaction of NSAIDs with other organic anions, such as prostaglandins[157] and penicillin.[158] The results of these studies suggest that NSAIDs may be transported via the renal organic anion transport. Consistent with these results, we have previously demonstrated the interaction of rOat1 with NSAIDs using an oocyte expression system.[159]

Using S_2-hOATs as well as S_2-hOCT1 and S_2-hOCT2, we elucidated the interaction of human OATs and OCTs with various NSAIDs (Table 15.2).[160] NSAIDs tested (i.e., acetaminophen, acetylsalicylate, salicylate, diclofenac, ibuprofen, indomethacin, ketoprofen, mefenamic acid, naproxen, piroxicam, phenacetin, and sulindac) inhibited organic anion uptake mediated by hOAT1, hOAT2, hOAT3, and OAT4. Although the organic cation uptake mediated by hOCT1 and hOCT2 was also inhibited by some NSAIDs, hOCT1 and hOCT2 did not mediate the uptake of NSAIDs. This suggests the interactions of hOATs with NSAIDs to be associated with the pharmacokinetics and the induction of adverse reactions of NSAIDs.

15.4.4. Antiviral Drugs

Both acyclovir (ACV) and ganciclovir (GCV) are acyclic guanosine derivatives.[161] ACV is used in the treatment of various forms of herpes simplex infections.[161] Valacyclovir (VACV) is the L-valyl ester of ACV, which is active against herpes simplex virus types 1 and 2, and varicella zoster virus.[162] GCV is used in the treatment of cytomegalovirus infections in acquired immunodeficiency syndrome and in transplant patients.[161] On the other hand, 3′-azido-3′-deoxythymidine (zidovudine, AZT) is widely used for the treatment of HIV infection.[162] Approximately 83% of ACV, 90% of GCV, and 80% of AZT are excreted in their unchanged forms by the kidney.[163–165] The renal excretion of ACV and AZT is reduced by probenecid, a typical inhibitor of organic anion transport.[163,166,167] Although neither possesses a typical anionic moiety, the results suggest that the renal organic anion transport system is responsible for the tubular secretion of these drugs. On the other hand, the involvement of an organic cation transport system has also been suggested in the tubular secretion of AZT because cimetidine, an organic cation, also reduces the renal clearance of AZT.[166]

Using S_2-hOATs as well as S_2-hOCT1 and S_2-hOCT2, we elucidated the interaction of human OATs and OCTs with antiviral agents such as ACV, GCV, and AZT (Table 15.2).[35] Time- and concentration-dependent uptake of ACV and GCV was

TABLE 15.2. Comparison of the Substrate Selectivities of hOAT1, hOAT2, hOAT3, hOAT4, hOCT1, and hOCT2, as Determined in Stable Transfected S_2 Cells

	Uptake K_m (μM) via:					
Labeled Compound	hOAT1	hOAT2	hOAT3	hOAT4	hOCT1	hOCT2
[^{14}C]TEA	−	−	−	−	+	+
[^{14}C]PAH	+	+	+	+	−	−
[^{14}C]Glutarate	+	+	+	+	−	−
[^{3}H]Estrone sulfate	+	+	+	+	−	−
[^{14}C]Urate	+	−	+	+	−	−
[^{14}C]Succinate	+	−	−	+	−	−
[^{14}C]Azidothymidine	+ (45.9)	+ (26.8)	+ (145.1)	+ (151.8)	−	−
[^{3}H]Acyclovir	+ (342.3)	−	−	−	+ (151.2)	−
[^{3}H]Ganciclovir	+ (895.5)	−	−	−	+ (516.2)	−
[^{3}H]Valaciclovir	−	−	+	−	−	−
[^{3}H]PGE$_2$	+	+	+	+	+	+
[^{3}H]PGF$_2$ α	+	+	+	+	+	+
[^{14}H]Indomethacin	+	−	−	−	−	−
[^{14}C]Salicylate	+	+	+	−	−	−
[^{3}H]Cimetidine	+	−	+	−	+	−
[^{3}H]Tetracycline	+	+	+	+	−	−
[^{3}H]Methotrexate	+	−	+	−	−	−
[^{3}H]cGMP	+	−	+	−	−	−
[^{3}H]Ochratoxin A	+	−	+	+	−	−
[^{3}H]DHEAS	−	−	+	+	−	−

Source: 13,15,19,27,33,35–37,40,45,46,49,51,160 and unpublished observation (Anzai and Endou).

observed in S_2-hOAT1 and S_2-hOCT1. In contrast, uptake of valacyclovir, L-valyl ester of ACV, was observed only in S_2-hOAT3. On the other hand, AZT uptake was observed in S_2-hOAT1, S_2-hOAT2, S_2-hOAT3, and S_2-OAT4.

15.5. MISCELLANEOUS ASPECTS OF RENAL DRUG TRANSPORTERS

15.5.1. Transporter-Mediated Drug–Drug Interactions

Drugs present in plasma could affect the transport of these drugs individually while mutually influencing the pharmacokinetics of the drugs. A notable example is the concomitant use of probenecid and penicillin G; the half-life of penicillin G is prolonged significantly when combined with probenecid compared with when it is administered alone. It has also been reported that methotrexate (MTX) administration with acidic drugs, such as NSAIDs, and β-lactam antibiotics, causes a severe suppression of bone marrow. NSAIDs and β-lactam antibiotics inhibit the tubular secretion of MTX, thereby reducing its renal clearance. As a consequence, unwanted side effects, such as bone marrow suppression, could occur as a result of the increase in plasma MTX levels.[46,168] These phenomena can be explained at the level of OAT1 and OAT3.

Similar OAT-mediated drug–drug interactions have been reported for diuretics,[169] nucleoside analogs,[170] and antiviral drugs.[171]

15.5.2. Transporter-Mediated Nephrotoxicity

Nephrotoxic Drugs OATs are involved in the development of organ-specific toxicity of drugs and their metabolites. For example, the nephrotoxic effects of β-lactam antibiotics (e.g., cephaloridine) and carbapenem antibiotics are closely associated with OATs,[172] and OATs are also responsible for the nephrotoxicity of antiviral drugs such as adefovir and cidofovir.[173] Indeed, β-lactam antibiotics[141] and antiviral drugs[174] exhibit a significantly high cytotoxicity in OAT1-transfected cell cultures.

The nephrotoxicity of all these compounds could be reduced by coadministration of other substrates of OATs or inhibitors of OATs. Indeed, recently, a new application of probenecid as a nephroprotectant in therapy with the antiviral drug cidofovir was determined.[175] When probenecid is coadministered with cidofovir, probenecid inhibits the tubular accumulation of cidofovir, thus reducing its potential risk.

Environmental Substances Ochratoxin A is a mycotoxin that contaminates cereals and is thought to be responsible for Balkan nephropathy,[176] an endemic nephropathy that exhibits characteristic chronic tubulointerstitial changes. In OAT1-expressing oocytes and OAT1-transfected cell cultures, the addition of ochratoxin A to the culture media decreases cell viability.[177] This decreased cell viability is abolished by the nontoxic substrates of OAT1, such as PAH.

Uremic Toxin The uremic toxins and their metabolites, produced during the catabolism process within the body, seem to be associated with the exacerbation of renal function in renal failure. Indoxyl sulfate, a uremic toxin derived from dietary proteins, is a substrate of OATs.[178–180] Immunohistochemical analyses revealed that 5/6-nephrectomized rats, an animal model of chronic renal failure, showed higher intensities of OAT1 and OAT3 proteins than did sham-operated rats.[180] These data suggest that OATs are also involved in the progression of chronic renal failure. In the body, uremic substances that cannot be eliminated by glomerular filtration during renal failure should be removed via tubular secretion mediated by OATs. In these cases, OAT protein expression levels increased, resulting in the accumulation of toxic substances in the tubules. OAT1 and OAT3 are also involved in the uptake of other uremic toxins, such as 3-carboxy-4-methyl-5-propyl-2-furanpropionate, indoleacetate, and hippurate.[181]

15.5.3. In Vitro and In Vivo Model Systems to Study Renal Drug Transport

The substrate specificities of drug transporters and drug discovery based on the transport mechanisms become increasingly important these days. The identification of compounds that are accepted by transporters can help the selection and optimization of novel drug candidates. For the screening of transport activities, high-throughput assays for transporters using their expression system seems necessary in the early stages of drug discovery. cDNA-transfected cells such as S_2-OATs/OCTs (Table 15.2) and/or cRNA-injected oocytes are commonly used gene expression systems. Recently, cultured cells stably transfected with both uptake (OATP-C or OATP8) and efflux transporters (MRP2) have become available.[182,183] These in vitro models, reproducing the polarity of transporter expression and the transport direction, will therefore contribute to predict the in vivo vectorial drug transport from blood to the lumen.

In addition, the generation of gene knockout animals could provide new information on the contribution of individual transporters in the intact organ. Knockout mice for OCT1,[184] OCT3,[185] OAT1,[186] and OAT3[187] revealed the importance of drug uptake of OCT1 in the liver, of OCT2 and OAT1 in kidney, and of OAT3 in kidney and choroids plexus. Therefore, the relative contribution of each transporter to overall drug uptake in vivo may be estimated using gene knockout animals.

15.6. CONCLUSIONS

Molecular identification of renal drug transporters has brought us a step further in the elucidation of the molecular mechanisms for drug elimination and distribution in the kidneys. Clarification of the physiological and pharmacological roles of each cloned transporter to the overall renal handling of drugs is essential. To better predict the in vivo kinetic profile of drugs from in vitro data, understanding of species differences in substrate selectivity, tissue distribution, and the expressed level of drug transporters is necessary. Furthermore, functionally relevant transporter SNPs as well as SNP mutations in the promoter–enhancer region of transporters should be explored to obtain insights into the structure–function relationship of drug transporters and the in vivo relevance of genetic heterogeneity in drug transporters. The prediction of pharmacokinetics in humans based on an understanding of such transport mechanisms should thus allow therapeutic agents to be used more safely.

REFERENCES

1. Marshall EK Jr, Vickers JL. 1923. The mechanism of the elimination of phenolsulphonephthalein by the kidney: a proof of secretion by the convoluted tubules. Bull Johns Hopkins Hosp 34:1–6.
2. Weiner IM. 1973. Transport of weak acids and bases. In: Orloff J, Berliner RW, editors. *Handbook of Physiology: Renal Physiology*, Section 8. Washington, DC: American Physiological Society, pp 521–554.
3. Ullrich KJ. 1997. Renal transporters for organic anions and organic cations: structural requirements for substrates. J Membr Biol 158:95–107.
4. Grantham JJ, Chonko AM. 1991. Renal handling of organic anions and cations: excretion of uric acid. In: Brenner BM, Rector FC, editors. *The kidney*, 4th ed. Philadelphia, PA: WB Saunders, pp 483–509.

5. Pritchard JB, Miller DS. 1993. Mechanisms mediating renal secretion of organic anions and cations. Physiol Rev 73:765–796.
6. Moller JV, Sheikh MI. 1982 Renal organic anion transport system: pharmacological, physiological, and biochemical aspects. Pharmacol Rev 34:315–358.
7. Inui KI, Masuda S, Saito H. 2000. Cellular and molecular aspects of drug transport in the kidney. Kidney Int 58:944–958.
8. Mol WE, Fokkema GN, Weert B, Meijer DK. 1988. Mechanisms for the hepatic uptake of organic cations: studies with the muscle relaxant vecuronium in isolated rat hepatocytes. J Pharmacol Exp Ther 244:268–275.
9. Steen H, Merema M, Meijer DK. 1992. A multispecific uptake system for taurocholate, cardiac glycosides and cationic drugs in the liver. Biochem Pharmacol 44:2323–2331.
10. Sica AD, Schoolwerth AC. 1996. Renal handling of organic anions and cations and renal excretion of uric acid. In: Brenner BM, Rector FC, editors. *The Kidney*, 5th ed. Philadelphia, PA: WB Saunders, pp. 607–626.
11. Burckhardt G, Pritchard JB. 2000. Organic anion and cation antiporters. In: Seldin DW, Giebisch G, editors. *The Kidney: Physiology and Pathophysiology*, 3rd ed. Philadelphia, PA: Lippincott Williams & Wilkins, pp. 193–222.
12. Roch-Ramel F. 1998. Renal transport of organic anions. Curr Opin Nephrol Hypertens 7:517–524.
13. Russel FG, Masereeuw R, van Aubel RA. 2002. Molecular aspects of renal anionic drug transport. Annu Rev Physiol 64:563–594.
14. Koepsell H, Schmitt BM, Gorboulev V. 2003. Organic cation transporters. Rev Physiol Biochem Pharmacol 150:36–90.
15. van Montfoort JE, Hagenbuch B, Groothuis GM, Koepsell H, Meier PJ, Meijer DK. 2003. Drug uptake systems in liver and kidney. Curr Drug Metab 4:185–211.
16. Burckhardt BC, Burckhardt G. 2003. Transport of organic anions across the basolateral membrane of proximal tubule cells. Rev Physiol Biochem Pharmacol 146:95–158.
17. Wright SH, Dantzler WH. 2004. Molecular and cellular physiology of renal organic cation and anion transport. Physiol Rev 84(3):987–1049.
18. You G. 2004. The role of organic ion transporters in drug disposition: an update. Curr Drug Metab 5:55–62.
19. Lee W, Kim RB. 2004. Transporters and renal drug elimination. Annu Rev Pharmacol Toxicol 44:137–66.
20. Anzai N, Kanai Y, Endou H. 2006. Organic anion transporter family: current knowledge. J Pharmacol Sci 100:411–426.
21. Sekine T, Watanabe N, Hosoyamada M, Kanai Y, Endou H. 1997. Expression cloning and characterization of a novel multispecific organic anion transporter. J Biol Chem 272:18526–18529.
22. Sweet DH, Wolff NA, Pritchard JB. 1997. Expression cloning and characterization of ROAT1: the basolateral organic anion transporter in rat kidney. J Biol Chem 272:30088–30095.
23. Wolff NA, Werner A, Burkhardt S, Burckhardt G. 1997. Expression cloning and characterization of a renal organic anion transporter from winter flounder. FEBS Lett 417:287–291.
24. Lopez-Nieto CE, You G, Bush KT, Barros EJ, Beier DR, Nigam SK. 1997. Molecular cloning and characterization of NKT, a gene product related to the organic cation

transporter family that is almost exclusively expressed in the kidney. J Biol Chem 272:6471–6478.

25. Shimada H, Moewes B, Burckhardt G. 1987. Indirect coupling to Na^+ of *p*-aminohippuric acid uptake into rat renal basolateral membrane vesicles. Am J Physiol 253:F795–F801.
26. Sekine T, Cha SH, Endou H. 2000. The multispecific organic anion transporter (OAT) family. Pflugers Arch 440:337–350.
27. Hosoyamada M, Sekine T, Kanai Y, Endou H. 1999. Molecular cloning and functional expression of a multispecific organic anion transporter from human kidney. Am J Physiol 276:F122–F128.
28. Bahn A, Prawitt D, Buttler D, Reid G, Enklaar T, Wolff NA, Ebbinghaus C, Hillemann A, Schulten HJ, Gunawan B, et al. 2000. Genomic structure and in vivo expression of the human organic anion transporter 1 (hOAT1) gene. Biochem Biophys Res Commun 275:623–630.
29. Bahn A, Ebbinghaus C, Ebbinghaus D, Ponimaskin EG, Füzesi L, Burckhardt G, Hagos Y. 2004. Expression studies and functional characterization of renal human organic anion transporter 1 isoforms. Drug Metab Dispos 32:424–430.
30. Simonson GD, Vincent AC, Roberg KJ, Huang Y, Iwanij V. 1994. Molecular cloning and characterization of a novel liver-specific transport protein. J Cell Sci 107:1065–1072.
31. Sekine T, Cha SH, Tsuda M, Apiwattanakul N, Nakajima N, Kanai Y, Endou H. 1998. Identification of multispecific organic anion transporter 2 expressed predominantly in the liver. FEBS Lett 429:179–182.
32. Morita N, Kusuhara H, Sekine T, Endou H, Sugiyama Y. 2001. Functional characterization of rat organic anion transporter 2 in LLC-PK1 cells. J Pharmacol Exp Ther 298:1179–1184.
33. Enomoto A, Takeda M, Shimoda M, Narikawa S, Kobayashi Y, Kobayashi Y, Yamamoto T, Sekine T, Cha SH, Niwa T, Endou H. 2002. Interaction of human organic anion transporters 2 and 4 with organic anion transport inhibitors. J Pharmacol Exp Ther 301: 797–802.
34. Kobayashi Y, Ohshiro N, Shibusawa A, Sasaki T, Tokuyama S, Sekine T, Endou H, Yamamoto T. 2002. Isolation, characterization and differential gene expression of multispecific organic anion transporter 2 in mice. Mol Pharmacol 62:7–14.
35. Takeda M, Khamdang S, Narikawa S, Kimura H, Kobayashi Y, Yamamoto T, Cha SH, Sekine T, Endou H. 2002. Human organic anion transporters and human organic cation transporters mediate renal antiviral transport. J Pharmacol Exp Ther 300:918–924.
36. Kimura H, Takeda M, Narikawa S, Enomoto A, Ichida K, Endou H. 2002. Human organic anion transporters and human organic cation transporters mediate renal transport of prostaglandins. J Pharmacol Exp Ther 301:293–298.
37. Babu E, Takeda M, Narikawa S, Kobayashi Y, Yamamoto T, Cha SH, Sekine T, Sakthisekaran D, Endou H. 2002. Human organic anion transporters mediate the transport of tetracycline. Jpn J Pharmacol 88:69–76.
38. Kusuhara H, Sekine T, Utsunomiya-Tate N, Tsuda M, Kojima R, Cha SH, Sugiyama Y, Kanai Y, Endou H. 1999. Molecular cloning and characterization of a new multispecific organic anion transporter from rat brain. J Biol Chem 274:13675–13680.
39. Brady KP, Dushkin H, Fornzler D, Koike T, Magner F, Her H, Gullans S, Segre GV, Green RM, Beier DR. 1999. A novel putative transporter maps to the osteosclerosis (oc) mutation and is not expressed in the oc mutant mouse. Genomics 56:254–261.

40. Cha SH, Sekine T, Fukushima JI, Kanai Y, Kobayashi Y, Goya T, Endou H. 2001. Identification and characterization of human organic anion transporter 3 expressing predominantly in the kidney. Mol Pharmacol 59:1277–1286.
41. Kojima R, Sekine T, Kawachi M, Cha SH, Suzuki Y, Endou H. 2002. Immunolocalization of multispecific organic anion transporters, OAT1, OAT2, and OAT3, in rat kidney. J Am Soc Nephrol 13:848–857.
42. Nagata Y, Kusuhara H, Endou H, Sugiyama Y. 2002. Expression and functional characterization of rat organic anion transporter 3 (rOat3) in the choroid plexus. Mol Pharmacol 61:982–988.
43. Sweet DH, Miller DS, Pritchard JB, Fujiwara Y, Beier DR, Nigam SK. 2002. Impaired organic anion transport in kidney and choroid plexus of organic anion transporter 3 (Oat3 (Slc22a8)) knockout mice. J Biol Chem 277:26934–26943.
44. Ohtsuki S, Kikkawa T, Mori S, Hori S, Takanaga H, Otagiri M, Terasaki T. 2004. Mouse reduced in osteosclerosis transporter functions as an organic anion transporter 3 and is localized at abluminal membrane of blood–brain barrier. J Pharmacol Exp Ther 309:1273–1281.
45. Jung KY, Takeda M, Kim DK, Tojo A, Narikawa S, Yoo BS, Hosoyamada M, Cha SH, Sekine T, Endou H. 2001. Characterization of ochratoxin A transport by human organic anion transporters. Life Sci 69:2123–2135.
46. Takeda M, Khamdang S, Narikawa S, Kimura H, Hosoyamada M, Cha SH, Sekine T, Endou H. 2002. Characterization of methotrexate transport and its drug interactions with human organic anion transporters. J Pharmacol Exp Ther 302:666–671.
47. Sweet DH, Chan LM, Walden R, Yang XP, Miller DS, Pritchard JB. 2003. Organic anion transporter 3 (Slc22a8) is a dicarboxylate exchanger indirectly coupled to the Na^+ gradient. Am J Physiol Renal Physiol 284:F763–F769.
48. Bakhiya A, Bahn A, Burckhardt G, Wolff N. 2003. Human organic anion transporter 3 (hOAT3) can operate as an exchanger and mediate secretory urate flux. Cell Physiol Biochem 13:249–256.
49. Cha SH, Sekine T, Kusuhara H, Yu E, Kim JY, Kim DK, Sugiyama Y, Kanai Y, Endou H. 2000. Molecular cloning and characterization of multispecific organic anion transporter 4 expressed in the placenta. J Biol Chem 275:4507–4512.
50. Ugele B, St-Pierre MV, Pihusch M, Bahn A, Hantschmann P. 2003. Characterization and identification of steroid sulfate transporters of human placenta. Am J Physiol Endocrinol Metab 284:E390–E398.
51. Babu E, Takeda M, Narikawa S, Kobayashi Y, Enomoto A, Tojo A, Cha SH, Sekine T, Sakthisekaran D, Endou H. 2002. Role of human organic anion transporter 4 in the transport of ochratoxin A. Biochim Biophys Acta 1590:64–75.
52. Anzai N, Enomoto A, Endou H. 2005. Renal urate handling: clinical relevance of recent advances. Curr Rheumatol Rep 7:227–234.
53. Ekaratanawong S, Anzai N, Jutabha P, Miyazaki H, Noshiro R, Takeda M, Kanai Y, Sophasan S, Endou H. 2004. Human organic anion transporter 4 is a renal apical organic anion/dicarboxylate exchanger in the proximal tubules. J Pharmacol Sci 94:297–304.
54. Enomoto A, Kimura H, Chairoungdua A, Shigeta Y, Jutabha P, Cha SH, Hosoyamada M, Takeda M, Sekine T, Igarashi T, et al. 2002. Molecular identification of a renal urate anion exchanger that regulates blood urate levels. Nature 417:447–452.

55. Youngblood GL, Sweet DH. 2004. Identification and functional assessment of the novel murine organic anion transporter Oat5 (Slc22a19) expressed in kidney. Am J Physiol Renal Physiol 287:F236–F244.
56. Anzai N, Jutabha P, Enomoto A, Yokoyama H, Nonoguchi H, Hirata T, Shiraya K, He X, Cha SH, Takeda M, et al. 2005. Functional characterization of rat organic anion transporter 5 (Slc22a19) at the apical membrane of renal proximal tubules. J Pharmacol Exp Ther 315:534–544.
57. Kwak JO, Kim HW, Oh KJ, Ko CB, Park H, Cha SH. 2005. Characterization of mouse organic anion transporter 5 as a renal steroid sulfate transporter. J Steroid Biochem Mol Biol 97:369–375.
58. Sun W, Wu RR, van Poelje PD, Erion MD. 2001. Isolation of a family of organic anion transporters from human liver and kidney. Biochem Biophys Res Commun 283:417–422.
59. Jacquemin E, Hagenbuch B, Stieger B, Wolkoff AW, Meier PJ. 1994. Expression cloning of a rat liver Na^+-independent organic anion transporter. Proc Natl Acad Sci U S A 91:133–137.
60. Hagenbuch B, Meier PJ. 2004. Organic anion transporting polypeptides of the OATP/ SLC21 family: phylogenetic classification as OATP/ SLCO superfamily, new nomenclature and molecular/functional properties. Pflugers Arch 447:653–665.
61. Mikkaichi T, Suzuki T, Onogawa T, Tanemoto M, Mizutamari H, Okada M, Chaki T, Masuda S, Tokui T, Eto N, et al. 2004. Isolation and characterization of a digoxin transporter and its rat homologue expressed in the kidney. Proc Natl Acad Sci U S A 101:3569–3574.
62. Saito H, Masuda S, Inui K. 1996. Cloning and functional characterization of a novel rat organic anion transporter mediating basolateral uptake of methotrexate in the kidney. J Biol Chem 271:20719–20725.
63. Masuda S, Ibaramoto K, Takeuchi A, Saito H, Hashimoto Y, Inui KI. 1999. Cloning and functional characterization of a new multispecific organic anion transporter, OAT-K2, in rat kidney. Mol Pharmacol 55:743–752.
64. Abe T, Kakyo M, Sakagami H, Tokui T, Nishio T, Tanemoto M, Nomura H, Hebert SC, Matsuno S, Kondo H, Yawo H. 1998. Molecular characterization and tissue distribution of a new organic anion transporter subtype (oatp3) that transports thyroid hormones and taurocholate and comparison with oatp2. J Biol Chem 273:22395–22401.
65. Choudhuri S, Cherrington NJ, Li N, Klaassen CD. 2003. Constitutive expression of various xenobiotic and endobiotic transporter mRNAs in the choroid plexus of rats. Drug Metab Dispos 31:1337–1345.
66. Ford JM, Hait WN. 1990. Pharmacology of drugs that alter multidrug resistance in cancer. Pharmacol Rev 42:155–199.
67. Reimer RJ, Edwards RH. 2004. Organic anion transport is the primary function of the SLC17/type I phosphate transporter family. Pflugers Arch 447:629–635.
68. Uchino H, Tamai I, Yamashita K, Minemoto Y, Sai Y, Yabuuchi H, Miyamoto K, Takeda E, Tsuji A. 2000. p-aminohippuric acid transport at renal apical membrane mediated by human inorganic phosphate transporter NPT1. Biochem Biophys Res Commun 270: 254–259.
69. Jutabha P, Kanai Y, Hosoyamada M, Chairoungdua A, Kim do K, Iribe Y, Babu E, Kim JY, Anzai N, Chatsudthipong V, Endou H. 2003. Identification of a novel voltage-driven organic anion transporter present at apical membrane of renal proximal tubule. J Biol Chem 278:27930–27938.

70. Terada T, Inui K. 2004. Peptide transporters: structure, function, regulation and application for drug delivery. Curr Drug Metab 5:85–94.
71. Daniel H, Kottra G. 2004. The proton oligopeptide cotransporter family SLC15 in physiology and pharmacology. Pflugers Arch 447:610–618.
72. Ogihara H, Saito H, Shin BC, Terado T, Takenoshita S, Nagamachi Y, Inui K, Takata K. 1996. Immuno-localization of H^+/peptide cotransporter in rat digestive tract. Biochem Biophys Res Commun 220:848–852.
73. Shen H, Smith DE, Yang T, Huang YG, Schnermann JB, Brosius FC 3rd. 1999. Localization of PEPT1 and PEPT2 proton-coupled oligopeptide transporter mRNA and protein in rat kidney. Am J Physiol 276:F658–F665.
74. Gray JH, Owen RP, Giacomini KM. 2004. The concentrative nucleoside transporter family, SLC28. Pflugers Arch 447:728–734.
75. Baldwin SA, Beal PR, Yao SY, King AE, Cass CE, Young JD. 2004. The equilibrative nucleoside transporter family, SLC29. Pflugers Arch 447:735–743.
76. Gottesman MM, Fojo T, Bates SE. 2002. Multidrug resistance in cancer: role of ATP-dependent transporters. Nat Rev Cancer 2:48–58.
77. Keppler D, König J, Büchler M. 1996. The canalicular multidrug resistance protein, cMRP/MRP2, a novel conjugate export pump expressed in the apical membrane of hepatocytes. Adv Enzyme Regul 37:321–333.
78. Leonard GD, Fojo T, Bates SE. 2003. The role of ABC transporters in clinical practice. Oncologist 8:411–424.
79. Smeets PH, van Aubel RA, Wouterse AC, van den Heuvel JJ, Russel FG. 2004. Contribution of multidrug resistance protein 2 (MRP2/ABCC2) to the renal excretion of *p*-aminohippurate (PAH) and identification of MRP4 (ABCC4) as a novel PAH transporter. J Am Soc Nephrol 15:2828–2835.
80. van Aubel RA, Smeets PH, van den Heuvel JJ, Russel FG. 2005. Human organic anion transporter MRP4 (ABCC4) is an efflux pump for the purine end metabolite urate with multiple allosteric substrate binding sites. Am J Physiol Renal Physiol 288:F327–F333.
81. Mao Q, Unadkat JD. 2005. Role of the breast cancer resistance protein (ABCG2) in drug transport. AAPS J 7:E118–E133.
82. Tanaka Y, Slitt AL, Leazer TM, Maher JM, Klaassen CD. 2005. Tissue distribution and hormonal regulation of the breast cancer resistance protein (Bcrp/Abcg2) in rats and mice. Biochem Biophys Res Commun 326:181–187.
83. Gründemann D, Gorboulev V, Gambaryan S, Veyhl M, Koepsell H. 1994. Drug excretion mediated by a new prototype of polyspecific transporter. Nature 372:549–552.
84. Koepsell H. 1998. Organic cation transporters in intestine, kidney, liver, and brain. Annu Rev Physiol 60:243–266.
85. Urakami Y, Okuda M, Masuda S, Saito H, Inui K. 1998. Functional characteristics and membrane localization of rat multispecific organic cation transporters, OCT1 and OCT2, mediating tubular secretion of cationic drugs. J Pharmacol Exp Ther 287:800–805.
86. Zhang L, Dresser MJ, Gray AT, Yost SC, Terashita S, Giacomini KM. 1997. Cloning and functional expression of a human liver organic cation transporter. Mol Pharmacol 51:913–921.
87. Gorboulev V, Ulzheimer JC, Akhoundova A, Ulzheimer-Teuber I, Karbach U, Quester S, Baumann C, Lang F, Busch AE, Koepsell H. 1997. Cloning and characterization of two human polyspecific organic cation transporters. DNA Cell Biol 16:871–881.

88. Zhang L, Gorset W, Dresser MJ, Giacomini KM. 1999. The interaction of n-tetraalkylammonium compounds with a human organic cation transporter, hOCT1. J Pharmacol Exp Ther 288:1192–1198.
89. Dresser MJ, Gray AT, Giacomini KM. 2000. Kinetic and selectivity differences between rodent, rabbit, and human organic cation transporters (OCT1). J Pharmacol Exp Ther 292:1146–1152.
90. Okuda M, Saito H, Urakami Y, Takano M, Inui K. 1996. cDNA cloning and functional expression of a novel rat kidney organic cation transporter, OCT2. Biochem Biophys Res Commun 224:500–507.
91. Karbach U, Kricke J, Meyer-Wentrup F, Gorboulev V, Volk C, Loffing-Cueni D, Kaissling B, Bachmann S, Koepsell H. 2000. Localization of organic cation transporters OCT1 and OCT2 in rat kidney. Am J Physiol Renal Physiol 279:F679-F687.
92. Sweet DH, Miller DS, Pritchard JB. 2000. Basolateral localization of organic cation transporter 2 in intact renal proximal tubules. Am J Physiol Renal Physiol 279:F826–F834.
93. Urakami Y, Akazawa M, Saito H, Okuda M, Inui K. 2002. cDNA cloning, functional characterization, and tissue distribution of an alternatively spliced variant of organic cation transporter hOCT2 predominantly expressed in the human kidney. J Am Soc Nephrol 13:1703–1710.
94. Kekuda R, Prasad PD, Wu X, Wang H, Fei YJ, Leibach FH, Ganapathy V. 1998. Cloning and functional characterization of a potential-sensitive, polyspecific organic cation transporter (OCT3) most abundantly expressed in placenta. J Biol Chem 273:15971–15979.
95. Gründemann D, Koster S, Kiefer N, Breidert T, Engelhardt M, Spitzenberger F, Obermuller N, Schömig E. 1998. Transport of monoamine transmitters by the organic cation transporter type 2, OCT2. J Biol Chem 273:30915–30920.
96. Wu X, George RL, Huang W, Wang H, Conway SJ, Leibach FH, Ganapathy V. 2000. Structural and functional characteristics and tissue distribution pattern of rat OCTN1, an organic cation transporter, cloned from placenta. Biochim Biophys Acta 1466:315–327.
97. Verhaagh S, Schweifer N, Barlow DP, Zwart R. 1999. Cloning of the mouse and human solute carrier 22a3 (Slc22a3/SLC22A3) identifies a conserved cluster of three organic cation transporters on mouse chromosome 17 and human 6q26–q27. Genomics 55:209–218.
98. Wu X, Kekuda R, Huang W, Fei YJ, Leibach FH, Chen J, Conway SJ, Ganapathy V. 1998. Identity of the organic cation transporter OCT3 as the extraneuronal monoamine transporter (uptake2) and evidence for the expression of the transporter in the brain. J Biol Chem 273:32776–32786.
99. Hayer-Zillgen M, Bruss M, Bonisch H. 2002. Expression and pharmacological profile of the human organic cation transporters hOCT1, hOCT2 and hOCT3. Br J Pharmacol 136:829–836.
100. Tamai I, Yabuuchi H, Nezu J, Sai Y, Oku A, Shimane M, Tsuji A. 1997. Cloning and characterization of a novel human pH-dependent organic cation transporter, OCTN1. FEBS Lett 419:107–111.
101. Yabuuchi H, Tamai I, Nezu J, Sakamoto K, Oku A, Shimane M, Sai Y, Tsuji A. 1999. Novel membrane transporter OCTN1 mediates multispecific, bidirectional, and pH-dependent transport of organic cations. J Pharmacol Exp Ther 289:768–773.

102. Gründemann D, Harlfinger S, Golz S, Geerts A, Lazar A, Berkels R, Jung N, Rubbert A, Schömig E. 2005. Discovery of the ergothioneine transporter. Proc Natl Acad Sci U S A 102:5256–5261.
103. 103. Wu X, Prasad PD, Leibach FH, Ganapathy V. 1998. cDNA sequence, transport function, and genomic organization of human OCTN2, a new member of the organic cation transporter family. Biochem Biophys Res Commun 246:589–595.
104. Wu X, Huang W, Prasad PD, Seth P, Rajan DP, Leibach FH, Chen J, Conway SJ, Ganapathy V. 1999. Functional characteristics and tissue distribution pattern of organic cation transporter 2 (OCTN2), an organic cation/carnitine transporter. J Pharmacol Exp Ther 290:1482–1492.
105. Tamai I, Ohashi R, Nezu J, Yabuuchi H, Oku A, Shimane M, Sai Y, Tsuji A. 1998. Molecular and functional identification of sodium ion-dependent, high affinity human carnitine transporter OCTN2. J Biol Chem 273:20378–20382.
106. Nezu J, Tamai I, Oku A, Ohashi R, Yabuuchi H, Hashimoto N, Nikaido H, Sai Y, Koizumi A, Shoji Y, et al. 1999. Primary systemic carnitine deficiency is caused by mutations in a gene encoding sodium ion-dependent carnitine transporter. Nat Genet 21: 91–94.
107. Tang NL, Hwu WL, Chan RT, Law LK, Fung LM, Zhang WM. 2002. A founder mutation (R254X) of SLC22A5 (OCTN2) in Chinese primary carnitine deficiency patients. Hum Mutat 20:232.
108. Wang Y, Ye J, Ganapathy V, Longo N. 1999. Mutations in the organic cation/carnitine transporter OCTN2 in primary carnitine deficiency. Proc Natl Acad Sci U S A 96:2356–2360.
109. Ohashi R, Tamai I, Inano A, Katsura M, Sai Y, Nezu J, Tsuji A. 2002. Studies on functional sites of organic cation/carnitine transporter OCTN2 (SLC22A5) using a Ser467Cys mutant protein. J Pharmacol Exp Ther 302:1286–1294.
110. Seth P, Wu X, Huang W, Leibach FH, Ganapathy V. 1999. Mutations in novel organic cation transporter (OCTN2), an organic cation/carnitine transporter, with differential effects on the organic cation transport function and the carnitine transport function. J Biol Chem 274:33388–33392.
111. Sekine T, Kusuhara H, Utsunomiya-Tate N, Tsuda M, Sugiyama Y, Kanai Y, Endou H. 1998. Molecular cloning and characterization of high-affinity carnitine transporter from rat intestine. Biochem Biophys Res Commun 251:586–591.
112. Thiebaut F, Tsuruo T, Hamada H, Gottesman MM, Pastan I, Willingham MC. 1987. Cellular localization of the multidrug-resistance gene product P-glycoprotein in normal human tissues. Proc Natl Acad Sci U S A 84:7735–7738.
113. Otsuka M, Matsumoto T, Morimoto R, Arioka S, Omote H, Moriyama Y. 2005. A human transporter protein that mediates the final excretion step for toxic organic cations. Proc Natl Acad Sci U S A 102:17923–17928.
114. Ljubojevic M, Herak-Kramberger CM, Hagos Y, Bahn A, Endou H, Burckhardt G, Sabolic I. 2004. Rat renal cortical OAT1 and OAT3 exhibit gender differences determined by both androgen stimulation and estrogen inhibition. Am J Physiol Renal Physiol 287:F124–F138.
115. Buist SC, Cherrington NJ, Choudhuri S, Hartley DP, Klaassen CD. 2002. Gender-specific and developmental influences on the expression of rat organic anion transporters. J Pharmacol Exp Ther 301:145–151.

116. Kobayashi Y, Hirokawa N, Ohshiro N, Sekine T, Sasaki T, Tokuyama S, Endou H, Yamamoto T. 2002. Differential gene expression of organic anion transporters in male and female rats. Biochem Biophys Res Commun 290:482–487.

117. Ljubojevic M, Balen D, Breljak D, Kusan M, Anzai N, Bahn A, Burckhardt G, Sabolic I. 2006. Renal expression of organic anion transporter OAT2 in rats and mice is regulated by sex hormones. Am J Physiol Renal Physiol (in press).

118. Hosoyamada M, Ichida K, Enomoto A, Hosoya T, Endou H. 2004. Function and localization of urate transporter 1 in mouse kidney. J Am Soc Nephrol 15:261–268.

119. Urakami Y, Okuda M, Saito H, Inui K. 2000. Hormonal regulation of organic cation transporter OCT2 expression in rat kidney. FEBS Lett 473:173-176.

120. Shu Y, Bello CL, Mangravite LM, Feng B, Giacomini KM. 2001. Functional characteristics and steroid hormone-mediated regulation of an organic cation transporter in Madin-Darby canine kidney cells. J Pharmacol Exp Ther 299:392–398.

121. Pavlova A, Sakurai H, Leclercq B, Beier DR, Yu AS, Nigam SK. 2000. Developmentally regulated expression of organic ion transporters NKT (OAT1), OCT1, NLT (OAT2), and Roct. Am J Physiol Renal Physiol 278:F635–F643.

122. Nakajima N, Sekine T, Cha SH, Tojo A, Hosoyamada M, Kanai Y, Yan K, Awa S, Endou H. 2000. Developmental changes in multispecific organic anion transporter 1 expression in the rat kidney. Kidney Int 57:1608–1616.

123. Lu R, Chan BS, Schuster VL. 1999. Cloning of the human kidney PAH transporter: narrow substrate specificity and regulation by protein kinase C. Am J Physiol 276:F295–F303.

124. Uwai Y, Okuda M, Takami K, Hashimoto Y, Inui K. 1998. Functional characterization of the rat multispecific organic anion transporter OAT1 mediating basolateral uptake of anionic drugs in the kidney. FEBS Lett 438:321–324.

125. You G, Kuze K, Kohanski RA, Amsler K, Henderson S. 2000. Regulation of mOAT-mediated organic anion transport by okadaic acid and protein kinase C in LLC-PK1 cells. J Biol Chem 275:10278–10284.

126. Wolff NA, Thies K, Kuhnke N, Reid G, Friedrich B, Lang F, Burckhardt G. 2003. Protein kinase C activation downregulates human organic anion transporter 1-mediated transport through carrier internalization. J Am Soc Nephrol 14:1959–1968.

127. Sauvant C, Hesse D, Holzinger H, Evans KK, Dantzler WH, Gekle M. 2004. Action of EGF and PGE2 on basolateral organic anion uptake in rabbit proximal renal tubules and hOAT1 expressed in human kidney epithelial cells. Am J Physiol Renal Physiol 286:F774–F783.

128. Mehrens T, Lelleck S, Cetinkaya I, Knollmann M, Hohage H, Gorboulev V, Boknik P, Koepsell H, Schlatter E. 2000. The affinity of the organic cation transporter rOCT1 is increased by protein kinase C–dependent phosphorylation. J Am Soc Nephrol 11:1216–1224.

129. Cetinkaya I, Ciarimboli G, Yalcinkaya G, Mehrens T, Velic A, Hirsch JR, Gorboulev V, Koepsell H, Schlatter E. 2003. Regulation of human organic cation transporter hOCT2 by PKA, PI3K, and calmodulin-dependent kinases. Am J Physiol Renal Physiol 284:F293–F302.

130. Anzai N, Jutabha P, Kanai Y, Endou H. 2005. Integrated physiology of proximal tubular organic anion transport. Curr Opin Nephrol Hypertens 14:472–479.

131. Anzai N, Miyazaki H, Noshiro R, Khamdang S, Chairoungdua A, Shin HJ, Enomoto A, Sakamoto S, Hirata T, Tomita K, Kanai Y, Endou H. 2004. The multivalent PDZ

domain–containing protein PDZK1 regulates transport activity of renal urate–anion exchanger URAT1 via its C-terminal. J Biol Chem 279:45942–45950.

132. Miyazaki H, Anzai N, Ekaratanawong S, Sakata T, Shin HJ, Jutabha P, Hirata T, He X, Nonoguchi H, Tomita K, Kanai Y, Endou H. 2005. Modulation of renal apical organic anion transporter 4 function by two PDZ domain–containing proteins. J Am Soc Nephrol 16:3498–3506.

133. Noshiro R, Anzai N, Sakata T, Miyazaki H, Terada T, Shin HJ, He X, Miura D, Inui K, Kanai Y, Endou H. 2006. The PDZ domain protein PDZK1 interacts with human peptide transporter PEPT2 and enhances its transport activity. Kidney Int 70:275–282.

134. Kato Y, Sai Y, Yoshida K, Watanabe C, Hirata T, Tsuji A. 2005. PDZK1 directly regulates the function of organic cation/carnitine transporter OCTN2. Mol Pharmacol 67:734–743.

135. Kuze K, Graves P, Leahy A, Wilson P, Stuhlmann H, You G. 1999. Heterologous expression and functional characterization of a mouse renal organic anion transporter in mammalian cells. J Biol Chem 274:1519–1524.

136. Tanaka K, Xu W, Zhou F, You G. 2004. Role of glycosylation in the organic anion transporter OAT1, J Biol Chem 279:14961–14966.

137. Hori R, Ishikawa Y, Takano M, Okano T, Kitazawa S, Inui K. 1982. The interaction of cephalosporin antibiotics with renal cortex of rats: accumulation to cortical slices and binding to purified plasma membrane. Biochem Pharmacol 31:2267–2272.

138. Takano M, Okano T, Inui K, Hori R. 1989. Transport of cephalosporin antibiotics in rat renal basolateral membranes. J Pharm Pharmacol 41:795–796.

139. Kasher JS, Holohan PD, Ross CR. 1983. Effect of cephaloridine on the transport of organic ions in dog kidney plasma membrane vesicles. J Pharmacol Exp Ther 225:606–610.

140. Jariyawat S, Sekine T, Takeda M, Apiwattanakul N, Kanai Y, Sophasan S, Endou H. 1999. The interaction and transport of beta-lactam antibiotics with the cloned rat renal organic anion transporter 1. J Pharmacol Exp Ther 290:672–677.

141. Jung KY, Takeda M, Shimoda M, Narikawa S, Tojo A, Kim do K, Chairoungdua A, Choi BK, Kusuhara H, Sugiyama Y, Sekine T, Endou H. 2002. Involvement of rat organic anion transporter 3 (rOAT3) in cephaloridine-induced nephrotoxicity: in comparison with rOAT1. Life Sci 70:1861–1874.

142. Takeda M, Babu E, Narikawa S, Endou H. 2002. Interaction of human organic anion transporters with various cephalosporin antibiotics. Eur J Pharmacol 438:137–142.

143. Ives HE. 2001. Diuretic agents. In: Katzung BG, editor. *Basic and Clinical Pharmacology*, 8th ed. New York: McGraw-Hill, pp 245–265.

144. Bekersky I, Popick A. 1986. Disposition of bumetanide in the isolated perfused rat kidney: effects of probenecid and dose response. Am J Cardiol 57:33A–37A.

145. Lee LJ, Cook JA, Smith DE. 1986. Renal transport kinetics of furosemide in the isolated perfused rat kidney. J Pharamacokinet Biopharm 14:157–174.

146. Odlind B, Lonnerholm G. 1982. Renal tubular secretion and effects of chlorothiazide, hydrochlorothiazide and clopamide: a study in the avian kidney. Acta Pharmacol Toxicol 51:187–197.

147. Gemba M, Taniguchi M, Matsushima Y. 1981. Effect of bumetanide on *p*-aminohippurate transport in renal cortical slices. J Pharmacobiodyn 4:162–166.

148. Bidiville J, Roch-Ramel F. 1986. Competition of organic anions for furosemide and *p*-aminohippurate secretion in the rabbit. J Pharmacol Exp Ther 237:636–643.

149. Hasannejad H, Takeda M, Taki K, Shin HJ, Babu E, Jutabha P, Khamdang S, Aleboyeh M, Onozato ML, Tojo A, et al. 2004. Interactions of human organic anion transporters with diuretics. J Pharmacol Exp Ther 308:1021–1029.

150. De Zeeuw D, Jacobson HR, Brater DC. 1988. Indomethacin secretion in the isolated perfused proximal straight rabbit tubule, incidence for two parallel transport mechanisms. J Clin Invest 81:1585–1592.

151. Cox PGF, Van Os CH, Russel GM. 1992. Accumulation of salicylic acid and indomethacin in isolated proximal tubular cells of the rat kidney. Pharmacol Res 27:241–252.

152. Ferrier B, Martin M, Roch-Ramel F. 1983. Effects of *p*-aminohippurate and pyrazinoate on the renal excretion of salicylate in the rat: a micropuncture study. J Pharmacol Exp Ther 224:451–458.

153. Schild L, Roch-Ramel F. 1988. Transport of salicylate in proximal tubule (S2 segment) isolated from rabbit kidney. Am J Physiol 254:F554–F561.

154. Despopoulos A. 1960. Renal metabolism of salicylate and salicylurate. Am J Physiol 198:230–232.

155. Putney JW Jr, Borzelleca JF. 1973. Active accumulation of [^{14}C]-salicylic acid by rat kidney cortex in vitro. J Pharmacol Exp Ther 186:600–608.

156. Chatton J, Roch-Ramel F. 1992. Transport of salicylic acid through monolayers of a kidney epithelial cell line (LLC-PK1). J Pharmacol Exp Ther 261:518–524.

157. Bito LZ, Salvador EV. 1976. Effects of anti-inflammatory agents and some other drugs on prostaglandin biotransport. J Pharmacol Exp Ther 198:481–488.

158. Nierenberg DW. 1987. Drug inhibition of penicillin tubular secretion: concordance between in vitro and clinical findings. J Pharmacol Exp Ther 240:712–716.

159. Apiwattanakul N, Sekine T, Chairoungdua A, Kanai Y, Nakajima N, Sophasan S, Endou H. 1999. Transport properties of nonsteroidal anti-inflammatory drugs by organic anion transporter 1 expressed in *Xenopus laevis* oocytes. Mol Pharmacol 55:847–854.

160. Khamdang S, Takeda M, Noshiro R, Narikawa S, Enomoto A, Anzai N, Piyachaturawat P, Endou H. 2002. Interactions of human organic anion transporters and human organic cation transporters with nonsteroidal anti-inflammatory drugs. J Pharmacol Exp Ther 303:534–539.

161. Safrin S. 2001. Antiviral agents. In: Katzung BG, editor. *Basic and Clinical Pharmacology*, 8th ed. New York: McGraw-Hill, pp 823–849.

162. de Miranda P, Godd SS, Yarchoan R, Thomas RV, Blum MR, Myers CE, Broder S. 1989. Alteration of zidovudine pharmacokinetics by probenecid in patients with AIDS or AIDS-related complex. Clin Pharmacol Ther 46:494–500.

163. Laskin OL, de Miranda P, King DH, Page DA, Longstreth JA, Rocco L, Lietman PS. 1982. Effects of probenecid on the pharmacokinetics and elimination of acyclovir in humans. Antimicrob Agents Chemother 21:804–807.

164. Yarchoan R, Mitsuya H, Myers CE, Broader S. 1989. Clinical pharmacology of 3′-azido-2′,3′-dideoxythymidine (Zidovudine) and related dideoxynucleosides. N Engl J Med 321:726–738.

165. Morse GD, Shelton MJ, O'Donnell AM. 1993. Comparative pharmacokinetics of antiviral nucleoside analogues. Clin Pharmacokinet 24:101–123.

166. Chatton J, Odone M, Besseghir K, Roch-Ramel F. 1990. Renal secretion of 3′-azido-3′-deoxythymidine by the rat. J Pharmacol Exp Ther 255:140–145.

167. Mays DC, Dixon KF, Balboa A, Pawluk LJ, Bauer MR, Nawoot S, Gerber N. 1991. A non-primate animal model applicable to zidovudine pharmacokinetics in humans: inhibition of glucuronidation and renal excretion of zidovudine by probenecid in rats. J Pharmacol Exp Ther 259:1261–1270.

168. Nozaki N, Kusuhara H, Endou H, Sugiyama Y. 2004. Quantitative evaluation of the drug–drug interactions between methotrexate and nonsteroidal anti-inflammatory drugs in the renal uptake process based on the contribution of organic anion transporters and reduced folate carrier. J Pharmacol Exp Ther 309:226–234.

169. Uwai Y, Saito H, Hashimoto Y, Inui KI. 2000. Interaction and transport of thiazide diuretics, loop diuretics, and acetazolamide via rat renal organic anion transporter rOAT1. J Pharmacol Exp Ther 295:261–265.

170. Hasegawa M, Kusuhara H, Endou H, Sugiyama Y. 2003. Contribution of organic anion transporters to the renal uptake of anionic compounds and nucleoside derivatives in rat. J Pharmacol Exp Ther 305:1087–1097.

171. Mulato AS, Ho ES, Cihlar T. 2000. Nonsteroidal anti-inflammatory drugs efficiently reduce the transport and cytotoxicity of adefovir mediated by the human renal organic anion transporter 1. J Pharmacol Exp Ther 295:10–15.

172. Endou H. 1998. Recent advances in molecular mechanisms of nephrotoxicity, Toxicol Lett 102–103:29–33.

173. Cihlar T, Lin DC, Pritchard JB, Fuller MD, Mendel DB, Sweet DH. 1999. The antiviral nucleotide analogs cidofovir and adefovir are novel substrates for human and rat renal organic anion transporter 1. Mol Pharmacol 56:570–580.

174. Ho ES, Lin DC, Mendel DB, Cihlar T. 2000. Cytotoxicity of antiviral nucleotides adefovir and cidofovir is induced by the expression of human renal organic anion transporter 1. J Am Soc Nephrol 11:383–393.

175. Lacy SA, Hitchcock MJ, Lee WA, Tellier P, Cundy KC. 1998. Effect of oral probenecid coadministration on the chronic toxicity and pharmacokinetics of intravenous cidofovir in cynomolgus monkeys. Toxicol Sci 44:97–106.

176. Pfohl-Leszkowicz A, Petkova-Bocharova T, Chernozemsky IN, Castegnaro M. 2002. Balkan endemic nephropathy and associated urinary tract tumours: a review on aetiological causes and the potential role of mycotoxins. Food Addit Contam 19:282–302.

177. Tsuda M, Sekine T, Takeda M, Cha SH, Kanai Y, Kimura M, Endou H. 1999. Transport of ochratoxin A by renal multispecific organic anion transporter 1. J Pharmacol Exp Ther 289:1301–1305.

178. Deguchi T, Ohtsuki S, Otagiri M, Takanaga H, Asaba H, Mori S, Terasaki T. 2002. Major role of organic anion transporter 3 in the transport of indoxyl sulfate in the kidney. Kidney Int 61:1760–1768.

179. Motojima M, Hosokawa A, Yamato H, Muraki T, Yoshioka T. 2002. Uraemic toxins induce proximal tubular injury via organic anion transporter 1–mediated uptake. Br J Pharmacol 135:555–563.

180. Enomoto A, Takeda M, Tojo A, Sekine T, Cha SH, Khamdang S, Takayama F, Aoyama I, Nakamura S, Endou H, Niwa T. 2002. Role of organic anion transporters in the tubular transport of indoxyl sulfate and the induction of its nephrotoxicity. J Am Soc Nephrol 13:1711–1720.

181. Deguchi T, Kusuhara H, Takadate A, Endou H, Otagiri M, Sugiyama Y. 2004. Characterization of uremic toxin transport by organic anion transporters in the kidney. Kidney Int 65:162–174.

182. Cui Y, König J, Keppler D. 2001. Vectorial transport by double-transfected cells expressing the human uptake transporter SLC21A8 and the apical export pump ABCC2. Mol Pharmacol 60:934-943.

183. Sasaki M, Suzuki H, Ito K, Abe T, Sugiyama Y. 2002. Transcellular transport of organic anions across a double–transfected Madin–Darby canine kidney II cell monolayer expressing both human organic anion–transporting polypeptide (OATP2/SLC21A6) and multidrug resistance–associated protein 2 (MRP2/ABCC2). J Biol Chem 277:6497–6503.

184. Jonker JW, Wagenaar E, Mol CA, Buitelaar M, Koepsell H, Smit JW, Schinkel AH. 2001. Reduced hepatic uptake and intestinal excretion of organic cations in mice with a targeted disruption of the organic cation transporter 1 (Oct1 [Slc22a1]) gene. Mol Cell Biol 21:5471–5477.

185. Jonker JW, Wagenaar E, Van Eijl S, Schinkel AH. 2003. Deficiency in the organic cation transporters 1 and 2 (Oct1/Oct2 [Slc22a1/Slc22a2]) in mice abolishes renal secretion of organic cations. Mol Cell Biol 23:7902–7908.

186. Eraly SA, Vallon V, Vaughn DA, Gangoiti JA, Richter K, Nagle M, Monte JC, Rieg T, Truong DM, Long JM, Barshop BA, Kaler G, Nigam SK. 2006. Decreased renal organic anion secretion and plasma accumulation of endogenous organic anions in OAT1 knockout mice. J Biol Chem 281:5072–5083.

187. Sweet DH, Miller DS, Pritchard JB, Fujiwara Y, Beier DR, Nigam SK. 2002. Impaired organic anion transport in kidney and choroid plexus of organic anion transporter 3 (Oat3 (Slc22a8)) knockout mice. J Biol Chem 277:26934–26943.

188. van de Water FM, Masereeuw R, Russel FG. 2005. Function and regulation of multidrug resistance proteins (MRPs) in the renal elimination of organic anions. Drug Metab Rev 37:443–471.

16

DRUG TRANSPORTERS IN THE INTESTINE

YAMING SU AND PATRICK J. SINKO
Rutgers, the State University of New Jersey, Piscataway, New Jersey

Drug Transporters: Molecular Characterization and Role in Drug Disposition, Edited by Guofeng You and Marilyn E. Morris
Copyright © 2007 John Wiley & Sons, Inc.

16.1. INTRODUCTION

Even though recent advances in drug delivery technologies have made it possible to deliver drugs by nonoral routes (e.g., to the lungs), oral administration is still preferred for many reasons. Ease of administration, avoiding vascular access complications, and reducing frequent hospital visits improve the quality of a patient's life. Additionally, the steady-state plasma drug concentrations following chronic oral administration mimic continuous intravenous injection or infusion and ensure therapeutic effectiveness of the drug.

The main absorption site of orally delivered drugs is the small intestine. The human small intestine is approximately 3 m long with an inner diameter of 3 to 4 cm. It consists of three sections: the duodenum, jejunum, and ileum, which comprise 5, 50, and 45% of the length, respectively. The primary function of colon is the reabsorption of fluid. Its role in drug absorption is typically very limited, even though it has the capacity to be an absorption site for certain types of drugs. Intestinal epithelial cells are a heterogeneous population of cells that include enterocytes or absorptive cells, goblet cells that secrete mucin, endocrine cells, Paneth cells, M-cells, tuft, and cup cells. Enterocytes are polarized with distinct apical and basolateral membranes that are separated by tight junctions. They dominate the cellular population of the villus epithelium and are responsible for the majority of nutrient and drug absorption from the small intestine.

Many factors influence the absorption of orally administered drugs including (1) physicochemical properties of the drug molecule (e.g., lipophilicity, solubility, stability, ionization, crystal form), (2) pharmaceutical factors (e.g., disintegration and dissolution rate, excipients, dosage form), and (3) physiological factors (e.g., gastric emptying rate, intestinal motility, metabolic enzymes, transporters). Once the drug compound is released from its dosage form, it exists as a soluble molecule. Its permeation from intestinal lumen to blood or lymph circulation includes the passage across a series of barriers, such as hydrodynamic boundary layer, glycocalyx (i.e., the carbohydrate-rich zone on the cell surface), mucus layer, microvillus mucosal brushborder membrane (apical membrane) in parallel with a tight junction barrier, basolateral membrane of enterocytes, and parallel lamina propria endothelial membranes of blood capillaries and lymphatic central lacteal. It is generally thought that apical and basolateral membranes of enterocytes as well as tight junctions represent the rate-limiting barriers to drug transport. However, some in vitro and in situ studies showed that the preepithelial barriers, such as the stagnant layer of water, mucus, and glycocalyx, might significantly affect the overall absorption rate of some rapidly transported drugs.[1]

16.2. INTESTINAL PERMEATION OF DRUGS

The transport of a drug molecule across the intestinal epithelial cell (specifically, enterocytes) monolayer can be divided into several routes, as depicted schematically in Figure 16.1. Transcellular pathways refer to the permeation across the apical and

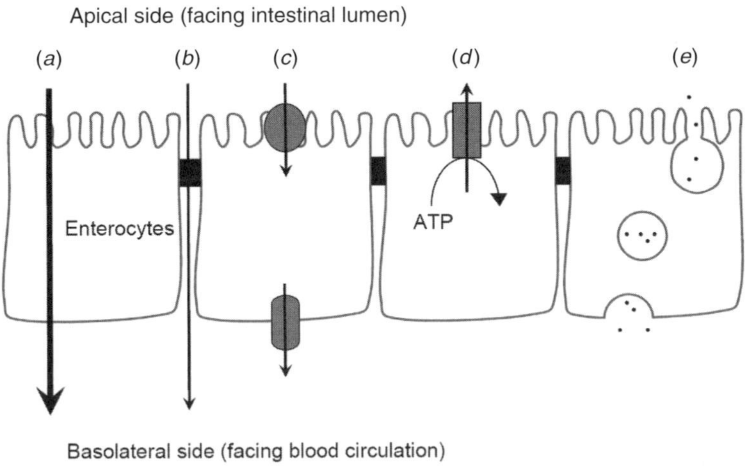

FIGURE 16.1. Drug transport routes across the intestinal epithelial cell monolayer: (*a*) passive transcellular diffusion; (*b*) paracellular diffusion; (*c*) transporter-mediated absorption; (*d*) transporter-mediated secretion; (*e*) transcytosis.

basolateral membrane of enterocytes, while paracellular permeation occurs through the gaps between adjacent epithelial cells. The rate-limiting step in paracellular transport is the barrier known as the *tight junctions*, which consists of large complexes of multiple different proteins that link adjacent cells together to form the intestinal mucosal membrane. Transcellular transport can be further divided into passive transcellular diffusion, carrier-mediated transport, and transcytosis.

16.2.1. Transcellular Diffusion

Transcellular passive diffusion is the route that many drugs take to permeate the intestinal epithelium. This type of transport does not require adenosine triphosphate (ATP) hydrolysis since its driving force is the concentration gradient of the drug, which makes the drug move toward regions with low drug concentration (e.g., from the intestinal lumen to the blood). Passive diffusion can be mathematically described by *Fick's first law of diffusion*:

$$J = \frac{dM}{dt\,S} = P(C_1 - C_2) \tag{1}$$

where J is the amount of drug flowing through a unit cross section of a barrier in unit time, known as flux, and P is the permeability coefficient (also called the permeability). C_1 and C_2 are the drug concentrations at the apical and basolateral side of enterocytes, respectively. It is noteworthy that drug transport within the enterocytes is ignored when using this equation. In fact, once the drug molecule transports across the enterocytes, it will rapidly be carried away from the basolateral side by the blood and/or lymphatic circulation. Therefore, C_2 becomes insignificant. This is known as *sink conditions* and serves as the driving force for drug absorption. Passive diffusion from the intestine is governed by the equation

$$J = PC_1 \tag{2}$$

The passive diffusive permeability is defined as

$$P = \frac{KD}{h} \tag{3}$$

where K is the partition coefficient between the aqueous phase and the membrane, D is the membrane diffusion coefficient (or membrane diffusivity), and h is the membrane thickness. It is very difficult to calculate a P value using this equation under physiological conditions because the determination of K, D, or h is difficult. Practically, permeability can be measured using the equation

$$P = \frac{dM}{dt}\frac{1}{SC_1} \tag{4}$$

where dM/dt is the slope of a linear region of transported mass versus time S is the surface area of membrane where transport takes place and C_1 is the concentration on the intestinal luminal side (also known as the donor side).

Numerous drug-transporting membrane proteins have been identified in the intestinal tissues of humans and various laboratory animal species. While the detailed working mechanisms of most transporters remains unclear to us at this time, it has been deduced and demonstrated experimentally that carrier-mediated drug transport is concentration dependent and saturable. Therefore, transporter-facilitated (i.e., carrier-mediated) drug transport is described by the following form of the *Michaelis–Menton equation*:

$$J_c = \frac{J_{\max}C}{K_m + C} \tag{5}$$

where J_c is transporter-facilitated drug flux, $J_{\max}$ is the maximal drug flux, K_m is the Michaelis constant (or drug-transporter affinity constant), and C is drug concentration. In a manner similar to enzymatic degradation reactions, transporter-facilitated drug flux can also be inhibited by other compounds in a competitive or noncompetitive manner. The equations governing these two types of inhibitions follow, and their derivations, are similar to typical enzymatic reaction equations.

Competitive inhibition:

$$J_c = \frac{J_{\max}C}{K_m(1 + 1/K_i) + C} \tag{6}$$

Noncompetitive inhibition:

$$J_c = \frac{J_{\max} C}{K_m + (1 + I/K_i)C} \tag{7}$$

where the inhibitor concentration and the inhibitor–transporter affinity constant are I and K_i, respectively.

It is noteworthy that the transport inhibition observed in vitro or in vivo may not be described by the two specific mechanisms listed previously. This is particularly true when the transporter has a broad spectrum of substrates and the binding sites are ambiguous. Mixed inhibition mechanisms are often involved in such cases. Additionally, recent studies suggest that some transporters, such as multidrug resistance protein 2 (ABCC2/MRP2) contains two substrate binding sites with one site (denoted as A) for transporting substrates and the other (denoted as B) for regulating the affinity of the transport site (i.e., binding site A) for the substrate.[2,3] This allosteric modulation of a drug transporter is exemplified by an in vitro study showing that the transport of saquinavir by MRP2 can be enhanced (or stimulated) by other compounds, such as probenecid or sulfanitran.[3]

16.2.2. Paracellular Transport

The paracellular route does not play a significant role in the small intestinal transport of modern drugs, for a couple of reasons. First, the surface area for paracellular spaces is estimated to be only 0.01% of the total surface area of the small intestine.[4] Second, the tight junctions between the adjacent enterocytes restricts the movement of large molecules (molecular weight > 200 Da). Over the past several years, using absorption enhancers to increase oral bioavailability has gained significant attention.[5] Some absorption enhancers can temporally loosen the intestinal tight junctions, thereby allowing large molecules to permeate the intestinal epithelium into the systemic circulation.[6]

16.2.3. Transcytosis

Endocytosis is a process by which a substance gains entry into the cells without passing through the phospholipid membrane bilayer. This type of cellular uptake occurs via several mechanisms: phagocytosis, pinocytosis, and receptor-mediated endocytosis. Endocytosis of a substance in the gut can potentially also lead to subsequent transcytosis across the gut epithelial cell (i.e., the substance enters differentiated epithelial cells from one side, then migrate through the cells to exit on the other side). This type of transport mechanistically describes the intestinal absorption of some biologically active macromolecules, such as immunoglobulin[7] and vitamin B_{12}.[8] Transcytosis is also utilized by many viruses, plant lectins, and plant toxins to gain entry into the human body. Several strategies have been pursed to take advantage of this mechanism for a targeted delivery and improved bioavailability of orally administered pharmaceuticals such as insulin.[9,10]

It is safe to say that transport of drug substances across the intestinal membrane is a complex and dynamic process. Overall, total drug absorption across a membrane can be expressed conceptually as the sum of the individual components that occur in parallel:

$$J_e = J_c + J_m + J_p + J_t \tag{8}$$

where J_e is the total absorption rate (also referred to as the apparent or effective flux), and J_c, J_m, J_p, and J_t are the transporter-mediated, passive, paracellular, and transcytosis-mediated absorption components, respectively. Transporter-mediated drug absorption might be a positive or negative value in order to reflect the direction of transport (e.g., efflux may be designated as a negative value). This equation becomes more complicated if one attempts to replace J_c by the product of P_c (carrier-mediated permeability) and C (drug concentration). This is due to the asymmetrical membrane localization of transporters and consequently different drug concentrations encountered by them.[11] For example, the drug concentration available for a brush border membrane–located absorptive transporter (e.g., proton-dependent peptide transporter PepT1) is the drug concentration in intestinal lumen, while drug concentration inside enterocytes is supplied for the efflux transporters at the basolateral membrane (e.g., basolaterally localized peptide transporters). The quantitative contribution of each component in equation (8) (i.e., the route of transport across intestinal membrane) to the total absorption rate is influenced by many factors, including the physicochemical properties of the drugs and physiological conditions of gastrointestinal (GI) tract. In the following sections we focus on the role of small intestinal transporters in drug absorption.

16.3. DRUG TRANSPORTERS IN THE SMALL INTESTINE

Numerous drug-transporting membrane proteins (Table 16.1) have been described in intestinal tissues, and most of them belong to two major transporter superfamilies. These are the ATP-binding cassette (ABC) and solute carrier (SLC) family.

The well-studied ABC transporters in intestine include ABCB1 [MDR1, P-glycoprotein (Pgp)], ABCC1, 2, and 3 (MRP1, 2, and 3), as well as ABCG2 [breast cancer resistance protein (BCRP)]. Available data demonstrate that the expression levels of ABC transporters vary along the GI tract. MDR1 expression gradually increases from duodenum to colon, and its message RNA level in colon is similar to that in ileum, which was approximately sixfold higher than in the duodenum.[12] Moreover, characterization of the regional intestinal kinetics of drug efflux in rat and human intestine revealed that the magnitude of Pgp-mediated efflux correlates with the expression levels of Pgp. The efflux ratios (B $\rightarrow$ A permeability/A $\rightarrow$ B permeability; B and A denote basolateral and apical membrane, respectively) in the ileum are typically higher than in other regions.[13] The relative abundance of other ABC transporters message RNA at different sites of intestine was also assessed by quantitative real-time polymerase chain reaction (PCR).[14] The ranking of transporter

TABLE 16.1. Well-Studied Drug Transporters in the Intestine

		Symbol	Alias	Subcellular Localization	Substrates
ABC	Subfamily B	ABCB1	MDR1, Pgp	Apical membrane of enterocytes[17]	Anti-cancer drugs: paclitaxel[45], vinblastine[70]
					Anti-HIV drugs: saquinavir[71]
					Immunosuppressants: tacrolimus[72]
					Antihistamine drugs: fexofenadine[73]
					Cardiovascular drugs: Digoxin[46]
	Subfamily C	ABCC1	MRP1	Basolateral membrane of crypt cells[16]	Anticancer drugs: Doxorubicin[74], Daunomycin,[62] vincristine[63]
		ABCC2	MRP2, CMOAT	Apical membrane	Anticancer drugs: PEITC,[75] EGCG,[76] etoposide[77]
					Anti-HIV drugs: saquinaivir[78]
		ABCC3	MRP3	Basolateral membrane[79]	
	Subfamily G	ABCG2	BCRP, MXR, ABCP	Apical membrane[80]	Anticancer drugs: Topotecan,[57] mitoxantrone[81]
SLC	SLCO subfamily	SLCO1A2	OATP, OATP1, OATP-A	Unknown	Anti-HIV drugs: saquinavir[36]
					Antihistamine drugs: fexofenadine[73]
		SLCO2B1	OATP-B	Apical membrane[82]	Lipid-lowering drugs: pravastatin[82]
	SLC15A subfamily	SLC15A1	PEPT1	Apical membrane of villus epithelial cells[83]	Cephalosporins, penicillins, angiotensin-converting enzyme (ACE) inhibitors, antiviral drugs (see review[84])

gene expression in the duodenum was MRP3 $\gg$ MDR1 $>$ MRP2 $>$ MRP1. In the terminal ileum the ranking order was MDR1 $>$ MRP3 $\gg$ MRP1 $>$ MRP2. In all segments of the colon (ascending, transverse, descending, and sigmoid colon) transporter gene expression increased in the order MRP3 $\gg$ MDR1 $>$ MRP1 $\gg$ MRP2. A similar study revealed that the jejunal ABC transporter expression levels have the following rank: BCRP $\cong$ MRP 2 $>$ MDR1 $\cong$ MRP3 $\cong$ MRP1.[15]

ABC transporters are often restricted to the specific cellular domains on particular cell types in the gastrointestinal tract. For example, it has been shown that MRP1 is present mainly at the basolateral membrane of crypt cells (e.g., Paneth cells) in the small intestine; however, its expression in the differentiated enterocytes at the tip of villi cannot be detected.[16] P-gp expression along the crypt–villus axis was also unevenly distributed in rat ileum, with the highest level in villus cells, but it is barely detectable in crypt epithelium.[17] Similarly, the subcellular localization of MRP2 in polarized intestinal epithelial cells was investigated using high-resolution microscopy techniques after immunostaining. The results unequivocally demonstrated apical expression in superficial columnar epithelial cells.[18]

Another transport family that is involved in drug absorption and disposition is the solute carrier (SLC) family. The members within the SLC superfamily are responsible for the transport of a variety of endogenous and exogenous substances, such as amino acids, glucose, oligopeptides, antibiotics, nonsteroid anti-inflammatory agents, and antitumor and anti-HIV drugs. Data currently available indicate that drug transport–relevant SLC members in intestine include (1) solute carrier organic anion transporter families (SLCO subfamilies) such as SLCO1A2 (OATP-A) and SLCO2B1 (OATP-B); and (2) solute carrier family 15 (SLC15A subfamily), such as SLC15A1 (PepT1). SLC22A members such as organic anion or cation transporters (OATs or OCTs) have also been identified in the intestine, but they seem to be of greater importance for transporting small endogenous molecules in kidney. In this chapter, only well-characterized and drug transport–relevant SLC members are described.

16.4. IMPACT OF SMALL INTESTINAL TRANSPORTERS ON ORAL ABSORPTION OF DRUGS

As discussed in Section 16.3, intestinal drug absorption is the sum of transport via several different pathways. Since paracellular transport and transcytosis do not play a significant role in the intestinal absorption of most modern new chemical entities,[19] these two components can be ignored in equation (8). In this case, the rate of drug absorption is controlled only by passive diffusion and carrier-mediated transport. If carrier-mediated transport (J_c) is replaced by the Michaelis–Menton equation, and passive transcellular diffusion (J_m) with the product of passive permeability (P_m) and drug concentration in intestinal lumen (C), apparent intestinal drug transport can be depicted as

$$J_e = J_c + J_m = \frac{J_{\max} C}{K_m + C} + P_m C \qquad (9)$$

In Section 16.3, the drug concentration to which a transporter is exposed was discussed and shown to be different depending on the localization and directionality of the transporter. However, the drug concentration in the intestinal lumen (C) is used for the calculation of transporter-mediated flux [i.e., $J_{\max}C/(K_m + C)$] in equation (9). Therefore, this equation only governs drug transport involving brush border membrane–localized transporters. In fact, it is generally believed and demonstrated by currently available data that apically situated transporters play the most significant role in the absorption of drug across the intestinal wall. So this simplified equation is valid under most situations. It is also noteworthy that drug concentrations in enterocytes were used to calculate apical efflux transporter (e.g., Pgp)-mediated flux in some modeling studies,[11] but it is justifiable to use lumen drug concentrations since the stepwise working mechanisms of efflux transporters such as Pgp are not completely known at this point in time.

Referring to equation (9), it is clear that in vivo drug transport across intestinal epithelial cells is governed by four independent variables: (1) transporter–substrate affinity (K_m), (2) capacity of transport by carrier ($J_{\max}$); (3) intestinal drug concentration (C), and (4) the drug's permeability from passive diffusion (P_m). The importance of intestinal transporters in the absorption of orally administered drugs can be quantitatively reflected by the percentage of J_c in J_e or the ratio of J_c to J_m. For a high-solubility/high-permeability drug [biopharmaceutical classification system (BCS) class I], passive diffusion will dominate drug transport since the value of P_mC is much higher than J_c. Additionally, the high solubility of such compounds allows large concentrations in the gut to saturate most transporters, and consequently, diminishes the influence of transporters on absorption. However, when a low-solubility/high-permeability drug (BCS class II) is dosed orally, or when a high-permeability drug is administered with low dose, it is very likely that efflux transporters at the intestinal brush border membrane could significantly influence the drug's absorption properties over clinically relevant concentration ranges. This conclusion is deduced based on the following two factors: (1) low available concentration and high permeability do not lead to a high value of P_mC, meaning that transcellular passive diffusion is low; and (2) the high permeability of these compounds permits ready access into the gut membranes, but the low solubility limits drug concentrations available in intestinal epithelial cells, thereby preventing saturation of the efflux transporters. For high-solubility/low-permeability drugs (BCS class III), apically located absorptive transporters (e.g., PepT1) may prove to play a significant role in intestinal absorption provided that the transporters are not saturated [i.e., substrate-binding affinity is low, or a high K_m value in equation (9).] In the following sections, some relatively well studied intestinal transporters are discussed in detail to illustrate the impact of transporters on oral drug absorption.

16.4.1. PepT1-Mediated Absorptive Transport

The existence of a proton–peptide symport with an electrogenic nature has been suggested in studies employing brush border membrane vesicles in the early 1980s.[20] Due to technological advances in molecular biology, the first proton-coupled mammalian peptide transporter (PepT1) was cloned from rabbit intestine in the early 1990s.[21] Subsequently, the orthologs of rabbit PepT1 were identified from other species, including humans.[22] The gene encoding human PepT1 (hPepT1) maps to human chromosome 13q33–34 and consists of 23 exons. As predicted by hydropathy analysis, the membrane topology model of PepT1 suggests 12 transmembrane domains (TMD) and a large extracellular loop between TMD 9 and 10 with the N- and C-termini facing the cytosol.[21,23] The expression of PepT1 was found primarily in small intestine, with low levels in the liver and kidney.[22] In the human GI tract, hPepT1 appears to have higher expression levels in the duodenum than in jejunum or ileum.[24] Immunohistochemical studies revealed that PepT1 protein is localized predominantly to the apical microvillous plasma membrane of the absorptive epithelial cells in rat small intestine.[25] The uptake of PepT1 substrates (i.e., di- and tripeptides or structurally related drugs) is mediated by a proton gradient and the membrane potential at the apical surface of epithelial cells. Briefly, an inward proton gradient is established at the brush border membrane by the Na^+/H^+ exchanger, and then the influx of protons back into the epithelial cells is coupled by PepT1 to transport its substrates, thus, the system is known as a proton-dependent cotransport system.[26]

PepT1 has generally been characterized as a low-affinity/high-capacity transporter with a wide variety of compounds as substrates. Drug molecules transported by PepT1 include β-lactam antibiotics such as penicillins and cephalosporins,[27] angiotensin-converting enzyme (ACE) inhibitors such as captopril, and the ester prodrugs enalapril and fosinopril.[28] Prodrugs of acyclovir (e.g., valacyclovir) and L-dopa (e.g., L-dopa-L-Phe) can also be recognized and transported by PepT1.[29,30] While conventional approaches to enhancing the bioavailability of orally administered drugs focused on the optimization of dissolution, solubility, and passive permeability of drugs, coupling of active drugs (e.g., acyclovir and L-dopa) with an amino acid (e.g., Val or Phe) to target PepT1 significantly improves the intestinal absorption of the drugs by recognition and uptake via PepT1.[29,31]

16.4.2. OATP-Mediated Absorptive Transport

Members of the OATP family that have been found in the human intestine and are relatively well studied are OATP-A and OATP-B. OATP-A was originally cloned from the human liver,[32] and its transcript is found predominantly in the blood–brain barrier in cerebral capillary endothelial cells, and to a lesser extent in the intestine.[33,34] Although immunohistochemical studies to identify the precise subcellular location of OATP-A at intestinal brush border membrane have not provided an unequivocal conclusion at the present time, some in vivo functional data suggest that OATP-A is localized at the apical side of intestinal epithelial cells as an absorptive transporter.[35] The drugs transported by OATP-A include fexofenadine.[35] and saquinavir.[36] Using OATP-A cRNA-injected *Xenopus laevis* oocytes, the affinity (K_m) between saquinavir and OATP-A was characterized as 36.4 ± 21.8 μM,[36] suggesting that the impact of OATP-A on intestinal transport of saquinavir after oral delivery is questionable since at typical luminal concentrations, OATP-A would be saturated for much of the drug's residence time in the intestine. Dresser et al.[35] studied the inhibitory activity of grapefruit, orange, and apple juice on OATP-A- or Pgp-mediated fexofenadine uptake or efflux, respectively, at the in vitro level and found that the inhibitory potency of fruit juices on OATP-A is much higher than that on Pgp. In a clinical study,[35] oral administration of fexofenadine in the presence of fruit juices led to a three- to fourfold decrease in the area under the plasma concentration–time curve (AUC) and a twofold decrease in $C_{\max}$, while the urinary clearance of fexofenadine remained unchanged compared to administration with water. These in vivo data suggest that fruit juices may decrease the oral bioavailability of fexofenadine by inhibition of OATP-A-mediated drug absorption.

Few transporters belonging to the OATP family have been identified in the GI tract to date; however, most functional characterization work has been carried out at the cell culture level. In vivo data to further support the role of OATPs on oral drug absorption are still very limited. One challenge to correlate in vitro findings with in vivo results is the ambiguous orthologous gene product in experimental animals such as rats or mice. For example, rat Oatp1, 2, and 3 share some substrates with human OATP-A, but none of them represents the ortholog of human OATP-A, because their amino acid sequence are only 67 to 73% identical to human OATP-A. Additionally, the tissue distribution patterns or cellular localizations of human OATP-A are quite different from rat Oatp1, 2, and 3.[37] In vivo assessment of the role of OATPs in drug absorption is also significantly impeded by the lack of specific inhibitors for these transporters. It may be feasible to use substrates as competitive inhibitors under certain in vitro situations (e.g., OATP-A gene transfected cells) where single transporters are present in the system, but the overlapping substrate specificity among OATP family members makes it impossible to use this strategy for in vivo functional assessment of individual OATPs. The clinical significance of intestinal OATPs in drug absorption needs to be established and confirmed.

16.4.3. Pgp-Mediated Secretory Transport

Pgp is probably the best known and most thoroughly characterized secretory drug transporter in the gut. It can transport a variety of drugs from many therapeutic classes with diverse structures and pharmacological activities. The number of substrates and inhibitors for Pgp is increasing continuously. Early studies demonstrating Pgp-mediated secretory transport in the intestine were carried out by Hunter et al.[38–40] using polarized human intestinal epithelial cell lines (e.g., Caco-2, HT29, and T84 cells), which were cultured on porous supporting materials and had Pgp expression at the apical domain. The vectorial transport of vinblastine, a typical Pgp substrate, was measured in this in vitro cell culture system. Their results showed that basolateral to apical (B $\rightarrow$ A, secretory) transport of vinblastine is markedly higher than apical to basolateral (A $\rightarrow$ B, absorptive) transport. Moreover, the secretory transport of vinblastine could be reduced significantly when Pgp was inhibited by verapamil and nifedipine. Using similar approaches described in this pioneering in vitro work, numerous studies have been performed, with the results demonstrating that intestinal

Pgp may transport a number of therapeutic agents and possibly lead to low oral absorption of some drugs.[41]

The important role of Pgp in intestinal absorption of orally administered drugs is also supported by a large body of in vivo evidence using Pgp knockout mice. In contrast to humans, who have only one gene (MDR1) encoding Pgp, mice have two genes (i.e., mdr1a and mdr1b) working together to fulfill the equivalent function of human Pgp. Further studies demonstrated that the relative abundance of mdr1a and mdr1b transcripts is tissue or organ specific, with mdr1a highly expressed in intestinal epithelial cells and at the blood–brain and blood–testis barriers, whereas mdr1b is highly expressed in the adrenal gland, pregnant uterus, and ovaries. Both mdr1a and mdr1b genes are expressed substantially in many other tissues, including the liver, kidney, lung, heart, and spleen.[42,43] Pgp knockout mice are now available as mdr1a(–/–), mdr1b(–/–), and mdr1a/1b(–/–) (double knockouts).[43,44] A study by Sparreboom et al.[45] assessed the effect of gut Pgp on the pharmacokinetics of paclitaxel using mdr1a(–/–) mice. The results showed that the AUC of paclitaxel was two- and sixfold higher in mdr1a(–/–) mice than in wild-type mice after intravenous and oral drug administration, respectively. Additionally, cumulative (0 to 96 hour) fecal drug excretion decreased from 87% (after oral administration) and 40% (after intravenous administration) of the administered dose in wild-type mice to less than 3% in mdr1a(–/–) mice. Since the biliary secretion of paclitaxel is similar between wild-type and mdr1a(–/–) mice, the reduced fecal excretion in mdr1a(–/–) mice was attributed to the lower intestinal Pgp-mediated secretion as compared to that of wild-type mice. Collectively, their results suggest that intestinal Pgp not only limit the oral absorption of paclitaxel but also mediate direct secretion of the drug from the systemic circulation into the intestinal lumen. The role of Pgp in intestinal drug secretion has also been demonstrated for digoxin, which is metabolized to a minor extent in humans. In bile duct–ligated mice (i.e., no biliary flux into the intestinal lumen), the intestinal secretion of digoxin was reduced from 16% of the administered dose in wild-type mice to 1.5% in mdr1a/1b(–/–) mice.[44] Using a recently developed perfusion catheter, which isolates human jejunal segments by inflating balloons at the desired point inside the lumen, Drescher et al.[46] measured the intestinal secretion of digoxin after intravenous administration. Within 3 hours of injection, 0.45% of the dosed digoxin (1 mg) was eliminated into a 20-cm-long segment of jejunum. Additionally, perfusion of the isolated segment with Pgp inhibitor (quinidine) reduced intestinal secretion of digoxin to 0.23%. When their findings were scaled up from one segment to the entire intestine, it was estimated that at least 11% of the intravenously administered digoxin was directly secreted into the intestinal lumen within 3 hours, and approximately half was mediated by Pgp.

The unequivocal involvement of Pgp in intestinal drug efflux was also assessed using excised intestinal segments mounted onto Ussing diffusion chambers. For example, studies on the vectorial transport of digoxin across the intestinal wall demonstrated that the efflux ratio was as high as 10-fold greater; however, this prominent directional difference of digoxin transport diminished in mdr1a(–/–) mice. Makhey et al.[13] investigated the mechanisms and kinetics of intestinal secretory transport process of the anticancer drug etoposide (a topoisomerase II inhibitor) in rat or human intestine. The results showed that etoposide secretion is regionally dependent with efflux ratios of 5.4, 9.7, and 4.8 in rat jejunum, ileum, and colon, respectively. Moreover, the presence of quinidine in the diffusion buffer could markedly reduce etoposide efflux ratio highlighting the critical role of intestinal Pgp in the absorption and secretion of etoposide. Diffusion studies using excised intestinal tissue have proved useful in mechanistically characterizing the drug transport process, however, kinetic parameters (e.g., permeability) are usually obtained at steady state. Therefore, better in vitro–in vivo correlations (IVIVCs) are necessary for accurate prediction of drug absorption and/or secretion in vivo.

Since the discovery of Pgp, a large number of in vitro and in vivo studies have been carried out to assess its detoxification function as a secretory transporter. Pgp-mediated transport reduces drug exposure by preventing the entry of xenobiotics into the human body and/or facilitating their elimination out of body. However, intestinal Pgp-mediated efflux may not play a significant role in oral bioavailability of certain drugs even though they have been demonstrated to be typical substrates for Pgp at the cell culture level. This is well demonstrated for high-permeability/high-solubility drugs such as verapamil.[11,12] As discussed in Section provided that the clinical dose is relatively high, high solubility and permeability will make simple diffusion [value of $P_m C$ in equation (9)] the dominant factor in drug transport, eclipsing the effect of Pgp-mediated efflux. Many pharmaceutical companies are now using several in vitro methods (cell- or enzyme-based) to screen Pgp substrate at the early development stage. However, the decision of discontinuing an in vitro–identified Pgp substrate for late drug development must be made with great caution, since the very low concentrations used for in vitro studies results in readily detected Pgp function, and the role of efflux transporter in the oral availability of certain drugs may not be substantial in actual clinical practice.[47]

Most Pgp-transported drugs are also substrates for phase I metabolic enzyme cytochrome P450 (CYP) 3A4. The coexpression of these two proteins in the intestine and the interplay between them complicates the demonstration of either protein's contribution to drug absorption. It is now generally accepted that intestinal metabolic enzymes and efflux transporters work coordinately as a detoxification system, resulting in the poor bioavailability of certain drugs.[48]

16.4.4. BCRP-Mediated Secretory Transport

BCRP was cloned independently by several research groups from drug-resistant cancer cells[49,50] or placenta.[51] Unlike other ABC transporters, BCRP contains only one membrane-spanning domain and one nucleotide-binding domain. Some available experimental evidence suggests that BCRP may function as a homodimer or homotetramer,[52–54] so it is also called a *half-transporter*.

BCRP is expressed abundantly at the apical membrane of the small intestine and colon epithelium, where it limits drug absorption and/or facilitates secretion back into lumen. This has been well demonstrated by several animal and clinical studies. Jonker et al.[55] compared the bioavailability of the anticancer drug topotecan (TPT, a topoisomerase I inhibitor) in Bcrp knockout mice and found that the AUC of orally administered TPT is about sixfold higher in Bcrp-deficient mice than in wild-type mice. Chemically knocking down mice Bcrp function by its inhibitor GF120918 (also

Pgp inhibitor) also results in a significant increase in TPT oral absorption.[56] Subsequently, TPT in combination with GF120918 demonstrated increased oral bioavailability in the clinical setting.[57] Since TPT is not transported extensively by Pgp and undergoes minimal CYP3A-catalyzed metabolism,[58] the pharmacokinetic interaction observed between TPT and GF120918 could be ascribed predominantly to the inhibition of the efflux transporter BCRP.

Other than limiting drug entry into the body, BCRP also restricts exposure to dietary carcinogens such as 2-amino-1-methyl-6-phenylimidazo[4,5-*b*]pyridine (PhIP). Pharmacokinetic studies by van Herwaarden et al.[59] demonstrated that at a dose of 1 mg/kg [^{14}C]PhIP, the AUC of oral and intravenous administration was 2.9- and 2.2-fold higher in Bcrp knockout mice than in wild-type mice, respectively. In mice with cannulated gallbladders, both biliary and direct intestinal secretion of [^{14}C]PhIP were greatly reduced in Bcrp knockout mice compared with wild-type mice. The data suggest that Bcrp effectively restricts the exposure of mice to ingested PhIP by decreasing its absorption from the small intestine and increasing biliary and intestinal secretion. Since PhIP is the most abundant heterocyclic amine present in various protein-containing foods, BCRP/bcrp1 is believed to play an important role in protection from the toxicity of normal food constituents. The importance of BCRP as a detoxification efflux transporter in the gut was also highlighted by Jonker et al.,[55] who found that Bcrp knockout mice are prone to developing phototoxic lesions on light-exposed areas of the skin when fed a diet containing large amounts of chlorophyll. Further studies showed that Bcrp efficiently limits the uptake of chlorophyll-breakdown product pheophorbide a, and the deficiency of Bcrp increases the exposure of mice to pheophorbide a, leading to the high risk of protoporphyria and diet-dependent phototoxicity.

16.4.5. Transporters Localized on the Basolateral Membrane of Enterocytes

Little is known about the transport process occurring at the basolateral membrane of the intestinal epithelium. Some functional studies provide strong evidence for the existence of basolateral peptide transporters which are distinguishable from PepT1[60,61]; however, these putative transporters have not yet been cloned. MRP1 is expressed primarily at the basolateral domain of the crypt cells in mouse and human small intestine.[16] In vitro studies showed that MRP1 can efflux a variety of compounds, including daunomycin[62] but and vincristine,[63] but its role in drug transport in the intestine has not yet been clearly demonstrated.

16.5. FUNCTIONAL MODULATION OF INTESTINAL TRANSPORTERS TO OPTIMIZE ORAL ABSORPTION OF DRUGS

In view of the importance of intestinal transporters in the absorption of certain drugs, tremendous efforts have been made to identify chemical inhibitors of secretory transporters, hoping to knock down the efflux activities of these gatekeepers, thereby increasing the bioavailability of some poorly absorbed drugs. Up to now, the characterization of Pgp inhibitors represents the most successful example in this research area and has generated a variety of specific and potent inhibitors, which can be classified into three generations based on the time sequence of their identification and the inhibitory mechanism. First-generation Pgp inhibitors include verapamil, cyclosporin A, tamoxifen, and several calmodulin antagonists. Indeed, these inhibitors are also Pgp substrates; therefore, the inhibitory mechanisms have been attributed to competitive binding or transport. Owing to widespread availability (most of them are commercially available), the first-generation Pgp inhibitors have been used extensively to delineate the role of Pgp in drug transport across the cell membrane (e.g., drug-resistant tumor cells) or tissue barriers (e.g., intestine, liver, kidney, and BBB), where Pgp protein resides. However, their clinical application for boosting drug bioavailability is limited because the desired inhibitory potency on Pgp requires a high blood concentration of inhibitors (at least three- to fivefold higher than their binding affinity K_m), at which concentrations side effects will be evoked. Subsequently, the second generation of Pgp inhibitors was synthesized with the goal of making the inhibitors more potent and less toxic. Representative second-generation Pgp inhibitors include biricodar (VX-710) and PSC833 (valspodar), a nonimmunosupressive derivative of cyclosporin D, which is 10- to 20-fold more potent than cyclosporin A in its ability to inhibit Pgp.[64] Coadministration of the anticancer drug paclitaxel with PSC833 resulted in a 10-fold increased oral bioavailability of paclitaxel in mice.[65] During the past several years, more potent and specific Pgp modulators, such as LY335979 (zosuquidar), GF120918 (elacridar), XR9576 (tariquidar), and R101933 (laniquidar), have been developed as third-generation Pgp inhibitors. A mass balance study demonstrated that concurrent administration of GF120918 results in almost complete oral absorption of paclitaxel.[66] Using intestinal and vascular access–ported (IVAP) rabbits, our laboratory found that in the presence of GF120918, the bioavailability of upper small intestine–administered saquinavir increased twofold.[67]

Using inhibitors of secretory transporters (e.g., Pgp) to boost the bioavailability of some poorly absorbed drugs has generated some encouraging results. However, it was noted recently that this strategy may be questionable for dual Pgp and CYP3A substrates. Wu and Benet examined the influence of GF120918 on the disposition of tacrolimus (a substrate for both Pgp and CYP3A) in isolated perfused rat liver and found that tacrolimus AUC could be reduced significantly in the presence of Pgp inhibitor.[68] In a clinical study,[69] patients receiving paclitaxel intravenous infusion with concurrent oral administration of PSC833 showed increased plasma concentrations of 6α-hydroxypaclitaxel, a major metabolite of paclitaxel, compared to patients treated with paclitaxel alone. These findings suggested that the Pgp inhibitor may sequester its substrate in hepatocytes by inhibiting its biliary excretion, resulting in greater metabolism by CYP3A.

16.6. CONCLUSIONS

Although our knowledge of physical pharmacy and advancements in manufacturing technology have enabled the production of oral medicines with better absorption profiles, low and erratic oral bioavailability still prevents many promising drug candidates from entering clinical trials. It has been demonstrated that drug transporters in the intestinal tract may play a critical role in the absorption and disposition of some drugs. However, they may also mediate significant drug–drug interactions. Recognition of the importance of intestinal transporters will certainly help the development of drugs with optimal absorption properties and benefit the design of individualized dosing regimens.

REFERENCES

1. Levitt MD, Furne JK, Levitt DG. 1992. Shaking of the intact rat and intestinal angulation diminish the jejunal unstirred layer. Gastroenterology 103(5):1460–1466.
2. Huisman MT, Smit JW, Crommentuyn KM, Zelcer N, Wiltshire HR, Beijnen JH, Schinkel AH. 2002. Multidrug resistance protein 2 (MRP2) transports HIV protease inhibitors, and transport can be enhanced by other drugs. AIDS 16(17):2295–2301.
3. Zelcer N, Huisman MT, Reid G, Wielinga P, Breedveld P, Kuil A, Knipscheer P, Schellens JH, Schinkel AH, Borst P. 2003. Evidence for two interacting ligand binding sites in human multidrug resistance protein 2 (ATP binding cassette C2). J Biol Chem 278(26):23538–23544.
4. Pappenheimer JR, Reiss KZ. 1987. Contribution of solvent drag through intercellular junctions to absorption of nutrients by the small intestine of the rat. J Membr Biol 100(2):123–136.
5. Aungst BJ. 2000. Intestinal permeation enhancers. J Pharm Sci 89(4):429–442.
6. Salama NN, Fasano A, Thakar M, Eddington ND. 2004. The effect of delta G on the transport and oral absorption of macromolecules. J Pharm Sci 93(5):1310–1319.
7. Mostov KE. 1994. Transepithelial transport of immunoglobulins. Annu Rev Immunol 12:63–84.
8. Dan N, Cutler DF. 1994. Transcytosis and processing of intrinsic factor-cobalamin in Caco-2 cells. J Biol Chem 269(29):18849–18855.
9. Shah D, Shen WC. 1996. Transcellular delivery of an insulin-transferrin conjugate in enterocyte-like Caco-2 cells. J Pharm Sci 85(12):1306–1311.
10. Xia CQ, Wang J, Shen WC. 2000. Hypoglycemic effect of insulin-transferrin conjugate in streptozotocin-induced diabetic rats. J Pharmacol Exp Ther 295(2):594–600.
11. Kwon H, Lionberger RA, Yu LX. 2004. Impact of P-glycoprotein-mediated intestinal efflux kinetics on oral bioavailability of P-glycoprotein substrates. Mol Pharm 1(6):455–465.
12. Cao X, Yu LX, Barbaciru C, Landowski CP, Shin HC, Gibbs S, Miller HA, Amidon GL, Sun D. 2005. Permeability dominates in vivo intestinal absorption of P-gp substrate with high solubility and high permeability. Mol Pharm 2(4):329–340.
13. Makhey VD, Guo A, Norris DA, Hu P, Yan J, Sinko PJ. 1998. Characterization of the regional intestinal kinetics of drug efflux in rat and human intestine and in Caco-2 cells. Pharm Res 15(8):1160–1167.
14. Zimmermann C, Gutmann H, Hruz P, Gutzwiller JP, Beglinger C, Drewe J. 2005. Mapping of MDR1 and MRP1-5 mRNA expression along the human intestinal tract. Drug Metab Dispos 33(2):219–224.

15. Taipalensuu J, Tornblom H, Lindberg G, Einarsson C, Sjoqvist F, Melhus H, Garberg P, Sjostrom B, Lundgren B, Artursson P. 2001. Correlation of gene expression of ten drug efflux proteins of the ATP-binding cassette transporter family in normal human jejunum and in human intestinal epithelial Caco-2 cell monolayers. J Pharmacol Exp Ther 299(1):164–170.
16. Peng KC, Cluzeaud F, Bens M, Van Huyen JP, Wioland MA, Lacave R, Vandewalle A. 1999. Tissue and cell distribution of the multidrug resistance–associated protein (MRP) in mouse intestine and kidney. J Histochem Cytochem 47(6):757–768.
17. Collett A, Higgs NB, Sims E, Rowland M, Warhurst G. 1999. Modulation of the permeability of H2 receptor antagonists cimetidine and ranitidine by P-glycoprotein in rat intestine and the human colonic cell line Caco-2. J Pharmacol Exp Ther 288(1):171–178.
18. Thiebaut F, Tsuruo T, Hamada H, Gottesman MM, Pastan I, Willingham MC. 1987. Cellular localization of the multidrug-resistance gene product P-glycoprotein in normal human tissues. Proc Natl Acad Sci U S A 84(21):7735–7738.
19. Stenberg P, Luthman K, Artursson P. 2000. Virtual screening of intestinal drug permeability. J Control Release 65(1–2):231–243.
20. Ganapathy V, Leibach FH. 1983. Role of pH gradient and membrane potential in dipeptide transport in intestinal and renal brush-border membrane vesicles from the rabbit. Studies with L-carnosine and glycyl-L-proline. J Biol Chem 258(23):14189–14192.
21. Fei YJ, Kanai Y, Nussberger S, Ganapathy V, Leibach FH, Romero MF, Singh SK, Boron WF, Hediger MA. 1994. Expression cloning of a mammalian proton-coupled oligopeptide transporter. Nature 368(6471):563–566.
22. Liang R, Fei YJ, Prasad PD, Ramamoorthy S, Han H, Yang-Feng TL, Hediger MA, Ganapathy V, Leibach FH. 1995. Human intestinal H^+/peptide cotransporter: cloning, functional expression, and chromosomal localization. J Biol Chem 270(12):6456–6463.
23. Covitz KM, Amidon GL, Sadée W. 1998. Membrane topology of the human dipeptide transporter, hPEPT1, determined by epitope insertions. Biochemistry 37(43):15214–15221.
24. Herrera-Ruiz D, Wang Q, Gudmundsson OS, Cook TJ, Smith RL, Faria TN, Knipp GT. 2001. Spatial expression patterns of peptide transporters in the human and rat gastrointestinal tracts, Caco-2 in vitro cell culture model, and multiple human tissues. AAPS PharmSci 3(1):E9.
25. Ogihara H, Saito H, Shin BC, Terado T, Takenoshita S, Nagamachi Y, Inui K, Takata K. 1996. Immuno-localization of H^+/peptide cotransporter in rat digestive tract. Biochem Biophys Res Commun 220(3):848–852.
26. Adibi SA. 1997. The oligopeptide transporter (Pept-1) in human intestine: biology and function. Gastroenterology 113(1):332–340.
27. Ganapathy ME, Brandsch M, Prasad PD, Ganapathy V, Leibach FH. 1995. Differential recognition of beta-lactam antibiotics by intestinal and renal peptide transporters, PEPT 1 and PEPT 2. J Biol Chem 270(43):25672–25677.
28. Zhu T, Chen XZ, Steel A, Hediger MA, Smith DE. 2000. Differential recognition of ACE inhibitors in *Xenopus laevis* oocytes expressing rat PEPT1 and PEPT2. Pharm Res 17(5):526–532.
29. Balimane PV, Tamai I, Guo A, Nakanishi T, Kitada H, Leibach FH, Tsuji A, Sinko PJ. 1998. Direct evidence for peptide transporter (PepT1)-mediated uptake of a nonpeptide prodrug, valacyclovir. Biochem Biophys Res Commun 250(2):246–251.

30. Tamai I, Nakanishi T, Nakahara H, Sai Y, Ganapathy V, Leibach FH, Tsuji A. 1998. Improvement of L-dopa absorption by dipeptidyl derivation, utilizing peptide transporter PepT1. J Pharm Sci 87(12):1542–1546.

31. Han H, de Vrueh RL, Rhie JK, Covitz KM, Smith PL, Lee CP, Oh DM, Sadée W, Amidon GL. 1998. 5′-Amino acid esters of antiviral nucleosides, acyclovir, and AZT are absorbed by the intestinal PEPT1 peptide transporter. Pharm Res 15(8):1154–1159.

32. Kullak-Ublick GA, Hagenbuch B, Stieger B, Schteingart CD, Hofmann AF, Wolkoff AW, Meier PJ. 1995. Molecular and functional characterization of an organic anion transporting polypeptide cloned from human liver. Gastroenterology 109(4):1274–1282.

33. Gao B, Hagenbuch B, Kullak-Ublick GA, Benke D, Aguzzi A, Meier PJ. 2000. Organic anion-transporting polypeptides mediate transport of opioid peptides across blood–brain barrier. J Pharmacol Exp Ther 294(1):73–79.

34. Tamai I, Nezu J, Uchino H, Sai Y, Oku A, Shimane M, Tsuji A. 2000. Molecular identification and characterization of novel members of the human organic anion transporter (OATP) family. Biochem Biophys Res Commun 273(1):251–260.

35. Dresser GK, Bailey DG, Leake BF, Schwarz UI, Dawson PA, Freeman DJ, Kim RB. 2002. Fruit juices inhibit organic anion transporting polypeptide-mediated drug uptake to decrease the oral availability of fexofenadine. Clin Pharmacol Ther 71(1):11–20.

36. Su Y, Zhang X, Sinko PJ. 2004. Human organic anion–transporting polypeptide OATP-A (SLC21A3) acts in concert with P-glycoprotein and multidrug resistance protein 2 in the vectorial transport of saquinavir in Hep G2 cells. Mol Pharm 1(1):49–56.

37. Kullak-Ublick GA, Ismair MG, Stieger B, Landmann L, Huber R, Pizzagalli F, Fattinger K, Meier PJ, Hagenbuch B. 2001. Organic anion-transporting polypeptide B (OATP-B) and its functional comparison with three other OATPs of human liver. Gastroenterology 120(2):525–533.

38. Hunter J, Hirst BH, Simmons NL. 1991. Epithelial secretion of vinblastine by human intestinal adenocarcinoma cell (HCT-8 and T84) layers expressing P-glycoprotein. Br J Cancer 64(3):437–444.

39. Hunter J, Hirst BH, Simmons NL. 1993. Drug absorption limited by P-glycoprotein-mediated secretory drug transport in human intestinal epithelial Caco-2 cell layers. Pharm Res 10(5):743–749.

40. Hunter J, Jepson MA, Tsuruo T, Simmons NL, Hirst BH. 1993. Functional expression of P-glycoprotein in apical membranes of human intestinal Caco-2 cells: kinetics of vinblastine secretion and interaction with modulators. J Biol Chem 268(20):14991–14997.

41. Adachi Y, Suzuki H, Sugiyama Y. 2001. Comparative studies on in vitro methods for evaluating in vivo function of MDR1 P-glycoprotein. Pharm Res 18(12):1660–1668.

42. Croop JM, Raymond M, Haber D, Devault A, Arceci RJ, Gros P, Housman DE. 1989. The three mouse multidrug resistance (mdr) genes are expressed in a tissue-specific manner in normal mouse tissues. Mol Cell Biol 9(3):1346–1350.

43. Schinkel AH, Smit JJ, van Tellingen O, Beijnen JH, Wagenaar E, van Deemter L, Mol CA, van der Valk MA, Robanus-Maandag EC, te Riele HP, et al. 1994. Disruption of the mouse mdr1a P-glycoprotein gene leads to a deficiency in the blood–brain barrier and to increased sensitivity to drugs. Cell 77(4):491–502.

44. Schinkel AH, Mayer U, Wagenaar E, Mol CA, van Deemter L, Smit JJ, van der Valk MA, Voordouw AC, Spits H, van Tellingen O, et al. 1997. Normal viability and altered

pharmacokinetics in mice lacking mdr1-type (drug-transporting) P-glycoproteins. Proc Natl Acad Sci U S A 94(8):4028–4033.

45. Sparreboom A, van Asperen J, Mayer U, Schinkel AH, Smit JW, Meijer DK, Borst P, Nooijen WJ, Beijnen JH, van Tellingen O. 1997. Limited oral bioavailability and active epithelial excretion of paclitaxel (Taxol) caused by P-glycoprotein in the intestine. Proc Natl Acad Sci U S A 94(5):2031–2035.
46. Drescher S, Glaeser H, Murdter T, Hitzl M, Eichelbaum M, Fromm MF. 2003. P-Glycoprotein-mediated intestinal and biliary digoxin transport in humans. Clin Pharmacol Ther 73(3):223–231.
47. Lin JH. 2004. How significant is the role of P-glycoprotein in drug absorption and brain uptake? Drugs Today (Barc) 40(1):5–22.
48. Benet LZ, Cummins CL, Wu CY. 2004. Unmasking the dynamic interplay between efflux transporters and metabolic enzymes. Int J Pharm 277(1–2):3–9.
49. Doyle LA, Yang W, Abruzzo LV, Krogmann T, Gao Y, Rishi AK, Ross DD. 1998. A multidrug resistance transporter from human MCF-7 breast cancer cells. Proc Natl Acad Sci U S A 95(26):15665–15670.
50. Miyake K, Mickley L, Litman T, Zhan Z, Robey R, Cristensen B, Brangi M, Greenberger L, Dean M, Fojo T, Bates SE. 1999. Molecular cloning of cDNAs which are highly overexpressed in mitoxantrone-resistant cells: demonstration of homology to ABC transport genes. Cancer Res 59(1):8–13.
51. Allikmets R, Schriml LM, Hutchinson A, Romano-Spica V, Dean M. 1998. A human placenta-specific ATP-binding cassette gene (ABCP) on chromosome 4q22 that is involved in multidrug resistance. Cancer Res 58(23):5337–5339.
52. Kage K, Tsukahara S, Sugiyama T, Asada S, Ishikawa E, Tsuruo T, Sugimoto Y. 2002. Dominant-negative inhibition of breast cancer resistance protein as drug efflux pump through the inhibition of S–S dependent homodimerization. Int J Cancer 97(5):626–630.
53. Nakanishi T, Doyle LA, Hassel B, Wei Y, Bauer KS, Wu S, Pumplin DW, Fang HB, Ross DD. 2003. Functional characterization of human breast cancer resistance protein (BCRP, ABCG2) expressed in the oocytes of *Xenopus laevis*. Mol Pharmacol 64(6):1452–1462.
54. Xu J, Liu Y, Yang Y, Bates S, Zhang JT. 2004. Characterization of oligomeric human half-ABC transporter ATP-binding cassette G2. J Biol Chem 279(19):19781–19789.
55. Jonker JW, Buitelaar M, Wagenaar E, van der Valk MA, Scheffer GL, Scheper RJ, Plosch T, Kuipers F, Elferink RP, Rosing H, Beijnen JH, Schinkel. AH 2002. The breast cancer resistance protein protects against a major chlorophyll-derived dietary phototoxin and protoporphyria. Proc Natl Acad Sci U S A 99(24):15649–15654.
56. Jonker JW, Smit JW, Brinkhuis RF, Maliepaard M, Beijnen JH, Schellens JH, Schinkel AH. 2000. Role of breast cancer resistance protein in the bioavailability and fetal penetration of topotecan. J Natl Cancer Inst 92(20):1651–1656.
57. Kruijtzer CM, Beijnen JH, Rosing H, Bokkel Huinink WW, Schot M, Jewell RC, Paul EM, Schellens JH. 2002. Increased oral bioavailability of topotecan in combination with the breast cancer resistance protein and P-glycoprotein inhibitor GF120918. J Clin Oncol 20(13):2943–2950.
58. Rosing H, van Zomeren DM, Doyle E, Bult A, Beijnen JH. 1998. O-glucuronidation, a newly identified metabolic pathway for topotecan and *N*-desmethyl topotecan. Anticancer Drugs 9(7):587–592.

59. van Herwaarden AE, Jonker JW, Wagenaar E, Brinkhuis RF, Schellens JH, Beijnen JH, Schinkel AH. 2003. The breast cancer resistance protein (Bcrp1/Abcg2) restricts exposure to the dietary carcinogen 2-amino-1-methyl-6-phenylimidazo[4,5-b]pyridine. Cancer Res 63(19):6447–6452.
60. Matsumoto S, Saito H, Inui K. 1994. Transcellular transport of oral cephalosporins in human intestinal epithelial cells, Caco-2: interaction with dipeptide transport systems in apical and basolateral membranes. J Pharmacol Exp Ther 270(2):498–504.
61. Terada T, Sawada K, Saito H, Hashimoto Y, Inui K. 1999. Functional characteristics of basolateral peptide transporter in the human intestinal cell line Caco-2. Am J Physiol 276(6 Pt 1):G1435–G1441.
62. Tseng E, Kamath A, Morris ME. 2002. Effect of organic isothiocyanates on the P-glycoprotein- and MRP1-mediated transport of daunomycin and vinblastine. Pharm Res 19(10):1509–1515.
63. Mao Q, Deeley RG, Cole SP. 2000. Functional reconstitution of substrate transport by purified multidrug resistance protein MRP1 (ABCC1) in phospholipid vesicles. J Biol Chem 275(44):34166–34172.
64. Twentyman PR, Bleehen NM. 1991. Resistance modification by PSC-833, a novel nonimmunosuppressive cyclosporin [corrected]. Eur J Cancer 27(12):1639–1642.
65. van Asperen J, van Tellingen O, Sparreboom A, Schinkel AH, Borst P, Nooijen WJ, Beijnen JH. 1997. Enhanced oral bioavailability of paclitaxel in mice treated with the P-glycoprotein blocker SDZ PSC 833. Br J Cancer 76(9):1181–1183.
66. Bardelmeijer HA, Ouwehand M, Beijnen JH, Schellens JH, van Tellingen O. 2004. Efficacy of novel P-glycoprotein inhibitors to increase the oral uptake of paclitaxel in mice. Invest New Drugs 22(3):219–229.
67. Sinko PJ, Kunta JR, Usansky HH, Perry BA. 2004. Differentiation of gut and hepatic first pass metabolism and secretion of saquinavir in ported rabbits. J Pharmacol Exp Ther 310(1):359–366.
68. Wu CY, Benet LZ. 2003. Disposition of tacrolimus in isolated perfused rat liver: influence of troleandomycin, cyclosporine, and gg918. Drug Metab Dispos 31(11):1292–1295.
69. Kang MH, Figg WD, Ando Y, Blagosklonny MV, Liewehr D, Fojo T, Bates SE. 2001. The P-glycoprotein antagonist PSC 833 increases the plasma concentrations of 6alpha-hydroxypaclitaxel, a major metabolite of paclitaxel. Clin Cancer Res 7(6):1610–1617.
70. van Asperen J, van Tellingen O, Beijnen JH. 2000. The role of mdr1a P-glycoprotein in the biliary and intestinal secretion of doxorubicin and vinblastine in mice. Drug Metab Dispos 28(3):264–267.
71. Kim RB, Fromm MF, Wandel C, Leake B, Wood AJ, Roden DM, Wilkinson GR. 1998. The drug transporter P-glycoprotein limits oral absorption and brain entry of HIV-1 protease inhibitors. J Clin Invest 101(2):289–294.
72. Yokogawa K, Takahashi M, Tamai I, Konishi H, Nomura M, Moritani S, Miyamoto K, Tsuji A. 1999. P-glycoprotein-dependent disposition kinetics of tacrolimus: studies in mdr1a knockout mice. Pharm Res 16(8):1213–1218.
73. Cvetkovic M, Leake B, Fromm MF, Wilkinson GR, Kim RB. 1999. OATP and P-glycoprotein transporters mediate the cellular uptake and excretion of fexofenadine. Drug Metab Dispos 27(8):866–871.

74. Cole SP, Bhardwaj G, Gerlach JH, Mackie JE, Grant CE, Almquist KC, Stewart AJ, Kurz EU, Duncan AM, Deeley. RG 1992. Overexpression of a transporter gene in a multidrug-resistant human lung cancer cell line. Science 258(5088):1650–1654.
75. Ji Y, Morris ME. 2005. Transport of dietary phenethyl isothiocyanate is mediated by multidrug resistance protein 2 but not P-glycoprotein. Biochem Pharmacol 70(4):640–647.
76. Hong J, Lambert JD, Lee SH, Sinko PJ, Yang CS. 2003. Involvement of multidrug resistance-associated proteins in regulating cellular levels of (−)-epigallocatechin-3-gallate and its methyl metabolites. Biochem Biophys Res Commun 310(1):222–227.
77. Guo A, Marinaro W, Hu P, Sinko PJ. 2002. Delineating the contribution of secretory transporters in the efflux of etoposide using Madin–Darby Canine kidney (MDCK) cells overexpressing P-glycoprotein (Pgp), multidrug resistance–associated Protein (MRP1), and canalicular multispecific organic anion transporter (cMOAT). Drug Metab Dispos 30(4):457–463.
78. Williams GC, Liu A, Knipp G, Sinko PJ. 2002. Direct evidence that saquinavir is transported by multidrug resistance-associated protein (MRP1) and canalicular multispecific organic anion transporter (MRP2). Antimicrob Agents Chemother 46(11):3456–3462.
79. Rost D, Mahner S, Sugiyama Y, Stremmel W. 2002. Expression and localization of the multidrug resistance-associated protein 3 in rat small and large intestine. Am J Physiol Gastrointest Liver Physiol 282(4):G720–G726.
80. Maliepaard M, Scheffer GL, Faneyte IF, van Gastelen MA, Pijnenborg AC, Schinkel AH, van de Vijver MJ, Scheper RJ, Schellens JH. 2001. Subcellular localization and distribution of the breast cancer resistance protein transporter in normal human tissues. Cancer Res 61(8):3458–3464.
81. Robey RW, Honjo Y, Morisaki K, Nadjem TA, Runge S, Risbood M, Poruchynsky MS, Bates SE. 2003. Mutations at amino-acid 482 in the ABCG2 gene affect substrate and antagonist specificity. Br J Cancer 89(10):1971–1978.
82. Kobayashi D, Nozawa T, Imai K, Nezu J, Tsuji A, Tamai I. 2003. Involvement of human organic anion transporting polypeptide OATP-B (SLC21A9) in pH-dependent transport across intestinal apical membrane. J Pharmacol Exp Ther 306(2):703–708.
83. Ziegler TR, Fernandez-Estivariz C, Gu LH, Bazargan N, Umeakunne K, Wallace TM, Diaz EE, Rosado KE, Pascal RR, Galloway JR, Wilcox JN, Leader LM. 2002. Distribution of the H^+/peptide transporter PepT1 in human intestine: up-regulated expression in the colonic mucosa of patients with short-bowel syndrome. Am J Clin Nutr 75(5):922–930.
84. Herrera-Ruiz D, Knipp GT. 2003. Current perspectives on established and putative mammalian oligopeptide transporters. J Pharm Sci 92(4):691–714.

17

REGULATION OF DRUG TRANSPORTER ACTIVITY

MEI HONG AND GUOFENG YOU

Rutgers, the State University of New Jersey, Piscataway, New Jersey

17.1. INTRODUCTION

Drug transporters are key determinants in the absorption, distribution, and excretion of a diverse array of environmental toxins and clinically important drugs, and are therefore critical for the survival of mammalian species. Alteration in the function of these drug transporters plays an important role in the intra- and interindividual variability of the therapeutic efficacy and the toxicity of the drugs. As a result, the activity of drug transporters must be under tight regulation so as to carry out their

Drug Transporters: Molecular Characterization and Role in Drug Disposition, Edited by Guofeng You and Marilyn E. Morris
Copyright © 2007 John Wiley & Sons, Inc.

normal duties. Key players involved in the regulation of transporters are hormones, protein kinases, nuclear receptors, scaffolding proteins, and disease conditions. These players may affect transporter activity at multiple levels, including (1) when and how often a gene encoding a given transporter is transcribed (transcriptional control), (2) how the primary RNA transcript is spliced or processed (RNA processing control), (3) which mRNA in the cytoplasm is translated by ribosomes (translational control), (4) which mRNA is destabilized in the cytoplasm (mRNA degradation control), and (5) how a transporter is modified and assembled after it has been made (posttranslational control). Regulation of transporter activity at the gene level usually occurs within hours and days and is therefore classified as *long-term* or *chronic regulation*. Long-term regulation usually occurs when the body undergoes massive change, such as during the development or occurrence of disease. Regulation at the posttranslational level usually occurs within minutes or hours and is therefore classified as *short-term* or *acute regulation*. Short-term regulation usually occurs when the body has to deal with rapidly changing amounts of substances as a consequence of variable intake of drugs, fluids, or meals as well as metabolic activity. In this chapter we describe the mechanisms for posttranslational regulation of drug transporters.

17.2. GLYCOSYLATION

Glycosylation is the most common and diverse form of posttranslational modification for newly synthesized proteins. It is a process in which sugars are added covalently to proteins. When sugars are added to the NH_2 group on the side chain of an asparagine (Asn) residue of the protein, the process is called *N-linked glycosylation*. When sugars are added to the OH group of serine (Ser) or threonine (Thr) side chains of the proteins, the process is called *O-linked glycosylation*. N-linked glycosylation occurs primarily in the endoplasmic reticulum (ER) (Figure 17.1).[1] Briefly, a dolichol pyrophosphate precursor ($Glc_3Man_9GlcNAc_2$) is at first transferred to an Asn side chain of Asn-*X*-Ser/Thr consensus sequence (X can be any amino acid except proline) for N-linked oligosaccharides in a nascent polypeptide in the ER. Processing is initiated by the removal of the three terminal glucose residues and at least one mannose residue in the endoplasmic reticulum, followed by transportation to Golgi apparatus, where mannose residues are further trimmed, and *N*-acetylglucosamine, galactose, and sialic acid residues are added sequentially. The newly synthesized glycoproteins then exit the Golgi and are transported to their final destination. O-linked glycosylation mainly happens in the Golgi and at a later stage during protein processing. The enzyme responsible for this process is called UDP-*N*-acetyl-D-galactosamine:polypeptide *N*-acetylgalactosaminyltransferase. It starts with the attachment of *N*-acetylgalactosamine to the hydroxyl groups of serine or threonine and is followed by the addition of other carbohydrates, such as galactose and sialic acid. Glycosylation has been demonstrated to play critical roles in the regulation of membrane targeting,[2,3] protein folding,[4,5] the maintenance of protein stability (resistance to proteolysis),[6,7] and providing recognition structures for interaction with diverse external ligands.[8,9]

FIGURE 17.1. Scheme of N-linked oligosaccharide biosynthetic pathway in the endoplasmic reticulum and the Golgi apparatus. Different forms of sugars are added by different types of transferases at various processing steps.

Many transporters are found to possess consensus sites for glycosylation in their amino acid sequences. This is also true for many of the drug transporters. For example, multiple potential sites for N-linked glycosylation were identified in all members of the organic anion transporter (OAT) family [solute carrier (SLC) 22A]. Mutagenesis studies on OAT1[2] and OAT4[4] in cultured cells revealed that although disruption of N-glycosylation at individual sites by replacing asparagine with glutamine had no effect on the transport activities of these transporters, simultaneous elimination of all the glycosylation sites caused retention of these transporters in an intracellular compartment, suggesting that addition of sugars to the transporters plays a critical role in the targeting of these transporters to the plasma membrane. Not only is acquisition of sugars, the first step in the glycosylation process, crucial for OAT activity, but modification of added sugars, the processing step, has also been proven to be important. With the use of mutant Chinese hamster ovary (CHO)-Lec cells lacking various enzymes required for glycosylation processing, it was shown that processing of glycosylation from a mannose-rich type to a complex type is associated with an increased affinity of OAT4 for its substrate.[4]

In contrast to the members of OAT family, where disruption of N-glycosylation at individual sites has no effect on transport activity, elimination of glycosylation at individual sites of organic cation transporter OCT2 showed a differential effect on its transport function.[10] Removal of the glycosylation site at position 112 of OCT2 impaired the trafficking of OCT2 to the plasma membrane, removal of glycosylation site at position 96 reduced the turnover number of the transporter, whereas removal each of the three glycosylation sites at positions 71, 96, and 112 all increased the affinity of the transporter for its substrate.

P-Glycoproteins are also heavily glycosylated plasma membrane proteins. Evaluation of the significance of N-glycosylation of human P-glycoprotein (MDR1)[11] revealed that transfection of cDNA encoding a N-glycosylation-deficient P-glycoprotein yielded drug-resistant clones with a much lower frequency than did transfection of wild-type cDNA, suggesting that N-glycosylation may prevent P-glycoprotein from ending up or getting stuck in the wrong subcellular compartments, and may improve the efficiency of P-glycoprotein routing or sorting.

The functional significance of N-glycosylation was also investigated with a naturally occurring glycosylation-defective mutant of organic anion–transporting polypeptide OATP1A2 (also known as human OATP-A or OATP1).[12] Genotypic analyses of subjects from various ethnic populations identified six nonsynonymous mutants within the coding region. One of the variants, A404T, has a mutation at a glycosylation site (N135I). In vitro assessment revealed that the A404T variant had a shift in the apparent molecular size, indicating an alteration in glycosylation status. This variant also had a markedly reduced capacity for mediating the cellular uptake of all the OATP1A2 substrate tested. Cell surface biotinylation and immunofluorescence confocal microscopy suggested that altered plasma membrane expression of the transporter might contribute to reduced transport activity associated with the A404T variant.[12]

17.3. UBIQUITINATION

Ubiquitination is a three-step process. In the first step, ubiquitin, an 8-kDa polypeptide, is activated by a ubiquitin-activating enzyme. The activated ubiquitin is subsequently transferred to an ubiquitin carrier protein. Finally, ubiquitin-protein ligase catalyzes the covalent binding of ubiquitin to the target protein. Ubiquitination of cellular proteins usually serves to tag them for rapid degradation and can therefore modulate their stability and activity (Figure 17.2). Cells degrade proteins through two major systems, the proteasome and the lysosome. The proteasome is involved in the degradation of most cytosolic and nuclear proteins as well as some membrane proteins[13–15] and removes misfolded or misaggregated proteins in the endoplasmic

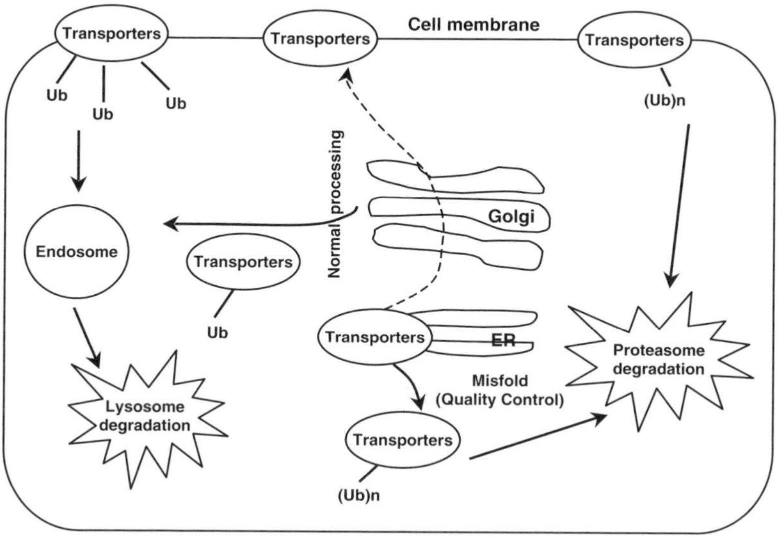

FIGURE 17.2. Ubiquitin-dependent degradation pathways of transporters. Polyubiquitinated transporters from the membrane will be degraded within proteasome. Misfolded proteins can also be ubiquitinated and removed by proteasome degradation. The mono-ubiquinated transporter proteins, on the other hand, would be degraded through the lysome pathway. Ub, ubiquitin; (Ub)n, ubiquitin polymers.

reticulum.[16] The lysosome degrades membrane proteins and extracellular materials that enter the cell via endocytosis.[15] Ubiquitin-mediated degradation usually occurs in proteasome. It was found[17] that in drug-resistant cancer cells, P-glycoprotein was ubiquitinated constitutively. Transfection of multidrug-resistant cells with wild-type ubiquitin increased the ubiquitination of P-glycoprotein and increased its degradation. Proteasome inhibitor MG-132 induced accumulation of ubiquitinated P-glycoprotein, suggesting involvement of the proteasome in turnover of the transporter. Enhanced ubiquitination of P-glycoprotein resulted in reduced function of the transporter, as demonstrated by increased intracellular drug accumulation and increased cellular sensitivity to drugs transported by P-glycoprotein.

In the cholangiocytes of liver, apical sodium-dependent bile acid transporter (ASBT) was found to be a short-lived protein and was associated with ubiquitin.[18] The inflammatory cytokine interleukin-1β (IL-1β) induced down-regulation of ASBT expression. Such down-regulation was accompanied by an increase in ABST polyubiquitin conjugates and a reduced ASBT half-life. However, in phosphorylation-deficient mutants, the ASBT half-life is prolonged markedly, IL-1β-induced ASBT ubiquitination is reduced significantly, and IL-1β failed to increase ASBT degradation. These results indicated that ASBT undergoes ubiquitin-mediated degradation under basal conditions, and such degradation is increased by IL-1β due to the serine/threonine phosphorylation of the transporter.

17.4. PHOSPHORYLATION

Phosphorylation is the covalent attachment of one or several phosphate groups to the hydroxyl side chains of serine, threonine, or tyrosine on proteins. Most phosphate on proteins of animal cells is on serine residues, less on threonine, with a very small amount on tyrosine residues. The phosphorylation process occurs in the cytosol. Phosphorylation influences the conformation and charge of the protein, thereby also its activity (either up or down), cellular location, or association with other proteins. Phosphorylation is catalyzed by protein kinases, which move a phosphate group from an adenosine triphosphate (ATP) molecule to the proteins. Tyrosine kinases phosphorylate proteins on tyrosine; serine/threonine kinases phosphorylate proteins on serine or threonine. However, the phosphate groups can also be removed from the protein by a process called *dephosphorylation*. This process is catalyzed by protein phosphatases. The amount of phosphate that is associated with the protein is thus determined by the relative activities of the kinase and phosphatase. Together, protein kinases and protein phosphatases act in an exactly opposite fashion to regulate a population of target proteins by controlling their phosphorylation state. Reversible protein phosphorylation is responsible for regulation of cellular processes as diverse as mobilization of glucose from glycogen,[19,20] prevention of transplant rejection by cyclosporine,[21] and development of a cancer form such as chronic myeloid leukemia.[22,23] An ion channel, may be closed when dephosphorylated but open when phosphorylated. Thus, a protein kinase would be responsible for opening the ion channel, and a protein phosphatase would be responsible for closing that channel.

Many drug transporters can also be regulated by reversible phosphorylation. It has been shown[24] that in HEK293 cells stably expressing rat organic cation transporter rOCT1, stimulation of protein kinase C (PKC) by *sn*-1,2-dioctanoyl glycerol results in a significant increase in the transport affinity of rOCT1 for its substrates tetraethylammonium, tetrapenthylammonium, and quinine. Such increase in transport affinity was accompanied by serine phosphorylation of the transporter. It was therefore proposed that the phosphorylation of rOCT1 by PKC results in conformational changes at the substrate-binding site.

Mouse organic anion transporter mOAT1 is another example whose transport activity is regulated by reversible phosphorylation. It was found[25] that treatment of mOAT1-expressing LLC-PK1 cells with okadaic acid resulted in an increase in the phosphorylation of mOAT1. Okadaic acid is a potent inhibitor of protein phosphatase 1 and protein phosphatase 2A, two of the four major serine/threonine protein phosphatases in the cytosol of mammalian cells. Okadaic acid readily enters cells, and numerous studies have demonstrated that it enhances the phosphorylation of many cellular proteins, presumably by preventing dephosphorylation. It was shown that okadaic acid–induced phosphorylation of mOAT1 paralleled in time and concentration the decrease of mOAT1-mediated transport of p-aminohippurate (PAH), a protypical organic anion. Phosphoamino acid analysis indicated that phosphorylation occurred primarily on serine residues. These results suggest that the increase in serine phosphorylation of mOAT1 by okadaic acid is, at least in part, responsible for the okadaic acid-induced decrease in basolateral PAH transport.

In addition to affecting the ability of a transporter to bind its substrate, phosphorylation may also affect the ability of the transporter to interact with its accessory proteins. It has been shown that the phosphorylation of multidrug resistance–associated protein MRP2 has profound effects on the binding of this transporter with PDZ proteins.[26] PDZ proteins contain specific domains that through binding to their targeting molecules are known to play important roles in scaffolding, protein trafficking, and the regulation of membrane transport activity. In MRP2, Ser-1542 forms part of the consensus sequence for phosphorylation by PKC. It is shown that when Ser-1542 was replaced by alanine (MRP2 S1542A), mimicking the dephosphorylation state of MRP2, the binding of MRP2 to the PDZ protein EBP50 was much less than that of the wild-type MRP2. In contrast, when Ser-1542 was replaced by glutamic acid (MRP2 S1542E), mimicking the phosphorylation state of MRP2, the binding of MRP2 to PDZ proteins EBP50 and IKEPP was much stronger than that of the wild type.

In rat hepatocyte, the organic anion transport mediated by organic anion–transporting polypeptide Oatp1 was down-regulated by extracellular ATP.[27] ATP was speculated to exert its effect through P2Y purinergic receptor, which then leads to the activation of intracellular signaling pathways. The down-regulation of Oatp1 function was also observed when hepatocytes were incubated with phosphatase inhibitors okadaic acid and calyculin A. Exposure of hepatocytes to both extracellular ATP and okadaic acid resulted in the phosphorylation of the transporter without affecting its cell surface presentation, suggesting that loss of transport activity is not caused by transporter internalization. Since Oatp1 functions as an organic anion exchanger, it is hypothesized that addition of a negatively charged phosphate group to the inner domain of the transporter may prevent this exchange from occurring.

17.5. DISULFIDE BONDS

A *disulfide bond* (SS bond), also called a *disulfide bridge*, is a strong covalent bond formed by oxidation of two sulfhydryl groups (—SH) present in cysteine residue. In eukaryotic cells, disulfide bonds are generally formed in the lumen of endoplasmic reticulum (ER) but not in the cytosol. This is due to the oxidative environment of the ER and the reducing environment of the cytosol, which is maintained by the high ratio of oxidized glutathione to reduced glutathione. Thus, disulfide bonds are found only in secretory proteins, lysosomal proteins, and the exoplasmic domains of membrane proteins such as transporters. The formation of disulfide bonds can be reversed by reducing conditions. These conditions may include the presence of agents with free sulfhydryl groups such as dithiothreitol (DTT), β-mercaptoethanol, or glutathione. A disulfide bond that links two peptide chains together is called an *intermolecular disulfide bond*, whereas a disulfide bond that links different parts of one peptide chain is called an *intramolecular disulfide bond*. Disulfide bonds are very important to the folding, subunit assembly, and functioning of proteins. The greater the number of disulfide bonds, the less susceptible the protein is to denaturation by forces such as detergents and heat.

Breast cancer resistance protein (BCRP/ABCG2) is believed to depend on both inter- and intramolecular disulfide bonds for its structural and functional integrity.[28] Unlike most of the other ABC transporters, which usually have two nucleotide-binding domains and two transmembrane domains, ABCG2 consists of only one nucleotide-binding domain followed by one transmembrane domain. Thus, ABCG2 has been thought to be a half-transporter that may function as a homodimer. Three extracellular cysteines (Cys-603, Cys-608, and Cys-592) were identified in this transporter. It was found that the transporter migrates as a dimer in SDS-PAGE under nonreducing conditions. Mutation of Cys-603 to Ala (C603A) caused the transporter to migrate as a single monomeric band. Therefore, Cys-603 forms an intermolecular disulfide bond. However, this mutation had no effect on efficient membrane targeting and the function of the transporter. In contrast to C603A, both C592A and C608A displayed impaired membrane targeting and function. Moreover, when only Cys-592 or Cys-608 were present (C592A/C603A and C603A/C608A), the transporter displayed impaired plasma membrane expression and function. These data suggest that Cys-592 and Cys-608 form an intramolecular disulfide bridge in ABCG2 that is critical for its function.

In contrast to the role of disulfide bond in BCRP, where its presence maintains the structural and functional integrity of the transporter, the disulfide bond in human P-glycoprotein (MDR1) has an inhibitory effect.[29] It was shown that the ATPase activity of human MDR1was stimulated by treatment with reducing reagent DTT, suggesting the presence of inhibitory disulfide bonds. The DTT treatment also resulted in a shift of molecular sizes on SDS gels from an oligomeric complex to a monomer. Using proteins containing different combination of naturally occurring cysteine residues, it was demonstrated that Cys-431 and Cys-1074, located in the Walker A sequences of nucleotide-binding sites 1 and 2 (NBS1 and NBS2), are involved in the formation of both intra- and intermolecular disulfide bonds. The intermolecular disulfide bonds were formed between Cys-431 and Cys-431 or between Cys-1074 and Cys-1074, which is responsible for dimeric complex of the protein, and the intramolecular disulfide bond was formed between Cys-431 and Cys-1074. The ATPase activity of proteins containing both Cys-431 and Cys-1074 (i.e., CL-4C and Cys-431/Cys-1074) was activated by DTT to about the same extent as the wild type, whereas proteins in which only one of these two Cys residues was present (single Cys-431, single Cys-1074) were activated to little or no extent. Thus, the major inhibitory disulfide is the intramolecular one that forms between Cys-431 and Cys-1074.

17.6. OLIGOMERIZATION

Single polypeptides can associate with each other through an intermolecular disulfide bond, as discussed above. A more common and widely occurring association of single polypeptides with one another to form larger protein complexes is through noncovalent bonding, such as hydrophobic interaction. Individual polypeptides in such complexes are referred to as *subunits*. The geometrically specific arrangements and stoichiometry of the composition of the complexes is crucial for the activity of these proteins. Oligomerization can be *homomeric* (self-association) or *heteromeric* (association with a different polypeptide). The heteromeric composition of most protein complexes gives the cells an additional level of variability and complexity that it can use for its activity. Often, heteromeric compositions of protein complexes are tissue specific or developmental specific, and multiple genes can control the activity of a single heteromeric protein complex.

17.6.1. Homo-oligomerization

In recent years, there have been numerous studies showing that transport proteins often exist as oligomers. Oligomerization plays critical role in various aspects of transporter function. Each subunit in the oligomer may form a pore itself and allows the translocation of its substrate, a mechanism mimicking water channel CHIP28.[30] On the other hand, several subunits in the oligomer may be required to form a single pore, as in K^+ channels.[31] In addition to the functional role mentioned above, oligomerization is also believed to play a role in the membrane trafficking and stability of transporters. After synthesis in the endoplasmic reticulum (ER), proteins undergo a strict process of quality control. Newly synthesized transporters may contain retention signals and are thereby retained in the ER. Oligomerization may shield or hide such signals and therefore is essential for the egress of transporters from the ER for subsequent targeting to the plasma membrane.[32–34]

One example of the homo-oligomerization of drug transporters is the human organic anion transporter OAT1.[35] Chemical cross-linking of intact membrane proteins from LLC-PK1 cells stably expressing hOAT1 as well as from rat kidney converted quantitatively OAT1 monomer to a putative trimer and a higher order of oligomer indicates that OAT1 is present in the membrane as multimeric complexes. The oligomers are not disulfide bonded, because removing the reducing reagent β-mercaptoethanol from SDS gels did not promote oligomerization of OAT1. When coexpressed in LLC-PK1 cells, FLAG-tagged hOAT1 coimmunoprecipitated with myc-tagged hOAT1. The hOAT1 oligomer was also detected in gel filtration chromatography of total membranes from hOAT1-expressing LLC-PK1 cells. Cell surface biotinylation with membrane-impermeable reagents and metabolic labeling with [^{35}S]methionine followed by immunoprecipitation showed that the oligomeric hOAT1 did not contain any other proteins. Therefore, hOAT1 exists in the plasma membrane of LLC-PK1 cells as a homo-oligomer, possibly trimer, and higher order of oligomer. However, the functional consequence of such oligomerization remains to be elucidated.

17.6.2. Hetero-oligomerization

Drug transporters are often seen to form hetero-oligomers with their associating proteins. The key organs for drug disposition, such as kidney, brain, intestine, liver, and placenta, are made of polarized epithelial cells. The capacity of any polarized epithelial cell type to mediate a specific transport process is dependent on its capacity to deliver the appropriate transport proteins to its apical and basolateral surfaces. The same transport proteins may be called upon to serve as apical or basolateral proteins in order to fulfill the transport missions assigned to each of the many different epithelial cell types. The actual destination to which a given transporter is directed is chosen by the epithelial cell and is determined through protein–protein interactions between the transporter and the components of the sorting machinery expressed by the epithelial cells. In this manner, each type of transporting epithelial cell can differentially distribute transport proteins so as to achieve the localizations required by its particular physiological role. Abnormal membrane sorting and trafficking of the transporters is the key cause for many clinical syndromes.[36–38] After the sorting machinery delivers the transporter to a specific cell surface, maintenance of the correct localization at that surface requires interactions of the transporters with structural proteins located at or near the plasma membrane. It has been shown that the protein–protein interactions that orchestrate the polarized distributions of transport proteins may also regulate their functions. Certain transport proteins are not constitutive components of a particular cell surface domain. Instead, these proteins commute or recycle between the cell surface and an intracellular storage compartment. In response to concentration changes in intracellular second messenger(s) concentrations, transporters are either inserted into or retrieved from the cell surface. By manipulating the surface populations of selected transport proteins, epithelial cells can precisely modulate their physiologic(al) properties. Therefore, during their journey from the endoplasmic reticulum through the secretary pathway to the cell surface, transporters interact with various accessory proteins. It is these interactions that determine their localization on the specific cell surface, domain, their stability at the specific cell surface, and their shuttling between the specific cell surface and the intracellular compartments when responding to stimuli.

Many drug transporters have been shown to be associated with other proteins to fulfill their function(s). PDZ proteins, for example, are one of the most common interacting partners with transporters. PDZ proteins contain multiple PDZ domains ranging from 80 to 90 amino acids in length and bind typically to proteins containing PDZ consensus binding sites, the tripeptide motif (S/T)X (X = any amino acid and = a hydrophobic residue) at their C-termini.[39] These multidomain molecules not only target and provide scaffolds for protein–protein interactions but also modulate the function of receptors and ion channels by which they associate.[40,41] The disruption of the association between PDZ proteins and their targets contributes to the pathogenesis of a number of human diseases, probably because of the failure of PDZ proteins to appropriately target and modulate the actions of associated proteins.[42,43]

Several known members of the organic anion transporter polypeptide (OATP) family have PDZ consensus–binding sites. Studies using protein mass fingerprinting and immunoprecipitation[44] showed that PDZ protein PDZK1 is the major interacting protein of Oatp1a1 both in 293T cells cotransfected with Oatp1a1 and PDZK1 and in native rat liver membrane extracts. Using PDZK1 knockout mouse liver to further examine the functional significance of the interaction between PDZK1 and Oatp1a1, it was found that Oatp1a1 was located predominantly in intracellular structures, in contrast to its normal basolateral plasma membrane distribution, suggesting a critical role for oligomerization of Oatp1a1 with PDZK1 for its proper subcellular localization and function.

PDZK1 was also found to interact with urate-anion exchanger URAT1 and hOAT4, two members of the organic anion transporter (OAT) family. Through a yeast two-hybrid screening, in vitro binding assay, coimmunoprecipitation, and surface plasmon resonance analysis, it was revealed[45] that the wild-type URAT1, but not its mutant lacking the PDZ consensus–binding site, interacts directly with PDZK1. The association of URAT1 with PDZK1 enhanced urate transport activity in HEK293 cells stably expressing URAT1 and cotransfected with PDZK1. Deletion of the URAT1 C-terminal PDZ consensus binding site abolished this effect. Augmentation of the transport activity was accompanied by a significant increase in the maximum transport velocity V_{max} of urate transport and was associated with increased surface expression level of URAT1 protein. By analyses similar to those used in the study of URAT1 and PDZK1 interaction, it was shown[46] that OAT4 wild-type but not a mutant lacking the PDZ consensus–binding site interacted directly with both PDZK1 and NHERF1. OAT4, PDZK1, and NHERF1 proteins were colocalized at the apical membrane of renal proximal tubules. The association with PDZK1 or NHERF1 enhanced OAT4-mediated transport activity in HEK293 cells stably expressing URAT1 transfected with PDZK1 or NHERF1. Deletion of the OAT4 C-terminal PDZ consensus–binding site abolished this effect. Augmentation of the transport activity was accompanied by an increase in maximum transport velocity V_{max} of estrone sulfate transport and was associated with an increased surface expression level of OAT4 protein.

Interaction of PDZ proteins with members of organic cation transporter families OCTN1 and OCTN2 was also demonstrated.[47] A pull-down study using recombinant C-terminal proteins of OCTN identified specific interaction of OCTN1 and OCTN2 with PDZK1, IKEPP, and NHERF2. Both yeast two-hybrid and pull-down studies suggested that the last four amino acids in OCTN1 and OCTN2 are required for the interaction. The interaction of PDZK1 with the C-terminus of OCTN2 was also confirmed in a pull-down study using kidney brush border membrane vesicles. Immunohistochemical analysis revealed that both PDZK1 and OCTN2 are colocalized in brush border membranes of the kidney. Double transfection of OCTN2 and PDZK1 stimulated the uptake by OCTN2 of its endogenous substrate carnitine. Such an increase was not observed for OCTN2 with deletion of the last four amino acids. In contrast to increased surface expression of OAT4 and URAT1 when associating with PDZK1, the surface expression of OCTN2 was not affected by the association with PDZK1.

In addition to the interaction with PDZ proteins as mentioned above, drug transporter activity can also be modulated through interacting with another class of proteins: caveolins. Caveolins are a major structural component of caveolae, the small flask-shaped and detergent-insoluble structures in the plasma membrane.[48,49] Various signaling molecules are found within caveolae, and their functional interaction with caveolins plays an important role in transmembrane signaling. Several lines of evidence indicate that caveolin may act as a scaffolding protein by direct interaction with and modulation of the activity of multiple signaling molecules. The compartmentation of various signaling molecules in caveolae and their direct and functional interaction with caveolin provides a paradigm by which these membrane microdomains are involved in regulating signal transduction pathways.[50]

The interaction of caveolins with members of organic anion transporter family OAT1 and OAT3 have been demonstrated.[51,52] By Western blot analysis using isolated caveolae-enriched membrane fractions or immunoprecipitates by respective antibodies from the rat kidney, it was shown that rOAT1 and caveolin-2 colocalized in the same fractions and formed complexes with each other. Similarly, rOAT3 and caveolin-1 colocalized in the same fractions and associated with each other. These results were confirmed by performing confocal microscopy with immunocytochemistry using primary cultured renal proximal tubular cells. When the synthesized cRNA of rOAT1 along with the antisense oligodeoxynucleotides of *Xenopus* caveolin-2 were coinjected into *Xenopus* oocytes, the uptakes of *p*-aminohippurate and methotrexate were decreased significantly. Similarly, when rOAT3's synthesized cRNA along with the antisense oligodeoxynucleotide of caveolin-1 were coinjected into *Xenopus* oocytes, the estrone sulfate uptake was reduced significantly. Therefore, the function of these transporters can be modulated by caveolins.

Caveolin has also been shown to modulate the function of P-glycoprotein (Pgp).[53] Pgp was identified in caveolar microdomains isolated from an in vitro model of the blood–brain barrier, formed by coculture of bovine brain capillary endothelial cells with astrocytes, and was found to interact with caveolin-1 and caveolin-2. Caveolae-disrupting agents filipin III and nystatin decrease Pgp transport activity. In addition, mutations in the caveolin-binding motif present in Pgp reduced the interaction of Pgp with caveolin-1 and increased the transport activity of Pgp.

17.7. CONCLUSIONS

Cells exploit many types of posttranslational modulation as tools with which to effect acute physiological responses. Although the development of heterologous expression systems has facilitated the identification of different posttranslational modifications on drug transporters, much remains to be learned about the effects of these modifications on transporter function in native cells. The mechanisms that regulate the activity of these transporters need to be analyzed to place these proteins into the context of cellular and organismal physiology and homeostasis. Modifications of different transporters, as well as the pathways involved in regulating these modifications, could vary depending on cell type and on the specific membrane domain (apical vs. basolateral) to which a transporter is localized. The interaction of transporters with different types of accessory proteins could also affect the posttranslational modifications of the transporters, potentially allowing further heterogeneity in the regulation of transporters.

REFERENCES

1. Kornfeld R, Kornfeld S (1985) Assembly of asparagine-linked oligosaccharides. Annu Rev Biochem 54:631–664.
2. Tanaka K, Xu W, Zhou F, You G (2004) Role of glycosylation in the organic anion transporter 1 (OAT1). J Biol Chem 279:14961–14966.

3. Lee TK, Koh AS, Cui Z, Pierce RH, Ballatori N (2003) N-Glycosylation controls functional activity of Oatp1, an organic anion transporter. Am J Physiol Gastrointest Liver Physiol 285:G371–G381.
4. Zhou FF, Xu W, Hong M, Pan Z, Sinko PJ, Ma J, You G (2005) The role of N-linked glycosylation in protein folding, membrane, targeting, and substrate binding of human organic anion transporter 4 hOAT4. Mol Pharmacol 67:868–876.
5. Kameh H, Landole-Marticorena C, Charuk JH, Schachter H, Reithmeier RA (1998) Structural and functional consequences of an N-glycosylation mutation (HEMPAS) affecting human erythrocyte membrane glycoproteins. Biochem Cell Biol 76:823–835.
6. Buck TM, Eledge J, Skach WR (2004) Evidence for stabilization of aquaporin-2 folding mutants by N-linked glycosylation in endoplasmic reticulum. Am J Physiol Cell Physiol 287:C1292–C1299.
7. Khanna R, Myers MP, Laine M, Papazian DM (2001) Glycosylation increases potassium channel stability and surface expression in mammalian cells. J Biol Chem 276:34028–34034.
8. Ott RJ, Hui AC, Giacominil KM (1992) Inhibition of N-linked glycosylation affects organic cation transport across the brush border membrane of opossum kidney (OK) cells. J Biol Chem 267:133–139.
9. Bernardo AA, Kear FT, Arruda JA (1997) The renal cortical Na^+/HCO_3-cotransporter VI: the effect of chemical modification in cotransporter activity. J Membr Biol 158:49–57.
10. Pelis RM, Suhre WM, Wright SH (2006) Functional influence of N-glycosylation in OCT2-mediated tetraethylammonium transport. Am J Physiol Renal Physiol 290:F1118–F1126.
11. Schinkel AH, Kemp S, Dolle M, Rudenko G, Wagenaar E (1992) N-Glycosylation and deletion mutants of the human MDR1P-glycoprotein. J Biol Chem 268:7474–7481.
12. Lee W, Glaeser H, Smith LH, Roberts RL, Moeckel GW, Gervasini G, Leake BF, Kim RB (2005) Polymorphisms in human organic anion-transporting polypeptide 1A2 (OATP1A2). J Biol Chem 280:9610–9617.
13. Jensen TJ, Loo MA, Pind D, Williams DB, Goldberg AL, Riordan JR (1995) Multiple proteolytic systems, including the proteasome, contribute to CFTR processing. Cell 83:129–135.
14. Sepp-Lorenzino L, Ma Z, Lebwohl DE, Vinitsky A, Rosen N (1995) Herbimycin A induces the 20 S proteasome- and ubiquitin-dependent degradation of receptor tyrosine kinases. J Biol Chem 270:16580–16587.
15. Ward CL, Omura S, Kopito RR (1995) Degradation of CFTR by the ubiquitin–proteasome pathway. Cell 83:121–127.
16. Kopito RR (1997) ER quality control: the cytoplasmic connection. Cell 88:427–430.
17. Zhang Z, Wu JY, Hait WN, Yang JM (2004) Regulation of the stability of P-glycoprotein by ubiquitination. Mol Pharmacol 66:395–403.
18. Xia X, Roundtree M, Merikhi A, Lu X, Shentu S, Lesage G (2004) Degradation of the apical sodium-dependent bile acid transporter by the ubiquitin–proteasome pathway in cholangiocytes. J Biol Chem 279:44931–44937.
19. Nuttall FQ, Gilboe DP, Gannon MC, Niewoehner CB, Tan AW (1988) Regulation of glycogen synthesis in the liver. Am J Med 85:77–85.
20. Longnus SL, Segalen C, Giudicelli J, Sahan MP, Farese RV, Van Obberghen E (2005) Insulin signalling downstream of protein kinase B is potentiated by 5′AMP-activated protein kinase in rat hearts in vivo. Diabetologia 48:2591–2601.

21. Erxleben C, Liao Y, Gentile S, Chin D, Gomez-Alegria C, Mori Y, Birnbaumer L, Armstrong DL (2006) Cyclosporin and Timothy syndrome increase mode 2 gating of CaV1.2 calcium channels through aberrant phosphorylation of S6 helices. Proc Natl Acad Sci U S A 103:3932–3927.
22. Eriksson S, Arner E, Spasokoukotskaja T, Wang L, Karlsson A, Brosjo O, Gunven P, Julusson G, Liliemark J (1994) Properties and levels of deoxynucleoside kinases in normal and tumor cells: implications for chemotherapy. Adv Enzyme Regul 34:13–25.
23. Coppo P, Flamant S, De Mas V, Jarrier P, Guillier M, Bonnet ML, Lacout C, Guilhot F, Vainchenker W, Turhan AG (2006) BCR-ABL activates STAT3 via JAK and MEK pathways in human cells. Br J Haematol 134:171–179.
24. Mehrens T, Lelleck S, Cetinkaya I, Knollmann M, Hohage H, Gorboulev V, Boknik P, Koepsell H, Schlatter E (2000) The affinity of the organic cation transporter rOCT1 is increased by protein kinase C–dependent phosphorylation. J Am Soc Nephrol 11:1216–1224.
25. You G, Kuze K, Kohanski RA, Amsler K, Henderson S (2000) Regulation of mOAT-mediated organic anion transport by okadaic acid and protein kinase C in LLC-PK(1) cells. J Biol Chem 275:10278–10284.
26. Hegedus T, Sessler T, Scott R, Thelin W, Bakos E, Varadi A, Szabo K, Homolya L, Milgram SL, Sarkadi B (2003) C-Terminal phosphorylation of MRP2 modulates its interaction with PDZ proteins. Biochem Biophys Res Commun 302:454–461.
27. Glavy JS, Wu SM, Wang PJ, Orr GA, Wolkoff AW (2000) Down-regulation by extracellular ATP of rat hepatocyte organic anion transport is mediated by serine phosphorylation of oatp1. J Biol Chem 275:1479–1484.
28. Henriksen U, Fog JU, Litman T, Gether U (2005) Identification of intra- and intermolecular disulfide bridges in the multidrug resistance transporter ABCG2. J Biol Chem 280:36926–36934.
29. Urbatsch IL, Gimi K, Wilke-Mounts S, Lerner-Marmarosh N, Rousseau ME, Gros P, Senior AE (2001) Cysteines 431 and 1074 are responsible for inhibitory disulfide cross-linking between the two nucleotide-binding sites in human P-glycoprotein. J Biol Chem 276:26980–26987.
30. Preston GM, Jung JS, Guggino WB, Agre P (1993) The mercury-sensitive residue at cysteine 189 in the CHIP28 water channel. J Biol Chem 268:17–20.
31. Mackinnon R (1990) Determination of the subunit stoichiometry of a voltage-activated potassium channel. Nature 350:232–235.
32. Ellgaard L, Helenius A (2003) Quality control in the endoplasmic reticulum. Nat Rev Mol Cell Biol 4:181–191.
33. Veenhoff LM, Heuberger EHML, Po B (2002) Quaternary structure and function of transport proteins. Trends Biochem Sci 27:242–249.
34. Scholze P, Freissmuth M, Sitte HH (2002) Mutations within an intramembrane leucine heptad repeat disrupt oligomer formation of the rat GABA transporter 1. J Biol Chem 277:43682–43690.
35. Hong M, Xu W, Yoshida T, Tanaka K, Wolff DJ, Zhou F, Inouye M, You G (2005) Human organic anion transporter hOAT1 forms homooligomers. J Biol Chem 280:32285–32290.
36. Dugani CB, Klip A (2005) Glucose transporter 4: cycling, compartments and controversies. EMBO Rep 6:1137–1142.

37. Muth TR, Caplan MJ (2003) Transport protein trafficking in polarized cells. Annu Rev Cell Dev Biol 19:333–366.
38. Wilson PD, Sherwood AC, Palla K, Du J, Watson R, Norman JT (1991) Reversed polarity of Na(+) -K(+) -ATPase: mislocation to apical plasma membranes in polycystic kidney disease epithelia. Am J Physiol 260:F420–F430.
39. Biber J, Gisler SM, Hernando N, Murer H (2005) Protein/protein interactions (PDZ) in proximal tubules. J Membr Biol 203:111–118.
40. Kato Y, Watanabe C, Tsuji A (2006) Regulation of drug transporters by PDZ adaptor proteins and nuclear receptors. Eur J Pharm Sci 27:487–500.
41. Perego C, Vanoni C, Villa A, Longhi R, Kaech SM, Frohli E, Hajnal A, Kim SK, Pietrini G (1999) PDZ-mediated interactions retain the epithelial GABA transporter on the basolateral surface of polarized epithelial cells. EMBO J 18:2384–2393.
42. Raghuram V, Hormuth H, Foskett JK (2003) A kinase-regulated mechanism controls CFTR channel gating by disrupting bivalent PDZ domain interactions. Proc Natl Acad Sci U S A 100:9620–9625.
43. Shenolikar S, Voltz JW, Minkoff CM, Wade JB, Weinman EJ (2002) Targeted disruption of the mouse NHERF-1 gene promotes internalization of proximal tubule sodium-phosphate cotransporter type IIa and renal phosphate wasting. Proc Natl Acad Sci U S A 99:11470–11475.
44. Wang P, Wang JJ, Xiao Y, Murray JW, Novikoff PM, Angeletti RH, Orr GA, Lan D, Silver DL, Wolkoff AW (2005) Interaction with PDZK1 is required for expression of organic anion transporting protein 1A1 on the hepatocyte surface. J Biol Chem 280:30143–30149.
45. Anzai N, Miyazaki H, Noshiro R, Khamdang S, Chairoungdua A, Shin HJ, Enomoto A, Sakamoto S, Hirata T, Tomita K, Kanai Y, Endou H (2004) The multivalent PDZ domain-containing protein PDZK1 regulates transport activity of renal urate-anion exchanger URAT1 via its C terminus. J Biol Chem 279:45942–45950.
46. Miyazaki H, Anzai N, Ekaratanawong S, Sakata T, Shin HJ, Jutabha P, Hirata T, He X, Nonoguchi H, Tomita K, Kanai Y, Endou H (2005) Modulation of renal apical organic anion transporter 4 function by two PDZ domain–containing proteins. J Am Soc Nephrol 16:3498–3506.
47. Kato Y, Sai Y, Yoshida K, Watanabe C, Hirata T, Tsuji A (2005) PDZK1 directly regulates the function of organic cation/carnitine transporter OCTN2. Mol Pharmacol 67:734–743.
48. Palade GE (1953) Fine structure of blood capillaries. J Appl Physiol 24:1424.
49. Yamada E (1955) The fine structure of the gall bladder epithelium of the mouse. J Biophys Biochem Cytol 1:445–458.
50. Schwencke C, Braun-Dullaeus RC, Wunderlich C, Strasser RH. (2006) Caveolae and caveolin in transmembrane signaling: implications for human disease. Cardiovasc Res 70:42–49.
51. Kwak JO, Kim HW, Song JH, Kim MJ, Park HS, Hyun DK, Kim DS, Cha SH (2005) Evidence for rat organic anion transporter 3 association with caveolin-1 in rat kidney. IUBMB Life 57:109–117.
52. Kwak JO, Kim HW, Oh KJ, Kim DS, Han KO, Cha SH (2005) Co-localization and interaction of organic anion transporter 1 with caveolin-2 in rat kidney. Exp Mol Med 37:204–212.
53. Jodoin J, Demeule M, Fenart L, Cecchelli R, Farmer S, Linton KJ, Higgins CF, Béliveau R (2003) P-Glycoprotein in blood–brain barrier endothelial cells: interaction and oligomerization with caveolins. J Neurochem 87:1010–1023.

18

EXPERIMENTAL APPROACHES TO THE STUDY OF DRUG TRANSPORTERS

YOSHIYUKI KUBO, YUKIO KATO, AND AKIRA TSUJI

Kanazawa University, Kakuma, Kanazawa, Japan

Drug Transporters: Molecular Characterization and Role in Drug Disposition, Edited by Guofeng You and Marilyn E. Morris
Copyright © 2007 John Wiley & Sons, Inc.

18.1. INTRODUCTION

Integration of chemical, physical, and engineering technologies has led to great advances in experimental methods in the biological sciences. For example, high-throughput sequencing technology has allowed us to establish the full sequence of the human genome, which has had enormous implications for understanding physiological systems and identifying the novel target proteins for drugs. Further, three-dimensional structure analysis of proteins will permit us to design novel drugs. Equally, in research on drug transporters, the use of modern integrated technology should enable the development of superior drugs and drug delivery systems, as well as eliminating side effects. In this chapter we describe the methods available to study drug transporters.

To understand the mechanisms of drug transport and to apply this knowledge to drug development and medical treatment, it is important to identify novel transporters that participate in pharmacological and physiological events and to clarify the functions and roles of known transporters. For these purposes, there are various in vivo and in vitro approaches (Figure 18.1).

18.2. IN VIVO EXPERIMENTS

In vivo study with model animals is an effective approach to elucidating the role of transporters in pharmacological and physiological events. Recently, many types of knockout or knockdown mice have become available. In addition, the RNA interference technology provides a convenient method for knockdown research in cells or animals. Here, we focus on comparative studies of normal and knockout or knockdown mice.

18.2.1. Preparation of Knockout Mice

To generate knockout mice, embryonic stem (ES) cells derived from mice are required.[1–3] In ES cells, the wild locus is mutated by homologous recombination (i.e., replacement of the wild locus with a targeting vector). Targeting vectors consist of a drug-resistance genes such as the neomycin-resistance gene, with homologous alignment of the target gene and a negative selection marker gene such as subunit

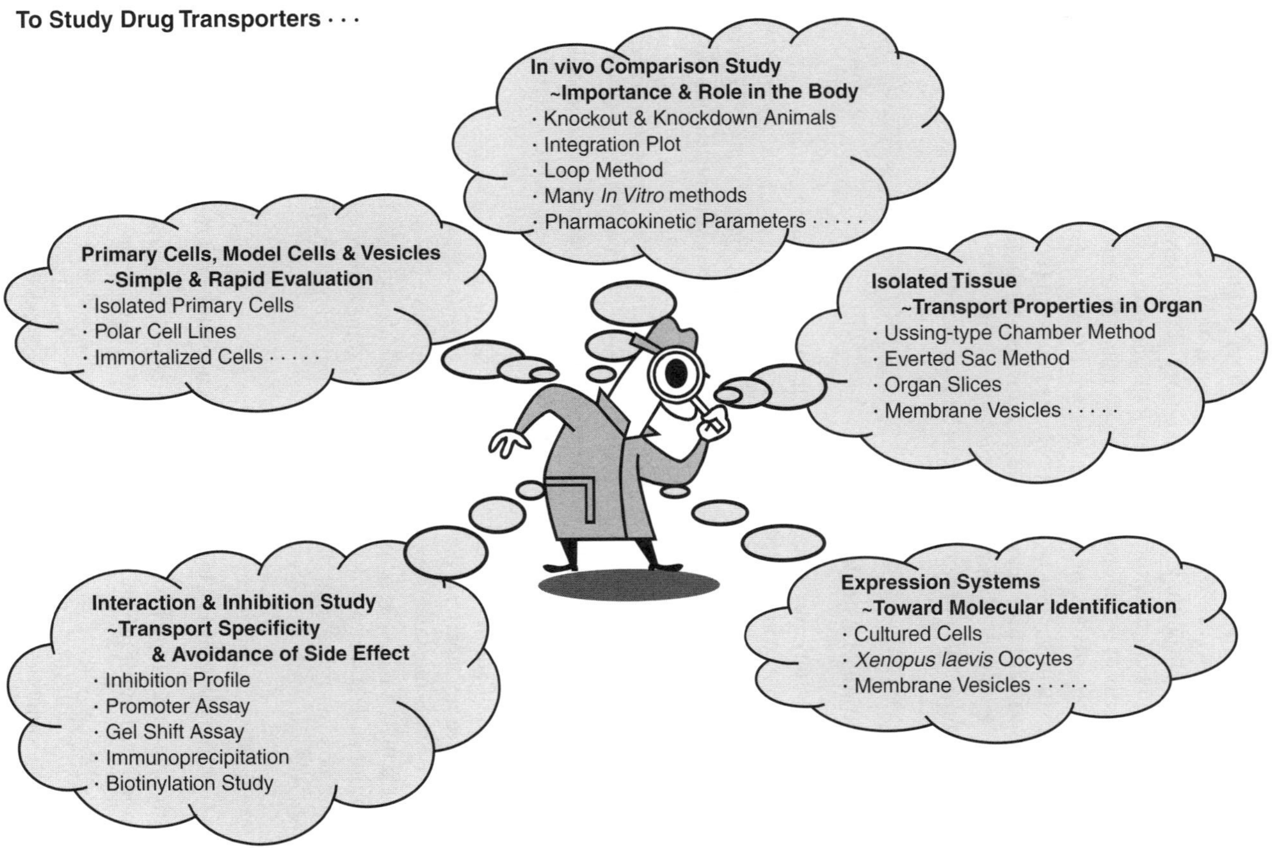

FIGURE 18.1. Methods for the study of drug transporters: summary of in vivo and in vitro approaches.

A gene of diphtheria toxin.[4–6] Appropriate design and construction of the targeting vector is the key to successful cloning of positive homologous recombinants. Cloning of homologous recombinants is also a critical step in the preparation of knockout mice, since it is technically difficult. In the authors' experiences, only 0 to 4 clones of positive homologous recombinant may be found on screening about 1000 clones.[7] After cloning, the cells are introduced into embryos by microinjection with suitable micromanipulation equipment, followed by transplanting of the ES cell-injected embryos into the uterus of pseudopregnant female mice. The chimeric progeny are born about 3 weeks later. Mating of chimeric and wild-type mice provides heterozygous (+/−) mice, and mating of these mice provides knockout (homozygous, −/−) mice. The genomic Southern blot analysis and polymerase chain reaction (PCR) are used for genotyping. Established knockout mice, such as *mdr1*$^{-/-}$, can be purchased from breeding companies, and chimeric mice can be prepared commercially by specialized companies.

18.2.2. Preparation of Knockdown Mice

Knockdown technology can be applied to all animals. Knockdown studies with RNA interference technology are an effective approach for in vitro research, and the technology has also been applied in vivo.[8–10] The hydrodynamics method allows effective suppression of gene expression in liver, kidney, lung, spleen, and heart. In addition, it is inexpensive, rapid, and does not require special equipment or techniques.[11–14]

In the hydrodynamics method, the design of siRNA is crucial. Prior to intravenous injection, cellular knockdown assay with the designed siRNA should be conducted to examine knockdown efficiency and cellular viability. The injection volume of RNAi solution, which is prepared by dissolving RNAi in phosphate-buffered saline (PBS), is determined by calculating the injection volume based on 0.08 mL of RNAi solution per 1 g of weight of animals, when the weight of the animals is between 11 and 33 g. The animals are anesthetized on a warm plate, and the RNAi solution is injected intravenously into the tail using a 26G needle. In the case of injection into 20-g animals, 1.6 mL of RNAi solution is injected within 5 seconds. In general, 48 hours after injection, a decrease in mRNA is observed, and the suppressing effect on protein expression persists for several days, although monitoring of protein expression may be required.[15]

18.2.3. Comparative Study of Normal and Knockout or Knockdown Mice

To identify pharmacological or physiological roles of drug transporters, and to evaluate the contributions of transporters to drug behavior in humans, comparative studies of normal and knockout or knockdown animals can provide persuasive evidence.[16,17] By comparing drug accumulation, concentration, distribution, bioavailablity, and other pharmacokinetic parameters in wild and gene-modified animals, the roles of transporters can often be deduced. For this purpose it is possible to use a range of methods, including in vivo methods, calculation of pharmacokinetic parameters, and transport assays with isolated tissues, primary cells, or membrane vesicles.

By means of the integration plot method, early-phase uptake of drugs by several organs can be accessed, throwing light on the roles of carrier-mediated transport in drug distribution and accumulation.[18,19] In this method, mice are anesthetized and a radiolabeled test compound is injected with or without 0.5 M unlabeled test compound in saline. In addition, [^{14}C]mannitol is often coinjected as a paracellular route marker. At appropriate times, blood is withdrawn from the right jugular vein and plasma is separated by centrifugation. After 1, 2, 3 or 5 minutes, mice are sacrificed, and the target organs are excised immediately. The organ are rinsed with ice-cold saline, blotted with dry filter paper, weighed, and solubilized. The solubilized organs are mixed with hydrogen peroxide and the solution is neutralized with 5 M HCl. The solubilized organs and plasma are mixed with scintillation liquid, and the associated radioactivity is measured with a liquid scintillation counter, followed by calculations of the tissue uptake clearance.

By using the integration plot method, we studied the distribution of an inhibitor of xanthine oxidase, Y-700, in the liver.[20] Following the intravenous administration of [^{14}C]Y-700 to rats, the liver and kidneys were removed and the amounts of [^{14}C]Y-700 taken up in these organs were measured. The values of tissue uptake clearance were 1.04 and 0.316 mL/minute per kilogram in the liver and kidney, respectively, suggesting efficient uptake of Y-700 by the liver in vivo.[21]

In comparative studies, alterations of pharmacokinetic parameters can be useful in evaluating drug behavior and the pharmacological roles of transporters. Here we focus on the measurement of drug disposition and tissue distribution. Animals are starved overnight, anesthetized, and bolus-injected with radiolabeled test compound via the jugular vein. [^{3}H]Inulin is injected simultaneously with the radiolabeled test compounds to evaluate the glomerular filtration rate (GFR). Serial blood samples are collected from the intraorbital venous plexus using heparinized capillary tubes at designated time intervals in individual mice during the experiment. Urine samples are collected by washing the bladder with saline (0.5 mL) at designated times through a catheter. Pharmacokinetic parameters, the area under the plasma concentration–time curve (AUC), the elimination rate constant (k), the steady-state distribution volume (V_{dss}), and the total body clearance ($\mathrm{CL_{tot}}$) can be estimated by means of model-independent moment analysis or compartment analysis using Win–Nonlin (SCI) or other computer-fitting techniques. GFR is estimated by inulin or creatinine clearance. The renal clearance (CL_r) and renal secretory clearance (CL_s) are calculated by using the equations $\mathrm{CL}_r = \mathrm{X/AUC}$ and $\mathrm{CL}_s = \mathrm{CL}_r - \mathrm{GFR}$, where X is the urinary excretion amount of the test compound.

To evaluate the tissue distribution of drugs, the tissue-to-plasma concentration ratio (K_p) is often adopted. Mice are decapitated at the appropriate time after a single intravenous injection of radiolabeled test compound. Tissues are quickly excised, rinsed well with ice-cold saline, blotted to dryness, and weighed. Plasma and urine samples are mixed with scintillation fluid. Tissue samples are dissolved, mixed with scintillation fluid, and neutralized with 1 N HCl, then the associated radioactivity is measured. If nonradiolabeled test compounds are used, quantification can be performed with high-performance liquid chromatography (HPLC) or other analytical techniques.

In our study of systemic carnitine defficiency (SCD) by using *jvs* mice lacking the organic cation transporter Octn2,[22] we compared the K_p values in *jvs* and wild-type mice at 4 hours after tetraethylammonium administration. In kidney, *jvs* mice showed significantly a higher K_p value of TEA (25.0 $\pm$ 4.6) than did wild-type mice (9.89 $\pm$ 1.16), and the K_p values in brain, lung, liver, and spleen of *jvs* mice were significantly lower. We proposed that the increase in kidney may be explained in terms of decreased renal apical secretory transport activity, and that the decrease of K_p values in other tissues may be explained in terms of decreased tissue-uptake activities.[23]

18.3. ISOLATED TISSUE METHODS

To predict the involvement of transporters in absorption and distribution, studies with isolated tissues are often performed. To investigate intestinal absorption mechanisms, the loop method, Ussing-type chamber method, or everted sac method can be adopted. The loop method is an in situ approach, whereas the Ussing-type chamber and everted sac methods are performed with isolated intestinal tissues. Sliced organs are often used, except in the case of intestine.

18.3.1. Loop Method

The loop method is an in situ method for studying the intestinal absorption mechanisms of drugs. Typically, mice are anesthetized, the intestine is exposed by midline abdominal incision, and the bile duct is ligated. A 5-cm closed loop of intestine is prepared by ligation at both ends after clearing the gut by passing warmed isotonic 2-morpholinoethanesulfonic (MES) acid buffer (pH 6.4, 290 mOsm/kg) slowly through it until the effluent is clear, and expelling the remaining solution with air pumped through a syringe. Each loop is filled with 0.2 mL of isotonic MES buffer. In the case of intestinal absorption study, the mouse is kept on a warm plate at 37°C for a 10-minute recovery period. The test compound is introduced into the loop. After 15 minutes the solution in the loop is collected and the loop is rinsed with isotonic MES buffer to give a total effluent volume of 5 mL. To estimate the remaining amount of the test compound, the concentration in the effluent is measured by HPLC or by other analytical techniques.[24]

Using the loop method, we compared the absorption of grepafloxacin in mouse small intestine of *mdr1a/1b* double-knockout and wild-type mice. The knockout mice showed significantly lower fractional absorption in the intestinal loop, suggesting that mdr1 restricts the uptake of grepafloxacin from mouse small intestine.[24,25]

18.3.2. Ussing-Type Chamber Method

A Ussing-type chamber consists of two buffer tanks with isolated tissues (intestine, skin, etc.) mounted between them.[26] Drug movements from the apical to the basolateral side and from the basolateral to the apical side can be measured to study directional drug transport in tissues. We have used the Ussing-type chamber method to investigate the intestinal absorption of a β-lactam antibiotic, cefadroxil, which is a representative substrate of Pept1. Rat intestinal tissue sheets consisting of the mucosa and most of the muscularis mucosa were mounted vertically in an Ussing-type chamber that provided an exposed area of 0.5 cm^2. The volume of bathing solution on each side was 5 mL, and the temperature was maintained at 37°C. The test solution (pH 7.4 or 6.0) was gassed with 95% O_2–5% CO_2 before and during the transport experiments. At intervals, the amount of cefadroxil transported was determined by HCPL with an ultraviolet detector.[25]

The drug transport can be estimated in terms of permeation (L/cm_2) obtained by dividing the amount transported (mol/cm_2 or g/cm_2) by the initial concentration of test compound on the donor side (mol/L or g/L). The permeability coefficient (cm/s) can be obtained from the slope of the linear portion of the permeation–time curve. In addition, the Ussing-type chamber method also permits us to evaluate accumulation of drugs and the regional dependence of transport in tissues.[24,25]

18.3.3. Everted Sac Method

The everted sac method is also used for studying drug absorption and distribution in the intestine.[27] The animals are anesthetized and an appropriate length of intestine is excised. The isolated portion is everted and divided into small segments of 4-cm length, which are fixed over polyethylene tubes. Each polyethylene tube carries two lines marked at a distance of 3 cm, and the everted tissue is tied with cotton threads at these marks. A 20-cm-length thread is ligated at the edge of the intestine so that the intestine can be removed from the medium immediately at the desired time. The everted segments are preincubated in buffer solution under 95% O_2–5% CO_2. After preincubation, the intestine is placed immediately in assay medium containing the test compound and gassed with 95% O_2–5% CO_2. After the desired period of incubation, the tissues are washed in ice-cold isotonic buffer and blotted on filter paper, followed by measurement with the appropriate analytical equipment.[28] We have used this approach to examine the time course, concentration dependence, and mode of inhibition of uptake of several β-lactam antibiotics by small intestinal tissues.[28]

18.3.4. Sliced Organs

Transport assays with sliced organs or tissues can throw light on the contribution of transporters to the uptake or distribution of drugs. Recently, we showed the involvement of a transporter in the renal excretion of a novel diuretic inhibitor, M17055, with this method.[29]

Rat kidneys were removed under anesthesia and soaked in ice-cold transport buffer (adjusted to pH 7.4). Two rat renal cortical slices were collected from a kidney using a Staddie–Riggs microtome (the weight of each slice was 20 to 40 mg). The slices were preincubated for 5 minutes in the transport buffer at the experimental temperature with bubbling of 95% O_2 and 5% CO_2. The transport buffer was then exchanged for fresh buffer containing [^{14}C]M17055. At the designated times, the slices were picked up and washed twice with ice-cold transport buffer. Finally, the slices were solubilized and neutralized for measurement of the radioactivity with a liquid scintillation counter.

The uptake of [^{14}C]M17055 by rat renal cortical slices showed a time-dependent increase. In addition, we have examined the effects of various compounds on the uptake of [^{14}C]M17055 by rat renal cortical slices, and the results suggested that the excretion of M17055 is at least partially mediated by the organic anion transporter OAT1. An Eadie–Hofstee plot of data obtained at various substrate concentrations provides the K_m value.[30]

18.4. PRIMARY CELL CULTURES AND ESTABLISHED MODEL CELL LINES

Primary cell cultures and established model cells may be the simplest systems that reflect the physiological conditions and can provide clues to the involvement of transporters in drug absorption, distribution, and excretion. In particular, primary cultured cells can permit us to evaluate carrier-mediated transport and to identify directly the molecules responsible. The introduction of established model cell lines has made it possible to conduct experiments routinely.

18.4.1. Isolated Hepatocytes

Isolated hepatocytes and enterocytes are often used as primary cultured cells. Here we focus on rat hepatocytes. Rat hepatocytes are isolated by means of the collagenase perfusion procedure.[31] After isolation they are suspended at 4°C in albumin-free Krebs–Henseleit buffer. In the preparation of isolated primary cells it is important to check cell viability using the trypan blue (0.4% w/v) exclusion test; we use only hepatocytes with more than 95% viability.[32] After preincubation of isolated hepatocytes for 5 minutes at 37°C, the cells and the transport buffer containing a radiolabeled drug are incubated separately at 37°C for 5 minutes and then mixed to initiate uptake. At designated times, 200-μL aliquots are taken and analyzed using the silicon layer method described later. Although primary cultured cells may require maturation, suitable combinations of cells and analytic methods permit us to access the effects on drug uptake, of ATP depressants, low temperature, and various compounds, as well as to determine the concentration dependences of uptake.[33,34] In addition, human hepatocytes are available from some organizations in the United States and China, and are useful to examine drug behavior in humans.[23]

18.4.2. Fibroblasts

Primary cultured cells can be an excellent tool in comparative studies with knockout or knockdown animals.[35] We have studied the loss of transport capability for cationic compounds in the *jvs* mouse using primary cultured fibroblasts. Transport measurements were performed by the dish method. The uptake of [^{14}C]TEA by fibroblasts from *jvs* mice was significantly lower than that by cells from normal mice, as was also the case for [^{3}H]carnitine. Furthermore, the efflux transport of [^{14}C]TEA was measured after preincubation of fibroblasts with [^{14}C]TEA, to demonstrate the bidirectional transport of TEA by Octn2.[23]

18.4.3. Established Cell Lines

In organs and tissues relevant to drug behavior, drugs are transported by polar cells such as intestinal and renal epithelial cells bearing directional drug transport systems. Therefore, established model cell lines with polarity provide useful model systems to study directional transport of drugs. Representative lines include Caco-2 and LLC-PK1 cells. Furthermore, conditionally immortalized cell lines are also available as new in vitro models for the study of transporters.[36]

Caco-2 Cells Caco-2 cells derived from human colon carcinoma are often used as a model for intestinal epithelial cells to predict drug behavior in the intestine.[37] For cultivation of Caco-2 cells, we adopt Dulbecco's modified Eagle's medium, containing 10% fetal calf serum, 1% nonessential amino acids, 2 mM L-glutamine, penicillin, and streptomycin in a humidified incubator at 37°C under an atmosphere of 5% CO_2 in air. For transport experiments, Caco-2 cells are grown on dishes and on Transwell microporous polycarbonate membranes. Caco-2 cells are cultured for 21 to 23 days before assays.[25,38,39]

LLC-PK1 Cells LLC-PK1 cells lines derived from pig kidney are often used for studies of MDR1 (ABCB1), since human MDR1 is expressed on the apical membrane of LLC-GA5-COL150 cells and confers upon them resistance to colchicines and other P-glycoprotein substrates. In particular, comparative studies between LLC-PK1 and LLC-GA5-COL150 cells can be performed to examine whether a test compound is a substrate of MDR1. Ueda et al. used this approach to show that steroidal hormones could be transported directionally by MDR1.[40]

LLC-PK1 cells are grown in M199 medium containing 10% fetal calf serum, 14.3 mM $NaHCO_3$ and 3% L-glutamine. For cultivation of LLC-GA5-COL150 cells, 150 ng/mL colchicine is added to the medium. For transport experiments, the cells are grown on Transwell microporous polycarbonate membranes and are cultured for 3 or 4 days before use.[25,41]

18.4.4. Transfected Cell Lines and *Xenopus laevis* Oocytes

For evaluation of the transport activity of drug transporters, gene expression systems, such as cultured cells and *Xenopus laevis* oocytes, provide simple experimental tools to evaluate substrate specificity and the driving force of transporters. Here we describe transport assay in transporter-expressing HEK293 cells and *X. laevis* oocytes.

Expression of Drug Transporters Expressed in HEK293 Cells For evaluation of the transport activity of drug transporters, several types of cultured mammalian cell expression systems can be used. We have adopted HEK293 cells derived from human embryonic kidney. HEK293 cells are routinely grown in Dulbecco's modified Eagle's medium containing 10% fetal calf serum, penicillin, and streptomycin in a humidified incubator at 37°C under an atmosphere of 5% CO_2 in air. After cloning or purchase of cDNAs encoding transporters, they are inserted into mammalian expression vectors such as the pcDNA series (Invitrogen Inc.) that include a CMV promoter and neomycin resistance gene (*Neo*^{r}). The constructs are transiently transfected into HEK293 cells. In the case of transient expression, a transport study is often performed at 48 hours after the transfection.[42,43]

We used the HEK293 cells expression system to examine the transport activity of OCTN1 as an organic cation transporter. After 24 hours of cultivation of HEK 293 cells in 10-cm dishes, 10 μg/dish of the expression vector with or without human OCTN1 cDNA was transfected into HEK293 cells by means of the calcium phosphate precipitation method. At 48 hours after transfection, the cells were harvested and suspended in transport medium. This suspension and a solution of a radiolabeled test compound in the transport medium were each incubated at 37°C for 10 minutes, and then transport was initiated by mixing them. At appropriate times, 200-μL aliquots of the assay mixture were sampled and analyzed by the silicon layer method. These experiments provided data concerning the time, pH, and concentration dependence of [^{14}C]TEA transport by OCTN1, suggesting that OCTN1 is the strongest candidate for the unknown renal H^+/cation antiporter.[44] For transport studies with cultured cells expressing transporters or with primary cultured cells, several assay methods, such as the dish method, and the silicon layer method, are available.

Expression in* Xenopus laevis *Oocytes *Xenopus laevis* oocyte expression systems have been used extensively for the study of membrane transporters because they offer the advantages of low background transport, high transporter expression level, and proper posttranslational modifications. Drugs transported into oocytes can be analyzed directly by measuring radioactivity. In the early stage of research, many members of the solute carrier (SLC) superfamily were identified by expression cloning methods utilizing *Xenopus laevis* oocytes.[45–48] In the postgenome era, the expression cloning method is being used less frequently, because scientists can utilize a gene database. However, *Xenopus laevis* oocytes are useful as an assay system for transporters, and drug transporter–expressing oocytes are commercially available. They are used, for example, for screening candidate drugs by pharmaceutical companies.

Capped cRNAs for transporters are synthesized in vitro by using T7 RNA polymerase from plasmids containing transporter cDNAs. Oocytes are harvested from oocyte-positive *X. laevis*, dissected and treated with collagenase in a calcium-free OR2 solution (82.5 mM NaCl, 2.0 mM KCl, 1.0 mM $MgCl_2 \cdot 6H_2O$, 5.0 mM HEPES, pH 7.5). The oocytes are washed with modified Barth's solution about 10 times and peeled, then injected individually with 50 μL of transporter cRNA (~500 ng/μL) and incubated at 18°C for 3 days in modified Barth's solution (88 mM NaCl, 1 mM KCl, 2.4 mM $NaHCO_3$, 0.82 mM $MgSO_4$, 0.33 mM $Ca(NO_3)_2$, 0.41 mM $CaCl_2$, 10 mM HEPES, pH 7.5) containing 50 μg/mL gentamicin. Uptake assays are performed by incubating transporter cRNA- or water-injected oocytes with 500 μL of transport buffer containing a radiolabeled test compound for 10, 30, 60, and 90 minutes. After uptake assay, oocytes are washed with ice-cold transport buffer. For uptake study, oocytes are lysed individually in 5% SDS, and the amount of the radiolabeled test compound transported into each oocyte can be determined by liquid scintillation counting.[42]

We have employed *X. laevis* oocytes for functional characterization of the organic cation transporter human OCTN1, using tetraethylammonium (TEA) as a model organic cation. The uptake of [^{14}C]TEA into OCTN1 cRNA-injected oocytes was significantly higher than that into water-injected oocytes, and the uptake of [^{14}C]TEA increased linearly until 90 minutes, and was saturable. Calculation of the kinetic parameters gave apparent K_m and V_{max} values of 0.195 ± 0.033 mM and 18.5 ± 1.45 pmol/60 minutes per oocyte, respectively. In addition, we examined the effects of pH and membrane potential on the uptake of [^{14}C]TEA by human OCTN1 by altering the transport buffer pH or replacing Na^+ with K^+ in the transport buffer. The results of these experiments suggested pH dependence and membrane potential independence of human OCTN1.[42]

18.4.5. Techniques to Study Cellular Uptake or Transport

Dish Method The dish method is suitable for transport study in adherent cells. Cultured cells are cultivated in an appropriate medium in a humidified incubator at 37°C under an atmosphere of 5% CO_2 in air. In the transport measurement, the cells cultured on plastic dishes are incubated with transport medium (Hanks' balanced salt solution: 136.7 mM NaCl, 5.36 mM KCl, 0.952 mM $CaCl_2$, 0.812 mM $MgSO_4$, 0.441 mM KH_2PO_4, 0.385 mM Na_2HPO_4, 25 mM D-glucose, 10 mM HEPES, pH 7.4) containing a radiolabeled test compound to initiate the uptake study. After an appropriate time, the dishes are washed three times with ice-cold Hanks' balanced salt solution and solubilized 1 M NaOH; then the radioactivity is measured in a liquid scintillation counter after neutralization with HCl and addition of liquid scintillation fluid.[23]

Silicon Layer Method Cultured cells are cultivated in an appropriate medium in a humidified incubator at 37°C under an atmosphere of 5% CO_2 in air. The cells are harvested with a rubber policeman, washed twice, and suspended in the transport buffer. After preincubation of cells for 5 minutes at 37°C, the cell suspension is mixed with transport medium containing the radiolabeled test compound to initiate the uptake study. At appropriate times, 200-μL aliquots of the mixture are placed in centrifuge tubes containing 50 μL of 3 M KOH, covered with a layer (100 μL) of silicon oil and liquid paraffin mixture (density = 1.03). Then the samples are centrifuged, and the resulting cell pellets are solubilized in 3 M KOH and neutralized with 30 μL of 5 M HCl. The associated radioactivity is measured in a liquid scintillation counter.[49,50]

In the case of *trans*-stimulated uptake experiments, the cells are preloaded with unlabeled compounds for 30 minutes, and subsequent uptake of radiolabeled compounds by the cells is measured. In addition, in the case of *trans*-stimulated efflux experiments, the cells are preloaded with radiolabeled compounds for 30 minutes at 37°C. Then the cells are spun down, suspended in ice-cold transport medium, and washed twice at 0°C. The cells are again suspended in the transport medium containing a test compound, and efflux is initiated at 37°C. At appropriate times, 100-mL aliquots of the mixture are sampled.[49]

Transwell Method The transwell method is useful to study the directional transport of test compounds in polar cells such as Caco-2 and LLC-PK1. Confluent cells are washed with Hanks' balanced salt solution (HBSS, pH 7.4, 325 mOsm/kg). Then 0.5- and 1.5-mL aliquots of HBSS are added on the apical and basolateral sides, respectively, of the cell insert. To study apical-to-basolateral or basolateral-to-apical flux, the test compound is added to the apical or basolateral solution, respectively. At appropriate times, samples (0.5 mL of basolateral-side or 0.2 mL of apical-side solution) are collected from the acceptor compartment and replaced with an equal volume of HBSS, followed by measurement of the drug concentration with suitable analytical equipment.[25,38,39]

We have used the Transwell method to study the transport of quinolones, grepafloxacin, sparfloxacin, and levofloxacin in Caco-2 and LLC-PK1 cells. Directional transport of [^{14}C]grepafloxacin was studied by using monolayers of Caco-2 cells grown on Transwells. The apical-to-basolateral permeability coefficient was found to be $2.60 \pm 0.32 \times 10^{-5}$ cm/s, which was 74 times higher than the value of mannitol transport (0.035×10^{-5} cm/s), representing paracellular transport. The basolateral-to-apical permeability coefficient of [^{14}C]grepafloxacin was $4.80 \pm 0.08 \times 10^{-5}$ cm/s and was significantly higher than that in the apical-to-basolateral directional, suggesting that the secretory-directional transport of grepafloxacin can be ascribed to carrier-mediated transport. We also used cells of the LLC-PK1 series to examine the transepithelial transport of [^{14}C]grepafloxacin, [^{14}C]sparfloxacin, and [^{14}C]levofloxacin. LLC-GA5-COL150 cells showed greater basolateral-to-apical and smaller apical-to-basolateral permeation of these compounds, indicating that grepafloxacin, sparfloxacin, and levofloxacin are substrates for MDR1.[25,41]

18.5. MEMBRANE VESICLES

Membrane vesicles provide an excellent experimental system for examining transport functions under various conditions. Generally, membrane vesicles can be prepared from animal tissues and cultured cells; those derived from tissues are used to investigate drug transport in organs. In kidney, small intestine, and liver, epithelial cells have polarity, and the plasma membrane has different transport properties on the apical side and basolateral sides. Methods have been established for preparing membrane vesicles derived from either apical or basolateral membrane, and they are often called brush-border membrane vesicles (BBMV) and basolateral membrane vesicles (BLMVs), respectively. These are powerful tools for the detailed study of directional transport systems.

18.5.1. Membrane Vesicles from Cultured Cells

Membrane vesicles can be used to investigate in detail the properties of transporters, including investigating the driving forces for transport, such as energy coupling, effects of *trans*- or *cis*-stimulation, and effects of intravesicular conditions. Membrane vesicles have been suggested to be a useful tool for pharmaceutical drug screening. Here we describe a method for preparation of vesicles from HEK293 cells with the expression vectors harboring transporter cDNAs.[51]

HEK293 cells expressing transporters are grown routinely in Dulbecco's modified Eagle's medium containing 10% fetal calf serum, penicillin, and streptomycin in a humidified incubator at 37°C under an atmosphere of 5% CO_2 in air. The transporter-expressing cell clone is cultured on 30 plastic dishes (150 mm in diameter) containing appropriate selection drugs, and when the cells reach confluence, they are harvested by scraping with a rubber policeman and washed twice by centrifugation in PBS. The cells are suspended in 25 mL of buffer 1 (10 mM NaCl, 1.5 mM $MgCl_2$, 0.02 mM phenylmethylsulfonyl fluoride, 10 mM Tris–HCl, pH 7.4) and placed on ice for 30 minutes. The cells are incubated in a cell disruption bomb (Parr 4635) and equilibrated with nitrogen gas at 700 psi with gentle stirring for 20 minutes. Then the pressure is released, ethylenediaminetetraacetic acid is added to a final concentration of 1 mM, and the content is homogenized with a Teflon-glass homogenizer (20 strokes) at 4°C. The cell homogenate is collected and centrifuged for 5 minutes at 1000 rpm and 4°C. The supernatant is layered on buffer 2 (35% sucrose, 10 mM Tris–HCl, pH 7.4) in a centrifuge tube and centrifuged for 90 minutes at 18,000*g* and 4°C. The white-colored fluffy layer at the boundary of buffer 2 is collected, suspended in buffer 3 (150 mM mannitol, 75 mM potassium gluconate, 10 mM HEPES–Tris, pH 7.4), and centrifuged at 100,000*g* for 2 hours at 4°C. The resulting pellet is again suspended in buffer 3 and centrifuged at 100,000*g* for 3 hours at 4°C. The final pellet is suspended in buffer 3 and stored at −80°C until used for transport experiments. The content of protein in each preparation can be measured by the method of Bradford.[52,53]

We employed membrane vesicles prepared from HEK293 expressing OCTN2 to examine the Na^+ dependence of OCTN2. Overshoot uptake of L-[^{3}H]carnitine by membrane vesicles from OCTN2-expressing cells was observed in the presence of extravesicular Na^+, whereas it was not observed in the absence of extravesicular Na^+, indicating that OCTN2 mediates the Na^+-coupled transport of L-carnitine.[52]

18.5.2. Membrane Vesicles from Tissues

By using membrane vesicles from tissues, we can investigate carrier-mediated drug transport in organs. Here we describe the preparation of renal BBMVs. The kidneys are rapidly removed under anesthesia and chilled in ice-cold isolation buffer. The cortex is separated from the medulla with scissors, minced weighed, suspended in 10 volumes of homogenization buffer (10 mM mannitol, 2 mM Tris–H_2SO_4, pH 7.4), and homogenized. A sample of homogenate is collected for protein and enzyme assays. After addition of 10 mM $MgCl_2$, the homogenate is stirred on ice for 15 minutes and centrifuged at 1900*g* for 12 minutes. The resulting supernatants are centrifuged at 20,000*g* for 12 minutes. The pellet is resuspended in 40 mL of homogenization buffer with a Dounce glass pestle homogenizer, and the $MgCl_2$ precipitation step is repeated. The suspension is centrifuged at 1900*g* for 12 minutes and the supernatant is further centrifuged at 20,000*g* for 12 minutes. This pellet is resuspended in 20 mL of intravesicular buffer, homogenized with a Dounce glass pestle homogenizer, and centrifuged at 30,000*g* for 20 minutes. This step is repeated once more and the final pellet is suspended in intravesicular buffer by aspirating the suspension with a syringe through a 25G needle. Prepared BBMVs are frozen and stored at −80°C in aliquots of 0.1 mL until the transport experiments. The purity of the BBMVs is confirmed by measuring alkaline phosphatase activity.[54–57]

For transport assays, frozen BBMVs are rapidly thawed under flowing water. The uptake study is performed by means of a rapid filtration method. Data analysis is performed after the subtraction of nonspecific absorption on the filter from the apparent uptake values. Basolateral membrane vesicles can be prepared by a modification of this method.[30]

18.5.3. Rapid Filtration Techique

Transport studies using membrane vesicles are performed by a rapid filtration technique.[58] Membrane vesicles stored at −80°C are thawed rapidly at 37°C and kept on ice. After preincubation for 30 minutes at 25°C, 10 μL of membrane vesicle suspension containing 20 μg of protein is mixed with 90 μL of drug solution containing the test compound in buffer 4 (150 mM NaCl, 20 mM mannitol, 10 mM HEPES–Tris, pH 7.4) to initiate uptake studies. The uptake reaction is terminated by diluting the mixture with 1 mL of buffer 5 (20 mM mannitol, 160 mM potassium chloride, 5 mM HEPES–Tris, pH 7.4) at 4°C. The diluted sample is filtered immediately through a membrane filter (Millipore, HAWP, 0.45-μm pore size), which is washed four times with buffer 5. The radioactivity retained on the membrane filter can be quantified with a liquid scintillation counter or other appropriate system.[54–57]

18.6. ANALYSIS OF DRUG INTERACTION MECHANISMS

For the prediction and clarification of drug interactions involving transporters, it is important to know the substrate specificity and the regulation mechanism of transporters, and this can also provide pharmacological and physiological information. Here we focus on inhibition studies and transcriptional and posttranscriptional regulation studies, which can clarify the substrate specificity of transporters and give insight into induction mechanisms and functional modification, respectively.

18.6.1. Inhibition Studies

Inhibition studies can give information about substrate specificity of transporters, as well as sensitivity to various inhibitors, which have implications for the behavior of drugs in tissues and the functional properties of transporters expressed in cells or vesicles. In general, inhibition studies involve examining the effects of a nonradiolabeled test compound on the transport of a radiolabeled probe compound. Compounds structurally analogous to substrates are often used as inhibitors. In addition, to allow the classification of transporters, TEA is characteristically used as a cationic compound; uridine and thymidine as nucleosides; and probenecid, estradiol-17-glucuronide, bromosulfophthalein (BSP), taurocholate, *p*-aminohippuric acid (PAH), indomethacin, and salicylate as anionic compounds. The experimental procedures for inhibition studies are similar to those used in transport assays with animals, tissues, cells, and oocytes.

We performed an inhibition study with rat renal cortical slices to characterize the transport mechanism of a novel diuretic drug, M17055. The effects of various compounds on [^{14}C]M17055 uptake were measured at 37°C. The concentration of [^{14}C]M17055 was 1 mM, and inhibitors used were 1 mM probenecid, 500 mM *p*-aminohippuric acid (PAH), 200 mM benzylpenicillin (PCG,) and 1 mM estradiol-17-glucuronide, 1 mM estrone 3-sulfate (E3S), 1 mM taurocholic acid (TCA), 1 mM tetraethylammonium (TEA), and 1 mM cimetidine. Probenecid, a well-known inhibitor of organic anion transport systems, decreased [^{14}C]M17055 uptake significantly. E3S also showed remarkable inhibition, whereas TCA was less effective. PAH inhibited [^{14}C]M17055 uptake, whereas PCG, which selectively inhibits OAT3 while having little effect on OAT1, did not. In addition, a transport study with OAT1-expressing HEK293 cells showed higher [^{14}C]M17055 uptake activity than did cells transfected with vector alone. PAH and probenecid inhibited the uptake of [^{14}C]M17055, whereas PCG had few effects. Overall, the results indicated that OAT1 contributes at least in part to the renal distribution of M17055.[30]

Thus, comparison of results obtained from tissues or primary cells with the known properties of transporters can allow identification of the responsible or major transporters involved in transport of particular drugs. When there are several transporters with similar substrate specificities, it will be useful to perform the inhibition study with tissues and cells devised by knockdown or knockout approaches. For example, in a comparative knockdown study using transporter-expressing cells and human model cell lines, the responsible molecule can be suggested by the loss of substrate transport activity or sensitivity to inhibitors.

18.6.2. Elucidation of Regulatory Mechanisms

The results of the human genome project indicate that approximately 50 ABC transporter genes and approximately 360 SLC transporter genes may exist in humans. So the question arises: "How is the overall expression pattern of transporters controlled?" To answer this it is necessary to clarify the mechanisms regulating the function and expression of individual transporters. In this section we describe methods for studying transcriptional and posttranscriptional regulation which can provide us with information about induction and functional modifications of transporters by administration of drugs.

Transcriptional Regulation

Reporter Gene Assay In initial transcriptional regulation studies, reporter gene assays are often performed to determine the minimal promoter and the region responsible for transcriptional regulation. The 5′-region upstream of the transcription start site in the transporter gene is PCR-amplified using genomic DNAs and high-fidelity DNA polymerase. The upstream and downstream primers are designed to include internal restriction enzyme sites, since the PCR product is ligated into a firefly luciferase reporter gene vector such as pGL3-Basic (Promega Co.). After subcloning, the 5′-deleted constructs are generated by digestion of the construct harboring the PCR product with appropriate restriction enzymes and self-ligation. After cloning of the deletion constructs, the nucleotide sequences are confirmed with a DNA sequencer.

Cultured cells are maintained in the appropriate culture medium supplemented with 10% fetal calf serum and other appropriate additives. Cells are plated into 24-well plates, followed by transfection of the reporter constructs and the *Renilla reniformis* vector pRL-TK (Promega Co.) on the following day. Six hours after transfection, the culture medium is changed to medium containing 2.5% fetal calf serum and 5% horse serum, and the cells are incubated for 48 hours at 37°C. The firefly and *Renilla* lucuferase activities are determined 48 hours after the transfection using a dual luciferase assay kit and a luminometer. The firefly activity is normalized to the *Renilla* activity, and changes in the ratio of firefly to *Renilla* luciferase activities can be used to identify the minimal promoter region.[59,60]

In addition to determination of the minimal promoter region, computational sequence analysis can reveal transcription factor recognition sites, which may contribute to the transcriptional regulation. Computational sequence analysis can be performed with TFSEARC, TRANSFAC, or similar software.

Gel Mobility Shift Assay To confirm the possible contribution of transcription factor recognition sites, gel shift assays are often performed to identify nuclear proteins that interact with the sites. By using cell-derived nuclear extracts and a putative binding motif oligonucleotide within the promoter, protein–DNA interactions can be detected. Confirmation can be obtained by using transcription factor–specific antibodies, oligonucleotide competitors, or an excess of unlabeled probe oligonucleotide.[61]

In the preparation of probes, oligonucleotides of the sense and antisense strands containing putative binding sites of target nucleotide-binding protein are synthesized and labeled with γ-[^{32}P]ATP. Commercial gel shift assays are available and are easy to use. The binding mixture consists of 1 μg of nuclear extract of cell, unlabeled competitor probes, and several required components in buffer solution. After preincubation at room temperature for 10 minutes, labeled probes are added, and the binding mixture is incubated for a further 20 minutes. For supershift assays, 1 μg of antibody is added to a target protein 10 minutes before addition of the labeled probes. The volume of the binding mixture is 10 μL throughout the experiment. The DNA–protein complex is then separated on a 4% polyacrylamide gel, and the gels are dried, followed by autoradiography.[62]

After the confirmation of transcription factor recognition sites, it is necessary to investigate the effects of the transcription factor(s). For this purpose, luciferase assay with cultured cells expressing the transcription factor is convenient. In the same experimental systems, site-directed mutagenesis of the transcription factor recognition sites can also provide useful confirmation.

Posttranscriptional Regulation Posttranscriptional regulation of transport proteins may involve kinations, interaction with regulation factors, modification, and so on, which may result in changes of specificity and localization. Here, we focus on interactions with regulation factors. To investigate protein–protein interactions between the transporter and regulatory factors and to identify putative regulation factors, immunoprecipitation, pull-down assay, and the yeast two-hybrid method are often employed.[63–65]

Immunoprecipitation Immunoprecipitation is initiated by mixing lysates of transporter- and the tag sequence–fused putative regulatory factor–expressing cells, or putative regulation protein-expressing cells and membrane fraction prepared from tissues. Here, we describe the use of transporter and putative regulatory protein-expressing cells. Cultured cells stably expressing the putative regulatory protein and cells transiently expressing the transporter are washed twice with phosphate-buffered saline (PBS) and collected with a rubber policeman followed by centrifugation. The pellets obtained are solubilized in RIPA-Y buffer. The cell lysates are mixed and then incubated with the antitag sequence antibody prebound to Protein G Sepharose at 4°C for 4 hours, followed by centrifugation and washing three times with ice-cold PBS. Samples are analyzed by Western blot analysis. If the construct of the putative regulatory protein is fused to c-myc, anti-c-myc antibody is prebound to Protein G Sepharose, and Western blot analysis is performed with antitransporter antibody. A specific signal of the transporter on the blotting membrane suggests the occurrence of a protein–protein interaction.[64] In general, to clarify interactions among proteins, a single approach is not sufficient.

Biotinylation Study of Surface Proteins To investigate transporter expression changes and localization, biotinylation is often performed. Although Western blot analysis of the membrane fraction indicates the amounts of transporters in the cell membrane including organelle membranes, a biotinylation study allows us to determine the amount of functional transporter localized on the plasma membrane.

At 48 hours after transfection of fluorescent protein-fused transporter, the cells are harvested and washed three times with PBS. Cells (2.5×10^7 cells/mL) are then incubated at 4°C with 0.5 mg/mL sulfo-NHS-LC-Biotin for 1 hour, washed with PBS, incubated again with sulfo-NHSLC-biotin in the same manner, washed, and solubilized in 20 mM phosphate buffer. The solubilized fraction is incubated at 4°C for 30 minutes with 50 μL of immobilized streptavidin, which is then washed three times with PBS containing 0.02% Tween 20 and subjected to Western blot analysis.[65]

To study interactions and change of localization under physiological conditions, immunocyto- or immunohistochemical analysis is performed with double staining using specific antibodies. In addition, posttranscriptional regulation may influence specificity and transport capacity, which can be accessed by transport and inhibition studies using cultured cell expression systems.

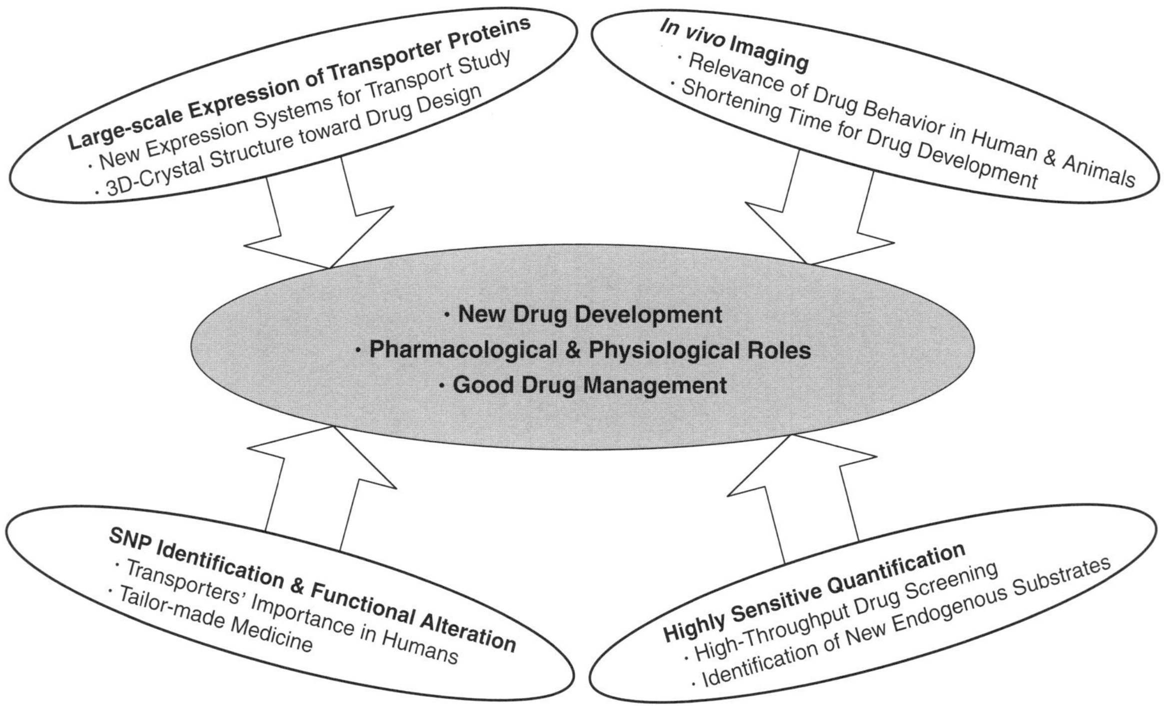

FIGURE 18.2. New technology for studying drug transporters.

18.7. CONCLUSIONS

In this chapter we have presented methodological aspects of popular approaches to the study of drug transporters. Application of these methods to study pharmacological issues is considered in detail in other chapters. We have focused on methods that we have employed ourselves, because descriptions based on actual examples should be easier to understand. In the case of transport studies, we have described experiments with radiolabeled probe compounds, but HPLC or LC/MS/MS systems are also suitable for this purpose.

As mentioned in our introduction, it is important for research in this field to apply state-of-the-art technology, so finally, we introduce some novel approaches for the study of transporters (Figure 18.2). Saitoh and others used a baculoviral expression system to obtain a large amount of purified PEPT1. This technology is likely to become a useful one to study the functions, activities, regulation, and crystal structures of drug transporters.[66–68] In vivo imaging systems such as positron emission tomography (PET) and computed tomography (CT) are used widely in clinical diagnostics and can be modified for use with small animals such as rodents. They can be used to visualize cellular and molecular events, in drug-treated living animals, and can be powerful tools for in vivo experiments under physiological conditions.[69,70] Single-nucleotide polymorphism (SNP) analyses with clinical samples can help to elucidate the physiological and pharmacological functions or importance of drug transporters. Recently, the involvement of transporters in drug–drug interaction has been reported, based on examination of SNPs and clinical samples. Understanding the SNPs of transporters and associated proteins may contribute to the establishment of "tailor-made medicine."[71–75]

Many methods are available for similar purposes, and the choice of methods may depend on the research environments, such as available equipments, experiences, funds, personalities, and so on. Nevertheless, we emphasize that evidence obtained form a single approach generally cannot be considered definitive and requires the support of data obtained by independent methods.

REFERENCES

1. Yagi T, Tokunaga T, Furuta Y, Nada S, Yoshida M, Tsukada T, Saga Y, Takeda N, Ikawa Y, Aizawa S. 1993. A novel ES cell line, TT2, with high germline-differentiating potency. Anal Biochem 214:70–76.
2. Tokunaga T, Tsunoda Y. 1992. Efficacious production of viable germ-line chimeras between embryonic stem (ES) cells and 8-cell stage embryos. Dev Growth Diff 34:561–566.
3. Nagy A, Rossant J, Nagy R, Abramow-Newerly W, Roder JC. 1993. Derivation of completely cell culture-derived mice from early-passage embryonic stem cells. Proc Natl Acad Sci USA 90:8424–8428.
4. Soriano P, Montogomery C, Geske R, Bradley A. 1991. Targeted disruption of the c-src proto-oncogene leads to osteopetrosis in mice. Cell 64:693–702.

5. Nada S, Yagi T, Takeda H, Tokunaga T, Nakagawa H, Ikawa Y, Okada M, Aizawa S. 1993. Constitutive activation of Src family kinases in mouse embryos that lack Csk. Cell 73:1125–1135.
6. Yagi T, Ikawa Y, Yoshida K, Shigetani Y, Takeda N, Mabuchi I, Yamamoto T, Aizawa S. 1990. Homologous recombination at c-fyn locus of mouse embryonic stem cells with use of diphtheria toxin A-fragment gene in negative selection. Proc Natl Acad Sci USA 87:9918–9922.
7. Kubo Y, Sekiya S, Ohigashi M, Takenaka C, Tamura K, Nada S, Nishi T, Yamamoto A, Yamaguchi A. 2005. ABCA5 resides in lysosomes, and ABCA5 knockout mice develop lysosome disease-like symptoms. Mol Cell Biol 25:4138–4149.
8. Hasuwa H, Kaseda K, Einarsdottir T, Okabe M. 2002. Small interfering RNA and gene silencing in transgenic mice and rats. FEBS Lett 532:227–230.
9. Tiscornia G, Singer O, Ikawa M, Verma IM. 2003. A general method for gene knockdown in mice by using lentiviral vectors expressing small interfering RNA. Proc Natl Acad Sci USA 100:1844–1848.
10. Liu F, Song Y, Liu D. 1999. Hydrodynamics-based transfection in animals by systemic administration of plasmid DNA. Gene Ther 6:1258–1266.
11. Kobayashi N, Nishikawa M, Takakura Y. 2005. The hydrodynamics-based procedure for controlling the pharmacokinetics of gene medicines at whole body, organ and cellular levels. Adv Drug Deliv Rev 57:713–731.
12. Kobayashi N, Nishikawa M, Hirata K, Takakura Y. 2004. Hydrodynamics-based procedure involves transient hyperpermeability in the hepatic cellular membrane: implication of a nonspecific process in efficient intracellular gene delivery. J Gene Med 6:584–592.
13. Kobayashi N, Hirata K, Chen S, Kawase A, Nishikawa M, Takakura Y. 2004. Hepatic delivery of particulates in the submicron range by a hydrodynamics-based procedure: implications for particulate gene delivery systems. J Gene Med 6:455–463.
14. Kobayashi N, Matsui Y, Kawase A, Hirata K, Miyagishi M, Taira K, Nishikawa M, Takakura Y. 2004. Vector-based in vivo RNA interference: dose- and time-dependent suppression of transgene expression. J Pharmacol Exp Ther 308:688–693.
15. Matsui Y, Kobayashi N, Nishikawa M, Takakura Y. 2005. Sequence-specific suppression of mdr1a/1b expression in mice via RNA interference. Pharm Res 22:2091–2098.
16. Schinkel AH, Smit JJ, van Tellingen O, Beijnen JH, Wagenaar E, van Deemter L, Mol CA, van der Valk MA, Robanus-Maandag EC, te Riele HP, et al. 1994. Disruption of the mouse mdr1a P-glycoprotein gene leads to a deficiency in the blood–brain barrier and to increased sensitivity to drugs. Cell 77:491–502.
17. Jonker JW, Buitelaar M, Wagenaar E, van der Valk MA, Scheffer GL, Scheper RJ, Plosch T, Kuipers F, Elferink RP, Rosing H, Beijnen JH, Schinkel AH. 2002. The breast cancer resistance protein protects against a major chlorophyll-derived dietary phototoxin and protoporphyria. Proc Natl Acad Sci USA 99:15649–15654.
18. Murata M, Tamai I, Sai Y, Nagata O, Kato H, Sugiyama Y, Tsuji A. 1998. Hepatobiliary transport kinetics of HSR-903, a new quinolone antibacterial agent. Drug Metab Dispos 26:1113–1119.
19. Sasabe H, Terasaki T, Tsuji A, Sugiyama Y. 1997. Carrier-mediated hepatic uptake of quinolone antibiotics in the rat. J Pharmacol Exp Ther 282:162–171.

20. Ishibuchi S, Morimoto H, Oe T, Ikebe T, Inoue H, Fukunari A, Kamezawa M, Yamada I, Naka Y. 2001. Synthesis and structure–activity relationships of 1-phenylpyrazoles as xanthine oxidase inhibitors. Bioorg Med Chem Lett 11:879–882.
21. Sai Y, Kato Y, Nakamura K, Kato S, Nishimura T, Kubo Y, Tamai I, Yang S, Hu Z, Yamada I, Tsuji A. 2006. Carrier-mediated hepatic uptake of a novel nonrenal excretion type uric acid generation inhibitor, Y-700. J Pharm Sci 95:336–347.
22. Nezu J, Tamai I, Oku A, Ohashi R, Yabuuchi H, Hashimoto N, Nikaido H, Sai Y, Koizumi A, Shoji Y, et al. 1999. Primary systemic carnitine deficiency is caused by mutations in a gene encoding sodium ion–dependent carnitine transporter. Nat Genet 21:91–94.
23. Ohashi R, Tamai I, Nezu J, Nikaido H, Hashimoto N, Oku A, Sai Y, Shimane M, Tsuji A. 2001. Molecular and physiological evidence for multifunctionality of carnitine/organic cation transporter OCTN2. Mol Pharmacol 59:358–366.
24. Naruhashi K, Tamai I, Inoue N, Muraoka H, Sai Y, Suzuki N, Tsuji A. 2002. Involvement of multidrug resistance–associated protein2 in intestinal secretion of grepafloxacin in rats. Antimicrob Agents Chemother 46:344–349.
25. Naruhashi K, Tamai I, Inoue N, Muraoka H, Sai Y, Suzuki N, Tsuji A. 2001. Active intestinal secretion of new quinolone antimicrobials and the partial contribution of P-glycoprotein. J Pharm Pharmacol 53:699–709.
26. Tamai I, Takanaga H, Maeda H, Yabuuchi H, Sai Y, Suzuki Y, Tsuji A. 1997. Intestinal brush-border transport of monocarboxylic acids mediated by proton-coupled transport and anion antiport mechanisms. J Pharm Pharmacol 49:108–112.
27. Tsuji A, Nakashima E, Kagami I, Honjo N, Yamana T. 1977. Effect of dose-concentration on the absorption of amoxicillin and ampicillin from the rat intestine. J Pharm Pharmacol 29:707–708.
28. Nakashima E, Tsuji A, Mizuo H, Yamana T. 1984. Kinetics and mechanism of in vitro uptake of amino-beta-lactam antibiotics by rat small intestine and relation to the intact-peptide transport system. Biochem Pharmacol 33:3345–3352.
29. Shinkawa T, Yamasaki F, Notsu T, Nakakuki M, Nishijima K, Yoshitomi K, Imai M. 1993. Loop and distal actions of a novel diuretic, M17055. Eur J Pharmacol 238:317–325.
30. Nishimura T, Kato Y, Sai Y, Ogihara T, Tsuji A. 2004. Characterization of renal excretion mechanism for a novel diuretic, M17055, in rats. J Pharm Sci 93:2558–2566.
31. Baur H, Kasperek S, Pfaff E. 1975. Criteria of viability of isolated liver cells. Hoppe–Seyler Z Physiol Chem 356: 827–838.
32. Moldeus P, Hogberg J, Orrenius S. 1978. Isolation and use of liver cells. Methods Enzymol 52:60–71.
33. Nozawa T, Tamai I, Sai Y, Nezu J, Tsuji A. 2003. Contribution of organic anion transporting polypeptide OATP-C to hepatic elimination of the opioid pentapeptide analogue [D-Ala2, D-Leu5]-enkephalin. J Pharm Pharmacol 55:1013–1020.
34. Tsuji A, Terasaki T, Tamai I, Nakashima E, Takanosu K. 1985. A carrier-mediated transport system for benzylpenicillin in isolated hepatocytes. J Pharm Pharmacol 37:55–57.
35. Hashimoto N, Suzuki F, Tamai I, Nikaido H, Kuwajima M, Hayakawa J, Tsuji A. 1998. Gene-dose effect on carnitine transport activity in embryonic fibroblasts of jvs mice as a model of human carnitine transporter deficiency. Biochem Pharmacol 55:1729–1732.
36. Terasaki T, Hosoya K. 2001. Conditionally immortalized cell lines as a new in vitro model for the study of barrier functions. Biol Pharm Bull 24:111–118.

37. Artursson P. 1990. Epithelial transport of drugs in cell culture. I: A model for studying the passive diffusion of drugs over intestinal absorptive (Caco-2) cells. J Pharm Sci 79:476–482.
38. Tamai I, Saheki A, Saitoh R, Sai Y, Yamada I, Tsuji A. 1997. Nonlinear intestinal absorption of 5-hydroxytryptamine receptor antagonist caused by absorptive and secretory transporters. J Pharmacol Exp Ther 283:108–115.
39. Tsuji A, Takanaga H, Tamai I, Terasaki T. 1994. Transcellular transport of benzoic acid across Caco-2 cells by a pH-dependent and carrier-mediated transport mechanism. Pharm Res 11:30–37.
40. Ueda K, Okamura N, Hirai M, Tanigawara Y, Saeki T, Kioka N, Komano T, Hori R. 1992. Human P-glycoprotein transports cortisol, aldosterone, and dexamethasone, but not progesterone. J Biol Chem 267:24248–24252.
41. Saito H, Yamamoto M, Inui K, Hori R. 1992. Transcellular transport of organic cation across monolayers of kidney epithelial cell line LLC-PK. Am J Physiol 262:C59–C66.
42. Yabuuchi H, Tamai I, Nezu J, Sakamoto K, Oku A, Shimane M, Sai Y, Tsuji A. 1999. Novel membrane transporter OCTN1 mediates multispecific, bidirectional, and pH-dependent transport of organic cations. J Pharmacol Exp Ther 289:768–773.
43. Tamai I, Ohashi R, Nezu J, Sai Y, Kobayashi D, Oku A, Shimane M, Tsuji A. 2000. Molecular and functional characterization of organic cation/carnitine transporter family in mice. J Biol Chem 275:40064–40072.
44. Tamai I, Yabuuchi H, Nezu J, Sai Y, Oku A, Shimane M, Tsuji A. 1997. Cloning and characterization of a novel human pH-dependent organic cation transporter, OCTN1. FEBS Lett 419:107–111.
45. Simonson GD, Vincent AC, Roberg KJ, Huang Y, Iwanij V. 1994. Molecular cloning and characterization of a novel liver-specific transport protein. J Cell Sci 107:1065–1072.
46. Fei YJ, Kanai Y, Nussberger S, Ganapathy V, Leibach FH, Romero MF, Singh SK, Boron WF, Hediger MA. 1994. Expression cloning of a mammalian proton-coupled oligopeptide transporter. Nature 368:563–566.
47. You G, Smith CP, Kanai Y, Lee WS, Stelzner M, Hediger MA. 1993. Cloning and characterization of the vasopressin-regulated urea transporter. Nature 365:844–847.
48. Kanai Y, Hediger MA. 1992. Primary structure and functional characterization of a high-affinity glutamate transporter. Nature 360:467–471.
49. Schwenk M. 1980. Transport systems of isolated hepatocytes studies on the transport of biliary compounds. Arch Toxicol 44:113–126.
50. Terasaki T, Tamai I, Takanosu K, Nakashima E, Tsuji A. 1986. Kinetic evidence for a common transport route of benzylpenicillin and probenecid by freshly prepared hepatocytes in rats: influence of sodium ion, organic anions, amino acids and peptides on benzylpenicillin uptake. J Pharmacobiodyn 9:18–28.
51. Lever JE. 1977. Active amino acid transport in plasma membrane vesicles from Simian virus 40-transformed mouse fibroblasts. Characteristics of electrochemical Na^+ gradient-stimulated uptake. J Biol Chem 252:1990–1997.
52. Tamai I, China K, Sai Y, Kobayashi D, Nezu J, Kawahara E, Tsuji A. 2001. Na(+)-coupled transport of L-carnitine via high-affinity carnitine transporter OCTN2 and its subcellular localization in kidney. Biochim Biophys Acta 6:1512:273–284.

53. Tamai I, Nakanishi T, Kobayashi D, China K, Kosugi Y, Nezu J, Sai Y, Tsuji A. 2004. Involvement of OCTN1 (SLC22A4) in pH-dependent transport of organic cations. Mol Pharmacol 1:57–66.

54. Rafizadeh C, Roch-Ramel F, Schali C. 1987. Tetraethylammonium transport in renal brush border membrane vesicles of the rabbit. J Pharmacol Exp Ther 240:308–313.

55. Malathi P, Preiser H, Fairclough P, Mallett P, Crane RK. 1979. A rapid method for the isolation of kidney brush border membranes. Biochim Biophys Acta 554:259–263.

56. Li Q, Sai Y, Kato Y, Muraoka H, Tamai I, Tsuji A. 2004. Transporter-mediated renal handling of nafamostat mesilate. J Pharm Sci 93:262–272.

57. Tamai I, Tsuji A, Kin Y. 1988. Carrier-mediated transport of cefixime, a new cephalosporin antibiotic, via an organic anion transport system in the rat renal brush-border membrane. J Pharmacol Exp Ther 246:338–344.

58. Rafizadeh C, Manganel M, Roch-Ramel F, Schali C. 1986. Transport of organic cations in brush border membrane vesicles from rabbit kidney cortex. Pflugers Arch 407:404–408.

59. Shimakura J, Terada T, Katsura T, Inui K. 2005. Characterization of the human peptide transporter PEPT1 promoter: Sp1 functions as a basal transcriptional regulator of human PEPT1. Am J Physiol Gastrointest Liver Physiol 289:G471–G477.

60. Maeda T, Hirayama M, Kobayashi D, Tamai I. Regulation of testis-specific carnitine transporter (octn3) gene by proximal *cis*-acting elements Sp1 in mice. Biochem Pharmacol 70:858–868.

61. Stewart AFR, Suzow J, Kubota T, Ueyama T, Chen HH. 1998. Transcription factor RTEF-1 mediates alpha-adrenergic reactivation of the fetal gene program in cardiac myocytes. Circ Res 83:43–49.

62. Maeda T, Mazzulli JR, Farrance AF, Stewart AFR. 2002. Mouse DTEF-1 (ETFR-1, TEF-5) is a transcriptional activator in alpha 1–adrenergic agonist-stimulated cardiac myocytes. J Biol Chem 277:24346–24352.

63. Kato Y, Misra S, Puertollano R, Hurley JH, Bonifacino JS. 2005. Phosphoregulation of sorting signal-VHS domain interactions by a direct electrostatic mechanism. Nat Struct Biol 9:532–536.

64. Kato Y, Yoshida K, Watanabe C, Sai Y, Tsuji A. 2004. Screening of the interaction between xenobiotic transporters and PDZ proteins. Pharm Res 21:1886–1894.

65. Kato Y, Sai Y, Yoshida K, Watanabe C, Hirata T, Tsuji A. 2005. PDZK1 directly regulates the function of organic cation/carnitine transporter OCTN2. Mol Pharmacol 67:734–743.

66. Saitoh R, Ohtomo T, Ito Y, Nezu J, Kimura N, Funahashi SI, Aso Y, Ohizumi I, Kodama T, Hamakubo T, Tsuchiya M. 2005. Recovery of functional peptide transporter PepT1 in budded baculovirus fraction. Protein Expr Purif 46:130–135.

67. Grisshammer R, Tate CG. 1995. Overexpression of integral membrane proteins for structural studies. Q Rev Biophys 28:315–422.

68. Loisel TP, Ansanay H, St-Onge S, Gay B, Boulanger P, Strosberg AD, Marullo S, Bouvier M. 1997. Recovery of homogeneous and functional beta 2-adrenergic receptors from extracellular baculovirus particles. Nat Biotechnol 15:1300–1304.

69. Rao VV, Dahlheimer JL, Bardgett ME, Snyder AZ, Finch RA, Sartorelli AC, Piwnica-Worms D. 1999. Choroid plexus epithelial expression of MDR1 P-glycoprotein and multidrug resistance–associated protein contribute to the blood–cerebrospinal fluid–drug permeability barrier. Proc Natl Acad Sci USA 96:3900–3905.

70. Dyszlewski M, Blake HM, Dahlheimer JL, Pica CM, Piwnica-Worms D. 2002. Characterization of a novel ^{99}mTc-carbonyl complex as a functional probe of MDR1 P-glycoprotein transport activity. Mol Imag 1:24–35.

71. Ishikawa T, Tsuji A, Inui K, Sai Y, Anzai N, Wada M, Endou H, Sumino Y. 2004. The genetic polymorphism of drug transporters: functional analysis approaches. Pharmacogenomics 5:67–99.

72. Iwai M, Suzuki H, Ieiri I, Otsubo K, Sugiyama Y. 2004. Functional analysis of single nucleotide polymorphisms of hepatic organic anion transporter OATP1B1 (OATP-C). Pharmacogenetics 14:749–757.

73. Kondo C, Suzuki H, Itoda M, Ozawa S, Sawada J, Kobayashi D, Ieiri I, Mine K, Ohtsubo K, Sugiyama Y. 2004. Functional analysis of SNPs variants of BCRP/ABCG2. Pharm Res 21:1895–1903.

74. Hirouchi M, Suzuki H, Itoda M, Ozawa S, Sawada J, Ieiri I, Ohtsubo K, Sugiyama Y. 2004. Characterization of the cellular localization, expression level, and function of SNP variants of MRP2/ABCC2. Pharm Res 21:742–748.

75. Kawasaki Y, Kato Y, Sai Y, Tsuji A. 2004. Functional characterization of human organic cation transporter OCTN1 single nucleotide polymorphisms in the Japanese population. J Pharm Sci 93:2920–2926.

19

IN VITRO–IN VIVO SCALE-UP OF DRUG TRANSPORT ACTIVITIES

KAZUYA MAEDA AND YUICHI SUGIYAMA

University of Tokyo, Tokyo, Japan

Drug Transporters: Molecular Characterization and Role in Drug Disposition, Edited by Guofeng You and Marilyn E. Morris
Copyright © 2007 John Wiley & Sons, Inc.

19.1. INTRODUCTION

Up to the present, many transporters have been identified and characterized in rodents and humans, as discussed in earlier chapters. Substrate specificities and transport kinetics of ligands for various transporters have been investigated thoroughly using transporter expression systems such as cRNA-injected *Xenopus laevis* oocytes, recombinant mammalian cells, and membrane vesicles expressing exogenous transporters. In addition, transport properties of substrates in organs and barriers have also been characterized by several experimental systems, such as in situ organ perfusion, primary cells isolated from organs, and organ slices. One of the major reasons for the discontinuation of the clinical development of new drugs is their unfavorable pharmacokinetics in humans. Examples of the importance of transporters in the pharmacokinetics of several drugs have been published.[1–3] To clarify the pharmacokinetic properties of new drugs during the early phase of drug development efficiently, it is important to be able to predict the in vivo pharmacokinetics from in vitro experimental results, and consideration needs to be given to the methodologies for in vitro–in vivo scale-up of transporter activities. Therefore, in this chapter we review and discuss various methods for the prediction of pharmacokinetics from in vitro results.

19.2. THEORETICAL BACKGROUND FOR THE PREDICTION OF IN VIVO PHARMACOKINETICS FROM IN VITRO DATA

We can estimate the intrinsic transport activity of substrates for each transporter by using gene expression systems. To extrapolate the intrinsic transport activities of substrates for individual transporters to their in vivo pharmacokinetics, we must consider several issues based on the pharmacokinetic concepts described below.[2–4]

19.2.1. Contribution of Each Transporter to Overall Membrane Transport

Several transporters are expressed on the same membrane of the same tissue. Due to the broad substrate specificities of transporters, one compound is often recognized by multiple transporters. Moreover, especially for lipophilic compounds, which can easily penetrate the plasma membrane, passive diffusion may not be negligible. Therefore, the intrinsic membrane transport clearance ($CL_{int,membrane}$) is described as

$$CL_{int,membrane} = \sum_i CL_{transport,i} + CL_{passive} \quad (1)$$

where $CL_{transport,i}$ and $CL_{passive}$ represent the active transport clearance mediated by transporter i and passive transport clearance. We can roughly distinguish the active transport clearance from passive clearance by measuring the reduction in transport activity in the presence of metabolic inhibitors [e.g., *p*-trifluoromethoxyphenylhydrazone (FCCP), rotenone, sodium azide] or at a low temperature.[5] The methods for determining the quantitative contribution of each transporter to the overall transport are discussed in the following sections.

19.2.2. Rate-Determining Process in Apparent Intrinsic Organ Clearance

Generally, the detoxification process consists of several parts: (1) uptake of compounds from blood to tissue cells, (2) metabolism in an intracellular compartment, (3) backflux from cells to blood, and (4) excretion from cells into another compartment, such as bile (liver) and urine (kidney). The "apparent" intrinsic organ clearance ($CL_{int,all}$) in such cases can be described as follows:

$$CL_{int,all} = CL_{u,influx} \frac{CL_{u,metab} + CL_{u,eff}}{(CL_{u,metab} + CL_{u,eff}) + CL_{u,back}} \quad (2)$$

where $CL_{u,influx}$, $CL_{u,metab}$, $CL_{u,eff}$, and $CL_{u,back}$ represent the influx clearance from blood, metabolic clearance, efflux clearance from cells, and clearance for backflux, respectively. If ($CL_{u,metab} + CL_{u,eff}$) is much greater than $CL_{u,back}$, the apparent intrinsic clearance is very close to $CL_{u,influx}$ and the influx clearance is solely determined by the overall intrinsic clearance:

$$CL_{int,all} \sim CL_{u,influx} \quad (3)$$

On the other hand, if ($CL_{u,metab} + CL_{u,eff}$) is much lower than $CL_{u,back}$, the apparent intrinsic clearance is determined by the activity of all the processes:

$$CL_{int,all} \sim CL_{u,influx} \times \frac{CL_{u,metab} + CL_{u,eff}}{CL_{u,back}} \quad (4)$$

When the membrane transport clearance is very rapid and the membrane transport on the blood side is symmetric ($CL_{u,influx} = CL_{u,back}$), equation (1) can be approximated by

$$CL_{int,all} \sim CL_{u,metab} + CL_{u,eff} \quad (5)$$

This equation is applicable to lipophilic drugs exhibiting rapid membrane permeability. Considering these equations, the important thing is that even if the major clearance mechanism of a drug is metabolism, the apparent intrinsic organ clearance is determined not by metabolic intrinsic clearance but by influx clearance under certain conditions, as discussed above when a drug is recognized by influx transporters.

19.2.3. Relationship Between Apparent Intrinsic Organ Clearance and Organ Clearance

According to pharmacokinetic theory, a number of models have been proposed to explain organ clearance from the apparent intrinsic clearance, organ blood flow, and protein unbound fraction. The simplest mathematical model is the well-stirred model, in which instantaneous and complete mixing within an organ is assumed. Based on this model, organ clearance ($\mathrm{CL_{org}}$) can be described by as

$$\mathrm{CL_{org}} = \frac{Q_{\mathrm{org}} f_B \mathrm{CL_{int,all}}}{Q_{\mathrm{org}} + f_B \mathrm{CL_{int,all}}} \tag{6}$$

where Q_{org}, f_B, and $\mathrm{CL_{int,all}}$ represent organ blood flow, protein unbound fraction in blood, and intrinsic organ clearance as discussed above, respectively. If Q_{org} is much smaller than $f_B\mathrm{CL_{int,all}}$, equation (6) is approximated by

$$\mathrm{CL_{org}} \sim Q_{\mathrm{org}} \tag{7}$$

In this situation, organ clearance is limited by blood flow. On the other hand, if Q_{org} is much larger than $f_B\mathrm{CL_{int,all}}$, equation (6) is approximated by

$$\mathrm{CL_{org}} \sim f_B \mathrm{CL_{int,all}} \tag{8}$$

In this case, organ clearance is affected directly by intrinsic clearance and protein unbound fraction.

19.3. PREDICTION OF HEPATIC TRANSPORT FROM IN VITRO DATA

Hepatic transport properties can be investigated by several in vitro methods. Isolated hepatocytes are very useful for evaluating the hepatic uptake of compounds. Hepatocytes are easily obtained from animals following perfusion of a collagenase-containing buffer. In addition, we can now obtain batches of human cryopreserved hepatocytes from several commercial sources. Shitara et al. have checked whether active transport systems are retained properly in cryopreserved human hepatocytes by comparing the estradiol-17β-glucuronide ($E_2$17βG) uptake (OATP substrate) and Na^+-dependent taurocholate (TCA) uptake (NTCP substrate) in freshly isolated hepatocytes with that in cryopreserved hepatocytes. It was found that there was a clear time-dependent saturable uptake of $E_2$17βG and TCA in cryopreserved hepatocytes, although the change in uptake clearance before and after cryopreservation exhibited a large interbatch variability among five different preparations of human hepatocytes.[6] This large difference was probably due to both the interindividual differences in intrinsic transport activity and the different conditions of isolation and cryopreservation of the hepatocytes. In our experience, when several batches of commercially available human cryoperserved hepatocytes are randomly screened by measuring the transport activity of $E_2 17\beta G$ and TCA, about 40% of the batches tested showed adequate transport activity, while the other batches exhibited no detectable transport activity. Also, the transport activity exhibited by each batch of hepatocytes did not always correlate with the metabolic activities described in the batch sheets. Therefore, before we check the transport characteristics of several compounds using human hepatocytes, we prescreen the uptake clearance of $E_2 17\beta G$ and TCA in many batches and select in advance at least three batches of hepatocytes with high transport activity.[7] Cultured hepatocytes can also be used, due to their ease of handling; however, we must keep in mind that several reports have indicated that long-term (>1 day) culture on collagen-coated dishes results in a drastic reduction of the mRNA and protein levels of several transporters and in the uptake activity of organic anions, such as pravastatin.[8–10]

Theoretically, the hepatic uptake intrinsic clearance can be estimated simply by scaling up the uptake clearance in hepatocytes to the in vivo level. By multiplying the uptake clearance per cell by the number of cells per gram liver, it is possible to extrapolate in vitro uptake data to obtain the in vivo intrinsic uptake clearance. Miyauchi et al. have demonstrated that the uptake clearance of 15 drugs calculated from isolated hepatocytes correlated well with the values estimated by the in situ multiple indicator dilution (MID) method, although the in situ clearance appeared to reach an upper limit possibly because the diffusion of compounds in the unstirred water layer became the rate-determining process.[11] Kato et al. have also shown that the uptake clearance of four types of endothelin antagonists obtained from integration plot analysis after intravenous administration of the compounds to rats is almost comparable with that calculated from the uptake clearance in isolated rat hepatocytes, assuming the well-stirred model[12] (Figure 19.1). This evidence suggests that isolated hepatocytes are a good model for predicting hepatic uptake clearance.

Regarding uptake transport in human liver, some SLC (solute carrier) transporters are expressed, such as NTCP (Na^+-taurocholate cotransporting polypeptide) and OATP (organic anion transporting polypeptide) family transporters, as discussed in other chapters. In particular, OATP1B1, OATP1B3, and OATP2B1 show very broad substrate specificities, and these overlap with one another, so that one compound is often recognized by multiple transporters. The transport properties of each transporter can be evaluated by using gene expression systems (e.g., mammalian cells, *Xenopus* oocytes). However, the transporters that accept one compound are not always important for overall hepatic uptake if the relative contribution is very minor compared with that of other transporters. Therefore, it is essential to determine the quantitative contribution to hepatic uptake of each transporter to show the importance of each transporter under in vivo conditions. When the function and/or expression level of one transporter is changed due to genetic polymorphisms, pathophysiological conditions and transporter-mediated drug–drug interactions information about the contribution is necessary to predict the change in the in vivo pharmacokinetics from in vitro data.

Several estimation methods have been developed for this purpose. Kouzuki et al. have proposed a method using reference compounds to determine the contribution of rat Oatp1a1 and Ntcp to the hepatic uptake of bile acids and organic anions.[13,14] This concept was originally established in the field of metabolic enzymes by Crespi and Penman,[15] who named it the *relative activity factor* (RAF) *method*. Using this method

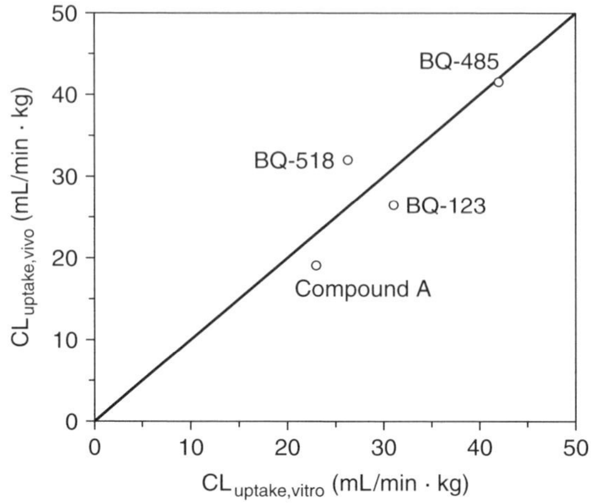

FIGURE 19.1. Comparison between in vivo hepatic uptake clearance of four types of endothelin antagonists and that predicted from in vitro data. The hepatic uptake clearance observed in vivo ($CL_{uptake,vivo}$) was obtained from the initial slope of the integration plot and plotted against that predicted from the results in isolated rat hepatocytes ($CL_{uptake,vitro}$). The straight line indicates 1 : 1 correlation. (From ref. 12, with the kind permission of the American Society for Pharmacology and Experimental Therapeutics.)

they checked the transport activity of both test and reference compounds, which should be specific substrates for single transporters, in short-term cultured rat hepatocytes and transporter-expressing COS-7 cells. Then they estimated the contribution from the equations

$$\text{contribution}(\%) = \frac{R_{COS}}{R_{hep}} \times 100 \tag{9}$$

$$R_{exp} = \frac{CL_{uptake,COS(test)}}{CL_{uptake,COS(reference)}} \times 100 \tag{10}$$

$$R_{hep} = \frac{CL_{uptake,hep(test)}}{CL_{uptake,hep(reference)}} \times 100 \tag{11}$$

where $CL_{uptake,COS(test)}$ and $CL_{uptake,COS(reference)}$ represent the uptake clearances of the test and reference compounds in transporter-transfected COS-7 cells, respectively, and $CL_{uptake,hep(test)}$ and $CL_{uptake,hep(reference)}$ represent the uptake clearances of the test and reference compounds in isolated rat hepatocytes, respectively. To estimate the contribution of rat Oatp1 and Ntcp, they used as reference compounds, TCA for Ntcp and $E_2 17\beta G$ for Oatp1a1. They found that rat Ntcp was responsible for the hepatic uptake of bile acids. On the other hand, some organic anions were partially taken up via Oatp1a1, but the hepatic uptake of other anions, such as pravastatin and DNP-SG, could not be explained by Oatp1a1-mediated transport, suggesting that uptake transporters other than Oatp1a1 are involved in their uptake. Now, other hepatic uptake transporters, such as Oatp1a4, Oatp1b2, and Oat2, have also been characterized, and they can accept a variety of anions.[16–18] Therefore, $E_2 17\beta G$ can no longer be used as a reference compound for Oatp1a1, but Crespi and Penman's concept is applicable to estimation of the relative contribution.

Hirano et al. have applied this concept to human hepatocytes to estimate the relative contribution of OATP1B1 and OATP1B3 to the hepatic uptake of $E_2 17\beta G$ and pitavastatin, a novel 3-hydroxy-3-methylglutarylcoenzyme A (HMG-CoA) reductase inhibitor, in cryopreserved human hepatocytes (Figure 19.2*a*).[7] As reference compounds, they used estrone 3-sulfate (E-sul) for OATP1B1 and cholecystokinin octapeptide (CCK-8) for OATP1B3. As with the previous method, they calculated the ratio of the uptake clearance of the reference compounds in human hepatocytes to that in the expression systems and defined R_{act} for OATP1B1 and 1B3. By multiplying the R_{act} value by the uptake clearance of test compounds (CL_{test}), it is possible to estimate the uptake clearance of the test compounds mediated by a specific transporter in human liver. Assuming that the hepatic uptake clearance (CL_{hep}) could be explained by OATP1B1- and OATP1B3-mediated transport, the following equation applies:

$$CL_{hep} = R_{act,OATP1B1} CL_{test,OATP1B1} + R_{act,OATP1B3} CL_{test,OATP1B3} \tag{12}$$

They demonstrated that both pitavastatin and $E_2 17\beta G$ were taken up mainly by OATP1B1 in three independent batches of human hepatocytes and that the uptake clearance observed in human hepatocytes was almost identical to the sum of the estimated clearance mediated by OATP1B1 and 1B3. They also confirmed their results by two different approaches.[7,19] One involves direct estimation of the ratio of the expression level of OATP1B1, 1B3, and 2B1 in human hepatocytes to that in expression systems by comparing the band density of Western blot analysis and estimating their contributions using that ratio instead of the R_{act} value shown above (Figure 19.2*b*).[7,19] The other approach is to estimate the inhibitable portion of the uptake of test compounds in human hepatocytes in the presence of a specific inhibitor for each transporter (Figure 19.2*c*).[19] E-sul can be used as a specific inhibitor as well as a specific substrate for OATP1B1 (Figure 19.3*a*), whereas CCK-8 inhibited both OATP1B1 and OATP1B3 to the same extent, indicating that CCK-8 could not be used as a specific inhibitor, although it is a specific substrate for OATP1B3. The uptake of pitavastatin and $E_2 17\beta G$ was completely inhibited by excess concentration of E-sul (Figure 19.3*b*). Therefore, the importance of OATP1B3 in their hepatic uptake was supported by these different approaches.

Shimizu et al. examined the uptake of fexofenadine in OATP1B1-, 1B3-, and 2B1-expressing cells, and significant uptake could be observed only via OATP1B3, although the uptake in OATP1B1- and 2B1-expressing cells was slightly higher compared with that in control cells.[20] They estimated the contribution of OATP1B3 to the uptake of fexofenadine based on the method using reference compounds and concluded that OATP1B3 rather than OATP1B1 is mainly involved in fexofenadine transport because its observed clearance in OATP1B1-expressing cells was much

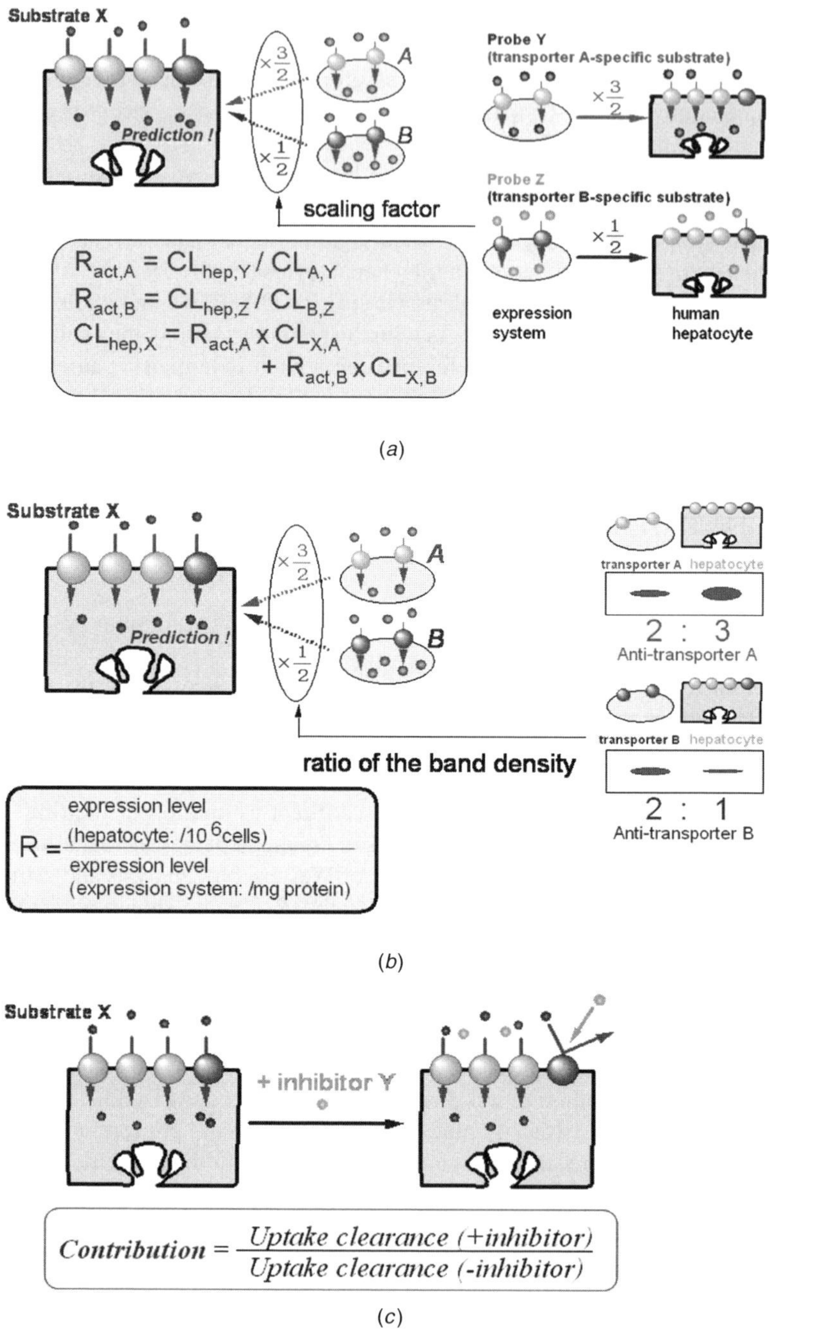

FIGURE 19.2. Schematic diagram of the method for estimating the contribution of each transporter to the overall hepatic uptake: (*a*) using reference compounds; (*b*) using the relative expression levels estimated from Western blot analysis; (*c*) using transporter-specific inhibitors. The details are described in the text. (*See insert for color representation of figure.*)

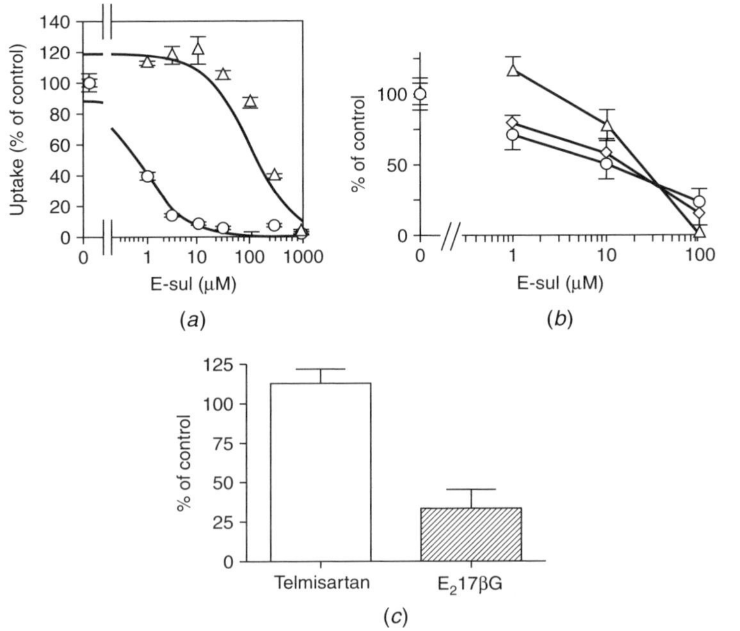

FIGURE 19.3. Specific inhibition of OATP1B1-mediated uptake by estrone 3-sulfate and its different inhibitory effects on the uptake of pitavastatin and telmisartan in human hepatocytes. (*a*) Inhibitory effect of E-sul on the OATP1B1-mediated uptake of $E_2 17\beta G$ (circle) and OATP1B3-mediated uptake of CCK-8 (triangle) in transporter-expressing HEK293 cells in the presence of 0.3% human serum albumin. (*b*) Inhibitory effect of E-sul on the uptake of pitavastatin in three batches of human hepatocytes. Open circles, triangles, and squares represent the uptake in human hepatocytes of lots OCF, 094, and ETR, respectively. (*c*) Inhibitory effect of 30 μM E-sul on the uptake of telmisartan in human hepatocytes in the presence of 0.3% human serum albumin. The *y*-axis represents the percentage of the saturable uptake of telmisartan and $E_2 17\beta G$ in the absence of E-sul. [(*a*, *c*) From ref. 21; (*b*) from ref. 19. With the kind permission of the American Society for Pharmacology and Experimental Therapeutics.]

lower than the value expected calculated from the assumption that OATP1B1 and 1B3 contribute equally to the fexofenadine uptake. Ishiguro et al. have demonstrated that telmisartan, a novel angiotensin II receptor antagonist, could be taken up by OATP1B3 but not by OATP1B1, and telmisartan uptake in human hepatocytes could not be inhibited by 30 μM E-sul (Figure 19.3*c*), indicating that OATP1B3 is the main transporter involved in its uptake.[21] On the other hand, in the case of valsartan, which is in the same category of drugs as telmisartan, both OATP1B1 and 1B3 are involved in its uptake, indicating that the relative importance of each transporter might be different even between compounds with a similar chemical structure.[22] Each approach has both advantages and disadvantages, so we recommend that users compare the results obtained from different methods and validate their results.

Gene-silencing techniques such as antisense, ribozyme, and RNA interference (RNAi) are also powerful tools for determining the transport activity of specific proteins. Hagenbuch et al. have investigated the effect of coinjection of transporter (Ntcp or Oatp1a1)-specific antisense oligonucleotide on the uptake of bromosulfophthalein (BSP) and TCA in *Xenopus* oocytes injected with total rat liver mRNA.[23] They succeeded in obtaining a significant reduction in the expression level of target transporter specifically and concluded that the Na^+-dependent and Na^+-independent uptakes of TCA were due almost entirely to Ntcp and Oatp1a1, respectively, whereas only half of the BSP uptake could be explained by Oatp1a1. Nakai et al. adopted the same approach to estimate the contribution of OATP1B1 to the hepatic uptake of pravastatin and $E_2 17\beta G$ in humans.[24] They demonstrated that oocytes microinjected with human liver poly(A) mRNA exhibited Na^+-independent uptake of pravastatin and $E_2 17\beta G$ and that a simultaneous injection of OATP1B1 antisense oligonucleotides abolished this uptake completely, suggesting that OATP1B1 is a major transporter involved in their uptake. In this approach we should note the assumption that the relative expression level of each uptake transporter in *Xenopus* oocytes is the same as that in rat liver. Currently, we cannot identify the transport activity mediated by a specific transporter only by using specific inhibitors. Takagi et al. have constructed 2′-*O*,4′-*C*-ethylene-bridged nucleic acid residues that can be incorporated into antisense oligonucleotides targeted to rat Oatp1a1, 1a4, and 1a5, and they checked the selective inhibition of Oatp subtypes.[25] Recently, some groups have succeeded in the construction of small interference RNAs (siRNAs), which can efficiently reduce the expression level of specific transporters such as MDR1 and MRP2.[26,27] However, it is fairly difficult to apply these gene-silencing techniques to primary cultured hepatocytes because long-term culture dramatically reduces the expression level of several transporters,[8–10] although generally it takes a few days to knock down the protein by depletion of mRNA expression, and optimization of the culture conditions is required for this analysis.

Regarding the evaluation of efflux transport in liver, one of the most popular experimental systems involves the use of canalicular membrane vesicles (CMVs). It is difficult to evaluate the transport activity of efflux transporters in cell systems because substrates cannot easily reach the intracellular compartment, so this system is often used for the rapid determination of adenosine triphosphate (ATP)-dependent efflux transport of substrates across the bile canalicular membrane. Aoki et al. have compared the in vitro transport clearance of nine substrates in rat CMVs defined as the initial velocity for the ATP-dependent uptake divided by the substrate concentration of the incubation medium with in vivo biliary excretion clearance, defined as the biliary excretion rate normalized by the unbound concentration in rat liver at steady state. They found a significant correlation between in vitro and in vivo clearance, suggesting that in vivo biliary excretion clearance can be predicted from in vitro transport studies using CMVs.[28]

To evaluate the biliary excretion clearance in humans, several methods have been created.[29] Niinuma et al. have shown that the transport activities of several substrates were observed in human CMVs prepared from six independent batches of human liver blocks, and interestingly, the initial uptake clearance of glucuronides was not so different between rat and human CMVs,[30] whereas that of other compounds in human CMVs was significantly lower than that in rat CMVs. This result suggests a species difference in the substrate specificities of common transporters such as MRP2, or the expression of multiple transporters with substrate specificities that differ from each other. Shilling et al. have demonstrated that the interspecies variation of the uptake clearance depends on the substrate by using CMVs prepared from rats, dogs, monkeys, and humans, implying the species difference of the expression and function of each efflux transporter.[31] Some specific inhibitors of efflux transporters have been proposed. For example, Ko143 preferentially inhibits BCRP-mediated transport,[32] while PSC833 and LY335979 inhibit the MDR1-mediated transport more potently than transport via other efflux transporters.[33,34] By evaluating the effect of transporter-specific inhibitors on the ATP-dependent transport of test compounds in human CMVs, the relative contribution of each transporter to the biliary excretion can be assessed.

Recently, LeCluyse et al. have demonstrated that a collagen-sandwich culture enabled the hepatocytes to form a bile canalicular "pocket" between the adjacent cells,[35] and depletion of Ca^{2+} from the incubation medium disrupted the bile canaliculi rapidly.[36] The advantage of this culture configuration is that the polarity and expression level of uptake and efflux transporters are fully retained for several days, unlike normal culture on rigid collagen, and the biliary excretion of compounds can be evaluated in intact cell systems by differential cumulative uptake in monolayers preincubated in Ca^{2+}-containing buffer and Ca^{2+}-free buffer.[37,38] Liu et al. have found that the in vitro biliary clearance of five compounds (inulin, salicylate, methotrexate, [D-pen^{2,5}]enkephalin, and taurocholate) in rat hepatocytes calculated from the amount excreted into a bile canalicular pocket divided by the area under the incubation medium concentration–time profile correlated well with their in vivo intrinsic biliary clearance, suggesting that this system is useful for prediction of the in vivo biliary excretion of many compounds[37] (Figure 19.4). Tian et al. succeeded in applying siRNAs targeted to Mrp2 and Mrp3 in sandwich-cultured rat hepatocytes.[39]

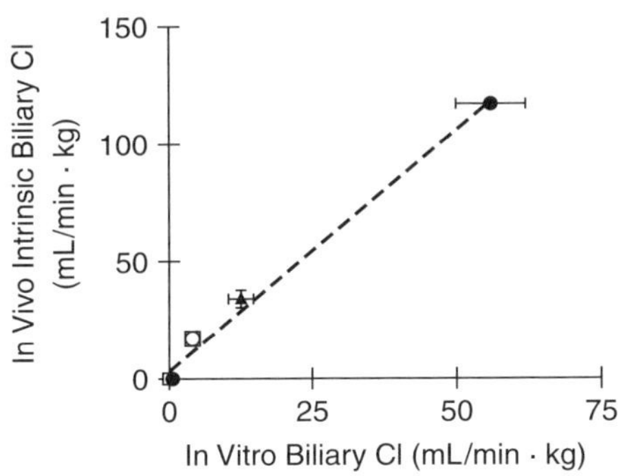

FIGURE 19.4. Relationship between in vivo intrinsic biliary clearance and in vitro estimated biliary clearance calculated from the result in 96-hour sandwich-cultured hepatocytes for model substrates, inulin (open square), salicylate (closed diamond), methotrexate (open circle), [D-pen2,5]enkephalin (closed triangle) and taurocholate (closed circle). The dashed line is the fit of a linear regression equation to the data ($r^2 = 0.99$). [From ref. 37, with the kind permission of the American Society for Pharmacology and Experimental Therapeutics.]

They observed that the knockdown of Mrp2 resulted in a reduction in the biliary excretion index of carboxydichlorofluorescein (CDF), whereas knockdown of Mrp3 caused an increase in the biliary excretion index, and the degree of the reduction of the protein expression level was in good agreement with the change in CDF disposition. Recently, Bi et al. have evaluated the biliary excretion of several substrates of a variety of efflux transporters in sandwich-cultured cryopreserved human hepatocytes.[40] By scaling up the in vitro biliary excretion clearance by multiplying the clearance per cell by the number of cells per gram liver, it is possible to estimate the in vivo clearance in humans from in vitro data.

Another approach to evaluating the efflux clearance is to use double-transfected cells, which express both uptake and efflux transporters. Originally, Cui et al. and Sasaki et al. established OATP1B3/MRP2 and OATP1B1/MRP2 double transfectants.[41,42] If a compound is a bisubstrate of uptake and efflux transporters, basal-to-apical transcellular transport is significantly greater than that in the opposite direction. Therefore, this system is suitable for high-throughput screening of bisubstrates. To extrapolate the in vivo biliary excretion clearance from in vitro experiments, Sasaki et al. have measured the basal-to-apical transcellular transport clearances of seven compounds in rat Oatp1b2/Mrp2 double transfectants and their in vivo biliary clearances ($CL_{bile,blood}$) calculated from the biliary excretion rate normalized by the blood concentration at steady state.[43] They found that the in vivo and in vitro clearances can be described by

$$CL_{bile,blood} = \frac{Q_H f_b PS_{b-a} \alpha}{Q_H + f_b PS_{b-a} \alpha} \tag{13}$$

where Q_H, f_b, and PS_{b-a} represent the hepatic blood flow rate, protein unbound fraction of the compounds in blood, and transcellular transport clearance in double transfectants corrected by the fact that 1 g of liver contains 160 mg of protein. Also, α is the scaling factor used for the quantitative prediction of the in vivo clearance from in vitro results. When α was 17.9, all data fitted the theoretical curve well as described by equation (13) (Figure 19.5). We and others have constructed several kinds of double and multiple transfectants, such as NTCP/BSEP and (OATP1B1 or OATP1B3)/(MRP2, MDR1, or BCRP) double-transfected cells, and the development and validation of the methodology for extrapolating the in vivo clearance from in vitro data are ongoing.[44–47]

19.4. PREDICTION OF RENAL TRANSPORT FROM IN VITRO DATA

The renal clearance consists of three components: (1) glomerular filtration, (2) tubular secretion, and (3) reabsorption. In particular, several renal transporters are known to be involved in (2) and (3). The renal uptake transport can be estimated using kidney slices. Hasegawa et al. have demonstrated that *p*-aminohippurate (PAH) was significantly transported by rat Oat1 but not Oat3, while pravastatin was taken up by Oat3 but not Oat1, indicating that PAH and pravastatin can be used as specific substrates of rat Oat1 and Oat3, respectively.[48] An inhibition study using gene-expression systems revealed

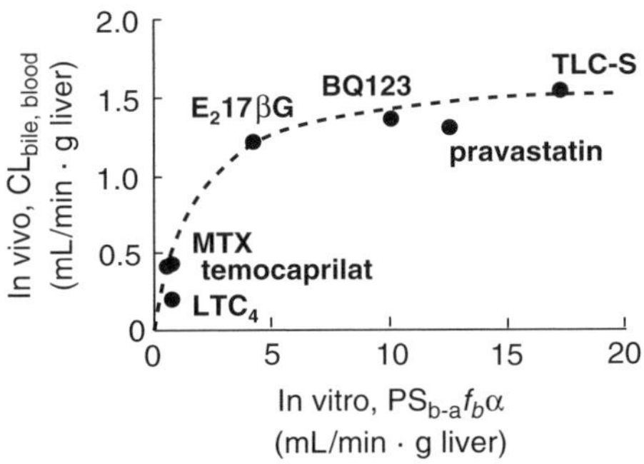

FIGURE 19.5. Comparison of the in vivo biliary excretion clearances with the in vitro transcellular transport clearance across an Oatp1b2/Mrp2 double transfectant. The in vivo biliary excretion clearance of seven compounds was compared with the transcellular transport clearance in a double transfectant. The *x*-axis represents the basal-to-apical transcellular transport clearance across the Oatp1b2/Mrp2 double transfectant (PS_{b-a}) multiplied by the protein-unbound fraction in blood (f_b) and a scaling factor (α). Closed circles represent the data for which the *x*-axis values were corrected for the scaling factor ($\alpha = 17.9$). The dotted line represents the fitted line based on equation (13) with an α value of 17.9. $E_2 17\beta G$, estradiol 17β-glucuronide; MTX, methotrexate; BQ123, cyclo-[D-Asp-Pro-D-Val-Leu-D-Trp]; TLC-S, taurolithocholate sulfate; LTC_4, leukotriene C_4. (From ref. 43, with the kind permission of the American Society for Pharmacology and Experimental Therapeutics.)

that pravastatin and benzylpenicillin (PCG) are more potent inhibitors of rat Oat3 than Oat1, whereas PAH inhibited Oat1-mediated uptake more potently than Oat3-mediated uptake. Also, the inhibitory effect of PCG and pravastatin on the pravastatin uptake in rat kidney slices was stronger than that of PAH, which also supports the major contribution of Oat3 to the renal uptake of pravastatin. Hasegawa et al. have also established methodology for determining the contribution of Oat1 and Oat3 to the overall renal uptake of test compounds in rats and predicting the renal uptake clearance in kidney slices using reference compounds (PAH and pravastatin).[49] This concept (RAF method) is similar to the one explained in Section 19.3. They have compared the uptake clearance of 10 organic anions as well as reference compounds in rat Oat1- and Oat3-expressing LLC-PK1 cells with that in rat kidney slices. As a result, for most compounds, the uptake clearance predicted per gram kidney estimated by the RAF method is almost identical to the uptake clearance observed in kidney slices (Figure 19.6).

In the case of salicylate (closed inverted triangle) and acyclovir (open diamond), the values predicted are lower than the values observed, indicating that transporters other than Oat1 and Oat3 may be involved in their renal uptake, whereas the clearance predicted for 2,4-dichlorophenoxyacetate (2,4-D) (open triangle) was greater than the value observed, suggesting that diffusion in kidney slices may be a rate-determining step in its renal uptake. Deguchi et al. have applied the same approach to the evaluation of the transport mechanism of uremic toxins.[50] They have investigated the contribution of Oat1 and Oat3 to the renal uptake of indoxyl sulfate (IS), 3-carboxy-4-methyl-5-propyl-2-furanpropionate (CMPF), indoleacetate (IA), and hippurate (HA) by the

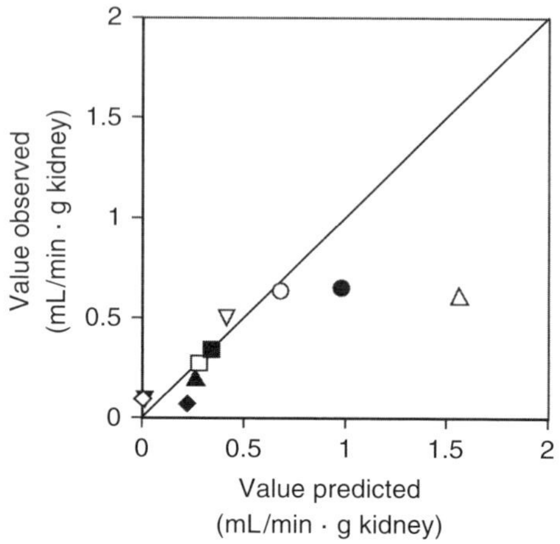

FIGURE 19.6. Relationship between the values observed and predicted for renal uptake clearance mediated by rat Oat1 and Oat3. The values predicted represent the sum of the rat Oat1- and Oat3-mediated uptake corrected by the relative transport activity of the test compounds with the reference compounds (PAH for Oat1 and pravastatin for Oat3). Each symbol represents an individual substrate. (From ref. 49, with the kind permission of the American Society for Pharmacology and Experimental Therapeutics.)

RAF method using PAH and pravastatin as reference compounds, and investigated the inhibitory effects of PAH, PCG, and pravastatin on their uptake in rat kidney slices. They concluded that OAT3 accounts primarily for CMPF uptake and OAT1 for IA and HA uptake, whereas OAT1 and OAT3 contribute almost equally to the renal uptake of IS.

To predict the uptake clearance in human kidney directly from in vitro analyses, Nozaki et al. characterized the transport properties of compounds by using human kidney slices prepared from the normal parts of a kidney carcinoma removed during surgery.[91] Regarding the efflux transporters in the kidney, some candidate transporters have been characterized, but it is still unclear which transporters are responsible for renal efflux transport. Hasegawa et al. have found that the basal-to-apical transcellular transport of some Oat ligands across the LLC-PK1 cell monolayer was enhanced by the transfection of rat Oat1 or Oat3, suggesting that LLC-PK1 cells express the efflux transporters endogenously (Hasegawa et al., submitted). They also found that substitution of Na^+ ions by K^+ ions in the incubation medium resulted in a reduction in the transcellular clearance and that the extent of this reduction depended on the substrates, implying that both voltage-dependent and voltage-independent systems are involved in the efflux of substrates in LLC-PK1 cells. They then excluded the involvement of urate transporter (UAT) in the PAH efflux, due to the fact that they found no inhibition of its transport by pyrazinoic acid and transfection of siRNA and suggested that MRP4 contributes partially to its efflux in LLC-PK1 cells since MRP4 siRNA reduced its transcellular transport. Brush border membrane vesicles (BBMVs) are also useful tools for the characterization of efflux transport systems,[51] but further analyses are required to investigate the in vitro–in vivo correlation of efflux transport in the kidney. Regarding the reabsorption process, we can qualitatively demonstrate the involvement of reabsorption in the renal clearance by observing the reduction in the reabsorption by diuresis induced by mannitol.[52] However, there is no appropriate in vitro experimental system for the quantitative prediction of transport activity involving reabsorption.

19.5. PREDICTION OF BLOOD–BRAIN BARRIER AND BLOOD–CSF BARRIER TRANSPORT FROM IN VITRO DATA

There are many transporters on the BBB and BCSFB that strictly regulate the entry of xenobiotics into the brain. In general, a change in the transport activity in BBB and BCSFB does not affect the body clearance and time profiles of the plasma concentration, but markedly affects the brain distribution, so it is difficult to predict the function of transporters in BBB and BCSFB from the in vivo pharmacokinetics of drugs, and several methods to assess the drug permeability into the brain have been created.[53] The role of Pgp in the brain disposition of drugs has been studied most extensively. Adachi et al. have investigated whether the in vivo brain distribution in mdr1a/1b knockout and normal mice can be predicted from in vitro transcellular transport studies using MDR1-expressing LLC-PK1 cells.[54] As shown in Figure 19.7*a*, apical-to-basal and basal-to-apical flux across the cell monolayer can be described by

$$\mathrm{PS}_{\text{a-to-b}} = \mathrm{PS}_{a,\mathrm{inf}} \frac{\mathrm{PS}_{b,\mathrm{eff}}}{\mathrm{PS}_{a,\mathrm{eff}} + \mathrm{PS}_{b,\mathrm{eff}}} \tag{14}$$

$$\mathrm{PS}_{\text{b-to-a}} = \mathrm{PS}_{b,\mathrm{inf}} \frac{\mathrm{PS}_{a,\mathrm{eff}}}{\mathrm{PS}_{a,\mathrm{eff}} + \mathrm{PS}_{b,\mathrm{eff}}} \tag{15}$$

where $\mathrm{PS}_{a,\mathrm{inf}}$ and $\mathrm{PS}_{a,\mathrm{eff}}$ represent the influx clearance from the apical compartment to the cells and the efflux clearance from cells to the apical compartment, respectively, and $\mathrm{PS}_{b,\mathrm{inf}}$ and $\mathrm{PS}_{b,\mathrm{eff}}$ represent the influx clearance and efflux clearance across the basal membrane. The flux ratio ($\mathrm{PS}_{\text{b-to-a}}/\mathrm{PS}_{\text{a-to-b}}$) in MDR1-expressing LLC-PK1 cells can be calculated as

$$\text{flux ratio} = \frac{\mathrm{PS}_{b,\mathrm{inf}}\,(\mathrm{PS}_{a,\mathrm{eff}} + \mathrm{PS}_{\mathrm{Pgp}})}{\mathrm{PS}_{a,\mathrm{inf}}\,\mathrm{PS}_{b,\mathrm{eff}}} \tag{16}$$

Also, the corrected flux ratio is defined as the ratio of the flux ratio in MDR1-expressing LLC-PK1 cells to that in parent LLC-PK1 cells:

$$\text{corrected flux ratio} = 1 + \frac{\mathrm{PS}_{\mathrm{Pgp}}}{\mathrm{PS}_{a,\mathrm{eff}}} \tag{17}$$

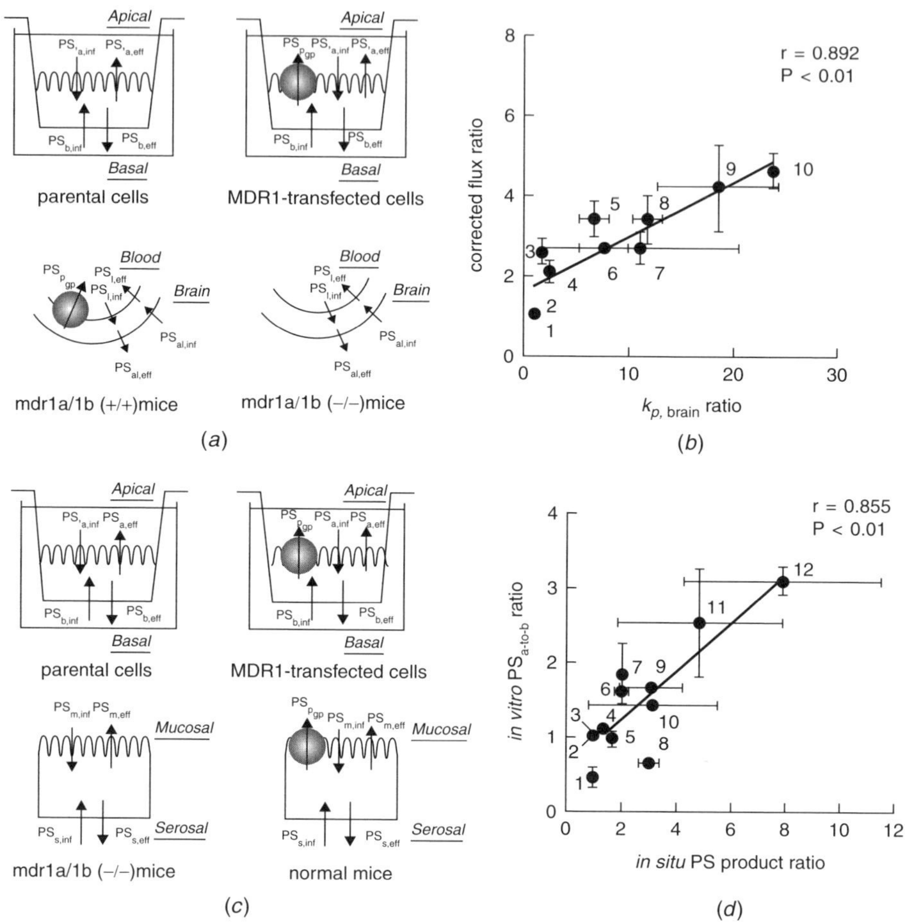

FIGURE 19.7. Prediction of the in vivo Pgp function in the intestine and in brain from in vitro transcellular transport studies in MDR1-expressing LLC-PK1 cells. (*a*) Schematic diagram illustrating the relationship between the transcellular transport clearance of compounds in MDR1-expressing and parent LLC-PK1 cells (upper figure) and the brain-to-plasma concentration ratio in mdr1a/1b(–/–) mice and normal mice (lower figure). The details are discussed in the text. (*b*) Correlation of the $K_{p,\mathrm{brain}}$ ratio calculated from the in vivo brain-to-plasma concentration ratio in mdr1a/1b(–/–) mice and normal mice and the in vitro corrected flux ratio calculated from the transcellular transport clearance across the LLC-PK1 cell monolayer. Each symbol corresponds to an individual substrate. (*c*) Schematic diagram illustrating the relationship between the transcellular transport clearance of compounds in MDR1-expressing and parent LLC-PK1 cells (upper figure) and the intestinal absorption clearance in mdr1a/1b(–/–) mice and normal mice (lower figure). The details are discussed in the text. (*d*) Correlation of the PS product ratio calculated from the results of in situ intestinal perfusion studies and the in vitro $PS_{\text{a-to-b}}$ ratio calculated from the transcellular transport clearance across the LLC-PK1 cell monolayer. Each symbol corresponds to an individual substrate. [(*a*, *b*) From ref. 54; (*c*, *d*) from ref. 69. With the kind permission of Springer Science and Business Media.]

Regarding the in vivo brain distribution, the transport clearance across the BBB from blood to brain and that in the opposite direction is given by

$$\mathrm{PS}_{\text{blood-to-brain}} = \mathrm{PS}_{l,\text{inf}} \frac{\mathrm{PS}_{al,\text{eff}}}{\mathrm{PS}_{l,\text{eff}} + \mathrm{PS}_{al,\text{eff}}} \tag{18}$$

$$\mathrm{PS}_{\text{brain-to-blood}} = \mathrm{PS}_{al,\text{inf}} \frac{\mathrm{PS}_{l,\text{eff}}}{\mathrm{PS}_{l,\text{eff}} + \mathrm{PS}_{al,\text{eff}}} \tag{19}$$

where $\mathrm{PS}_{l,\text{inf}}$ and $\mathrm{PS}_{l,\text{eff}}$ represent the clearance for the influx and efflux clearance across the luminal membrane, and $\mathrm{PS}_{\text{al,inf}}$ and $\mathrm{PS}_{\text{al,eff}}$ represent the influx and efflux clearance across the abluminal membrane, respectively (Figure 19.7*a*). Then, because the brain-to-plasma concentration ratio ($K_{p,\text{brain}}$) is calculated by the ratio $\mathrm{PS}_{\text{blood-to-brain}}/\,\mathrm{PS}_{\text{brain-to-blood}}$, $K_{p,\text{brain}}$ is given by

$$K_{p,\text{brain}} = \frac{\mathrm{PS}_{l,\text{inf}}\mathrm{PS}_{al,\text{eff}}}{\mathrm{PS}_{al,\text{inf}}\mathrm{PS}_{l,\text{eff}}} \tag{20}$$

So the $K_{p,\text{brain}}$ ratio, defined as the ratio of $K_{p,\text{brain}}$ in mdr1a/1b knockout mice to that in normal mice, can be described by

$$K_{p,\text{brain}}\ \text{ratio} = 1 + \frac{\mathrm{PS}_{\text{Pgp}}}{\mathrm{PS}_{l,\text{eff}}} \tag{21}$$

We should note that equations (17) and (21) have the same form and we can compare the in vitro corrected flux ratio and in vivo $K_{p,\text{brain}}$ ratio directly. They found a significant correlation between these two parameters among 10 compounds, whereas ATP hydrolysis activity mediated by P-glycoprotein (Pgp) was not significantly correlated with the $K_{p,\text{brain}}$ ratio (Figure 19.7*b*). The good correlation between the corrected flux ratio and the $K_{p,\text{brain}}$ ratio among seven compounds was also shown by Yamazaki et al.[55]

Several transporters other than Pgp are also expressed on the BBB. To clarify the contribution of each transporter to the total efflux of $E_217\beta G$ from the brain, Sugiyama et al. have established a method for estimating the contribution of Oatp1, 2 and Oat1, 3 by examining the effect of specific inhibitors on the elimination rate calculated using the in situ brain efflux index (BEI) method.[56] They confirmed that probenecid is a nonselective inhibitor of the efflux transport and that TCA, PAH, and digoxin can be used as selective inhibitors of the Oatp family, Oat family, and Oatp2, respectively. Considering the inhibitory effects of these four compounds on the $E_217\beta G$ efflux rate from brain, they concluded that Oatp2 and the Oat family (mainly, Oat3) account for about 40% and 20% of the overall efflux of $E_217\beta G$ across the BBB, respectively. Using the same approach, Kikuchi et al. have investigated the relative importance of Oat3 and Oatp2 in the efflux transport of pravastatin and pitavastatin across the BBB.[57] The pravastatin efflux measured by the BEI method was partially inhibited by all the inhibitors (PAH, TCA, and digoxin), whereas the pitavastatin efflux was inhibited almost completely by digoxin, an Oatp2-specific inhibitor. This suggests that the relative contribution of Oatp2 and Oat3 to the efflux transport of pravastatin and pitavastatin is different, even in the same category of drugs.

The same type of approach was applied to the functional analyses of drug transport across BCSFB. Nagata et al. have investigated the uptake mechanisms of PAH and PCG in isolated rat choroid plexus, especially concentrating on rat Oat3.[58] The inhibition study revealed that PAH and PCG shared the same transport pathway because the K_i values of six different inhibitors for the uptake of PCG correlates well with those for the uptake of PAH in the isolated choroid plexus. Moreover, they have shown that the K_i values for the uptake of PCG in the isolated choroid plexus were almost identical to those found in rat Oat3-expressing LLC-PK1 cells, suggesting that rat Oat3 is responsible for the uptake of PCG and PAH in the rat choroid plexus. Kuroda et al. have investigated the efflux mechanisms of cefaclor and cephalexin, differing by only one functional group via the choroid plexus by using transporter-specific inhibitors.[59] The elimination of cefaclor from the cerebrospinal fluid after intracerebroventricular administration was inhibited by PCG but not by glycylsarcosine (GlySar), which is used as a PEPT2 inhibitor, while GlySar, but not PCG, inhibited the elimination of cephalexin. Also, the uptake of cefaclor in the rat isolated choroid plexus was not inhibited by other Oat3 substrates (cimetidine and PAH). These results suggest that the efflux of cefaclor is mediated by a PCG-sensitive mechanism distinct from Oat3, while that of cephalexin is thought to be accounted for mainly by PEPT2. Comparison of the inhibitory effects of some specific inhibitors on the transport of test compounds in gene expression systems of individual transporters and in situ experiments allows us to gain an insight into the role of specific transporters in transport across the BBB and BCSFB.

Recently, conditionally immortalized brain endothelial cell lines from transgenic mice and rats harboring the temperature-sensitive simian virus 40 large T-antigen, named TM-BBB and TR-BBB, have been established by Hosoya et al.[60,61] These cell lines maintain the expression of several types of transporters, such as GLUT1, MCT1, neurotransmitter transporters, and ABC transporters.[62] Terasaki et al. have shown that there is a good correlation between the predicted BBB permeability clearance estimated by transcellular transport clearance across TM-BBB or TR-BBB monolayers and that observed from in vivo analyses, suggesting that these types of cell lines can be used to predict the BBB permeability of drugs.[62] Also, recently, Hino et al. have succeeded in partially reducing the Oat3-mediated brain-to-blood efflux of PCG by rapid injection of siRNA targeted to mouse Oat3 into the tail vein 36 hours before the BEI experiment.[63] Although the gene-silencing effect was not investigated in that report, the siRNAs with an efficient delivery system may enable us to directly clarify the role of specific transporters in transport via BBB and BCSFB in in vivo situations.

19.6. PREDICTION OF INTESTINAL TRANSPORT FROM IN VITRO DATA

The role of MDR1 in the intestinal absorption of drugs has long been discussed, and transcellular transport across the Caco-2 monolayer is frequently used to predict intestinal absorption. Some reports have shown that Caco-2 cells possess multiple drug transporters, whose expression levels are similar to those of the human intestine,[64,65] although some of them, such as BCRP and PEPT1, have been reported to be expressed to a lower degree than in the human intestine,[65] and Chong et al. have shown that substrates of dipeptide transporters (β-lactam antibiotics and ACE inhibitors) do not readily cross Caco-2 monolayers, in contrast to their complete absorption in vivo.[66] Moreover, it is known that the transcellular transport clearance of the same series of drugs across Caco-2 monolayers exhibits large interlaboratory differences, due to differences in cell culture conditions, passage numbers, and protocols.[67,68] Some reports have also shown that the expression level of transporters depends largely on the culture conditions.[67,68] Such evidence makes it difficult to predict quantitatively the in vivo human intestinal permeability from the transcellular transport of drugs in Caco-2 cells. Adachi et al. have measured the membrane permeability clearance determined by in situ intestinal perfusion using mdr1a/1b knockout and normal mice and the transcellular transport clearance in MDR1-expressing and parent LLC-PK1 cells.[69]

Based on pharmacokinetic theory, the in situ permeability clearance can be described as (Figure 19.7*c*)

$$\mathrm{PS}_{\text{in situ}} = \mathrm{PS}_{m,\text{inf}} \frac{\mathrm{PS}_{s,\text{eff}}}{\mathrm{PS}_{m,\text{eff}} + \mathrm{PS}_{s,\text{eff}} + \mathrm{PS}_{\text{Pgp}}} \tag{22}$$

Accordingly, the PS product ratio, defined as the ratio of the $\mathrm{PS}_{\text{in situ}}$ of mdr1a/1b knockout mice to that of normal mice, can be expressed as

$$\text{PS product ratio} = 1 + \frac{\mathrm{PS}_{\text{Pgp}}}{\mathrm{PS}_{m,\text{eff}} + \mathrm{PS}_{s,\text{eff}}} \tag{23}$$

On the other hand, the apical-to-basal transcellular transport clearance across the LLC-PK1 monolayer is given by (Figure 19.7*c*)

$$\mathrm{PS}_{a\text{-}to\text{-}b} = \mathrm{PS}_{a,\text{inf}} \frac{\mathrm{PS}_{b,\text{eff}}}{\mathrm{PS}_{a,\text{eff}} + \mathrm{PS}_{b,\text{eff}} + \mathrm{PS}_{\text{Pgp}}} \tag{24}$$

The $\mathrm{PS}_{a\text{-}to\text{-}b}$ ratio, defined as the ratio of the $\mathrm{PS}_{a\text{-}to\text{-}b}$ in parent LLC-PK1 cells to that in MDR1-expressing LLC-PK1 cells is given by

$$\mathrm{PS}_{a\text{-}to\text{-}b}\ \text{ratio} = 1 + \frac{\mathrm{PS}_{\text{Pgp}}}{\mathrm{PS}_{a,\text{eff}} + \mathrm{PS}_{b,\text{eff}}} \tag{25}$$

It has been shown that there is a clear correlation between the in situ PS product ratio and the in vitro $\mathrm{PS}_{a\text{-}to\text{-}b}$ ratio, suggesting that MDR1 function in the human small intestine can be estimated from in vitro transcellular transport studies using MDR1-expressing LLC-PK1 cells (Figure 19.7*d*). On the apical membrane, efflux transporters other than MDR1 are expressed, so even if vectorial apical-to-basal transport of compounds, is observed, they are not always a substrate of MDR1 and other transporters are involved in the backflux of compounds to the apical compartment. Watanabe et al. have constructed stable MDR1 knockdown Caco-2 cells to examine the specific impact of MDR1 on the transcellular transport of compounds, and symmetrical transcellular transport of several MDR1 substrates (digoxin, vincristine, rhodamine 123, and daunomycin) was observed in MDR1 knockdown Caco-2 cells. This suggests that MDR1 really plays an important role in restricting the membrane permeation of these ligands in Caco-2 monolayers.[70]

Several uptake and efflux transporters other than MDR1 are expressed in the human intestine. However, currently, the relative importance of each transporter in terms of intestinal absorption remains to be determined. For example, some reports have suggested that there are uptake transporters for organic anions such as fexofenadine in the intestine.[71,72] Kikuchi et al. have reported the saturation of the intestinal absorption of fexofenadine in rats using the Ussing-type chamber method in the presence of an MDR1 specific inhibitor, ketoconazole.[73] They suggested the possible involvement of Oatp3 due to the similar K_m values for fexofenadine transport obtained from the results of Oatp3-expressing *Xenopus* oocytes and the Ussing chamber, although no direct evidence has yet been found.

19.7. PREDICTION OF DRUG–DRUG INTERACTIONS FROM IN VITRO DATA

Clinical examples in which transporter-mediated drug–drug interactions are involved are being reported increasingly.[4,74] To avoid a critical clinical drug interaction in advance of launching a drug it is necessary to predict in vivo drug–drug interactions from in vitro experiments in a quantitative manner. Based on pharmacokinetic concepts, when an inhibitor blocks the transporter-mediated transport of a substrate, the intrinsic transport clearance (PS_{int}) in the presence of the inhibitor is given by equation (26), assuming that the inhibitor competitively or noncompetitively inhibits the transport of the substrate and that the substrate concentration is much lower than the K_m value:

$$PS_{int} = \frac{V_{max}}{K_m(1 + I_u/K_i)} \tag{26}$$

where V_{max}, K_m, I_u, and K_i represent the maximum transport activity and Michaelis constant for a substrate, and the protein unbound concentration and inhibition constant for an inhibitor, respectively. Then the degree of inhibition (R) is defined as

$$R = \frac{PS_{int}(+\text{inhibitor})}{PS_{int}(-\text{inhibitor})} = \frac{1}{1 + I_u/K_i} \tag{27}$$

Due to the broad substrate specificities of transporters, a compound is often recognized by multiple transporters. When these transporters function in parallel, if an inhibitor can inhibit the multiple transporters with different inhibition potencies, the net degree of inhibition (R) is given by

$$R = \frac{\sum_n \mathrm{PS}_{\mathrm{int},n}(+I)}{\sum_n \mathrm{PS}_{\mathrm{int},n}(-I)} = \sum_n R_n \frac{\mathrm{PS}_{\mathrm{int},n}(-I)}{\sum_n \mathrm{PS}_{\mathrm{int},n}(-I)} = \sum_n R_n f_n = \sum_n \frac{f_n}{1 + I_u/K_{i,n}} \tag{28}$$

where $\mathrm{PS}_{\mathrm{int},n}(+I)$, $\mathrm{PS}_{\mathrm{int},n}(-I)$, R_n, and f_n represent the intrinsic membrane transport clearance mediated by transporter n in the presence and absence of an inhibitor, the R value for transporter n, and the contribution of transporter n to the overall transport clearance, respectively. Therefore, considering this equation, when we predict a clinical drug–drug interaction, we must obtain information about the contribution of each transporter to the overall membrane transport (f_n), the protein unbound concentration of the inhibitor (I_u), and its inhibition constant for each transporter ($K_{i,n}$). Also, f_n can be obtained by several methods described in Section 19.3 and $K_{i,n}$ can be estimated from an in vitro inhibition experiment using suitable gene-expression systems.

On the other hand, it is fairly difficult to estimate the exact I_u value because I_u is defined as the unbound concentration at the capillaries and inside the cells for estimating drug–drug interaction involving uptake and efflux processes, respectively, and it cannot be measured directly in vivo. Therefore, to avoid any false-negative predictions, especially during the early phase of drug development, when an inhibitor is administered by the intravenous route, the maximum concentration in the circulating blood is generally used as the I_u value. In the case of oral administration of an inhibitor, it is possible that the inhibitor concentration at the inlet to the liver is higher than the maximum concentration in the circulating blood. Ito et al. have proposed an equation for estimating the unbound maximum concentration at the inlet to the liver[75]:

$$I_{u,\mathrm{in,max}} = f_u \left(I_{\mathrm{max}} + \frac{k_a D F_a}{Q_h} \right) \tag{29}$$

where f_u, I_{max}, k_a, D, F_a, and Q_h represent the blood protein unbound fraction, the maximum concentration in the circulating blood, the absorption rate constant, the dose, and fraction absorbed into intestinal cells and the hepatic blood flow, respectively. To consider the maximum concentration of an inhibitor, k_a and F_a values are usually set at 0.1 per minute (maximum gastric emptying time = 10 minutes) and 1, respectively. In interpretating the results obtained from this approach, we need to keep in mind that even if the R value for the combination of drugs is calculated to be less than 1, a drug–drug interaction does not always occur. On the other hand, if the R value is nearly equal to 1, we can disregard a drug–drug interaction for that combination of drugs.

Shitara et al. have investigated the mechanism of the drug–drug interaction between cerivastatin and cyclosporin A by using several in vitro experimental systems.[76] They have shown that the K_i values of cerivastatin for the uptake of cerivastatin both in human hepatocytes and OATP1B1-expressing MDCKII cells are less than 1 μM, while the metabolism of cerivastatin in pooled human microsomes is inhibited by cerivastatin with K_i values of more than 30 μM. Considering that the maximum plasma unbound concentration of cerivastatin in clinical situations is around 0.1 μM, it is still possible that cyclosporin A inhibits the hepatic uptake of cerivastatin via OATP1B1 because after oral administration of cyclosporin A, the maximum concentration at the inlet to the liver is much higher than that in the circulating blood, as discussed above. The clinical reports showing that coadministration of cyclosporin A causes an increase in the plasma concentration of OATP substrates such as statins, repaglinide, and bosentan also support the inhibitory effect of cyclosporin A on OATP transporters.[2,77]

To identify the clinically relevant drug–drug interaction for pitavastatin from in vitro experiments, Hirano et al. have measured the inhibitory effects of several drugs on the OATP1B1-mediated transport of pitavastatin by using OATP1B1-expressing HEK293 cells, and they estimated the ratio of the uptake clearance in the absence of inhibitor to that in its presence (I/R).[19] As a result, several drugs, with the ratios greater than 2.5, have the potential to interfere with the OATP1B1-mediated uptake of pitavastatin. Campbell et al. have examined the inhibition potencies of several drugs on the OATP1B1-mediated uptake to screen candidate drugs producing drug-induced hyperbilirubinemia.[78] They measured the K_i values for OATP1B1-mediated $E_2 17\beta G$ uptake and compared the clinical maximum unbound concentration in the blood with the K_i values. As a result, indinavir, cyclosporin A, and rifamycin SV inhibited human OATP1B1 at therapeutically relevant concentrations and had the potential to induce hyperbilirubinemia, whereas saquinavir, which is also a potent inhibitor of OATP1B1, failed to inhibit OATP1B1 in clinical situations. These results are in good agreement with the clinical reports of drug-induced hyperbilirubinemia.

Transporter-mediated drug–drug interactions can occur during biliary excretion as well as during hepatic uptake. Horikawa et al. have searched for drugs that can be used as inhibitors of MRP2 in clinical situations using the same approach and assuming that the protein unbound concentration inside the cells is equal to that in the circulating plasma.[79] They found that some drugs (probenecid, sulfobromophthalein, and cefodizime) had relatively smaller R values, but considering that the plasma half-life of sulfobromophthalein and cefodizime is shorter than that of probenecid and that probenecid is an orally administered drug, they finally proposed that probenecid is a candidate inhibitor of MRP2 which can be used in clinical situations. They also investigated the potential cholestatic activity of drugs in the same way and concluded that the majority of cholestasis-inducing drugs have only a minimal inhibitory effect on rat BSEP and MRP2.[80]

Ueda et al. have established a quantitative prediction method for examining alterations in the pharmacokinetics of drugs caused by inhibition of uptake as well as efflux.[81] They tried to predict the drug–drug interaction between methotrexate and probenecid. In this strategy, the inhibitory effect of probenecid on the hepatic uptake of methotrexate was evaluated using isolated rat hepatocytes, and the effects on its biliary excretion were examined using bile canalicular membrane vesicles (CMVs). The degree of inhibition of the uptake and efflux in vivo was comparable with that predicted from in vitro experiments. This allowed the R values for uptake (R_{uptake}) and efflux

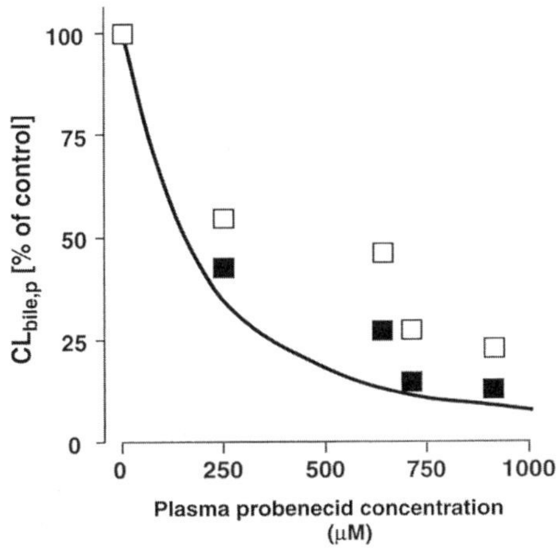

FIGURE 19.8. Extrapolation of the drug interaction between methotrexate and probenecid involving both hepatic uptake and biliary excretion from in vitro data. The y-axis represents the biliary excretion clearance of methotrexate in the presence of several concentrations of probenecid with respect to the circulating plasma. The open squares show the $CL_{bile,p}$ values of methotrexate observed in vivo. The closed squares and solid line represent the predicted $CL_{bile,p}$ values derived from the intrinsic biliary excretion clearance estimated using the unbound concentration of inhibitor in the liver and plasma, respectively. (From ref. 81, with the kind permission of the American Society for Pharmacology and Experimental Therapeutics.)

($R_{excretion}$) to be calculated. The net degree of inhibition (R_{net}) can be described as

$$R_{net} \leq R_{uptake} R_{excretion} \tag{30}$$

They showed that the degree of reduction in the hepatic clearance was overestimated by a simple calculation of the product of the reduction in the hepatic uptake and biliary excretion [equation (30)]. This method is useful for avoiding false-negative predictions (Figure 19.8).

The transporters in kidney are also main targets of drug–drug interactions. For example, coadministration of probenecid with famotidine results in a reduction in the renal clearance of famotidine in humans but not in rats.[82,83] To clarify this interspecies difference, Tahara et al. examined the effect of probenecid on the uptake of famotidine mediated by renal transporters.[84] They showed that the transport activity of famotidine in human OAT3 was relatively greater than that in rat Oat3 and that the renal uptake of organic cations in humans is accounted for solely by human OCT2, which transports famotidine to a lesser degree than does rat Oct1. In vitro inhibition studies suggested that probenecid could not inhibit the OCT-mediated uptake. Therefore, they proposed that in the case of rats, the relative contribution of Octs to the overall renal uptake of famotidine is thought to be high and inhibition of Oat3-mediated uptake by probenecid may not have a significant effect on the renal clearance of famotidine, whereas in the case of humans, the relative contribution of OAT3 is high compared with that in rats,

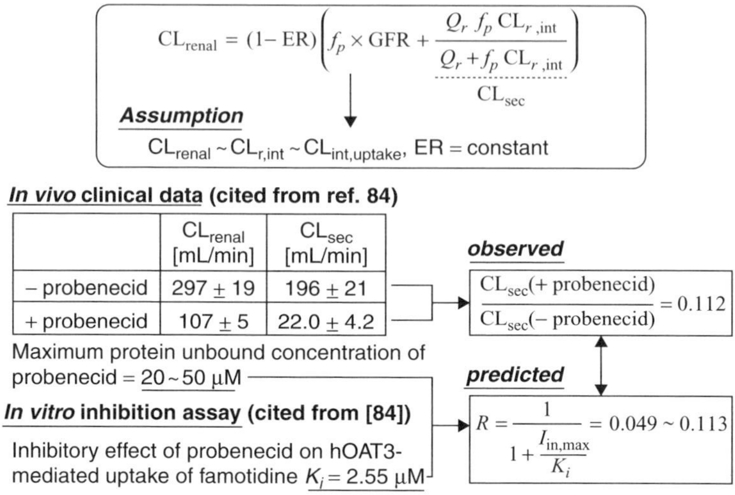

FIGURE 19.9. Prediction of the drug–drug interaction between famotidine and probenecid in humans from in vitro data. Renal clearance (CL_{renal}) is described by this equation. ER, reabsorption ratio; Q_r, renal blood flow; f_p, plasma protein unbound fraction; $CL_{r,int}$, intrinsic secretion clearance; GFR, glomerular filtration rate. Based on the pharmacokinetic parameters, renal clearance can be approximated by $CL_{r,int}$ in this case, and assuming that renal uptake is the rate-determining process of overall renal secretion, $CL_{r,int}$ is thought to be equal to intrinsic uptake clearance ($CL_{int,\ uptake}$). The extent of decrease in the renal secretion clearance in the clinical study was well predicted from the information of the maximum protein unbound concentration of probenecid in the clinical study and the K_i value of probenecid for the human OAT3-mediated uptake of famotidine. (From ref. 84.)

and the inhibition of OAT3-mediated uptake of famotidine by probenecid causes a drug–drug interaction (Figure 19.9).

Tahara et al. also examined the interspecies differences in the transport activities of OAT1 and OAT3 in rats, monkeys, and humans using gene expression systems.[85] In the case of OAT1-mediated uptake, a good correlation of the relative uptake clearances of 11 compounds was observed between humans and monkeys and between humans and rats. On the other hand, the relative clearances of nine substrates mediated by human OAT3 did not correlate well with those mediated by rat Oat3. They also showed that the tubular secretion clearance of famotidine was reduced to one-tenth when coadministered with probenecid, and the drug–drug interaction between famotidine and probenecid was observed in both monkeys and humans.[86] Moreover, OCT1 is not expressed in monkey kidney. These results suggest that the monkey is a more appropriate animal model for predicting OAT3-mediated drug–drug interactions.

Nozaki et al. have examined the contribution of transporters [Oat1, Oat3, and reduced folate carrier (RFC)-1] to the overall renal uptake of methotrexate and the inhibitory effects of several nonsteroidal anti-inflammatory drugs (NSAIDs) on the transporter-mediated uptake of methotrexate to investigate the mechanism of the drug–drug interaction between methotrexate and NSAIDs.[87] The uptake of methotrexate was partly inhibited by PCG (Oat3 inhibitor) and folate (RFC-1 inhibitor) but not PAH (Oat1 inhibitor), suggesting that Oat3 and RFC-1 contribute equally to methotrexate uptake in the rat kidney. Also, the inhibitory effects of NSAIDs on the Oat3-mediated uptake of methotrexate were more potent than the effects on Oat1 and RFC-1, and they discussed how Oat3-mediated methotrexate uptake could be inhibited by salicylate, indomethacin, and phenylbutazone in clinical situations, whereas other NSAIDs, which were reported to cause severe drug–drug interactions, were unlikely to inhibit the renal uptake of methotrexate.

Human-derived organ samples are now available for the evaluation of drug transport in humans. However, there are few reports of the prediction of the in vivo pharmacokinetics from in vitro data, and more examples are needed as well as validation of the prediction methods for drug–drug interactions using human samples.

19.8. CONCLUSIONS

We have discussed some examples of the prediction of in vivo pharmacokinetics from in vitro experimental results. Recently, several clinical reports have demonstrated that many factors, such as genetic polymorphisms, drug–drug interactions, and pathophysiological conditions, affect the transport activities of multiple transporters and alter in vivo pharmacokinetics and subsequent pharmacological effects. In particular, in humans, it is impossible to measure the concentration in each tissue and the site of drug action. To predict the time-dependent whole-body pharmacokinetics in vivo, physiologically based pharmacokinetic (PBPK) models are powerful tools.[88] Integrating the kinetic parameters for each transporter and metabolic enzyme and physiological parameters such as blood flow rate and protein unbound fraction into PBPK models will enable us to simulate the time-dependent concentrations in plasma and tissues. Once we have constructed a good PBPK model, we can easily understand the importance of each enzyme and transporter by simulation using modified kinetic parameters for each molecule (called *sensitivity analysis*).

The methodology for constructing an accurate PBPK model needs to be investigated in detail to provide information about transporters and metabolic enzymes at molecular levels to help with drug development and individualized drug therapy. Using positron emission tomography (PET) or single photon emission computed tomography (SPECT), we can now obtain detailed data on tissue distribution directly in humans. Some reports have even succeeded in quantifying drug concentrations in the human brain.[89,90] In future, by combining these data and a variety of in vitro data, we should be able to construct more accurate pharmacokinetic models for humans.

REFERENCES

1. Ho, R.H. and Kim, R.B. (2005). Transporters and drug therapy: implications for drug disposition and disease. Clin Pharmacol Ther 78, 260–277.
2. Shitara, Y., Horie, T. and Sugiyama, Y. (2006). Transporters as a determinant of drug clearance and tissue distribution. Eur J Pharm Sci 27, 425–446.

3. Giacomini, K.M. and Sugiyama, Y. (2006). Membrane transporters and drug response. In: *Goodman & Gilman's The Pharmacological Basis of Therapeutics,* 11th ed., ed. by Brunton, L.L., McGraw-Hill, New York, pp. 41–70.
4. Kusuhara, H. and Sugiyama, Y. (2001). Drug–drug interactions involving the membrane transport process. In: *Drug–Drug-Interactions*, ed. by Rodrigues, A.D., Marcel Dekker, Basel, Switzerland, pp. 123–188.
5. Yamazaki, M., Suzuki, H., Hanano, M. and Sugiyama, Y. (1993). Different relationships between cellular ATP and hepatic uptake among taurocholate, cholate, and organic anions. Am J Physiol 264, G693–G701.
6. Shitara, Y., Li, A.P., Kato, Y., Lu, C., Ito, K., Itoh, T. and Sugiyama, Y. (2003). Function of uptake transporters for taurocholate and estradiol 17beta-D-glucuronide in cryopreserved human hepatocytes. Drug Metab Pharmacokinet 18, 33–41.
7. Hirano, M., Maeda, K., Shitara, Y. and Sugiyama, Y. (2004). Contribution of OATP2 (OATP1B1) and OATP8 (OATP1B3) to the hepatic uptake of pitavastatin in humans. J Pharmacol Exp Ther 311, 139–146.
8. Jigorel, E., Le Vee, M., Boursier-Neyret, C., Bertrand, M. and Fardel, O. (2005). Functional expression of sinusoidal drug transporters in primary human and rat hepatocytes. Drug Metab Dispos 33, 1418–1422.
9. Ishigami, M., Tokui, T., Komai, T., Tsukahara, K., Yamazaki, M. and Sugiyama, Y. (1995). Evaluation of the uptake of pravastatin by perfused rat liver and primary cultured rat hepatocytes. Pharm Res 12, 1741–1745.
10. Rippin, S.J., Hagenbuch, B., Meier, P.J. and Stieger, B. (2001). Cholestatic expression pattern of sinusoidal and canalicular organic anion transport systems in primary cultured rat hepatocytes. Hepatology 33, 776–782.
11. Miyauchi, S., Sawada, Y., Iga, T., Hanano, M. and Sugiyama, Y. (1993). Comparison of the hepatic uptake clearances of fifteen drugs with a wide range of membrane permeabilities in isolated rat hepatocytes and perfused rat livers. Pharm Res 10, 434–440.
12. Kato, Y., Akhteruzzaman, S., Hisaka, A. and Sugiyama, Y. (1999). Hepatobiliary transport governs overall elimination of peptidic endothelin antagonists in rats. J Pharmacol Exp Ther 288, 568–574.
13. Kouzuki, H., Suzuki, H., Ito, K., Ohashi, R. and Sugiyama, Y. (1998). Contribution of sodium taurocholate co-transporting polypeptide to the uptake of its possible substrates into rat hepatocytes. J Pharmacol Exp Ther 286, 1043–1050.
14. Kouzuki, H., Suzuki, H., Ito, K., Ohashi, R. and Sugiyama, Y. (1999). Contribution of organic anion transporting polypeptide to uptake of its possible substrates into rat hepatocytes. J Pharmacol Exp Ther 288, 627–634.
15. Crespi, C.L. and Penman, B.W. (1997). Use of cDNA-expressed human cytochrome P450 enzymes to study potential drug–drug interactions. Adv Pharmacol 43, 171–188.
16. Sekine, T., Cha, S.H., Tsuda, M., Apiwattanakul, N., Nakajima, N., Kanai, Y. and Endou, H. (1998). Identification of multispecific organic anion transporter 2 expressed predominantly in the liver. FEBS Lett 429, 179–182.
17. Noé, B., Hagenbuch, B., Stieger, B. and Meier, P.J. (1997). Isolation of a multispecific organic anion and cardiac glycoside transporter from rat brain. Proc Natl Acad Sci USA 94, 10346–10350.
18. Cattori, V., Hagenbuch, B., Hagenbuch, N., Stieger, B., Ha, R., Winterhalter, K.E. and Meier, P.J. (2000). Identification of organic anion transporting polypeptide 4 (Oatp4) as a

major full-length isoform of the liver-specific transporter-1 (rlst-1) in rat liver. FEBS Lett 474, 242–245.

19. Hirano, M., Maeda, K., Shitara, Y. and Sugiyama, Y. (2006). Drug–drug interaction between pitavastatin and various drugs via OATP1B1. Drug Metab Dispos 34, 1229–1236.
20. Shimizu, M., Fuse, K., Okudaira, K., Nishigaki, R., Maeda, K., Kusuhara, H. and Sugiyama, Y. (2005). Contribution of OATP (organic anion–transporting polypeptide) family transporters to the hepatic uptake of fexofenadine in humans. Drug Metab Dispos 33, 1477–1481.
21. Ishiguro, N., Maeda, K., Kishimoto, W., Saito, A., Harada, A., Ebner, T., Roth, W., Igarashi, T. and Sugiyama, Y. (2006). Predominant contribution of OATP1B3 to the hepatic uptake of telmisartan, an angiotensin II receptor antagonist, in humans. Drug Metab Dispos 34, 1109–1115.
22. Yamashiro, W., Maeda, K., Hirouchi, M., Adachi, Y., Hu, Z. and Sugiyama, Y. (2006). Involvement of transporters in the hepatic uptake and biliary excretion of valsartan, a selective antagonist of the angiotensin II AT1-receptor, in humans. Drug Metab Dispos 34, 1247–1254.
23. Hagenbuch, B., Scharschmidt, B.F. and Meier, P.J. (1996). Effect of antisense oligonucleotides on the expression of hepatocellular bile acid and organic anion uptake systems in *Xenopus laevis* oocytes. Biochem J 316(Pt 3), 901–904.
24. Nakai, D., Nakagomi, R., Furuta, Y., Tokui, T., Abe, T., Ikeda, T. and Nishimura, K. (2001). Human liver-specific organic anion transporter, LST-1, mediates uptake of pravastatin by human hepatocytes. J Pharmacol Exp Ther 297, 861–867.
25. Takagi, M., Morita, K., Nakai, D., Nakagomi, R., Tokui, T. and Koizumi, M. (2004). Enhancement of the inhibitory activity of oatp antisense oligonucleotides by incorporation of 2′-*O*,4′-*C*-ethylene-bridged nucleic acids (ENA) without a loss of subtype selectivity. Biochemistry 43, 4501–4510.
26. Materna, V., Stege, A., Surowiak, P., Priebsch, A. and Lage, H. (2006). RNA interference-triggered reversal of ABCC2-dependent cisplatin resistance in human cancer cells. Biochem Biophys Res Commun 348, 153–157.
27. Nieth, C., Priebsch, A., Stege, A. and Lage, H. (2003). Modulation of the classical multidrug resistance (MDR) phenotype by RNA interference (RNAi). FEBS Lett 545, 144–150.
28. Aoki, J., Suzuki, H. and Sugiyama, Y. (2000). Quantitative prediction of in vivo biliary excretion clearance across the bile canalicular membrane from in vitro transport studies with isolated membrane vesicles. In: *Proceedings of the Millennial World Congress of Pharmaceutical Sciences*, April 16–20, San Francisco, CA, p. 92.
29. Ghibellini, G., Leslie, E.M. and Brouwer, K.L. (2006). Methods to evaluate biliary excretion of drugs in humans: an updated review. Mol Pharmacol 3, 198–211.
30. Niinuma, K., Kato, Y., Suzuki, H., Tyson, C.A., Weizer, V., Dabbs, J.E., Froehlich, R., Green, C.E. and Sugiyama, Y. (1999). Primary active transport of organic anions on bile canalicular membrane in humans. Am J Physiol 276, G1153–G1164.
31. Shilling, A.D., Azam, F., Kao, J. and Leung, L. (2006). Use of canalicular membrane vesicles (CMVs) from rats, dogs, monkeys and humans to assess drug transport across the canalicular membrane. J Pharmacol Toxicol Methods 53, 186–197.
32. Allen, J.D., van Loevezijn, A., Lakhai, J.M., van der Valk, M., van Tellingen, O., Reid, G., Schellens, J.H., Koomen, G.J. and Schinkel, A.H. (2002). Potent and specific inhibition of

the breast cancer resistance protein multidrug transporter in vitro and in mouse intestine by a novel analogue of fumitremorgin C. Mol Cancer Ther 1, 417–425.

33. Dantzig, A.H., Shepard, R.L., Cao, J., Law, K.L., Ehlhardt, W.J., Baughman, T.M., Bumol, T.F. and Starling, J.J. (1996). Reversal of P-glycoprotein-mediated multidrug resistance by a potent cyclopropyldibenzosuberane modulator, LY335979. Cancer Res 56, 4171–4179.

34. Kusunoki, N., Takara, K., Tanigawara, Y., Yamauchi, A., Ueda, K., Komada, F., Ku, Y., Kuroda, Y., Saitoh, Y. and Okumura, K. (1998). Inhibitory effects of a cyclosporin derivative, SDZ PSC 833, on transport of doxorubicin and vinblastine via human P-glycoprotein. Jpn J Cancer Res 89, 1220–1228.

35. LeCluyse, E.L., Audus, K.L. and Hochman, J.H. (1994). Formation of extensive canalicular networks by rat hepatocytes cultured in collagen-sandwich configuration. Am J Physiol 266, C1764–C1774.

36. Liu, X., LeCluyse, E.L., Brouwer, K.R., Lightfoot, R.M., Lee, J.I. and Brouwer, K.L. (1999). Use of Ca^{2+} modulation to evaluate biliary excretion in sandwich-cultured rat hepatocytes. J Pharmacol Exp Ther 289, 1592–1599.

37. Liu, X., Chism, J.P., LeCluyse, E.L., Brouwer, K.R. and Brouwer, K.L. (1999). Correlation of biliary excretion in sandwich-cultured rat hepatocytes and in vivo in rats. Drug Metab Dispos 27, 637–644.

38. Hoffmaster, K.A., Turncliff, R.Z., LeCluyse, E.L., Kim, R.B., Meier, P.J. and Brouwer, K.L. (2004). P-Glycoprotein expression, localization, and function in sandwich-cultured primary rat and human hepatocytes: relevance to the hepatobiliary disposition of a model opioid peptide. Pharm Res 21, 1294–1302.

39. Tian, X., Zamek-Gliszczynski, M.J., Zhang, P. and Brouwer, K.L. (2004). Modulation of multidrug resistance-associated protein 2 (Mrp2) and Mrp3 expression and function with small interfering RNA in sandwich-cultured rat hepatocytes. Mol Pharmacol 66, 1004–1010.

40. Bi, Y.A., Kazolias, D. and Duignan, D.B. (2006). Use of cryopreserved human hepatocytes in sandwich culture to measure hepatobiliary transport. Drug Metab Dispos 34, 1658–1665.

41. Cui, Y., Konig, J. and Keppler, D. (2001). Vectorial transport by double-transfected cells expressing the human uptake transporter SLC21A8 and the apical export pump ABCC2. Mol Pharmacol 60, 934–943.

42. Sasaki, M., Suzuki, H., Ito, K., Abe, T. and Sugiyama, Y. (2002). Transcellular transport of organic anions across a double-transfected Madin–Darby canine kidney II cell monolayer expressing both human organic anion–transporting polypeptide (OATP2/SLC21A6) and multidrug resistance–associated protein 2 (MRP2/ABCC2). J Biol Chem 277, 6497–6503.

43. Sasaki, M., Suzuki, H., Aoki, J., Ito, K., Meier, P.J. and Sugiyama, Y. (2004). Prediction of in vivo biliary clearance from the in vitro transcellular transport of organic anions across a double-transfected Madin–Darby canine kidney II monolayer expressing both rat organic anion transporting polypeptide 4 and multidrug resistance associated protein 2. Mol Pharmacol 66, 450–459.

44. Kopplow, K., Letschert, K., König, J., Walter, B. and Keppler, D. (2005). Human hepatobiliary transport of organic anions analyzed by quadruple-transfected cells. Mol Pharmacol 68, 1031–1038.

45. Spears, K.J., Ross, J., Stenhouse, A., Ward, C.J., Goh, L.B., Wolf, C.R., Morgan, P., Ayrton, A. and Friedberg, T.H. (2005). Directional trans-epithelial transport of organic anions in

porcine LLC-PK1 cells that co-express human OATP1B1 (OATP-C) and MRP2. Biochem Pharmacol 69, 415–423.

46. Mita, S., Suzuki, H., Akita, H., Hayashi, H., Onuki, R., Hofmann, A.F. and Sugiyama, Y. (2006). Vectorial transport of unconjugated and conjugated bile salts by monolayers of LLC-PK1 cells doubly transfected with human NTCP and BSEP or with rat Ntcp and Bsep. Am J Physiol Gastrointest Liver Physiol 290, G550–G556.

47. Matsushima, S., Maeda, K., Kondo, C., Hirano, M., Sasaki, M., Suzuki, H. and Sugiyama, Y. (2005). Identification of the hepatic efflux transporters of organic anions using double-transfected Madin–Darby canine kidney II cells expressing human organic anion-transporting polypeptide 1B1 (OATP1B1)/multidrug resistance-associated protein 2, OATP1B1/multidrug resistance 1, and OATP1B1/breast cancer resistance protein. J Pharmacol Exp Ther 314, 1059–1067.

48. Hasegawa, M., Kusuhara, H., Sugiyama, D., Ito, K., Ueda, S., Endou, H. and Sugiyama, Y. (2002). Functional involvement of rat organic anion transporter 3 (rOat3; Slc22a8) in the renal uptake of organic anions. J Pharmacol Exp Ther 300, 746–753.

49. Hasegawa, M., Kusuhara, H., Endou, H. and Sugiyama, Y. (2003). Contribution of organic anion transporters to the renal uptake of anionic compounds and nucleoside derivatives in rat. J Pharmacol Exp Ther 305, 1087–1097.

50. Deguchi, T., Kusuhara, H., Takadate, A., Endou, H., Otagiri, M. and Sugiyama, Y. (2004). Characterization of uremic toxin transport by organic anion transporters in the kidney. Kidney Int 65, 162–174.

51. Han, Y.H., Kato, Y. and Sugiyama, Y. (1999). Binding and transport of methotrexate and its derivative, MX-68, across the brush-border membrane in rat kidney. Biopharm Drug Dispos 20, 361–367.

52. Koren, G., Klein, J., Bentur, Y. and Giesbrecht, E. (1989). The effects of mannitol diuresis on digoxin and phenobarbital handling by the kidney: implications for tubular reabsorption and secretion of the cardiac glycoside. Clin Invest Med 12, 279–284.

53. Nicolazzo, J.A., Charman, S.A. and Charman, W.N. (2006). Methods to assess drug permeability across the blood–brain barrier. J Pharm Pharmacol 58, 281–293.

54. Adachi, Y., Suzuki, H. and Sugiyama, Y. (2001). Comparative studies on in vitro methods for evaluating in vivo function of MDR1 P-glycoprotein. Pharm Res 18, 1660–1668.

55. Yamazaki, M., Neway, W.E., Ohe, T., Chen, I., Rowe, J.F., Hochman, J.H., Chiba, M. and Lin, J.H. (2001). In vitro substrate identification studies for P-glycoprotein-mediated transport: species difference and predictability of in vivo results. J Pharmacol Exp Ther 296, 723–735.

56. Sugiyama, D., Kusuhara, H., Shitara, Y., Abe, T., Meier, P.J., Sekine, T., Endou, H., Suzuki, H. and Sugiyama, Y. (2001). Characterization of the efflux transport of 17beta-estradiol-D-17beta-glucuronide from the brain across the blood–brain barrier. J Pharmacol Exp Ther 298, 316–322.

57. Kikuchi, R., Kusuhara, H., Abe, T., Endou, H. and Sugiyama, Y. (2004). Involvement of multiple transporters in the efflux of 3-hydroxy-3-methylglutaryl-CoA reductase inhibitors across the blood–brain barrier. J Pharmacol Exp Ther 311, 1147–1153.

58. Nagata, Y., Kusuhara, H., Endou, H. and Sugiyama, Y. (2002). Expression and functional characterization of rat organic anion transporter 3 (rOat3) in the choroid plexus. Mol Pharmacol 61, 982–988.

59. Kuroda, M., Kusuhara, H., Endou, H. and Sugiyama, Y. (2005). Rapid elimination of cefaclor from the cerebrospinal fluid is mediated by a benzylpenicillin-sensitive mechanism distinct from organic anion transporter 3. J Pharmacol Exp Ther 314, 855–861.

60. Hosoya, K., Tetsuka, K., Nagase, K., Tomi, M., Saeki, S., Ohtsuki, S., Takanaga, H., Yanai, N., Obinata, M., Kikuchi, A., Okano, T. and Terasaki, T. (2000). Conditionally immortalized brain capillary endothelial cell lines established from a transgenic mouse harboring temperature-sensitive simian virus 40 large T-antigen gene. AAPS PharmSci 2, E27.

61. Hosoya, K.I., Takashima, T., Tetsuka, K., Nagura, T., Ohtsuki, S., Takanaga, H., Ueda, M., Yanai, N., Obinata, M. and Terasaki, T. (2000). mRna expression and transport characterization of conditionally immortalized rat brain capillary endothelial cell lines; a new in vitro BBB model for drug targeting. J Drug Target 8, 357–370.

62. Terasaki, T., Ohtsuki, S., Hori, S., Takanaga, H., Nakashima, E. and Hosoya, K. (2003). New approaches to in vitro models of blood–brain barrier drug transport. Drug Discov Today 8, 944–954.

63. Hino, T., Yokota, T., Ito, S., Nishina, K., Kang, Y.S., Mori, S., Hori, S., Kanda, T., Terasaki, T. and Mizusawa, H. (2006). In vivo delivery of small interfering RNA targeting brain capillary endothelial cells. Biochem Biophys Res Commun 340, 263–267.

64. Taipalensuu, J., Tornblom, H., Lindberg, G., Einarsson, C., Sjoqvist, F., Melhus, H., Garberg, P., Sjostrom, B., Lundgren, B. and Artursson, P. (2001). Correlation of gene expression of ten drug efflux proteins of the ATP-binding cassette transporter family in normal human jejunum and in human intestinal epithelial Caco-2 cell monolayers. J Pharmacol Exp Ther 299, 164–170.

65. Englund, G., Rorsman, F., Ronnblom, A., Karlbom, U., Lazorova, L., Grasjo, J., Kindmark, A. and Artursson, P. (2006). Regional levels of drug transporters along the human intestinal tract: Co-expression of ABC and SLC transporters and comparison with Caco-2 cells. Eur J Pharm Sci 29, 269–277.

66. Chong, S., Dando, S.A., Soucek, K.M. and Morrison, R.A. (1996). In vitro permeability through caco-2 cells is not quantitatively predictive of in vivo absorption for peptide-like drugs absorbed via the dipeptide transporter system. Pharm Res 13, 120–123.

67. Balimane, P.V., Han, Y.H. and Chong, S. (2006). Current industrial practices of assessing permeability and P-glycoprotein interaction. AAPS J 8, E1–E13.

68. Balimane, P.V. and Chong, S. (2005). Cell culture-based models for intestinal permeability: a critique. Drug Discov Today 10, 335–343.

69. Adachi, Y., Suzuki, H. and Sugiyama, Y. (2003). Quantitative evaluation of the function of small intestinal P-glycoprotein: comparative studies between in situ and in vitro. Pharm Res 20, 1163–1169.

70. Watanabe, T., Onuki, R., Yamashita, S., Taira, K. and Sugiyama, Y. (2005). Construction of a functional transporter analysis system using MDR1 knockdown Caco-2 cells. Pharm Res 22, 1287–1293.

71. Dresser, G.K., Bailey, D.G., Leake, B.F., Schwarz, U.I., Dawson, P.A., Freeman, D.J. and Kim, R.B. (2002). Fruit juices inhibit organic anion transporting polypeptide-mediated drug uptake to decrease the oral availability of fexofenadine. Clin Pharmacol Ther 71, 11–20.

72. Kamath, A.V., Yao, M., Zhang, Y. and Chong, S. (2005). Effect of fruit juices on the oral bioavailability of fexofenadine in rats. J Pharm Sci 94, 233–239.

73. Kikuchi, A., Nozawa, T., Wakasawa, T., Maeda, T. and Tamai, I. (2006). Transporter-mediated intestinal absorption of fexofenadine in rats. Drug Metab Pharmacokinet 21, 308–314.

74. Endres, C.J., Hsiao, P., Chung, F.S. and Unadkat, J.D. (2006). The role of transporters in drug interactions. Eur J Pharm Sci 27, 501–517.

75. Ito, K., Iwatsubo, T., Kanamitsu, S., Ueda, K., Suzuki, H. and Sugiyama, Y. (1998). Prediction of pharmacokinetic alterations caused by drug–drug interactions: metabolic interaction in the liver. Pharmacol Rev 50, 387–412.

76. Shitara, Y., Itoh, T., Sato, H., Li, A.P. and Sugiyama, Y. (2003). Inhibition of transporter-mediated hepatic uptake as a mechanism for drug–drug interaction between cerivastatin and cyclosporin A. J Pharmacol Exp Ther 304, 610–616.

77. Shitara, Y., Sato, H. and Sugiyama, Y. (2005). Evaluation of drug–drug interaction in the hepatobiliary and renal transport of drugs. Annu Rev Pharmacol Toxicol 45, 689–723.

78. Campbell, S.D., de Morais, S.M. and Xu, J.J. (2004). Inhibition of human organic anion transporting polypeptide OATP 1B1 as a mechanism of drug-induced hyperbilirubinemia. Chem Biol Interact 150, 179–187.

79. Horikawa, M., Kato, Y., Tyson, C.A. and Sugiyama, Y. (2002). The potential for an interaction between MRP2 (ABCC2) and various therapeutic agents: probenecid as a candidate inhibitor of the biliary excretion of irinotecan metabolites. Drug Metab Pharmacokinet 17, 23–33.

80. Horikawa, M., Kato, Y., Tyson, C.A. and Sugiyama, Y. (2003). Potential cholestatic activity of various therapeutic agents assessed by bile canalicular membrane vesicles isolated from rats and humans. Drug Metab Pharmacokinet 18, 16–22.

81. Ueda, K., Kato, Y., Komatsu, K. and Sugiyama, Y. (2001). Inhibition of biliary excretion of methotrexate by probenecid in rats: quantitative prediction of interaction from in vitro data. J Pharmacol Exp Ther 297, 1036–1043.

82. Inotsume, N., Nishimura, M., Nakano, M., Fujiyama, S. and Sato, T. (1990). The inhibitory effect of probenecid on renal excretion of famotidine in young, healthy volunteers. J Clin Pharmacol 30, 50–56.

83. Lin, J.H., Los, L.E., Ulm, E.H. and Duggan, D.E. (1988). Kinetic studies on the competition between famotidine and cimetidine in rats: evidence of multiple renal secretory systems for organic cations. Drug Metab Dispos 16, 52–56.

84. Tahara, H., Kusuhara, H., Endou, H., Koepsell, H., Imaoka, T., Fuse, E. and Sugiyama, Y. (2005). A species difference in the transport activities of H2 receptor antagonists by rat and human renal organic anion and cation transporters. J Pharmacol Exp Ther 315, 337–345.

85. Tahara, H., Shono, M., Kusuhara, H., Kinoshita, H., Fuse, E., Takadate, A., Otagiri, M. and Sugiyama, Y. (2005). Molecular cloning and functional analyses of OAT1 and OAT3 from cynomolgus monkey kidney. Pharm Res 22, 647–660.

86. Tahara, H., Kusuhara, H., Chida, M., Fuse, E. and Sugiyama, Y. (2006). Is the monkey an appropriate animal model to examine drug–drug interactions involving renal clearance? Effect of probenecid on the renal elimination of H2 receptor antagonists. J Pharmacol Exp Ther 316, 1187–1194.

87. Nozaki, Y., Kusuhara, H., Endou, H. and Sugiyama, Y. (2004). Quantitative evaluation of the drug–drug interactions between methotrexate and nonsteroidal anti-inflammatory drugs in the renal uptake process based on the contribution of organic anion transporters and reduced folate carrier. J Pharmacol Exp Ther 309, 226–234.

88. Shitara, Y. and Sugiyama, Y. (2006). Pharmacokinetic and pharmacodynamic alterations of 3-hydroxy-3-methylglutaryl coenzyme A (HMG-CoA) reductase inhibitors: drug–drug interactions and interindividual differences in transporter and metabolic enzyme functions. Pharmacol Ther 112, 71–105.

89. Takano, A., Kusuhara, H., Suhara, T., Ieiri, I., Morimoto, T., Lee, Y.J., Maeda, J., Ikoma, Y., Ito, H., Suzuki, K. and Sugiyama, Y. (2006). Evaluation of in vivo P-glycoprotein function at the blood–brain barrier among MDR1 gene polymorphisms by using ^{11}C-verapamil. J Nucl Med 47, 1427–1433.

90. Sasongko, L., Link, J.M., Muzi, M., Mankoff, D.A., Yang, X., Collier, A.C., Shoner, S.C. and Unadkat, J.D. (2005). Imaging P-glycoprotein transport activity at the human blood–brain barrier with positron emission tomography. Clin Pharmacol Ther 77, 503–514.

91. Nozaki, Y., Kusuhara, H., Kondo, T., Hasegawa, H., Shiroyanagi, Y., Nakazawa, H., Okano, T., and Sugiyama, Y. (2007). Characterization of the uptake of organic anion transporter (OAT)1 and OAT3 substrates by human kidney slices. J Pharm Exp Ther 321, 362–369.

20

AGE- AND GENDER-RELATED DIFFERENCES IN XENOBIOTIC TRANSPORTER EXPRESSION

CURTIS D. KLAASSEN AND XINGGUO CHENG
University of Kansas Medical Center, Kansas City, Kansas

Drug Transporters: Molecular Characterization and Role in Drug Disposition, Edited by Guofeng You and Marilyn E. Morris
Copyright © 2007 John Wiley & Sons, Inc.

20.1. INTRODUCTION

For decades it has been known that infants cannot simply be regarded as miniature adults when describing their response to drugs. However, the reasons for the differences between infants and adults have not been well delineated. Although variations in pharmacokinetics and pharmacodynamics are assumed to be responsible for producing these differences, their relative importance in producing these effects is not understood. The purpose of this review is to illustrate the differences in expression of drug transporters in young animals, as well as gender differences.

In laboratory animals the pharmacokinetics of numerous chemicals is different in newborns than in adults. In general, chemicals are eliminated from the body more slowly in young animals. In the 1960s it was determined that the expression of some drug-metabolizing enzymes is not fully expressed at birth, which is responsible for the delayed elimination of some xenobiotics from the body.[1,2]

In the mid-1980s, when P-glycoprotein (multidrug resistance protein, Mdr) was first cloned,[3,4] we began to understand the function of xenobiotic transporters in eliminating xenobiotics from the body. Engineering of the first transporter-null mice in the mid-1990s[5] enabled the scientific community to understand more thoroughly how transporters affect the pharmacokinetics of xenobiotics. It is now generally accepted that transporters play a critical role in the absorption, distribution, and elimination of xenobiotics. It thus follows that variations in transporter expression can lead to differences in pharmacokinetics, and therefore, differences observed in the age- and gender-specific pharmacokinetics of drugs may be explained by differences in the expression of xenobiotic transporters.

Mouse is the species in which the most thorough studies have been performed to examine the gender- and age-related differences in transporters. Therefore, we emphasize the data from mice, but data from other species are included where available. Whereas it would be desirable to quantify the transporter protein and function at various ages, the availability of antibodies and specific substrates for the various transporters are not available, and thus most of the data presented in this chapter relate to mRNA.

20.2. HEPATOBILIARY TRANSPORTERS

20.2.1. Hepatobiliary Excretion in Newborns

Many but not all drugs and other foreign compounds are more toxic in newborn than in adult animals, and some of these compounds, which are excreted primarily into bile, are more toxic in newborn than in adult rats. For example, the following ratios of the LD_{50} of drugs in adults over those in newborn rats demonstrate increased toxicity. Probenecid is 2.8 times more toxic in newborn than in adult rats; indocyanine green, 3.3; iopanoic acid, 3.8; digoxin, 3.8; digitoxin, 4.6; colchicine, 21; and ouabain, 40 times.[6,7] These observations suggest that the increased sensitivity of newborn rats to these chemicals may be due to immaturity of their hepatic excretory mechanisms.

Because ouabain is 40 times more toxic in newborn than in adult rats,[6,7] is not biotransformed prior to its excretion,[8,9] and is excreted from the body almost entirely into bile,[8] it is an ideal prototype to use to determine whether the hepatic excretory mechanism is immature in newborns.[6] The toxicity of ouabain decreased gradually in rats from 3 to 12 days of age, with a marked decrease in toxicity observed between 12 and 21 days of age. After 30 days of age, the toxicity of ouabain was relatively constant. Ouabain was removed very slowly from the plasma of 7-day-old rats, with a half-time of 30 minutes compared to a half-life in an adult of approximately 5 minutes. The reason for this longer half-life of ouabain in newborn rats is the inability of its liver to remove ouabain from the plasma. The concentration of ouabain in the liver of an adult rat reaches 50 times that of plasma, whereas the liver of newborns cannot concentrate ouabain at all.[6] Ouabain uptake into hepatocytes isolated from 12-day-old rats is much slower than hepatocytes isolated from adult rats.[10] The ability of liver to extract ouabain from plasma and concentrate it in liver develops concurrently with the decrease in toxicity. It appears that this inability of immature liver to extract ouabain from the plasma and to excrete it into bile results in a prolonged high concentration of ouabain in plasma, which is associated with a higher toxicity.[6]

Lower hepatic removal rates in newborn rats have also been demonstrated for organic acids. For example, sulfobromophthalein (BSP) and indocyanine green (ICG) are extracted from plasma and excreted into bile less efficiently in newborn than in adult rats.[11] This difference is not as striking as that observed with ouabain, but the time course of development of increasing clearance of the organic acids parallels that observed with ouabain. Decreased clearance of BSP in newborns does not appear to be due to a decreased conjugation of BSP, as similar results are obtained when the conjugate (BSP-glutathione) is administered.[11] It has also been demonstrated that the increased toxicity of colchicine in newborn rats is also due largely to the immaturity of hepatic excretory function.[12] Whereas the immaturity in hepatic excretion of ouabain in newborns appears to be in the transfer of ouabain from plasma to liver, for colchicine. it is the transfer from liver to bile.[12,13]

Biliary excretion is not mature in newborn rats,[7,14] dogs, rabbits,[14] and guinea pigs.[15] Indirect evidence indicates that newborn humans also have a decreased capacity to excrete foreign compounds into bile.[16]

20.2.2. Hepatic Uptake

There are a number of transporters in liver responsible for the uptake and efflux of chemicals (Figure 20.1). As noted in earlier chapters, the organic anion–transporting polypeptide (Oatp) is an important family of uptake transporters, consisting of 15 transporters in mice.[17] The uptake transporter expressed most specifically in liver is

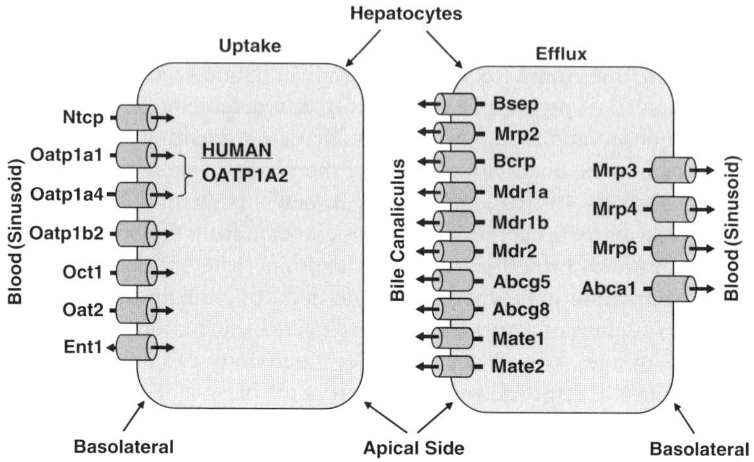

FIGURE 20.1. Cellular localization and transport direction of known transporters in liver. Uptake transporters from sinusoidal blood are depicted in the hepatocyte on the left. Efflux transporters in the sinusoidal (basolateral) or canalicular (apical) membrane are shown in the hepatocyte on the right.

Oatp1b2[17,18]; however, as noted in Figure 20.2, six Oatp transporters are expressed in mouse liver.

The developmental and gender differences in Oatp expression in mice are shown in Figure 20.2. Oatp1a1, 1a4, and 1a6 are rodent organic anion transporters.[17,19] There is only one human ortholog for these three transporters: OATP1A2. Expression is low at birth for all three transporters. Expression of Oatp1a1 reaches mature levels at 1 month of age, whereas Oatp1a4 and 1a6 reach adult levels earlier. The developmental expression of Oatp1b2 is extremely low before birth, increases abruptly at birth, and reaches adult levels at 3 weeks of age.[17] The ontogenic expression of Oatp1b2 in rats is similar to that in mice.[20] In mice, most of the other Oatps have a developmental pattern similar to that of Oatp1b2, with the exception of Oatp2a1, also known as the prostaglandin transporter.[21] Mouse Oatp2a1 is expressed consistently at similar levels from 2 days before birth to adulthood.[17] Figure 20.2 indicates that gender differences exist in the expression of some Oatps in mouse liver: male-predominant expression of Oatp1a1 and a female-predominant expression of Oatp1a4. The gender-divergent expression of the Oatps in mouse liver becomes apparent at approximately 1 month of age.

The equilibrative nucleoside transporter 1 (Ent1) mRNA is expressed at similar levels at all ages in mouse livers,[22] except that its expression is highest at birth. In general, organic anion transporters (Oats) are expressed primarily in kidney; however, Oat2 is also expressed in liver,[23] with an unusual ontogenic pattern, with very low expression at birth, reaching peak expression at day 15 and decreasing thereafter to adult levels. In rat livers, Oat2 expression also occurs after birth.[24] Organic cation transporter 1 (Oct1) increases slowly to mature levels over the first 3 weeks of life.[25]

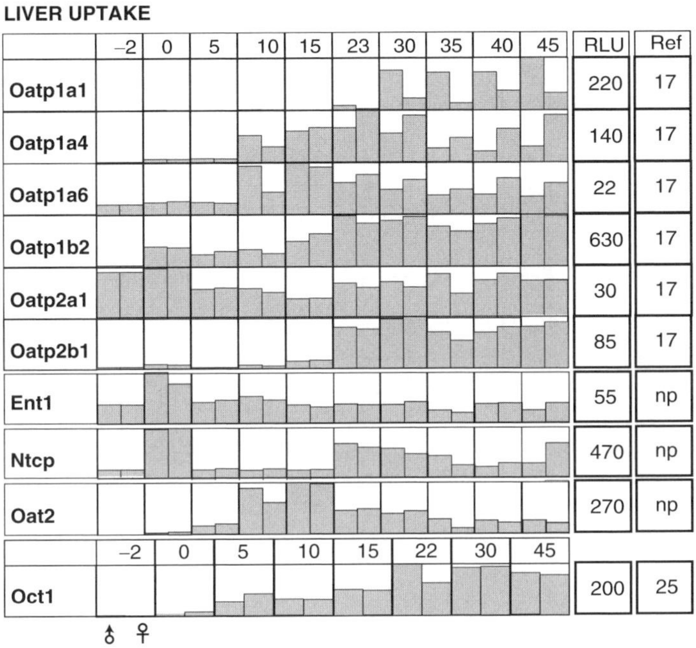

FIGURE 20.2. Ontogenic expression of uptake transporters Oatp1a1, 1a4, 1a6, 1b2, 2a1, and 2b1, Ent1, Ntcp, Oat2, and Oct1 in livers of male and female mice. The total RNA from C57BL/6 mice at each age ($n = 5$/gender) was analyzed by the bDNA method. Data are presented as a ratio of mRNA expression at that age to the highest observed in mouse liver. RLU (relative light unit) represents the highest mRNA expression observed in mouse liver. ♂ represents mRNA expression levels in male mouse liver; ♀ represents mRNA expression levels in female mouse liver. np, not published.

Sodium taurocholate cotransporting polypeptide (Ntcp) is a bile acid–uptake transporter[26] whose ontogenic expression in mice is quite unusual. Ntcp expression is very low 2 days before birth, attains maximal levels at birth, and then decreases by 5 days of age to fetal levels, remaining low until 23 days of age, when a second sharp increase in its expression occurs. The Ntcp mRNA gradually decreases afterward, and then increases in females but not in males at 45 days of age, when the gender difference in Ntcp mRNA expression becomes apparent. In rats, Ntcp expression appears to reach adult levels by 1 week of age.[27]

20.2.3. Hepatic Efflux into Bile

Many drugs and xenobiotics are metabolized in liver to more water-soluble chemicals. Therefore, these metabolites do not readily pass membranes; most chemicals exit the liver via transporters. Glutathione conjugates are often transported into bile, and sulfate conjugates are usually transported back into blood, whereas glucuronide conjugates are transported from the liver into both bile and blood.

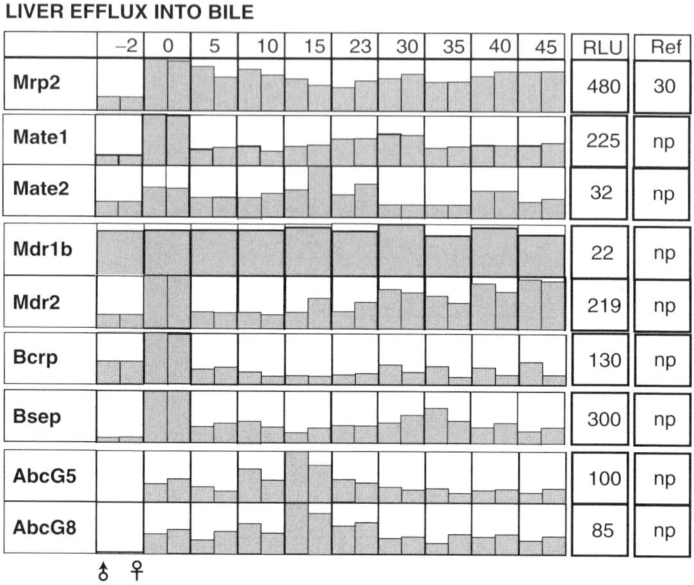

FIGURE 20.3. Ontogenic expression of canalicular (apical) efflux transporters Abcg5 and g8, Bcrp, Bsep, Mate1, Mate2, Mdr1b, Mdr2, and Mrp2 in livers of male and female mice. The total RNA from C57BL/6 mice at each age ($n = 5$/gender) was analyzed by the bDNA method. Data are presented as a ratio of mRNA expression at that age to the highest observed in mouse liver. RLU (relative light unit) represents the highest mRNA expression observed in mouse liver. ♂ represents mRNA expression levels in male mouse liver; ♀ represents mRNA expression levels in female mouse liver.

The multidrug resistance–associated protein (Mrp) family has the major role of transporting chemicals out of cells.[28] Mrp2 is highly expressed in liver and is responsible for hepatobiliary excretion of reduced and oxidized glutathione as well as glutathione and glucuronide conjugates. Hepatobiliary disposition of conjugated bilirubin is impaired in Dubin–Johnson syndrome patients, who have nonfunctional Mrp2.[29] As noted in Figure 20.3, the mRNA expression of Mrp2 in livers is very low in mice before birth. At birth, Mrp2 reaches its highest expression.[30] In rats, there is also no marked increase in Mrp2 mRNA between 1 day of age and maturity.[31,32]

The multidrug and toxin extrusion (Mate) family consists of two transporters, and both are localized in the bile canaliculus and excrete cationic drugs.[33,34] As noted in Figure 20.3, the expression of Mate1 and of Mate2 were relatively the same at all ages except for a "surge" at birth.

The multidrug resistance protein (Mdr) family of transporters also transport chemicals across the canaliculus into bile. Whereas neither Mdr1a nor 1b is highly expressed in liver, Mdr1b appears to be more abundant than Mdr1a in the liver of mice. These transporters often transport bulky neutral and cationic chemicals.[35] As noted in Figure 20.3, Mdr1b is expressed similarly in mice of all ages, and no gender differences are observed. Mdr2 is a flippase, functioning to move phospholipids from the inner to the outer leaflet of the canalicular membrane.[36,37] Mdr2 mRNA expression markedly increases at birth, decreases rapidly after birth, and gradually increases to adult levels at 1 month of age.

The breast cancer resistance protein (Bcrp/Abcg2) has high expression in mouse liver[38] before birth, peaks at birth, and after birth, decreases rapidly. At approximately 30 days of age, Bcrp expression in male mice increases, whereas in females it does not, thus creating a gender difference in expression in adult mice.[38]

Bile salt export pump (Bsep) is the main transporter for the hepatobiliary excretion of bile acids. Targeted inactivation of mouse Bsep gene results in nonprogressive but persistent intrahepatic cholestasis.[39] Mutation of human BSEP gene is linked to the etiology of progressive familial intrahepatic cholestasis subtype 2.[40] Bsep mRNA expression is low before birth in mice, reaches the highest level at birth, decreases to adult levels by 5 days of age, and remains relatively constant thereafter. In rats, Bsep is also expressed at relatively high levels at 1 day of age.[32]

The adenosine triphosphate (ATP)-binding cassette proteins (Abcg5 and g8) are expressed predominantly in intestine, where they are important for limiting the absorption of plant steroids.[41] Abcg5 and Abcg8 are also located in the bile canaliculus and excrete these compounds into bile. The ontogenic expression of Abcg5 and Abcg8 in liver is very similar. Expression of both is extremely low before birth, increases markedly at parturition, continuing to increase until day 15, and then decreases to adult levels that are approximately one-half of that observed on day 15.

20.2.4. Hepatic Efflux into Blood

In contrast to Mrp2, Mrp3, 4, and 6 transport chemicals from liver back into blood (Figure 20.1). Mrp3 mRNA is lowly expressed until 30 days of age, when expression peaks (Figure 20.4). Mrp4 is expressed at similar levels at all ages, except at day 0, the age at which it is most highly expressed.

Mutation in MRP6 is associated with the disease pseudoxanthomia elasticum, characterized by progressive calcification of elastic fibers in many tissues, especially the skin.[42,43] Ontogenic expression of Mrp6 in mouse liver is nondetectable until day 10, when it reaches peak expression, and decreases thereafter to less than one-half that in adults.

The organic solute transporters (Ostα and β) heterodimerize to form a transporter that moves bile acids from enterocytes into blood.[44] Ostα and β are also expressed in liver but at lower levels, and are thought to transport bile acids from liver back into blood.[44] The ontogenic expression of Ostα and Ostβ is quite different in mouse liver, even though they heterodimerize to produce a functional transporter. Ostα is expressed at a low level before birth and is expressed most highly on the day of birth. Ostα decreases after birth, with an intermediate increase on day 15 and further increases at days 40 and 45, reaching adult expression levels. In contrast, Ostβ expression decreases during the first month of life, although it also depicts the highest expression at birth.

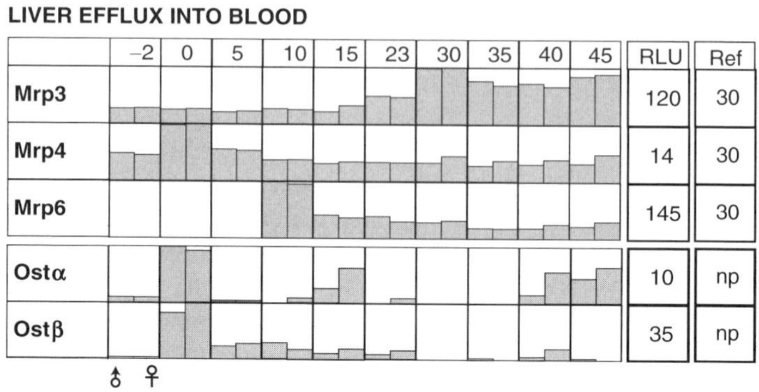

FIGURE 20.4. Ontogenic expression of sinusoidal (basolateral) efflux transporters Mrp3, 4, and 6, Ostα, and Ostβ in livers of male and female mice. The total RNA from C57BL/6 mice at each age ($n = 5$/gender) was analyzed by the bDNA method. Data are presented as a ratio of mRNA expression at that age to the highest observed in mouse liver. RLU (relative light unit) represents the highest mRNA expression observed in mouse liver. ♂ represents mRNA expression levels in male mouse liver; ♀ represents mRNA expression levels in female mouse liver.

20.2.5. Hepatic Transporter Ontogeny Patterns

Table 20.1 indicates the approximate age when the mRNA of each transporter reaches adult levels in mice. Most transporters are expressed at low levels before birth. Only four transporters (Oatp2a1, Ent1, Abcg5, and Abcg8) are expressed at adult levels before birth. Probably the most surprising data show a surge in mRNA of one-half of the transporters at the day of birth. The mechanism of this marked increase at birth is not known. After this surge, the mRNA of most transporters remains high; however, Mdr2 decreases and then reaches adult levels at an older age. Four of the transporters in liver (Oatp1a4, Oatp1a6, Oat2, and Mrp6) reach adult levels at 5 to 10 days of age, four (Oatp1b2, Oatp2b1, Oct1, and Mdr2) at 15 to 23 days of age, and two (Oatp1a1 and Mrp3) at 30 days of age or older.

20.3. RENAL TRANSPORTERS

20.3.1. Renal Excretion in Newborns

The kidneys of newborn animals are functionally immature. The renal clearance of *p*-aminohippuric acid (PAH), an organic acid, has been shown to be low in humans,[45] dogs,[46] rats,[47] rabbits,[48] and sheep.[49] The renal transport of tetraethylammonium (TEA), an organic base, is also underdeveloped in newborns.[50] Because PAH and TEA are transported by proximal tubule cells into the filtrate, the decrease in renal clearance of these chemicals in the newborn indicates that both organic acid and base transporters are immature.[50] The studies described below indicate that the mRNA expression of many of the renal transporters is responsible for immaturity of renal excretory function.

TABLE 20.1. Hepatic Transporter Ontogeny Patterns

Function	Transporters	Fetal (−2)	Birth (0)	Early (5–10)	Middle (15–23)	Late (30→)
Liver uptake	Oatp1a1					*
	Oatp1a4			*		
	Oatp1a6			*		
	Oatp1b2				*	
	Oatp2a1	*				
	Oatp2b1				*	
	Ent1	*	*			
	Ntcp		*			
	Oat2			*		
	Oct1				*	
Liver efflux into bile	Mrp2		*			
	Mate1		*			
	Mate2		*			
	Mdr1b					
	Mdr2		*		*	
	Bcrp		*			
	Bsep		*			
	Abcg5	*	*			
	Abcg8	*	*			
Liver efflux into blood	Mrp3					*
	Mrp4		*			
	Mrp6			*		
	Ostα		*			
	Ostβ		*			

*Approximately adult levels.

20.3.2. Renal Uptake from Blood

The kidney contains numerous transporters for the uptake of chemicals into proximal tubule cells, efflux into filtrate, reabsorption from the filtrate, and efflux back into blood (Figure 20.5). In kidney, organic anion transporters (Oats), in general, are most important for uptake of organic anions from blood.[51,52] As noted in Figure 20.6, Oat1 and 3 mRNA in mouse kidney are less at birth than in adults. Oat1 begins to increase at 30 days of age and reaches adult levels by 40 days of age. Oat1 exhibits gender differences in mRNA expression; higher expression is observed in kidneys of male mice at 30 days of age and older. Oat3 reaches adult levels of expression in mice at 35 days of age. In rats, Oat3 matured at the youngest age, Oat1 at an intermediate age, and Oat2 at a later age.[23] The expression of Oat1 is male-predominant in rat kidneys as in mice.[23,51]

Oatp4c1 transports xenobiotics, such as digoxin, from blood into proximal tubule cells.[53] Its expression is relatively constant from before birth until 2 weeks of age, when it more than doubles, to adult levels.[17]

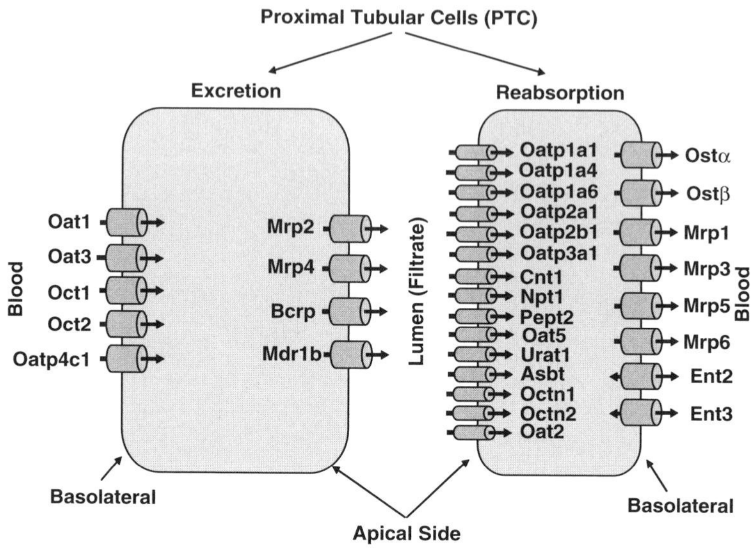

FIGURE 20.5. Cellular localization and transport direction of known transporters in kidney. Excretion transporters from basolateral and apical membrane are depicted in the proximal tubular cells (PTCs) of the kidney on the left. Reabsorption transporters from basolateral and apical membrane are shown in the PTCs of the kidney on the right.

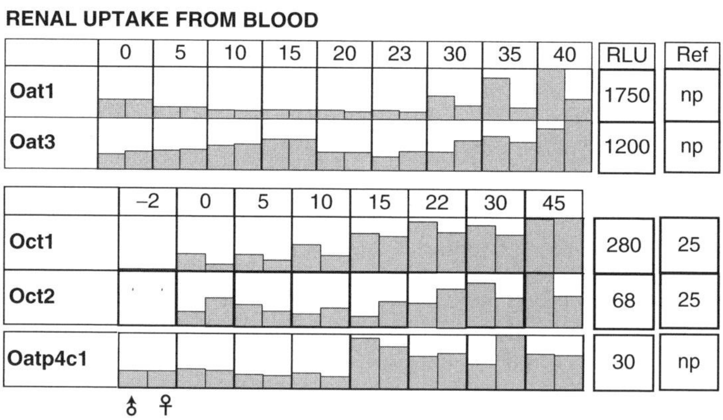

FIGURE 20.6. Ontogenic expression of basolateral uptake transporters Oat1, Oat3, Oct1, Oct2, and Oatp4c1 in proximal tubular cells (PTCs) of kidneys of male and female mice. The total RNA from C57BL/6 mice at each age ($n = 5$/gender) was analyzed by the bDNA method. Data are presented as a ratio of mRNA expression at that age to the highest observed in mouse kidney. RLU (relative light unit) represents the highest mRNA expression observed in mouse kidney. ♂ represents mRNA expression levels in male mouse kidney; ♀ represents mRNA expression levels in female mouse kidney.

The organic cation transporters (Oct and Octn) are important for the transport of organic cations from blood into kidney.[54] All four mouse Octs have very low expression before birth (Figure 20.6), and their expression increases gradually to adult levels at 3 weeks of age.[25,54] In rats, the developmental expression of Oct1 and 2 also increase after birth, with consistent increases observed for the first 4 to 6 weeks of age.[55]

20.3.3. Renal Uptake from Filtrate (Reabsorption)

There are seven Oatps that are expressed in mouse kidneys.[17] Of these, Oatp1a1, 1a4, and possibly 1a6 transport xenobiotics from the lumen into proximal tubules, thus facilitating reabsorption. The ontogenic pattern of Oatp1a1 is most unusual (Figure 20.7). The renal expression of Oatp1a1 is very low in young mice. Not until 30 days of age is there any detectable expression of Oatp1a1 in male mouse kidneys. Between 30 and 45 days of age, a marked increase in Oatp1a1 is observed in male but not in female mice. Thus, a major gender difference in Oatp1a1 is observed in kidneys of mice. Oatp1a4 is expressed at similar levels from before birth to adulthood. Oatp1a6 exhibits a relatively linear increase in expression during the first 45 days of life. Oatp2a1 and 2b1 expression are relatively constant from birth to adulthood. The renal expression of Oatp3a1 before birth is about one-third of that observed in adult male mice. Oatp3a1 expression remains relatively constant, but at day 30 it increases in males, resulting in gender differences in expression.

The development of other renal reabsorption transporters in mice has distinctive ontogenic patterns (Figure 20.7). The concentrative nucleotide transporter 1 (Cnt1) has a gradual increase in expression during the first 45 days of life. The peptide transporter (Pept) 2 is relatively low until it reaches its peak expression on day 15. Oat2 mRNA in mouse kidneys is less at birth than in adults. Oat2 is expressed at very low levels in mouse kidneys until 30 days of age. After 30 days of age, Oat2 begins to increase and reaches adult levels by 40 days of age. Oat5 has low fetal expression in kidney, with a major increase observed at birth. Oat5 expression is relatively constant until day 22, when it increases in female mice, resulting in a gender difference in expression. Urat1, the uric acid transporter,[56] is expressed at low levels in kidneys before birth, and gradually increases over the first 6 weeks of age. A male-predominant gender difference is observed in adult mice.[57] The apical sodium bile acid transporter (Asbt) is expressed at very low levels in the kidney until 3 weeks of age, and by 6 weeks of age it is more highly expressed in female mice.

20.3.4. Renal Efflux into Filtrate

The multidrug resistance–associated protein (Mrp) family of transporters are extremely important in the efflux of chemicals from cells. In kidney, as in liver, Mrp2 plays a major role in the efflux of chemicals across the apical membrane, and thus from proximal tubule cells into the filtrate. In mice, Mrp2 expression is very low before birth and throughout the first 10 days of life, thereafter increasing to adult levels by 15 days of age[30] (Figure 20.8). Mrp4 expression gradually increases in kidney over the first 6 weeks of life.[30]

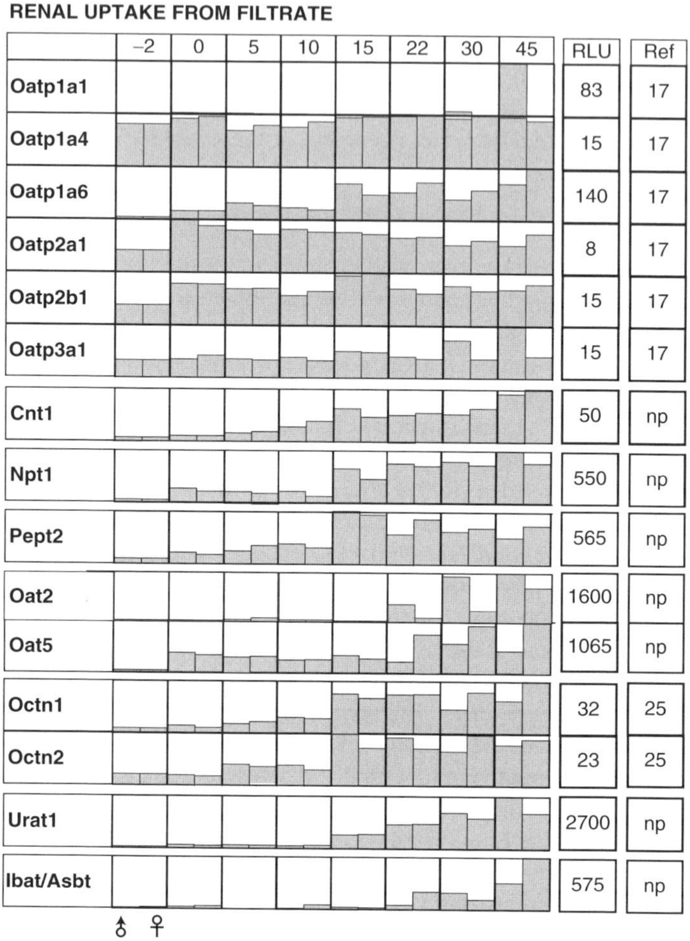

FIGURE 20.7. Ontogenic expression of apical reabsorption transporters Oatp1a1, 1a4, 1a6, 2a1, 2b1, and 3a1, Oat2, Oat5, Octn1, Octn2, Cnt1, Npt1, Pept2, Urat1, and Ibat/Asbt in proximal tubular cells (PTCs) of kidneys of male and female mice. The total RNA from C57BL/6 mice at each age (n = 5/gender) was analyzed by the bDNA method. Data are presented as a ratio of mRNA expression at that age to the highest observed in mouse kidney. RLU (relative light unit) represents the highest mRNA expression observed in mouse kidney. ♂ represents mRNA expression levels in male mouse kidney; ♀ represents mRNA expression levels in female mouse kidney.

The breast cancer resistance protein (Bcrp/Abcg2) is expressed in kidney before birth, but gradually increases fivefold before reaching adult levels at 22 days of age. The multidrug resistance protein 1b (Mdr1b) effluxes basic drugs from the proximal tubule cells into filtrate. Mdr1b is expressed in kidney before birth at about one-fourth the level in adult female mice. The developmental expression of Mdr1a and 1b mRNA and protein in rats appear to be similar to that in mice.[31,58] From days

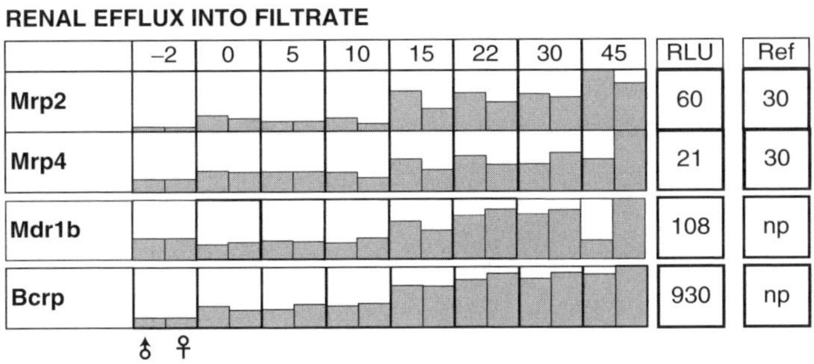

FIGURE 20.8. Ontogenic expression of apical efflux transporters Bcrp, Mdr1b, Mrp2, and Mrp4 in proximal tubular cells (PTCs) of kidneys of male and female mice. The total RNA from C57BL/6 mice at each age (n = 5/gender) was analyzed by the bDNA method. Data are presented as a ratio of mRNA expression at that age to the highest observed in mouse kidney. RLU (relative light unit) represents the highest mRNA expression observed in mouse kidney. ♂ represents mRNA expression levels in male mouse kidney; ♀ represents mRNA expression levels in female mouse kidney.

10 to 22, Mdr1b expression increases, whereas expression decreases between days 30 and 45 in male mice, resulting in a gender difference in expression of Mdr1b. The equilibrative nucleoside transporter (Ent) 2 and 3 are expressed in kidney.[22] Ent2 is expressed similarly at various ages in mouse kidneys, whereas Ent3 shows a continuous increase in expression from birth through day 45.

20.3.5. Renal Efflux into Blood

The Mrps in kidneys, not only efflux chemicals into the filtrate but can also transport xenobiotics into blood (Figures 20.5 and 20.9). Whereas Mrp2 and 4 are localized in the apical portion of renal tubular cell, Mrp1, 3, 5, and 6 exist in the basolateral portion of renal tubular cell membranes and efflux chemicals back into blood.[59,60] Figure 20.9 illustrates the developmental pattern of expression of these four Mrps. Mrp1 expression is relatively constant at various ages in mice[30] and rats;[31] however, its expression is higher at days 0 and 15 than at other ages. A later decrease in expression in male mouse kidneys results in a female-predominant gender difference. Mrp3 expression in kidney is low until 15 days of age, when the expression triples.[30] At 30 days of age, Mrp3 expression in male kidneys decreases, also resulting in a female-predominant gender difference in expression. Mrp5 expression in kidney is similar before birth as in adult mice, but its expression is at least twice as high at birth, but then it gradually decreases to adult expression levels. Mrp6 is expressed at high levels before birth and gradually increases to levels observed in adults.

Organic solute transporters (Ost) α and β are important for effluxing bile acids from enterocytes back into blood.[61] In kidneys, they also probably transfer chemicals back into blood.[44] Ostα and β both have low expression before birth, increase to about one-fourth of adult expression at birth, and reach maturity between 10 and 15 days of age (Figure 20.9).

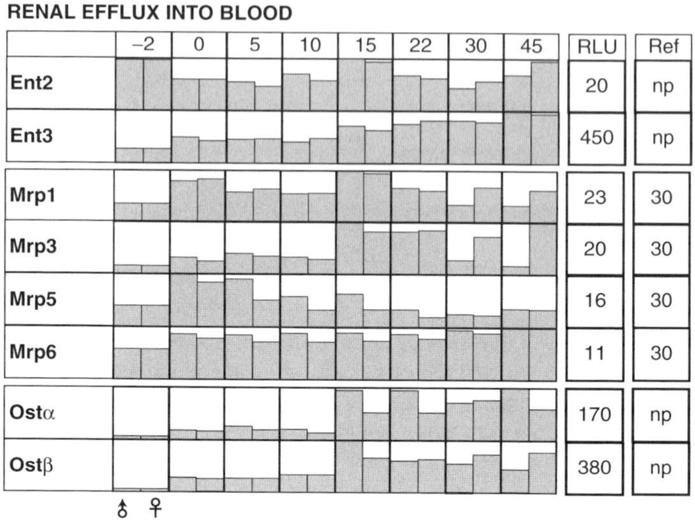

FIGURE 20.9. Ontogenic expression of basolateral efflux transporters Ent2 and 3, Mrp1, 3, 5, and 6, Ostα, and Ostβ in proximal tubular cells (PTCs) of kidneys of male and female mice. The total RNA from C57BL/6 mice at each age ($n = 5$/gender) was analyzed by the bDNA method. Data are presented as a ratio of mRNA expression at that age to the highest observed in mouse kidney. RLU (relative light unit) represents the highest mRNA expression observed in mouse kidney. ♂ represents mRNA expression levels in male mouse kidney; ♀ represents mRNA expression levels in female mouse kidney.

20.3.6. Renal Transporter Ontogeny Patterns

Similar to the ontogenic patterns of transporters in liver, only a few of the transporters are expressed at adult levels before birth (Table 20.2). None of the transporters that are responsible for elimination of chemicals into urine (i.e., those that are either responsible for uptake of xenobiotics from blood or efflux into filtrate) are expressed at adult levels before birth. However, a few transporters responsible for the reabsorption of chemicals are at adult levels of expression at birth. For example, four of the six Oatp transporters that probably reabsorb chemicals from the filtrate (only Oatp1a1 has been shown to be localized to the lumenal membrane) are expressed before birth, as well as Cnt2 and Mrp6, which probably function to transport chemicals from kidney back into blood.

Whereas a major surge in expression of half of the transporters are observed in the liver at birth, only 5 of 32 transporters expressed in kidney exhibit this phenomenon: Oatp2a1, Oatp2b1, Oat5, Mrp1, and Mrp5. Very few of the transporters in kidneys reach maturity at 5 to 10 days of age (Mrp1 and 5). Most of the transporters (19 of 32) in kidneys reach adult expression levels at 15 to 23 days of age. A few of the transporters (6 of 32) reach mature levels in kidneys at 30 days of age and thereafter.

TABLE 20.2. Kidney Transporter Ontogeny Patterns

Function	Transporters	Fetal (−2)	Birth (0)	Early (5–10)	Middle (15–23)	Late (30→)
Renal uptake from blood	Oat1					*
	Oat3					*
	Oct1				*	
	Oct2				*	
	Octn1				*	
	Octn2				*	
	Oatp4c1				*	
Renal uptake from filtrate	Oatp1a1					*
	Oatp1a4	*				
	Oatp1a6				*	
	Oatp2a1	*	*			
	Oatp2b1	*	*			
	Oatp3a1	*				*
	Cnt1				*	
	Npt1				*	
	Pept2				*	
	Oat2					*
	Oat5		*		*	
	Urat1				*	
	Ibat/Asbt					*
Renal efflux into filtrate	Mrp2				*	
	Mrp4				*	
	Mdr1b				*	
	Bcrp				*	
Renal efflux into blood	Ent2	*				
	Ent3				*	
	Mrp1		*			
	Mrp3				*	
	Mrp5	*	*			
	Mrp6	*				
	Ostα				*	
	Ostβ				*	

*Approximately adult levels.

20.4. INTESTINAL TRANSPORTERS

20.4.1. Intestinal Uptake from Lumen

Transporters in the intestine are important in the absorption of some chemicals and in preventing the absorption of others (Figure 20.10). The concentrative nucleoside transporters (Cnts) are important for the absorption of nucleosides from the intestine.[25,55] Cnt1 and 3 have a similar ontogenic expression profile, being relatively lowly expressed during the first 3 weeks of life and then increasing to maximal expression at 6 weeks of age (Figure 20.11). Cnt2 has a very different expression pattern: is

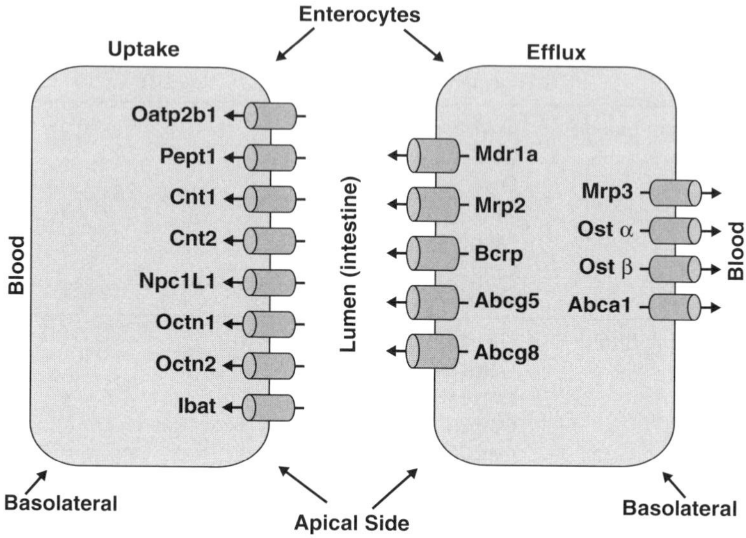

FIGURE 20.10. Cellular localization and transport direction of known transporters in intestine. Uptake transporters in apical membrane are depicted in the enterocytes on the left. Efflux transporters in the basolateral and apical membrane are shown in the enterocytes on the right.

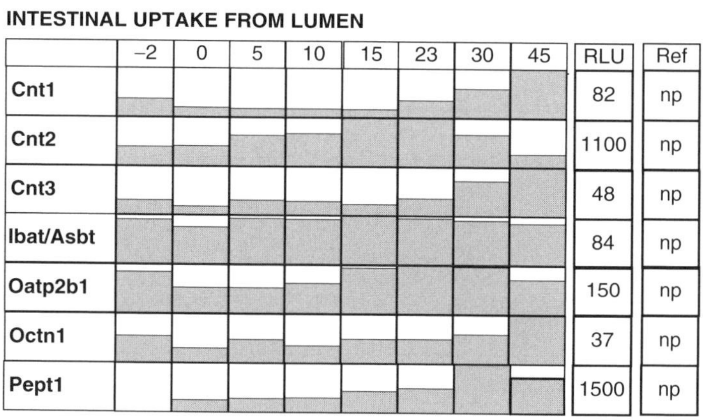

FIGURE 20.11. Ontogenic expression of apical uptake transporters Cnt1, 2, and 3, Ibat/Asbt, Oatp2b1, Octn1, and Pept1 in enterocytes of male mice. Previous studies showed that there is no gender difference in mRNA expression of these transporters in adult mouse duodenum. Therefore, the total RNA from male C57BL/6 mouse duodenum at each age ($n = 5$) was analyzed by the bDNA method. Data are presented as a ratio of mRNA expression at that age to the highest observed in mouse duodenum. RLU (relative light unit) represents the highest mRNA expression observed in mouse duodenum.

expressed in the intestine of the fetus, reaches its maximum expression at 2 to 3 weeks of age, and decreases to an adult level lower than that in the fetus.

The intestinal bile acid transporter, also known as the apical sodium bile acid transporter (Ibat/Asbt), is expressed at similar levels in all ages examined. In rats, the Asbt protein is very low before 3 weeks of age.[62] Oatp2b1 mRNA is expressed at a similar level in enterocytes of mice of all ages. The organic cation transporter (Octn1) is expressed from 2 days before birth through the first month of life at about one-half that observed in 45-day-old mice.[25] The peptide transporter (pept) 1 expression is nonexistent two days before birth, but expression increases gradually over the first month to adult levels.

20.4.2. Intestinal Efflux into Lumen

The intestinal efflux transporters are important in preventing the absorption of some chemicals into the systemic circulation. For example, Mdr1a decreases the absorption of many bulky cationic organic compounds. As noted in Figure 20.12, Mdr1a expression in the intestine is low 2 days before birth but is expressed at nearly adult levels at birth, continuing through adulthood. Mrp2 is expressed in mouse intestine before birth, maximally expressed at days 0 to 15, and decreases by more than 50% to adult levels. Mrp7 is expressed at constant levels from 2 days before birth to adulthood.[30] Abcg5 and g8 prevent the absorption of plant sterols.[63,64] The ontogenic patterns for Abcg5 and g8 are quite different (Figure 20.12). Abcg5 reaches its maximum expression at day 15 and then decreases, whereas Abcg8 reaches its maximum expression earlier (days 5 and 10), and then decreases markedly.

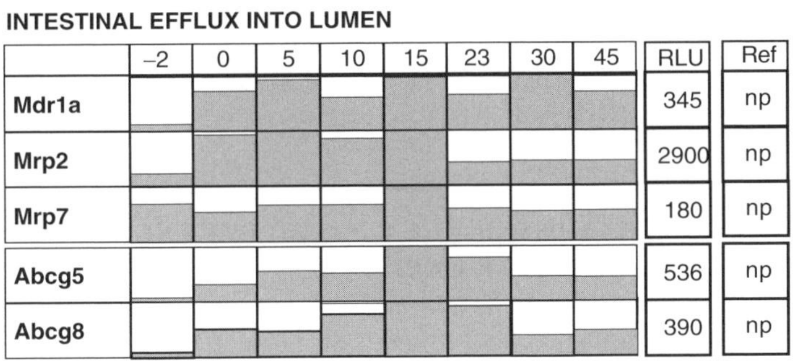

FIGURE 20.12. Ontogenic expression of apical efflux transporters Abcg5 and g8, Mdr1a, Mrp2, and Mrp7 in enterocytes of male mice. Previous studies showed that there is no gender difference in mRNA expression of these transporters in the adult mouse duodenum. Therefore, the total RNA from male C57BL/6 mouse duodenum at each age ($n = 5$) was analyzed by the bDNA method. Data are presented as the ratio of mRNA expression at that age to the highest observed in mouse duodenum. RLU (relative light unit) represents the highest mRNA expression observed in mouse duodenum.

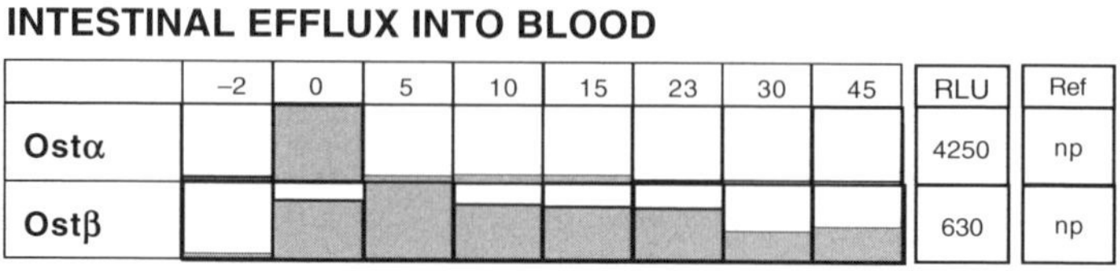

FIGURE 20.13. Ontogenic expression of basolateral efflux transporters Ostα and Ostβ in enterocytes of male mice. Previous studies showed that there is no gender difference in mRNA expression of these transporters in adult mouse duodenum. Therefore, the total RNA from male C57BL/6 mouse duodenum at each age ($n = 5$) was analyzed by the bDNA method. Data are presented as the ratio of mRNA expression at that age to the highest observed in mouse duodenum. RLU (relative light unit) represents the highest mRNA expression observed in mouse duodenum.

20.4.3. Intestinal Efflux into Blood

The organic solute transporters Ostα and β form heterodimers and their main function is to transport bile acids out of enterocytes into blood.[61] Their expression in intestine is very low before birth, and reaches a maximum on the day of birth (Figure 20.13). Surprisingly, their expression then decreases. Ostα expression decreases more than 95%, and Ostβ decreases more than 50%.

20.5. TRANSPORTERS IN THE BRAIN

As noted in Figure 20.14, there are a number of transporters that are expressed in brain. Figure 20.14 illustrates that Mdr1a, Mdr1b, Ent2, Oatp1a4, Oatp1c1, Mrp4, and Mrp5 are all expressed in brain. These seven transporters are expressed in brain before birth. Mdr1b, Ent2, Mrp4, and Mrp5 expression does not vary markedly with age. In contrast, Mdr1a, Oatp1a4, and 1c1 are low at birth. Mdr1a reaches adult levels at 3 weeks of age. Oatp1a4 reaches maturity at 30 days of age, whereas Oatp1c1 reaches adult levels at 10 days of age.

20.6. MECHANISMS FOR AGE-RELATED DIFFERENCES IN XENOBIOTIC TRANSPORTER EXPRESSION

It is clear from the information above that most transporters in liver and kidney are not expressed at adult levels at birth. This is not surprising because excretory function before birth is performed by the mother. After birth there is a need to develop processes for elimination of chemicals.

Little is known about the molecular mechanisms responsible for the developmental expression of transporters. As noted previously, there is a surge of some transporters at birth. Might this be due to hormones or the stress of birth? It is not known. Might the later increases be due to change in diet from milk to chow, puberty, or other changes?

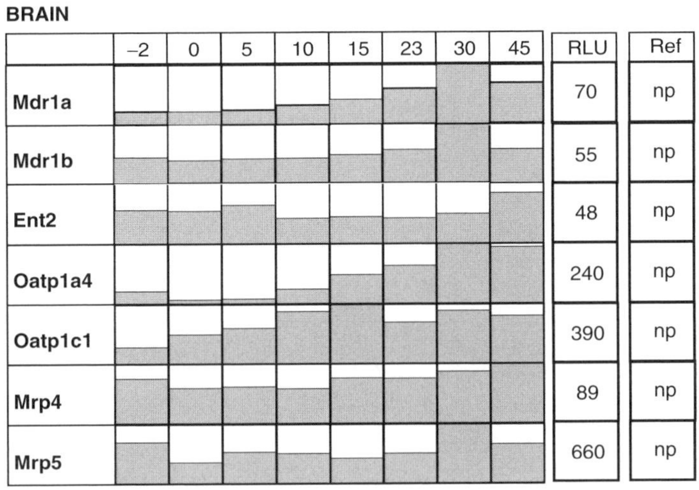

FIGURE 20.14. Ontogenic expression of transporters Mdr1a, Mdr1b, Ent2, Mrp4, Mrp5, Oatp1a4, and Oatp1c1 in the brain of male mice. Previous studies showed that there is no gender difference in mRNA expression of these transporters in the brain of adult mice. Therefore, the total RNA from male C57BL/6 mouse brain at each age ($n = 5$) was analyzed by the bDNA method. Data are presented as the ratio of mRNA expression at that age to the highest observed in the brain. RLU (relative light unit) represents the highest mRNA expression observed in the brain.

This is not known either. Little is known about the transcription factors that regulate the basal developmental expression of transporters.

The importance of the transcription factor hepatic nuclear factor 1 alpha (HNF1α) on the constitutive expression of transporters in liver and kidney has been examined in HNF1α-null mice.[65] HNF1α is a positive regulatory factor for expression of the Oat family of transporters in both liver and kidney and most of the Oatp family members in these two organs. In contrast, HNF1α is a negative regulatory factor for most Mdr and Mrp transporters in mouse liver and kidney. Very little information is available on other transcription factors that regulate the constitutive expression of transporters in adult animals, and essentially nothing is known about the transcription factors responsible for the development pattern of transporters in newborn animals. This is a fertile research area.

20.7. GENDER DIFFERENCES IN TRANSPORTER EXPRESSION

Over the years we have learned that there are gender differences in the disposition of xenobiotics, especially in rodents. Some of these differences are due to gender variations in expression of phase I and II drug-metabolizing enzymes. However, gender differences in transporters might be equally important.

TABLE 20.3. Gender Differences in Transporter Expression in Mice

	Tissue	Gender Difference[a] M/F	Gender Difference[a] F/M	Refs.
Oatp1a1	Liver	2.2		17,87
Oatp1a4	Liver		2.4	17,87
Oatp1a1	Kidney	19.0		17,87
Oatp3a1	Kidney	2.8		17,87
Bcrp	Liver	2.4		38
Mdr1b	Kidney		2.9	N.P[b]
Map3	Kidney		25.0	30,88
Map4	Kidney		2.4	30,88
Mate1	Kidney	10		N.P.
Mate2	Kidney		2.2	N.P.
Ntcp	Liver		1.7	N.P.
Ent1	Kidney		1.5	22
Ent2	Kidney		1.4	22
Ibat/Asbt	Kidney		1.9	N.P.
Ostα	Liver		1.5	N.P.
Ostα	Kidney	1.7		N.P.
Ostβ	Kidney		1.8	N.P.
Oat1	Kidney	2.4		51
Oat5	Kidney		2.4	N.P.
Urat1	Kidney	1.5		57
Oct2	Kidney	1.8		25
Octn1	Kidney		1.6	25

[a]Gender difference is the ratio of mRNA expression of a transporter in adult male and female mice.
[b]N.P., not published.

Table 20.3 illustrates numerous examples of gender differences in the transporter expression in mice. Gender differences in the expression of five transporters were noted in mouse liver: Oatp1a1, Oatp1a4, Bcrp, Ntcp, and Ostα.[17,38] Meanwhile 15 transporters exhibit gender-divergent expression in mouse kidneys: Oatp1a1, Oatp3a1, Mdr1b, Mrp3, Mrp4, Ent1, Ent2, Asbt, Ostα, Ostβ, Oat1, Oat5, Urat1, Oct2, and Octn1.[22,25,30,51] Eight of the transporters have higher expression in male mice, and 12 transporters are more highly expressed in female mice. The magnitude of gender differences for most of the transporters is about two fold in mRNA expression; however, the expression of Oatp1a1 is 19-fold higher in male mouse kidneys, whereas Mrp3 has a 25-fold higher expression in female mouse kidneys.

As shown in Table 20.4, gender differences in a specific transporter are not observed consistently in all species. For example, whereas Oat2 expression in mice is higher in kidneys of male mice, in rats it is higher in kidneys of females.[23,66] Ntcp is more highly expressed in livers of female mice, but in male rats.[67] In rabbit kidneys, no gender difference is observed in Oats and Octs, indicating caution in extrapolating transport-related gender differences between species.[68] However, the expression of

TABLE 20.4. Gender Differences in Transporter Expression in Mice and Rats

	Tissue	Gender Difference[a]	
		Mice	Rats
Oatp1a1	Liver	M>F	
Oatp1a4	Liver	F>M	M>F
Oatp1a1	Kidney	M>F	M>F
Oatp3a1	Kidney	M>F	M>F
Bcrp	Liver	M>F	M>F
Mdr1b	Kidney	F>M	
Mrp3	Kidney	F>M	
Mrp4	Kidney	F>M	
Mate1	Kidney	M>F	
Mate2	Kidney	F>M	
Ntcp	Liver	F>M	M>F
Ent1	Kidney	F>M	
Ent2	Kidney	F>M	
Ibat/Asbt	Kidney	F>M	
Ostα	Liver	F>M	
Ostα	Kidney	M>F	
Ostβ	Kidney	F>M	
Oat1	Kidney	M>F	
Oat2	Kidney		F>M
Oat5	Kidney	F>M	
Urat1	Kidney	M>F	
Oct2	Kidney	M>F	M>F
Octn1	Kidney	F>M	

[a]Gender difference is the ratio of mRNA expression of a transporter in adult male and female mice.

Oatp1a1 is male-predominant in kidneys of rats[20,69,70] and mice,[17,71] and Oct2 is male-predominant in both rats and mice.[25,55,72]

20.8. MECHANISMS OF GENDER DIFFERENCES

Gender differences in transporter gene expression may be the result of regulation by sex hormones and/or gender-dimorphic growth-hormone (GH) secretory patterns. Androgens and estrogens alter gene expression by stimulating gene transcription directly or stabilizing the mRNA of certain genes.[73–75] Growth hormone is also an important regulator of gender-divergent gene expression. Gender-divergent secretion patterns of GH lead to differential patterns in gene expression. In rats, males secrete GH in high-amplitude pulses with a regular frequency. Between pulses, serum GH levels are nondetectable.[76] In contrast, female rats secrete GH in low-amplitude pulses with greater frequency and higher trough levels than males, resulting in a continuously detectable baseline of serum GH.[77] These GH-secretory patterns are responsible for male-specific expression of rat Cyp2c11 and female-specific Cyp2c12, respectively.[78] GH-secretion pattern in male mice is similar to that in male rats.[79] In female mice, GH is secreted at regular intervals with a nondetectable baseline between pulses, however, the pulses are more frequent (1 to 1.5 hours) than those in male mice (2.5 hours).[79] The GH-secretory pattern in male mice is responsible for induction of male-predominant Cyp2d9 and repression of female-predominant Cyp2a4 in liver.[80,81]

Several animal models are used to investigate the effects of hormones on gene expression. Hypophysectomy (HPX) is surgical removal of the pituitary, which obliterates the production of several hormones, including luteinizing hormone, follicle-stimulating hormone, adrenocorticotropic hormone, and prolactin. Gonadectomy (GNX) is the surgical removal of testes or ovaries, the organs primarily responsible for sex hormone production. A mutant mouse model, the lit/lit mouse, has a spontaneous mutation in the GH-releasing hormone receptor (GHRH-R), which leads to impaired GH secretion.[82-85] Unlike hypophysectomy, the lit/lit mouse model circumvents the loss of other pituitary hormones and is still responsive to GH therapy.[81,86]

Table 20.5 indicates the results of experiments in mice to determine whether sex hormones or the growth-hormone secretory pattern is responsible for the gender differences in transporters. The three experimental protocols described above were used: (1) hypophysectomy, followed by testosterone, estrogen, male-pattern growth hormone, or female-pattern growth hormone; (2) gonadectomy, followed by testosterone or estrogen; and (3) the lit/lit mouse, followed by male- or female-pattern growth hormone.

Using these experimental designs, the male-predominant expression of Oatp1a1 in mouse livers is due to the stimulatory effect of male-pattern growth hormone (GH) secretion, whereas the female-predominant expression of Oatp1a4 in mouse livers is due to the inhibitory effect of male-pattern GH secretion. In contrast, the male-predominant expression of Oatp1a1 and 3a1 in mouse kidneys appears to be due to testosterone. The male-predominant expression of Oatp1a1 in liver and kidney of mice is due to different mechanisms; in liver it is due to male-pattern GH secretion, whereas in kidney it is testosterone.[87]

The female-predominant expression of the bile acid transporter Ntcp in mouse livers is due to the inhibitory effect of male-pattern GH secretion. The male-predominant expression of Bcrp in mouse livers appears to be due to the stimulatory effect of testosterone,[38] whereas the female-predominant expression of Mrp3 in mouse kidneys is due to estrogen stimulation.[88] Whereas the gender difference in expression of all the transporters above appears to be due to one factor, the female-predominant expression of Mrp4 in kidney of mice appears to be due to inhibitory effects of both testosterone and male-pattern growth-hormone secretion.[88]

20.9. CONCLUSIONS

The data described in this chapter indicate that many of the transporters involved in transport of chemicals into and out of cells are observed to have age- and

TABLE 20.5. Regulation of Gender-Dimorphic Expression of Transporters in Mice[a]

			HPX				GNX		Lit/Lit			
	Tissue	Phenotype	Testo	Estro	GH_M	GH_F	Testo	Estro	GH_M	GH_F	Conclusions	Ref.
Oatp1a1	Liver	M/F = 2.2			↑		↑		↑		GH_M: ↑	87
Oatp1a4	Liver	F/M = 2.4		↑	↓		↓		↓		GH_M: ↓	87
Oatp1a1	Kidney	M/F = 19	↑				↑				Testo: ↑	87
Oatp3a1	Kidney	M/F = 2.8	↑				↑				Testo: ↑	87
Ntcp	Liver	F/M = 1.7			↓				↓		GH_M: ↓	N.P.
Bcrp	Liver	M/F = 2.4	N.A.	N.A.	N.A.	N.A.	↑				Testo: ↑	38
Mrp3	Kidney	F/M = 25		↑			↓	↑			Testo: ↑	88
Mrp4	Kidney	F/M = 2.4	↓		↓		↓		↓	↓	Testo: ↓ GH_M: ↓	88

[a] ↑, up-regulation; ↓, down-regulation; N.A, not available; HPX, hypophysectomy; GNX, gonadectomy; lit/lit, mice with mutation in GH-releasing-hormone receptor gene; GH_M, male-pattern GH administration; GH_F, female-pattern GH administration; N.P, not published; data from the laboratory of the authors.

gender-specific differences in mRNA expression in mice. The implication of these observations is that age- and gender-related differences in pharmacokinetics may be ascribed in part to differences in transporter expression. Therefore, when attempting to explain age- or gender-related discrepancies in xenobiotic pharmacokinetics, functional studies of transporters that display variations in expression should be considered to exclude or confirm a contribution of transporter expression patterns to the differences observed in pharmacokinetics.

Acknowledgments

The authors wish to acknowledge present and previous predoctoral and postdoctoral follows of this laboratory for their laboratory work in obtaining much of the data in the chapter and for making suggestions for its improvement.

REFERENCES

1. Jondorf WR, Parke DV, Williams RT. 1958. Studies in detoxication. The metabolism of halogenobenzenes: 1:2:3:4-, 1:2:3:5- and 1:2:4:5-tetrachlorobenzenes. Biochem J 69:181–189.
2. Fouts JR, Adamson RH. 1959. Drug metabolism in the newborn rabbit. Science 129:897–898.
3. Riordan JR, Deuchars K, Kartner N, Alon N, Trent J, Ling V. 1985. Amplification of P-glycoprotein genes in multidrug-resistant mammalian cell lines. Nature 316:817–819.
4. Gros P, Croop J, Housman D. 1986. Mammalian multidrug resistance gene: complete cDNA sequence indicates strong homology to bacterial transport proteins. Cell 47:371–380.
5. Schinkel AH, Smit JJ, van Tellingen O, Beijnen JH, Wagenaar E, van Deemter L, Mol CA, van der Valk MA, Robanus-Maandag EC, te Riele HP, et al. 1994. Disruption of the mouse mdr1a P-glycoprotein gene leads to a deficiency in the blood–brain barrier and to increased sensitivity to drugs. Cell 77:491–502.
6. Klaassen CD. 1972. Immaturity of the newborn rat's hepatic excretory function for ouabain. J Pharmacol Exp Ther 183:520–526.
7. Klaassen CD. 1973. Comparison of the toxicity of chemicals in newborn rats to bile duct-ligated and sham-operated rats and mice. Toxicol Appl Pharmacol 24:37–44.
8. Kupferberg HJ, Schankl LS. 1968. Biliary secretion of ouabain-^{3}H and its uptake by liver slices in the rat. Am J Physiol 214:1048–1053.
9. Russell JQ, Klaassen CD. 1972. Species variation in the biliary excretion of ouabain. J Pharmacol Exp Ther 183:513–519.
10. Stacey NH, Klaassen CD. 1979. Uptake of ouabain by isolated hepatocytes from livers of developing rats. J Pharmacol Exp Ther 211:360–363.
11. Klaassen CD. 1973. Hepatic excretory function in the newborn rat. J Pharmacol Exp Ther 184:721–728.
12. Hunter AL, Klaassen CD. 1975. Biliary excretion of colchicine in newborn rats. Drug Metab Dispos 3:530–535.

13. Klaassen CD. 1977. Biliary excretion In: *Handbook of Physiology: Reactions to Environmental Agents*. Lee, D.H.K. (Ed.), Am Physiol Soc, Washington, D.C., sec 9, ch 34, pp. 537–553.
14. Klaassen CD. 1975. Hepatic uptake of cardiac glycosides in newborn rats, rabbits and dogs. Biochem Pharmacol 24:923–925.
15. Whelan G, Hoch J, Combes B. 1970. A direct assessment of the importance of conjugation for biliary transport of sulfobromophthalein sodium. J Lab Clin Med 75:542–557.
16. Vest MF. 1962. Conjugation of sulfobromophthalein in newborn infants and children. J Clin Invest 41:1013–1020.
17. Cheng X, Maher J, Chen C, Klaassen CD. 2005. Tissue distribution and ontogeny of mouse organic anion–transporting polypeptides (Oatps). Drug Metab Dispos 33:1062–1073.
18. Ogura K, Choudhuri S, Klaassen CD. 2000. Full-length cDNA cloning and genomic organization of the mouse liver–specific organic anion transporter-1 (lst-1). Biochem Biophys Res Commun 272:563–570.
19. Hagenbuch B, Meier PJ. 2003. The superfamily of organic anion transporting polypeptides. Biochim Biophys Acta 1609:1–18.
20. Li N, Hartley DP, Cherrington NJ, Klaassen CD. 2002. Tissue expression, ontogeny, and inducibility of rat organic anion transporting polypeptide 4. J Pharmacol Exp Ther 301:551–560.
21. Lu R, Kanai N, Bao Y, Schuster VL. 1996. Cloning, in vitro expression, and tissue distribution of a human prostaglandin transporter cDNA (hPGT). J Clin Invest 98:1142–1149.
22. Lu H, Chen C, Klaassen C. 2004. Tissue distribution of concentrative and equilibrative nucleoside transporters in male and female rats and mice. Drug Metab Dispos 32:1455–1461.
23. Buist SC, Cherrington NJ, Choudhuri S, Hartley DP, Klaassen CD. 2002. Gender-specific and developmental influences on the expression of rat organic anion transporters. J Pharmacol Exp Ther 301:145–151.
24. Simonson GD, Vincent AC, Roberg KJ, Huang Y, Iwanij V. 1994. Molecular cloning and characterization of a novel liver–specific transport protein. J Cell Sci 107(Pt 4):1065–1072.
25. Alnouti Y, Petrick JS, Klaassen CD. 2006. Tissue distribution and ontogeny of organic cation transporters in mice. Drug Metab Dispos 34:477–482.
26. Hagenbuch B, Meier PJ. 1994. Molecular cloning, chromosomal localization, and functional characterization of a human liver Na^+/bile acid cotransporter. J Clin Invest 93:1326–1331.
27. Hardikar W, Ananthanarayanan M, Suchy FJ. 1995. Differential ontogenic regulation of basolateral and canalicular bile acid transport proteins in rat liver. J Biol Chem 270:20841–20846.
28. Haimeur A, Conseil G, Deeley RG, Cole SP. 2004. The MRP-related and BCRP/ABCG2 multidrug resistance proteins: biology, substrate specificity and regulation. Curr Drug Metab 5:21–53.
29. Keitel V, Nies AT, Brom M, Hummel-Eisenbeiss J, Spring H, Keppler D. 2003. A common Dubin–Johnson syndrome mutation impairs protein maturation and transport activity of MRP2 (ABCC2). Am J Physiol Gastrointest Liver Physiol 284:G165–G174.

30. Maher JM, Slitt AL, Cherrington NJ, Cheng X, Klaassen CD. 2005. Tissue distribution and hepatic and renal ontogeny of the multidrug resistance–associated protein (Mrp) family in mice. Drug Metab Dispos 33:947–955.
31. Rosati A, Maniori S, Decorti G, Candussio L, Giraldi T, Bartoli F. 2003. Physiological regulation of P-glycoprotein, MRP1, MRP2 and cytochrome P450 3A2 during rat ontogeny. Dev Growth Differ 45:377–387.
32. Tomer G, Ananthanarayanan M, Weymann A, Balasubramanian N, Suchy FJ. 2003. Differential developmental regulation of rat liver canalicular membrane transporters Bsep and Mrp2. Pediatr Res 53:288–294.
33. Otsuka M, Matsumoto T, Morimoto R, Arioka S, Omote H, Moriyama Y. 2005. A human transporter protein that mediates the final excretion step for toxic organic cations. Proc Natl Acad Sci U S A 102:17923–17928.
34. Hiasa M, Matsumoto T, Komatsu T, Moriyama Y. 2006. Wide variety of locations for rodent MATE1, a transporter protein that mediates the final excretion step for toxic organic cations. Am J Physiol Cell Physiol 291:C678–C686.
35. Smit JW, Weert B, Schinkel AH, Meijer DK. 1998. Heterologous expression of various P-glycoproteins in polarized epithelial cells induces directional transport of small (type 1) and bulky (type 2) cationic drugs. J Pharmacol Exp Ther 286:321–327.
36. Smit JJ, Schinkel AH, Oude Elferink RP, Groen AK, Wagenaar E, van Deemter L, Mol CA, Ottenhoff R, van der Lugt NM, van Roon MA, et al. 1993. Homozygous disruption of the murine mdr2 P-glycoprotein gene leads to a complete absence of phospholipid from bile and to liver disease. Cell 75:451–462.
37. Ruetz S, Gros P. 1994. Phosphatidylcholine translocase: a physiological role for the mdr2 gene. Cell 77:1071–1081.
38. Tanaka Y, Slitt AL, Leazer TM, Maher JM, Klaassen CD. 2005. Tissue distribution and hormonal regulation of the breast cancer resistance protein (Bcrp/Abcg2) in rats and mice. Biochem Biophys Res Commun 326:181–187.
39. Wang R, Salem M, Yousef IM, Tuchweber B, Lam P, Childs SJ, Helgason CD, Ackerley C, Phillips MJ, Ling V. 2001. Targeted inactivation of sister of P-glycoprotein gene (spgp) in mice results in nonprogressive but persistent intrahepatic cholestasis. Proc Natl Acad Sci U S A 98:2011–2016.
40. Strautnieks SS, Bull LN, Knisely AS, Kocoshis SA, Dahl N, Arnell H, Sokal E, Dahan K, Childs S, Ling V, et al. 1998. A gene encoding a liver–specific ABC transporter is mutated in progressive familial intrahepatic cholestasis. Nat Genet 20:233–238.
41. Dieter MZ, Maher JM, Cheng X, Klaassen CD. 2004. Expression and regulation of the sterol half-transporter genes ABCG5 and ABCG8 in rats. Comp Biochem Physiol C Toxicol Pharmacol 139:209–218.
42. Bergen AA, Plomp AS, Schuurman EJ, Terry S, Breuning M, Dauwerse H, Swart J, Kool M, van Soest S, Baas F, ten Brink JB, de Jong PT. 2000. Mutations in ABCC6 cause pseudoxanthoma elasticum. Nat Genet 25:228–231.
43. Ringpfeil F, Lebwohl MG, Christiano AM, Uitto J. 2000. Pseudoxanthoma elasticum: mutations in the MRP6 gene encoding a transmembrane ATP-binding cassette (ABC) transporter. Proc Natl Acad Sci U S A 97:6001–6006.
44. Ballatori N, Christian WV, Lee JY, Dawson PA, Soroka CJ, Boyer JL, Madejczyk MS, Li N. 2005. OSTalpha-OSTbeta: a major basolateral bile acid and steroid transporter in human intestinal, renal, and biliary epithelia. Hepatology 42:1270–1279.

45. Calcagno PL, Rubin MI. 1963. Renal extraction of *para*-aminohippurate in infants and children. J Clin Invest 42:1632–1639.
46. Horster M, Valtin H. 1971. Postnatal development of renal function: micropuncture and clearance studies in the dog. J Clin Invest 50:779–795.
47. Horster M, Lewy JE. 1970. Filtration fraction and extraction of PAH during neonatal period in the rat. Am J Physiol 219:1061–1065.
48. Lewy JE. 1974. *Proceeding of the 3rd International Symposium on Pediatic Nephrology*, Washington, DC.
49. Phelps DL, Omori K, Oh W. 1976. PAH clearance, sodium excretion, and PAH extraction ratio in acidotic near-term lambs treated with hypertonic sodium bicarbonate. Biol Neonate 28:57–64.
50. Rennick B, Hamilton B, Evans R. 1961. Development of renal tubular transports of TEA and PAH in the puppy and piglet. Am J Physiol 201:743–746.
51. Buist SC, Klaassen CD. 2004. Rat and mouse differences in gender-predominant expression of organic anion transporter (Oat1–3; Slc22a6–8) mRNA levels. Drug Metab Dispos 32:620–625.
52. Anzai N, Kanai Y, Endou H. 2006. Organic anion transporter family: current knowledge. J Pharmacol Sci 100:411–426.
53. Mikkaichi T, Suzuki T, Onogawa T, Tanemoto M, Mizutamari H, Okada M, Chaki T, Masuda S, Tokui T, Eto N, et al. 2004. Isolation and characterization of a digoxin transporter and its rat homologue expressed in the kidney. Proc Natl Acad Sci U S A 101:3569–3574.
54. Wright SH. 2005. Role of organic cation transporters in the renal handling of therapeutic agents and xenobiotics. Toxicol Appl Pharmacol 204:309–319.
55. Slitt AL, Cherrington NJ, Hartley DP, Leazer TM, Klaassen CD. 2002. Tissue distribution and renal developmental changes in rat organic cation transporter mRNA levels. Drug Metab Dispos 30:212–219.
56. Enomoto A, Kimura H, Chairoungdua A, Shigeta Y, Jutabha P, Cha SH, Hosoyamada M, Takeda M, Sekine T, Igarashi T, et al. 2002. Molecular identification of a renal urate anion exchanger that regulates blood urate levels. Nature 417:447–452.
57. Hosoyamada M, Ichida K, Enomoto A, Hosoya T, Endou H. 2004. Function and localization of urate transporter 1 in mouse kidney. J Am Soc Nephrol 15:261–268.
58. Pinto N, Halachmi N, Verjee Z, Woodland C, Klein J, Koren G. 2005. Ontogeny of renal P-glycoprotein expression in mice: correlation with digoxin renal clearance. Pediatr Res 58:1284–1289.
59. Sekine T, Miyazaki H, Endou H. 2006. Molecular physiology of renal organic anion transporters. Am J Physiol Renal Physiol 290:F251–F261.
60. Shitara Y, Sato H, Sugiyama Y. 2005. Evaluation of drug–drug interaction in the hepatobiliary and renal transport of drugs. Annu Rev Pharmacol Toxicol 45:689–723.
61. Dawson PA, Hubbert M, Haywood J, Craddock AL, Zerangue N, Christian WV, Ballatori N. 2005. The heteromeric organic solute transporter alpha-beta, Ostalpha-Ostbeta, is an ileal basolateral bile acid transporter. J Biol Chem 280:6960–6968.
62. Shneider BL, Dawson PA, Christie DM, Hardikar W, Wong MH, Suchy FJ. 1995. Cloning and molecular characterization of the ontogeny of a rat ileal sodium-dependent bile acid transporter. J Clin Invest 95:745–754.

63. Duan LP, Wang HH, Wang DQ. 2004. Cholesterol absorption is mainly regulated by the jejunal and ileal ATP-binding cassette sterol efflux transporters Abcg5 and Abcg8 in mice. J Lipid Res 45:1312–1323.

64. Berge KE, Tian H, Graf GA, Yu L, Grishin NV, Schultz J, Kwiterovich P, Shan B, Barnes R, Hobbs HH. 2000. Accumulation of dietary cholesterol in sitosterolemia caused by mutations in adjacent ABC transporters. Science 290:1771–1775.

65. Maher JM, Slitt AL, Callaghan TN, Cheng X, Cheung C, Gonzalez FJ, Klaassen CD. 2006. Alterations in transporter expression in liver, kidney, and duodenum after targeted disruption of the transcription factor HNF1alpha. Biochem Pharmacol 72:512–522.

66. Kobayashi Y, Ohshiro N, Shibusawa A, Sasaki T, Tokuyama S, Sekine T, Endou H, Yamamoto T. 2002. Isolation, characterization and differential gene expression of multispecific organic anion transporter 2 in mice. Mol Pharmacol 62:7–14.

67. Simon FR, Fortune J, Iwahashi M, Qadri I, Sutherland E. 2004. Multihormonal regulation of hepatic sinusoidal Ntcp gene expression. Am J Physiol Gastrointest Liver Physiol 287:G782–G794.

68. Groves CE, Suhre WB, Cherrington NJ, Wright SH. 2006. Sex differences in the mRNA, protein, and functional expression of organic anion transporter (Oat) 1, Oat3, and organic cation transporter (Oct) 2 in rabbit renal proximal tubules. J Pharmacol Exp Ther 316:743–752.

69. Lu R, Kanai N, Bao Y, Wolkoff AW, Schuster VL. 1996. Regulation of renal oatp mRNA expression by testosterone. Am J Physiol 270:F332–F337.

70. Rost D, Kopplow K, Gehrke S, Mueller S, Friess H, Ittrich C, Mayer D, Stiehl A. 2005. Gender-specific expression of liver organic anion transporters in rat. Eur J Clin Invest 35:635–643.

71. Isern J, Hagenbuch B, Stieger B, Meier PJ, Meseguer A. 2001. Functional analysis and androgen-regulated expression of mouse organic anion transporting polypeptide 1 (Oatp1) in the kidney. Biochim Biophys Acta 1518:73–78.

72. Asaka J, Terada T, Okuda M, Katsura T, Inui K. 2006. Androgen receptor is responsible for rat organic cation transporter 2 gene regulation but not for rOCT1 and rOCT3. Pharm Res 23:697–704.

73. Beato M. 1989. Gene regulation by steroid hormones. Cell 56:335–344.

74. Kimura N, Arai K, Sahara Y, Suzuki H. 1994. Estradiol transcriptionally and posttranscriptionally up-regulates thyrotropin-releasing hormone receptor messenger ribonucleic acid in rat pituitary cells. Endocrinology 134:432–440.

75. Paul SJ, Ortolano GA, Haisenleder DJ, Stewart JM, Shupnik MA, Marshall JC. 1990. Gonadotropin subunit messenger RNA concentrations after blockade of gonadotropin-releasing hormone action: testosterone selectively increases follicle-stimulating hormone beta-subunit messenger RNA by posttranscriptional mechanisms. Mol Endocrinol 4:1943–1955.

76. Tannenbaum GS, Martin JB. 1976. Evidence for an endogenous ultradian rhythm governing growth hormone secretion in the rat. Endocrinology 98:562–570.

77. Saunders A, Terry LC, Audet J, Brazeau P, Martin JB. 1976. Dynamic studies of growth hormone and prolactin secretion in the female rat. Neuroendocrinology 21:193–203.

78. Waxman DJ, Pampori NA, Ram PA, Agrawal AK, Shapiro BH. 1991. Interpulse interval in circulating growth hormone patterns regulates sexually dimorphic expression of hepatic cytochrome P450. Proc Natl Acad Sci U S A 88:6868–6872.

79. MacLeod JN, Pampori NA, Shapiro BH. 1991. Sex differences in the ultradian pattern of plasma growth hormone concentrations in mice. J Endocrinol 131:395–399.

80. Aida K, Negishi M. 1993. A *trans*-acting locus regulates transcriptional repression of the female-specific steroid 15 alpha-hydroxylase gene in male mice. J Mol Endocrinol 11:213–222.

81. Noshiro M, Negishi M. 1986. Pretranslational regulation of sex-dependent testosterone hydroxylases by growth hormone in mouse liver. J Biol Chem 261:15923–15927.

82. Beamer WH, Eicher EM. 1976. Stimulation of growth in the little mouse. J Endocrinol 71:37–45.

83. Cheng TC, Beamer WG, Phillips JA 3rd, Bartke A, Mallonee RL, Dowling C. 1983. Etiology of growth hormone deficiency in little, Ames, and Snell dwarf mice. Endocrinology 113:1669–1678.

84. Jansson JO, Downs TR, Beamer WG, Frohman LA. 1986. Receptor-associated resistance to growth hormone-releasing factor in dwarf "little" mice. Science 232:511–512.

85. Lin SC, Lin CR, Gukovsky I, Lusis AJ, Sawchenko PE, Rosenfeld MG. 1993. Molecular basis of the little mouse phenotype and implications for cell type–specific growth. Nature 364:208–213.

86. Kasukawa Y, Baylink DJ, Guo R, Mohan S. 2003. Evidence that sensitivity to growth hormone (GH) is growth period and tissue type dependent: studies in GH-deficient lit/lit mice. Endocrinology 144:3950–3957.

87. Cheng X, Maher JM, Lu H, Klaassen CD. 2006. Endocrine regulation of gender-divergent mouse organic anion transporting polypeptide (Oatp) expression. Mol Pharmacol 70:1291–1297.

88. Maher JM, Cheng X, Tanaka Y, Scheffer GL, Klaassen CD. 2006. Hormonal regulation of renal multidrug resistance-associated proteins 3 and 4 (Mrp3 and Mrp4) in mice. Biochem Pharmacol 71:1470–1478.

21

POLYMORPHISMS OF DRUG TRANSPORTERS AND CLINICAL RELEVANCE

LESLIE W. CHINN, JASON M. GOW, AND DEANNA L. KROETZ
University of California–San Francisco, San Francisco, California

21.1. GENETIC VARIATION AND DRUG RESPONSE

Numerous environmental and genetic factors influence interindividual variation in drug response. Differences in age, disease state, alcohol and recreational drug use, concomitant drug and natural-product administration, and organ function are all associated with variability in specific drug responses.[1] Pharmacogenetics is the study of the contribution of genetic variability in drug response genes to variability in

Drug Transporters: Molecular Characterization and Role in Drug Disposition, Edited by Guofeng You and Marilyn E. Morris
Copyright © 2007 John Wiley & Sons, Inc.

drug efficacy and toxicity.[2–4] Until recently, most of the efforts in pharmacogenetics focused on drug-metabolizing enzymes and transporters. For example, genetic variability in thiopurine methyltransferase is known to dramatically influence an individual patient's response rate and likelihood of toxicity for the anticancer agent 6-mercaptopurine.[5] Similarly, patients with functional variants of cytochrome P450 2C9 (CYP2C9) and vitamin K epoxide reductase (VKORC1) are at increased risk of breakthrough bleeding and require lower doses of warfarin.[6,7] These examples highlight the importance of genetic variability in drug-metabolizing enzymes and drug targets in determining drug response and toxicity. It is also well recognized that membrane transporters play an important role in determining plasma and intracellular drug levels. A growing interest in understanding the molecular mechanisms of regulation of drug transporters and in defining their in vivo roles in drug disposition makes it of great importance to understand how genetic variability in membrane transporters contributes to interindividual variation in drug response.

The importance of genetic variation in determining a given phenotype can be investigated in two ways: (1) using a phenotype-to-genotype approach, in which an observable phenotype is associated with a gene and polymorphisms in that gene, and (2) using a genotype-to-phenotype approach, in which a systematic examination of naturally occurring polymorphisms takes place, followed by functional and/or clinical examination of the effect of these variants on protein function and a clinical or cellular phenotype. Both of these approaches have been in widespread use only since the development of techniques that enabled the sequencing of the human genome in a time- and cost-effective manner. In phenotype-to-genotype studies the genetic basis for a given phenotype is mapped to a pathogenic locus in the genome by comparing patterns of linkage disequilibrium in affected and unaffected individuals. Once a putative locus is identified, candidate genes most likely involved in a physiological pathway associated with the phenotype are identified and polymorphism screening is carried out. The genotype-to-phenotype approach has been more widely used for drug transporters. For example, the *ABCB1* gene encoding P-glycoprotein has been screened for genetic variation in the coding region, and cell-based systems have been used to characterize the effect of amino acid changes on transport function.[8–10] The best characterized variants of *ABCB1*, 2677G>T/A and 3435C>T, have subsequently been examined for an association with drug pharmacokinetics, response, and toxicity.[11–14]

As the sequencing of the human genome neared completion, the nascent field of pharmacogenomics began to draw interest. The vast majority of the human genome is identical in all individuals—less than 1% of the genome is variable.[15] This raises the intriguing question of whether the small number of interindividual variations in the genome can be used by scientists and clinicians to predict biological outcomes and tailor drug therapies to reduce toxicities while maximizing efficacy. Although this is, of course, the hope of researchers in the field of pharmacogenetics, a significant amount of research is still necessary before these goals are fulfilled. In particular, the field of drug transporter pharmacogenetics is still in its infancy. Large-scale efforts to identify the nature and extent of genetic variation in adenosine triphosphate (ATP)-binding cassette (ABC) and solute carrier (SLC) transporters is well under way, although most data available cover only the coding region of these genes (see www.pharmgkb.org).

Functional analysis of membrane transporter variants lags further behind and clinical studies are largely inconclusive. However, the increased understanding of important roles for drug transporters in determining intracellular concentrations and drug response makes this an exciting and vibrant area of research. Presented below is the current state of understanding of the pharmacogenetics of membrane transporters, including results from genetic, functional, and clinical studies. Future directions and challenges are also discussed.

21.2. GENETIC VARIATION IN MEMBRANE TRANSPORTERS

Systematic examination of genetic variation has been carried out for a number of membrane transporter genes, including members of the *ABCB*, *ABCC*, *ABCG*, *SLC10*, *SLC15*, *SLCO*, *SLC22*, *SLC28*, and *SLC29* families.[8–10,16–59] Most of these single nucleotide polymorphisms (SNPs) can be found in several publicly available databases, including the Pharmacogenetics Knowledge Base (PharmGKB), which catalogs genotypic and phenotypic pharmacogenetic data (www.pharmgkb.org). In most cases, only the coding region and surrounding intron–exon boundaries were examined, based on the hypothesis that nonsynonymous changes leading to an amino acid change would be most likely to influence transport function.[8–10,16–22,24–26,29,33,36–45,47,48,50,51,53,55–59] A limited number of data are available regarding genetic variation in the untranslated and promoter regions of membrane transporter genes.[23,27,28,32,34,35,49,52,54] Polymorphisms in untranslated regions may influence mRNA stability and translation, while promoter region variants may influence transcription and gene/protein expression. In light of the relative paucity of common coding region variants that affect transport function (see Section 21.3), it is increasingly interesting to determine whether genetic variation in these noncoding regions may have a greater impact on transporter function. Many studies have focused on a single ethnic population and have examined a sufficient number of representative DNA samples to provide reasonable estimates of minor allele frequencies for common variants (>5% minor allele frequency) in that population.[8,16,20,23,28,29,31–35,40,41,49,54] Others have carried out deep resequencing in multiple ethnic groups, including Caucasians, African Americans, Mexican Americans, and Asian Americans.[9,10,17,18,21,22,24,25,36,38,43–45,48,51] The latter approach is important since genetic variation will differ across ethnic groups, and population-specific information is necessary for application to the clinical setting.

A recent analysis of 24 membrane transporter genes that were screened for polymorphisms in exonic and flanking intronic regions revealed some interesting trends in genetic variation.[60] A total of 680 single-nucleotide polymorphisms (SNPs) were identified in 96 kb of sequence that was screened in DNA samples from 247 subjects, including 100 Caucasians, 100 African Americans, and 30 Asians. Similar numbers of coding- and non-coding-region SNPs were found, and only 12 SNPs were identified in intron–exon boundaries. The deep resequencing across the Caucasian and African-American populations in the Leabman et al. analysis identified a large number of population-specific SNPs.[60] Of the 680 SNPs that were identified, more than two-thirds were population specific, and more than half of these were rare mutations found on only one of 494 chromosomes. A larger number of common population-specific SNPs (>5% minor allele frequency) were found in the African-American population compared to the Caucasian population.

Interestingly, the number of synonymous (silent) and nonsynonymous (resulting in an amino acid change) SNPs identified in the 24 membrane transporter genes was also similar (175 and 155, respectively).[60] However, statistical genetic analysis of the variation observed, which takes into account the frequency of the SNP and the number of alleles that were screened (denoted as the average heterozygosity, π), revealed that variation was about three- to fourfold more common at synonymous positions than at nonsynonymous positions (Figure 21.1*a*). This suggests that there is some selective pressure on membrane transporter genes to suppress dramatic changes in transporter function. Genetic variation was strikingly low at nonsynonymous sites for

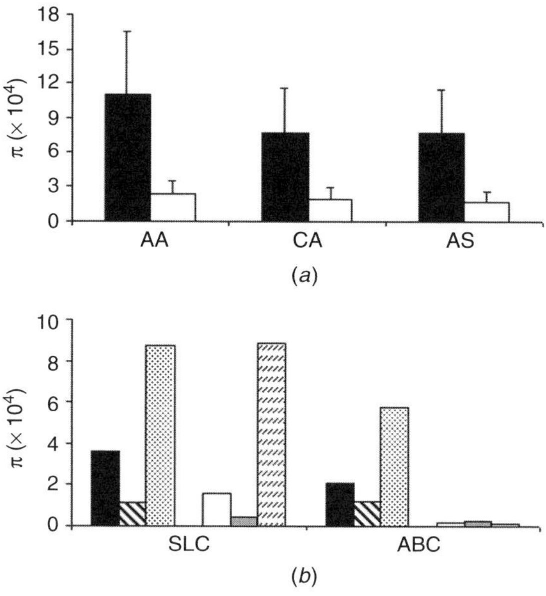

FIGURE 21.1. (*a*) Nucleotide diversity (π) is greater at synonymous than at nonsynonymous sites. The π values are shown for synonymous (black bars), and nonsynonymous (open bars) sites and were calculated separately for a population of 100 African Americans (AA), 100 Caucasians (CA), and 30 Asian Americans (AS). The values shown are the mean ± S.D. (*b*) Nucleotide diversity is generally higher in the loops than in the transmembrane domains and is particularly low for evolutionarily conserved residues. The π values are shown for all nonsynonymous loop residues (black bars), evolutionarily conserved loop residues (diagonal lines), evolutionarily unconserved loop residues (dotted bars), all nonsynonymous transmembrane domain residues (open bars), evolutionarily conserved transmembrane domain residues (gray bars), and evolutionarily unconserved transmembrane domain residues (stippled bars). The values are shown separately for 18 SLC and 5 ABC transporters and are the mean ± S.D. (Data from ref. 60.)

the *SLC29A1* gene encoding ENT1, *ABCC1* encoding MRP1, and *SLC22A3* encoding OCT3.[60] Amino acid diversity was also much lower in the transmembrane regions of the transporters compared to the loops (Figure 21.1*b*). This was particularly true for the ABC transporters, with calculated π values for nonsynonymous sites varying more than 13-fold between loops and transmembrane domains. In the loop regions, nonsynonymous SNPs were much more common at evolutionarily unconserved sites than at conserved sites. A similar relationship held for the transmembrane domains of the SLC, but not the ABC transporters. Of the five ABC transporters analyzed in this study, average heterozygosity in the transmembrane regions was extremely low at all sites, irrespective of sequence conservation across species.[60] A number of algorithms were evaluated that parsed nonsynonymous SNPs into various categories based on chemical similarities and evolutionary relatedness. Assuming that deleterious amino acid changes would be selected against and thus found at low frequency, consideration of conservation at the variant site across orthologous species is predicted to be the best indicator of a detrimental effect on transporter function.[51,60] In fact, functional analysis of 15 amino acid variants of OCT1 revealed that evolutionary conservation is a strong predictor of function, although consideration of the degree of chemical change is also informative.[51]

In general, insertion and deletion (indel) mutations are relatively rare.[61–63] In the 24 membrane transporter genes that were studied by Leabman and colleagues, a total of 29 indels were found, including eight that affected the coding region.[60] The majority of these coding indels resulted in the addition or deletion of a single amino acid with no disruption of the reading frame. A common insertion of a Leu at position 140 in CNT1 is found at a frequency of ~30% in Caucasians and African Americans, and a frameshift deletion at codon 385 in this same transporter occurs at a frequency of 3% in African Americans.[24]

Haplotypes define the combination of genotypes across a given gene or a multigenic region and are expected to more accurately predict functional consequences of genetic variation than is consideration of single SNPs. The near completion of the HapMap project to define haplotype blocks across all human genes provides a wealth of information regarding genetic variation in persons of African, European, and Asian descent, and this information is being used increasingly in the design of genetic association studies.[64] Haplotype structure has been determined for several membrane transporter genes based on the variants identified during population screening.[9,10,24,41,43,45,46,48] In some cases, a relatively low number of haplotypes describes the genetic variation in a given population; for example, a single haplotype (*SLC29A1**1) accounted for more than 90% of the 494 chromosomes screened in one study.[43] In contrast, almost 60 coding region haplotypes were estimated for *SLC28A1*.[24] Numerous haplotypes have been estimated for *ABCB1*, although relatively few contain nonsynonymous variants that would result in altered protein sequence.[9,10] Of these, *ABCB1**13 is the most common haplotype in Caucasians and Asian Americans and contains three intronic variants: two synonymous changes (1236C>T and 3435C>T) and a single nonsynonymous site (2677G>T/A, Ala893Ser).[10] The *SLC22A4* and *SLC22A5* genes are located in tandem on chromosome 5q31 and have a functional haplotype that spans both genes and includes a missense mutation in *SLC22A4* and a promoter region variant in *SLC22A5*.[46] Interestingly, this haplotype has been shown to alter transport function and transcriptional activity and is associated with an increased risk of Crohn's disease.[46]

21.3. FUNCTIONAL ANALYSIS OF TRANSPORTER VARIANTS

In contrast to genetic variation data that can be amassed in large quantities using state of the art high-throughput methods, characterization of the functional consequences of genetic variation in membrane transporters is significantly more challenging. Since initial resequencing efforts to identify transporter variants were exoncentric, functional studies have focused largely on the effects of amino acid changes on transporter function. The transmembrane domains probably confer substrate specificity for SLC and ABC transporters, while specific transmembrane and loop domains may control proper trafficking to the plasma membrane.[65–68] The large intracellular loops of ABC transporters contain ATP-binding domains that are required for energy-dependent transport.[69] Amino acid changes can potentially affect transport kinetics and substrate/inhibitor selectivity, properties that can be examined in heterologous expression systems in vitro. The basic approach in examining changes in transporter function is to express the reference and variant proteins in a heterologous expression system and to measure the transport of known substrates and interaction with known inhibitors. Determination of transport kinetics provides additional mechanistic data regarding the effect of a SNP or haplotype on transport function. The functional consequences of untranslated region (UTR) and promoter variants can also be examined in vitro. Variants in the UTR may affect mRNA stability and promoter region variants can influence transcription and ultimately the expression and function of the transporters. Cell-based systems are used to compare mRNA stability and transcriptional activity of reference and variant transporter sequences. Such in vitro studies of the functional consequences of genetic variation in membrane transporters can serve as the basis for future studies aimed at determining the clinical consequences of functionally significant variants.

21.3.1. SLC Transporters

Functional analysis of coding-region changes have been reported for many of the SLC transporters, and a summary of these studies is presented in Table 21.1. The function of members of the SLC family as uptake transporters greatly facilitates their characterization relative to the ABC efflux transporters, thus explaining our increased understanding of the impact of amino acid changes on SLC transporter activity. In a recent analysis, data were culminated from multiple studies to elucidate general principles related to the functional effects of amino acid variants on SLC membrane transporters.[70] Eleven transporters (OCT1, OCT2, OCTN1, OCTN2, OAT1, OAT3, CNT1, CNT2, CNT3, ENT1, and ENT2) were analyzed for the number of variants that caused a change in transport function. A total of 88 variants across the 11 transporters

TABLE 21.1. Summary of Functional Consequences of Nonsynonymous Polymorphisms in SLC and ABC Membrane Transporters

Transporter	Variant	Function	Substrates	Refs.
CNT1	V189I	↓	Gemcitabine	24
	V385del, S546P	↓↓	Thymidine	
CNT3	S5N	↔	Adenosine, cladribine, fludarabine, inosine, thymidine, cytidine, uridine	18,20
	G367R	↓	Inosine, thymidine	
ENT2	D5Y	↓	Gemcitabine, fludarabine, hypoxanthine, inosine, uridine	45
	S282del	↓↓		
OAT1	R50H	↓	Adefovir, cidofovir, tenofovir	19
	K525I	↔		
	R454Q	↓↓	*p*-Aminohippurate, methotrexate, ochratoxin A	22
OAT3	R149S, Q239del, I260R	↓↓	Estrone 3-sulfate, cimetidine	21,70
	R277W, I305F	↓		
OATP1B1	F73L, V82A, I353T, G488A	↓↓	Estrone sulfate, estradiol 17β-glucuronide, atorvastatin, cerivastatin, pravastatin	55,75–78
	N130D	↔		
	V174A	↔, ↓, ↓↓		
	N130D + V174A	↓		
	N432D	↔, ↓		
OCT1	S14F	↑	MPP^+, TEA	51,73
	R61C, P341L	↓		
	F160L	↔		
	G220V, P283L, R287G, G401S, G465R	↓↓		
OCT2	F45lns, M165I, R400C	↓	MPP^+	74
OCTN1	D165G, M205I, R282del	↓	Tetraethylammonium	46,70,172
	L503F	↑		
	T306I	↔		
	G462E	↓↓		
OCTN2	F17L	↓	Carnitine	70,173,174
	E452K, S467C	↓↓		

(Continued)

TABLE 21.1. ***(Continued)***

Transporter	Variant	Function	Substrates	Refs.
P-gp	N21D, F103L, A998T S400N A893S	↔ ↔, ↓ ↔, ↑	Bisantrene, calcein-AM, daunorubicin, bodipy-FL forskolin, bodipy-FL prazosin, bodipy-FL verapamil, bodipy-FL vinblastine, rhodamine 123, vinblastine, vincristine, doxorubicin, digoxin, amprenavir, indinavir, lopinavir, ritonavir, saquinavir, calcein-AM, cyclosporin A, verapamil	9,10,80,81
MRP1	T73I, S92F, T117M, R230Q, R633Q, G671V, R723Q, C1047S, R1058Q, S1512L R433S, A989T	↔ ↓, ↔	Leukotriene C_4, estradiol 17β-glucuronide, glutathione	84,85
MRP2	R412G, R1150H V417I, A1450T R768W, I1173F S789F Q1382R	↓ ↔ ↓, ↓↓ ↑ ↓↓	Methotrexate, glutathione-methylfluorescein, 2,4-dinitrophenyl-S-glutathione, leukotriene C_4, estradiol 17β-glucuronide, gluthothione-monochlorobimane, carboxyfluorescein	90,91,93,94
MXR	V12M, Q141K A149P, R163K, Q166E, P269S I206L S441N, N590Y, D620N	↓, ↔ ↔ ↑ ↓	Mitoxantrone, topotecan, doxorubicin, SN-38, esterone 3-sulfate, DHEAS, PAH, bodipy FL-prazosin, pheophorbide a	97,175,176
BSEP	E297G R432T D482G	↔, ↓ ↓ ↔	Glycocholate, taurocholate	177,178

were tested functionally in heterologous expression systems with model substrates, and 22 variants reduced function. Interestingly, those variants that were detrimental to function were more likely to be amino acid changes at evolutionarily conserved positions. Another intriguing result from this multitransporter analysis is that some protein-altering variants can change substrate specificity. In a survey of nine SLC transporters, 14% of the variants exhibited altered substrate specificity. Such a finding will have important implications on clinical studies to investigate how SLC transporter polymorphisms influence pharmacokinetic and pharmacodynamic properties of therapeutic compounds. In general, the SLC transporter variants discussed in this section are highlighted because of their common allele frequency and/or functional effect.

Using a *Xenopus laevis* oocyte expression system, 15 nonsynonymous variants of *SLC22A1* (OCT1) were evaluated with respect to MPP^+ uptake.[51] Of the six OCT1 variants with altered function, two showed reduced MPP^+ uptake (Arg61Cys and Pro341Leu), three had complete loss of function (Gly220Val, Gly401Ser, and Gly465Arg), and one variant increased accumulation of MPP^+ (Ser14Phe). In addition to decreased MPP^+ transport, the Cys88Arg, Pro341Leu, and Gly401Ser variants have decreased transport of tetraethylammonium (TEA), and the latter two variants have diminished ability to transport serotonin relative to the reference OCT1.[71–73] The Pro341Leu variant is of particular interest since it is found at a frequency of greater than 8% in African Americans and Asian Americans.[51] Kinetic analysis of the Arg61Cys and Pro341Leu variants with MPP^+ showed a simultaneous decrease in V_{max} and increase in K_m.[51] While the relevance of V_{max} changes measured in heterologous expression systems is unclear, the increased K_m suggests that the Cys61 and Leu341 variants of OCT1 have decreased affinity for MPP^+ compared to the reference transporter. In contrast, the nonfunctional Gly465Arg OCT1 variant had impaired localization to the basolateral membrane relative to the reference transporter.[51] Analysis of a related transporter, OCT2, showed a reduced K_m for MPP^+ with the Lys432Gln variant relative to the reference transporter, suggesting increased affinity for this substrate.[74]

OATP1B1 (*SLCO1B1*) has multiple protein-altering variants that have been functionally studied in vitro and many show reduced transport of model substrates (Table 21.1). The most common nonsynonymous polymorphisms are at residues 130 (Asn130Asp) and 174 (Val174Ala), which are located in the second extracellular loop and fourth transmembrane domain, respectively. Asn130Asp is a common variant (>30%) in African Americans, European Americans, and Japanese while Val174Ala is common in European Americans (14%) and is much less frequent in African Americans and Japanese (<3%).[55,75] Heterologous expression of the Asn130Asp and Val174Ala variants generally shows no effect on the transport of estrone sulfate and estradiol 17β-glucuronide.[75–77] Similarly, these variants had no effect on the transport of pravastatin or the active irinotecan metabolite SN-38.[76] However, in some systems the Val174Ala variant shows reduced uptake of both estrone sulfate and estradiol 17β-glucuronide.[55,78] A natural haplotype of these two variants (*SLCO1B1**15) is found in Japanese subjects at a frequency of 3%.[75] When the two variants were expressed together, there was decreased uptake of estradiol 17β-glucuronide, estrone 3-sulfate, pravastatin, and SN-38.[76–78] In several studies, cellular localization experiments showed that the Val174Ala variant was expressed at lower levels on the plasma membrane and was also found in intracellular compartments.[55,78] It is possible that *SLCO1B1**15 alters membrane trafficking and that the reduction in OATP1B1 transporters on the plasma membrane decreases the V_{max}. Additional nonsynonymous variants of OATP1B1 (Phe73Leu, Val82Ala, Ile353Thr, and Gly488Ala) show reduced function for either estrone 3-sulfate or estradiol 17β-glucuronide (Table 21.1), but their allele frequency is generally less than 2% in European Americans and African Americans.[55]

The concentrative and equilibrative nucleoside transporters have a relatively small number of functionally significant variants. In a study of 14 amino acid variants of CNT1 only two (Val385del and Ser546Pro) were unable to transport thymidine.[24] Despite normal levels of thymidine uptake by the Val189Ile CNT1 variant, differential interaction with gemcitabine was noted. Similarly, only one of 10 CNT3 variants (Gly367Arg) showed loss of thymidine and inosine uptake.[18,20] Neither of the two nonsynonymous ENT1 variants had altered transport, but a deletion variant of ENT2 (Δ845–846) was unable to transport inosine, guanosine, uridine, hypoxanthine, fludarabine or gemcitabine.[43,45] Reduced transport of these substrates was also noted for the Asp5Tyr variant of ENT2.[45]

Of the organic anion transporters, nonsynonymous variants have been characterized functionally for OAT1 and OAT3. The majority of OAT1 variants are able to transport *p*-aminohippurate, ochratoxin A, and methotrexate to a degree similar to that of the reference transporter, although the Arg454Gln variant of OAT1 has complete loss of transport of these three substrates.[22] The transport of the antiretrovirals adefovir, cidofovir, and tenofovir is impaired significantly with the Arg50His variant of OAT1.[19] The Arg50His variant is found at 17% frequency in African Americans, suggesting that it could have a significant effect on the disposition of these antiretrovirals in this ethnic population. The uptake of estrone sulfate and cimetidine is abolished in the Arg149Ser, Gln239del, and Ile305Phe variants of OAT3.[21,70] However, only the Ile305Phe variant is found at a frequency of >1% (3.5% in Asian Americans), so the clinical significance of these changes might be minimal.

The majority of polymorphic sites with functional consequences have negative effects on transport function; however, there are rare examples where variants increase function. A nonsynonymous variant in OCT1 causes a Ser-to-Phe change at residue 14 (Ser14Phe) which was associated with a 75% increased uptake of MPP^+ and TEA and a ~twofold increase in Cl_{int} (V_{max}/K_m) compared to the reference OCT1.[51] One general assumption is that if an amino acid position is necessary for optimal protein function, it should be conserved across species. An alignment with other mammalian sequences orthologous to OCT1, as well as human OCT2 and OCT3, showed that Phe14 is evolutionarily conserved except for in human OCT1.[51] Although the mechanism for this increased function is unclear, the variation at a highly conserved residue is consistent with the functional change observed.

The in vitro functional effects of a variant may not always be clear if a limited number of substrates are tested. As discussed earlier, a variant may show only functional effects with specific substrates. One such example is OCTN1. Mammalian cell lines were transiently transfected with the reference and Leu503Phe OCTN1 to see how multiple unrelated substrates affected transport.[46] Enhanced function with Phe503 OCTN1 was observed with tetrabutylammonium, tetraethylammonium, and tetrapentylammonium; however, reduced function was seen with carnitine, choline, cimetidine, lidocaine, N-methylnicotinamide, and verapamil. Significant differences were seen in K_m values, suggesting that residue 503 influences substrate affinity. These studies nicely illustrate how detailed functional analysis of membrane transporter variants can provide important information regarding transporter biology.

In some cases, substrate selectivity of genetic variants is assessed indirectly by inhibition studies. Four OCT2 variants were analyzed in *Xenopus laevis* oocytes using the model substrate MPP^+, and only the Lys432Gln variant had altered transport.[36] Interestingly, tetramethylammonium, tetraethylammonium, and tetrapropylammonium each inhibited the reference OCT2 and the Met165Ile, Ala270Ser, Arg400Cys, and Lys432Gln variants to a similar degree. In contrast, tetrabutylammonium was a more potent inhibitor of the Arg400Cys and Lys432Gln variants and a weaker inhibitor of the Ala270Ser variant than was the reference.

21.3.2. ABC Transporters

Amino acid variants of multiple ABC transporters have been characterized functionally in vitro, and many of the results from these studies are summarized in Table 21.1. The most widely studied ABC transporter is P-glycoprotein (Pgp), which is encoded by *ABCB1*. There are dozens of polymorphic sites in *ABCB1* that have been identified through multiple SNP discovery efforts.[8–10] Functional analyses of all of the nonsynonymous variants are slowly accumulating but the most frequently studied SNP is the triallelic variant at amino acid position 893 that is located in the sixth intracellular loop near the C-terminus. The reference Ala can change into a common Ser or a lower frequency Thr.[9,10] The Ala-to-Ser variant was first seen in the drug-resistant cell line MCF-7/Adr, which is a breast cancer cell line that overexpresses ABCB1.[79] ABCB1 was cloned from MCF-7/Adr and transfected into different drug-sensitive cell types. The cells acquired the drug resistance phenotype but no comparison was made with the reference Ala at codon 893. The functional effects of the Ala893Ser polymorphism were first examined in a specialized mouse fibroblast cell line (NIH3T3-GP+E86).[9] In this study the Ser893 Pgp showed enhanced efflux of the model substrate digoxin. In contrast, analysis of the function of five nonsynonymous variants, including Ala893Ser, in HeLa cells (human cervical cancer cell line) using seven different substrates showed no differences in transport compared to the reference Pgp.[80] Two subsequent studies looking at Ala893Ser also concluded that there was no in vitro functional difference using established Pgp substrates, such as calcein-AM and verapamil.[10,81] It is possible that the varying results for the Ser893 variant reflect differences between the heterologous expression systems used in these analyses or result from substrate-dependent effects of this polymorphism. Other Pgp variants have been characterized in vitro, including Asn21Asp, Phe103Leu, Ser400Asn, and Ala998Thr.[80,82,83] In most cases there were no differences in function between the reference and variant Pgps; however, Ser400Asn shows increased resistance to vinblastine and vincristine and decreased transepithelial flux of rhodamine 123.[83] The Ser400Asn variant also has altered affinity to several protease inhibitors.[82]

Limited in vitro functional data exist for polymorphisms of *ABCC1* encoding MRP1. In a single study the function of ten nonsynonymous variants were examined using three different MRP1 substrates. There appeared to be minor substrate-dependent changes for some of the variants; however, only the Ala989Thr was determined to have a significant functional effect. Specifically, the 989Thr MRP1 showed a 50% decrease in function when transporting estradiol 17β-glucuronide but had normal transport function with leukotriene C_4 and glutathione.[84] Amino acid 989 is at the membrane interface of the seventh extracellular loop and may play an important role in substrate interactions. The functional effects of a variant of amino acid 433, located at the interface between the cytoplasm and the plasma membrane, have also been investigated.[85] The Arg433Ser polymorphism occurs in the fourth cytoplasmic loop close to the membrane interface of transmembrane domain 8. MRP1 was expressed transiently in HEK293 and HeLa cells, and membrane vesicles were used to measure transport of leukotriene C_4, estradiol 17β-glucuronide, and estrone sulfate. The Ser433 MRP1 showed a 50% decrease in transport of leukotriene C_4 and estrone sulfate as well as a decrease in V_{max} for both substrates relative to the reference transporter.[85] It is possible that the conversion from a positively charged amino acid (Arg) to a neutral one (Ser) disrupted substrate affinity.[86]

A deficiency in MRP2 function can lead to altered transport of conjugated bilirubin in the liver, resulting in the hyperbilirubinemic disease known as Dubin–Johnson syndrome.[87,88] Certain mutations in *ABCC2* are thought to form an inactive protein product and are regarded as the molecular basis of Dubin–Johnson syndrome.[89–91] In some cases, including the Arg768Trp and I1173Phe variants, loss of MRP2 transport function is a result of defects in protein maturation and sorting to the apical membrane.[90,92] In contrast, the Gln1382Arg and Arg1150His polymorphisms have no effect on localization but disrupt the nucleotide-binding domain and ATP-dependent transport of MRP2 substrates.[90,91] A rare mutation resulting in the Arg412Gly variant MRP2 has recently been associated with loss of methotrexate function in vitro and higher plasma methotrexate levels in vivo.[93] The most common nonsynonymous variant of MRP2 described to date is a Val417Ile change found at a frequency of 12 to 17% in major ethnic populations (see www.pharmgkb.org). A comparison of reference and Ile174 MRP2 transport of estradiol 17β-glucuronide, leukotriene C_4, and 2,4-dinitrophenol-*S*-glutathione showed no significant changes in function.[94]

The ATP-binding domains for ABC transporters are important for function as is evident in their high degree of conservation across subfamilies and species.[95] A priori any modifications made to these regions would be predicted to alter energy-dependent transport of substrates. Pseudoxanthoma elasticum is a rare heritable disorder defined by the calcification of elastic fibers, and dozens of mutations in *ABCC6* (encoding MRP6) have been associated with pseudoxanthoma elasticum.[96] Many of the mutations cause nonsynonymous changes in or around the nucleotide-binding domains (NBDs) of MRP6, which most likely produce an aberrant MRP6 transporter. Three variant amino acid positions located in the second nucleotide-binding domain (NBD2) of MRP6—Val1298Phe, Gly1302Arg and Gly1321Ser—are conserved residues thought to be important for hydrogen bonding with ATP and/or proper side-chain folding.[96] These variants were expressed individually in Sf9 insect cells and transport of *N*-ethylmaleimide *S*-glutathione and leukotriene C_4 was measured in membrane vesicles. There was no difference in expression between the reference and variant MRP6 transporters, but each variant showed an 80% decrease in transport activity for both substrates compared to reference MRP6. Val1298Phe and Gly1302Arg MRP6 variants also showed impaired ATP binding using a nucleotide-trapping technique.[96] The amino acid changes in NBD2 of MRP6 have loss-of-function consequences that demonstrate the importance of the ATP binding domains in ABC transporters.

Most of the functional data on ABCG2 polymorphisms focuses on the Gln141Lys variant in the cytoplasmic loop of the protein. This polymorphism is found at a frequency of approximately 9% in Caucasians and less than 2% in African Americans.[97andwww.pharmgkb.org] Initial studies showed that Lys141 MXR had lower in vitro protein expression as well as decreased in vitro drug resistance to mitoxantrone, topotecan, and SN-38.[29] More recently, the Gln141Lys variant has also been linked with a defect in ATPase activity associated with an increased K_m for ATP.[97] This is somewhat surprising since the polymorphism of note is not located in the nucleotide-binding domain. Changes in ATPase activity were also associated with increased sensitivity to mitoxantrone, doxorubicin, and topotecan.

Limited studies have been reported to date on the effects of UTR and promoter variants on ABC transporter expression and function. *ABCC1* has a G-to-C transversion 260 bp upstream (−260G>C) of the transcriptional start site that is located in the core promoter region. Based on an interpopulation genome analysis, it was predicted that −260G>C underwent recent positive selection, and as a result this promoter variant may be functionally relevant.[98] The transcriptional activity of the reference and −260G>C variant of *ABCC1* was studied in four different mammalian cell lines using reporter constructs, and in each case the variant C allele had at least a twofold greater level of promoter activity. Bioinformatic analysis indicates that this SNP resides in a putative c-ETS-1 transcription factor binding site.[98] The variant C-allele disrupts the c-ETS-1 binding sequence and may negate the repressive properties c-ETS-1 has on transcription.

A recently recognized mechanism for changes in transporter mRNA and protein expression is allelic imbalance. The highly studied 3435C>T polymorphism in *ABCB1* was shown to be associated with the expression of ABCB1 mRNA in liver.[99] Specifically, the expression of the 3435C allele was 10 to 60% higher than that of the 3435T allele, which translated into differences in ABCB1 hepatic mRNA levels that were 3435 genotype dependent. Cell-based studies showed that the mRNA transcripts containing the 3435T allele had decreased stability relative to the reference allele. The possibility that this is a more general mechanism for altered expression and function of transporters is currently being explored.

21.4. CLINICAL SIGNIFICANCE OF DRUG TRANSPORTER VARIANTS

All genotype–phenotype association studies require the collection or availability of clearly defined and well-documented phenotypes. Pharmacogenetic phenotypes include drug response expressed either in a continuous (e.g., decrease in blood pressure) or categorical (e.g., 5-year disease-free survival) manner and the presence of a drug-related toxicity such as peripheral neuropathy, nephrotoxicity, or specific CNS disturbances. Appropriate phenotypes for association studies with membrane transporter variants can be any pharmacokinetic parameter that drug transporters influence (e.g., CL or F) as well as measures of drug efficacy or toxicity. In the latter case, drug transporters may be implicated as a major determinant of drug concentration at the target or toxicity site. In this section we describe a number of examples of SLC and ABC transporter polymorphisms that have been associated with clinical endpoints, ranging from pharmacokinetics and pharmacodynamics to congenital disorders; a comprehensive summary of SLC and ABC transporter genotypes that have been linked to clinical phenotypes can be found in Table 21.2.

21.4.1. SLC Transporters

Proteins in the SLC superfamily transport small molecules in a process that does not directly utilize the energy of the cell. Most, if not all, SLC transporters have both endogenous and xenobiotic substrates that are transported with varying degrees of efficiency. SLC family members often have overlapping substrate specificity, although their tissue distribution may show isoform-specific patterns. Variation in SLC genes may cause defects in the function of physiologically important pathways such as bile acid metabolism or autoimmune function as well as drug disposition and response. The investigation of several serious disorders led to the discovery and characterization of SLC gene polymorphisms.

One early example of a phenotype–genotype association study began with the publication of a case report in 1975 in which a boy with underdeveloped musculature and weakness as well as recurring episodes of hepatic and cerebral dysfunction was found to have marked carnitine deficiency in skeletal muscle, plasma, and liver.[100] Oral administration of carnitine returned plasma levels to normal and symptoms improved, although muscle and liver carnitine levels did not change. This disorder was classified as primary systemic carnitine deficiency (SCD, OMIM 212140).[100] The pathogenic locus for SCD was mapped by linkage analysis to chromosome 5q in a Japanese family in 1998.[101] The *SLC22A5* (OCTN2) gene was cloned in 1999 and was found to transport carnitine in a sodium-dependent manner in vitro. Analysis of three SCD family pedigrees uncovered a number of mutations that resulted in a truncated OCTN2: a 113-bp deletion containing the start codon encoded a transporter that lacked the first two transmembrane domains, a cytosine insertion just after the start codon led to a frameshift, a SNP in the first codon of exon 2 resulted in the creation of a premature stop codon, and a splice site mutation removed exon 9 and caused a frameshift. All of these mutations abolish transporter function and cause the carnitine deficiency phenotype.[102] At least 20 additional polymorphisms in *SLC22A5*

TABLE 21.2. Summary of Clinical Associations with SLC and ABC Membrane Transporter Polymorphisms[a]

HUGO Name	SNP	Population	Outcome Marker	Effect	Ref.
SLC22A4, *SLC22A5*	1672C>T (Leu503Phe), −207G>C	625 CD, 363 UC patients	CD or UC susceptibility and severity	1672T/−207C haplotype associated with ↑ CD risk, earlier age of disease onset, and ↓ need for surgery	179
		241 CD, 247 UC Japanese patients	CD or UC susceptibility	No statistical association	180
		120 Caucasian patients	CD susceptibility	1672T/−207C haplotype associated with CD	181
		769 CD, 186 UC Caucasian patients	CD or UC susceptibility and severity	1672T/−207C haplotype associated with penetrating disease but not susceptibility	182
		679 patients	CD or UC susceptibility	1672T/−207C haplotype not associated with CD in the absence of IBD risk haplotype; no association with UC	183
		229 Caucasian children, 502 parents	CD or UC susceptibility	1672T/−207C haplotype not associated with CD and UC in the absence of IBD risk haplotype	107
		503 patients	CD susceptibility	1672T/–207C haplotype associated with CD	46
	11 SNPs	909 Caucasian patients	RA susceptibility	No statistical association	184
SLC22A4	Intronic SNPs slc2F1 and slc2F2, 1672C>T (Leu503Phe)	918 RA, 507 CD Caucasian patients	CD or RA susceptibility	No statistical association	185

(*Continued*)

TABLE 21.2. (*Continued*)

HUGO Name	SNP	Population	Outcome Marker	Effect	Ref.
SLCO1B1	*1a, *1b, *5	30 healthy Caucasian males	Pravastatin PK	Compared to *1a/*1a: *1b carriers AUC(0–6 h) ↓; *5 carriers AUC(0–6 h) ↑; *1b A_e(0–12 h) ↓	111
	*1b, *15	23 healthy Japanese males	Pravastatin PK	Compared to *1b/*1b: *15 carriers CL and CL_{nr} ↓	41
SLC16A1	2 missense mutations	5 patients	Subnormal erythrocyte lactate transport; muscle injury following exercise	Mutations are at conserved site and are not found in healthy controls	112
SLC10A2	Ala171Ser, 169C>T	458 colorectal adenoma patients	Colorectal cancer susceptibility	169CT genotype associated with risk of colorectal cancer compared to CC genotype (no TT group); Ala171Ser not associated with colorectal cancer risk	186
SLC2A1	−2841A>T, intron 2 +22999G>T	92 CCRC patients undergoing nephrectomy	CCRC susceptibility	−2841T and +22999G alleles associated with clear-cell renal carcinoma	187
ABCB1	3435C>T	14–21 healthy Caucasians	Digoxin PK, Pgp duodenal expression	3435TT genotype associated with ↓ duodenal MDR1 expression and ↑ digoxin levels than CC or CT	8

24 healthy volunteers	Digoxin PK, Pgp duodenal expression	3435T genotype associated with ↑ AUC(0–4 h) and C_{max} than 3435C	125
50 healthy Caucasian subjects	Digoxin PK	No statistical association	126
metaanalysis	Digoxin AUC, C_{max}, baseline intestinal MDR1 and Pgp expression	No statistical association	127
14 healthy subjects	Cyclosporine PK	No statistical association	129
44 Caucasian patients	Chemotherapy response, MDR1 expression, Pgp expression, Pgp activity in lymphoblast cells	No statistical association	188
113 Caucasian children	Occurrence and outcome of ALL	TT associated with occurrence of ALL; CC associated with ↓ probability of event-free survival and overall survival	189
102 children	Etoposide PK	CC associated with ↑ day 29 CL	190
53 patients	MDR1 lymphoblast expression; overall survival	No statistical association	191
123 patients	Plasma antiretroviral concentration, change in CD4 cell count	3435TT>CT>CC ↑ in CD4 cells; 3435TT<CT<CC median plasma drug concentration	136

(*Continued*)

TABLE 21.2. ***(Continued)***

HUGO Name	SNP	Population	Outcome Marker	Effect	Ref.
		340 HIV+ patients	Virologic failure, efavirenz PK	3435TT associated with decreased likelihood of virologic failure but not efavirenz exposure	137
		149 drug-naive HIV+ patients	CD4 cell recovery	No statistical association	140
		31 drug-naive HIV+ patients	Viral decay, CD4 cell recovery, ritonavir PK	No statistical association	139
		461 drug-naive HIV+ Caucasians	Time to virological success or failure or immunological failure	Trend toward earlier virological failure in 3435CC group	141
		32–43 HIV+ patients	LPV plasma trough levels, 12-hEFV levels	No statistical association	192
		200 drug-resistant, 155 drug-responsive patients	Refractory epilepsy	3435CC associated with drug-resistant epilepsy	147
		230 drug-resistant, 170 drug-responsive patients	Refractory epilepsy	No statistical association	151
		401 drug-resistant; 208 drug-responsive patients	Refractory epilepsy	No statistical association	150
		63 drug-resistant; 108 drug-responsive Koreans	Refractory epilepsy	No statistical association	131
		30 early-onset, 77 late-onset (59 pesticide-exposed) patients	Onset of Parkinson's disease	Frequency of 3435TT ↑ in exposed vs. nonexposed patients (fivefold ↑ disease risk; CT associated with threefold ↑ disease risk)	157

		149 UC, 126 CD patients	UC, CD susceptibility	3435TT genotype associated with UC, no association with Crohn's disease	160
		135 CD, 123 UC Caucasian patients	UC, CD susceptibility	No statistical association	163
		66 Japanese UC patients	Early- and late-onset UC	3435T associated with susceptibility to late-onset UC, but not early-onset UC	161
		97 Caucasian cancer patients	Paclitaxel PK	No statistical association	193
		45 matched tumor and normal colon tissue samples	Colorectal cancer susceptibility	No statistical association	194
		68 patients treated preop with anthracyclines with or without taxanes	Response to preoperative chemotherapy	3435TT associated with clinical complete response to chemotherapy	195
		179 patients	CCRC susceptibility; renal Pgp expression	3435T allele associated with occurrence of tumors and ↓ Pgp expression in normal renal tissue	196
	2677G>T	30 patients	Lymphoblast Pgp expression and function; response to chemotherapy; survival	2677GG and TT associated with poor survival; no association between genotype and Pgp expression or function	197
	3435C>T, 2677G>T/A (Ala893Ser/Thr)	32 healthy subjects	Digoxin PK	3435TT genotype associated with ↑ AUC and A_e (48 h) than CC; trend toward ↑ AUC with 2677TT than 2677GG	124

(*Continued*)

TABLE 21.2. (*Continued*)

HUGO Name	SNP	Population	Outcome Marker	Effect	Ref.
		15 healthy subjects	Digoxin PK	F lowest in 2677GG/3435CC, intermediate in 2677GT/3435CT, and highest in 2677TT/3435TT subjects; renal CL highest in 2677GG/3435CC subjects and lowest in 2677TT/3435TT subjects	198
		33 healthy Korean males	Fexofenadine PK	2677AA/3435CC ↓ AUC(0–24 h) compared to other genotypes; 3435TT genotype ↑ AUC(0–24 h) and C_{max} compared to CC	128
		101 Korean patients	Lymphoblast Pgp function, remission, 3-year event-free survival	3435CC associated with ↓ Pgp activity; 3435CC, 2677GG, and 3435C/2677G haplotype associated with complete remission and 3-year event-free survival but not overall survival	199
		52 children	Vincristine PK and side effects	No statistical association	200
		411 drug-naive HIV+ patients	Disease progression	No statistical association	201
	1236C>T, 2677G>T (Ala893Ser), 3435C>T	37 Caucasians, 23 African Americans	Fexofenadine PK	1236TT/2677TT/3435TT haplotype associated with ↓ AUC compared to reference	9

	405 AML patients (136 for expression)	Lymphoblast MDR1 expression, overall survival, remission rate, risk of relapse	3435CC associated with ↓ MDR1 expression and ↓ overall survival with a ↑ probability of risk of relapse	130
	Retrospective study of HIV+ patients	CD4 cell recovery	Marginal association with 1236 genotype	202
	210 epileptic patients	Degree of drug resistance	1236C/2677G/3435C haplotype increased risk of drug resistance	152
	66 cancer patients	Hepatic elimination over 90 min	2677T and 3435T alleles associated with ↓ elimination constant (k_H) for hepatic elimination of ^{99m}Tc-sestamibi	203
	28 Caucasian patients with solid tumors	Tipifarnib PK	1236C>T associated with ↑ AUC	204
	65 cancer patients	Irinotecan and SN-38 PK	1236C>T associated with higher irinotecan and SN-38 AUC	205
−129T>C, 2677G>T/A (Ala893Ser/Thr), 3435C>T	25 drug-resistant, 20 drug-sensitive Turkish patients	Drug resistance or sensitivity	No statistical association	206
	25 early-onset, 70 late-onset Parkinson's disease patients	Early vs. late onset	No statistical association	156
3435C>T, intron 26, +80T>C	28 HIV+ patients	Cellular AUC	3435TT>CT>CC nelfinavir cellular AUC; +80CC>CT>TT nelfinavir cellular AUC	138

(*Continued*)

TABLE 21.2. (*Continued*)

HUGO Name	SNP	Population	Outcome Marker	Effect	Ref.
	Intron 1, −41G>A, −145G>C, −129C>T, 1236C>T, 2677G>T/A, 3435C>T, 4036G>A	206 Chinese patients	Parkinson's disease	1236, 2677, 3435 genotypes associated with Parkinson's disease; 2677T and 3435T associated with late onset	207
	5′UTR SNP, −129T>C, 5 intronic SNPs, 1236C>T, 2677G>T, 3435C>T	144 UC, 163 CD patients	Refractory CD, CD, UC	1236T/2677T/3435T (and two intronic SNPs in LD with this haplotype) alleles associated with ↑ risk for refractory CD and UC (significant only as haplotype)	162
	Intronic SNP rs3789243 (intron 3)	249 UC, 179 CD patients	UC, CD	rs3789243 G allele associated with susceptibility to UC but not CD	171
	Promoter, +8T>C	139 patients	Osteogenic sarcoma susceptibility	+8CT genotype more frequent in patients than in controls	208
ABCB1, ABCC2	1249G>A (ABCC2)	14 brain tissue samples	Protein expression	1249 associated with different labeling; no effect of MDR1 polymorphisms on Pgp expression	209
ABCB11	4 nonsense mutations; 6 missense mutations	PFIC families	PFIC susceptibility	Mutations associated with PFIC	114
	10 mutations	PFIC patients	BSEP protein expression in liver sections	Mutations associated with lack of hepatic BSEP expression	115

ABCC3	15 SNPs	84 (RNA) and 50 (protein) Caucasian liver samples	mRNA and protein levels	−211C>T associated with lower MRP3 mRNA levels (affects binding of nuclear factors)	35
ABCG2	421C>A (141Glu>Lys)	12 patients	Topotecan bioavailability	Two patients with CA genotype had increased F compared to 10 patients with CC genotype	164
ABCG2	421C>A (141Glu>Lys)	200 patients	Nonpapillary RCC susceptibility	421A carriers are at higher risk of developing nonpapillary RCC	210
ABCG2	34G>A (12Val>Met), 421C>A (141Gln>Lys), 376C>T (126Gln>stop)	100 placentas	mRNA and protein levels	421A allele associated with decreased protein levels compared to 421C	211
	421 C>A (141Gln>Lys)	84 Caucasian cancer patients	Irinotecan and SN-38 PK	No statistical association	212
	9 SNPs	91 liver, 53 intestinal samples	mRNA and protein levels	No statistical association	58
ABCB1, *ABCG2*	3435C>T, 195769-72del	29 Asian nasopharyngeal carcinoma patients	Irinotecan metabolism and PK	3435CC had lower irinotecan C_{max}; *ABCG2* deletion associated with lower extent of irinotecan conversion to SN-38	213

[a]CD, Crohn's disease; UC, ulcerative colitis; IBD, inflammatory bowel disease; RA, rheumatoid arthritis; CCRC, clear-cell renal carcinoma; RCC, renal cell carcinoma; PFIC, progressive familial intrahepatic cholestatis; PK, pharmacokinetics.

have been discovered, all of which lead to nonfunctional OCTN2 and presentation of primary SCD.[103] Knowledge of the genetic and molecular causes of SCD allows earlier diagnosis and more effective treatment; in the majority of cases, clinical carnitine administration alleviates symptoms of SCD.[104]

The *SLC22A4* and *SLC22A5* (encoding OCTN1 and OCTN2, respectively) genes are located in tandem on chromosome 5q23.3.[105] At the amino acid level, OCTN1 and OCTN2 are 77% identical and 88% similar.[106] OCTN1 and OCTN2 are both widely expressed in human tissues, although OCTN1 is not expressed in adult liver, whereas OCTN2 is found in hepatocytes.[106] Despite the high level of sequence identity and overlapping tissue distribution, OCTN1 and OCTN2 transport different substrates; therefore, polymorphisms in these genes result in different clinical phenotypes. For example, polymorphisms in OCTN1 have not been associated with SCD, although OCTN2 polymorphisms clearly lead to the development of SCD. However, both the OCTN1 and OCTN2 genes have been tentatively linked to several autoimmune disorders, such as Crohn's disease and rheumatoid arthritis.[107,108] Using linkage disequilibrium mapping, a common multigenic haplotype encompassing both OCTN1 (a missense substitution) and OCTN2 (a promoter polymorphism) was associated with Crohn's disease susceptibility and severity.[46] The molecular mechanism that leads to these autoimmune disorders is not yet known but may involve the thiol ergothionine, which was recently characterized as a physiological substrate of OCTN1.[109]

Several members of the SLCO (OATP) family of transporters have been characterized as xenobiotic uptake transporters, perhaps playing a complementary role to the xenobiotic efflux transporters in the ABCB and ABCC gene families. OATP1B1, in particular, has gained attention because of its ability to facilitate uptake of drugs into hepatocytes, which suggests that this transporter may affect drug disposition and pharmacokinetics. One recent study identified naturally occurring polymorphisms in a population of Africans and European Americans and demonstrated the detrimental effects of several polymorphisms on OATP1B1 transport or expression in vitro.[55] This initial report and related findings[76,77] led to the development of several clinical studies that investigated the impact of OATP1B1 polymorphisms on pravastatin pharmacokinetics in vivo.

The cholesterol synthesis inhibitor pravastatin, used widely in the treatment of hypercholesterolemia, is known to be a substrate for OATP1B1 (encoded by *SLCO1B1*).[110] In one clinical study, the *SLCO1B1**5 haplotype (consisting of the nonsynonymous SNP 521T>C, Val174Ala) increased pravastatin AUC after a single oral dose, indicating reduced hepatocellular uptake.[111] In another study, the *SLCO1B1**15 haplotype, containing the Val174Ala and Asn130Asp amino acid changes, was compared to the *SLCO1B1**1b haplotype, which consists of only the Asn130Asp amino acid change. After a single oral dose of pravastatin, carriers of the *SLCO1B1**15 allele had reduced clearance compared to those with the *SLCO1B1**1b allele.[41] Previous studies have shown that the *SLCO1B1**1b allele has no effect on OATP1B1 function in vitro,[55,75] so these data suggest that the reduced pravastatin clearance exhibited by *SLCO1B1**15 carriers is due primarily to the Val174Ala change. This correlates with the in vitro findings that the 521T>C (Val174Ala) polymorphism significantly reduced uptake of the OATP1B1 model substrates estrone sulfate and estradiol

17β-D-glucuronide, which may be attributed to the reduced surface expression of the OATP1B1*5 protein.[55]

The *SLCO1B1* 521 T>C variation is an example of a transporter polymorphism that does not have a phenotype except during situations in which the transporter is challenged by an environmental stress: in this case, pravastatin administration. Another example of a transporter with a "hidden" phenotype is *SLC16A1*, the gene that encodes the monocarboxylate transporter MCT1. In 2000, five patients were identified with signs of abnormal muscle injury upon exercise or heat exposure due to reduced lactate transport in erythrocytes. Upon examination of these patients, a number of missense mutations were identified in *SLC16A1* that presumably led to muscle damage during exercise or heat exposure.[112] Although the functional consequences of these polymorphisms have yet to be confirmed in vitro, it was recently shown that exercise rapidly increases MCT1 expression in rat muscle,[113] indicating that MCT1 has a prominent role in lactate transport during exercise.

In the liver, as in muscle, numerous tissue-specific transporters with specialized functions are expressed. Hepatic ABC and SLC transporters are involved in the movement of bile acid. Defects in these transporters are commonly manifested as cholestasis, or the suppression of biliary flow from the liver. There are several types of cholestatic disorders, including heritable disorders such as progressive familial intrahepatic cholestasis (PFIC). In one cohort of patients with PFIC, positional cloning led to the implication of *ABCB11*, encoding the bile salt export pump (BSEP), in a more specific disorder classified as PFIC-2 (OMIM 601847).[114] A mutation in this gene eliminates expression of BSEP, as shown by antibody staining in liver samples from patients with PFIC-2.[115] In another form of PFIC, PFIC-3, the MDR3 transporter is absent or defective in liver samples due to polymorphisms in *ABCB4*.[116]

The discovery that the ABC transporters BSEP and MDR3 play such a crucial role in bile acid homeostasis led to a study in which the expression of other hepatic transporters was examined in patients with PFIC-2 and PFIC-3. Besides investigating the mRNA levels of ABCB4 and ABCB11, expression of SLC10A1 (NTCP), SLCO1B1 (OATP1B1), and SLCO1B3 (OATP1B3) was also measured. This study highlighted the interplay between the SLC uptake transporters and the ABC efflux transporters, finding that in liver samples from patients with PFIC-2 or PFIC-3, the SLC bile salt uptake transporters also had decreased expression, thus compensating for the reduced hepatic elimination of bile salts by the ABC efflux transporters BSEP or MDR3.[117] It is plausible that polymorphisms in multiple hepatic transporters would have an additive effect; therefore, the severity of cholestasis may be increased significantly.

21.4.2. ABC Transporters

Although some members of the ABC superfamily are involved in crucial physiological functions such as biliary homeostasis, the ABC transporters that cause multidrug resistance have been more widely studied with respect to their role in drug disposition and response. Homologous ABC transporters can be found in evolutionarily lower organisms such as yeast, suggesting that ABC transporters play a crucial part in the existence and survival of an organism.[118,119] Presumably, the transport activity of these proteins has a protective effect that has been widely conserved throughout evolution.

The majority of multidrug resistance transporter research has focused on P-glycoprotein (Pgp), a xenobiotic efflux transporter that is encoded by *ABCB1*. Pgp was identified in 1979 as the protein that reduced the permeability of Chinese hamster ovary (CHO) cells, thereby limiting the intracellular accumulation of the anti-inflammatory drug colchicine.[120] In the 1980s, Pgp was found to cause resistance to a wide spectrum of drugs, including anticancer agents, and overexpression of the transporter in tumor samples and cell lines was reported.[121] Characterization of the substrate specificity and elucidation of the molecular mechanism of Pgp has progressed, but many aspects of this transporter remain to be investigated.

A number of associations with SNPs in ABC transporters and various pharmacokinetic or pharmacodynamic phenotypes have been reported in the last five years. The genotype–phenotype approach was commonly used to study the effects of *ABCB1* polymorphisms and response to drug therapy. The coding region of the gene was first sequenced in 2000 in a small number of healthy subjects in whom Pgp expression and function had also been characterized. A number of polymorphisms were found, one of which, the synonymous 3435C>T SNP, was associated with decreased intestinal Pgp function and expression.[8] More comprehensive SNP discovery efforts followed, in larger cohorts and in ethnically diverse populations.[9,10,122] Those SNPs that occurred at high frequencies or which were predicted to have a functional effect were examined in vitro. The most widely studied SNPs included 3435C>T, 2677G>A/T (Ala893Ser/Thr), and 1236C>T, which were found to be in significant linkage disequilibrium.[9,10,123] The common haplotype containing 3435T, 2677T (893Ser), and 1236T was named *ABCB1**13 and had a frequency greater than 30% in Caucasians, Asian Americans, Mexican Americans, and Pacific Islanders.[10]

The in vitro analyses of the functional consequences of some of these SNPs were suggestive of a clinical effect; for example, one study reported a significant increase in digoxin efflux associated with the *ABCB1**13 haplotype, although protein expression was not influenced.[9] Clinical pharmacogenetics studies quickly followed, with results that were often conflicting. Complicating the matter was the fact that most studies looked at each SNP individually instead of considering them in their naturally occurring haplotypes. For instance, several studies found that the 3435T genotype increased digoxin AUC significantly but found no significant correlation between 2677 genotype and digoxin pharmacokinetics.[124,125] Another study found no association between 3435 genotype and digoxin pharmacokinetics.[126] It is probable that multiple haplotypes were included in each genotype group, and for this reason, the effects of these haplotypes may have been obscured.

While there are many reports on *ABCB1* polymorphisms and pharmacokinetics of clinically used Pgp substrates, including digoxin,[127] fexofenadine,[128] and cyclosporine,[129] there are also a number of studies that examine the effects of *ABCB1* polymorphisms on clinical endpoints such as drug response or survival. The predictive value of *ABCB1* genotype has been studied with respect to rates of remission and survival in patients with acute myeloid leukemia (AML). One study found that the wild-type 1236C, 2677G, and 3435C alleles were associated with significantly decreased survival rate and increased probability of relapse, although conversely, these genotypes were correlated with low ABCB1 mRNA expression.[130] Contrary to the first study, a second study showed that the wild-type 3435C and 2677G alleles were strongly associated with an increased probability of AML remission.[131]

The role of ABC transporter pharmacogenetics has also been examined in HIV. The HIV virus targets CD4 T-lymphocytes and the site of action for HIV antiretroviral drugs are lymphoblastoid cells. mRNA expression of several ABC transporters, including ABCB1, is measurable in lymphocytes,[132] and the addition of ABC transporter inhibitors in vitro increases the activity of the HIV antiretroviral drugs AZT and indinavir,[133] suggesting that ABC transporters limit drug entry in lymphocytes. Consistent with these findings, Pgp transport of HIV protease inhibitors has been demonstrated in cultured cells.[134] All of these data have contributed to the hypothesis that genetic polymorphisms in *ABCB1*, and possibly other ABC transporters, affect the pharmacokinetics and pharmacodynamics of HIV antiretroviral drugs.[135]

The first study testing this hypothesis was reported by Fellay et al. in 2002. Patients with the *ABCB1* 3435TT genotype receiving protease inhibitor–based therapy had higher increases in CD4 count than other genotypes after 6 months, and the *ABCB1* 3435TT genotype was a strong predictor of immunologic response.[136] Fellay et al. also associated the *ABCB1* 3435TT genotype with lower nelfinavir plasma concentrations, which may be explained by more nelfinavir being retained in lymphocytes.[136] The *ABCB1* 3435T genotype was associated with increased efavirenz efficacy, but not with efavirenz plasma concentrations,[137] and with higher lymphocyte nelfinavir concentrations in HIV seropositive patients.[138] In follow-up studies, one group reported an association between the *ABCB1* 3435TT genotype and increased efavirenz efficacy;[139] however, the majority of studies have not been able conclusively to validate an association between *ABCB1* 3435 genotype and HIV antiretroviral pharmacokinetics or efficacy.[139–141] It is important to note that unlike the original report, these studies were performed in antiretroviral-naive subjects.

As with HIV antiretroviral efficacy, the data concerning the impact of ABC transporter pharmacogenetics on the field of drug-resistant epilepsy are discordant. In 1995, an exploratory study found that in 11 of 19 brain specimens removed from patients with refractory epilepsy, ABCB1 mRNA levels were significantly higher than in normal brain; these samples also had increased Pgp expression, as determined by immunohistochemistry.[142] Potschka et al. demonstrated that Pgp transports the antiepileptic drugs phenytoin and carbamazepine in a rat microdialysis model,[143,144] although the transport properties of these compounds were challenged by several in vitro reports.[145,146] More recently, the association between the *ABCB1* 3435C>T polymorphism and refractory epilepsy has been explored. In a study of 315 epileptic patients (200 drug-resistant, 115 drug-responsive), the frequency of the *ABCB1* 3435CC genotype was higher in patients with drug-resistant epilepsy than in those with drug-responsive epilepsy,[147] consistent with previous findings that the 3435C allele was associated with higher Pgp expression than was the 3435T allele.[8,148,149] In a second study that used the same criteria for classification of drug resistance, however, the frequency of the *ABCB1* 3435C>T genotype was not different between drug-responsive or drug-resistant patients.[150] Several replication studies have also failed to reproduce the correlation between *ABCB1* 3435 genotype and refractory epilepsy,[131,151] although another study demonstrated an association between the *ABCB1* 1236C/2677G/3435C haplotype and increased pharmacoresistance in patients with temporal lobe epilepsy.[152] The conflicting data in the field speak to the lack of knowledge about the causal mechanism of refractory epilepsy, although it is likely that transporters play a role.

ABCB1 genetic variability has also recently been considered in Parkinson's disease, another debilitating neurological disorder. Because it has been shown that Pgp located on the blood–brain barrier limits the brain's exposure to xenobiotics in both mice[153,154] and humans,[155] it was hypothesized that the level of Pgp expression and/or function at the blood–brain barrier modulates the amount of neurotoxic agents that enters the brain and therefore susceptibility to neurological disorders. A small pilot study examined the effect of the *ABCB1* 2677G>T/3435C>T haplotype on patient age at the onset of Parkinson's disease. Although there was no statistically significant association between *ABCB1* genotype and Parkinson's disease, the frequency of the 2677T/3435T haplotype was highest in the patient group with early-onset Parkinson's disease (36%), second highest in the group with late-onset Parkinson's disease (23%), and lowest in the control group (19%).[156] This haplotype has previously been linked with decreased Pgp expression,[8,148,149] providing a plausible hypothesis for the association with earlier development of Parkinson's disease: lower Pgp expression at the blood–brain barrier leads to higher brain exposure to neurotoxic xenobiotics, thus increasing susceptibility to Parkinson's. The trends in this study have been reproduced,[157] suggesting that *ABCB1* genotype may be a risk factor for the development of Parkinson's disease.

The connection between Pgp function and Parkinson's disease is still being established, but *ABCB1* is a logical candidate gene for this disease, given that the passage of neurotoxic xenobiotics across the blood–brain barrier is likely controlled to some degree by membrane transporters. At times, however, xenobiotic membrane transporters are not obvious candidate genes, and the association between transporter pharmacogenetics and clinical phenotype is somewhat surprising. For example, ulcerative colitis is a disorder attributed to an abnormal immune response to bacteria in the digestive tract.[158] *mdr1a* knockout mice were reported to be susceptible to a form of colitis similar to ulcerative colitis,[159] prompting an investigation into the effect of the *ABCB1* 3435C>T polymorphism on predisposition to ulcerative colitis. Because the 3435T allele has been associated with lower Pgp expression,[8,148,149] it was hypothesized that *ABCB1* 3435T carriers were predisposed to ulcerative colitis. In a study with 149 patients with ulcerative colitis and 998 controls, the 3435TT genotype was associated significantly with ulcerative colitis, suggesting a previously uncharacterized antibacterial role for Pgp.[160] This association was replicated in a recent study in which the 3435T allele was correlated with susceptibility to late-onset ulcerative colitis.[161] However, this genotype–phenotype relationship was not replicated in two other studies;[162,163] subsequently, questions were raised about how the composition of patient and control populations can affect the reproducibility of clinical findings.

Through SNP discovery efforts, polymorphisms in other ABC transporter genes have been identified, and several have been singled out as causing altered transport function in vitro, including the 421C>A (Gln141Lys) polymorphism in *ABCG2*, which encodes MXR. This polymorphism caused reduced MXR expression and transport of the anticancer agent topotecan in vitro,[29] and in a small preliminary study, two carriers of the 421A allele had increased topotecan bioavailability compared to persons with the wild-type allele, indicating reduced intestinal MXR efflux activity.[164] Pharmacogenetic studies of other ABC transporters that are thought to be involved in multidrug resistance are ongoing. High expression of several members of the ABCC family of transporters has been demonstrated in numerous tissues and tumor types,[165] suggesting that polymorphisms in these genes may affect the development and degree of multidrug resistance.

Although genetic variation in the vast majority of ABC transporters implicated in multidrug resistance have been studied using a genotype–phenotype approach, there are a few ABC transporters in which polymorphisms were identified through a phenotype–genotype approach. One prominent example is *ABCC7*, which was mapped in 1989 as the pathogenic locus of cystic fibrosis (OMIM 219700). A common ΔF508 amino acid deletion was discovered in the cystic fibrosis transmembrane conductance regulator (CFTR) protein of patients with cystic fibrosis.[166] This three-nucleotide deletion in *ABCC7* causes a defective chloride channel in approximately 70% of cystic fibrosis patients; other polymorphisms are responsible for the remaining 30% of cases. Additional examples of phenotype–genotype studies include the polymorphisms in *ABCC2* (MRP2) that lead to a rare liver disorder called *Dubin–Johnson syndrome* (DJS) (OMIM 237500),[89] and polymorphisms in *ABCC6* (MRP6) that have been associated with pseudoxanthoma elasticum (PXE) (OMIM 264800), a disorder of the connective tissue.[167]

21.5. CONCLUSIONS

To date, most of the polymorphisms in transporter genes that result in a clinical effect have been identified through phenotype-to-genotype studies. These polymorphisms generally result in a decrease in expression and/or function of the transporter, leading to an obvious and observable clinical phenotype. Genotype-to-phenotype studies provide a wealth of information about genetic diversity but have been less successful at identifying polymorphisms that result in substantial clinical effects. This is in part because of the multiplicity built into the genome and the ability of one gene to compensate for the altered function of another, and in part because most clinical phenomena, including drug response and toxicity, are complex events that involve multiple genes. The consideration of genetic variation in a group of genes implicated in the response and toxicity to a given drug—a pathway approach—may assist in the elucidation of the effects of transporter polymorphisms in more complex clinical phenotypes. An increasing number of drug response pathways are available at www.pharmgkb.org and serve as an important resource for the design of pharmacogenetic studies.

Appropriate study design is another important consideration for pharmacogenetics studies. In many of the early transporter genotype–phenotype association studies, sample sizes were small and sufficient power to detect meaningful differences was often absent. In addition to power considerations, the issue of haplotypic diversity must also be incorporated into study design. In many cases there is a lack of well-documented functional polymorphisms for a given transporter that is implicated in the phenotype of interest. In this case, the use of tagging SNPs that incorporate genetic diversity data from the HapMap project and reflect variation across the entire gene should be used.[168,169] Such an approach has recently been used in several *ABCB1* association studies.[170,171]

Information regarding the functional consequences of genetic variation in transporter genes is increasing, but there will be a continuing need for studies in this area. A shift in focus from amino acid-altering variants to polymorphisms in the untranslated promoter and intronic regions of genes is currently underway. It remains important to understand the molecular basis for an association with any transporter polymorphism and a clinical phenotype. In many cases, initial association studies do not reveal the causative SNP, and additional studies are necessary to identify the molecular mechanism for a change in expression and/or function.

Research in the area of drug transporter pharmacogenetics has already increased our understanding of interindividual variability in drug response and toxicity. The ultimate goal of this work is to be able to use patient-specific genotype data to guide the proper selection of drug, dose, and regimen. In the future, a comprehensive understanding of the genetic regulation of drug-metabolizing enzymes, transporters, and targets will be applied in a rational manner to guide drug therapy. Incorporation of emerging principles in human genetics and the adoption of molecular techniques for studying protein function and expression will aid in the advancement of this exciting field.

Acknowledgments

The work from the authors' laboratory discussed in this chapter was funded by NIH grant GM61390.

REFERENCES

1. Vesell ES. 1977. Genetic and environmental factors affecting drug disposition in man. Clin Pharmacol Ther 22:659–679.
2. McLeod HL. 2004. Drug pathways: moving beyond single gene pharmacogenetics. Pharmacogenomics 5:139–141.
3. Eichelbaum M, Ingelman-Sundberg M, Evans WE. 2006. Pharmacogenomics and individualized drug therapy. Annu Rev Med 57:119–137.
4. Evans WE, Relling MV. 2004. Moving towards individualized medicine with pharmacogenomics. Nature 429:464–468.
5. Krynetskiy EY, Evans WE. 2004. Closing the gap between science and clinical practice: the thiopurine *S*-methyltransferase polymorphism moves forward. Pharmacogenetics 14:395–396.
6. Sconce EA, Khan TI, Wynne HA, Avery P, Monkhouse L, King BP, Wood P, Kesteven P, Daly AK, Kamali F. 2005. The impact of CYP2C9 and VKORC1 genetic polymorphism

and patient characteristics upon warfarin dose requirements: proposal for a new dosing regimen. Blood 106:2329–2333.

7. Rieder MJ, Reiner AP, Gage BF, Nickerson DA, Eby CS, McLeod HL, Blough DK, Thummel KE, Veenstra DL, Rettie AE. 2005. Effect of VKORC1 haplotypes on transcriptional regulation and warfarin dose. N Engl J Med 352:2285–2293.
8. Hoffmeyer S, Burk O, von Richter O, Arnold HP, Brockmöller J, Johne A, Cascorbi I, Gerloff T, Roots I, Eichelbaum M, Brinkmann U. 2000. Functional polymorphisms of the human multidrug-resistance gene: multiple sequence variations and correlation of one allele with P-glycoprotein expression and activity in vivo. Proc Natl Acad Sci U S A 97:3473–3478.
9. Kim RB, Leake BF, Choo EF, Dresser GK, Kubba SV, Schwarz UI, Taylor A, Xie HG, McKinsey J, Zhou S, et al. 2001. Identification of functionally variant *MDR1* alleles among European Americans and African Americans. Clin Pharmacol Ther 70:189–199.
10. Kroetz DL, Pauli-Magnus C, Hodges LM, Huang CC, Kawamoto M, Johns SJ, Stryke D, Ferrin TE, DeYoung J, Taylor T, et al. 2003. Sequence diversity and haplotype structure in the human ABCB1 (MDR1, multidrug resistance transporter) gene. Pharmacogenetics 13:481–494.
11. Marzolini C, Paus E, Buclin T, Kim RB. 2004. Polymorphisms in human MDR1 (P-glycoprotein): recent advances and clinical relevance. Clin Pharmacol Ther 75:13–33.
12. Pauli-Magnus C, Kroetz DL. 2004. Functional implications of genetic polymorphisms in the multidrug resistance gene MDR1 (ABCB1). Pharm Res 21:904–913.
13. Schwab M, Eichelbaum M, Fromm MF. 2003. Genetic polymorphisms of the human MDR1 drug transporter. Annu Rev Pharmacol Toxicol 43:285–307.
14. Sakaeda T, Nakamura T, Okumura K. 2003. Pharmacogenetics of MDR1 and its impact on the pharmacokinetics and pharmacodynamics of drugs. Pharmacogenomics 4:397–410.
15. Sachidanandam R, Weissman D, Schmidt SC, Kakol JM, Stein LD, Marth G, Sherry S, Mullikin JC, Mortimore BJ, Willey DL, et al. 2001. A map of human genome sequence variation containing 1.42 million single-nucleotide polymorphisms. Nature 409:928–933.
16. Allabi AC, Horsmans Y, Issaoui B, Gala JL. 2005. Single nucleotide polymorphisms of ABCB1 (MDR1) gene and distinct haplotype profile in a West Black African population. Eur J Clin Pharmacol 61:97–102.
17. Anderle P, Nielsen CU, Pinsonneault J, Krog PL, Brodin B, Sadée W. 2006. Genetic variants of the human dipeptide transporter PEPT1. J Pharmacol Exp Ther 316:636–646.
18. Badagnani I, Chan W, Castro RA, Brett CM, Huang CC, Stryke D, Kawamoto M, Johns SJ, Ferrin TE, Carlson EJ, Burchard EG, Giacomini KM. 2005. Functional analysis of genetic variants in the human concentrative nucleoside transporter 3 (CNT3; SLC28A3). Pharmacogenom J 5:157–165.
19. Bleasby K, Hall LA, Perry JL, Mohrenweiser HW, Pritchard JB. 2005. Functional consequences of single-nucleotide polymorphisms in the human organic anion transporter hOAT1 (SLC22A6). J Pharmacol Exp Ther 314:923–931.
20. Damaraju S, Zhang J, Visser F, Tackaberry T, Dufour J, Smith KM, Slugoski M, Ritzel MW, Baldwin SA, Young JD, Cass CE. 2005. Identification and functional characterization of variants in human concentrative nucleoside transporter 3, hCNT3 (SLC28A3), arising from single-nucleotide polymorphisms in coding regions of the hCNT3 gene. Pharmacogenet Genom 15:173–182.

21. Erdman AR, Mangravite LM, Urban TJ, Lagpacan LL, Castro RA, de la Cruz M, Chan W, Huang CC, Johns SJ, Kawamoto M, et al. 2006. The human organic anion transporter 3 (OAT3; SLC22A8): genetic variation and functional genomics. Am J Physiol Renal Physiol 290:F905–F912.
22. Fujita T, Brown C, Carlson EJ, Taylor T, de la Cruz M, Johns SJ, Stryke D, Kawamoto M, Fujita K, Castro R, et al. 2005. Functional analysis of polymorphisms in the organic anion transporter, SLC22A6 (OAT1). Pharmacogenet Genom 15:201–209.
23. Fukushima-Uesaka H, Maekawa K, Ozawa S, Komamura K, Ueno K, Shibakawa M, Kamakura S, Kitakaze M, Tomoike H, Saito Y, Sawada J. 2004. Fourteen novel single-nucleotide polymorphisms in the SLC22A2 gene encoding human organic cation transporter (OCT2). Drug Metab Pharmacokinet 19:239–244.
24. Gray JH, Mangravite LM, Owen RP, Urban TJ, Chan W, Carlson EJ, Huang CC, Kawamoto M, Johns SJ, Stryke D, Ferrin TE, Giacomini KM. 2004. Functional and genetic diversity in the concentrative nucleoside transporter, CNT1, in human populations. Mol Pharmacol 65:512–519.
25. Ho RH, Leake BF, Roberts RL, Lee W, Kim RB. 2004. Ethnicity-dependent polymorphism in Na^+-taurocholate cotransporting polypeptide (SLC10A1) reveals a domain critical for bile acid substrate recognition. J Biol Chem 279:7213–7222.
26. Honda T, Dan Y, Koyabu N, Ieiri I, Otsubo K, Higuchi S, Ohtani H, Sawada Y. 2002. Polymorphism of MDR1 gene in healthy Japanese subjects: a novel SNP with an amino acid substitution (Glu108Lys). Drug Metab Pharmacokinet 17:479–481.
27. Honjo Y, Morisaki K, Huff LM, Robey RW, Hung J, Dean M, Bates SE. 2002. Single-nucleotide polymorphism (SNP) analysis in the ABC half-transporter ABCG2 (MXR/BCRP/ABCP1). Cancer Biol Ther 1:696–702.
28. Iida A, Saito S, Sekine A, Mishima C, Kondo K, Kitamura Y, Harigae S, Osawa S, Nakamura Y. 2001. Catalog of 258 single-nucleotide polymorphisms (SNPs) in genes encoding three organic anion transporters, three organic anion-transporting polypeptides, and three NADH:ubiquinone oxidoreductase flavoproteins. J Hum Genet 46:668–683.
29. Imai Y, Nakane M, Kage K, Tsukahara S, Ishikawa E, Tsuruo T, Miki Y, Sugimoto Y. 2002. C421A polymorphism in the human breast cancer resistance protein gene is associated with low expression of Q141K protein and low-level drug resistance. Mol Cancer Ther 1:611–616.
30. Ito S, Ieiri I, Tanabe M, Suzuki A, Higuchi S, Otsubo K. 2001. Polymorphism of the ABC transporter genes, MDR1, MRP1 and MRP2/cMOAT, in healthy Japanese subjects. Pharmacogenetics 11:175–184.
31. Itoda M, Saito Y, Soyama A, Saeki M, Murayama N, Ishida S, Sai K, Nagano M, Suzuki H, Sugiyama Y, Ozawa S, Sawada J. 2002. Polymorphisms in the ABCC2 (cMOAT/MRP2) gene found in 72 established cell lines derived from Japanese individuals: an association between single nucleotide polymorphisms in the 5′-untranslated region and exon 28. Drug Metab Dispos 30:363–364.
32. Itoda M, Saito Y, Komamura K, Ueno K, Kamakura S, Ozawa S, Sawada J. 2002. Twelve novel single-nucleotide polymorphisms in ABCB1/MDR1 among Japanese patients with ventricular tachycardia who were administered amiodarone. Drug Metab Pharmacokinet 17:566–571.
33. Itoda M, Saito Y, Shirao K, Minami H, Ohtsu A, Yoshida T, Saijo N, Suzuki H, Sugiyama Y, Ozawa S, Sawada J. 2003. Eight novel single-nucleotide polymorphisms in

ABCG2/BCRP in Japanese cancer patients administered irinotacan. Drug Metab Pharmacokinet 18:212–217.

34. Itoda M, Saito Y, Maekawa K, Hichiya H, Komamura K, Kamakura S, Kitakaze M, Tomoike H, Ueno K, Ozawa S, Sawada J. 2004. Seven novel single-nucleotide polymorphisms in the human SLC22A1 gene encoding organic cation transporter 1 (OCT1). Drug Metab Pharmacokinet 19:308–312.
35. Lang T, Hitzl M, Burk O, Mornhinweg E, Keil A, Kerb R, Klein K, Zanger UM, Eichelbaum M, Fromm MF. 2004. Genetic polymorphisms in the multidrug resistance–associated protein 3 (ABCC3, MRP3) gene and relationship to its mRNA and protein expression in human liver. Pharmacogenetics 14:155–164.
36. Leabman MK, Huang CC, Kawamoto M, Johns SJ, Stryke D, Ferrin TE, DeYoung J, Taylor T, Clark AG, Herskowitz I, Giacomini KM. 2002. Polymorphisms in a human kidney xenobiotic transporter, OCT2, exhibit altered function. Pharmacogenetics 12:395–405.
37. Lee YM, Cui Y, König J, Risch A, Jäger B, Drings P, Bartsch H, Keppler D, Nies AT. 2004. Identification and functional characterization of the natural variant MRP3-Arg1297His of human multidrug resistance protein 3 (MRP3/ABCC3). Pharmacogenetics 14:213–223.
38. Lee W, Glaeser H, Smith LH, Roberts RL, Moeckel GW, Gervasini G, Leake BF, Kim RB. 2005. Polymorphisms in human organic anion–transporting polypeptide 1A2 (OATP1A2): implications for altered drug disposition and central nervous system drug entry. J Biol Chem 280:9610–9617.
39. Materna V, Lage H. 2003. Homozygous mutation Arg768Trp in the ABC-transporter encoding gene MRP2/cMOAT/ABCC2 causes Dubin–Johnson syndrome in a Caucasian patient. J Hum Genet 48:484–486.
40. Michalski C, Cui Y, Nies AT, Nuessler AK, Neuhaus P, Zanger UM, Klein K, Eichelbaum M, Keppler D, König J. 2002. A naturally occurring mutation in the SLC21A6 gene causing impaired membrane localization of the hepatocyte uptake transporter. J Biol Chem 277:43058–43063.
41. Nishizato Y, Ieiri I, Suzuki H, Kimura M, Kawabata K, Hirota T, Takane H, Irie S, Kusuhara H, Urasaki Y, et al. 2003. Polymorphisms of OATP-C (SLC21A6) and OAT3 (SLC22A8) genes: consequences for pravastatin pharmacokinetics. Clin Pharmacol Ther 73:554–565.
42. Oelkers P, Kirby LC, Heubi JE, Dawson PA. 1997. Primary bile acid malabsorption caused by mutations in the ileal sodium-dependent bile acid transporter gene (SLC10A2). J Clin Invest 99:1880–1887.
43. Osato DH, Huang CC, Kawamoto M, Johns SJ, Stryke D, Wang J, Ferrin TE, Herskowitz I, Giacomini KM. 2003. Functional characterization in yeast of genetic variants in the human equilibrative nucleoside transporter, ENT1. Pharmacogenetics 13:297–301.
44. Owen RP, Gray JH, Taylor TR, Carlson EJ, Huang CC, Kawamoto M, Johns SJ, Stryke D, Ferrin TE, Giacomini KM. 2005. Genetic analysis and functional characterization of polymorphisms in the human concentrative nucleoside transporter, CNT2. Pharmacogenet Genom 15:83–90.
45. Owen RP, Lagpacan LL, Taylor TR, de la Cruz M, Huang CC, Kawamoto M, Johns SJ, Stryke D, Ferrin TE, Giacomini KM. 2006. Functional characterization and haplotype analysis of polymorphisms in the human equilibrative nucleoside transporter, ENT2. Drug Metab Dispos 34:12–15.

46. Peltekova VD, Wintle RF, Rubin LA, Amos CI, Huang Q, Gu X, Newman B, Van Oene M, Cescon D, Greenberg G, et al. 2004. Functional variants of OCTN cation transporter genes are associated with Crohn disease. Nat Genet 36:471–475.

47. Perdu J, Germain DP. 2001. Identification of novel polymorphisms in the pM5 and MRP1 (ABCC1) genes at locus 16p13.1 and exclusion of both genes as responsible for pseudoxanthoma elasticum. Hum Mutat 17:74–75.

48. Pinsonneault J, Nielsen CU, Sadée W. 2004. Genetic variants of the human H^+/dipeptide transporter PEPT2: analysis of haplotype functions. J Pharmacol Exp Ther 311:1088–1096.

49. Saito S, Iida A, Sekine A, Miura Y, Ogawa C, Kawauchi S, Higuchi S, Nakamura Y. 2002. Three hundred twenty-six genetic variations in genes encoding nine members of ATP-binding cassette, subfamily B (ABCB/MDR/TAP), in the Japanese population. J Hum Genet 47:38–50.

50. Shoda J, Suzuki H, Sugiyama Y, Hirouchi M, Utsunomiya H, Oda K, Kawamoto T, Matsuzaki Y, Tanaka N. 2003. Novel mutations identified in the human multidrug resistance–associated protein 2 (MRP2/ABCC2) gene in a Japanese patient with Dubin–Johnson syndrome. Hepatol Res 27:323–326.

51. Shu Y, Leabman MK, Feng B, Mangravite LM, Huang CC, Stryke D, Kawamoto M, Johns SJ, DeYoung J, Carlson E, et al. 2003. Evolutionary conservation predicts function of variants of the human organic cation transporter, OCT1. Proc Natl Acad Sci U S A 100:5902–5907.

52. Takane H, Kobayashi D, Hirota T, Kigawa J, Terakawa N, Otsubo K, Ieiri I. 2004. Haplotype-oriented genetic analysis and functional assessment of promoter variants in the MDR1 (ABCB1) gene. J Pharmacol Exp Ther 311:1179–1187.

53. Tang NL, Ganapathy V, Wu X, Hui J, Seth P, Yuen PM, Wanders RJ, Fok TF, Hjelm NM. 1999. Mutations of OCTN2, an organic cation/carnitine transporter, lead to deficient cellular carnitine uptake in primary carnitine deficiency. Hum Mol Genet 8:655–660.

54. Taniguchi S, Mochida Y, Uchiumi T, Tahira T, Hayashi K, Takagi K, Shimada M, Maehara Y, Kuwano H, Kono S, et al. 2003. Genetic polymorphism at the 5′ regulatory region of multidrug resistance 1 (MDR1) and its association with interindividual variation of expression level in the colon. Mol Cancer Ther 2:1351–1359.

55. Tirona RG, Leake BF, Merino G, Kim RB. 2001. Polymorphisms in OATP-C: identification of multiple allelic variants associated with altered transport activity among European- and African-Americans. J Biol Chem 276:35669–35675.

56. Wang H, Hao B, Zhou K, Chen X, Wu S, Zhou G, Zhu Y, He F. 2004. Linkage disequilibrium and haplotype architecture for two ABC transporter genes (ABCC1 and ABCG2) in Chinese population: implications for pharmacogenomic association studies. Ann Hum Genet 68:563–573.

57. Xu G, Bhatnagar V, Wen G, Hamilton BA, Eraly SA, Nigam SK. 2005. Analyses of coding-region polymorphisms in apical and basolateral human organic anion transporter (OAT) genes [OAT1 (NKT), OAT2, OAT3, OAT4, URAT (RST)]. Kidney Int 68:1491–1499.

58. Zamber CP, Lamba JK, Yasuda K, Farnum J, Thummel K, Schuetz JD, Schuetz EG. 2003. Natural allelic variants of breast cancer resistance protein (BCRP) and their relationship to BCRP expression in human intestine. Pharmacogenetics 13:19–28.

59. Zhang EY, Fu DJ, Pak YA, Stewart T, Mukhopadhyay N, Wrighton SA, Hillgren KM. 2004. Genetic polymorphisms in human proton-dependent dipeptide transporter PEPT1: implications for the functional role of Pro586. J Pharmacol Exp Ther 310:437–445.

60. Leabman MK, Huang CC, DeYoung J, Carlson EJ, Taylor TR, de la Cruz M, Johns SJ, Stryke D, Kawamoto M, Urban TJ, et al. 2003. Pharmacogenetics of Membrane Transporters. I. Natural variation in human membrane transporter genes reveals evolutionary and functional constraints. Proc Natl Acad Sci U S A 100:5896–5901.

61. Cargill M, Altshuler D, Ireland J, Sklar P, Ardlie K, Patil N, Lane CR, Lim EP, Kalayanaraman N, Nemesh J, et al. 1999. Characterization of single-nucleotide polymorphisms in coding-regions of human genes. Nat Genet 22:231–238.

62. Halushka MK, Fan JB, Bentley K, Hsie L, Shen N, Weder A, Cooper R, Lipshutz R, Chakravarti A. 1999. Patterns of single-nucleotide polymorphisms in candidate genes for blood-pressure homeostasis. Nat Genet 22:239–247.

63. Stephens JC, Schneider JA, Tanguay DA, Choi J, Acharya T, Stanley SE, Jiang R, Messer CJ, Chew A, Han JH, et al. 2001. Haplotype variation and linkage disequilibrium in 313 human genes. Science 293:489–493.

64. Altshuler D, Brooks LD, Chakravarti A, Collins FS, Daly MJ, Donnelly P. 2005. A haplotype map of the human genome. Nature 437:1299–1320.

65. Feng B, Dresser MJ, Shu Y, Johns SJ, Giacomini KM. 2001. Arginine 454 and lysine 370 are essential for the anion specificity of the organic anion transporter, rOAT3. Biochemistry 40:5511–5520.

66. Konno T, Ebihara T, Hisaeda K, Uchiumi T, Nakamura T, Shirakusa T, Kuwano M, Wada M. 2003. Identification of domains participating in the substrate specificity and subcellular localization of the multidrug resistance proteins MRP1 and MRP2. J Biol Chem 278:22908–22917.

67. Litman T, Druley TE, Stein WD, Bates SE. 2001. From MDR to MXR: new understanding of multidrug resistance systems, their properties and clinical significance. Cell Mol Life Sci 58:931–959.

68. Wang J, Giacomini KM. 1999. Serine 318 is essential for the pyrimidine selectivity of the N2 Na^+-nucleoside transporter. J Biol Chem 274:2298–2302.

69. Ambudkar SV, Kim IW, Sauna ZE. 2006. The power of the pump: mechanisms of action of P-glycoprotein (ABCB1). Eur J Pharm Sci 27:392–400.

70. Urban TJ, Sebro R, Hurowitz EH, Leabman MK, Badagnani I, Lagpacan LL, Risch N, Giacomini KM. 2006. Functional genomics of membrane transporters in human populations. Genome Res 16:223–230.

71. Takeuchi A, Motohashi H, Okuda M, Inui K. 2003. Decreased function of genetic variants, Pro283Leu and Arg287Gly, in human organic cation transporter hOCT1. Drug Metab Pharmacokinet 18:409–412.

72. Kerb R, Brinkmann U, Chatskaia N, Gorbunov D, Gorboulev V, Mornhinweg E, Keil A, Eichelbaum M, Koepsell H. 2002. Identification of genetic variations of the human organic cation transporter hOCT1 and their functional consequences. Pharmacogenetics 12:591–595.

73. Sakata T, Anzai N, Shin HJ, Noshiro R, Hirata T, Yokoyama H, Kanai Y, Endou H. 2004. Novel single-nucleotide polymorphisms of organic cation transporter 1 (SLC22A1) affecting transport functions. Biochem Biophys Res Commun 313:789–793.

74. Leabman MK, Huang CC, Kawamoto M, Johns SJ, Stryke D, Ferrin TE, DeYoung J, Taylor T, Clark AG, Herskowitz I, Giacomini KM. 2002. Polymorphisms in a human kidney xenobiotic transporter, OCT2, exhibit altered function. Pharmacogenetics 12:395–405.

75. Nozawa T, Nakajima M, Tamai I, Noda K, Nezu J, Sai Y, Tsuji A, Yokoi T. 2002. Genetic polymorphisms of human organic anion transporters OATP-C (SLC21A6) and OATP-B (SLC21A9): allele frequencies in the Japanese population and functional analysis. J Pharmacol Exp Ther 302:804–813.

76. Nozawa T, Minami H, Sugiura S, Tsuji A, Tamai I. 2005. Role of organic anion transporter OATP1B1 (OATP-C) in hepatic uptake of irinotecan and its active metabolite, 7-ethyl-10-hydroxycamptothecin: in vitro evidence and effect of single nucleotide polymorphisms. Drug Metab Dispos 33:434–439.

77. Iwai M, Suzuki H, Ieiri I, Otsubo K, Sugiyama Y. 2004. Functional analysis of single-nucleotide polymorphisms of hepatic organic anion transporter OATP1B1 (OATP-C). Pharmacogenetics 14:749–757.

78. Kameyama Y, Yamashita K, Kobayashi K, Hosokawa M, Chiba K. 2005. Functional characterization of SLCO1B1 (OATP-C) variants, SLCO1B1*5, SLCO1B1*15 and SLCO1B1*15^{+}C1007G, by using transient expression systems of HeLa and HEK293 cells. Pharmacogenet Genom 15:513–522.

79. Fairchild CR, Moscow JA, O'Brien EE, Cowan KH. 1990. Multidrug resistance in cells transfected with human genes encoding a variant P-glycoprotein and glutathione S-transferase-pi. Mol Pharmacol 37:801–809.

80. Kimchi-Sarfaty C, Gribar JJ, Gottesman MM. 2002. Functional characterization of coding polymorphisms in the human MDR1 gene using a vaccinia virus expression system. Mol Pharmacol 62:1–6.

81. Morita N, Yasumori T, Nakayama K. 2003. Human MDR1 polymorphism: G2677T/A and C3435T have no effect on MDR1 transport activities. Biochem Pharmacol 65:1843–1852.

82. Woodahl EL, Yang Z, Bui T, Shen DD, Ho RJ. 2005. MDR1 G1199A polymorphism alters permeability of HIV protease inhibitors across P-glycoprotein-expressing epithelial cells. AIDS 19:1617–1625.

83. Woodahl EL, Ho RJ. 2004. The role of MDR1 genetic polymorphisms in interindividual variability in P-glycoprotein expression and function. Curr Drug Metab 5:11–19.

84. Letourneau IJ, Deeley RG, Cole SP. 2005. Functional characterization of nonsynonymous single-nucleotide polymorphisms in the gene encoding human multidrug resistance protein 1 (MRP1/ABCC1). Pharmacogenet Genom 15:647–657.

85. Conrad S, Kauffmann HM, Ito K, Leslie EM, Deeley RG, Schrenk D, Cole SP. 2002. A naturally occurring mutation in MRP1 results in a selective decrease in organic anion transport and in increased doxorubicin resistance. Pharmacogenetics 12:321–330.

86. Haimeur A, Conseil G, Deeley RG, Cole SP. 2004. Mutations of charged amino acids in or near the transmembrane helices of the second membrane spanning domain differentially affect the substrate specificity and transport activity of the multidrug resistance protein MRP1 (ABCC1). Mol Pharmacol 65:1375–1385.

87. Paulusma CC, Oude Elferink RP. 1997. The canalicular multispecific organic anion transporter and conjugated hyperbilirubinemia in rat and man. J Mol Med 75:420–428.

88. Buchler M, König J, Brom M, Kartenbeck J, Spring H, Horie T, Keppler D. 1996. cDNA cloning of the hepatocyte canalicular isoform of the multidrug resistance protein, cMrp,

reveals a novel conjugate export pump deficient in hyperbilirubinemic mutant rats. J Biol Chem 271:15091–15098.

89. Toh S, Wada M, Uchiumi T, Inokuchi A, Makino Y, Horie Y, Adachi Y, Sakisaka S, Kuwano M. 1999. Genomic structure of the canalicular multispecific organic anion–transporter gene (MRP2/cMOAT) and mutations in the ATP-binding-cassette region in Dubin–Johnson syndrome. Am J Hum Genet 64:739–746.

90. Hashimoto K, Uchiumi T, Konno T, Ebihara T, Nakamura T, Wada M, Sakisaka S, Maniwa F, Amachi T, Ueda K, Kuwano M. 2002. Trafficking and functional defects by mutations of the ATP-binding domains in MRP2 in patients with Dubin–Johnson syndrome. Hepatology 36:1236–1245.

91. Mor-Cohen R, Zivelin A, Rosenberg N, Shani M, Muallem S, Seligsohn U. 2001. Identification and functional analysis of two novel mutations in the multidrug resistance protein 2 gene in Israeli patients with Dubin–Johnson syndrome. J Biol Chem 276:36923–36930.

92. Keitel V, Nies AT, Brom M, Hummel-Eisenbeiss J, Spring H, Keppler D. 2003. A common Dubin–Johnson syndrome mutation impairs protein maturation and transport activity of MRP2 (ABCC2). Am J Physiol Gastrointest Liver Physiol 284:G165–G174.

93. Hulot JS, Villard E, Maguy A, Morel V, Mir L, Tostivint I, William-Faltaos D, Fernandez C, Hatem S, Deray G, et al. 2005. A mutation in the drug transporter gene ABCC2 associated with impaired methotrexate elimination. Pharmacogenet Genom 15:277–285.

94. Hirouchi M, Suzuki H, Itoda M, Ozawa S, Sawada J, Ieiri I, Ohtsubo K, Sugiyama Y. 2004. Characterization of the cellular localization, expression level, and function of SNP variants of MRP2/ABCC2. Pharm Res 21:742–748.

95. Dean M, Rzhetsky A, Allikmets R. 2001. The human ATP-binding cassette (ABC) transporter superfamily. Genome Res 11:1156–1166.

96. Ilias A, Urban Z, Seidl TL, Le Saux O, Sinko E, Boyd CD, Sarkadi B, Varadi A. 2002. Loss of ATP-dependent transport activity in pseudoxanthoma elasticum–associated mutants of human ABCC6 (MRP6). J Biol Chem 277:16860–16867.

97. Mizuarai S, Aozasa N, Kotani H. 2004. single-nucleotide polymorphisms result in impaired membrane localization and reduced atpase activity in multidrug transporter ABCG2. Int J Cancer 109:238–246.

98. Wang Z, Wang B, Tang K, Lee EJ, Chong SS, Lee CG. 2005. A functional polymorphism within the MRP1 gene locus identified through its genomic signature of positive selection. Hum Mol Genet 14:2075–2087.

99. Wang D, Johnson AD, Papp AC, Kroetz DL, Sadée W. 2005. Multidrug resistance polypeptide 1 (MDR1, ABCB1) variant 3435C>T affects mRNA stability. Pharmacogenet Genom 15:693–704.

100. Karpati G, Carpenter S, Engel AG, Watters G, Allen J, Rothman S, Klassen G, Mamer OA. 1975. The syndrome of systemic carnitine deficiency: clinical, morphologic, biochemical, and pathophysiologic features. Neurology 25:16–24.

101. Shoji Y, Koizumi A, Kayo T, Ohata T, Takahashi T, Harada K, Takada G. 1998. Evidence for linkage of human primary systemic carnitine deficiency with D5S436: a novel gene locus on chromosome 5q. Am J Hum Genet 63:101–108.

102. Nezu J, Tamai I, Oku A, Ohashi R, Yabuuchi H, Hashimoto N, Nikaido H, Sai Y, Koizumi A, Shoji Y, et al. 1999. Primary systemic carnitine deficiency is caused by mutations in a gene encoding sodium ion–dependent carnitine transporter. Nat Genet 21:91–94.

103. Lahjouji K, Mitchell GA, Qureshi IA. 2001. Carnitine transport by organic cation transporters and systemic carnitine deficiency. Mol Genet Metab 73:287–297.

104. Lamhonwah AM, Olpin SE, Pollitt RJ, Vianey-Saban C, Divry P, Guffon N, Besley GT, Onizuka R, De Meirleir LJ, et al. 2002. Novel OCTN2 mutations: no genotype–phenotype correlations: early carnitine therapy prevents cardiomyopathy. Am J Med Genet 111:271–284.

105. Eraly SA, Hamilton BA, Nigam SK. 2003. Organic anion and cation transporters occur in pairs of similar and similarly expressed genes. Biochem Biophys Res Commun 300:333–342.

106. Wu X, Prasad PD, Leibach FH, Ganapathy V. 1998. cDNA sequence, transport function, and genomic organization of human OCTN2, a new member of the organic cation transporter family. Biochem Biophys Res Commun 246:589–595.

107. Russell RK, Drummond H, Nimmo E, Anderson N, Noble C, Wilson D, Gillett P, McGrogan P, Hassan K, Weaver L, et al. 2006. Analysis of the influence of OCTN1/2 variants within the IBD5 locus on disease susceptibility and growth parameters in early-onset inflammatory bowel disease. Gut 55:1114–1123.

108. Tokuhiro S, Yamada R, Chang X, Suzuki A, Kochi Y, Sawada T, Suzuki M, Nagasaki M, Ohtsuki M, Ono M, et al. 2003. An intronic SNP in a RUNX1 binding site of SLC22A4, encoding an organic cation transporter, is associated with rheumatoid arthritis. Nat Genet 35:341–348.

109. Gründemann D, Harlfinger S, Golz S, Geerts A, Lazar A, Berkels R, Jung N, Rubbert A, Schömig E. 2005. Discovery of the ergothioneine transporter. Proc Natl Acad Sci U S A 102:5256–5261.

110. Tokui T, Nakai D, Nakagomi R, Yawo H, Abe T, Sugiyama Y. 1999. Pravastatin, an HMG-CoA reductase inhibitor, is transported by rat organic anion transporting polypeptide, oatp2. Pharm Res 16:904–908.

111. Mwinyi J, Johne A, Bauer S, Roots I, Gerloff T. 2004. Evidence for inverse effects of OATP-C (SLC21A6) 5 and 1b haplotypes on pravastatin kinetics. Clin Pharmacol Ther 75:415–421.

112. Merezhinskaya N, Fishbein WN, Davis JI, Foellmer JW. 2000. Mutations in MCT1 cDNA in patients with symptomatic deficiency in lactate transport. Muscle Nerve 23:90–97.

113. Coles L, Litt J, Hatta H, Bonen A. 2004. Exercise rapidly increases expression of the monocarboxylate transporters MCT1 and MCT4 in rat muscle. J Physiol 561:253–261.

114. Strautnieks SS, Bull LN, Knisely AS, Kocoshis SA, Dahl N, Arnell H, Sokal E, Dahan K, Childs S, Ling V, et al. 1998. A gene encoding a liver-specific ABC transporter is mutated in progressive familial intrahepatic cholestasis. Nat Genet 20:233–238.

115. Jansen PL, Strautnieks SS, Jacquemin E, Hadchouel M, Sokal EM, Hooiveld GJ, Köning JH, De Jager-Krikken A, Kuipers F, Stellaard F, et al. 1999. Hepatocanalicular bile salt export pump deficiency in patients with progressive familial intrahepatic cholestasis. Gastroenterology 117:1370–1379.

116. Deleuze JF, Jacquemin E, Dubuisson C, Cresteil D, Dumont M, Erlinger S, Bernard O, Hadchouel M. 1996. Defect of multidrug-resistance 3 gene expression in a subtype of progressive familial intrahepatic cholestasis. Hepatology 23:904–908.

117. Keitel V, Burdelski M, Warskulat U, Kuhlkamp T, Keppler D, Haussinger D, Kubitz R. 2005. Expression and localization of hepatobiliary transport proteins in progressive familial intrahepatic cholestasis. Hepatology 41:1160–1172.

118. Croop JM. 1993. P-Glycoprotein structure and evolutionary homologies. Cytotechnology 12:1–32.
119. Michaelis S, Berkower C. 1995. Sequence comparison of yeast ATP-binding cassette proteins. Cold Spring Harbor Symp Quant Biol 60:291–307.
120. Riordan JR, Ling V. 1979. Purification of P-glycoprotein from plasma membrane vesicles of Chinese hamster ovary cell mutants with reduced colchicine permeability. J Biol Chem 254:12701–12705.
121. Lum BL, Gosland MP, Kaubisch S, Sikic BI. 1993. Molecular targets in oncology: implications of the multidrug resistance gene. Pharmacotherapy 13:88–109.
122. Cascorbi I, Gerloff T, Johne A, Meisel C, Hoffmeyer S, Schwab M, Schäffeler E, Eichelbaum M, Brinkmann U, Roots I. 2001. Frequency of single-nucleotide polymorphisms in the P-glycoprotein drug transporter MDR1 gene in white subjects. Clin Pharmacol Ther 69:169–174.
123. Tang K, Ngoi SM, Gwee PC, Chua JM, Lee EJ, Chong SS, Lee CG. 2002. Distinct haplotype profiles and strong linkage disequilibrium at the MDR1 multidrug transporter gene locus in three ethnic Asian populations. Pharmacogenetics 12:437–450.
124. Verstuyft C, Schwab M, Schäffeler E, Kerb R, Brinkmann U, Jaillon P, Funck-Brentano C, Becquemont L. 2003. Digoxin pharmacokinetics and MDR1 genetic polymorphisms. Eur J Clin Pharmacol 58:809–812.
125. Johne A, Kopke K, Gerloff T, Mai I, Rietbrock S, Meisel C, Hoffmeyer S, Kerb R, Fromm MF, Brinkmann U, et al. 2002. Modulation of steady-state kinetics of digoxin by haplotypes of the P-glycoprotein MDR1 gene. Clin Pharmacol Ther 72:584–594.
126. Gerloff T, Schaefer M, Johne A, Oselin K, Meisel C, Cascorbi I, Roots I. 2002. MDR1 genotypes do not influence the absorption of a single oral dose of 1 mg digoxin in healthy white males. Br J Clin Pharmacol 54:610–616.
127. Chowbay B, Li H, David M, Cheung YB, Lee EJ. 2005. Meta-analysis of the influence of MDR1 C3435T polymorphism on digoxin pharmacokinetics and MDR1 gene expression. Br J Clin Pharmacol 60:159–171.
128. Yi SY, Hong KS, Lim HS, Chung JY, Oh DS, Kim JR, Jung HR, Cho JY, Yu KS, Jang IJ, Shin SG. 2004. A variant 2677A allele of the MDR1 gene affects fexofenadine disposition. Clin Pharmacol Ther 76:418–427.
129. Min DI, Ellingrod VL. 2002. C3435T mutation in exon 26 of the human MDR1 gene and cyclosporine pharmacokinetics in healthy subjects. Ther Drug Monit 24:400–404.
130. Illmer T, Schuler US, Thiede C, Schwarz UI, Kim RB, Gotthard S, Freund D, Schakel U, Ehninger G, Schaich M. 2002. MDR1 gene polymorphisms affect therapy outcome in acute myeloid leukemia patients. Cancer Res 62:4955–4962.
131. Kim DW, Kim M, Lee SK, Kang R, Lee SY. 2006. Lack of association between C3435T nucleotide MDR1 genetic polymorphism and multidrug-resistant epilepsy. Seizure 15:344–347.
132. Albermann N, Schmitz-Winnenthal FH, Z'Graggen K, Volk C, Hoffmann MM, Haefeli WE, Weiss J. 2005. Expression of the drug transporters MDR1/ABCB1, MRP1/ABCC1, MRP2/ABCC2, BCRP/ABCG2, and PXR in peripheral blood mononuclear cells and their relationship with the expression in intestine and liver. Biochem Pharmacol 70:949–958.
133. Jorajuria S, Dereuddre-Bosquet N, Becher F, Martin S, Porcheray F, Garrigues A, Mabondzo A, Benech H, Grassi J, Orlowski S, Dormont D, Clayette P. 2004. ATP binding

cassette multidrug transporters limit the anti-HIV activity of zidovudine and indinavir in infected human macrophages. Antivir Ther 9:519–528.

134. Kim RB. 2003. Drug transporters in HIV therapy. Top HIV Med 11:136–139.

135. Quirk E, McLeod H, Powderly W. 2004. The pharmacogenetics of antiretroviral therapy: a review of studies to date. Clin Infect Dis 39:98–106.

136. Fellay J, Marzolini C, Meaden ER, Back DJ, Buclin T, Chave JP, Decosterd LA, Furrer H, Opravil M, Pantaleo G, et al. 2002. Response to antiretroviral treatment in HIV-1-infected individuals with allelic variants of the multidrug resistance transporter 1: a pharmacogenetics study. Lancet 359:30–36.

137. Haas DW, Smeaton LM, Shafer RW, Robbins GK, Morse GD, Labbe L, Wilkinson GR, Clifford DB, D'Aquila RT, De Gruttola V, et al. 2005. Pharmacogenetics of long-term responses to antiretroviral regimens containing efavirenz and/or nelfinavir: an Adult AIDS Clinical Trials Group study. J Infect Dis 192:1931–1942.

138. Colombo S, Soranzo N, Rotger M, Sprenger R, Bleiber G, Furrer H, Buclin T, Goldstein D, Decosterd L, Telenti A. 2005. Influence of ABCB1, ABCC1, ABCC2, and ABCG2 haplotypes on the cellular exposure of nelfinavir in vivo. Pharmacogenet Genom 15:599–608.

139. Haas DW, Wu H, Li H, Bosch RJ, Lederman MM, Kuritzkes D, Landay A, Connick E, Benson C, Wilkinson GR, Kessler H, Kim RB. 2003. MDR1 gene polymorphisms and phase 1 viral decay during HIV-1 infection: an Adult AIDS Clinical Trials Group study. J Acquir Immune Defic Syndr 34:295–298.

140. Nasi M, Borghi V, Pinti M, Bellodi C, Lugli E, Maffei S, Troiano L, Richeldi L, Mussini C, Esposito R, Cossarizza A. 2003. MDR1 C3435T genetic polymorphism does not influence the response to antiretroviral therapy in drug-naive HIV-positive patients. AIDS 17:1696–1698.

141. Brumme ZL, Dong WW, Chan KJ, Hogg RS, Montaner JS, O'Shaughnessy MV, Harrigan PR. 2003. Influence of polymorphisms within the CX3CR1 and MDR-1 genes on initial antiretroviral therapy response. AIDS 17:201–208.

142. Tishler DM, Weinberg KI, Hinton DR, Barbaro N, Annett GM, Raffel C. 1995. MDR1 gene expression in brain of patients with medically intractable epilepsy. Epilepsia 36: 1–6.

143. Potschka H, Fedrowitz M, Loscher W. 2001. P-Glycoprotein and multidrug resistance–associated protein are involved in the regulation of extracellular levels of the major antiepileptic drug carbamazepine in the brain. Neuroreport 12:3557–3560.

144. Potschka H, Loscher W. 2001. In vivo evidence for P-glycoprotein-mediated transport of phenytoin at the blood–brain barrier of rats. Epilepsia 42:1231–1240.

145. Maines LW, Antonetti DA, Wolpert EB, Smith CD. 2005. Evaluation of the role of P-glycoprotein in the uptake of paroxetine, clozapine, phenytoin and carbamazapine by bovine retinal endothelial cells. Neuropharmacology 49:610–617.

146. Owen A, Pirmohamed M, Tettey JN, Morgan P, Chadwick D, Park BK. 2001. Carbamazepine is not a substrate for P-glycoprotein. Br J Clin Pharmacol 51:345–349.

147. Siddiqui A, Kerb R, Weale ME, Brinkmann U, Smith A, Goldstein DB, Wood NW, Sisodiya SM. 2003. Association of multidrug resistance in epilepsy with a polymorphism in the drug-transporter gene ABCB1. N Engl J Med 348:1442–1448.

148. Tanabe M, Ieiri I, Nagata N, Inoue K, Ito S, Kanamori Y, Takahashi M, Kurata Y, Kigawa J, Higuchi S, Terakawa N, Otsubo K. 2001. Expression of P-glycoprotein in human

placenta: relation to genetic polymorphism of the multidrug resistance (MDR)-1 gene. J Pharmacol Exp Ther 297:1137–1143.

149. Hitzl M, Drescher S, van der Kuip H, Schäffeler E, Fischer J, Schwab M, Eichelbaum M, Fromm MF. 2001. The C3435T mutation in the human MDR1 gene is associated with altered efflux of the P-glycoprotein substrate rhodamine 123 from $CD56^+$ natural killer cells. Pharmacogenetics 11:293–298.

150. Tan NC, Heron SE, Scheffer IE, Pelekanos JT, McMahon JM, Vears DF, Mulley JC, Berkovic SF. 2004. Failure to confirm association of a polymorphism in ABCB1 with multidrug-resistant epilepsy. Neurology 63:1090–1092.

151. Sills GJ, Mohanraj R, Butler E, McCrindle S, Collier L, Wilson EA, Brodie MJ. 2005. Lack of association between the C3435T polymorphism in the human multidrug resistance (MDR1) gene and response to antiepileptic drug treatment. Epilepsia 46:643– 647.

152. Zimprich F, Sunder-Plassmann R, Stogmann E, Gleiss A, Dal-Bianco A, Zimprich A, Plumer S, Baumgartner C, Mannhalter C. 2004. Association of an ABCB1 gene haplotype with pharmacoresistance in temporal lobe epilepsy. Neurology 63:1087–1089.

153. Kim RB, Fromm MF, Wandel C, Leake B, Wood AJ, Roden DM, Wilkinson GR. 1998. The drug transporter P-glycoprotein limits oral absorption and brain entry of HIV-1 protease inhibitors. J Clin Invest 101:289–294.

154. Chen C, Hanson E, Watson JW, Lee JS. 2003. P-Glycoprotein limits the brain penetration of nonsedating but not sedating H1–antagonists. Drug Metab Dispos 31:312–318.

155. Sasongko L, Link JM, Muzi M, Mankoff DA, Yang X, Collier AC, Shoner SC, Unadkat JD. 2005. Imaging P-glycoprotein transport activity at the human blood–brain barrier with positron emission tomography. Clin Pharmacol Ther 77:503–514.

156. Furuno T, Landi MT, Ceroni M, Caporaso N, Bernucci I, Nappi G, Martignoni E, Schäffeler E, Eichelbaum M, Schwab M, Zanger UM. 2002. Expression polymorphism of the blood–brain barrier component P-glycoprotein (MDR1) in relation to Parkinson's disease. Pharmacogenetics 12:529–534.

157. Drozdzik M, Bialecka M, Mysliwiec K, Honczarenko K, Stankiewicz J, Sych Z. 2003. Polymorphism in the P-glycoprotein drug transporter MDR1 gene: a possible link between environmental and genetic factors in Parkinson's disease. Pharmacogenetics 13: 259–263.

158. Sartor RB. 1997. Pathogenesis and immune mechanisms of chronic inflammatory bowel diseases. Am J Gastroenterol 92:5S–11S.

159. Panwala CM, Jones JC, Viney JL. 1998. A novel model of inflammatory bowel disease: mice deficient for the multiple drug resistance gene, mdr1a, spontaneously develop colitis. J Immunol 161:5733–5744.

160. Schwab M, Schäffeler E, Marx C, Fromm MF, Kaskas B, Metzler J, Stange E, Herfarth H, Schoelmerich J, Gregor M, et al. 2003. Association between the C3435T MDR1 gene polymorphism and susceptibility for ulcerative colitis. Gastroenterology 124: 26–33.

161. Osuga T, Sakaeda T, Nakamura T, Yamada T, Koyama T, Tamura T, Aoyama N, Okamura N, Kasuga M, Okumura K. 2006. MDR1 C3435T polymorphism is predictive of later onset of ulcerative colitis in Japanese. Biol Pharm Bull 29:324–329.

162. Potocnik U, Ferkolj I, Glavac D, Dean M. 2004. Polymorphisms in multidrug resistance 1 (MDR1) gene are associated with refractory Crohn disease and ulcerative colitis. Genes Immun 5:530–539.

163. Glas J, Torok HP, Schiemann U, Folwaczny C. 2004. MDR1 gene polymorphism in ulcerative colitis. Gastroenterology 126:367.

164. Sparreboom A, Loos WJ, Burger H, Sissung TM, Verweij J, Figg WD, Nooter K, Gelderblom H. 2005. Effect of ABCG2 genotype on the oral bioavailability of topotecan. Cancer Biol Ther 4:650–658.

165. Haimeur A, Conseil G, Deeley RG, Cole SP. 2004. The MRP-related and BCRP/ABCG2 multidrug resistance proteins: biology, substrate specificity and regulation. Curr Drug Metab 5:21–53.

166. Riordan JR, Rommens JM, Kerem B, Alon N, Rozmahel R, Grzelczak Z, Zielenski J, Lok S, Plavsic N, Chou JL, et al. 1989. Identification of the cystic fibrosis gene: cloning and characterization of complementary DNA. Science 245:1066–1073.

167. Ringpfeil F, Lebwohl MG, Christiano AM, Uitto J. 2000. Pseudoxanthoma elasticum: mutations in the MRP6 gene encoding a transmembrane ATP-binding cassette (ABC) transporter. Proc Natl Acad Sci U S A 97:6001–6006.

168. Goldstein DB. 2005. Haplotype tagging in pharmacogenetics. Novartis Found Symp 267:14–19; discussion, 19–30.

169. Ke X, Miretti MM, Broxholme J, Hunt S, Beck S, Bentley DR, Deloukas P, Cardon LR. 2005. A comparison of tagging methods and their tagging space. Hum Mol Genet 14:2757–2767.

170. Bercovich D, Friedlander Y, Korem S, Houminer A, Hoffman A, Kleinberg L, Shochat C, Leitersdorf E, Meiner V. 2006. The association of common SNPs and haplotypes in the CETP and MDR1 genes with lipids response to fluvastatin in familial hypercholesterolemia. Atherosclerosis 185:97–107.

171. Ho GT, Soranzo N, Nimmo ER, Tenesa A, Goldstein DB, Satsangi J. 2006. ABCB1/MDR1 gene determines susceptibility and phenotype in ulcerative colitis: discrimination of critical variants using a gene-wide haplotype tagging approach. Hum Mol Genet 15:797–805.

172. Kawasaki Y, Kato Y, Sai Y, Tsuji A. 2004. Functional characterization of human organic cation transporter OCTN1 single-nucleotide polymorphisms in the Japanese population. J Pharm Sci 93:2920–2926.

173. Ohashi R, Tamai I, Inano A, Katsura M, Sai Y, Nezu J, Tsuji A. 2002. Studies on functional sites of organic cation/carnitine transporter OCTN2 (SLC22A5) using a Ser467Cys mutant protein. J Pharmacol Exp Ther 302:1286–1294.

174. Wang Y, Kelly MA, Cowan TM, Longo N. 2000. A missense mutation in the OCTN2 gene associated with residual carnitine transport activity. Hum Mutat 15:238–245.

175. Kondo C, Suzuki H, Itoda M, Ozawa S, Sawada J, Kobayashi D, Ieiri I, Mine K, Ohtsubo K, Sugiyama Y. 2004. Functional analysis of SNPs variants of BCRP/ABCG2. Pharm Res 21:1895–1903.

176. Vethanayagam RR, Wang H, Gupta A, Zhang Y, Lewis F, Unadkat JD, Mao Q. 2005. Functional analysis of the human variants of breast cancer resistance protein: I206L, N590Y, and D620N. Drug Metab Dispos 33:697–705.

177. Hayashi H, Takada T, Suzuki H, Akita H, Sugiyama Y. 2005. Two common PFIC2 mutations are associated with the impaired membrane trafficking of BSEP/ABCB11. Hepatology 41:916–924.

178. Noé J, Kullak-Ublick GA, Jochum W, Stieger B, Kerb R, Haberl M, Mullhaupt B, Meier PJ, Pauli-Magnus C. 2005. Impaired expression and function of the bile salt export

pump due to three novel ABCB11 mutations in intrahepatic cholestasis. J Hepatol 43: 536–543.

179. Torok HP, Glas J, Tonenchi L, Lohse P, Muller-Myhsok B, Limbersky O, Neugebauer C, Schnitzler F, Seiderer J, Tillack C, et al. 2005. Polymorphisms in the DLG5 and OCTN cation transporter genes in Crohn's disease. Gut 54:1421–1427.

180. Tosa M, Negoro K, Kinouchi Y, Abe H, Nomura E, Takagi S, Aihara H, Oomori S, Sugimura M, Takahashi K, et al. 2006. Lack of association between IBD5 and Crohn's disease in Japanese patients demonstrates population-specific differences in inflammatory bowel disease. Scand J Gastroenterol 41:48–53.

181. Gazouli M, Mantzaris G, Archimandritis AJ, Nasioulas G, Anagnou NP. 2005. single-nucleotide polymorphisms of OCTN1, OCTN2, and DLG5 genes in Greek patients with Crohn's disease. World J Gastroenterol 11:7525–7530.

182. Vermeire S, Pierik M, Hlavaty T, Claessens G, van Schuerbeeck N, Joossens S, Ferrante M, Henckaerts L, Bueno de Mesquita M, Vlietinck R, Rutgeerts P. 2005. Association of organic cation transporter risk haplotype with perianal penetrating Crohn's disease but not with susceptibility to IBD. Gastroenterology 129:1845–1853.

183. Noble CL, Nimmo ER, Drummond H, Ho GT, Tenesa A, Smith L, Anderson N, Arnott ID, Satsangi J. 2005. The contribution of OCTN1/2 variants within the IBD5 locus to disease susceptibility and severity in Crohn's disease. Gastroenterology 129:1854–1864.

184. Barton A, Eyre S, Bowes J, Ho P, John S, Worthington J. 2005. Investigation of the SLC22A4 gene (associated with rheumatoid arthritis in a Japanese population) in a United Kingdom population of rheumatoid arthritis patients. Arthritis Rheum 52:752–758.

185. Newman B, Wintle RF, van Oene M, Yazdanpanah M, Owen J, Johnson B, Gu X, Amos CI, Keystone E, Rubin LA, Siminovitch KA. 2005. SLC22A4 polymorphisms implicated in rheumatoid arthritis and Crohn's disease are not associated with rheumatoid arthritis in a Canadian Caucasian population. Arthritis Rheum 52:425–429.

186. Wang W, Xue S, Ingles SA, Chen Q, Diep AT, Frankl HD, Stolz A, Haile RW. 2001. An association between genetic polymorphisms in the ileal sodium-dependent bile acid transporter gene and the risk of colorectal adenomas. Cancer Epidemiol Biomark Prev 10:931–936.

187. Page T, Hodgkinson AD, Ollerenshaw M, Hammonds JC, Demaine AG. 2005. Glucose transporter polymorphisms are associated with clear-cell renal carcinoma. Cancer Genet Cytogenet 163:151–155.

188. Jamroziak K, Balcerczak E, Cebula B, Kowalczyk M, Panczyk M, Janus A, Smolewski P, Mirowski M, Robak T. 2005. Multi-drug transporter MDR1 gene polymorphism and prognosis in adult acute lymphoblastic leukemia. Pharmacol Rep 57:882–888.

189. Jamroziak K, Mlynarski W, Balcerczak E, Mistygacz M, Trelinska J, Mirowski M, Bodalski J, Robak T. 2004. Functional C3435T polymorphism of MDR1 gene: an impact on genetic susceptibility and clinical outcome of childhood acute lymphoblastic leukemia. Eur J Haematol 72:314–321.

190. Kishi S, Yang W, Boureau B, Morand S, Das S, Chen P, Cook EH, Rosner GL, Schuetz E, Pui CH, Relling MV. 2004. Effects of prednisone and genetic polymorphisms on etoposide disposition in children with acute lymphoblastic leukemia. Blood 103:67–72.

191. Efferth T, Sauerbrey A, Steinbach D, Gebhart E, Drexler HG, Miyachi H, Chitambar CR, Becker CM, Zintl F, Humeny A. 2003. Analysis of single nucleotide polymorphism

C3435T of the multidrug resistance gene MDR1 in acute lymphoblastic leukemia. Int J Oncol 23:509–517.

192. Winzer R, Langmann P, Zilly M, Tollmann F, Schubert J, Klinker H, Weissbrich B. 2003. No influence of the P-glycoprotein genotype (MDR1 C3435T) on plasma levels of lopinavir and efavirenz during antiretroviral treatment. Eur J Med Res 8:531–534.

193. Henningsson A, Marsh S, Loos WJ, Karlsson MO, Garsa A, Mross K, Mielke S, Vigano L, Locatelli A, Verweij J, Sparreboom A, McLeod HL. 2005. Association of CYP2C8, CYP3A4, CYP3A5, and ABCB1 polymorphisms with the pharmacokinetics of paclitaxel. Clin Cancer Res 11:8097–8104.

194. Humeny A, Rodel F, Rodel C, Sauer R, Füzesi L, Becker C, Efferth T. 2003. MDR1 single-nucleotide polymorphism C3435T in normal colorectal tissue and colorectal carcinomas detected by MALDI-TOF mass spectrometry. Anticancer Res 23:2735–2740.

195. Kafka A, Sauer G, Jaeger C, Grundmann R, Kreienberg R, Zeillinger R, Deissler H. 2003. Polymorphism C3435T of the MDR-1 gene predicts response to preoperative chemotherapy in locally advanced breast cancer. Int J Oncol 22:1117–1121.

196. Siegsmund M, Brinkmann U, Schäffeler E, Weirich G, Schwab M, Eichelbaum M, Fritz P, Burk O, Decker J, Alken P, et al. 2002. Association of the P-glycoprotein transporter MDR1(C3435T) polymorphism with the susceptibility to renal epithelial tumors. J Am Soc Nephrol 13:1847–1854.

197. van den Heuvel-Eibrink MM, Wiemer EA, de Boevere MJ, van der Holt B, Vossebeld PJ, Pieters R, Sonneveld P. 2001. MDR1 gene-related clonal selection and P-glycoprotein function and expression in relapsed or refractory acute myeloid leukemia. Blood 97:3605–3611.

198. Kurata Y, Ieiri I, Kimura M, Morita T, Irie S, Urae A, Ohdo S, Ohtani H, Sawada Y, Higuchi S, Otsubo K. 2002. Role of human MDR1 gene polymorphism in bioavailability and interaction of digoxin, a substrate of P-glycoprotein. Clin Pharmacol Ther 72:209–219.

199. Kim DH, Park JY, Sohn SK, Lee NY, Baek JH, Jeon SB, Kim JG, Suh JS, Do YR, Lee KB. 2006. Multidrug resistance-1 gene polymorphisms associated with treatment outcomes in de novo acute myeloid leukemia. Int J Cancer 118:2195–2201.

200. Plasschaert SL, Groninger E, Boezen M, Kema I, de Vries EG, Uges D, Veerman AJ, Kamps WA, Vellenga E, de Graaf SS, de Bont ES. 2004. Influence of functional polymorphisms of the MDR1 gene on vincristine pharmacokinetics in childhood acute lymphoblastic leukemia. Clin Pharmacol Ther 76:220–229.

201. Bleiber G, May M, Suarez C, Martinez R, Marzolini C, Egger M, Telenti A. 2004. MDR1 genetic polymorphism does not modify either cell permissiveness to HIV-1 or disease progression before treatment. J Infect Dis 189:583–586.

202. Zhu D, Taguchi-Nakamura H, Goto M, Odawara T, Nakamura T, Yamada H, Kotaki H, Sugiura W, Iwamoto A, Kitamura Y. 2004. Influence of single-nucleotide polymorphisms in the multidrug resistance-1 gene on the cellular export of nelfinavir and its clinical implication for highly active antiretroviral therapy. Antivir Ther 9:929–935.

203. Wong M, Evans S, Rivory LP, Hoskins JM, Mann GJ, Farlow D, Clarke CL, Balleine RL, Gurney H. 2005. Hepatic technetium Tc ^{99}m-labeled sestamibi elimination rate and ABCB1 (MDR1) genotype as indicators of ABCB1 (P-glycoprotein) activity in patients with cancer. Clin Pharmacol Ther 77:33–42.

204. Sparreboom A, Marsh S, Mathijssen RH, Verweij J, McLeod HL. 2004. Pharmacogenetics of tipifarnib (R115777) transport and metabolism in cancer patients. Invest New Drugs 22:285–289.

205. Mathijssen RH, Marsh S, Karlsson MO, Xie R, Baker SD, Verweij J, Sparreboom A, McLeod HL. 2003. Irinotecan pathway genotype analysis to predict pharmacokinetics. Clin Cancer Res 9:3246–3253.

206. Kaya P, Gunduz U, Arpaci F, Ural AU, Guran S. 2005. Identification of polymorphisms on the MDR1 gene among Turkish population and their effects on multidrug resistance in acute leukemia patients. Am J Hematol 80:26–34.

207. Lee CG, Tang K, Cheung YB, Wong LP, Tan C, Shen H, Zhao Y, Pavanni R, Lee EJ, Wong MC, Chong SS, Tan EK. 2004. MDR1, the blood–brain barrier transporter, is associated with Parkinson's disease in ethnic Chinese. J Med Genet 41:e60.

208. Rund D, Azar I, Shperling O. 1999. A mutation in the promoter of the multidrug resistance gene (MDR1) in human hematological malignancies may contribute to the pathogenesis of resistant disease. Adv Exp Med Biol 457:71–75.

209. Vogelgesang S, Kunert-Keil C, Cascorbi I, Mosyagin I, Schroder E, Runge U, Jedlitschky G, Kroemer HK, Oertel J, Gaab MR, et al. 2004. Expression of multidrug transporters in dysembryoplastic neuroepithelial tumors causing intractable epilepsy. Clin Neuropathol 23:223–231.

210. Korenaga Y, Naito K, Okayama N, Hirata H, Suehiro Y, Hamanaka Y, Matsuyama H, Hinoda Y. 2005. Association of the BCRP C421A polymorphism with nonpapillary renal cell carcinoma. Int J Cancer 117:431–434.

211. Kobayashi D, Ieiri I, Hirota T, Takane H, Maegawa S, Kigawa J, Suzuki H, Nanba E, Oshimura M, Terakawa N, et al. 2005. Functional assessment of ABCG2 (BCRP) gene polymorphisms to protein expression in human placenta. Drug Metab Dispos 33:94–101.

212. de Jong FA, Marsh S, Mathijssen RH, King C, Verweij J, Sparreboom A, McLeod HL. 2004. ABCG2 pharmacogenetics: ethnic differences in allele frequency and assessment of influence on irinotecan disposition. Clin Cancer Res 10:5889–5894.

213. Zhou Q, Sparreboom A, Tan EH, Cheung YB, Lee A, Poon D, Lee EJ, Chowbay B. 2005. Pharmacogenetic profiling across the irinotecan pathway in Asian patients with cancer. Br J Clin Pharmacol 59:415–424.

22

DIET/NUTRIENT INTERACTIONS WITH DRUG TRANSPORTERS

XIAODONG WANG AND MARILYN E. MORRIS
University at Buffalo, State University of New York, Amherst, New York

22.1. INTRODUCTION

Dietary effects on drug pharmacokinetics have been well documented.[1,2] A number of mechanisms are responsible for food–drug interactions, including alterations in physiological conditions (e.g., gastric pH, gastric emptying, intestinal motility, hepatic blood or bile flow rate), complexation of drugs with dietary components, and modulation of drug-metabolizing enzymes by dietary constituents.[3,4] Over the past decade, numerous food–drug interaction studies have focused primarily on the effects of diet on drug-metabolizing enzymes. Alterations in activities of drug-metabolizing enzymes can subsequently change the pharmacokinetics of drugs that are substrates of these enzymes.[4–6] Foods that contain complex mixtures of phytochemicals, such as

Drug Transporters: Molecular Characterization and Role in Drug Disposition, Edited by Guofeng You and Marilyn E. Morris
Copyright © 2007 John Wiley & Sons, Inc.

fruits, vegetables, herbs, and teas, have great potential to modulate the activities of both phase I and phase II enzymes.[4] Many clinically significant metabolic food–drug interactions have been reported, as exemplified by the interactions between cytochrome P450 (CYP) enzymes and diet and dietary supplements, such as grapefruit juice and St. John's wort.[4,6,7] In recent years, great advances have been made in elucidating the roles of drug transporters in drug disposition.[8–10] It is now recognized increasingly that drug transporters can also contribute significantly to food–drug interactions.

Drug transporters can be generally classified into two major groups: efflux and uptake transporters. Most efflux transporters belong to a superfamily of adenosine triphosphate (ATP)-binding cassette (ABC) transporters, which utilize energy derived from ATP hydrolysis to export substrates out of cells against a concentration gradient. Some major members of this group of transporters include P-glycoprotein (Pgp; MDR1, *ABCB1*), multidrug resistance–associated proteins (MRPs, *ABCC*), and breast cancer resistance protein (BCRP, *ABCG2*).[11] In contrast, uptake transporters mediate the translocation of drugs into cells. Included in this group of transporters are the organic anion–transporting polypeptide (OATP, *SLCO*) family, the organic anion transporter (OAT, *SLC22A*) family, and the organic cation transporter (OCT, *SLC22A*) family.[12–14] Many of these efflux and uptake transporters are expressed in various major organs, such as intestine, liver, kidney, blood–brain barrier, and placenta, suggesting their essential roles in governing the oral absorption, intestinal, hepatobiliary, and renal excretion of a variety of endogenous and exogenous compounds.[9–11,15–18] In addition, overexpression of efflux transporters (e.g., Pgp, MRP, and BCRP) in tumor cells limits intracellular accumulation of a broad range of structurally and functionally unrelated anticancer agents and leads to inefficient cell killing, a phenomenon known as multidrug resistance (MDR), which remains the primary obstacle to successful cancer chemotherapy.[18–20] The transport activities of these efflux transporters have been shown to confer cellular resistance to many widely used and clinically important anticancer agents, such as anthracyclines, *Vinca* alkaloids, camptothecin derivatives, palitaxel, and mitoxantrone.[18,21] Due to the important roles of these transporters in drug disposition and cancer therapy, modulation of these drug transporters by inhibitors or inducers may significantly alter the pharmacokinetics and therapeutic efficacy of many anticancer drugs, resulting in beneficial or adverse drug–drug interactions. The importance of transporter-mediated drug interactions is being recognized increasingly and has been indicated in a number of preclinical and clinical studies.[22–27]

In recent years, dietary supplements (e.g., St. John's wort, ginkgo, garlic, ginseng) have been widely used in Western countries due to an increased public interest in alternative medicine and disease prevention.[7,28,29] Emerging evidence to date has suggested that many of these dietary supplements, as well as nutrients contained in these dietary supplements (e.g., flavonoids), are potent inhibitors or inducers of several major efflux and uptake drug transporters.[30–35] In addition, significant or even life-threatening pharmacokinetic interactions between diet/nutrients and drug transporters have been observed in a number of animal and clinical studies.[25,37,38] Given the growing consumption of dietary supplements and the essential roles of drug transporters in drug disposition, it is important to understand the interactions of diet and nutrients with drug transporters as well as the potential clinical consequences of these interactions. In this review we focus on the interactions of diet and nutrients with drug transporters arising from in vitro and in vivo studies, with specific emphasis on several major efflux transporters (e.g., Pgp, MRP, and BCRP) and uptake transporter (e.g., OATP).

22.2. DIET/NUTRIENT INTERACTIONS WITH DRUG TRANSPORTERS

22.2.1. Interactions of Diet and Dietary Supplements with Drug Transporters

Dietary supplements are products (other than tobacco) intended to supplement the diet. Many top-selling herbal products, such as St. John's wort, garlic, green tea, ginseng, and milk thistle, are classified as dietary supplements.[29,39] Since their marketing does not require Food and Drug Administration (FDA) approval, the potential interactions of these herbal products with conventional drugs, in general, have not been carefully evaluated, raising a serious concern about the safety of using these products. Recent in vitro and in vivo studies have shown that many dietary items and supplements can modulate the activities of both efflux and uptake transporters and alter the pharmacokinetics of various therapeutic agents, resulting in clinically important drug interactions (Table 22.1).

St. John's Wort St. John's wort is widely used in the treatment of depression as an over-the-counter herbal medicine.[40] As a complex mixture, St. John's wort contains multiple constituents, such as hypericin, pseudohypericin, hyperforin, and the flavonoids quercetin and its methylated form, isorhamnetin.[41,42] A large number of clinically relevant St. John's wort–drug interactions have been reported, many of which are drug transporter–mediated interactions.

Effects on Drug Transporters Several studies have shown that St. John's wort induced Pgp both in vitro and in vivo. In LS-180 intestinal carcinoma cells, Pgp expression was induced significantly after 3 days of exposure to St. John's wort (a fourfold increase at 300 μg/mL) or hypericin (a sevenfold increase at 3 μM) in a dose-dependent fashion. The induction of Pgp in LS-180 cells resulted in enhanced efflux and decreased accumulation of rhodamine 123, a fluorescent molecule that acts as a probe of Pgp.[43] In an in vivo induction study, 14 days of administration of St. John's wort increased the Pgp level significantly: the intestinal Pgp 3.8-fold in rats and the duodenal Pgp level 1.4-fold in humans.[31] Similarly, chronic treatment with St. John's wort (3 × 600 mg/day for 16 days) caused a 4.2-fold increase in Pgp expression level in the peripheral blood lymphocytes from healthy volunteers, resulting in reduced accumulation of rhodamine 123.[44] The underlying mechanism of Pgp induction by St. John's wort is presumably due to the activation of pregnane X receptor (PXR), a nuclear receptor that regulates Pgp expression. It has been shown that hyperforin, a major constituent of St. John's wort, is a potent agonist for PXR with a K_i value of 27 nM.[45,46]

TABLE 22.1. Effects of Dietary Supplements on Drug Transporters

Dietary Supplement	In Vitro/ In Vivo	Substrate	Effects	Mechanisms	Refs.
St. John's wort	In vitro	Rhodamine 123	$\uparrow$ Accumulation	Inhibition of Pgp after acute exposure	43
			$\downarrow$ Accumulation due to $\uparrow$ Pgp protein level	Induction of Pgp after chronic exposure	43,44
	In vitro	Ritonavir	$\downarrow$ Efflux	Inhibition of Pgp	48
	In vitro	Daunorubicin/ calcein-AM	$\downarrow$ Efflux	Inhibition of Pgp	49
	In vitro		$\uparrow$ MRP2 mRNA level		50
	In vivo		$\uparrow$ Pgp protein level	Activation of PXR	31
	In vivo		$\uparrow$ MRP2 protein level		51
	In vivo	Digoxin	$\downarrow$ AUC, $\downarrow C_{max}$	Induction of Pgp	24,31,53
	In vivo	Fexofenadine	$\uparrow C_{max}$, $\downarrow$ clearance	Inhibition of Pgp (single dose)	57
			$\leftrightarrow C_{max}$, $\leftrightarrow$ AUC	Induction of Pgp (chronic treatment)	
	In vivo	Indinavir	$\downarrow$ AUC, $\downarrow C_{trough}$	Induction of Pgp/CYP3A4	59
	In vivo	Cyclosporine	$\downarrow$ AUC, $\downarrow C_{plasma}$	Induction of Pgp/CYP3A4	60,61
	In vivo	Tacrolimus	$\downarrow$ AUC, $\uparrow$ clearance	Induction of Pgp/CYP3A4	64
Grapefruit juice	In vitro	Talinolol/digoxin	$\uparrow$ Absorptive transport	Inhibition of Pgp	72,74
	In vitro	Rhodamine 123/ fexofenadine/ saquinavir	$\downarrow$ Efflux	Inhibition of Pgp	73,77,78
	In vitro	Vinblastine	$\downarrow$ Efflux or $\uparrow$ uptake	Inhibition of Pgp	71,79
	In vitro	Vincristine	$\uparrow$ Uptake	Inhibition of Pgp	175
	In vitro		$\downarrow$ Pgp protein and mRNA levels		80

	In vitro	Vinblastine/ saquinavir	↓ Efflux	Inhibition of MRP2/Pgp	75
	In vitro	Fexofenadine	↓ Uptake	Inhibition of OATP/oatp	25
	In vitro	Estrone 3-sulfate	↓ Uptake	Inhibition of OATP-B	81
	In vivo	Cyclosporine	↑ AUC, ↑ C_{max}	Inhibition of Pgp/CYP3A4	82–84
	In vivo	Talinolol	↑ AUC, ↑ C_{max}	Inhibition of Pgp	72
			↓ AUC, ↓ C_{max}	See the text	86
	In vivo	Fexofenadine	↓ AUC, ↓ C_{max}	Inhibition of OATP	25,91
	In vivo	Digoxin	↔ C_{max}, ↔ AUC	See the text	87,88
Garlic	In vitro	Ritonavir	↓ Efflux	Inhibition of Pgp	48
	In vitro		↓ Pgp protein level		96
	In vivo		↓ Pgp protein level		96
	In vivo		↑ Mrp2 protein level		97
	In vivo	Saquinavir/ ritonavir	↓ AUC, ↓ C_{max}	Induction of Pgp/CYP3A4	98,99
Green tea	In vitro	Rhodamine 123	↑ Accumulation	Inhibition of Pgp	104
	In vitro	Doxorubicin	↓ Efflux	Inhibition of Pgp	176
	In vitro		↓ Pgp gene expression ↓ ATPase activity		105
	In vitro	Methotrexate	↓ Efflux	Inhibition of MRP2	106
	In vitro	Estrone 3-sulfate	↓ Uptake	Inhibition of OATP-B	110

(*Continued*)

TABLE 22.1. (*Continued*)

Dietary Supplement	In Vitro/ In Vivo	Substrate	Effects	Mechanisms	Refs.
Ginseng	In vitro	Rhodamine 123/ vinblastine	↑ Accumulation ↔ Pgp mRNA or protein levels	Inhibition of Pgp	114
		Azidopine	↓ Photoaffinity labeling		
	In vitro	Daunorubicin	↑ Accumulation ↔ Pgp protein level	Inhibition of Pgp	115
		Azidopine	↓ Photoaffinity labeling		
Milk thistle	In vitro	Daunomycin	↑ Accumulation ↓ ATPase activity	Inhibition of Pgp	120
		Azidopine	↓ Photoaffinity labeling		
	In vitro	Digoxin/ vinblastine	↓ Efflux/↑ accumulation	Inhibition of Pgp	121
	In vitro	Daunomycin/ vinblastine	↑ Accumulation	Inhibition of MRP1	32
	In vivo	Digoxin	↔ C_{max}, ↔ AUC		122
	In vivo	Indinavir	↔ C_{max}, ↔ AUC		123–125
	In vivo	Irinotecan	↔ C_{max}, ↔ AUC		127
Kava	In vitro	Calcein-AM	↑ Accumulation	Inhibition of Pgp	132
	In vivo	Kawain	↑ AUC, ↑ C_{max}	Inhibition of Pgp/CYP3A4	133

In addition to altering Pgp expression, St. John's wort was also shown to interact with Pgp via modulating Pgp-mediated efflux. Using calcein-AM as a fluorescent marker of Pgp, Weber et al.[47] reported that St. John's wort extracts, as well as some constituents such as quercetin and hyperforin, potently modulated the transport by Pgp in VLB cells (a human lymphocytic leukemia cell line expressing Pgp) and in porcine brain capillary endothelial cells (PBCECs). In Caco-2 cells or canine kidney cells stably expressing Pgp (MDCK-MDR1), hypericin and quercetin significantly inhibited Pgp-mediated efflux of ritonavir, resulting in increased intracellular uptake or a decreased basal-to-apical/apical-to-basal transport ratio of ritonavir.[48] Similarly, Wang et al.[49] showed that hyperforin and hypericin significantly inhibited Pgp activity with IC_{50} values of approximately 30 μM by using daunorubicin and calcein-AM as fluorescent substrates.

Recently, several studies suggested that St. John's wort could also induce MRP2 expression both in vitro and in vivo. In human heptocellular carcinoma HepG2 cells, a 24-hour exposure of St. John's wort or hyperforin to HepG2 cells resulted in 1.55- to 1.72-fold increases in MRP2 mRNA levels.[50] In rats, when St. John's wort was given at a dose of 400 mg/kg per day for 10 days, the amount of MRP2 in the liver was increased significantly, to 304% of control. The increase in MRP2 was maximal at 10 days after St. John's wort treatment and lasted for at least 30 days.[51]

In Vivo Drug Interactions In a single-blind, placebo-controlled parallel study, it was shown that the area under the plasma concentration–time curve (AUC) and peak plasma concentration (C_{max}) of digoxin (a well-known Pgp substrate[52]) were decreased by 25% and 26%, respectively, after the ingestion of St. John's wort for 10 days (3 × 300 mg/day) by healthy volunteers.[24] Similarly, administration of St. John's wort to eight healthy male volunteers at a dose of 300 mg three times a day for 2 weeks decreased the bioavailability of digoxin.[31] In another study, coadministration of the hyperforin-rich extract to healthy volunteers for 14 days significantly reduced the AUC and C_{max} of digoxin: by 24.8% and 37%, respectively.[53] Given the fact that digoxin undergoes limited metabolic transformation, it is likely that induction of Pgp might be the underlying mechanism responsible for the altered pharmacokinetic profiles of digoxin.[24,54]

Interestingly, the effects of St. John's wort on Pgp activity appear to be exposure duration–dependent. Using fexofenadine, a Pgp substrate with minimal metabolic transformation,[55,56] Wang et al.[57] reported that a single oral dose (900 mg) of St. John's wort to healthy volunteers increased the C_{max} of fexofenadine by 45% and decreased the oral clearance by 20%, indicating inhibition of intestinal Pgp. However, long-term administration (3 × 300 mg/day for 14 days) reversed the changes in fexofenadine disposition observed with single-dose administration. The mechanism underlying this biphasic effect of St. John's wort is possibly due to the initial inhibition of Pgp after acute exposure, followed by a significant induction of intestinal Pgp after long-term use.[58]

In addition, coadministration of St. John's wort (3 × 300 mg/day for 14 days) significantly decreased indinavir (a protease inhibitor) AUC by 54% and the plasma trough concentration (C_{trough}) by 81% in healthy subjects.[59] Moreover, significant reductions in AUC and plasma concentration of cyclosporine, an immunosuppressant, were observed in renal transplant patients after prolonged administration of St. John's wort.[60,61] It was suggested in several case studies that the decreased cyclosporine concentration during treatment with St. John's wort was the possible cause of acute heart and kidney transplant rejection.[62,63] A similar pharmacokinetic interaction was observed between St. John's wort and another immunosuppressant, tacrolimus. Coadministration of St. John's wort (3 × 300 mg/day for 18 days) to 10 healthy volunteers significantly decreased tacrolimus AUC and increased its apparent oral clearance.[64] Since indinavir, cyclosporine, and tacrolimus are dual substrates for both Pgp and CYP3A4, it is likely that these interactions might involve induction of Pgp or CYP3A4, or both.[44,64,65]

Grapefruit Juice The interaction of grapefruit juice with some drugs was discovered accidentally when grapefruit juice was used to mask the taste of ethanol in a study using the calcium channel blocker felodipine. Coadministration of grapefruit juice increased the bioavailability of felodipine threefold.[66] Since then, a number of studies have shown that grapefruit juice can enhance the bioavailability of many clinically important drugs.[67] The predominant mechanism for grapefruit juice–drug interactions appears to be the inhibition of CYP3A4 in the small intestine, resulting in reduced presystemic metabolism and enhanced drug oral bioavailability.[68] Another important mechanism of grapefruit juice-drug interactions, as revealed in many recent studies, is through the modulation of drug transporters, especially Pgp and OATP.[25,69–75]

Effects on Drug Transporters The major constituents in grapefruit juice include flavonoids (e.g., naringin, naringenin, quercetin, kaempferol) and furanocoumarins (e.g., bergamottin, 6′, 7′-dihydroxybergamottin). Many studies have shown that both grapefruit juice and its major constituents can modulate the activities of several drug transporters.

An early report of the grapefruit juice–drug interaction indicated that grapefruit juice activated the transport of several Pgp substrates across MDCK-MDR1 cells.[76] However, the majority of the more recent studies have indicated that grapefruit juice has an inhibitory effect on Pgp-mediated transport. In Caco-2 cells using talinolol and digoxin as specific, yet metabolically stable Pgp substrates, the apical-to-basolateral (absorptive) transport of talinolol and digoxin was increased significantly.[72,74] Similarly, in Caco-2 cells and a rat everted sac model, grapefruit juice extracts or some major components such as naringin and 6′,7′-dihydroxybergamottin decreased significantly the efflux or increased intracellular accumulation of several Pgp substrates, such as rhodamine 123, fexofenadine, and saquinavir.[73,75,77,78] Using vinblastine as a Pgp substrate, a number of studies indicated that several extracts of grapefruit juice inhibited the activity of Pgp in Caco-2 cells with different potency. The ethyl acetate extracts of grapefruit juice, followed by diethyl ethyl and methylene chloride extracts, exhibited the greatest potency in increasing the permeability coefficient of the apical-to-basolateral transport of vinblastine.[71,79]

In addition to modulating Pgp activity, grapefruit juice has been shown to interact with Pgp via other mechanisms. Using an ATP-hydrolysis assay, Wang et al.[70] reported that bergamottin increased ATP hydrolysis approximately 2.3-fold, with a K_m value of 8 μM, suggesting that grapefruit juice components might also modulate Pgp ATPase activity. In an in vitro study, exposure of grapefruit juice as well as kaempferol and naringenin to human proximal tubular HK-2 cells for 4 days decreased Pgp mRNA and protein levels significantly in a dose-dependent manner.[80]

Recently, it was shown that grapefruit juice could also interact with other drug transporters. In human MRP2 transfected porcine kidney epithelial cells (LLC-MRP2), ethyl acetate extracts of grapefruit juice or its components (bergamottin and 6′,7′-dihydroxybergamottin) inhibited MRP2-mediated transport of vinblastine and saquinavir significantly.[75] Interestingly, recent interaction studies on grapefruit juice and fexofenadine (a substrate for both Pgp and human OATP or rat oatp[55]) indicated that grapefruit juice preferentially inhibited human OATP and rat oatp rather than Pgp in cell culture studies at a concentration of 5% of normal strength. Several constituents, including bergamottin, 6′,7′-dihydroxybergamottin, naringenin, and hesperidin, at a concentration of 50 μM significantly inhibited fexofenadine uptake mediated by rat Oatp3. Moreover, 6′,7′-dihydroxybergamottin potently inhibited rat oatp1 with an IC_{50} value of 0.28 μM, whereas no significant inhibitory effect on Pgp was observed.[25] Moreover, in human embryonic kidney 293 cells stably expressing OATP-B, grapefruit juice at a concentration of 5% significantly inhibited OATP-B-mediated uptake of estrone 3-sulfate by 82%. Major grapefruit juice constituents, including naringin, naringenin, quercetin, bergamottin, and 6′,7′-dihydroxybergamottin, at a concentration of 10 μM, also significantly inhibited OATP-B-mediated uptake of estrone 3-sulfate by 39, 28, 21, 60, and 43%, respectively.[81]

In Vivo Drug Interactions Several studies have indicated that grapefruit juice can produce significant increases in the AUC (45 to 60%) and $C_{\max}$ (35 to 43%) of cyclosporine in healthy subjects.[82–84] Compared with CYP3A4, intestinal Pgp was suggested to be a more important determinant of cyclosporine bioavailability, presumably by being a rate-limiting step in drug absorption.[85]

In rats, grapefruit juice caused a nearly twofold increase in $C_{\max}$ and more than a 35% increase in AUC of talinolol, indicating an important role of Pgp in talinolol disposition.[72] In a recent clinical study, however, intake of grapefruit juice significantly decreased the AUC, $C_{\max}$, and urinary excretion of talinolol to 56, 57, and 56%, respectively.[86] One possible explanation for this discrepancy is the difference in experimental protocols (adjusted and unadjusted pH for grapefruit juice) since talinolol has a pH-dependent intermediate lipid solubility and low water solubility. Another explanation suggested by the authors is that other intestinal uptake transporters might be responsible for talinolol uptake, and grapefruit juice may have species-different effects on these transporters.[86]

On the other hand, inhibition of Pgp by grapefruit juice had no significant effects on digoxin pharmacokinetics. In two clinical studies, grapefruit juice did not affect AUC, $C_{\max}$, elimination half-life, or renal clearance of digoxin, although it decreased the digoxin absorption rate constant significantly.[87,88] The lack of in vivo effect of grapefruit juice on digoxin is probably due to the fact that digoxin has a high bioavailability after oral administration, and therefore Pgp does not have a significant role in determining its absorption and bioavailability after oral administration.[89,90]

Consistent with the in vitro data of OATP inhibition, grapefruit juice decreased the AUC, C_{max}, and urinary excretion values of fexofenadine to 30 to 40% of those with water, with no change in the time to C_{max}, elimination half-life ($t_{1/2}$), renal clearance, or urinary volume in humans.[25] Presumably, the mechanism underlying this interaction is related to the reduced intestinal fexofenadine absorption due to the inhibition of the uptake transporter OATP. A recent clinical study indicated that the volume of grapefruit juice also had a significant effect on fexofenadine bioavailability. A 300-mL volume of grapefruit juice decreased the AUC and C_{max} of fexofenadine to 58 and 53%, respectively, to that following ingestion of the same volume of water. With the total dose being the same, a 1200-mL volume of grapefruit juice had a more pronounced effect and decreased these parameters to 36 and 33%, respectively, of those with the corresponding volume of water. It was concluded that grapefruit juice, at a commonly consumed volume, decreased the oral bioavailability of fexofenadine, probably due to direct inhibition of uptake by intestinal OATP-A. A much higher volume caused an additional modest effect, possibly from reduced intestinal concentration and transit time of fexofenadine.[91]

Garlic Garlic is one of the best-selling herbal supplements in the United States and has long been used as an herbal medicine for its lipid-lowering, cardioprotective, antioxidant, antineoplastic, antimicrobial, and antiplatelet effects.[29,92] Allicin (diallyl thiosulfinate) is believed to be one of the major active ingredients in garlic, which can be further converted to a number of organic sulfur compounds, such as allyl, diallyl, and methyl sulfides.[93]

Recent studies indicated that garlic and its components could interact with drug transporters via multiple mechanisms. In an in vitro study, allicin inhibited the Pgp-mediated efflux of ritonavir significantly, increasing uptake of ritonavir in MDCK-MDR1 cells with an IC_{50} value of 119 μM. Moreover, allicin (50 μM) remarkably inhibited Pgp-mediated ritonavir transport across Caco-2 cell monolayers, resulting in an efflux ratio of ritonavir close to unity.[48] Using purified Pgp cell membranes and a colorimetric ATPase assay, Foster et al.[94] showed that garlic extracts inhibited Pgp activity significantly, although the potency was low to moderate compared with verapamil, a positive control. However, in Pgp-overexpressing human carcinoma KB-C2 cells, diallyl sulfide or diallyl trisulfide (50 μM) had no significant effects on daunorubicin intracellular accumulation.[95] Interestingly, in Pgp-overexpressing K562-resistant cells, exposure of a nontoxic concentration of diallyl sulfide (8.75 mM) resulted in a time-dependent reduction of Pgp expression levels, with a maximum effect observed at 72 hours. In addition, oral administration of diallyl sulfide (5 mg/kg body weight) effectively reduced *Vinca* alkaloid–induced Pgp overexpression in mouse hepatocyte.[96] In contrast, oral administration of diallyl disulfide (200 mg/kg for 3 days) to rats induced Mrp2 expression sevenfold in renal brush border membranes, while no Pgp induction was observed.[97]

Two clinical trials have indicated that administration of garlic supplements decreased AUC and C_{max} of HIV protease inhibitors saquinavir and ritonavir, substrates for both CYP3A4 and Pgp.[98,99] As suggested by the authors, one mechanism is probably due to the induction of CYP3A4 in the gut mucosa. However, the contribution from induction of Pgp cannot be excluded since saquinavir and ritonavir are also Pgp substrates.[98,100,101]

Green Tea Green tea is widely consumed as a beverage. Tea polyphenols, known as catechins, are the major constituents in green tea. The most abundant catechins in a typical brewed green tea include 10 to 15% (−)-epigallocatechin gallate (EGCG), 6 to 10% (−)-epigallocatechin (EGC), 2 to 3% (−)-epicatechin gallate (ECG), and 2% (−)-epicatechin (EC).[102] The extracts of green tea were reported to exhibit a variety of beneficial health effects, especially chemopreventive, anticarcinogenic, and antioxidant effects.[103]

Several recent studies indicated that green tea components could interact with Pgp and inhibit its transport activity. In a multidrug-resistant cell line CH^RC5, green tea polyphenols (30 μg/mL) inhibited the photolabeling of Pgp by 75% and increased the accumulation of rhodamine 123 threefold. Among the catechins present in green tea, EGCG, ECG, and (−)-catechin gallate (CG) were the major determinants responsible for inhibiting Pgp. In addition, EGCG was able to potentiate the cytotoxicity of vinblastine in CH^RC5 cells.[104] In carcinoma KB-A1 cells, green tea polyphenols (40 μg/mL) and EGCG (10 μg/mL) enhanced doxorubicin cytotoxicity 5.2- and 2.5-fold, respectively. More detailed studies revealed that green tea polyphenols were able to inhibit Pgp ATPase activity and down-regulate Pgp expression.[105]

The effects of green tea extracts on MRP2 were also investigated. In human gastrointestinal epithelial LS-180 cells, green tea extracts, at a concentration of 0.01 mg/mL, had no effect on either activity or expression of MRP2. However, at a concentration of 0.1 mg/mL, green tea extracts significantly inhibited MRP2-mediated efflux of methotrexate (a MRP2 substrate) in MRP2-overexpressing MDCK (MDCK-MRP2) cells, resulting in increased cellular accumulation of methotrexate. In contrast to the inhibitory effects on Pgp, the green tea components EGCG and EGC did not contribute to the MRP2 inhibition activity.[106] Using Caco-2 and MDCK-MRP cells, green tea polyphenols such as EGCG, ECG, EGC, and EC were shown to be potential substrates of MRPs, suggesting the importance of MRPs in determining cellular concentrations of green tea components.[107–109]

In a recent study involving 15 herbal extracts, green tea extracts were shown to potently inhibit OATP-B-mediated uptake of estrone 3-sulfate at the putative gastrointestinal concentration (400 μg/mL). Moreover, OATP-B inhibition by green tea extracts exhibited concentration dependence with an IC_{50} value of 22.1 ± 4.9 μg/mL.[110]

In in vivo xenograft studies, green tea extracts were shown to be effective in reversal of cancer multidrug resistance.[105,111,112] For example, in mice bearing doxorubicin-resistant human carcinoma KB-A-1 cells, the combination of doxorubicin with EGCG (40 mg/kg) increased the doxorubicin concentration in the tumors by 51% and potentiated doxorubicin-induced apoptosis of tumor cells. Compared with the mice given either doxorubicin or EGCG alone, there was a considerable reduction of tumor weights in mice given both EGCG and doxorubicin.[112]

Ginseng Ginseng is a well-known traditional Oriental herbal medicine and has gained popularity as a dietary supplement in the United States during the last decade. Ginsenosides are considered the major active constituents and contribute to the many pharmacological activities of ginseng, including cognition-enhancing, neuroprotective, antioxidant, antineoplastic, and immunomodulatory effects.[113]

The effects of ginseng constituents on Pgp-mediated multidrug resistance were investigated both in vitro and in vivo. In drug-resistant KBV20C cells, ginsenoside Rg(3) increased cellular accumulation of rhodamine 123 and vinblastine significantly in a dose-dependent manner. In addition, Rg(3), at the concentration of 20 μM, greatly enhanced cytotoxicity and reversed multidrug resistance to several anticancer drugs, such as doxorubicin, colchicine, vincristine, and etoposide. However, Rg(3) did not affect either mRNA or protein levels of Pgp in KBV20C cells. A photo-affinity labeling study with [^{3}H]azidopine revealed that 100 μM Rg(3) completely inhibited [^{3}H]azidopine binding to Pgp, indicating that inhibition of drug efflux by Rg(3) was probably due to the competition for the common substrate binding sites on Pgp. In mice implanted with doxorubicin-resistant murine leukemia P388 cells, combining 4 mg/kg doxorubicin with10 mg/kg Rg(3) enhanced the cytotoxicity and efficacy of doxorubicin, resulting in a significant increase in life span and suppression in tumor growth in the mice.[114]

Similarly, in acute myelogenous leukemia cell sublines AML-2/D100 (overexpressing Pgp) and AML-2/DX100 (overexpressing MRP), protopanaxatriol ginsenosides (PTG) were able to reverse Pgp (but not MRP)-mediated resistance in a concentration-dependent manner. PTG (100 μg/mL) increased daunorubicin accumulation in the AML-2/D100 subline twofold higher than that observed in the presence of verapamil (5 μg/mL) but 1.5-fold less than that by cyclosporin A (3 μg/mL). In contrast to verapamil and cyclosporin A, the maximum noncytotoxic concentrations of PTG had no effects on Pgp expression. Moreover, PTG, at a concentration of 200 μg/mL or more, completely inhibited the [^{3}H]azidopine binding to Pgp.[115] These results indicated that PTG probably reversed Pgp-mediated multidrug resistance via direct interaction with Pgp at the azidopine site, resulting in increased intracellular accumulation of other Pgp substrates. Although ginseng constituents are potent multidrug resistance chemosensitizers, no studies to date have been conducted to determine the potential in vivo drug interactions.

Milk Thistle Milk thistle is one of the most popular herbal medicines, used as a hepatoprotectant in the therapy of various liver diseases, including hepatitis and alcoholic cirrhosis.[116,117] The major constituent of milk thistle is silymarin, a mixture of flavonolignans such as silybin, isosilybin, silychristin, and silydianin.[118,119]

In a Pgp-overexpressing human breast cancer cell line (MDA435/LCC6MDR1), silymarin increased daunomycin cellular accumulation significantly in a concentration-dependent manner. At a concentration of 50 μM, silymarin enhanced daunomycin accumulation approximately fourfold. Moreover, silymarin (100 μM) potentiated doxorubicin cytotoxicity threefold in resistant cells. In the presence of 100 μM silymarin, the verapamil-induced Pgp ATPase activity was inhibited completely and [^{3}H]azidopine photoaffinity labeling of Pgp was decreased significantly. However, preincubation cells with 50 μM silymarin did not change Pgp protein levels.[120] In Caco-2 cells, silymarin significantly inhibited Pgp-mediated transport of digoxin across the cell monolayers in a concentration-dependent manner, resulting in decreased efflux and increased cellular accumulation of digoxin.[121] These results indicated that silymarin had inhibitory effects on Pgp-mediated drug efflux and the mechanism of the interaction involved, at least in part, a direct interaction.

In addition to interacting with Pgp, silymarin also modulates MRP1 activity. In MRP1 overexpressing human pancreatic adenocarcinoma Panc-1 cells, silymarin (50 μM) increased the cellular accumulation of vinblastine and daunomycin approximately threefold. However, the inhibition of MRP1 by silymarin was not due to the inhibition of glutathione *S*-transferase activity or depletion of intracellular glutathione.[32]

In vivo drug interactions with milk thistle have been reported recently. However, the results are less compelling compared with the in vitro results. Using digoxin as a selective probe of Pgp, the effects of milk thistle on digoxin pharmacokinetics were studied in 16 healthy subjects. However, no statistically significant changes in AUC, $C_{\max}$, and oral clearance (CL/F) were observed after administration of milk thistle (900 mg/day) for 2 weeks.[122] Similar to the case of grapefruit juice, this lack of in vivo effect of milk thistle on digoxin is probably due to the high inherent bioavailability of digoxin.[90] In addition, the administration of milk thistle for 14 days,[123] 21 days,[124] and 28 days[125] to healthy subjects had no significant effects on the pharmacoknetics of indinavir. In cancer patients, short-term (4 days) or more prolonged intake of milk thistle (12 days) did not change the pharmacokinetics of irinotecan, a substrate for CY3A4, Pgp, and BCRP.[126,127] Given the fact that indinavir and irinotecan are substrates for both drug-metabolizing enzymes and transporters, the lack of drug interactions with milk thistle is probably due to the complicated in vivo regulation of metabolizing enzymes and drug transporters, which are highly susceptible to both induction and inhibition by xenobiotics. Therefore, further studies with better substrate candidates are necessary to characterize in vivo drug interactions with milk thistle.

Kava Kava has been used for many centuries as a traditional intoxicating beverage in the Pacific islands. During the past few decades, kava has gained popularity in Western countries as an herbal supplement for its anxiolytic, antistress, and sedative properties.[128] The major constituents in kava are kavalactones, a mixture of more than 18 different α-pyrones, including kawain, methysticin, yangonin, dihydrokawain, desmethoxyyangonin, and dihydromethysticin.[129,130] However, in 2002, due to the reported cases of liver toxicity, kava extracts were withdrawn from the market in several countries, which prompted wide discussion on kava's relative benefits and risks as an herbal remedy.[29,131]

In a Pgp-overexpressing cell line P388/dx, a kava crude extract and six main kavalactones (kawain, dihydrokawain, methysticin, dihydromethysticin, yangonin, and desmethoxyyangonin) exhibited moderate to potent inhibitory activities on

Pgp-mediated efflux of calcein-AM. The crude extract and the kavalactones increased cellular accumulation of calcein-AM significantly, causing increased intracellular fluorescence intensity. The concentrations needed to double baseline fluorescence were 170 μg/mL and 17 to 90 μM for crude extract and six kavalactones, respectively.[132]

A recent in vivo study indicated that oral coadministration of kava extract (256 mg/kg) significantly changed the pharmacokinetcs of kawain, resulting in a tripling of kawain AUC and a doubling of C_{max}. Moreover, kava extract and kavalactones significantly inhibited CYP enzymes and modestly modulated Pgp ATPase activities in vitro. It was concluded that mechanisms by which kava extract altered the pharmacokinetics of kawain might include inhibition of CYP450 and/or Pgp.[133]

22.2.2. Interactions of Flavonoids with Drug Transporters

Flavonoids are a class of polyphenolic compounds widely present in fruits, vegetables, and plant-derived beverages, and in many herbal products marketed as over-the-counter dietary supplements, such as St. John's wort, green tea, and milk thistle. Concentrations of some flavonoids, such as naringin and hesperidin, abundant in fruit juices, have been reported to be as high as 145 to 638[134] and 200 to 450 mg/L,[135,136] respectively. The average daily intake of total flavonoids from the U.S. diet was estimated to be 200 mg to 1 g[137–141]. The structures of a number of flavonoid subclasses are shown in Table 22.2. Flavonoids have long been associated with a variety of biochemical and pharmacological properties, including antioxidant, antiviral, anticarcinogenic, and anti-inflammatory activities, with no or low toxicity.[142,143] These health-promoting activities indicate that flavonoids may play a protective role in cancer prevention and cardiovascular diseases, as well as other age-related degenerative diseases.[143–145] Recently, numerous studies have indicated that flavonoids could interact with several efflux and uptake transporters such as Pgp, MRP1, BCRP, and OATP, suggesting the potential roles of flavonoids for in vivo drug interactions.[30,34,120,146]

Interactions with Pgp The effects of flavonoids on Pgp have been studied extensively during the past decade (Table 22.3). It was shown clearly from these studies that many of these flavonoids demonstrated Pgp-modulating activities. However, the effects of flavonoids, especially for some flavonols, were cell line-dependent, concentration-dependent, and substrate-dependent. For example, in Pgp-expressing HCT-15 colon cells, flavonols such as quercetin, kaempferol, and galangin were shown to stimulate adriamycin efflux.[147] In contrast, in MCF-7-ADR-resistant breast cancer cells, quercetin was able to restore sensitivity to adriamycin and inhibit rhodamine 123 efflux.[148] Interestingly, in mouse brain capillary endothelial cells (MBEC4), quercetin and kaempferol exhibited biphasic effects by activating Pgp at a low concentration (10 μM) and inhibiting Pgp at a high concentration (50 μM).[149] In rat hepatocytes, the effects of quercetin, kaempferol, and galangin on rhodamine 123 and doxorubicin efflux were shown to be substrate-dependent.[150] Using a purified and reconstituted Pgp system, Shapiro and Ling reported that quercetin inhibited Pgp-mediated Hoechst 33342 efflux and enhanced its accumulation in resistant CH^RC5 cells. This effect was, at least partly, caused by the inhibition of Pgp ATPase activity by quercetin.[151] More

TABLE 22.2. Chemical Structures of Subclasses of Flavonoids

Structural Formula	Representative Flavonoids	Substitutions[a]					
		5	7	2′	3′	4′	5′
Flavones 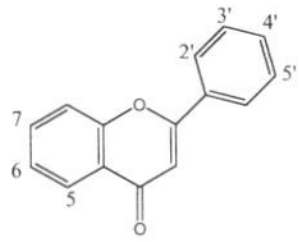	Apigenin	OH	OH	H	H	OH	H
	Chrysin	OH	OH	H	H	H	H
	Luteolin	OH	OH	H	OH	OH	H
	Diosmetin	OH	OH	H	OH	O-Me	H
Flavonols 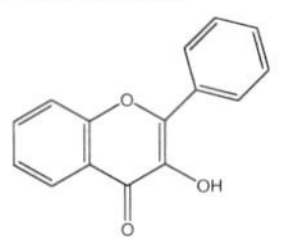	Fisetin	H	OH	H	OH	OH	H
	Galangin	OH	OH	H	H	H	H
	Kaempferol	OH	OH	H	H	OH	H
	Morin	OH	OH	OH	H	OH	H
	Myricetin	OH	OH	H	OH	OH	OH
	Quercetin	OH	OH	H	OH	OH	H
Flavanones 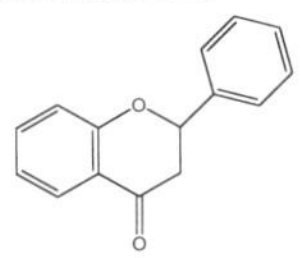	Hesperitin	OH	OH	H	OH	O-Me	H
	Naringenin	OH	OH	H	H	OH	H
Flavanols 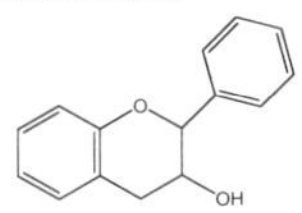	Epicatechin	OH	OH	H	OH	OH	H
	Epigallocatechin	OH	OH	H	OH	OH	OH
Flavanolols 	Silibinin	OH	OH	H	H	Selane	H
Isoflavones 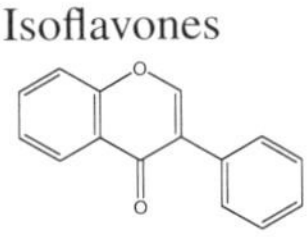	Biochanin A	OH	OH	H	H	O-Me	H
	Genistein	OH	OH	H	H	OH	H
	Daidzein	H	OH	H	H	OH	H

[a] O-Me, methoxy.

recently, in Pgp-overexpressing KB-C2 carcinoma cells, quercetin and kaempferol significantly inhibited Pgp activity and increased cellular accumulation of rhodamine 123 and daunorubicin.[152] The reason(s) for these observed disparate results is still unclear. One possible explanation might be the existence of multiple drug-binding sites on Pgp and various allosteric effects of flavonoids on these sites. It has been shown that quercetin could preferentially bind to the Hoechst 33342–binding site on Pgp and inhibited Hoechst 33342 transport, presumably via competitive inhibition. In contrast, binding of quercetin to the Hoechst 33342 site resulted in increased binding

TABLE 22.3. Interactions of Flavonoids with Pgp

Flavonoid	Cell Line Used	Substrates	Effect on Pgp	Refs.
Flavones				
Apigenin	MCF-7 breast cancer cells	Daunomycin	↔ Activity	177
Chrysin	MCF-7 breast cancer cells	Daunomycin	↓ Activity	177
	MBEC4 endothelial cells	Vincristine	↓ Activity	149
Luteolin	MCF-7 breast cancer cells	Daunomycin	↔ Activity	177
Diosmin	MCF-7 breast cancer cells	Daunomycin	↔ Activity	177
Flavonols				
Fisetin	KB-C2 carinoma cells	Daunorubicin	↓ Activity	152
	MCF-7 breast cancer cells	Daunomycin	↔ Activity	177
Galangin	MCF-7 breast cancer cells	Daunomycin	↑ Expression	177
	Rat hepatocytes	Rhodamine 123/ doxorubicin	↑ or ↓ Activity (substrate-dependent)	150
	HCT-15 colon cells	Adriamycin	↑ Activity	147
Kaempferol	KB-C2 carinoma cells	Rhodamine 123	↓ Activity	152
		Daunorubicin	↓ Activity	
	MCF-7 breast cancer cells	Daunomycin	↔ Activity	177
	MBEC4 endothelial cells	Vincristine	Biphasic effect	149
	Rat hepatocytes	Rhodamine 123/ doxorubicin	↑ or ↓ Activity (substrate-dependent)	150
	HCT-15 colon cells	Adriamycin	↑ Activity	147
Morin	KB-C2 carinoma cells	Daunorubicin	↓ Activity	152
	MCF-7 and MDA435/LCC6 breast cancer cells	Daunomycin	↓ Activity	120
	MCF-7 breast cancer cells	Daunomycin	↓ Activity	177

Myricetin	KB-C2 carinoma cells	Daunorubicin	↓ Activity	152
	MCF-7 breast cancer cells	Daunomycin	↔ Activity	177
Quercetin	KB-C2 carinoma cells	Rhodamine 123	↓ Activity	152
		Daunorubicin	↓ Activity	
	MCF-7 breast cancer cells	Daunomycin	↓ Activity	177,178
		Adriamycin	↓ Activity	148
		Rhodamine 123		
	MBEC4 endothelial cells	Vincristine	Biphasic effect	149
	CH^RC5	Hoechst 33342	↓ Activity	151
	Rat hepatocytes	Rhodamine 123/ doxorubicin	↑ or ↓ Activity (substrate-dependent)	150
	HCT-15 colon cells	Adriamycin	↑ Activity	147
Rutin	KB-C2 carinoma cells	Rhodamine 123	↔ Activity	152
		Daunorubicin	↔ Activity	
	MCF-7 breast cancer cells	Daunomycin	↔ Activity	177
	MBEC4 endothelial cells	Vincristine	↔ Activity	149
Flavanones				
Hesperitin	MBEC4 endothelial cells	Vincristine	↓ Activity	177
	MCF-7 breast cancer cells	Daunomycin	↔ Activity	149
Hesperidin	MBEC4 endothelial cells	Vincristine	↔ Activity	149

(Continued)

TABLE 22.3. (*Continued*)

Flavonoid	Cell Line Used	Substrates	Effect on Pgp	Refs.
Naringenin	KB-C2 carinoma cells	Daunorubicin	↔ Activity	152
	MCF-7 breast cancer cells	Daunomycin	↓ Activity	178
			↔ Activity	177
	MBEC4 endothelial cells	Vincristine	↓ Activity	149
Naringin	MCF-7 breast cancer cells	Daunomycin	↔ Activity	177
	MBEC4 endothelial cells	Vincristine	↔ Activity	149
Flavanols				
Catechin	CH^RC5 cells	Rhodamine 123	↔ Activity	104
CG	CH^RC5 cells	Rhodamine 123	↓ Activity	104
Epicatechin	KB-C2 carinoma cells	Rhodamine 123	↔ Activity	179
		Daunorubicin	↔ Activity	
	CH^RC5 cells	Rhodamine 123	↔ Activity	104
ECG	KB-C2 carinoma cells	Rhodamine 123	↓ Activity	179
		Daunorubicin	↓ Activity	
	CH^RC5 cells	Rhodamine 123	↓ Activity	104
EGC	KB-C2 carinoma cells	Rhodamine 123	↓ Activity	179
		Daunorubicin	↓ Activity	
	CH^RC5 cells	Rhodamine 123	↔ Activity	104
	MCF-7 breast cancer cells	Daunomycin	↔ Activity	177

EGCG	KB-A1 carcinoma cells	Doxorubicin	↓ Activity	180
	KB-C2 carinoma cells	Rhodamine 123	↓ Activity	179
		Daunorubicin	↓ Activity	
	CH^RC5 cells	Rhodamine 123	↓ Activity	104
Flavanolols				
Silibinin/ silymarin	MCF-7 and MDA435/LCC6 breast cancer cells	Daunomycin	↓ Activity	120,178
	Caco-2 cells	Digoxin	↓ Activity	121
		Vinblastine	↓ Activity	
	MCF-7 breast cancer cells	Daunomycin	↓ Activity	177
Isoflavones				
Biochanin A	MCF-7 and MDA435/LCC6 breast cancer cells	Daunomycin	↓ Activity	120,178
	Caco-2 cells	Digoxin	↓ Activity	121
		Vinblastine	↓ Activity	
	MCF-7 breast cancer cells	Daunomycin	↓ Activity	177
Genistein	MCF-7 breast cancer cells	Daunomycin	↓ Activity	177
	BC19/3 breast cancer cells	Rhodamine 123 Daunorubicin	↓ Activity	181

of rhodamine 123 to another site on Pgp and stimulated rhodamine 123 transport.[153] Alternatively, the difference in experimental conditions, such as the different cell lines, the different substrates and their concentrations, and the concentrations of flavonoids used, might contribute to the discrepancies as well. As shown in the MBEC4 study, low concentrations of quercetin and kaempferol activated Pgp possibly via enhancing the phosphorylation (and hence activity) of Pgp, whereas high concentrations of quercetin and kaempferol inhibited Pgp directly.[149] Despite these earlier controversial observations, the majority of recent studies have shown that many flavonoids aglycones have an inhibitory activity on Pgp-mediated drug transport (Table 22.3).

Using purified C-terminal nucleotide-binding domain (NBD2) from mouse Pgp, Conseil et al. studied the structure–activity relationship of flavonoids interacting with Pgp based on their binding affinity. As a suitable tool for the rapid screening of Pgp modulators, NBD2 contains an ATP-binding site as well as a close but distinct hydrophobic steroid RU486-binding site.[30] It was shown in this study that flavones (apigenin) and flavonols (quercetin) had higher binding affinity than flavanones (naringenin), isoflavones (genistein), or glycosylated derivatives (rutin). Interestingly, the flavonol kaempferide exibited bifunctional interactions with both the ATP-binding and hydrophobic steroid–binding sites.[30] It was shown in other studies that among a total of 29 flavonoids tested, flavonols were able to bind to the ATP-binding site but flavones did not, suggesting the essential roles of hydroxyl at position 3 on the C ring in interacting with ATP-binding site.[154] Moreover, hydrophobic substitution by prenylation at either position 6 or 8 on the A ring increased the binding affinity of flavonoids for Pgp NBD2 significantly and abolished the interactions of flavonols with the ATP-binding site.[154] Based on these findings, a tentative mechanism for the interaction of flavonoids with Pgp was proposed by Di Pietro et al.[154] In this model, flavonols appear to interact with the ATP-binding site via hydroxyl groups at positions 3 and 5 on the A or C ring, whereas the rest of the molecule would interact with the hydrophobic steroid-binding region. Increasing hydrophobicity by prenylation will shift flavonol binding from the ATP-binding site to the vicinal steroid-binding and transmembrane domain (TMD) (Figure 22.1).

Additional structure–activity relationship studies using NBD2 and cell lines suggest that high Pgp-modulating activities are associated with molecules containing a 2–3 double bond (planar structure), 3- and 5-hydroxyl groups, and hydrophobic groups on the A or B ring. Moreover, glycosylation would dramatically decrease flavonoid Pgp-modulating activity, as exemplified by rutin, heperidin, and naringin (Table 22.3). With regard to flavanols (e.g., catechin, catechin gallate, EGC, EGCG), the presence of a galloyl moiety on the C ring increases their activities markedly.[152,154–156]

The pharmacokinetic interactions of flavonoids with drug transporters have been reported in a number of studies, most of which were focused on the interactions between Pgp and quercetin. Using paclitaxel as a Pgp substrate, Choi et al.[157] investigated the effects of quercetin on paclitaxel pharmacokinetics in rats. It was shown in this study that the oral administration of quercetin increased the AUC and $C_{\max}$ of paclitaxel (po dose, 40 mg/kg) in a dose-dependent manner. At a dose of 20 mg/kg, quercetin administration resulted in a 3.1- and 2.7-fold increase in paclitaxel AUC and

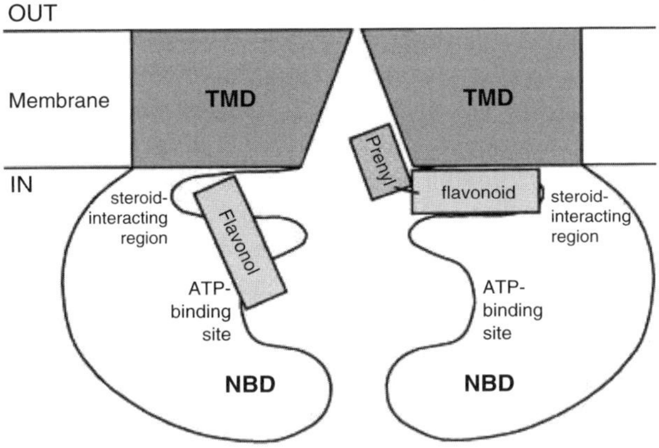

FIGURE 22.1. Interactions of flavonoids with P-glycoprotein and related multidrug resistance transporters. Flavonols such as kaempferide, quercetin, and galangin display bifunctional interactions with NBDs at both the ATP-binding site and the hydrophobic steroid-binding region. Prenylation of flavonoids would greatly increase the hydrophobicity of flavonoids, shifting the flavonol binding outside the ATP-binding site to vicinal steroid-binding region and TMD drug-binding site. (From ref. 154, with permission.)

C_{max}, respectively. Moreover, the absolute bioavailability of paclitaxel was improved from 2% (control group) to 6.2% (20 mg/kg quercetin treatment group). The $t_{1/2}$ and MRT of paclitaxel were also greatly prolonged compared to those of the control group. Similarly, the oral administration of 40 mg/kg quercetin to pigs increased the AUC of digoxin (p.o. dose, 0.02 mg/kg) by 170% and increased the C_{max} by 413%. Unexpectedly, increasing the dose of quercetin to 50 mg/kg resulted in sudden death of two of three pigs within 30 minutes after digoxin administration.[37] The adverse interaction observed in this study raised a serious safety concern of concomitant use of dietary supplements and clinically important drugs, especially those with a narrow therapeutic index. In another study, quercetin (100 μM) significantly inhibited Pgp-mediated transport of moxidectin and enhanced cellular accumulation of moxidectin in rat hepatocytes. In addition, subcutaneous administration of quercetin (10 mg/kg) increased the AUC of moxidectin (p.o. dose, 0.2 mg/kg) by 83.8%.[158] Interestingly, in a study by Hsiu et al.,[159] the oral administration of quercetin (50 mg/kg) to pigs and rats resulted in decreases in the AUC of cyclosporine by 56 and 43%, respectively. However, in a recent clinical study, oral administration of quercetin (5 mg/kg, 30 minutes or 3 days pretreatment) increased cyclosporine AUC by 36 and 47%, respectively.[38] Possible explanations for the contradictory results of cyclosporine might be species differences, methods of administration (concomitant or separate administration), and doses (low or high doses). All these studies indicated that flavonoid–Pgp interaction could occur in vivo. However, the pharmacokinetic interactions observed might result from the interactions with Pgp and/or CYP3A or both, since most of these Pgp substrates are also CYP3A substrates.

Interactions with MRPs A growing number of reports have been published regarding the modulating effects of flavonoids on MRPs. Many flavonoids have been shown to interact with MRPs, and the potential mechanisms of MRPs reversal might involve (1) decreasing intracellular glutathione (GSH) concentrations via stimulating GSH transport, (2) altering the expression of MRPs, (3) influencing drug transport possibly via direct or indirect binding interactions with MRPs at substrate or allosteric binding sites, and (4) affecting ATPase activity, ATP binding, and ADP release (Table 22.4). Using recombinant nucleotide-binding domain (NBD1) from human MRP1, Trompier et al.[160] studied the direct interactions of flavonoids with NBD1 and revealed the presence of multiple flavonoid-binding sites. In this study, dehydrosilybin was found to bind to a ATP-binding site and to inhibit leukotriene C_4 (LTC_4) transport. Similar to the case of Pgp, hydrophobic C-isoprenylation of dehydrosilybin increased the binding affinity for NBD1 but shifted the flavonoid binding outside the ATP site and decreased the inhibition of LTC_4 transport.

A number of structure–activity relationship studies indicated that flavones and flavonols were more potent in modulating MRP1 activities than were isoflavones, flavanols, flavanones, and flavanolols. Glycosylation of flavonoids resulted in a decrease in the inhibitory activity.[146,161,162] The structural features necessary for high MRP1 inhibitory potency include (1) a planar molecular structure due to the presence of a 2–3 double bond, (2) the presence of both 3′- and 4′-hydroxyl groups on the B ring, and (3) hydrophobic substitution of 4′-hydroxyl group on the B ring.[161–163] In a recent study including 29 flavonoids, diosmetin (3′,5,7-trihydroxy-4′-methoxyflavone) was identified as the most potent MRP1 inhibitor, with an IC_{50} value of 2.7 ± 0.6 μM.[162] In contrast to the wide variety of flavonoids that can inhibit MRP1, MRP2 displays higher selectivity for flavonoid inhibition. Among 29 flavonoids tested, only robinetin and myricetin inhibited MRP2-mediated calcein efflux, with IC_{50} values of 15.0 ± 3.5 and 22.2 ± 3.9 μM, respectively. The presence of a pyrogallol group on the B ring of flavonols was an important structural characteristic of flavonoids for MRP2 inhibition.[162]

Interactions with BCRP Several recent studies have demonstrated that many naturally occurring flavonoids can inhibit BCRP. Among all the subclasses of flavonoids tested, flavones seemed to be the most potent BCRP inhibitors[164,165] (Table 22.5). The EC_{50} values of the flavones chrysin and apigenin for BCRP inhibition (measured as the concentration of flavonoids for producing 50% of the maximal increase in mitoxantrone accumulation in BCRP-overexpressing MCF-7 MX100 cells) were shown to be within the sub- or low-micromolar range (0.39 ± 0.13 and 1.66 ± 0.55 μM, respectively).[165] Glycosylation dramatically decreased the BCRP-inhibiting activities of some flavonoids.[34,164] Recent structure–activity relationship studies indicated that high BCRP inhibitory potency was associated with flavonoids with the following characteristics: (1) a planar molecular structure due to the presence of a 2–3 double bond, (2) hydroxylation at position 5, (3) lack of hydroxylation at position 3; (4) B ring attached at position 2, and (4) hydrophobic substitution of a 6, 7, 8, or 4′-hydroxyl group.[165,166] Interestingly, it was shown in a recent study that prenylation at position 6 strongly enhanced both inhibitory potency and specificity of flavones. Compared

TABLE 22.4. Interactions of Flavonoids with MRPs

Flavonoid	Cell Line Used	Substrates	Effect on Pgp	Refs.
Flavones				
Apigenin	GLC_4/ADR lung cancer cells (MRP1)	Daunorubicin	↑ Accumulation	182
	GLC_4/ADR lung cancer cells/ membrane vesicles (MRP1)	Daunorubicin	↑ Accumulation ↑ ATPase activity	183
	Vesicles from HeLa-MRP1 cells	Glutathione	↑ GSH transport	146,184
	Vesicles from HeLa-MRP1 cells	$LTC_4/E_2 17\beta$ G	↓ Transport	146
	Panc-1 cells (MRP1)	DNM	↓ Accumulation	32
		VBL	↑ Accumulation	
	Human erythrocytes	BCPCF	↓ Efflux via MRP1	163
	MDCK-MRP1	Calcein	↓ Efflux	162
	MDCK-MRP2	Calcein	↔ Efflux	
Chrysin	Panc-1 cells (MRP1)	DNM/VBL	↑ Accumulation	32
	MDCK-MRP1	Calcein	↔ Efflux	162
	MDCK-MRP2	Calcein	↔ Efflux	
Luteolin	Panc-1 cells (MRP1)	DNM	↓ Accumulation	32
		VBL	↔ Accumulation	
	MCF7-pMTG5 cells	DNP-SG	↓ Efflux via MRP1	161
	MDCK-MRP1	Calcein	↓ Efflux	162
	MDCK-MRP2	Calcein	↔ Efflux	
Diosmetin	MDCK-MRP1	Calcein	↓ Efflux	162
	MDCK-MRP2	Calcein	↔ Efflux	
Diosmin	Panc-1 cells (MRP1)	DNM/VBL	↔ Accumulation	32

(Continued)

TABLE 22.4. *(Continued)*

Flavonoid	Cell Line Used	Substrates	Effect on Pgp	Refs.
Flavonols				
Fisetin	Panc-1 cells (MRP1)	DNM/VBL	↓ Accumulation	32
	MDCK-MRP1	Calcein	↔ Efflux	162
	MDCK-MRP2	Calcein	↔ Efflux	
Galangin	Panc-1 cells (MRP1)	DNM	↓ Accumulation	32
		VBL	↔ Accumulation	
	MCF7-pMTG5 cells	DNP-SG	↓ Efflux via MRP1	161
	MDCK-MRP1	Calcein	↓ Efflux	162
	MDCK-MRP2	Calcein	↔ Efflux	
	Vesicles from GLC_4/ADR cells		↑ ATPase activity	185
Kaempferol	GLC_4/ADR lung cancer cells/ membrane vesicles (MRP1)	Daunorubicin	↑ Accumulation ↑ ATPase activity	183
	Vesicles from HeLa-MRP1 cells	$LTC_4/E_217\beta G$	↓ Transport	146
		Glutathione	↔ GSH transport ↑ ATPase activity ↑ ADP trapping	
	Panc-1 cells (MRP1)	DNM/VBL	↑ Accumulation	32
	MCF7-pMTG5 cells	DNP-SG	↓ Efflux via MRP1	161
	MDCK-MRP1	Calcein	↓ Efflux	162
	MDCK-MRP2	Calcein	↔ Efflux	
Morin	Panc-1 cells (MRP1)	DNM/VBL	↑ Accumulation ↑ GSH level	32
	MCF7-pMTG5 cells	DNP-SG	↓ Efflux via MRP1	161
	Human erythrocytes	BCPCF	↓ Efflux via MRP1	163
	MDCK-MRP1	Calcein	↓ Efflux	162
	MDCK-MRP2	Calcein	↔ Efflux	

Myricetin	Vesicles from HeLa-MRP1 cells	LTC_4	↓ Transport	146
		$E_2 17\beta$ G	↔ Transport	
		Glutathione	↑ GSH transport	
	Panc-1 cells (MRP1)	DNM	↓ Accumulation	32
		VBL	↔ Accumulation	
	MCF7-pMTG5 cells	DNP-SG	↓ Efflux via MRP1	161
	MDCK-MRP1/2	Vincristine	↓ Efflux	186
	MDCK-MRP1/2	Calcein	↓ Efflux	162
Quercetin	GLC_4/ADR lung cancer cells (MRP1)	Daunorubicin	↑ Accumulation	182
	Vesicles from HeLa-MRP1 cells	Glutathione	↑ GSH transport	146,184
	Vesicles from HeLa-MRP1 cells	$LTC_4/E_2 17\beta G$	↓ Transport	146
			↓ ATPase activity	
			↑ ADP trapping	
	MCF-7 breast cancer cells		↑ MRP1 mRNA level	187
	Panc-1 cells (MRP1)	DNM/VBL	↑ Accumulation	32
			↓ GSH level	
	MCF7-pMTG5 cells	DNP-SG	↓ Efflux via MRP1	161
	Vesicles from human erythrocytes	DNP-SG	↓ Transport via MRP1	188
		cGMP	↓ Transport via MRP4	
	HEK293-MRP1	Calcein	↑ Accumulation	
	HEK293-MRP5	BCECF	↑ Accumulation	
	MDCK-MRP1	Calcein	↓ Efflux	162
	MDCK-MRP2	Calcein	↔ Efflux	
Rutin	Panc-1 cells (MRP1)	DNM	↔ Accumulation	32
		VBL	↓ Accumulation	

(*Continued*)

TABLE 22.4. ***(Continued)***

Flavonoid	Cell Line Used	Substrates	Effect on Pgp	Refs.
Flavanones				
Hesperitin	Panc-1 cells (MRP1)	DNM/VBL	↔ Accumulation	32
	Vesicles from human erythrocytes	DNP-SG	↓ Transport via MRP1	188
		cGMP	↓ Transport via MRP4	
	HEK293-MRP1	Calcein	↔ Accumulation	
	HEK293-MRP5	BCECF	↔ Accumulation	
Naringenin	Vesicles from HeLa-MRP1 cells	Glutathione	↑ GSH transport	146,184
	Vesicles from HeLa-MRP1 cells	$LTC_4/E_2 17\beta G$	↓ Transport	146
			↑ ATPase activity	
			↑ ADP trapping	
	Panc-1 cells (MRP1)	DNM/VBL	↔ Accumulation	32
	Vesicles from human erythrocytes	DNP-SG	↓ Transport via MRP1	188
		cGMP	↓ Transport via MRP4	
	HEK293-MRP1	Calcein	↔ Accumulation	
	HEK293-MRP5	BCECF	↔ Accumulation	
Naringin	Vesicles from HeLa-MRP1 cells	LTC_4	↔ Transport	146
	Panc-1 cells (MRP1)	DNM/VBL	↔ Accumulation	32
Flavanols				
Catechin	MCF7-pMTG5 cells	DNP-SG	↔ Efflux via MRP1	161
	MDCK-MRP1	Calcein	↔ Efflux	162
	MDCK-MRP2	Calcein	↔ Efflux	
EGC	Panc-1 cells (MRP1)	DNM	↔ Accumulation	32
		VBL	↑ Accumulation	

Flavanolols				
Silibinin/ silymarin	Panc-1 cells (MRP1)	DNM/VBL	↑ Accumulation	32
			↓ GSH level	
	Vesicles from BHK-21-MRP1 cells	LTC_4	↓ Transport	160
	Human erythrocytes	BCPCF	↓ Efflux via MRP1	163
	Vesicles from human erythrocytes	DNP-SG	↓ Transport via MRP1	188
		cGMP	↓ Transport via MRP4	
	HEK293-MRP1	Calcein	↑ Accumulation	
	HEK293-MRP5	BCECF	↑ Accumulation	
Isoflavones				
Biochanin A	GLC_4/ADR lung cancer cells (MRP1)	Daunorubicin	↑ Accumulation	182
	Panc-1 cells (MRP1)	DNM/VBL	↑ Accumulation	32
			↓ GSH level	
Genistein	GLC_4/ADR lung cancer cells (MRP1)	Daunorubicin	↑ Accumulation	182
			↓ Transport	189
	K562/TPA leukemia cells (MRP1)	Adriamycin	↑ Accumulation	190
	Vesicles from GLC_4/ADR cells		↑ ATPase activity	185
	GLC_4/ADR lung cancer cells/ membrane vesicles (MRP1)	Daunorubicin	↑ Accumulation	183
			↑ ATPase activity	
	Vesicles from HeLa-MRP1 cells	LTC_4	↓ Transport	146
	Vesicles from HeLa-MRP1 cells	Glutathione	↑ GSH transport	184
	Panc-1 cells (MRP1)	DNM/VBL	↑ Accumulation	32
			↓ GSH level	
	Human erythrocytes	BCPCF	↓ Efflux via MRP1	163
Genistin	Vesicles from GLC_4/ADR cells		↔ ATPase activity	185
	GLC_4/ADR lung cancer cells/ membrane vesicles (MRP1)	Daunorubicin	↔ Accumulation	183
			↔ ATPase activity	
	Vesicles from HeLa-MRP1 cells	LTC_4	↔ Transport	146

TABLE 22.5. Interactions of Flavonoids with BCRP

Flavonoid	Cell Line Used	Substrates	Effect on Pgp	Refs.
Flavones				
Apigenin	MCF-7/MX100 and NCI-H460 MX20	Mitoxantrone	↑ Accumulation	34
	K562/BCRP	SN38	Reverse resistance	164
Chrysin	MCF-7/MX100 and NCI-H460 MX20	Mitoxantrone	↑ Accumulation	34
	K562/BCRP	SN38	Reverse resistance	164
Luteolin	MCF-7/MX100 and NCI-H460 MX20	Mitoxantrone	↑ Accumulation	34
	K562/BCRP	SN38	Reverse resistance	164
Diosmetin	K562/BCRP	SN38	Reverse resistance	164
Diosmin	K562/BCRP	SN38	No effects	164
Flavonols				
Fisetin	MCF-7/MX100 and NCI-H460 MX20	Mitoxantrone	↑ Accumulation	34
	K562/BCRP	SN38	No effects	164
Galangin	K562/BCRP	SN38	Reverse resistance	164
Kaempferol	MCF-7/MX100 and NCI-H460 MX20	Mitoxantrone	↑ Accumulation	34
	K562/BCRP	Mitoxantrone/SN38	Reverse resistance	164
Morin	MCF-7/MX100 and NCI-H460 MX20	Mitoxantrone	↔ Accumulation	34
Myricetin	MCF-7/MX100 and NCI-H460 MX20	Mitoxantrone	↔ Accumulation	34
	K562/BCRP	SN38	No effects	164
Quercetin	MCF-7/MX100 and NCI-H460 MX20	Mitoxantrone	↑ Accumulation	34
	MCF-7/MR and K562/BCRP	Mitoxantrone/prazosin	↑ Accumulation	33
	K562/BCRP	SN38	Moderate reversal	164
Rutin	K562/BCRP	SN38	No effects	164
Flavanones				
Hesperitin	MCF-7/MX100 and NCI-H460 MX20	Mitoxantrone	↑ Accumulation	34
	MCF-7/MR and K562/BCRP	Mitoxantrone/prazosin	↑ Accumulation	33
	K562/BCRP	SN38	Reverse resistance	164

Naringenin	MCF-7/MX100 and NCI-H460 MX20	Mitoxantrone	↑ Accumulation	34
	K562/BCRP	Mitoxantrone/SN38	Reverse resistance	164
		Topotecan	↑ Accumulation	
Naringin	MCF-7/MX100 and NCI-H460 MX20	Mitoxantrone	↔ Accumulation	34
Flavanols				
Catechin	K562/BCRP	SN38	No effects	164
EGC	MCF-7/MX100 and NCI-H460 MX20	Mitoxantrone	↔ Accumulation	34
EGCG	MCF-7/MX100 and NCI-H460 MX20	Mitoxantrone	↔ Accumulation	34
Flavanolols				
Silibinin/	MCF-7/MX100 and NCI-H460 MX20	Mitoxantrone	↑ Accumulation	34
silymarin	MCF-7/MR and K562/BCRP	Mitoxantrone/ prazosin	↑ Accumulation	33
	K562/BCRP	SN38	Moderate reversal	164
Isoflavones				
Biochanin A	MCF-7/MX100 and NCI-H460 MX20	Mitoxantrone	↑ Accumulation	34
Genistein	MCF-7/MX100 and NCI-H460 MX20	Mitoxantrone	↑ Accumulation	34
	K562/BCRP	Mitoxantrone/SN38	Reversal resistance	164
		Topotecan	↑ Accumulation	
Daidzein	MCF-7/MX100 and NCI-H460 MX20	Mitoxantrone	↑ Accumulation	34
	MCF-7/MR and K562/BCRP	Mitoxantrone/ prazosin	↑ Accumulation	33
	K562/BCRP	SN38	Moderate reversal	164

with chrysin ($IC_{50} = 4.6 \pm 0.5\mu M$), the inhibitory potency of 6-prenylchrysin was enhanced significantly ($IC_{50} = 0.29 \pm 0.06\mu M$). Moreover, 6-prenylchrysin seemed to represent a specific BCRP inhibitor since no interaction was detected with either Pgp or MRP1.[166]

Interactions with OATP In a recent study, 20 naturally occurring flavonoids and some of their corresponding glycosides were investigated for their modulatory effects on OATP-C by using [^{3}H]dehydroepiandrosterone sulfate (DHEAS) as a probe substrate. Many of the flavonoids tested (including biochanin A, genistein, and epigallocatechin-3-gallate) inhibited [^{3}H]DHEAS uptake significantly in a concentration-dependent manner, with biochanin A being one of the most potent inhibitors, with an IC_{50} of 11.3 $\pm$ 3.22 μM. A kinetic study revealed that biochanin A inhibited [^{3}H]DHEAS uptake in a noncompetitive manner with a K_i value of 10.2 ± 1.89 μM. Four of the eight pairs of tested flavonoids and their glycosides [i.e., genistein/genistin, diosmetin/diosmin, epigallocatechin (EGC)/epigallocatechin-3-gallate (EGCG), and quercetin/rutin] exhibited distinct effects on [^{3}H]DHEAS uptake. For example, genistin did not inhibit DHEAS uptake whereas genistein did, and rutin stimulated uptake whereas quercetin had no effect[36] (Table 22.6). In another study using HEK293 cells stably transfected with OATP-B, Fuchikami et al. identified that some flavanols from green tea extracts (e.g., catechin, EC, EGC, ECG, EGCG) were potent inhibitors of OATP-B. At a concentration as low as 10 μM, EC, ECG, and EGCG significantly inhibited OATP-B-mediated uptake of estrone-3 sulfate.[110] Similar to Pgp inhibition, the presence of a galloyl moiety on the C ring of green tea catechins (e.g., EGCG vs. EGC, ECG vs. EC) markedly enhanced their potency in both OATP-B and OATP-C inhibition.[36,110]

22.2.3. Interactions of Organic Isothiocyanates with Drug Transporters

Organic isothiocyanates (ITCs, R—N=C=S), also known as mustard oils, are widely present in cruciferous vegetables, such as broccoli, watercress, cabbage, and cauliflower. Human consumption of glucosinolates, the precursors of ITCs in plants, has been estimated as high as 300 mg/day, and milligram quantities of ITCs can be released from consumption of normal amounts of vegetables.[167,168] Recently, dietary supplements containing ITCs have been used increasingly, due to the beneficial health effects of ITCs, especially cancer chemopreventive effects.[169,170] However, the mechanisms of ITCs in cancer chemoprevention have not been fully characterized.

Recently, a number of studies have indicated that ITCs might reverse resistance to anticancer drugs via interacting with drug transporters. Using Pgp-overexpressing MCF-7/ADR and MRP1-overexpressing Panc-1 cells, Tseng et al.[171] evaluated the effects of ITCs on Pgp- and MRP1-mediated transport of chemotherapeutic agents. Among all the ITCs tested, 1-naphthyl-isothiocyanate (NITC) increased the accumulation of daunomycin (DNM) and vinblastine (VBL) significantly in both resistant cell lines. Benzylisothiocyanate (BITC) and phenylhexylisothiocyanate (PHITC) increased the accumulation of DNM and/or VBL in MCF-7/ADR cells significantly, whereas phenethylisothiocyanate (PEITC), erysolin, PHITC, and phenylbutylisothiocyanate (PBITC) increased the accumulation of DNM and/or VBL in Panc-1

TABLE 22.6. Interactions of Flavonoids with OATPs

Flavonoid	Cell Line Used	Substrates	Effect on OATPs	Refs.
Flavones				
Apigenin	HeLa/OATP-C	DHEAS	↓ Uptake	36
Chrysin	HeLa/OATP-C	DHEAS	↔ Uptake	36
Luteolin	HeLa/OATP-C	DHEAS	↓ Uptake	36
Diosmetin	HeLa/OATP-C	DHEAS	↓ Uptake	36
Diosmin	HeLa/OATP-C	DHEAS	↔ Uptake	36
Flavonols				
Fisetin	HeLa/OATP-C	DHEAS	↓ Uptake	36
Galangin	HeLa/OATP-C	DHEAS	↔ Uptake	36
Kaempferol	HeLa/OATP-C	DHEAS	↔ Uptake	36
Morin	HeLa/OATP-C	DHEAS	↓ Uptake	36
Myricetin	HeLa/OATP-C	DHEAS	↓ Uptake	36
Quercetin	HeLa/OATP-C	DHEAS	↔ Uptake	36
Rutin	HeLa/OATP-C	DHEAS	↑ Uptake	36
	HEK/OATP-B	Estrone 3-sulfate	↓ Uptake	110
Flavanones				
Hesperitin	HeLa/OATP-C	DHEAS	↓ Uptake	36
Hesperidin	HeLa/OATP-C	DHEAS	↓ Uptake	36
Naringenin	HeLa/OATP-C	DHEAS	↓ Uptake	36
Naringin	HeLa/OATP-C	DHEAS	↓ Uptake	36
Flavanols				
Catechin	HEK/OATP-B	Estrone 3-sulfate	↓ Uptake	110
Epicatechin	HEK/OATP-B	Estrone 3-sulfate	↓ Uptake	110
EGC	HeLa/OATP-C	DHEAS	↔ Uptake	36
	HEK/OATP-B	Estrone 3-sulfate	↓ Uptake	110
ECG	HEK/OATP-B	Estrone 3-sulfate	↓ Uptake	110
EGCG	HeLa/OATP-C	DHEAS	↓ Uptake	36
	HEK/OATP-B	Estrone 3-sulfate	↓ Uptake	110
Flavanolols				
Silibinin/ silymarin	HeLa/OATP-C	DHEAS	↓ Uptake	36
Isoflavones				
Biochanin A	HeLa/OATP-C	DHEAS	↓ Uptake	36
Genistein	HeLa/OATP-C	DHEAS	↓ Uptake	36
Genistin	HeLa/OATP-C	DHEAS	↔ Uptake	36
Daidzein	HeLa/OATP-C	DHEAS	↔ Uptake	36
Daidzin	HeLa/OATP-C	DHEAS	↔ Uptake	36

cells. In another study, Hu et al.[172] reported that BITC and PEITC were able to deplete cellular concentrations of glutathione (GSH) in Panc-1 and Caco-2 cells after 2- and 24-hour ITC treatments. However, no significant changes in glutathione-*S*-transferase activity were found in the presence of BITC, PEITC, or NITC. In addition, PEITC and/or its metabolites was shown to be transported by MRP1 and MRP2 but not by Pgp.[172,173] These results indicate that certain dietary ITCs inhibit the Pgp- and MRP1-mediated efflux of anticancer drugs in MDR cancer cells, and the interactions probably involve multiple mechanisms.

In a recent study, the effects of 12 ITCs on the cellular accumulation of mitoxantrone (MX) were measured in BCRP-overexpressing human breast cancer (MCF-7) and large cell lung carcinoma (NCI-H460) cells. At a concentration of 10 or 30 μM, seven ITCs increased MX accumulation significantly in both cell lines and reversed MX cytotoxicity, indicating that ITCs could also modulate BCRP activities.[174]

22.3. CONCLUSIONS

In recent years, a wealth of evidence has been generated from in vitro and in vivo studies showing that many diets and nutrients interact extensively with drug transporters and play critical roles in multidrug resistance reversal and drug disposition. The importance of transporter-mediated diet–drug interactions has been recognized increasingly and reported in a number of preclinical and clinical studies. Some nutrients rich in fruits and vegetables, such as flavonoids and isothiocyanates, have been identified as potent inhibitors or inducers of major efflux or uptake transporters. The structural preferences of flavonoids for some efflux transporters have been described in several structure–activity relationship studies. However, diet and nutrient interactions with drug transporters still remain largely unknown. Most in vivo interaction studies reported to date are focused primarily on the interactions between dietary supplements and Pgp. Given the facts that MRPs and BCRP, and OATP are also essential in drug disposition, it is important to appreciate pharmacokinetic interactions of diet and nutrients with these transporters. The concentrations of many flavonoids required to produce significant modulation on activities of these transporters appear to be, in general, within the micromolar range, which is achievable in the intestine after intake of food and especially, dietary supplements. Therefore, altered disposition of MRP, BCRP, and OATP substrates following ingestion of a regular diet is likely to occur and needs further characterization. Moreover, structure–activity relationship (QSAR) studies will be crucial to better understand the potential of dietary components for inhibiting or inducing drug transport and the potential for significant in vivo diet/dietary supplement–drug interactions.

REFERENCES

1. Singh BN. 1999. Effects of food on clinical pharmacokinetics. Clin Pharmacokinet 37(3):213–255.

2. Anderson KE. 1998. Influences of diet and nutrition on clinical pharmacokinetics. Clin Pharmacokinet 14(6):325–346.
3. Evans AM. 2000. Influence of dietary components on the gastrointestinal metabolism and transport of drugs. Ther Drug Monit 22(1):131–136.
4. Harris RZ, Jang GR, Tsunoda S. 2003. Dietary effects on drug metabolism and transport. Clin Pharmacokinet 42(13):1071–1088.
5. Walter-Sack I, Klotz U. 1996. Influence of diet and nutritional status on drug metabolism. Clin Pharmacokinet 31(1):47–64.
6. Ioannides C. 2002. Pharmacokinetic interactions between herbal remedies and medicinal drugs. Xenobiotica 32(6):451–478.
7. Ohnishi N, Yokoyama T. 2004. Interactions between medicines and functional foods or dietary supplements. Keio J Med 53(3):137–150.
8. Kim RB. 2002. Transporters and xenobiotic disposition. Toxicology 181–182:291–297.
9. Ho RH, Kim RB. 2005. Transporters and drug therapy: implications for drug disposition and disease. Clin Pharmacol Ther 78(3):260–277.
10. Beringer PM, Slaughter RL. 2005. Transporters and their impact on drug disposition. Ann Pharmacother 39(6):1097–1108.
11. Leslie EM, Deeley RG, Cole SP. 2005. Multidrug resistance proteins: role of P-glycoprotein, MRP1, MRP2, and BCRP (ABCG2) in tissue defense. Toxicol Appl Pharmacol 204(3):216–237.
12. Hagenbuch B, Meier PJ. 2004. Organic anion transporting polypeptides of the OATP/SLC21 family: phylogenetic classification as OATP/ SLCO superfamily, new nomenclature and molecular/functional properties. Pflugers Arch 447(5):653–665.
13. You G. 2004. Towards an understanding of organic anion transporters: structure–function relationships. Med Res Rev 24(6):762–774.
14. Koepsell H, Schmitt BM, Gorboulev V. 2003. Organic cation transporters. Rev Physiol Biochem Pharmacol 150:36–90.
15. Allen JD, Schinkel AH. 2002. Multidrug resistance and pharmacological protection mediated by the breast cancer resistance protein (BCRP/ABCG2). Mol Cancer Ther 1(6):427–434.
16. Borst P, Evers R, Kool M, Wijnholds J. 2000. A family of drug transporters: the multidrug resistance-associated proteins. J Natl Cancer Inst 92(16):1295–1302.
17. König J, Nies AT, Cui Y, Leier I, Keppler D. 1999. Conjugate export pumps of the multidrug resistance protein (MRP) family: localization, substrate specificity, and MRP2-mediated drug resistance. Biochim Biophys Acta 1461(2):377–394.
18. Litman T, Druley TE, Stein WD, Bates SE. 2001. From MDR to MXR: new understanding of multidrug resistance systems, their properties and clinical significance. Cell Mol Life Sci 58(7):931–959.
19. Gottesman MM, Pastan I. 1993. Biochemistry of multidrug resistance mediated by the multidrug transporter. Annu Rev Biochem 62:385–427.
20. Leonard GD, Fojo T, Bates SE. 2003. The role of ABC transporters in clinical practice. Oncologist 8(5):411–424.
21. Doyle LA, Ross DD. 2003. Multidrug resistance mediated by the breast cancer resistance protein BCRP (ABCG2). Oncogene 22(47):7340–7358.

22. Westphal K, Weinbrenner A, Giessmann T, Stuhr M, Franke G, Zschiesche M, Oertel R, Terhaag B, Kroemer HK, Siegmund W. 2000. Oral bioavailability of digoxin is enhanced by talinolol: evidence for involvement of intestinal P-glycoprotein. Clin Pharmacol Ther 68(1):6–12.
23. Westphal K, Weinbrenner A, Zschiesche M, Franke G, Knoke M, Oertel R, Fritz P, von Richter O, Warzok R, Hachenberg T, et al. 2000. Induction of P-glycoprotein by rifampin increases intestinal secretion of talinolol in human beings: a new type of drug/drug interaction. Clin Pharmacol Ther 68(4):345–355.
24. Johne A, Brockmöller J, Bauer S, Maurer A, Langheinrich M, Roots I. 1999. Pharmacokinetic interaction of digoxin with an herbal extract from St John's wort (*Hypericum perforatum*). Clin Pharmacol Ther 66(4):338–345.
25. Dresser GK, Bailey DG, Leake BF, Schwarz UI, Dawson PA, Freeman DJ, Kim RB. 2002. Fruit juices inhibit organic anion transporting polypeptide-mediated drug uptake to decrease the oral availability of fexofenadine. Clin Pharmacol Ther 71(1):11–20.
26. Kruijtzer CM, Beijnen JH, Rosing H, ten Bokkel Huinink WW, Schot M, Jewell RC, Paul EM, Schellens JH. 2002. Increased oral bioavailability of topotecan in combination with the breast cancer resistance protein and P-glycoprotein inhibitor GF120918. J Clin Oncol 20(13):2943–2950.
27. Jonker JW, Smit JW, Brinkhuis RF, Maliepaard M, Beijnen JH, Schellens JH, Schinkel AH. 2000. Role of breast cancer resistance protein in the bioavailability and fetal penetration of topotecan. J Natl Cancer Inst 92(20):1651–1656.
28. Izzo AA. 2005. Herb–drug interactions: an overview of the clinical evidence. Fundam Clin Pharmacol 19(1):1–16.
29. Sparreboom A, Cox MC, Acharya MR, Figg WD. 2004. Herbal remedies in the United States: potential adverse interactions with anticancer agents. J Clin Oncol 22(12):2489–2503.
30. Conseil G, Baubichon-Cortay H, Dayan G, Jault JM, Barron D, Di Pietro A. 1998. Flavonoids: a class of modulators with bifunctional interactions at vicinal ATP- and steroid-binding sites on mouse P-glycoprotein. Proc Natl Acad Sci USA 95(17):9831–9836.
31. Durr D, Stieger B, Kullak-Ublick GA, Rentsch KM, Steinert HC, Meier PJ, Fattinger K. 2000. St John's wort induces intestinal P-glycoprotein/MDR1 and intestinal and hepatic CYP3A4. Clin Pharmacol Ther 68(6):598–604.
32. Nguyen H, Zhang S, Morris ME. 2003. Effect of flavonoids on MRP1-mediated transport in Panc-1 cells. J Pharm Sci 92(2):250–257.
33. Cooray HC, Janvilisri T, van Veen HW, Hladky SB, Barrand MA. 2004. Interaction of the breast cancer resistance protein with plant polyphenols. Biochem Biophys Res Commun 317(1):269–275.
34. Zhang S, Yang X, Morris ME. 2004. Flavonoids are inhibitors of breast cancer resistance protein (ABCG2)-mediated transport. Mol Pharmacol 65(5):1208–1216.
35. Zhou S, Lim LY, Chowbay B. 2004. Herbal modulation of P-glycoprotein. Drug Metab Rev 36(1):57–104.
36. Wang X, Wolkoff AW, Morris ME. 2005. Flavonoids as a novel class of human organic anion-transporting polypeptide OATP1B1 (OATP-C) modulators. Drug Metab Dispos 33(11):1666–1672.

37. Wang YH, Chao PD, Hsiu SL, Wen KC, Hou YC. 2004. Lethal quercetin-digoxin interaction in pigs. Life Sci 74(10):1191–1197.
38. Choi JS, Choi BC, Choi KE. 2004. Effect of quercetin on the pharmacokinetics of oral cyclosporine. Am J Health Syst Pharm 61(22):2406–2409.
39. Huang SM, Hall SD, Watkins P, Love LA, Serabjit-Singh C, Betz JM, Hoffman FA, Honig P, Coates PM, Bull J, et al. 2004. Drug interactions with herbal products and grapefruit juice: a conference report. Clin Pharmacol Ther 75(1):1–12.
40. Bilia AR, Gallori S, Vincieri FF. 2002. St. John's wort and depression: efficacy, safety and tolerability—an update. Life Sci 70(26):3077–3096.
41. Schulz HU, Schurer M, Bassler D, Weiser D. 2005. Investigation of the bioavailability of hypericin, pseudohypericin, hyperforin and the flavonoids quercetin and isorhamnetin following single and multiple oral dosing of a hypericum extract containing tablet. Arzneimittelforschung 55(1):15–22.
42. Obach RS. 2000. Inhibition of human cytochrome P450 enzymes by constituents of St. John's wort, an herbal preparation used in the treatment of depression. J Pharmacol Exp Ther 294(1):88–95.
43. Perloff MD, von Moltke LL, Stormer E, Shader RI, Greenblatt DJ. 2001. Saint John's wort: an in vitro analysis of P-glycoprotein induction due to extended exposure. Br J Pharmacol 134(8):1601–1608.
44. Hennessy M, Kelleher D, Spiers JP, Barry M, Kavanagh P, Back D, Mulcahy F, Feely J. 2002. St John's wort increases expression of P-glycoprotein: implications for drug interactions. Br J Clin Pharmacol 53(1):75–82.
45. Geick A, Eichelbaum M, Burk O. 2001. Nuclear receptor response elements mediate induction of intestinal MDR1 by rifampin. J Biol Chem 276(18):14581–14587.
46. Moore LB, Goodwin B, Jones SA, Wisely GB, Serabjit-Singh CJ, Willson TM, Collins JL, Kliewer SA. 2000. St. John's wort induces hepatic drug metabolism through activation of the pregnane X receptor. Proc Natl Acad Sci USA 97(13):7500–7502.
47. Weber CC, Kressmann S, Fricker G, Muller WE. 2004. Modulation of P-glycoprotein function by St John's wort extract and its major constituents. Pharmacopsychiatry 37(6):292–298.
48. Patel J, Buddha B, Dey S, Pal D, Mitra AK. 2004. In vitro interaction of the HIV protease inhibitor ritonavir with herbal constituents: changes in Pgp and CYP3A4 activity. Am J Ther 11(4):262–277.
49. Wang EJ, Barecki-Roach M, Johnson WW. 2004. Quantitative characterization of direct P-glycoprotein inhibition by St John's wort constituents hypericin and hyperforin. J Pharm Pharmacol 56(1):123–128.
50. Krusekopf S, Roots I. 2005. St. John's wort and its constituent hyperforin concordantly regulate expression of genes encoding enzymes involved in basic cellular pathways. Pharmacogenet Genom 15(11):817–829.
51. Shibayama Y, Ikeda R, Motoya T, Yamada K. 2004. St. John's wort (*Hypericum perforatum*) induces overexpression of multidrug resistance protein 2 (MRP2) in rats: a 30-day ingestion study. Food Chem Toxicol 42(6):995–1002.
52. Tanigawara Y, Okamura N, Hirai M, Yasuhara M, Ueda K, Kioka N, Komano T, Hori R. 1992. Transport of digoxin by human P-glycoprotein expressed in a porcine kidney epithelial cell line (LLC-PK1). J Pharmacol Exp Ther 263(2):840–845.

53. Mueller SC, Uehleke B, Woehling H, Petzsch M, Majcher-Peszynska J, Hehl EM, Sievers H, Frank B, Riethling AK, Drewelow B. 2004. Effect of St John's wort dose and preparations on the pharmacokinetics of digoxin. Clin Pharmacol Ther 75(6):546–557.

54. Lacarelle B, Rahmani R, de Sousa G, Durand A, Placidi M, Cano JP. 1991. Metabolism of digoxin, digoxigenin digitoxosides and digoxigenin in human hepatocytes and liver microsomes. Fundam Clin Pharmacol 5(7):567–582.

55. Cvetkovic M, Leake B, Fromm MF, Wilkinson GR, Kim RB. 1999. OATP and P-glycoprotein transporters mediate the cellular uptake and excretion of fexofenadine. Drug Metab Dispos 27(8):866–871.

56. Molimard M, Diquet B, Benedetti MS. 2004. Comparison of pharmacokinetics and metabolism of desloratadine, fexofenadine, levocetirizine and mizolastine in humans. Fundam Clin Pharmacol 18(4):399–411.

57. Wang Z, Hamman MA, Huang SM, Lesko LJ, Hall SD. 2002. Effect of St John's wort on the pharmacokinetics of fexofenadine. Clin Pharmacol Ther 71(6):414–420.

58. Xie HG, Kim RB. 2005. St John's wort–associated drug interactions: short-term inhibition and long-term induction? Clin Pharmacol Ther 78(1):19–24.

59. Piscitelli SC, Burstein AH, Chaitt D, Alfaro RM, Falloon J. 2000. Indinavir concentrations and St John's wort. Lancet 355(9203):547–548.

60. Bauer S, Stormer E, Johne A, Kruger H, Budde K, Neumayer HH, Roots I, Mai I. 2003. Alterations in cyclosporin A pharmacokinetics and metabolism during treatment with St. John's wort in renal transplant patients. Br J Clin Pharmacol 55(2):203–211.

61. Mai I, Bauer S, Perloff ES, Johne A, Uehleke B, Frank B, Budde K, Roots I. 2004. Hyperforin content determines the magnitude of the St John's wort–cyclosporine drug interaction. Clin Pharmacol Ther 76(4):330–340.

62. Ruschitzka F, Meier PJ, Turina M, Luscher TF, Noll G. 2000. Acute heart transplant rejection due to Saint John's wort. Lancet 355(9203):548–549.

63. Barone GW, Gurley BJ, Ketel BL, Lightfoot ML, Abul-Ezz SR. 2000. Drug interaction between St. John's wort and cyclosporine. Ann Pharmacother 34(9):1013–1016.

64. Hebert MF, Park JM, Chen YL, Akhtar S, Larson AM. 2004. Effects of St. John's wort (Hypericum perforatum) on tacrolimus pharmacokinetics in healthy volunteers. J Clin Pharmacol 44(1):89–94.

65. Dresser GK, Schwarz UI, Wilkinson GR, Kim RB. 2003. Coordinate induction of both cytochrome P4503A and MDR1 by St John's wort in healthy subjects. Clin Pharmacol Ther 73(1):41–50.

66. Bailey DG, Spence JD, Munoz C, Arnold JM. 1991. Interaction of citrus juices with felodipine and nifedipine. Lancet 337(8736):268–269.

67. Bailey DG, Malcolm J, Arnold O, Spence JD. 1998. Grapefruit juice–drug interactions. Br J Clin Pharmacol 46(2):101–110.

68. Dahan A, Altman H. 2004. Food–drug interaction: grapefruit juice augments drug bioavailability—mechanism, extent and relevance. Eur J Clin Nutr 58(1):1–9.

69. Kane GC, Lipsky JJ. 2000. Drug–grapefruit juice interactions. Mayo Clin Proc 75(9):933–942.

70. Wang EJ, Casciano CN, Clement RP, Johnson WW. 2001. Inhibition of P-glycoprotein transport function by grapefruit juice psoralen. Pharm Res 18(4):432–438.

71. Ohnishi A, Matsuo H, Yamada S, Takanaga H, Morimoto S, Shoyama Y, Ohtani H, Sawada Y. 2000. Effect of furanocoumarin derivatives in grapefruit juice on the uptake of vinblastine by Caco-2 cells and on the activity of cytochrome P450 3A4. Br J Pharmacol 130(6):1369–1377.
72. Spahn-Langguth H, Langguth P. 2001. Grapefruit juice enhances intestinal absorption of the P-glycoprotein substrate talinolol. Eur J Pharm Sci 12(4):361–367.
73. Tian R, Koyabu N, Takanaga H, Matsuo H, Ohtani H, Sawada Y. 2002. Effects of grapefruit juice and orange juice on the intestinal efflux of P-glycoprotein substrates. Pharm Res 19(6):802–809.
74. Xu J, Go ML, Lim LY. 2003. Modulation of digoxin transport across Caco-2 cell monolayers by citrus fruit juices: lime, lemon, grapefruit, and pummelo. Pharm Res 20(2):169–176.
75. Honda Y, Ushigome F, Koyabu N, Morimoto S, Shoyama Y, Uchiumi T, Kuwano M, Ohtani H, Sawada Y. 2004. Effects of grapefruit juice and orange juice components on P-glycoprotein- and MRP2-mediated drug efflux. Br J Pharmacol 143(7):856–864.
76. Soldner A, Christians U, Susanto M, Wacher VJ, Silverman JA, Benet LZ. 1999. Grapefruit juice activates P-glycoprotein-mediated drug transport. Pharm Res 16(4):478–485.
77. Lim SL, Lim LY. 2006. Effects of citrus fruit juices on cytotoxicity and drug transport pathways of Caco-2 cell monolayers. Int J Pharm 307(1):42–50.
78. Eagling VA, Profit L, Back DJ. 1999. Inhibition of the CYP3A4-mediated metabolism and P-glycoprotein-mediated transport of the HIV-1 protease inhibitor saquinavir by grapefruit juice components. Br J Clin Pharmacol 48(4):543–552.
79. Takanaga H, Ohnishi A, Matsuo H, Sawada Y. 1998. Inhibition of vinblastine efflux mediated by P-glycoprotein by grapefruit juice components in Caco-2 cells. Biol Pharm Bull 21(10):1062–1066.
80. Romiti N, Tramonti G, Donati A, Chieli E. 2004. Effects of grapefruit juice on the multidrug transporter P-glycoprotein in the human proximal tubular cell line HK-2. Life Sci 76(3):293–302.
81. Satoh H, Yamashita F, Tsujimoto M, Murakami H, Koyabu N, Ohtani H, Sawada Y. 2005. Citrus juices inhibit the function of human organic anion–transporting polypeptide OATP-B. Drug Metab Dispos 33(4):518–523.
82. Yee GC, Stanley DL, Pessa LJ, Dalla Costa T, Beltz SE, Ruiz J, Lowenthal DT. 1995. Effect of grapefruit juice on blood cyclosporin concentration. Lancet 345(8955):955–956.
83. Ducharme MP, Warbasse LH, Edwards DJ. 1995. Disposition of intravenous and oral cyclosporine after administration with grapefruit juice. Clin Pharmacol Ther 57(5):485–491.
84. Edwards DJ, Fitzsimmons ME, Schuetz EG, Yasuda K, Ducharme MP, Warbasse LH, Woster PM, Schuetz JD, Watkins P. 1999. 6′,7′-Dihydroxybergamottin in grapefruit juice and Seville orange juice: effects on cyclosporine disposition, enterocyte CYP3A4, and P-glycoprotein. Clin Pharmacol Ther 65(3):237–244.
85. Lown KS, Mayo RR, Leichtman AB, Hsiao HL, Turgeon DK, Schmiedlin-Ren P, Brown MB, Guo W, Rossi SJ, Benet LZ, Watkins PB. 1997. Role of intestinal P-glycoprotein (mdr1) in interpatient variation in the oral bioavailability of cyclosporine. Clin Pharmacol Ther 62(3):248–260.

86. Schwarz UI, Seemann D, Oertel R, Miehlke S, Kuhlisch E, Fromm MF, Kim RB, Bailey DG, Kirch W. 2005. Grapefruit juice ingestion significantly reduces talinolol bioavailability. Clin Pharmacol Ther 77(4):291–301.
87. Parker RB, Yates CR, Soberman JE, Laizure SC. 2003. Effects of grapefruit juice on intestinal P-glycoprotein: evaluation using digoxin in humans. Pharmacotherapy 23(8):979–987.
88. Becquemont L, Verstuyft C, Kerb R, Brinkmann U, Lebot M, Jaillon P, Funck-Brentano C. 2001. Effect of grapefruit juice on digoxin pharmacokinetics in humans. Clin Pharmacol Ther 70(4):311–316.
89. Dresser GK, Bailey DG. 2003. The effects of fruit juices on drug disposition: a new model for drug interactions. Eur J Clin Invest 33(Suppl 2):10–16.
90. Magnani B, Malini PL. 1995. Cardiac glycosides. Drug interactions of clinical significance. Drug Saf 12(2):97–109.
91. Dresser GK, Kim RB, Bailey DG. 2005. Effect of grapefruit juice volume on the reduction of fexofenadine bioavailability: possible role of organic anion transporting polypeptides. Clin Pharmacol Ther 77(3):170–177.
92. Tattelman E. 2005. Health effects of garlic. Am Fam Physician 72(1):103–106.
93. Lawson LD, Gardner CD. 2005. Composition, stability, and bioavailability of garlic products used in a clinical trial. J Agric Food Chem 53(16):6254–6261.
94. Foster BC, Foster MS, Vandenhoek S, Krantis A, Budzinski JW, Arnason JT, Gallicano KD, Choudri S. 2001. An in vitro evaluation of human cytochrome P450 3A4 and P-glycoprotein inhibition by garlic. J Pharm Pharm Sci 4(2):176–184.
95. Nabekura T, Kamiyama S, Kitagawa S. 2005. Effects of dietary chemopreventive phytochemicals on P-glycoprotein function. Biochem Biophys Res Commun 327(3):866–870.
96. Arora A, Seth K, Shukla Y. 2004. Reversal of P-glycoprotein-mediated multidrug resistance by diallyl sulfide in K562 leukemic cells and in mouse liver. Carcinogenesis 25(6):941–949.
97. Demeule M, Brossard M, Turcotte S, Regina A, Jodoin J, Béliveau R. 2004. Diallyl disulfide, a chemopreventive agent in garlic, induces multidrug resistance-associated protein 2 expression. Biochem Biophys Res Commun 324(2):937–945.
98. Piscitelli SC, Burstein AH, Welden N, Gallicano KD, Falloon J. 2002. The effect of garlic supplements on the pharmacokinetics of saquinavir. Clin Infect Dis 34(2):234–238.
99. Gallicano K, Foster B, Choudhri S. 2003. Effect of short-term administration of garlic supplements on single-dose ritonavir pharmacokinetics in healthy volunteers. Br J Clin Pharmacol 55(2):199–202.
100. Kim AE, Dintaman JM, Waddell DS, Silverman JA. 1998. Saquinavir, an HIV protease inhibitor, is transported by P-glycoprotein. J Pharmacol Exp Ther 286(3):1439–1445.
101. Lee CG, Gottesman MM, Cardarelli CO, Ramachandra M, Jeang KT, Ambudkar SV, Pastan I, Dey S. 1998. HIV-1 protease inhibitors are substrates for the MDR1 multidrug transporter. Biochemistry 37(11):3594–3601.
102. Fujiki H. 2005. Green tea: health benefits as cancer preventive for humans. Chem Rec 5(3):119–132.
103. Yang CS, Maliakal P, Meng X. 2002. Inhibition of carcinogenesis by tea. Annu Rev Pharmacol Toxicol 42:25–54.

104. Jodoin J, Demeule M, Beliveau R. 2002. Inhibition of the multidrug resistance P-glycoprotein activity by green tea polyphenols. Biochim Biophys Acta 1542(1–3):149–159.

105. Mei Y, Qian F, Wei D, Liu J. 2004. Reversal of cancer multidrug resistance by green tea polyphenols. J Pharm Pharmacol 56(10):1307–1314.

106. Netsch MI, Gutmann H, Luescher S, Brill S, Schmidlin CB, Kreuter MH, Drewe J. 2005. Inhibitory activity of a green tea extract and some of its constituents on multidrug resistance-associated protein 2 functionality. Planta Med 71(2):135–141.

107. Zhang L, Zheng Y, Chow MS, Zuo Z. 2004. Investigation of intestinal absorption and disposition of green tea catechins by Caco-2 monolayer model. Int J Pharm 287(1–2):1–12.

108. Hong J, Lambert JD, Lee SH, Sinko PJ, Yang CS. 2003. Involvement of multidrug resistance-associated proteins in regulating cellular levels of (−)-epigallocatechin-3-gallate and its methyl metabolites. Biochem Biophys Res Commun 310(1):222–227.

109. Vaidyanathan JB, Walle T. 2003. Cellular uptake and efflux of the tea flavonoid (−)epicatechin-3-gallate in the human intestinal cell line Caco-2. J Pharmacol Exp Ther 307(2):745–752.

110. Fuchikami H, Satoh H, Tsujimoto M, Ohdo S, Ohtani H, Sawada Y. 2006. Effects of herbal extracts on the function of human organic anion transporting polypeptide, OATP-B. Drug Metab Dispos 34(4):577–582.

111. Sartippour MR, Heber D, Ma J, Lu Q, Go VL, Nguyen M. 2001. Green tea and its catechins inhibit breast cancer xenografts. Nutr Cancer 40(2):149–156.

112. Zhang Q, Wei D, Liu J. 2004. In vivo reversal of doxorubicin resistance by (−)-epigallocatechin gallate in a solid human carcinoma xenograft. Cancer Lett 208(2):179–186.

113. Attele AS, Wu JA, Yuan CS. 1999. Ginseng pharmacology: multiple constituents and multiple actions. Biochem Pharmacol 58(11):1685–1693.

114. Kim SW, Kwon HY, Chi DW, Shim JH, Park JD, Lee YH, Pyo S, Rhee DK. 2003. Reversal of P-glycoprotein-mediated multidrug resistance by ginsenoside Rg(3). Biochem Pharmacol 65(1):75–82.

115. Choi CH, Kang G, Min YD. 2003. Reversal of P-glycoprotein-mediated multidrug resistance by protopanaxatriol ginsenosides from Korean red ginseng. Planta Med 69(3):235–240.

116. Jacobs BP, Dennehy C, Ramirez G, Sapp J, Lawrence VA. 2002. Milk thistle for the treatment of liver disease: a systematic review and meta-analysis. Am J Med 113(6):506–515.

117. Rambaldi A, Jacobs BP, Iaquinto G, Gluud C. 2005. Milk thistle for alcoholic and/or hepatitis B or C liver diseases: a systematic cochrane hepato-biliary group review with meta-analyses of randomized clinical trials. Am J Gastroenterol 100(11):2583–2591.

118. Bilia AR, Bergonzi MC, Gallori S, Mazzi G, Vincieri FF. 2002. Stability of the constituents of calendula, milk-thistle and passionflower tinctures by LC-DAD and LC-MS. J Pharm Biomed Anal 30(3):613–624.

119. Kvasnicka F, Biba B, Sevcik R, Voldrich M, Kratka J. 2003. Analysis of the active components of silymarin. J Chromatogr A 990(1–2):239–245.

120. Zhang S, Morris ME. 2003. Effects of the flavonoids biochanin A, morin, phloretin, and silymarin on P-glycoprotein-mediated transport. J Pharmacol Exp Ther 304(3):1258–1267.

121. Zhang S, Morris ME. 2003. Effect of the flavonoids biochanin A and silymarin on the P-glycoprotein-mediated transport of digoxin and vinblastine in human intestinal Caco-2 cells. Pharm Res 20(8):1184–1191.

122. Gurley BJ, Barone GW, Williams DK, Carrier J, Breen P, Yates CR, Song PF, Hubbard MA, Tong Y, Cheboyina S. 2006. Effect of milk thistle (*Silybum marianum*) and black cohosh (*Cimicifuga racemosa*) supplementation on digoxin pharmacokinetics in humans. Drug Metab Dispos 34(1):69–74.

123. DiCenzo R, Shelton M, Jordan K, Koval C, Forrest A, Reichman R, Morse G. 2003. Coadministration of milk thistle and indinavir in healthy subjects. Pharmacotherapy 23(7):866–870.

124. Piscitelli SC, Formentini E, Burstein AH, Alfaro R, Jagannatha S, Falloon J. 2002. Effect of milk thistle on the pharmacokinetics of indinavir in healthy volunteers. Pharmacotherapy 22(5):551–556.

125. Mills E, Wilson K, Clarke M, Foster B, Walker S, Rachlis B, DeGroot N, Montori VM, Gold W, Phillips E, Myers S, Gallicano K. 2005. Milk thistle and indinavir: a randomized controlled pharmacokinetics study and meta-analysis. Eur J Clin Pharmacol 61(1):1–7.

126. Smith NF, Figg WD, Sparreboom A. 2006. Pharmacogenetics of irinotecan metabolism and transport: an update. Toxicol In Vitro 20(2):163–175.

127. van Erp NP, Baker SD, Zhao M, Rudek MA, Guchelaar HJ, Nortier JW, Sparreboom A, Gelderblom H. 2005. Effect of milk thistle (*Silybum marianum*) on the pharmacokinetics of irinotecan. Clin Cancer Res 11(21):7800–7806.

128. Singh YN. 2005. Potential for interaction of kava and St. John's wort with drugs. J Ethnopharmacol 100(1–2):108–113.

129. Bilia AR, Scalise L, Bergonzi MC, Vincieri FF. 2004. Analysis of kavalactones from *Piper methysticum* (kava-kava). J Chromatogr B Anal Technol Biomed Life Sci 812(1–2):203–214.

130. Bilia AR, Bergonzi MC, Lazari D, Vincieri FF. 2002. Characterization of commercial kava-kava herbal drug and herbal drug preparations by means of nuclear magnetic resonance spectroscopy. J Agric Food Chem 50(18):5016–5025.

131. Ulbricht C, Basch E, Boon H, Ernst E, Hammerness P, Sollars D, Tsourounis C, Woods J, Bent S. 2005. Safety review of kava (*Piper methysticum*) by the Natural Standard Research Collaboration. Expert Opin Drug Saf 4(4):779–794.

132. Weiss J, Sauer A, Frank A, Unger M. 2005. Extracts and kavalactones of *Piper methysticum* G. Forst (kava-kava) inhibit P-glycoprotein in vitro. Drug Metab Dispos 33(11):1580–1583.

133. Mathews JM, Etheridge AS, Valentine JL, Black SR, Coleman DP, Patel P, So J, Burka LT. 2005. Pharmacokinetics and disposition of the kavalactone kawain: interaction with kava extract and kavalactones in vivo and in vitro. Drug Metab Dispos 33(10):1555–1563.

134. Ross SA, Ziska DS, Zhao K, ElSohly MA. 2000. Variance of common flavonoids by brand of grapefruit juice. Fitoterapia 71(2):154–161.

135. Erlund I, Meririnne E, Alfthan G, Aro A. 2001. Plasma kinetics and urinary excretion of the flavanones naringenin and hesperetin in humans after ingestion of orange juice and grapefruit juice. J Nutr 131(2):235–241.

136. Manach C, Morand C, Gil-Izquierdo A, Bouteloup-Demange C, Remesy C. 2003. Bioavailability in humans of the flavanones hesperidin and narirutin after the ingestion of two doses of orange juice. Eur J Clin Nutr 57(2):235–242.

137. Kuhnau J 1976. The flavonoids. A class of semi-essential food components: their role in human nutrition. World Rev Nutr Diet 24:117–191.

138. Harborne JB, Williams CA. 2000. Advances in flavonoid research since 1992. Phytochemistry 55(6):481–504.

139. Scalbert A, Williamson G. 2000. Dietary intake and bioavailability of polyphenols. J Nutr 130(8S Suppl):2073S–2085S.

140. Vinson JA, Hao Y, Su X, Bose P. 1998. Phenol antioxidant quantity and quality in foods: vegetables. J Agric Food Chem 46(9):3630–3634.

141. Vinson JA, Su X, Zubik L, Bose P. 2001. Phenol antioxidant quantity and quality in foods: fruits. J Agric Food Chem 49(11):5315–5321.

142. Middleton E, Jr., Kandaswami C, Theoharides TC. 2000. The effects of plant flavonoids on mammalian cells: implications for inflammation, heart disease, and cancer. Pharmacol Rev 52(4):673–751.

143. Havsteen BH. 2002. The biochemistry and medical significance of the flavonoids. Pharmacol Ther 96(2–3):67–202.

144. Hertog MG, Feskens EJ, Hollman PC, Katan MB, Kromhout D. 1993. Dietary antioxidant flavonoids and risk of coronary heart disease: the Zutphen Elderly Study. Lancet 342(8878):1007–1011.

145. Kohno H, Tanaka T, Kawabata K, Hirose Y, Sugie S, Tsuda H, Mori H. 2002. Silymarin, a naturally occurring polyphenolic antioxidant flavonoid, inhibits azoxymethane-induced colon carcinogenesis in male F344 rats. Int J Cancer 101(5):461–468.

146. Leslie EM, Mao Q, Oleschuk CJ, Deeley RG, Cole SP. 2001. Modulation of multidrug resistance protein 1 (MRP1/ABCC1) transport and ATPase activities by interaction with dietary flavonoids. Mol Pharmacol 59(5):1171–1180.

147. Critchfield JW, Welsh CJ, Phang JM, Yeh GC. 1994. Modulation of adriamycin accumulation and efflux by flavonoids in HCT-15 colon cells. Activation of P-glycoprotein as a putative mechanism. Biochem Pharmacol 48(7):1437–1445.

148. Scambia G, Ranelletti FO, Panici PB, De Vincenzo R, Bonanno G, Ferrandina G, Piantelli M, Bussa S, Rumi C, Cianfriglia M, et al. 1994. Quercetin potentiates the effect of adriamycin in a multidrug-resistant MCF-7 human breast-cancer cell line: P-glycoprotein as a possible target. Cancer Chemother Pharmacol 34(6):459–464.

149. Mitsunaga Y, Takanaga H, Matsuo H, Naito M, Tsuruo T, Ohtani H, Sawada Y. 2000. Effect of bioflavonoids on vincristine transport across blood–brain barrier. Eur J Pharmacol 395(3):193–201.

150. Chieli E, Romiti N, Cervelli F, Tongiani R. 1995. Effects of flavonols on P-glycoprotein activity in cultured rat hepatocytes. Life Sci 57(19):1741–1751.

151. Shapiro AB, Ling V. 1997. Effect of quercetin on Hoechst 33342 transport by purified and reconstituted P-glycoprotein. Biochem Pharmacol 53(4):587–596.

152. Kitagawa S, Nabekura T, Takahashi T, Nakamura Y, Sakamoto H, Tano H, Hirai M, Tsukahara G. 2005. Structure–activity relationships of the inhibitory effects of flavonoids on P-glycoprotein-mediated transport in KB-C2 cells. Biol Pharm Bull 28(12):2274–2278.

153. Shapiro AB, Ling V. 1997. Positively cooperative sites for drug transport by P-glycoprotein with distinct drug specificities. Eur J Biochem 250(1):130–137.
154. Di Pietro A, Conseil G, Perez-Victoria JM, Dayan G, Baubichon-Cortay H, Trompier D, Steinfels E, Jault JM, de Wet H, Maitrejean M, et al. 2002. Modulation by flavonoids of cell multidrug resistance mediated by P-glycoprotein and related ABC transporters. Cell Mol Life Sci 59(2):307–322.
155. Boumendjel A, Di Pietro A, Dumontet C, Barron D. 2002. Recent advances in the discovery of flavonoids and analogs with high-affinity binding to P-glycoprotein responsible for cancer cell multidrug resistance. Med Res Rev 22(5):512–529.
156. Kitagawa S. 2006. Inhibitory effects of polyphenols on P-glycoprotein-mediated transport. Biol Pharm Bull 29(1):1–6.
157. Choi JS, Jo BW, Kim YC. 2004. Enhanced paclitaxel bioavailability after oral administration of paclitaxel or prodrug to rats pretreated with quercetin. Eur J Pharm Biopharm 57(2):313–318.
158. Dupuy J, Larrieu G, Sutra JF, Lespine A, Alvinerie M. 2003. Enhancement of moxidectin bioavailability in lamb by a natural flavonoid: quercetin. Vet Parasitol 112(4):337–347.
159. Hsiu SL, Hou YC, Wang YH, Tsao CW, Su SF, Chao PD. 2002. Quercetin significantly decreased cyclosporin oral bioavailability in pigs and rats. Life Sci 72(3):227–235.
160. Trompier D, Baubichon-Cortay H, Chang XB, Maitrejean M, Barron D, Riordon JR, Di Pietro A. 2003. Multiple flavonoid-binding sites within multidrug resistance protein MRP1. Cell Mol Life Sci 60(10):2164–2177.
161. van Zanden JJ, Geraets L, Wortelboer HM, van Bladeren PJ, Rietjens IM, Cnubben NH. 2004. Structural requirements for the flavonoid-mediated modulation of glutathione S-transferase P1-1 and GS-X pump activity in MCF7 breast cancer cells. Biochem Pharmacol 67(8):1607–1617.
162. van Zanden JJ, Wortelboer HM, Bijlsma S, Punt A, Usta M, Bladeren PJ, Rietjens IM, Cnubben NH. 2005. Quantitative structure activity relationship studies on the flavonoid mediated inhibition of multidrug resistance proteins 1 and 2. Biochem Pharmacol 69(4):699–708.
163. Lania-Pietrzak B, Michalak K, Hendrich AB, Mosiadz D, Grynkiewicz G, Motohashi N, Shirataki Y. 2005. Modulation of MRP1 protein transport by plant, and synthetically modified flavonoids. Life Sci 77(15):1879–1891.
164. Imai Y, Tsukahara S, Asada S, Sugimoto Y. 2004. Phytoestrogens/flavonoids reverse breast cancer resistance protein/ABCG2-mediated multidrug resistance. Cancer Res 64(12):4346–4352.
165. Zhang S, Yang X, Coburn RA, Morris ME. 2005. Structure activity relationships and quantitative structure activity relationships for the flavonoid-mediated inhibition of breast cancer resistance protein. Biochem Pharmacol 70(4):627–639.
166. Ahmed-Belkacem A, Pozza A, Munoz-Martinez F, Bates SE, Castanys S, Gamarro F, Di Pietro A, Perez-Victoria JM. 2005. Flavonoid structure–activity studies identify 6-prenylchrysin and tectochrysin as potent and specific inhibitors of breast cancer resistance protein ABCG2. Cancer Res 65(11):4852–4860.
167. Hecht SS. 1995. Chemoprevention by isothiocyanates. J Cell Biochem Suppl 22:195–209.
168. Talalay P, Fahey JW. 2001. Phytochemicals from cruciferous plants protect against cancer by modulating carcinogen metabolism. J Nutr 131(11 Suppl):3027S–3033S.

169. Hecht SS. 2000. Inhibition of carcinogenesis by isothiocyanates. Drug Metab Rev 32(3–4):395–411.

170. Kelloff GJ, Crowell JA, Steele VE, Lubet RA, Malone WA, Boone CW, Kopelovich L, Hawk ET, Lieberman R, Lawrence JA, et al. 2000. Progress in cancer chemoprevention: development of diet-derived chemopreventive agents. J Nutr 130(2S Suppl):467S–471S.

171. Tseng E, Kamath A, Morris ME. 2002. Effect of organic isothiocyanates on the P-glycoprotein- and MRP1-mediated transport of daunomycin and vinblastine. Pharm Res 19(10):1509–1515.

172. Hu K, Morris ME. 2004. Effects of benzyl-, phenethyl-, and alpha-naphthyl isothiocyanates on P-glycoprotein- and MRP1-mediated transport. J Pharm Sci 93(7):1901–1911.

173. Ji Y, Morris ME. 2005. Transport of dietary phenethyl isothiocyanate is mediated by multidrug resistance protein 2 but not P-glycoprotein. Biochem Pharmacol 70(4):640–647.

174. Ji Y, Morris ME. 2004. Effect of organic isothiocyanates on breast cancer resistance protein (ABCG2)-mediated transport. Pharm Res 21(12):2261–2269.

175. Ikegawa T, Ushigome F, Koyabu N, Morimoto S, Shoyama Y, Naito M, Tsuruo T, Ohtani H, Sawada Y. 2000. Inhibition of P-glycoprotein by orange juice components, polymethoxyflavones in adriamycin-resistant human myelogenous leukemia (K562/ADM) cells. Cancer Lett 160(1):21–28.

176. Sadzuka Y, Sugiyama T, Sonobe T. 2000. Efficacies of tea components on doxorubicin induced antitumor activity and reversal of multidrug resistance. Toxicol Lett 114(1–3):155–162.

177. Tseng E, Liang W, Wallen C, Morris ME. 2001. Effect of flavonoids on Pgp mediated transport in a human breast cancer cell line. AAPS PharmSci 3:S2081.

178. Chung SY, Sung MK, Kim NH, Jang JO, Go EJ, Lee HJ. 2005. Inhibition of P-glycoprotein by natural products in human breast cancer cells. Arch Pharm Res 28(7):823–828.

179. Kitagawa S, Nabekura T, Kamiyama S. 2004. Inhibition of P-glycoprotein function by tea catechins in KB-C2 cells. J Pharm Pharmacol 56(8):1001–1005.

180. Qian F, Wei D, Zhang Q, Yang S. 2005. Modulation of P-glycoprotein function and reversal of multidrug resistance by (−)-epigallocatechin gallate in human cancer cells. Biomed Pharmacother 59(3):64–69.

181. Castro AF, Altenberg GA. 1997. Inhibition of drug transport by genistein in multidrug-resistant cells expressing P-glycoprotein. Biochem Pharmacol 53(1):89–93.

182. Versantvoort CH, Schuurhuis GJ, Pinedo HM, Eekman CA, Kuiper CM, Lankelma J, Broxterman HJ. 1993. Genistein modulates the decreased drug accumulation in non-P-glycoprotein mediated multidrug resistant tumour cells. Br J Cancer 68(5):939–946.

183. Hooijberg JH, Broxterman HJ, Scheffer GL, Vrasdonk C, Heijn M, de Jong MC, Scheper RJ, Lankelma J, Pinedo HM. 1999. Potent interaction of flavopiridol with MRP1. Br J Cancer 81(2):269–276.

184. Leslie EM, Deeley RG, Cole SP. 2003. Bioflavonoid stimulation of glutathione transport by the 190-kDa multidrug resistance protein 1 (MRP1). Drug Metab Dispos 31(1):11–15.

185. Hooijberg JH, Broxterman HJ, Heijn M, Fles DL, Lankelma J, Pinedo HM. 1997. Modulation by (iso)flavonoids of the ATPase activity of the multidrug resistance protein. FEBS Lett 413(2):344–348.

186. van Zanden JJ, de Mul A, Wortelboer HM, Usta M, van Bladeren PJ, Rietjens IM, Cnubben NH. 2005. Reversal of in vitro cellular MRP1 and MRP2 mediated vincristine resistance by the flavonoid myricetin. Biochem Pharmacol 69(11):1657–1665.

187. Kauffmann HM, Pfannschmidt S, Zoller H, Benz A, Vorderstemann B, Webster JI, Schrenk D. 2002. Influence of redox-active compounds and PXR-activators on human MRP1 and MRP2 gene expression. Toxicology 171(2–3):137–146.

188. Wu CP, Calcagno AM, Hladky SB, Ambudkar SV, Barrand MA. 2005. Modulatory effects of plant phenols on human multidrug-resistance proteins 1, 4 and 5 (ABCC1, 4 and 5). FEBS J 272(18):4725–4740.

189. Versantvoort CH, Broxterman HJ, Lankelma J, Feller N, Pinedo HM. 1994. Competitive inhibition by genistein and ATP dependence of daunorubicin transport in intact MRP overexpressing human small cell lung cancer cells. Biochem Pharmacol 48(6):1129–1136.

190. Takeda Y, Nishio K, Niitani H, Saijo N. 1994. Reversal of multidrug resistance by tyrosine-kinase inhibitors in a non-P-glycoprotein-mediated multidrug-resistant cell line. Int J Cancer 57(2):229–239.

23

INTERPLAY OF DRUG TRANSPORTERS AND ENZYMES ON HEPATIC DRUG PROCESSING

K. SANDY PANG, HUADONG SUN, AND SHANJUN LIU
Leslie Dan Faculty of Pharmacy, University of Toronto, Toronto, Ontario, Canada

23.1. INTRODUCTION

The liver is a powerhouse of drug-metabolizing enzymes and is of paramount importance in first-pass drug removal since the organ connects the gastrointestinal tract to the general circulation. Removal of the orally administered dose due to first-pass removal reduces the fraction of dose absorbed or the systemic bioavailability.[1] Upon reaching the sinusoid, the unbound drug enters the hepatocyte by passive diffusion or

Drug Transporters: Molecular Characterization and Role in Drug Disposition, Edited by Guofeng You and Marilyn E. Morris
Copyright © 2007 John Wiley & Sons, Inc.

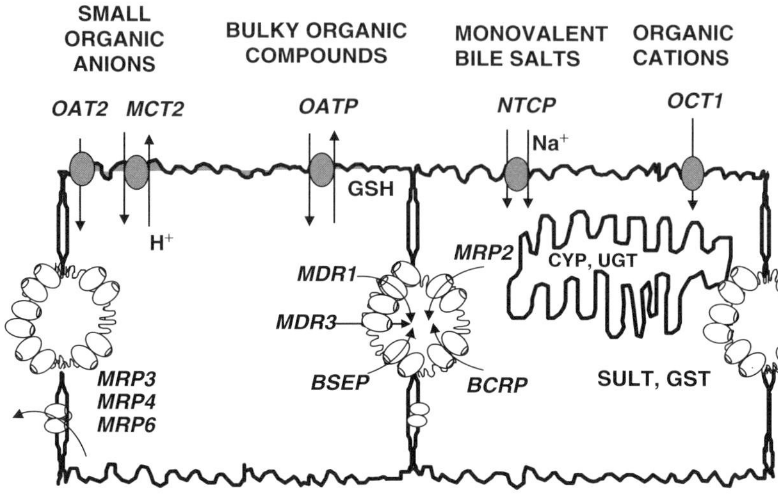

FIGURE 23.1. Schematic diagram of transport and metabolism of drugs in the hepatocyte which shows influx transporters such as OATP, NTCP, OAT2, OCT1, and MCT2 at the sinusoidal membrane; efflux transporters such as MRP3, MRP4, and MRP6 at the basolateral membrane; and canalicular efflux transporters such as Pgp or MDR1, MDR3, MRP2, BSEP, and BCRP at the canalicular membrane. Enzymes such as CYP, UGT, SULT and GST are present to mediate intracellular metabolism. (From ref. 84, with permission.)

is taken up by transporters. Within the cell, the drug may be effluxed immediately back to the blood or eliminated irreversibly by metabolism or secretion into the bile. To date, blood flow, vascular [plasma protein and red blood cell (RBC)] binding, transporters, and metabolic enzymes are regarded as important determinants that affect hepatic drug extraction.[2,3]

Most drug removal activities have been attributed to the cytochrome P450s (CYPs), UDP-glucuronosyltransferases (UGTs), glutathione *S*-transferases (GSTs), and the sulfotransferases (SULTs) in accounting for irreversible loss of drug due to metabolism.[4] Uptake transporters on the sinusoidal membrane include the OATPs (organic anion transporting polypeptides), NTCP (sodium-dependent taurocholate–cotransporting polypeptide), OCTs (organic cation transporters), and MCT2 (monocarboxylic acid transporter 2), and the ABC transporters, Pgp (P-glycoprotein or MDR1, multidrug resistance protein 1), MRP2 (multidrug resistance–associated protein 2), BSEP (bile salt export pump), and BCRP (breast cancer resistance protein) on the canalicular membrane (Figure 23.1).[5–9] Moreover, it is recognized that hepatic uptake transporters may rate-limit metabolism and excretion in the liver.[10–12] Zonal heterogeneities of transporters and metabolic enzymes further modulate drug disposition in the liver.[13,14] For phase II reactions, rate-limiting roles for cosubstrates[15,16] in conjugation reactions have been described. Complicating factors include the interconversion or futile cycling between drug and metabolite, as this presence tends to alter the simple relationships of clearance with respect to transporters, enzymes, and flow.[17–22]

The hepatic microcirculation consists of the dual perfusion from the portal vein (PV, 75%) and hepatic artery (HA, 25%). The flow pattern to the liver and the micromixing of the circulation are of paramount importance. The incompleteness of micromixing of PV and HA and bypass of the HA will affect drug clearance in liver.[23–26] If flow to the region is absent, metabolism and excretion would not occur. Hence, drugs in the circulation are delivered to the sinusoidal membrane and must recruit transporters for entry or enter passively before they are acted on by the enzymes intracellularly. The flow pattern is expected to be more important in the modulation of clearances of drugs with high hepatic extraction ratios.[27]

23.2. MODELS OF HEPATIC DRUG CLEARANCE

The efficiency of hepatic removal is based on clearance concepts. Hepatic clearance ($CL_{liver,tot}$) is the volume of perfusing blood cleared of the drug per unit time. $CL_{liver,tot}$ is a proportionality constant that equals the rate of elimination divided by the concentration of drug entering the liver and is the product of the hepatic blood flow (Q) and the extraction ratio (E).[28] The relationship that $CL_{liver,tot} = QE$ is simple but misleading, since E is modulated by common factors such as flow rate, unbound fraction, the enzymes, and transporters.

Various hepatic clearance models have been developed to elucidate the removal mechanisms and predict drug extraction in the liver. Two conventionally used models are the *well-stirred model*, which views the liver as a well-stirred compartment with concentration of drug in the liver in equilibrium with that in the emergent blood, and the *parallel tube model*, which regards the liver as a series of parallel tubes with enzymes distributed evenly around the tubes, with the concentration of drug declining along the length of the tube.[27–31] These models reflect extreme flow conditions of complete and rapid mixing (well-stirred model) or lack of mixing (parallel tube model) (Figure 23.2). The reality is an intermediate degree of mixing, the *dispersion*

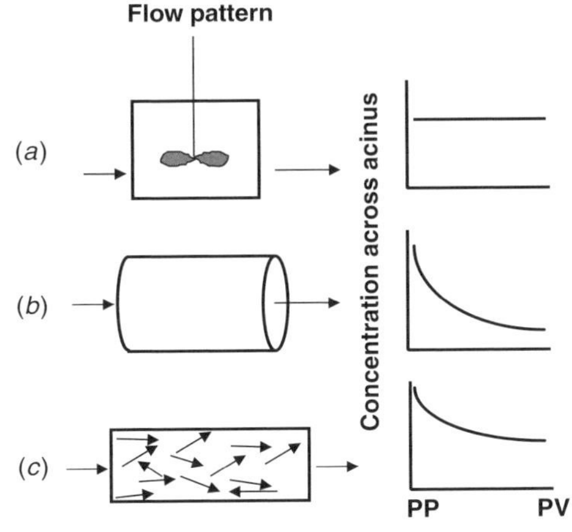

FIGURE 23.2. Hepatic clearance models based on flow patterns: (*a*) bulk flow, infinite mixing: well-stirred model (dispersion number $D_N = \infty$); (*b*) plug flow, no mixing: parallel tube model ($D_N = 0$); (*c*) dispersion flow, intermediate mixing: dispersion model ($D_N > 0$).

model, imparting some finite value for the D_N number that yields asymmetric outflow dilution curves.[32,33] The three models have been extended to include the sinusoidal transmembrane barrier[34–38] for a description of the hepatic clearances of solutes. Other hepatic clearance models include the enzyme-distributed model,[39] the series-compartment model,[40] the zonal-compartment model,[14] and Goresky's model that describes variable capillary transit times and barrier-limited transport.[41] Zonal heterogeneities of enzymes and transporters[13,42,43] are additional variables that need to be considered in hepatic drug clearance model. This is particularly important for highly extracted compounds in which cellular substrate concentration is maintained low and enzymes and transporters compete for the substrate.

In the absence of a barrier and futile cycling, the important equations detailing the steady-state extraction of the models are for the well-stirred model,

$$E = \frac{f_u \mathrm{CL_{int}}}{f_u \mathrm{CL_{int}} + Q} \tag{1}$$

for the parallel-tube model,

$$E = 1 - e^{-f_u \mathrm{CL_{int}}/Q} \tag{2}$$

and for the dispersion model,

$$E = 1 - e^{-[f_u \mathrm{CL_{int}}/Q + (f_u \mathrm{CL_{int}}/Q)^2 D_N]} \tag{3}$$

where f_u is the unbound fraction in the vasculature (plasma or blood with rapid drug equilibration), $\mathrm{CL_{int}}$ the total intrinsic clearance (sum of $V_{\max}/K_m$ or intrinsic clearances for metabolism and excretion under first-order conditions), Q the blood flow rate and D_N the dispersion number. The equations have been extended to include transport barriers. For the well-stirred model,

$$E = \frac{f_u \mathrm{CL_{influx}} \mathrm{CL_{int}}}{Q\mathrm{CL_{efflux}} + \mathrm{CL_{int}}(Q + f_u \mathrm{CL_{influx}})} \tag{4}$$

for the parallel tube model,

$$E = 1 - e^{-f_u \mathrm{CL_{int}} \mathrm{CL_{influx}}/[Q(\mathrm{CL_{int}} + \mathrm{CL_{efflux}})]} \tag{5}$$

and for the dispersion model,

$$E = 1 - e^{-[(1-\sqrt{(1+4\mathrm{D_N R_N})})/(2\mathrm{D_N})]} \tag{6}$$

where the efficiency number R_N contains terms pertaining to unbound fraction f_u, the total intrinsic clearance $\mathrm{CL_{int}}$, and the influx and efflux clearance parameters $\mathrm{CL_{influx}}$ and $\mathrm{CL_{efflux}}$, respectively:

$$R_N = \frac{f_u \mathrm{CL_{int}} \mathrm{CL_{influx}}}{Q(\mathrm{CL_{efflux}} + \mathrm{CL_{int}})}$$

23.3. IN VITRO SYSTEMS

Questions are often posed as to how to interrelate in vitro data to in vivo data: namely, how to model the data in order to answer questions mechanistically and make predictions. Vital information is provided from in vitro systems to arrive at transport data to predict events in vivo. The in vitro systems for the study of drug transport include use of cell lines transfected with transporter genes[44,45] and hepatocyte uptake studies.[1,46,47] Transfected cells and freshly prepared isolated hepatocytes are often used to determine, within the linear uptake period, the extent of cellular accumulation to yield the initial velocity of uptake (amount/time). Beyond these time points, efflux and metabolism become dominant and drug accumulation is no longer proportional to time. The velocity data derived may be fitted to the Michaelis–Menten equation or similar relationships to provide the maximum transport capacity (V_{max}) and affinity constant (Michaelis-Menten constant K_m or concentration of drug at half V_{max}). Because of the time-involvement and lack of availability of human livers for the preparation of fresh hepatocytes, cryopreservation has been used to retain and preserve functions of the hepatocyte.[48,49] Zonal hepatocytes, prepared according to the digitonin–collagenase method of Lindros and Pentïlla,[50] have been used for the study of acinar drug uptake. Vesicles from either basolateral or canalicular membranes may be used for study. The transmembrane transport often aims at identification of the driving forces of uptake.[51,52] Inside-out vesicles are needed for examination of canalicular transport, especially for drugs that are too polar to traverse the basolateral membrane. With molecular cloning, in vitro gene expression systems in *Xenopus laevis* oocytes or mammalian cells have been applied extensively to drug transport studies, in either transiently or stably transfected systems.[44,45,53,54] Double transfection systems that express rat transporters such as Ntcp/Bsep[55] have been developed. Moreover, double-transfected MDCK II or LLC-PK1 cells expressing human OATP2/MRP2,[57,58] OATP2/MDR1, OATP2/BCRP,[59] or OATP8/MRP2,[60,61] or even quadruple-transfected cells[62] have been utilized to study vectorial transport of solutes in cell monolayers. Transgenic animals bearing a human transporter gene have succeeded to show that the Eisai hyperbilirubinemic rat that is deficient in Mrp2 could regain its ability to excrete Mrp2-substrates into bile.[63]

Analogously, many systems exist at the subcellular level: the S9 fraction, cytosol, and microsomes purified enzyme systems[64,65] for the assessment of metabolic activities. In other systems, transgenic animals,[66–68] antibodies,[69] and reconstituted or cDNA-expressed systems[70] are used routinely to identify the metabolic system as well as the kinetic parameters (K_m and V_{max}). However, one drawback of these in vitro studies is that the cofactors or cosubstrates are usually added in excess, and this type of study rarely divulges the rate-limiting role of cofactors or cosubstrates that may be depleted in vivo.[15,16] In other instances, the in vitro system is not able to fully reflect the activity of the enzyme in vivo.[71,72]

The isolated hepatocyte that retains the intact cellular membrane houses both transporters and enzymes and encompasses the pertinent information for both transporter and enzymes.[11,21] Similar types of information may be obtained in the sandwich-cultured hepatocytes[73–75] and bioreactors composed of cultured hepatocytes.[76,77] However, caution is needed with cultured systems since enzyme levels and regulators and coactivators may be dependent on the culture system. Some of the models may be cholestatic in nature, due to the inability to clear bile acids from the canaliculi when cells are in culture. Liver slices, standardized and improved with precision cut,[78,79] are also useful in the investigation of drug biotransformation and toxicity. Others have used gene-knockout or mutant animals.[80–83]

In this chapter we highlight the use of in vitro data in unraveling the mechanisms and rate-limiting steps underlying the hepatic clearances of drugs, some of high extraction ratios and some of poor extraction ratios. Data from the perfused liver preparation were used since the intact organ consists of flow perfusing the hepatocytes in their native cellular architecture and environment, and the system provides a platform to investigate interplay of the determinants of drug clearance.[84] We demonstrate how in vitro data on transport and metabolism may be used within simple (PBPK) and zonal (ZPBPK) physiologically based pharmacokinetic models to provide an integrated approach to study the hepatic drug extraction. We found that there is a good correlation between data in vitro from hepatocytes for transport and those in the perfusion system. The resulting model may be utilized for the study and prediction of the interplay between transporters and enzymes.

23.4. SIMPLE AND ZONAL PHYSIOLOGICALLY BASED PHARMACOKINETIC MODELS

The zonal model is based on extending the simple physiologically based pharmacokinetic model (PBPK) (Figure 23.3*a*).[85,86] The recirculating system includes three zones in the liver to describe heterogeneities of transporters and enzymes (Figure 23.3*b*).[14] The simple PBPK model consists of the reservoir or the central (blood) compartment, the liver plasma or blood (PL), liver tissue (L), and bile compartment (bile). Upon expansion, the new ZPBPK model comprises the periportal (PP, around the portal region, at the inlet), midzonal, and perivenous (PV, around the central vein, at the outlet) regions.[14] Transport clearances across the sinusoidal membrane for drug from hepatic plasma to tissue and from tissue to hepatic plasma are characterized by influx (CL_{influx}) and efflux (CL_{efflux}) intrinsic clearances, respectively, and these are designated with the subscript, i to denote each ith zonal region. Drug within tissue is metabolized by enzymes of intrinsic clearance, $CL_{int,met}$(V_{max}/K_m for first-order condition); the $V_{max,i}$ for each zonal region may vary to reflect enzyme zonation, whereas K_m is constant among the zones. Biliary excretion of drug is a function of the biliary intrinsic clearance, $CL_{int,sec}$, and these may be subdivided into $CL_{int,sec,i}$ for the canalicular membrane for each respective zonal region (Figure 23.3*b*). The total intrinsic clearance of the liver is the $\sum CL_{int,i}$ for $i = 1$ to 3. Through our experience with modeling and simulation with any given intrinsic clearance, the repetitive removal of drug by enzymes or canalicular proteins that are dispersed among the zonal regions would furnish a higher extraction efficiency than that when the enzymes or canalicular transporters are confined within one compartment only.

Much was learned from the simpler physiologically based pharmacokinetic (PBPK) model, since solutions for the biliary ($CL_{liver,ex}$), metabolic ($CL_{liver,met}$), and

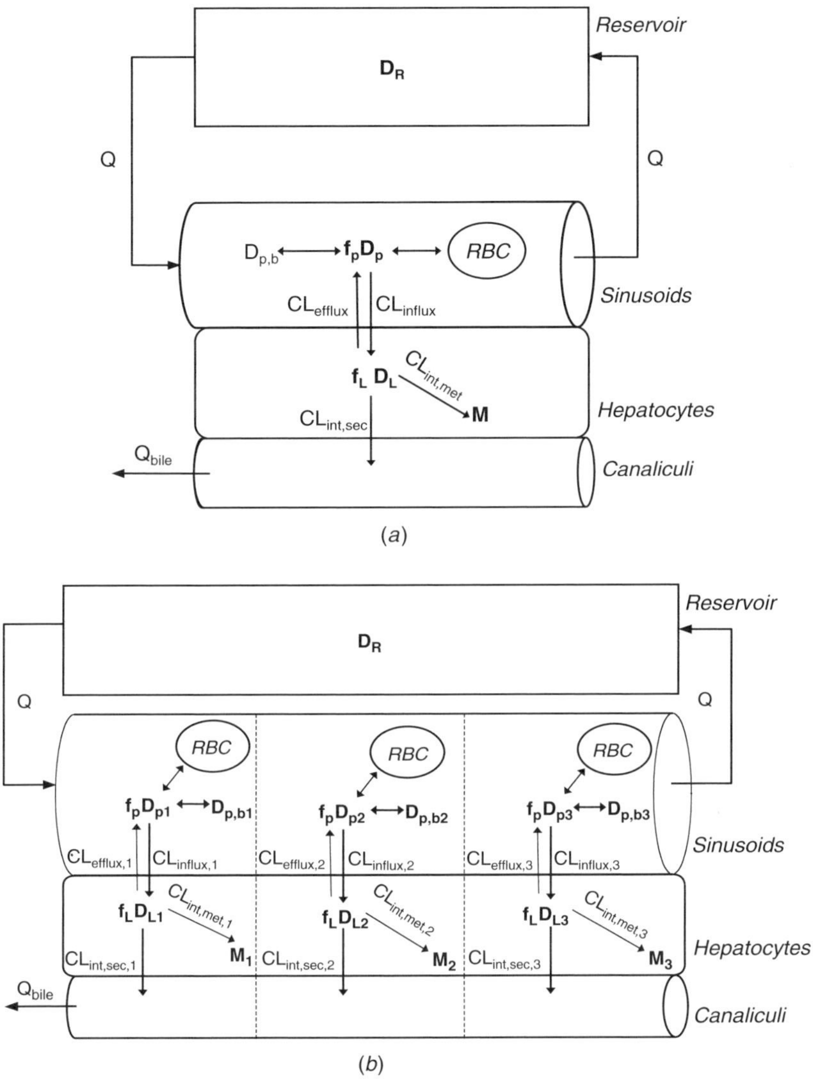

FIGURE 23.3. Simple (PBPK) (*a*) and zonal (ZPBPK) (*b*) physiologically based pharmacokinetic models. In the PBPK model (*a*), the liver compartment consists of the liver blood, tissue, and bile compartments. Influx (CL_{influx}) and efflux (CL_{efflux}) clearances, as well as the metabolic ($CL_{int,met}$) and secretory ($CL_{int,sec}$) intrinsic clearances, regulate levels of the drug in tissue (D_L), plasma (D_P), and reservoir (D_R). M is the total amount of metabolite formed, including the amounts of metabolite in reservoir, bile, and liver tissue as well as those for subsequently formed metabolites; Q and Q_{bile} are the liver blood flow and bile flow rates, respectively. In the ZPBPK model (*b*), the liver is divided into three zones (PP zone or zone 1, middle zone or zone 2, and PV zone or zone 3) according to the hepatic microcirculation. A subscript i denotes the influx ($CL_{influx,i}$), efflux ($CL_{efflux,i}$), and metabolic ($CL_{int,met,i}$) and secretory ($CL_{int,sec,i}$) intrinsic clearances in different zones. The models may be further modified by RBC partitioning and unbound fraction in plasma, f_p.

total ($CL_{liver,tot}$) clearances existed with the physiological variables: the hepatic flow rate (Q), unbound fraction (denoted by f_u and f_L, the unbound fractions in plasma and tissue, respectively), $CL_{int,met}$, $CL_{int,sec}$, and the basolateral influx (CL_{influx}) and efflux (CL_{efflux}) clearances.[85,86] Modeling readily considers the binding of drugs that partition rapidly or slowly into RBCs. All of the above named parameters would influence clearances in a complex fashion (Table 23.1).[85] As shown, the unbound fraction in tissue is absent under first-order conditions, showing that tissue binding is irrelevant for first-order conditions. It is immediately recognized from these mathematical relations that the secretory intrinsic clearance ($CL_{int,sec}$) is present in the determination of the metabolic clearance ($CL_{liver,met}$), and the metabolic intrinsic clearance term ($CL_{int,met}$) is present in the determination of the excretory clearance ($CL_{liver,ex}$). Upon dividing the equations, it may be shown that

$$\frac{CL_{liver,ex}}{CL_{liver,met}} = \frac{CL_{int,sec}}{CL_{int,met}} \tag{7}$$

Equation (7) may be used to ascertain $CL_{int,sec}$, which normally remains obscure in the intact liver. Since $CL_{liver,ex}$ and $CL_{liver,met}$ are measured parameters, and $CL_{int,met}$ may be estimated from in vitro studies, $CL_{int,sec}$ may be estimated from equation (7).

The next strategy was to express heterogeneity from in vitro data to the ZPBPK model. Normally, we obtain direct measurements of the biliary clearance, $CL_{liver,ex}$. The metabolic clearance, $CL_{liver,met}$, is obtained either directly or indirectly as the difference between total and the biliary clearances. Since $CL_{int,met}$ may be estimated from in vitro studies (S9, microsomal or cytosolic incubations), $CL_{int,sec}$ may be approximated by equation (7), which will hold true when all of the transporters and enzymes are evenly distributed among the zones (unpublished simulations). But equation (7) will not hold when transporter/enzyme heterogeneity and futile cycling exist, as will be divulged later with the example on estradiol 17β-D-glucuronide ($E_2$17G). Nonetheless, the estimate of $CL_{int,sec}$ from equation (7) provides an initial estimate for fitting purposes. The acinar transporter activity for the canalicular (described by $CL_{int,sec}$) or sinusoidal (CL_{influx} or CL_{efflux}) membrane may be assessed qualitatively from confocal microscopy or from Western blotting on the relative expression of transporter proteins in preparations derived from PP and PV hepatocytes.[47,87] Direct verification of acinar transport may be secured from uptake studies involving zonally enriched isolated hepatocytes.[46,47,87]

23.5. INTERPLAY BETWEEN TRANSPORTERS AND ENZYMES

After verification of the appropriateness of the PBPK and ZPBPK models with various drug examples, these models may then be used for examination of the interplay between enzyme and transporter. The first account was from Sirianni and Pang,[85] who solved the equations and examined the competition between metabolism and excretion in the PBPK model. Based on sound modeling and pharmacokinetic theory,

TABLE 23.1. Solved Equations on the Biliary Excretion Clearance, Hepatic Metabolic Clearance, and Total Hepatic Clearance

Clearance Term	Solutions for the Simple PBPK Model[a]: CL_{influx} and CL_{efflux} are comparable to Q, $CL_{int,met}$, and $CL_{int,sec}$	Solutions for the Simple PBPK Model[a]: $CL_{influx} = CL_{efflux} \gg Q$, $CL_{int,met}$, and $CL_{int,sec}$
$CL_{liver,ex}$	$\frac{Q f_u CL_{influx} CL_{int,sec}}{Q(CL_{efflux} + CL_{int,met} + CL_{int,sec}) + f_u CL_{influx}(CL_{int,sec} + CL_{int,met})}$	$\frac{Q; f_u CL_{int,sec}}{Q + f_u(CL_{int,sec} + CL_{int,met})}$
$CL_{liver,met}$	$\frac{Q f_u CL_{influx} CL_{int,met}}{Q(CL_{efflux} + CL_{int,met} + CL_{int,sec}) + f_u CL_{influx}(CL_{int,sec} + CL_{int,met})}$	$\frac{Q; f_u CL_{int,met}}{Q + f_u(CL_{int,sec} + CL_{int,met})}$
$CL_{liver,tot}$	$\frac{Q f_u CL_{influx}(CL_{int,sec} + CL_{int,met})}{Q(CL_{efflux} + CL_{int,met} + CL_{int,sec}) + f_u CL_{influx}(CL_{int,sec} + CL_{int,met})}$	$\frac{Q; f_u(CL_{int,sec} + CL_{int,met})}{Q + f_u(CL_{int,sec} + CL_{int,met})}$

Source: Ref. 85; with permission.

[a]These equations are reduced in complexity when $CL_{influx} = CL_{efflux} \gg Q_L$, $CL_{int,met}$ and $CL_{int,sec}$. Q is the plasma flow rate to the liver.

it may be deduced that suppression of one pathway should elevate the rate of drug removal by compensatory pathways, whereas total clearance would be decreased.[85,87]

23.5.1. Enalapril

Enalapril, an angiotensin-converting enzyme (ACE) inhibitor, is taken up by rat Oatp1a1 and human OATP1B2 into the liver[53,88] and is metabolized by the carboxylesterases to enalaprilat in the hepatocyte, demonstrating a moderately high extraction ratio ($E \sim 0.7$) (Figure 23.4).[90] The precursor–product pair is excreted via Mrp2/MRP2 into bile in the rat and human livers.[88] In the rat, the metabolic intrinsic clearance ($V_{\mathrm{max,met}}/K_{\mathrm{m,met}}$ under the linear conditions) was estimated from S9 fractions.[14] Further probing with zonal cells showed acinar metabolism with zonal S9 fractions, but there was a lack of heterogeneity in the transporters, Oatp1a1 and Mrp2 (Figure 23.5) (Table 23.2).

Taking the view that the ratio of the excretory to metabolic clearances equals the ratio of the intrinsic clearances ($\mathrm{CL_{int,sec}}/\mathrm{CL_{int,met}}$) under linear conditions [equation (7)] for the PBPK model, estimate of the $\mathrm{CL_{int,sec}}$ may be obtained with known excretion and metabolic clearances, and the experimentally determined value of $\mathrm{CL_{int,met}}$ from either S9 or microsomal–cytosolic preparations. The only unknown left is $\mathrm{CL_{efflux}}$, which may be estimated according to the equations for the well-stirred, parallel tube, or dispersion models [equations (4) to (6)]. The estimates may be used as initial estimates for fitting with the simple PBPK model to the data (Table 23.3).

The parameters derived may be utilized to simulate various conditions to test the effects of each of the variables on hepatic drug extraction. To study the interplay of transporters and enzymes in hepatic drug processing, $\mathrm{CL_{int,sec}}$ may be changed from 100% to 200%, then 50% with the PBPK model; it is noted that when $\mathrm{CL_{int,sec}}$ equals zero, the condition mimics the EHBR liver that lacks Mrp2, the transporter responsible

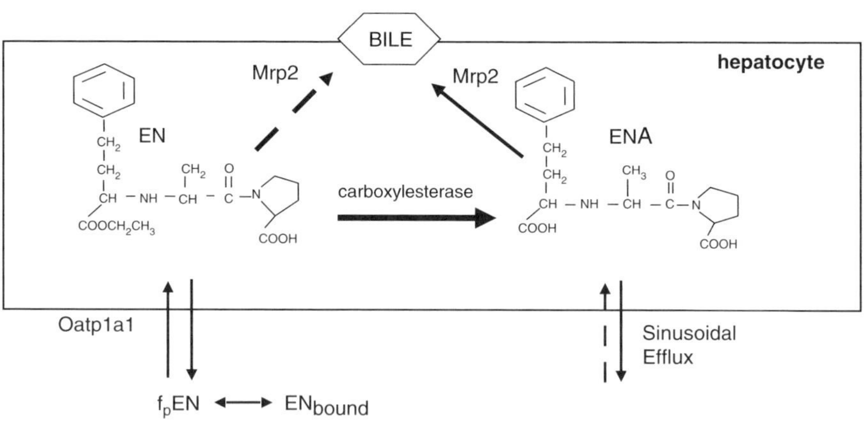

FIGURE 23.4. Biological fate of enalapril (EN) and its metabolite, enalaprilat (ENA), in rat liver. Enalapril but not its diacid metabolite enalaprilat is taken up by Oatp1a1, and both compounds are excreted into bile via Mrp2; enalapril is metabolized to enalaprilat by the carboxylesterases. (From ref. 87, with permission.)

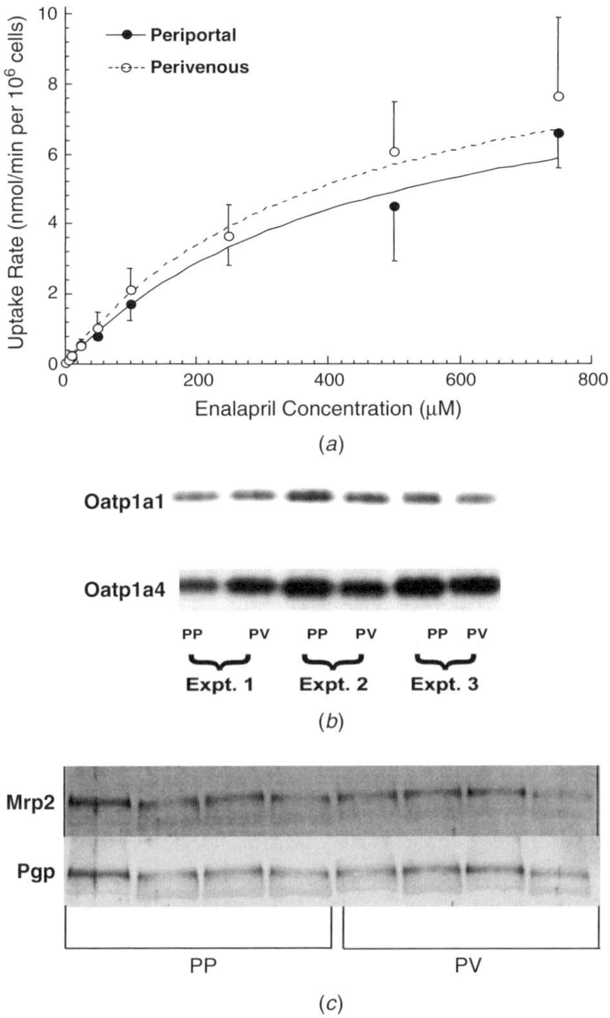

FIGURE 23.5. Uptake of enalapril by PP (solid circle) and PV (open circle) rat hepatocytes with increasing concentrations (mean $\pm$ S.D., $n = 4$). The lines are based on the average fitted constants shown in Table 23.2 for a single saturable uptake component (*a*). Protein expression of Oatp1a1 and Oatp1a4 in sinusoidal membrane fragments (*b*) and of Mrp2 and Pgp in canalicular membrane fragments (*c*) was similar to that for PP and PV hepatocytes. (From refs. 12, 47, and 89, with permission.)

for the canalicular transport of enalapril into bile.[88] The process is then repeated for $CL_{int,met}$, which is changed from 100% to 200%, then 50%; the $CL_{int,sec}$ value and other parameters are kept unchanged for the simulation. One conclusion that may be made from the simulation results is that a change in CL_{influx} or CL_{efflux} would result in similar trends for the biliary, metabolic, and total clearances. When the influx clearance is

TABLE 23.2. Kinetic Parameters for Metabolism of Enalapril by S9 Enzymes Derived from Homogeneous or Zonal Rat Hepatocytes and the Transport Constants Derived from Initial Uptake Velocities by Homogeneous or Zonal Rat Hepatocytes

	K_m (μM)	V_{max} (nmol/min/g liver)
Transport[a]		
Periportal hepatocytes	461 ± 117	1235 ± 260
Perivenous hepatocytes	441 ± 50	1352 ± 260
Homogeneous hepatocytes	344 ± 52	1430 ± 195
S9 metabolism[b]		
Periportal hepatocytes	1049 ± 335	547 ± 310
Perivenous hepatocytes	2612 ± 236	2096 ± 600
Homogeneous hepatocytes	1308 ± 419	800 ± 300

Source: Refs. 14 and 47; with permission.

[a]Based on 130 × 10^6 cells per gram of liver.

[b]Based on 100 mg of S9 protein per gram of liver.

increased or decreased, all of the clearances increase or decrease concomitantly, but the ratio of the biliary–metabolic clearances remains unchanged (Figure 23.6*a*), since $CL_{int,sec}/CL_{int,met}$ is constant [equation (7)]. The reverse trend holds for the basolateral efflux clearance (Figure 23.6*b*); a decrease of CL_{efflux} would bring about increases in the clearances, whereas an increase in CL_{efflux} evokes decreases in the clearances. The ratio of drug metabolized to excreted would remain unchanged, since $CL_{int,sec}/CL_{int,met}$ is constant [equation (7)]. These changes are well predicted by the equations shown in Table 23.1, and the results are summarized in Table 23.4.

Changes in $CL_{int,sec}$ or $CL_{int,met}$ will modulate the clearances differentially. As may be envisioned, many scenarios may be simulated to demonstrate the competing nature of transporters and metabolic enzymes. For enalapril that is excreted only to a minor extent, exhibiting a low $CL_{int,sec}$, the metabolic and total clearances would be modulated only very slightly with changes in $CL_{int,sec}$ (Figure 23.6*d*), whereas changes in $CL_{int,met}$ would evoke large changes in the total, metabolic, and biliary clearances (Figure 23.6*c*; Table 23.4). Generally speaking, for a drug that undergoes metabolism and biliary secretion, an increase in the intrinsic clearance of one pathway will decrease the observed or "apparent" clearance of the alternative pathway, and the total clearance is increased. By contrast, a decrease in intrinsic clearance of one pathway will increase clearance of the alternative pathway and decrease the total hepatic clearance. These compensatory mechanisms (seesaw phenomenon) are due to changes in the intracellular substrate concentration, increasing upon inhibition and decreasing with activation of the elimination pathways. The ratio of the biliary and metabolic clearances will change when the intrinsic metabolic or secretory clearance is altered (Figure 23.6*c* and *d*).

With the added complexity of zonation in enzymes, the ZPBPK model that encompasses zonal metabolism of enalapril has been added as an illustration. The total $CL_{int,met}$, being the sum of the zonal intrinsic clearances, $CL_{int,met,i}$ (PP, midzonal, and PV activities represented 27, 32, and 41% of the total $CL_{int,met}$ from S9 studies), is the

TABLE 23.3. Interrelationships Between the Metabolic Intrinsic Clearance ($CL_{int,met}$) and Secretory Intrinsic Clearance ($CL_{int,sec}$), the Influx and Efflux (CL_{influx} and CL_{efflux}) Clearances of Enalapril Based on the Equations Shown for the Simple PBPK (Table 23.1) for the Single-Pass Perfused Rat Liver Preparation, and in the ZPBPK Model in Which Metabolism in Zonal Regions Are 27, 32, and 41% of the Total $CL_{int,met}$ (ZPBPK1) or Evenly Distributed (ZPBPK2)

		Predictions[a]		
	Data Observed	PBPK	ZPBPK1	ZPBPK2
CL_{influx} from hepatocyte uptake studies in vitro (mL/min/g)	4.00[b]			
$CL_{int,met}$ from S9 in vitro metabolic studies (mL/min/g)	0.61[c]			
Unbound fraction in plasma, f_p	0.55			
Plasma flow rate, Q (mL/min/g)	0.8			
V_R (mL/g)[b]	20			
V_S (mL/g)[c]	0.149			
V_L (mL/min/g)	0.6			
Hepatic metabolic clearance $CL_{liver,met}$ (mL/min/g)	0.345[d]	0.340	0.386	0.388
Biliary clearance $CL_{liver,ex}$ in perfused liver preparation (mL/min/g)	0.022[d]	0.022	0.026	0.025
Total hepatic clearance $CL_{liver,tot}$ in perfused liver preparation (mL/min/g)	0.366[d]	0.362	0.411	0.413
$CL_{liver,met}/CL_{liver,ex} = CL_{int,met}/CL_{int,sec}$ [equation (7)]	15.7	15.7	14.8	15.7
$CL_{int,sec}$ calculated (mL/min/g)	0.039[e]			
CL_{efflux} (mL/min/g)	1.46[f]			

[a] Based on simulated AUC from intravenous dosing from Scientist.
[b] In vitro hepatocyte data of Abu-Zahra and Pang[14] (from Table 23.2).
[c] In vitro S9 data of Abu-Zahra et al.[47] (from Table 23.2).
[d] Observation of de Lannoy et al.[86] from liver perfusion studies.
[e] According to equation (7).
[f] Fit to simple PBPK model.

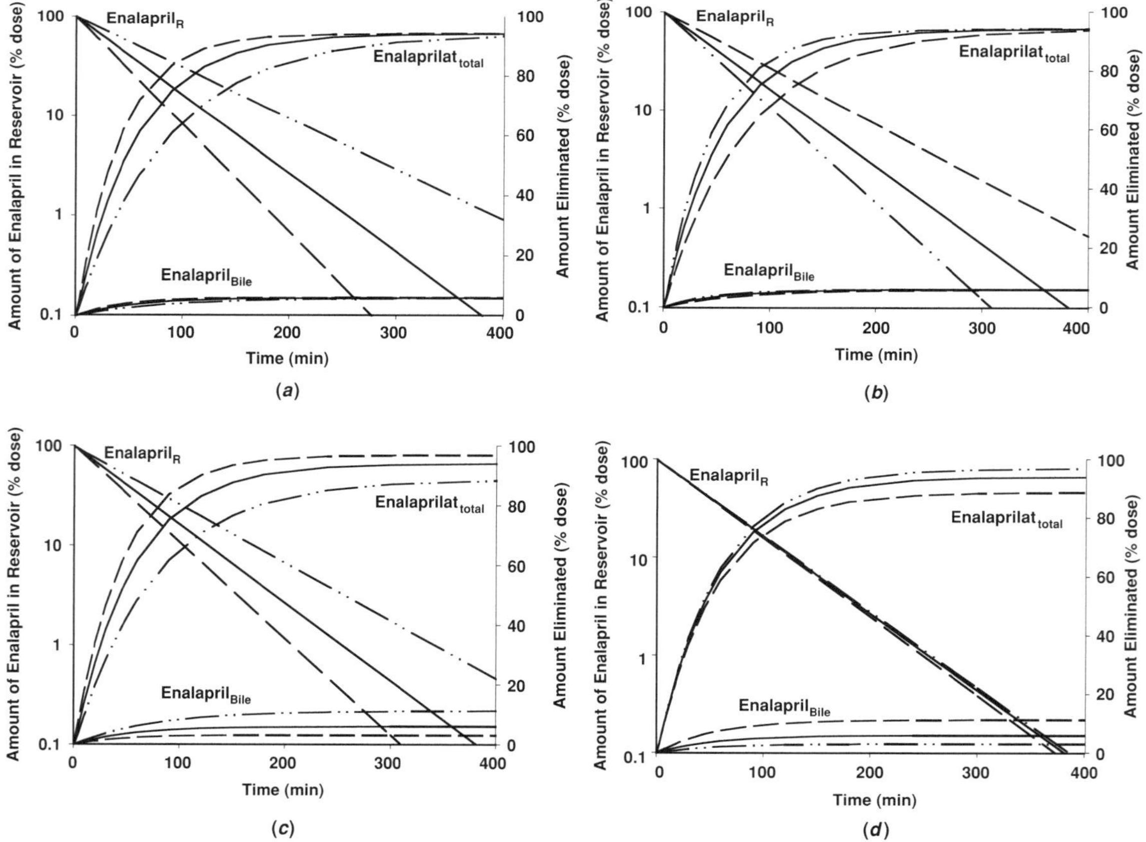

FIGURE 23.6. Simulations on enalapril, based on the simple PBPK model. The CL_{influx} (*a*), CL_{efflux} (*b*), $CL_{int,met}$ (*c*) or $CL_{int,sec}$ (*d*) were altered from 100% (solid line) to 200% (short dashed line) and 50% (dashed–dotted–dotted line); all other parameters were kept constant in the simulation to examine the decay of enalapril in reservoir (enalapril$_R$) and the total amounts of enalapril excreted into bile (enalapril$_{bile}$) and enalaprilat formed (enalaprilat$_{total}$.)

TABLE 23.4. Similar Trends in the Total Hepatic, Metabolic, and Excretion Clearance When the Influx, Efflux, Metabolic, or Secretory Intrinsic Clearances Are Changed Under Linear Conditions for Both the Single Liver Compartment PBPK Model (Figure 23.3*a*) and the ZPBPK Model (Figure 23.3*b*)

	Predictions for Clearances		
Changes	$CL_{liver,tot}$	$CL_{liver,met}$	$CL_{liver,ex}$
$\uparrow CL_{influx}$	↑	↑	↑
$\downarrow CL_{influx}$	↓	↓	↓
$\uparrow CL_{efflux}$	↓	↓	↓
$\downarrow CL_{efflux}$	↑	↑	↑
$\uparrow CL_{int,met}$	↑	↑	↓
$\downarrow CL_{int,met}$	↓	↓	↑
$\uparrow CL_{int,sec}$	↑	↓	↑
$\downarrow CL_{int,sec}$	↓	↑	↓
$\uparrow f_P$ or f_u	↑	↑	↑
$\downarrow f_P$ or f_u	↓	↓	↓

basis of the simulations. With the heterogeneity of enzyme included in the ZPBPK model, clearances and trends in the transporter–enzyme interplay are similar to those for the simple PBPK model (cf. Figures 23.7 and 23.6). With the $CL_{int,met}$ unevenly dispersed among the zones, slightly higher clearances result for the ZPBPK model than that predicted by the PBPK model. The ratio $CL_{liver,met}/CL_{liver,ex}$ is not $CL_{int,met}/CL_{int,sec}$ for the ZPBPK model (Table 23.4). Similar to the results obtained for the PBPK model (Table 23.4), decreasing the intrinsic clearance of one elimination pathway will increase the apparent clearances of alternative removal pathways, and the total hepatic clearance is decreased, whereas increasing the intrinsic clearance of one pathway will decrease the apparent clearances of alternative pathways and total clearance will be increased (Figure 23.7*c* and *d*).

There are various unexpected results in the literature: namely, the inhibition of P-glycoprotein (Pgp or Mdr1) appeared to produce an increase in $CL_{liver,tot}$.[92,93] The observation is tenable only when efflux is also reduced in the presence of inhibitors.[87] It must be borne in mind that inhibitors or inducers exert multiple actions in addition to their actions on certain pathways. Simulation studies are of great utility to show the underlying mechanism for the changes in clearance.

23.5.2. Digoxin

The second example is digoxin (Dg3), a commonly used cardiotonic drug. In the rat liver, digoxin is metabolized primarily by cytochrome P450 3a2 (Cyp3a2) that is enriched in the PV zone[94] to di- and monodigitoxosides as well as digoxigenin (Figure 23.8).[95,96] In addition, digoxin is excreted unchanged into bile by Pgp.[80] The transport of digoxin into rat hepatocytes is by passive diffusion and Oatp1a4, shown in transport studies involving Oatp1a4-transfected *Xenopus laevis* oocytes[97–99] or LCC-PK1 cells.[54] Digoxin uptake into rat zonal (PP and PV) hepatocytes showed similar

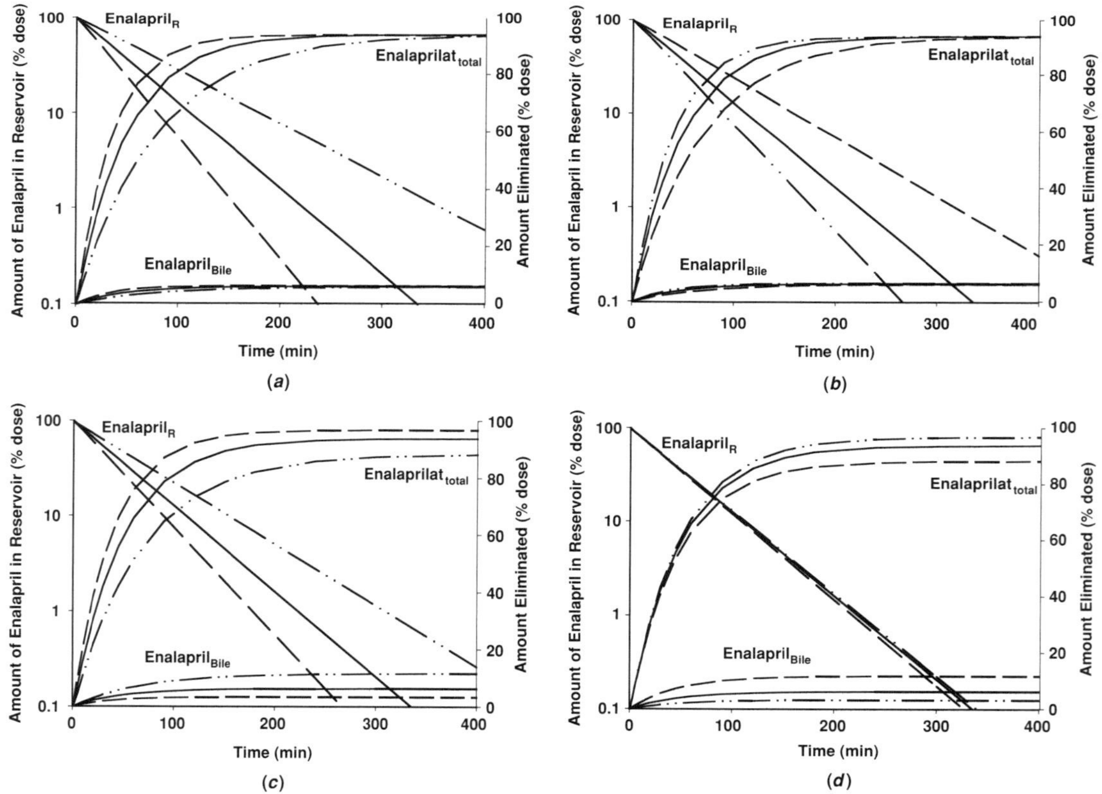

FIGURE 23.7. Simulations of enalapril based on the ZPBPK model. The CL_{influx} (*a*), CL_{efflux} (*b*), $CL_{int,met}$ (*c*) or $CL_{int,sec}$ (*d*) were altered from 100% (solid line) to 200% (short dashed line) and 50% (dashed–dotted–dotted line); all other parameters were kept constant in the simulation to examine the decay of enalapril in reservoir (enalapril$_R$), and the total amounts of enalapril excreted into bile (enalapril$_{bile}$) and enalaprilat formed (enalaprilat$_{total}$). The $CL_{int,met,i}$ for the PP, midzonal, and PV zones were assigned values of 0.16, 0.20, and 0.25 mL/min/g, respectively, based on the data of Abu-Zahra and Pang.[14]

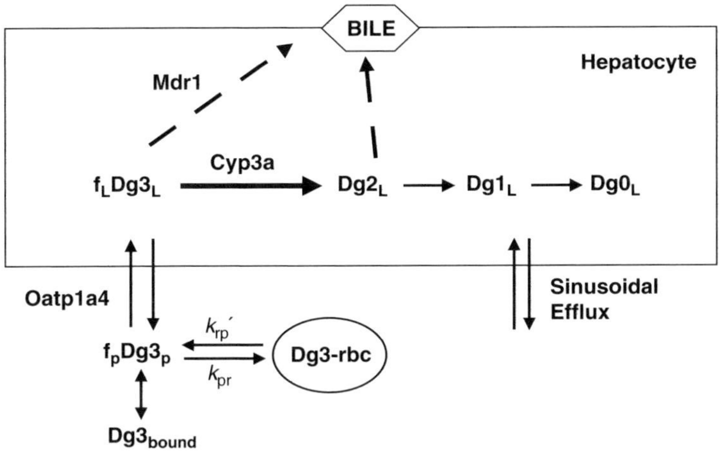

FIGURE 23.8. Schematic presentation of the fates of digoxin (Dg3) and its metabolites, Dg2 (di-digitoxoside), Dg1 (mono-digitoxoside), and Dg0 (digoxigenin) in the rat liver. Dg3 is taken up by Oatp1a4 and is excreted via Pgp/Mdr1 into bile. Dg3 is metabolized by Cyp3a to Dg2, Dg1, and Dg0. (From ref. 96, with permission.)

uptake clearances when estimated as the sum of saturable (CL_{uptake} or V_{max}/K_m) and nonsaturable (PS_{uptake}) components (Figure 23.9 and Table 23.5). The findings contrasted with those of others who showed a perivenous abundance of Oatp1a4.[100] Western blotting (Figure 23.5*b*) and confocal immunofluorescent microscopy verified the lack of zonal distribution of the Oatp1a4.[89] These, when scaled up, provided the influx clearance (CL_{influx}) for the transport of digoxin into rat liver. The distribution of Pgp/Mdr1 in zones 1, 2, and 3 was similar (Figure 23.5*c*).[89]

Modeling of the digoxin data (reservoir and bile) from recirculating rat liver perfusion studies in the absence of RBCs and albumin at a 40-mL/min flow rate (KHB–perfused liver), and with 20% RBC–1% albumin at 10 mL/min (RBC–alb–perfused liver) was successful (Figure 23.10).[89] Normally, the free drug concentration may be estimated as the unbound fraction multiplied by the total drug concentration. However, the slow exchange of digoxin between plasma and RBCs (Figure 23.8) necessitated consideration of these binding–debinding rate constants in the ZPBPK model (Figure 23.3*b*). The exchange-rate constants from plasma to RBCs (k_{pr} or $0.468 \pm 0.021\ \text{min}^{-1}$) and from RBCs to plasma (k'_{rp} or $1.81 \pm 0.12\ \text{min}^{-1}$) of digoxin demonstrate a slow (RBC) distribution and modest plasma protein binding (an unbound plasma fraction of 0.64).[89] The slow efflux from RBCs may constitute an impediment to drug removal.[101] To enable fitting, the uptake clearance CL_{influx}, provided by scale-up of hepatocyte uptake data (Table 23.5), and the parameters on RBC partitioning (k_{pr} and k'_{rp}) and digoxin binding in plasma (an unbound fraction in f_p) were assigned. The values of $CL_{influx,i}$, $CL_{efflux,i}$, and $CL_{int,sec,i}$ were partitioned evenly for each zone (one-third of the value of the liver), whereas that for $CL_{int,met,i}$ for metabolism by Cyp3a2 for zones 1, 2, and 3, respectively, was assigned arbitrarily

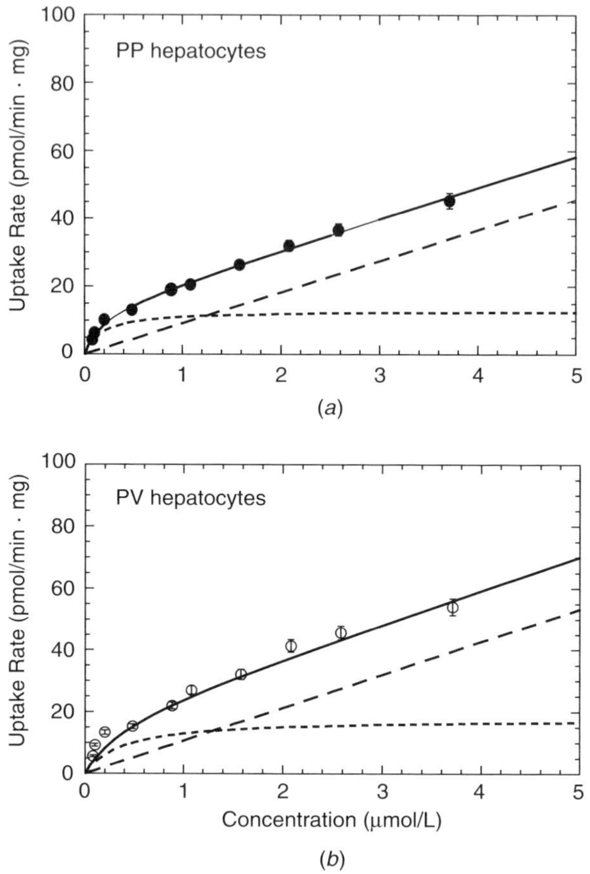

FIGURE 23.9. Rates of digoxin uptake by PP and PV isolated rat hepatocytes ($n = 4$): the fitted lines are the sum of saturable (small dashed lines) and nonsaturable (medium dashed lines) components. (From ref. 89, with permission.)

TABLE 23.5. In Vitro Metabolism and Transport Data on Digoxin in the Rat Hepatic Microsomes and Isolated Zonal Hepatocytes

	K_m^a (nM)	V_{max}^a (pmol/min/mg)	PS_{uptake}^b (μL/min/mg)	CL_{int}^c (μL/min/mg)
Metabolism[d]	125,000	362	—	2.90
Transport				
Periportal hepatocytes	180 ± 112	13 ± 8	9.2 ± 1.3	83 ± 29
Perivenous hepatocytes	390 ± 406	18 ± 4.9	10.7 ± 2.5	111 ± 80

Source: ref. 89; with permission.

[a] Saturable component.

[b] Nonsaturable or linear uptake clearance.

[c] Total hepatocyte uptake clearance is equal to $V_{max}/K_m + PS_{uptake}$ under linear conditions; PS_{uptake} is the permeability–surface area product by passive diffusion.

[d] Metabolic data from Salphati and Benet.[96]

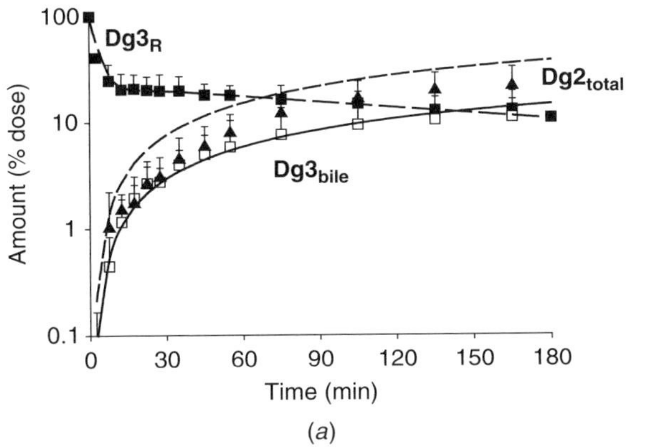

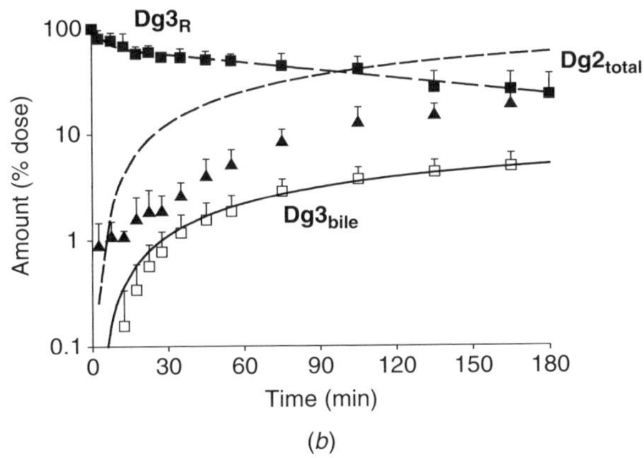

FIGURE 23.10. Fits of the Dg3 data in reservoir ($Dg3_R$) and bile ($Dg3_{bile}$) to the ZPBPK model (Figure 23.3*b*) for the KHB-perfused (without binding) (*a*) and RBC-Alb-perfused (with binding) (*b*) livers. The solid triangle symbols are observations on the summed amounts of Dg2 in reservoir and bile (% dose); however, the total amount of Dg2 formed ($Dg2_{total}$) predicted, was underestimated since Dg2 amounts in liver was not accessible for all time points. (From ref. 89, with permission.)

as 10, 30, and 60% of the total $CL_{int,met}$ (Table 23.6). Again, the fit to the ZPBPK model provided reasonable estimates of CL_{efflux}, $CL_{int,met}$, $CL_{int,sec}$, and f_L and was consistent with the data (Figure 23.10) and the observed clearances (Table 23.6).

Simulations are then made to illustrate the various relevant points pertaining to the removal of digoxin, a poorly cleared compound. When the blood flow rate is increased from 10 mL/min to 40 mL/min, it is apparent that the flow would not perturb the concentration profile and decay processes of digoxin, a poorly extracted compound (Figure 23.11*a*). Whereas when red cell binding is set as nonexistent ($k_{pr} = k'_{rp} = 0$), the decay profile of digoxin is hastened, and all clearance values, excretory and metabolic, are increased (Figure 23.11*b*). When albumin binding is further set as

TABLE 23.6. Assigned and Fitted Parameters to ZPBPK Based on Average Data of Digoxin

	Observed/Assigned	Fitted/Predicted	
	RBC-Alb-Perfused Livers (20% RBC–1% Albumin)	ZPBPK1[a]	ZPBPK2[b]
Q (mL/min/g)	0.811		
Hct	0.142		
V_R (mL/g)[c]	12.2		
V_S (mL/g)[d]	0.149		
V_L (mL/g)[e]	0.663		
k'_{rp} (min^{-1})[f]	1.81		
k_{pr} (min^{-1})[f]	0.468		
f_p[g]	0.64		
CL_{influx} (mL/min/g)[h]	12.1		
CL_{efflux} (mL/min/g)		14.3 ± 388	14.1 ± 204
$CL_{int,met}$ (mL/min/g)		0.202 ± 5.49	0.197 ± 2.84
$CL_{int,sec}$ (mL/min/g)		0.017 ± 0.469	0.017 ± 0.250
f_L		0.051 ± 1.40	0.052 ± 0.749
$CL_{liver,ex}$ (mL/min/g)	0.011 ± 0.007	0.008	0.008
$CL_{liver,met}$ (mL/min/g)	0.103 ± 0.051	0.093	0.094
$CL_{liver,tot}$ (mL/min/g)	0.123 ± 0.061	0.101	0.102
E	0.144 ± 0.054	0.125	0.126

Source: ref. 89.

[a]Fitted parameters when the metabolic activity, $CL_{int,met}$, was distributed heterogeneously (10% : 30% : 60% of total intrinsic clearance in zones 1, 2, and 3, respectively).

[b]Fitted parameters when $CL_{int,met}$ was distributed homogeneously in zones 1, 2, and 3.

[c]Mean of reservoir volume normalized to averaged liver weight.

[d]Sinusoidal blood volume of liver.

[e]Cellular water space of liver.

[f]Partition rate constants of digoxin between plasma and RBCs.

[g]Unbound fraction in plasma (1 and 2% albumin) for 0.01 to 100 μM of digoxin.

[h]Calculated from data of Table 23.5, the in vitro, hepatocyte uptake data were scaled up with the scaling factor (α/β, where α is 1.25×10^8 cells/g liver and β is 1×10^6 cells/mg protein).[11]

zero ($f_p = 1$), steeper changes in decay are observed, and all of the clearance values are increased further (Figure 23.11*b*). Trends simulated for the interplay between transporters and enzymes (Figure 23.12) are identical to those of enalapril (Figures 23.6 and 23.7). The simulated results from digoxin with the ZPBPK model (with and without heterogeneous distribution of Cyp3a2, ZPBPK1, and ZPBPK2, respectively; Table 23.6) again confirm identical trends on the competing nature of transporters and metabolic enzymes, as observed previously for enalapril (Table 23.4). The lack of effect of flow on the clearances of digoxin is consistent with the fact that digoxin is a poorly extracted compound. Due to the low E, the presence of heterogeneity of enzyme $CL_{int,met,i}$ is also devoid of effect since the intrahepatic concentration gradient is shallow, and the enzyme is unable to perturb the substrate concentration within the cell.

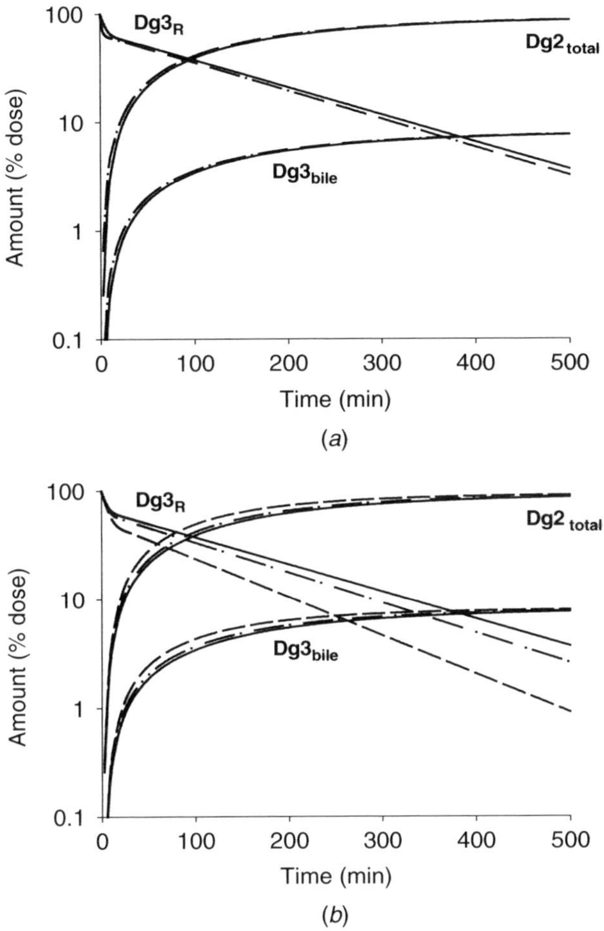

FIGURE 23.11. Simulation of the effect of flow rate (*a*) or binding (*b*) on Dg3 in reservoir ($Dg3_R$), bile ($Dg3_{bile}$), and total Dg2 formation ($Dg2_{total}$) with the ZPBPK model (Figure 3*b*). (*a*) The flow rate was increased from 10 mL/min (solid line) to 40 mL/min (dashed–dotted line); (*b*) The on and off binding constants for digoxin with red cells were set as zero ($k'_{rp} = k_{pr} = 0$) (dashed–dotted line), suggesting lack of RBC binding, and binding to BSA was further set to zero ($f_p = 1$) (dashed line). (From ref. 89, with permission.)

23.5.3. Estradiol 17β-D-Glucuronide

Estradiol 17β-D-glucuronide (E_217G) is taken up into rat hepatocytes at the sinusoidal membrane by members of the sodium-independent organic anion transporting polypeptide: Oatp1a1, Oatp1a4, and Oatp1b2 (Figure 23.13).[102–104] In the rat liver, futile cycling between E_217G and estradiol 3-sulfate-17β-D-glucuronide (E_23S17G) occurs via estrogen sulfotransferase or sulfotransferase 1e1 (Sult1e1) and arylsulfatase C.[105] Since the metabolite, E_23S17G, is not transported across the sinusoidal

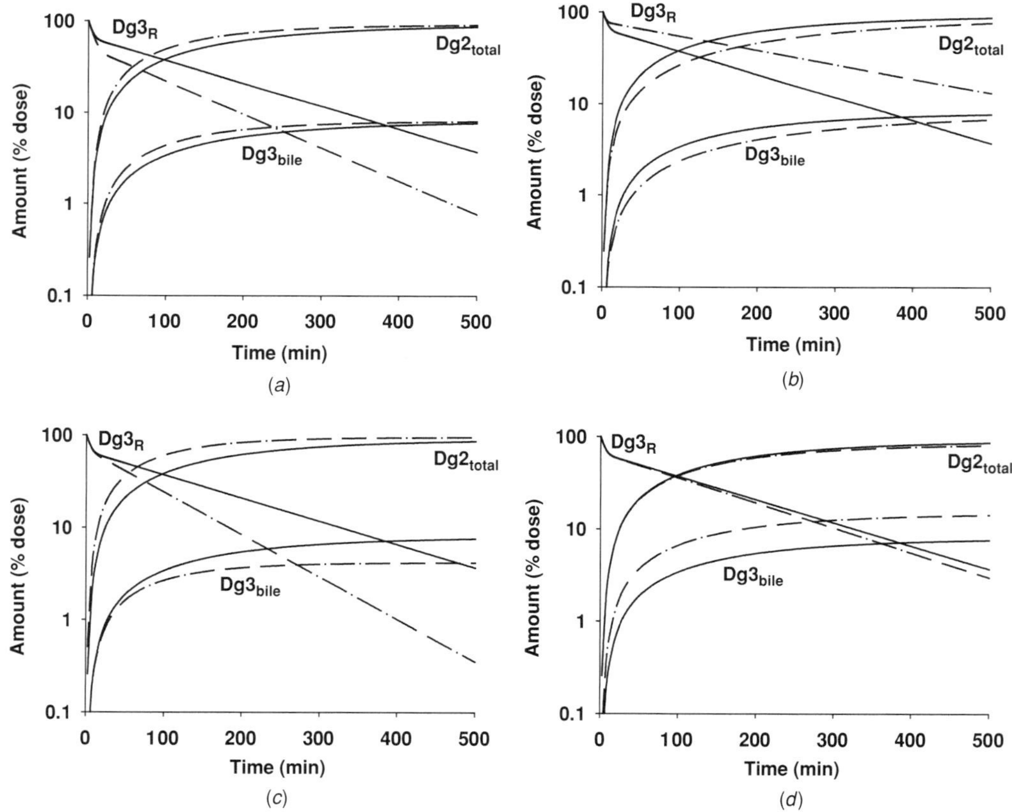

FIGURE 23.12. Simulation of the effect of doubling the CL_{influx} (*a*), CL_{efflux} (*b*), $CL_{int,met}$ (*c*), and $CL_{int,sec}$ (*d*) on Dg3 in reservoir ($Dg3_R$), bile ($Dg3_{bile}$), and total Dg2 formation ($Dg2_{total}$) with the ZPBPK model (Figure 22.3*b*). Changes from the original, controlled condition (solid line) were denoted as (dashed–dotted lines). (From ref. 89, with permission.)

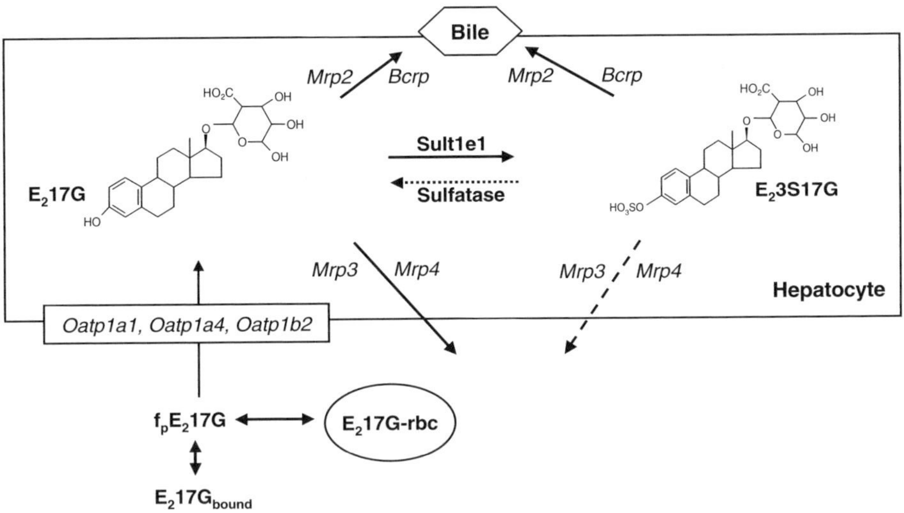

FIGURE 23.13. Schematic of the biological fate of $E_2$17G in rat liver. $E_2$17G is sulfated by Sult1e1 to $E_2$3S17G, which may be desulfated by arylsulfatases back to $E_2$17G. Influx transporters (Oatp1a1, Oatp1a4, and Oatp1b2) bring $E_2$17G into the liver. The conjugated estrogens, $E_2$17G and $E_2$3S17G, may be excreted into bile by Mrp2 and Bcrp, or efflux out by Mrp3 and Mrp4. (From ref. 105, with permission.)

membrane, the metabolite appears exclusively in bile.[105] Both $E_2$17G and $E_2$3S17G are excreted by Mrp2, and to a lesser extent Bcrp, into bile.[106–109] The efflux of $E_2$17G at the basolateral membrane may be mediated by Mrp3 and Mrp4,[110,113] whose levels are very low at physiological conditions unless cholestasis occurs.[9,111,112,114]

In a metastatic liver tumor model in which microcirculatory changes were absent in male Wag/Rij rats,[115] enzymes and transporters in the peritumor tissue were found to be altered with tumor progression. Specifically, Oatp1a1 and Oatp1b2 were decreased significantly (40%) compared to those in sham-operated rat liver; sulfatase activity was unchanged, whereas sulfotransferase activity by Sult1e1 was increased significantly (Figure 23.14).[105] These in turn altered the removal kinetics $E_2$17G (Figure 23.15). The data from both sham-operated and metastatic tumor liver have been fit to a PBPK model that includes futile cycling. The fit revealed a decreased influx clearance (CL_{influx}, 40% decrease) and increased sulfation activity ($CL_{int,met}$, 40%) (Table 23.7). The model is able to predict the perfusion data (Figure 23.15), with and without tumor, and altered protein levels of enzyme and uptake transporters (Figure 23.14). For $E_2$17G that is taken up rapidly, however, the decrease in CL_{influx} in tumor liver would not reduce the overall clearance of $E_2$17G, since the CL_{influx} remains to be very high. When the decrease in CL_{influx} exceeds 40% (Figure 23.16), the reduction in clearance (total, biliary, and metabolic) would then become apparent. For this flow-limited and highly cleared substrate, increased Sult1e1 levels increases sulfation and the net metabolic clearance, bringing about a reduction in the biliary excretion clearance (Table 23.7).

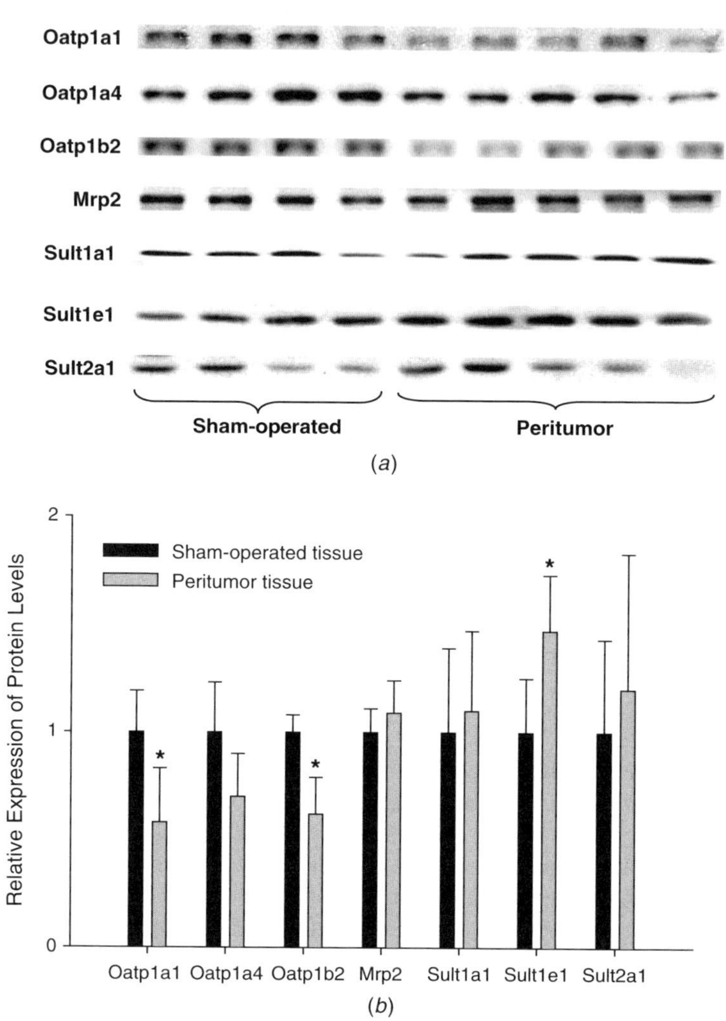

FIGURE 23.14. Western blots (relative protein expression) (*a*) and integrated densitometric analysis (*b*) of crude membrane fractions (for the Oatps and Mrp2) and cytosolic fractions (for Sult's) prepared from sham-operated rat liver tissue ($n = 4$) and peritumor tissue of metastatic rat livers ($n = 5$). An asterisk denotes that $p < 0.05$ between sham-operated and peritumor tissue. (From ref. 105, with permission.)

Due to futile cycling, equation (7) no longer holds even for the PBPK model; the ratio $CL_{liver,met}/CL_{liver,sec}$ fails to reflect $CL_{int,sult}/CL_{int,sec}$ (2.56/1.55 or 1.65) (Table 23.8). Due to the high extraction ratio for $E_2 17G$, clearance values would change more dramatically between the PBPK and ZPBPK models that engender heterogeneity even when the same given intrinsic clearance is used for prediction (Table 23.8). The ZPBPK model consists of variable $CL_{int,sult,i}$ for sulfation in the PP, midzonal,

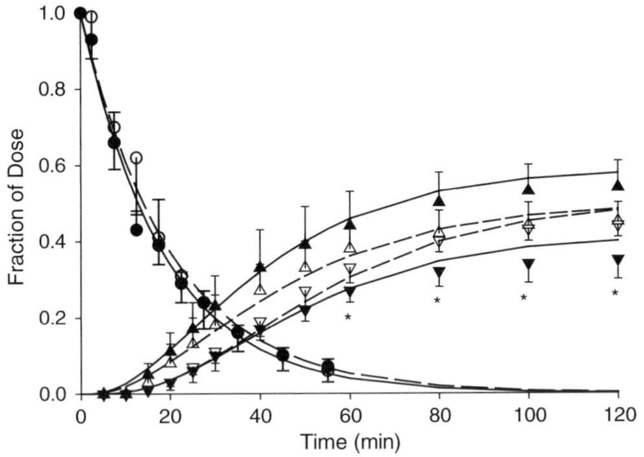

FIGURE 23.15. Fits of liver perfusion data: E_217G in reservoir perfusate and E_217G and E_23S17G in bile from metastatic tumor-bearing livers ($n = 5$, open symbols) and sham-operated livers ($n = 4$, solid symbols) at 4 weeks' postdevelopment of tumor to a PBPK model. Perfusate E_217G (circle) and the cumulative amounts of E_217G (triangle up) and E_23S17G (triangle down) in bile were adequately predicted by the PBPK model (solid line for sham-operated and dashed line for metastatic tumor group). A statistical difference (*, $p < 0.05$) was noted for the cumulative biliary excreted amounts of $[^3H]E_23S17G$ (and the biliary clearances) between sham-operated and metastatic tumor groups. (From ref. 105, with permission.)

and PV regions that were assigned 13, 33, and 54% of the $CL_{int,sult}$, respectively, according to Tan et al., who reported the acinar distribution of Sult1e1 and evenness of the sulfatases in the futile cycling of estrone sulfate.[116] With the added complexity of futile cycling, different patterns would surface for predictions according to the PBPK and ZPBPK models in describing changes in CL_{influx} on the clearances of E_217G. While the patterns of change predicted for E_217G with the PBPK model would show similarity with those for enalapril and digoxin—an increase in CL_{influx} increased all clearances—patterns distinctly different from those of the PBPK model are predicted for the ZPBPK model. According to the ZPBPK, an increase in CL_{influx} increases $CL_{liver,ex}$ and $CL_{liver,tot}$ but decreases the metabolic clearance, $CL_{liver,met}$. This is attributed to the perivenous zonation of Sult1e1 and even distribution of the Mrp2 and is expected for E_217G, a highly cleared compound (Table 23.8). Rapid entry into hepatocytes and absence of basolateral efflux for E_217G would trap larger amounts of the substrate in zone 1, the first accessible zone after onset of perfusion, for both sulfation and excretion. Due to less efficiency of sulfation for zone 1, more is excreted. The perivenous abundance of Sult1e1 therefore renders a lower apparent metabolic clearance in comparison to the scenario when Sult1e1 is evenly distributed (Table 23.8). With changes in CL_{influx}, the $CL_{liver,met}/CL_{liver,ex}$ ratio would remain constant for the PBPK model but is altered in the ZPBPK model, as would also be expected for drugs undergoing futile cycling or drugs that are highly extracted (Table 23.8).

TABLE 23.7. Assigned or Fitted Parameters from Force Fitting of $E_2$17G and $E_2$3S17G Data from In Situ Recirculating Perfused Metastatic and Sham-Operated Rat Liver Preparations at 4 Weeks' Postinoculation with a Simple Physiological (PBPK) Model

Parameters	Sham-Operated Livers		Tumor Livers
Observations			
Net $CL_{liver,ex}$ (mL/min)	6.33		4.90
Net $CL_{liver,met}$ (mL/min)	4.45		4.95
Net $CL_{liver,tot}$ (mL/min)	10.8		9.85
Net extraction ratio, E_H	0.91		0.82
$CL_{liver,met}/CL_{liver,ex}$	0.70		1.01
Fitted parameters			
Influx clearance of $E_2$17G, CL_{influx}(mL/min)	546 ± 172[a]		283 ± 50
Efflux clearance of $E_2$17G, CL_{efflux}(mL/min)		0	
Sulfation intrinsic clearance of $E_2$17G, $CL_{int,sult}$(mL/min)	2.56 ± 1.94		3.69 ± 2.79
Desulfation intrinsic clearance of $E_2$3S17G, $CL_{int,desult}\{M\}$(mL/min)		1.19 ± 0.62	
Biliary intrinsic clearance of $E_2$17G, $CL_{int,sec}$(mL/min)		1.55 ± 0.61	
Biliary intrinsic clearance of $E_2$3S 17G, $CL_{int,sec}\{M\}$ (mL/min)		0.88 ± 0.27	
$CL_{int,sult}/CL_{int,sec}$	1.65		2.38

Source: ref. 105; with permission.

[a]Fitted parameter ± SD.

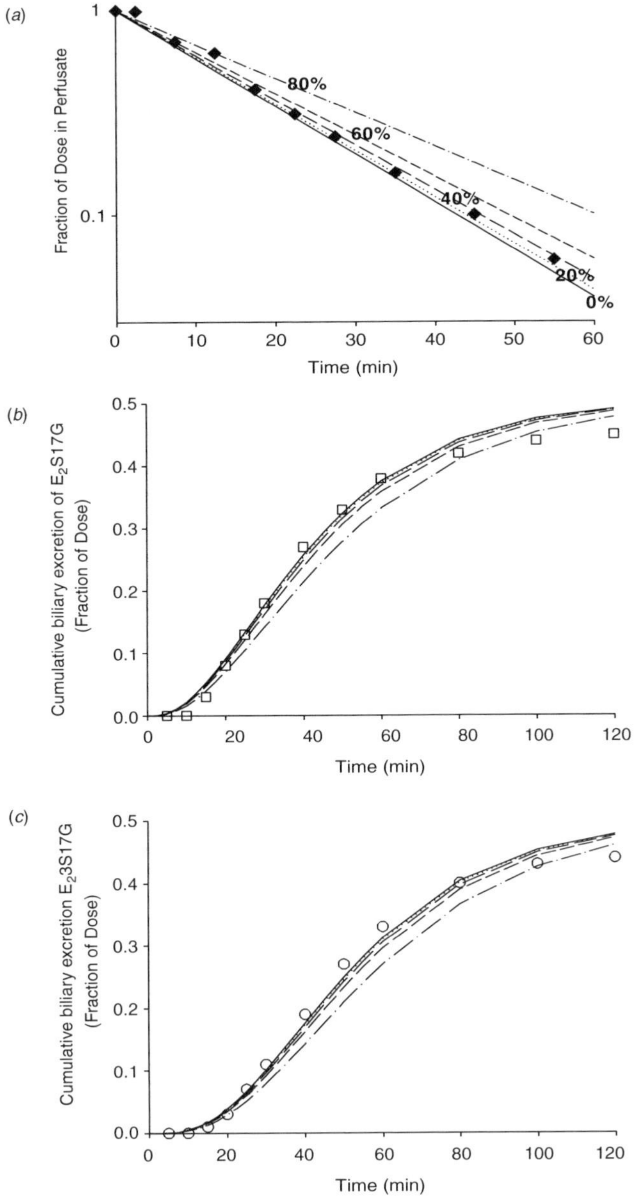

FIGURE 23.16. Profiles simulated for perfusate $E_2$17G (*a*), $E_2$17G (*b*), and $E_2$3S17G (*c*) in bile. Parameters for simulation were based on the intrinsic clearances reported for the sham-operated liver preparations (Table 23.7; $CL_{efflux} = 0$) except that a higher $CL_{int,sult}$ value (41% increase, as in tumor group) was used. CL_{influx} was decreased (down-regulated by 0%, 20%, 40%, 60%, 80%) to simulate the profiles for $E_2$17G and $E_2$3S17G. The solid line, unchanged CL_{influx}, and the dotted, medium-dashed, short-dashed, and dashed-dotted lines are associated with reductions in CL_{influx} of 0, 20, 40, 60, and 80%, respectively. The total clearance of $E_2$17G remained apparently unchanged when CL_{influx} retained 60% of its original value. (From ref. 105, with permission.)

TABLE 23.8. Simulation Results in Total Hepatic, Metabolic, and Excretion Clearances of $E_2$17G when the Influx, Sulfation, Secretory Intrinsic Clearances for $E_2$17G and Desulfation and Secretory Intrinsic Clearances for $E_2$3S17G Are Changed Under Linear Conditions for Both Single Liver Compartment PBPK and ZPBPK Models

Intrinsic Clearance (mL/min)		Clearances (mL/min) for PBPK				Clearances (mL/min) for ZPBPK[a]			
		$CL_{liver,ex}$	$CL_{liver,met}$	$CL_{liver,tot}$	$CL_{liver,met}/CL_{liver,ex}$	$CL_{liver,ex}$	$CL_{liver,met}$	$CL_{liver,tot}$	$CL_{liver,met}/CL_{liver,ex}$
Influx intrinsic clearance of $E_2$17G, CL_{influx}	273	5.8	4.0	9.8	0.70	7.9	3.3	11.2	0.42
	546	6.3	4.5	10.8	0.70	8.6	3.2	11.8	0.37
	1092	6.7	4.7	11.4	0.70	9.0	2.9	12.0	0.32
Sulfation intrinsic clearance of $E_2$17G, $CL_{int,sult}$	1.28	8.0	2.8	10.8	0.35	9.9	1.9	11.8	0.19
	2.56	6.3	4.5	10.8	0.70	8.6	3.2	11.8	0.37
	5.12	4.5	6.3	10.8	1.40	6.9	4.9	11.8	0.71
Biliary intrinsic clearance of $E_2$17G, $CL_{int,sec}$	0.78	4.5	6.3	10.8	1.40	6.9	4.9	11.8	0.71
	1.55	6.3	4.5	10.8	0.70	8.6	3.2	11.8	0.37
	3.10	8.0	2.8	10.8	0.35	9.9	1.9	11.8	0.19
Desulfation intrinsic clearance of $E_2$3S17G, $CL_{int,desult}\{M\}$	0.60	5.4	5.4	10.8	0.99	7.8	4.0	11.8	0.51
	1.19	6.3	4.5	10.8	0.70	8.6	3.2	11.8	0.37
	2.38	7.5	3.3	10.8	0.45	9.5	2.3	11.8	0.24
Biliary intrinsic clearance of $E_2$3S 17G, $CL_{int,sec}\{M\}$	0.44	7.5	3.3	10.8	0.45	9.5	2.3	11.8	0.24
	0.88	6.3	4.5	10.8	0.70	8.6	3.2	11.8	0.37
	1.76	5.4	5.4	10.8	0.99	7.8	4.0	11.8	0.51

[a]The ZPBPK model was based on uneven sulfation activities of Sult1e1 of 13, 33, and 54%, respectively, in zones 1, 2, and 3 and even distribution of Oatps and Mrp2 for uptake and excretion and evenly distributed desulfation activity of $E_2$3S17G.

The trends predicted for the biliary, metabolic, and total clearance of $E_2$17G with varying $CL_{int,sec}$ or $CL_{int,sult}$ are similar to those for enalapril and digoxin, in which futile cycling is absent. Patterns for the PBPK and ZPBPK models are also similar. An increase in the CL_{int} of one pathway would decrease the clearance of the alternative pathway but increase the total clearance, whereas a decrease in the CL_{int} of one pathway would increase the clearance of the alternative pathway but decrease the total clearance (Table 23.8). The existence of futile cycling further empowers the metabolite, $E_2$3S17G, to exert considerable influence on the rate and extent of hepatic removal of the parent drug via desulfation intrinsic clearance ($CL_{int,desult}\{M\}$) and the secretory intrinsic clearance of the metabolite, $CL_{int,sec}\{M\}$, by altering the intracellular concentration of $E_2$17G. According to both the PBPK and ZPBPK models, an increase in desulfation clearance of $E_2$3S17G ($CL_{int,desult}\{M\}$) decreases the net metabolic clearance and increases the apparent biliary clearance of $E_2$17G (Table 23.8). Biliary excretion of the metabolite ($CL_{int,sec}\{M\}$), which removes the metabolite irreversibly and prevents futile cycling, effectively increases the net sulfation clearance and results in attenuated biliary excretion of the parent drug (Table 23.8). These patterns are unique for $E_2$17G and $E_2$3S17G since the parent drug $E_2$17G fails to reenter the circulation at the basolateral membrane, and the metabolite $E_2$3S17G does not cross the basolateral membrane at all.

23.6. CONCLUSIONS

The predictive power of the simple and zonal physiologically based pharmacokinetic liver models has been showcased in the chapter. The drug examples above reveal that PBPK liver model has exquisite properties in presenting integrative views on hepatic drug extraction. The ZPBPK liver model, which is based on the acinar distribution of transporters and enzymes, is superior, since it is able to describe the attendant heterogeneities of transporters and enzymes. To use this integrated approach, in vitro hepatocyte transport and metabolic data and information on the zonal distribution of transporters and enzymes are building blocks of the ZPBPK model. Enalapril and digoxin have been chosen as the examples because of the richness of in vitro data on transport and metabolism and ex vivo liver perfusion data in our laboratory. The data and the simulations aptly demonstrate the interplay of transporters and enzymes. The concepts developed are pertinent to all drugs and allow us to explain quantitatively the effects of blood flow, vascular (plasma and RBC) binding, transport, and metabolism on hepatic drug disposition. The models are able to predict changes in the biliary, metabolic, and total hepatic clearances upon alteration of one or multiple parameters. More important, the PBPK and ZPBPK models share common views on the interplay of enzymes and transporters (Table 23.4). These concepts have been based solidly on mass transfer principles and will be unyielding in terms of the predictions.

However, the scenario is altered with futile cycling, as exemplified by estradiol 17β-D-glucuronide that is sulfated to its 3-sulfate conjugate, which in turn is desulfated back to the parent glucuronide. Due to the absence of $E_2$17G basolateral efflux and the complication due to futile cycling, the interplay between transporter and enzyme becomes more complex. A new pattern of change, depending on the distributions of the enzyme and secretory transporter, surfaces with changes in CL_{influx} (Table 23.8). Nonetheless, the seesaw patterns in the clearances with changes in CL_{int} would persist for both PBPK and ZPBPK for drugs that undergo futile cycling (Table 23.8) and are readily discerned with modeling and simulations. These types of ZPBPK models lend great usefulness to decipher drug–drug interactions in liver and the type of changes accompanying the use of inhibitors and inducers.

Acknowledgments

The work was supported by CIHR (MOP 64350).

REFERENCES

1. Gibaldi M, Boyes RN, Feldman S. 1971. Influence of first-pass effect on availability of drugs on oral administration. J Pharm Sci 60:1338–1340.
2. Pang KS, Chiba M. 1994. Metabolism: Scaling up from in vitro to organ and whole body. In: Welling PG, Balant LP, editors. *Handbook of Experimental Pharmacology*. Stuttgart, Germany: Springer-Verlag, pp 101–187.
3. Pang KS, Geng W, Schwab AJ, Goresky CA. 1998. Probing the structure and function of the liver with the multiple indicator dilution technique. In: Bassingthwaigthe J, Goresky CA, Lenihan JN, editors. *Whole Organ Approach to Cellular Metabolism: Capillary Permeation, Cellular Transport and Reaction Kinetics*. Berlin: Springer-Verlag, pp 325–368.
4. Parkinson A. 2001. Biotransformation of xenobiotics. In: Klaassen C, editor. *Casarett & Doull's Toxicology*. Columbus, OH: McGraw-Hill, pp 133–224.
5. Keppler D, Arias IM. 1997. Hepatic canalicular membrane. Introduction: transport across the hepatocyte canalicular membrane. FASEB J 11:15–18.
6. Shitara Y, Horie T, Sugiyama Y. 2006. Transporters as a determinant of drug clearance and tissue distribution. Eur J Pharm Sci 27:425–446.
7. Eloranta JJ, Kullak Ublick GA. 2005. Coordinate transcriptional regulation of bile acid homeostasis and drug metabolism. Arch Biochem Biophys 433:397–412.
8. Tirona RG, Kim RB. 2005. Nuclear receptors and drug disposition gene regulation. Pharm Sci 94:1169–1186.
9. Zamek-Gliszczynski MJ, Hoffmaser KA, Nesasa K-I, Tallman MN, Brouwer KLR. 2006. Integration of hepatic transporters and phase II metabolizing enzymes: mechanisms of hepatic excretion of sulfate, glucuronide, and glutathione metabolites. Eur J Pharm Sci 27:447–486.
10. Goresky CA, Bach GC, Nadeau BE. 1973. On the uptake of materials by the intact liver: the transport and net removal of galactose. J Clin Invest 52:991–1009.
11. Tirona RG, Pang KS. 1999. Bimolecular glutathione conjugation of ethacrynic acid and efflux of the glutathione adduct by periportal and perivenous rat hepatocytes. J Pharmacol Exp Ther 290:1230–1241.

12. Tirona, RG, Tan E, Meier G, Pang KS. 1999. Uptake and glutathione conjugation kinetics of ethacrynic acid in rat liver: in vitro and perfusion studies. J Pharmacol Exp Ther 291:1210–1219.
13. Pang KS, Terrell JA. 1981. Retrograde perfusion to probe the heterogeneous distribution of hepatic drug metabolizing enzymes in rats. J Pharmacol Exp Ther 216:339–346.
14. Abu-Zahra TN, Pang KS. 2000. Effect of zonal transport and metabolism on hepatic removal: enalapril hydrolysis in zonal, isolated rat hepatocytes in vitro and correlation with perfusion data. Drug Metab Dispos 28:807–813.
15. Hjelle JJ, Hazelton GA, Klaassen CD. 1985. Acetaminophen decreases adenosine 3′-phosphate 5′-phosphosulfate and uridine diphosphoglucuronic acid in rat liver. Drug Metab Dispos 13:35–41.
16. Dills RL, Howell SR, Klaassen CD. 1987. Hepatic UDP-glucose and UDP-glucuornic acid synthesis rates in rats during a reduced energy state. Drug Metab Dispos 15:281–288.
17. Wagner JG, DiSanto AR, Gillespie WR, Albert KS. 1981. Reversible metabolism and pharmacokinetics: application to prednisone–prednisolone. Res Commun Chem Pathol Pharmacol 32:387–405.
18. Ebling WF, Szefler SJ, Jusko WJ. 1986. Methylprednisolone disposition in rabbits: analysis, prodrug conversion, reversible metabolism, and comparison with man. Drug Metab Dispos 13:296–304.
19. Cheng HY, Jusko WJ. 1990. Mean interconversion times and distribution rate parameters for drugs undergoing reversible metabolism. Pharm Res 7:1003–1110.
20. Ratna S, Chiba M, Bandyopadhyay L, Pang KS. 1993. Futile cycling between 4-methylumbelliferone and its conjugates in perfused rat liver. Hepatology 17:838–853.
21. Tan E, Pang KS. 2001. Sulfation is rate limiting in the futile cycling between estrone and estrone sulfate in enriched periportal and perivenous rat hepatocytes. Drug Metab Dispos 29:335–346.
22. Tan E, Lu T, Pang KS. 2001. Futile cycling of estrone sulfate and estrone in the recirculating, perfused rat liver. J Pharmacol Exp Ther 297:423–436.
23. Pang KS, Sherman IA, Schwab AJ, Geng W, Barker F III, Dlugosz JA, Cuerrier G, Goresky CA. 1994. Role of the hepatic artery in the metabolism of phenacetin and acetaminophen: an intravital microscopic and multiple indicator dilution study in perfused rat liver. Hepatology 20:672–683.
24. Sherman IA, Barker F III, Dugloz J, Sadeghi FM, Pang KS. 1996. Dynamics of arterial and portal flow interactions in perfused rat liver: an intravital microscopic study. Am J Physiol 271:G201–G210.
25. Takasaki S, Hano H. 2001. Three-dimensional observations of the human hepatic artery (arterial system in the liver). J Hepatol 34:455–466.
26. Foley DP, Ricciardi R, Traylor AN, McLaughlin TJ, Donohue SE, Wheeler SM, Meyers WC, Quarfordt SH. 2003. Effect of hepatic artery flow on bile secretory function after cold ischemia. Am J Transplant 3:148–155.
27. Pang KS, Rowland M. 1977. Hepatic clearance of drugs: II. Experimental evidence for acceptance of the "well-stirred" model over the "parallel tube" model using lidocaine in the perfused rat liver in situ preparation. J Pharmacokinet Biopharm 5:655–680.
28. Rowland M, Benet LZ, Graham GG. 1973. Clearance concepts in pharmacokinetics. J Pharmacokinet Biopharm 1:123–136.

29. Winkler K, Keiding S, Tygstrup N. 1973. Clearance as a quantitative measure of liver function. In: Paumgartner P, Presig R, editors. *The Liver: Quantitative Aspects of Structure and Functions*. Basel, Switzerland, Karger, pp 144–155.

30. Pang, KS, Rowland M. 1977. Hepatic clearance of drugs. I. Theoretical considerations of a "well-stirred" model and a "parallel tube" model. Influence of hepatic blood flow, plasma and blood cell binding, and the hepatocellular enzymatic activity on hepatic drug clearance. J Pharmacokinet Biopharm 5:625–653.

31. Pang, KS, Rowland M. 1977. Hepatic clearance of drugs. III. Additional experimental evidence for supporting the "well-stirred" model, using metabolite (MEGX) data generated from lidocaine under varying hepatic blood flow rates and linear conditions in the perfused rat liver in situ preparation. J Pharmacokinet Biopharm 5:681–690.

32. Perl W, Chinard FP. 1968. A convection–diffusion model of indicator transport through an organ. Circ Res 22:273–298.

33. Roberts MS, Rowland M. 1985. Hepatic elimination–dispersion model. J Pharm Sci 74:585–587.

34. Pang, KS, Gillette JR. 1978. Theoretical relationships between area under the curve and route of administration of drugs and their precursors for evaluating sites and pathways of metabolism. J Pharm Sci 67:703–704.

35. Sato H, Sugiyama Y, Miyauchi S, Sawada Y, Iga T, Hanano M. 1986. A simulation study on the effect of a uniform diffusional barrier across hepatocytes on drug metabolism by evenly or unevenly distributed uni-enzyme in the liver. J Pharm Sci 75:3–8.

36. de Lannoy IAM, Pang KS. 1986. A commentary: presence of a diffusional barrier on metabolite kinetics. Enalaprilat as a generated versus preformed metabolite. Drug Metab Dispos 14:513–520.

37. de Lannoy IAM, Pang KS. 1987. Effect of diffusional barriers on drug and metabolite kinetics. Drug Metab Dispos 15:51–58.

38. Roberts MS, Donaldson JD, Rowland M. 1988. Models of hepatic elimination: comparison of stochastic models to describe residence time distributions and to predict the influence of drug distribution, enzyme heterogeneity, and systemic recycling on hepatic elimination. J Pharmacokinet Biopharm 16:41–83.

39. Bass L, Robinson P, Bracken AJ. 1978. Hepatic elimination of flowing substrates: the distributed model. J Theor Biol 72:161–184.

40. Gray MR, Tam YK. 1987. The series-compartment model for hepatic elimination. Drug Metab Dispos 15:27–31.

41. Goresky CA, Bach GC, Nadeau BE. 1973. On the uptake of materials by the intact liver: the transport and net removal of galactose. J Clin Invest 52:991–1009.

42. Xu X, Selick P, Pang KS. 1993. Nonlinear protein binding and enzyme heterogeneity: effects on hepatic drug removal. J Pharmacokinet Biopharm 21:43–74.

43. Kwon Y, Morris ME. 1997. Membrane transport in hepatic clearance of drugs: II. Zonal distribution patterns of concentration-dependent transport and elimination processes. Pharm Res 14:780–785.

44. Hagenbuch B, Meier PJ. 1994. Molecular cloning, chromosomal localization, and functional characterization of a human liver Na^+/bile acid cotransporter. J Clin Invest 93: 1326–1331.

45. Shi X, Bai S, Ford AC, Burk RD, Jacquemin E, Hagenbuch B, Meier PJ, Wolkoff AW. 1995. Stable inducible expression of a functional rat liver organic anion transport protein in HeLa cells. J Biol Chem 270:25591–25595.

46. Tan E, Tirona RG, Pang KS. 1999. Lack of zonal uptake of estrone sulfate in enriched periportal and perivenous isolated rat hepatocytes. Drug Metab Dispos 27:336–341.
47. Abu-Zahra TN, Wolkoff A, Kim RB, Pang KS. 2000. Uptake of enalapril and expression of organic anion transporting polypeptide 1 in zonal, isolated rat hepatocytes. Drug Metab Dispos 28:801–806.
48. Shitara Y, Li AP, Kato Y, Lu C, Ito K, Itoh T, Sugiyama Y. 2003. Function of uptake transporters for taurocholate and estradiol 17β-D-glucuronide in cryopreserved human hepatocytes. Drug Metab Pharmacokinet 18:33–41.
49. Houle R, Raoul J, Levesque JF, Pang KS, Nicoll-Griffith DA, Silva JM. 2003. Retention of transporter and metabolic activities in cryopreserved, isolated rat hepatocytes. Drug Metab Dispos 31:447–451.
50. Lindros KO, Penttïla KE. 1985. Digitonin-collagenase perfusion for efficient separation of periportal or perivenous hepatocytes. Biochem J 228:757–760.
51. Meier PJ, Boyer JL. 1990. Preparation of basolateral (sinusoidal) and canalicular plasma membrane vesicles for the study of hepatic transport processes. Methods Enzymol 192:534–545.
52. Shilling AF, Azam F, Kao J, Leung L. 2005. Use of canalicular membrane vesicles (CMVs) from rats, dogs, monkeys and humans to assess drug transport across the canalicular membrane. J Pharmacol Toxicol Methods 53:186–197.
53. Pang KS, Wang PJ, Chung AY, Wolkoff AW. 1998. The modified dipeptide, enalapril, an angiotensin-converting enzyme inhibitor, is transported by the rat liver organic anion transport protein. Hepatology 28:341–1346.
54. Shitara Y, Sugiyama D, Kusuhara H, Kato Y, Abe T, Meier PJ, Itoh T, Sugiyama Y. 2002. Comparative inhibitory effects of different compounds on rat oatpl (slc21a1)- and Oatp2 (Slc21a5)-mediated transport. Pharm Res 19:147–153.
55. Mita S, Suzuki H, Akita H, Stieger B, Meier PJ, Hofmann AF, Sugiyama Y. 2005. Vectorial transport of bile salts across MDCK cells expressing both rat Na^+-taurocholate cotransporting polypeptide and rat bile salt export pump. Am J Physiol 288: G159–G167.
56. Sasaki M, Suzuki H, Aoki J, Ito K, Meier PJ, Sugiyama Y. 2004. Prediction of in vivo biliary clearance from the in vitro transcellular transport of organic anions across a double-transfected Madin–Darby canine kidney II monolayer expressing both rat organic anion transporting polypeptide 4 and multidrug resistance associated protein 2. Mol Pharmacol 66:450–459.
57. Sasaki M, Suzuki H, Ito K, Abe T, Sugiyama Y 2002. Transcellular transport of organic anions across a double-transfected Madin–Darby canine kidney II cell monolayer expressing both human organic anion-transporting polypeptide (OATP2/SLC21A6) and multidrug resistance–associated protein 2 (MRP2/ABCC2). J Biol Chem 277:6497–6503.
58. Spears KJ, Ross J, Stenhouse A, Ward CJ, Goh LB, Wolf CR, Morgan P, Ayrton A, Friedberg, TH. 2005. Directional trans-epithelial transport of organic anions in porcine LLC-PK1 cells that co-express human OATP1B1 (OATP-C) and MRP2. Biochem Pharmacol 69:415–423.
59. Matsushima S, Maeda K, Kondo C, Hirano M, Sasaki M, Suzuki H, Sugiyama Y. 2005. Identification of the hepatic efflux transporters of organic anions using double-transfected Madin–Darby canine kidney II cells expressing human organic anion-transporting

polypeptide1B1(OATP1B1)/multidrug resistance-associated protein 2, OATP1B1/multidrug resistance 1, and OATP1B1/breast cancer resistance protein. J Pharmacol Exp Ther 314:1059–1067.

60. Cui Y, König J, Keppler D. 2001. Vectorial transport by double-transfected cells expressing the human uptake transporter SLC21A8 and the apical export pump ABCC2. Mol Pharmacol 60:934–943.

61. Letschert K, Komatsu M, Hummel-Eisenbeiss J, Keppler D. 2005. Vectorial transport of the peptide CCK-8 by double-transfected MDCKII cells stably expressing the organic anion transporter OATP1B3 (OATP8) and the export pump ABCC2. J Pharmacol Exp Ther 313:549–556.

62. Kopplow K, Letschert K, König J, Walter B, Keppler D. 2005. Human hepatobiliary transport of organic anions analyzed by quadruple-transfected cells. Mol Pharmacol 68:1031–1038.

63. Hirouchi M, Suzuki H, Sugiyama Y. 2005. Treatment of hyperbilirubinemia in Eisai hyperbilirubinemic rat by transfection human MRP2/ABCC2 gene. Pharm Res 22:661–666.

64. Miller RE, Guengerich FP. 1983. Metabolism of trichloroethylene in isolated hepatocytes, microsomes, and reconstituted enzyme systems containing cytochrome P-450. Cancer Res 43:1145–1152.

65. Yu A, Dong H, Lang D, Haining RL. 2001. Characterization of detromethorphan O- and N-demethylation catalyzed by highly purified recombinant human CYP2D6. Drug Metab Dispos 29:1362–1365.

66. Gonzalez FJ. 2003. Role of gene knockout and transgenic mice in the study of xenobiotic metabolism. Drug Metab Rev 35:319–335.

67. Gonzalez FJ. 2004. Cytochrome P450 humanised mice. Human Genom 4:300–306.

68. Gonzalez FJ, Yu AM. 2006. Cytochrome P450 and xenobiotic receptor humanized mice. Annu Rev Pharmacol Toxicol 46:41–64.

69. Mei Q, Tang C, Lin Y, Rushmore TH, Shou M. 2002. Inhibition kinetics of monoclonal antibodies against cytochromes P450. Drug Metab Dispos 30:701–708.

70. Kobyashi K, Urashima K, Shimada N, Chiba K. 2002. Substrate specificity for rat cytochrome P450 (CYP) isoforms: screening with cDNA-expressed systems in the rat. Biochem Pharmacol 63:889–896.

71. Lin JH, Wong BK. 2002. Complexities of glucuronidation affecting in vitro–in vivo extrapolation. Curr Drug Metab 3:623–646.

72. Miners JO, Knights KM, Houston JB, Mackenzie PI. 2006. In vitro–in vivo correlation for drugs and other compounds eliminated by glucuronidation in humans: pitfalls and promises. Biochem Pharmacol 71:1531–1539.

73. Liu X, Chism P, LeCluyse EL, Brouwer KR, Brouwer KL. 1999. Correlation of biliary excretion in sandwich-cultured rat hepatocytes and in vivo in rats. Drug Metab Dispos 27:637–644.

74. Hoffmaster KA, Turncliff RZ, LeCluyse EL, Kim RB, Meier PJ, Brouwer KL. 2004. P-glycoprotein expression, localization, and function in sandwich-cultured primary rat and human hepatocytes: relevance to the hepatobiliary disposition of a model opioid peptide. Pharm Res 21:1294–1302.

75. Hoffmaster KA, Zamek-Gliszczynski MJ, Pollack GM, Brouwer KL. 2004. Hepatobiliary disposition of the metabolically stable opioid peptide [D-Pen2, D-Pen5]-enkephalin

(DPDPE): pharmacokinetic consequences of the interplay between multiple transport systems. J Pharmacol Exp Ther 311:1203–1210.

76. Zeilinger K, Sauer IM, Pless G, Strobel C, Rudzitis J, Wang A, Nussler AK, Grebe A, Mao L, Auth SH, et al. 2002. Three-dimensional co-culture of primary human liver cells in bioreactors for in vitro drug studies: effects of the initial cell quality on the long-term maintenance of hepatocyte-specific functions. Altern Lab Anim 30:525–538.
77. Allen JW, Khetani SR, Bhatia SN. 2005. In vitro zonation and toxicity in a hepatocyte bioreactor. Toxicol Sci 84:110–119.
78. de Kanter R, Monshouwer M, Meijer DK, Groothuis GM. 2002. Precision cut organ slices as a tool to study toxicity and metabolism of xenobiotics with special reference to non-hepatic tissues. Curr Drug Metab 3:39–59.
79. van de Bovenkamp M, Groothuis GM, Draaisma AL, Merema MT, Bezuijen JI, van Gils MJ, Meijer DK, Friedman SL, Olinga P. 2005. Precision-cut liver slices as a new model to study toxicity-induced hepatic stellate cell activation in a physiologic milieu. Toxicol Sci 85:632–638.
80. Schinkel AH, Wagenaar E, van Deemter L, Mol CA, Borst P. 1995. Absence of mdr1a P-glycoprotein in mice affects tissue distribution and pharmacokinetics of dexamethasone, digoxin and cyclosporine A. J Clin Invest 96:1698–1705.
81. Adachi Y, Suzuki H, Schikel AH, Sugiyama Y. 2005. Role of breast cancer resistance protein (Bcrp1/Abcg2) in the extrusion of glucuronide and sulfate conjugates from enterocytes to intestinal lumen. Mol Pharmacol 67:923–928.
82. Nezasa K, Tian X, Zamek-Gliszczynski MJ, Patel NJ, Raub TJ, Brouwer KL. 2006. Altered hepatobiliary disposition of 5 (and 6)-carboxy-2′,7′-dichlorofluorescein in Abcg2 (Bcrp1) and Abcc2 (Mrp2) knockout mice. Drug Metab Dispos 34:718–723.
83. Zamek-Gliszczynski MJ, Hoffmaster KA, Humphreys JE, Tian X, Nezasa KI, Brouwer KL. 2006. Differential involvement of Mrp2 (Abcc2) and Bcrp (Abcg2) in biliary excretion of 4-methylumbelliferyl glucuronide and sulfate in the rat. J Pharmacol Exp Ther 319:459–467.
84. Liu L, Pang KS. 2006. An integrated approach to model hepatic drug clearance. Eur J Pharm Sci 29:215–230.
85. Sirianni GL, Pang KS. 1997. Organ clearance concepts: new perspectives on old principles. J Pharmacokinet Biopharm 25:449–470.
86. de Lannoy IA, Barker F 3rd, Pang KS. 1993. Formed and preformed metabolite excretion clearances in liver, a metabolite formation organ: studies on enalapril and enalaprilat in the single-pass and recirculating perfused rat liver. J Pharmacokinet Biopharm 21: 395–422.
87. Liu L, Pang KS. 2005. The roles of transporters and enzymes in hepatic drug processing. Drug Metab Dispos 33:1–9.
88. Liu L, Cui Y, Chung AY, Shitara Y, Sugiyama Y, Keppler D, Pang KS. 2006. Vectorial transport of enalapril by rOatp1a1/rMrp2 and OATP1B1 and OATP1B3/MRP2 in rat and human livers. J Pharmacol Exp Ther 318:395–402.
89. Liu L, Mak E, Tirona RG, Tan E, Novikoff PM, Wang P, Wolkoff AW, Pang KS. 2005. Vascular binding, blood flow, transporter, and enzyme interactions on the processing of digoxin in rat liver. J Pharmacol Exp Ther 315:433–448.

90. Tocco DJ, deLuna FA, Duncan AE, Vassil TC, Ulm EH. 1982. The physiological disposition and metabolism of enalapril maleate in laboratory animals. Drug Metab Dispos 10:15–19.

91. Liu L, Cui Y, Chung AY, Shitara Y, Sugiyama Y, Keppler D, Pang KS. 2006. Vectorial transport of enalapril by rOatp1a1/rMrp2 and OATP1B1 and OATP1B3/MRP2 in rat and human livers. J Pharmacol Exp Ther 318:395–402.

92. Wu CY, Benet LZ. 2003. Disposition of tacrolimus in isolated perfused rat liver: influence of troleandomycin, cyclosporine, and GG918. Drug Metab Dispos 31:1292–1295.

93. Lam JL, Benet LZ. 2004. Hepatic microsome studies are insufficient to characterize in vivo hepatic metabolic clearance and metabolic drug–drug interactions: studies of digoxin metabolism in primary rat hepatocytes versus microsomes. Drug Metab Dispos 32:1311–1316.

94. Oinonen T, Lindros KO. 1995. Hormonal regulation of the zonated expression of cytochrome P-450 3A in rat liver. Biochem J 309:55–61.

95. Harrison LI, Gibaldi M. 1976. Pharmacokinetics of digoxin in the rat. Drug Metab Dispos 4:88–93.

96. Salphati L, Benet LZ. 1999. Metabolism of digoxin and digoxigenin digitoxosides in rat liver microsomes: involvement of cytochrome P4503A. Xenobiotica 29:171–185.

97. Noé B, Hagenbuch B, Stieger B, Meier PJ. 1997. Isolation of a multispecific organic anion and cardiac glycoside transporter from rat brain. Proc Natl Acad Sci U S A 94:10346–10350.

98. Guo GL, Klaassen CD. 2001. Protein kinase C suppresses rat organic anion transporting polypeptide 1- and 2-mediated uptake. J Pharmacol Exp Ther 299:551–557.

99. Hagenbuch N, Reichel C, Stieger B, Cattori V, Fattinger KE, Landmann L, Meier PJ, Kullak-Ublick GA. 2001. Effect of phenobarbital on the expression of bile salt and organic anion transporters of rat liver. J Hepatol 34:881–887.

100. Reichel C, Gao B, van Montfoort J, Cattori V, Rahner C, Hagenbuch B, Stieger B, Kamisako T, Meier PJ. 1999. Localization and function of the organic anion–transporting polypeptide Oatp2 in rat liver. Gastroenterology 117:688–695.

101. Goresky CA, Bach GG, Nadeau BE. 1975. Red cell carriage of label: its limiting effect on the exchange of materials in the liver. Circ Res 36:328–351.

102. Bossuyt X, Müller M, Hagenbuch B, Meier PJ. 1996. Polyspecific drug and steroid clearance by an organic anion transporter of mammalian liver. J Pharmacol Exp Ther 276:891–896.

103. Cattori V, van Montfoort JE, Stieger B, Landmann L, Meijer DK, Winterhalter KH, Meier PJ, Hagenbuch B. 2001. Localization of organic anion transporting polypeptide 4 (Oatp4) in rat liver and comparison of its substrate specificity with Oatp1, Oatp2 and Oatp3. Pflugers Arch 443:188–195.

104. Noé B, Hagenbuch B, Stieger B, Meier PJ. 1997. Isolation of a multispecific organic anion and cardiac glycoside transporter from rat brain. Proc Natl Acad Sci U S A 94:10346–10350.

105. Sun H, Liu L, Pang KS. 2006. Increased estrogen sulfation of estradiol 17β-D glucuronide in metastatic tumor rat livers. J Pharmacol Exp Ther 319:818–831.

106. Takikawa H, Yamazaki R, Sano N, Yamanaka M. 1996. Biliary excretion of estradiol-17β-glucuronide in the rat. Hepatology 23:607–613.

107. Morikawa A, Goto Y, Suzuki H, Hirohashi T, Sugiyama Y. 2000. Biliary excretion of 17β-estradiol 17β-D-glucuronide is predominantly mediated by cMOAT/MRP2. Pharm Res 17:546–552.

108. Suzuki M, Suzuki H, Sugimoto Y, Sugiyama Y. 2003. ABCG2 transports sulfated conjugates of steroids and xenobiotics. J Biol Chem 278:22644–22649.

109. Imai Y, Asada S, Tsukahara S, Ishikawa E, Tsuruo T, Sugimoto Y. 2003. Breast cancer resistance protein exports sulfated estrogens but not free estrogens. Mol Pharmacol 64:610–618.

110. Zeng H, Liu G, Rea PA, Kruh GD. 2000. Transport of amphipathic anions by human multidrug resistance protein 3. Cancer Res 60:4779–4784.

111. Hirohashi T, Suzuki H, and Sugiyama Y. 1999. Characterization of the transport properties of cloned rat multidrug resistance–associated protein 3 (MRP3). J Biol Chem 274:15181–15185.

112. Ninomiya M, Ito K, Horie T. 2005. Functional analysis of dog multidrug resistance–associated protein 2 (Mrp2) in comparison with rat Mrp2. Drug Metab Dispos 33:225–232.

113. Chen ZS, Lee K, Kruh GD. 2001. Transport of cyclic nucleotides and estradiol 17β-D-glucuronide by multidrug resistance protein: 4. Resistance to 6-mercaptopurine and 6-thioguanine. J Biol Chem 276:33747–33754.

114. Denk GU, Soroka CJ, Takeyama Y, Chen WE, Schuetz JD, Boyer JL. 2004. Multidrug resistance–associated protein 4 is up-regulated in liver but down-regulated in kidney in obstructive cholestasis in the rat. J Hepatol 40:585–591.

115. Liu L, Sun H, Valji W, Pang KS. 2006. Changes in liver spaces, transporters, enzymes, and enalapril disposition in a rat metastatic liver tumor model. Am J Physiol (submitted).

116. Tan E, Pang KS. 2001. Sulfation is rate limiting in the futile cycling between estrone and estrone sulfate in enriched periportal and perivenous rat hepatocytes. Drug Metab Dispos 29:335–346.

24

CLINICAL RELEVANCE: DRUG–DRUG INTERACTION, PHARMACOKINETICS, PHARMACODYNAMICS, AND TOXICITY

SERENA MARCHETTI

Netherlands Cancer Institute, Amsterdam, The Netherlands; and University of Florence, Azienda Ospedaliero-Universitaria Careggi, Florence, Italy

ROBERTO MAZZANTI

University of Florence, Azienda Ospedaliero-Universitaria Careggi, Florence, Italy

JOS H. BEIJNEN AND JAN H. M. SCHELLENS

Netherlands Cancer Institute, Amsterdam, The Netherlands; and Utrecht University, Utrecht, The Netherlands

Drug Transporters: Molecular Characterization and Role in Drug Disposition, Edited by Guofeng You and Marilyn E. Morris
Copyright © 2007 John Wiley & Sons, Inc.

24.1. INTRODUCTION

In patients, drug–drug interactions can result in unexpected life-threatening and even lethal toxicities. Up to 10% of all hospital admissions in general hospitals are caused by improper use of drugs and combinations of drugs, resulting in potentially severe drug–drug interactions.[1,2] Adverse drug reactions can be especially severe when these interactions involve cytotoxic anticancer agents.[3,4] Anticancer drugs are dosed close to the maximum dose tolerated, and factors affecting the pharmacokinetics may therefore greatly increase the likelihood of development of life-threatening toxicities. Thus far, drug–drug interactions have been thought to result from inhibition of drug metabolism, displacement out of the protein binding, or pharmaceutical interactions. However, interference at the level of the ATP-binding cassette (ABC) and other transporters is increasingly being identified as the mechanism behind clinically important drug–drug interactions. These factors are the subject of this chapter.

24.2. INTERACTIONS MEDIATED BY ABC DRUG TRANSPORTERS

24.2.1. ABCB1 (MDR1, P-Glycoprotein)

Impact of Polymorphism on Function Currently, at least 105 variants of the ABCB1 gene have been identified, with significant differences in their frequencies among different ethnic groups. The majority of these single-nucleotide polymorphisms (SNPs) involve intronic or noncoding regions, thus do not affect the P-glycoprotein (Pgp) amino acid sequence, whereas several variants in the ABCB1 coding regions result in amino acid change and potentially affect Pgp expression and activity. Hoffmeyer et al. reported an association between an SNP in exon 26 (C3435T) of ABCB1, reduction in duodenal Pgp levels, and higher peak plasma concentrations of the Pgp substrate digoxin in healthy volunteers.[5] Confirming and contradicting studies have subsequently been published about the influence of SNPs in ABCB1 on the disposition of digoxin and on other Pgp substrate drugs (such as fexofenadine, tacrolimus, irinotecan, SN-38, paclitaxel, and cyclosporin A) and on Pgp expression and activity (for reviews, see refs. 6 to 13).

Moreover, by potentially altering the physiologic protective role of Pgp, genetic variation in ABCB1, has recently been assessed in the etiology of several human pathophysiological conditions. An increasing number of studies have associated certain SNPs in ABCB1 with susceptibility to such diseases as pharmacoresistant epilepsy, Parkinson's disease, inflammatory bowel diseases (e.g., ulcerative colitis, Crohn's disease), colorectal cancer, and renal carcinoma.[14–19]

In addition, C3435T polymorphism has been linked to an increased immunological response (CD4 count) to the anti-HIV protease inhibitor nelfinavir in HIV-positive patients,[20] although some later studies confirmed and others contradicted the association of certain ABCB1 SNPs with the efficacy of antiviral therapy in HIV patients.[21–26] Recently, ABCB1 SNP C3435T has been associated with antiemetic treatment efficacy with 5-hydroxytryptamine type 3 receptor antagonists (e.g., granisetron, ondansetron, tropisetron) in patients with cancer,[27] whereas in patients affected by depression, the same polymorphism has been linked to the development of postural hypotension induced by the antidepressant nortriptyline.[28] The 2677T SNP has been recognized as a positive predictor of tacrolimus-induced neurotoxicity.[29] Furthermore, in a recent study, hypercholesterolemic patients carrying the 1236T variant allele showed higher lipid-lowering efficacy of simvastatin treatment than for homozygotes with the wild-type allele. 1236T, 2677A/T, and 3435T alleles were also associated with reduced incidence of adverse reaction to the 3-hydroxy-3-methylglutarylcoenzyme A (HMG-CoA) synthase inhibitor simvastatin.[30]

MDR1 gene polymorphism is also suggested to affect the outcome of patients with several malignancies. Goreva et al. reported an association between C3435T and G3677T SNPs in ABCB1 and the risk of drug resistance in patients with lymphoproliferative diseases.[31] A correlation between several commonly occurring ABCB1 SNPs and overall survival and risk of relapse has been reported in patients affected by acute myeloid leukemia treated with etoposide, mitoxantrone, or daunorubicin (well-known Pgp substrates).[32] Moreover, ABCB1 SNP C3435T has been suggested as a significant predictor of the treatment outcome in children affected by acute lymphoblastic leukemia, although these findings have not been confirmed in adults.[33–35] Another study showed an increased response to preoperative chemotherapy in breast cancer patients homozygous for the C3435T genotype,[36] whereas conflicting results have been reported regarding the impact of genetic variation of the MDR1 gene (in particular, G2677T/A) on the response to paclitaxel chemotherapy in patients with ovarian carcinoma.[37] Taken together, all these findings suggest that despite the numerous conflicting results, an interethnic difference in ABCB1 SNPs exists, and several ABCB genetic variants, resulting in altered Pgp function, may contribute to the interindividual variability in toxicity and pharmacokinetics of drugs and to the susceptibility to certain diseases. Additional studies to clarify the clinical implications of ABCB1 polymorphisms are needed.[13,38]

Main Substrate Classes (Clinically Applied) Pgp presents high transport capacity and broad substrate specificity. A wide number of clinically relevant drugs with structurally different features belonging to different classes [e.g., several anticancer drugs, some HIV protease inhibitors, H2-receptor antagonists, antiarrythmics (cardiac glycosides and calcium channel blockers), immunosuppressive agents, corticosteroids, antiemetic and antidiarrheal agents, analgesics, antibiotics, anthelmintics, antiepileptics, sedatives, antidepressants] can be transported by Pgp (for a review, see ref. 39); in general, they are hydrophobic and amphipatic molecules in nature, uncharged or basic, although zwitterionic and negatively charged compounds can also be transported.

Inhibitors (Competitive and Noncompetitive) Some Pgp drug substrates are able to inhibit Pgp-mediated transport of other substrates. The discovery by Tsuruo and colleagues[40] that verapamil (a weak Pgp substrate) could reverse Pgp-mediated multidrug resistance in leukemia cells was followed by the identification of several other Pgp inhibitors[41,42] that can block Pgp activity by competition for drug-binding sites (competitive inhibitors) or by blockade of the adenosine triphosphate (ATP) hydrolysis process (noncompetitive inhibitors). The first agents identified as Pgp inhibitors were drugs (e.g., verapamil and cyclosporin A) already used in the clinic, which were themselves transported by Pgp (called first-generation inhibitors). Because of their low substrate selectivity and the concomitant inhibition of the drug-metabolizing cytochrome P4503A4 enzyme (CYP3A4), second (cyclosporin A analog PSC833)- and third (LY335979, VX710, GF120918, XR9576)-generation Pgp inhibitors were developed. These and other selective Pgp inhibitors have been tested extensively preclinically and in patients to reverse multidrug resistance. Of interest is the fact that GF120918 (elacridar), originally developed as a Pgp inhibitor, was also identified as an effective breast cancer resistance protein (BCRP)- (ABCG2) inhibitor.[43]

Recently, it has been reported that the benzimidazole gastric H^+,K^+-ATPase proton pump inhibitors (PPIs) omeprazole, pantoprazole, lansoprazole, and rabeprazole, which are used by up to 50% of patients with cancer, are effective inhibitors of Pgp,[44] although their potency toward BCRP inhibition is even greater.[45] Drug interactions with benzimidazoles are reported increasingly.[1,2,46–49]

In addition, several widely used drugs have been described to inhibit Pgp function, thus potentially leading to relevant drug–drug interactions. They include various antimicrobial agents (e.g., ceftriaxone, cefoperazone, clarithromycin, erythromycin, itraconazole, ketoconazole), Ca^{2+} antagonists (verapamil, diltiazem, quinidine, quinine, nifedipine, nicardipine), HIV protease inhibitors (ritonavir, indinavir, saquinavir, nelfinavir), and other compounds, such as amiodarone, propanolol, dipyridamole, tacrolimus, hydrocortisone, progesterone, and tamoxifen, to name a few.[50,51]

Inducers Clinical and preclinical findings reveal that the expression of Pgp (as in some of the CYP isoenzymes) is inducible. Recent in vitro studies have demonstrated that several drugs, including rifampicin, paclitaxel, reserpine, and hyperforin (one of the components of St John's wort), can induce CYP3A4 and MDR1 gene expression through a similar mechanism.[52–54] Other Pgp inducers are clotrimazole, phenobarbital, phenytoin, troglitazone, and the flavonoids kaempferol and quercetin.[50] Clinically relevant drug–drug interactions between Pgp substrates and inducers have been reported (see below).

Pharmacological and Toxicological Function The pharmacological functions of Pgp have been studied extensively in vitro and in vivo models. Pgp was first described as a plasma membrane protein that could cause multidrug resistance in tumor cells by actively extruding a wide range of structurally diverse compounds, thus contributing to the resistance against chemotherapy that occurs in several cancers. In addition, the strategically physiological distribution of Pgp in organs that play key roles in processes of drug absorption, distribution, and elimination of compounds suggests that Pgp has a relevant impact on limiting cellular uptake of drugs out of the blood circulation into the brain, placenta, and testis and from the intestinal lumen into epithelial cells lining the gut. In addition, Pgp may also mediate excretion of drugs out of hepatocytes into the bile canaliculi and out of renal tubules into the urine. The effect of Pgp on the pharmacokinetics of substrate drugs has been demonstrated in vivo using *Mdr1a* and *Mdr1a/1b* knockout mice. Mice lacking one or both *Mdr* genes showed significant alterations in drug absorption, distribution, and elimination.[55–58] Compared to wild-type mice, $Mdr1a^{(-/-)}$ and $Mdr1a/1b^{(-/-)}$ mice displayed increased sensitivity to the centrally neurotoxic anthelmintic ivermectin and other known Pgp substrates, such as vinblastin, digoxin, and cyclosporin A. Compared with wild-type mice, $Mdr1a^{(-/-)}$ and $Mdr1a/1b^{(-/-)}$ knockout mice also presented higher concentrations of drugs in many tissues (especially in the brain) and a reduction in the rate of drug elimination. Other in vitro and in vivo studies documented the effect of Pgp expression on the apparent oral bioavailability of substrate drugs. Hunter et al. reported the apical efflux of vinblastin across Caco-2 cells layers, and the efflux was inhibited in the presence of the Pgp inhibitor verapamil.[59] In in vivo experiments, the apparent bioavailability of the Pgp substrate paclitaxel after oral administration was 11% in $Mdr1a^{(+/+)}$ and 35% in $Mdr1a^{(-/-)}$ mice.[60] Bardelmeijer et al.[61] reported an apparent bioavailability after oral administration of docetaxel, another Pgp substrate, of 3.6% in wild-type mice and 22.7% in $Mdr1a/1b^{(-/-)}$ knockout mice. In other studies, oral administration of Pgp the inhibitors valspodar (PSC833), cyclosporin A, and elacridar (GF120918), followed by oral paclitaxel[62–64] or oral docetaxel,[61] resulted in a significant increase in the apparent oral bioavailability in wild-type mice treated with a Pgp inhibitor contrasted with those without (Figure 24.1*a*). These findings lead to important potential clinical implications. Drug–drug interaction between substrates and Pgp inhibitors can modify the drug's pharmacokinetics by increasing bioavailability and organ uptake, leading to more adverse drug reactions and toxicities. Possibly, coadministration of substrates for Pgp and Pgp inducing agents may lead to reduction of plasma drug levels and consequently, undertreatment.

Furthermore, the localization of Pgp in the placenta has been shown to play a key role in preventing fetal exposure to various potentially harmful or therapeutic compounds. Fetal exposure to an isomer of the pesticide ivermectin, L-652,280, a well-known Pgp substrate reported to cause cleft palate in mice, was significantly higher in *Mdr1a* knockout fetuses than in wild-type fetuses. Accordingly, the incidence of cleft palate was 100% in $Mdr1a^{(-/-)}$ litters, 30% in $Mdr1a^{(+/-)}$ litters, and 0% in $Mdr1a^{(+/+)}$ litters.[65] Moreover, Pgp has been shown to reduce fetal penetration of Pgp substrate drugs: Administration of digoxin, saquinavir, or paclitaxel (well-known Pgp substrates and clinically used drugs) to heterozygous $Mdr1a^{(+/-)}/1b^{(+/-)}$ pregnant

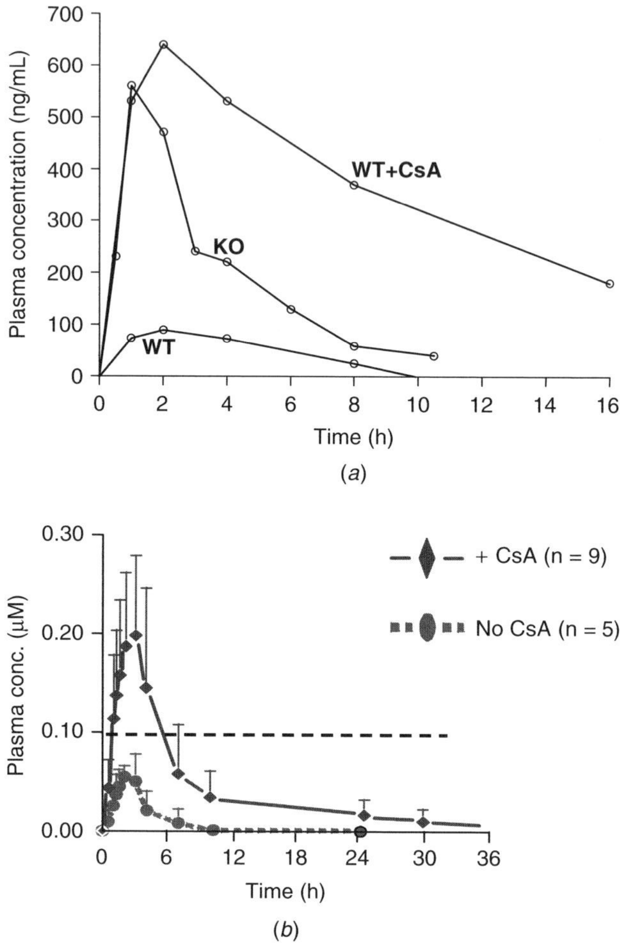

FIGURE 24.1. (*a*) Coadministration of oral paclitaxel and cyclosporin A (CsA) in wild-type (WT) mice resulted in a significant increase in the area under the curve (AUC) of paclitaxel in plasma. The effect was even greater than the AUC of paclitaxel when given to Pgp-deficient mdr1a/b knockout (KO) mice. The results indicate that Pgp effectively prevents oral uptake of paclitaxel from the gut. CsA may also have inhibited CYP3A to explain the additional increase in the AUC compared with the experiment in KO mice. (*b*) Coadministration of oral paclitaxel and cyclosporin A in patients with advanced cancer resulted in a significant increase in the AUC of paclitaxel in plasma. These results are the clinical proof of the concept that inhibition of Pgp (and possibly also CYP3A in the gut epithelium) results in a significant increase in the uptake of paclitaxel from the gut, leading to a significant increase in the systemic exposure to paclitaxel. [(*a*) From ref. 60, with permission; (*b*) from ref. 162, with permission.]

mice resulted in a significant increase in *Mdr1a*$^{(-/-)}$/*1b*$^{(-/-)}$ fetal drug concentrations compared with those in *Mdr1a*$^{(+/+)}$/*1b*$^{(+/+)}$ fetuses. Oral coadministration of a Pgp inhibitor (GF120918 or PSC833) to pregnant *Mdr1a*$^{(+/-)}$/*1b*$^{(+/-)}$ mice led to an increase in fetal drug distribution similar to that reported in *Mdr1a*$^{(-/-)}$/*1b*$^{(-/-)}$ fetuses.[66] In humans, studies with in vitro perfused term placenta showed that the fetal-to-maternal clearance index of the Pgp substrates indinavir and vinblastin was two- to threefold higher than the maternal-to-fetal clearance.[67] Moreover, treatment with the Pgp inhibitors PSC833 or GF120918 significantly increased the maternal-to-fetal clearance index of the Pgp substrate saquinavir,[68] suggesting that inhibition of Pgp activity in the placenta can affect the distribution and consequently, the fetal toxicity and/or efficacy of Pgp substrate drugs.

Drug–Drug Interactions In the literature, several drug–drug interactions mediated by Pgp have been described (see Table 24.1). In general, study of the involvement of Pgp in drug–drug interactions is difficult to prove in humans because due to the overlapping substrate specificity of inhibitors and inducers between CYP3A4 and Pgp, many drug interactions may involve both CYP3A4 enzymes and Pgp.[69] Moreover, Pgp and CYP3A4 may be linked functionally, and several potential mechanisms by which the functions of Pgp and CYP3A4 could be complementary have been proposed.[70] Furthermore, drug–drug interactions may involve additional ATP-binding cassette transporters as well.

Clinical drug–drug interactions were reported in the literature between digoxin and other drugs, such as quinidine,[71–73] verapamil,[74] propafenone,[75,76] talinolol,[77] clarithromycin,[78] itraconazole,[79] and erythromycin.[80] Coadministration of quinidine and digoxin increased digoxin plasma concentrations, resulting in clinical toxicity. One possible explanation proposed for this drug–drug interaction is the inhibition of Pgp by quinidine with two main mechanisms: reduction of the renal secretion of digoxin by blocking Pgp activity in the renal tubule, and direct inhibition (mediated by quinidine) of the intestinal elimination of digoxin. In vitro and in vivo experiments have shown that digoxin is a Pgp substrate[56] with only a minimal contribution of metabolism to its disposition.[81] In vitro experiments demonstrated that the Pgp-mediated transport of digoxin can be inhibited by low concentrations of quinidine,[82] and a decreased renal excretion of digoxin by quinidine and verapamil has been demonstrated in in vitro studies using isolated perfused rat and dog kidney.[83,84] In vivo experiments performed in wild-type and *Mdr1a* $^{(-/-)}$ knockout mice reported that coadministration of quinidine with digoxin in wild-type mice increased plasma digoxin levels by more 70%, whereas there was no effect on plasma digoxin levels for the combination in knockout mice. Furthermore, other in vivo experiments, in rats revealed a possible interaction between quinidine and the intestinal absorption of digoxin.[85,86] Finally, Drescher et al.[87] reported in humans that the direct elimination of digoxin into the intestinal lumen can be inhibited by intraluminal quinidine. In in vitro experiments, verapamil was found to be an efficient Pgp inhibitor[40] and has been used in vivo in several animal and human studies as a Pgp modulator. Similar mechanisms involved in the digoxin–verapamil interaction can be postulated for the interaction described between digoxin and propafenone, clarithromycin, or

TABLE 24.1. Examples of the Possible Involvement of MDR1 in Clinical Drug–Drug Interactions

Drug	Inhibitor/ Inducer	Measured Effect/Toxicity	Putative Mechanism	Refs.
Digoxin	Quinidine	Increased plasma levels, decreased renal clearance	Inhibition of MDR1	71,72
	Verapamil	Increased plasma levels, decreased renal clearance	Inhibition of MDR1	74,89
	Talinolol	Increased plasma levels and AUC, decreased renal clearance	Inhibition of MDR1	77
	Clarithromycin	Increased plasma levels, decreased renal clearance	Inhibition of MDR1	78
	Erithromycin	Increased plasma levels, decreased renal clearance	Inhibition of MDR1	80,274
	Itraconazole	Increased plasma levels, decreased renal clearance	Inhibition of MDR1	79,275
	Statins (e.g., atorvastatin)	Increased plasma levels, decreased renal clearance	Inhibition of MDR1	277
	Ritonavir	Increased plasma AUC and terminal half-life, toxicity effects	Inhibition of MDR1	90,103
Paclitaxel	Cyclosporin A	Increased apparent bioavailability	Inhibition of MDR1, CYP3A	162
	Elacridar	Increased bioavailability	Inhibition of MDR1, CYP3A	278
Docetaxel	Cyclosporin A	Increased bioavailability	Inhibition of MDR1, CYP3A	163
Saquinavir	Ritonavir	Increased apparent oral bioavailability	Inhibition of MDR1, CYP3A	92,95
Tacrolimus	Verapamil	Increased plasma levels and toxicity	Inhibition of MDR1, CYP3A	107
	Ketoconazole	Increased apparent oral bioavailability	Inhibition of MDR1, CYP3A	279
Talinolol	Verapamil	Increased plasma levels[a]	Inhibition of MDR1	114,115
	Erythromycin	Increased AUC	Inhibition of MDR1	112
Cyclosporin A	Erythromicin	Increased plasma AUC and peak plasma concentrations	Inhibition of MDR1, CYP3A	280,281

TABLE 24.1. (*Continued*)

Drug	Inhibitor/ Inducer	Measured Effect/Toxicity	Putative Mechanism	Refs.
Loperamide	Quinidine	Increased CNS adverse effects	Inhibition of MDR1	181
Digoxin	Rifampin	Decreased plasma levels and AUC	Induction of MDR1, CYP3A	118
Talinolol	Rifampin	Decreased AUC	Induction of MDR1	119
Tacrolimus	Rifampin	Decreased apparent oral bioavailability, decreased total clearance	Induction of MDR1, CYP3A	282
Fexofenadine	Rifampin	Decreased AUC, peak plasma concentrations, increased oral clearance	Induction of MDR1, CYP3A	120
Cyclosporin A	Rifampin	Decreased oral bioavailability	Induction of MDR1, CYP3A	121
Digoxin	St John's Worth	Decreased AUC and peak plasma concentrations	Induction of MDR1	125,283
Cyclosporin A	St John's Worth	Decreased plasma levels	Induction of MDR1	123
Indinavir	St John's Worth	Decreased plasma levels	Induction of MDR1, CYP3A	124
Tacrolimus	St John's Worth	Decreased plasma levels	Induction of MDR1, CYP3A	131
Topotecan	Elacridar (GF120918)	Increased apparent oral availability	Inhibition of BCRP, MDR1	164
Methotrexate	Omeprazole/ Pantoprazole	Increased AUC, decreased clearance	Inhibition of BCRP, MDR1	45,111

[a]The supposed molecular mechanism of this drug–drug interaction came from preclinical findings; in a clinical study, coadministration of talinolol and verapamil resulted in a decrease in talinolol oral AUC (area under the plasma concentration–time curve) compared with placebo[112] (see the text).

erythromycin, as in vitro experiments reported that all these drugs could reduce the renal secretion of digoxin by inhibiting Pgp activity in the renal tubule.[88,89]

Recently, Phillips et al. reported a case of a clinical drug–drug interaction between the HIV protease inhibitor ritonavir and digoxin in a 61-year-old woman undergoing antiretroviral triple therapy combined with digoxin, in whom digoxin toxicity developed 3 days after coadministration of ritonavir.[90] Although it has been demonstrated that ritonavir is a potent inhibitor of the metabolizing CYP3A4 and other cytochrome P450 isoforms[91–94] and that the reported ritonavir–drug interaction is likely to occur mainly at the metabolic level,[93–96] this mechanism cannot explain this drug–drug interaction, as the contribution of metabolism to digoxin elimination is negligible in humans.[81] On the other hand, it has been demonstrated that digoxin is a good Pgp substrate[83,84] that lacks Pgp inhibitor properties in vitro, and several studies reported that ritonavir (like other HIV protease inhibitors) is a substrate[97–100] and an inhibitor of Pgp in vitro,[97,99,101] including Pgp in renal tubules.[102] Furthermore, in a clinical study in 12 healthy volunteers, repeated oral administration of ritonavir increased the area under the plasma concentration–time curve (AUC) of digoxin by 86% and its volume of distribution by 77%. Ritonavir decreased the nonrenal and renal digoxin clearance by 48 and 35%, respectively, and resulted in an increase in digoxin terminal half-life in plasma by 156%.[103] Although the additional inhibition of other digoxin-transporting systems by ritonavir cannot be excluded completely, these findings support the hypothesis that therapeutic doses of ritonavir also inhibit drug transport in humans in addition to the well-known and marked inhibitory effects on drug metabolism.

A clinical drug–drug interaction has also been described between ritonavir and another HIV protease inhibitor, saquinavir. When saquinavir was coadministered with ritonavir, its oral bioavailability increased dramatically in both animals and humans.[92,93,95,104,105] In vitro, animal and clinical experiments have shown that the main mechanism involved in this interaction is probably ritonavir-mediated inhibition of CYP3A-mediated drug metabolism.[95,96,101,106] However, a marginal role of Pgp modulation cannot be excluded completely, because as reported above, saquinavir and ritonavir are both substrates and inhibitors of Pgp and CYP3A, and Pgp may be linked functionally.[51,70]

Drug–drug interaction is also documented between the immunosuppressive tacrolimus and verapamil. Hebert and Lam described a patient with dramatically increased tacrolimus plasma levels and toxicity symptoms during coadministration of verapamil.[107] Since tacrolimus and verapamil are well-known substrates and/or inhibitors for Pgp and CYP3A, a contribution of Pgp modulation could be postulated in this interaction. In addition, in vitro studies showed Pgp-mediated efflux of tacrolimus from intestinal epithelial cells (Caco-2 cells),[108] and in vivo studies demonstrated a contribution of Pgp to oral bioavailability of tacrolimus in rats.[109] Recently, clinical reports regarding the possible involvement of proton pump inhibitors in drug–drug interactions have been reported. Sipe et al. described a serious case of rhabdomyolysis causing atrioventricular block in a patient taking atorvastatin, esomeprazole, and clarithromycin.[110] In patients receiving high-dose methotrexate therapy, concurrent administration of benzimidazoles was associated with a significant decrease in clearance and significantly higher plasma concentrations of methotrexate.[111] Although proton pump inhibitors are known to interact with drug-metabolizing enzymes, since a recent in vitro study has demonstrated that omeprazole, lansoprazole, and pantoprazole are substrates and inhibitors of Pgp and are able to down-regulate digoxin efflux,[44] the authors speculated that a possible contribution to this reaction could be Pgp inhibition by esomeprazole altering atorvastatin's normal significant first-pass clearance.

Drug–drug interaction has also been reported between the β_1-adrenergic antagonist talinolol and erythromycin or verapamil in humans. An increase in oral bioavailability of talinolol after concomitant administration of erythromycin is observed in humans.[112] In vitro studies showed that talinolol, a drug that is eliminated primarily via the urine without any significant systemic metabolism, is a Pgp substrate actively transported across intestinal Caco-2 cells. Verapamil was able to block this transport.[113] In rats, coadministration of talinolol and verapamil resulted in increased oral AUC of talinolol.[114] In vivo experiments in humans demonstrated that talinolol is actively secreted into the small intestine, and the secretion is reduced by the intraluminal supply of the Pgp inhibitor verapamil.[115] These findings were not confirmed in a clinical study: Coadministration of talinolol with verapamil decreased the oral bioavailability of talinolol without affecting the renal clearance of the drug in healthy volunteers. However, the authors attributed this unexpected result to intestinal metabolism. Therefore, the relatively low dose of verapamil that was administered may have been insufficient to inhibit Pgp significantly.[113,116,117]

Other clinically relevant drug interactions described in the literature involve the antimicrobial drug rifampicin, a well-known inducer of intestinal CYP3A4. However, recent findings indicated that it can also induce Pgp expression. In a clinical study the oral bioavailability of digoxin in eight healthy volunteers was decreased by 30% during rifampicin therapy.[118] Intestinal biopsies obtained from the same patients before and after administration of rifampicin showed a significant increase in intestinal Pgp expression after administration of the antimicrobial drug, which correlated inversely with the oral AUC of digoxin. In addition, pretreatment with rifampicin had little effect on the AUC and renal clearance of digoxin. These results suggest that the digoxin-rifampicin interaction occurs primarily at the intestinal level. Chronic exposure to rifampicin can thus result in Pgp induction.[118] Similar interactions with rifampicin have been reported for talinolol,[119] fexofenadine,[120] and cyclosporin A.[121] Administration of rifampicin and talinolol to eight healthy male volunteers resulted in a significant reduction in the AUC after intravenous and oral talinolol and substantially increased expression of duodenal Pgp.[119] Hamman et al.[120] reported reduction of peak plasma concentration and increased oral clearance of fexofenadine in 20 healthy volunteers after oral coadministration of rifampicin. The investigators concluded that this interaction has been caused by induction of intestinal Pgp. However, in view of the recent findings indicating that fexofenadine is also an inducer of CYP3A4 and that organic anion–transporting polypeptides (OATPs) are involved in the hepatic uptake of fexofenadine,[122] it is likely that the interaction is due to a combination of factors. Hebert et al. reported a reduction in the oral bioavailability of cyclosporin A from 27% to 10% after oral administration of rifampicin to six healthy volunteers.[121] Since cyclosporin A is a substrate for both CYP3A4 and Pgp, and since rifampicin is an inducer of both CYP3A4 and Pgp, the interaction is probably due to a combination of CYP3A4 and Pgp induction.

Finally, clinically relevant drug–drug interactions have been reported between the over-the-counter antidepressant herbal St John's wort (SJW) and a wide range of drugs. Chronic administration of SJW together with cyclosporin A has been associated with a significant reduction of cyclosporin plasma levels and increased risk of acute organ rejection in transplanted patients.[123] In healthy volunteers, administration of SJW together with the HIV1 protease inhibitor indinavir produced an approximately 57% reduction in the plasma AUC of indinavir.[124] Coadministration of SJW with digoxin produced a 18% reduction in the plasma AUC of digoxin and a 40% increase in the expression of intestinal Pgp.[125] Other clinical studies confirmed that coadministration of SJW significantly reduced plasma concentrations of drugs such as oral contraceptives, cyclosporin A, tacrolimus, warfarin, omeprazole, verapamil, fexofenadine, and some others, leading to important clinical implications (i.e., undertreatment and failure of therapies). Induction of CYP3A4 and enhanced Pgp expression have been demonstrated to be responsible for these drug–drug interactions [for reviews, see refs. 126 to 136].

Interactions mediated by Pgp that have clinically relevant consequences have also been reported for some excipients used in pharmaceutical formulations. In in vitro experiments, Tween 80 was able to inhibit Pgp activity and to increase daunorubicin intracellular levels in cell cultures.[137] Cremophor EL and Tween 80 (substances used in drug formulations to dissolve some lipophylic and/or poorly soluble drugs) were reported to increase the oral absorption of saquinavir and digoxin, respectively, through interaction with Pgp activity.[138,139] In addition, food and dietary constituents are possible Pgp modulators. A recent report described the interaction between fexofenadine and grapefruit, orange, and apple juice,[140] although these complex interactions are supported by rather contradictory results from in vivo and in vitro studies. For example, grapefruit juice has been reported to enhance intestinal absorption and to increase the plasma levels of the Pgp substrate talinolol in rats,[141] whereas decreased oral bioavailability of talinolol was observed in humans after ingestion of grapefruit juice.[142] Species differences, multiple mechanisms, and other transporters may be the cause of this apparent discrepancy; indeed, grapefruit juice is reported to be an enzymatic inhibitor of CYP3A, a moderate inhibitor of Pgp, and a potent inhibitor of OATP drug-uptake transporters.[122,141,143–145]

Possible Clinical Benefit of Drug–Drug Interactions On the other hand, the study of drug–drug interactions with Pgp modulators is an interesting research field, as Pgp was discovered and described for its ability to confer the multidrug (MDR) phenotype to cancer cells. The modulation of Pgp activity was at first seen as a useful strategy for increasing the penetration and retention of anticancer drugs in resistant tumor cells, thus overcoming the intrinsic or acquired resistance against chemotherapy that occurs in several cancers. Since the mid-1980s, various clinical trials with anticancer drugs in combination with Pgp modulators (calcium channel blockers—nifedipine or verapamil—or cyclosporin A) have been performed.[146–148] Unfortunately, with only a few exceptions,[149–153] these studies did not show any survival benefit for the combination of anticancer drug plus Pgp inhibitor.[154–158] In addition, because the Pgp inhibitors used in those trials presented overlap in substrate specificity with CYP3A4 inhibitors, pharmacokinetic interactions occurred, resulting in increased toxicity. To date, some clinical trials using second- and third-generation Pgp inhibitors with the aim to reverse MDR in tumor cells have been performed, and others are ongoing.[159–161]

Modulation of Pgp activity with selective inhibitors could also be a useful strategy to increase the oral bioavailability of Pgp substrate drugs, in particular to develop oral formulations of anticancer drugs transported by Pgp. Several in vivo studies in animals (see above) and clinical trials in humans have been performed to evaluate the feasibility and the safety of this approach (coadministration of a substrate drug and a Pgp inhibitor). In a clinical study, cyclosporin A, an effective Pgp blocker, followed by oral paclitaxel (a well-known Pgp substrate), increased systemic exposure to paclitaxel eightfold[162] (Figure 24.1*b*). Cyclosporin A also effectively increased the oral bioavailability of docetaxel from 8% to 91%.[163] Elacridar, an effective inhibitor of BCRP as well as of Pgp, increased the oral bioavailability of topotecan from 40% to 97%.[164] These oral strategies turned out to be safe and clinically active.[164–166]

Modulation of Pgp activity to improve the oral bioavailability of drug substrates is actually a field of research for HIV protease inhibitors (HPIs), as it has been demonstrated that almost all of them (in particular, indinavir, saquinavir, nelfinavir, and ritonavir) are good substrates of Pgp.[97–99,167] Inhibition of intestinal and possibly also hepatobiliary Pgp would increase HPI oral bioavailability; retard HPI elimination, avoiding undertreatment and development of HPI-resistant HIV variants; and reduce the necessary frequency of drug dosing. This may result in increased compliance of patients to the therapy.[167] The inhibition of Pgp in subclasses of lymphocytes might increase the intracellular HPI concentration, thereby enhancing the drug's effectiveness.[167] In addition, blockade of Pgp in the blood–brain, blood–testis, and blood–nerve barriers would increase HPI penetration and retention in the putative pharmacological sanctuaries that are thought to harbor reservoirs of poorly tractable HIV copies that can reinvade the circulation, thus avoiding eradication of the HIV. In in vivo experiments, the concentrations of indinavir, nelfinavir, and saquinavir in the brain were increased significantly in Pgp knockout mice compared with wild-type mice.[98,168] Edwards et al. reported a significantly increased accumulation of amprenavir in brain tissue after coadministration of the Pgp/BCRP inhibitor GF120918 (elacridar).[169]

Indeed, an interesting clinical application of selective modulation of Pgp activity might lead to an increase in the passage of certain drugs across the blood–brain barrier, which might considerably extend the range of drugs available for treatment of brain disorders.[170] These include primary and metastatic tumors, microbial infections, HIV infections, mood disorders, and neurological treatment-resistant diseases (e.g., refractory epilepsy and schizophrenia). Furthermore, preclinical studies have shown that the brain penetration of anticancer drugs which are transported by Pgp, such as paclitaxel, docetaxel, and imatinib, can be improved by concomitant use of Pgp inhibitors, such as cyclosporin A, valspodar, elacridar, and zosuquidar.[171–175] A clinical study determining the brain penetration of paclitaxel in combination with elacridar in patients with primary brain tumors is ongoing, and the preliminary results are reported to be promising.[176] Similarly, clinical trials are exploring the activity of imatinib (Gleevec) against central nervous system (CNS) tumor glioblastoma,[177] based on promising preclinical results. However, taking into account that imatinib is a good Pgp and BCRP substrate drug with limited distribution to the brain,[178,179] and that preclinical studies reported that combination of imatinib with an effective

Pgp inhibitor resulted in improved CNS accumulation,[174,180] modulation of Pgp as well as BCRP activity can be a useful strategy to improve CNS penetration of imatinib.[176,180]

However, the safety of this approach should be explored carefully, as modulation of Pgp in the blood–brain barrier may lead to increased CNS accumulation of unwanted potentially toxic xenobiotics and endogenous compounds. Preclinical studies in wild-type mice and *Mdr1a/b* knockout mice demonstrated that *Mdr1a/1b* knockout mice are fertile and viable, but they are more sensitive than wild-type mice to a range of drugs and toxins.[55,57,58] Moreover, absence or inhibition of Pgp activity can alter the specific pharmacodynamic activity of some Pgp substrate drugs, leading to CNS toxicity and adverse drug effects. For instance, the safe clinical use of the antidiarrheal drug loperamide may also be critically dependent on the presence of Pgp in the blood–brain barrier. Loperamide is a potent opiate which demonstrates nearly exclusively peripheral opiatelike effects, on the gastrointestinal tract and no central effects, because it is a Pgp substrate. Thus, normally, it cannot accumulate in the CNS. In *Mdr1a* knockout mice, however, loperamide showed pronounced opiatelike effects and sometimes lethal effects at doses that are safe in wild-type mice.[57] In humans, coadministration of loperamide with the Pgp inhibitor quinidine produced opiate-induced respiratory depression, a clear central opiate effect that is normally not seen in humans.[181] Along the same line, blocking of placental Pgp in HIV-infected pregnant women might be used to enhance HPI levels in the unborn child shortly before and during delivery, thereby reducing the risk of HIV infection of the fetus. However, the safety of this approach needs to be studied in greater detail. For instance, preclinical data using *Mdr1a/1b* knockout mice demonstrated significantly increased fetal penetration of the HPIs indinavir and saquinavir, indicating that placental Pgp might have a protective role for the fetus.[66,167]

24.2.2. ABCG2 (BCRP)

Impact of Polymorphism on Function Currently, more than 80 naturally occurring SNPs in the BCRP gene have been identified with different ethnic distribution.[182–186] Few commonly occurring SNPs with presumed clinical consequences have been studied: the G34A (V12M, replacing valine by methionine), C376T (with a stop codon for glutamine 126), and C421A (Q141K, a transversion changing glutamine to lysine at codon 141).[183] Recently, 19 ABCG2 SNPs were found in a Dutch population, of which seven were unknown previously.[186] The Q141K polymorphism located in exon 5 showed an allele frequency ranging between 1 and 35%, with the highest allele frequencies in Japanese and Chinese populations and rare detection in African and African-American subjects.[185,187] In vitro functional characterization and in vivo protein expression in human tissue samples of certain ABCG2 allelic variants (e.g., V12M, Q141K, C34A) have been performed, but contradicting results were reported.[183,185,188–190] However, a recent clinical study has documented a correlation between Q141K polymorphism and significant pharmacokinetics changes of the BCRP substrate diflomotecan when administered intravenously, suggesting that

ABCG2 SNPs may affect the expression of BCRP protein (in this case reducing its functional expression), thus changing the pharmacokinetics of BCRP substrate drugs.[191] In another pilot study, increased oral bioavailability of topotecan was observed in two patients heterozygous for the nonsynonymous variant C421A (which displays a significantly reduced transport of topotecan in vitro), compared to patients with the wild-type allele.[192] On this basis, Özvegy-Laczka et al. hypothesized that the high frequency of the allele C421A in the Japanese population (35%) might be one explanation for the unexpectedly high toxicity of the BCRP substrate tyrosine kinase inhibitor gefitinib observed in the Japanese population.[193–195] Similarly, it was proposed that ABCG2 SNPs could explain, at least in part, the large interindividual variability observed in the pharmacokinetics of BCRP substrate drugs, in particular irinotecan and its metabolites SN-38 and SN-38 glucuronide and topotecan.[164,196,198,199] Although in a clinical study performed in cancer patients the pharmacokinetics of irinotecan and SN-38 were not significantly different in wild-type patients and in those carrying at least one ABCG2 variant allele (K141), one of the two homozygous subjects showed marked accumulation of SN-38 and SN-38 glucuronide and severe toxicity.[187] In another recent study performed in Asian patients, genetic, polymorphisms in ABCG2, as well as in ABCB1, has been reported to potentially affect the pharmacokinetics of irinotecan and its metabolites.[8] In addition, the BCRP SNP C421A has been shown to influence the disposition of 9-aminocamptothecin in patients with advanced solid tumors.[200]

A number of mutants of wild-type BCRP (detected in normal tissues) found in some drug-selected resistant cell lines were shown to affect the substrate specificity of BCRP. In particular, the amino acid at position 482 is reported to be an important determinant of substrate recognition by BCRP.[201–205] In contrast to wild-type BCRP (with an Arg at position 482), mutants with a threonine or glycine (R482T or R482G) were reported to transport daunorubicin and rhodamine 123 but not methotrexate.[201,206] Recently, several BCRP mutants with different drug resistance profiles have been generated in vitro and characterized functionally.[207] However, mutants with amino acid changes at position 482 found in some drug-selected resistant cell lines have so far not been identified in normal populations or in DNA samples of cancer patients.

Main Substrate Classes (Clinically Applied) Functional characterization studies have demonstrated that BCRP can transport a wide range of substrates, ranging from chemotherapeutic agents to conjugated and unconjugated organic anions and chemical toxicants.[208] Thus, BCRP displays a broad spectrum of substrate specificity that is overlapping, but distinct from that of Pgp and multidrug resistance protein 1 (MRP1). Clinical and preclinical findings suggest that mitoxantrone, epipodophyllotoxins (etoposide), and several camptothecin derivatives (e.g., irinotecan and its active metabolite SN-38, topotecan, homocamptothecins) are substrates of BCRP. In particular, camptothecin analogs with high polarity are transported by BCRP.[199,208–210] Flavopiridol, a new cyclin-dependent kinase inhibitor currently in clinical trials, appears to be a BCRP substrate.[202] Methotrexate and methotrexate polyglutamates are reported to be substrates of wild-type BCRP only.[45,203,204,206] In contrast, anthracyclines (e.g., doxorubicin, daunorubicin, epirubicin) are transported by the BCRP mutants R482G and R482T only.[201,207]

BCRP also actively transports sulfated conjugates (e.g., estrone 3-sulfate, dehydroepiandosterone)[211,212] and, with less affinity, GSH and glucuronide conjugates.[203] Other BCRP substrates include some chemical toxicants and carcinogenic substances [e.g., pheophorbide a and 2-amino-1-methyl-6-phenylimidazol[4,5-*b*]pyridine (PhIP)],[213–215] the tyrosine kinase inhibitors CI1033 (canertinib)[216] and STI571 (imatinib or Gleevec),[180] phosphatidylserine,[217] proton pump inhibitors (pantoprazole, omeprazole, lansoprazole),[45] the antibiotic nitrofurantoin,[218] phytoestrogens and flavonoids,[219] prazosin,[220] several fluoroquinolone antibiotics,[221] the HMG-CoA reductase inhibitor pitavastatin (and possibly other statins),[222] and the HIV nucleoside reverse transcriptase inhibitors lamivudine and zidovudine.[223–225] In addition, fluorescent compounds, such as BODIPY-prazosin and Hoechst 33342 are BCRP substrates.[220,226,227] However, rhodamine 123 and Lyso-Tracker Green appear to be transported by mutants R482G and R482T only.[202]

Inhibitors (Competitive and Noncompetitive) Recently, a large number of BCRP inhibitors with diverse chemical structures have been described, of which some show overlap in substrate selectivity with other ABC transporters, whereas others are substrates of BCRP, thus inhibiting BCRP in a competitive manner. The acridone carboxamide derivative GF120918 (elacridar) is an effective Pgp and BCRP inhibitor.[43,164,228,229] The pipecolinate derivative VX-710 (biricodar) is reported to be a BCRP as well as a Pgp and a MRP1 inhibitor.[230] The HIV protease inhibitors ritonavir, saquinavir, and nelfinavir are effective inhibitors (but not substrates) of BCRP,[231] whereas ritonavir is also described to be a weak Pgp inhibitor.[97,99,101,103] Recently, proton pump inhibitors (omeprazole, pantoprazole, and lansoprazole) have been reported to be Pgp and BCRP substrates and inhibitors.[44,45] Other BCRP inhibitors are the natural product fumitremorgin C and its derivatives Ko132, Ko134, and K143, which appear to reverse BCRP, but not Pgp- or MRP1-mediated drug resistance.[232–234] The coumermycin antibiotic novobiocin,[210,235] tryptostatin A,[236] reserpine, tamoxifen, estrone, and several food dietary flavonoids (in particular, biochanin A and chrysin)[219] have been shown to block BCRP activity. Recently, the tyrosine kinase inhibitors CI1033 (canertinib), STI571 (imatinib), EKI-785, and ZD1839 (gefitinib, Iressa) were reported to be potent BCRP inhibitors in vitro and in vivo.[216,237–240] Finally, in in vitro studies the immunosuppressants cyclosporin A, tacrolimus, and sirolimus were revealed to be effective inhibitors (but not substrates) of BCRP at clinically relevant concentrations.[241]

Pharmacological and Toxicological Function The strategic localization of BCRP in placenta, small intestine, colon, liver, blood–brain barrier, and to a lesser extent in human kidney suggests that BCRP, as well as other ABC transporters, play a crucial protective role for the fetus, for the body as a whole, and especially for the brain. BCRP limits the oral absorption of substrates and increases biliary elimination of substrate xenobiotics, thus affecting the pharmacological behavior of these compounds. In preclinical studies, the oral bioavailability of topotecan, a much better

BCRP than Pgp substrate, was increased dramatically by oral coadministration of the BCRP/Pgp inhibitor elacridar in both wild-type and Pgp knockout mice. Treatment with elacridar only marginally decreased the plasma clearance and hepatobiliary excretion of topotecan.[229] In a recent clinical study, coadministration of elacridar increased the apparent oral bioavailability of topotecan in cancer patients significantly, from 40% to 97%. Elacridar reduced the plasma clearance of topotecan administered intravenously by only 10%, indicating that increased systemic exposure after oral administration of topotecan when given in combination with elacridar is due almost entirely to increased uptake of topotecan from the gut.[164] Polli et al. also described an important role of BCRP in the bioavailability and disposition of GV196771, a potent antagonist of the NMDA receptor developed for the treatment of neuropathic pain and reported to be a good BCRP and poor Pgp substrate. Coadministration of elacridar in Pgp knockout mice resulted in a significantly increased AUC of GV196771.[242] Moreover, in in vivo pharmacokinetic studies using wild-type and *Bcrp1* knockout mice, the plasma concentration of ciprofloxacin (a commonly used fluoroquinolone antibiotic) after oral administration was increased more than twofold in *Bcrp1*$^{(-/-)}$ compared with wild-type mice.[221] In the same animal model, Bcrp1 (the homolog of human BCRP) effectively reduced the intestinal absorption and increased biliary and intestinal elimination of the Bcrp1 substrate PhIP (a dietary carcinogen), supporting the hypothesis that BCRP has a protective role in the body and can affect the pharmacokinetics of substrate drugs.[215] Recently, Cisternino et al. reported that Bcrp1 reduced brain penetration of the BCRP substrates mitoxantrone and prazosin.[243] All these findings suggest that BCRP may play an important role in regulating drug absorption and disposition. Clinically, no increased or unexpected toxicities have been seen in studies modulating the oral uptake of BCRP substrate drugs.[165]

Drug–Drug Interactions Clinically important drug–drug interactions have recently been reported between the antifolate drug methotrexate (MTX); its main metabolite, 7-OH-methotrexate (7-OH-MTX); and the benzimidazoles omeprazole and pantoprazole, resulting in long-lasting extensive myelosuppression associated with systemic infections and severe mucositis. In a case report, omeprazole was shown to inhibit MTX clearance, resulting in sustained highly toxic levels[46] (Figure 24.2*a*). In another case report, coadministration of pantoprazole resulted in a 70% increase in the serum concentration of the metabolite 7-OH-MTX.[48] In a clinical study, patients receiving benzimidazoles and high-dose MTX therapy displayed reduced MTX clearance and increased MTX plasma levels.[111] Although it was challenging to assess the contribution of BCRP in these drug–drug interactions, as several transporters are involved in MTX uptake and efflux out of cells (reduced folate carrier 1; MRP1-4[244–249]; human organic anion transporters hOAT1, hOAT3, and hOAT4[529]), several preclinical findings support the hypothesis that high-dose MTX–benzimidazoles interactions are mainly BCRP mediated. MTX is a BCRP substrate, and Volk[204,206] and Chen[203] showed that besides MTX, the active di- and triglutamylated forms of MTX could be transported by BCRP. In in vitro studies the proton pump inhibitors omeprazole and pantoprazole have been demonstrated to be BCRP substrates and effective BCRP and Pgp inhibitors.[44,45] In in vivo experiments, the AUC of MTX after intravenous

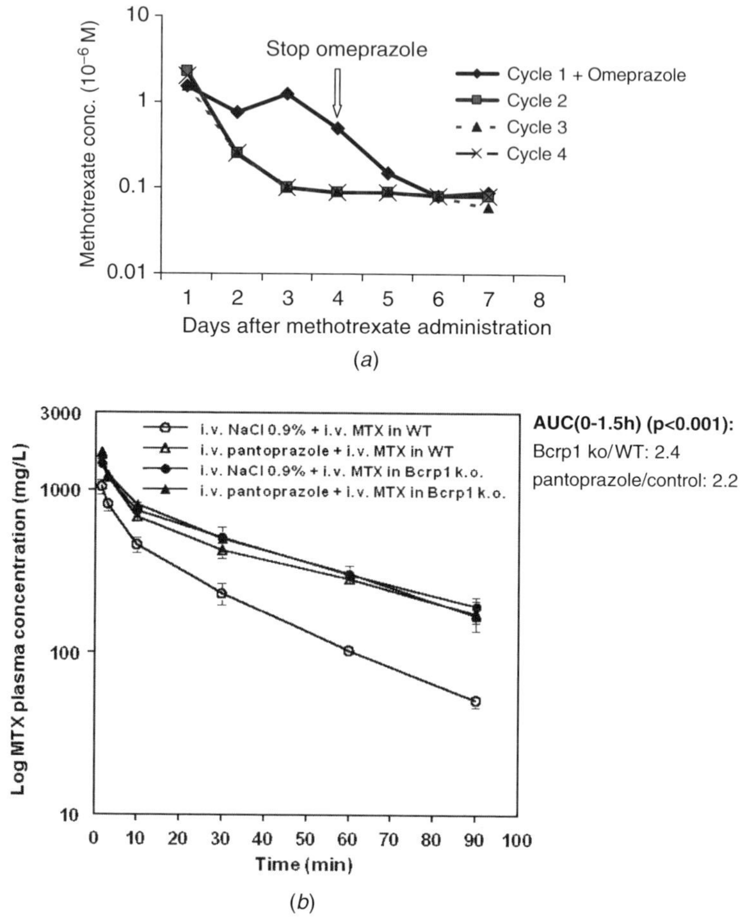

FIGURE 24.2. (*a*) Omeprazole increased the plasma exposure to high-dose methotrexate. The mechanism of this important drug–drug interaction is explained in (*b*). (*b*) In vivo studies with wild-type and Bcrp1 knockout (k.o.) mice, revealing that absence of Bcrp1 and coadministration of the BCRP and Pgp inhibitor pantoprazole effectively increased plasma exposure to high-dose methotrexate (MTX). The aim of these studies was to unravel the mechanism of the drug–drug interaction reported between high-dose methotrexate and benzimidazole proton pump inhibitors in patients [see (*a*)]. [(*a*) From ref. 46; (*b*) from ref. 45. With permission from the American Association for Cancer Research.]

administration was increased 1.8-fold, and its clearance was reduced about twofold by coadministration of pantoprazole in wild-type mice, reaching similar levels as in *Bcrp1* knockout mice, while the AUC and the clearance of MTX were not affected by pantoprazole in *Bcrp1* knockout mice[45] (Figure 24.2*b*). Together, these findings suggest that inhibition of BCRP may explain the clinical interaction between high-dose methotrexate and benzimidazoles.

Preclinical findings showed that some tyrosine kinase inhibitors (TKIs) are also BCRP substrates and/or inhibitors. In in vitro experiments the HER TKI CI1033 enhanced the cytotoxicity of SN-38 and topotecan by inhibiting BCRP-mediated drug efflux from resistant cells.[216] Furthermore, in several in vitro assay systems, Hegedus et al. and Özvegy-Laczka et al. found that imatinib mesylate, gefitinib, and EKI-785 interacted with BCRP at submicromolar concentrations, whereas the other multidrug transporters, MDR1 and MRP1, were much less sensitive to these agents. Moreover, at higher concentrations of the three TKIs reported, inhibition of BCRP, MDR1, and MRP1 was observed in each case.[237,250] In other studies, gefitinib was able to overcome BCRP-mediated drug resistance by inhibiting the pump function of BCRP in BCRP-transfected human myelogenous and murine lymphocytic leukemia cells and in human colon cancer cells expressing BCRP endogenously. Mice transplanted with lymphocytic leukemia cells BCRP-transfected and treated with a combination of irinotecan and gefitinib presented significantly longer survival than those treated with irinotecan or gefitinib alone.[239] Recently, Stewart et al. reported that the combination of gefitinib and irinotecan resulted in greater than additive antitumor activity in tumor models and that activity was independent of tumor Erb-B1 status. Furthermore, gefitinib enhanced the oral bioavailability of irinotecan and topotecan in mice.[238] This BCRP-mediated drug–drug interaction could be relevant for clinical applications, as it has been demonstrated that protracted schedules of administration increase the efficacy of several camptothecin analogs, including irinotecan and topotecan.[251–253] This can best be achieved by oral dosing. However, oral dosing of irinotecan and topotecan is associated with significant limitations because of, in part, poor oral bioavailability and large interpatient variability.[196,254] Potentially, gefitinib might enhance oral absorption, decrease interpatient variability, and at the cellular level, might reverse tumor resistance mediated by BCRP, thereby increasing the effectiveness of the anticancer therapy.

Moreover, as gefitinib was also reported to be a BCRP substrate,[193] modulation of BCRP activity by coadministration of BCRP inhibitors might also improve the brain penetration of gefitinib, thereby increasing its activity against brain metastases of non-small-cell lung cancer patients.[255] Furthermore, due to the extensive expression of both BCRP and Pgp in the blood–brain barrier and overlap between BCRP and Pgp in substrate specificity, modulation of BCRP, as well as Pgp, by BCRP and Pgp inhibitors could be a good strategy to improve brain penetration of substrate drugs. Preclinical in vitro and in vivo studies reported significant activity of imatinib mesylate against glioblastoma,[177] but limited brain penetration of imatinib has been observed in humans and nonhuman primates.[178,179,256] Preclinical in vitro and animal studies have shown that imatinib is a Pgp substrate. Pgp limits the distribution of imatinib to the brain and the Pgp inhibitors cyclosporin A and LY335979 could improve the brain penetration of imatinib in mice.[174,257] Recently, imatinib has been reported to be a good BCRP substrate and inhibitor,[180] and coadministration of the BCRP and Pgp inhibitors pantoprazole or elacridar increased the CNS concentration of imatinib significantly in mice. This might be of interest for the treatment of patients with primary or metastatic brain tumors sensitive to anticancer drugs with high affinity for Pgp and/or BCRP.[176]

Potentially clinically relevant BCRP-mediated drug–drug interactions can also be predicted between flavonoids or flavonoid-containing food and herbal products and BCRP-substrate drugs. Flavonoids are the most abundant polyphenols present in fruits, in plant-derived beverages, and to a lesser extent, in vegetables.[258] Due to their recognized protective effect on cancer, cardiovascular diseases, and osteoporosis,[259] flavonoid-containing dietary supplements and herbal products are widely used and their consumption will probably increase. Preclinical studies showed that flavonoids are Pgp and CYP3A4 inhibitors.[260,261] Recently, it has been reported that flavonoids effectively inhibit BCRP-mediated drug resistance at relatively low concentrations.[219,262–264] These concentrations could well be achieved in the gut after the ingestion of flavonoid-containing foods or dietary supplements. Moreover, when coadministered, some flavonoids were reported to inhibit BCRP.[264] Recently, the flavonoids chrysin and 7,8-benzoflavone were reported to inhibit human BCRP-mediated transport of topotecan with relatively high potency in BCRP-overexpressing cells. However, oral coadministration of topotecan with chrysin and 7,8-benzoflavone failed to alter topotecan pharmacokinetics significantly in rats or *Mdr1a/1b* knockout mice.[265] Although there is an incomplete in vitro–in vivo correlation, these findings point in the direction of possibly clinically relevant flavonoid–drug interactions.

Other relevant BCRP-mediated drug–drug interactions can be predicted between some steroid drugs and BCRP-substrate drugs. Recently, several glucocorticoid drugs, including prednisone and dexamethasone, were shown to inhibit BCRP efficiently in vitro.[266] Due to the fact that many of the BCRP substrates identified are anticancer drugs, and that prednisone and dexamethasone are frequently used in chemotherapeutic regimens in the treatment of lymphoid leukemias and solid tumors, either for their anticancer activity or for their ability to reduce adverse side effects of chemotherapy, pharmacokinetic interactions could occur. Moreover, it can be speculated that high systemic levels of glucocorticoids contribute to reverse BCRP-mediated multidrug resistance. Similarly, as several commercially available estrogen antagonists and agonists (in particular, diethylstilbestrol, tamoxifen, and tamoxifen derivatives) have been reported to reverse drug resistance effectively in BCRP-overexpressing cells,[267] clinical drug–drug interactions can be predicted in patients taking these BCRP inhibitors and BCRP-substrate drugs. Whether the modulation of BCRP activity by these drugs is clinically relevant and can be used to improve the oral bioavailability of BCRP substrate drugs and potentiation of their cytotoxic activity is still an open question.

Drug–drug interactions might also take place in patients taking highly active antiretroviral therapy (HAART) for the treatment of HIV type 1 (HIV-1) infection, as recently the HIV protease inhibitors ritonavir, saquinavir, and nelfinavir have been reported to block BCRP activity effectively in vitro, although ritonavir, saquinavir, nelfinavir, indinavir, and amprenavir were not substrates of BCRP.[231] Preclinical experiments revealed that the nucleoside reverse-transcriptase inhibitors (NRTIs) zidovudine (AZT) and lamivudine (3TC) are probably transported by BCRP,[223] and preclinical studies have shown that antiretroviral drugs used in HAART are substrates and/or inhibitors of other ABC transporters, in particular Pgp and MRP4, MRP5, and MRP8.[268–270] Further studies are warranted to evaluate if the modulation of BCRP,

Pgp, and MRP activity is a useful strategy to increase the intracellular concentrations, the oral bioavailability, and the brain penetration of HIV antiretroviral drug substrates when given in combination with potent inhibitors of these transporters. Drug–drug interaction has been described between the anticancer drug irinotecan and the immunosuppressive tacrolimus in a patient with hepatocellular carcinoma after liver transplantation. Concomitant administration of irinotecan and tacrolimus resulted in increased plasma concentrations of SN-38 and severe diarrhea.[271] The mechanism of this interaction remains unknown, but as tacrolimus has recently been reported to inhibit BCRP efficiently in vitro[241], the inhibition of BCRP may explain, at least in part, this clinically relevant feature. Other clinically relevant drug–drug interactions involving the anticancer drug irinotecan and its metabolites have been described, but as irinotecan displays an extremely complex pharmacokinetics that involves several metabolizing enzymes and membrane transporters, it is likely that various mechanisms and different transporters are involved in these interactions (see the section "MRP2").

Modulation of BCRP activity could also be a useful strategy to modify the pharmacokinetic property of anthelmintic benzimidazole drugs such as albendazole and fenbendazole. This is a class of drugs used for the treatment of intestinal luminal parasites because they are poorly absorbed from the gut.[272] As the sulfoxide derivatives of albendazole and fenbendazole can be actively transported by BCRP (but not by Pgp or MRP2) in vitro, clinical use of efficacious BCRP inhibitors might improve their oral bioavailability by reducing their intestinal elimination and hepatobiliary secretion. Therefore, this strategy, resulting in increased plasma levels and duration of systemic exposure to sulfoxide derivatives of methylcarbamate albendazoles, might improve the therapeutic efficacy of these compounds against extraintestinal (systemic) infections.[273]

24.2.3. ABCC Family (Multidrug Resistance–Associated Proteins, MRP1 to MRP9)

MRP1 (ABCC1)

Impact of Polymorphism on Function MRP1 (ABCC1) SNPs have been described in various ethnic populations, but they are not associated with any known genetic disorder or phenotype, or with altered pharmacokinetics or toxicity of substrate drugs, in particular, irinotecan.[7,284–288] A mutation of R433S in MRP1 has been identified and displayed decreased organic anion transport and increased resistance to doxorubicin.[289] In vitro functional characterization studies have shown that some MRP1 allelic variants display altered substrate specificity.[289] For this reason we speculate that some MRP1 variants could be associated with changes in drug disposition, but further studies are awaited.

Main Substrate Classes (Clinically Applied) MRP1 is primarily an organic anion transporter. MRP1 can also transport compounds that are conjugated or complexed to glutathione (GSH), sulfate, glutamate, or glucuronide.[244,248,290–296] In addition,

MRP1 is able to cotransport neutral/basic drugs with reduced GSH: In in vitro experiments it has been reported that the ATP-dependent vesicular transport of vincristine, daunorubicin, and etoposide by MRP1 required reduced GSH, and GSH-depleting agents such as buthionine sulfoximine (BSO) could sensitize MRP1-overexpressing cells to a number of anticancer agents.[292,297–303] Antimonial and arsenical oxianions are also cotransported by MRP1 with GSH[304] or complexed to GSH.[303] Thus, MRP1 transports a broad range of physiological compounds, xenobiotics, and drugs. Leukotriene C_4 (LTC_4) and its metabolites, D_4 (LTD_4) and E_4 (LTE_4), are good substrates of MRP1, as well as other prostaglandin derivatives, the cholestatic 17β-estradiol-glucuronide and the sulfated bile salt sulfatolithocholate.[290,292,295,305] A variety of chemical toxicants and their metabolites with potential carcinogenic activity are also transported by MRP1. In vitro data show that the GSH-conjugated stereoisomers of aflatoxin B1,[306] the glucuronide conjugate of some nicotine metabolites,[296,307] the GSH conjugates of the herbicide metolachlor, and the toxicants 1-Cl-2,4-dinitrobenzene and 4-nitroquinoline 1-oxide are all substrates of MRP1.[308–310] Clinically relevant substrates of MRP1 include several classes of anticancer agents, such as anthracenedione (mitoxantrone); epipodophyllotoxins; *Vinca* alkaloids; anthracyclines; camptothecins (topotecan, irinotecan, and the unconjugated and conjugated forms of its active metabolite SN-38)[311]; conjugates of alkylating agents (thiotepa,[312] cyclophosphamide, chlorambucil, and melphalan); and the antiandrogen flutamide and its metabolite, hydroxyflutamide.[313] MRP1 can also confer resistance to short-term exposure to methotrexate, a folate antimetabolite.[244,248] Recently, the HIV protease inhibitors ritonavir and saquinavir, as well as conjugates of the diuretic ethacrynic acid,[316] were found to be transported by MRP.[314,315]

Inhibitors (Competitive and Noncompetitive) Various classes of compounds with MRP-inhibiting activity have been described. Sulfinpyrazone, benzbromarone, probenecid, and indomethacin can modulate MRP1 activity, but they are nonspecific inhibitors, as these molecules modulate the activity of many transporters, and based on in vitro findings, they should be used at relatively high concentrations to inhibit MRP1 activity in vivo.[317–320] Relatively specific modulators of MRP1 are MK571 (a LTD_4 receptor antagonist),[321] ONO-1078 (a peptide leukotriene receptor antagonist),[322] glibenclamide [a sulfonylurea derivative that inhibits MRP1 as well as sulfonylurea receptor 1 (SUR1)][323] and several peptidomimetic GSH-conjugate analogs.[324] Several tricyclic isoxazole derivatives (e.g., LY475776, LY329146, LY402913) are highly specific and potent MRP1 inhibitors, as they block the MRP1-mediated LTC_4 transport in a GSH-dependent manner.[294,325–327] Some Pgp inhibitors, such as VX-710 (Biricodar), PSC 833 (valspodar), verapamil,[328] cyclosporin A, agosterol A,[329] PAK-105P,[330,331] S9788 [332] as well as several bioflavonoids (e.g., genistein, quercetin),[333,334] nonsteroidal anti-inflammatory drugs (NSAIDs),[335] steroid derivatives (e.g., RU486, budesonide), and imidathiazole derivatives have been shown to inhibit MRP1, but with low affinity and poor specificity.[296] Recently, MRP1-specific antisense oligonucleotides and cDNA, ribozymes, and small interfering RNA molecules have been developed. For instance, ISIS 7597 (a MRP1-specific antisense oligonucleotide) has been shown to down-regulate MRP1 successfully in a xenograft model of human neuroblastoma.[336]

Inducers In some cell systems MRP1 expression can be induced by pro-oxidant compounds such as quercetin, sulindac, menadione, pyrrolidinedithiocarbamate, and *tert*-butylhydroquinone.[337–339] Several compounds that generate reactive oxygen species (e.g., TNFα)[340] and nitric oxide donors[341] are also reported to induce MRP1 expression.

Pharmacological and Toxicological Function The pharmacological and toxicological functions of MRP1 have been studied using *Mrp1*$^{(-/-)}$ mice. Increased sensitivity to the MRP1 substrate etoposide after intravenous administration in *Mrp1*$^{(-/-)}$ mice has been described as well as dramatically increased vincristine and etoposide toxicity after intraperitoneal administration to *Mrp1*$^{(-/-)}$ and *Mdr1a/1b*$^{(-/-)}$ (triple knockout) mice compared with *Mdr1a/1b*$^{(-/-)}$ and wild-type mice. In particular, the *Mrp1*$^{(-/-)}$ and/or *Mdr1a/1b*$^{(-/-)}$ mice showed increased sensitivity of bone marrow precursor cells and some epithelia containing high levels of Mrp1 (e.g., oropharyngeal mucosa and collecting tubules of the kidney) to Mrp1 substrate drugs (etoposide and vincristine). They also showed increased anticancer drug–induced destruction of sperm cells in the testis and increased drug (etoposide) levels in the cerebrospinal fluid (CSF).[342–346] Recently, MRP1 has been reported to participate in the brain uptake and accumulation of substrate drugs. Indeed, although saquinavir has been described as a good Pgp substrate and it has been reported that affinity for Pgp affects its brain distribution significantly, in a preclinical study in mice the brain uptake of saquinavir increased more than fourfold in the presence of the selective MRP inhibitor MK571.[347] Other in vitro experiments confirmed that at the cellular level, the endogenous expression of MRP1 (together with Pgp) can contribute to the basal resistance of cell lines to a variety of anticancer drugs. Murine fibroblast and embryonic stem cell lines lacking the expression of *Mrp1, Mdr1a*, and *Mdr1b* genes displayed significantly increased sensitivity to anthracyclines, topotecan, SN-38, epipodophyllotoxins, and arsenite.[348,349] All these findings together suggest that MRP1 plays an important role in protection of the body and that administration of MRP1 modulators can alter the pharmacokinetics of coadministered drugs, thus leading to potentially clinically relevant drug–drug interactions.

MRP2 (cMOAT, ABCC2)

Impact of Polymorphism on Function Several mutations of MRP2 (cMOAT, ABCC2) have been described in humans and animals. In humans some mutations (missense, nonsense, splice site, and deletion mutations) in the MRP2 gene, resulting in the absence or functional inactivity of MRP2 in the bile canalicular membrane, have been reported in persons affected by the Dubin–Johnson syndrome, an autosomal recessive inheritable disorder characterized by conjugated hyperbilirubinemia and increased urinary coproporphyrin I fraction.[15,350] Although the hepatic function of patients with Dubin–Johnson syndrome is reported to be normal (probably due to a compensation by other transporters), it has been suggested that those patients, however, are at increased risk of drug-induced toxicity.[351] In rats, two differently occurring mutations in the Mrp2 gene generate the Mrp2-deficient GY/TR$^-$ or Eisai hyperbilirubinemic (EHBR) rat strains.[352–356] Recently, in a study performed in 64 Caucasian cancer patients, a functional MRP2 SNP (C3972T) able to affect irinotecan pharmacokinetics has been found. Patients carrying the 3972T allele showed higher AUC of irinotecan and its metabolite, SN-38 glucuronide. SNP analysis of MRP2 has also been performed in healthy Japanese subjects, in a Dutch population, and in several cell lines established from surgically dissected tumors from Japanese patients. Several SNPs have been identified, and some of them have been characterized functionally, but their clinical consequences and impact on drug disposition need to be clarified.[15,286,350,357,358] Interindividual differences in the expression level of MRP2 in liver and gut have also been found, and it is supposed that together with single-nucleotide polymorphism, the induction/down-regulation of this transporter can be affected by different disease states (e.g., cholestatic conditions, hepatitis) and food and/or drug intake.[352,359–362] Clinical implications of this observation are not yet known.

Main Substrate Classes (Clinically Applied) There are many similarities between compounds transported by MRP2 and MRP1, although there is not complete overlap. Anticancer drugs transported by MRP2 include anthracyclines (doxorubicin and epirubicin), camptothecin derivatives (irinotecan/SN-38), *Vinca* alkaloids (vinblastin and vincristine), mitoxantrone, cisplatin, and probably, etoposide.[244,296,363–367] Recently, it has been reported that MRP2 transports taxanes (paclitaxel and docetaxel) and that this transport is stimulated by probenecid.[368] Like MRP1 and MRP3, MRP2 is able to confer resistance after short-term (brief) exposure to high concentrations of methotrexate.[244,248,369] In addition, as MRP2 is primarily an organic anion transporter, it seems very likely that basic drugs (e.g., vinblastin) are cotransported with GSH.[363,367,370] Substrates of MRP2 also include many amphipathic anion drugs and endogenous compounds and GSH, glucuronide, and sulfate conjugates. MRP2 is the principal transporter of bilirubin mono- and bisglucuronides into the bile.[371,372] Other MRP2 endogenous substrates include LTC_4, reduced and oxidized GSH (GSSG), LTD_4, LTE_4, estradiol-17β-glucuronide, L-thyronine, and glucuronide conjugates of such drugs as diclofenac and acetaminophen.[350,373–375] Taurolithocholate sulfate and taurochenodeoxycholate sulfate are also substrates of MRP2.[376,377] Other clinically important drugs transported by MRP2 are pravastatin; temocaprilat; ampicillin; ceftriaxone; grepafloxacin and its glucuronide conjugate, BQ-123; sulfinpyrazone; the HIV protease inhibitors saquinavir, ritonavir, and indinavir; *p*-aminohippurate; and possibly, arsenic trioxide.[296,367,378–381] Furthermore, MRP2 has been shown to mediate the transport of some carcinogens, such as PhIP (2-amino-1-methyl-6-phenylimidazo[4,5-*b*]pyridine),[382,383] the glucuronide conjugate of the nicotine metabolite 4-(methylnitrosamino)-1-(3-pyridyl)-1-butanol (NNAL),[307] the fungal toxin ochratoxin A,[384] arsenite, cadmium, and α-naphthylisothiocyanate.[370,381]

Inhibitors and Inducers Inhibitors of MRP2 have been described, but they are not highly specific. MK571 inhibits MRP2 but also MRP1 and MRP3, and cyclosporin

A, PSC 833, PAK-104P, sulfinpyrazone, benzbromarone, and probenecid, in addition to Pgp, are reported to block MRP2 activity in intact cells. Many of the anionic substrates of MRP2 (e.g., LTC_4, phenolphtalein glucuronide, fluorescein, methotrexate) can act as competitive inhibitors in in vitro systems.[319,385] Interestingly, certain MRP2 modulators can inhibit MRP2-mediated transport of a number of substrates (e.g., probenecid inhibits methotrexate transport by MRP2) but stimulate transport of others (e.g., probenecid stimulates transport of taxanes, etoposide, vinblastin, and HPIs).[244,367,368,386,387]

Zelcer et al. demonstrated that a large variety of clinically used compounds could stimulate MRP2 transport.[388] Fromm et al. reported that rifampicin is able to induce intestinal MRP2 as well as Pgp in humans.[362] Regulation of MRP2 activity can also take place at the cellular level. Dynamic endocytic retrieval and exocytic insertion of MRP2 between the canalicular membrane and an intracellular pool of vesicles are involved. Treatment of rats with lipopolysaccharide (which induces cholestasis) also induces endocytic retrieval of MRP2.[389,390] Implications for the disposition of substrate drugs of MRP2 are currently unknown.

In vitro, transcriptional regulation of MRP2 has also been characterized, and various MRP2-inducing agents have been found. In primary cultures of human and/or rat hepatocytes, dexamethasone, rifampicin, tamoxifen, cisplatin, cycloheximide, phenobarbital, 2-acetaminofluorene, cholic acid, chenodeoxycholic acid, clotrimazole, pregnenolone, sodium arsenite, and oltipraz, serving here as examples, induced MRP2 expression.[350,391–400] Induction of liver Mrp2 expression has also been described in Wistar rats treated with St. John's wort.[400] Similarly, treatment of Sprague–Dawley rats with dexamethasone induced Mrp2 expression in liver and kidney.[401] However, further investigations are needed, as discrepancies have been reported between in vitro and in vivo findings. For instance, in contrast with results observed in cultured hepatocytes, phenobarbital treatment of rats did not increase Mrp2 expression.[402–404]

Pharmacological and Toxicological Function The pharmacological and toxicological functions of MRP2 have been studied extensively in in vitro and in vivo models, using rat strains defective for Mrp2 (GY/TR- and EHBR) and cells from humans suffering from the Dubin–Johnson syndrome. In these models the pharmacokinetics of MRP2 substrate compounds are altered significantly, suggesting that MRP2 can affect the pharmacological behavior of these substrate drugs. MRP2 is involved in the hepatobiliary excretion of clinically important anionic drugs and intracellularly formed glucuronide and glutathione conjugates of many drugs,[352,354,372,405] thus playing a key role in the phase III xenobiotic detoxification system, the biliar excretion of conjugated metabolites produced by phase I and II hepatic enzymes. MRP2 probably has a central role in the disposition and hepatobiliary elimination of substrate drugs. In Mrp2-deficient rats, a reduction was found in the biliary excretion of cefodizime, ceftriaxone, and grepafloxacin, antimicrobials used clinically in the treatment of infections of the biliary tract, indicating that efficient excretion of these drugs into the bile is probably mediated by MRP2.[379,406–408] Similarly, the biliary elimination of the HMG-CoA reductase inhibitor pravastatin, a known MRP2 substrate, has been reported to be threefold lower in EHBR than in normal rats.[409] Biliary excretion of methotrexate, mediated by MRP2 as well as BCRP, has been proposed to be responsible for its intestinal toxic effects.[45,410] Furthermore, secretion into the bile via MRP2 of the glucuronide derivative of the nonsteroidal anti-inflammatory drug diclofenac is supposed to cause toxic effects to the bile canalicular membrane.[376]

Moreover, MRP2, together with BCRP and Pgp, may contribute to the gastrointestinal excretion of the anticancer drug irinotecan and its metabolites SN-38, and SN-38 glucuronide. Irinotecan, SN-38, and SN-38 glucuronide are transported by BCRP and to a lesser extent by MRP2 and Pgp.[197,199,411,412] The intestinal excretion of irinotecan and its metabolites has been linked to severe unpredictable diarrhea, a frequent dose-limiting effect of irinotecan that shows wide interpatient variability.[310,413–419] Several strategies to prevent this important toxicity using efficient inhibitors of the gastrointestinal transport of irinotecan and its metabolites have been investigated. In rats, coadministration of probenecid (a MRP2 inhibitor) reduced SN-38 biliary excretion and decreased irinotecan-induced late-onset diarrhea.[378,420] In other studies, coadministration of irinotecan with cyclosporin A (a Pgp, BCRP, and MRP2 inhibitor)[240,421] resulted in significantly reduced excretion of irinotecan and SN-38 via the biliary route and/or the intestinal route in rats.[422,423] In patients affected by fluorouracil-refractory metastatic colorectal cancer, a clinical phase I study of intravenous irinotecan and oral cyclosporin A revealed an increased plasma AUC_{0-24h} of irinotecan, SN-38, and SN-38 glucuronide, and a reduction of irinotecan clearance and of diarrhea with preservation of antitumor activity.[424] Similar results were found in another study performed in a larger patient population with refractory solid tumors or lymphoma.[425]

MRP2-mediated biliary excretion of drugs as part of a multiple elimination pathway (such as biliary and urinary) has been proposed to confer pharmacokinetic advantage to substrate drugs (e.g., temocapril), thus avoiding potentially toxic effects. In patients with renal failure, the relatively stable pharmacokinetics of temocaprilat [the active metabolite of the angiotensin-converting enzyme (ACE) inhibitor temocapril], compared with other ACE inhibitors (such as enalapril, captopril, and cilazapril), is thought to be due to its dual excretion via both bile and urine,[426–429] whereas the other ACE inhibitors are eliminated primarily via renal excretion. Preclinical findings show that temocaprilat is transported by MRP2 and that MRP2 is the predominant factor in determining its biliary excretion. The biliary clearance after intravenous administration of temocaprilat was significantly lower in EHBR than in Sprague–Dawley rats. In contrast, the other ACE inhibitors do not appear to be good MRP2 substrates.[430] In addition, intestinal MRP2 limited the oral bioavailability of substrate compounds. In preclinical studies, absorption after oral administration and the tissue levels in several organs (liver, kidney, lung, and colon) of the food-derived carcinogen PhIP (a proven MRP2 substrate) were approximately twofold higher in TR^- than in Wistar rats. Furthermore, after intravenous administration, the levels of PhIP and its metabolites in bile and the direct intestinal excretion of unmetabolized PhIP were significantly reduced in TR^- compared with Wistar rats.[370,382] In vitro experiments using reverted sacs prepared from jejunum and the ussing chamber have shown an important reduction of clearance by intestinal excretion and vectorial transcellular transport of 2,4-dinitrophenyl-*S*-glutathione, a MRP2 substrate, in EHBR compared with normal rats.[430,431]

Extrapolating these findings to the known MRP2 drug substrates, it is likely that these compounds are also affected by intestinal MRP2 activity in their oral bioavailability and direct intestinal excretion. It is also possible that in some studies the overlapping transport activity of the highly expressed intestinal transporters Pgp and BCRP has masked the relevance of MRP2 in this function.[432] Recently, in preclinical studies using Wistar and TR^- rats, MRP2 has been reported to limit the oral bioavailability of the HIV protease inhibitor saquinavir [433] and of the taxanes paclitaxel and docetaxel, substrates for which both Pgp and MRP2 have affinity.[368,380,433] In addition, recent research demonstrates that MRP2 modulates the brain penetration of clinically important substrate drugs. In TR^- rats, brain extracellular levels and the anticonvulsant effect of the antiepileptic drug phenytoin were significantly increased compared with normal rats. Moreover, using in vivo microdialysis in rats, a significantly enhanced brain extracellular concentration of phenytoin has been reported with local perfusion of the anion transporter inhibitor probenecid.[434–436] A significant increase of drug penetration into the brain by probenecid was reported previously for valproate[437] and carbamazepine[438] and has been attributed to inhibition of MRP2 in the blood–brain barrier, although probenecid is not a specific inhibitor of MRP2.[439] Subsequently, a study using TR^- rats reported increased anticonvulsant response to carbamazepine in the kindling model of temporal lobe epilepsy, but no difference in brain drug levels.[435] These findings have important clinical relevance, as recently it has been hypothesized that overexpression of Pgp and MRPs may be involved in the generation of drug resistance in epileptic patients.[14,440–442] Therefore, the development of active and more specific MRPs and/or Pgp inhibitors might overcome drug resistance to antiepileptic drugs.[443] Furthermore, due to the broad MRP2 substrate specificity, its pharmacological role and the wide range of clinically used drugs that can modulate MRP2 activity, several clinically relevant drug–drug interactions mediated by MRP2 could be anticipated.

Clinically relevant drug–drug interactions have been described when rifampicin is coadministered with morphine or propafenone, resulting in loss of analgesic effect of morphine or in reduced bioavailability of propafenone, respectively.[444–446] These interactions can be attributed in part to induction of intestinal Pgp and phase I drug-metabolizing enzymes. However, rifampicin has also been demonstrated to induce the intestinal expression and activity of MRP2.[362] Therefore, induction of MRP2 expression mediated by rifampicin could contribute to the reduction in intestinal absorption, plasma concentrations, and urinary recovery of morphine and propafenone that has been observed.

A drug–drug interaction has also been described between phenobarbital and acetaminophen glucuronide: Pretreatment with phenobarbital significantly increased acetaminophen glucuronide (AP-G) formation but significantly reduced AP-G biliary excretion in rats,[447] whereas acute phenobarbital treatment decreased the biliary excretion of AP-G together with a moderate reduction in AP-G production.[448] The decreased biliary excretion of AP-G due to phenobarbital treatment can be attributed to impairment in biliary excretion or up-regulation in basolateral transport of AP-G. Although the impaired biliary excretion of AP-G after phenobarbital pretreatment may be attributed primarily to the induction of hepatic MRP3 by phenobarbital (see below), it could also be explained by inhibition of Mrp2-mediated AP-G transport by the phenobarbital metabolite *p*-hydroxyphenobarbital glucuronide (*p*-OHPBG).[374] Pharmacokinetic modeling and simulation studies suggest that the biliary excretion of AP-G is mediated almost exclusively by Mrp2.[373] Furthermore, the AP-G biliary clearance was markedly decreased and the AP-G basolateral clearance significantly increased in TR^- rat liver. In plasma membrane vesicles, *p*-OHPBG significantly inhibited Mrp2.[374]

Another drug–drug interaction has been reported between probenecid and the HIV protease inhibitor (HPI) saquinavir and the anticancer drug paclitaxel. In in vitro experiments MRP2 appears to transport saquinavir and paclitaxel efficiently and the transport of both drugs is stimulated by probenecid.[314,368,380] In wild-type and Pgp deficient mice (used because saquinavir and ritonavir are also good Pgp substrates), coadministration of saquinavir, ritonavir (to inhibit saquinavir metabolism), and probenecid resulted in strongly decreased saquinavir plasma levels.[435] Although the same experiment performed in Mrp2-deficient rats showed that Mrp2 transport function was not the main cause for the decreased HPI plasma levels, it cannot be excluded that in vivo there may be MRP2 stimulation that contributes to this drug–drug interaction. As probenecid is still used in general clinical practice in some countries, in particular in HIV/AIDS patients, for treatment of gout or in adjunct to antimicrobial therapy (e.g., with penicillin and/or cephalosporins) to boost antibiotic plasma concentrations, this reported drug–drug interaction possibly has clinical consequences. It may lead to a drop in HPI plasma levels, resulting in suboptimal therapy; selection of mutant, resistant HIV strains; and subsequently, to failure of therapy. Similarly, coadministration of probenecid and other MRP2-stimulating drugs in cancer patients during chemotherapy with MRP2 substrate drugs (e.g., taxanes, etoposide) might enhance MRP2-mediated drug resistance, increase the elimination of parenterally administered drugs, and decrease the oral bioavailability of anticancer drugs, with potential effects on therapeutic efficacy.[368] It is noteworthy that recently, considerably high levels of MRP2 have been reported in cancers originating from lung, breast, ovarian, renal, and colon carcinomas.[449] To date there is no clinical evidence of the aforementioned drug interaction with probenecid; however, selected combinations are worth investigating.

Furthermore, Zelcer et al. and Bakos et al. recently reported a broad range of clinically applied drugs that can stimulate and/or inhibit MRP2 activity in vitro, thus potentially leading to clinically relevant (beneficial or adverse) drug–drug interactions.[319,388] Indomethacin, penicillin G, pantoprazole (as well as other proton pumps inhibitors, such as omeprazole and lansoprazole), furosemide, glibenclamide and sulfanitran stimulated MRP2 transport in in vitro models. However, based on the complex model of the MRP2 interaction proposed (two distinguishable binding sites, one site from which the drug is transported and a second site that regulates the former allosterically), the effect of a transport modulator appeared to depend on the substrate transported. For example, probenecid significantly stimulated transport of

17β-estradiol 17β-D-glucuronide ($E_2$17βG) but inhibited the transport of methotrexate, two well-known MRP2 substrates.[388] In addition, other widely used drugs and herbal medicines, such as rifampicin, tamoxifen, dexamethasone, and St John's wort, which are often coadministered with other MRP2 substrate drugs (in particular, in anticancer therapy) have been demonstrated to induce intestinal or liver MRP2 in humans, rodents, and primates.[362,392,400] However, it is important to note that the stimulation of intestinal and hepatic MRP2 could lead to differential clinical effects. On the one hand, it may result in decreased bioavailability of drugs, which may lead to reduced treatment efficacy, but on the other hand, it could be a beneficial strategy to stimulate the secretion of toxic metabolites (e.g., to improve the clearance of HPIs in case of an accidental overdose or HPI-mediated toxicity). The exact clinical implications have, however, not been documented, and before attributing consequences to these in vitro observations, careful interaction studies should be performed with humans.

MRP3 (ABCC3)

Impact of Polymorphism on Function Little information is available about MRP3 (ABCC3) genetic polymorphism.[284,450,451] By screening of a Japanese population, Saito et al. recently identified several intron variations of ABCC3 and three exon SNPs, leading to synonymous amino acid changes in MRP3.[284] Screening 103 Caucasians, Lang et al. identified multiple synonymous and nonsynonymous mutations in the human MRP3 gene, one of which affected hepatic MRP3 expression.[451] In another study performed using several Caucasian patients with lung cancer or other respiratory diseases, a naturally occurring variant of MRP3 was described (MRP3-Arg 1297 His), with cellular location and transport characteristics similar to those of MRP3.[450] To date, no clinical phenotype has been attributed to a functionally deficient MRP3. However, nonsynonymous polymorphic MRP3 variants with altered transport characteristics are likely to exist.

Main Substrate Classes (Clinically Applied) MRP3 is an organic anion transporter with substrate specificity in part overlapping with MRP1 and MRP2. MRP3 transports several anticancer drugs, such as the epipodophyllotoxin derivatives etoposide and teniposide, methotrexate, and possibly vincristine, probably in a non-GSH-dependent manner.[368,452] Substrates of MRP3 include $E_2$17βG, a variety of bile salts, LTC_4, dinitrophenyl-*S*-GSH (GS-DNP) (but not GSH) and several glucuronide and sulfate conjugates of steroids,[452–454] prostaglandins [in particular, the glutathione conjugate of 15-*d*-prostaglandin J_2, an important product of prostaglandin D_2 (PGD_2) metabolism],[455] bile acids,[456,457] and commonly used drugs (e.g., acetaminophen and morphine).[376,458,459]

Inhibitors (Competitive and Noncompetitive) In vitro, several organic anion transport inhibitors, such as indomethacin, probenecid, benzbromarone, and sulfinpyrazone, could reverse MRP3-mediated decreased accumulation of substrate drugs, suggesting a possible inhibiting effect on MRP3 activity.[452]

Inducers Several groups have reported that liver MRP3 expression can be induced in response to a number of toxins, such as β-naphtoflavone[460] and 2-acetylaminofluorene,[461] oltipraz, diallyl sulfide, and *trans*-stilbene oxide,[462] and to drugs such as phenobarbital,[374,463,464] acetaminophen,[465] the proton pump inhibitor omeprazole,[460] and several nonsteroidal anti-inflammatory drugs.[339] Certain bile salts (e.g., chenodeoxycholic acid) can also act as inducers of MRP3 expression.[466] In humans, increased MRP3 expression has been detected in several hereditary and acquired liver diseases, including Dubin–Johnson syndrome,[467] primary biliary cirrhosis,[468] type 3 progressive familial intrahepatic cholestasis,[469] and obstructive cholestasis.[359,470]

Pharmacological and Toxicological Function Recently, the role of MRP3 in drug disposition has been investigated. In Mrp3 knockout mice, administration of the commonly used analgesic and antipyretic acetaminophen (AP) resulted in dramatically increased AP-glucuronide (AP-G) levels in the liver and bile and in a reduction of plasma AP-G levels compared with $Mrp3^{(+/+)}$ mice. In the isolated perfused liver, a strong decrease in AP-G secretion was found in the perfusate of $Mrp3^{(-/-)}$ livers, suggesting that in the absence of Mrp3, AP-G accumulates sufficiently in the liver to redirect a fraction of its excretion through low-affinity canalicular transport involving Mrp2.[471] This hypothesis is supported by previous preclinical studies investigating the hepatobiliary disposition of AP in Mrp2 (TR^-)-deficient rats, in which Mrp3 up-regulation was demonstrated. In this model, administration of AP resulted in a reduction in the biliary elimination of AP-G without retention of this metabolite in hepatocytes, and increased secretion of AP-G into the urine.[472] The same concept was demonstrated by Xiong et al.[373] using isolated perfused livers from TR^- and wild-type rats.

Similarly, Zelcer et al.[458] recently reported pharmacokinetic alterations of injected morphine and morphine-glucuronides in $Mrp3^{(-/-)}$ mice with a reduction in antinociceptive potency of injected morphine-6-glucuronide in $Mrp3^{(-/-)}$ mice. In addition, MRP3 has been demonstrated to affect the disposition of substrate drugs, thus possibly changing their pharmacological and toxicological profile. For instance, several findings support the hypothesis that induction of liver expression of MRP3 by AP may contribute to AP-decreased hepatotoxicity, as overexpression of MRP3 might minimize AP liver exposure, thus reducing AP activation to its toxic reactive metabolite. Although speculative, this hypothesis could contribute to the explanation of clinical cases of patients who took high-doses of AP without developing liver damage.[473] In rats and mice, treatment with an acute toxic dose of AP increased the hepatic expression of Mrp2 and increased the biliary excretion of its model substrates dinitrophenyl-*S*-GSH and GSSG. Repeated AP dosages correlated with a significant Mrp3 induction in the liver (but not to further MRP2 up-regulation), a shift from canalicular to basolateral efflux of AP-G and a decrease in its enterohepatic recirculation.[465,474–476] Moreover, in other preclinical studies, Xiong et al.[374] and Slitt et al.[462] used different Mrp3 inducers and demonstrated increased urinary excretion of the AP-G metabolite. This modulation of the route of drug excretion by MRP3 may lead to clinically relevant drug–drug interactions.

As reported above, induction of hepatic Mrp3 by phenobarbital is one of the most important mechanisms involved in the reduction of the biliary excretion of acetaminophen glucuronide after phenobarbital pretreatment of rats.[374] Phenobarbital is a well-known enzyme inducer that also induces the hepatocyte basolateral transporters Mrp3 and organic anion–transporting polypeptide 2 (Oatp2) in Sprague–Dawley rats.[374,470,477] Studies of the recirculating isolated perfused rat liver model have suggested that Mrp3 is involved in the basolateral transport of AP-G. Experiments with plasma inside-out membrane vesicles showed that AP-G is a low-affinity substrate for Mrp2 and Mrp3 and that *p*-hydroxyphenobarbital glucuronide (*p*-OHPBG), a major phenobarbital metabolite, significantly inhibited Mrp2- and Mrp3-mediated transport processes.[373] These findings support the hypothesis that the impaired biliary excretion of AP-G after phenobarbital pretreatment can be attributed primarily to the induction of hepatic Mrp3.

MRP4 MRP4 substrates include cyclic nucleotides and nucleotide analogs widely used in antiviral and anticancer therapy [e.g., 9-2-phosphonylmethoxyethyladenine (PMEA), azydothymidine (AZT) lamivudine, 2′,3′-dideoxy-3′-thiacytidine, ganciclovir, 6-mercaptopurine, thioguanine, cladribine],[478–483] cyclic GMP (guanosine 3′-5′-cyclic monophosphate) and AMP (adenosine 3′-5′-cyclic monophosphate), methotrexate, leucovorin,[249,478] cyclophosphamide,[484] several camptothecins (topotecan, irinotecan, its active metabolite SN-38, and rubitecan),[484–486] GSH, and folate.[246,249,484,487,488] The involvement of GSH in MRP4 cotransport of certain substrates is still controversial, as different groups reported contradicting results.[479,487,489] Several drugs have been reported to inhibit MRP4 activity efficiently in vitro, but their specificity and clinical applicability are still unclear. Buthionine-sulfoximine (BSO), MK571, celecoxib, and diclofenac as well as dipyridamole, dilazep, nitrobenzyl, mercaptopurine ribozide, probenecid, and sulfinpyrazone have been described as MRP4 blockers.[481,484] Little is known about induction of MRP4 expression. Recently, a study in infected human macrophages indicated that AZT treatment induced MRP4 mRNA expression.[490]

The pharmacological role of MRP4 is not yet completely known. However, due to the broad tissue distribution of MRP4, it is likely that its modulation by coadministered drugs (inhibitors or inducers) may have important pharmacokinetic implications for substrate drugs. In this respect, a recent and interesting study from Leggas et al.[491] reported enhanced accumulation after intravenous administration of topotecan (a MRP4 substrate) in brain tissue and cerebrospinal fluid (CSF) of Mrp4$^{(-/-)}$ mice compared with Mrp4$^{(+/+)}$ mice. Moreover, in a previous study,[492] the transport of intraventricularly injected methotrexate from the choroid plexus into the blood circulation was inhibited by probenecid, a reported Mrp4 inhibitor. Therefore, Mrp4 appears to play a significant role in brain penetration of selected cytotoxins and other therapeutic agents. Modulation of MRP4 activity and expression may enhance brain penetration of drugs, thus possibly improving the treatment of primary brain tumors and other cerebral diseases. At the cellular level, inhibition of MRP4 activity may be a useful strategy to increase the intracellular concentration of drugs, thus overcoming drug resistance. Interestingly, the expression level of MRP4 was associated with poor prognosis in patients with neuroblastoma.[486] Moreover, in a preclinical study, coadministration of celecoxib enhanced the antitumor activity of irinotecan and reduced the gastrointestinal side effects in mice bearing HT-29 and colon-26 tumor. One of the possible mechanisms involved in this beneficial drug–drug interaction might be the modulation of MRP4 function, but further studies are needed to explore the role of MRP4 in resistance, toxicity, and pharmacokinetics of irinotecan and its metabolites.[493] To date, several clinical phase I/II studies exploring the efficacy and safety of the association of irinotecan and celecoxib, frequently in combination with other drugs (e.g., docetaxel, 5 fluorouracil/leucovorin, thalidomide), in colorectal cancer, lung cancer, and malignant glioma have been performed, but the results are inconclusive.[494–499]

The multidrug resistance proteins **MRP5** to **MRP9** have not yet been completely characterized. Thus, their physiological and pharmacological role and possible involvement in drug–drug interactions are not clear at this moment.

24.3. INTERACTIONS MEDIATED BY ORGANIC ANION AND CATION TRANSPORTERS

24.3.1. Organic Anion Transporters

Impact of Polymorphism on Function Genetic variation in OATs has been studied in several different populations, but currently little is known about the impact of these genetic variants on drug disposition, toxicity, and disease.[500,501] A total of 25 single-nucleotide polymorphisms (SNPs) were identified from OAT1–OAT3 gene loci in healthy Japanese volunteers and involved the 5′ flanking regions, the 5′ untranslated regions, the coding regions, and introns, but their functional consequences were not explored.[502] More recently, SNPs in OAT family members were analyzed in an ethnically diverse sample of 96 individuals as well as combinations of OATs1 to 4 and URAT1 in particular ethnic groups. A total of 29 SNPs were identified, 14 of which were nonsynonymous and most of which were located in the OAT4 gene. Subjects with nonsynonymous SNPs in OATs localized in both apical and basolateral membranes, as well as with combinations of synonymous SNPs in OAT1 and OAT3, were found. Certain ethnic groups displayed a high prevalence of nonsynonymous SNPs in particular OATs (such as OAT4 in sub-Saharan Africans), and this finding may support the hypothesis of an association between ethnic group and risk to develop toxic and adverse drug reactions to several commonly prescribed drugs.[501] Furthermore, polymorphism in OAT4 has been suggested to potentially affect the fetal in utero risk because of toxicity by endogenous and/or exogenous compounds.[501]

Recently, sequencing of the SLC22A12 gene in 32 Japanese subjects affected by idiopathic renal hypouricemia, it was demonstrated that URAT1 was responsible for most cases of renal hypouricemia, especially severe renal hypouricemia. In these Japanese patients, eight new mutations and two previously identified mutations of SLC22A12 gene were found, resulting in loss of URAT1 function and clinically renal hypouricemia. In particular mutation G774A, changing tryptophan 258 (TGG) to a stop codon (TGA) and resulting in a truncated nonfunctional URAT1 protein, which was associated with the development of hypouricaemia, was reported as the major SLC22A12 renal hypouricemia allele in Japan.[501,503,504] In addition, two other homozygous missense mutations in SLC22A12 gene, leading the substitution of threonine by methionine (T217M) and glutamic acid by aspartic acid (E298D), respectively, were found in subjects affected by renal hypouricaemia.[505] Very recently, a SNP in the SLC22CA12 gene was associated with elevated serum uric acid levels among Japanese subjects. This SNP might be an independent genetic marker for predicting hyperuricemia, but further investigations are needed.[506]

In general, OAT1 has been shown low genetic and functional diversity suggesting its important role in body homeostasis. However, lately, 20 SNPs in human OAT1 were observed in an ethnically diverse population, of which one nonsynonymous SNP (R50H, encoding change from arginine to histidine) was associated with in vitro impaired transport of the nucleoside phosphonate analogs adefovir, cidofovir, and tenofovir.[507] Analogously, in another study, only one OAT1 genetic variant (R454Q) between six nonsynonymous SNPs identified in an ethnically diverse population was reported to be nonfunctional, but the subsequent clinical study performed in a single African-American family failed to demonstrate a significant difference in renal clearance of the antiviral drug adefovir, a well-known OAT1 substrate, between subjects heterozygous for the nonfunctional variant OAT1-R454Q and subjects not carrying this variant.[508] In another study performed in a Japanese population, polymorphisms in the OAT3 gene was not associated with changes in renal handling of the HMG-CoA reductase inhibitor pravastatin, an expected OAT3 substrate.[509,510] Finally, SNPs in the 5′ regulatory regions of the SLC22A6 and SLC22A8 genes, potentially affecting the transcription of OAT1 and OAT3, respectively, have been described, but their importance in affecting the human variation in organic anionic drug handling as well as the possibility of a reciprocal influence of the nucleotide polymorphisms in OAT1 and OAT3 remain to be established.[511]

These studies evaluating the potential relationship between OAT SNPs and the therapeutic effect of drugs have a significant clinical relevance. Indeed, they can help us understand and predict altered drug responses and toxicities, in particular ethnic subgroups that carry certain SNPs in OATs. It has been proposed that SNPs in OATs could be responsible, in part, for the different therapeutic activity of diuretics and ACE inhibitors, widely used antihypertensive drugs, and well-known OAT substrates. This mechanism could at least partly explain that about two-thirds of hypertensive patients are inadequately treated.[512] Testing for variants in OATs could also be a useful strategy to predict a predisposition to adverse events during treatment of a large group of the population. For instance, study of the impact of OATs SNPs on the incidence of nephrotoxicity caused by the nucleoside phosphonate analog cidofovir (an OATs substrate) could be clinically relevant, as cidofovir has recently been proposed for the treatment of smallpox infections in the case of bioterrorist attacks.[512]

Main Substrate Classes (Clinically Applied) and Inhibitors Generally, OATs mediate the transport of structurally diverse organic anions with broad overlapping substrate specificities. Interspecies differences in OAT substrates have also been described. OAT1 is an organic anion/dicarboxylate exchanger reported to transport a wide range of organic anions but also uncharged molecules such as steroid hormones. Typically, OAT1 substrates are compounds with a hydrophobic domain of 4 to 10 Å. In addition, functional studies suggested that substrates with an increase in the negative charge and/or with electron-attracting side groups (e.g., Cl, Br, NO_2) interact with more affinity with OAT1.

OAT1 substrates include the model organic anion *p*-aminohippurate (PAH) and endogenous anionic compounds such as the cyclic nucleotides cAMP and cGMP, folate,α-ketoglutarate, prostaglandin E_2, urate, indole acetate, and indoxyl sulfate. In addition, several classes of widely used drugs are transported by OAT1. Antibiotic agents (e.g., β-lactam antibiotics—penicillins and cephalosporines, tetracyclines, and quinolone gyrase inhibitors) and diuretics (e.g., furosemide, acetazolamide, bumetanide, hydrochlorothiazide, ethacrynate, tienilate) are transported by OAT1. Moreover, OAT1 substrates comprise several antiviral drugs (e.g., acyclovir, ganciclovir, lamivudine, zidovudine, stavudine, trifluridine adefovir, cidofovir, tenofovir, zalcitabine), nonsteroidal anti-inflammatory drugs (NSAIDs: indomethacin, acetylsalicylate, ketoprofen, salicylate), cimetidine, methotrexate, neurotransmitter metabolites (e.g., vanilmandelic acid), heavy metal chelators [2,3-dimercaptopropane sulfonate (DMPS)], test agents (fluorescein and 6-carboxyfluorescein), and toxins (in particular, ochratoxin A) [for a review, see refs. 514 to 517]. In addition to the human substrates, rat OAT1 has been shown to interact with several ACE inhibitors (captopril, enalapril, delapril, quinapril, ramipril), angiotensin II antagonists (telmisartan, candesartan, valsartan, losartan), the antiepileptic valproate, and several neurotransmitter metabolites (e.g., 5-methoxytryptophol, 5-hydroxyindole-3-acetic acid, D,L-4-hydroxyl-3-methoxymandelic acid).[514–516]

Inhibitors of human OAT1 have also been reported. Probenecid, the classic inhibitor of the renal organic anion secretion system, has been shown to block OAT1 activity with high affinity but low specificity, as it also inhibited sat-1, OAT2, and OAT3. Furthermore, in in vitro studies PAH cellular uptake OAT1-mediated was inhibited by benzylpenicillin with low affinity and by cephaloridine, cephradine, doxycycline, minocycline, oxytetracycline, and tetracycline. Human OAT1 inhibitors include some antiviral drugs (e.g., acyclovir, adefovir, cidofovir and its prodrug, zidovudine, and ganciclovir) as well as certain diuretics (furosemide, bumetanide) and NSAIDs (diclofenac, ibuprofen, flurbiprofen, indomethacin, ketoprofen, naproxen, etodolac, diflunisal, phenacetin, piroxicam, and salicylate). Cimetidine was shown to inhibit OAT1 and OAT2 in a noncompetitive manner and OAT3 and OAT4 in a competitive way. Pravastatin was able to block OAT1 and OAT3 activity in vitro.[514,517–521] A wider range of OAT1 inhibitors comes from in vitro experiments using the rat Oat1 isoform. In these studies several β-lactam antibiotics (penicillins and cephalosporines such as piperacillin, cloxacillin, nafcillin, cefazolin, cephalexin, cephaloridine, ceftriaxone, cefoperazone, and others) were found to inhibit rat Oat1 activity in a competitive manner as well as a wide range of clinically used antiviral drugs, diuretics (acetazolamide, bumetanide, hydrochlorothiazide), and NSAIDs (in addition to the compounds tested with human isoform, also salicylurate, tolmetin, benzydamine, antipyrine, aminopyrine, acetylsalicylate, and paracetamol).[522–525] Substrate specificity of human OAT2 has not yet been characterized completely, although based on in vitro studies, it is currently accepted that OAT2 transports PAH, α-ketoglutarate, cAMP, and prostaglandins E_2 and $F_{2\alpha}$.[515,526] Clinically applied drug substrates of OAT2 are, for example, the antiviral zidovudine, the cytostatic methotrexate, and the antimicrobial tetracyclines and cephalosporines. Due to the low transport rates found in in vitro experiments for ochratoxin A, valproate, methotrexate, and tetracyclines, it has been suggested that OAT2 does not contribute significantly to proximal tubular transport of these compounds. Probenecid was shown to inhibit human OAT2 in vitro but with low affinity; tetracycline, oxytetracycline, and minocycline were reported to inhibit the uptake of prostaglandin $F_{2\alpha}$.[527–532]

Human OAT3 was demonstrated to translocate PAH, cAMP, glutarate, sulfate, or glucuronide conjugates of steroid hormones (e.g., dehydroepiandrosterone sulfate, estrone sulfate, estradiol glucuronide), prostaglandins E_2 and $F_{2\alpha}$, taurocholate, and urate, as well as clinically relevant drugs such as tetracycline, methotrexate, salicylate, cimetidine, zidovudine, valacyclovir, and ochratoxin A. Several cephalosporins (cefadroxil, cefamandole, cefazolin, cefoperazone, cefotaxime, ceftriaxone, cephaloridine, and cephalothin) have also been reported to interact with human OAT3.[514,515,528–530,533] In addition to probenecid, benzylpenicillin, quinidine, the loop diuretics furosemide and bumetanide, and the NSAIDs diclofenac, ibuprofen, and indomethacin have been reported to inhibit the cellular uptake of estrone 3-sulfate by human OAT3 in vitro. Furthermore, piroxicam was observed to inhibit OAT3-mediated ochratoxin A transport in a competitive manner.[520]

Less is known regarding the substrate specificity and inhibitors of the other OAT members: OAT4 showed affinity to various sulfate but not to glucuronide conjugates. Indeed, OAT4 has been reported to interact with estrone sulfate with high affinity and with prostaglandins E_2 and $F_{2\alpha}$, dehydroepiandrosterone sulfate, and ochratoxin A. Very little transport of PAH was described.[515,529,531,534,535] Functional characterization has been evaluated only for rat OAT5, whereas data on the human isoform are lacking. When expressed in *Xenopus* oocytes, rat Oat5 was able to transport sulfoconjugated steroids (e.g., estrone sulfate) and ochratoxin A. Rat Oat5 was also shown to interact with penicillin G, several diuretics, certain NSAIDs, sulfobromophtalein, and some substrate conjugates (e.g., β-estradiol sulfate).[536,537]

Finally, human URAT1 is reported to act as an antiporter exchanging urate against certain organic anions (e.g., lactate, nicotinate, acetoacetate, hydroxybutirate) and chloride. Orotate, pyrazinecarboxylic acid, and nicotinate have been reported to inhibit URAT1 as well as several uricosuric compounds that have been shown to reduce hyperuricemia in experimental and clinical conditions (e.g., benzbromarone, fenylbutazone, sulfinpyrazone, probenecid, NSAIDs, and diuretic). In addition, the angiotensin II receptor antagonist losartan (DuP7-53) has been described to determine a transient uricosuria with a reduction in blood urate levels in healthy volunteers, probably due to inhibition of urate reabsorption in renal proximal tubule.[505,538,539]

Inducers Little is reported about induction of OAT expression and activity. Steroid hormones, in particular androgens, have been suggested to affect OAT expression, leading to the sexual dimorphism in OATs reported in rodents and supposed in humans. Indeed, in rats, testosterone has been shown to increase OAT3 and to decrease OAT2 messenger RNA levels, respectively.[540–542] Moreover, phosphorylation has been hypothesized to affect OAT activity posttranslationally: The epidermal growth factor (EGF) has been reported to induce OAT activity, whereas protein kinase C and certain protein kinase C activators (e.g., phorbol esters) led to an inhibition/internalization of human OAT1.[543,544]

Pharmacological and Toxicological Function Given the increasing number of drugs discovered to be OAT substrates, study of the pharmacological and toxicological roles of OATs has acquired clinical relevance. The recent generation of OAT gene knockout mice has provided in vivo models for this purpose, although the presence of species differences in substrate specificity and tissue distribution between OATs always has to be considered. Moreover, due to the functional redundancy of OATs, generation of knockout of multiple OAT genes (e.g., $OAT1^{-/-}/OAT3^{-/-}$ or $OAT4^{-/-}/URAT1^{-/-}$) is expected to help further in the elucidation of the in vivo function of these clinically relevant transporters. Both Oat1 and Oat3 knockout mice recently developed appeared healthy and viable and did not exhibit any obvious phenotype. Oat1 knockout mice showed a significant loss of renal organic anion secretion and markedly increased plasma levels of a wide range of endogenous (e.g., benzoate, *N*-acetylaspartate, 4-hydroxyphenyllactate, 4-hydroxyphenylpyruvate, 4-hydroxyphenylacetate) and specific exogenous compounds, some of which have been demonstrated to interact with OAT1 in vitro. For instance, in Oat1 knockout mice a reduced renal excretion of the loop diuretic furosemide (a well-known OAT1 substrate) was demonstrated, resulting in altered diuretic effect of this drug. Similarly, OAT1 can be predicted to affect the pharmacokinetics of other clinically used substrate drugs (β-lactam antibiotics, NSAIDs, antiviral nucleoside analogs), and the concomitant treatment with OAT substrates and inhibitors may lead to a reduction in renal clearance and to an increase in plasma levels and the half-lives of substrate compounds, thus potentially increasing the risk of their extrarenal toxicity.[545] For instance, drug–drug interactions between the antifolate methotrexate (a well-known OATs substrate) and penicillin, probenecid, and NSAIDs, have been reported to determine life-threatening complications due to severe myelosuppression caused by a decreased methotrexate renal secretion.[528]

Oat3 knockout mice have recently been generated. They display an altered organic anion transport physiological phenotype in renal and choroid plexus epithelia. In ex vivo experiments, uptake of Oat1 substrates, in particular taurocholate, but also estrone sulfate and to a lesser extent PAH, was reduced significantly in renal slices prepared from Oat3 knockout mice compared with slices from wild-type mice. Similarly, using intact choroid plexus, uptake of the Oat substrate fluorescein was markedly reduced in Oat3 knockout mice compared with wild-type mice. These findings suggest that at least in mice, Oat3, together with other members of the OAT family and other organic anion transporters, may play an important role in drug distribution and in brain protection from toxic injury by regulating the composition of the extracellular fluid of the central nervous system.[546]

Moreover, renal toxicity of several clinically used drugs (tetracyclines, cephaloridine, adefovir, cidofovir) as well as nephrotoxins (ochratoxin A, citrinin, mercuric conjugates) and uremic toxins (indoxyl sulfate, indoleacetic acid, *p*-hydroxyhippuric acid, and *o*-hydroxyhippuric acid) has been suggested to be caused by their accumulation in the renal proximal tubular cells via OATs, in particular OAT1.[547] This hypothesis is further supported by the protective role observed for OAT inhibitors or competitive substrates such as probenecid and several NSAIDs against nephrotoxicity from ochratoxin A, adefovir, cidofovir, cephaloridine, and mercury. In this view, drug–drug interactions due to competition for OAT transport is expected to determine a decrease in the renal secretion, thus leading to decreased nephrotoxicity, increased half-life, and potentially enhanced extrarenal toxicity of substrates.

For instance, ochratoxin A, a mycotoxin recently identified as an etiological factor of the endemic Balkan nephropathy, is transported by OAT1, OAT2, and OAT4. Accumulation of ochratoxin A, in proximal tubular cells via OATs has been reported as a primary event in the development of ochratoxin A nephrotoxicity. The OAT inhibitor probenecid has been observed to decrease the renal clearance of ochratoxin A, whereas piroxicam and octanoate, which inhibit OAT1-mediated uptake of ochratoxin A, have been reported to prevent the nephrotoxicity of ochratoxin A.[548–551] Therefore, the administration of OAT inhibitors or competitive substrates has been evaluated as a useful strategy to decrease nephrotoxicity of clinically used nephrotoxic OAT substrate drugs such as β-lactam antibiotics and antiviral nucleotide analogs: in particular, cidofovir, adefovir, and its orally available prodrug adefovir dipivoxil.

Organic anion transporter activity has also been proposed to be involved in the chelation therapy of the enviromental neurotoxicant methylmercury. *N*-acetylcysteine and dimercaptopropanesulfonate have been reported to increase dramatically the urinary methylmercury excretions in animals and humans, probably by forming mercaptide complexes, which are high-affinity substrates for Oat1. Therefore, organic anion transporters may facilitate the urinary methylmercury excretion.[552] Furthermore, based on the recent detection of OAT3 at the basolateral side of the blood–brain barrier and at the brush border membrane of the choroid plexus in rats and mice, it has been suggested that OAT3, together with other transporters belonging to the MRP family, localized at the luminal membrane of capillary endothelial cells of the blood–brain barrier, may participate in the efflux of metabolites of monoamine neurotransmitters, uremic toxins, and endogenous compounds (e.g., *p*-aminohippuric acid) but also of several drugs from the brain. In particular, OAT3 and members of MRPs (especially MRP4 and MRP5) have been proposed to reduce the brain penetration of the thiopurines nucleobase analogs 6-mercaptopurine and 6-thioguanine, drugs used widely in maintaining remission in the treatment for acute lymphoblastic leukemia (ALL), especially in pediatric patients. It has recently been suggested that coadministration of 6-mercaptopurine together with OAT3-specific inhibitors could be a useful strategy to increase thiopurine concentrations in the brain interstitial and cerebrospinal fluid, thus enhancing their pharmacological effect to prevent CNS relapses during treatment of patients with ALL.[553] Indeed, it was demonstrated in animal models that 6-mercaptopurine efflux from the brain was inhibited by OATs and/or MRPs inhibitors, such as benzylpenicillin, cimetidine, and sulfinpyrazone.[553]

Drug–Drug Interactions Several clinically relevant drug–drug interactions mediated by OAT family members have been described and others can be predicted. Indeed, due to the multispecific substrate recognition of OATs, OAT substrate drugs coexisting in the plasma may compete for the transport processes, thus modulating a drug's pharmacokinetics. Drug–drug interactions have been reported between methotrexate and NSAIDs, probenecid, and penicillin G, resulting in severe and even life-threatening effects, including bone marrow suppression, hepatitis, and acute renal failure.[554–560] These interactions may be caused at least in part by protein-binding displacement, inhibition of prostaglandin synthesis with reduction of glomerular filtration, and decrease of renal secretion of methotrexate due to competition for renal organic anion transporters. Several drug transporters are involved in methotrexate elimination and disposition, such as OATs, MRP4, MRP2, reduced folate carrier (RFC-1), oatp1, oatp3, and NPT1 (human type I sodium-dependent inorganic phosphate transporter). OATs have recently been suggested to contribute to interactions with methotrexate. Lately, in vitro studies using renal slices and mouse proximal tubular cells stably expressing human OAT1, OAT3, and OAT4 have shown that these transporters are the site of drug interactions between methotrexate and NSAIDs, probenecid, and penicillin G. In addition, it has been suggested that human OAT3 mediates the drug interactions between methotrexate and salicylate, phenylbutazone, indomethacin, or probenecid in vivo, as the OAT3 inhibition by these NSAIDs and probenecid took place at therapeutically relevant concentrations in vitro. In contrast, NSAIDs were reported to inhibit methotrexate transport mediated by RFC-1 in vitro at concentration levels higher than those clinically relevant.[523,528,561]

Interactions between methotrexate and the NSAID loxoprofen, a propionic acid derivative widely used in Japan, have been suggested. One Japanese patient with Hodgkin's disease treated with methotrexate at high-doses and loxoprofen experienced a delay in the elimination of plasma methotrexate. Subsequent in vitro studies showed that loxoprofen and its *trans*-OH metabolite inhibited methotrexate transport by human OAT1 and OAT3 at clinically relevant concentrations.[562] Severe drug–drug interactions between methotrexate and other NSAIDs (e.g., diclofenac, ketoprofen, ibuprofen, piroxicam, naproxen) have been reported, but the mechanism is still unclear.[556,563]

Drug–drug interactions have also been described between cephalosporin antibiotics and probenecid. As several cephalosporins are substrates of OATs and probenecid is recognized as a good organic anion transporter inhibitor, some of these drug–drug interactions may be due to an OAT-mediated uptake process. For instance, coadministration of probenecid has been shown to alter the pharmacokinetics of cefadroxil, cefamandole, and ceftriaxone, resulting in increased peak plasma concentration and half-life of these antibiotics.[564–567] As reported above, clinically beneficial drug–drug interactions may involve OAT substrates and OAT inhibitors, such as probenecid and NSAIDs. These inhibitors, by blocking OAT activity, may reduce the tubular secretion of OAT substrate drugs, thus decreasing their nephrotoxicity. Indeed, probenecid has been reported to decrease the nephrotoxicity of cephaloridine, adefovir, and cidofovir and of the toxic compounds ochratoxin A and mercury.[568] In addition, betamipron and cilastatin have been suggested to prevent nephrotoxicity of the antimicrobial panipenem and imipenem, respectively, via inhibition of OATs.[577]

Analogously, in in vitro studies, several NSAIDs (e.g., ketoprofen, diflunisal, flurbiprofen, indomethacin, ibuprofen, naproxen, diclofenac, and diflunisal) were reported to reduce significantly the cytotoxicity of adefovir and its orally available prodrug, adefovir dipivoxil (antiviral OAT substrate drugs widely used in HIV therapy and currently in clinical evaluation against chronic hepatitis B virus infections and herpes viruses and retroviruses) at clinically relevant plasma concentrations, indicating that they may exhibit in vivo nephroprotective effects when coadministered with adefovir or other nephrotoxic OAT1 substrate drugs.[518] Moreover, the concomitant administration of OAT substrate drugs and inhibitors may result in the impaired efficacy of drugs at the renal level. Indeed, most diuretics are actively secreted by renal OATs, thus reaching their target molecules in the kidney; coadministration of these diuretics with inhibitors of renal organic anion transporters is predicted to reduce their diuretic effect. In Oat1 knockout mice a reduction in tubular secretion and a decrease in diuretic effect of the loop diuretic furosemide was demonstrated. In humans, after treatment with probenecid, the renal clearance of intravenously administered furosemide was decreased significantly.[570] Drug–drug interactions probably mediated by organic anion transporters have been described between the antiviral zidovudine (AZT) and cimetidine. In humans, cimetidine has been shown to inhibit the renal clearance of AZT. In vitro studies have shown that AZT is an OAT1, OAT2, OAT3, and OAT4 substrate but it is not transported by OCTs, and cimetidine has been identified as a potent inhibitor of OAT3 (as well as OCT1 and OCT2). Thus, this clinically relevant drug–drug interaction may be mediated by OATs.[529,533,571,572]

In contrast, contradicting results have been reported regarding the mechanism of interaction between the antiviral acyclovir and probenecid. In humans, after oral administration of probenecid, the mean terminal half-life and the AUC of intravenously administered acyclovir were increased significantly, whereas the urinary excretion of the antiviral drug was reduced.[573] Although probenecid showed significant inhibition of organic anion uptake mediated by OAT1 and OAT3 in in vitro models, it exerted only a weak inhibitory effect on the acyclovir transport mediated by OAT1, thus suggesting that OAT1 is not responsible for the drug interaction between acyclovir and probenecid.[529] A drug–drug interaction probably mediated by OAT3 has also been described in humans between famotidine and probenecid. Coadministration of famotidine, and probenecid resulted in increased plasma concentrations and decreased renal clearance of famotidine, probably due to inhibition of OAT3-mediated transport of famotidine by probenecid.[574]

Finally, due to the large number of compounds that have been shown to interact with urate transporter 1 (URAT1), modulations of URAT1 activity by drugs used clinically is predicted to affect the urate plasma levels. It has been observed that several anti-inflammatory drugs may have uricosuric or antiuricosuric effects by inhibiting or stimulating URAT1. Drugs that stimulate the expression and the activity of URAT1 may be used in the treatment of hyperuricemia, a pathophysiological condition that leads to problems such as gout, hypertension, and cardiovascular disease, whereas the use of drugs that inhibit URAT1 may increase urate plasma levels, thus enhancing the risk of gout. A pharmacokinetic interaction has been described between the uricosuric compound benzbromarone and oxypurinol, an active metabolite of allopurinol, a xanthine oxidase inhibitor widely used for the treatment of hyperuricemia and gout. Although in a previous study no significant effect of benzbromarone on the oxypurinol pharmacokinetics were found after a single dose of the drug in healthy volunteers,[575] in a subsequent study evaluating pharmacokinetic parameters at steady state, concomitant administration of benzbromarone increased the renal clearance and decreased plasma levels of oxypurinol in healthy subjects.[576] It was suggested that this drug–drug interaction could potentially lead to a reduction of uricosuric effect of allopurinol, but another study performed in 14 patients with gout showed that despite the pharmacokinetic interaction between oxypurinol and benzbromarone, which resulted in a lower oxypurinol plasma level, the combination therapy consisting of Allomaron (100 mg of allopurinol plus 20 mg of benzbromarone twice daily) was superior to allopurinol alone (Zyloprim: 100 mg allopurinol twice daily) in lowering plasma acid uric concentrations, probably because of the added uricosuric effect of benzbromarone.[577] In any case, it has been suggested that the effects of such combined therapy may be less than additive when compared with the separate effects of similar doses of the two drugs. In preclinical studies, oxypurinol has been shown to be a substrate of URAT1, and its uptake in oocytes expressing URAT1 was inhibited by benzbromarone at clinically relevant concentrations.[578] These findings support the hypothesis that benzbromarone increases the clearance of oxypurinol by interacting at the URAT1 level, thus reducing oxypurinol renal tubular reabsorption. Analogously, the URAT1-mediated transport of oxypurinol was inhibited by probenecid: at least in part explaining the drug–drug interaction between probenecid and oxypurinol in humans, resulting in enhanced renal clearance of oxypurinol.[579] In contrast, concomitant treatment with pyrazinamide, an antituberculous agent with also antiuricosuric properties, has been reported to increase the renal clearance of oxypurinol through activation of URAT1.[579]

24.3.2. Organic Anion–Transporting Polypeptides

Impact of Polymorphism on Function Recently, SNPs of OATPs have been identified,[580] several of which have shown a reduction in transport activity in vitro and in vivo. These SNPs may contribute to the interindividual variability in the pharmacological/toxicological profile of OATP substrate drugs that have the liver as one of their pharmacological targets (e.g., HMG-CoA reductase inhibitors) and/or have a plasma clearance affected by OATP1B1-mediated hepatic uptake. Indeed, a recent study performed in a Japanese population showed that the T521C (Val174Ala) polymorphism in the OATP1B1 gene was linked with increased systemic exposure to pravastatin, a highly hydrophilic semisynthetic inhibitor of HMG-CoA reductase.[509] In subsequent studies, the haplotypes 130Asp/174Ala and −11187A/130Asp/174Ala and the SNP 174Ala were associated with altered transport activity in vitro, reduced renal and total clearance, and significantly increased AUC of pravastatin in healthy Caucasian, Japanese, European, and African-American populations.[509,581–584] Subsequent increased exposure to pravastatin due to a decreased uptake of the statins from blood into hepatocytes in subjects carrying these haplotypes has been suggested to enhance the risk of myotoxic effects. On the other hand, as statins reduce cholesterol levels primarily by inhibiting the hepatic HMG-CoA reductase, a reduction of the cholesterol-lowering efficacy of pravastatin could be predicted.[581] In contrast, the haplotype 130AspVal174, was reported to increase the hepatic OATP1B1-dependent uptake of pravastatin, thus putatively contributing to lower oral bioavailability of this statin (by reducing its AUC after oral administration), other analog drugs, and possibly valsartan and temocapril, two widely used drugs transported by OATP1B1.[585,586] Analogously, certain OATP1B1 variant haplotypes have been found to reduce the uptake from blood into hepatocytes of pitavastatin, a new HMG-CoA reductase inhibitor in healthy Korean volunteers.[982] Allelic variants of OATP1B1 have been reported to affect the pharmacokinetics of repaglinide, an antidiabetic drug transported by OATP1B1.[587] These findings support the hypothesis that SNPs in OATPs may influence the interindividual variability in disposition and drug response in the population.

Genetic polymorphism in the SLCO1B3 and SLCO2B1 genes has been studied, and several SNPs have been identified. A complete functional characterization of these variants is required to assess their possible implications for drug disposition.[583,588] In addition, several OATP1A2 SNPs were identified in the Japanese population, and some of them were located in regulatory regions of the SLCO1A2 gene. As OATP1A2 is localized predominantly in the capillary endothelial cells of the brain, allelic variants of OATP1A2 could affect the brain distribution and toxicity of several substrate drugs.[502,589]

Main Substrate Classes (Clinically Applied) In general, OATPs mediate the transport of structurally diverse compounds such as organic anions, cations, neutral or zwitterionic substances, and certain peptidomimetic agents. The principal OATP physiological substrates are bile acids, bilirubin, steroids, thyroid hormones, prostaglandins, and cholecystokinine. An increasing number of clinically used drugs has been recognized to be transported by OATPs. OATP substrate drugs include several HMG-CoA reductase inhibitors, and in particular, pravastatin, the antihistaminergic compound fexofenadine, the angiotensin-converting enzyme inhibitors enalapril and temocaprilat, the cardiac glycoside digoxin, the antidiabetic repaglinide, the folate analog methotrexate, the endothelin receptor antagonist BQ-123, and several antimicrobial agents, such as benzylpenicillin and rifampin.

In particular, OATP1A2 has a broad substrate specificity and has been reported to transport bile salts and bromosulfophtalein (BSP), steroid sulfates, thyroid hormones [triiodothyronine (T3), thyroxine (T4), and reverse T3], prostaglandin E_2, fexofenadine, opioid peptides [e.g., deltorphin II and (*d*-penicillamine)enkephalin], rocuronium, *N*-methylquinine and *N*-methylquinidine, ouabain, the endothelin receptor antagonist BQ-123, the thrombin inhibitor CRC-220, and certain magnetic resonance imaging contrast agents.[122,590–594] OATP1B1 substrates include bile salts, conjugated and unconjugated bilirubin, BSP, steroid conjugates, T3 and T4, peptides, natural toxins such as microcystin and phalloidin, and drugs such as pravastatin, methotrexate, benzylpenicillin, and rifampicin.[595–599] OATP1B3 displays similar broad substrate specificity as OATP1B1, but there are some differences in affinity.[600] Moreover, in contrast to OATP1B1, unconjugated bilirubin is not a substrate for OATP1B3.[601] In addition, OATP1B3 has also been reported to transport leukotriene C_4, linear and cyclic peptides, the intestinal peptide cholecystokinin 8 (CCK-8), digoxin, ouabain, and deltorphin II.[602,603]

OATP1C1 has narrower substrate specificity: T4 (thyroxine) and reverse T3 (reverse triiodothyronine) are the highest affinity substrates, whereas BSP, steroid sulfates and conjugates, and T3 are transported to a lesser extent.[591] Analogously, OATP4A1 exerts high affinity for T4, T3, and reverse T3, but it was also found to transport taurocholate, steroid sulfates and conjugates, prostaglandin E_2, and benzylpenicillin.[599,604] Although there is wide tissue distribution, the substrate specificity of OATP2B1 is rather limited. OATP2B1 substrates include BSP and steroid sulfates (e.g., estrone 3-sulfate and dehydroepiandrosterone sulfate), whereas contrasting results have been reported for steroid conjugates and prostaglandin E_2.[591,602] To date, less is known about OATP4C1 substrate specificity. OATP4C1 has been reported to transport cardiac glycosides (digoxin and ouabain), thyroid hormones (triiodothyronine and thyroxine), cAMP, and methotrexate in a sodium-independent manner.[605]

Inhibitors (Competitive, Noncompetitive) Several compounds have been reported to inhibit OATP activity. Cyclosporin has been shown to inhibit OATP1B1 in vivo, thus modulating the pharmacokinetics of repaglinide, a novel antidiabetic drug.[606] The antimicrobial rifamycin SV was able to strongly block OATP1B1-, OATP1B3-, OATP2B1-, and OATP1A2-mediated transport of BSP in vitro and in vivo, whereas rifampicin, a drug structurally related to rifamycin SV, was shown to inhibit in a competitive manner OATP1B3 primarily, and to a lesser extent, OATP1B1, in preclinical and clinical studies.[607,608] Rifamycin SV and rifampicin decreased BSP clearance in humans significantly and were shown to interfere with OATPs-mediated estradiol-17β-glucuronide transport, probably through inhibition of OATPs-mediated hepatic uptake. Moreover, rifamycin SV and rifampicin were shown to interfere with hepatic organic anion uptake by inhibition of Oatp1a1 and Oatp1a4 in rats. Inhibition of OATPs transport activity by rifamycin antibiotics has been suggested to determine the reduced hepatic bilirubin/organic anion elimination observed during initial treatment with these antimicrobial agents.[607–610]

Recently, grapefruit, orange, and apple juices as well as several of their furanocoumarin, bioflavonoid, and bergamottin constituents (such as 6′,7′-dihydroxybergamottin, bergamottin, naringin, hesperidin, methoxypsoralen) were reported to reduce human OATP and rat Oatp activity in vitro and in vivo, thus leading to clinically relevant drug–food interactions.[139,611] For instance, in a clinical study performed in 12 healthy volunteers, grapefruit juice at a commonly consumed volume diminished the oral bioavailability of coadministered fexofenadine sufficiently to be clinically relevant. The main molecular mechanism of this interaction could be the direct inhibition mediated by grapefruit juice on the intestinal OATP1A2 uptake of fexofenadine.[612] Analogously, the antimicrobials ketoconazole and erythromycin are described as OATP inhibitors.[122] Recently, troglitazone sulfate, the metabolite of troglitazone (an insulin-sensitizing drug developed for the treatment of type 2 diabetes mellitus) as well as other thiazolidinediones (e.g., pioglitazone and rosiglitazone) have been demonstrated to inhibit OATP1B1 and/or OATP1B3 in vitro, thus suggesting a possible involvement of troglitazone sulfate in troglitazone hepatotoxicity, the rare side effect that caused withdrawal of the drug from the market.[613]

Finally, in in vitro studies, several Pgp substrates and inhibitors were shown to inhibit OATP transport activity to various extents. The HMG-CoA reductase inhibitor lovastatin and the HIV-1 protease inhibitors ritonavir, saquinavir, and nelfinavir, as well as quinidine, ketoconazole, and verapamil,[122] were reported to reduce OATP-mediated transport activity.

Inducers Several studies have demonstrated that OATP expression can be induced by various compounds that are well-known ligands and activators of the nuclear receptors pregnane X receptor (PXR) and constitutive androstane receptor (CAR). Furthermore, the expression of certain OATPs (such as OATP1B1 and OATP1B3) has been suggested to be under transcriptional control by hepatocyte nuclear factor 1α (NF1α). Indeed, in a recent study, rat Oatp1b1 expression was shown to be modulated by several classes of drug-metabolizing enzyme inducers. In particular, Oatp1b1 expression was increased moderately by CAR ligands and increased dramatically by PXR ligands: phenobarbital, diallyl sulfide, and polychlorinated biphenyl were reported to increase Oatp1b1 expression. Other compounds, such as spironolactone, dexamethasone, and diethylhexylphthalate, were shown to enhance Oatp1b1 expression by a different mechanism.[614]

Previous studies in humans revealed that the barbiturate phenobarbital increased the hepatic clearance of organic anions BSP and bilirubin from plasma.[615] In preclinical studies, phenobarbital treatment enhanced the transport maximum for the excretion of bilirubin into bile in rats and significantly stimulated (more than sixfold) the maximum uptake velocity of BSP in isolated rat hepatocytes.[616,617] Analogously, pretreatment of rats with phenobarbital and the synthetic steroid pregnenolone-16α-carbonitril resulted in a reduced plasma half-life, increased hepatic clearance, and decreased toxicity of certain cardiac glycosides (e.g., digoxin, digitoxin, ouabain) that are well-known OATP substrates.[618–620] Increased hepatic uptake of cardiac glycosides by the two compounds was also induced in isolated hepatocytes.[621] More recently, phenobarbital and pregnenolone-16a-carbonitril were shown to enhance the expression of Oatp1B1 in rats, and consequently, the uptake of digoxin was increased.[403] However, in primary human and mouse hepatocytes, phenobarbital decreased OATP1B3, OATP2B1, and Oatp1a1 mRNA expression, respectively.[622,623] These findings support the hypothesis that modulation of Oatp/OATP expression in rats, as well as in humans, by phenobarbital and other compounds may affect the pharmacokinetics of coadministered OATP substrate drugs, leading to potentially clinically relevant drug–drug interactions.[624–626] On the other hand, as the hepatic expression of OATP1B3 has been reported to depend on hepatocyte nuclear factor 1α (HNF1α) and on the bile acid nuclear receptor farnesoid X-activated receptor

(FXR)/BAR (bile acid receptor), it has been suggested that the induction of SLCO1B3 gene expression by bile acids could maintain the hepatic elimination of xenobiotics and peptides under cholestatic conditions.[591,627,628] Finally, testosterone has been reported to induce renal rat Oatp1a2.[629]

Pharmacological and Toxicological Function Organic anion transporter polypeptides play a significant role in mediating the transport of anionic, neutral, and cationic organic endogenous and exogenous compounds. Given their expression in organs such as the liver, intestine, and brain, and the increasing number of known substrate drugs, OATPs are expected to affect drug bioavailability and tissue distribution. Furthermore, as some OATPs are coexpressed in several tissues with efflux carriers of the ABC family (such as Pgp and MRPs), the coordinated activity of both drug-uptake and efflux transporters may be important in drug disposition. Moreover, the liver-specific expression of some OATPs suggests a significant role for them in detoxification reactions, to prevent the accumulation of xenobiotics and toxic endogenous compounds in certain tissues. The expression of OATP1A2 in the capillary endothelial cells of brain supports the hypothesis that OATP may modulate the brain penetration of substrate drugs. However, to date little is known about the impact of OATP activity on the pharmacokinetics of substrate drugs, due to the lack of animal models, because even if the generation of mice with targeted disruption of OATPs/Oatps could be a useful model, there is no orthology between rodents and humans. Therefore, the physiological and pharmacological role of OATPs may be studied by identification of naturally occurring null or functionally deficient alleles in the population and their association with particular phenotypes, or a particular impairment in the pharmacokinetics of substrate drugs. However, as discussed above, OATP activity has been reported to affect the disposition of substrate drugs such as the HMG-CoA reductase inhibitors pravastatin and rosuvastatin, the cardiac glycosides digoxin and ouabain, the antihistaminergic compound fexofenadine, and the antidiabetic repaglinide. Similarly, it is expected that the pharmacokinetics of numerous other OATP substrate drugs are influenced by OATP activity.

Drug–Drug Interactions Several clinically relevant drug–drug and food–drug interactions mediated by OATPs have been reported; other drug–drug interactions can be predicted by the concomitant use of OATP substrates and inhibitors/inducers, and some of them can be clinically beneficial. As reported above, a clinically relevant drug–drug interaction has been described between the immunosuppressive cyclosporin A (a well-known Pgp, CYP3A4, and OATP1B1 inhibitor) and the antidiabetic repaglinide: Coadministration of the two drugs resulted in increased repaglinide plasma levels, with less effect in subjects with the SLCO1B1 521TC gene variant.[606]

A drug–drug interaction was also observed between the HMG-CoA reductase inhibitor rosuvastatin and cyclosporin A. Heart transplant patients receiving cyclosporin A for antirejection presented a significant increase in rosuvastatin AUC compared with historical controls.[630] A clinical interaction has also been described between cerivastatin and cyclosporin A: The plasma AUC and maximum plasma concentration of cerivastatin increased by four- and fivefold, respectively, when administered concomitantly with cyclosporin A. Although cerivastatin is metabolized by cytochrome P450 2C8 (CYP2C8) and 3A4 (CYP3A4), the interaction is probably better explained by a transporter-mediated process, in particular by cyclosporin A inhibition of cerivastatin OATP1B1 transport.[631,632] Similarly, the interactions reported in the literature between pravastatin, atorvastatin, and pitavastatin (other HMG-CoA reductase inhibitors and well-known OATP1B1 substrates) and cyclosporin A may be ascribed at least in part to a transporter-based mechanism and in particular to cyclosporin A inhibition of OATP1B1 activity rather than a metabolically mediated process.[633–635]

Recently, the widely clinically applied antihypertriglyceridemic drug gemfibrozil has been reported to interact when coadministered with certain statins, such as cerivastatin, pravastatin, simvastatin, lovastatin, pitavastatin and rosuvastatin.[636–642] In particular, the interaction with cerivastatin has been considered responsible for the side effect of myotoxicity, including lethal rhabdomyolysis.[637,643] Anyway these findings must be interpreted cautiously, as gemfibrozil may inhibit other transporters, such as MRP2 and BCRP or metabolic enzymes (e.g., CYP2C8 and CYP2C9). Indeed, although gemfibrozil has been reported to inhibit OATP1B1 transport activity and that the interaction with rosuvastatin was ascribed to gemfibrozil-mediated OATP1B1 inhibition of rosuvastatin transport, other authors suggested that this interaction is probably metabolism mediated.[639,644–646]

Clinically relevant drug–drug interaction has been described between digoxin and amiodarone, resulting in increased digoxin plasma levels and toxicity.[647–649] Recent findings suggest that other mechanisms besides the modulation of Pgp (classically considered the main interaction site) take part in this interaction. Indeed, in preclinical and clinical studies, amiodarone appeared to increase digoxin plasma levels in humans and in rats by decreasing the hepatic elimination of digoxin rather than by decreasing its urinary secretion.[648,650,651] Recently, digoxin has been described to be transported by human OATP1B3, OATP4C1, and rat Oatp1a4 (previously named Oatp2).[602,605,618,652] Moreover, at therapeutic concentrations, amiodarone was able to inhibit digoxin uptake significantly by oatp2-expressing *Xenopus* oocytes and by isolated rat hepatocytes, whereas only slight inhibition of Pgp-mediated digoxin transport was documented in cell monolayers overexpressing Pgp after incubation with amiodarone.[276] In addition, using the isolated-perfused liver system, amiodarone increased the plasma levels of digoxin by inhibiting its uptake into rat hepatocytes.[653] Therefore, the inhibitory effect of amiodarone on the hepatic uptake of digoxin by rat oatp2 appears greater than that on the tubular and biliary secretion of digoxin mediated by Pgp. These findings suggest that modulation of OATP may be one of the main molecular mechanisms involved in the digoxin–amiodarone interaction, at least in rats. Additional studies are needed to further confirm this hypothesis in humans. In particular, it might be interesting to investigate whether OATP1B3 and/or OATP4C1 could be considered the functional counterparts of rat Oatp1a4 and to elucidate the effect of amiodarone on human OATPs-mediated transport of digoxin.

A significant food–drug interaction has been reported between fexofenadine and several fruit juices. In humans, pretreatment with grapefruit juice caused a significant reduction in the maximum plasma concentration and AUC of fexofenadine after oral administration.[654] Analogously, orange and apple juices have been observed to reduce the bioavailability of fexofenadine. Although several fruit juice components have been reported to affect Pgp activity, the fruit juice–fexofenadine interactions above described appeared to be mediated by a net inhibition of OATP activity by citrus bioflavonoids and bergamottin, substances with which these fruit juices are enriched.[140,612]

Clinically relevant drug–drug interactions have also been described between the antimicrobials ketoconazole or erythromycin and fexofenadine, but although these antimicrobials have been reported to inhibit OATP activity, the interactions appear more likely to be mediated by inhibition of Pgp activity.[122] Drug–drug interactions mediated by OATP have also been documented between the barbiturate phenobarbital and cardiac glycosides, in particular digoxin. Pretreatment with phenobarbital resulted in a reduced half-life, toxic and the therapeutic activities of cardiac glycosides, digoxin and ouabain, in dogs and rats, respectively, an effect probably mediated by hepatic OATP induction. Indeed, cardiac glycosides are OATP substrates, and enhanced Oatp1B1 expression was observed in livers of rats treated with phenobarbital and pregnenolone-16α-carbonitrile.[403,477,618,655,656] Based on these preclinical findings, a similar interaction can be predicted in humans. The study of OATP inhibitors and/or inducers is important to explain, predict, and thus avoid clinically disadvantageous drug–drug or drug–food interactions. Moreover, the development and the clinical use of selective OATP inhibitors may have relevant implications for drug therapy.

The oral bioavailability of an OATP substrate drug with significant OATP-mediated hepatic first-pass elimination (such as the thrombin inhibitor CRC 220) could be improved by coadministration of an OATP-specific inhibitor.[657] In contrast, concomitant treatment with an OATP inhibitor and an intrahepatically active drug (such as pravastatin) could reduce the hepatic uptake of the drug, thus leading to a decreased drug efficacy and therapeutic failure.[596] Administration of a selective OATP inhibitor could also be useful in reducing hepatotoxicity in patients affected by an OATP substrate intoxication (e.g., microcystin); the OATP inhibitor could reduce the hepatic toxin uptake and the enterohepatic toxins circulation.[658]

24.3.3. Organic Cation Transporters

Impact of Polymorphism on Function Recently, different groups have described several polymorphic genetic variations in organic cation transporters, but little is still known about the effect of these genetic variants on drug disposition and disease.[659–663] Kerb et al. reported the identification of 25 SNPs in the hOCT1 gene in 57 Caucasians, and functional characterization studies revealed that three of these variants (Arg61Cys, Cys88Arg, and Gly401Ser) resulted in decreased transport function.[661] From a screening study for variants performed in 247 subjects with different ethnic backgrounds, Leabman et al. identified 15 variants of OCT1 and 28 polymorphisms in OCT2.[662,663] Some of these variants displayed a significantly altered (decreased or increased) transport activity (or substrate selectivity).[659,662] Further studies in humans are awaited to elucidate the relationship between genetic variation in organic cation transporters and drug disposition and toxicities.

Main Substrate Classes (Clinically Applied) and Inhibitors In general, the OCTs mediate the transport of structurally diverse small hydrophilic organic cations, and they display a broadly but not completely overlapping substrate specificity. Inter-species differences in the function of the OCTs have also been described. OCT substrates include the model substrate tetraethylammonium (TEA), the neurotoxin 1-methyl-4-phenylpyridinium (MPP+), and clinically applied drugs such as antidiabetics (biguanides)[664] antiparkinson drugs (amantadine and memantine),[665,666] β-blockers (acebutolol),[665] the H2-receptor antagonist cimetidine,[667,668] skeletal muscle relaxants (e.g., vecuronium),[665,668,669] and several other endogenous compounds, such as dopamine,[666,668] noradrenalin, serotonin, histamine,[666,667,670] creatinine, and choline.[665,671] Moreover, OCT1 and OCT2 have been demonstrated to transport some anionic prostanoids (i.e., prostaglandins and their derivatives), indicating that a positive charge is not an absolute prerequisite for being an OCT substrate.[526] Lately, the tyrosine kinase inhibitor imatinib mesylate was shown to be transported into cells by OCT1, thus confirming previous studies suggesting that influx proteins in general (and OCT1 in particular) might play an important role in imatinib mesylate cell accumulation and response during treatment.[672,673] In a group of patients affected by chronic myeloid leukemia, the cell expression/function of OCT1 was correlated significantly with patient response to imatinib mesylate treatment.[671]

Similarly, increased accumulation and cytotoxicity of the anticancer drug oxaliplatin (but not of cisplatin or carboplatin) was reported in human OCT1 and OCT2 transfected cells, indicating that oxaliplatin is a substrate of these transporters. These findings suggest that OCT1 and OCT2 tumor expression might be a determinant of the anticancer activity of oxaliplatin and might contribute to its antitumor specificity. Additional studies are needed to elucidate the role of tumor OCT expression as a marker for selecting specific platinum-based therapies in individual patients.[675] Moreover, the consequences of the modulation of OCTs on pharmacokinetics and anticancer activity of oxaliplatin should be explored.

Several compounds have been shown to inhibit uptake of the prototypes TEA or MPP+ in a competitive or noncompetitive manner, suggesting that they can interact with OCTs. Cimetidine appears to produce a noncompetitive inhibition of OCTs, whereas cisplatin has been reported to inhibit TEA uptake by mouse Oct2 competitively.[676] The antiarrhytmic procainamide is reported to inhibit OCT1 as well as OCT2 and OCT3 efficiently; quinine and quinidine appear to alter the activity of OCT1 and OCT2.[665,669,671,677] In a mammalian expression system for murine OCT2 and OCT3, several steroids (in particular, β-estradiol, corticosterone, deoxycorticosterone, papaverine, testosterone, and progesterone) were found to inhibit TEA uptake.[678] In addiction, the HIV protease inhibitors ritonavir, saquinavir, indinavir, and nelfinavir have been reported to be potent inhibitors but poor substrates for human OCT1.[669,679] Finally, several flavonoids (such as quercetin, kaempferol, naringenin, isoquercitrin, spiraeoside, and others) have been shown to inhibit OCT2-mediated transport in vitro.[680]

Pharmacological and Toxicological Function The pharmacological and toxicological roles of OCT1-3 have been investigated in vivo using knockout mice generated for all three organic cation transporters.[681–683] Oct1$^{(-/-)}$, Oct2$^{(-/-)}$, Orct3$^{(-/-)}$, and Oct1/2$^{(-/-)}$ mice are viable, fertile and do not display any obvious physiological defect, but present altered pharmacokinetics of substrate organic cations. In Oct 1$^{(-/-)}$ mice accumulation of the model OCT1 substrate TEA (tetraethylammonium) in liver was significantly lower than in wild-type mice after intravenous administration, whereas direct intestinal excretion of TEA was about twofold reduced.[681] Similarly, decreased liver accumulation but no difference in intestinal secretion was found in Oct1$^{(-/-)}$ mice for the neurotoxin 1-methyl-4-phenylpyridium (MPP$^+$) and [^{131}I]metaiodobenzylguanidine (MIBG), an anticancer drug clinically used in detection and treatment of tumors of neuroadrenergic origin, such as neuroblastoma and pheochromocytoma.[684] These findings supported the idea that OCT1 can affect the liver distribution of substrate drugs. Extrapolating to humans, modulation of OCT1 expression and/or activity might have clinically important consequences in drug therapy. Reduced liver uptake of drugs may result in reduction of efficacy for drugs that have the target or undergo to metabolic activation in the liver, but it can also be positive for limiting the toxicity of hepatotoxic drugs. Indeed, Wang et al. demonstrated that Oct1 affects the distribution of the biguanide metformin and plays a key role in biguanide-induced lactic acidosis, a life-threatening adverse effect of biguanides. Distribution in liver and small intestine of the intravenously administered metformin, as well as the blood lactate concentration, were decreased significantly in Oct1$^{(-/-)}$ compared with wild-type mice.[664,685] In all these animal experiments no clear difference was found in the renal distribution and elimination of the OCT1 substrate compounds tested, probably due to the shift from hepatobiliary toward renal elimination and to the functional redundancy of OCT1 with OCT2 in mice.[664,681] Therefore, the impact of OCTs activity on renal distribution of substrates was elucidated by generating Oct2 single- and Oct1/2 double-knockout mice. Using this model, Jonker et al. reported that although the absence of Oct2 had little effect on the renal distribution of substrate compounds (such as TEA), the concomitant deficiency of Oct1 and Oct2 in mice resulted in a complete abolishment of the renal secretion of TEA.[686] By extrapolating from rodents to humans (where only OCT2 and not OCT1 has been reported to be expressed in the kidney), it can be expected that OCT2 deficiency in humans may result in altered renal elimination and consequently, increased exposure to some drugs.

The pharmacological and physiological functions of Oct3 were partly elucidated by generation of Oct3 knockout mice.[683] Oct3 knockout mice do not present obvious physiological or neural phenotype and also no imbalance of monoamine neurotransmitters (noradrenalin and dopamine). However, Oct3$^{(-/-)}$ mice showed a significant reduction in heart uptake of MPP$^+$, a neurotoxin involved in the etiology of Parkinson's disease, suggesting that Oct3 is an essential component in vivo for the transport activity of the extraneuronal monoamine clearance system named uptake-2.[683]

Indeed, the extraneural uptake-2 system was first discovered in myocardiocytes of the heart in rats, and high expression of Oct3 has been reported in heart of rodents and humans.[687] In addition, the placenta was identified as a novel site of the extraneuronal uptake-2 system: in pregnant females of an Oct3 heterozygous cross, a threefold-reduced MPP^+ accumulation after intravenous administration was found in $Oct3^{(-/-)}$ compared with wild-type fetuses.[683] No difference in MPP^+ uptake was found in other Oct3-expressing organs, probably due to the functional redundancy within the different OCTs and other transporters.

Drug–Drug Interactions A number of drug–drug interactions probably mediated by OCTs have been described. A drug–drug interaction has been observed between the H2-receptor inhibitor cimetidine and the antiarrythmic procainamide. Treatment with cimetidine is reported to increase the AUC significantly and to reduce the renal clearance of coadministered procainamide.[688] Similarly, coadministration of cimetidine with the biguanide metformin has been shown to increase the plasma levels and to reduce the renal excretion of metformin, thus leading to clinically relevant consequences.[689] Recently, a clinically relevant interaction was reported between the antiarrythmic drug pilsicainide and the H1-receptor antagonist cetirizine in a Japanese patient with moderate renal insufficiency. In a subsequent study using healthy volunteers, when these two drugs were coadministered, they mutually inhibited renal clearance and decreased their elimination constant. Moreover, as in vitro studies both cetirizine and pilsicainide were able to inhibit the transport of substrates mediated by MDR1 and OCT2, and pilsicainide was reported to be excreted into urine probably via both MDR1 and OCT2, it has been suggested that this drug interaction may be mediated by both MDR1 and OCT2.[690] In a previous study in healthy volunteers coadministration of pilsicainide and cimetidine resulted in an increased AUC of pilsicainide by on average 33%, a prolonged elimination half-life, and a reduced apparent renal clearance of pilsicainide.[691]

24.3.4. Organic Cation/Carnitine Transporters

Impact of Polymorphism on Function Recently, single-nucleotide polymorphisms (SNPs) of OCTN1 have been identified in the Japanese population. Two SNPs (G462E and T306I) have been functionally characterized: The single amino acid mutation T306I does not seem to affect TEA transport activity, whereas the G462E mutation altered OCTN1 transport activity, presumably affecting the physiological function of OCNT1 and the pharmacokinetics of its substrate drugs.[692] Several mutations of OCTN2 causing loss of uptake activity and certain silent polymorphisms of OCTN2 have been reported in ethnically diverse subjects, in patients affected by primary carnitine deficiency, and in some of their relatives; their impact on the pharmacological profile of substrate drugs needs to be investigated further.[693–699] However, it is likely that mutations causing loss of carnitine transport ability may result in reduced transport of the fourth-generation β-lactam antibiotics (such as cefepime and cefoselis) that are reported to interact with OCTN2. Therefore, it has been proposed that patients with systemic carnitine deficiency may have reduced distribution and altered pharmacological profile (in particular, increased renal clearance and therefore decreased systemic half-life) of β-lactam antibiotics and other OCTN2 substrate drugs.[700]

Main Substrates Classes (Clinically Applied) and Inhibitors (Competitive and Noncompetitive) Carnitine and carnitine esters (acetyl- and propionylcarnitine), together with other important organic cations, are the main physiological substrates of OCTN1 and OCTN2. OCTN1 and OCTN2 substrates include TEA and the Ca^{2+} antagonists verapamil and quinidine. OCTN2 has been reported to transport betaine, choline, lysine, and methionine.[701] In in vitro models of several clinically applied drugs, such as the β-lactam antibiotics cephaloridine, cefepime, cefoselis, and cefluprenam, were reported to inhibit OCTN2-mediated carnitine transport in a competitive manner, suggesting themselves being substrates of OCTN2.[700] Analogously, widely used compounds such as quinidine, procainamide, desipramine, cimetidine, clonidine, emetine, and the hormones aldosterone and corticosterone compete with carnitine and/or acetylcarnitine for OCTN2 transport.[698,702,703] OCTN1 is inhibited by structurally diverse compounds, including cephaloridine, quinine, cimetidine, procainamide, and pyrilamine.[703]

Valproic acid seems to interfere with the regulation or synthesis of carnitine transporters, and it or its metabolites are reported to be responsible for inhibition of L-carnitine transport in cell cultures.[700,702,704] Emetine (an antiamebic and emetic compound) and pivalic acid (a substance contained in several antibiotics used in the treatment of respiratory and urinary tract infections) have been reported to inhibit OCTN2 activity. In humans, long-term treatment with emetine and pivalic acid, as well as therapy with valproic acid and betaine, may lead to a secondary or acquired systemic carnitine deficiency.[701,704–709]

Pharmacological and Toxicological Function, and Interactions Organic cation/carnitine transporters are physiologically important in mediating the transport of carnitine, carnitine ester derivatives, and several organic cations. OCTN substrates include also a number of clinically relevant drugs. These findings, together with the wide tissue distribution of OCTNs and the observation that OCTN activity may be inhibited by a variety of xenobiotics, suggest a significant pharmacological and toxicological role for OCTNs. At the renal level OCTN2 mediates the reabsorption of substrate drugs, thus affecting their disposition and clearance. Consequently, by increasing the systemic half-life of substrate drugs, OCTN2 may enhance their therapeutic efficacy and modulate their toxicological properties. For instance, one of the reported mechanisms of nephrotoxicity caused by the β-lactam antibiotic cephaloridine is competition with carnitine transport at the level of OCTN2 in renal tubular cells and mitochondria, possibly leading to renal mitochondrial damage.[710,711]

OCTN1 and OCTN2 have also been proposed to affect the intestinal absorption and pharmacokinetic behavior of a number of drugs. In rats, oral administration of sulpiride, a dopamine D_2 receptor antagonist, with OCTN1 and OCTN2 substrates and/or inhibitors decreased the plasma sulpiride concentration.[712,713] Moreover, systemic carnitine deficiency has been associated with long-term treatment with the antiepileptic valproic acid, the β-lactam antibiotic pivampicillin, the emetic and antibiotic emetine, the nucleoside analog zidovudine, and the compound betaine. Although the mechanisms responsible for this drug toxicity have not been characterized completely, the competitive inhibition of OCTN2-mediated carnitine transport by at least some of these drugs (valproic acid, emetine) has been reported to cause the drug–induced carnitine deficiency. These findings suggest that the drug–drug interactions between carnitine and other OCTN2 substrate drugs mediated by OCTN2 could be clinically relevant.

Analogously, the inhibitory effect of anticonvulsants on carnitine transport by the human placental carnitine transporter may cause the fetal anticonvulsant syndrome, of which some symptoms are similar to the fetal carnitine deficiency phenotype. Indeed, fetuses and neonates may not be able to synthesize adequate amounts of carnitine that is essential in fetal metabolic functions and tissue development and maturation. Recent in vitro experiments performed with vesicles derived from placental brush border membranes showed that several clinically used anticonvulsants, in particular valproic acid, phenytoin, and tiagabine, were able to inhibit carnitine uptake competitively.[714,715] Moreover, carnitine supplementation has been described to reduce the adverse reactions caused by long-term valproic acid treatment.[716,717] These findings suggest a possible beneficial role of carnitine supplementation during pregnancy in women treated with certain anticonvulsants. However, considering that carnitine transporters have recently been involved in carnitine transport across the blood–brain barrier, carnitine administration could competitively inhibit the uptake of anticonvulsants in the brain.[714,718] The drug–drug interactions at the level of OCTN1 and 2 outlined may have clinical implications; however, it is warranted to investigate this in greater detail. In Table 24.2, examples of clinical drug–drug interactions probably mediated by organic anion and cation transporters are reported.

24.4. INTERACTIONS MEDIATED BY PEPTIDE TRANSPORTERS

Impact of Polymorphism on Function To date, no polymorphisms in human peptide transporter genes have been described. However, based on the relevant (patho)physiological and pharmacological role of peptide transporters (PEPTs), it can be predicted that genetic polymorphism in PEPT and peptide/histidine transporter (PHT) genes may influence the function of the transporters, affect drug disposition, and contribute to interindividual variability in drug therapy.

Main Substrates Classes (Clinically Applied) There are some differences in substrate specificity and affinity as well as in transport capacity between PEPT1 and PEPT2. PEPT1 is considered a low-affinity, high-capacity transporter, whereas PEPT2 is a high-affinity, low-capacity system. The physiological substrates of PEPT1 and PEPT2 include all 400 dipeptides and about 8000 possible tripeptides derived from the proteogenic L-α-amino acids. These transporters are stereoselective, as they

TABLE 24.2. Examples of the Possible Involvement of Organic Anion and Cation Transporters in Clinical Drug–Drug Interactions

Drug	Inhibitor/ Inducer	Measured Effect/Toxicity	Putative Mechanism	Refs.
Penicillin	Probenecid	Decreased renal clearance, prolonged half-life	Inhibition of OATs	720,721
ACE inhibitors	Probenecid	Decreased renal clearance, prolonged half-life	Inhibition of OATs	721–726
Methotrexate	Penicillin	Decreased renal clearance, increased toxicity	Inhibition of OATs	721,727–730
	NSAIDs	Decreased renal clearance	Inhibition of OATs	529,557–559, 563,564
	Probenecid	Decreased renal clearance	Inhibition of OATs	529,555, 730
Adefovir/ cidofovir	Probenecid	Reduced nephrotoxicity	Inhibition of OATs	569,731–733
Cephalosporins	Probenecid	Increased peak plasma concentration and terminal half-life	Inhibition of OATs	565,734–736
Furosemide	Probenecid	Decreased renal clearance	Inhibition of OATs	571
Zidovudine	Cimetidine	Decreased renal clearance	Inhibition of OCTs, OATs	737,738
	Probenecid	Decreased renal clearance	Inhibition of OATs	738–740
Famotidine	Probenecid	Increased plasma concentration, decreased renal clearance	Inhibition of OATs	575
Oxypurinol	Benzbromarone	Decreased plasma level, increased renal clearance	Inhibition of URAT1	577,579
Digoxin	Amiodarone	Increased plasma levels and toxicity	Inhibition of OATPs, MDR1	276,649, 652,654, 741
Fexofenadine	Fruit juices (grapefruit, orange, apple)	Decreased plasma AUC and maximum plasma concentration	Inhibition of OATPs, MDR1	128,140
	Erythromycin	Increased plasma levels	Inhibition of OATPs, MDR1	123,742
Repaglinide	Cyclosporin A	Increased plasma AUC	Inhibition of OTAPs, MDR1	607,743, 744

TABLE 24.2. (*Continued*)

Drug	Inhibitor/ Inducer	Measured Effect/Toxicity	Putative Mechanism	Refs.
Statins	Cyclosporin A	Increased plasma AUC and maximum plasma concentration	Inhibition of OATPs, CYP3A4, CYP2C8, CYP2C9	631
Pravastatin	Cyclosporin A	Increased AUC and plasma levels	Inhibition of OATPs	643
Cerivastatin	Cyclosporin A	Increased AUC and plasma levels	Inhibition of OATPs, CYP2C8/3A4	632,633
Atorvastatin	Cyclosporin A	Increased AUC and plasma levels	Inhibition of OATPs, CYP3A4	636
Pitavastatin	Cyclosporin A	Increased AUC and plasma levels	Inhibition of OATPs	745
Statins	Gemfibrozil	Increased myotoxicity	Inhibition of OATPs, MRP2, BCRP, CYP2C8, CYP2C9	645,647
Simvastatin	Gemfibrozil	Increased AUC and plasma levels	Inhibition of OATPs	637
Cerivastatin	Gemfibrozil	Increased AUC and plasma levels	Inhibition of OATPs, CYP2C8/3A4	638,644, 645
Lovastatin	Gemfibrozil	Increased AUC and plasma levels	Inhibition of OATPs	641
Rosuvastatin	Gemfibrozil	Increased AUC and plasma levels	Inhibition of OATPs, CYP2C9	640
Metformin	Cimetidine	Decreased renal clearance, increased AUC	Inhibition of OCTs, OATs, OATPs	690
Procainamide	Cimetidine	Decreased renal clearance, increased AUC	Inhibition of OCTs, OATs, OATPs	689,720, 721, 746
Dofetilide	Cimetidine	Decreased renal clearance, increased AUC	Inhibition of OCTs, OATs, OATPs	721,747
Pilsicainide	Cimetidine	Decreased renal clearance, increased AUC and half life	Inhibition of OCTs, OATs, OATPs	692
Levofloxacin	Cimetidine	Decreased renal clearance, increased AUC	Inhibition of OCTs, OATs, OATPs	721
Pilsicainide	Cetirizime	Mutually reduction of renal clearance	Inhibition of OCTs, MDR1	691

display higher affinity for oligopeptides containing L-enantiomers of amino acids residues than for peptides with D-enantiomers. Moreover, as PEPT1 has recently been reported to transport simple ω-amino fatty acids that do not contain a peptide bond, several studies were performed to identify the essential features of PEPT substrates. The minimal structure of PEPT substrates consists of two charged moieties at opposite ends (carboxyl and amino groups) separated by at least four methylene groups.[747]

Furthermore, PEPT1 and PEPT2 effectively transport a significant number of peptidomimetic drugs, including penicillin β-lactam antibiotics (e.g., ampicillin, cyclacillin) and cephalosporins (e.g., cefadroxil, cefixime, ceftibuten, cephalexin, cepharadine),[748,749] ACE inhibitors (e.g., captopril and the ester prodrugs fosinopril and enalapril),[750,751] and the aminopeptidase inhibitor bestatin (also named ubenimex).[752] PEPT substrates also include some prodrugs, such as 3,4-dihydroxy-L-phenylalanine (L-dopa, used in the treatment of Parkinson disease), the antiviral L-valyl ester of acyclovir (called valaciclovir), and the L-valine ester of zidovudine. Several nonpeptidic compounds are also transported by PEPTs, such as the photosensitizing agent 5-aminolevulinic acid (δ-ALA), widely used in photodynamic therapy,[753] and sulpiride, a selective dopamine D_2 receptor antagonist.[712] Less is known about PHT1 and PHT2 substrates. Both are reported to mediate the proton-dependent uptake of histidine and several di- and tripeptides. The rat PHT1 isoform showed a high affinity for carnosine in various in vitro studies.[754,755]

Inhibitors Several inhibitors of peptide transporters were identified during screening for substrates of these transporters, whereas others were synthesized more recently by a rational approach. 4-Aminomethylbenzoic acid (4-AMBA) is the first competitive nontransported PEPT1 inhibitor described[756]; PEPT1 was reported to be blocked efficiently by diethylpyrocarbonate.[757] Other clinically relevant noncompetitive PEPT1 inhibitors are the orally active sulfonylurea antidiabetic drugs nateglinide and glibenclamide as well as tolbutamide and chlorpropamide.[758,759] Recently, several high-affinity competitive inhibitors of PEPT1 and PEPT2 have been developed. They are lysine-containing dipeptide derivatives starting from lysyl-4-nitro-benzyloxycarbonylproline (Lys[Z(NO_2)]-Pro).[760,761] Moreover, certain compounds with a structure related to cephalosporins and ACE inhibitor-ester prodrugs, such as quinalapril and quinalaprilat, fosinoprilat, and enalaprilat, have been reported to block peptide transporter activity with low affinity.[762–764]

Inducers The activity of peptide transporters can be modulated by several substances and in different pathophysiological conditions. In particular, PEPT1 may be modulated at the transcriptional level or at the level of translocation of transporter proteins to the plasma membrane. Treatment with insulin, leptin, human growth hormone, pentazocine, clonidine, or Ca^{2+} channel blockers as well as addition of high amounts of dipeptides in a medium of cultured cells (Caco-2 cells) expressing PEPT1 were reported to induce PEPT1 expression.[765–770] In contrast, cell exposure to the immunosuppressive drugs tacrolimus and cyclosporin A, to thyroid hormone T3, or epidermal growth factor are reported to inhibit PEPT1 activity.[766,771,772]

Dietary conditions and several pathophysiological states can modulate PEPT expression: In rats, a brief period of fasting or starvation or administration of a diet enriched in certain free amino acids and peptides was associated with up-regulation of PEPT1 expression.[773–776] A circadian regulation of intestinal PEPT1 expression has also been described in rats fed ad libitum and was able to affect the pharmacokinetics of the oral antibiotic ceftibuten, a peptidelike drug. However, the diurnal rhythm of intestinal rat PEPT1 activity was disrupted by fasting.[777,778] Diabetes, induced by treatment of rats with streptozotocin, resulted in an increased expression of PEPT1 in the small intestine.[779] Unusually high intestinal colonic levels of PEPT1 have also been identified in patients affected by short-bowel syndrome, chronic ulcerative colitis, Crohn's disease, or acute cryptosporidiosis, whereas PEPT1 is virtually absent under normal conditions of the gastrointestinal tract.[780–783] A potential role of PEPT1 in inflammatory bowel diseases has recently been suggested, as PEPT1 was reported to transport formyl-Met-Leu-Phe (fMLP), a neutrophil chemotactic factor produced by *Escherichia coli*.[784,785] In contrast, endotoxin and lipopolysaccharide are associated with down-regulation of PEPT1 in animal models.[786] Less is known regarding modulation of PEPT2 expression: Up-regulation of PEPT2 has been described in rat remnant kidney after unilateral nephrectomy, which could possibly modulate the pharmacokinetics of drugs transported by PEPT2 in patients affected by chronic renal failure.[787]

Pharmacological and Toxicological Function The physiological activity, the wide substrate specificity, and the tissue distribution of peptide transporters suggest that they could affect the oral bioavailability and the pharmacokinetics of substrate drugs. In general, compounds that are structurally similar to oligopeptides and are well-known PEPT1 substrates have good oral bioavailability, whereas their class analogs not transported by PEPT1 display slower and less complete absorption after oral administration. Indeed, in a study comparing several penicillins and cephalosporins, a significant correlation was found between the in vitro affinity for PEPT1 and their reported oral availability. In particular, aminopenicillins, and aminocephalosporin that displayed higher affinity and higher in vitro transport rates than other β-lactam antibiotics showed higher absorption rates in vivo.[788–790] Moreover, it has been reported that the expression level of PEPT1 in the rat intestine correlated significantly with intestinal absorptive transport of cephalexin and cefadroxil.[776,791]

Based on these findings, a strategy was developed to improve the oral bioavailability of drugs. It consists of coupling an amino acid residue to a drug to obtain a peptide or a peptidomimetic compound transported by peptide transporters. For example, L-dopa-L-Phe, a peptide derivative developed as a prodrug of L-dopa, displayed an intestinal uptake around 40-fold higher than that of free L-dopa, a drug widely used in treatment of Parkinson's disease.[792] Analogously, valacyclovir, the L-valyl ester of acyclovir, a well-known antiviral drug, shows three- to fivefold higher oral bioavailability than that of free acyclovir, and in in vitro models, valacyclovir, but not acyclovir, appears to be a PEPT1 substrate.[793–796] The examples of L-valine ester of zidovudine, a compound commonly used in the treatment of HIV infection and the amino acid esterification of other antiviral nucleosides have led to the development of other compounds that become substrates of peptide transporters after modification by amino acid esterification, thereby improving their pharmacological profile.[796] Furthermore, the peptidic prodrugs of a series of novel aminomethyl tetrahydrofuranyl (THF)-1β-methylcarbapenems showed higher efficacy and intestinal uptake than that of the active parent drug molecules.[797]

Moreover, both PEPT1 and PEPT2 may, by mediating the renal tubular reabsorption of substrates, affect the half-life, the clearance, and thereby the pharmacokinetics of substrate drugs. β-Lactam antibiotics transported by peptide transporters have more rapid intestinal uptake, increased availability after oral administration, and a longer half-life than that of antimicrobials that are not PEPT substrates.[789,790] Peptide transporters may also affect the tissue distribution of substrates in the body, thus determining a specific pattern of tissue drug delivery. This characteristic could be exploited in the treatment of selected diseases. For instance, δ-ALA, a precursor of cellular porphyrin synthesis used widely as a photosensitizer in the photodynamic therapy of tumors, is a well-known PEPT1 and PEPT2 substrate. This may affect ALA's rapid intestinal absorption and renal clearance but may also determine its accumulation in other tissues with PEPT2 expression, such as the brain, mammary gland, and lung, thus possibly improving under well-selected conditions the response of these organs to photodynamic therapy.[753,798,799] In addition, due to the proposed role of PEPT2 in mediating the influx and efflux of peptides from cerebrospinal fluid, PEPT2 may affect the disposition of peptidomimetic drugs (e.g., aminocephalosporins and penicillins) and xenobiotics in the cerebrospinal fluid and the brain. Therefore, the development of drugs with limited PEPT2 affinity or the design of selective PEPT2 inhibitors could be a useful strategy to improve brain delivery of drugs aimed to treat brain disorders.[800,801] Finally, PEPTs have been found on the cell membrane of a range of cancer cells; however, it is currently unknown whether this expression affects the outcome of anticancer drug therapy.

Drug–Drug Interactions Clinically relevant drug–drug interactions mediated by peptide transporters have been reported between different β-lactam antibiotics and/or inhibitors of these transporters. Oral coadministration of cefadroxil and cephalexin has been shown to delay and decrease the time to maximal plasma concentration and the AUC, respectively, of cefadroxil, presumably due to competitive inhibition by cephalexin of the intestinal PEPT1-mediated transport of cefadroxil.[802] Analogously, altered pharmacokinetics of the β-lactam ampicillin and amoxycillin were described after oral administration, together with cyclacillin, to healthy volunteers. These findings, together with evidence obtained previously in rats, suggested competition for absorption between these penicillins in the human gut at the level of peptide drug transporters.[803,804]

Via a different mechanism, nifedipine, a widely used Ca^{2+} antagonist, has been observed to increase the bioavailability of orally administered amoxicillin and cefixime in humans and cephalexin in rats. One of the mechanisms proposed to explain these drug–drug interactions involves a change in the intestinal surface pH mediated by nifedipine (due to a decreased concentration of intracellular Ca^{2+}), thereby increasing the driving force for β-lactam drug transport mediated by PEPT1.[770,791,805–807]

Clinically relevant drug–drug interactions mediated by PEPT at the intestinal absorption level may be postulated between certain sulfonylureas and peptidelike coadministered drugs. As several sulfonylureas, such as nateglinide, glibenclamide, and tolbutamide, appear to be good inhibitors of PEPT1 activity, oral coadministration of glibenclamide or nateglinide and β-lactam antibiotics to diabetic patients may lead to a reduction in the oral bioavailability of the β-lactam compounds, thereby affecting their plasma levels and possibly also their therapeutic efficacy.[758,759]

On the other hand, taken in consideration the postulated involvement of PEPT1 in intestinal inflammation via transport of the neutrophil chemotactic factor formyl-Met-Leu-Phe, the activity of PEPT2 at the renal and brain level and the affinity of both peptide transporters for δ-ALA, long-term administration of PEPT inhibitors, such as glibenclamide or nateglinide, might affect states of intestinal inflammation, porphyrin metabolism, and brain tissue drug delivery. It is of interest to explore these concepts further preclinically and eventually, test the clinical feasibility in patients.

24.5. INTERACTIONS MEDIATED BY MONOCARBOXYLATE TRANSPORTERS

Impact of Polymorphism on Function Mutations of monocarboxylate transporters (MCTs) have been reported to cause lactate transport deficiency with important pathophysiological effects. Some decades ago an otherwise healthy patient who developed severe chest pain and elevated serum creatine kinase levels after heavy exercise has been described. He was reported to have a defective red cell lactate efflux and a delayed reduction in muscle lactate after exercise.[808] In a subsequent study, genetic analysis of this patient and other subjects with reported defective lactate transport led to the identification of missense mutations of MCT1 gene with altered protein function.[809] Recently, mutations of MCT8 (a MCT that has been shown to participate in thyroid hormone transport and metabolism) have been identified, and several of them have been associated with the Allan–Herndon–Dudley syndrome (AHDS), an X-linked condition characterized by severe mental retardation, neurological dysfunction, and elevated serum triiodothyronine (T3) levels.[810–812]

Main Substrates Classes (Clinically Applied) MCT1 substrates consist of a broad range of short-chain monocarboxylates, especially those substituted in the C2 and C3 positions such as lactate, pyruvate, acetoacetate, β-hydroxybutyrate, some branched-chain keto acids (e.g., α-ketoisocaproate), and acetate, proprionate, and butyrate.[813,814] MCT1, as well as other MCTs, is also able to transport some exogenous or pharmacologically active compounds, usually consisting of monovalent weak organic acids with the carboxyl bound to a lateral small hydrophobic or hydrophilic group. Putative clinically relevant MCT1 substrates include salicylic acid, benzoic acid, nicotinic acid, foscarnet, and *R*- and *S*-mandelic acid.[815–817] MCTs in general, and MCT1 in particular, have been implicated in the transport of some β-lactam antibiotics (e.g., phenethicillin, propicillin, carindacillin)[818,819] and HMG-CoA reductase inhibitors (e.g., simvastatin, lovastatin, pravastatin, atorvastatin) at the blood–brain barrier.[820–822] MCT2 has been found to transport a wide range of monocarboxylates with substantially higher affinity than MCT1, in particular for pyruvate,[823,824] whereas MCT4 displayed lower affinity for most MCT1 substrates.[825,826] To date, the substrate specificity of the other MCT isoforms has not been described fully, with the exceptions of T-type-amino acid transporter 1 (TAT-1), which appears to transport aromatic amino acids (phenylalanine, tyrosine, tryptophan, L-dopa), and of MCT8, which is reported to transport the thyroid hormones T4 and T3.[827,828]

Inhibitors (Competitive and Noncompetitive) Thus far, several MCT1 inhibitors have been described, but none of them is specific. MCT1 inhibitors can be divided into several groups: The first is composed of substituted bulky or aromatic monocarboxylates (including 2-oxo-4-methylpentanoate and phenylpyruvate) and cyanocinnamate derivatives [e.g., α-cyano-4-hydrocinnamate (CHC)] that act as competitive inhibitors. The second group consists of amphiphilic compounds with different structures that also inhibit the anion exchanger AE1 and other membrane transporters. This group includes bioflavonoids, such as phloretin and quercetin, and anion transport inhibitors such as niflumic acid and 5-nitro-2-(3-phenyl-propylamino)benzoate (NPPB).

Stilbenedisulfonates [e.g., 4,4′-diisothiocyanostilbene-2,2′-disulfonate (DIDS), 4,4′-dibenzamidostilbene-2,2′-disulfonate (DBDS)] are reversible weak inhibitors of MCT1, although on prolonged exposure, DBDS appears to be able to block irreversibly MCT1 activity.[984] Other irreversible inhibitors are a broad range of amino reagents (e.g., phenylglyoxal, pyridoxal phosphate) and the organomercurial thiol reagent *p*-chloromercuribenzene sulfonate (pCMBS).[814,827] MCT2 is more sensitive to inhibition by a range of MCT blockers, including CHC, DBDS, and DIDS, but is insensitive to pCMBS.[823,824,829] MCT3 is reported to be insensitive to CHC, p-CMBS, and phloretin.[830] MCT4 exhibits less affinity for most inhibitors than MCT1 and is reported to be insensitive to DIDS.[826]

Inducers Several studies suggested that in human muscles the lactate/H^+ transport capacity and MCT expression could be enhanced by training. In one study, 7 days of bicycle training increased the content of MCT1 in the vastus lateralis muscle with an increase in the femoral venous lactate concentration during exercise.[985] In another study, high-intensity knee-extensor exercise was reported to increase the MCT1 and MCT4 content by 76 and 36%, respectively, with an increase in sarcolemmal transport of lactate measured in vesicles prepared from muscle tissue obtained by needle biopsies.[831] Subsequently, in another study, lon-gterm leg cycle endurance training has been associated with an increase in MCT1 and MCT4 in the human vastus lateralis muscle.[832] Also, MCTs have been reported to be rapidly up-regulated by exercise.[833] Indeed, in human muscles, MCT1 and MCT4 proteins were increased 2 and 4 days after a prolonged exercise bout.[834] This suggests that up-regulation of MCT1 and MCT4 involves both posttranscriptional and transcriptional mechanisms. Conversely, lactate transport capacity is reduced and MCTs are down-regulated after denervation of muscle or spinal injury.[835–837] Moreover, in rat heart, MCT1 is increased after ligation of a branch of the left coronary artery, whereas in rats with streptozotocin-induced diabetes, MCT1 expression is reported to be reduced in the heart, in adipose tissue, and together with MCT4, in skeletal muscle,[838,839] although previous studies reported contrasting results.[985] In another study, MCT1 content in skeletal muscle of patients with type 2 diabetes was lower than in the muscle of healthy men. In addition, training increased MCT1 content in both groups, whereas in healthy people, but not in patients with type 2 diabetes, MCT4 increased in response to training.[840]

In brain tissue, MCT expression levels were found to vary in endothelial and parenchymal cells during peri- and postnatal development in rats. For instance, it was observed that endothelial mature astrocytes and parenchymal cells in rat brain have increased levels of MCT1 postnatally, an increase after weaning, and quite high levels at adulthood. These modifications in MCT expression seem to reflect formation of the blood–brain barrier and the switch in the postnatal period from a prevalent use of monocarboxylates to glucose utilization by the adult brain.[841] Furthermore, in adult rat a ketogenic diet as well as ischemic conditions were reported to induce brain MCT1 expression.[842,843] Moreover, in cultured cortical neurons, MCT2 expression was enhanced by noradrenalin, suggesting that MCT expression may be regulated by synaptic activity.[844,845] In addition, treatment of rat brain endothelial cells by cobalt chloride increased MCT1 expression.[817] In retinal explants, MCT1 was induced by hypoxia and vascular endothelial growth factor,[817,846] whereas in cultured macrophages, lipopolysaccharides and tumor necrosis factor-α determine increased expression of MCT1.[847]

In human colonic epithelial cells, treatment with butyrate has been reported to up-regulate MCT1 expression, and exposure to leptin, a hormone involved in regulation of cellular metabolism, was able to further increase butyrate uptake.[848,849] In rat, treatment with testosterone has recently been associated with an increase in MCT1 and MCT4 proteins in skeletal muscles but not in the heart.[850] Finally, as MCT1 and MCT4 activity has been linked to the expression of some membrane glycoproteins (e.g., GP-70 in humans, CD147, OX-47, basigin, etc., in other species), overexpression of these glycoproteins by several factors can modulate the expression and activity of MCT transporters.[851,852]

Pharmacological and Toxicological Function In view of their cellular location and tissue distribution, it has been proposed that MCTs, in particular MCT1, can be involved in intestinal absorption, blood–brain transport, and liver delivery of some weak organic acid drugs. MCTs may be involved in the intestinal absorption of some β-lactam antibiotics and nonsteroidal anti-inflammatory drugs, as they are reported to be good MCT substrates in vitro.[817] In addition, as monocarboxylate activity is proton dependent, the transport of substrate drugs by MCTs across the epithelial layer of the small intestine may be facilitated by the presence, at the luminal side of the brush border membrane, of a low pH. This creates a proton gradient that could facilitate the monocarboxylate transport.[853] MCT1 expression at the blood–brain barrier may play a crucial role in the efflux of certain drugs from the brain. Indeed, the low distribution of probenecid and 6-mercaptopurine in the brain has been proposed to depend in part on their MCT-mediated efflux.[854,855]

The tissue-specific expression of MCTs may also explain the development of certain side effects that follow the intake of a range of MCT substrate drugs. For instance, the lipophilic HMG-CoA reductase inhibitors lovastatin and simvastatin, which present a carboxylic acid moiety and are used clinically for the treatment of hypercholesterolemia and mixed dyslipidemia, have been suspected to be MCT1 substrate drugs. In particular, it has been suggested that MCTs (especially, MCT1) may contribute to the transport of simvastatin across the blood–brain barrier into the brain, thus determining some of the CNS side effects (predominantly sleeplessness) associated with the intake of these drugs. In vivo brain perfusion studies confirm that penetration of these drugs across the blood–brain barrier correlates with these side effects, and in vitro studies reported a possible role of MCT1 in the uptake of simvastatin.[820,821] A possible contribution of MCTs to the intestinal absorption of atorvastatin, another HMG-CoA reductase inhibitor, has been speculated, but other mechanisms and/or transporters may be involved as well.[822]

Drug–Drug Interactions To date, no clinically relevant drug–drug interactions mediated by MCTs have been described. However, because MCTs are involved in important pathophysiological conditions (hypoglycemia, diabetes, starvation, cerebral and heart ischemia), modulation of MCTs expression and activity may be a useful strategy to protect some tissues from ischemic or metabolic damage. Recently, transfection of MCT2 in cultured neurons has been reported to confer neuroprotection against toxicity.[856,857] Further studies need to be undertaken to unravel the role of MCTs in clinically relevant drug–drug interactions.

24.6. INTERACTIONS MEDIATED BY THE NUCLEOSIDE TRANSPORTERS

Impact of Polymorphism on Function Recently, genetic variability in nucleoside transporter family members have been described, although no pathophysiological conditions have been associated with this family to date. However, large interindividual differences in response to anticancer and antiviral nucleoside analogs and substrates of nucleoside transporters have been observed, suggesting that genetic variability in nucleoside transporters could lead to reduced function or nonfunctional transporter proteins. This could result in interpatient variability in systemic and intracellular levels of nucleoside analog drugs. In general, equilibrative nucleoside transporters (ENTs) appear to have less genetic and functional diversity than do concentrative nucleoside transporters (CNTs). No significant differences in the functional characterization were found between the two recently identified nonsynonymous variants of ENT1 and the reference ENT1.[858] Five protein-altering variants were identified for ENT2, but they were found at very low frequency, making the hypothesis unlikely that genetic variation of ENT2 significantly affects the variability in drug response that is observed clinically.[859]

Regarding the CNT family, CNT1 and CNT2 display more genetic diversity than CNT3. Recently, Gray et al. have observed in an ethnically diverse population a high degree of genetic and functional variation in CNT1. In particular, two nonfunctional variants and one variant (CNT1-Val189Ile) with reduced affinity for the anticancer pyrimidine nucleoside analog gemcitabine were described.[860,861] In addition, a non-synonymous CNT2 variant (CNT2-F355S) with significant allelic frequency was identified in the African-American population; it displayed altered specificity for inosine and uridine.[860] In contrast, CNT3 has less genetic and functional diversity, suggesting that CNT3 is more critical for viable human life, tolerating less genetic and functional diversity.[858] Possibly, more genetic variation can be tolerated in CNT1 and CNT2, because together with CNT3, they display overlapping tissue distribution and substrate specificity.

Main Substrate Classes (Clinically Applied) In general, most nucleosides, nucleobases, and their analogs used in anticancer and antiviral therapy are substrates of the nucleoside transport systems. There are many similarities between the spectrum of compounds transported by ENTs and CNTs, although there is no complete overlap. ENT1 displays broad substrate specificity for purine and pyrimidine nucleosides, with higher affinity for adenosine and lower affinity for cytidine, but it is reported not to transport nucleobases.[862–866] Anticancer and antiviral drugs transported by ENT1 include gemcitabine, cytarabine, fludarabine, cladribine, and ribavirin, whereas the nucleoside analogs 2′,3′-dideoxycytidine (zalcitabine, ddC) and 2′,3′-dideoxyinosine (didanosine, ddI) are only weak substrates. The antiviral drug 3′-azido-3′deoxythymidine (zidovudine, AZT) is reported not to be a substrate for ENT1.[862,864,867–871]

ENT2 transports a broad range of substrates, including purine and pyrimidine nucleosides (although with a lower apparent affinity than ENT1, with the exception of inosine),[863,864,872] nucleobases (hypoxantine, adenine, guanine, uracil, thymine, and cytosine),[873] and possibly, cyclic nucleotides.[874] In contrast to ENT1, ENT2 can transport AZT and displays much greater affinity for ddC and ddI.[864,870] Recently, it has been reported that the anticancer drugs gemcitabine and fludarabine are substrates of ENT2.[875–877]

Characterization of the substrate specificity of ENT3 and ENT4 is currently in progress. ENT3 appears to be able to transport several nucleosides (with the exception of hypoxantine) and antiviral nucleoside analogs such as AZT, ddC, and ddI. Adenosine is reported to be a weak substrate for ENT4.[878,879] CNT1 selectively transports pyrimidine nucleosides (cytidine, thymidine, uridine) and adenosine.[880,881] CNT1 substrates also include the antiviral nucleoside analogs zidovudine (3′-azido-3′-deoxythymidine, AZT), lamivudine (2′,3′-dideoxy-3′-thiacytidine, 3TC), zalcitabine (2′,3′-dideoxycytidine, ddC), the anticancer drugs gemcitabine (2′,2′-difluorodeoxycytidine, dFdC), cytarabine [(1-β-D-arabinofuranosyl)cytosine, AraC], and 5′-deoxy-5-fluorouridine (5′-DFUR, which is the active metabolite of capecitabine).[869,880,882–884] In contrast, CNT2 transports purine nucleosides (adenosine, guanosine, inosine), uridine, and formycin B selectively.[881,885–886] CNT2 substrates also include didanosine (2′,3′-dideoxyinosine, ddI) and ribavirin,[880,886–888] but CNT2 does not appear to transport other antiviral drugs, such as zidovudine, zalcitabine, or the anticancer nucleoside analogs currently used in anticancer chemotherapy.[870,886,888] CNT3 is widely selective for nucleoside substrates, accepting both purine and pyrimidine nucleosides. CNT3 also transports several nucleoside analogs, such as fludarabine, cladribine, zebularine, 5-fluorouridine, gemcitabine, ddI, ddC, and AZT, although the latter weakly.[889–891]

Inhibitors (Competitive and Noncompetitive) Historically, inhibition activity of nitrobenzylthioinosine (NBTI) was used to classify ENTs. ENT1 is reported to be highly sensitive to inhibition by NBTI (K_i for ENT1 is in the nanomolar range), whereas ENT2 is resistant to inhibition by NBTI up to micromolar concentrations.[892,893] Coronary vasodilators such as dipyridamole, dilazep, and the lidoflazine-related compounds lidoflazine, mioflazine, soluflazine, R75231, and draflazine are reported to be potent ENT1 competitive inhibitors, whereas ENT2 is much less susceptible to inhibition by these compounds.[863,878,894] Recently, several dipyridamole analogs with potent inhibiting activity against nucleoside transporters and with a better pharmacological profile have been identified.[895] Moreover, a large number of protein kinase inhibitors, including some inhibitors of serine/threonine or tyrosine kinases, such as STI-571 (imatinib mesylate, Gleevec), protein kinase C inhibitors (e.g., staurosporine, arcyriarubin A), rapamycin, kinase inhibitors against p38 MAPK (mitogen-activated protein kinase) EGF receptor kinase, and others, have been reported to inhibit ENT1 and/or ENT2 in a human erythroleukemia cell line.[896–898] The antidiabetic troglitazone, a thiazolidinedione, was able to inhibit ENT1 competitively, whereas it did not have any effect on ENT2 activity.[899] In vitro, ethanol is suggested to inhibit ENT1 selectively after acute administration and to down-regulate ENT1 after chronic exposure.[900,901] Lately, a range of 6-benzylthioinosine analogs, currently studied as potential agents against *Toxoplasma gondii*, have been described as inhibitors of ENT1.[902] Potential clinical consequences of these recent observations warrant further investigations.

Inducers Expression of nucleoside transporters appears to be regulated by several cellular and systemic events, as expression of nucleoside transporter is cell cycle-dependent and varies with cell differentiation and with the cellular deoxynucleotide levels.[903–905] In in vivo studies, after 48 hours of starvation, rats showed an increase in CNT1 protein levels in jejunum and liver. Moreover, nucleotide supplemented diet increased the CNT1 level in the liver and decreased it in the small intestine.[906] In rodents, after partial hepatectomy a rapid up-regulation of CNT1 and CNT2 with no significant modification in ENT1 mRNA levels has been described.[905,907] ENT1-dependent nucleoside transport appears to be regulated by protein kinase C (PKC) isoforms, in particular δ and/or ε. In human cultured cells, long treatment with phorbol 12-myristate 13-acetate (PMA) is associated with a decrease in ENT1 transport, whereas acute stimulation of PKC increases the ENT1 nucleoside uptake. In human BLS1 and Raji B lymphocytes, treatment with PMA and lipopolysaccharide (LPS) up-regulated CNTs and led to a down-regulation of ENT1 in a PKC-dependent manner.[908–910] In murine bone marrow–derived macrophages, induction of proliferation with M-CSF is reported to up-regulate es-type transport activity (presumably mediated by ENT1), whereas macrophage activation mediated by interferon γ (IFN-γ) determined the induction of CNT1 and CNT2. Moreover, IFN-γ was able to down-regulate the M-CSF-induced expression of ENT1.[910,912] In addition, hormonal regulation of nucleoside transporters has been reported in various tissues and cell types. In a study using rat liver, glucagon was reported to up-regulate the nucleoside transport, whereas insulin was able to stimulate Na^+-dependent uridine uptake by stimulation of synthesis of the transporters.[912] In cultures of rat B and T lymphocytes, nucleoside transporter expression levels were reported to be regulated independently and differentially by glucose and insulin.[913,914] Elevated levels of glucose were also reported to induce the expression of ENT1 in human aortic smooth muscle cells.[915] High concentrations of glucose reduced the expression level and transport activity of ENT1 and affected adenosine transport in human umbilical vein endothelial cells (HUVECs) with a mechanism that involved endothelial nitric oxide synthase, PKC, and MAP kinases.[916,917] The thyroid hormone T3 and dexamethasone were able to induce selective up-regulation of CNT2 and down-regulation of ENT1 and ENT2 in rat fetal hepatocytes.[918] T3 has also been reported to induce the expression of nucleoside transporters and to stimulate adenosine transport in cultured chromaffin cells.[919] Possible clinical implications of these findings need to be explored further.

Pharmacological and Toxicological Function The cellular location of nucleoside transporters in several normal and cancer tissues and in organs, which play a key role in drug distribution, suggest that they may influence the intracellular bioavailability and the pharmacokinetic, therapeutic, and toxicological profiles of affected nucleoside analog substrate drugs. Variation in nucleoside transporter expression may contribute to variation in drug efficacy and drug resistance: Transport of nucleoside analogs is the first step in tumor cytotoxicity. In general, tumors display highly variable patterns of ENT and CNT expression, which affects drug bioavailability and action. Decreased nucleoside transporter expression and/or activity in tumor tissues and in virally infected cells that are targets for nucleoside antiviral drugs have been correlated with reduced drug-uptake and the development of drug resistance. Chemosensitivity and ex vivo cytotoxicity of cytarabine, fludarabine, and other nucleoside analogs have been correlated to expression of subtypes of nucleoside transporters in several in vitro studies.[875,876,884,920–927] Furthermore, gemcitabine cytotoxicity was correlated to nucleoside transporter activity in cell lines derived from a variety of lymphoid and human pancreatic adenocarcinomas with defined different patterns of nucleoside transporter expression.[868,875]

The location of nucleoside transporters (in particular, CNTs) in the intestine, kidney, and liver suggests that these transporters may affect the systemic absorption and disposition of nucleoside analog substrate drugs. The significant expression of CNT2 in the gastrointestinal tract is supposed to be a major factor in absorption of orally administered ribavirin and other natural purine and purine analogs.[887,928] In addition, the broad tissue distribution of ENT1 appears to be a major determinant of the unusually large distribution volume of ribavirin.[867,929] Furthermore, the renal disposition of natural nucleosides and nucleoside analogs appears to be affected by nucleoside transporters widely expressed in the kidney.[930] In contrast, although some nucleoside transporters have been found in brain capillary endothelial cells that form the blood–brain barrier, it is still not known whether they mediate the vectorial transfer of nucleosides and nucleoside analogs across the blood–brain barrier. However, due to the lack of animal models knocked out for the various nucleoside transporter genes, together with the overlap in substrate specificity between the various nucleoside transporters and between other membrane protein pumps able to translocate nucleoside analogs, it is difficult to evaluate the contribution of individual nucleoside transporters to in vivo disposition of nucleoside analogs. To date, only ENT1-null mice have been generated.[900]

At this moment, no clinically relevant drug–drug interactions mediated by nucleoside transporters have been described, although the possible therapeutic benefits resulting from the concomitant use of nucleoside analogs and nucleoside transporter inhibitors are being evaluated. In particular, by modulating the extracellular concentrations of adenosine in several tissues (they block the intracellular uptake and reuptake of adenosine), ENT inhibitors can potentiate adenosine-mediated protective cellular processes. For this reason, ENT inhibitors such as dipyridamole, dilazep and draflazine have been used in the treatment of cardiovascular diseases.[931,932] Dipyridamole appears to inhibit the nucleoside/adenosine uptake into endothelial cells, thus prolonging and enhancing the cardioprotective effects of adenosine. This mechanism might explain the therapeutic benefit of dipyridamole during percutaneous transluminal coronary angioplasty in humans.[933] Indeed, the infusion of the ENT inhibitor R-75231 before coronary artery occlusion has been reported to increase myocardial adenosine levels, to reduce infarct size, and to enhance postischemic recovery in swine.[934] Administration of nucleoside transporter inhibitors has also been reported to have promising beneficial effects in the treatment of ischemic neuronal damage: The xanthine derivative propentofylline has been studied as a neuroprotective drug for brain ischemic injury.[935] One of the proposed mechanisms by which propentofylline appears to protect neurons from ischemic cerebral damage is the inhibition of the cellular uptake of adenosine (due to the ability of propentofylline to block ENT1, ENT2, and CNT1), resulting in increased extracellular concentrations of endogenous adenosine and enhanced stimulation of adenosine receptors.[935–938] Indeed, preischemic administration of the phosphorylated prodrug form of nitrobenzylthioinosine increased brain adenosine levels and reduced ischemic neuronal death in rats.[939,940]

Inhibition of nucleoside transporter activity may also be a potentially useful strategy in the treatment of chronic pain. Adenosine A1 receptor activation in the spinal cord has antinociceptive effects,[941] and in mice, administration of nucleoside transport inhibitors have been described to enhance the opioid-mediated antinociception.[942] Moreover, ENT1 was found at significant levels in the dorsal horn of the rat spinal cord, and administration of the ENT1 inhibitor nitrobenzylthioinosine resulted in increased extracellular adenosine levels that modulated glutamate release at the presynaptic A1 receptors. This would finally lead to a suppression of nociceptive neurotransmission.[943]

Another potentially beneficial drug–drug interaction mediated by nucleoside transporters involves nucleoside transporter inhibitors and nonnucleoside antimetabolite drugs: In vitro studies reported that dipyridamole was able to potentiate the cytotoxic effect of methotrexate, pemetrexed disodium (Alimta), lometrexol (which inhibits de novo synthesis of purine and pyrimidine nucleotides), and certain thymidylate synthase inhibitors by preventing the nucleoside and/or base cellular salvage mediated by nucleoside transporters.[895,944–949] However, clinical trials that evaluate whether coadministration of dipyridamole and antimetabolite drugs can be a useful strategy to improve the efficacy of antimetabolite therapy have been unsuccessful. This is probably due to the presence of nucleobase transporters in cancer cells that are insensitive to nucleoside transport inhibitors and to low concentrations of free dipyridamole in the plasma, due to its avid binding to serum proteins (in particular to the $\alpha 1$ acid glycoprotein).[892,895,950]

Finally, given the recent finding that several kinase inhibitors widely used in the treatment of certain cancer types can significantly block the activity of ENT1 and/or ENT2, clinically relevant drug–drug interactions between kinase inhibitors and nucleoside analogs can be speculated and should be explored to avoid potential antagonistic interactions between these drugs. On the other hand, as ENTs can mediate both cellular uptake and efflux of therapeutic nucleoside analogs, the chronomodulated coadministration of selective ENT inhibitors with nucleoside analogs (in particular, if they are also transported into cells by CNTs) may improve the cytotoxicity of nucleoside analogs. In vitro, dipyridamole has been reported to increase intracellular ara-CTP levels by blocking cellular efflux of cytosine arabinoside in human leukemic myeloblasts.[951] In the same way, in cultured human leukemic lymphoblasts the cytotoxicity of cladribine (2-chlorodeoxyadenosine) has been shown to be improved by subsequent treatment with nitrobenzylthioinosine or dipyridamole.[952]

24.7. INTERACTIONS MEDIATED BY THE FOLATE UPTAKE TRANSPORTER

Impact of Polymorphism on Function Recently, a common SNP at position 80 in exon 2 of the RFC1 gene has been identified. This mutation substitutes a histidine (CAG codon) at residue 27 of the protein with an arginine (CGG codon), leading to the production of a RFC1 protein with reduced activity.[953] Recently, several epidemiological studies have analyzed the possible role of the RFC1 A80G polymorphism, independently and/or in association with maternal periconceptional folate intake and other genetic variants of cellular enzymes involved in folate metabolism (e.g., methylenetetrahydrofolate reductase), as a significant risk factor for the development of neural tube defects, Down's syndrome, and congenital heart and orofacial defects, all congenital diseases linked to impaired folate homeostasis. Although the results of these studies are not yet conclusive (or even contrasting) a contribution of the RFC1 A80G variant to the susceptibility of the population to these and other pathological conditions cannot be excluded.[953–959] Moreover, the G80A RFC1 polymorphism, via the modulation of extra- and intracellular folate levels, has been proposed to be a protective factor against thrombosis [960] and to contribute to interindividual variability in response and toxicities to high-dose therapy with methotrexate.[961] In contrast with recent epidemiological findings supporting an association between polymorphisms in folate-metabolizing enzymes and risk of colorectal and other cancers, RFC1 G80A polymorphism was not associated with altered colon cancer risk.[962]

Main Substrate Classes (Clinically Applied) RFC1 mediates transport of folate and its derivatives, with high affinity for reduced folates (e.g., *N*5-formyltetrahydrofolate, *N*5-methyltetrahydrofolate). RFC1 substrates also include thiamine derivatives (in particular, active thiamin metabolites such as thiamine monophosphate) [963] and several folate analog drugs used in cancer chemotherapy and for the treatment of rheumatic diseases. Indeed, RFC1 has been reported to contribute to the cellular transport of methotrexate and other classical antifolate drugs, such as raltitrexed (Tomudex, ZD1694), pemetrexed (Alimta, LY231514), the thymidylate synthase inhibitor ZD9331, and the structural analog of aminopterin N^{α}-(4-amino-4-deoxypteroyl)-N^{δ}-hemiphthaloyl-L-ornithine (PT523).[964–967]

Inhibitors Specific RFC1 inhibitors have not yet been identified completely. Probenecid and other anion transporter inhibitors have been reported to block RFC1 activity, whereas recently, sulfasalazine, a drug widely used in the treatment of rheumatoid arthritis and inflammatory bowel disease, has been shown to act as a potent noncompetitive inhibitor of RFC1.[968] Prostaglandin A1 has also been described as a noncompetitive inhibitor of the RFC1-mediated transport of methotrexate.[969] Nitric oxide (NO) and several NO donors (e.g., 3-nitroso-*N*-acetylpenicillamine, sodium nitroprusside) were recently reported to inhibit, specifically and reversibly, the activity of RFC1 in cultured human retinal pigment epithelial cells. In addition, hyperglycemic conditions have been shown to reduce the expression and activity of RFC1 in the same cell system and in retinal pigment epithelium of diabetic mice.[970,971]

Inducers In contrast, dietary folate deficiency has been reported to induce RFC1 expression significantly and specifically at the rat intestinal brush border membrane, suggesting regulation of the transport process by the dietary substrate level.[972] Moreover, 5-amino-4-imidazolecarboxamide riboside has been shown to potentiate the influx of methotrexate in the human T-lymphoblastic leukemia cell line (CCRF-CEM) by induction of RFC1, suggesting a possible role of 5-amino-4-imidazolecarboxamide riboside in enhancing the therapeutic effect of methotrexate against childhood acute lymphoblastic leukemia.[973]

Pharmacological and Toxicological Function The contribution of RFC1 to pharmacokinetic, therapeutic, and toxicological profiles of folate analog drug substrates for this transporter is difficult to evaluate in in vivo animal models, as the homozygous deletion of the RFC gene in the mouse is embryonically lethal. However, for methotrexate and other classical antifolate drugs, the transport mediated by RFC1 plays an essential role in cellular uptake and tissue drug delivery: Impaired RFC1 expression and/or activity has been reported to contribute to the development of resistance to antifolate drugs used in the treatment of cancer. On the other hand, the ubiquitous expression of the RFC1 in normal tissues reduces the specific tissue distribution of antifolate drugs in cancer cells and the patient tolerability to these drug substrates for RFC1.

Drug–Drug Interactions Clinically relevant drug–drug and drug–nutrient interactions mediated by RFC1 have been evaluated. Sulfasalazine has been reported to inhibit RFC1 activity, thus interfering with intestinal uptake of folate and other coadministered RFC1 substrate drugs. Indeed, administration of sulfasalazine to patients with inflammatory bowel diseases and/or rheumatoid arthritis has been associated with macrocytic anemia and other pathophysiological conditions related to folic acid deficiency.[974–977] Sulfasalazine was able to reduce the cellular uptake of methotrexate partially at clinically relevant plasma concentrations in several in vitro models.[968,978] In patients with rheumatoid arthritis, the combination therapy of sulfasalazine–methotrexate did not display the additive/synergistic effects expected from the well-known efficacy of monotherapy with either sulfasalazine or methotrexate alone.[979,980] Coadministration of sulfasalazine and methotrexate resulted in a significant loss in cellular methotrexate uptake and in elevated homocysteine levels (revealing folate deficiency). In contrast, sulfasalazine-induced cellular folate deficiency has been suggested as a useful strategy to increase the cellular sensitivity to trimetrexate, a folate antagonist with high activity in cells with low intracellular folate levels due to impaired RFC1expression and/or activity.[968,981] These findings suggest a potential synergistic combination of sulfasalazine with trimetrexate.

Another potentially clinically relevant therapeutic strategy may be postulated in patients with retinal diseases due to impaired RFC1 activity, in particular in the presence of elevated levels of nitric oxide at the retinal pigment epithelium. As ascorbate and glutathione have been reported to reverse the inhibitory effect of nitric oxide on RFC1 activity, these antioxidant compounds could be used in patients with diabetes mellitus, where circulating levels of nitric oxide are elevated.[971] The cellular biochemistry of folates and antifolates is intriguing: It has already been studied for several decades and in different disease conditions. It is clear that a proficient folate carrier is also essential in the cellular uptake of natural folate and folate antagonists. As outlined, it is at this level that folate–drug and drug–drug interactions take place that can have clinical implications.

24.8. CONCLUSIONS

More and more it becomes clear that endogenously expressed transport proteins mediate clinically important drug–drug, food–drug, and herb–drug interactions. Drug–drug interactions may involve drugs from different therapeutic classes with widely different chemical structures and physicochemical characteristics. Up to 10% of all hospital admissions in general hospitals are caused by improper use of drugs and combinations of drugs, resulting in potentially severe drug–drug interactions. Adverse drug reactions can be especially severe when these interactions involve cytotoxic anticancer agents. Anticancer drugs are dosed close to the maximum dose tolerated, and factors affecting the pharmacokinetics may therefore greatly increase the likelihood of development of life-threatening toxicities.

Thus far, drug–drug interactions have been thought to result from inhibition of drug metabolism, inhibition of renal drug excretion, displacement out of the protein binding, or pharmaceutical interactions. However, interference at the level of ABC transporters and other drug transporters is increasingly being identified as the mechanism behind clinically important drug–drug interactions. In the field of cancer, unwanted clinically important anticancer drug–drug interactions may comprise interactions when two or more anticancer agents are given together, resulting in increased toxicity of the combination, or when a non-anticancer drug is coadministered with an anticancer agent, thereby increasing the toxicity of the anticancer agent.

The importance of ABC transporters in drug–drug interactions is increasingly being identified. Pgp is involved in the interactions between cyclosporin A or verapamil and oral digoxin. Azole antifungals, such as fluconazole and itraconazole, interact with Pgp, explaining drug interactions with digoxin and other drugs. Benzimidazoles are transported by and inhibit Pgp. Pgp regulates oral bioavailability and tissue distribution of the immunosuppressant tacrolimus. Pgp mediates drug interactions between antiretroviral drugs and comedication. Also, genetic variability in the MDR1 gene affects absorption and tissue distribution of Pgp substrate drugs. At the renal level, ABC and other transporters, including human organic anion transporters (hOAT1, 3, and 4), play significant roles in drug secretion. Interactions at this level (e.g., between MTX and NSAIDs) can result in enhanced toxicity of MTX.

After absorption, many drugs are detoxified by conjugation with glutathione, glucuronic acid, or sulfate, which results in acidic charged conjugates that cannot diffuse through cell membranes. ABC transporters can have a function in the detoxification pathway by mediating the export of these conjugates. Also, in renal clearance of drugs, ABC transporters play an important role in contributing to the active secretion of substrate drugs. Drug–drug interactions mediated by BCRP may also take place by coadministration of HIV protease inhibitors, which were recently identified as BCRP inhibitors. However, the possible role of a number of transporter proteins in mediating drug–drug interactions needs to be explored further in vivo and in humans. This concerns, for example, the ABCC family members MRP4 to MRP9, the organic cation transporters, the organic cation/carnitine transporters, the peptide transporters, the monocarboxylate transporters, and the nucleoside transporters. In addition, possible consequences of genetic variability on the expression and activity of these transporters need to be explored in clinical studies.

Furthermore, complementary and alternative medicine (CAM) use, such as herbs, food, and vitamins, by patients has increased significantly in recent years. Surveys have shown that the prevalence of CAM use among cancer patients receiving conventional therapy is 54 to 77%, and that about 72% do not inform their treating physician. CAM use significantly increases the risk of interactions with anticancer drugs, especially because of the small therapeutic range and steep dose–toxicity curve of these drugs. Clinically relevant problems are seen with St. John's wort and grapefruit juice. St. John's wort decreases significantly the plasma levels of SN38 and the active metabolite of irinotecan, and increases imatinib clearance. Grapefruit juice reduces oral bioavailability of etoposide. However, it is expected that CAM–drug interactions are responsible for more of the so-far-unresolved interindividual variation and clinical problems seen in cancer and noncancer patients.

The main causes of interactions are changes in the pharmacokinetics of drugs, although interactions at the pharmacodynamic level are also possible. Many drugs are cleared by biotransformation and are subsequently transported by Pgp and BCRP (ABCG2) or other transporters. Altered expression or activity of these enzymes can lead to decreased therapeutic efficacy or increased toxicity. Expression levels of these proteins are regulated by nuclear receptors such as the pregnane X receptor, constitutive androstane receptor, and the vitamin D–binding receptor. Some (active constituents of) herbs, such as hyperforin from St. John's wort, can activate one or more of these receptors, thereby increasing the expression of metabolizing enzymes and transporters. Garlic, ginseng, milk thistle, and kava inhibit CYP activity directly. Pgp activity can be inhibited by curcumin, milk thistle, and quercetin, and its expression induced by St. John's wort. However, for most CAM, it is unknown whether they affect metabolizing enzymes and/or drug transporters, leading to unwanted PK interactions with drug therapy. Increased knowledge of drug–drug, food–drug, and herb–drug interactions and of genetic factors affecting pharmacokinetics and pharmacodynamics is expected to improve drug safety and will enable drug therapy tailored to individual patients' needs.

REFERENCES

1. Fattinger K, Roos M, Vergeres P, Holenstein C, Kind B, Masche U, Stocker DN. 2000. Epidemiology of drug exposure and adverse drug reaction in two Swiss departments of internal medicine. Br J Clin Pharmacol 499: 158–167.
2. Zoppi M, Braunschweig S, Kuenzi UP, Maibach R, Hoigne R. 2000. Incidence of lethal adverse drug reactions in the comprehensive hospital drug monitoring, a 20-year survey, 1974–1993, based on the data of Berne/St.Gallen. Eur J Clin Pharmacol 56: 427–430.
3. van Meerten E, Verweij J, Schellens JH.1995. Antineoplastic agents: drug interactions of clinical significance. Drug Saf 12: 168–182.
4. Beijnen JH, Schellens JHM. 2004. Drug interactions in oncology. Lancet Oncol 5: 489–496.
5. Hoffmeyer S, Burk O, von Richter O, Arnold HP, Brockmöller J, Johne A, Cascorbi I, Gerloff T, Roots I, Eichelbaum M, Brinkmann U. 2000. Functional polymorphisms of the human multidrug resistance gene: multiple sequence variations and correlation of one allele with P-glycoprotein expression and activity in vivo. Natl Acad Sci U S A 97: 3473–3478.
6. Marzolini C, Paus E, Buclin T, Kim RB. 2004. Polymorphisms in human MDR1 (P-glycoprotein): recent advances and clinical relevance. Clin Pharmacol Ther 75: 13–33.
7. Mathijssen RHJ, Marsh S, Karlsson MO, Xie R, Baker SD, Verweij J, Sparreboom A, McLeod HL. 2003. Irinotecan pathway genotype analysis to predict pharmacokinetics. Clin Cancer Res 9: 3246–3253.

8. Zhou Q, Sparreboom A, Tan EH, Cheung YB, Lee A, Poon D, Lee EJ, Chowbay B. 2005. Pharnacogenetic profiling across the irinotecan pathway in Asian patients with cancer. Br J Clin Pharmacol 59: 415–424.
9. de Jong FA, de Jonge MJ, Verweij J, Mathijssen RH. 2006. Role of pharmacogenetics in irinotecan therapy. Cancer Lett 234: 90–106.
10. Yamaguchi H, Hishinuma T, Endo N, Tsukamoto H, Kishikawa Y, Sato M, Murai Y, Hiratsuka M, Ito K, Okamura C, et al. 2006. Genetic variation in ABCB1 influences paclitaxel pharmacokinetics in Japanese patients with ovarian cancer. Int J Gynecol Cancer 16: 979–985.
11. Henningsson A, Marsh S, Loos WJ, Karlsson MO, Garsa A, Mross K, Mielke S, Vigano L, Locatelli A, Verweij J, Sparreboom A, McLeod HL. 2005. Association of CYP2C8, CYP3A4, CYP3A5, and ABCB1 polymorphisms with the pharmacokinetics of paclitaxel. Clin Cancer Res 11: 8097–8104.
12. Verstuyft C, Schwab M, Schäffeler E, Kerb R, Brinkmann U, Jaillon P, Funck-Brentano C, Becquemont L. 2003. Digoxin pharmacokinetics and MDR1 genetic polymorphisms. Eur J Clin Pharmacol 58: 809–812.
13. Schwab M, Eichelbaum M, Fromm MF. 2003. Genetic polymorphisms of the human MDR1 drug transporter. Annu Rev Pharmacol Toxicol 43: 285–307.
14. Siddiqui A, Kerb R, Weale ME, Brinkmann U, Smith A, Goldstein DB, Wood NW, Sisodiya SM. 2003. Association of multidrug resistance in epilepsy with a polymorphism in the drug-transporter gene ABCB1. N Engl J Med 348: 1442–1448.
15. Wada M. 2006. Single nucleotide polymorphisms in ABCC2 and ABCB1 genes and their clinical impact in physiology and drug response. Cancer Lett 234: 40–50.
16. Siegsmund M, Brinkmann U, Schäffeler E, Weirich G, Schwab M, Eichelbaum M, Fritz P, Burk O, Decker J, Alken P, et al. 2002. Association of the P-glycoprotein transporter MDR1(C3435T) polymorphism with the susceptibility to renal epithelial tumors. J Am Soc Nephrol 13: 1847–1854.
17. Drozdzik M, Bialecka M, Mysliwiec K, Honczarenko K, Stankiewicz J, Sych Z. 2003. Polymorphism in the P-glycoprotein drug transporter MDR1 gene: a possible link between enviromental and genetic factors in Parkinson's disease. Pharmacogenetics 13: 259–263.
18. Furuno T, Landi MT, Ceroni M, Caporaso N, Bernucci I, Nappi G, Martignoni E, Schäffeler E, Eichelbaum M, Schwab M, Zanger UM. 2002. Expression polymorphism of the blood–brain barrier component P-glycoprotein (MDR1) in relation to Parkinson's disease. Pharmacogenetics 12: 529–534.
19. Schwab M, Schäffeler E, Marx C, Fromm MF, Kaskas B, Metzler J, Stange E, Herfarth H, Schoelmerich J, Gregor M, et al. 2003. Association between the C3435T MDR1 gene polymorphism and susceptibility for ulcerative colitis. Gastroenterology 124: 26–33.
20. Fellay J, Marzolini C, Meaden ER, Back DJ, Buclin T, Chave JP, Decosterd LA, Furrer H, Opravil M, Pantaleo G, et al. 2002. Response to antiretroviral treatment in HIV-1-infected individuals with allelic variants of the multidrug resistance transporter 1: a pharmacogenetics study. Lancet 359: 30–36.
21. Nasi M, Borghi V, Pinti M, Bellodi C, Lugli E, Maffei S, Troiano L, Richeldi L, Mussini C, Esposito R, Cossarizza A. 2003. MDR1 C3435T genetic polymorphism does not influence the response to antiretroviral therapy in drug-naive HIV-positive patients. AIDS 17: 1696–1698.

22. Brumme ZL, Dong WW, Chan KJ, Hogg RS, Montaner JS, O'Shaughnessy MV, Harrigan PR. 2003. Influence of polymorphisms whitin thr CX3CR1 and MDR-1 genes on initial antiretroviral therapy response. AIDS 17: 201–208.

23. Zhu D, Taguchi-Nakamura H, Goto M, Odawara T, Nakamura T, Yamada H, Kotaki H, Sugiura W, Iwamoto A, Kitamura Y. 2004. Influence of single-nucleotide polymorphisms in the multidrug resistance-1 gene on the cellular export of nelfinavir and its clinical implication for highly active antiretroviral therapy. Antivir Ther 9: 929–935.

24. Winzer R, Langmann P, Zilly M, Tollmann F, Schubert J, Klinker H, Weissbrich B. 2005. No influence of the P-glycoprotein polymorphisms MDR1 G2677T/A and C3435T on the virological and immunological response in treatment naive HIV-positive patients. Ann Clin Microbiol Antimicrob 4: 3.

25. Saitoh A, Singh KK, Powell CA, Fenton T, Fletcher CV, Brundage R, Starr S, Spector SA. 2005. An MDR1-3435 variant is associated with higher plasma nelfinavir levels and more rapid virologic response in HIV-1 infected children. AIDS 19: 371–380.

26. Verstuyft C, Marcellin F, Morand-Joubert L, Launay O, Brendel K, Mentre F, Peytavin G, Gerard L, Becquemont L, Aboulker JP. 2005. Absence of association between MDR1 genetic polymorphisms, indinavir pharmacokinetics and response to highly active antiretroviral therapy. AIDS 19: 2127–2131.

27. Babaoglu MO, Bayar B, Aynacioglu AS, Kerb R, Abali H, Celik I, Bozkurt A. 2005. Association of the ABCB1 3435C>T polymorphism with antiemetic efficacy of 5-hydroxytryptamine type 3 antagonists. Clin Pharmacol Ther 78: 619–626.

28. Roberts RL, Joyce PR, Mulder RT, Begg EJ, Kennedy MA. 2002. A common P-glycoprotein polymorphism is associated with nortriptyline-induced postural hypotension in patients treated for major depression. Pharmacogenom J 2: 191–196.

29. Yamauchi A, Ieiri I, Kataoka Y, Tanabe M, Nishizaki T, Oishi R, Higuchi S, Otsubo K, Sugimachi K. 2002. Neurotoxicity induced by tacrolimus after liver transplantation: relation to genetic polymorphisms of the ABCB1 (MDR1) gene. Transplantation 74: 571–572.

30. Fiegenbaum M, de Silveira FR, van der Sand CR, van der Sand LC, Ferreira ME, Pires RC, Hutz MH. 2005. The role of common variants of ABCB1, CYP3A4 and CYP3A5 genes in lipid-lowering efficacy and safety of simvastatin treatment. Clin Pharmacol Ther 78: 551–558.

31. Goreva OB, Grishanova AY, Mukhin OV, Domnikova NP, Lyakhovich VV. 2003. Possible prediction of the efficiency of chemotherapy in patients with with lymphoproliferative diseases based on MDR1 gene G2677T and C3435T polymorphisms. Bull Exp Biol Med 136: 183–185.

32. Illmer T, Schuler US, Thiede C Schwarz UI, Kim RB, Gotthard S, Freund D, Schakel U, Ehninger G, Schaich M. 2002. MDR1 gene polymorphisms affect therapy outcome in acute myeloid leukemia patients. Cancer Res 62: 4955–4962.

33. Jamroziak K, Mlynarski W, Balcerczak E, Mistygacz M, Trelinska J, Mirowski M, Bodalski J, Robak T. 2004. Functional C3435T polymorphism of MDR1 gene: an impact on genetic susceptibility and clinical outcome of childhood acute lymphoblastic leukemia. Eur J Haematol 72: 314–321.

34. Jamroziak K, Robak T. 2004. Pharmacogenomics of MDR1/ABCB1 gene: the influence on risk and clinical outcome of haematological malignancies. Hematology 9: 91–105.

35. Jamroziak K, Balcerczak E, Cebula B, Kowalczyk M, Panczyk M, Janus A, Smolewski P, Mirowski M, Robak T. 2005. Multi-drug transporter MDR1 gene polymorphism and prognosis in adult acute lymphoblastic leukemia. Pharmacol Rep 57: 882–888.

36. Kafka A, Sauer G, Jaeger C, Grundmann R, Kreienberg R, Zeillinger R, Deissler H. 2003. Polymorphism C3435T of the MDR-1 gene predicts response to preoperative chemotherapy in locally advanced breast cancer. Int J Oncol 22: 1117–1121.

37. Green H, Soderkvist P, Rosenberg P, Horvath G, Peterson C. 2006. mdr-1 single-nucleotide polymorphisms in ovarian cancer tissue: G2677T/A correlates with response to paclitaxel chemotherapy. Clin Cancer Res 12: 854–859.

38. Kurata Y, Ieiri I, Kimura M, Morita T, Irie S, Urae A, Ohdo S, Ohtani H. 2002. Role of human MDR1 gene polymorphisms in bioavailability: interaction of digoxin, a substrate of P-glycoprotein. Clin Pharmacol Ther 72: 209–219.

39. Schinkel AH, Jonker JW. 2003. Mammalian drug efflux transporters of the ATP binding cassette (ABC) family: an overview. Adv Drug Deliv Rev 55: 3–29.

40. Tsuruo T, Lida H, Tsukagoshi S, Sakurai Y. 1981. Overcoming of vincristine resistance in P388 leukemia in vivo and in vitro through enhanced cytotoxicity of vincristine and vinblastin by verapamil. Cancer Res 41: 1967–1972.

41. Sikic BI. 1997. Pharmacological approaches to reversing multidrug resistance. Semin Hematol 34: 40–47.

42. Sikic BI. 1999. New approaches in cancer treatment. Ann Oncol 10: 149–153.

43. De Bruin M, Miyake K, Litman T, Robey R, Bates SE. 1999. Reversal of resistance by GF120918 in cell lines expressing the ABC half-transporter, MXR. Cancer Lett 146: 117–126.

44. Pauli-Magnus C, Rekersbrink S, Kloz U, Gromm MF. 2001. Interaction of omeprazole, lansoprazole and pantoprazole with P-glycoprotein. Arch Pharmacol 364: 551–557.

45. Breedveld P, Zelcer N, Pluim D, Sonmezer O, Tibben MM, Beijnen JH, Schinkel AH, van Tellingen O, Borst P, Schellens JHM. 2004. Mechanism of the pharmacokinetic interaction between methotrexate and benzimidazoles: potential role for breast cancer resistance protein in clinical drug–drug interactions. Cancer Res 16: 5804–58011.

46. Reid T, Yuen A, Catolico M, Carlson RW. 1993. Impact of omeprazole on the plasma clearance of methotrexate. Cancer Chemother Pharmacol 33: 82–84.

47. Mannesse CK, Derkx FH, de Ridder MA, Man in't veld AJ, van der Cammen TJ. 2000. Contribution of adverse drug reactions to hospital admission of older patients. Age Ageing 29: 35–39.

48. Tröger U, Stotzel B, Martens-Lobenhoffe J, Gollnick H, Meijer FP. 2002. Drug Points: Severe myalgia from an interaction between treatments with pantoprazole and methotrexate. Br Med J 22: 1497.

49. De Maat MM, Ekhart GC, Huitema AD, Koks CH, Mulder JW, Beijnen JH. 2003. Drug interactions between antiretroviral drugs and comedication. Clin Pharmacokinet 42: 223–282.

50. DuBuske LM. 2005. The role of P-glycoprotein and organic anion-transporting polypeptides in drug interactions. Drug Saf 28: 789–801.

51. Sankatsing SUC, Beijnen JH, Schinkel AH, Lange JMA, Prins JM. 2004. P-Glycoprotein in human immunodeficiency virus type 1 infection and therapy. Antimicrob Agents Chemother 48: 1073–1081.

52. Schuetz EG, Schinkel AH, Relling MV, Schuetz JD. 1996. P-Glycoprotein: a major determinant of rifampicin-inducible expression of cytochrome P4503A in mice and humans. Proc Natl Acad Sci U S A 93: 4001–4005.

53. Geick A, Eichelbaum M, Burk O. 2001. Nuclear receptor response elements mediate induction of intestinal MDR1 by rifampin. J Biol Chem 276: 14581–14587.

54. Synold TW, Dussault I, Forman BM. 2001. The orphan nuclear receptor SXR coordinately regulates drug metabolism and efflux. Nat Med 7: 584–590.

55. Schinkel AH, Smit JJM, van Tellingen O, Beijnen JH, Wagenaar E, van Deemter L, Mol CAAM, van der Valk MA, Robanus Maandag EC, te Riele HPJ, Berns AJM, Borst P. 1994. Disruption of the mouse mdr1a P-glycoprotein gene leads to a deficiency in the blood–brain barrier and to increased sensitivity to drugs. Cell 77: 491–502.

56. Schinkel AH, Wagenaar E, van Deemter L, Mol CAAM, Borst P. 1995. Absence of the mdr1a P-glycoprotein in mice affects tissue distribution and pharmacokinetics of dexamethasone, digoxin and cyclosporin A. J Clin Invest 96: 1698–1705.

57. Schinkel AH, Wagenaar E, Mol CAAM, Van Deemter L. 1996. P-Glycoprotein in the blood–brain barrier of mice influences the brain penetration and pharmacological activity of many drugs. J Clin Invest 97: 2517–2524.

58. Schinkel AH, Mayer U, Wagenaar E, Mol CA, van Deemter L, Smit JJ, van der Valk MA, Voordouw AC, Spits H, van Tellingen O, Zijlmans JM, Fibbe WE, Borst P. 1997. Normal viability and altered pharmacokinetics in mice lacking mdr1-type (drug-transporting) P-glycoproteins. Proc Natl Acad Sci U S A 94: 4028–4033.

59. Hunter J, Hirst BH, Simmons NL. 1993. Drug absorption limited by P-glycoprotein-mediated secretory drug transport in human intestinal epithelial Caco-2 cell layers. Pharm Res 10: 743–749.

60. Sparreboom A, van Asperen J, Mayer U, Schinkel AH, Smit JW, Meijer DK, Borst P, Nooijen WJ, Beijnen JH, van Tellingen O. 1997. Limited oral bioavailability and active epithelial excretion of paclitaxel (Taxol) caused by P-glycoprotein in the intestine. Proc Natl Acad Sci U S A 94: 2031–2035.

61. Bardelmeijer HA, Ouwehand M, Buckle T, Huisman MT, Schellens JH, Beijnen JH, van Tellingen O. 2002. Low systemic exposure of oral docetaxel in mice resulting from extensive first-pass metabolism is boosted by ritonavir. Cancer Res 62: 6158–6164.

62. van Asperen J, van Tellingen O, Sparreboom A, Schinkel AH, Borst P, Nooijen WJ, Beijnen JH. 1997. Enhanced oral bioavailability of paclitaxel in mice treated with the P-glycoprotein blocker SDZ PSC 833. Br J Cancer 76: 1181–1183.

63. van Asperen J, van Tellingen O, van der Valk MA, Rozenhart M, Beijnen JH. 1998. Enhanced oral absorption and decreased elimination of paclitaxel in mice cotreated with cyclosporin A. Clin Cancer Res 4: 2293–2297.

64. Bardelmeijer HA, Beijnen JH, Brouwer KR, Rosing H, Nooijen WJ, Schellens JHM, Van Tellingen O. 2000. Increased oral bioavailability of paclitaxel by GF120918 in mice through selective modulation of P-glycoprotein. Clin Cancer Res 6: 4416–4421.

65. Lankas GR, Wise LD, Cartwright ME, Pippert T, Umberhauer DR. 1998. Placental P-glycoprotein deficiency enhances susceptibility to chemically induced birth defects in mice. Reprod Toxicol 12: 457–463.

66. Smith JW, Huisman MT, van Tellingen O, Wiltshire HR, Schinkel AH. 1999. Absence or pharmacological blocking of placental P-glycoprotein profoundly increases fetal drug exposure. J Clin Invest 104: 1441–1447.

67. Sudhakaran S, Ghabrial H, Nation RL, Kong DC, Gude NM, Angus PW, Rayner CR. 2005. Differential bidirectional transfer of indinavir in the isolated perfused human placenta. Antimicrob Agents Chemother 49: 1023–1028.
68. Molsa M, Heikkinen T, Hakkola J, Hakala K, Wallerman O, Wadelius M, Wadelius C, Laine K. 2005. Functional role of P-glycoprotein in the human blood–placental barrier. Clin Pharmacol Ther 78: 123–131.
69. Wacher VJ, Wu CY, Benet LZ. 1995. Overlapping substrate specificities and tissue distribution of cytochrome P450 3A and P-glycoprotein: implications for drug delivery and activity in cancer chemotherapy. Mol Carcinog 13: 129–134.
70. Watkins PB. 1997. The barrier function of CYP3A4 and P-glycoprotein in the small bowel. Adv Drug Deliv Rev 27: 161–170.
71. Bigger JT, Leahey EB. 1982. Quinidine and digoxin. An important interaction. Drugs 24: 229–239.
72. Leahey EB, Reiffel JA, Heissenbuttel RH, Drusin RE, Lovejoy WP, Bigger JT Jr. 1979. Enhanced cardiac effect of digoxin during quinidine treatment. Arch Intern Med 139: 519–521. Abstract only.
73. Doering W. 1979. Quinidine–digoxin interaction: pharmacokinetics, underlying mechanism and clinical implications. N Engl J Med 301: 400–404.
74. Klein HO, Lang R, Weiss E, Di Segni E, Libhaber C, Guerrero J, Kaplinsky E. 1982. The influence of verapamil on serum digoxin concentration. Circulation 65: 998–1003.
75. Belz GG, Doering W, Munkes R, Matthews J. 1983. Interaction between digoxin and calcium antagonists and antiarrhythmic drugs. Clin Pharmacol Ther 33: 410–417.
76. Calvo MV, Martin-Suarez A, Martin Luengo C, Avila C, Cascon M, Dominguez-Gil Hurle A. 1989. Interaction between digoxin and propafenone. Ther Drug Monit 11: 10–15.
77. Westphal K, Weinbrenner A, Giessmann T, Stuhr M, Franke G, Zschiesche M, Oertel R, Terhaag B, Kroemer HK, Siegmund W. 2000. Oral bioavailability of digoxin is enhanced by talinolol: evidence for involvement of intestinal P-glycoprotein. Clin Pharmacol Ther 68: 6–12.
78. Wakasugi H, Yano I, Ito T, Hashida T, Futami T, Nohara R, Sasayama S, Inui KI. 1998. Effect of clarithromycin on renal excretion of digoxin: Interaction with P-glycoprotein. Clin Pharmacol Ther 64: 123–128.
79. Jalava KM, Partanen J, Neuvonen PJ. 1997. Itraconazole decreases renal clearance of digoxin. Ther Drug Monit 19: 609–613.
80. Maxwell DL, Gilmour-White SK, Hall MR. 1989. Digoxin toxicity due to interaction of digoxin with erythromycin. BMJ 298: 572.
81. Hinderling PH, Hartmann D. 1991. Pharmacokinetics of digoxin and main metabolites/derivatives in healthy humans. Ther Drug Monit 13: 381–401.
82. Fromm MF, Kim RB, Stein CM, Wilkinson GR, Roden DM. 1999. Inhibition of P-glycoprotein-mediated drug transport: a unifying mechanism to explain the interaction between digoxin and quinidine. Circulation 99: 552–557.
83. Hori R, Okamura N, Aiba T, Tanigawara Y. 1993. Role of P-glycoprotein in renal tubular secretion of digoxin in the isolated perfused rat kidney. J Pharmacol Exp Ther 266: 1620–1625.
84. De Lannoy IA, Koren G, Klein J, Charuk J, Silverman M. 1992. Cyclosporin and quinidine inhibition of renal digoxin excretion: evidence for luminal secretion of digoxin. Am J Physiol 263: F613–F622.

85. Su SF, Huang JD. 1996. Inhibition of the intestinal digoxin absorption and exsorption by quinidine. Drug Metab Dispos 24: 142–147.
86. Sababi M, Borga O, Hultkvist-Bengtsson U. 2001. The role of P-glycoprotein in limiting intestinal regional absorption of digoxin in rats. Eur J Pharm Sci 14: 21–27.
87. Drescher S, Glaeser H, Murdter T, Hitzl M, Eichelbaum M, Fromm MF. 2003. P-Glycoprotein-mediated intestinal and biliary digoxin transport in humans. Clin Pharmacol Ther 73: 223–231.
88. Woodland C, Verjee Z, Giesbrecht E, Koren G, Ito S. 1997. The digoxin–propafenone interaction: characterization of a mechanism using renal tubular cell monolayers. J Pharmacol Exp Ther 283: 39–45.
89. Verschraagen M, Koks CHW, Schellens JHM, Beijnen JH. 1999. P-glycoprotein system as a determinant of drug interactions: the case of digoxin–verapamil. Pharmacol Res 40: 301–306.
90. Phillips EJ, Rachlis AR, Ito S. 2003. Digoxin toxicity and ritonavir: A drug interaction mediated through P-glycoprotein? AIDS 17: 1577–1578.
91. Eagling VA, Back DJ, Barry MG. 1997. Differential inhibition of cytochrome P450 isoforms by the protease inhibitors, ritonavir, saquinavir and indinavir. Br J Clin Pharmacol 44: 190–194.
92. Kempf DJ, Marsh KC, Kumar G, Rodriques AD, Denissen JF, McDonald E, Kukulka MJ, Hsu A, Granneman GR, Baroldi PA, et al. 1997. Pharmacokinetic enhancement of inhibitors of the human immunodeficiency virus protease by coadministration with ritonavir. Antimicrob Agents Chemother 41: 654–660.
93. Koudriakova T, Iatsimirskaia E, Utkin I, Gangl E, Vouros P, Storozhuk E, Orza D, Marinina J, Gerber N. 1998. Metabolism of the human immunodeficiency virus protease inhibitors indinavir and ritonavir by human intestinal microsomes and expressed cytochrome P4503A4/3A5: mechanism-based inactivation of cytochrome P4503A by ritonavir. Drug Metab Dispos 26: 552–561.
94. Kumar GN, Dykstra J, Roberts EM, Jayanti VK, Hickman D, Uchic J, Yao Y, Surber B, Thomas S, Granneman GR. 1999. Potent inhibition of the cytochrome P-450 3A-mediated human liver microsomal metabolism of a novel HIV protease inhibitor by ritonavir: a positive drug–drug interaction. Drug Metab Dispos 27: 902–908.
95. Hsu A, Granneman GR, Cao G, Carothers L, El-Shourbagy T, Baroldi P, Erdman K, Brown F, Sun E, Leonard JM. 1998. Pharmacokinetic interactions between two human immunodeficiency virus protease inhibitors, ritonavir and saquinavir. Clin Pharmacol Ther 63: 453–464.
96. Hsu A, Granneman R, Cao G, Carothers L, Japour A, El-Shourbagy T, Dennis S, Berg J, Erdman K, Leonard JM, Sun E. 1998. Pharmacokinetic interaction between ritonavir and indinavir in healthy volunteers. Antimicrob Agents Chemother 42: 2784–2791.
97. Alsenz J, Steffen H, Alex R. 1998. Active apical secretory efflux of the HIV protease inhibitors saquinavir and ritonavir in Caco-2 cell monolayers. Pharm Res 15: 423–428.
98. Kim RB, Fromm MF, Wandel C, Leake B, Wood AJJ, Roden DM. 1998. The drug transporter P-glycoprotein limits oral absorption and brain entry of HIV-1 protease inhibors. J Clin Invest 101: 289–294.
99. Lee CG, Gottesman MM, Cardarelli CO, Ramachandra M, Jeang KT, Ambudkar SV, Pastan I, Dey S. 1998. HIV-1 protease inhibitors are substrates for the MDR1 multidrug transporter. Biochemistry 37: 3594–3601.

100. Polli JW, Jarrett JL, Studenberg SD, Humphreys JE, Dennis SW, Brouwer KR, Woolley JL. 1999. Role of P-glycoprotein on the CNS disposition of amprenavir (141W94), an HIV protease inhibitor. Pharm Res 16: 1206–1212.

101. Huisman MT, Smit JW, Wiltshire HR, Hoetelmans RMW, Beijnen JH, Schinkel AH. 2001. P-Glycoprotein limits oral availability, brain, and fetal penetration of saquinavir even with high-doses of ritonavir. Mol Pharmacol 59: 806–813.

102. Gutmann H, Fricker G, Drewe J, Toeroek M, Miller DS. 1999. Interactions of HIV protease inhibitors with ATP-dependent drug export proteins. Mol Pharmacol 56: 383–389.

103. Ding R, Tayrouz Y, Riedel KD, Burhenn J, Weiss J, Mikus G, Haefeli WE. 2004. Substantial pharmacokinetic interaction between digoxin and ritonavir in healthy volunteers. Clin Pharmacol Ther 76: 73–84.

104. Cameron DW, Japour AJ, Xu Y, Hsu A, Mellors J, Farthing C, Cohen C, Poretz D, Markowitz M, Follansbee S, et al. 1999. Ritonavir and saquinavir combination therapy for the treatment of HIV infection. AIDS 13: 213–224.

105. Kaufmann GR, Duncombe C, Cunningham P, Beveridge A, Carr A, Sayer D, French M, Cooper DA. 1998. Treatment response and durability of a double protease inhibitor therapy with saquinavir and ritonavir in an osbervational cohort of HV-1-infected individuals. AIDS 12: 1625–1630.

106. Kumar GN, Rodrigues AD, Buko AM, Denissen JF. 1996. Cytochrome P450-mediated metabolism of the HIV-1 protease inhibitor ritonavir (ABT-538) in human liver microsomes. J Pharmacol Exp Ther 277: 423–431.

107. Hebert MF, Lam AY. 1999. Diltiazem increases tacrolimus concentrations. Ann Pharmacother 33: 680–682.

108. Arima H, Yunomae K, Hirayama F, Uekama K. 2001. Contribution of P-glycoprotein to the enhancing effects of dimethyl-β-cyclodextrin on oral bioavailability of tacrolimus. J Pharmacol Exp Ther 297: 547–555.

109. Hashimoto Y, Sasa H, Shimomura M, Inui K. 1998. Effects of intestinal and hepatic metabolism on the bioavailability of tacrolimus in rats. Pharm Res 15: 1609–1613.

110. Sipe BE, Jones RJJ, Bokhart GH. 2003. Rhabdomyolysis causing AV blockade due to possible atorvastatin, esomeprazole, and clarithromycin interaction. Ann Pharmacother 37: 808–811.

111. Joerger M, Huitema AD, van den Bongard HJ, Baas P, Schornagel JH, Schellens JH, Beijnen JH. 2006. Determinants of the elimination of methotrexate and 7-hydroxy-methotrexate following high-dose infusional therapy to cancer patients. Br J Clin Pharmacol 62(1): 71–80.

112. Schwarz UI, Gramatte T, Krappweis J, Oertel R, Kirch W. 2000. P-Glycoprotein inhibitor erythromycin increases oral bioavailability of talinolol in humans. Int J Clin Pharmacol Ther 38: 161–167.

113. Wetterich U, Spahn-Langguth H, Mutschler E, Terhaag B, Rösch W, Langguth P. 1996. Evidence for intestinal secretion as an additional clearance pathway of talinolol enantiomers: concentration- and dose-dependent absorption in vitro and in vivo. Pharm Res 13: 514–522.

114. Spahn-Langguth H, Baktir G, Radschuweit A, Okyar A, Terhaag B, Ader P, Hanafy A, Langguth P. 1998. P-Glycoprotein transporters and the gastrointestinal tract: evaluation of the potential in vivo relevance of in vitro data employing talinolol as model compound. Int J Clin Pharmacol Ther 36: 16–24.

115. Gramatté T, Oertel R. 1999. Intestinal secretion of intravenous talinolol is inhibited by luminal R-verapamil. Clin Pharmacol Ther 66: 239–245.

116. Schwarz UI, Gramatté T, Krappweis J, Berndt A, Oertel R, von Richter O, Kirch W. 1999. Unexpected effect of verapamil on oral bioavailability of the β-blocker talinolol in humans. Clin Pharmacol Ther 65: 283–290.

117. Gramatté T, Oertel R, Terhaag B, Kirch W. 1996. Direct demonstration of small intestinal secretion and site-dependent absorption of the β-blocker talinolol in humans. Clin Pharmaocol Ther 59: 541–549.

118. Greiner B, Eichelbaum M, Fritz P, Kreichgauer HP, von Richter O, Zundler J, Kroemer HK. 1999. The role of intestinal P-glycoprotein in the interaction of digoxin and rifampin. J Clin Invest 104: 147–153.

119. Westphal K, Weinbrenner A, Zschiesche M, Franke G, Knoke M, Oertel R, Fritz P, von Richter O, Warzok R, Hachenberg T, et al. 2000. Induction of P-glycoprotein by rifampin increases intestinal secretion of talinolol in human beings: a new type of drug/drug interaction. Clin Pharmacol Ther 68: 345–355.

120. Hamman MA, Bruce MA, Haehner-Daniels BD, Hall SD. 2001. The effect of rifampin administration on the disposition of fexofenadine. Clin Pharmacol Ther 69: 114–121.

121. Hebert MF, Roberts JP, Prueksaritanont T, Benet LZ. 1992. Bioavailability of cyclosporine with concomitant rifampin administration is markedly less than predicted by hepatic enzyme induction. Clin Pharmacol Ther 52: 453–457.

122. Cvetkovic M, Leake B, Fromm MF, Wilkinson GR, Kim RB 1999. OATP and P-glycoprotein transporters mediate the cellular uptake and excretion of fexofenadine. Drug Metab Dispos 27: 866–871.

123. Ruschitzka F, Mejer PJ, Turina M, Luscher TF, Noll G. 2000. Acute heart transplant rejection due to Saint John's wort. Lancet 355: 548–549.

124. Piscitelli SC, Burstein AH, Chaitt D, Alfaro RM, Falloon J. 2000. Indinavir concentrations and St John's wort. Lancet 355: 547–548.

125. Durr D, Stieger B, Kullak-Ublick GA, Rentsch KM, Steinert HC, Meier PJ, Fattinger K. 2000. St. John's wort induces intestinal P-glycoprotein/MDR1 and intestinal and hepatic CYP3A4. Clin Pharmacol Ther 68: 598–604.

126. Zhou S, Chan E, Pan SQ, Huang M, Lee EJ. 2004. Pharmacokinetic interactions of drugs with St. John's wort. J Psychopharmacol 18: 262–276.

127. Schwarz UI, Buschel B, Kirch W. 2003. Unwanted pregnancy on self-medication with St. John's wort despite hormonal contraception. Br J Clin Pharmacol 55: 112–113.

128. Dresser GK, Schwarz UI, Wilkinson GR, Kim RB. 2003. Coordinate induction of both cytochrome P4503A and MDR1 by St. John's wort in healthy subjects. Clin Pharmacol Ther 73: 41–50.

129. Wang Z, Hamman MA, Huang SM, Lesko LJ, Hall SD. 2002. Effect of St. John's wort on the pharmacokinetics of fexofenadine. Clin Pharmacol Ther 71: 414–420.

130. Mai I, Stormer E, Bauer S, Kruger H, Budde K, Roots I. 2003. Impact of St. John's wort treatment on the pharmacokinetics of tacrolimus and mycophenolic acid in renal transplant patients. Nephrol Dial Transplant 18: 819–822.

131. Hebert MF, Park JM, Chen YL, Akhtar S, Larson AM. 2004. Effects of St. John's wort (*Hypericum perforatum*) on tacrolimus pharmacokinetics in healthy volunteers. J Clin Pharmacol 44: 89–94.

132. Wang LS, Zhou G, Zhu B, Wu J, Wang JG, Abd El-Aty AM, Li T, Liu J, Yang TL, Wang D, Zhong XY, Zhou HH. 2004. St. John's wort induces both cytochrome P450 3A4-catalyzed sulfoxidation and 2C19-dependent hydroxylation of omeprazole. Clin Pharmacol Ther 75: 191–197.

133. Tannergren C, Engman H, Knutson L, Hedeland M, Bondesson U, Lennernas H. 2004. St. John's wort decreases the bioavailability of R- and S-verapamil through induction of the first-pass metabolism. Clin Pharmacol Ther 75: 298–309.

134. Frye RF, Fitzgerald SM, Lagattuta TF, Hruska MW, Egorin MJ. 2004. Effect of St. John's wort on imatinib mesylate pharmacokinetics. Clin Pharmacol Ther 76: 323–329.

135. Bauer S, Stormer E, Johne A, Kruger H, Budde K, Neumayer HH, Roots I, Mai I. 2003. Alterations in cyclosporin A pharmacokinetics and metabolism during treatment with St. John's wort in renal transplant patients. Br J Clin Pharmacol 55: 203–211.

136. Hall SD, Wang Z, Huang SM, Hamman MA, Vasavada N, Adigun AQ, Hilligoss JK, Miller M, Gorski JC. 2003. The interaction between St. John's wort and an oral contraceptive. Clin Pharmacol Ther 74: 525–535.

137. Woodcock DM, Linsenmeyer ME, Chojnowski G, Kriegler AB, Nink V, Webster LK, Sawyer WH. 1992. Reversal of multidrug resistance by surfactants. Br J Cancer 66: 62–68.

138. Martin-Facklam M, Burhenne J, Ding R, Fricker R, Mikus G, Walter-Sack I, Haefeli WE. 2002. Dose-dependent increase of saquinavir bioavailability by the pharmaceutic aid cremophor EL. Br J Clin Pharmacol 53: 576–581.

139. Tayrouz Y, Ding R, Burhenne J, Riedel KD, Weiss J, Hoppe-Tichy T, Haefeli WE, Mikus G. 2003. Pharmacokinetic and pharmaceutic interaction between digoxin and Cremophor RH40. Clin Pharmacol Ther 73: 397–405.

140. Dresser GK, Bailey DG, Leake BF, Schwarz UI, Dawson PA, Freeman DJ, Kim RB. 2002. Fruit juices inhibit organic anion transporting polypeptide-mediated drug uptake to decrease the oral availability of fexofenadine. Clin Pharmacol Ther 71: 11–20.

141. Spahn-Langguth H, Langguth P. 2001. Grapefruit juice enhances intestinal absorption of the P-glycoprotein substrate talinolol. Eur J Pharm Sci 12: 361–367.

142. Schwarz UI, Seemann D, Oertel R, Miehlke S, Kuhlisch E, Fromm MF, Kim RB, Bailey DG, Kirch W. 2005. Grapefruit juice ingestion significantly reduces talinolol bioavailability. Clin Pharmacol Ther 77: 291–301.

143. Wang EJ, Casciano CN, Clement RP, Johnson WW. 2001. Inhibition of P-glycoprotein transport function by grapefruit juice psoralen. Pharm Res 18: 432–438.

144. Takanaga H, Ohnishi A, Matsuo H, Sawada Y. 1998. Inhibition of vinblastin efflux mediated by P-glycoprotein by grapefruit juice components in Caco-2 cells. Biol Pharm Bull 21: 1062–1066.

145. Becquemont L, Verstuyft C, Kerb R, Brinkmann U, Lebot M, Jaillon P, Funck-Brentano C. 2001. Effect of grapefruit juice on digoxin pharmacokinetics in humans. Clin Pharmacol Ther 70: 311–316.

146. Rogan AM, Hamilton TC, Young RC, Klecker RW Jr, Ozols RF. 1984. Reversal of adriamycin resistance by verapamil in human ovarian cancer. Science 224: 994–996.

147. Fisher GA, Sikic BI. 1995. Clinical studies with modulators of multidrug resistance. Hematol Oncol Clin North Am 9: 363–382.

148. List AF, Spier C, Greer J, Wolff S, Hutter J, Dorr R, Salmon S, Futscher B, Baier M, Dalton W. 1993. Phase I/II trial of cyclosporine as a chemotherapy-resistance modifier in acute leukemia. J Clin Oncol 11: 1652–1660.

149. Chan HS, DeBoer G, Thiessen JJ, Budning A, Kingston JE, O'Brien JM, Koren G, Giesbrecht E, Haddad G, Verjee Z, et al. 1996. Combining cyclosporin with chemotherapy controls intraocular retinoblastoma without requiring radiation. Clin Cancer Res 2: 1499–1508.

150. List AF. 1993. Multidrug resistance: clinical relevance in acute leukemia. Oncology 7: 23–28.

151. List AF, Kopecky KJ, Willman CL, Head DR, Persons DL, Slovak ML, Dorr R, Karanes C, Hynes HE, Doroshow JH, Shurafa M, Appelbaum FR. 2001. Benefit of cyclosporine modulation of drug resistance in patients with poor-risk acute myeloid leukemia: a Southwest Oncology Group study. Blood 98: 3212–3220.

152. Belpomme D, Gauthier S, Pujade-Lauraine E, Facchini T, Goudier MJ, Krakowski I, Netter-Pinon G, Frenay M, Gousset C, Marie FN, Benmiloud M, Sturtz F. 2000. Verapamil increases the survival of patients with anthracycline-resistant metastatic breast carcinoma. Ann Oncol 11: 1471–1476.

153. Millward MJ, Cantwell BM, Munro NC, Robinson A, Corris PA, Harris AL. 1993. Oral verapamil with chemotherapy for advanced non-small cell lung cancer: a randomised study. Br J Cancer 67: 1031–1035.

154. Dalton WS, Crowley JJ, Salmon SS, Grogan TM, Laufman LR, Weiss GR, Bonnet JD. 1995. A phase III randomized study of oral verapamil as a chemosensitizer to reverse drug resistance in patients with refractory myeloma. Cancer 75: 815–820.

155. Milroy R. 1993. A randomised clinical study of verapamil in addition to combination chemotherapy in small cell lung cancer. Br J Cancer 68: 813–818.

156. Wishart GC, Bissett D, Paul J, Jodrell D, Harnett A, Habeshaw T, Kerr DJ, Macham MA, Soukop M, Leonard RC. 1994. Quinidine as a resistance modulator of epirubicin in advanced breast cancer: mature results of a placebo-controlled randomized trial. J Clin Oncol 12: 1771–1777.

157. Sonneveld P, Suciu S, Weijermans P, Beksac M, Neuwirtova R, Solbu G, Lokhorst H, van der Lelie J, Dohner H, Gerhartz H, et al. 2001. Cyclosporin A combined with vincristine, doxorubicin and dexamethasone (VAD) compared with VAD alone in patients with advanced refractory multiple myeloma: an EORTC-HOVON randomized phase III study (06914). Br J Haematol 115: 895–902.

158. Solary E, Drenou B, Campos L, De Crémoux P, Mugneret F, Moreau P, Lioure B, Falkenrodt A, Witz B, Bernard M, et al. 2003. Quinine as a multidrug resistance inhibitor: a phase 3 multicentric randomized study in adult de novo acute myelogenous leukemia. Blood 102: 1202–1210.

159. Friedenberg WR, Rue M, Blood EA, Dalton WS, Shustik C, Larson RA, Sonneveld P, Greipp PR. 2006. Phase III study of PSC-833 (valspodar) in combination with vincristine, doxorubicin, and dexamethasone (valspodar/VAD) versus VAD alone in patients with recurring or refractory multiple myeloma (E1A95): a trial of the Eastern Cooperative Oncology Group. Cancer 106: 830–838.

160. Greenberg PL, Lee SJ, Advani R, Tallman MS, Sikic BI, Letendre L, Dugan K, Lum B, Chin DL, Dewald G, et al. 2004. Mitoxantrone, etoposide, and cytarabine with or without valspodar in patients with relapsed or refractory acute myeloid leukemia and high-risk myelodysplastic syndrome: a phase III trial (E2995). J Clin Oncol 22(6): 1078–1086.

161. Fracasso PM, Goldstein LJ, de Alwis DP, Rader JS, Arquette MA, Goodner SA, Wright LP, Fears CL, Gazak RJ, Andre VA. 2004. Phase I study of docetaxel in combination

with the P-glycoprotein inhibitor, zosuquidar, in resistant malignancies. Cancer Res 10: 7220–7228.

162. Meerum Terwogt JM, Beijnen JH, Ten Bokkel Huinink WW, Rosing H, and Schellens JHM. 1999. Coadministration of oral cyclosporin A enables oral therapy with paclitaxel. Clin Cancer Res 5: 3379–3384.
163. Malingrè MM, Richel DJ, Beijnen JH, Rosing H, Koopman FJ, Ten Bokkel Huinink WW, Schot ME, Schellens JH. 2001. Coadministration of cyclosporine strongly enhances the oral bioavailability of docetaxel. J Clin Oncol 19, 1160–1166.
164. Kruijtzer CMF, Beijnen JH, Rosing H, ten Bokkel Huinink WW, Schot M, Jewell RC, Paul EM, Schellens JHM. 2002. Increased oral bioavailability of topotecan in combination with the breast cancer resistance protein (BCRP) and P-glycoprotein (Pgp) inhibitor GF120918. J Clin Oncol 20: 2943–2950.
165. Kruijtzer CMF, Beijnen JH, Schellens JHM. 2002. Improvement of oral drug treatment by temporary inhibition of drug transporters and/or cytochrome P450 in the gastrointestinal tract and liver: an overview. Oncologist 7: 516–530.
166. Kruijtzer CMF, Schellens JHM, Mezger J, Scheulen ME, Keilholz U, Beijnen JH, Rosing H, Mathôt RAA, Marcus S, Van Tinteren H, Baas P. 2002. Phase II and pharmacologic study of weekly oral paclitaxel plus cyclosporine in patients with advanced non-small-cell lung cancer. J Clin Oncol 20: 4508–4516.
167. Huisman MT, Smit JW, Schinkel AH. 2000. Significance of P-glycoprotein for the pharmacology and clinical use of HIV protease inhibitors. AIDS 14: 237–242.
168. Washington CB, Wiltshire HR, Man M, Moy T, Harris SR, Worth E, Weigl P, Liang Z, Hall D, Marriott L, Blaschke TF. 2000. The disposition of saquinavir in normal and P-glycoprotein deficient mice, rats, and in cultured cells. Drug Metab Dispos 28: 1058–1062.
169. Edwards JE, Brouwer KR, McNamara PJ. 2002. GF120918, a P-glycoprotein modulator, increases the concentration of unbound amprenavir in the central nervous system in rats. Antimicrob Agents Chemother 46: 2284–2286.
170. van Asperen J, Mayer U, van Tellingen O, Beijnen JH. 1997. The functional role of P-glycoprotein in the blood–brain barrier. J Pharm Sci 86: 881–884.
171. Kemper EM, van Zandbergen AE, Cleypool C, Mos HA, Boogerd W, Beijnen JH, van Tellingen O. 2003. Increased penetration of paclitaxel into the brain by inhibition of P-glycoprotein. Clin Cancer Res 9: 2849–2855.
172. Kemper EM, Boogerd W, Thuis I, Beijnen JH, van Tellingen O. 2004. Modulation of the blood–brain barrier in oncology: Therapeutic opportunities for the treatment of brain tumours? Cancer Treat Rev 30: 415–423.
173. Kemper EM, Verheij M, Boogerd W, Beijnen JH, van Tellingen O. 2004. Improved penetration of docetaxel into the brain by coadministration of inhibitors of P-glycoprotein. Eur J Cancer 40: 1269–1274.
174. Dai H, Marbach P, Lemaire M, Hayes M, Elmquist WF. 2003. Distribution of STI-571 to the brain is limited by P-glycoprotein-mediated efflux. J Pharmacol Exp Ther 304: 1085–1092.
175. Fellner S, Bauer B, Miller DS, Schaffrik M, Fankhanel M, Spruss T, Bernhardt G, Graeff C, Farber L, Gschaidmeier H, Buschauer A, Fricker G. 2002. Transport of paclitaxel (Taxol) across the blood–brain barrier in vitro and in vivo. J Clin Invest 110: 1309–1318.

176. Breedveld P, Beijnen JH, Schellens JHM. 2006. Use of P-glycoprotein and BCRP inhibitors to improve oral bioavailability and CNS penetration of anticancer drugs. Trends Pharmacol Sci 27: 17–24.

177. Kilic T, Alberta JA, Zdunek PR, Acar M, Iannarelli P, O'Reilly T, Buchdunger E, Black PM, Stiles CD. 2000. Intracranial inhibition of platelet-derived growth factor–mediated glioblastoma cell growth by an orally active kinase inhibitor of the 2-phenylaminopyrimidine class. Cancer Res 60: 5143–5150.

178. Petzer AL, Gunsilius E, Hayes M, Stockhammer G, Duba HC, Schneller F, Grunewald K, Poewe W, Gastl G. 2002. Low concentrations of STI571 in the cerebrospinal fluid: a case report. Br J Haematol 117: 623–625.

179. Takayama N, Sato N, O'Brien SG, Ikeda Y, Okamoto S. 2002. Imatinib mesylate has limited activity against the central nervous system involvement of Philadelphia chromosome-positive acute lymphoblastic leukaemia due to poor penetration into cerebrospinal fluid. Br J Haematol 119: 106–108.

180. Breedveld P, Pluim D, Cipriani G, Wielinga P, van Tellingen O, Schinkel AH, Schellens JH. 2005. The effect of Bcrp1 (Abcg2) on the in vivo pharmacokinetics and brain penetration of imatinib mesylate (Gleevec): implications for the use of breast cancer resistance protein and P-glycoprotein inhibitors to enable the brain penetration of imatinib in patients. Cancer Res 65: 2577–2582.

181. Sadeque AJ, Wandel C, He H, Shah S, Wood AJ. 2000. Increased drug delivery to the brain by P-glycoprotein inhibition. Clin Pharmacol Ther 68: 231–237.

182. Honjo Y, Morisaki K, Huff LM, Robey RW, Hung J, Dean M, Bates SE. 2002. Single-nucleotide polymorphism (SNP) analysis in the ABC half-transporter ABCG2 (MXR/BCRP/ABCP1). Cancer Biol Ther 1: 696–702.

183. Imai Y, Nakane M, Kage KM, Tsukahara S, Ishikawa E, Tsuruo T, Miki Y, Sugimoto Y. 2002. C421A polymorphism in the human breast cancer resistance protein gene is associated with low expression of Q141K protein and low-level drug resistance. Mol Cancer Ther 1: 611–616.

184. Iida A, Saito S, Sekine A, Mishima C, Kitamura Y, Kondo K, Harigae S, Osawa S, Nakamura Y. 2002. Catalog of 605 single-nucleotide polymorphisms (SNPs) among 13 genes encoding human ATP-binding cassette transporters: ABCA4, ABCA7, ABCA8, ABCD1, ABCD3, ABCD4, ABCE1, ABCF1, ABCG1, ABCG2, ABCG4, ABCG5, and ABCG8. J Hum Genet 47: 285–310.

185. Zamber CP, Lamba JK, Yasuda K, Farnum J, Thummel K, Schuetz JD, Schuetz EG. 2003. Natural allelic variants of breast cancer resistance protein (BCRP) and their relationship to BCRP expression in human intestine. Pharmacogenetics 13: 19–28.

186. Bosch TM, Kjellberg LM, Bouwers A, Koeleman BP, Schellens JH, Beijnen JH, Smits PH, Meijerman I. 2005. Detection of SNPs in the ABCG2 gene in a Dutch population. Am J Pharmacogenom 5: 123–131.

187. de Jong FA, Marsh S, Mathijssen RH, King C, Verweij J, Sparreboom A, McLeod HL. 2004. ABCG2 pharmacogenetics: ethnic differences in allele frequency and assessment of influence on irinotecan disposition. Clin Cancer Res 10: 5889–5894.

188. Kobayashi D, Ieiri I, Hirota T, Takane H, Maegawa S, Kigawa J, Suzuki H, Nanba E, Oshimura M, Terakawa N, et al. 2005. Functional assessment of abcg2 (bcrp) gene polymorphisms to protein expression in human placenta. Drug Metab Dispos 33: 94–101.

189. Mizuarai S, Aozasa N, Kotani H. 2004. Single nucleotide polymorphisms result in impaired membrane localization and reduced atpase activity in multidrug transporter ABCG2. Int J Cancer 109: 238–246.

190. Morisaki K, Robey RW, Ozvegy-Laczka CS, Honjo Y, Polgar O, Steadman K, Sarkadi B, Bates SE. 2005. Single nucleotide polymorphisms modify the transporter activity of ABCG2. Cancer Chemother Pharm 56: 161–172.

191. Sparreboom A, Gelderblom H, Marsh S, Ahluwalia R, Obach R, Principe P, Twelves C, Verweij J, McLeod HL. 2004. Diflomotecan pharmacokinetics in relation to ABCG2 421C>A genotype. Clin Pharmacol Ther 76: 38–44.

192. Sparreboom A, Loos WJ, Burger H, Sissung TM, Verweij J, Figg WD, Nooter K, Gelderblom H. 2005. Effect of ABCG2 genotype on the oral bioavailability of topotecan. Cancer Biol Ther 4: 650–658.

193. Özvegy-Laczka C, Cserepes J, Elkind NB, Sarkadi B. 2005. Tyrosine kinase inhibitor resistance in cancer: role of ABC multidrug transporters. Drug Resist Updates 8: 15–26.

194. Elkind NB, Szentpétery Z, Apáti A, Özvegy-Laczka C, Várady G, Ujhelly O, Szabó K, Homolya L, Váradi A, Buday L. et al. 2005. Multidrug transporter ABCG2 prevents tumor cell death induced by the epidermal growth factor receptor inhibitor Iressa (ZD1839, Gefitinib). Cancer Res 65: 1770–1777.

195. Shih YN, Chiu CH, Tsai CM, Perng RP. 2005. Interstitial pneumonia during gefitinib treatment of non-small-cell lung cancer. J Chin Med Assoc 68: 183–186.

196. Schellens JHM, Creemers GJ, Beijnen JH, Rosing H, de Boer-Dennert M, McDonald M, Davies B, Verweij J. 1996. Bioavailability and pharmacokinetics of oral topotecan: a new topoisomerase I inhibitor. Br J Cancer 73: 1268–1271.

197. Schellens JHM, Maliepaard M, Scheper RJ, Scheffer GL, Jonker JW, Smit JW, Beijnen JH, Schinkel AH. 2000. Transport of topoisomerase I inhibitors by the breast cancer resistance protein: potential clinical implication. Ann Acad Sci 922: 188–194.

198. Zamboni WC, Bowman LC, Tan M, Santana VM, Houghton PJ, Meyer WH, Pratt CB, Heideman RL, Gajjar AJ, Pappo AS, Stewart CF. 1999. Interpatient variability in bioavailability of the intravenous formulation of topotecan given orally to children with recurrent solid tumors. Cancer Chemother Pharmacol 6: 454–460.

199. Nakatomi K, Yshikawa M, Oka M, Ikegami Y, Hayasaka S, Sano K, Shiozawa K, Kawabata S, Soda H, Ishikawa T, Tanabe S, Kohno S. 2001. Transport of 7-ethyl-10-hydroxycamptothecin (SN-38) by breast cancer resistance protein ABCG2 in human lung cancer cells. Biochem Biophys Res Commun 288: 827–832.

200. Zamboni WC, Ramanathan RK, McLeod HL, Mani S, Potter DM, Strychor S, Maruca LJ, King CR, Jung LL, Parise RA, et al. 2006. Disposition of 9-nitrocamptothecin and its 9-aminocamptothecin metabolite in relation to ABC transporters genotypes. Invest New Drugs 24: 393–401.

201. Honjo Y, Hrycyna CA, Yan QW, Medina-Perez WY, Robey RW, van de Laar A, Litman T, Dean M, Bates SE. 2001. Acquired mutations in the MXR/BCRP/ABCP gene alter substrate specificity in MXR/BCRP/ABCP-overexpressing cells. Cancer Res 61: 6635–6639.

202. Robey RW, Honjo Y, Morisaki K, Nadjem TA, Runge S, Risbood M, Poruchynsky MS, Bates SE. 2003. Mutations at amino-acid 482 in the ABCG2 gene affect substrate and antagonist specificity. Br J Cancer 89: 1971–1978.

203. Chen ZS, Robey RW, Belinsky MG, Shchaveleva I, Ren XQ, Sugimoto Y, Ross DD, Bates SE, Kruh GD. 2003. Transport of methotrexate, methotrexate polyglutamates, and 17β-estraiol 17-(β-D-glucuronide) by ABCG2 (BCRP/MXR): effects of acquired mutations at R482 on methotrexate transport. Cancer Res 63: 4048–4054.

204. Volk EL, Schneider E. 2003. wild-type breast cancer resistance protein (BCRP/ABCG2) is a methotrexate polyglutamate transporter. Cancer Res 63: 5538–5543.

205. Allen JD, Schinkel AH. 2002. Multidrug resistance and pharmacological protection mediated by the breast cancer resistance protein (BCRP/ABCG2). Mol Cancer Ther 1: 427–434.

206. Volk EL, Farley KM, Wu Y, Li F, Robey RW, Scheider E. 2002. Overexpression of wild-type breast cancer resistance protein mediates methotrexate resistance. Cancer Res 62: 5035–5040.

207. Miwa M, Tsukahara S, Ishikawa E, Asada S, Imai Y, Sugimoto Y. 2003. Single amino acid substitutions in the transmembrane domains of breast cancer resistance protein (BCRP) alter cross resistance patterns in transfectants. Int J Cancer 107: 757–763.

208. Mao Q, Unadkat JD. 2005. Role of the breast cancer resistance protein (ABCG2) in drug transport. AAPS J 7: 118–133.

209. Maliepaard M, van Gastelen MA, Tohgo A, Hausheer FH, Van Waardenburg RCAM, De Jong LA, Pluim D, Beijnen JH, Schellens JHM. 2001. Circumvention of breast cancer resistance protein (BCRP)–mediated resistance to camptothecins in vitro using non-substrate drugs or the BCRP inhibitor GF120918. Clin Cancer Res 7: 935–941.

210. Shiozawa K, Oka M, Soda H, Yoshikawa M, Ikegami Y, Tsurutani J, Nakatomi K, Namakura Y, Dor S, Kitazaki T, et al. 2004. Reversal of breast cancer resistance protein (BCRP/ABCG2)-mediated drug resistance by novobiocin, a coumermycin antibiotic. Int J Cancer 108: 146–151.

211. Suzuki M, Suzuki H, Sugimoto Y, Sugiyama Y. 2003. ABCG2 transports sulfated conjugates of steroids and xenobiotics. J Biol Chem 278: 22644–22649.

212. Imai Y, Asada S, Tsukahara S, Ishikawa E, Tsuruo T, Sugimoto Y. 2003. Breast cancer resistance protein exports sulfated estrogens but not free estrogens. Mol Pharmacol 64: 610–618.

213. Robey RW, Steadman K, Polgar O, Morisaki K, Blayney M, Mistry P, Bates SE. 2004. Pheophorbide a is a specific probe for ABCG2 function and inhibition. Cancer Res 64: 1242–1246.

214. Jonker JW, Buitelaar M, Wagenaar E, van der Valk, MA, Scheffer GL, Scheper RJ, Plösch T, Kuipers F, Oude Elferink RPJ, Rosing H, Beijnen JH, Schinkel AH. 2002. The breast cancer resistance proteine protects against a major chlorophyll-derived dietary phototoxin and protoporphyria. Proc Natl Acad Sci 99: 15649–15654.

215. van Herwaarden AE, Jonker JW, Wagenaar E, Brinkhuis RF, Schellens JHM, Beijnen JH, Schinkel AH. 2003. The breast cancer resistance protein (Bcrp1/Abcg2) restricts exposure to the dietary carcinogen 2-amino-1-methyl-6-phenylimidazo[4,5-*b*]pyridine. Cancer Res 63: 6447–6452.

216. Erlichman C, Boerner SA, Hallgren CG, Spieker R, Wang XY, James CD, Scheffer GL, Maliepaard M, Ross DD, Bible KC, Kaufmann SH. 2001. The HER tyrosine kinase inhibitor CI1033 enhances cytotoxicity of 7-ethyl-10-hydroxycamptothecin and topotecan by inhibiting breast cancer resistance protein–mediated drug efflux. Cancer Res 61: 739–748.

217. Woehlecke H, Pohl A, Alder-Baerens N, Lage H, Herrmann A. 2003. Enhanced exposure of phosphatidylserine in human gastric carcinoma cells overexpressing the half-size ABC transporter BCRP (ABCG2). Biochem J 376(Pt 2): 489–495.

218. Merino G, Jonker JW, Wagenaar E, Van Herwaarden AE, Schinkel AH. 2005. The breast cancer resistance protein (BCRP/ABG2) affects pharmacokinetics, heptobiliary excretion, and milk secretion of the antibiotic nitrofurantoin. Mol Pharmacol 67: 1758–1764.

219. Imai Y, Tsukahara S, Asada S, Sugimoto Y. 2004. Phytoestrogens/flavonoids reverse breast cancer resistance protein/ABCG2–mediated multidrug resistance. Cancer Res 64: 4346–4352.

220. Robey RW, Honjo Y, van de Laar A, Miyake K, Regis JT, Litman T, Bates SE. 2001. A functional assay for detection of the mitoxantrone resistance protein, MXR (ABCG2). Biochim Biophys Acta 1512: 171–182.

221. Merino G, Alvarez AI, Pulido MM, Molina AJ, Schinkel AH, Prieto JG. 2006. Breast Cancer Resistance Protein (BCRP/ABCG2) transports fluoroquinolone antibiotics and affects their oral availability, pharmacokinetics and milk secretion. Drug Metab Dispos 34: 690–695.

222. Hirano M, Maeda K, Matsushima S, Nozaki Y, Kusuhara H, Sugiyama Y. 2005. Involvement of BCRP (ABCG2) in the biliary excretion of pitavastatin. Mol Pharmacol 68: 800–807.

223. Wang X, Furukawa T, Nitanda T, Okamoto M, Sugimoto Y, Akiyama SC, Baba M. 2003. Breast cancer resistance protein (BCRP/ABCG2) induces cellular resistance to HIV-1 nucleoside reverse transcriptase inhibitors. Mol Pharmacol 63: 65–72.

224. Wang X, Nitanda T, Shi M, Okamoto M, Furukawa T, Sugimoto Y, Akiyama S, Baba M. 2004. Induction of cellular resistance to nucleoside reverse transcriptase inhibitors by the wild-type breast cancer resistance protein. Biochem Pharmacol 68: 1363–1370.

225. Wang X, Baba M. 2005. The role of breast cancer resistance protein (BCRP/ABCG2) in cellular resistance to HIV-1 nucleoside reverse transcriptase inhibitors. Antivir Chem Chemother 16: 213–216.

226. Kim M, Turnquist H, Jackson J, Sgagias M, Yan Y, Gong M, Dean M, Sharp JG, Cowan K. 2002. The multidrug resistance transporter ABCG2 (breast cancer resistance protein 1) effluxes Hoechst 33342 and is overexpressed in hematopoietic stem cells. Clin Cancer Res 8: 22–28.

227. Scharenberg CW, Harkey MA, Torok-Storb B. 2002. The ABCG2 transporter is an efficient Hoechst 33342 efflux pump and is preferentially expressed by immature human hematopoietic progenitors. Blood 99: 507–512.

228. Allen JD, van Dort S, Buitelaar M, van Tellingen O, Schinkel AH. 2003. Mouse breast cancer resistance protein (Bcrp1/Abcg2) mediates etoposide resistance and transport, but etoposide oral availability is limited primarily by P-glycoprotein. Cancer Res 63: 1339–1344.

229. Jonker JW, Smit JW, Brinkhuis RF, Maliepaard M, Beijnen JH, Schellens JHM, Schinkel AH. 2000. Role of breast cancer resistance protein in the bioavailability and fetal penetration of topotecan. J Natl Cancer Inst 92: 1628–1629.

230. Minderman H, O'Loughlin KL, Pendyala L, Baer MR. 2004. VX-710 (biricodar) increases drug retention and enhances chemosensitivity in resistant cells overexpressing P-glycoprotein, multidrug resistance protein, and breast cancer resistance protein. Clin Cancer Res 10: 1826–1834.

231. Gupta A, Zhang Y, Unadkat JD, Mao Q. 2004. HIV protease inhibitors are inhibitors but not substrates of the human breast cancer resistance protein (BCRP/ABCG2). J Pharmacol Exp Ther 310: 334–341.

232. Rabindran SK, Ross DD, Doyle LA, Yang W, Greenberger LM. 2000. Fumitremorgin C reverses multidrug resistance in cells transfected with the breast cancer resistance protein. Cancer Res 60: 47–50.

233. Allen JD, van Loevezijn A, Lakhai JM, van der valk M, van Tellingen O, Reid G, Schellens JHM, Koomen GJ, Schinkel AH. 2002. Potent and specific inhibition of the breast cancer resistance protein multidrug transporter in vitro and in mouse intestine by a novel analogue of fumitremorgin C. Mol Cancer Ther 1: 417–425.

234. van Loevezijn A, Allen JD, Schinkel AH, Koomen GJ. 2001. Inhibition of BCRP-mediated drug efflux by fumitremorgin-type indolyl diketopiperazines. Bioorg Med Chem Lett 11: 29–32.

235. Yang CH, Chen YC, Kuo ML. 2003. Novobiocin sensitizes BCRP/MXR/ABCP overexpressing topotecan-resistant human breast carcinoma cells to topotecan and mitoxantrone. Anticancer Res 23: 2519–2523.

236. Woehlecke H, Osada H, Herrmann A, Lage H. 2003. Reversal of breast cancer resistance protein-mediated drug resistance by tryprostatin A. Int J Cancer 107: 721–728.

237. Özvegy-Laczka C, Hegedus T, Várady G, Ujhelly O, Schuetz JD, Váradi A, Kéri G, Orfi L, Német K, Sarkadi B. 2004. High-affinity interaction of tyrosine kinase inhibitors with the ABCG2 multidrug transporter. Mol Pharmacol 65: 1485–1495.

238. Stewart CF, Leggas M, Schuetz JD, Panetta JC, Cheshire PJ, Peterson J, Daw N, Jenkins JJ, Gilbertson R, Germain GS, Harwood FC, Houghton PJ. 2004. Gefitinib enhances the antitumor activity and oral bioavailability of irinotecan in mice. Cancer Res 64: 7491–7499.

239. Yanase K, Tsukahara S, Asada S, Ishikawa E, Imai Y, Sugimoto Y. 2004. Gefitinib reverses breast cancer resistance protein–mediated drug resistance. Mol Cancer Ther 3: 1119–1125.

240. Houghton PJ, Germain GS, Harwood FC, Schuetz JD, Stewart CF, Buchdunger E, Traxler P. 2004. Imatinib mesylate is a potent inhibitor of the ABCG2 (BCRP) transporter and reverses resistance to topotecan and SN-38 in vitro. Cancer Res 64: 2333–2337.

241. Gupta A, Dai Y, Vethanayagam RR, Hebert MF, Thummel KE, Unadkat JD, Ross DD, Mao Q. 2006. Cyclosporin A, tacrolimus and sirolimus are potent inhibitors of the human breast cancer resistance protein (ABCG2) and reverse resistance to mitoxantrone and topotecan. Cancer Chemother Pharmacol 58: 374–383.

242. Polli JW, Baughman TM, Humphreys JE, Jordan KH, Mote AL, Webster LO, Barnaby RJ, Vitulli G, Bertolotti L, Read KD, Serabjit-Singh CJ. 2004. The systemic exposure of an *N*-methyl-D-aspartate receptor antagonist is limited in mice by the P-glycoprotein and breast cancer resistance protein efflux transporters. Drug Metab Dispos 32: 722–726.

243. Cisternino S, Mercier C, Bourasset F, Roux F, Scherrmann JM. 2004. Expression, upregulation, and transport activity of the multidrug-resistance protein Abcg2 at the mouse blood–brain barrier. Cancer Res 64: 3296–3301.

244. Hooijberg JH, Broxterman HJ, Kool M, Assaraf YG, Peters GJ, Noordhuis P, Scheper RJ, Borst P, Pinedo HM, Jansen G. 1999. Antifolate resistance mediated by the multidrug resistance proteins MRP1 and MRP2. Cancer Res 59: 2532–2535.

245. Borst P, Evers R, Kool M, Wijnholds JA. 2000. A family of drug transporters: the multidrug resistance–associated proteins. J Natl Cancer Inst 16: 1295–1302.

246. Borst P, Oude Elferink R. 2002. Mammalian ABC transporters in health and disease. Annu Rev Biochem 71: 537–593.

247. Ito KI, Oleschuk CJ, Westlake C, Vasa MZ, Deeley RG, Cole SPC. 2001. Mutation of Trp1254 in the multispecific organic anion transporter, multidrug resistance protein 2 (MRP2), alters substrate specificity and results in loss of methotrexate transport activity. J Biol Chem 41: 38108–38114.

248. Zeng H, Chen ZS, Belinsky MG, Rea A, Kruh GD. 2001. Transport of methotrexate (MTX) and folates by multidrug resistance protein (MRP) 3 and MRP1: effect of polyglutamylation on MTX transport. Cancer Res 61: 7225–7232.

249. Chen ZS, Lee K, Walther S, Blanchard-Raftogianis R, Kuwano M, Zeng H, Kruh GD. 2002. Analysis of methotrexate and folate transport by multidrug resistance protein 4 (ABCC4): MRP4 is a component of the methotrexate efflux system. Cancer Res 62: 3144–3150.

250. Hegedus T, Orfi L, Seprodi A, Varadi A, Sarkadi B, Keri G. 2002. Interaction of tyrosine kinase inhibitors with the human multidrug transporter proteins, MDR1 and MRP1. Biochim Biophys Acta 1587: 318–325.

251. Houghton PJ, Stewart CF, Zamboni WC, Thompson J, Luo X, Danks MK, Houghton JA. 1996. Schedule-dependent efficacy of camptothecins in models of human cancer. Ann N Y Acad Sci 803: 188–201.

252. Furman WL, Stewart CF, Poquette CA, Pratt CB, Santana VM, Zamboni WC, Bowman LC, Ma MK, Hoffer FA, Meyer WH, et al. 1999. Direct translation of a protracted irinotecan schedule from a xenograft model to a phase I trial in children. J Clin Oncol 17: 1815–1824.

253. Furman WL, Stewart CF, Kirstein M, Kepner JL, Bernstein ML, Kung F, Vietti TJ, Steuber CP, Becton DL, Baruchel S, Pratt C. 2002. Protracted intermittent schedule of topotecan in children with refractory acute leukemia: a pediatric oncology group study. J Clin Oncol 20: 1617–1624.

254. Drengler RL, Kuhn JG, Schaaf LJ, Rodriguez GI, Villalona-Calero MA, Hammond LA, Stephenson JA Jr, Hodges S, Kraynak MA, Staton BA, et al. 1999. Phase I and pharmacokinetic trial of oral irinotecan administered daily for 5 days every 3 weeks in patients with solid tumors. J Clin Oncol 17: 685–696.

255. Ceresoli GL, Cappuzzo F, Gregorc V, Bartolini S, Crino L, Villa E. 2004. Gefitinib in patients with brain metastases from non-small-cell lung cancer: a prospective trial. Ann Oncol 15: 1042–1047.

256. Neville K, Parise RA, Thompson P, Aleksic A, Egorin MJ, Balis FM, McGuffey L, McCully C, Berg SL, Blaney SM. 2004. Plasma and cerebrospinal fluid pharmacokinetics of imatinib after administration to nonhuman primates. Clin Cancer Res 10: 2525–2529.

257. Hamada A, Miyano H, Watanabe H, Saito H. 2003. Interaction of imatinib mesylate with human P-glycoprotein. J Pharmacol Exp Ther 307: 824–828.

258. Scalbert A, Williamson G. 2000. Dietary intake and bioavailability of polyphenols. J Nutr 130(8S Suppl): 2073S–2085S.

259. Arts IC, Hollman PC. 2005. Polyphenols and disease risk in epidemiologic studies. Am J Clin Nutr 81(1 Suppl): 317S–325S.

260. Conseil G, Baubichon-Cortay H, Dayan G, Jault JM, Barron D, Di Pietro A. 1998. Flavonoids: a class of modulators with bifunctional interactions at vicinal ATP- and steroid-binding sites on mouse P-glycoprotein. Proc Natl Acad Sci U S A 95: 9831–9836.

261. Ho PC, Saville DJ, Wanwimolruk S. 2001. Inhibition of human CYP3A4 activity by grapefruit flavonoids, furanocoumarins and related compounds. J Pharm Pharm Sci 4: 217–227.

262. Yoshikawa M, Ikegami Y, Sano K, Yoshida H, Mitomo H, Sawada S, Ishikawa T. 2004. Transport of SN-38 by the wild-type of human ABC transporter ABCG2 and its inhibition by quercetin, a natural flavonoid. J Exp Ther Oncol 4: 25–35.

263. Zhang S, Yang X, Morris ME. 2004. Combined effects of multiple flavonoids on breast cancer resistance protein (ABCG2)–mediated transport. Pharm Res 21: 1263–1273.

264. Zhang S, Yang X, Morris ME. 2004. Flavonoids are inhibitors of breast cancer resistance protein (ABCG2)–mediated transport. Mol Pharmacol 65: 1208–1216.

265. Zhang S, Wang X, Sagawa K, Morris ME. 2005. Flavonoids chrysin and benzoflavone, potent breast cancer resistance protein inhibitors, have no significant effect on topotecan pharmacokinetics in rats or mdr1a/1b (–/–) mice. Drug Metab Dispos 33: 34134–3418.

266. Pavek P, Merino G, Wagenaar E, Bolscher E, Novotna M, Jonker JW, Schinkel AH. 2005. Human breast cancer resistance protein: interactions with steroid drugs, hormones, the dietary carcinogen 2-amino-1-methyl-6-phenylimidazo(4,5-*b*)pyridine, and transport of cimetidine. J Pharmacol Exp Ther 312: 144–152.

267. Sugimoto Y, Tsukahara S, Imai Y, Sugimoto Y, Ueda K, Tsuruo T. 2003. Reversal of breast cancer resistance protein–mediated drug resistance by estrogen antagonists and agonists. Mol Cancer Ther 2: 105–112.

268. Guo Y, Kotova E, Chen ZS, Lee K, Hopper-Borge E, Belinsky MG, Kruh GD. 2003. MRP8, ATP-binding cassette C11 (ABCC11), is a cyclic nucleotide efflux pump and a resistance factor for fluoropyrimidines 2′,3′-dideoxycytidine and 9′-(2′-phosphonylmethoxyethyl)adenine. J Biol Chem 278: 29509–29514.

269. Wijnholds J, Mol CA, van Deemter L, de Haas M, Scheffer GL, Baas F, Beijnen JH, Scheper RJ, Hatse S, De Clercq E, Balzarini J, Borst P. 2000 Multidrug-resistance protein 5 is a multispecific organic anion transporter able to transport nucleotide analogs. Proc Natl Acad Sci U S A 97: 7476–7481.

270. Wijnholds J. 2002. Drug resistance caused by multidrug resistance-associated proteins. Novartis Found Symp 243: 69–79.

271. Gornet JM, Lokiec F, Dclos-Vallee JC, Azoulay D, Goldwasser F. 2001. Severe CPT-11-induced diarrhea in presence of FK-506 following liver transplantation for hepatocelluler carcinoma. Anticancer Res 21: 4203–4206.

272. Capece BP, Castells G, Perez F, Arboix M, Cristofol C. 2000. Pharmacokinetic behaviour of albendazole sulphoxide enantiomers in male and female sheep. Vet Res Commun 24: 339–348.

273. Merino G, Jonker JW, Wagenaar E, Pulido MM, Molina AJ, Alvarez AI, Schinkel AH. 2005. Transport of anthelmintic benzimidazole drugs by breast cancer resistance protein (BCRP/ABCG2). Drug Metab Dispos 33: 614–618.

274. Morton MR, Cooper JW. 1989. Erythromycin-induced digoxin toxicity. DICP 23: 668–670. Abstract only.

275. Woodland C, Ito S, Koren G. 1998. A model for the prediction of digoxin-drug interactions at the renal tubular cell level. Ther Drug Monit 20: 134–138.

276. Kodawara T, Masuda S, Wakasugi H, Uwai Y, Futami T, Saito H, Abe T, Inui KI. 2002. Organic anion transporter oatp2-mediated interaction between digoxin and amiodarone in the liver. Pharm Res 19: 738–743.

277. Boyd RA, Stern RH, Stewart BH, Wu X, Reyner EL, Zegarac EA, Randinitis EJ, Whitfield L. 2000. Atorvastatin coadministration may increase digoxin concentrations by inhibition of intestinal P-glycoprotein-mediated secretion. J Clin Pharmacol 40: 91–98.

278. Malingrè MM, Beijnen JH, Rosing H, Koopman FJ, Jewell RC, Paul EM, Ten Bokkel Huinink WW, Schellens JH. 2001. Co-administration of GF120918 significantly increases the systemic exposure to oral paclitaxel in cancer patients. Br J Cancer 84: 42–47.

279. Floren LC, Bekersky I, Benet LZ, Mekki Q, Dressler D, Lee JW, Roberts JP, Hebert MF. 1997. Tacrolimus oral bioavailability doubles with coadministration of ketoconazole. Clin Pharmacol Ther 62: 41–49.

280. Gupta SK, Bakran A, Johnson RW, Rowland M. 1989. Cyclosporin-erythromycin interaction in renal transplant patients. Br J Clin Pharmacol 27: 475–481.

281. Gupta SK, Bakran A, Johnson RW, Rowland M. 1988. Erythromycin enhances the absorption of cyclosporin. Br J Clin Pharmacol 25: 401–402.

282. Hebert MF, Fisher RM, Marsh CL, Dressler D, Bekersky I. 1999. Effects of rifampin on tacrolimus pharmacokinetics in healthy volunteers. J Clin Pharmacol 39: 91–96.

283. Johne A, Brockmöller J, Bauer S, Maurer A, Langheinrich M, Roots I. 1999. Pharmacokinetic interaction of digoxin with an herbal extract from St. John's wort (Hypericum perforatum). Clin Pharmacol Ther 66: 338–345.

284. Saito S, Iida A, Sekine A, Miura Y, Ogawa C, Kawauchi S, Higuchi S, Nakamura Y. 2002. Identification of 779 genetic variations in eight genes encoding members of the ATP-binding cassette, subfamily C (ABCC/MRP/CFTR. J Hum Genet 47: 147–171.

285. Conrad S, Kauffmann HM, Ito K, Deeley RG, Cole SP, Schrenk D. 2001. Identification of human multidrug resistance protein 1 (MRP1) mutations and characterization of a G671V substitution. J Hum Genet 46: 656–663.

286. Ito S, Ieiri I, Tanabe M, Suzuki A, Higuchi S, Otsubo K. 2001. Polymorphism of the ABC transporter genes, MDR1, MRP1 and MRP2/cMOAT, in healthy Japanese subjects. Pharmacogenetics 11: 175–184.

287. Moriya Y, Nakamura T, Horinouchi M, Sakaeda T, Tamura T, Aoyama N, Shirakawa T, Gotoh A, Fujimoto S, Matsuo M, Kasuga M, Okumura K. 2002. Effects of polymorphisms of MDR1, MRP1, and MRP2 genes on their mRNA expression levels in duodenal enterocytes of healthy Japanese subjects. Biol Pharm Bull 25: 1356–1359.

288. Perdu J, Germain DP. 2001. Identification of novel polymorphisms in the pM5 and MRP1 (ABCC1) genes at locus 16p13.1 and exclusion of both genes as responsible for pseudoxanthoma elasticum. Hum Mutat 17: 74–75.

289. Conrad S, Kauffmann HM, Ito K, Leslie EM, Deeley RG, Schrenk D, Cole SP. 2002. A naturally occurring mutation in MRP1 results in a selective decrease in organic anion transport and in increased doxorubicin resistance. Pharmacogenetics 4: 321–330.

290. Jedlitschky G, Leier I, Buchholz U, Barnouin K, Kurz G, Keppler D. 1996. Transport of glutathione, glucuronate, and sulfate conjugates by the MRP gene-encoded conjugate export pump. Cancer Res 56: 988–994.

291. Loe DW, Almquist KC, Cole SP, Deeley RG. 1996. ATP-dependent 17 beta-estradiol 17-(beta-D-glucuronide) transport by multidrug resistance protein (MRP). Inhibition by cholestatic steroids. J Biol Chem 271: 9683–9689.

292. Loe DW, Almquist KC, Deeley RG, Cole SP. 1996. Multidrug resistance protein (MRP)–mediated transport of leukotriene C4 and chemotherapeutic agents in membrane vesicles: demonstration of glutathione-dependent vincristine transport. J Biol Chem 271: 9675–9682.

293. Mao Q, Deeley RG, Cole SP. 2000. Functional reconstitution of substrate transport by purified multidrug resistance protein MRP1 (ABCC1) in phospholipid vesicles. J Biol Chem 275: 34166–34172.

294. Mao Q, Qiu W, Weigl KE, Lander PA, Tabas LB, Shepard RL, Dantzig AH, Deeley RG, Cole SP. 2002. GSH-dependent photolabeling of multidrug resistance protein MRP1 (ABCC1) by [125I]LY475776. Evidence of a major binding site in the COOH-proximal membrane spanning domain. J Biol Chem 277: 28690–28699.

295. Jedlitschky G, Keppler D. 2002. Transport of leukotriene C4 and structurally related conjugates. Vitam Horm 64: 153–184.

296. Haimeur A, Conseil G, Deeley RG, Cole SP. 2004. The MRP-related and BCRP/ABCG2 multidrug resistance proteins: biology, substrate specificity and regulation. Curr Drug Metab 5: 21–53.

297. Loe DW, Deeley RG, Cole SP. 1998. Characterization of vincristine transport by the M(r) 190,000 multidrug resistance protein (MRP): evidence for cotransport with reduced glutathione. Cancer Res 58: 5130–5136.

298. Salerno M, Garnier-Suillerot A. 2001. Kinetics of glutathione and daunorubicin efflux from multidrug resistance protein overexpressing small-cell lung cancer cells. Eur J Pharmacol 421: 1–9.

299. Renes J, de Vries EG, Nienhuis EF, Jansen PL, Müller M. 1999. ATP- and glutathione-dependent transport of chemotherapeutic drugs by the multidrug resistance protein MRP1. Br J Pharmacol 126: 681–688.

300. Chuman Y, Chen ZS, Seto K, Sumizawa T, Furukawa T, Tani A, Haraguchi M, Niwa K, Yamada K, Aikou T, Akiyama S. 1998. Reversal of MRP-mediated vincristine resistance in KB cells by buthionine sulfoximine in combination with PAK-104P. Cancer Lett 129: 69–76.

301. Rappa G, Lorico A, Flavell RA, Sartorelli AC. 1997. Evidence that the multidrug resistance protein (MRP) functions as a co-transporter of glutathione and natural product toxins. Cancer Res 57: 5232–5237.

302. Rappa G, Gamcsik MP, Mitina RL, Baum C, Fodstad O, Lorico A. 2003. Retroviral transfer of MRP1 and gamma-glutamyl cysteine synthetase modulates cell sensitivity to L-buthionine-*S*,*R*-sulphoximine (BSO): new rationale for the use of BSO in cancer therapy. Eur J Cancer 39: 120–128.

303. Zaman GJ, Lankelma J, van Tellingen O, Beijnen JH, Dekker H, Paulusma C, Oude Elferink RPJ, Baas F, Borst P. 1995. Role of glutathione in the export of compounds from cells by the multidrug-resistance–associated protein. Proc Natl Acad Sci U S A 92: 7690–7694.

304. Salerno M, Petroutsa M, Garnier-Suillerot A. 2002. The MRP1-mediated effluxes of arsenic and antimony do not require arsenic–glutathione and antimony–glutathione complex formation. J Bioenerg Biomembr 34: 135–145.

305. Jedlitschky G, Leier I, Buchholz U, Center M, Keppler D. 1994. ATP-dependent transport of glutathione S-conjugates by the multidrug resistance-associated protein. Cancer Res 54: 4833–4836.

306. Loe DW, Stewart RK, Massey TE, Deeley RG, Cole SP. 1997. ATP-dependent transport of aflatoxin B1 and its glutathione conjugates by the product of the multidrug resistance protein (MRP) gene. Mol Pharmacol 51: 1034–1041.

307. Leslie EM, Ito K, Upadhyaya P, Hecht SS, Deeley RG, Cole SP. 2001. Transport of the beta -*O*-glucuronide conjugate of the tobacco-specific carcinogen 4-(methylnitrosamino)-1-(3-pyridyl)-1-butanol (NNAL) by the multidrug resistance protein 1 (MRP1): requirement for glutathione or a non-sulfur-containing analog. J Biol Chem 276: 27846–27854.

308. Leslie EM, Deeley RG, Cole SP. 2001. Toxicological relevance of the multidrug resistance protein 1, MRP1 (ABCC1) and related transporters. Toxicology 167: 3–23.

309. Diah SK, Smitherman PK, Townsend AJ, Morrow CS. 1999. Detoxification of 1-chloro-2,4-dinitrobenzene in MCF7 breast cancer cells expressing glutathione *S*-transferase P1-1 and/or multidrug resistance protein 1. Toxicol Appl Pharmacol 157: 85–93.

310. Morrow CS, Diah S, Smitherman PK, Schneider E, Townsend AJ. 1998. Multidrug resistance protein and glutathione *S*-transferase P1-1 act in synergy to confer protection from 4-nitroquinoline 1-oxide toxicity. Carcinogenesis 19: 109–115.

311. Chu XY, Suzuki H, Ueda K, Kato Y, Akiyama S, Sugiyama Y. 1999. Active efflux of CPT-11 and its metabolites in human KB-derived cell lines. J Pharmacol Exp Ther 288: 735–741.

312. Cnubben NH, Rommens AJ, Oudshoorn MJ, van Bladeren PJ. 1998. Glutathione-dependent biotransformation of the alkylating drug thiotepa and transport of its metabolite monoglutathionylthiotepa in human MCF-7 breast cancer cells. Cancer Res 58: 4616–4623.

313. Grzywacz MJ, Yang JM, Hait WN. 2003. Effect of the multidrug resistance protein on the transport of the antiandrogen flutamide. Cancer Res 63: 2492–2498.

314. Williams GC, Liu A, Knipp G, Sinko PJ. 2002. Direct evidence that saquinavir is transported by multidrug resistance–associated protein (MRP1) and canalicular multispecific organic anion transporter (MRP2). Antimicrob Agents Chemother 46: 3456–3462.

315. Olson DP, Scadden DT, D'Aquila RT, De Pasquale MP. 2002. The protease inhibitor ritonavir inhibits the functional activity of the multidrug resistance related-protein 1 (MRP-1). AIDS 16: 1743–1747.

316. Zaman GJ, Cnubben NH, van Bladeren PJ, Evers R, Borst P. 1996. Transport of the glutathione conjugate of ethacrynic acid by the human multidrug resistance protein MRP. FEBS Lelt 391: 126–130.

317. Gollapudi S, Kim CH, Tran BN, Sangha S, Gupta S. 1997. Probenecid reverses multidrug resistance in multidrug resistance–associated protein-overexpressing HL60/AR and H69/AR cells but not in P-glycoprotein-overexpressing HL60/Tax and P388/ADR cells. Cancer Chemother Pharmacol 40: 150–158.

318. Draper MP, Martell RL, Levy SB. 1997. Indomethacin-mediated reversal of multidrug resistance and drug efflux in human and murine cell lines overexpressing MRP, but not P-glycoprotein. Br J Cancer 75: 810–815.

319. Bakos E, Evers R, Sinko EM, Varadi A, Borst P, Sarkadi B. 2000. Interactions of the human multidrug resistance proteins MRP1 and MRP2 with organis anions. Mol Pharmacol 57: 760–768.

320. Roller A, Bahr OR, Streffer J, Winter S, Heneka M, Deininger M, Meyermann R, Naumann U, Gulbins E, Weller M. 1999. Selective potentiation of drug cytotoxicity by NSAID in human glioma cells: the role of COX-1 and MRP. Biochem Biophys Res Commun 259: 600–605.

321. Gekeler V, Ise W, Sanders KH, Ulrich WR, Beck J. 1995. The leukotriene LTD4 receptor antagonist MK571 specifically modulates MRP associated multidrug resistance. Biochem Biophys Res Commun 208: 345–352.

322. Nakano R, Oka M, Nakamura T, Fukuda M, Kawabata S, Terashi K, Tsukamoto K, Noguchi Y, Soda H, Kohno S. 1998. A leukotriene receptor antagonist, ONO-1078, modulates drug sensitivity and leukotriene C4 efflux in lung cancer cells expressing multidrug resistance protein. Biochem Biophys Res Commun 251: 307–312.

323. Payen L, Delugin L, Courtois A, Trinquart Y, Guillouzo A, Fardel O. 2001. The sulphonylurea glibenclamide inhibits multidrug resistance protein (MRP1) activity in human lung cancer cells. Br J Pharmacol 132 (Feb): 778–784.

324. Burg D, Wielinga P, Zelcer N, Saeki T, Mulder GJ, Borst P. 2002. Inhibition of the multidrug resistance protein 1 (MRP1) by peptidomimetic glutathione-conjugate analogs. Mol Pharmacol 62: 1160–1166.

325. Norman BH, Dantzig AH, Kroin JS, Law KL, Tabas LB, Shepard RL, Palkowitz AD, Hauser KL, Winter MA, Sluka JP, Starling JJ. 1999. Reversal of resistance in multidrug resistance protein (MRP1)–overexpressing cells by LY329146. Bioorg Med Chem Lett 9: 3381–3386.

326. Norman BH, Gruber JM, Hollinshead SP, Wilson JW, Starling JJ, Law KL, Self TD, Tabas LB, Williams DC, Paul DC, Wagner MM, Dantzig AH. 2002. Tricyclic isoxazoles are novel inhibitors of the multidrug resistance protein (MRP1). Bioorg Med Chem Lett 12: 883–886.

327. Norman BH, Lander PA, Gruber JM, Kroin JS, Cohen JD, Jungheim LN, Starling JJ, Law KL, Self TD, Tabas LB, et al. 2005. Cyclohexyl-linked tricyclic isoxazoles are potent and selective modulators of the multidrug resistance protein (MRP1). Bioorg Med Chem Lett 15: 5526–5530.

328. Loe DW, Deeley RG, Cole SPC. 2000. Verapamil stimulates glutathione transport by the 190-kDa multidrug resistance protein 1 (MRP1). J Pharmacol Exp Ther 296: 530–538.

329. Aoki S, Chen ZS, Higasiyama K, Setiawan A, Akiyama S, Kobayashi M. 2001. Reversing effect of agosterol A, a spongean sterol acetate, on multidrug resistance in human carcinoma cells. Jpn J Cancer Res 92: 886–895. Abstract only.

330. Sumizawa T, Chen ZS, Chuman Y, Seto K, Furukawa T, Haraguchi M, Tani A, Shudo N, Akiyama SI. 1997. Reversal of multidrug resistance–associated protein-mediated drug resistance by the pyridine analog PAK-104P. Mol Pharmacol 51: 399–405.

331. Marbeuf-Gueye C, Salerno M, Quidu P, Garnier-Suillerot A. 2000. Inhibition of the P-glycoprotein- and multidrug resistance protein–mediated efflux of anthracyclines and calceinacetoxymethyl ester by PAK-104P. Eur J Pharmacol 391: 207–216.

332. Bichat F, Solis-Recendez G, Poullain MG, Poupon MF, Khayat D, Bastian G. 1998. S9788 modulation of P-glycoprotein- and multidrug-related protein-mediated multidrug resistance by Servier 9788 in doxorubicin-resistant MCF7 cells. Biochem Pharmacol 56: 497–502.

333. Hooijberg JH, Broxterman HJ, Heijn M, Fles DL, Lankelma J, Pinedo HM. 1997. Modulation by (iso)flavonoids of the ATPase activity of the multidrug resistance protein. FEBS Lett 413: 344–348.

334. Leslie EM, Mao Q, Oleschuk CJ, Deeley RG, Cole SPC. 2001. Modulation of multidrug resistance protein 1 (MRP1/ABCC1) transport and atpase activities by interaction with dietary flavonoids. Mol Pharmacol 59: 1171–1180.

335. Duffy CP, Elliott CJ, O'Connor RA, Heenan MM, Coyle S, Cleary IM, Kavanagh K, Verhaegen S, O'Loughlin CM, NicAmhlaoibh R, Clynes M. 1998. Enhancement of chemotherapeutic drug toxicity to human tumour cells in vitro by a subset of non-steroidal anti-inflammatory drugs (NSAIDs). Eur J Cancer 34: 1250–1259.

336. Kuss BJ, Corbo M, Lau WM, Fennell DA, Dean NM, Cotter FE. 2002. In vitro and in vivo downregulation of MRP1 by antisense oligonucleotides: a potential role in neuroblastoma therapy. Int J Cancer 98: 128–133.

337. Kurz EU, Cole SP, Deeley RG. 2001. Identification of DNA–protein interactions in the 5′ flanking and 5′ untranslated regions of the human multidrug resistance protein (MRP1) gene: evaluation of a putative antioxidant response element/AP-1 binding site. Biochem Biophys Res Commun 285: 981–990.

338. Yamane Y, Furuichi M, Song R, Van NT, Mulcahy RT, Ishikawa T, Kuo MT. 1998. Expression of multidrug resistance protein/GS-X pump and gamma-glutamylcysteine synthetase genes is regulated by oxidative stress. J Biol Chem 273: 31075–31085.

339. Kauffmann HM, Pfannschmidt S, Zoller H, Benz A, Vorderstemann B, Webster JI, Schrenk DC. 2002. Influence of redox-active compounds and PXR-activators on human MRP1 and MRP2 gene expression. Toxicology 171: 137–146.

340. Stein U, Walther W, Laurencot CM, Scheffer GL, Scheper RJ, Shoemaker RH.1997. Tumor necrosis factor-alpha and expression of the multidrug resistance-associated genes LRP and MRP. J Natl Cancer Inst 89: 807–813.

341. Ikegami Y, Tatebe S, Lin-Lee YC, Xie QW, Ishikawa T, Kuo MT. 2000. Induction of MRP1 and gamma-glutamylcysteine synthetase gene expression by interleukin 1beta is mediated by nitric oxide–related signalings in human colorectal cancer cells. J Cell Physiol 185: 293–301.

342. Wijnholds J, Evers R, van Leusden MR, Mol CAAM, Zaman GJR, Mayer U, Beijnen JH, van der Valk M, Krimpenfort P, Borst P. 1997. Increased sensitivity to anticancer drugs and decreased inflammatory response in mice lacking the multidrug resistance–associated protein. Nat Med 11: 1275–1279.

343. Wijnholds J, Scheffer GL, van der Valk M, van der Valk P, Beijnen JH, Scheper RJ, Borst P. 1998. Multidrug resistance protein 1 protects the oropharyngeal mucosal layer and the testicular tubules against drug-induced damage. J Exp Med 188: 797–808.

344. Wijnholds J, deLange EC, Scheffer GL, van den Berg DJ, Mol CA, van der Valk M, Schinkel AH, Scheper RJ, Breimer DD, Borst P. 2000. Multidrug resistance protein 1 protects the choroid plexus epithelium and contributes to the blood-cerebrospinal fluid barrier. J Clin Invest 105: 279–285.

345. Johnson DR, Finch RA, Lin ZP, Zeiss CJ, Sartorelli AC. 2001. The pharmacological phenotype of combined multidrug-resistance mdr1a/1b- and mrp1-deficient mice. Cancer Res 61: 1469–1476.

346. Lorico A, Rappa G, Finch RA, Yang D, Flavell RA, Sartorelli AC. 1997. Disruption of the murine MRP (multidrug resistance protein) gene leads to increased sensitivity to etoposide (VP-16) and increased levels of glutathione. Cancer Res 57: 5238–5242.

347. Park S, Sinko PJ. 2005. P-Glycopotein and multidrug resistance–asociated proteins limit the brain uptake of saquinavir in mice. J Pharmacol Exp Ther 312: 1249–1256.

348. Lorico A, Rappa G, Flavell RA, Sartorelli AC. 1996. Double knockout of the MRP gene leads to increased drug sensitivity in vitro. Cancer Res 56: 5351–5355.

349. Allen JD, Brinkhuis RF, van Deemter L, Wijnholds J, Schinkel AJ. 2000. Extensive contribution of the multidrug transporters P-glycoprotein and Mrp1 to basal drug resistance. Cancer Res 60: 5761–5766.

350. Suzuki H, Sugiyama Y. 2002. Single nucleotide polymorphisms in multidrug resistance associated protein 2 (MRP2/ABCC2): its impact on drug disposition. Adv Drug Deliv Rev 54: 1311–1331.

351. Elferink RO, Groen AK. 2002. Genetic defects in hepatobiliary transport. Biochim Biophys Acta 1586: 129–145.

352. Paulusma CC, Bosma PJ, Zaman GJ, Bakker CT, Otter M, Scheffer GL, Scheper RJ, Borst P, Oude Elferink RP. 1996. Congenital jaundice in rats with a mutation in a multidrug resistance-associated protein gene. Science 271: 1126–1128.

353. Paulusma CC, Kool M, Bosma PJ, Scheffer GL, ter Borg F, Scheper RJ, Tytgat GN, Borst P, Baas F, Oude Elferink RP. 1997. A mutation in the human canalicular multispecific organic anion transporter gene causes the Dubin–Johnson syndrome. Hepatology 25: 1539–1542.

354. Paulusma CC, Oude Elferink RP. 1997. The canalicular multispecific organic anion transporter and conjugated hyperbilirubinemia in rat and man. J Mol Med 75: 420–428.

355. Ito K, Suzuki H, Hirohashi T, Kume K, Shimizu T, Sugiyama Y. 1997. Molecular cloning of canalicular multispecific organic anion transporter defective in EHBR. Am J Physiol 272: G16–G22.

356. Jansen PL, Peters WH, Lamers WH. 1985. Hereditary chronic conjugated hyperbilirubinemia in mutant rats caused by defective hepatic anion transport. Hepatology 5: 573–579.

357. Itoda M, Saito Y, Soyama A, Saeki M, Murayama N, Ishida S, Sai K, Nagano M, Suzuki H, Sugiyama Y, Ozawa S, Sawada J. 2002. Polymorphisms in the ABCC2 (cMOAT/MRP2) gene found in 72 estabilished cell lines derived from Japanese individuals: an association between single-nucleotide polymorphisms in the 5′-untranslated region and exon 28. Drug Metab Dispos 30: 363–364.

358. Bosch TM, Doodeman VD, Smits PH, Meijerman I, Schellens JH, Beijnen JH. 2006. Pharmacogenetic screening for polymorphisms in drug-metabolizing enzymes and drug transpoters in a Dutch population. Mol Diagn Ther 10: 175–185.

359. Shoda J, Kano M, Oda K, Kamiya J, Nimura Y, Suzuki H, Sugiyama Y, Miyazaki H, Todoroki T, Stengelin S, et al. 2001. The expression levels of plasma membrane transporters in the cholestatic liver of patients undergoing biliary drainage and their association with the impairment of biliary secretory function. Am J Gastroenterol 96: 3368–3378.

360. Zollner G, Fickert P, Zenz R, Fuchsbichler A, Stumptner C, Kenner L, Ferenci P, Stauber RE, Krejs GJ, Denk H, Zatloukal K, Trauner M. 2001. Hepatobiliary transporter expression in percutaneous liver biopsies of patients with cholestatic liver diseases. Hepatology 33: 633–646.

361. Hinoshita E, Taguchi K, Inokuchi A, Uchiumi T, Kinukawa N, Shimada M, Tsuneyoshi M, Sugimachi K, Kuwano M. 2001. Decreased expression of an ATP-binding cassette transporter, MRP2, in human livers with hepatitis C virus infection. J Hepatol 35: 765–773.

362. Fromm MF, Kauffmann HM, Fritz P, Burk O, Kroemer HK, Warzok RW, Eichelbaum M, Siegmund W, Schrenk D. 2000. The effect of rifampin treatment on intestinal expression of human MRP transporters. Am J Pathol 157: 1575–1580.

363. Cui Y, König J, Buchholz JK, Spring H, Leier I, Keppler D. 1999. Drug resistance and ATP-dependent conjugate transport mediated by the apical multidrug resistance protein, MRP2, permanently expressed in human and canine cells. Mol Pharmacol 55: 929–937.

364. Koike K, Kawabe T, Tanaka T, Toh S, Uchiumi T, Wada M, Akiyama S, Ono M, Kuwano M. 1997. A canalicular multispecific organic anion transporter (cMOAT) antisense cDNA enhances drug sensitivity in human hepatic cancer cells. Cancer Res 57: 5475–5479.

365. Borst P, Evers R, Kool M, Wijnholds J. 1999. The multidrug resistance protein family. Biochim Biophys Acta 1461: 347–357.

366. Evers R, Kool M, van Deemter L, Janssen H, Calafat J, Oomen LC, Paulusma CC, Oude Elferink RP, Baas F, Schinkel AH, Borst P. 1998. Drug export activity of the human canalicular multispecific organic anion transporter in polarized kidney MDCK cells expressing cMOAT (MRP2) cDNA. J Clin Invest 101: 1310–1319.

367. Evers R, de Haas M, Sparidans R, Beijnen J, Wielinga PR, Lankelma J, Borst P. 2000. Vinblastin and sulfinpyrazone export by the multidrug resistance protein MRP2 is associated with glutathione export. Br J Cancer 83: 375–383.

368. Huisman MT, Chhatta A, van Tellingen O, Beijnen JH, Schinkel AH. 2005. MRP2 (ABCC2) transports taxanes and confers paclitaxel resistance and both processes are stimulated by probenecid. Int J Cancer 116: 824–829.

369. Kool M, van der Linden M, de Haas M, Scheffer GL, De Vree JML, Smith AJ, Jansen G, Peters GI, Ponne N, Scheper RJ, et al. 1999. MRP3, an organic anion transporter able to transport anti-cancer drugs. Proc Natl Acad Sci U S A 96: 6914–6919.

370. Dietrich CG, Ottenhoff R, de Waart DR, Oude Elferink RP. 2001. Role of MRP2 and GSH in intrahepatic cycling of toxins. Toxicology 167: 73–81.

371. Konno T, Ebihara T, Hisaeda K, Uchiumi T, Nakamura T, Shirakusa T, Kuwano M, Wada M. 2003. Identification of domains participating in the substrate specificity and subcellular localization of the multidrug resistance proteins MRP1 and MRP2. J Biol Chem 278: 22908–22917.

372. Keppler D, Leier I, Jedlitschky G. 1997. Transport of glutathione conjugates and glucuronides by the multidrug resistance proteins MRP1 and MRP2. Biol Chem 378: 787–791.

373. Xiong H, Turner KC, Ward ES, Jansen PL, Brouwer KL. 2000. Altered hepatobiliary disposition of acetaminophen glucuronide in isolated perfused livers from multidrug resistance-associated protein 2-deficient TR(-) rats. J Pharmacol Exp Ther 295: 512–518.

374. Xiong H, Suzuki H, Sugiyama Y, Meier PJ, Pollack GM, Brouwer KLR. 2002. Mechanisms of impaired biliary excretion of acetaminophen glucuronide after acute phenobarbital treatment or phenobarbital pretreatment. Drug Metab Dispos 30: 962–969.

375. Seitz S, Kretz-Rommel A, Oude Elferink RP, Boelsterli UA. 1998. Selective protein adduct formation of diclofenac glucuronide is critically dependent on the rat canalicular conjugate export pump (Mrp2). Chem Res Toxicol 11: 513–519.

376. Sasaki M, Suzuki H, Ito K, Abe T, Sugiyama Y. 2002. Transcellular transport of organic anions across a double-transfected Madin–Darby canine kidney II cell monolayer expressing both human organic anion-transporting polypeptide (OATP2/SLC21A6) and multidrug resistance-associated protein 2 (MRP2/ABCC2). J Biol Chem 277: 6497–6503.

377. Akita H, Suzuki H, Ito K, Kinoshita S, Sato N, Takikawa H, Sugiyama Y. 2001. Characterization of bile acid transport mediated by multidrug resistance associated protein 2 and bile salt export pump. Biochim Biophys Acta 1511: 7–16.

378. Horikawa M, Kato Y, Tyson CA, Sugiyama Y. 2002. The potential for an interaction between MRP2 (ABCC2) and various therapeutic agents: probenecid as a candidate inhibitor of the biliary excretion of irinotecan metabolites. Drug Metab Pharmacokinet 17: 23–33.

379. Naruhashi K, Tamai I, Inoue N, Muraoka Hm Sai Y, Suzuki N, Tsuji A. 2002. Involvement of multidrug resistance-associated protein 2 in intestinal secretion of grepafloxacin in rats. Antimicrob Agents Chemother 46: 344–349.

380. Huisman MT, Smit JW, Crommentuyn KM, Zelcer N, Wiltshire HR, Beijnen JH, Schinkel AH. 2002. Multidrug resistance protein 2 (MRP2) transports HIV protease inhibitors, and transport can be enhanced by other drugs. AIDS 16: 2295–2301.

381. Kala SV, Neely MW, Kala G, Prater CI, Atwood DW, Rice JS, Lieberman MW. 2000. The MRP2/cMOAT transporter and arsenic–glutathione complex formation are required for biliary excretion of arsenic. J Biol Chem 275: 33404–33408.

382. Dietrich CG, de Waart DR, Ottenhoff R, Bootsma AH, Van Gennip AH, Oude Elferink RPJ. 2001. Mrp2-deficiency in the rat impairs biliary and intestinal excretion and influences metabolism and disposition of the food-derived carcinogen 2-amino-1-methyl-6-phenylimidazo[4,5-*b*]pyridine. Carcinogenesis 22: 805–811.

383. Dietrich CG, de Waart DR, Ottenhoff R, Schoots IG, Oude Elferink RPJ. 2001. Increased bioavailability of the food-derived carcinogen 2-amino-1-methyl-6-phenylimidazo[4,5-*b*]pyridine in MRP2-deficient rats. Mol Pharmacol 59: 974–980.

384. Leier I, Hummel-Eisenbeiss J, Cui Y, Keppler D. 2000. ATP-dependent *para*-aminohippurate transport by apical multidrug resistance protein MRP2. Kidney Int 57: 1636–1642.

385. van Aubel RA, van Kuijck MA, Koenderink JB, Deen PM, van Os CH, Russel FG. 1998. Adenosine triphosphate-dependent transport of anionic conjugates by the rabbit multidrug resistance–associated protein Mrp2 expressed in insect cells. Mol Pharmacol 53: 1062–1067.

386. Chen ZS, Furukawa T, Sumizawa T, Ono K, Ueda K, Seto K, Akiyama SI. 1999. ATP-Dependent efflux of CPT-11 and SN-38 by the multidrug resistance protein (MRP) and its inhibition by PAK-104P. Mol Pharmacol 55: 921–928.

387. Chen ZS, Kawabe T, Ono M, Aoki S, Sumizawa T, Furukawa T, Uchiumi T, Wada M, Kuwano M, Akiyama SI. 1999. Effect of multidrug resistance–reversing agents on transporting activity of human canalicular multispecific organic anion transporter. Mol Pharmacol 56: 1219–1228.

388. Zelcer N, Huisman MT, Reid G, Wielinga P, Breedveld P, Kuil A, Knipscher P, Schellens JHM, Schinkel AH, Borst P. 2003. Evidence for two interacting ligand binding sites in human multidrug resistance protein 2 (ATP binding cassette C2). J Biol Chem 278: 23538–23544.

389. Dombrowski F, Kubitz R, Chittattu A, Wettstein M, Saha N, Haussinger D. 2000. Electron-microscopic demonstration of multidrug resistance protein 2 (Mrp2) retrieval from the canalicular membrane in response to hyperosmolarity and lipopolysaccharide. Biochem J 348 (Pt 1): 183–188.

390. Kipp H, Arias IM. 2000. Intracellular trafficking and regulation of canalicular ATP-binding cassette transporters. Semin Liver Dis 20: 339–351.

391. Kauffmann HM, Keppler D, Kartenbeck J, Schrenk D. 1997. Induction of cMrp/cMoat gene expression by cisplatin, 2-acetylaminofluorene, or cycloheximide in rat heptocytes. Hepatology 26: 980–985.

392. Kauffmann HM, Keppler D, Gant TW, Schrenk D. 1998. Induction of hepatic mrp2 (cmrp/cmoat) gene expression in nonhuman primates treated with rifampicin or tamoxifen. Arch Toxicol 72: 763–768.

393. Courtois A, Payen L, Guillouzo A, Fardel O. 1999. Up-regulation of multidrug resistance–associated protein 2 (MRP2) expression in rat hepatocytes by dexamethasone. FEBS Lett 459: 381–385.

394. Courtois A, Payen L, Le Ferrec E, Scheffer GL, Trinquart Y, Guillouzo A, Fardel O. 2002. Differential regulation of multidrug-resistance associated protein 2 (MRP2) and cytochromes P450 2B1/2 and 3A1/2 in phenobarbital-treated hepatocytes. Biochem Pharmacol 63: 333–341.

395. Kubitz R, Warskulat U, Schmitt M, Haussinger D. 1999. Dexamethasone- and osmolarity-dependent expression of the multidrug-resistance protein 2 in cultured rat hepatocytes. Biochem J 340: 585–591.

396. Payen L, Sparfel L, Courtois A, Vernhet L, Guillouzo A, Fardel O. 2002. The drug efflux pump MRP2: regulation of expression in physiopathological situations and by endogenous and exogenous compounds. Cell Biol Toxicol 18: 221–233.

397. Kast HR, Goodwin B, Tarr PT, Jones SA, Anisfeld AM, Stoltz CM, Tontonoz P, Kliewer S, Willson TM, Edwards PA. 2002. Regulation of multidrug resistance–associated protein 2 (ABCC2) by the nuclear receptors pregnane X receptor, farnesoid X–activated receptor, and constitutive androstane receptor. J Biol Chem 277: 2908–2915.

398. Vernhet L, Seite MP, Allain N, Guillouzo A, Fardel O. 2001. Arsenic inducers expression of the multidrug resistance–associated protein 2 (MRP2) gene in primary rat and human hapatocytes. J Pharmacol Exp Ther 298: 234–239.

399. Schrenk D, Baus PR, Ermel N, Klein C, Vorderstemann B, Kauffmann HM. 2001. Up-regulation of transporters of the MRP family by drugs and toxins. Toxicol Lett 120: 51–57.

400. Shibayama Y, Ikeda R, Motoya T, Yamada K. 2004. St. John's wort (hypericum perforatum) induces overexpression of multidrug resistance protein 2 (MRP2) in rats: a 30-days ingestion study. Food Chem Toxicol 42: 995–1002.

401. Demeule M, Jodoin J, Beaulieu E, Brossard M, Béliveau. 1999. Dexamethasone modulation of multidrug transporters in normal tissues. FEBS Lett 442: 208–214.

402. Gerk PM, Vore M. 2002. Regulation of expression of the multidrug resistance–associated protein 2 (MRP2) and its role in drug disposition. J Pharmacol Exp Ther 302: 407–415.

403. Hagenbuch N, Reichel C, Stieger B, Cattori V, Fattinger KE, Landmann L, Meier PJ, Kullak-Ublick GA. 2001. Effect of phenobarbital on the expression of bile salt and organic anion transporters of rat liver. J Hepatol 34: 881–887.

404. Johnson DR, Habeebu SS, Klaassen CD. 2002. Increase in bile flow and biliary excretion of glutathione-derived sulfhydryls in rats by drug-metabolizing enzyme inducers is mediated by multidrug resistance protein 2. Toxicol Sci 66: 16–26.

405. König J, Nies AT, Cui Y, Leier I, Keppler D. 1999. Conjugate export pumps of the multidrug resistance protein (MRP) family: localization, substrate specificity, and MRP2-mediated drug resistance. Biochim Biophys Acta 1461: 377–394.

406. Sathirakul K, Suzuki H, Yasuda K, Hanano M, Tagaya O, Horie T, Sugiyama Y. 1993. Kinetic analysis of hepatobiliary transport of organic anions in Eisai hyperbilirubinemic mutant rats. J Pharmacol Exp Ther 265: 1301–1312.

407. Kusuhara H, Suzuki H, Sugiyama Y. 1998. The role of P-glycoprotein and canalicular multispecific organic anion transporter in the hepatobiliary excretion of drugs. J Pharm Sci 87: 1025–1040.

408. Sasabe H, Tsuji A, Sugiyama Y. 1998. Carrier-mediated mechanism for the biliary excretion of the quinolone antibiotic grepafloxacin and its glucuronide in rats. J Pharmacol Exp Ther 284: 1033–1039.

409. Yamazaki M, Akiyama S, Ni'inuma K, Nishigaki R, Sugiyama Y. 1997. Biliary excretion of pravastatin in rats: contribution of the excretion pathway mediated by canalicular multispecific organic anion transporter. Drug Metab Dispos 25: 1123–1129.

410. Masuda M, I'izuka Y, Yamazaki M, Nishigaki R, Kato Y, Ni'inuma K, Suzuki H, Sugiyama Y. 1997. Methotrexate is excreted into the bile by canalicular multispecific organic anion transporter in rats. Cancer Res 57: 3506–3510.

411. Lalloo AK, Luo FR, Guo A, Paranjpe PV, Lee SH, Vyas V, Rubin E, Sinko PJ. 2004 Membrane transport of camptothecin: facilitation by human P-glycoprotein (ABCB1) and multidrug resistance protein 2 (ABCC2). BMC Med 2: 16.

412. Luo FR, Paranjpe PV, Guo A, Rubin E, Sinko P. 2002. Intestinal transport of irinotecan in Caco-2 cells and MDCKII cells overexpessing efflux transporters Pgp, cMOAT, and MRP1. Drug Metab Dispos 30: 763–770.

413. Rowinsky EK, Grochow LB, Ettinger DS, Sartorius SE, Lubejko BG, Chen TL, Rock MK, Donehower RC. 1994. Phase I and pharmacological study of the novel topoisomerase I inhibitor 7-ethyl-10-[4-(1-piperidino)-1-piperidino]carbonyloxycamptothecin (CPT-11) administered as a ninety-minute infusion every 3 weeks. Cancer Res 54: 427–436.

414. Araki E, Ishikawa M, Iigo M, Koide T, Itabashi M, Hoshi A. 1993. Relationship between development of diarrhea and the concentration of SN-38, an active metabolite of CPT-11, in the intestine and the blood plasma of athymic mice following intraperitoneal administration of CPT-11. Jpn J Cancer Res 84: 697–702.

415. Sugiyama Y, Kato Y, Chu X. 1998. Multiplicity of biliary excretion mechanisms for the camptothecin derivative irinotecan (CPT-11), its metabolite SN-38, and its glucuronide: role of canalicular multispecific organic anion transporter and P-glycoprotein. Cancer Chemother Pharmacol 42 (Suppl): S44–S49.

416. Chu XY, Kato Y, Niinuma K, Sudo KI, Hakusui H, Sugiyama Y. 1997. Multispecific organic anion transporter is responsible for the biliary excretion of the camptothecin derivative irinotecan and its metabolites in rats. J Pharmacol and Exp Ther 1: 304–314.

417. Chu XY, Kato Y, Sugiyama Y. 1997. Multiplicity of biliary excretion mechanisms for irinotecan, CPT-11, and its metabolites in rats. Cancer Res 57: 1934–1938.

418. Chu XY, Kato Y, Ueda K, Suzuki H, Niinuma K, Tyson CA, Weizer V, Dabbs JE, Froehlich R, Green CE, Sugiyama Y. 1998. Biliary excretion mechanism of CPT-11 and its metabolites in humans: involvement of primary active transporters. Cancer Res 58: 5137–5143.

419. Chu XY, Kato Y, Sugiyama Y. 1999. Possible involvement of P-glycoprotein in biliary excretion of CPT-11 in rats. Short communication. Drug Metab Dispos 4: 440–441.

420. Horikawa M, Kato Y, Sugiyama Y. 2002. Reduced gastrointestinal toxicity following inhibition of the biliary excretion of irinotecan and its metabolites by probenecid in rats. Pharm Res 19: 1345–1353.

421. Qadir M, O'Loughlin KL, Fricke SM, Williamson NA, Greco WR, Minderman H, Baer MR. 2005. Cyclosporin A is a broad-spectrum multidrug resistance modulator. Clin Cancer Res 11: 2320–2326.

422. Arimori K, Kuroki N, Hidaka M, Iwakiri T, Yamsaki K, Okumura M, Ono H, Takamura N, Kikuchi M, Nakano M. 2003. Effect of P-glycoprotein modulator, cyclosporin A, on the gastrointestinal excretion of irinotecan and its metabolite SN-38 in rats. Pharm Res 20: 910–917.

423. Gupta E, Safa AR, Wang X, Ratain MJ. 1996. Pharmacokinetic modulation of irinotecan and metabolites by cyclosporin A. Cancer Res 56: 1309–1314.

424. Chester JD, Joel SP, Cheeseman SL, Hall GD, Braun MS, Perry J, Davis T, Button CJ, Seymour MT. 2003. Phase I and pharmacokinetic study of intravenous irinotecan plus oral ciclosporin in patients with fuorouracil-refractory metastatic colon cancer. J Clin Oncol 21: 1125–1132.

425. Innocenti F, Undevia SD, Ramirez J, Mani S, Schilsky RL, Vogelzang NJ, Prado M, Ratain MJ. 2004. A phase I trial of pharmacologic modulation of irinotecan with cyclosporine and phenobarbital. Clin Pharmacol Ther 76: 490–502.

426. Oguchi H, Miyasaka M, Koiwai T, Tokunaga S, Hora K, Sato K, Yoshie T, Shioya H, Furuta S. 1993. Pharmacokinetics of temocapril and enalapril in patients with various degrees of renal insufficiency. Clin Pharmacokinet 24: 421–427.

427. Nakashima M, Yamamoto J, Shibata M, Uematsu T, Shinjo H, Akahori T, Shioya H, Sugiyama K, Kawahara Y. 1992. Pharmacokinetics of temocapril hydrochloride, a novel angiotensin converting enzyme inhibitor, in renal insufficiency. Eur J Clin Pharmacol 43: 657–659.

428. Hoyer J, Schulte KL, Lenz T. 1993. Clinical pharmacokinetics of angiotensin converting enzyme (ACE) inhibitors in renal failure. Clin Pharmacokinet 24: 230–254.

429. Puchler K, Eckl KM, Fritsche L, Renneisen K, Neumayer HH, Sierakowski B, Lavrijssen AT, Thomsen T, Roots I. 1997. Pharmacokinetics of temocapril and temocaprilat after 14 once daily oral doses of temocapril in hypertensive patients with varying degrees of renal impairment. Br J Clin Pharmacol 44: 531–536.

430. Ishizuka H, Konno K, Naganuma H, Sasahara K, Kawahara Y, Niinuma K, Suzuki H, Sugiyama Y. 1997. Temocaprilat, a novel angiotensin-converting enzyme inhibitor, is excreted in bile via an ATP-dependent active transporter (cMOAT) that is deficient in Eisai hyperbilirubinemic mutant rats (EHBR). J Pharmacol Exp Ther 280: 1304–1311.

431. Gotoh Y, Suzuki H, Kinoshita S, Hirohashi T, Kato Y, Sugiyama Y. 2000. Involvement of an organic anion transporter (canalicular multispecific organic anion transporter/multidrug resistance–associated protein 2) in gastrointestinal secretion of glutathione conjugates in rats. J Pharmacol Exp Ther 292: 433–439.

432. Taipalensuu J, Tornblom H, Lindberg G, Einarsson C, Sjoqvist F, Melhus H, Garberg P, Sjostrom B, Lundgren B, Artursson P. 2001.Correlation of gene expression of ten drug efflux proteins of the ATP-binding cassette transporter family in normal human jejunum and in human intestinal epithelial Caco-2 cell monolayers. J Pharmacol Exp Ther 299: 164–170.

433. Huisman MT, Looije NA, Crommentuyn KML, Elferink RO, Rosing H, Beijnen JH, Schinkel AH. 2003. Probenecid decreases the oral availability of MRP2 (ABCC2) substrates drugs. Multidrug transporters and the pharmacokinetics of HIV protease inhibitors. Ph.D. dissertation, University of Utrecht, The Netherlands, Chapter 5, pp 96–115.

434. Potschka H, Löscher W. 2001. Multidrug resistance–associated protein is involved in the regulation of extracellular levels of phenytoin in the brain. Neuroreport 12: 2387–2389.

435. Potschka H, Fedrowitz M, Löscher W. 2003. Brain access and anticonvulsant efficacy of carbamazepine, lamotrigine, and felbamate in ABCC2/MRP2-deficient TR^-rats. Epilepsia 44: 1479–1487.

436. Potschka H, Fedrowitz M, Löscher W. 2003. Multidrug resistance protein MRP2 contributes to blood–brain barrier function and restricts antiepileptic drug activity. J Pharmacol Exp Ther 306: 124–131.

437. Frey HH, Löscher W. 1978. Distribution of valproate across the interface between blood and cerebrospinal fluid. Neuropharmacology 17: 637–642.

438. Potschka H, Fedrowitz, Löscher W. 2001. P-Glycoprotein and multidrug resistance associated protein are involved in the regulation of extracellular levels of the major antiepileptic drug carbamazepine in the brain. Neuroreport 12: 3557–3560.

439. Löscher W, Potschka H. 2002. Role of multidrug transporters in pharmacoresistance to antiepileptic drugs. J Pharmacol Exp Ther 301: 7–14.

440. Tishler DM, Weinberg KI, Hinton DR, Barbaro N, Annett GM, Raffel C. 1995. MDR1 gene expression in brain of patients with medically intractable epilepsy. Epilepsia 36: 1–6.

441. Sisodiya SM, Lin WR, Harding BN, Squier MV, Thom M. 2002. Drug resistance in epilepsy: expression of drug resistance proteins in common causes of refractory epilepsy. Brain 125: 22–31.

442. Dombrowski SM, Desai SY, Marroni M, Cucullo L, Goodrich K, Bingaman W, Mayberg MR, Bengez L, Janigro D. 2001. Overexpression of multiple drug resistance genes in endothelial cells from patients with refractory epilepsy. Epilepsia 42: 1501–1506.

443. Löscher W, Potschka H. 2005. Drug resistance in brain diseases and the role of drug efflux transporters. Nat Rev Neurosci 6: 591–602.

444. Fromm MF, Eckhardt K, Li S, Schänzle G, Hofmann U, Mikus G, Eichelbaum M. 1997. Loss of analgesic effect of morphine due to coadministration of rifampin. Pain 72: 261–267.

445. Dilger K, Greiner B, Fromm MF, Hofmann U, Kroemer HK, Eichelbaum M. 1999. Consequences of rifampicin treatment on propafenone disposition in extensive and poor metabolizers of CYP2D6. Pharmacogenetics 9: 551–559.

446. Strayhorn VA, Baciewicz AM, Self TH. 1997. Update on rifampin drug interactions, III. Arch Intern Med 157: 2453–2458.

447. Brouwer KL, Jones JA. 1990. Altered hepatobiliary disposition of acetaminophen metabolites after phenobarbital pretreatment and renal ligation: evidence for impaired biliary excretion and a diffusional barrier. J Pharmacol Exp Ther 252: 657–664.

448. Studenberg SD, Brouwer KL. 1992. Impaired biliary excretion of acetaminophen glucuronide in the isolated perfused rat liver after acute phenobarbital treatment and in vivo phenobarbital pretreatment. J Pharmacol Exp Ther 261: 1022–1027.

449. Sandusky GE, Mintze KS, Pratt SE, Dantzig AH. 2002. Expression of multidrug resistance-associated protein 2 (MRP2) in normal human tissues and carcinomas using tissue microarrays. Histopathology 41: 65–74.

450. Lee YMA, Cui Y, König J, Risch A, Jäger B, Drings P, Bartsch H, Keppler D, Nies AT. 2004. Indentification and functional characterization of the natural vairant MRP3-Arg^{1297} His of human multidrug resistance protein 3 (MRP3/ABCC3). Pharmacogenetics 14: 213–223.

451. Lang T, Hitzl M, Burk O, Mornhinweg E, Keil A, Kerb R, Klein K, Zanger UM, Eichelbaum M, Fromm MF. 2004. Genetic polymorphisms in the multidrug resistance-associated protein 3 (ABCC3, MRP3) gene and relationship to its mRNA and protein expression in human liver. Pharmacogenetics 14: 155–164.

452. Zelcer N, Saeki T, Reid G, Beijnen JH, Borst P. 2001. Characterization of drug transport by the human multidrug resistance protein 3 (ABCC3). J Biol Chem 276: 46400–46407.

453. Hirohashi T, Suzuki H, Sugiyama Y. 1999. Characterization of the transport properties of cloned rat multidrug resistance–associated protein 3 (MRP3). J Biol Chem 274: 15181–15185.

454. Zeng H, Liu G, Rea PA, Kruh GD. 2000. Transport of amphipathic anions by human multidrug resistance protein 3. Cancer Res 60: 4779–4784.

455. Paumi CM, Wright M, Townsend AJ, Morrow CS. 2003. Multidrug resistance protein (MRP) 1 and MRP3 attenuate cytotoxic and transactivating effects of the cyclopentenone prostaglandin, 15-deoxy-delta(12,14)prostaglandin J2 in MCF7 breast cancer cells. Biochemistry 42: 5429–5437.

456. Hirohashi T, Suzuki H, Takikawa H, Sugiyama Y. 2000. ATP-dependent transport of bile salts by rat multidrug resistance–associated protein 3 (Mrp3). J Biol Chem 275: 2905–2910.

457. Zelcer N, Saeki T, Bot I, Kuil A, Borst P. 2003. Transport of bile acids in multidrug-resistance-protein 3-overexpressing cells co-transfected with the ileal Na^+-dependent bile-acid transporter. Biochem J 369(Pt 1): 23–30.

458. Zelcer N, Wetering KV, Hillebrand M, Sarton E, Kuil A, Wielinga PR, Tephly Th, Dahan A, Beijnen JH, Borst P. 2005. Mice lacking multidrug resistance protein 3 show altered morphine pharmacokinetics and morphine-6-glucuronide antinociception. Proc Natl Acad Sci U S A 102: 7274–7279.

459. Zelcer N, Wetering KV, Waart RD, Scheffer GL, Marschall HU, Wielinga PR, Kuil A, Kunne C, Smith A, Valk MV, et al. 2006. Mice lacking Mrp3 (Abcc3) have normal bile salt transport, but altered hepatic transport of endogenous glucuronides. J Hepatol 44: 768–775.

460. Hitzl M, Klein K, Zanger UM, Fritz P, Nüssler AK, Neuhaus P, Fromm MF. 2003. Influence of omeprazole on multidrug reistance protein 3 expression in human liver. J Pharmacol Exp Ther 304: 524–530.

461. Stockel B, König J, Nies AT, Cui Y, Brom M, Keppler D. 2000. Characterization of the 5′-flanking region of the human multidrug resistance protein 2 (MRP2) gene and its regulation in comparison withthe multidrug resistance protein 3 (MRP3) gene. Eur J Biochem 267: 1347–1358.

462. Slitt AL, Cherrington NJ, Maher JHM, Klaassen CD. 2003. Induction of multidrug resistance protein 3 in rat liver is associated with altered vectorial excretion of acetaminophen metabolites. Drug Metab Dispos 31: 1176–1186.

463. Kiuchi Y, Suzuki H, Hirohashi T, Tyson CA, Sugiyama Y. 1998. cDNA cloning and inducible expression of human multidrug resistance associated protein 3 (MRP3). FEBS Lett 433: 149–152.

464. Cherrington NJ, Hartley DP, Li N, Johnson DR, Klaassen CD. 2002. Organ distribution of multidrug resistance proteins 1, 2, and 3 (Mrp1, 2, and 3) mRNA and hepatic induction of Mrp3 by constitutive androstane receptor activators in rats. J Pharmacol Exp Ther 300: 97–104.

465. Ghanem CI, Ruiz ML, Villanueva SSM, Luquita MG, Catania VA, Jones B, Bengochea LA, Vore M, Mottino AD. 2005. Shift from biliary to urinary elimination of acetaminophen–glucuronide in acetaminophen-pretreated rats. J Pharmacol Exp Ther 315: 987–995.

466. Inokuchi A, Hinoshita E, Iwamoto Y, Kohno K, Kuwano M, Uchiumi T. 2001. Enhanced expression of the human multidrug resistance protein 3 by bile salt in human enterocytes: A transcriptional control of a plausible bile acid transporter. J Biol Chem 276: 46822–46829.

467. König J, Rost D, Cui Y, Keppler D. 1999. Characterization of the human multidrug resistance protein isoform MRP3 localized to the basolateral hepatocyte membrane. Hepatology 29: 1156–1163.

468. Zollner G, Fickert P, Silbert D, Fuchsbichler A, Marschall HU, Zatloukal K, Denk H, Trauner M. 2003. Adaptive changes in hepatobiliary transporter expression in primary biliary cirrhosis. J Hepatol 38: 717–727.

469. Scheffer GL, Kool M, de Haas M, de Vree JM, Pijnenborg AC, Bosman DK, Elferink RP, van der Valk P, Borst P, Scheper RJ. 2002. Tissue distribution and induction of human multidrug resistant protein 3. Lab Invest 82: 193–201.

470. Ogawa K, Suzuki H, Hirohashi T, Ishikawa T, Meier PJ, Hirose K, Akizawa T, Yoshioka M, Sugiyama Y. 2000. Characterization of inducible nature of MRP3 in rat liver. Am J Physiol Gastrointest Liver Physiol 278: G438–G446.

471. Manautou JE, de Waart DR, Kunne C, Zelcer N, Goedken M, Borst P, Elferink RO. 2005. Altered disposition of acetaminophen in mice with a disruption of the Mrp3 gene. Hepatology 42: 1091–1098.

472. Chen C, Hennig GE, Manautou JE. 2003. Hepatobiliary excretion of acetaminophen glutathione conjugate and its derivatives in transport-deficient (TR-) hyperbilirubinemic rats. Drug Metab Dispos 31: 798–804.

473. Tredger JM, Thuluvath P, Williams R, Murray-Lyon IM. 1995. Metabolic basis for high paracetamol dosage without hepatic injury: a case study. Hum Exp Toxicol 14: 8–12.

474. Ghanem CI, Gomez PC, Arana MC, Perassolo M, Ruiz ML, Villanueva SS, Ochoa EJ, Catania VA, Bengochea LA, Mottino AD. 2004. Effect of acetaminophen on expression and activity of rat liver multidrug resistance–associated protein 2 and P-glycoprotein. Biochem Pharmacol 68: 791–798.

475. Aleksunes LM, Slitt AM, Cherrington NJ, Thibodeau MS, Klaassen CD, Manautou JE. 2005. Differential expression of mouse hepatic transporter genes in response to acetaminophen and carbon tetrachloride. Toxicol Sci 83: 44–52.

476. Aleksunes LM, Scheffer GL, Jakowski AB, Pruimboom-Brees IM, Manautou JE. 2006. Coordinated expression of multidrug resistance-associated proteins (Mrps) in mouse liver during toxicant-induced injury. Toxicol Sci 89: 370–379.

477. Rausch-Derra LC, Hartley DP, Meier PJ, Klaassen CD. 2001. Differential effects of microsomal enzyme-inducing chemicals on the hepatic expression of rat organic anion transporters, OATP1 and OATP2. Hepatology 33: 1469–1478.

478. Lee K, Klein-Szanto AJ, Kruh GD. 2000. Analysis of the MRP4 drug resistance profile in transfected NIH3T3 cells. J Natl Cancer Inst 92: 1934–1940.

479. Lai L, Tan TMC. 2002. Role of glutathione in the multidrug resistance protein 4 (MRP4/ABCC4)-mediated efflux of cAMP and resistance to purine analogues. Biochem J 361: 497–503.

480. Reid G, Wielinga P, Zelcer N, de Haas M, van Deemster L, Wijnholds J, Balzarini J, Borst P. 2003. Characterization of the transport of nucleoside analog drugs by the human multidrug resistance proteins MRP4 and MRP5. Mol Pharmacol 63: 1094–1103.

481. Reid G, Wielinga P, Zelcer N, van der Heijden I, Kuil A, de Haas M, Wijnholds J, Borst P. 2003. The human multidrug resistance protein MRP4 functions as a prostaglandin efflux transporter and is inhibited by nonsteroidal antiinflammatory drugs. Proc Natl Acad Sci U S A 100: 9244–9249.

482. Schuetz JD, Connelly M, Sun D, Paibir SG, Flynn PM, Srinivas RV, Kumar A, Fridland A. 1999. MRP4: a previously unidentified factor in resistance to nucleoside-based antiviral drugs. Nat Med 5: 1048–1051.

483. Chen ZS, Lee K, Kruh GD. 2001. Transport of cyclic nucleotides and estradiol 17-beta-D-glucuronide by multidrug resistance protein 4: resistance to 6-mercaptopurine and 6-thioguanine. J Biol Chem 276: 33747–33754.

484. Tian Q, Zhang J, Chin Tan TM, Chan E, Duan W, Chan SY, Boelsterli UA, Ho PCL, Yang H, Bian JS, et al. 2005. Human multidrug resistance associated protein 4 confers resistance to camptothecins. Pharm Res 22: 1837–1853.

485. Tian Q, Zhang J, Chan SY, Tan TM, Duan W, Huang M, Zhu YZ, Chan E, Yu Q, Nie YQ, et al. 2006. Topotecan is a substrate for multidrug resistance associated protein 4. Curr Drug Metab 7: 105–118.

486. Norris MD, Smith J, Tanabe K, Tobin P, Flemming C, Scheffer GL, Wielinga P, Cohn SL, London WB, Marshall G, Allen JD, Haber M. 2005. Expression of multidrug transporter MRP4/ABCC4 is a marker of poor prognosis in neuroblastoma and confers resistance to irinotecan in vitro. Mol Cancer Ther 4: 547–553.

487. Rius M, Nies AT, Hummel-Eisenbeiss J, Jedlitschky G, Keppler D. 2003. Cotransport of reduced glutathione with bile salts by MRP4 (ABCC4) localized to the basolateral hepatocyte membrane. Hepatology 38: 374–384.

488. Rius M, Hummel-Eisenbeiss J, Hofmann AF, Keppler D. 2006. Substrate specificity of human ABCC4 (MRP4)–mediated cotransport of bile acids and reduced glutathione. Am J Physiol Gastrointest Liver Physiol 290: G640–G649.

489. Wielinga PR, van der Heijden I, Reid G, Beijnen JH, Wijnholds J, Borst P. 2003. Characterization of the MRP4- and MRP5-mediated transport of cyclic nucleotides from intact cells. J Biol Chem 278: 17664–17671.

490. Jorajuria S, Dereuddre-Bosquet N, Becher F, Martin S, Porcheray F, Garrigues A, Mabondzo A, Benech H, Grassi J, Orlowski S, Dormont D, Clayette P. 2004. ATP

binding cassette multidrug transporters limit the anti-HIV activity of zidovudine and indinavir in infected human macrophages. Antivir Ther 9: 519–528.

491. Leggas M, Adachi M, Scheffer GL, Sun D, Wielinga P, Du G, Mercer KE, Zhuang Y, Panetta JC, Johnston B, et al. 2004. Mrp4 confers resistance to topotecan and protects the brain from chemotherapy. Mol Cell Biol 24: 7612–7621.

492. Spector R. 1976. Inhibition of methotrexate transport from cerebrospinal fluid by probenecid. Cancer Treat Rep 60: 913–916.

493. Trifan OC, Durham WF, Salazar VS, Horton J, Levine BD, Zweifel BS, Davis TW, Masferrer JL. 2002. Cyclooxygenase-2 inhibition with celecoxib enhances antitumor efficacy and reduces diarrhea side effect of CPT-11. Cancer Res 62: 5778–5784.

494. Pan CX, Loehrer P, Seitz D, Helft P, Juliar B, Ansari R, Pletcher W, Vinson J, Cheng L, Sweeney C. 2005. A phase II trial of irinotecan, 5-fluorouracil and leucovorin combined with celecoxib and glutamine as first-line therapy for advanced colorectal cancer. Oncology 69: 63–70.

495. Villalona-Calero M, Schaaf L, Phillips G, Otterson G, Panico K, Duan W, Kleiber B, Shah M, Young D, Wu WH, Kuhn J. 2007. Thalidomide and celecoxib as potential modulators of irinotecan's activity in cancer patients. Cancer Chemother Pharmacol 59: 23–33.

496. Maiello E, Giuliani F, Gebbia V, Di Renzo N, Pezzella G, Romito S, Mallamaci R, Lopez M, Colucci G. 2006. FOLFIRI with or without celecoxib in advanced colorectal cancer: a randomized phase II study of the Gruppo Oncologico dell'Italia Meridionale (GOIM). Ann Oncol 17 (Suppl 7): vii55-vii59.

497. Argiris A, Kut V, Luong L, Avram MJ. 2006. Phase I and pharmacokinetic study of docetaxel, irinotecan, and celecoxib in patients with advanced non-small cell lung cancer. Invest New Drugs 24: 203–212.

498. Dy GK, Mandrekar S, Peethambaram PP, Okuno SH, Croghan GC, Hanson LJ, Furth A, Adjei AA. 2005. A phase I trial of celecoxib in combination with docetaxel and irinotecan in patients with advanced cancer. Cancer Chemother Pharmacol 56: 623–628.

499. Reardon DA, Quinn JA, Vredenburgh J, Rich JN, Gururangan S, Badruddoja M, Herndon JE 2nd, Dowell JM, Friedman AH, Friedman HS. 2005. Phase II trial of irinotecan plus celecoxib in adults with recurrent malignant glioma. Cancer 103: 329–338.

500. Vormfelde SV, Schirmer M, Hagos Y, Toliat MR, Engelhardt S, Meineke I, Burckhardt G, Nürnberg P, Brockmöller J. 2006. Torsemide renal clearance and genetic variation in luminal and basolateral organic anion transporters. Br J Clin Pharmacol 62: 323–335.

501. Xu G, Bhatnagar V, Wen G, Hamilton BA, Eraly SA, Nigam SK. 2005. Analyses of coding region polymorphisms in apical and basolateral human organic anion transporter (OAT) genes [OAT1 (NKT), OAT2, OAT3, OAT4, URAT (RST)]. Kidney Int 68: 1491–1499.

502. Iida A, Saito S, Sekine A, Mishima C, Kondo K, Kitamura Y, Harigae S, Osawa S, Nakamura Y. 2001. Catalog of 258 single-nucleotide polymorphisms (SNPs) in genes encoding three organic anion transporters, three organic anion-transporting polypeptides, and three NADH:ubiquinone oxidoreductase flavoproteins. J Hum Genet 46: 668–683.

503. Ichida K, Hosoyamada M, Hisatome I, Enomoto A, Hikita M, Endou H, Hosoya T. 2004. Clinical and molecular analysis of patients with renal hypouricemia in Japan: influence of URAT1 gene on urinary urate excretion. J Am Soc Nephrol 15: 164–173.

504. Komoda F, Sekine T, Inatomi J, Enomoto A Endou H, Ota T, Matsuyama T, Ogata T, Ikeda M, Awazu M, et al. 2004. The W258X mutation in SLC22A12 is the predominant cause of Japanese renal hypouricemia. Pediatr Nephrol 19: 728–733.

505. Enomoto A, Kimura H, Chairoungdua A, Shigeta Y, Jutabha P, Cha SH, Hosoyamada M, Takeda M, Sekine T, Igarashi T, et al. 2002. Molecular identification of a renal urate anion exchanger that regulates blood urate levels. Nature 417: 447–452.

506. Shima Y, Teruya K, Ohta H. 2006. Association between intronic SNP in urate-anion exchanger gene, SLC22A12, and serum uric acid levels in Japanese. Life Sci 79: 2234–2237.

507. Bleasby K, Hall LA, Perry JL, Mohrenweiser HW, Pritchard JB. 2005. Functional consequences of single-nucleotide polymorphisms in the human organic anion transporter hOAT1 (SLC22A6). J Pharmacol Exp Ther 314: 923–931.

508. Fujita T, Brown C, Carlson EJ, Taylor T, de la Cruz M, Johns SJ, Stryke D, Kawamoto M, Fujita K, Castro R, et al. 2005. Functional analysis of polymorphisms in the organic anion transporter, SLC22A6 (OAT1). Pharmacogenet Genom 15: 201–209.

509. Nishizato Y, Ieiri I, Suzuki H, Kimura M, Kawabata K, Hirota T, Takane H, Irie S, Kusushara H, Urasaki Y, et al. 2003. Polymorphisms of OATP-C (SLC21A6) and OAT3 (SLC22A8) genes: consequences for pravastatin pharmacokinetics. Clin Pharmacol Ther 73: 554–565.

510. Hasegawa M, Kusuhara H, Sugiyama D, Ito K, Ueda S, Endou H, Sugiyama Y. 2002. Functional involvement of rat organic anion transporter 3 (rOAT3; SLC22a8) in the renal uptake of organic anions. J Pharmacol Exp Ther 300: 746–753.

511. Bhatnagar V, Xu G, Hamilton BA, Truong DM, Eraly SA, Wu W, Nigam SK. 2006. Analyses of 5′ regulatory region polymorphisms in human SLC22A6 (OAT1) and SLC22A8 (OAT3). J Hum Genet 51: 575–580.

512. Eraly SA, Blantz RC, Bhatnagar V, Nigam SK. 2003. Novel aspects of renal organic anion transporters. Curr Opin Nephrol Hypertens 12: 551–558.

513. Burckhardt BC, Burckhardt G. 2003. Transport of organic anions across the basolateral membrane of proximal tubule cells. Rev Physiol Biochem Pharmacol 146: 95–158.

514. Dresser MJ, Leabman MK, Giacomini KM. 2001. Transporters involved in the elimination of drugs in the kidney: organic anion transporters and organic cation transporters. J Pharm Sci 90: 397–421.

515. Sekine T, Cha SH, Endou H. 2000. The multispecific organic anion transporter (OAT) family. Pflugers Arch 440: 337–350.

516. Anzai N, Kanai Y, Endou H. 2006. Organic anion transporter family: current knowledge. J Pharmacol Sci 100: 411–426.

517. Race JE, Grassl SM, Williams WJ, Holtzman EJ. 1999. Molecular cloning and characterization of two novel human renal organic anion transporters (hOAT1 and hOAT3). Biochem Biophys Res Commun 255: 508–514.

518. Mulato AS, Ho ES, Cihlar T. 2000. Nonsteroidal anti-inflammatory drugs efficiently reduce the transport and cytotoxicity of adefovir mediated by the human renal organic anion transporter 1. J Pharmacol Exp Ther 295: 10–15.

519. Hosoyamada M, Sekine T, Kanai Y, Endou H. 1999. Molecular cloning and functional expression of a multispecific organic anion transporter from human kidney. Am J Physiol 276: F122–F128.

520. Khamdang S, Takeda M, Noshiro R, Narikawa S, Enomoto A, Anzai N, Piyachaturawat P, Endou H. 2002. Interactions of human organic anion transporters and human organic cation transporters with nonsteroidal anti-inflammatory drugs. J Pharmacol Exp Ther 303: 534–539.

521. Khamdang S, Takeda M, Shimoda M, Noshiro R, Narikawa S, Huang XL, Enomoto A, Piyachaturawat P, Endou H. 2004. Interactions of human- and rat-organic anion transporters with pravastatin and cimetidine. J Pharmacol Sci 94: 197–202.

522. Uwai Y, Saito H, Hashimoto Y, Inui KI. 2000. Interaction and transport of thiazide diuretics, loop diuretics, and acetazolamide via rat renal organic anion transporter rOAT1. J Pharmacol Exp Ther 295: 261–265.

523. Uwai Y, Saito H, Inui K. 2000. Interaction between methotrexate and nonsteroidal anti-inflammatory drugs in organic anion transporter. Eur J Pharmacol 409: 31–36.

524. Jariyawat S, Sekine T, Takeda M, Apiwattanakul N, Kanai Y, Sophasan S, Enou H. 1999. The interaction and transport of β-lactam antibiotics with the cloned rat renal organic anion transporter 1. J Pharmacol Exp Ther 290: 673–677.

525. Hasegawa M, Kusuhara H, Endou H, Sugiyama Y. 2003. Contribution of organic anion transporters to the renal uptake of anionic compounds and nucleoside derivatives in rat. J Pharmacol Exp Ther 305: 1087–1097.

526. Kimura H, Takeda M, Narikawa S, Enomoto A, Ichida K, Endou H. 2002. Human organic anion transporters and human organic cation transporters mediate renal transport of prostaglandins. J Pharmacol Exp Ther 301: 293–298.

527. Babu E, Takeda M, Narikawa S, Kobayashi Y, Yamamoto T, Cha SH, Sekine T, Sakthisekaran D, Endou H. 2002. Human organic anion transporters mediate the transport of tetracycline. Jpn J Pharmacol 88: 69–76. Abstract only.

528. Takeda M, Khamdang S. Narikawa S, Kimura H, Hosoyamada M, Cha SH, Sekine T, Endou H. 2002. Characterization of methotrexate transport and its drug interactions with human organic anion transporters. J Pharmacol Exp Ther 302: 666–671.

529. Takeda M, Khamdang S, Narikawa S, Kimura H, Kobayashi Y, Yamamoto T, Cha SH, Sekine T, Endou H. 2002. Human organic anion transporters and human organic cation transporters mediate renal antiviral transport. J Pharmacol Exp Ther 300: 918–924.

530. Takeda M, Babu E, Narikawa S, Endou H. 2002. Interaction of human organic anion transporters with various cephalosporin antibiotics. Eur J Pharmacol 438: 137–142.

531. Enomoto A, Takeda M, Shimoda M, Narikawa S, Kobayashi Y, Kobayashi Y, Yamamoto T, Sekine T, Cha SH, Niwa T, Endou H. 2002. Interaction of human organic anion transporters 2 and 4 with organic anion transport inhibitors. J Pharmacol Exp Ther 301: 797–802.

532. Khamdang S, Takeda M, Babu E, Noshiro R, Onozato ML, Tojo A, Enomoto A, Huang XL, Narikawa S, Anzai N, Piyachaturawat P, Endou H. 2003. Interaction of human and rat organic anion transporter 2 with various cephalosporin antibiotics. Eur J Pharmacol 465: 1–7.

533. Cha SH, Sekine T, Fukushima JI, Kanai Y, Kobayashi Y, Goya T, Endou H. 2001. Identification and characterization of human organic anion transporter 3 expressing predominantly in the kidney. Mol Pharmacol 59: 1277–1286.

534. Cha SH, Sekine T, Kusuhara H, Yu E, Kim JY, Kim DK, Sugiyama Y, Kanai Y, Endou H. 2000. Molecular cloning and characterization of multispecific organic anion transporter 4 expressed in the placenta. J Biol Chem 275: 4507–4512.

535. Babu E, Takeda M, Narikawa S, Kobayashi Y, Enomoto A, Tojo A, Cha SH, Sekine T, Sakthisekaran D, Endou H. 2002. Role of human organic anion transporter 4 in the transport of ochratoxin A. Biochim Biophys Acta 1590: 64–75.

536. Youngblood GL, Sweet DH. 2004. Identification and functional assessment of the novel murine organic anion transporter Oat5 (Slc22a19) expressed in kidney. Am J Physiol Renal Physiol 287: F236–F244.

537. Anzai N, Jutabha P, Enomoto A, Yokoyama H, Nonoguchi H, Hirata T, Shiraya K, He X, Cha SH, Takeda M, et al. 2005. Functional characterization of rat organic anion transporter 5 (Slc22a19) at the apical membrane of renal proximal tubules. J Pharmacol Exp Ther 315: 534–544.

538. Nakashima M, Uematsu T, Kosuge K, Kanamaru M. 1992. Pilot study of the uricosuric effects of DuP-753, a new angiotensin II receptor antagonist, in healthy subjects. Eur J Clin Pharmacol 42: 333–335.

539. Enomoto A, Endou H. 2005. Roles of organic anion transporters (OATs) and a urate transporter (URAT1) in the pathophysiology of human disease. Clin Exp Nephrol 9:195–205.

540. Kobayashi Y, Hirokawa N, Ohshiro N, Sekine T, Sasaki T, Tokuyama S, Endou H, Yamamoto T. 2002. Differential gene expression of organic anion transporters in male and female rats. Biochem Biophys Res Commun 290: 482–487.

541. Kudo N, Katakura M, Sato Y, Kawashima Y. 2002. Sex hormones regulated renal transport of perfluorooctanoic acid. Chem Biol Interact 139: 301–316.

542. Buist SC, Cherrington NJ, Choudhuri S, Hartley DP, Klaassen CD. 2002. Gender-specific and developmental influences on the expression of rat organic anion transporters. J Pharmacol Exp Ther 301: 145–151.

543. Sauvant C, Holzinger H, Gekle M. 2001. Modulation of the basolateral and apical step of transepithelial organic anion secretion in proximal tubular opossum kidney cells, acute effects of eppidermal growth factor and mitogen-activated protein kinase. J Biol Chem 276: 14695–14703.

544. Lu R, Chan BS, Schuster VL. 1999. Cloning of the human kidney PAH transporter: narrow substrate specificity and regulation by protein kinase C. Am J Physiol 276: F295–F303.

545. Eraly SA, Vallon V, Vaughn DA, Gangoiti JA, Richter K, Nagle M, Monte JC, Rieg T, Truong DM, Long JM, et al. 2006. Decreased renal organic anion secretion and plasma accumulation of endogenous organic anions in OAT1 knock-out mice. J Biol Chem 281: 5072–5083.

546. Sweet DH, Miller DS, Pritchard JB, Fujiwara Y, Beier DR, Nigam SK. 2002. Impaired organic anion transport in kidney and choroid plexus of organic anion transporter 3 (Oat3 (Slc22a8)) knockout mice. J Biol Chem 277: 26934–26943.

547. Ho ES, Lin DC, Mendel DB, Cihlar T. 2000. Cytotoxicity of antiviral nucleotides adefovir and cidofovir is induced by the expression of human renal organic anion transporter 1. J Am Soc Nephrol 11: 383–393.

548. Tsuda M, Sekine T, Takeda M, Cha SJ, Kanai Y, Kimura M, Endou H. 1999. Transport of ochratoxin A by renal multispecific organic anion transporter 1. J Pharmacol Exp Ther 289: 1301–1305.

549. Stein AF, Phillips TD, Kubena LF, Harvey RB. 1985. Renal tubular secretion and reabsorption as factors in ochratoxicosis: effects of probenecid on nephrotoxicity. J Toxicol Environ Health 16: 593–605.

550. Creppy EE, Baudrimont I, Betbeder A-M. 1995. Prevention of nephrotoxicity of ochratoxin A, a food contaminant. Toxicol Lett 82–83: 869–877.

551. Baudrimont I, Murn M, Betbeder AM, Guilcher J, Creppy EE. 1995. Effect of piroxicam on the nephrotoxicity induced by ochratoxin A in rats. Toxicology 95: 147–154.

552. Koh AS, Simmons-Willis TA, Pritchard JB, Grassl SM, Ballatori N. 2002. Identification of a mechanism by which the methylmercury antidotes *N*-acetylcysteine and dimercaptopropanesulfonate enhance urinary metal excretion: transport by the renal organic anion transporter-1. Mol Pharmacol 62: 921–926.
553. Mori S, Ohtsuki S, Takanaga H, Kikkawa T, Kang YS, Terasaki T. 2004. Organic anion transporter 3 is involved in the brain-to-blood efflux transport of thiopurine nucleobase analogs. J Neurochem 90: 931–941.
554. Basin KS, Escalante A, Beardmore TD. 1991. Severe pancytopenia in a patient taking low dose methotrexate and probenecid. J Rheumatol 18: 609–610.
555. Frenia ML, Long KS. 1992. Methotrexate and nonsteroidal antiinflammatory drug interactions. Ann Pharmacother 26: 234–237.
556. Thyss A, Milano G, Kubar J, Namer M, Achneider M. 1986. Clinical and pharmacokinetic evidence of a life-threatening interaction between methotrexate and ketoprofen. Lancet 1: 256–258.
557. Maiche AG. 1986. Acute renal failure due to concomitant action of methotrexate and indomethacin. Lancet 1: 1390.
558. Tracy TS, Krohn K, Jones DR, Bradley JD, Hall SD, Brater DC. 1992. The effects of salicylate, ibuprofen and naproxen on the disposition of methotrexate in patients with reumathoid arthritis. Eur J Clin Pharmacol 42: 121–125.
559. Furst DE. 1995. Practical clinical pharmacology and drug interactions of low-dose methotrexate therapy in rheumatoid arthritis. Br J Rheumatol 34(Suppl 2): 20–25.
560. Sathi N, Ackah J, Dawson J. 2006. Methotrexate induced neutropenia associated with coprescription of penicillins: serious and under-reported? Rheumatology (Oxford) 45: 361–362.
561. Nozaki Y, Kusuhara H, Endou H, Sugiyama Y. 2004. Quantitative evaluation of the drug–drug interactions between methotrexate and nonsteroidal anti-inflammatory drugs in the renal uptake process based on the contribution of organic anion transporters and reduced folate carrier. J Pharmacol Exp Ther 309: 226–234.
562. Uwai Y, Taniguchi R, Motohashi H, Saito H, Okuda M, Inui KI. 2004. Methotrexate-loxoprofen interaction: involvement of human organic anion transporters hOAT1 and hOAT3. Drug Metab Pharmacokinet 19: 369–374.
563. Ng HW, MacFarlane AW, Graham RM, Verbov JL. 1987. Near fatal drug interactions with methotrexate given for psoriasis. Br Med J 295: 752–753.
564. Brown GR. 1993. Cephalosporin–probenecid drug interactions. Clin Pharmacokinet 24: 289–300.
565. Marino EL, Dominguez-Gil A. 1981. The pharmacokinetics of cefadroxil associated with probenecid. Int J Clin Pharmacol Ther Toxicol 19: 506–508.
566. Griffiths RS, Black HR, Brier GL, Wolny JD. 1977. Effect of probenecid on the blood levels and urinary excretion of cefamandole. Antimicrob Agents Chemother 11: 809–812.
567. Stoeckel K, Trueb V, Dubach UC, McNamara PJ. 1988. Effect of probenecid on the elimination and protein binding of ceftriaxone. Eur J Clin Pharmacol 34: 151–156.
568. Lacy SA, Hitchcock MJM, Lee WA, Tellier P, Cundy KC. 1998. Effect of oral probenecid coadministration on the chronic toxicity and pharmacokinetics of intravenous cidofovir in cynomolgus monkeys. Toxicol Sci 4: 97–106.
569. Goa KL, Noble S. 2003. Panipenem/betamipron. Drugs 63: 913–925.

570. Chennavasin P, Seiwell R, Brater DC, Liang WM. 1979. Pharmacodynamic analysis of the furosemide–probenecid interaction in man. Kidney Int 16: 187–195.

571. Urakami Y, Okuda M, Masuda S, Saito H, Inui KI. 1998. Functional characteristics and membrane localization of rat multispecific organic cation transporters, OCT1 and OCT2, mediating tubular secretion of cationic drugs. J Pharmacol Exp Ther 287: 800–805.

572. Chatton J, Odone M, Besseghir K, Roch-Ramel F. 1990. Renal secretion of 3′-azido-3′-deoxythymidine by the rat. J Pharmacol Exp Ther 255: 140–145.

573. Laskin OL, de Miranda P, King DH, Page DA, Longstreth JA, Rocco L, Lietman PS. 1982. Effects of probenecid on the pharmacokinetics and elimination of acyclovir in humans. Antimicrob Agents Chemother 21: 804–807.

574. Inotsume N, Nishimura M, Nakano M, Fujiyama S, Sato T. 1990. The inhibitory effect of probenecid on renal excretion of famotidine in young, healthy volunteers. J Clin Pharmacol 30(1): 50–56.

575. Breithaupt B, Tittel M. 1982. Kinetics of allopurinol after single intravenous and oral doses: noninteraction with benzbromarone and hydrochlorothiazide. Eur J Clin Pharmacol 22: 77–84.

576. Colin JN, Farinotti R, Fredj G, Tod M, Clavel JP, Vignon E, Dietlin F. 1986. Kinetics of allopurinol and oxypurinol after chronic oral administration. Interaction with benzbromarone. Eur J Clin Pharmacol 30: 75–80.

577. Muller FO, Schall R, Groenewoud G, Hundt HKL, Merwe JC, Dyk M. 1993. The effect of benzbromarone on allopurinol/oxypurinol kinetics in patients with gout. Eur J Clin Pharmacol 44: 69–72.

578. Iwanaga T, Kobayashi D, Hirayama M, Maeda T, Tamai I. 2005. Involvement of uric acid transporter in increased renal clearance of the xanthine oxidase inhibitor oxypurinol induced by a uricosuric agent, benzbromarone. Drug Metab Dispos 33: 1791–1795.

579. Yamamoto T, Moriwaki Y, Takahashi Y, Suda M, Higashino K. 1991. Effects of pyrazinamide, probenecid and benzbromarone on renal excretion of oxypurinol. Ann Rheum Dis 50: 631–633.

580. König J, Seithel A, Gradhand U, Fromm MF. 2006. Pharmacogenomics of human OATP transporters. Naunyn Schmiedebergs Arch Pharmacol 372: 432–443.

581. Niemi M, Schäffeler E, Lang T, Fromm MF, Neuvonen M, Kyrklund C, Backman JT, Kerb R, Schwab M, Neuvonen PJ, Eichelbaum M, Kivisto KT. 2004. High plasma pravastatin concentrations are associated with single-nucleotide polymorphisms and haplotypes of organic anion transporting polypeptide-C (OATP-C, SLCO1B1). Pharmacogenetics 14: 429–440.

582. Tirona RG, Leake BF, Merino G, Kim RB. 2001. Polymorphisms in OATP-C: identification of multiple allelic variants associated with altered transport activity among European- and African-Americans. J Biol Chem 276: 35669–35675.

583. Nozawa T, Nakajima M, Tamai I, Noda K, Nezu J, Sai Y, Tsuji A, Yokoi T. 2002. Genetic polymorphisms of human organic anion transporters OATP-C (SLC21A6) and OATP-B (SLC21A9): allele frequencies in the Japanese population and functional analysis. J Pharmacol Exp Ther 302: 804–813.

584. Michalski C, Cui Y, Nies AT, Nuessler AK, Neuhaus P, Zanger UM, Klein K, Eichelbaum M, Keppler D, König J. 2002. A naturally occurring mutation in the SLC21A6 gene causing impaired membrane localization of the hepatocyte uptake trasnporter. J Biol Chem 277: 43058–43063.

585. Mwinyi J, Johne A, Bauer S, Roots I, Gerloff T. 2004. Evidence for inverse effects of OATP-C (SLC21A6) 5 and 1b haplotypes on pravastatin kinetics. Clin Pharmacol Ther 75: 415–421.

586. Maeda K, Ieiri I, Yasuda K, Fujino A, Fujiwara H, Otsubo K, Hirano M, Watanabe T, Kitamura Y, Kusuhara H, Sugiyama Y. 2006. Effects of organic anion transporting polypeptide 1B1 haplotype on pharmacokinetics of pravastatin, valsartan, and temocapril. Clin Pharmacol Ther 79: 427–439.

587. Niemi M, Backman JT, Kajosaari LI, Leathart JB, Neuvonen M, Daly AK, Eichelbaum M, Kivisto KT, Neuvonen PJ. 2005. Polymorphic organic anion transporting polypeptide 1B1 is a major determinant of repaglinide pharmacokinetics. Clin Pharmacol Ther 77: 468–478.

588. Letschert K, Keppler D, Konig J. 2004. Mutations in the SLCO1B3 gene affecting the substrate specificity of the hepatocellular uptake transporter OATP1B3 (OATP8). Pharmacogenetics 14: 441–452.

589. Lee W, Glaeser H, Smith LH, Roberts RL, Moeckel GW, Gervasini G, Leake BF, Kim RB. 2005. Polymorphisms in human organic anion transporting polypeptide 1A2 (OATP1A2): implications for altered drug disposition and central nervous system drug entry. J Biol Chem 280: 9610–9617.

590. Hagenbuch B, Meier PJ. 2003. The superfamily of organic anion transporting polypeptides. Biochim Biophys Acta 1609: 1–18.

591. Hagenbuch B, Meier PJ. 2004. Organic anion–transporting polypeptides of the OATP/ SLC21 family: phylogenetic classification as OATP/ SLCO superfamily, new nomenclature and molecular/functional properties. Pflugers Arch 447: 653–665.

592. Kullak-Ublick GA, Hagenbuch B, Stieger B, Schteingart CD, Hofmann AF, Wolkoff AW, Meier PJ. 1995. Molecular and functional characterization of an organic anion transporting polypeptide cloned from human liver. Gastroenterology 109: 1274–1282.

593. Kullak-Ublick GA, Fisch T, Oswald M, Hagenbuch B, Meier PJ, Beuers U, Paumgartner G. 1998. Dehydroepiandrosterone sulfate (DHEAS): identification of a carrier protein in human liver and brain. FEBS Lett 424: 173–176.

594. Gao B, Hagenbuch B, Kullak-Ublick GA, Benke D, Aguzzi A, Meier PJ. 2000. Organic anion-transportingpolypeptides mediate transport of opioid peptides across blood–brain barrier. J Pharmacol Exp Ther 294: 73–79.

595. Hsiang B, Zhu Y, Wang Z, Wu Y, Sasseville V, Yang W-P, Kirchgessner TG. 1999. A novel human hepatic organic anion transporting polypeptide (OATP2). J Biol Chem 274: 37161–37168.

596. Nakai D, Nakagomi R, Furuta Y, Tokui T, Abe T, Ikeda T, Nishimura K. 2001. Human liver–specific organic anion transporter, LST-1, mediates uptake of pravastatin by human hepatocytes. J Pharmacol Exp Ther 297: 861–867.

597. Abe T, Kakyo M, Tokui T, Nakagomi R, Nishio T, Nakai D, Nomura H, Unno M, Suzuki M, Naitoh T, Matsuno S, Yawo H. 1999. Identification of a novel gene family encoding human liver–specific organic anion transporter LST-1. J Biol Chem 274: 17159–17163.

598. König J, Cui Y, Nies AT, Keppler D. 2000. A novel human organic anion transporting polypeptide localized to the basolateral hepatocyte membrane. Am J Physiol Gastrointest Liver Physiol 278: G156–G164.

599. Tamai I, Nezu JI, Uchino H, Sai Y, Oku A, Shimane M, Tsjuji A. 2000. Molecular identification and characterization of novel members of the human organic anion transporter (OATP) family. Biochem Biophys Res Commun 273: 251–260.

600. König J, Cui Y, Nies AT, Keppler D. 2000. Localization and genomic organization of a new hepatocellular organic anion transporting polypeptide. J Biol Chem 275: 23161–23168.

601. Cui Y, König J, Leier I, Buchholz U, Keppler D. 2001. Hepatic uptake of bilirubin and its conjugates by the human organic anion transporter SLC21A6. J Biol Chem 276: 9626–9630.

602. Kullak-Ublick GA, Ismair MG, Stieger B, Landmann L, Huber R, Pizzagalli F, Fattinger K, Meier PJ, Hagenbuch B. 2001. Organic anion-transporting polypeptide B (OATP-B) and its functional comparison with three other OATPs of human liver. Gastroenterology 120: 525–533.

603. Ismair MG, Stieger B, Cattori V, Hagenbuch B, Fried M, Meier PJ, Kullak-Ublick GA. 2001. Hepatic uptake of cholecystokinin octapeptide by organic anion-transporting polypeptides OATP4 and OATP8 of rat and human liver. Gastroenterology 121: 1185–1190.

604. Fujiwara K, Adachi H, Nishio T, Unno M, Tokui T, Okabe M, Onogawa T, Suzuki T, Asano N, Tanemoto M, et al. 2001. Identification of thyroid hormones transporters in humans: different molecules are involved in a tissue-specific manner. Endocrinology 142: 2005–2012.

605. Mikkaichi T, Suzuki T, Onogawa T, Tanemoto M, Mizutamari H, Okada M, Chaki T, Masuda S, Tokui T, Eto N, et al. 2004. Isolation and characterization of a digoxin transporter and its rat homologue expressed in the kidney. Proc Natl Acad Sci U S A 101: 3569–3574.

606. Kajosaari LI, Niemi M, Neuvonen M, Laitila J, Neuvonen PJ, Backman JT. 2005. Cyclosporin markedly raises the plasma concentrations of repaglinide. Clin Pharmacol Ther 78: 388–399.

607. Vavricka SR, van Montfoort J, Ha HR, Meier PJ, Fattinger K. 2002. Interactions of rifamycin SV and rifampicin with organic anion uptake systems of human liver. Hepatology 36: 164–172.

608. Fattinger K, Cattori V, Hagenbuch B, Meier PJ, Stieger B. 2000. Rifamycin SV and rifampicin exhibit differential inhibition of the hepatic rat organic anion transporting polypeptides, Oatp1 and Oatp2. Hepatology 32: 82–86.

609. Acocella G, Nicolis FB, Tenconi LT. 1965. The effect of an intravenous infusion of rifamycin SV on the excretion of bilirubin, bromsulphophthalein, and indocyanine green in man. Gastroenterology 49: 521–525.

610. Laudano OM. 1972. Effects of rifampicin on the blood clearance and biliary excretion of sulfobromophthalein in man. Farmaco 27: 622–627. Abstract only.

611. Dresser GK, Schwartz U, Leake B, Schwarz UI, Dawson PA, Freeman DJ, Kim RB. 2000. Grapefruit juice selectively inhibits OATP not P-glycoprotein. Drug Metab Rev 32(Suppl 2): 193.

612. Dresser GK, Kim RB, Bailey DG. 2005. Effect of grapefruit juice volume on the reduction of fexofenadine bioavailability: possible role of organic anion transporting polypeptides. Clin Pharmacol Ther 77: 170–177.

613. Nozawa T, Sugiura S, Nakajima M, Goto A, Yokoi T, Nezu J, Tsjuji A, Tamai I. 2004. Involvement of organic anion transporting polypeptides in the transport of troglitazone

sulfate: implications for understanding troglitazone hepatotoxicity. Drug Metab Dispos 32: 291–294.

614. Guo GL, Choudhuri S, Klaassen CD. 2002. Induction profile of rat organic anion transporting polypeptide 2 (oatp2) by prototypical drug-metabolizing enzyme inducers that activate gene expression through ligand-activated transcription factor pathways. J Pharmacol Exp Ther 300: 206–212.
615. Blaschke TF, Berk PD, Rodkey FL, Sharschmidt BF, Collison HA, Waggoner JG. 1974. Drugs and the liver. I. Effects of glutethimide and phenobarbital on hepatic bilirubin clearance, plasma bilirubin turn-over and carbon monoxide production in man. Biochem Pharmacol 23: 2795–2806.
616. Whelan G, Combes B. 1975. Phenobarbital-enhanced biliary excretion of administered unconjugated and conjugated sulfobromophthalein (BSP) in the rat. Biochem Pharmacol 24: 1283–1286.
617. Potter BJ, Ni JZ, Wolfe K, Stump D, Berk PD. 1994. Induction of a dose-related increase in sulfobromophthalein uptake velocity in freshly isolated rat hepatocytes by phenobarbital. Hepatology 20: 1078–1085.
618. Noè B, Hagenbuch B, Stieger B, Meier PJ. 1997. Isolation of a multispecific organic anion and cardiac glycoside transporter from rat brain. Proc Natl Acad Sci U S A 94: 10346–10350.
619. Klassen CD. 1974. Effects of microsomal enzyme inducers on the biliary excretion of cardiac glycosides. J Pharmacol Exp Ther 191: 201–211.
620. Klassen CD. 1974. Stimulation of the development of the hepatic excretory mechanism for ouabain in newborns rats with microsomial enzyme inducers. J Pharmacol Exp Ther 191: 212–218.
621. Eaton DL, Klassen CD. 1979. Effects of microsomal enzyme inducers on carrier-mediated transport systems in isolated rat hepatocytes. J Pharmacol Exp Ther 208: 381–385.
622. Cheng X, Maher J, Dieter MZ, Klaassen CD. 2005. Regulation of mouse organic anion-transporting polypeptides (Oatps) in liver by prototypical microsomal enzyme inducers that activate distinct transcription factor pathways. Drug Metab Dispos 33: 1276–1282.
623. Jigorel E, Le Vee M, Boursier-Neyret C, Parmentier Y, Fardel O. 2006. Differential regulation of sinusoidal and canalicular hepatic drug transporter expression by xenobiotics activating drug-sensing receptors in primary human hepatocytes. Drug Metab Dispos 34: 1756–1763.
624. Greenberger NJ, Thomas FB. 1973. Biliary excretion of 3H-digitoxin: modification by bile salts and phenobarbital. J Lab Clin Med 81: 241–251.
625. McClain RM, Levin AA, Posch R, Downing JC. 1989. The effect of phenobarbital on the metabolism and excretion of thyroxine in rats. Toxicol Appl Pharmacol 99: 216–228.
626. Nevasaari K, Alakare B, Karki NT. 1976. Different effects of microsomal enzyme inducers on the biliary excretion of digoxin. Acta Pharmacol Toxicol 39: 442–448.
627. Jung D, Hagenbuch B, Gresh L, Pontoglio M, Meier PJ, Kullak-Ublick GA. 2001. Characterization of the human OATP-C (SLC21A6) gene promoter and regulation of liver-specific OATP genes by hepatocyte nuclear factor 1 alpha. J Biol Chem 276: 37206–37214.
628. Jung D, Podvinec M, Meyer UA, Mangelsdorf DJ, Fried M, Meier PJ, Kullak-Ublick GA. 2002. Human organic anion–transporting polypeptide OATP8 (SLC21A8) promoter is transactivated by the farnesoid X receptor/bile acid receptor. Gastroenterology 122: 1954–1966.

629. Lu R, Kanai N, Bao Y, Wolkoff AW, Schuster VL. 1996. Regulation of renal oatp mRNA expression by testosterone. Am J Physiol 270: F332–F337.
630. Simonson SG, Raza A, Martin PD, Mitchell PD, Jarcho JA, Brown CD, Windass AS, Schneck DW. 2004. Rosuvastatin pharmacokinetics in heart transplant recipients administered administered an antirejection regimen including cyclosporine. Clin Pharmacol Ther 76: 167–177.
631. Mück W, Mai I, Fritsche L, Ochmann K, Rohde G, Unger S, Johne A, Bauer S, Budde K, Roots I, Neumayer HH. 1999. Increase in cerivastatin systemic exposure after single and multiple dosing in cyclosporine-treated kidney transplant recipients. Clin Pharmacol Ther 65: 251–261.
632. Shitara Y, Itoh T, Sato H, Li AP, Sugiyama Y. 2003. Inhibition of transporter-mediated hepatic uptake as a mechanism for drug-drug interaction between cerivastatin and cyclosporin A. J Pharmacol Exp Ther 304: 610–616.
633. Regazzi MB, Campana IC, Raddato V, Lesi C, Perani G, Gavazzi A, Vigano M. 1993. Altered disposition of pravastatin following concomitant drug therapy with cyclosporin A in transplant recipients. Transplant Proc 25: 2732–2734.
634. Hirano M, Maeda K, Shitara Y, Sugiyama Y. 2006. drug–drug interaction between pitavastatin and various drugs via OATP1B1. Drug Metab Dispos 34: 1229–1236.
635. Asberg A, Hartmann A, Fjeldsa E, Bergan S, Holdaas H. 2001. Bilateral pharmacokinetic interaction between cyclosporin A and atorvastatin in renal transplant recipients. Am J Transplant 1: 382–386.
636. Backman JT, Kyrklund C, Kivisto KT, Wang JS, Neuvonen PJ. 2000. Plasma concentrations of active simvastatin acid are increased by gemfibrozil. Clin Pharmacol Ther 68: 122–129.
637. Backman JT, Kyrklund C, Neuvonen M, Neuvonen PJ. 2002. Gemfibrozil greatly increases plasma concentrations of cerivastatin. Clin Pharmacol Ther 72: 685–691.
638. Mikkaichi T, Suzuki T, Tanemoto M, Ito S, Abe T. 2004. The organic anion transporter (OATP) family. Drug Metab Pharmacokinet 19: 171–179.
639. Schneck DW, Birmingham BK, Zalikowski JA, Mitchell PD, Wang Y, Martin PD, Lasseter KC, Brown CD, Windass AS, Raza A. 2004. The effect of gemfibrozil on the pharmacokinetics of rosuvastatin. Clin Pharmacol Ther 75: 455–463.
640. Kyrklund C, Backman JT, Kivisto KT, Neuvonen M, Laitila J, Neuvonen PJ. 2001. Plasma concentrations of active lovastatin acid are markedly increased by gemfibrozil but not by bezafibrate. Clin Pharmacol Ther 69: 340–345.
641. Kyrklund C, Backman JT, Neuvonen M, Neuvonen PJ. 2003. Gemfibrozil increases plasma pravastatin concentrations and reduces pravastatin renal clearance. Clin Pharmacol Ther 73: 538–544.
642. Endres CJ, Hsiao P, Chung FS, Unadkat JD. 2006. The role of transporters in drug interactions. Eur J Pharm Sci 27: 501–517.
643. Bruce-Joyce J, Dugas JM, MacCausland OE. 2001. Cerivastatin and gemfibrozil-associated rhabdomyolysis. Ann Pharmacother 35: 1016–1019.
644. Shitara Y, Hirano M, Sato H, Sugiyama Y. 2004. Gemfibrozil and its glucuronide inhibit the OATP2(OATP1B1:SLC21A6)-mediated hepatic uptake and CYP2C8-mediated metabolism of cerivastatin-analysis of the mechanism of the clinically relevant drug–drug interaction between cerivastatin and gemfibrozil. J Pharmacol Exp Ther 311: 228–236.

645. Shitara Y, Sato H, Sugiyama Y. 2005. Evaluation of drug–drug interaction in the hepatobiliary and renal transport of drugs. Annu Rev Pharmacol Toxicol 45: 689–723.

646. Wang JS, Neuvonen M, Wen X, Backman JT, Neuvonen PJ. 2002. Gemfibrozil inhibits CYP2C8-mediated cerivastatin metabolism in human liver microsomes. Drug Metab Dispos 30: 1352–1356.

647. Moysey JO, Jaggarao NSV, Grundy EN, Chamberlain DA. 1981. Amiodarone increases plasma digoxin concentration. Br Med J 282: 272.

648. Nademanee K, Kannan R, Hendrickson J, Ookhtens M, Kay I, Singh BN. 1984. Amiodarone–digoxin interaction: clinical significance, time course of development, potential pharmacokinetic mechanisms and therapeutic implications. J Am Coll Cardiol 4: 111–116.

649. Oetgen WJ, Sobol SM, Tri TB, Heydorn WH, Rakita L. 1984. Amiodarone–digoxin interaction. Clinical and experimental observations. Chest 86: 75–79.

650. Robinson K, Johnston A, Walker S, Mulrow JP, McKenna WJ, Holt DW. 1989. The digoxin–amiodarone interaction. Cardiovasc Drugs Ther 3: 25–28. Abstract only.

651. Braunschweig J, Stäubli M, Studer H. 1987. Interactions of amiodarone with digoxin in rats. Br J Pharmacol 92: 553–559.

652. Abe T, Unno M, Onogawa T, Tokui T, Kondo TN, Nagakomi R, Adachi H, Fujiwara K, Okabe M, Suzuki T, et al. 2001. LST-2, a human liver–specific organic anion transporter, determines methotrexate sensitivity in gastrointestinal cancers. Gastroenterology 120: 1689–1699.

653. Lambert C, Lamontage D, Hottlet H, Souich PD. 1989. Amiodarone–digoxin interaction in rats. A reduction in hepatic uptake. Drug Metab Dispos 17: 704–708.

654. Banfield C, Gupta S, Marino M, Lim J, Affrime M. 2002. Grapefruit juice reduces the oral bioavailability of fexofenadine but not desloratadine. Clin Pharmacokinet 41: 311–318.

655. Thompson TN, Klaassen CD. 1995. The effects of hepatic microsomal enzyme inducers on the pharmacokinetics of ouabain after portal and systemic administration to rats. J Pharm Pharmacol 47: 1041–1047.

656. Pedersoli WM, Ganjam VK, Nachreiner RF. 1980. Serum digoxin concentrations in dogs before, during, and after concomitant treatment with phenobarbital. Am J Vet Res 41: 1639–1642.

657. Eckhardt U, Horz JA, Petzinger E, Stuber W, Reers M, Dickneite G, Daniel H, Wagener M, Hagenbuch B, Stieger B, Meier PJ. 1996. The peptide-based thrombin inhibitor CRC 220 is a new substrate of the basolateral rat liver organic anion–transporting polypeptide. Hepatology 24: 380–384.

658. Fischer WJ, Altheimer S, Cattori V, Meier PJ, Dietrich DR, Hagenbuch B. 2005. Organic anion–transporting polypeptides expressed in liver and brain mediate uptake of microcystin. Toxicol Appl Pharmacol 203: 257–263.

659. Shu Y, Leabman MK, Feng B, Mangravite LM, Huang CC, Stryke D, Kawamoto M, Johns SJ, DeYoung J, Carlson E, et al. 2003. Evolutionary conservation predicts function of variants of the human organic cation transporter, OCT1. Proc Natl Acad Sci U S A 100: 5902–5907.

660. Saito S, Lida A, Sekine A, Ogawa C, Kawauchi S, Higuchi S, Nakamura Y. 2002. Catalog of 238 variations among six human genes encoding solute carriers (hSLCs) in the Japanese population. J Hum Genet 47: 576–584.

661. Kerb R, Brinkmann U, Chatskaia N, Gorbunov D, Gorboulev V, Mornhinweg E, Keil A, Eichelbaum M, Koepsell H. 2002. Identification of genetic variations of the human organic cation transporter hOCT1 and their functional consequences. Pharmacogenetics 12: 591–595.

662. Leabman MK, Huang CC, Kawamoto M, Johns SJ, Stryke D, Ferrin TE, DeYoung J, Taylor T, Clark AG, Herskowitz I, Giacomini KM. 2002. Polymorphisms in a human kidney xenobiotic transporter, OCT2, exhibit altered function. Pharmacogenetics 12: 395–405.

663. Leabman MK, Huang CC, DeYoung J, Carlson EJ, Taylor TR, de la Cruz M, Johns SJ, Stryke D, Kawamoto M, Urban TJ, et al. 2003. Natural variation in human membrane transporter genes reveals evolutionary and functional constraints. Proc Natl Acad Sci U S A 100: 5896–58901.

664. Wang DS, Jonker JW, Kato Y, Kusuhara H, Schinkel AH, Sugiyama Y. 2002. Involvement of organic cation transporter 1 in hepatic and intestinal distribution of metformin. J Pharmacol Exp Ther 302: 510–515.

665. Wang DS, Jonker JW, Kato Y, Kusuhara H, Schinkel AH, Sugiyama Y. 2002. Involvement of organic cation transporter 1 in hepatic and intestinal distribution of metformin. J Pharmacol Exp Ther 302: 510–515.

666. Busch AE, Karbach U, Miska D, Gorboulev V, Akhoundova A, Volk C, Arndt P, Ulzheimer JC, Sonders MS, Baumann C, et al. 1998. Human neurons express the polyspecific cation transporter hOCT2, which translocates monoamine neurotransmitters, amantadine, and memantine. Mol Pharmacol 54: 342–352.

667. Gründemann D, Liebich G, Kiefer N, Köster S, Schömig E. 1999. Selective substrates for non-neuronal monoamine transporters. Mol Pharmacol 56: 1–10.

668. Zhang L, Brett CM, Giacomini KM. 1998. Role of organic cation transporters in drug absorption and elimination. Annu Rev Pharmacol Toxicol 38: 431–460.

669. Zhang L, Gorset W, Dresser MJ, Giacomini KM. 1999. The interaction of *n*-tetraalkylammonium compounds with a human organic cation transporter, hOCT1. J Pharmacol Exp Ther 288: 1192–1198.

670. Gründemann D, Köster S, Kiefer N, Breidert T, Engelhardt M, Spitzenberger F, Obermuller N, Schömig E. 1998. Transport of monoamine transmitters by the organic cation transporter type 2, OCT2. J Biol Chem 273: 30915–30920.

671. Gorboulev V, Ulzheimer JC, Akhoundova A, Ulzheimer-Teuber I, Karbach U, Quester S, Baumann C, Lang F, Busch AE, Koepsell H. 1997. Cloning and characterization of two human polyspecific organic cation transporters. DNA Cell Biol 16: 871–881.

672. Crossman LC, Druker BJ, Deininger MW, Pirmohamed M, Wang L, Clark RE. 2005. hOCT 1 and resistance to imatinib. Blood 106: 1133–1134.

673. Thomas J, Wang L, Clark RE, Pirmohamed M. 2004. Active transport of imatinib into and out of cells: implications for drug resistance. Blood 104: 3739–3745.

674. White DL, Saunders VA, Dang P, Engler J, Zannettino AC, Cambareri AC, Quinn SR, Manley PW, Hughes TP. 2006. OCT-1-mediated influx is a key determinant of the intracellular uptake of imatinib but not nilotinib (AMN107): reduced OCT-1 activity is the cause of low in vitro sensitivity to imatinib. Blood 108: 697–704.

675. Zhang S, Lovejoy KS, Shima JE, Lagpacan LL, Shu Y, Lapuk A, Chen Y, Komori T, Gray JW, Chen X, Lippard SJ, Giacomini KM. 2006. Organic cation transporters are determinants of oxaliplatin cytotoxicity. Cancer Res 66: 8847–8857.

676. Pan BF, Sweet DH, Pritchard JB, Chen R, Nelson JA. 1999. A transfected cell model for the renal toxin transporter, rOCT2. Toxicol Sci 47: 181–186.

677. Wu X, Huang W, Ganapathy M, Wang H, Kekuda R, Conway SJ, Leibach FH, Ganapathy V. 2000. Structure, function, and regional distribution of the organic cation transporter OCT3 in the kidney. Am J Physiol Renal Physiol 279: 449–458.

678. Wu X, Kekuda R, Huang W, Fei YJ, Leibach FH, Chen J, Conway SJ, Ganapathy V. 1998. Identity of the organic cation transporter OCT3 as the extraneuronal monoamine transporter (uptake2) and evidence for the expression of the transporter in the brain. J Biol Chem 273: 32776–32786.

679. Zhang L, Gorset W, Washington CB, Blaschke TF, Kroetz DL, Giacomini KM. 2000. Interactions of HIV protease inhibitors with a human organic cation transporter in a mammalian expression system. Drug Metab Dispos 28: 329–334.

680. Ofer M, Wolffram S, Koggel A, Spahn-Langguth H, Langguth P. 2005. Modulation of drug transport by selected flavonoids: Involvement of Pgp and OCT? Eur J Pharm Sci 25: 263–271.

681. Jonker JW, Wagenaar E, Mol CA, Buitelaar M, Koepsell H, Smit JW, Schinkel AH. 2001. Reduced hepatic uptake and intestinal excretion of organic cations in mice with a targeted disruption of the organic cation transporter 1 (Oct1 [Slc22a1]) gene. Mol Cell Biol 21: 5471–5477.

682. Jonker JW, Schinkel AH. 2004. Pharmacological and physiological functions of the polyspecific organic cation transporters: OCT1, 2, and 3 (SLC22A1-3). J Pharmacol Exp Ther 308: 2–9.

683. Zwart R, Verhaagh S, Buitelaar M, Popp-Snijders C, Barlow DP. 2001. Impaired activity of the extraneuronal monoamine transporter system known as uptake-2 in Orct3/Slc22a3-deficient mice. Mol Cell Biol 21: 4188–4189.

684. Troncone L, Rufini V. 1997. 131I-MIBG therapy of neural crest tumours (review). Anticancer Res 17: 1823–1831.

685. Wang DS, Kusuhara H, Kato Y, Jonker JW, Schinkel AH, Sugiyama Y. 2003. Involvement of organic cation transporter 1 in the lactic acidosis caused by metformin. Mol Pharmacol 63: 844–848.

686. Jonker JW, Wagenaar E, Van Eijl S, Schinkel AH. 2003. Deficiency in the organic cation transporters 1 and 2 (Oct1/Oct2 [Slc22a1/Slc22a2]) in mice abolishes renal secretion of organic cations. Mol Cell Biol 23: 7902–7908.

687. Grundemann D, Schechinger B, Rappold GA, Schömig E. 1998. Molecular identification of the corticosterone-sensitive extraneuronal catecholamine transporter. Nat Neurosci 1: 349–351.

688. Somogyi A, McLean A, Heinzow B. 1983. Cimetidine–procainamide pharmacokinetic interaction in man: evidence of competition for tubular secretion of basic drugs. Eur J Clin Pharmacol 25: 339–345.

689. Somogyi A, Stockley C, Keal J, Rolan P, Bochner F. 1987. Reduction of metformin renal tubular secretion by cimetidine in man. Br J Clin Pharmacol 23: 545–551.

690. Tsuruoka S, Ioka T, Wakaumi M, Sakamoto K, Ookami H, Fujimura A. 2006. Severe arrhythmia as a result of the interaction of cetirizine and pilsicainide in a patient with renal insufficiency: first case presentation showing competition for excretion via renal multidrug resistance protein 1 and organic cation transporter 2. Clin Pharmacol Ther 79: 389–396.

691. Shiga T, Hashiguchi M, Urae A, Kasanuki H, Rikihisa T. 2000. Effect of cimetidine and probenecid on pilsicainide renal clearance in humans. Clin Pharmacol Ther 67: 222–228.

692. Kawasaki Y, Kato Y, Sai Y, Tsuji A. 2004. Functional characterization of human organic cation transporter OCTN1 single-nucleotide polymorphisms in the Japanese population. J Pharm Sci 93: 2920–2926.

693. Urban TJ, Gallagher RC, Brown C, Castro RA, Lagpacan LL, Brett CM, Taylor TR, Carlson EJ, Ferrin TE, Burchard EG, Packman S, Giacomini KM. 2006. Functional genetic diversity in the high-affinity carnitine transporter OCTN2 (SLC22A5). Mol Pharmacol, Aug 24 [Epub ahead of print].

694. Tang NLS, Ganapathy V, Wu X, Hui J, Seth Pm Yuen PMP, Fok TF, Hjelm NM. 1999. Mutations of OCTN2, an organic cation/carnitine transporter, lead to deficient cellular carnitine uptake in primary carnitine deficiency. Hum Mol Genet 8: 655–660.

695. Lahjouji K, Mitchell GA, Qureshi IA. 2001. Carnitine transport by organic cation transporters and systemic carnitine deficiency. Mol Genet Metab 73: 287–297.

696. Wang Y, Taroni F, Garavaglia B, Longo N. 2000. Functional analysis of mutations in the OCTN2 transporter causing primary carnitine deficiency: lack of genotype-phenotype correlation. Hum Mutat 16: 401–407.

697. Wang Y, Korman SH, Ye J, Gargus JJ, Gutman A, Taroni F, Garavaglia B, Longo N. 2001. Phenotype and genotype variation in primary carnitine deficiency. Genet Med 3: 387–392.

698. Wu X, Huang W, Prasad PD, Seth P, Rajan DP, Leibach FH, Chen J, Conway SJ, Ganapathy V. 1999. Functional characteristics and tissue distribution pattern of organic cation transporter 2 (OCTN2), an organic cation/carnitine transporter. J Pharmacol Exp Ther 290: 1482–1492.

699. Koizumi A, Nozaki J, Ohura T, Kayo T, Wada Y, Nezu J, Ohashi R, Tamai I, Shoji Y, Takada G, et al. 1999. Genetic epidemiology of the carnitine transporter OCTN2 gene in a Japanese population and phenotypic characterization in Japanese pedigrees with primary systemic carnitine deficiency. Hum Mol Genet 8: 2247–2254.

700. Ganapathy ME, Huang W, Rajan DP, Carter AL, Sugawara M, Iseki K, Leibach FH, Ganapathy V. 2000. Beta-lactam antibiotics as substrates for OCTN2, an organic cation/carnitine transporter. J Biol Chem 275: 1699–1707.

701. Wagner CA, Lükewille U, Kaltenbach S, Moschen I, Bröer A, Risler T, Bröer S, Lang F. 2000. Functional and pharmacological characterization of human Na(+)-carnitine cotransporter hOCTN2. Am J Physiol Renal Physiol 279: F584–F591.

702. Ohashi R, Tamai I, Yabuuchi H, Nezu JI, Oku A, Yoshimichi S, Shimane M, Tsuji A. 1999. Na(+)-dependent carnitine transport by organic cation transporter (OCTN2): its pharmacological and toxicological relevance. J Pharmacol Exp Ther 91: 778–784.

703. Yabuuchi H, Tamai I, Nezu J, Sakamoto K, Oku A, Shimane M, Sai Y, Tsuji A. 1999. Novel membrane transporter OCTN1 mediates multispecific, bidirectional and pH-dependent transport of organic cations. J Pharmacol Exp Ther 289: 768–773.

704. Yabuuchi H, Tamai I, Nezu J, Sakamoto K, Oku A, Shimane M, Sai Y, Tsuji A. 1999. Novel membrane transporter OCTN1 mediates multispecific, bidirectional and pH-dependent transport of organic cations. J Pharmacol Exp Ther 289: 768–773.

705. Tein I. 2003. Carnitine transport: pathophysiology and metabolism of known molecular defects. J Inherit Metab Dis 26: 147–169.

706. Abrahamsson K, Mellander M, Eriksson BO, Holme E, Jodal U, Lindstedt S. 1997. Cardiac effects of carnitine deficiency induced by antibiotics containing pivalic acid in children. Cardiol Young 7: 178–182.

707. Holme E, Greter J, Jacobson CE, Lindstedt S, Nordin I, Kristiansson B, Jodal U. 1989. Carnitine deficiency induced by pivampicillin and pivmecillinam therapy. Lancet 2: 469–473.

708. Kuntzer T, Reichmann H, Bogousslavsky J, Regli F. 1990. Emetine-induced myopathy and carnitine deficiency. J Neurol 237: 495–496.

709. Pons R, De Vivo DC. 1995. Primary and secondary carnitine deficiency syndromes. J Child Neurol 10(Suppl 2): S8–S24.

710. Tune BM. 1994. Renal tubular transport and nephrotoxicity of beta lactam antibiotics: structure–activity relationships. Miner Electrolyte Metab 20: 221–231.

711. Tune BM, Hsu CY. 1994. Toxicity of cephaloridine to carnitine transport and fatty acid metabolism in rabbit renal cortical mitochondria: structure–activity relationships. J Pharmacol Exp Ther 270: 873–880.

712. Watanabe K, Sawano T, Endo T, Sakata M, Sato J. 2002. Studies on intestinal absorption of sulpiride (2): transepithelial transport of sulpiride across the human intestinal cell line Caco-2. Biol Pharm Bull 25: 1345–1350.

713. Watanabe K, Sawano T, Jinriki T, Sato J. 2004. Studies on intestinal absorption of sulpiride (3): intestinal absorption of sulpiride in rats. Biol Pharm Bull 27: 77–81.

714. Wu SP, Shyu MK, Liou HH, Gau CS, Lin CJ. 2004. Interaction between anticonvulsants and human placental carnitine transporter. Epilepsia 45: 204–210.

715. Verrotti A, Greco R, Morgese G, Chiarelli F. 1999. Carnitine deficiency and hyperammonemia in children receiving valproic acid with and without other anticonvulsant drugs. Int J Clin Lab Res 29: 36–40.

716. Raskind JY, El-Chaar GM. 2000. The role of carnitine supplementation during valproic acid therapy. Ann Pharmacother 34: 630–638.

717. Bohan TP, Helton E, McDonald I, Konig S, Gazitt S, Sugimoto T, Scheffner D, Cusmano L, Li S, Koch G. 2001. Effect of L-carnitine treatment for valproate-induced hepatotoxicity. Neurology 56: 1405–1409.

718. Kido Y, Tamai I, Ohnari A, Sai Y, Kagami T, Nezu J, Nikaido H, Hashimoto N, Asano M, Tsuji A. 2001. Functional relevance of carnitine transporter OCTN2 to brain distribution of L-carnitine and acetyl-L-carnitine across the blood–brain barrier. J Neurochem 79: 959–969.

719. Somogyi A. 1996. Renal transport of drugs: specificity and molecular mechanisms. Clin Exp Pharmacol Physiol 23: 986–989.

720. Ayrton A, Morgan P. 2001. Role of transport proteins in drug absorption, distribution and excretion. Xenobiotica 31: 469–497.

721. Shionoiri H. 1993. Pharmacokinetic drug interactions with ACE inhibitors. Clin Pharmacokinet 25: 20–58.

722. Lin JH, Chen IW, Ulm EH, Duggan DE. 1988. Differential renal handling of angiotensin-converting enzyme inhibitors enalaprilat and lisinopril in rats. Drug Metab Dispos 16: 392–396.

723. Noormohamed FH, McNabb WR, Lant AF. 1990. Pharmacokinetic and pharmacodynamic actions of enalapril in humans: effect of probenecid pretreatment. J Pharmacol Exp Ther 253: 362–368.

724. Drummer OH, Thompson J, Hooper R, Jarrott B. 1985. Effect of probenecid on the disposition of captopril and captopril dimer in the rat. Biochem Pharmacol 34: 3347–3351.

725. Sinhvi SM, Duchin KL, Willard DA, McKinstry DN, Migdalof BH. 1982. Renal handling of captopril: effect of probenecid. Clin Pharmacol Ther 32: 182–189.

726. Zarychanski R, Wlodarczyk K, Ariano R, Bow E. 2006. Pharmacokinetic interaction between methotrexate and piperacillin/tazobactam resulting in prolonged toxic concentrations of methotrexate. J Antimicrob Chemother 58: 228–230.

727. Titier K, Lagrange F, Pehourq F, Edno-Mcheid L, Moore N, Molimard M. 2002. Pharmacokinetic interaction between high-dose methotrexate and oxacillin. Ther Drug Monit 24: 570–572.

728. Ronchera CL, Hernandez T, Peris JE, Torres F, Granero L, Jimenez NV, Pla JM. 1993. Pharmacokinetic interaction between high-dose methotrexate and amoxycillin. Ther Drug Monit 15: 375–379.

729. Iven H, Brasch H. 1986. Influence of the antibiotics piperacillin, doxycycline, and tobramycin on the pharmacokinetics of methotrexate in rabbits. Cancer Chemother Pharmacol 17: 218–222.

730. Cundy KC. 1999. Clinical pharmacokinetics of the antiviral nucleotide analogues cidofovir and adefovir. Clin Pharmacokinet 36: 127–143.

731. Servais A, Lechat P, Zahr N, Urien S, Aymard G, Jaudon MC, Deray G, Isnard Bagnis C. 2006. Tubular transporters and clearance of adefovir. Eur J Pharmacol 540: 168–174.

732. Cihlar T, Lin DC, Pritchard JB, Fuller MD, Mendel DB, Sweet DH. 1999. The antiviral nucleotide analogs cidofovir and adefovir are novel substrates for human and rat renal organic anion transporter 1. Mol Pharmacol 56: 570–580.

733. Gimeno MJ, Martinez M, Granero L, Torres-Molina F, Peris JE. 1996. Influence of probenecid on the renal excretion mechanisms of cefadroxil. Drug Metab Dispos 24: 270–272.

734. Spina SP, Dillon EC Jr. 2003. Effect of chronic probenecid therapy on cefazolin serum concentrations. Ann Pharmacother 37: 621–624.

735. Garton AM, Rennie RP, Gilpin J, Marrelli M, Shafran SD. 1997. Comparison of dose doubling with probenecid for sustaining serum cefuroxime levels. J Antimicrob Chemother 40: 903–906.

736. Fletcher CV, Henry WK, Noormohamed SE, Rhame FS, Balfour HH Jr. 1995. The effect of cimetidine and ranitidine administration with zidovudine. Pharmacotherapy 15: 701–708.

737. Aiba T, Sakurai Y, Tsukada S, Koizumi T. 1995. Effects of probenecid and cimetidine on the renal excretion of 3′-azido-3′-deoxythymidine in rats. J Pharmacol Exp Ther 272: 94–99.

738. de Miranda P, Good SS, Yarchoan R, Thomas RV, Blum MR, Myers CE. 1989. Alteration of zidovudine pharmacokinetics by probenecid in patiets with AIDS and AIDS-related complex. Clin Pharmacol Ther 46: 494–500.

739. Hedaya MA, Elmquist WF, Sawchuk RJ. 1990. Probenecid inhibits the metabolic and renal clearances of zidovudine (AZT) in human volunteers. Pharm Res 7: 411–417.

740. Fenster PE, White NW Jr, Hanson CD. 1985. Pharmacokinetic evaluation of the digoxin–amiodarone interaction. J Am Coll Cardiol 5: 108–112.

741. Milne RW, Larsen LA, Jorgensen KL, Bastlund J, Stretch GR, Evans AM. 2000. Hepatic disposition of fexofenadine: influence of the transport inhibitors erythromycin and dibromosulphothalein. Pharm Res 17: 1511–1515.

742. Backman JT, Kajosaari LI, Niemi M, Neuvonen PJ. 2006. Cyclosporine A increases plasma concentrations and effects of repaglinide. Am J Transplant 6: 2221–2222.

743. Turk T, Witzke O. 2006. Pharmacological interaction between cyclosporine A and repaglinide. Is it clinically relevant? Am J Transplant 6: 2223.

744. Hasunuma T, Nakamura M, Yachi T, Arisawa N, Fukushima K. 2003. The drug-drug interactions of pitavastatin (NK-104), a novel HMG-CoA reductase inhibitor and cyclosporin. J Clin Ther Med 19: 381–389.

745. Christian CD Jr, Meredith CG, Speeg KV Jr. 1984. Cimetidine inhibits renal procainamide clearance. Clin Pharmacol Ther 36: 221–227.

746. Abel S, Nichols DJ, Brearley CJ, Eve MD. 2000. Effect of cimetidine and ranitidine on pharmacokinetics and pharmacodynamics of a single dose of dofetilide. Br J Clin Pharmacol 49: 64–71.

747. Doring F, Will J, Amasheh S, Clauss W, Ahlbrecth H, Daniel H. 1998. Minimal molecular determinants of substrates for recognition by the intestinal peptide transporter. J Biol Chem 273: 23211–23218.

748. Terada T, Saito H, Mukai M, Inui K. 1997. Recognition of beta-lactam antibiotics by rat peptide transporters, PEPT1 and PEPT2, in LLC-PK1 cells. Am J Physiol 273(5 Pt 2): F706–F711.

749. Ganapathy ME, Prasad PD, Mackenzie B, Ganapathy V, Leibach FH. 1997. Interaction of anionic cephalosporins with the intestinal and renal peptide transporters PEPT 1 and PEPT 2. Biochim Biophys Acta 1324: 296–308.

750. Zhu T, Chen XZ, Steel A, Hediger MA, Smith DE. 2000. Differential recognition of ACE inhibitors in *Xenopus laevis* oocytes expressing rat PEPT1 and PEPT2. Pharm Res 17: 526–532.

751. Shu C, Shen H, Hopfer U, Smith DE. 2001. Mechanism of intestinal absorption and renal reabsorption of an orally active ace inhibitor: uptake and transport of fosinopril in cell cultures. Drug Metab Dispos 29: 1307–1315.

752. Saito H, Terada T, Okuda M, Sasaki S, Inui K. 1996. Molecular cloning and tissue distribution of rat peptide transporter PEPT2. Biochim Biophys Acta 1280: 173–177.

753. Doring F, Walter J, Will J, Focking M, Boll M, Amasheh S, Clauss W, Daniel H. 1998. Delta-aminolevulinic acid transport by intestinal and renal peptide transporters and its physiological and clinical implications. J Clin Invest 101: 2761–2767.

754. Yamashita T, Shimada S, Guo W, Sato K, Kohmura E, Hayakawa T, Takagi T, Tohyama M. 1997. Cloning and functional expression of a brain peptide/histidine transporter. J Biol Chem 272: 10205–10211.

755. Sakata K, Yamashita T, Maeda M, Moriyama Y, Shimada S, Tohyama M. 2001. Cloning of a lymphatic peptide/histidine transporter. Biochem J 356: 53–60.

756. Meredith D, Boyd CA, Bronk JR, Biley PD, Morgan KM, Collier ID, and Temple CS. 1998. 4-Aminomethylbenzoic acid is a non-translocated competitive inhibitor of the epithelial peptide transporter PepT1. J Physiol 512: 629–634.

757. Terada T, Saito H, Mukai M, Inui K. 1996. Identification of the histidine residues involved in the substrate recognition by a rat H^+/peptide cotransporter, PEPT1. FEBS Lett 394: 196–200.

758. Sawada K, Terada T, Saito H, Hashimoto Y, Inui K. 1999. Effects of glibenclamide on glycylsarcosine transport by the rat peptide transporters PEPT1 and PEPT2. Br J Pharmacol 128: 1159–1164.

759. Terada T, Sawada K, Saito H, Hashimoto Y, Inui K. 2000. Inhibitory effect of novel oral hypoglycemic agent nateglinide (AY4166) on peptide transporters PEPT1 and PEPT2. Eur J Pharmacol 392: 11–17.

760. Knutter I, Theis S, Hartrodt B, Born I, Brandsch M, Daniel H, Neubert K. 2001. A novel inhibitor of the mammalian peptide transporter PEPT1. Biochemistry 40: 4454–4458.

761. Theis S, Knutter I, Hartrodt B, Brandsch M, Kottra G, Neubert K, Daniel H. 2002. Synthesis and characterization of high affinity inhibitors of the H^+/peptide transporter PEPT2. J Biol Chem 277: 7287–7292.

762. Akarawut W, Lin CJ, Smith DE. 1998. Noncompetitive inhibition of glycylsarcosine transport by quinapril in rabbit renal brush border membrane vesicles: effect on high-affinity peptide transporter. J Pharmacol Exp Ther 287: 684–690.

763. Lin CJ, Akarawut W, Smith DE. 1999. Competitive inhibition of glycylsarcosine transport by enalapril in rabbit renal brush border membrane vesicles: interaction of ACE inhibitors with high-affinity H^+/peptide symporter. Pharm Res 16: 609–615.

764. Swaan PW, Stehouwer MFC, Tukker JJ. 1995. Molecular mechanism for the relative binding affinity to the intestinal peptide carrier. Comparison of three ACE inhibitors: enalapril, enalaprilat, and lisinopril. Biochim Biophys Acta 1236: 31–38.

765. Thamotharan M, Bawani SZ, Zhou X, Adibi SA. 1999. Hormonal regulation of oligopeptide transporter PepT-1 in a human intestinal cell line. Am J Physiol Cell Physiol 276: C821–C826.

766. Sun BW, Zhao XC, Wang GJ, Li N, Li JS. 2003. Hormonal regulation of dipeptide transporter (PepT1) in Caco-2 cells with normal and anoxia/reoxygenation management. World J Gastroenterol 9: 808–812.

767. Berlioz F, Maoret JJ, Paris H, Laburthe M, Farinotti R, Rozé C. 2000. α2-Adrenergic receptors stimulate oligopeptide transport in a human intestinal cell line. J Pharmacol Exp Ther 294: 466–472.

768. Buyse M, Berlioz F, Guilmeau S, Tsocas A, Voisin T, Peranzi G, Merlin D, Laburthe M, Lewin MJ, Rozé C, Bado A. 2001. PepT1 mediates epithelial transport of dipeptides and cephalexin is enhanced by luminal leptin in the small intestine. J Clin Invest 108: 1483–1494.

769. Fujita T, Majikawa Y, Umehisa S, Okada N, Yamamoto A, Ganapathy V, Leibach FH. 1999. Sigma receptor ligand-induced up-regulation of the H^+/peptide transporter PEPT1 in the human intestinal cell line Caco-2. Biochem Biophys Res Commun 261: 242–246.

770. Wenzel U, Kuntz S, Diestel S, Daniel H. 2002. PEPT1-mediated cefixime uptake into human intestinal epithelial cells is increased by $Ca2^+$ channel blockers. Antimicrob Agents Chemother 46: 1375–1380.

771. Ashida K, Katsura T, Motohashi H, Saito H, Inui K. 2002. Thyroid hormone regulates the activity and expression of the peptide transporter PepT1 in Caco-2 cells. Am J Physiol Gastrointest Liver Physiol 282: G617–G623.

772. Nielsen CU, Amstrup J, Steffansen B, Frokjaer S, Brodin B. 2001. Epidermal growth factor inhibits glycylsarcosine transport and hPepT1 expression in a human intestinal cell line. Am J Physiol Gastrointest Liver Physiol 281: G191–G199.

773. Ihara T, Tsujikawa T, Fujiyama Y, Bamba T. 2000. Regulation of PepT1 peptide transporter expression in the rat small intestine under malnourished conditions. Digestion 61: 59–67.

774. Thamotharan M, Bawani SZ, Zhou X, Adibi SA. 1999. Functional and molecular expression of intestinal oligopeptide transporter (Pept-1) after a brief fast. Metabolism 48: 681–684.

775. Shiraga T, Miyamoto K, Tanaka H, Yamamoto H, Taketani Y, Morita K, Tamai I, Tsuji A, Takeda E. 1999. Cellular and molecular mechanisms of dietary regulation on rat intestinal H^+/peptide transporter PepT1. Gastroenterology 116: 354–362.

776. Naruhashi K, Sai Y, Tamai I, Suzuki N, Tsuji A. 2002. Pept1 mRNA expression is induced by starvation and its level correlates with absorptive transport of cefadroxil longitudinally in the rat intestine. Pharm Res (N Y) 19: 1417–1423.

777. Pan X, Terada T, Irie M, Saito H, Inui K. 2002. Diurnal rhythm of H^+-peptide cotransporter in rat small intestine. Am J Physiol Gastrointest Liver Physiol 283: G57–G64.

778. Pan X, Terada T, Okuda M, Inui K. 2003. Altered diurnal rhythm of intestinal peptide transporter by fasting and its effects on the pharmacokinetics of ceftibuten. J Pharmacol Exp Ther 307: 626–632.

779. Gangopadhyay A, Thamotharan M, Adibi SA. 2002. Regulation of oligopeptide transporter (Pept-1) in experimental diabetes. Am J Physiol Gastrointest Liver Physiol 283: G133–G138.

780. Barbot L, Windsor E, Rome S, Tricottet V, Reynes M, Topouchian A, Huneau JF, Gobert JG, Tomé D, Kapel N. 2003. Intestinal peptide transporter PepT1 is over-expressed during acute cryptosporidiosis in suckling rats as a result of both malnutrition and experimental parasite infection. Parasitol Res 89: 364–370.

781. Ziegler TR, Fernandez-Estivariz C, Gu LH, Bazargan N, Umeakunne K, Wallace TM, Diaz EE, Rosado KE, Pascal RR, Galloway JR, Wilcox JN, Leader LM. 2002. Distribution of the H^+/peptide transporter PepT1 in human intestine: upregulated expression in the colonic mucosa of patients with short-bowel syndrome. Am J Clin Nutr 75: 922–930.

782. Merlin D, Si-Tahar M, Sitaraman SV, Eastburn K, Williams I, Liu X, Hediger MA, Madara JL. 2001. Colonic epithelial hPepT1 expression occurs in inflammatory bowel disease: transport of bacterial peptides influences expression of MHC class I molecules. Gastroenterology 120: 1666–1679.

783. Ford D, Howard A, Hirst BH. 2003. Expression of the peptide transporter hPepT1 in human colon: a potential route for colonic protein nitrogen abd drug absorption. Cell Biol 119: 37–43.

784. Merlin D, Steel A, Gewirtz AT, Hediger MA, Madara JL. 1998. hPepT1-mediated epithelial transport of bacteria-derived chemotactic peptides enhances neutrophil–epithelial interactions. J Clin Invest 1: 2011–2018.

785. Buyse M, Toscas A, Walker F, Merlin D, Bado A. 2002. PepT1-mediated fMLP transport induces intestinal inflammation in vivo. Am J Physiol Cell Physiol 283: C1795–1800.

786. Shu HJ, Takeda H, Shinzawa H, Takahashi T, Kawata S. 2002. PepT1-mediated fMLP transport induces intestinal inflammation in vivo. Am J Physiol Cell Physiol 283: C1795–C1800.

787. Takahashi K, Masuda S, Nakamura N, Saito H, Futami T, Doi T, Inui K. 2001. up-regulation of H^+-peptide cotransporter PEPT2 in rat remnant kidney. Am J Physiol 281: F1109–F1116.

788. Bretschneider B, Brandsch M, Neubert R. 1999. Intestinal transport of beta-lactam antibiotics at the H^+/peptide symporter (PEPT1), the uptake into Caco-2 cell monolayers and the transepithelial flux. Pharm Res 16 : 55–61.

789. Dantzig AH. 1997. Oral absorption of beta-lactams by intestinal peptide transport proteins. Adv Drug Deliv Rev 23: 63–76.

790. Ganapathy ME, Brandsch M, Prasad PD, Ganapathy V, Leibach FH. 1995. Differential recognition of beta-lactam antibiotics by intestinal and renal peptide transporters, PEPT 1 and PEPT 2. J Biol Chem 270: 25672–25677.

791. Berlioz F, Julien S, Tsocas A, Chariot J, Carbon C, Farinotti R, Rozé C. 1999. Neural modulation of cephalexin intestinal absorption through the di- and tripeptide brush border transporter of rat jejunum in vivo. J Pharmacol Exp Ther 288: 1037–1044.

792. Tamai I, Nakanishi T, Nakahara H, Sai Y, Ganapathy V, Leibach FH, Tsuji A. 1998. Improvement of L-dopa absorption by dipeptidyl derivation, utilizing peptide transporter PepT1. J Pharm Sci 87: 1542–1546.

793. Ganapathy ME, Huang W, Wang H, Ganapathy V, Leibach FH. 1998. Valacyclovir: a substrate for the intestinal and renal peptide transporters PEPT1 and PEPT2. Biochem Biophys Res Commun 246: 470–475.

794. Soul-Lawton J, Seaber E, On N, Wootton R, Rolan P, Posner J. 1995. Absolute bioavailability and metabolic disposition of valacyclovir, the L-valyl ester of acyclovir, following oral administration to humans. Antimicrob Agents Chemother 39: 2759–2764.

795. Balimane PV, Tamai I, Guo A, Nakanishi T, Kitada H, Leibach FH, Tsuji A, Sinko PJ. 1998. Direct evidence for peptide transporter (PepT1)-mediated uptake of a nonpeptide prodrug valacyclovir. Biochem Biophys Res Commun 250: 246–251.

796. Han H, de Vrueh RL, Rhie JK, Covitz KM, Smith PL, Lee CP, Oh DM, Sadée W, Amidon GL. 1998. 5′-Aminoacid esters of antiviral nucleosides, acyclovir, and AZT are absorbed by the intestinal PEPT1 peptide transporter. Pharm Res 15: 1154–1159.

797. Weiss WJ, Mikels SM, Petersen PJ, Jacobus NV, Bitha P, Lin YI, Testa RT. 1999. In vivo activities of peptidic prodrugs of novel aminomethyl tetrahydrofuranyl-1β-methylcarbapenems. Antimicrob Agents Chemother 43: 460–464.

798. Kelty CJ, Brown NJ, Reed MW, Ackroyd R. 2002. The use of 5-aminolevulinic acid as a photosensitiser in photodynamic therapy and photodiagnosis. Photochem Photobiol Sci 1: 158–168.

799. Neumann J, Brandsch M. 2003. Delta-aminolevulinic acid transport in cancer cells of the human extrahepatic biliary duct. J Pharmacol Exp Ther 305: 219–224.

800. Shen H, Smith DE, Keep RF, Xiang J, Brosius FC III. 2003. Targeted disruption of the PEPT2 gene markedly reduces dipeptide uptake in choroid plexus. J Biol Chem 278: 4786–4791.

801. Shu C, Shen H, Teuscher NS, Lorenzi PJ, Keep RF, Smith DE. 2002. Role of PEPT2 in peptide/mimetic trafficking at the blood-cerebrospinal fluid barrier: studies in rat choroid plexus epithelial cells in primary culture. J Pharmacol Exp Ther 301: 820–829.

802. Garrigues TM, Martin U, Peris-Ribera JE, Prescott LF. 1991. Dose-dependent absorption and elimination of cefadroxil in man. Eur J Clin Pharmacol 41: 179–183.

803. Sjövall J, Alvan G, Westerlund D. 1985. Oral cyclacillin interacts with the absorption of oral ampicillin, amoxycillin, and bacampicillin. Eur J Clin Pharmacol 29: 495–502. Abstract only.

804. Kimura T, Endo H, Yoshikawa M, Muranishi S, Sezaki H. 1978. Carrier-mediated transport systems for aminopenicillins in rat small intestine. J Pharmacobiodyn 1: 262–267. Abstract only.

805. Westphal JF, Trouvin JH, Deslandes A, Carbon C. 1990. Nifedipine enhances amoxicillin absorption kinetics and bioavailability in humans. J Pharmacol Exp Ther 255: 312–317.

806. Berlioz F, Lepere-Prevot B, Julien S, Tsocas A, Carbon C, Rozé C, Farinotti R. 2000. Chronic nifedipine dosing enhances cephalexin bioavailability and intestinal absorption in conscious rats. Drug Metab Dispos 28: 1267–1269.

807. Duverne C, Bouten A, Deslandes A, Westphal JF, Trouvin JH, Farinotti R, Carbon C. 1992. Modification of cefixime bioavailability by nifedipine in humans: involvement of the dipeptide carrier system. Antimicrob Agents Chemother 36: 2462–2467.

808. Fishbein WN. 1986. Lactate transporter defect: a new disease of muscle. Science 234: 1254–1256.

809. Merezhinskaya N, Fishbein WN, Davis JI, Foellmer JW. 2000. Mutations in MCT1 cDNA in patients with symptomatic deficiency in lactate transport. Muscle Nerve 23: 90–97.

810. Maranduba CM, Friesema EC, Kok F, Kester MH, Jansen J, Sertie AL, Passos-Bueno MR, Visser TJ. 2006. Decreased cellular uptake and metabolism in Allan–Herndon–Dudley syndrome (AHDS) due to a novel mutation in the MCT8 thyroid hormone transporter. J Med Genet 43: 457–460.

811. Friesema EC, Grueters A, Biebermann H, Krude H, von Moers A, Reeser M, Barrett TG, Mancilla EE, Svensson J, Kester MH, et al. 2004. Association between mutations in a thyroid hormone transporter and severe X-linked psychomotor retardation. Lancet 364: 1435–1437.

812. Friesema EC, Jansen J, Heuer H, Trajkovic M, Bauer K, Visser TJ. 2006. Mechanisms of disease: psychomotor retardation and high T3 levels caused by mutations in monocarboxylate transporter 8. Nat Clin Pract Endocrinol Metab 2: 512–523.

813. Poole RC, Halestrap AP. 1993. Transport of lactate and other monocarboxylates across mammalian plasma membranes. Am J Physiol 264: C761–C782.

814. Halestrap AP, Price NT. 1999. The proton-linked monocarboxylate transporter (MCT) family: structure, function and regulation. Biochem J 343: 281–299.

815. Kido Y, Tamai I, Okamoto M, Suzuki F, Tsuji A. 2000. Functional clarification of MCT1-mediated transport of monocarboxylic acids at the blood–brain barrier using in vitro cultured cells and in vivo BUI studies. Pharm Res 17: 55–62.

816. Tamai I, Takanaga H, Maeda H, Sai Y, Ogihara T, Higashida H, Tsuji A. 1995. Participation of a proton-cotransporter, MCT1, in the intestinal transport of monocarboxylic acids. Biochem Biophys Res Commun 214: 482–489.

817. Enerson BE, Drewes LR. 2003. Molecular features, regulation, and function of monocarboxylate transporters: implications for drug delivery. J Pharm Sci 92: 1531–1544.

818. Li YH, Tanno M, Itoh T, Yamada H. 1999. Role of the monocarboxylic acid transport system in the intestinal absorption of an orally active beta-lactam prodrug: carindacillin as a model. Int J Pharm 191: 151–159.

819. Itoh T, Tanno M, Li Y-H, Yamada H. 1998. Transport of phenethicillin into rat intestinal brush border membrane vescicles: Role of the monocarboxylic acid transport system. Int J Pharm 172: 103–112.

820. Saheki A, Terasaki T, Tamai I, Tsuji A. 1994. In vivo and in vitro blood-brain barrier transport of 3-hydroxy-3-methylglutaryl coenzyme A (HMG-CoA) reductase inhibitors. Pharm Res 11: 305–311.

821. Tsuji A, Saheki A, Tamai I, Terasaki T. 1993. Transport mechanism of 3-hydroxy-3-methylglutaryl coenzyme A reductase inhibitors at the blood–brain barrier. J Pharmacol Exp Ther 267: 1085–1090.

822. Wu X, Whitfield LR, Stewart BH. 2000. Atorvastatin transport in the Caco-2 cell model: contributions of P-glycoprotein and the proton-monocarboxylic acid co-transporter. Pharm Res 17: 209–215.

823. Lin RY, Vera JC, Chaganti RS, Golde DW. 1998. Human monocarboxylate transporter 2 (MCT2) is a high affinity pyruvate transporter. J Biol Chem 273: 28959–28965.

824. Bröer S, Bröer A, Schneider HP, Stegen C, Halestrap AP, Deitmer JW. 1999. Characterization of the high-affinity monocarboxylate transporter MCT2 in *Xenopus laevis* oocytes. Biochem J 341: 529–535.

825. Dimmer KS, Friedrich B, Lang F, Deitmer JW, Bröer S. 2000. The low-affinity monocarboxylate transporter MCT4 is adapted to the export of lactate in highly glycolytic cells. Biochem J 350(Pt 1): 219–227.

826. Manning Fox JE, Meredith D, Halestrap AP. 2000. Characterisation of human monocarboxylate transporter 4 substantiates its role in lactic acid efflux from skeletal muscle. J Physiol (Lond) 529: 285–293.

827. Halestrap AP, Meredith D. 2004. The SLC16 gene family: from monocarboxylate transporters (MCTs) to aromatic amino acid transporters and beyond. Pflugers Arch 447: 619–628.

828. Friesema EC, Jansen J, Milici C, Visser TJ. 2005. Thyroid hormone transporters. Vitam Horm 70: 137–167.

829. Garcia CK, Brown MS, Pathak RK, Goldstein JL. 1995. cDNA cloning of MCT2, a second monocarboxylate transporter expressed in different cells than MCT1. J Biol Chem 270: 1843–1849.

830. Grollman EF, Philp NJ, McPhie P, Ward RD, Sauer B. 2000. Determination of transport kinetics of chick MCT3 monocarboxylate transporter from retinal pigment epithelium by expression in genetically modified yeast. Biochemistry 39: 9351–9357.

831. Pilegaard H, Domino K, Noland T, Juel C, Hellsten Y, Halestrap AP, Bangsbo J. 1999. Effect of high-intensity exercise training on lactate/H^+ transport capacity in human skeletal muscle. Am J Physiol 276(2 Pt 1): E255–E261.

832. Dubouchaud H, Butterfield GE, Wolfel EE, Bergman BC, Brooks GA. 2000. Endurance training, expression, and physiology of LDH, MCT1, and MCT4 in human skeletal muscle. Am J Physiol Endocrinol Metab 278: E571–E579.

833. Baker SK, McCullagh KJA, Bonen A. 1998. Training intensity dependent and tissue specific increases lactate uptake and MCT1 in heart and muscle. J Appl Physiol 84: 987–994.

834. Green H, Halestrap A, Mockett C, O'Toole D, Grant S, Ouyang J. 2002. Increases in muscle MCT are associated with reductions in muscle lactate after a single exercise session in humans. Am J Physiol Endocrinol Metab 282: E154–E160.

835. Juel C, Halestrap AP. 1999. Lactate transport in skeletal muscle: role and regulation of the monocarboxylate transporter. J Physiol 517: 633–642.

836. Pilegaard H, Juel C. 1995. Lactate transport studied in sarcolemmal giant vesicles from rat skeletal muscles: effect of denervation. Am J Physiol 269: E679–E682.

837. Pilegaard H, Mohr T, Kjaer M, Juel C. 1998. Lactate/H+ transport in skeletal muscle from spinal-cord-injured patients. Scand J Med Sci Sports 8: 98–101.

838. Enoki T, Yoshida Y, Hatta H, Bonen A. 2003. Exercise training alleviates MCT1 and MCT4 reductions in heart and skeletal muscles of STZ-induced diabetic rats. J Appl Physiol 94: 2433–2438.

839. Hajduch E, Heyes RR, Watt PW, Hundal HS. 2000. Lactate transport in rat adipocytes: identification of monocarboxylate transporter 1 (MCT1) and its modulation during streptozotocin-induced diabetes. FEBS Lett 479: 89–92.

840. Juel C, Holten MK, Dela F. 2004. Effects of strength training on muscle lactate release and MCT1 and MCT4 content in healthy and type 2 diabetic humans. J Physiol 556: 297–304.

841. Pierre K, Pellerin L. 2005. Monocarboxylate transporters in the central nervous system: distribution, regulation and function. J Neurochem 94: 1–14.

842. Tseng MT, Chan SA, Schurr A. 2003. Ischemia-induced changes in monocarboxylate transporter 1 reactive cells in rat hippocampus. Neurol Res 25: 83–86.

843. Leino RL, Gerhart DZ, Duelli R, Enerson BE, Drewes LR. 2001. Diet-induced ketosis increases monocarboxylate transporter (MCT1) levels in rat brain. Neurochem Int 38: 519–527.

844. Pierre K, Debernardi R, Magistretti PJ, Pellerin L. 2003. Noradrenaline enhances monocarboxylate transporter 2 expression in cultured mouse cortical neurons via a translational regulation. J Neurochem 86: 1468–1476.

845. De Bernardi R, Pierre K, Lengacher S, Magistretti PJ, Pellerin L. 2003. Cell-specific expression pattern of monocarboxylate transporters in astrocytes and neurons observed in different mouse brain cortical cell cultures. J Neurosci Res 73: 141–155.

846. Knott RM, Robertson M, Muckersie E, Folefac VA, Fairhurst FE, Wileman SM, Forrester JV. 1999. A model system for the study of human retinal angiogenesis: activation of monocytes and endothelial cells and the association with the expression of the monocarboxylate transporter type 1 (MCT-1). Diabetologia 42: 870–877.

847. Hahn EL, Halestrap AP, Gamelli RL. 2000. Expression of the lactate transporter MCT1 in macrophages. Shock 13: 253–260.

848. Cuff Ma, Lambert DW, Shirazi-Beechey SP. 2002. Substrate-induced regulation of the human colonic monocarboxylate transporter MCT1. J Physiol 539: 361–371.

849. Buyse M, Sitaraman SV, Liu X, Bado A, Merlin D. 2002. Luminal leptin enhances CD147/MCT-1-mediated uptake of butyrate in the human intestinal cell line Caco 2-BBE. J Biol Chem 277: 28182–28190.

850. Enoki T, Yoshida Y, Lally J, Hatta H, Bonen A. 2006. Testosterone increases lactate transport, MCT1 and MCT4 in rat skeletal muscle. J Physiol 577: 433–443.

851. Nehme CL, Fayos BE, Bartles JR. 1995. Distribution of the integral plasma membrane glycoprotein CE9 (MRC OX-47) among rat tissues and its induction by diverse stimuli of metabolic activation. Biochem J 310: 693–698.

852. Kirk P, Wilson MC, Heddle C, Brown MH, Barclay AN, Halestrap AP. 2000. CD147 is tightly associated with lactate transporters MCT1 and MCT4 and facilitates their cell surface expression. EMBO J 19: 3896–3904.

853. Tamai I, Sai Y, Ono A, Kido Y, Yabuuchi H, Takanaga H, Satoh E, Ogihara T, Amano O, Izeki S, Tsuji A. 1999. Immunohistochemical and functional characterization of pH-dependent intestinal absorption of weak organic acids by the monocarboxylic acid transporter MCT1. J Pharm Pharmacol 51: 1113–1121.

854. Deguchi Y, Nozawa K, Yamada S, Yokoyama Y, Kimura R. 1997. Quantitative evaluation of brain distribution and blood–brain barrier efflux transport of probenecid in rats by microdialysis: possible involvement of the monocarboxylic acid transport system. J Pharmacol Exp Ther 280: 551–560.

855. Deguchi Y, Yokoyama Y, Sakamoto T, Hayashi H, Naito T, Yamada S, Kimura R. 2000. Brain distribution of 6-mercaptopurine is regulated by the efflux transport system in the blood–brain barrier. Life Sci 66: 649–662.

856. Bliss TM, Ip M, Cheng E, Minami M, Pellerin L, Magistretti P, Sapolsky RM. 2004. Dual-gene, dual-cell type therapy against an excitotoxic insult by bolstering neuroenergetics. J Neurosci 24: 6202–6208.

857. Sapolsky RM. 2003. Neuroprotective gene therapy against acute neurological insults. Nat Rev Neurosci 4: 61–69.

858. Osato DH, Huang CC, Kawamoto M, Johns SJ, Stryke D, Wang J, Ferrin TE, Herskowitz I, Giacomini KM. 2003. Functional characterization in yeast of genetic variants in the human equilibrative nucleoside transporter, ENT1. Pharmacogenetics 13: 297–301.

859. Owen RP, Lagpacan LL, Taylor TR, de la Cruz M, Huang CC, Kawamoto M, Johns SJ, Stryke D, Ferrin TE, Giacomini KM. 2006. Functional characterization and haplotype analysis of polymorphisms in the human equilibrative nucleoside transporter, ENT2. Drug Metab Dispos 34: 12–15.

860. Owen RP, Gray JH, Taylor TR, Carlson EJ, Huang CC, Kawamoto M, Johns SJ, Stryke D, Ferrin TE, Giacomini KM. 2005. Genetic analysis and functional characterization of polymorphisms in the human concentrative nucleoside transporter, CNT2. Pharmacogenet Genom 15: 83–90.

861. Gray JH, Mangravite LM, Owen RP, Urban TJ, Chan W, Carlson EJ, Huang CC, Kawamoto M, Johns SJ, Stryke D, Ferrin TE, Giacomini KM. 2004. Functional and genetic diversity in the concentrative nucleoside transporter, CNT1, in human populations. Mol Pharmacol 65: 512–519.

862. Griffiths M, Yao SY, Abidi F, Phillips SE, Cass CE, Young JD, Baldwin SA. 1997. Molecular cloning and characterization of a nitrobenzylthioinosine- insensitive (*ei*) equilibrative nucleoside transporter from human placenta. Biochem J 328: 739–743.

863. Ward JL, Sherali A, Mo ZP, Tse CM. 2000. Kinetic and pharmacological properties of cloned human equilibrative nucleoside transporters, ENT1 and ENT2, stably expressed in nucleoside transporter-deficient PK15 cells. J Biol Chem 275: 8375–8381.

864. Vickers MF, Kumar R, Visser F, Zhang J, Charania J, Raborn RT, Baldwin SA, Young JD, Cass CE. 2002. Comparison of the interaction of uridine, cytidine, and other pyrimidine nucleoside analogues with recombinant human equilibrative nucleoside transporter 2 (hENT2) produced in *Saccharomyces cerevisiae*. Biochem Cell Biol 80: 639–644.

865. Lum PY, Ngo LY, Baggen AH, Unadkat JD. 2000. Human intestinal es nucleoside transporter: molecular characterization and nucleoside inhibitory profiles. Cancer Chemother Pharmacol 45: 273–278.

866. Yao SY, Ng AM, Vickers MF, Sundaram M, Cass CE, Baldwin SA, Young JD. 2002. Functional and molecular characterization of nucleobase transport by recombinant human and rat equilibrative nucleoside transporters 1 and 2. Chimeric constructs reveal a role for the ENT2 helix 5–6 region in nucleobase translocation. J Biol Chem 277: 24938–24948.

867. Jarvis SM, Thorm JA, Glue P. 1998. Ribavirin uptake by human erythrocytes and the involvement of nytrobenzylthioinosine-sensitive (es)-nucleoside transporters. Br J Pharmacol 123: 1587–1592.

868. Mackey JR, Mani RS, Selner M, Mowles D, Young JD, Belt JA, Crawford CR, Cass CE. 1998. Functional nucleoside transporters are required for gemcitabine influx and manifestation of toxicity in cancer cell lines. Cancer Res 58: 4349–4357.

869. Mackey JR, Yao SY, Smith KM, Karpinski E, Baldwin SA, Cass CE, Young JD. 1999. Gemcitabine transport in xenopus oocytes expressing recombinant plasma membrane mammalian nucleoside transporters. J Natl Cancer Inst 91: 1876–1881.

870. Yao SY, Ng AM, Sundaram M, Cass CE, Baldwin SA, Young JD. 2001. Transport of antiviral 3′-deoxy-nucleoside drugs by recombinant human and rat equilibrative, nitrobenzylthioinosine (NBMPR)-insensitive (ENT2) nucleoside transporter proteins produced in *Xenopus* oocytes. Mol Membr Biol 18: 161–167.

871. Patil SD, Ngo LY, Unadkat JD. 2000. Structure-inhibitory profiles of nucleosides for the human intestinal N1 and N2 Na^+-nucleoside transporters. Cancer Chemother Pharmacol 46: 394–402.

872. Crawford CR, Patel DH, Naeve C, Belt JA. 1998. Cloning of the human equilibrative, nitrobenzylmercaptopurine riboside (NBMPR)–insensitive nucleoside transporter *ei* by functional expression in a transport-deficient cell line. J Biol Chem 273: 5288–5293.

873. Ward JL, Leung GP, Toan SV, Tse CM. 2003. Functional analysis of site directed glycosylation mutants of the human equilibrative nucleoside transporter-2. Arch Biochem Biophys 411: 19–26.

874. Guida L, Bruzzone S, Sturla L, Franco L, Zocchi E, De Flora A. 2002. Equilibrative and concentrative nucleoside transporters mediate influx of extracellular cyclic ADP-ribose into 3T3 murine fibroblasts. J Biol Chem 277: 47097–47105.

875. Garcia-Manteiga J, Molina-Arcas M, Casado FJ, Mazo A, Pastor-Anglada M. 2003. Nucleoside transporter profiles in human pancreatic cancer cells: role of hCNT1 in 2′,2′-difluorodeoxycytidine- induced cytotoxicity. Clin Cancer Res 9: 5000–5008.

876. Molina-Arcas M, Bellosillo B, Casado FJ, Mentserrat J, Gil J, Colomer D, Pastor-Anglada M. 2003. Fludarabine uptake mechanisms in B-cell chronic lymphocytic leucemia. Blood 101: 2328–2334.

877. Molina-Arcas M, Marce S, Villamor N, Huber-Ruano I, Casado FJ, Bellosillo B, Montserrat E, Gil J, Colomer D, Pastor-Anglada M. 2005. Equilibrative nucleoside transporter-2 (hENT2) protein expression correlates with ex vivo sensitivity to fludarabine in chronic lymphocytic leukemia (CLL) cells. Leukemia 19: 64–68.

878. Baldwin SA, Beal PR, Yao SY, King AE, Cass CE, Young JD. 2004. The equilibrative nucleoside transporter family, SLC29. Pflugers Arch 447: 735–743.

879. Baldwin SA, Yao SY, Hyde RJ, Ng AM, Foppolo S, Barnes K, Ritzel MW, Cass CE, Young JD. 2005. Functional characterization of novel human and mouse equilibrative nucleoside transporters (hENT3 and mENT3) located in intracellular membranes. J Biol Chem 280: 15880–15887.

880. Ritzel MW, Yao SY, Huang MY, Elliott JF, Cass CE, Young JD. 1997. Molecular cloning and functional expression of cDNAs encoding a human Na^+-nucleoside cotransporter (hCNT1). Am J Physiol 272: C707–C714.

881. Wang J, Su SF, Dresser MJ, Schaner ME, Washington CB, Giacomini KM. 1997. Na^+-dependent purine nucleoside transporter from human kidney: cloning and functional characterization. Am J Physiol 273: F1058–F1065.

882. Graham KA, Leithoff J, Coe IR, Mowles D, Mackey JR, Young JD, Cass CE. 2000. Differential transport of cytosine-containing nucleosides by recombinant human concentrative nucleoside transporter protein hCNT1. Nucleosides Nucleotides Nucleic Acids 19: 415–434.

883. Lostao MP, Mata JF, Larrayoz IM, Inzillo SM, Casado FJ, Pastor-Anglada M. 2000. Electrogenic uptake of nucleosides and nucleoside-derived drugs by the human nucleoside transporter 1 (hCNT1) expressed in *Xenopus laevis* oocytes. FEBS Lett 481: 137–140.

884. Mata JF, Garcia-Manteiga JM, Lostao MP, Fernandez-Veledo S, Guillen-Gomez E, Larrayoz IM, Lloberas J, Casado FJ, Pastor-Anglada M. 2001. Role of the human concentrative nucleoside transporter (hCNT1) in the cytotoxic action of 5′-deoxy-5-fluorouridine, an active intermediate metabolite of capecitabine, a novel oral anticancer drug. Mol Pharmacol 59: 1542–1548.

885. Wang J, Schaner ME, Thomassen S, Su SF, Piquette-Miller M, Giacomini KM. 1997. Functional and molecular characteristics of $Na(^+)$-dependent nucleoside transporters. Pharm Res 14: 1524–1532.

886. Schaner ME, Wang J, Zhang L, Su SF, Gerstin KM, Giaomini KM. 1999. Functional characterization of a human purine-selective, Na^+-dependent nucleoside transporter (hSPNT1) in a mammalian expression system. J Pharmacol Exp Ther 289: 1487–1491.

887. Patil SD, Ngo LY, Glue P, Unadkat JD. 1998. Intestinal absorption of ribavirin is preferentially mediated by the Na^+ nucleoside purine (N1) transporter. Pharm Res 15: 950–952.

888. Gerstin KM, Dresser MJ, Giacomini KM. 2002. Specificity of human and rat orthologs of the concentrative nucleoside transporter, SPNT. Am J Physiol Renal Physiol 283: F344–F349.

889. Ritzel MW, Ng AM, Yao SY, Graham K, Loewen SK, Smith KM, Ritzel G, Mowles DA, Carpenter P, Chen XZ, et al. 2001. Molecular identification and characterization of novel human and mouse concentrative Na+-nucleoside cotransporter proteins (hCNT3 and mCNT3) broadly selective for purine and pyrimidine nucleosides. J Biol Chem 276: 2914–2927.

890. Hu H, Endres CJ, Chang C, Umapathy NS, Lee EW, Fei YJ, Itagaki S, Swaan PW, Ganapathy V, Unadkat JD. 2006. Electrophysiological characterization and modeling of the structure activity relationship of the human concentrative nucleoside transporter 3 (hCNT3). Mol Pharmacol 69: 1542–1553.

891. Toan SV, To KK, Leung GP, de Souza MO, Ward JL, Tse CM. 2003. Genomic organization and functional characterization of the human concentrative nucleoside transporter-3 isoform (hCNT3) expressed in mammalian cells. Pflugers Arch 447: 195–204.

892. Baldwin SA, Mackey JR, Cass CE, Young JD. 1999. Nucleoside transporters: molecular biology and implications for therapeutic development. Mol Med Today 5: 216–224.

893. Griffiths M, Beaumont N, Yao SY, Sundaram M, Boumah CE, Davies A, Kwong FYP, Coe I, Cass CE, Young JD, Baldwin SA. 1997. Cloning of a human nucleoside transporter implicated in the cellular uptake of adenosine and chemotherapeutic drugs. Nat Med 3: 89–93.

894. Sundaram M, Yao SY, Ng AM, Griffiths M, Cass CE, Baldwin Sa, Young JD. 1998. Chimaeric constructs between human and rat equilibrative nucleoside transporters (hENT1 and rENT1) reveal hENT1 structural domains interacting with coronary vasoactive drugs. J Biol Chem 273: 21519–21525.

895. Smith PG, Thomas HD, Barlow HC, Griffin RJ, Golding BT, Calvert AH, Newell DR, Curtin NJ. 2001. In vitro and in vivo properties of novel nucleoside transport inhibitors with improved pharmacological properties that potentiate antifolate activity. Clin Cancer Res 7: 2105–2113.

896. Huang M, Wang Y, Collins M, Gu JJ, Mitchell BS, Graves LM. 2002. Inhibition of nucleoside transport by p38 MAPK inhibitors. J Biol Chem 277: 28364–28367.

897. Huang M, Wang Y, Cogut SB, Mitchell BS, Graves LM. 2003. Inhibition of nucleoside transport by protein kinase inhibitors. J Pharmacol Exp Ther 304: 753–760.

898. Huang M, Wang Y, Mitchell BS, Graves LM. 2004. Regulation of equilibrative nucleoside uptake by protein kinase inhibitors. Nucleosides Nucleotides Nucleic Acids 23: 1445–1450.

899. Leung GP, Man RY, Tse CM. 2005. Effect of thiazolidinediones on equilibrative nucleoside transporter-1 in human aortic smooth muscle cells. Biochem Pharmacol 70: 355–62.

900. Choi DS, Cascini MG, Mailliard W, Young H, Paredes P, McMahon T, Diamond I, Bonci A, Messing RO. 2004. The type 1 equilibrative nucleoside transporter regulates ethanol intoxication and preference. Nat Neurosci 7: 855–861.

901. Nagy LE, Diamond I, Casso DJ, Franklin C, Gordon AS. 1990. Ethanol increases extracellular adenosine by inhibiting adenosine uptake via the nucleoside transporter. J Biol Chem 265: 1946–1951.

902. Gupte A, Buolamwini JK, Yadav V, Chu CK, Naguib FN, el Kouni MH. 2005. 6-Benzylthioinosine analogues: promising anti-toxoplasmic agents as inhibitors of the mammalian nucleoside transporter ENT1 (es). Biochem Pharmacol 71: 69–73.

903. Valdes R, Casado FJ, Pastor-Anglada M. 2002. Cell-cycle-dependent regulation of CNT1, a concentrative nucleoside transporter involved in the uptake of cell-cycle-dependent nucleoside-derived anticancer drugs. Biochem Biophys Res Commun 296: 575–579.

904. Pressacco J, Wiley JS, Jamieson GP, Erlichman C, Hedley DW. 1995. Modulation of the equilibrative nucleoside transporter by inhibitors of DNA-synthesis. Br J Cancer 72: 939–942.

905. Pastor-Anglada M, Casado FJ, Valdes R, Mata J, Garcia-Manteiga J, Molina M. 2001. Complex regulation of nucleoside transporter expression in epithelial and immune system cells. Mol Membr Biol 18: 81–85.

906. Valdes R, Ortega MA, Casado FJ, Felipe A, Gil A, Sanchez-Pozo A, Pastor-Anglada M. 2000. Nutritional regulation of nucleoside transporter expression in rat small intestine. Gastroenterology 119: 1623–1630.

907. Felipe A, Valdes R, Santo B, Lloberas J, Casado J, Pastor-Anglada M. 1998. Na^+-dependent nucleoside transport in liver: two different isoforms from the same gene family are expressed in liver cells. Biochem J 330: 997–1001.

908. Coe I, Zhang Y, McKenzie T, Naydenova Z. 2002. PKC regulation of the human equilibrative nucleoside transporter, hENT1. FEBS Lett 517: 201–205.

909. Soler C, Felipe A, Mata JF, Casado FJ, Celada A, Pastor-Anglada M. 1998. Regulation of nucleoside transport by lipopolysaccharide, phorbol esters, and tumor necrosis factor-alpha in human B-lymphocytes. J Biol Chem 273: 26939–26945.

910. Soler C, Garcia-Manteiga J, Valdes R, Xaus J, Comalada M, Casado FJ, Pastor-Anglada M, Celada A, Felipe A. 2001. Macrophages require different nucleoside transport systems for proliferation and activation. FASEB J 15: 1979–1988.

911. Soler C, Felipe A, Garcia-Manteiga J, Serra M, Guillen-Gomez E, Casado FJ, MacLeod C, Modolell M, Pastor-Anglada M, Celada A. 2003. Interferon-gamma regulates nucleoside transport systems in macrophages through signal transduction and activator of transduction factor 1 (STAT1)-dependent and -independent signalling pathways. Biochem J 375: 777–783.

912. Gomez-Angelats M, del Santo B, Mercader J, Ferrer-Martinez A, Felipe A, Casado J, Pastor-Anglada M. 1996. Hormonal regulation of concentrative nucleoside transport in liver parenchymal cells. Biochem J 313: 915–920.

913. Sakowicz M, Szutowicz A, Pawelczyk T. 2004. Insulin and glucose induced changes in expression level of nucleoside transporters and adenosine transport in rat T lymphocytes. Biochem Pharmacol 68: 1309–1320.

914. Sakowicz M, Szutowicz A, Pawelczyk T. 2005. Differential effect of insulin and elevated glucose level on adenosine transport in rat B lymphocytes. Int Immunol 17: 145–154.

915. Leung GP, Man RY, Tse CM. 2005. D-Glucose upregulates adenosine transport in cultured human aortic smooth muscle cells. Am J Physiol Heart Circ Physiol 288: H2756–H2762.

916. Aguayo C, Casado J, Gonzalez M, Pearson JD, Martin RS, Casanello P, Pastor-Anglada M, Sobrevia L. 2005. Equilibrative nucleoside transporter 2 is expressed in human umbilical vein endothelium, but is not involved in the inhibition of adenosine transport induced by hyperglycaemia. Placenta 26: 641–653.

917. Montecinos VP, Aguayo C, Flores C, Wyatt AW, Pearson JD, Mann GE, Sobrevia L. 2000. Regulation of adenosine transport by D-glucose in human fetal endothelial cells: involvement of nitric oxide, protein kinase C and mitogen-activated protein kinase. J Physiol 529: 777–790.

918. del Santo B, Tarafa G, Felipe A, Casado FJ, Pastor-Anglada M. 2001. Developmental regulation of the concentrative nucleoside transporters CNT1 and CNT2 in rat liver. J Hepatol 34: 873–880.

919. Fideu MP, Miras-Portugal MT. 1992. Long term regulation of nucleoside transport by thyroid hormone (T3) in cultured chromaffin cells. Neurochem Res 17: 1099–1104.

920. Belt JA, Noel LD. 1988. Isolation and characterization of a mutant of L1210 murine leukemia deficient in nitrobenzylthioinosine-insensitive nucleoside transport. J Biol Chem 263: 13819–13822.

921. Galmarini CM, Thomas X, Calvo F, Rousselot P, Rabilloud M, El Jaffari A, Cros E, Dumontet C. 2002. In vivo mechanisms of resistance to cytarabine in acute myeloid leukaemia. Br J Haematol 117: 860–868.

922. Galmarini CM, Thomas X, Calvo F, Rousselot P, El Jafaari A, Cros E, Dumontet C. 2002. Potential mechanisms of resistance to cytarabine in AML patients. Leuk Res 26: 621–629.

923. Wright AM, Paterson AR, Sowa B, Akabutu JJ, Grundy PE, Gati WP. 2002. Cytotoxicity of 2-chlorodeoxyadenosine and arabinosylcytosine in leukaemic lymphoblasts from paediatric patients: significance of cellular nucleoside transporter content. Br J Haematol 116: 528–537.

924. Stam RW, den Boer ML, Meijerink JP, Ebus ME, Peters GJ, Noordhuis P, Janka-Schaub GE, Armstrong SA, Korsmeyer SJ, Pieters R. 2003. Differential mRNA expression of

Ara-C-metabolizing enzymes explains Ara-C sensitivity in MLL gene-rearranged infant acute lymphoblastic leukemia. Blood 101: 1270–1276.

925. Gati WP, Paterson AR, Larratt LM, Turner AR, Belch AR. 1997. Sensitivity of acute leukemia cells to cytarabine is a correlate of cellular es nucleoside transporter site content measured by flow cytometry with SAENTA-fluorescein. Blood 90: 346–353.

926. Gati WP, Paterson AR, Belch AR, Chlumecky V, Larratt LM, Mant MJ, Turner AR. 1998. Es nucleoside transporter content of acute leukemia cells: role in cell sensitivity to cytarabine (araC). Leuk Lymphoma 32: 45–54.

927. Hubeek I, Stam RW, Peters GJ, Broekhuizen R, Meijerink JP, van Wering ER, Gibson BE, Creutzig U, Zwaan CM, Cloos J, et al. 2005. The human equilibrative nucleoside transporter 1 mediates in vitro cytarabine sensitivity in childhood acute myeloid leukaemia. Br J Cancer 93: 1388–1394.

928. Pennycooke M, Chaudary N, Shuralyova I, Zhang Y, Coe IR. 2001. Differential expression of human nucleoside transporters in normal and tumor tissue. Biochem Biophys Res Commun 280: 951–959.

929. Glue P. 1999. The clinical pharmacology of ribavirin. Semin Liver Dis 19(Suppl 1): 17–24.

930. Mangravite LM, Xiao G, Giacomini KM. 2003. Localization of human equilibrative nucleoside transporters, hENT1 and hENT2, in renal epithelial cells. Am J Physiol Renal Physiol 284: F902–F910.

931. VerDonck K. 1994. Purine metabolism in the heart. Strategies for protection against myocardial ischaemia. Pharm World Sci 16: 69–76.

932. Dennis DM, Raatikainen P, Martens JR, Belardinelli L. 1996. Modulation of atrioventricular nodal function by metabolic and allosteric regulators of endogenous adenosine in guinea-pig heart. Circulation 94: 2551–2559.

933. Mubagwa K, Flameng W. 2001. Adenosine, adenosine receptors and myocardial protection: an updated overview. Cardiovasc Res 52: 25–39.

934. Martin BJ, Lasley RD, Mentzer RM Jr. 1997. Infarct size reduction with the nucleoside transport inhibitor R-75231 in swine. Am J Physiol 272: H1857–H1865.

935. Parkinson FE, Rudolphi KA, Fredholm BB. 1994. Propentofylline: a nucleoside transport inhibitor with neuroprotective effects in cerebral ischemia. Gen Pharmacol 25: 1053–1058.

936. Parkinson FE, Paterson AR, Young JD, Cass CE. 1993. Inhibitory effects of propentofylline on [3H]adenosine influx. A study of three nucleoside transport systems. Biochem Pharmacol 46: 891–896.

937. Andinè P, Rudolphi KA, Fredholm BB, Hagberg H. 1990. Effect of propentofylline (HWA 285) on extracellular purines and excitatory amino acids in CA1 of rat hippocampus during transient ischaemia. Br J Pharmacol 100: 814–818.

938. Dux E, Fastbom J, Ungerstedt U, Rudolphi K, Fredholm BB. 1990. Protective effect of adenosine and a novel xanthine derivative propentofylline on the cell damage after bilateral carotid occlusion in the gerbil hippocampus. Brain Res 516: 248–256.

939. Parkinson FE, Zhang YW, Shepel PN, Greenway SC, Peeling J, Geiger JD. 2000. Effects of nitrobenzylthioinosine on neuronal injury, adenosine levels, and adenosine receptor activity in rat forebrain ischemia. J Neurochem 75: 795–802.

940. Zhang YW, Shepel PN, Peeling J, Geiger JD, Parkinson FE. 2002. Effects of nitrobenzylthioinosine on adenosine levels and neuronal injury in rat forebrain ischemia. Neurosci Res Commun 30: 80–89.

941. Sawynok J. 1998. Adenosine receptor activation and nociception. Eur J Pharmacol 347: 1–11.

942. Keil GJ, DeLander GE. 1995. Time-dependent antinociceptive interactions between opioids and nucleoside transport inhibitors. J Pharmacol Exp Ther 274: 1387–1392.

943. Ackley MA, Governo RJ, Cass CE, Young JD, Baldwin SA, King AE. 2003. Control of glutamatergic neurotransmission in the rat spinal dorsal horn by the nucleoside transporter ENT1. J Physiol 548: 507–517.

944. Smith PG, Marshman E, Newell DR, Curtin NJ. 2000. Dipyridamole potentiates the in vitro activity of MTA (LY231514) by inhibition of thymidine transport. Br J Cancer 82: 924–930.

945. Chan TC, Howell SB. 1990. Role of hypoxanthine and thymidine in determining methotrexate plus dipyridamole cytotoxicity. Eur J Cancer 26: 907–911.

946. Hughes JM, Tattersall MH. 1989. Potentiation of methotrexate lymphocytotoxicity in vitro by inhibitors of nucleoside transport. Br J Cancer 59: 381–384.

947. Goel R, Sanga R, Howell SB. 1989. Pharmacologic basis for the use of dipyridamole to increase the selectivity of intraperitoneally delivered methotrexate. Cancer Chemother Pharmacol 25: 167–172.

948. Warlick CA, Sweeney CL, McIvor RS. 2000. Maintenance of differential methotrexate toxicity between cells expressing drug-resistant and wild-type dihydrofolate reductase activities in the presence of nucleosides through nucleoside transport inhibition. Biochem Pharmacol 59: 141–151.

949. Marshman E, Newell DR, Calvert AH, Dickinson AM, Patel HR, Campbell FC, Curtin NJ. 1998. Dipyridamole potentiates antipurine antifolate activity in the presence of hypoxanthine in tumor cells but not in normal tissues in vitro. Clin Cancer Res 4: 2895–2902.

950. Wadler S, Subar M, Green MD, Wiernik PH, Muggia FM. 1987. Phase II trial of oral methotrexate and dipyridamole in colorectal carcinoma. Cancer Treat Rep 71: 821–824.

951. Yang JL, White JC, Capizzi RL. 1992. Modulation of the cellular pharmacokinetics of ara-CTP in human leukemic blasts by dipyridamole. Cancer Chemother Pharmacol 29: 236–240.

952. Wright AM, Gati WP, Paterson AR. 2000. Enhancement of retention and cytotoxicity of 2-chlorodeoxyadenosine in cultured human leukemic lymphoblasts by nitrobenzylthioinosine, an inhibitor of equilibrative nucleoside transport. Leukemia 14: 52–60.

953. Chango A, Emery-Fillon N, de Courcy GP, Lambert D, Pfister M, Rosenblatt DS, Nicolas JP. 2000. A polymorphism (80G-A) in the reduced folate carrier gene and its associations with folate status and homocysteinemia. Mol Genet Metab 70: 310–315.

954. Chango A, Emery-Fillon N, Mircher C, Blehaut H, Lambert D, Herbeth B, James SJ, Rethore MO, Nicolas JP. 2005. No association between common polymorphisms in genes of folate and homocysteine metabolism and the risk of Down's syndrome among French mothers. Br J Nutr 94: 166–169.

955. Coppede F, Marini G, Bargagna S, Stuppia L, Minichilli F, Fontana I, Colognato R, Astrea G, Palka G, Migliore L. 2006. Folate gene polymorphisms and the risk of Down syndrome pregnancies in young Italian women. Am J Med Genet A 140: 1083–1091.

956. De Marco P, Calevo MG, Moroni A, Merello E, Raso A, Finnell RH, Zhu H, Andreussi L, Cama A, Capra V. 2003. Reduced folate carrier polymorphism (80A→G) and neural tube defects. Eur J Hum Genet 11: 245–252.

957. Barkai G, Abruzova S, Berkenstadt M, Heifetz S, Cuckle H. 2003. Frequency of Down syndrome and neural-tube defects in the same family. Lancet 361: 1331–1335.

958. Pei L, Zhu H, Ren A, Li Z, Hao L, Finnell RH, Li Z. 2005. Reduced folate carrier gene is a risk factor for neural tube defects in a Chinese population. Birth Defects Res A Clin Mol Teratol 73: 430–433. Abstract only.

959. O'Leary VB, Pangilinan F, Cox C, Parle-McDermott A, Conley M, Molloy AM, Kirke PN, Mills JL, Brody LC, Scott JM. 2006. Reduced folate carrier polymorphisms and neural tube defect risk. Mol Genet Metab 87: 364–369.

960. Yates Z, Lucock M. 2005. G80A reduced folate carrier SNP modulates cellular uptake of folate and affords protection against thrombosis via a non homocysteine related mechanism. Life Sci 77: 2735–2742.

961. Shimasaki N, Mori T, Samejima H, Sato R, Shimada H, Yahagi N, Torii C, Yoshihara H, Tanigawara Y, Takahashi T, Kosaki K. 2006. Effects of methylentetrahydrofolate reductase and reduced folate carrier 1 polymorphisms on high-dose methotrexate-induced toxicities in children with acute lymphoblastic leukemia or lymphoma. J Pediatr Hematol Oncol 28: 64–68.

962. Ulrich CM, Curtin K, Potter JD, Bigler J, Caan B, Slattery ML. 2005. Polymorphisms in the reduced folate carrier, thymidylate synthase, or methionine synthase and risk of colon cancer. Cancer Epidemiol Biomark Prev 14(11 Pt 1): 2509–2516.

963. Zhao R, Gao F, Goldman ID. 2002. Reduced folate carrier transports thiamine monophosphate: an alternative route for thiamine delivery into mammalian cells. Am J Physiol Cell Physiol 282: C1512–C1517.

964. Rothem L, Berman B, Stark M, Jansen G, Assaraf YG. 2005. The reduced folate carrier gene is a novel selectable marker for recombinant protein overexpression. Mol Pharmacol 68: 616–624.

965. Wright JE, Yurasek GK, Chen YN, Rosowsky A. 2003. Further studies on the interaction of nonpolyglutamatable aminopterin analogs with dihydrofolate reductase and the reduced folate carrier as determinants of in vitro antitumor activity. Biochem Pharmacol 65: 1427–1433.

966. Jackman AL, Boyle FT, Harrap KR. 1996. Tomudex (ZD1694): from concept to care, a programme in rational drug discovery. Invest New Drugs 14: 305–316.

967. Chattopadhyay S, Zhao R, Krupenko SA, Krupenko N, Goldman ID. 2006. The inverse relationship between reduced folate carrier function and pemetrexed activity in a human colon cancer cell line. Mol Cancer Ther 5:438–449.

968. Jansen G, van der Heijden J, Oerlemans R, Lems WF, Ifergan I, Scheper RJ, Asseraf YG, Dijkmans BAC. 2004. Sulfasalazine is a potent inhibitor of the reduced folate carrier: implications for combination therapies with methotrexate in rheumatoid arthritis. Arthritis Rheum 50: 2130–2139.

969. Assaraf YG, Sierra EE, Babani S, Goldman ID. 1999. Inhibitory effects of prostaglandin A1 on membrane transport of folates mediated by both the reduced folate carrier and ATP-driven exporters. Biochem Pharmacol 58: 1321–1327.

970. Naggar H, Ola MS, Moore P, Huang W, Bridges CC, Ganapathy V, Smith SB. 2002. Downregulation of reduced-folate transporter by glucose in cultured RPE cells and in RPE of diabetic mice. Invest Ophthalmol Vis Sci 43: 556–563.

971. Smith SB, Huang W, Chancy C, Ganapathy V. 1999. Regulation of the reduced-folate transporter by nitric oxide in cultured human retinal pigment epithelial cells. Biochem Biophys Res Commun 257: 279–283.

972. Chiao JH, Roy K, Tolner B, Yang C, Sirotnak FM. 1997. RFC-1 Gene expression regulates folate absorption in mouse small intestine. J Biol Chem 272: 11165–11170.

973. McGuire JJ, Haile WH, Yeh CC. 2006. 5-Amino-4-imidazolecarboxamide riboside potentiates both transport of reduced folates and antifolates by the human reduced folate carrier and their subsequent metabolism. Cancer Res 66: 3836–3844.

974. Prouse PJ, Shawe D, Gumpel JM. 1986. Macrocytic anaemia in patients treated with sulphasalazine for rheumatoid arthritis. Br Med J 293:1407.

975. Swinson CM, Perry J, Lumb M, Levi AJ. 1981. Role of sulfasalazine in the aetiology of folate deficiency in ulcerative colitis. Gut 22: 456–461.

976. Shiroky JB. 1993. Unsubstantiated "risk" of antifolate toxicity with combination methotrexate and sulfasalazine therapy. Arthritis Rheum 12: 1757–1758.

977. Longstreth GF, Green R. 1983. Folate status in patients receiving maintenance doses of sulfasalazine. Arch Intern Med 143: 902–904.

978. Zimmerman J. 1992. Drug interactions in intestinal transport of folic acid and methotrexate: further evidence for the heterogeneity of folate transport in the human small intestine. Biochem Pharmacol 44: 1839–1842.

979. Haagsma CJ, Russel FG, Vree TB, van Riel PL, van de Putte LB. 1996. Combination of methotrexate and sulfasalazine in patients with rheumatoid arthritis: pharmacokinetic analysis and relationship to clinical response. Br J Clin Pharmacol 42: 195–200.

980. Haagsma CJ, van Riel PL, da Jong AJ, van de Putte LB. 1997. Combination of sulfasalazine and methotrexate versus the single components in early rheumatoid arthritis: a randomized, controlled, double-blind, 52 week clinical trial. Br J Rheumatol 36: 1082–1088.

981. Haagsma CJ, Blom HJ, van Riel PL, van't Hof MA, Giesendorf BA, Oppenraaij-Emmerzaal D, van de Putte LB. 1999. Influence of sulphasalazine, methotrexate, and the combination of both on plasma homocysteine concentrations in patients with rheumatoid arthritis. Ann Rheum Dis 58: 79–84.

982. Chung JY, Cho JY, Yu KS, Kim JR, Oh DS, Jung HR, Lim KS, Moon KH, Shin SG, Jang IJ. 2005. Effect of OATP1B1 (SLCO1B1) variant alleles on the pharmacokinetics of pitavastatin in healthy volunteers. Clin Pharmacol Ther 78: 342–350.

983. Poole RC, Halestrap AP. 1991. Reversible and irreversible inhibition, by stilbenedisulphonates, of lactate transport into rat erythrocytes: identification of some new high-affinity inhibitors. Biochem J 275: 307–312.

984. Bonen A, McCullagh KJ, Putman CT, Hultman E, Jones NL, Heigenhauser GJ. 1998. Short-term training increases human muscle MCT1 and femoral venous lactate in relation to muscle lactate. Am J Physiol 274: E102–E107.

985. Py G, Lambert K, Milhavet O, Eydoux N, Prefaut C, Mercier J. 2002. Effects of streptozotocin-induced diabetes on markers of skeletal muscle metabolism and monocarboxylate transporter 1 to monocarboxylate transporter 4 transporters. Metabolism 51: 807–813.

INDEX

Drug Transporters: Molecular Characterization and Role in Drug Disposition, Edited by Guofeng You and Marilyn E. Morris

Copyright © 2007 John Wiley & Sons, Inc.